D0918742

Mobile Radio Networks

Second Edition

To

Antonie,
Thomas and Christoph

Mobile Radio Networks
Networking, Protocols and Traffic Performance

Second Edition

Bernhard H. Walke
Aachen University of Technology (RWTH), Germany

JOHN WILEY & SONS, LTD

First published under the title Mobilfunknetze und ihre Protokolle, Band 1 und 2

Baffins Lane, Chichester,
West Sussex, PO19 1UD, England

National 01243 779777
International (+44) 1243 779777
e-mail (for orders and customer service enquiries): cs-books@wiley.co.uk

Visit our Home Page on http://www.wiley.co.uk or http://www.wiley.com

Other Wiley Editorial Offices

John Wiley & Sons, Inc., 605 Third Avenue,
New York, NY 10158-0012, USA

WILEY-VCH Verlag GmbH
Pappelallee 3, D-69469 Weinheim, Germany

John Wiley & Sons Australia Ltd, 33 Park Road, Milton,
Queensland 4064, Australia

John Wiley & Sons (Canada) Ltd, 22 Worcester Road
Rexdale, Ontario, M9W 1L1, Canada

John Wiley & Sons (Asia) Pte Ltd, 2 Clementi Loop #02-01,
Jin Xing Distripark, Singapore 129809

British Library Cataloguing in Publication Data

A catalogue record for this book is available from the British Library

ISBN 0471 49902 1

Produced from PostScript files supplied by the author.
Printed and bound in Great Britain by Antony Rowe Ltd, Chippenham, Wiltshire.
This book is printed on acid-free paper responsibly manufactured from sustainable forestry, in which at least two trees are planted for each one used for paper production.

Contents

Preface

Until the late 1980s cellular mobile radio networks for public and private users in Europe were company-specific solutions and not intended for the mass market. The broader technical–scientific world therefore limited its interest to familiarizing itself with the systems and their concepts, without involving itself in the details.

Since the development of European standards for digital systems in the late 1980s and the subsequent introduction of these systems around 1990, mobile radio has become a mass market commodity. Digital mobile radio has evolved from being an add-on business to being a key sales sector for certain large telecommunications companies, making them market leaders worldwide. This has resulted in the technical–scientific world taking a greater interest in mobile radio.

The success of mobile radio is due to the great advances made in information technology, such as the microminiaturization of integrated circuits and components and the dramatic increase in the integration density of semiconductor devices on chips, which have been particularly important in the development of hand-portable mobile radio devices (handhelds): a mobile terminal essentially consists of a very powerful signal processor that incorporates as programs all the algorithms of transmission technology required for receiving and transmitting and for electric signalling. *Field Programmable Gate Arrays* (FPGAs) are used instead to reduce the power consumption.

On the other hand, the advances in information technology are also evident in the development of these algorithms for signal (de)modulation, synchronization of communicating parties, channel coding and channel equalization, i.e., receiver technology that allows the reliable reception of signals with a few microvolts of amplitude over the radio channel, which perceptually can be described as an intermittent electrical contact.

Another just as important contribution to information technology has been the development of services and protocols for the organization and operation of communications networks. Along with multiplexing functions that enable a large number of mobile terminals to communicate quasi-simultaneously over the radio interface of the mobile radio system, these networks comprise a telecommunications network that contains intelligent network functions for mobility management, as well as cryptographic procedures for data protection and data security of a level never previously available in any network.

Thinking about the *Global System for Mobile Communication* (GSM) (a standard of the *European Telecommunications Standards Institute* (ETSI)) which has been successful worldwide, one forgets that, in addition to this cellular system, many other concepts exist for new digital mobile radio systems that are attempting to repeat the success of GSM, with some of them aiming for applications other than narrowband voice communication, e.g., paging, trunked radio, cordless communication, *Wireless Local Area Networks* (WLANs) and satellite-supported personal communication. This book explores all these systems.

Since 1983, my research group has specialized in the development of services and protocols for private and public mobile radio systems, and has produced an extensive set of tools for software design, modelling and stochastic simulation of mobile radio systems. Through these tools, the mobile radio systems now being used in Europe, under discussion or in the process of being introduced, and described in this book, have been reproduced in a highly accurate form as large simulation program packages at my chair. These tools allow us to study the existing or forthcoming systems in their natural environments with the appropriate radio coverage, mobility and typical traffic volumes of their subscribers, and, based on this, to test our own approaches to the improvements and introduction of new services and protocols. Our proposals and the results of our work have successfully influenced the standardization of various systems at the ETSI, e.g., *General Packet Radio Service* (GPRS) and *HIPERLAN/2* (H/2).

The tools referred to are the outcome of the work of on average about 25 scientific assistants and 60 students working for their Master's theses per year, without whom it would not have been possible to incorporate so many details of so many systems. The work involved was not limited to the implementation of protocols for the respective systems, but ranged from the development of radio planning tools on the basis of empirical and ray-tracing techniques for given scenarios to Markov chain-based modelling of the radio channel, exemplary research into the modelling of receivers, studies on the effectiveness of adaptive channel coding, prototypic implementation of equalizers, development of models for bit error characteristics of different systems, development of procedures for dynamic channel allocation in large-scale systems and for the decentral organization of systems with wireless base stations, etc., all the way to the development of value-added services. This supplementary work proved to be necessary in order to establish with sufficient realism the difficult process of modelling real systems. It would not have been possible to present a description of the systems with the desired degree of detail without actually having implemented the services and protocols in realistic models for simulation of the systems.

The text has been gradually expanded from a first comprehensive presentation of GSM [1]. The text and many of the figures in this book are based on the input of many students whose names it would not be possible to mention individually. All I can do is convey my gratitude to all of them for their enthusiasm and for the thoroughness of their work in this collaboration. Their contribution was in modelling and evaluating the different systems and their modifications, and their input has helped my research assistants and me to develop a better understanding of the characteristics of the systems which have been considered.

The individual chapters of the book have been written in close cooperation with the research assistants responsible for the respective system models and they have been named. The chapters reflect the results of extensive research and development and in some cases incorporate material from final or earlier versions of twelve cycles' worth of lecture notes. I should like to take this opportunity to give my warmest thanks to them for the thoroughness of their contributions on the respective topics, for their assistance in dealing with the relevant Master's theses that they have co-supervised and for their role in creating such an excellent work atmosphere. I should particularly like to mention Peter Decker and Christian Wietfeld, who in earlier years helped to provide some of the background information for the lecture notes and later provided the about 150 pages nucleus around which the book crystallized by integrating existing text modules. Contributions to individual sections of the book have been made by the following (former) members of my research group:

- Branko Bjelajac (Chapter 16)
- Götz Brasche (Sections 3.11, 4.1)
- Peter Decker (Chapter 3)
- Matthias Fröhlich (Chapter 17)
- Eckhardt Geulen (Section 3.10.5)
- Alexander Guntsch (Chapter 16)
- Andreas Hettich (Chapter 13, Section 14.3)
- Martin Junius (Section 3.6)
- Arndt Kadelka (Chapter 6, 13)
- Matthias Lott (Chapter 6)
- Dietmar Petras (Section 2.8, Chapter 13)
- Christian Plenge (Chapter 10)
- Markus Scheibenbogen (Section 3.6, Chapter 10)
- Peter Seidenberg (Chapter 6)
- Matthias Siebert (Chapter 12)
- Martin Steppler (Sections 2.8, 7.2)
- Christian Wietfeld (Chapter 3, Section 6.13)

I especially want to thank my assistant Dirk Kuypers, who has shown so much dedication in the preparation of the manuscript. In addition to taking great care in incorporating years' worth of corrections and additions to the manuscript, he has made an effort to ensure that there is a homogeneity to the content and presentation—a task that particularly entailed extensive and frequent revisions to the tables and illustrative material. The precise correction work of Frank Mueller and Thomas Lammert and the careful reading of the final version of the manuscript by Carmelita Goerg are greatly appreciated.

I also want to convey my warm thanks to Mrs Jourdan von Schmoeger for the careful translation performed on the basis of a German version of the book into English. This work contributed very much to establish this book, although a lot remained to be done by the author, especially to translate the text in the figures, identify the misunderstandings introduced and align some wordings to the technical terms used by mobile radio specialists. I sincerely hope that I have not corrupted too much the style of presentation introduced by the professional translator.

Aachen, December 1998 Bernhard Walke

Addresses:

Homepage for chair: `http://www.comnets.rwth-aachen.de`
Errata: `http://www.comnets.rwth-aachen.de/~mfn/errata.html`
E-Mail address for corrections: `walke@comnets.rwth-aachen.de`
Address of chair: Communication Networks
RWTH Aachen University of Technology
D-52074 Aachen, Germany

Preface to the Second Edition

The positive response to the first edition, the rapid further development of mobile and wireless radio technology and systems and the introduction of new standards have been the motivators to update the book to a substantial extend in many chapters and to extent the number of systems described to those soon expected to become of great importance. As a result, the following systems descriptions have been included in the 2nd edition according to their state of development in early 2001: *General Packet Radio Service* (GPRS), *Enhanced Data Rates for GSM Evolution* (EDGE), *Universal Mobile Telecommunication System* (UMTS) FDD- and TDD-mode, *HIPERLAN/2* (H/2), Bluetooth, Selforganizing WLAN, of which only *General Packet Radio Service* (GPRS) was discussed with a basic description in the first edition.

The subtitle of the book has been extended from 'Networking and Protocols' to 'Networking, Protocols and Traffic Performance' to make visible a unique feature introduced in this edition: We present the traffic performance behaviour of the most important systems under realistic traffic load conditions and in addition are able to compare the performance seen by the user of some competing systems.

The basis for these results is a modelling technique developed at my chair that is based on the formal specification of the air-interfaces by means of the *Specification and Description Language* (SDL), resulting in a prototype implementation of the respective systems down to the bit-level that is sometimes also called emulation. The implementation of the system is then embedded in a simulation environment allowing us to load each mobile terminal with an individual traffic class derived from, say, a speech or e-mail traffic generator. The communication between terminal and base station, including the fixed network, is studied by stochastic simulation under a precise modelling of the interference conditions and resulting block error rates in a multi-cellular scenario to extract the traffic performance parameters of interest. What is called simulation is in fact the repeated loading of a system implementation by traffic generators and the statistical evaluation of the system's performance.

To be able to derive distribution functions of, e.g., the packet delay, we have used a very precise method for this developed at my chair. Examples of the simulation technique used are given in Sections 4.3 and 14.2.9.

The traffic performance results presented are examples only to characterize the systems according to the current status of development and should not be taken as the latest method of comparing systems against each other. Standards and related systems are continously developing, exploiting dormant potentials of systems to make them effective in the field. It can be expected that all of the newer systems will experience a substantial improvement in performance over time by introducing functions that are not available to date.

I would like to mention the many contributions to this edition by my PhD students who are named next to the Sections to which they have contributed. Besides those named in the Preface to the first edition, the following list names those who in particular have contributed to realize the book and introduce the traffic performance results—many thanks to all of them:

- Marc Althoff (Section 6.12)
- Norbert Esseling (Section 14.2.12)
- Ian Herwono (Section 14.4)
- Dirk Kuypers (Sections 7.2.5, 7.2.6)
- Stefan Mangold (Sections 14.2.10, 14.3)
- Michael Meyer (Section 6.14)
- Jörg Peetz (Section 14.2.13)
- Peter Seidenberg (Chapter 6)
- Peter Sievering (Sections 7.2.5–7.2.7)
- Peter Stuckmann (Chapter 4, Section 6.14)
- Ulrich Vornefeld (Section 14.2.11)
- Christoph Walke (Section 14.2.11)
- Bangnan Xu (Section 15.2)

Mrs. Jourdan von Schmoeger professionally perfromed the translation of German texts into English. Dirk Kuypers, as with the first edition, has again devoted months of work to transform the text contributions into a homogeneous style and took care for harmonized appearences of the tables and figures throughout the book. Many thanks to him, who in our group earned the title of a 'Dr. TEX'. Without his deep devotion to this book project it would have been much more difficult, if not impossible, to finish this work.

Aachen, May 2001 Bernhard Walke

Bibliography

[1] B. Walke. *Technik des Mobilfunks, in: Zellularer Mobilfunk, J. Kruse (Hrsg)*, pp. 17–63. net-Buch, Telekommunikation edition, 1990.

1
Introduction

During the first half of this century, the transmission of human voice through the telephone was the dominant means of communication next to telegraphy. Radio-supported mobile communication has constantly grown in importance during the last few decades and particularly the last few years to technical advances in transmission and switching technology as well as in microelectronics. Table 1.1 presents an overview of the chronological development of mobile radio systems.

In contrast to wireline networks, mobile radio networks that comply with the wish for geographically unrestricted communication can be used anywhere where it is not economic or possible to install cabling. Whereas the limiting factor with wireline networks is the network infrastructure that has to be created, the capacity of radio networks is determined by the frequency spectrum available and the physical attributes of radio waves in the earth's atmosphere.

The development of radio systems is influenced considerably by the scarcity of an important resource—frequencies. For instance, spectral efficiency can be improved through the digitalization of speech and the use of source and channel coding. Analogue radio systems have therefore been replaced more and more by digital mobile radio networks. Modern digital techniques used in modulation, coding and equalization enable bandwidth-efficient transmission and offer better interference behaviour and lower susceptibility to noise than analogue-modulated signals. Digital voice and data can be processed and stored before being transmitted and output to a receiver.

This allows the use of multiplexing methods such as *Time Division Multiplex* (TDM), *Frequency Division Multiplex* (FDM) and *Code Division Multiplex* (CDM) that enable services to be provided to many users. For example, with TDM a large number of users in a specific frequency bandwidth are able to exchange information without extremely high selectivity of the receiver. This means that fewer steep-edged filters and resonating elements are needed, thereby resulting in a cost reduction, whereas modems transmit in burst mode and therefore are more costly. Digital modulation techniques often produce a higher level of transmission quality and are also more compatible with existing digital fixed networks.

Mobile communication today is available in a broad spectrum of technological and service-specific forms. The aim of this book is to provide the reader with an overview of the digital communications networks that have been introduced over the last few years, along with the services these networks offer, their protocols and their traffic performance behaviour. Special emphasis is

Table 1.1: Chronological development of mobile radio systems

Year	Paging system standards	Cordless phone system standards	Mobile terrestrial system standards	Mobile satellite system standards
1980	POCSAG	CT0	NMT 450 Nordic Mobile Telephone	Inmarsat-A
1985		CT1	AMPS (USA) Advanced Mobile Phone System RC2000(F) Radio Communication C450 (D,P) Cellular TACS (UK) Total Access Communication System	Inmarsat-C
1990		CT2	GSM Global System for Mobile Communication DCS 1800 Digital Cellular System at 1800 MHz	Inmarsat-B Inmarsat-M Inmarsat-Paging
1994	ERMES		TFTS Terrestrial Flight Telephone System	
1998				Inmarsat-P21, Iridium, Aries, Odyssey, Globalstar, Ellipso
2000			UMTS Universal Mobile Telecommunication System FPLMTS Future Public Land Mobile Telecommunication System	
2001			Mobile Broadband System/ Wireless LANs	

given to systems in Europe that are currently being used, are being standardized or whose introduction is imminent.

Deregulation and liberalization of the telecommunications market, along with the various agreements on standardization, are having a major effect on the development of mobile radio systems. Detailed specifications are necessary in order to achieve compatibility between the products of different system and terminal suppliers. International and European standards bodies are defining mobile radio systems that can be used and operated across country boundaries. This will enable users to be reachable wherever they are roaming and will result in the cost-effective production of terminals per unit, thereby opening up the market to different types of customers. The most important standards organizations active in the mobile radio area are covered in Appendix B.

Physical connections over a radio channel are far more complex than those in a fixed network. Some of the main characteristics of radio transmission are therefore presented in Chapter 2.

Digital radio networks that currently exist at the planning stage or are in the process of being installed are described in the chapters that follow.

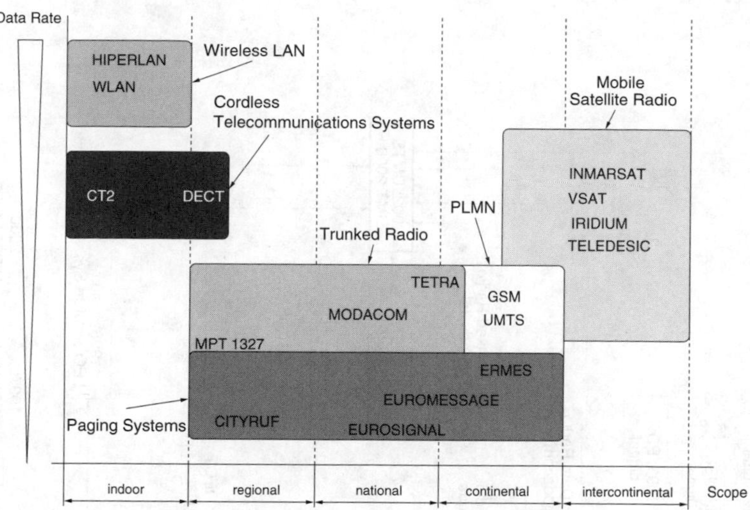

Figure 1.1: Different types of mobile radio systems

Some of the systems that were described in the first edition of this book have been dropped owing to their reduced importance from today's point of view. The respective chapters are still available on `www.comnets.rwth-aachen.de/~mfn`. Networks are differentiated according to the services they offer their users, their technical design and how they are used (see Figure 1.1). Significant differences exist between them in how and to what extent they support user mobility (see Figure 13.1).

Public cellular mobile radio systems extend the telephone service of wireline networks area-wide to mobile users. Older mobile radio networks were designed for pure voice transmission and use analogue transmission techniques. In 1958 the first public mobile radio network, the A-Netz, was installed in Germany and was subsequently followed by the cellular B and C networks (see Figure 1.2). Whereas the B-Netz is restricted to telephony, the C-Netz also offers data services. These networks have stepwise been replaced by digital mobile radio networks based on the ETSI/GSM standard, e.g., the most modern analogue system C-Netz has been closed down in 2001. Other analogue networks in Europe (see Table 1.2 and Figure 1.3) likewise are currently being replaced by GSM.

Third-Generation (3G) systems like UMTS are currently being introduced and expected to open public service in 2001 (Japan) and 2002 (Europe) (see Chapter 6).

Trunked mobile radio systems are optimized for commercial application and provide services like closed user group, e.g., for companies with mobile field staff. These systems enable voice and/or data communication over a half-duplex channel with a central site, typically between mobile users in an area geographically restricted owing to the transmitter power of the base

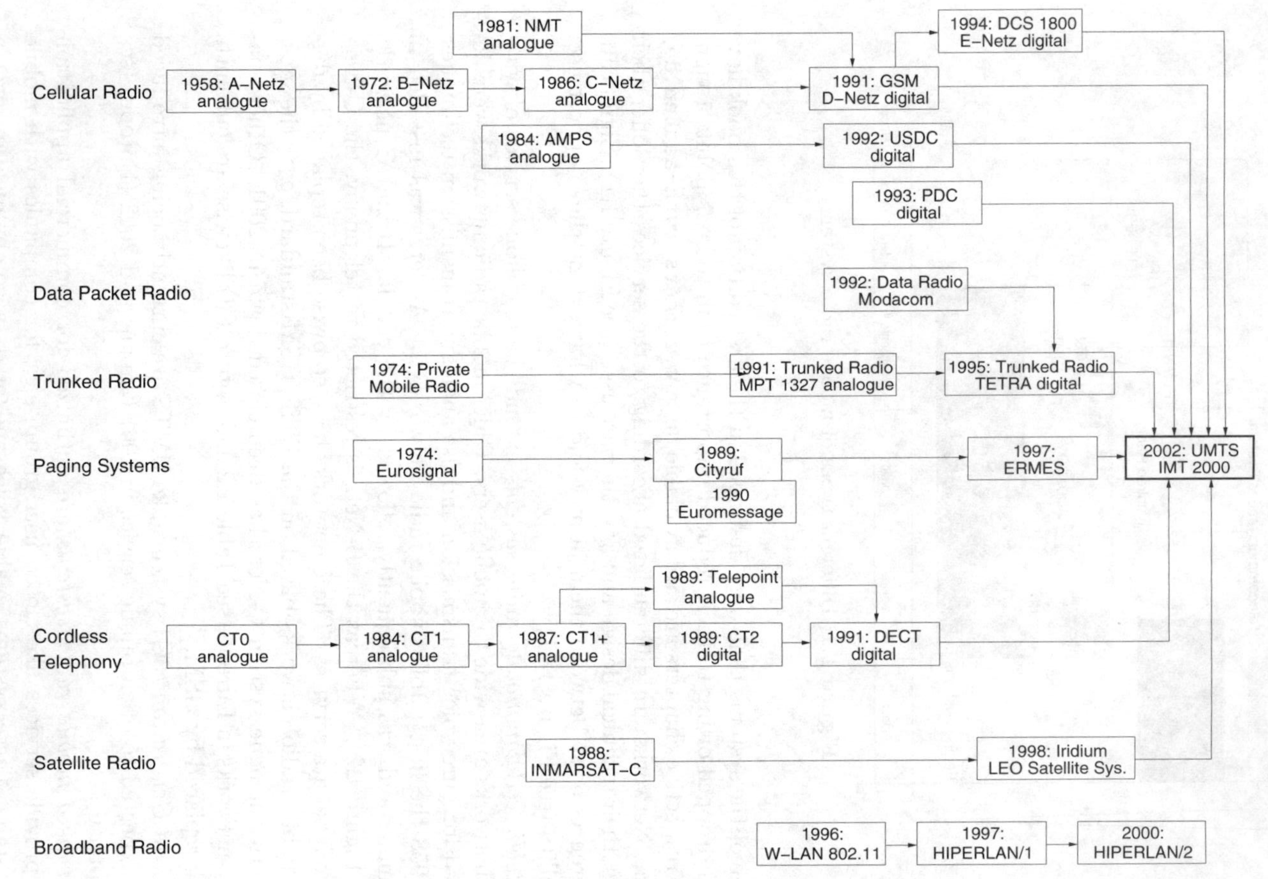

Figure 1.2: Evolution of mobile radio systems (dates refer to introduction in the field)

Table 1.2: An overview of analogue cellular mobile radio

Parameter	C 450	NMT 450	NMT 900	TACS	E-TACS	AMPS
Original country	Germany	Scandinavia	Scandinavia	GB	GB	USA
Standardized by	DBP Telekom			CRAG	CRAG	FCC
Introduced in	1985	1981	1986	1984		1983
Uplink [MHz]	450.3–454.74	453–457.5	890–915	890–915	872–905	824–849
Downlink [MHz]	461.3–465.74	463–467.5	935–960	935–960	917–950	869–894
Channel spacing [kHz]	20	25 (20)	25 (12.5)	25	25	30
Duplex range [MHz]	11	10	45	45	45	45
Access method	FDMA	FDMA	FDMA	FDMA	FDMA	FDMA
Modulation	FSK	FFSK	FFSK	PSK	PSK	PSK
MAH[a]	Yes	No	No	No	No	No
Cell diameter [km]		15–40	2–20			
Frequencies [#]	222	180 (220)	1000 (1999)	1000	1320	833
Data services [kbit/s]	2.4	–	–	–		2.4 (no HO)
Traffic capacity [Erl./km^2] (3 km distance)				14	14	12

[a]Mobile Assisted Handover

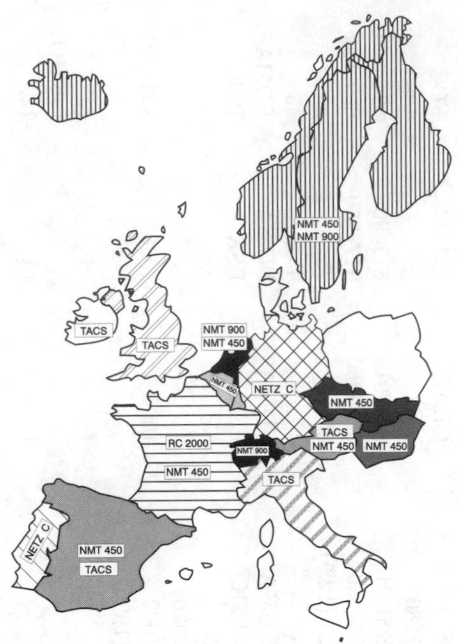

Figure 1.3: Distribution of analogue cellular systems in Europe

station; see Chapter 7. These systems are in broad usage also by public safety organizations.

Paging systems allow the direct paging of subscribers with mobile, pocket-sized receivers through the transmission of a signal or short message. Users with the appropriate pager can receive a call but not use the receiver to respond; see Chapter 8.

The concept of *cordless communication systems* summarizes services and applications based on cordless telephony. In principle, with cordless telephones, the cable between the telephone terminal and the handset was merely replaced by a radio path that allows a radio connection of up to 300 m/50 m (outdoors/indoors); see Chapters 9, 10 and 12.

Fixed wireless access systems also called *Wireless Local Loop* (WLL) systems are in use to bypass the fixed lines of local loop operators allowing comparable and even higher transmission speeds at the subscriber terminal, e.g. $n \cdot 64$ kbit/s, $m \cdot 2$ Mbit/s up to 8 and under specific conditions up to 34 Mbit/s.

Wireless Local Area Networks (WLANs) take into account the growing demand to avoid cabling of workstation computers; see Chapters 13 and 14 and *Personal Area Networks* (PANs) like Bluetooth help to avoid cabling close to the user's terminals; see Section 14.4. *Mobile Satellite Radio Systems* provide global communication and accessibility; see Chapter 16.

The mobile communications market is currently developing at a rapid pace, and it is anticipated that the next few years will bring a dramatic growth in the number of users and an increased demand for quality. As a result, the standardization bodies are continously developing new standards with the aim of providing a universal, services-integrating mobile telecommunications system in the near future; see Chapter 18.

1.1 Existing and New Networks and Services

1.1.1 GSM System

The spectacular growth of the GSM-based cellular mobile radio networks convey the impression that the essential development needed in this area has been accomplished through the introduction of these cellular mobile radio networks. What one forgets is that these networks have been designed as an "extension" of the *Integrated Services Digital Network* (ISDN) to the mobile user, but only address the needs to a limited degree: instead of two B-channels per user, only one with a considerably lower user data rate (13/6.5 kbit/s for voice and 9.6 kbit/s for data) is available. Likewise the ISDN-D channel has only been reproduced to a point: an X.25 packet service (X.31) on the D_m channel is not possible in GSM. The primary rate connection (2.048 Mbit/s) available with ISDN does not exist. The situation is a similar one with competing *Second-Generation* (2G) systems in the USA and Japan (see Table 1.3 and Figure 1.4).

A number of concepts of GSM have been developed further in order to head off the competitive pressure of 3G concepts for cellular networks (see Chapter 6) and to provide better support to mobile image and data services. The anticipated demand for ISDN-compatible mobile data services (64 kbit/s) is pressuring for rapid continued development of the radio interface. Relevant work has been carried out by ETSI GSM/2+. Examples include:

- *General Packet Radio Service* (GPRS) for multiplex data transmission of a large number of virtual connections over one or more *Traffic Channels* (TCHs).
- Multipoint, voice and data services for group communications as usual with trunked radio systems.
- Higher bit rate voice and services for images and data over parallel TCHs (*High-Speed Circuit-Switched Data*, HSCSD).
- Use of the total 200 kHz bandwidth per carrier with advanced modulation to provide high-bit-rate services up to 384 kbit/s (*Enhanced Data Rates for GSM Evolution*, EDGE) now a 3G system family member (see Section 4.2).

Table 1.3: An overview of 2nd generation digital cellular mobile radio

Parameter	GSM 900	GSM 1800	D-AMPS	CDMA	PDC
Orig. country	Europe	Europe	US	US, Korea	Japan
Standardized by	ETSI	ETSI	TIA 54	TIA 95	MPT
Introduced in	1992	1993	1991	Test 1991	
Uplink [MHz]	890–915	1710–1785	824–849	824–849	810–826
Downlink [MHz]	935–960	1805–1855	869–894	869–894	940–956
Channel spacing [kHz]	200	200	10	1230	25
Duplex range [MHz]	45	95	45	45	130
Access method	TDMA/FDD	TDMA/FDD	TDMA/FDD	CDMA	TDMA/FDD
Modulation	GMSK	GMSK	$\frac{\pi}{4}$-PQJPSK	QPSK/DQPSK	$\frac{\pi}{4}$-PQJPSK
Min. C/I [dB]	> 9	> 9	> 12		> 13
Speech codec	RPE-LPC-LTP	RPE-LPC-LTP	VSELP	ADPCM	VSELP
Speech data rate [kbit/s]	13 (6.5)	13	7.25	8.55	6.7
Frequencies [#]	124	374	832	10	1600
Time slots/freq.	8 (16)	8	3	–	3
Data service [kbit/s]	9.6	9.6	8	?	9.6–14.4
Max. speed MS [km/h]	250	250	100	100	100
Output power HU [W]	2–5–8–20	0.25–1	0.6–1.2–3–6	0.5–2–6.3	0.3–0.8–2
Traffic cap. [Erl./km^2] (3 km site dist.)	40 (84)		41		91
Spectr. efficiency $\left[\frac{\text{Erl.}}{\text{km}^2\cdot\text{MHz}}\right]$	1.1 ... 1.6	1.1 ... 1.6	1.64 ... 2.99		2 ... 3.63
Diversity meth.	Interleaving	Interleaving Slow FH	Interleaving Slow FH		Interleaving Antenna Div.

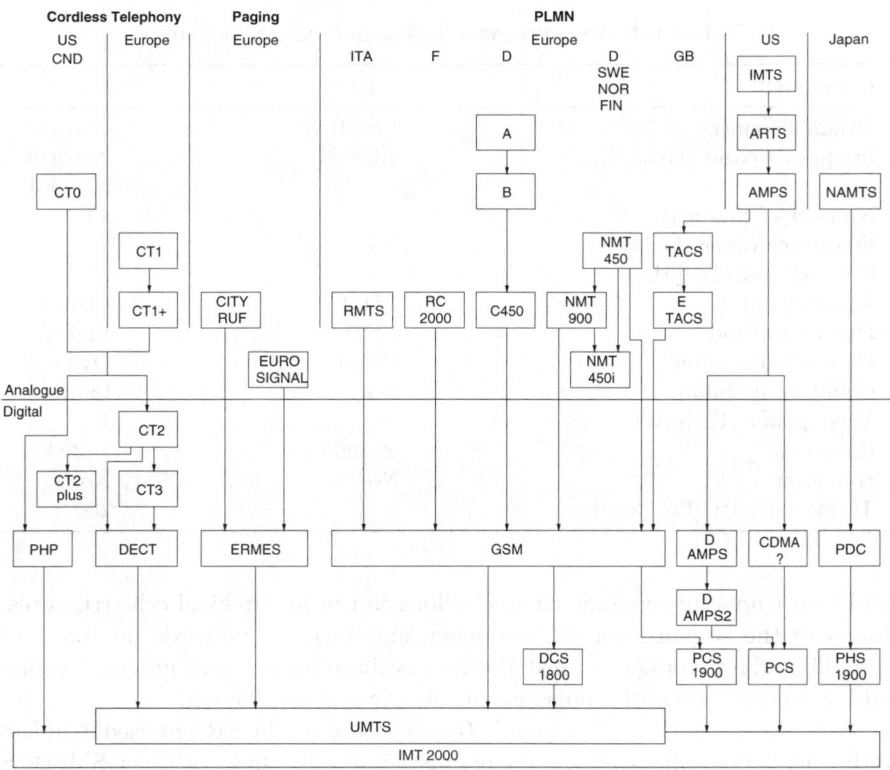

Figure 1.4: Overview of worldwide standards for mobile radio systems

These further developments resulted from research in areas such as radio planning, adaptive antennas, propagation modelling, modulation, source and channel coding, combined coding and modulation (*Codulation*) for the improvement of spectral efficiency, data compression, signal processing, microelectronics and switching technology, and electrical power storage. After the basic services have been agreed on and introduced on the network side, new services and applications have been provided through modifications to the radio interface and the incorporation of relevant options for terminals, e.g., mobile computing; limited support of multimedia applications are also available now with these GSM-based systems.

This development of cellular networks should not be viewed in isolation but must be seen from the standpoint of the continuing development of other mobile radio systems that have been introduced at the same time or later and in some cases represent competition (see Sections 1.1.2–1.1.3). Another point to be made is that cellular networks still do not offer a satisfactory solution for the radio coverage required within buildings and that there is still a great deal of scope for improving the planning tools for radio coverage prediction

Table 1.4: An overview of analogue cordless telephony

Parameter	CT0	CT1
Original country	US, JPN	Europe
Frequency band [MHz]	1.6 / 47	914–916 959–961
Number of channels	8	40
Frequency range [MHz]	0.4	4
Channel spacing [kHz]		25
Access method	FDMA	FDMA
Duplex method	FDD	FDD
Channel allocation	Fixed	Dynamic
Cellular capability	No	Limited
Max. power HU [mW]		10
Range [m]	< 1000	< 300
Handover	No	No
Traffic capacity [Erl./km^2]	1	200

and algorithms for dynamic channel allocation in hierarchical cell structures. Likewise the organization of intelligent mobile radio networks in terms of signalling, data storage and distribution, availability of (cross-network) value-added services, network management, etc., is still developing.

It is expected that *Terrestrial Trunked Radio* (TETRA)-based trunked radio networks will experience demands similar to those of the GSM/DCS 1800 systems, with the added complication of the algorithms and protocols for direct communication between mobile stations (*direct mode*) optionally available from the standard.

1.1.2 DECT

Digital cordless systems based on the ETSI standard *Digital Enhanced Cordless Telecommunications* (DECT) already in use (as "small" systems) have conquered the market as "large systems" providing partial coverage to the application areas of larger private branch exchange systems, thereby replacing the analogue predecessor systems (see Table 1.4). These systems are extremely suitable for mobile applications indoors and in close proximity to the respective base station (up to approximately 300 m outdoors).

Because only a fraction of future mobile users will be using communications services outside metropolitan areas, DECT systems have been used as personal communications systems in densely built-up areas to reach a high percentage of all mobile radio subscribers, e.g., in cities, with the function mobility management implemented. The *Personal Handyphone System* (PHS) build up in large cities in Japan (see Table 1.5 and Chapter 12) is another example of this.

The inherent strength of cordless systems lies in their suitability for providing coverage indoors, which easily lends itself to providing a limited range outdoors also.

Compared with GSM, DECT systems are less restrictive in terms of the services that can be specified at the radio interface, and therefore allow a higher level of flexibility, e.g., for use in a *Local Area Network* (LAN) or as the technology for wireless access to the telecommunications network (*Radio in the Local Loop*, RLL) (see Figure 10.64).

On one hand, DECT systems are preventing users from migrating from fixed network operators to mobile radio network operators, but, on the other hand, they promote a migration to mobile radio network operators because users typically already have a fixed network access point.

1.1.3 Radio Networks as a Bypass to the Local Loop

Deregulation in Europe ends the monopoly on voice services of the incumbent operator, and is resulting in the expansion of former corporate networks by new network operators in competition with the respective incumbent (some of them also using lines leased from the incumbent), who are providing services to large customers and (eventually) all conveniently located corporate and private customers. This expansion is being accompanied by the development and establishment of local cellular radio networks that use point-to-multipoint radio relay or fixed radio user connections (see Figure 11.6), offer ISDN-based and primary rate multiplex interfaces and can be used as access networks (*Fixed Wireless Access*, FWA) to fixed networks of the incumbent's competitors.

GSM only has limited application in this area because of its noticeably lower transmission rate compared with ISDN. In multichannel operation DECT can offer ISDN interfaces; the appropriate standards were drawn up by ETSI/RES 03 in 1996. FWA networks are closely related to the systems described in Sections 1.1.1 and 1.1.2 but require further development to enable them to make better use of frequencies and operate more cost-effectively.

Along with cellular networks that provide sectorial or radial coverage in the proximity of a base station, chains of base stations (DECT relay) and tree-like arrangements of radio links, starting from the fixed network access point, are also expected to bridge the "last mile" between fixed networks and customers in the local network area. The same frequency band used by cellular systems (e.g., with DECT) or public radio relay bands (e.g., 3.4/10/17/23/27/40/60 GHz) will be used.

ETSI *Broadband Radio Access Network* (BRAN) started in 1999 the specification of FWA systems under the name HiperACCESS and various versions of air-interfaces, each with its own name, are currently being discussed for the frequency bands mentioned.

Table 1.5: An overview of digital cordless telephony

Parameter	CT2	DECT	PHS
Original country	GB	Europe	Japan
Standardized by	BT (ETSI)	ETSI	TTC/RCR
Introduced in	1992	1993	
Frequency range [MHz]	864.1–868.1	1880–1990	1895–1918
Radio carrier spacing [MHz]	0.1	1.728	0.3
Channels [#]	40	120	
Transmitted data rate [kbit/s]	72	1152	
Channel allocation method	DCA	DCA	DCA
Speech data rate [kbit/s]	32	32	32
Speech coding	ADPCM G.721	ADPCM G.721	ADPCM
Control channels	In-call emb.	In-call emb. (logical channels: C, P, Q, N)	Fixed control carriers
In-call control channel data rate [kbit/s]	MUX 1.2, 0.75, 1.4, 1.5	4.8 (+ 1.6 CRC)	
Total channel data rate [kbit/s]	33/34	41.6	
Duplexing technique	TDD	TDD	TDD
Multiple access TDMA [Timeslots]	1 TDD	12 TDD	4 TDD
Carrier Usage FDMA/Multicarrier [# Carriers]	40	10	
Bits/TDMA timeslot (speech/data+emb. ctl.) [bit]	66/68	420 (4 bit Z-field)	
Timeslot duration (incl. guard time) [µs]	1	417	
TDMA frame period [ms]		10	
Modulation technique	GFFSK	GFFSK	$\pi/4$-DQPSK
Modulation index	0.4–0.7	0.45–0.55	
Traffic capacity [Erl./km^2]	250	10,000	
Handover	No	Yes	
Cellular capability	Limited	Yes	
Peak output power [mW]	10	250	10
Mean output power [mW]	5	10	

1.1.4 Wireless Local Area Networks (IEEE 802.11 WLAN, ETSI BRAN/HIPERLAN/2)

There is a considerable demand today for the wireless connection of (*movable*) workstation computers to provide flexibility in how and where equipment is installed in order to use standard Internet applications, which today are often accessed over a *Local Area Network* (LAN). Standardization has just produced solutions that constitute the first fast step in this direction. So-called single-hop solutions are currently possible; these tend to require the connection of a base station to a fixed network (e.g., LAN) for each office room served at the frequencies 2.4/5.3/40/60 GHz. Further development is possible and necessary to reduce the cabling required. The most successful system is IEEE 802.11 (b) WLAN operating at 2.4 GHz and offering a transmission rate of up to 11 Mbit/s. Another standardized system is ETSI BRAN *HIPERLAN/1* (H/1) operating at 5.3 GHz to provide 25 Mbit/s.

Owing to the increasing introduction and growing use of broadband services over fibre optic networks based on ATM transmission technology (broadband ISDN) with transmission rates of 34 (E3), 155, 622 and 2400 Mbit/s, a broadband option is required for connecting movable or mobile terminals, similar to GSM to connect to the narrowband ISDN. The current state of technology enables the implementation of radio-supported, cellular mobile broadband systems with up to 54 Mbit/s user data rates.

Since these networks permit data transfer rates comparable to LANs (typically up to 25 Mbit/s), they are more suitable for replacing LANs and less appropriate for supporting new multimedia services. These new services place real-time demands on a transmission system that in principle cannot be supported by the Internet, or at least not until considerable further development has been carried out in this area. Movable workstations along with mobile terminals can be supported. In addition to radio, media such as infrared and visible light are also being considered for wireless LANs. Terminal mobility (or movability) is placing new demands on Internet protocols.

Consequently, considerable efforts are being spent by the *Internet Engineering Task Force* (IETF) to complement the mobile Internet protocol (*mobile IP*) that was designed to be able to attach a computer at various access points to the Internet by new designs better suited for wireless systems. Cellular *Internet Protocol* (IP) and Hawai are similar protocols developed in 2000 by the IETF *Mobile Access Network* (MANET) project to better support wireless mobile terminals, see also Section 14.2.6.6.

1.1.5 Universal Mobile Telecommunications System UMTS

The MoU UMTS (*Memorandum of Understanding for the Introduction of UMTS*) group promoted a revolutionary (non-evolutionary) approach to the design of a *Third-Generation* (3G) system. The critical factor with current (2G) mobile communications systems is the bit rate, which is not sufficient for the new applications of the future and should be allocated flexibly as required. 3G was regarded by some people less as a totally new system but more as a further development of GSM. The EDGE system as one member of the *International Mobile Telecommunications at 2000 MHz* (IMT-2000) family of 3G standards followed this line. Besides that, a further development of the 2G system IS-95 *Code Division Multiple Access* (CDMA) used in the USA has been standardized with the name cdma2000. Other members of the IMT-2000 family are a further developed DECT system and three versions of UMTS (see Chapter 6).

1.1.6 Mobile Satellite Radio

Geostationary satellites are preferred for providing coverage to slowly moving (ships) and fixed stations because of the large receiving antennas required owing to high signal attenuation. Various groups of companies have established global mobile radio networks on the basis of low (700–1700 km height, *Low Earth Orbit* (LEO)) and medium–high (10000 km height, *Intermediate Circular Orbit* (ICO)) flying satellites, e.g. Iridium, Globalstar, ICO and are planning other systems (see Tables 1.6 and 1.7). The aim to guarantee radio coverage, e.g., at 1.6 GHz for hand-portable satellite receivers (300 g) has been reached with Iridium and Globalstar. These systems are primarily geared to providing coverage to rural and suburban areas. This means that, in addition to the efforts involved in the development and evaluation of these systems, issues concerning the cooperation with terrestrial mobile radio and fixed networks had also have to be resolved. Handover procedures in hierarchical cell structures, from picocells to satellite umbrella cells, have been developed, too.

In addition to the switching functions on board broadband satellites for the connection of mobile stations to a suitable ground base station, other problems still need to be resolved, such as routing between mobile satellites and the control of the radio links between the satellites. Satellite networks, like terrestrial mobile radio networks, will endeavour to pick up traffic close to the source and deliver it close to the destination, with no use or minimal use of other fixed networks. Research interest is focused on the problem of interference between space segments of the same or of different satellite systems and between space and ground segments.

1.1.7 Universal Personal Mobility

In addition to radio and transmission-related functions, mobile communication requires special services from the fixed networks. Mobile radio systems usually consist of a radio and a fixed part. The mobility management of users is essentially implemented through functions in the fixed network based on functions of *Common Channel Signalling System No. 7* (CCS7).

Architectures for *Universal Personal Telecommunication* (UPT) and *Intelligent Networks* (INs) have been developed worldwide for fixed networks and

Table 1.6: An overview of mobile satellite telecommunications, Part 1

Name	IRIDIUM	Project 21	Globalstar	The Calling Network (Brilliant Pebbles)	Odyssey
Prime company, country	Motorola, US	Inmarsat, GB	Qualcom, US (LQSS)	Global Com. Inc.	TRW, US
Services offered	Voice, Data, Fax, GPS, Paging	Voice, Data, GPS, Paging	Voice, Data, Fax, GPS, Paging	Voice	Voice, Data, GPS
Coverage	Global	Global	Global	Underdev. Regions	Global
Orbit type	LEO	ICO	LEO	LEO	MEO
Orbit height	778 km	10000 km	1389 km	600 km	10370 km
No. of orbits	6		8	21	3
Sats. per orbit	11		6	40	4
Total no. of sats.	66	12–15	48	840 (+80 spares)	12
Cells/sat.	48		6		37
Channels/sat.	4070		2700		2300
1st sat. launched	1996		1997	1996	
Full operation	1998	2001	1999	1999	
Costs	3.4–3.7 bill. $	2 bill. $	0.82–1.5 bill. $	6.5–7 bill. $	1.3–1.4 bill. $
Charge for voice service per min.	3 $		0.35–0.45 $		0.65 $
Terminal mode	Dual-mode	Dual-mode	Dual-mode		
Access method	FDMA/TDMA	Not yet decided	CDMA	FDMA/TDMA	

Table 1.7: An overview of mobile satellite telecommunications, Part 2

Name	Inmarsat A	NAME	Telesat	Arles	Ellipsat
Prime company, country	Inmarsat, GB	American Mobile Sat. Corp., US	Telesat Mobile, CND	Constellation Communication, US	Ellipsat Corp., US, GB, ISR
Services offered	Voice, Data, Telex	Voice, Data, Fax	Voice, Data, Fax	Voice, Data, Fax	Voice
Coverage	Global	Global	Global		Global (North and South Zone)
Orbit type	GEO	GEO	GEO		
Orbit height	36000 km	36000 km	36000 km		
Number of orbits	1	1	1		3 (North)
Sats. per orbit	3	3	3		5 (North)
Total no. of sats.		3	3	48	15 (North) + 6 (South)
Cells/sat.				7 . . . 19	
Channels/sat.				50	
1st sat. launched	1976	1995	1996		
Full operation	1979	1995	1996		
Costs				290 bill. $	219 bill. $
Charge for voice service per min.					0.5 $
Terminal mode	Sat. only	Dual-mode	Dual-mode		Dual-mode

standardized by the *Telecommunication Standardization Sector of ITU* (ITU-T). This means that it will eventually be possible to reach a person anywhere in the world under one personal subscriber number, for all services and over fixed and mobile radio networks, independently of the network service provider. The concepts for mobility support across network domains still need to be further developed.

Advantages are to be exploited and disadvantages avoided, with users given control in each specific situation over which callers are allowed to reach them, which services can be used and what should be done with other calls or incoming messages (concept of the subscriber's role).

All services marked as not to be switched to the subscriber will be dealt with according to his instructions, e.g., transferred into another form of service, routed to a storage device or diverted to a third party. These types of services will initially be primarily available for subscribers of mobile radio networks, because only they will have universal access to the network. Consequently, these services will be implemented and introduced within the context of mobile communications. The *Virtual Home Environment* (VHE) of UMTS is covering these aspects (see Section 6.13).

Finally, there should be some mention of the work involved in the *Telecommunication Infrastructure Networking Architecture* (TINA) that has been developed by the international consortium TINA-C to increase the flexibility in using communications networks.

1.2 Systems with Intelligent Antennas

Studies have recently been conducted into all types of mobile radio systems to examine possibilities for increasing spectrum efficiency [(bit/s)/(MHz·km^2)] through the use of *smart antenna arrays*, see also Section 16.4. For obvious reasons (dimensions, complexity, ability to use existing mobile devices without requiring changes), this technology is initially being discussed for use in the base stations of cellular systems. The range (and consequently the cell radius) can be increased or the transmitter power (and consequently the interference) reduced owing to the array gain achieved through adaptive forming of the antenna diagram called beamforming. This might result, at the end of development, in dynamic, radio-relay-like point-to-multipoint mobile communication.

Over and above this increase in efficiency through a reduction in the transmitting power and/or an increase in coverage range, it appears possible to implement true *Space Division Multiple Access* (SDMA) for a dramatic increase in spectral efficiency and network capacity [Erl./(MHz·km^2)]. This access protocol is not to be seen as an alternative to the established procedures *Code/Time/Frequency-Division Multiple Access* (C/T/FDMA) but instead as a compatible extension to them. The idea is that a receiver with an *antenna array* receives the signals of several users who are using the same time/frequency/code channel, and from this calculates the geographical *Directions of Arrival* (DoA). This directional information is used for spatial filtering on the uplink and beam forming on the downlink, which can be imagined as a simultaneous adaptive forming of the antenna directional diagram for each user with only one antenna array.

It is obvious that the protocols for the radio interfaces of existing mobile radio systems will have to be adapted to these new concepts. One example of that is given in Section 14.2.11. Radio resource management in the network will derive considerable advantages from a dynamic channel allocation procedure that minimizes the transmit power while guaranteeing a minimum interference level. This optimized channel allocation within a cell or beyond cell boundaries creates the expectation of a considerable increase in spectral efficiency and network capacity, and in fact irrespective of whether the basic system is a F/TDMA type (GSM), a CDMA type (IS-95) or a hybrid form (such as the UMTS under discussion). However, the results for the different basic systems will vary depending on the detailed formulation and optimization of the overall concept.

1.3 Mobile Radio Systems with Dynamic Channel Allocation and Multiple Use of Frequency Spectrum

Dynamic channel allocation is an intelligent method for allocating radio resources as required for wireless communication between a terminal and a base station. This measure on its own can increase the capacity of an ETSI/DECT system (one with standardized dynamic channel allocation) in indoor application cost-effectively by a factor of two to four compared with an ETSI/GSM system (one that uses fixed channel allocation). Some publications indicate it could be possible to achieve comparable or somewhat lower capacity increases with mobile radio systems.

Dynamic channel allocation produces higher capacity, which in turn enables a larger number of communications relationships to be implemented simultaneously in the available frequency ranges. This would be possible in the existing GSM if the respective procedures for channel allocation were implemented.

Owing to the scarcity of frequency spectrum for mobile radio applications, the *Federal Communications Commission* (FCC) (USA) and *European Radio Office* (ERO) have made inital allocations for joint use of the same spectrum for public mobile radio services.

Little is known at present about the compatibility of mobile radio systems in adjacent frequency bands and the compatibility of systems operating in the same frequency band. Here too it obviously comes down to improving spectral efficiency through measures allowing the competitive use of the same

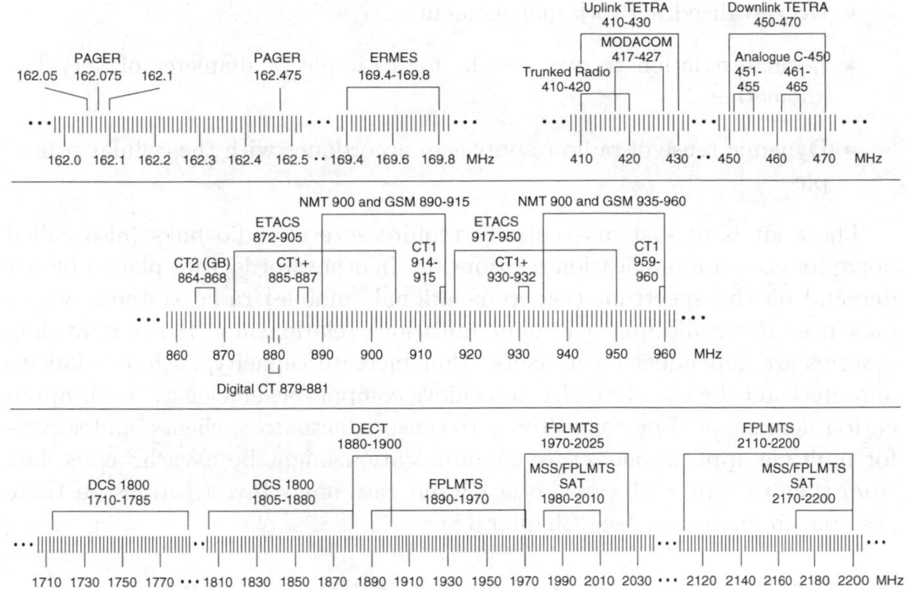

Figure 1.5: Frequency use for mobile radio services

frequency band. Figure 1.5 shows the neighbouring relationships between some of the radio systems affected.

1.4 Other Aspects

1.4.1 Self-Organizing Systems

Decentralized organizational forms (eliminating a centralized base station) would appear to have an advantage for applications with a high local density from wireless communicating stations that operate in frequency bands above 2.5 GHz and therefore require a line-of-sight connection between each other. Ad hoc networks are being discussed, the key feature of which is fully decentralized self-oganization. Other characteristics of these systems include:

- Use of some or all stations as relays on the multi-hop route between communicating stations.
- Support of synchronous and asynchronous transmission services, such as those customary with ISDN and local networks.
- Autonomous route selection and operation of the stations, including sleep mode.
- Installation of gateway stations for the link to fixed network.

- Decentralized network management.
- Local restriction to areas with, for example, a diameter of very few kilometres.
- Dynamic reuse of radio resources in accordance with the cellular principle.

These kinds of systems typically require several radio links (also called hops) for each communication relationship; in other words, they place a higher demand on the spectrum than conventional (mobile) radio systems, which only need one radio hop per communications relationship. Hence multi-hop systems are dependent on measures that increase capacity, such as adaptive antennas and the use of SDMA, to achieve comparable efficiency. Their application domains will be connectivity to sensors, actuators, clients, and servers for multiple applications where communication might be asynchronous, i.e., from time to time with possibly a low bit rate only. Investigations on these systems are underway (see Chapter 15).

1.4.2 Electromagnetic Environmental Compatibility

Conventional mobile radio systems use omnidirectional antennas, which adversely affect the environment because of the electromagnetic field produced ("electrosmog"). Intelligent antennas deliberately direct the transmitter power towards the receiver, which, compared with omnidirectional antennas at the same range, reduces transmitter power considerably.

The effects of electromagnetic waves on biological systems have been examined scientifically for many years with no indications of a negative impact on human health when the equipment is operated according to the regulations defined in the respective standards. The best knowledge on the use and control of radio waves is being taken into account in the development of new technologies for mobile radio systems.

In many places people are afraid of seeing the base station antennas of a mobile radio network close to their home or office and presume that the radio waves generated there will negatively affect their health. These people are not aware that most of them are exposed to the radio signal power of broadcast radio base stations that are much more distant, but the transmit power is two or three orders of magnitude higher than that of cellular radio systems, measured on the human body.

If a person wants to avoid the most serious exposure to mobile radio waves then they should not use a handheld, because the transmit antenna is close to the head then.

1.5 Historical Development

Communications networks began their triumphal march in 1843 when approval was granted by the American Congress for the first test section for Morse telegraphy along a rail route between Washington and Baltimore. Wired voice transmission was first possible through the invention of the telephone by Alexander Graham Bell in 1876.

1879 *Hughes* presented the phenomenon of electromagnetic waves to the Academy of Natural Sciences in London. Because the Maxwellian rules on the propagation of electromagnetic waves had not yet been recognized at that time, Hughes' results were rejected [1, 2].

1881 The first public telephone network in Berlin was installed. Point-to-point voice transmission was made possible with the help of the switchboard operators called *Fräuleins vom Amt*, who switched calls from one terminal to another terminal. Previously this service had only been offered by telegraphists.

During the following century, telephone networks were installed and constantly extended. Then they were equipped with automatic exchanges and expanded into regional, national and, finally, worldwide networks. The telephone became a part of daily life, although its use was restricted to fixed wired networks.

1888 *Hertz* was successfully able to reproduce and confirm the Maxwellian theory. He demonstrated that a spark produced from a transmitter at a nearby receiver produced a voltage. During the 1890s, *Tesla* extended the bridgeable distance.

1897 *Marconi* developed the first usable system for wireless telegraphic transmission over large distances. A Morse key was used to produce a spark in the transmitter. The receiver contained a coherer—a tube filled with ferrous powder that was connected to a direct current supply. The voltage was set so that the electric circuit, including an electromagnetic printer, did not close.

The received electromagnetic wave created by a spark in the transmitter causes the receiver circuit to be closed. A so-called Wagner hammer (see Figure 1.6) ensures, through shaking the coherer (C), that the conducting receiver circuit is opened again. Antennas A and B are adjusted to the oscillator frequency of the resonant circuit of the transmitter.

1901 *Marconi* succeeded in transmitting wireless signals over the Atlantic. However, the transmitting and receiving equipment used was so large that it could only be used in stationary locations.

1902 A radio from the company Telefunken (Germany) used in the military is shown in Figure 1.7. The top cart is carrying a 3 kW gas motor, which

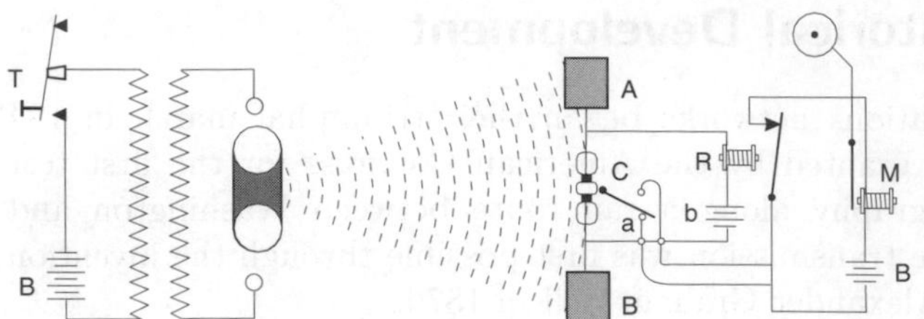

Figure 1.6: Wagner hammer

is driving a 1 kW alternating-current generator. The transmitting and receiving device is mounted on the cart below, and the Morse key is recognizable in the foreground. The antennas, which are not pictured, were very large because of the short-wave frequencies used.

1903 At this time, the first ships were equipped with radio facilities to provide shipping companies and the military with wireless communication.

Braun, Slaby and *v. Arco* of Telefunken developed a closed resonance circuit that improved the adjustment to a given frequency. At the same time, this provided a way of bypassing the Marconi patents.

1906 It became necessary to coordinate radio frequencies as more and more ships were being equipped with radio systems and it was possible to filter the spectrum occupied by a radio transmitter. At the first *World Administrative Radio Conference* (WARC) specific frequency bands were allocated to different services in order to limit reciprocal radio interference.

With the invention of the triode by *von Lieben* in 1910, transmitters based on sparking were very quickly replaced by smaller and lighter devices.

1912 The use of frequency bands up to 3 MHz was regulated at the second WARC. Higher frequencies were judged as being commercially not viable, and were therefore released for private use by radio amateurs. However, this decision was reviewed, and during the following years the commercial use of the spectrum was quickly regulated:

1927	up to 30 MHz	1947	up to 10.5 GHz
1932	up to 60 MHz	1959	up to 40 GHz
1938	up to 200 MHz	1979	up to 275 GHz

1935 The first transmitting and receiving equipment for private users (e.g., taxis) appeared on the market. They used electronic tubes and could be installed in vehicles (see Figure 1.8). The problem was that they completely filled up all the space in the boot.

Figure 1.7: Power and equipment cart for the Telefunken telegraphy system

Figure 1.8: Car telephone 1935

1952 It was now possible to call a user of a mobile terminal from a fixed network connection. Local radio systems started to be used more and more [2]:

- A single *Base Transceiver Station* (BTS) was used in an area 20–100 km in diameter.
- Around 20 participants shared one voice channel.
- In the beginning manual switching was still required to establish a connection to the fixed network; later this was replaced by automatic switching.
- Calls to a mobile device from the fixed network were supported.
- A file with user-specific data (*Home Location Register*, HLR) was established, identifying the local switching centre as the gateway to the BTS.
- Voice communication was carried out semi- or full-duplex.

1958 Isolated systems typically supplied coverage to city areas, and were restricted to their terminals. When users left the city, they could not use the mobile radio devices in another city, even if the same system was installed there (the frequency used was a different one and there was no roaming agreement between the individual operators).

Countrywide mobile radio systems allowed subscribers to establish a connection with any BTS of the system. Instead of one frequency channel, a whole bundle of frequencies was available to the mobile terminal. Any free channel could be used for transmission. Through the utilization of trunking gain, it was possible for more traffic to be carried with the same blocking probability. Base stations were either linked individually over gateway nodes to the fixed network or intermeshed together and connected over a central gateway to the fixed network.

The rest of this chapter describes the development of cellular radio networks in Germany. The A-Netz in Germany is an example of an early *Public Land Mobile Network* (PLMN). An operator supported the switching, and every BTS was available to radio telephones.

1972 The B-Netz was introduced in Germany, Austria, the Netherlands and Luxemburg. It supported fully automatic switching of the incoming and outgoing calls of *Mobile Stations* (MSs) and roaming between the four participating countries. A caller from the fixed network has to know the number of the base station where the mobile user is currently located. The mobile user's complete number consists of the location area code of the BTS, the number of the gateway node and the user identification. The MS is paged on a system-wide frequency, and receives a radio channel when it responds to the call.

A subscriber can restrict outgoing calls from an MS to certain frequency channels in order to optimize operating costs. A radio channel is used for inband signalling to establish a connection. If the mobile station leaves the coverage area of its BTS, the connection is broken off. There is no handover—either from the frequency channel of the BTS to another one of the same BTS or to a neighbouring cell.

1989 The C-Netz was the first mobile radio network in Germany in which automatic interruption-free *handover* was executed for mobile users changing from one radio supply zone (cell) to another cell. The network has fully automatic mobility management so that the location areas of the switched-on terminals are constantly updated and a user can automatically be located over the corresponding database when there is an incoming call, without any operator support. The network had a maximum of 850 000 subscribers in 1995 that has then been continuously decreasing in favour of the GSM systems. In 2001 the C-Netz was closed down.

1992 The D1-Netz based on the European ETSI/GSM standard was introduced. It transmits digitally, and eliminates the incompatibility that previously existed between national mobile radio networks in Europe. The first operator was T-Mobil, a subsidiary of Deutsche Telekom AG.

1993 As a result of the deregulation of mobile radio in Europe, the D2-Netz began operating as an area-wide GSM network in Germany. The operator is Mannesmann Mobilfunk GmbH.

1995 The E1-Netz based on the ETSI/GSM standard was launched as another area-wide mobile radio network. The operator is E-Plus Mobilfunk. Since the spectrum available at 900 MHz is occupied already by the two networks D1 and D2, the E1-Netz was assigned 1800 MHz as operating band.

1997 A licence was granted for the operation of another GSM1800 network E2, which must achieve 75 % area coverage in Germany and began operations in October 1998.

2000 Six licences were granted by auction for 3G systems operation. The licence fee per operator for the 10 MHz paired band plus 5 MHz unpaired band amounted to 7.7 billion USD, an incredibly high cost.

Many other mobile radio systems have been successfully introduced along with these public cellular systems (see Figure 1.2 and Tables 1.1–1.7).

Bibliography

[1] R. Gööck. *Die großen Erfindungen.* Siegloch Edition, Künzelsau, Deutschland, 1988.

[2] R. Klingler. *Die Entwicklung des öffentlichen Mobilfunks.* In *FIBA Kongress* EUROPÄISCHER MOBILFUNK, pp. 11–27, Munich, February 1989.

2
System Aspects

2.1 Fundamentals of Radio Transmission

In mobile radio systems, unlike wired networks, electromagnetic signals are transmitted in free space (see Figure 2.1). Therefore a total familiarity with the propagation characteristics of radio waves is a prerequisite in the development of mobile radio systems. In principle, the Maxwell equations explain all the phenomena of wave propagation. However, when used in the mobile radio area, this method can result in some complicated calculations or may not be applicable at all if the geometry or material constants are not known exactly. Therefore special methods were developed to determine the characteristics of radio channels, and these consider the key physical effects in different models. The choice of model depends on the frequency and range of the radio waves, the characteristics of the propagation medium and the antenna arrangement.

The propagation of electromagnetic waves in free space is extremely complex. Depending on the frequency and the corresponding wavelength, electromagnetic waves propagate as ground waves, surface waves, space waves or direct waves. The type of propagation is correlated with the range, or distance, at which a signal can be received (see Figure 2.2). The general rule is that the higher the frequency of the wave to be transmitted, the shorter the range.

Based on the curvature of the earth, waves of a lower frequency, i.e., larger wavelength, propagate as ground or surface waves. These waves can still be received from a great distance and even in tunnels.

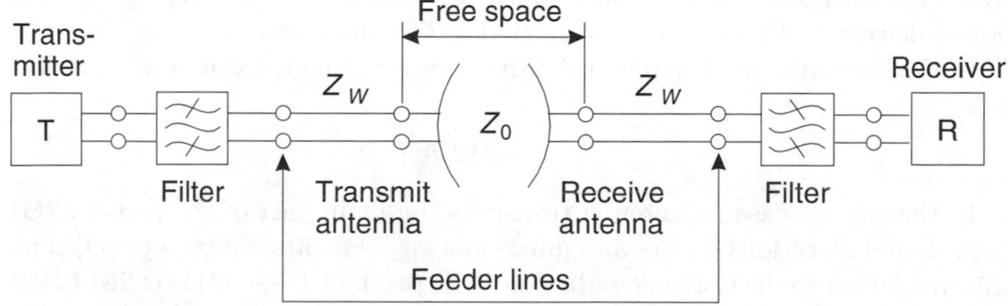

Figure 2.1: Radio transmission path: transmitter–receiver. Z_0 and Z_W are the radio wave resistances in free space and on the antenna feeder link

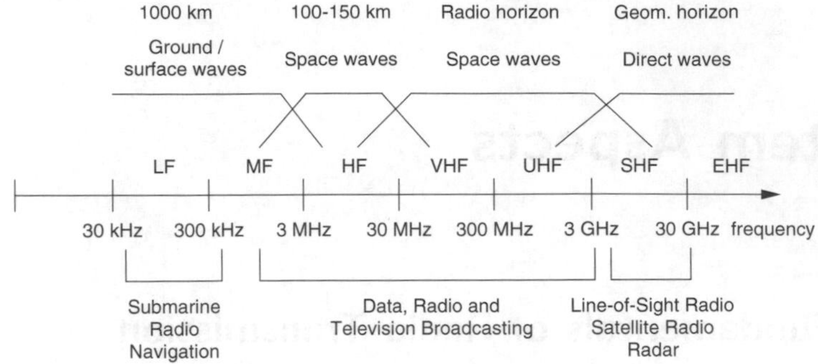

Figure 2.2: Propagation and range of electromagnetic waves in free space

In the higher frequencies it is usually space waves that form. Along with direct radiation, which, depending on the roughness and the conductivity of the earth's surface, is quickly attenuated, these waves are diffracted and reflected based on their frequency in the troposphere or in the ionosphere.

The range for lower frequencies lies between 100 and 150 km, whereas it decreases with higher frequencies because of the increasing transparency of the ionosphere, referred to as the radio horizon. When solar activity is intense, space waves can cover a distance of several thousand kilometres owing to multiple reflection on the conductive layers of the ionosphere and the earth's surface.

Waves with a frequency above 3 GHz propagate as direct waves, and consequently can only be received within the geometric (optical) horizon.

Another factor that determines the range of electromagnetic waves is their power. The field strength of an electromagnetic wave in free space decreases in inverse proportion to the distance to the transmitter, and the receiver input power therefore fades with the square of the distance. The received power for omnidirectional antennas can be described on the basis of the law of free-space propagation.

An ideal point-shaped source, a so-called isotropic radiator of signal energy, transmits its power P_0 uniformly in all directions Θ. The constant spatial power density is $P_T(\Theta) = \frac{P_0}{4\pi} = P_{iso}$ (index T: Transmitter).

In the isotropic case the power density flow F through a sphere with radius d is

$$F = \frac{P_0}{4\pi d^2}[W/m^2]. \tag{2.1}$$

In the normal case an antenna transmits the main part of the power $P(\Theta)$ in preferred directions (main- and minor lobes). The *antenna gain* $g_T(\Theta)$ puts this in relation to the isotropic radiation. The product $P_T g_T(\Theta)$ is called EIRP (*Effective Isotropically Radiated Power*). It expresses the power an isotropic source must radiate to transmit an equivalent power $P_T(\Theta)$ in the direction Θ.

An antenna with $g_T(\Theta)$, which transmits in the mean the total power P_0, transmits into the direction Θ the power density

$$P_T(\Theta) = g_T(\Theta)\frac{P_0}{4\pi}. \tag{2.2}$$

Integrating $P_T(\Theta)$ over all directions results in P_0.
The corresponding power flow density through a sphere with radius d is

$$F(\Theta) = \frac{P_T(\Theta)}{4\pi d^2} = g_T(\Theta)\frac{P_0}{(4\pi d)^2}. \tag{2.3}$$

The power P_R (Index R: Receiver) an antenna can take from the electromagnetic waves is the product of F and the effective antenna area which can be expressed as followed by the wavelength λ and the gain g_R of the receiver antenna

$$P_R = \frac{P_0 \cdot g_T(\Theta) \cdot g_R}{(4\pi d)^2} \cdot \lambda^2. \tag{2.4}$$

The term

$$L = \left(\frac{\lambda}{4\pi d}\right)^2$$

is referenced as free-space path loss because it describes the spatial diffusion of the transmitted energy over the path d.

In a logarithmic representation the difference $P_T - P_R$ corresponds to an expression $-10\log\frac{P_R}{P_T}$. In this representation the free-space loss L_F results (with $c = \lambda \cdot f$) in:

$$L_F = -10\log g_T - 10\log g_R + 20\log f + 20\log d - 20\log\frac{c}{4\pi} \tag{2.5}$$

In the case of an isotropic antennas the last expression reduces to

$$L_F = -20\log\frac{\lambda}{4\pi d} = -20\log(\frac{P_R}{P_T}). \tag{2.6}$$

2.1.1 Attenuation

Weather conditions cause changes to the atmosphere, which in turn affect the propagation conditions of waves. Attenuation is frequency-dependent and has a considerable effect on some frequencies, and a lesser one on others. For example, in the higher-frequency ranges above about 12 GHz attenuation is strong when it is foggy or raining because of the scattering and absorption of electromagnetic waves on drops of water.

Figure 2.3 shows the frequency-dependent attenuation of radio waves with horizontal free-space propagation in which, as applicable, the appropriate attenuation values for fog (B) or rain of different intensity (A) still need to be added to the gaseous attenuation (curve C). What is remarkable are the

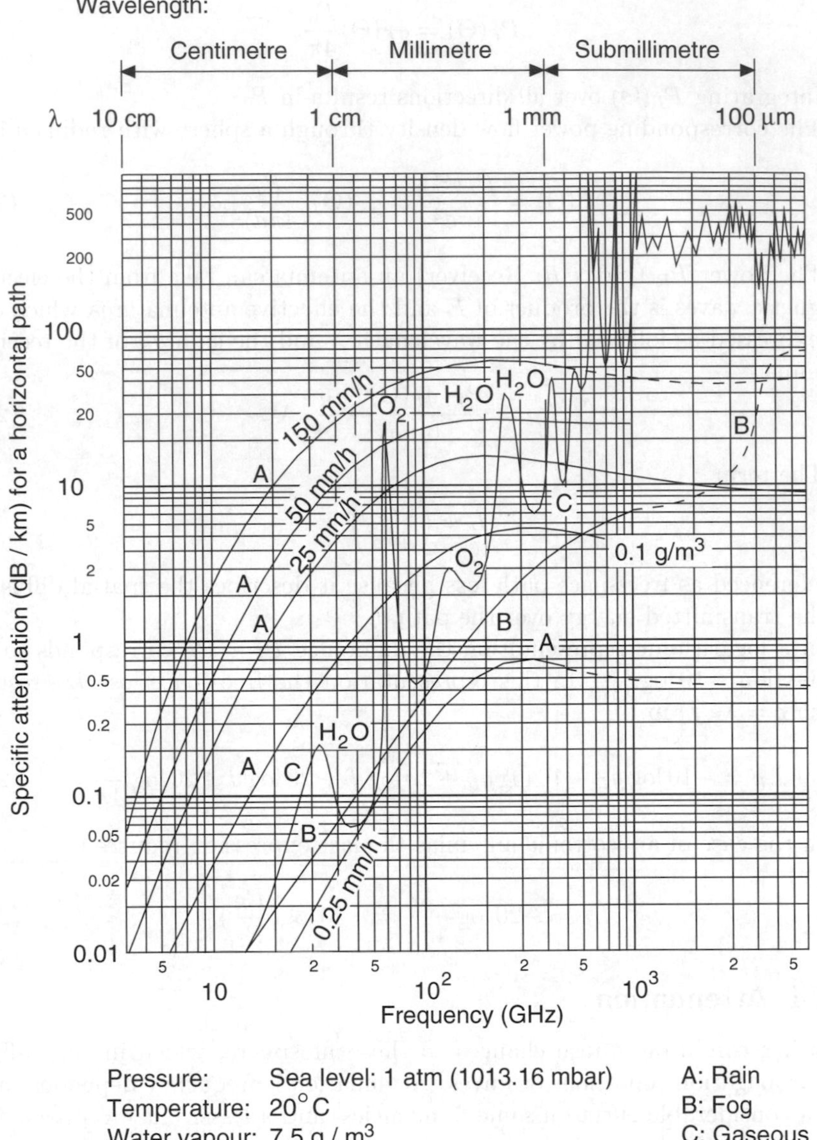

Figure 2.3: Attenuation of radio propagation depending on the frequency due to gaseous constituents and precipitation for transmission through the atmosphere, (from CCIR Rep. 719, 721)

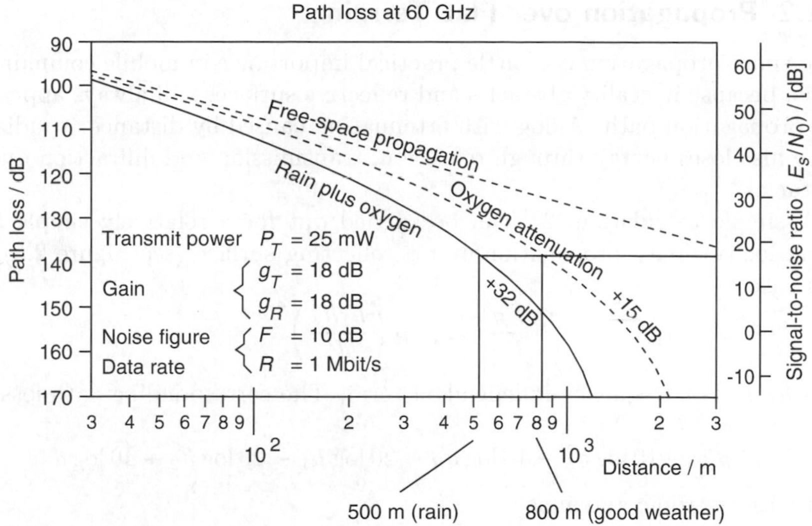

Figure 2.4: Attenuation due to weather conditions

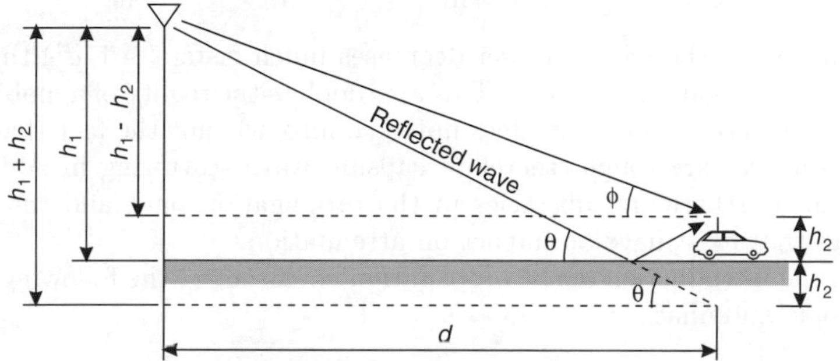

Figure 2.5: Model for two-path propagation due to reflection

resonant local attenuation maxima caused by water vapour (at 23, 150, etc., GHz) or oxygen (at 60 and 110 GHz).

Based on 60 GHz as an example, Figure 2.4 shows the propagation attenuation and the energy per symbol E_s related to N_0 (noise power), referred to as the signal-to-noise ratio, for antenna gain of $g_T = g_R = 18$ dB. These gains are achieved with directional antennas with approximately $20° \cdot 20°$ beam angles. The electric transmit power in the example is 25 mW, thereby producing the value 2 dBW = 1.6 W for the radiated microwave power (EIRP). The ranges which can be achieved are 800 m in good weather conditions and 500 m in rainy conditions (50 mm/h).

2.1.2 Propagation over Flat Terrain

Free-space propagation is of little practical importance in mobile communications, because in reality obstacles and reflective surfaces will always appear in the propagation path. Along with attenuation caused by distance, a radiated wave also loses energy through reflection, transmission and diffraction due to obstacles.

A simple calculation [27] can be carried out for a relatively simple case scenario: two-path propagation over a reflecting surface (see Figure 2.5). In this case

$$\frac{P_R}{P_T} = g_T g_R \left(\frac{h_1 h_2}{d^2}\right)^2$$

$d \gg h_1, h_2$ is a frequency-independent term. The corresponding path loss L_P is

$$L_P = -10\log g_T - 10\log g_R - 20\log h_1 - 20\log h_2 + 40\log d$$

and with isotropic antennas

$$\frac{L_F}{\text{dB}} = 120 - 20\log\frac{h_1}{\text{m}} - 20\log\frac{h_2}{\text{m}} + 40\log\frac{d}{\text{km}} \tag{2.7}$$

In this model the receive power decreases much faster ($\sim 1/d^4$) than with free-space propagation ($\sim 1/d^2$). This also depicts the reality of a mobile radio environment more closely but does not take into account the fact that actual ground surfaces are rough, therefore causing wave scattering in addition to reflection. Furthermore, obstacles in the propagation path and the type of buildings that exist have an impact on attenuation.

With the introduction of the propagation coefficent γ, the following applies to isotropic antennas:

$$P_R = P_T g_T g_R \left(\frac{\lambda}{4\pi}\right)^2 \frac{1}{d^\gamma} \tag{2.8}$$

Realistic values for γ are between 2 (free-space propagation) and 5 (strong attenuation, e.g., because of city buildings).

Different models can be used for calculating the path loss based on these parameters, and are presented in Section 2.2.

Figure 2.6 compares the resulting propagation attenuation at 1 GHz and at 60 GHz, taking into account O_2 absorption and interference caused by two-path propagation. This interference leads to signal fading in sharply defined geographical areas, and this is also relevant within the transmission range.

2.1.3 Fading in Propagation with a Large Number of Reflectors (Multipath Propagation)

Fading refers to fluctuations in the amplitude of a received signal that occur owing to propagation-related interference. Multipath propagation caused

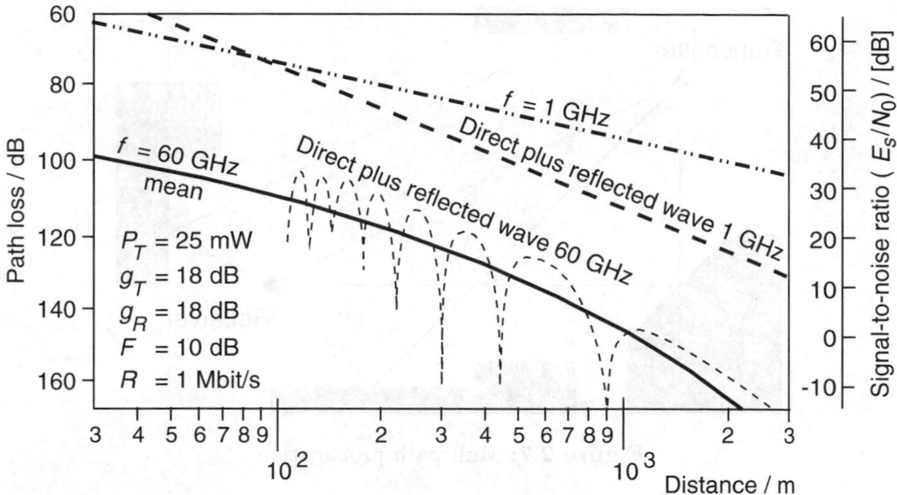

Figure 2.6: Propagation attenuation in two-path model taking into account O_2 absorption

by reflection and the scattering of radio waves lead to a situation in which transmitted signals arrive phase-shifted over paths of different lengths at the receiver and are superimposed there. This interference can strengthen, distort or even eliminate the received signal. There are many conditions that cause fading, and these will be covered below.

In a realistic radio environment waves reach a receiver not only over a direct path but also on several other paths from different directions (see Figure 2.7). A typical feature of multipath propagation (frequency-selective with broadband signals) is the existence of drops and boosts in level within the channel bandwidth that sometimes fall below the sensitivity threshold of the receiver or modulate it beyond its linear range.

The individual component waves can thereby superimpose themselves constructively or destructively and produce a stationary signal profile, referred to as multipath fading, which produces a typical signal profile on a path when the receiver is moving, referred to as *short-term fading* (see Figure 2.8).

The different time delays of component waves result in the widening of a channel's impulse response. This dispersion (or *delay spread*) can cause interference between transmitted symbols (*intersymbol interference*).

Furthermore, depending on the direction of incidence of a component wave, the moving receiver experiences either a positive or a negative Doppler shift, which results in a widening of the frequency spectrum.

In general the time characteristics of a signal envelope pattern can be described as follows:

$$r(t) = m(t)r_0(t) \tag{2.9}$$

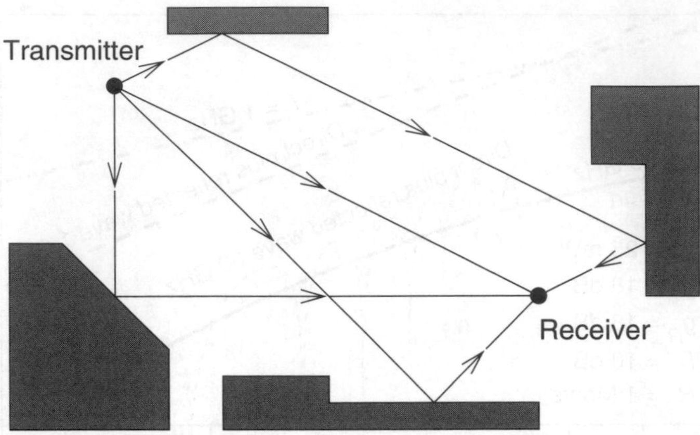

Figure 2.7: Multipath propagation

Here $m(t)$ signifies the current mean value of the signal level and $r_0(t)$ refers to the part caused by short-term fading. The local mean value $m(t)$ can be deduced from the overall signal level $r(t)$ by averaging $r(t)$ over a range of 40–200 λ [21].

The receive level can sometimes be improved considerably through the use of a diversity receiver with two antennas positioned in close proximity to each other ($n \cdot \lambda/2$; $n = 1, 2, \ldots$). Because of the different propagation paths of the radio waves, the receiving minima and maxima affected by fading of both antennas occur at different locations in the radio field, thereby always enabling the receiver to pick up the strongest available receive signal. See Figure 2.9, which shows the signal profile $r_i(t)$ of two antennas and the receive signal $r(t)$. With *scanning diversity* an antenna is replaced by a prevalent antenna when its signal level drops below a threshold A. With *selection diversity* it is always the antenna with the highest signal level that is used.

2.1.4 A Statistical Description of the Transmission Channel

It is only possible to provide a generic description of a transmission channel on the basis of a real-life scenario. In the frequency range of mobile radio being considered, changes such as the movement of reflectors alter propagation conditions. Signal statistics is another way of developing a mathematical understanding of the propagation channel.

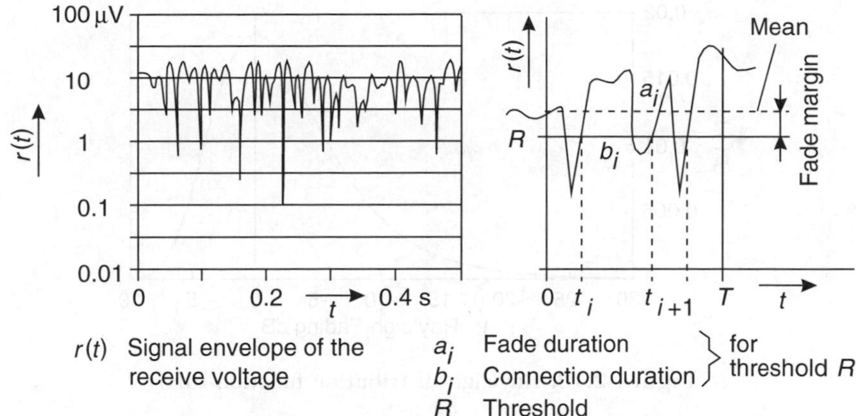

Figure 2.8: Receive signal voltage at a moving terminal under multipath fading (overall and in detail)

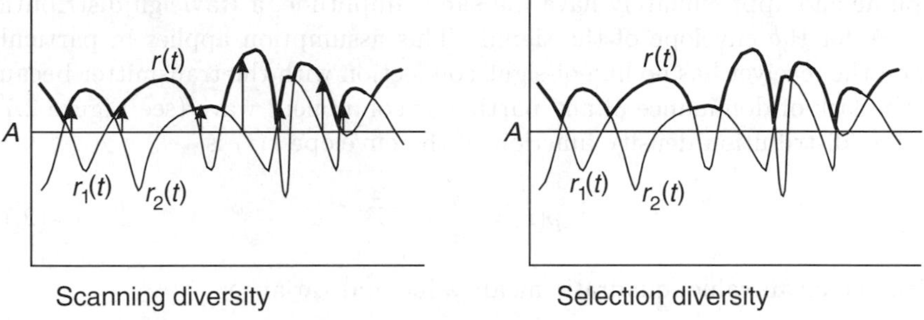

Figure 2.9: Diversity reception

2.1.4.1 Gaussian Distribution

The distribution function resulting from the superposition of an infinite number of statistically independent random variables is, based on the central limit theorem, a Gaussian function:

$$p(x) = \frac{1}{\sqrt{2\pi}\sigma} e^{-\frac{(x-m)^2}{2\sigma^2}} \tag{2.10}$$

No particular distribution function is required for the individual overlaid random variables, and they can even be uniformly distributed. The only prerequisite is that the variances of the individual random variables should be small in comparison with the overall variance.

A complete description of the Gaussian distribution is provided through its mean value m and the variance σ^2.

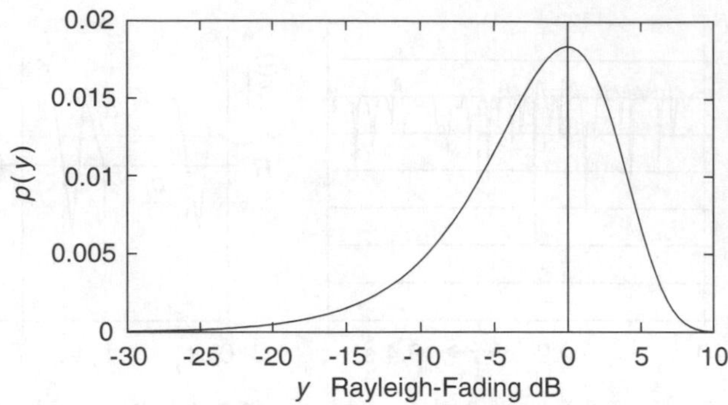

Figure 2.10: Rayleigh distribution function (dB)

2.1.4.2 Rayleigh Distribution

On the assumption that all component waves are approximately incident at a plane and approximately have the same amplitude, a Rayleigh distribution occurs for the envelope of the signal. This assumption applies in particular when the receiver has no line-of-sight connection with the transmitter because of the lack of dominance of any particular component wave (see Figure 2.7).

The distribution density function of the envelope $r(t)$ is

$$p(r) = \frac{r}{\sigma^2} e^{-\frac{r^2}{2\sigma^2}} \tag{2.11}$$

with the mean value, quadratic mean value and variance

$$E\{r\} = \sigma\sqrt{\frac{\pi}{2}}, \qquad E\{r^2\} = 2\sigma^2, \qquad \sigma_r^2 = \sigma^2\left(\frac{4-\pi}{2}\right)$$

This function is shown in Figure 2.11, see $r_s = 0$.

For the representation with $r(t) = m(t) \cdot r_0(t)$ a normalization of $E\{r_0^2\} = 1$ is common and useful. The logarithmic representation with $y = 20 \log r_0$ therefore produces

$$p(y) = \frac{10^{y/10}}{20 \log e} e^{-10^{y/10}}$$

with the mean value, variance and standard deviation ($C = 0.5772\ldots$ is Euler's constant)

$$E\{y\} = -C \cdot 10 \log e = -2.51 \text{ dB}$$

$$\sigma_y^2 = (10 \log e)^2 \pi^2/6 = 31.03 \text{ dB}, \qquad \sigma_y = 5.57 \text{ dB}$$

Figure 2.10 illustrates the distribution in half-logarithmic scaling.

Fading frequency The frequency of fading, which can be of the order of about 30 to 40 dB in depth, is dependent on the speed at which the receiver is moving, and can be described on the basis of the Doppler shift of the transmit frequency. The rate N_R at which the prescribed field strength level is exceeded is therefore calculated from

$$N_R = \sqrt{2\pi} f_m \rho e^{-\rho^2} \tag{2.12}$$

with f_m standing for the quotient arising from the vehicle speed v and wavelength λ

$$f_m = v/\lambda \tag{2.13}$$

and ρ indicating the relationship between the received signal level and the mean level.

Because the quadrature and in-phase components of the transmitted signal are Gaussian-distributed and the field strength follows a Rayleigh distribution, the signal fluctuations that arise due to multipath propagation are also referred to as *Rayleigh fading*.

The propagation paths are all of different lengths and have different reflection and transmission coefficients on the respective obstacles. This causes phase shifts on the individual incoming paths.

Signal fading due to Rayleigh fading occurs at intervals of the order of half the wavelength, $\lambda/2$.

Taking into account the attenuation and the multipath propagation with the complex elements of all the paths, the following attenuation can be observed in buildings according to [29]:

$$L = -20 \log \left(\frac{\lambda}{4\pi} \left| \sum_{i=0}^{n} \frac{\Gamma_i}{d_i} e^{\frac{2\pi d_i}{\lambda}} \right| \right) \tag{2.14}$$

L	overall attenuation in dB	n	number of incoming paths
d_i	length of ith path	λ	wavelength

Γ_i takes account of the reflections and transmissions experienced by the ith ray on the path between transmitter and receiver:

$$\Gamma_i = \prod_{j=0}^{r} R_j \prod_{k=0}^{t} T_k \tag{2.15}$$

R_j	jth reflection factor of ith route	r	number of reflections on jth path
T_k	kth transmission factor of ith route	t	number of reflections on kth path

2.1.4.3 The Rice Distribution

There are many cases in which the assumption of component waves having the same amplitude does not apply, especially when a line-of-sight connection dominates. The envelope is then described on the basis of a Rice distribution.

Table 2.1: Parameter values for the Rice distribution

	$d \leq 6\,\text{km}$		$d > 6\,\text{km}$	
Environment	r_s	$K = r_s^2/\sigma^2$	r_s	$K = r_s^2/\sigma^2$
Woodland	0.40	0.25	0.16	0.04
Small town	0.63	0.76	0.39	0.27
Village	0.74	1.15	0.40	0.24
Hamlet	0.81	1.61	0.77	1.35
Minor road	0.77	1.19	0.75	0.96
B-road	0.78	1.23	0.74	0.92
A-road	0.86	1.37	0.55	0.55

The distribution density function for the envelope $r(t)$ produces

$$p(r) = \frac{r}{\sigma^2} e^{-\frac{r^2+r_s^2}{2\sigma^2}} \mathrm{I}_0\left(\frac{r r_s}{\sigma^2}\right) \tag{2.16}$$

with I_0 is the Bessel function of 1st type and 0th order. The Rayleigh distribution is a special case of the Rice distribution for $r_s = 0$.

In concrete terms, r_s^2 represents the power of the direct, dominant component wave, and σ^2 that of the randomly distributed multipath component waves. Signal fades occur at longer intervals the further away the receiver is from the transmitter; see Figure 2.6.

Reference [24] contains parameter values for several measurements in rural areas (see Table 2.1). The values relate to the normalized signal envelope $r(t) = m(t) r_0(t)$ and a dB-mean value of 0 for r_0. Depending on the environment, σ^2 is clearly less than with Rayleigh fading. $K = r_s^2/\sigma^2 = 0$ corresponds to $r_s = 0$, i.e., there is no line-of-sight connection. $K \to \infty$ means that no multipath signals are being received. Figure 2.11 shows the Rice distribution for $\sigma = 1$.

There is no closed-form solution for the mean value and variance for the Rice distribution density function. These parameters can only be determined using approximation formulas and tables.

2.1.5 Reflection

Waves are completely reflected on smooth surfaces, but otherwise they are only partially reflected because of partial absorption—something that results in undesirable phase shifts.

If a propagating wave hits a wall, part of it is reflected and part transmitted, as is shown in Figure 2.12. The reflected part is a result of direct reflection and a multitude of multiple reflections on the inside of the wall. In this same way the entire transmitted part consists of one direct continuous wave and many component waves reflected in the wall; see [19].

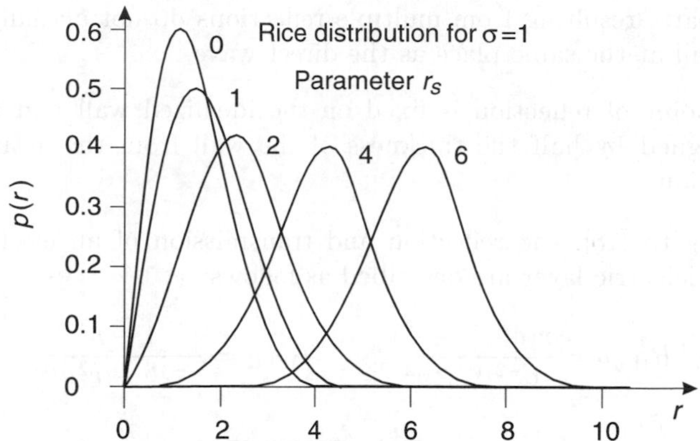

Figure 2.11: Rice distribution density function

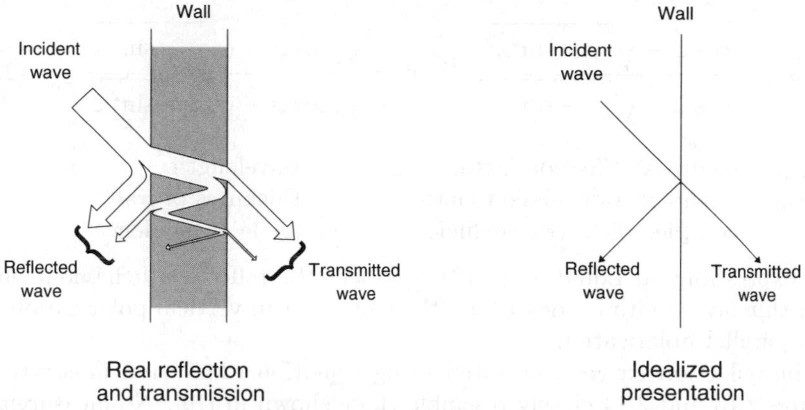

Figure 2.12: Reflection at a wall

The sum total of the reflected and the transmitted wave differs from the incident wave because the multiple reflections within the wall cause attenuation loss.

In the prediction of actual radio propagation (e.g., using ray tracing techniques) it is usually the geometric conditions of reflection and transmission on a wall—albeit in the idealized form presented in Figure 2.12—that are taken into account.

Geometric errors can occur for the following reasons:

1. Owing to refraction, the exit point of the transmitted wave on the inside of the wall is shifted vertically from the exit point in the simplified representation.

2. The parts resulting from multiple reflections do not actually exit from the wall at the same place as the direct wave.
3. The point of reflection is fixed on the idealized wall and is therefore misaligned by half the thickness of the wall from the actual point of reflection.

According to [18], the reflection and transmission of an electromagnetic wave on a dielectric layer are described as follows:

$$R_{Wall} = \frac{r(e^{-2j\psi} - 1)}{e^{-2j\psi} - r^2}, \qquad T_{Wall} = \frac{1 - r^2}{e^{-j\psi} - r^2 e^{j\psi}} \tag{2.17}$$

with

$$\psi = \frac{2\pi d}{\lambda}\sqrt{\underline{\varepsilon}_r - \sin^2\varphi} \tag{2.18}$$

and

$$r_\perp = \frac{\cos\varphi - \sqrt{\underline{\varepsilon}_r - \sin^2\varphi}}{\cos\varphi + \sqrt{\underline{\varepsilon}_r - \sin^2\varphi}}, \qquad r_\| = \frac{\underline{\varepsilon}_r\cos\varphi - \sqrt{\underline{\varepsilon}_r - \sin^2\varphi}}{\underline{\varepsilon}_r\cos\varphi + \sqrt{\underline{\varepsilon}_r - \sin^2\varphi}} \tag{2.19}$$

R_{Wall}	complex reflection factor	λ	wavelength
T_{Wall}	complex transmission factor	d	thickness of wall
$\underline{\varepsilon}_r$	complex dielectric coefficient	φ	angle of incidence

The expressions in Equation (2.19) represent the reflection behaviour on an ideal thin layer, with $r_\perp$ describing the behaviour in vertical polarization and $r_\|$ in parallel polarization.

The reflection curves calculated using Equation (2.17) and illustrated in Figures 2.13 and 2.14 closely resemble those shown in [19]. No measurement results are available for the transmission values (see Figures 2.15 and 2.16); they are deduced from the reflection coefficients.

The figures show the attenuation of the reflection or the transmission over the angle of incidence φ, with the attenuation being $20\log|R_{Wall}|$ [dB] and $20\log|T_{Wall}|$ [dB] respectively.

The results for the different polarization directions as a function of the angle of incidence indicate a sharp drop in the *Brewster angle* area. Otherwise the reflection factor increases from a minimum value of 0° to a maximum value of almost 90°. The minimum value, the gradient of the curve and the *Brewster angle* are dependent on the thickness and material of the wall.

The reflection characteristics of different materials in the area of 1–20 GHz are presented as attenuation curves in [19].

2.1.6 Diffraction

Diffraction describes the modification of propagating waves when obstructed. A wave is diffracted into the shadow space of an obstruction, thereby enabling

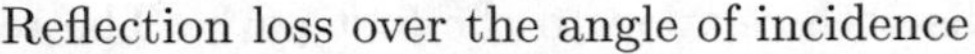

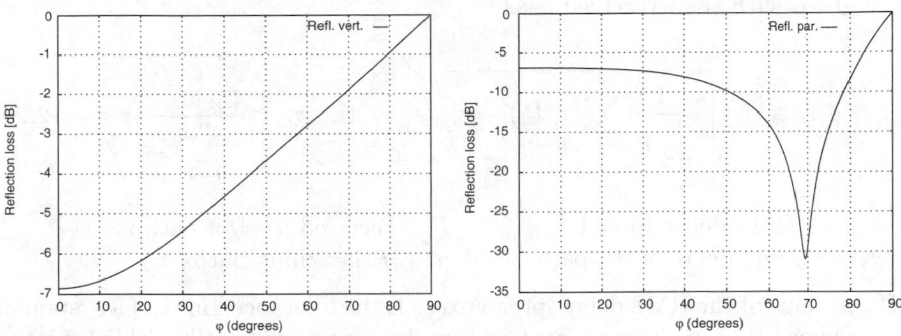

Figure 2.13: Concrete wall (wall thickness 150 mm), vertical polarization

Figure 2.14: Concrete wall (wall thickness 150 mm), horizontal polarization

Transmission loss over the angle of incidence

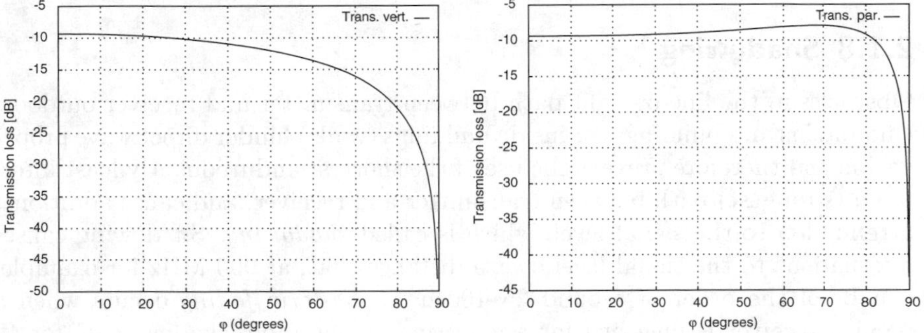

Figure 2.15: Concrete wall (wall thickness 150 mm), vertical polarization

Figure 2.16: Concrete wall (wall thickness 150 mm), horizontal polarization

it to reach an area that it could ordinarily only reach along a direct path through transmission.

The effect of diffraction becomes greater as the ratio of the wavelength to the dimension of the obstacle increases. Diffraction is negligible at frequencies above around 5 GHz.

2.1.7 RMS Delay Spread

The RMS (*root mean square*) delay spread describes the dispersion of a signal through multipath propagation and takes into account the time delays of all incoming paths with relation to the first path. The respective paths are weighted with their received level:

$$\tau_{rms} = \sqrt{\frac{1}{\sum_{i=1}^{n} P_i} \sum_{i=1}^{n} (\tau_i^2 P_i) - \tau_d^2}, \qquad \text{with} \quad \tau_d = \frac{\sum_{i=1}^{n} (\tau_i P_i)}{\sum_{i=1}^{n} P_i} \tag{2.20}$$

τ_{rms}	RMS delay spread	P_i	received level of path i
τ_i	time delay of ith path	n	# incoming paths

If the value of the RMS delay spread exceeds the tolerance limits of a system, it is assumed that error-free reception is no longer possible. When this happens, the waves travel over considerably different long paths, the levels of which are not negligible. If the resultant time dispersion of the signal is greater than the symbol duration during transmission then the receiver experiences intersymbol interference and bit errors.

2.1.8 Shadowing

Obstacles in the line-of-sight path between transmitter and receiver outdoors (mountains and buildings) or inside buildings (walls) hinder direct wave propagation and therefore prevent the use of the shortest and frequently least interfered (strongest) path between transmitter and receiver, and cause additional attenuation to the signal level, which is called *shadowing*. Shadowing causes fluctuations to the signal level over a distance that, at 900 MHz for example, can be of the order of around 25–100 m. *Long-term fading* occurs when a moving receiver is lingering for a long time in the radio shadow, e.g., for 10 to 40 s.

Measurements have revealed that the local mean value $m(t)$ in Equation (2.9) follows a lognormal distribution, i.e., $L_m = \log m(t)$ is normally distributed with a standard deviation of approximately 4 dB [21, 27]. This is also called *lognormal fading*. This approximation applies to statistics for large built-up areas.

2.1.9 Interference Caused by Other Systems

In addition to the interference caused by radio wave propagation, which has already been discussed, there is also secondary interference, such as the reciprocal effects of neighbouring radio systems on adjacent channels in the spectrum and electromagnetic impulses caused by other systems, car starters, generators and PCs—in other words *man-made noise*.

2.2 Models to Calculate the Radio Field

Reliable models for the calculation of expected signal levels are needed in the planning of radio networks, establishing of supply areas and siting of base stations. Data on terrain structure (topography) and buildings and vegetation (morphology) are required for these calculations.

Radio propagation in a mobile radio environment can be described on the basis of three components: long-term mean value, shadowing and short-term fading. The sum total of these components $L_P = L_l + L_m + L_s$ describes the resultant overall path loss between transmitter and receiver; see Figure 3.46.

Another factor to be considered is that mobile stations usually move at different speeds. The level, e.g., for determining GSM radio measurement data, is measured on a time-related basis, so that the level is also affected by the speed at which the mobile station is moving.

In the measurements by Okumura [25] the long-term mean value describes the level value averaged over a large physical area of 1–1.5 km. The effects of shadowing and short-term fading disappear through the averaging. This long-term mean value can be calculated using approximate models.

A description of the most common models used in calculating the mean value of the expected radio levels follows. A distinction is made between empirical models, which are based on measurement data, and theoretical models, which are based on the use of wave diffraction.

2.2.1 Empirical Models

The empirical approach is based on measurement data that when plotted as regression curves or analytical expressions can be used to calculate signal levels. The advantage of these models is that because of their measurement basis, they all take into account known and unknown factors of radio propagation. The disadvantage is that the models only cover certain frequencies and scenarios and sometimes have to be revalidated for other areas.

Reference [27] offers an overview of the different measurements and the models derived from them.

2.2.2 Diffraction Models

Diffraction theory can be used to obtain a description of radio propagation. In this case obstacles in uneven terrain are modelled as diffraction edges. A section of terrain of the line of sight, which can usually be obtained from a topographical database, is required for calculating diffraction loss. Figure 2.17 illustrates the principle for an edge.

For less steep forms of terrain, such as hills, cylinder diffraction can be used as a model to produce better results. All types of terrain must be represented using several diffraction edges. Many different methods are available for calculating the resultant diffraction loss (see the overview in [27]).

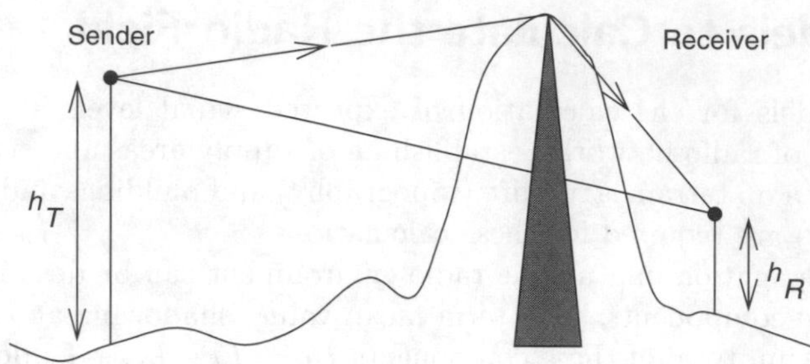

Figure 2.17: Obstacle in terrain as diffraction edge

Diffraction models have the advantage that they can be calculated without reference to any particular frequency or scenario, and consequently, in comparison with empirical models, can be used in a wider range of application (frequencies, distances). The disadvantages are that the accuracy of the calculation depends strongly on the accuracy of the topographical database and that the different approaches produce widely different results for terrains with several obstacles.

Because morphology plays an important role in the calculation of radio propagation, empirical correction factors are also required for the diffraction models. In practice, therefore, hybrid calculation methods are used with radio network planning tools.

2.2.3 Ray Tracing Techniques

The long-term mean value of a signal level can be calculated using empirical models and diffraction models. Some applications, such as the calculation of radio propagation in networks with microcells (<1 km radio range), require calculations with a more accurate resolution. The surroundings of the mobile and base station (such as the geometry of the buildings) must be taken into account if there is an interest in more than just the mean value over a large area.

There are empirical and theoretical approaches to carrying out calculations. Good results can be achieved using *ray tracing* techniques, which can, however, prove to be complicated for larger scenarios.

2.2.4 The Okumura/Hata Model

In 1962/63 and in 1965 Okumura carried out extensive measurements in and around Tokyo for the frequency range from 500 MHz to 2 GHz. The results of these measurements were published as regression curves [25]. To simplify radio field predictions, Hata linearized some of these curves and approximated them through analytical equations [12].

The basis for the calculation is an equation for the path loss in relatively flat terrain with city dwellings (see Section 2.2.4.1) and isotropic antennas:

$$\begin{aligned}\frac{L_P}{\text{dB}} &= 69.55 + 26.16 \log \frac{f}{\text{MHz}} - 13.82 \log \frac{h_T}{\text{m}} \\ &\quad - a(h_R) + \left(44.9 - 6.55 \log \frac{h_T}{\text{m}}\right) \log \frac{d}{\text{km}} \qquad (2.21)\end{aligned}$$

This equation applies to frequencies f from 150 to 1500 MHz, (effective) transmitting antenna heights h_T from 30 to 200 m and distances d from 1 to 20 km. $a(h_R)$ is a correction factor for the height of the receiving antenna h_R.

Based on a frequency of 900 MHz, a 30 m base station antenna height and a 1.5 m mobile station antenna height in a medium-sized city ($a(h_R) \approx 0$), Equation (2.21) gives

$$\frac{L_P}{\text{dB}} = 126.42 + 35.22 \log \frac{d}{\text{km}}$$

A comparison with free-space propagation (also see Section 2.1)

$$\frac{L_P}{\text{dB}} = 91.52 + 20 \log \frac{d}{\text{km}}$$

and with propagation over a flat area (Equation (2.7))

$$\frac{L_P}{\text{dB}} = 86.94 + 40 \log \frac{d}{\text{km}}$$

shows the path loss to be clearly higher than in these two theoretical models; with a value of 3.5, the propagation exponent is somewhat lower than it is with propagation over a flat area.

2.2.4.1 Terrain Profile

Okumura distinguishes between two types of terrain:

- *Quasi-flat terrain*
  Flat terrain with a maximum 20 m altitude difference.
- *Irregular terrain*
  All other irregular types of terrain, divided into:
  - *Rolling hilly terrain*
  - *Isolated mountain*
  - *General sloping terrain*
  - *Mixed land–sea path*

Diagrams with correction factors for calculating the radio propagation for each of these categories are provided.

2.2.4.2 Types of Morphology

Okumura refers to three types of morphology:

- *Open area*
  Open area without any major obstacles: farmland, etc.
- *Suburban area*
  City or town with houses and trees, low-density housing
- *Urban area*
  City with high rises or at least two-storey high buildings

The diagrams with the correction factors for the *open* and *suburban* morphology types in Okumura's work have been approximated into formulas by Hata. In each case the correction factors must be added to the basic path loss Equation (2.21):

Suburban

$$\frac{K_{\text{suburban}}}{\text{dB}} = -2\left(\log \frac{f}{28\text{MHz}}\right)^2 - 5.4$$

Open

$$\frac{K_{\text{open}}}{\text{dB}} = -4.78\left(\log \frac{f}{\text{MHz}}\right)^2 + 18.33 \log \frac{f}{\text{MHz}} - 40.94$$

The values for a frequency of 900 MHz are $K_{\text{suburban}} = -10$ dB, $K_{\text{open}} = -28.5$ dB, and for 1800 MHz they are $K_{\text{suburban}} = -12$ dB, $K_{\text{open}} = -32$ dB.

2.2.5 Radio Propagation in Microcells

Today's radio networks use hierarchical cell structures with small microcells below the level of conventional macrocells. This increases network capacity in areas with high traffic volumes (city centres, trade fairs, etc.).

Microcells cover areas extending up to several 100 m, with radio illumination greatly affected by the geometry of the base station surroundings. The Okumura/Hata model only lends itself to the calculation of the mean value for large areas and cell sizes with a radius of several kilometres. Other methods are required for calculating radio illumination in microcells [26, 36].

2.3 Cellular Systems

Traditional radio networks (1st-generation) that try to provide coverage to large areas by increasing the transmitting power of the individual base stations are only able to serve a limited number of subscribers because of the bandwidth used. In these radio networks an allocated radio channel is retained as long as possible, even if a receiver has already moved into another

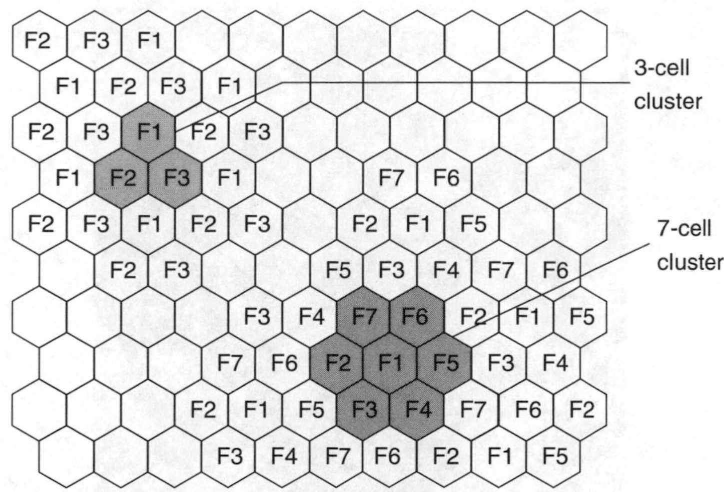

Figure 2.18: A cellular network supplying an extended area

supply area. Because the physical boundaries of a supply area are not precisely defined, neighbouring supply areas must use different radio channels in order to avoid interference. Where user density is high, this results in a heavy demand for frequencies, which, however, are restricted because of the scarcity of available spectrum.

The poor utilization of frequency spectrum in these radio networks and the increase in the number of mobile radio users, which these systems could no longer handle, led to the development of cellular networks.

Cellular networks are based on dividing the entire area over which a network is to be operated into radio cells, each of which is served by a base station. Each base station is only allowed to use a certain number of the total available frequency channels, which can only be reused after a sufficient interval to avoid any interference from neighbouring cells (see Figure 2.18). F_i $(i = 1, 2, \ldots, n)$ indicates the groups of frequency channels used in the respective cells.

In cellular networks the low transmitting power of the base stations enables them to use allocated frequencies only in a strictly defined area of the radio cell, thereby allowing these frequencies to be reused after a predetermined reuse distance.

Cells are generally represented as idealized regular hexagons, but because of topographical and environmental conditions, this is only an approximation of what actually occurs; see Figure 2.19. In reality radio cells are very irregular in their external shape and furthermore are designed to overlap with one another by approximately 10–15%. This enables mobile stations operating near the boundary of a cell to choose with which base station they set up a call.

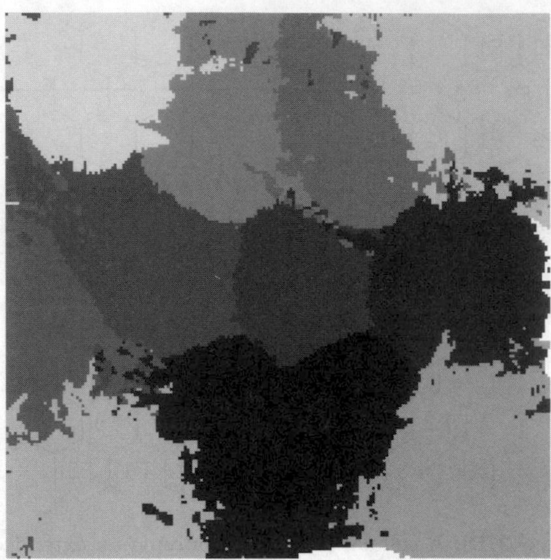

Figure 2.19: Simulated best server cells of a metropolitan scenario

2.3.1 Cluster Patterns and Carrier-to-Interference Ratio

Radio cells are combined into *clusters*, and each frequency is used once per cluster. Cells of a cluster neighbouring any other cluster may reuse all frequencies occupied in the cluster in a regular way and thereby repeat the cluster. Because the clusters must supply an entire area with coverage, only certain cluster patterns consisting of, e.g., 3, 4, 7, 9, 12 or 13 cells, are possible. The lower the number of cells per cluster, the greater the number of radio-frequency channels that can be used per cell for the entire mobile radio system in a given frequency band. However, frequencies reused within a short distance increase the co-channel interference across clusters. The carrier-to-interference ratio is the ratio of the received *carrier signal* C to the received *interference signal* I from the co-channels.

Figure 2.20 shows an example of frequency planning using a 3-cell cluster (frequency groups 1, 2, 3) and a regular hexagonal cell structure. This kind of uniform cluster structure always has six co-channel cells in close proximity to any base station. The interference usually originates from these cells; co-channel cells further away will be of less harm.

The distribution of interference level is caused by the overlapping of many independent sources of interference and therefore can be regarded as a normal distribution. On the downlink the base stations are the source of interference; on the uplink it is the mobile stations from the co-channel cells.

The number N of cells found in a cluster is called Cluster arrangement or also *Reuse* factor. The size of the cluster is limited towards the bottom by the C/I required. Networks with a small cluster size have a larger number

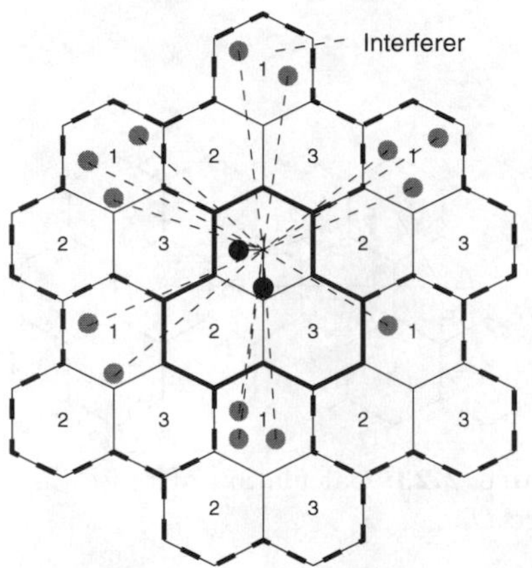

Figure 2.20: Interference by co-channel cells

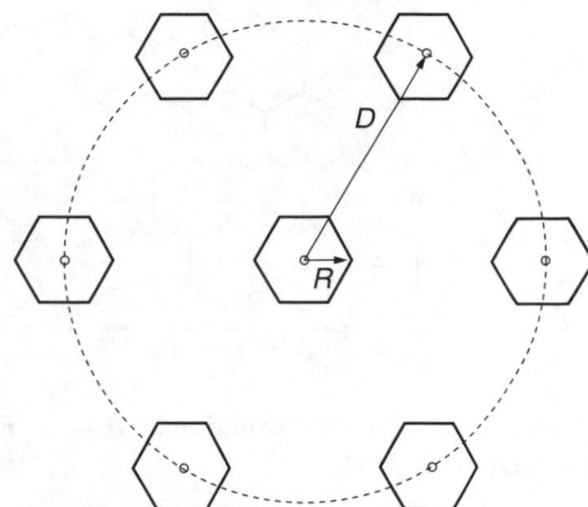

Figure 2.21: Co-channel cells in a cellular network

of frequencies per cell than those with large cluster sizes, i.e., the network capacity is greater.

In hexagonal systems the cluster size is calculated as

$$N = i^2 + ij + j^2 \quad i, j \varepsilon \mathbf{N_0} \tag{2.22}$$

The parameters i and j are determined by the steps that have to be counted in both coordinate directions in order to reach a co-channel cell. A hexagonal network with a cluster size 7 and parameters $i = 1$ and $j = 2$ is used in Figure 2.22 to clarify this connection.

Two cells that use the same subfrequency are called co-channel cells. The minimum distance between two co-channel cells is called reuse distance D and can be calculated as follows (see Figure 2.23):

$$D^2 = i(2h) + j^2(2h)^2 - 2ij(2h)^2 \cdot cos(120°) \tag{2.23}$$

$$D = \sqrt{3}\sqrt{i^2 + ij + j^2}R \tag{2.24}$$

$$= \sqrt{3N} \cdot R \tag{2.25}$$

2.3.2 C/I Ratio and Interference-Reduction Factor

Depending on the modulation technique employed and the technical equipment of the receiver, the smallest acceptable C/I ratio for a cellular network is based on the number of co-channel cells, the distance between these cells, the transmitting power and the terrain characteristics.

The C/I ratio required for cell planning is calculated below on the basis of an idealized concept for the model. The calculation considers a system such as the one in Figure 2.21. In the central cell with radius R a mobile station

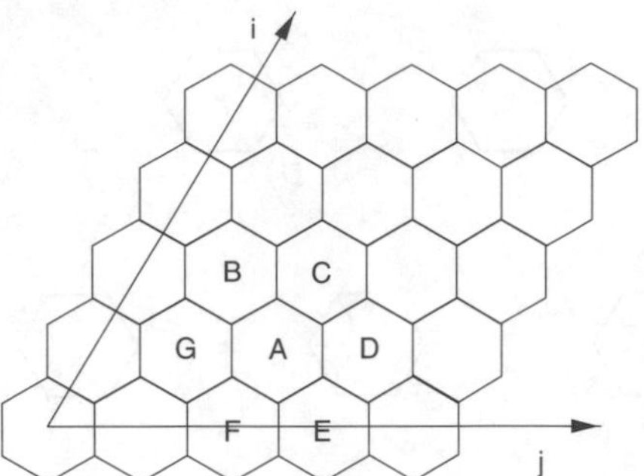

Figure 2.22: Cluster size and reuse distance $D(i = 1, j = 2)$

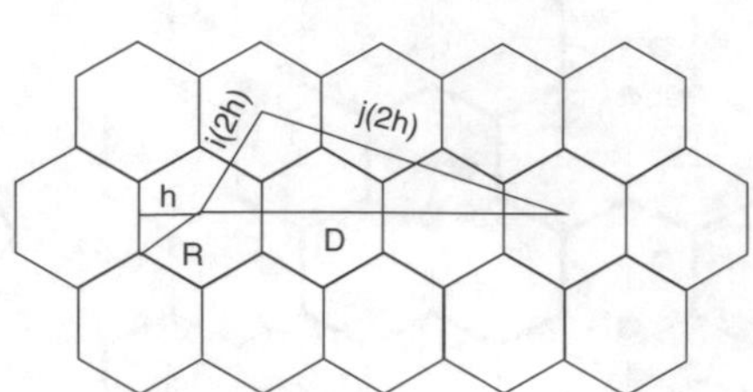

Figure 2.23: Calculation of reuse distance D

is in radio contact with its base station. The interference cells in which the mobile and base stations are transmitting on the same frequency and therefore can cause interference are arranged in a circle around the cell at distance D. Co-channel interference by cells further away can be ignored because of the larger associated path loss. On the assumption that statistically all mobile stations transmit independently of one another, the total interference power for a base station with K_I interfering mobile stations is calculated on the basis of a combination of the individual interference contribution I_k and the noise power N:

$$I = \sum_{k=1}^{K_I} I_k + N \tag{2.26}$$

The noise power N in the receiver is

$$N = FK\Theta v \tag{2.27}$$

F stands for the noise factor of the receiver (which is between 5 and 10), K for Boltzmann's constant ($1.38 \cdot 10^{-23}$ W·s/K), Θ for the temperature in K and v for the bit rate in bit/s. The ratios in the model produce a noise power of approximately $5.6 \cdot 10^{-12}$ mW, or a level of -112 dBm.

If one ignores the noise power in a homogeneous system in which all mobile stations are transmitting at the same power then the co-channel carrier-to-interference ratio only depends on where the mobile stations are located in relationship to one another:

$$\frac{C}{I} = \frac{R^{-\gamma}}{\sum_{k=1}^{K_I} D_k^{-\gamma}} \tag{2.28}$$

In Equation (2.28) D_k indicates the distance to the kth mobile station and γ indicates the path loss parameter for the receive power. Because $K_I = 6$

Table 2.2: Decrement factor and C/I on the downlink of different cluster sizes($\gamma = 4$), C/I calculated with a more refined version of Eq. (2.29)

Cluster size	3	4	7	9	12	13
D/R	3	3.464	4.58	5.20	6	6.24
C/I [dB]	11.3	13.8	18.66	20.85	23.34	24.04

cells are located at the same distance from the central cell in the model, and for purposes of simplification, all distances can be considered as equal, and it follows that

$$\frac{C}{I} = \frac{R^{-\gamma}}{6D^{-\gamma}} = \frac{1}{6}q^{\gamma} \tag{2.29}$$

The ratio $q = D/R$ is called the *co-channel interference reduction factor* and is

$$q = \frac{D}{R} = \left(6\frac{C}{I}\right)^{\frac{1}{\gamma}} \tag{2.30}$$

Aiming towards a co-channel carrier-to-interference ratio C/I of, e.g., 11 dB, using a realistic value of, e.g., $\gamma = 4$, produces a reduction factor of $q = 2.85$.

The next step in cell planning is the selection of a suitable cluster type to meet the requirement for a specific ratio of cell radius R to reuse distance D. If large clusters are used, the distance between the interfering cells will also be great; however, each cell will be able to carry less traffic because the trunk of frequency channels of the available band must be distributed among the clusters. Therefore, for maximum spectrum utilization efficiency, there has to be a balance between the number of cells per cluster and achievable transmission quality. See Table 2.2 for reduction factors and corresponding cluster sizes.

The minimum C/I ratio required for adequate connection quality depends on the channel access technique, the modulation technique and the coding selected. However, due to fading processes a more or less larger security distance is needed in addition to the minimum distance. Table 2.2 lists the scaled reuse distances (decrement factor) and co-channel carrier-to-interference ratios for the different cluster sizes calculated on the cell radius.

2.3.3 Traffic Load and Cell Radius

Another parameter in radio network planning is the size of the cell radii. By reducing the size of the cells, a system can be adapted to cope with a higher level of user density (see Figure 2.24). However, the system dictates the minimum possible size of a cell derived from cost/benefit considerations. In practice networks are usually initially set up with relatively large cell radii. If it turns out that this structure is no longer able to carry existing traffic, the cells with high traffic density are reduced in size through the process of

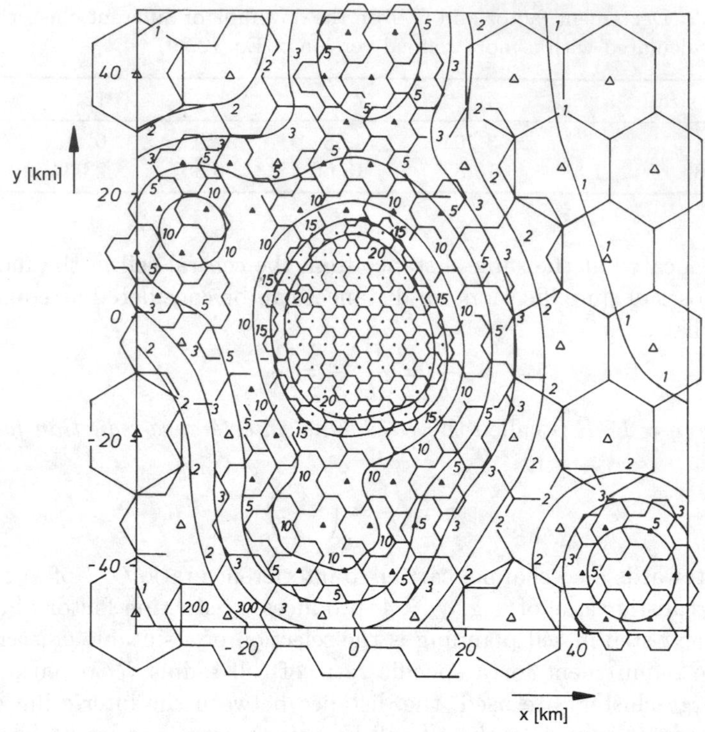

Figure 2.24: Adapting cell size to traffic density

cell splitting and equipped with additional base transceiver stations. This approach always provides an optimal solution in terms of network utilization, necessary signalling for handover, service quality for the user, as well as costs of the network infrastructure. The number of base station locations can be reduced by setting up three neighbouring cells, with the cells serving three 120° sectors (= cells) with different channel groups (also see Section 2.4).

The main advantage of cellular systems is that the same radio channels can be reused, and consequently coverage can be provided to areas of any size. The following requirements must be met to ensure that the system functions without any problems:

- Continual measurement of the field strength of the received signal ensures that a communicating mobile station is using the serving base station of the corresponding cell. The network is alerted as soon as a mobile station leaves a cell and the connection is automatically switched to the neighbouring base station. Cellular systems must therefore be capable of carrying out a change in radio channel as well as in base station during a connection. This process is called *handover* (see Section 3.6).

- A mobile radio network must be aware of which mobile radio users are currently roaming in its radio coverage area to enable it to page them if necessary. Mobile stations are therefore always assigned to a location area and reassigned to another area by the network when they change the service area. This service characteristic of cellular networks, which is called support of *roaming* (see Section 3.7.1), ensures that users can be reached at all times.

Complicated signalling protocols are required for operating cellular networks.

2.4 Sectorization and Spectral Efficiency

Cellular systems such as GSM are considered to be spectrum-efficient. Spectral efficiency relates the traffic capacity to frequency unit and surface element:

$$\text{Spectral efficiency}\left[\frac{\text{bit/s}}{\text{MHz}\cdot\text{km}^2}\right] = \frac{\text{Traffic [Erlang]}}{\text{Bandwidth}\times\text{Area}} \tag{2.31}$$

Traffic refers to all the traffic to be handled in a cellular system (i.e., the number of simultaneous voice and data connections), bandwidth is the total available channel capacity, and area refers to the total system area A [km^2].

The efficiency is dependent on the following parameters:

- Number of required radio channels per cell
- Cluster size or size of interference group

The sectorization of a given cluster reduces co-channel interference, because the power radiated backward from a directional base station antenna is very small and the number of interfering cells is reduced, i.e., the C/I ratio increases, owing to the directivity. On the other hand, the number of channels/sector drops and along with it the trunking gain.

2.4.1 Efficiency and Traffic Capacity

The following abbreviations are used in the next section:

V	average number of MS per km^2	n	number of channels per cell
t	traffic of a MS in Erlang	x	number of channels per sector
f_s	channel grid (30 kHz* in GSM)	S	number of sectors per cell
N	number of cells per cluster	A_s	sector area
Z	number of cells in system (A/A_h)	R	cell radius

*A GSM frequency channel is 200 kHz wide and carries 8 TDMA channels. With 25 kHz per TDMA traffic channel and an additional 12 % of channel capacity for signalling, the result is approximately 30 kHz/TDMA channel.

The area of a hexagon with side length R is

$$A_h = \frac{3\sqrt{3}}{2} R^2 \tag{2.32}$$

This produces the following areas for three or six-sector cells:

$$A_3 = \frac{\sqrt{3}}{2} R^2, \qquad A_6 = \frac{\sqrt{3}}{4} R^2 \tag{2.33}$$

2.4.1.1 Measures for the Spectrum Utilization Efficiency

Two measures of spectrum utilization efficiency are to distinguish:

- Efficiency of a modulation method, which is usually expressed by the quotient of transmission rate [bit/s] and frequency bandwidth [Hz] used:

$$\eta_{Mod} = \frac{\text{Transmission rate of a channel}}{\text{Bandwidth per channel used}} \left[\frac{\text{bit/s}}{\text{Hz}}\right] \tag{2.34}$$

- Efficiency of a cellular system seen from the transmission specialist's viewpoint, where the transmission rate [bit/s] of the radio channel is normalized to the system bandwidth:

$$\eta_{cell} = \frac{\text{Transmission rate} \times \text{No. of channels in system}}{\text{System bandwidth}} \left[\frac{\text{bit/s}}{\text{Hz}}\right] \tag{2.35}$$

In both cases a bit error ratio results that is characteristic for the modulation scheme and other receiver parameters selected. The two views are more or less identical.

The teletraffic engineer has a different view on spectrum utilization efficiency. He relies on additional parameters to calculate the traffic capacity per square kilometre for a given system concept under some quality of service constraints, and is then able to compare different cellular concepts.

According to Section 2.3.2, any calculations of the traffic capacity are possible only under the constraints of some C/I ratio. In [33] the spectrum utilization efficiency is claimed to be best characterized by calculation of the local traffic intensity per bandwidth unit; see Equation (2.31). The traffic intensity is exactly one Erlang if the radio channel is continuously being used. A channel being used up to $x\,\%$ only is carrying a traffic of $x/100$ Erlang. From this, the spectrum utilization efficiency can be defined as follows:

$$\eta = \frac{\text{No. of channels/Cell} \times \text{Offered traffic/Channel}}{\text{System bandwidth} \times \text{Area of a cell}} \left[\frac{\text{Erlang}}{\text{kHz} \cdot \text{km}^2}\right] \tag{2.36}$$

Here the system bandwidth is the product of the cluster size and the bandwidth per cell. The offered traffic differs from the carried traffic by the amount of traffic lost according to blocking of the system; see Appendix A.1.2.

Using the abbreviations introduced above, the spectrum efficiency for systems with sectorized and non-sectorized antennas can be defined, see [6]:

- With omnidirectional antennas, the maximum traffic capacity is the result of the traffic of a cell multiplied by the number of cells in the system:

$$T_{max} = VtA_hZ \tag{2.37}$$

The efficiency is then

$$\eta_{Omni} = \frac{VtA_hZ}{2f_snNA} = \frac{Vt}{2f_snN}\left[\frac{\text{Erlang}}{\text{km}^2 \cdot \text{kHz}}\right] \tag{2.38}$$

- With sector antennas, the maximum traffic capacity is calculated as

$$T_{max} = VtA_sSZ \tag{2.39}$$

The efficiency is then

$$\eta_{dir} = \frac{Vt}{2f_s(xS)N} \tag{2.40}$$

2.4.2 The Effect of Sectorization with a Given Cluster Size

Figures 2.25–2.27 show different clusters with 3-site and 6-site sectorization. The effects of sectorization can be examined on the basis of the following factors:

- Only the first ring of co-channel interference cells M is taken into account (see Figure 2.28). The effect of the second ring of interference cells is negligible.
- The propagation coefficient or attenuation factor is approximately $\gamma = 4$, so the received power is $P_R \sim P_T/R^4$.
- The interference reduction factor, per Equation (2.30), is $q = 4.6$ for $N = 7$ (see the table in Section 2.3.2).

According to Figure 2.28, the relationship between the number of sectors and the number of interference cells for a 7-sector cluster is

$S \hat{=}$ number of sectors	$M \hat{=}$ number of co-channel cells
1 (omni)	6
3	2
6	1

The same applies to other cluster sizes (see Section 2.3). Analogously to Equation (2.29), the C/I ratio for different clusters is:

$$C/I = \frac{q^4}{M} = \frac{(3N)^2}{M} \tag{2.41}$$

Without the simplifications made in Equation (2.29) C/I ratio for the sectorizations being considered is [20]

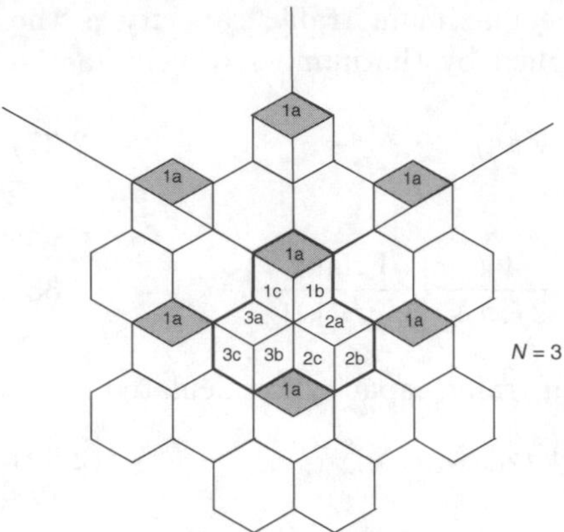

Figure 2.25: 3-sector cluster with 3-site sectorization

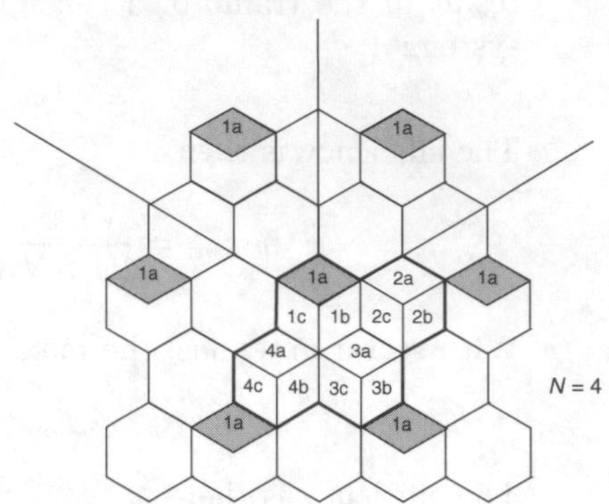

Figure 2.26: 4-sector cluster with 3-site sectorization

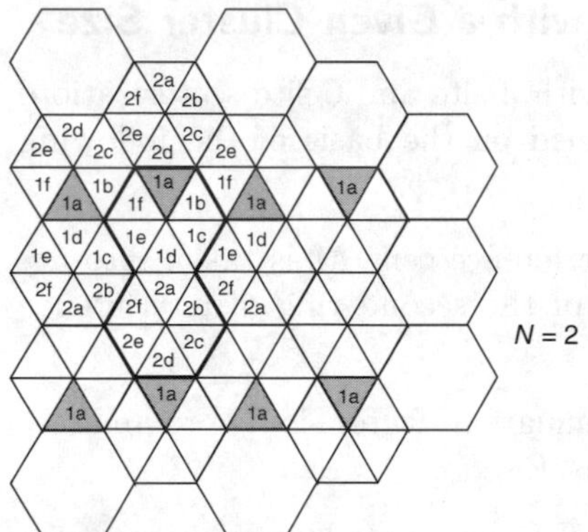

Figure 2.27: 2-sector cluster with 6-site sectorization

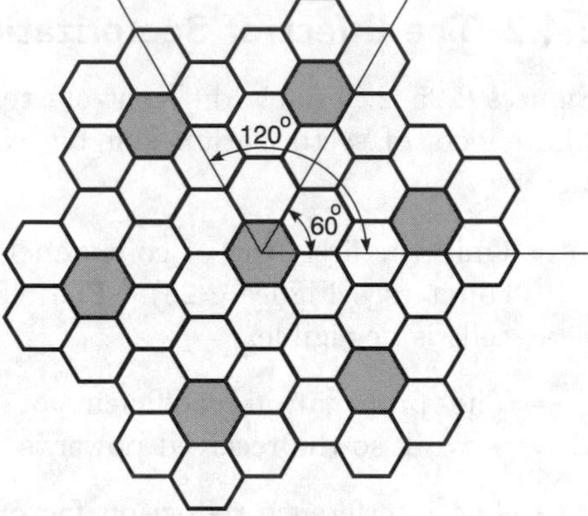

Figure 2.28: Distribution of co-channel interference cells

$N = 7$	Omni	3-sector	6-sector
C/I [dB]	17	25.3	29

The spectral efficiency can therefore be calculated from Equations (2.38) and (2.40). Figure 2.29 shows the spectral efficiency over the cell radius for a traffic value of $Vt = 0.2$ Erlang/km^2, which corresponds to a traffic density of 10 MS/km^2 with a traffic volume of 0.02 Erlang per MS (in other words, each MS talks an average of 1.2 minutes per hour on the mobile telephone).

Figure 2.30 reveals a negative percentage increase in efficiency for sectorized systems found for the cell radii shown. According to Appendix A.1.2, the traffic capacity per cell $Y = f(n, p_l)$ can be calculated analytically with given loss probability p_l:

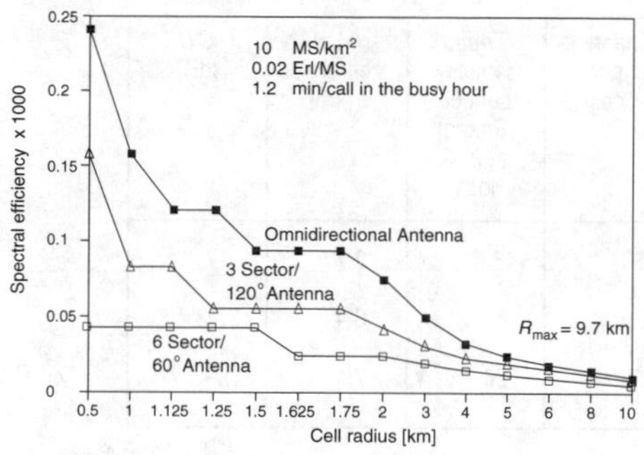

Figure 2.29: Spectral efficiency for $Vt = 0.2$ Erlang

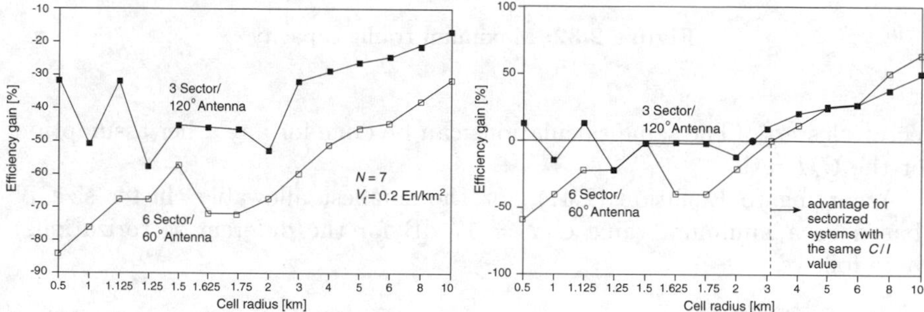

Figure 2.30: Efficiency gain with sectorized systems

Figure 2.31: Efficiency gain with unchanged/reduced clusters

$$Y = f(n, p_l) = VtA_sSp_l, \qquad \text{whereby } n = f(Vt, S, A_s)$$

For example, with a loss probability of 2 % it is possible to calculate the overall maximum traffic capacity for a system with a 2·12.5 MHz bandwidth or with 416 duplex channels; see Figure 2.32. The maximum cell radius for a traffic load per user of 0.2 Erlang/km^2 is also given.

In a 3-sector cell $n = 19$ channels result per sector, for which at $p_l = 2$ % a channel utilization of $\rho = 0.65$ can be determined from Appendix A.2.2, Figure A.7. The traffic capacity per cell (3 sectors) is then $\rho n \cdot 3 = 37$ Erl.

It is noticeable that sectorization in a given cluster size reduces spectral efficiency and maximum traffic capacity because of a decrease in trunking gain; the main advantage of sectorization is an improved C/I ratio!

2.4.3 Efficiency and Traffic Capacity with Sectorization and a Well-Chosen Cluster Size

Efficiency can be increased at the same time that sectorization occurs if the sectorized cluster is reduced in size but the C/I ratio is not allowed to drop below the value of, say 17 dB, achieved with omnidirectional antennas and 7-

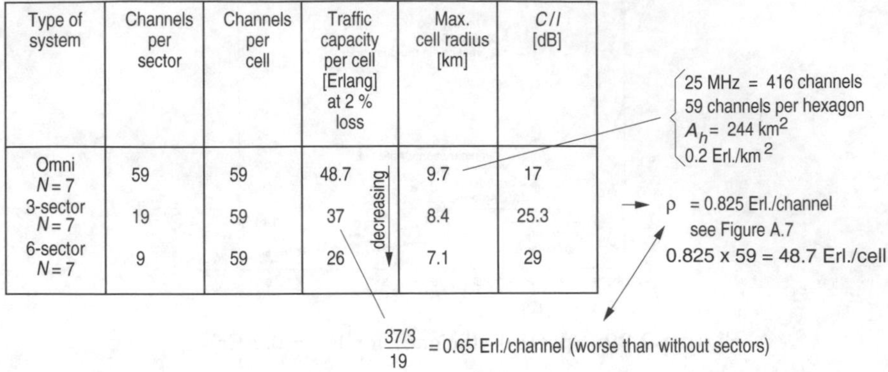

Type of system	Channels per sector	Channels per cell	Traffic capacity per cell [Erlang] at 2 % loss	Max. cell radius [km]	C/I [dB]
Omni N = 7	59	59	48.7	9.7	17
3-sector N = 7	19	59	37	8.4	25.3
6-sector N = 7	9	59	26	7.1	29

Figure 2.32: Maximum traffic capacity

sector clusters. (The same calculations can be done for any other assumption for the C/I ratio.)

According to Equation (2.41), for the smallest allowable cluster size N (based on a minimum value $C/I = 17$ dB for the different sectorizations) there follows:

- Without sectorization, the cluster size must be $N = 7$, so that $C/I = 17$ dB.
- For $S = 3$ ($M = 2$), $N = 4$ is allowable, with $C/I = 18.5$ dB resulting.
- For $S = 6$ ($M = 1$), $N = 3$ is allowable, with $C/I = 19$ dB resulting.

The spectral efficiency described in Equation (2.40) produces the result shown in Figure 2.31:

- From a cell radius of $R = 3$ km (at 0.2 Erl./km^2 traffic volume), sectorized systems offer advantages over omnidirectional systems.
- From a 6 km radius, six sectors tend to be better than three sectors.
- With higher traffic volumes, 3-sector antennas have an advantage with cell radii from:

Vt [Erlang/km^2]	0.5	1
R [km]	2	1.7

The values for maximum traffic capacity clearly indicate the advantages of sectorized systems (see the first table in Figure 2.33). Despite the loss in trunking gain, sector systems prove to be superior to non-sectorized systems when a comparison is made on the basis of equal values for the C/I ratio.

Maximum cell radius at 0.2 Erl./km^2 and various sectorization

7-sector Cluster: with $p_I = 0.02$ (without shadowing)

Type of system	Channels per sector	Channels per cell	Traffic capacity per cell [Erlang]	Max. cell radius [km]	C/I [dB]
Omni $N=7$	59	59	48.7	9.7	17
3-sector $N=4$	34	104	76.6	12.1	18.5
6-sector $N=3$	23	138	94.6	13.5	19

(traffic capacity: increasing ↓)

27-sector Cluster: with $p_I = 0.02$ (with a 13 dB shadowing reserve)

Type of system	Channels per sector	Channels per cell	Traffic capacity per cell [Erlang]	Max. cell radius [km]
Omni $N=27$	-	15	9.0 ↓	4.2
3-sector $N=16$	8	26	10.9	4.5
6-sector $N=12$	5	34	9.9 ↑	4.4

Necessary cluster size with a 13 dB shadowing reserve

Type of system	Interfering channels M	Reuse distance ratio q	Cluster size N	C/I [dB]
Omni	6	8.7	27	30.4
3-sector	2	6.8	16	30.6
6-sector	1	5.69	12	31.1

Figure 2.33: Traffic capacity with shadowing reserve

2.4.4 Sectorization with Shadowing

The necessary C/I values can be maintained in spite of radio shadowing if there is a large overlapping of cells and larger clusters with, e.g., $N = 12$, i.e., the interference reduction factor $q = 5.6$. The cluster size with sectorization shown at the bottom of Figure 2.33 is necessary in order to achieve a C/I of 17 dB with an interference probability of 0.1 for $\sigma = 6$ dB (σ is the standard deviation of the lognormal distribution).

The percentage of efficiency increase with sectorization for these systems can be displayed in a similar way as in Figure 2.31 [22]. This shows that at 0.2 Erl./km^2, even with a 13 dB shadowing reserve, sectorized systems are better than omnidirectional ones starting from a radius of approximately $R = 3$ km. At 0.5 or 1 Erl./km^2, even sector systems with a 2 or 1.7 km cell radius are better than omnidirectional systems. The traffic capacity per cell follows from Figure 2.33 (table on right), calculated on the same basis as before.

2.5 The ISO/OSI Reference Model

Since the exchange of information between communicating partners is complex in structure and difficult to understand, the entire communications process has been universally standardized and organized into individual well-defined hierarchical *layers*.

Each layer, with the exception of the top one, offers *services* to the layer directly above it. The way these services are implemented is through the passing of information between the *peer entities* of the respective layer of the communicating systems by means of *protocols*. In this process a layer

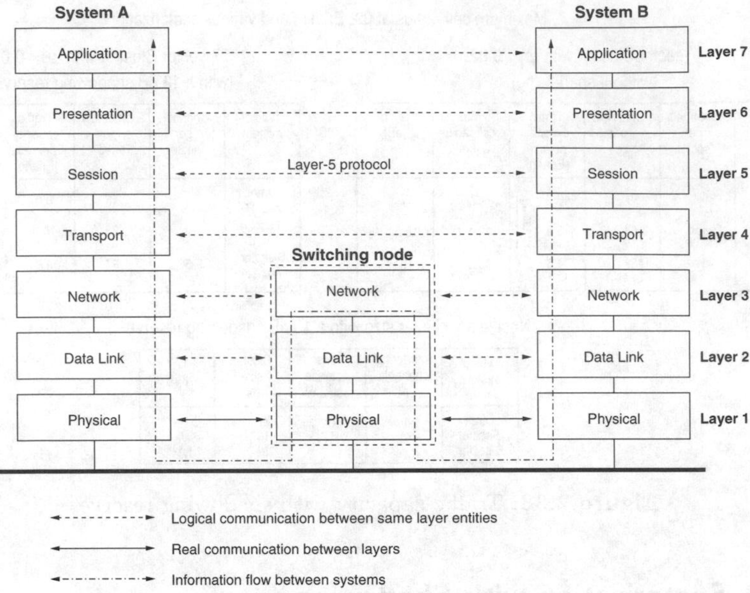

Figure 2.34: The ISO/OSI Reference Model

uses the services of the next layer below it. Therefore within a process each entity communicates directly only with the entity immediately above it or below it. The higher-ranking layer is referred to as the service user and the lower-ranking layer as the service provider.

The hierarchical model facilitates communication between developers, suppliers and users of communications systems. If a change is undertaken in one of the layers, it does not affect the others. Furthermore, the structure of the layers makes it easier for protocols to be implemented and standardized.

Taking these aspects into account, the International Standardization Organization (ISO) specified a generally accepted layered model, the *ISO/OSI Reference Model*, for *Open Systems Interconnection*, OSI, a description that refers to almost all the communications systems in use today. This model is called OSI, because it describes the connection of open digital systems compliant with the respective ISO standards.

The OSI model is based on different principles. Each layer carries out a precisely defined function, and each function has been stipulated in line with internationally standardized protocols. The boundary lines between the individual layers have been established to minimize the information flow over the interfaces. Each higher layer represents a new level of abstraction from the layers below it. To keep the number of layers and interfaces to a minimum, several different functions have also been added to the same layer. A seven-layer model (see Figure 2.34) was created as a result of these considerations.

The following is a brief description of the different tasks of the seven layers of the OSI Reference Model [13]:

Physical layer: Layer 1 The bit transmission layer (*physical layer*) provides the basis for communication and facilitates the transmission of bits over a communications medium. Layer 1 describes the electrical and mechanical characteristics, e.g., standardized plugs, synchronized transmission over cable or radio channels, synchronizing techniques, signal coding, and signal levels for the interface between terminal equipment and line termination.

Data link layer: Layer 2 The task of the *data link layer* is to interpret the bit stream of layer 1 as a sequence of data blocks and to forward them error-free to the network layer. Error-detection or correction codes are used to protect data from transmission errors. Thus, for example, systematic redundancy that is used at the receiving side for error detection is added by the transmitter to the data, which is transmitted in blocks (frames).

These frames are transmitted sequentially between peer entities of layer 2. If a transmission error is detected then an acknowledgement mechanism initiates a retransmission of the block and guarantees that the sequence will be maintained.

The data link layer adds special bit patterns to the start and to the end of blocks to ensure they are recognized. Because of flow control on both sides, the logical channel can be used individually by the communicating partner entities.

Layer 2 contains the access protocol for the medium and the functions for call set-up and termination with regard to the operated link.

Network layer: Layer 3 The *network layer* is responsible for the setting up, operation and termination of network connections between open systems. In particular, this includes *routing* and address interpretation, and optimal path selection when a connection is established or during a connection.

Layer 3 also has the task of multiplexing connections onto the channels of the individual subnets between the network nodes.

Transport layer: Layer 4 The *transport layer* has responsibility for end-to-end data transport. It controls the beginning and the end of a data communication, carries out the segmentation and reassembly of messages, and controls data flow. Error handling and data security, coordination between logical and physical equipment addresses and optimization of information transport paths also fall within the range of this layer's tasks.

The transport layer represents the connecting link between the network-dependent layers 1–3 and the totally network-independent overlaid layers 5–7, and provides the higher layers with a network-independent interface. The transport layer provides a service with a given quality to the communicating applications processes, regardless of the type of network used.

Session layer: Layer 5 The *session layer* controls communication between participating terminals, and contains functions for exchange of terminal identification, establishing the form of data exchange, dialogue management, tariff accounting and notification, resetting to an initialized logical checkpoint after dialogue errors have occurred, and dialogue synchronization.

Presentation layer: Layer 6 The *presentation layer* offers services to the application layer that transform data structures into a standard format for transmission agreed upon and recognized by all partners.

It also provides services such as data compression as well as encryption to increase the confidentiality and authenticity of data.

Application layer: Layer 7 The *application layer* forms the interface to the user or an applications process needing communications support. It contains standard services for supporting data transmission between user processes (e.g. file transfer), providing distributed database access, allowing a process to be run on different computers, and controlling and managing distributed systems.

2.6 Allocation of Radio Channels

Utilization of the capacity of a transmission medium can be improved through different methods that involve transmitting several connections simultaneously in multiplex mode. Multiplexing is a technique permitting multiple use of the transmission capacity of a medium. The following techniques are used in radio systems:

- *Frequency-division multiplexing* (FDM)
- *Time-division multiplexing* (TDM)
- *Code-division multiplexing* (CDM)
- *Space-division multiplexing* (SDM)

In addition to these multiplexing methods, which facilitate the multiple use of the capacity of a transmission medium by many communications channels, there are access techniques to the respective frequency, time, code and space channels, which are abbreviated as follows:

- *Frequency-division multiple access* (FDMA)
- *Time-division multiple access* (TDMA)

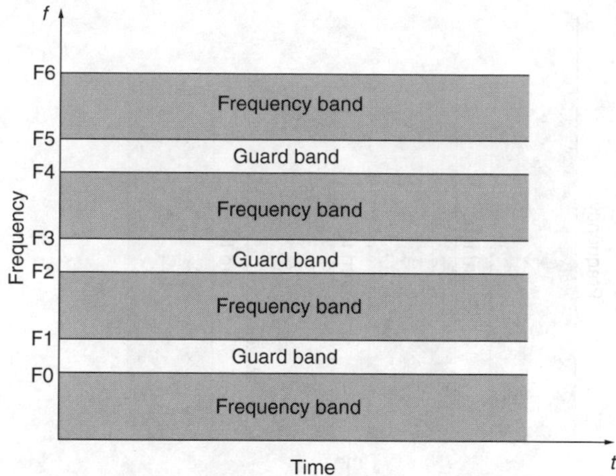

Figure 2.35: Frequency-division multiplexing technique, FDM

- *Code-division multiple access* (CDMA)
- *Space-division multiple access* (SDMA)

As layer 2 protocols based on the ISO/OSI Reference Model, these access techniques are only indicators of the respective class of protocols, and are specified in each individual case (for each system).

2.6.1 Frequency-Division Multiplexing, FDM

With frequency-division multiplexing, the spectrum available to the radio system is divided into several frequency bands that can be used simultaneously (see Figure 2.35).

Each frequency band is regarded as a physical channel that is allocated to two or more stations for communication. Each station is able to send or receive using the entire available transmission rate of the frequency band.

The frequency spectrum is divided into frequency bands through the modulation of different carrier frequencies when messages are being transmitted. On the receiving end the signals are separated through appropriate filtering. Because real filters have a finite slope characteristic, *guard bands* are necessary to prevent interference (cross-talk).

Therefore full utilization of the available frequency band is not possible. Many public land mobile radio systems, such as the GSM network and trunked radio systems, use frequency-division multiplexing techniques.

2.6.2 Time-Division Multiplexing, TDM

An FDM channel can sometimes offer more capacity than is required for a communications link. Periodically the frequency channel can then be allocated

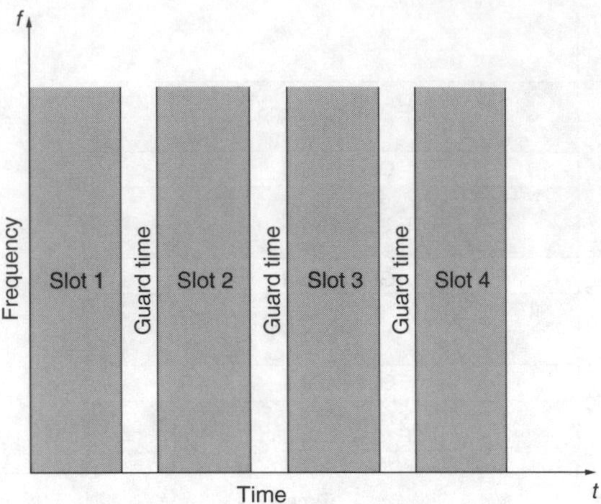

Figure 2.36: Time-division multiplexing, TDM

alternately to different communications links. This is the concept behind time-division multiplexing, in which the entire bandwidth of a radio channel is used but is divided into *time slots* that are periodically allocated to each station for the duration of a call (see Figure 2.36). The transmitting station can accommodate a certain number of data bits in a slot. The sequence of slots used by a station forms a time channel.

In some applications this regular allocation of time slots to stations, which results in constant engagement of the transmission medium, can be a disadvantage, especially when long transmission pauses occur. When this happens, the time slots are allocated centrally or decentrally to individual users as required. Instead of synchronous, asynchronous time slots are then allocated.

Access to a transmission medium using the TDM technique requires a multiplexer and, on the receiver's side, a demultiplexer, which must be capable of working together in perfect synchronization to enable transmitted messages to be allocated to the right time channels. Similarly to FDM, TDM systems must also have a *guard time*, in this case between the individual slots, to prevent synchronization errors and intersymbol interference resulting from signal propagation time differences. This guard time prevents the use of very short time slots and therefore reduces the utilization of the theoretically available capacity. Further, a delay is introduced to collect the data to be transmitted in a time slot. The radio modems must transmit in a burst mode now, since the frequency channel is available only during time slots with intermittent silence periods, and time compression of the message transmitted is necessary.

Although the TDM method is more frequency-economic than FDM, it requires very precise synchronization between the communicating parties, and therefore is technically more complex than FDM. Most of the digitally transmitting mobile radio systems also use TDM techniques in addition to FDM.

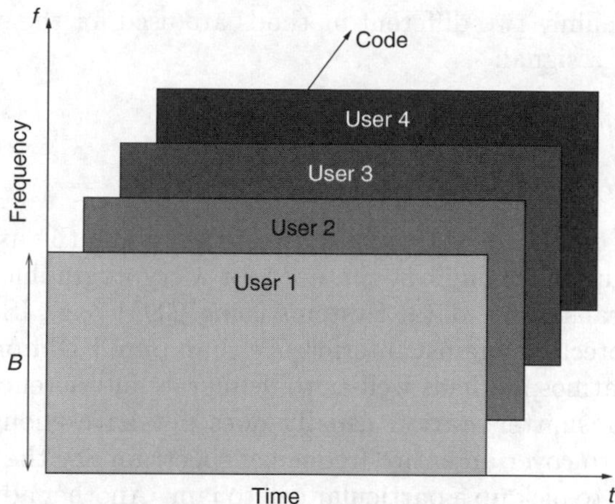

Figure 2.37: Code-division multiplexing, CDM

2.6.3 Code-Division Multiplexing, CDM

Code-division multiplexing separates the transmission channels neither into frequency areas nor into time slots. What is characteristic about this method is that it transmits narrowband signals in a wide frequency spectrum in which the narrowband signal is spread to a broadband signal through the use of a unique user code. This is called code spreading (see Figure 2.37). By this method a communications channel is defined by an unique code.

Each user in the radio communications systems is allocated a unique user code, which is used to spread the signal spectrum to be transmitted into multiples of the original bandwidth. These signals are then transmitted by the transmitters of the communicating terminals at the same time in the same frequency band, typically to/from the same base station. The user codes used by the transmitters must be selected in such a way that the receivers experience minimal interference despite the fact that transmission is taking place simultaneously. The use of an orthogonal *pseudo-noise* (PN) code for carrier modulation of the information being transmitted fulfils this condition.

The receiver, which must know the transmitter's user code, searches the broadband signal for the bit pattern of the transmitter's PN sequence. Through an autocorrelation function (ACF), the receiver is able to synchronize with the transmitter's code channel and de-spread the signal back to its original bandwidth. The signals of the other transmitters, the codes of which are different from the PN sequence selected, are not de-spread by this procedure, and therefore only contribute to the noise level of the received signal.

With a certain number of code channels on the same frequency channel the (*signal-to-noise ratio*, SNR) can fall below the value required for reception by the correlator. The CDM method thereby restricts the number of users who can use the same frequency channel.

In practice mainly two different methods are used for the spread spectrum transmission of a signal:

- *Direct sequencing* (DS)
- *Frequency hopping* (FH)

An advantage of CDM is that because of the coding the user data remains confidential, thus cancelling out the need for a cryptographic method for the protection of transmitted data. Systems using CDM (e.g., IS-95 and UMTS) have better protection against interference than pure FDM or TDM systems: this applies to atmospheric as well as to deliberate interference to the system. A jamming transmitter station usually does not have enough transmission power in order to cover an entire frequency spectrum nor the necessary information in order to pick up a particular call to jam. Another advantage over the TDM method is that the different transmitters in a CDM system do not need to synchronize their time. Because of the codes, they are self-synchronizing.

A system-inherent disadvantage of CDM is that transmitters and receivers must generate pseudo-random code sequences. When several stations are transmitting simultaneously (this is the normal way of operation), random statistical superimposition can occur, causing errors and creating the need for error-detection and correction measures. Furthermore, each receiver must utilize fast power control to prevent transmitters with strong signals from causing interference to the signals of weaker transmitters.

2.6.3.1 Direct Sequencing

Direct sequencing is a spreading technique in which the binary signals to be transmitted are added modulo two to the binary output signal of a pseudo-noise generator and then used to phase-modulate the carrier signals. A combination of the data bits with the pseudo-random bit sequence (*chip sequence*) converts the narrowband information signal to the large bandwidth of the PN signal, thereby producing a code channel [30]. The PN signal can use a Barker, Walsh or Gold code.

2.6.3.2 Frequency Hopping

In *frequency hopping* transmitters and receivers change transmission frequency synchronously and in quick succession (see Figure 2.38). The already-modulated information signal is added modulo two to the signal of a code generator, which controls a frequency synthesizer. This causes a considerable widening of the original bandwidth. Frequency hopping takes place either quickly (many *hops* per information bit) or slowly (one *hop* for a large number of information bits).

It is possible for several transmissions to take place at the same time in the available frequency range, but this can lead to collisions if two or more transmitters happen to be using the same frequency simultaneously. These

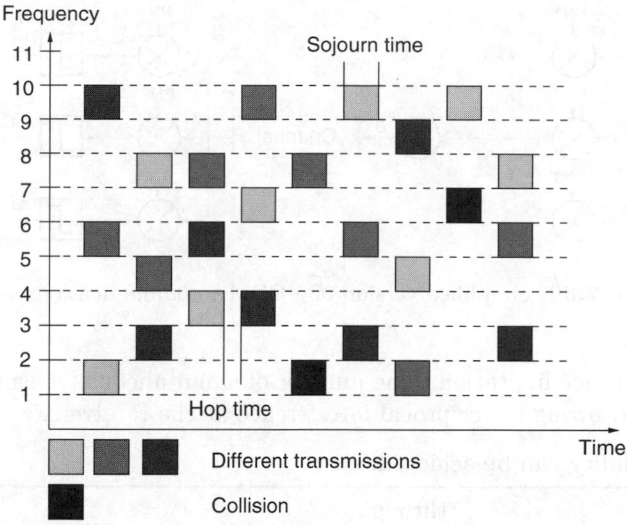

Figure 2.38: Frequency hopping spread spectrum technique

collisions, which can reduce the carrier to interference ratio to an unacceptable low level, are preventable through the use of orthogonal hopping codes.

The advantage of this method is that interference in the transmission channel, e.g., caused by co-channel interference, will usually only affect a small part of a message. The defective part of the communication can, if necessary, be compensated for on the receiving side through error-protection methods.

2.6.4 CDMA Technique for 2nd-Generation PLMNs

Since the early 1980s, CDMA has been used in experiments involving wireless communication in buildings and outdoors as well as over satellites [3, 30].

This technology has generated special interest since early 1991 because of an announcement by QUALCOMM USA in San Diego about a cellular CDMA system claiming to offer a particularly high level of spectral efficiency in cellular radio technology. Telecommunications Industry Association (TIA) Interim Standard 95 (IS-95) has been produced since then (see Section 5.3).

Its characteristics are:

- A frequency band B for all code channels.
- A large number of simultaneous digital signals in the same band B.
- Use of (almost) orthogonal signal sequences.
- Each signal sequence in a channel can be detected and decoded in the receiver if the code sequence is known.

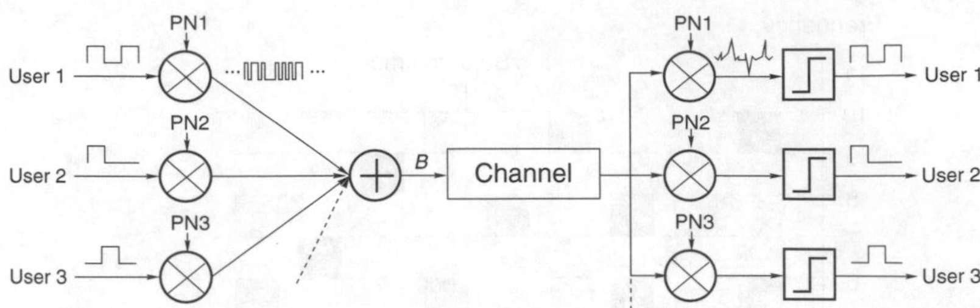

Figure 2.39: Simplified version of a CDMA communications system

- Interference limitation: The number of simultaneous transmitters is restricted owing to reciprocal interference at the receivers.

Orthogonality can be achieved:

with	through
TDM	Signal separation in the time domain
FDM	Signal transmission in different frequency bands
CDM	Channel-specific code for signal sequences

The bandwidth B of a user's CDMA signal is considerably wider than his channel bit rate R. The larger B/R is, the greater the number of simultaneous code channels in the same band B. The performance per bandwidth unit is reduced through spreading.

Example: If a 1 MHz signal is spread 100 times to $B = 100$ MHz and is transmitted with 1 W power then

- a receiver with a bandwidth of 1 MHz only receives 10 mW of interference power (attenuated because of the distance).
- a transmitter with a 1 MHz bandwidth causes interference to a receiver with a 100 MHz bandwidth with only 1/100 of its power.

Figure 2.39 shows a diagram of a CDMA communications system (without a modulator or a demodulator):

- User signal sequences are multiplied by individual spreading sequences. PN_1, PN_2, ... and transmitted over the same band.
- The signal sequences with errors are multiplied by the channel-specific codes in the (correlation) receiver and restored by an equalizer
- The overlaid signals of multiple users resembles Gaussian noise.

CDMA signals are not sensitive to narrowband interference. The method claims to offer spectral efficiency several times greater than what is possible with FDMA/TDMA.

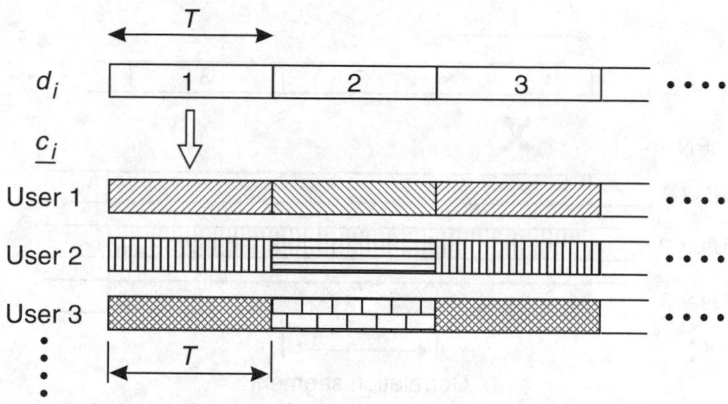

Figure 2.40: Orthogonal spreading of symbols

Problem with CDMA: All signal sequences received by the receiver must be of the same strength within approximately 1 dB, otherwise stronger signals will suppress weaker signals (*near/far problem*). An adaptive (fast) *power control* for the transmitter is required by the receiver.

2.6.4.1 Different Popular CDMA Techniques

Orthogonal spreading of symbols The data symbol d_i with the duration T of each user is individually mapped onto a set of orthogonal code sequences each of length T [3]:

$$\underline{c_k} = \{c_{k1}, c_{k2}, \ldots, c_{kL}\}, \tag{2.42}$$

The duration of each chip c_{kj} in the sequence is T/L, with L indicating the length of the code sequence (see Figure 2.40).

If data symbols for the basis m (m-ary) are used and purely orthogonal coding of the symbols is applied then an individual quantity of m orthogonal code sequences $\underline{c_k}$ is required for each user (channel).

The main advantages of symbol-by-symbol code spreading are as follows. The orthogonality and cross-correlation (i.e., channel interference) of the code sequences depend strongly on the code selected and therefore are controllable. If transmission and the reception of symbols of different channels occur simultaneously, e.g., within a cell, then capacity (channels/bandwidth) is increased substantially (i.e., the orthogonality is fully effective).

Initially, there was no satisfactory solution for establishing extensive code sequences with good cross-correlation properties on the basis of limited code lengths L, as required for cellular networks. Recently it has been shown that there exist large code families with the codes being nearly perfectly orthogonal to each other [31].

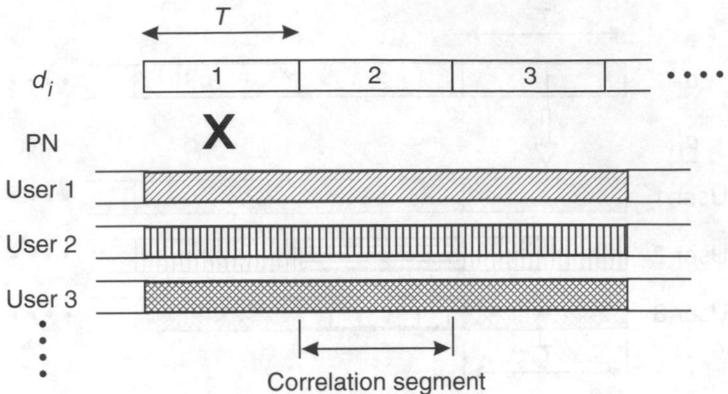

Figure 2.41: Code spreading based on the DS method

Spreading based on the direct sequencing (DS) method This involves the DS multiplication of each user's data signals with the user-specific PN binary chip sequence having a period clearly longer than that of the data symbol T (see Figure 2.41). The chip length is then $T/L \ll T$.

Figure 2.42 shows the effect of code spreading in the baseband. The interference caused by spread signals (*chips*) during transmission can be detected and suppressed after the despreading.

Advantage: The selection of code sequences and their allocation to users is greatly simplified (e.g., IS-95 uplink: a common PN sequence of length $2^{42} - 1$ for all users in all cells, with a user-specific starting sequence taken from this PN sequence; see Section 5.3.1).

Disadvantage: The selection of code sequence has little influence on the cross-correlation of the different code channels. *Reason*: Random cross-correlation of spread code segments of symbol length T, which only amount to a small part of the whole PN sequence.

Hybrid methods A PN sequence is symbolically multiplied by a Walsh-function (with period T) that is user-specific. A PN sequence plus a complete set of orthogonal Walsh functions produce a finite set of periodic DS spreading functions with guaranteed orthogonality with regard to cross-correlation during the symbol length T (see Figure 2.43). This method is applied by IS-95 on the downlink for each cell, see Section 5.3.2.

Frequency hopping techniques This is a code spreading system in which transmission is spread over different frequency bands, which are used sequentially: *fast frequency hopping* with many hops per symbol (bit).

Advantage: The system is robust with respect to different receiving levels—better than DS-CDMA.

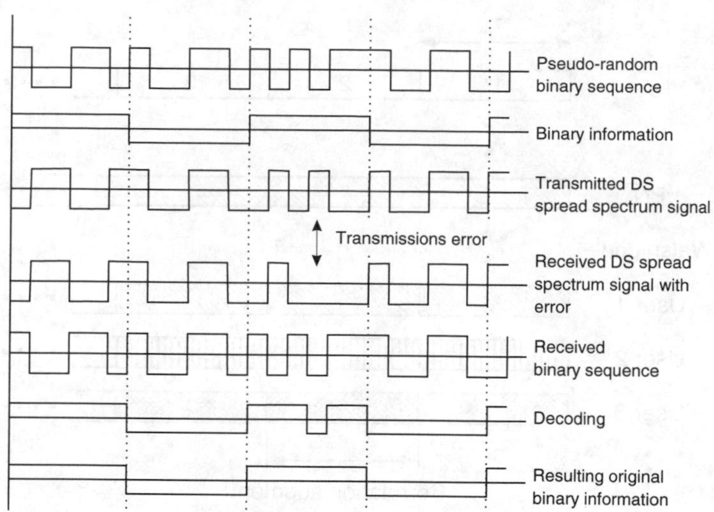

Figure 2.42: Example of DS spreading method

Disadvantage: The code allocation is complicated when there are a large number of users per cell. The technical complexity of frequency changing of up to 1 Mhop/s is required, and the complexity of the FH receiver should also be mentioned.

Example: In GSM (see Section 3.3.2), *slow frequency hopping* is used to avoid the consequences of signal fading on a small number of frequencies of the hopping sequence. The hopping frequency has 1 hop/burst (burst$\hat{=}$157 bit $\hat{=}$4.6 ms) $\rightarrow$ 217 hop/s.

CDMA combined with FDMA and TDMA The effectiveness of this method is underlined by its use with the UMTS; see Section 6.9.1.

The assumption that the spectral efficiency (Erlang per km^2 and MHz bandwidth) is as high with TDMA as it is with CDMA has been widely discussed in the literature, with apples being compared with pears; there does not seem to be a fairer mathematical comparison. In Section 6.14 we are able to show by large-scale systems simulation that the spectral efficiency of UMTS is not superior to EDGE, both being implemented according to the respective standards.

2.6.4.2 CDMA in Cellular Systems

- All frequency bands in each cell can be used.
- Because of the resultant interference ratio C/I, two channels (one each per BS) are usually operated simultaneously in the neighbouring area of two cells.

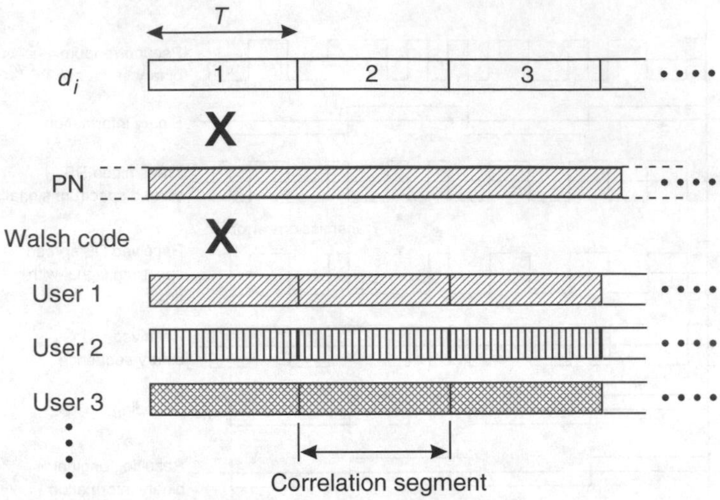

Figure 2.43: Combined DS and orthogonal symbolic spreading

- Power control must be performed quickly: a 1 kbit/s channel is required with IS-95; each 1 bit transmitted means 1 more level of power, each 0 bit means 1 power level less. There are many levels.
- Speech transmission with variable bit rate: 1 kbit/s up to approximately 6 kbit/s, depending on the speech activity.
- As the traffic load increases, the service quality decreases because of system limitations, e.g., *graceful degradation* of speech quality.

2.6.4.3 TDMA in Cellular Systems (GSM)

Per GSM network:

$$\begin{aligned} 61 \text{ channels} \cdot 200 \text{ kHz} &= 12.2 \text{ MHz} \\ N = 7 &\rightarrow 9 \text{ frequency channels/cell} \\ (N = 12 &\rightarrow 5 \text{ frequency channels/cell}) \end{aligned}$$

With CDMA:

$$N = 1 \text{ with } B = 12.2 \text{ MHz} \rightarrow 7\text{-fold spreading possible}$$

On the basis of past experience with CDMA systems, a spreading factor of 7 is clearly too low. A factor of 100 to 200 is required instead, B-CDMA (B$\hat{=}$broadband). Therefore a different channel structure is needed, e.g., not 200 kHz per channel but 10 kHz instead, which will produce a spreading factor of 140 (corresponding to the IS-95 approach, see Section 5.3).

2.6.5 Space-Division Multiplexing, SDM

With space-division multiplexing the frequencies used for transmission are reused at appropriate geometric intervals so that theoretically a huge volume of traffic can be transported over an enormous extended area despite the limited frequency spectrum.

This method is facilitated by the circumstance that the field strength of a radio signal decreases as the distance from the transmitter increases. Therefore the space-division multiplexing method uses the propagation attenuation of electromagnetic waves. If the distance from the transmitter is great enough, the signal becomes so weak that the noise due to interference that occurs when this frequency is reused by another transmitter can be tolerated (cluster principle).

Space-division multiplexing describes the basic concept of cluster formation that helps to provide cellular systems with an almost unlimited amount of traffic capacity restricted only by cost considerations. Sectorization of the supply area of a base station is another related measure.

There has been recent discussion about using phased-array antennas to set up multiple antenna beams that swivel electronically and can focus sharply, thereby allowing a base station's various *spot beams*, which have been angled towards different directions, to use the same frequency channel at the same time. Space-division multiplexing would be occurring in the same cell, thereby contributing to a dramatic increase in capacity. Spot beam antennas reduce the angle of incidence at the receiver for signal phasing, thereby reducing signal interference, i.e., dispersion, and allowing the use of simple (low-power) equalizers.

2.6.6 Hybrid Methods

In practice, hybrid multiplexing methods are usually applied because they offer efficient means of utilizing the frequency spectrum. Therefore a combination of two or several multiplexing techniques are applied in one system. By using a combination of technologies and incorporating the respective advantages of the different multiplexing techniques, it is possible to make good economic use of the radio channels (see Figure 2.44).

In the public mobile radio networks GSM, USDC and PDC, for example, frequency-division multiplexing as well as time-division multiplexing are used alongside the respective cellular structure. The systems DECT, TETRA, HIPERLAN and UMTS are other examples of the use of a combination of the two methods, FDM and TDM. All these systems are introduced later and described in detail.

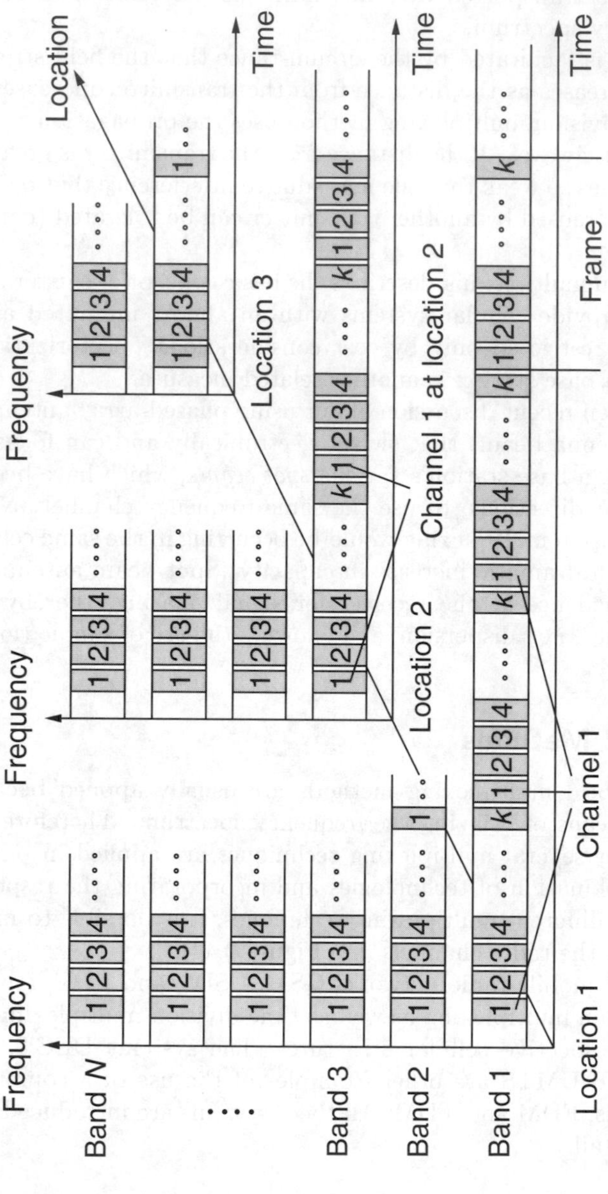

Figure 2.44: Using a hybrid multiplexing technique with a combination of FDM, TDM and SDM

2.7 Fundamentals of Error Protection

2.7.1 Error Protection in Radio Channels

Channel coding protects transmitted data from transmission errors. Because of fluctuations in signal level *(fading)* that characteristically occur in mobile radio transmission, the bit-error ratio can fluctuate considerably owing to distance, speed and the radio signal shadowing of stations communicating with each other.

2.7.1.1 Effects of Fading

A distinction is made between *long-term fading* and *short-term fading* of a received signal. Long-term fading is caused by shadowing, such as by buildings; therefore it has a greater chance of occurring in built-up urban areas than in flat countryside. In GSM networks this type of fading occurs at intervals of about 12–60 m; at a speed of 36 km/h this produces intervals in the range of 1.2–6 s.

Short-term fading is mainly caused by the multipath propagation of radio waves that are reflected or diffracted on obstacles such as buildings and hills (see Figure 2.7). The transmitted signal reaches the receiver with time delays and on paths with different lengths. Depending on the phase position of the signals of the individual paths, this results in interference (loss or boosts) of the received signal at the receiver. Fades are more or less at fixed locations in a given environment. The distances of the fading minima are frequency-dependent and about half a wavelength; thus in GSM at approximately 15 cm (in DCS 1800 at approximately 8 cm), which explains the term short-term fading seen from a mobile receiver.

2.7.1.2 Necessity of Error-Protection Methods

Error-protection methods reduce or eliminate the effects of interference to the receiving signal described above, and hence the bit-error ratio of the received data stream. Special codes are selected, which are used in conjunction with *interleaving* and *frequency hopping*. Short-term fading is much easier to combat than long-term fading.

Error-protection methods for data transmission have much higher requirements than those for speech. Whereas with speech a residual bit-error ratio of 10^{-2} is still considered tolerable, with data it should be lower than 10^{-7} if possible. This explains why different error-protection methods are used for speech and for data transmission in digital mobile radio networks.

A distinction is made between three different methods of error protection, namely error *detection*, error *correction* and error *handling*.

2.7.2 Error Detection

Through error detection it is possible to establish whether a received data word is a valid code word. However, data words that are recognized as being incorrect cannot be corrected.

In GSM checksums of cyclic codes (CRC $\hat{=}$ *Cyclic Redundancy Check*) are used for error detection. This usually involves BCH codes (*Bose-Chaudhuri-Hocquenhem-Codes*), which are described in detail in [32].

An (n, k)-BCH code is characterized by the following values:

k			block length *before* coding (data word length)
n	$=$	$2^m - 1$	block length *after* coding (code word length)
$m \cdot t$	$=$	$n - k$	number of redundant parity bits
d_{min}	$\geq$	$2t + 1$	Hamming (minimum code words) distance

In RLP, the data link control protocol for GSM (see Section 3.9.4), a (240,216) BCH code is used, i.e., a 24-bit long frame check sequence is attached to the original data word, which consists of $k = 216$ bits. The data word that is created after the coding has a length of $n = 240$ bits.

This code has a Hamming distance of $d_{min} = 7$, so that all defective code words with a maximum $d_{min} - 1 = 6$ errors (at random bit positions) are detected. More than 6 errors in a code word can result in the creation of another valid code word, which will be accepted as correct. If this happens, the code has failed in this particular case.

The residual bit-error ratio (RBER) P_e of an (n, k)-BCH code is

$$P_e = \sum_{i=d_{min}}^{n} r(i)p(i) \tag{2.43}$$

$$r(i) = \frac{A(i)}{\binom{i}{n}} \tag{2.44}$$

$$p(i) = \sum_{j=0}^{i} \binom{j}{n} BER^j (1 - BER)^{n-j} \tag{2.45}$$

The values represent:

- $r(i)$ probability (reduction function) with which an error pattern with the Hamming weight i corresponds to a valid code word $A(i)$ and therefore is not recognized as incorrect
- $A(i)$ weight distribution of a code
- $p(i)$ probability of the occurrence of an error based on weight i

The RBER for the range $d_{min} \leq i \leq n - d_{min}$ can then be estimated using the following formula:

$$P_e = \frac{\text{number of used code words} - 1}{\text{number of all possible code words}} = \frac{2^k - 1}{2^n - 1} \approx 2^{-(n-k)} \tag{2.46}$$

Accordingly, the residual bit-error ratio depends on the number of parity bits. Therefore it is irrelevant whether a data word with 50 bits or one with 3000 bits is protected by a 24-bit long CRC sequence; the residual error probability is equally high in both cases.

For the (240,216)-BCH code used in the RLP mentioned above the RBER approximates

$$P_e = 2^{-(240-216)} = 2^{-24} \approx 5.96 \cdot 10^{-8} \qquad (2.47)$$

An RBER of this magnitude is sufficient for most applications in mobile data transmission.

2.7.3 Error Correction

Forward error correction (FEC) also occurs in the context of error-protection coding. With forward error correction the transmitter adds so much redundancy to a data word that the receiver is able to correct a certain number of errors. In contrast to the ARQ methods described in Section 2.7.4, a reverse channel from the receiver to the transmitter is not required.

A distinction is made between two code families that are suitable for forward errror correction, namely the linear block codes and the convolutional codes. Linear block codes are systematic codes, i.e., a certain number of redundant bits are calculated from the data word to be protected and are transmitted with it. The coded data word can thus be divided into a redundant and a non-redundant part. On the other hand, the convolutional codes described later are non-systematic codes.

2.7.3.1 Reed–Solomon Codes

Reed–Solomon (RS) codes constitute a special class of BCH codes that use q-ary instead of binary symbols. Because of the binary character of the data being transmitted, RS codes are usually selected over the Galois field GF(2^m), thus at $q = 2^m$.

According to [22], an RS code can be described based on the following values:

k				k	block length before coding
n	$=$	$2^m - 1$	with	n	block length *after* coding
$d_{min} - 1$	$=$	$n - k$		d_{min}	Hamming distance
$2t + e$	$=$	$n - k$		t	number of correctable errors
				e	number of losses

RS codes are particularly suitable for the correction of errors that occur in bursts (bursty errors). This makes them especially useful in mechanical storage systems (e.g., hard disks) as well as with compact disks (CD), because these frequently have large blocks that are garbled (e.g., due to scratches on the surface). Because strongly correlated errors likewise occur in radio transmission owing to signal fading, RS codes are also highly suitable for this area.

Table 2.3: Block lengths of Reed–Solomon codes over GF(2^m)

GF(2^m)	Block length n in	
m	q-ary symbols	bit
3	7	21
4	15	60
5	31	155
6	63	378
7	127	889
8	255	2040

A particular feature of the RS code in comparison with the BCH code is that with given block length n the data word length k can be selected, and this selection always produces an optimal distance d_{min}.

Table 2.3 presents calculations for the block lengths of some RS codes. It can be seen that the block length with the factor m of the Galois field increases quickly, which results in very redundant codes. Abbreviated RS codes are used to counteract this problem (see [32]). This involves stuffing a given data block length with so many zero symbols that a new length is produced with the desired redundancy and consequently the required correction quality. The zero sequence itself is only used to form the check bits, but is not transmitted because before decoding the receiver can add these zero symbols to the same positions again.

2.7.3.2 Convolutional codes

In comparison with block codes, convolutional codes have a memory, i.e., a code word position is not only dependent on the actual data bit but on several preceding data word positions. An (n, k, m) convolutional code has an influence length m, k input and n output bits per data word. The direct result is a code ratio of $R = k/n$. Convolutional codes are explained in detail in [22].

Figure 2.45 presents a schematic structure of a simple (2,1,3) convolutional code with code ratio equal to 1/2. This coder has two internal shift registers T and three exclusive-or functional units. The input bit x_i is coded into two output bits $y_i^{(1)}$ and $y_i^{(2)}$. Together these two output bits form the code word.

Because of the two shift registers, the convolutional code can assume $2^2 = 4$ different states. In the literature these states are labelled as a, b, c and d.

The convolutional coder changes its state depending on the bit occurring at the input (0 or 1). A number of $(m-1)$ 0 bits, called tail bits, are added to the bit stream to ensure a return to the initial state a after the coding of any length of bit stream.

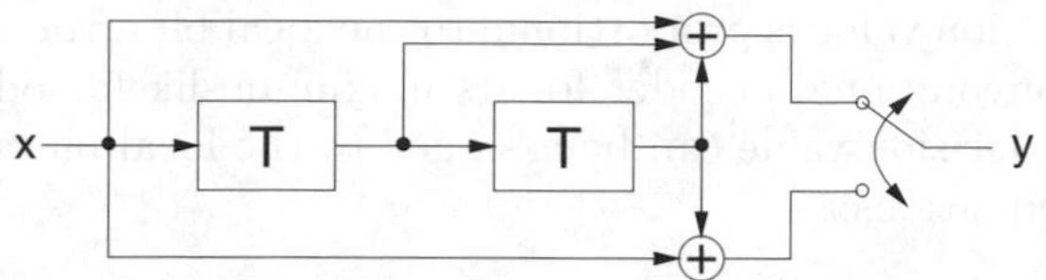

Figure 2.45: Structure of a (2,1,3) convolutional code

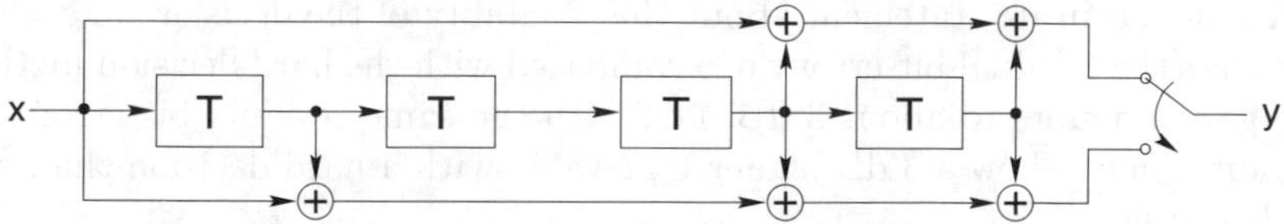

Figure 2.46: (2,1,5) convolutional coders in GSM

Viterbi algorithms and *Maximum Likelihood Decision* (MLD) are used in convolutional decoding. The path generating the bit sequence that comes closest to that of the received sequence is selected from the Trellis diagram [22, 32]. This path is then used for decoding the entire received bit sequence.

The residual bit-error ratio (RBER) remaining after convolutional decoding decreases in line with the number of internal shift registers of the coder. As an example, the RLP of GSM uses a (2,1,5) convolutional code. This convolutional coder referred to as GSM-96 has four registers (see Figure 2.46) and a lower RBER than the (2,1,3) convolutional coder shown in Figure 2.45.

The decoding complexity increases disproportionately to the number of memories for the convolutional coder. Convolutional coders with 2–10 shift registers are common today in quasi-real-time operations with decoding in the millisecond range.

Convolutional decoders are very suitable for the correction of uncorrelated errors, but are extremely sensitive when it comes to bursty distributed errors such as those which frequently occur in radio channels. Consequently, in mobile radio systems convolutional coders are almost exclusively used in combination with *interleaving* (see Section 3.3.7.1), a technique that is able to let bursty errors appear as single-bit errors.

2.7.3.3 Soft Decision

With the decoding method called *soft decision* the convolutional decoder receives additional information from the *equalizer* about the bit just received. In contrast to the *hard decision* method, in addition to the actual value of the bit (0 or 1) a statement about the reliability of the decision designated as a soft decision value is provided based on the received signal level by the equalizer after quantization (e.g., coded with 2–8 bits in GSM).

The soft decision value is proportional to the local bit error ratio and can be used by the convolutional decoder for its maximum-likelihood decision. The quantized soft decision value can be assigned to the local bit error ratio using the following equations:

$$p_e = \frac{e^{-sd}}{1+e^{-sd}}, \qquad sd = -\ln\left(\frac{p_e}{1-p_e}\right) \tag{2.48}$$

A more accurate statement about the reliability of the decision on each bit decreases the residual bit-error ratio compared with the hard decision method. The gain amounts to about 3 dB, i.e., with the same residual bit error ratio a channel must show a 3 dB higher C/I value with a hard decision than with a soft decision.

2.7.3.4 Punctured Convolutional Codes

Puncturing is a method for shortening a convolutional code. This involves deleting one or more bits from the output stream of the convolutional coder according to a prescribed scheme called a puncture table.

A puncture table consists of the elements 0 and 1 and is periodically readout. A 0 signifies that the corresponding bit is not being transmitted in the output bit stream; a 1 means that the bit is being sent. This shortens the coded sequence and weakens the convolutional code. So the puncturing removes part of the redundancy added with the convolution process.

The advantage of a punctured code is that different code ratios can be implemented. Based on a mother code with the ratio $1/n$, codes with higher code ratios can be developed through periodic puncturing.

Punctured codes can be decoded by the receiver using the same convolutional decoder as for the mother code. To do so, the receiver, also familiar with the puncture table, adds bits with a random value (e.g., 0) at the appropriate positions. This process is called repuncturing. To the Viterbi decoder this 'unknown bit' means that the probability of a 0 or a 1 is the same.

2.7.4 Error-Handling Methods by ARQ-Protocols

Unlike forward error correction, when *Automatic Repeat Request protocols* are employed, corrupted data words are not corrected by the receiver and then accepted, but are instead checked for correctness on the basis of an error-detection code, which may also be used for correcting them (see Section 2.7.2). If a data word is not recognized as being correct, the receiver requests the transmitter to transmit it again.

The prerequisite for this method is a reverse channel between transmitter and receiver over which the results of the error evaluation are transmitted to the transmitter as an *acknowledgement*. For this purpose all the transmitted data words are provided with a sequence number NS. A positive acknowledgement ACK[NS] is sent when the word received is correct. If it is defective then a negative acknowledgement NAK[NS] is sent along with a request that the data word should be resent. A data word has the structure of a protocol data unit of the ARQ-protocol used, and is mostly called a block, a frame or a packet.

The transmitter must maintain a copy of the packet sent until the acknowledgement indicates that it has been received error-free.

To ensure that a faulty transmission can be detected in the first place, the transmitter must transmit the checksum of an error-detecting code (CRC $\hat{=}$ *Cyclic Redundancy Check* or FCS $\hat{=}$ *Frame Check Sequence*) in the packet.

2.7.4.1 Frame Types and Flow Control

Flow control is implemented in order to regulate the data flow from the data source in such a way that the data sink is not overloaded. The mechanism is based on numbering in sequence all frames carrying data from higher layers. This numbering helps to monitor whether a frame is missing. Furthermore, it provides a means for confirming explicit receipt of frames and requesting that frames be resent when a layer 2 connection is in the acknowledged mode of transmission. Section 3.4.1 explains how this mode is reached. But first of all, here is an overview of the different types of frames that are usually sent between two peer entities in a data link layer in acknowledged mode of operation.

I-Frame, Information This frame transports the information fields provided by layer 3. The frames are numbered in sequence (NS, *Send Sequence Number*), and *piggy-back* acknowledgement is possible through the inclusion of an NR (*Number Received*) number.

RR-Frame, Receive Ready This indicates that a layer-2 entity is ready to receive I-frames. All frames that have already been received are acknowledged through the parameter NR contained in the RR frame.

RNR-Frame, Receive Not Ready This frame occurs when the transmitter of this frame is unable to receive further frames. This frame also contains an NR.

REJ-Frame, Reject This frame is used when a sequence error occurs while I-frames are being received. It indicates that the other side should repeat the I-frames. The repeat process begins with the I-frame in which the NS is equal to the parameter NR contained in the REJ-frame.

SREJ-Frame, Selective Reject An SREJ is an alternative to the REJ-frame. The difference is that a single I-frame is requested on a selective basis. The side that receives an SREJ should continue with normal operations after this requested I-frame has been sent.

Sequential numbering All I-frames transmitted are provided with a continuous number NS modulo n. If an I-frame is lost then its number can be used to confirm this fact.

All the frame types listed above contain a Received Number (NR). The NR in a frame that has been received represents an acknowledgement of all I-frames that have already been sent with the sequence numbers including NR−1. NR is the sequence number of the next I-frame to be sent. Both sequence numbers, NS and NR, are normally coded in bit fields of length n.

Sliding window protocol Protocols of the data link layer must be able to deal with special cases in which frames have been damaged or lost. The sliding window technique helps to control the information flow in such a way that communication between entities is kept synchronous.

Transmit window The transmitter maintains a list of successive sequence numbers (NS) that match the corresponding I-frames. These I-frames are in the transmit window. The lower edge of the window is marked by VA (*Acknowledge Variable*) and the upper edge by VS (*Send Variable*). The lower edge is updated through the NR frames. The upper edge is increased when a new I-frame is sent. I-frames remain in the window until receipt has been confirmed by the other side. If VA = VS applies, then the window is open. If all the possible window places are occupied (VS = VA + k, Window k), then no further I-frames are allowed to be sent; the window is closed. I-frames must be sent by the transmitter in the sequence corresponding to the information supplied by layer 3.

Receive window The receiver keeps a list of all the sequence numbers (NS) that correspond to the I-frames waiting to be received. VR (*Receive Variable*) indicates the lower edge of this window. The receive window has a fixed size k. I-frames can be accepted in this window in any sequence, as long as their sequence number (NS) fits in. It is important that the layer-2 entity passes on the information received from these I-frames to its layer 3 in the sequence in which it was sent.

Figure 2.47 presents an example of how the sliding window technique functions [34]. Each partner entity has a transmitting window as well as a receiving window. The transmitting and receiving windows can each be of a different size.

2.7.4.2 Time Monitoring

The confirmation of a transmitted frame that is part of a transmit window is monitored by a *timer*. If confirmation for the I-frame arrives on time, the timer is stopped. If no confirmation is received, the I-frame is resent after the timer runs out.

With a size-1 transmit window it is possible either to set up a separate timer for each window space or to pursue the principle of timer extension.

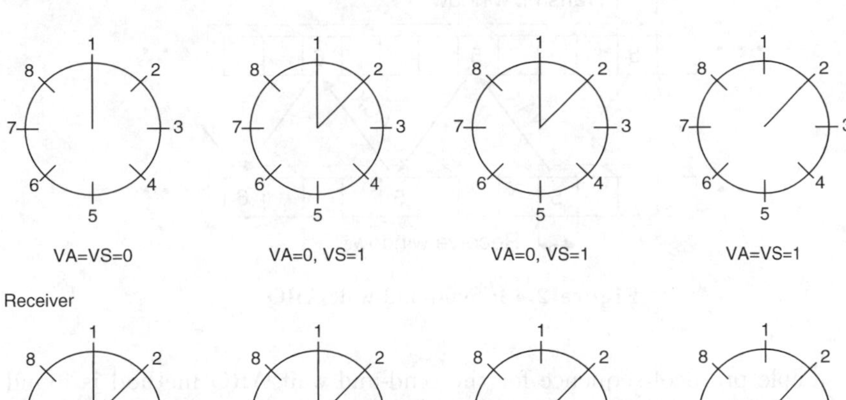

Figure 2.47: Size-1 sliding windows

It is also possible to use the timer to influence the window size. On the assumption that a size-4 transmit window is planned, if the same premise applies that a timer runs out for an I-frame before the next possible transmitting cycle for the next I-frame then only one window space is used in the entire window. In effect the protocol runs its course with window size 1.

2.7.4.3 Send-and-Wait ARQ Protocol

Protocols with a window size of 1 are referred to as *send-and-wait* protocols. This means that a transmitter sends an I-frame and then is required to wait for (positive or negative) acknowledgement that this I-frame has been received before the next I-frame can be sent. The $\mathrm{LAPD_m}$ belongs to this group of data link control protocols (see Section 3.4.1).

The transmitter responds to the receipt of acknowledgement NAK by retransmitting the packet, and the subsequent packet is then sent after an ACK has been received. The throughput with this method is minimal, especially with long signal propagation delays and with short data packets.

The minimal waiting time for an acknowledgement corresponds to the length of t_{rd} (*round-trip propagation delay*). The period of time t_{rd} is equal to double the transfer time t_f of the transmission channel.

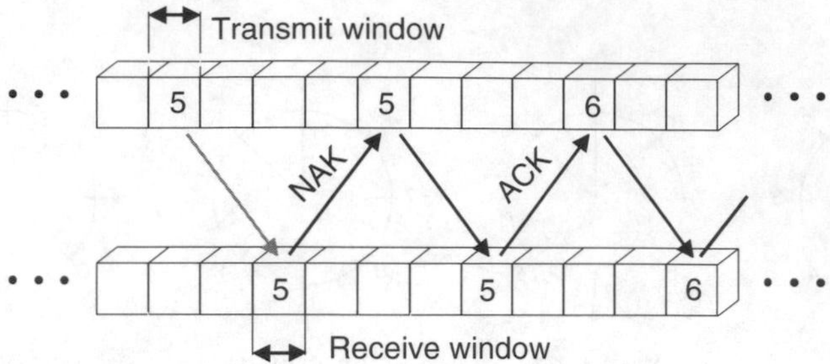

Figure 2.48: Send-and-wait ARQ

A sample protocol sequence for the send-and-wait ARQ method is found in Figure 2.48; the numbers refer to specific packets.

With

$$\begin{aligned} n &= \text{packet length in bits} \\ PER &= \text{packet error ratio} \\ c &= \text{delay } t_{rd} \text{ between end of transmission of last} \\ &\quad\ \text{packet and start of next packet in seconds} \\ v &= \text{transmission rate in bit/s} \end{aligned} \tag{2.49}$$

the throughput D (more accurately the utilization of the channel) is

$$D = \frac{n(1 - PER)}{n + cv} \tag{2.50}$$

If an ideally fast and interference-free channel is used in a model, a window size of 1 will certainly be sufficient. If a delay of the confirmation occurs on the route between the transmitting layer-2 entity and the receiving partner entity, the pipelining principle can be advantageous. Delays occur when the transmission time over the medium being used is very high, interleaving is implemented in layer 1 or an ACK/NAK frame is lost. With pipelining, a window size >1 is required. Pipelining means that a limited number of other I-frames are being sent, even if confirmation is not provided.

Pipelining over a seriously unreliable channel presents some problems. What happens, for example, if an I-frame is lost in the middle of a long sequence? The subsequent I-frames will arrive at the receiver before the transmitter is able to establish that an error has occurred. There are two methods for mastering these situations effectively.

2.7.4.4 Go-Back-N-ARQ Protocol

The Go-Back-N-ARQ method is frequently called the REJ method or cumulative ARQ. With this ARQ method data flow between transmitter and receiver on a duplex channel is more or less continuous.

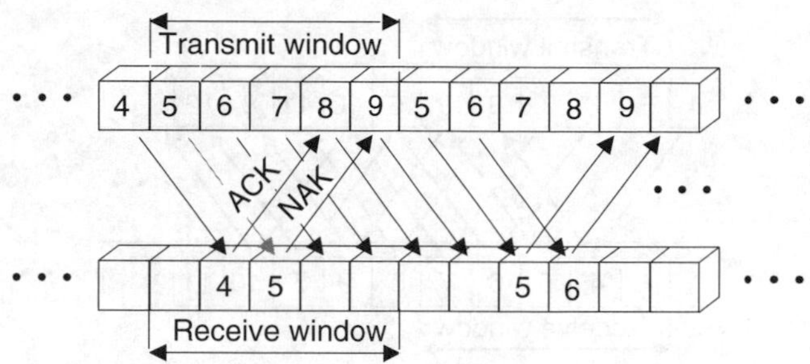

Figure 2.49: Go-Back-N-ARQ

With a transmit window that is unlimited in size, data packets with consecutive packet numbers NS are transmitted until the transmitter receives a NAK[NS_e] for a packet that the receiver has recognized as being defective. When this occurs, the transmitter interrupts transmission and switches back to the defective packet with the number NS_e. The packet with packet number NS_e and all subsequent packets are then resent. The receiver ignores all other packets until packet NS_e is received.

With the REJ, the transmitter requires sufficient buffer in which to store all the unacknowledged data packets. This is why a transmit window of finite size is usually agreed upon. If the transmitter receives a positive acknowledgement ACK[NS] for a packet already sent then all the packets with sequence numbers less than or equal to NS are considered to have been transmitted correctly. The receive window selected is often larger than 1 to enable positive and negative acknowledgements to be sent (*piggy-back*) with an information packet from receiver to transmitter. Acknowledgements may also be sent before the receive window is completely filled up. A section of an REJ protocol process is shown in Figure 2.49.

The throughput with REJ-ARQ methods can be determined on the basis of the expressions introduced in Section 2.7.4.3:

$$D = \frac{n(1 - PER)}{n + PER \cdot cv} \tag{2.51}$$

2.7.4.5 Selective-Reject ARQ-Protocol

As with REJ, the objective of the SREJ method is to transmit packets between transmitters and receivers as continuously as possible. With this method, when the receipt of a defective packet is detected, only the packet that is defective is selectively requested. In contrast to the REJ method, the packets that arrive in the meantime are not rejected but are filed in a storage buffer. Therefore the transmitter only resends the data packets that were transmitted with errors.

At the same time the receiver must store a packet with the number NS_i until all packets with lower packet numbers have been correctly received. This

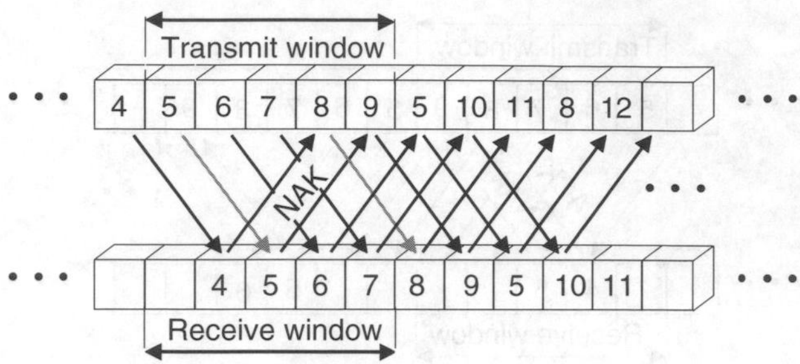

Figure 2.50: Selective-Reject-ARQ

is the only possible way to forward the packets in the correct sequence to the next highest ISO/OSI layer. For its part, the transmitter must keep the transmitted data packets in a storage buffer until a positive acknowledgement has been received for them, just as is the case with REJ. SREJ methods are not able to make full use of the transmitting window, because otherwise the uniqueness of the packet that has been requested again can be lost. Figure 2.50 shows the fundamentals of the SREJ process.

Assuming the availability of unlimited receiving storage capacity, the throughput with an SREJ-ARQ method amounts to

$$D = 1 - PER \tag{2.52}$$

Consequently the highest throughput of the three ARQ methods presented is achieved with SREJ.

2.7.4.6 Comparison Between FEC and ARQ Methods

The advantages and disadvantages of FEC and ARQ methods are summarized and compared here.

FEC methods

Advantages:
- No reverse channel required
- Constant throughput irrespective of channel quality
- Constant delay time between transmitter and receiver

Disadvantages:
- Low throughput due to high redundancy (low code ratio)
- Residual bit error ratio strongly dependent on channel quality
- Complicated coder and decoder algorithms required

ARQ methods

Advantages:
- High throughput when channel quality is good
- Guarantee of very low residual bit error ratio
- Less complex coder and decoder unit required

Disadvantages:
- Reverse channel required and additional signalling
- Throughput fluctuates depending on channel conditions
- Variable delay time, less suitable for real-time applications
- Additional storage needed by transmitter and receiver

2.7.4.7 Adaptive Coding

A mobile terminal experiences different channel qualities counted by the bit error ratio, depending on the current location, speed, etc., which is mainly characterized by the current receive signal level and C/I value at the mobile's receiver.

Since the channel coding defined for a mobile radio system has to take into account the worst possible situation in which the system should still be able to operate satisfactorily, the coding is overdimensioned in all the situations, where a terminal is operating in an excellent, good or acceptable receive situation.

Clearly the channel coding should be adaptive. This requires that the code ratio be dependent on the current receive conditions so that a given quality of service (typically measured by parameters like bit-error ratio, throughput, delay, etc.) can be guaranteed.

The result of an adaptive coding is that the net bit rate of the channel is dependent on the receive situation of the mobile terminal, since the redundant parity bits would depend on the current needs.

With speech transmission where the bit rate is continuous in time, during time intervals where a small code ratio is adaptively chosen, the channel throughput would be too high. Silence periods could then be inserted for the transmission of speech codec data and the interference generated by use of the channel to other communicating terminals would be reduced. The use of intermittent speech transmission is already practised, e.g., when using packet-data-based voice, as is usual with CDMA systems, or using discontinous transmission, as is usual with the GSM system.

With data services a variable bit rate resulting from adaptive channel coding might be of advantage, too. This especially is the case with packet-oriented data services.

2.7.5 Hybrid ARQ/FEC Methods

ARQ methods provide a high level of transmission security, which can also be maintained when channels are severely disrupted albeit at the cost of throughput.

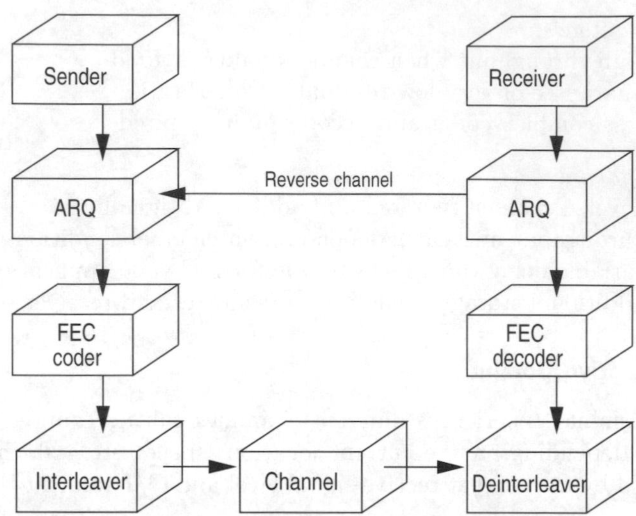

Figure 2.51: Hybrid ARQ/FEC system

FEC methods provide continuous channel throughput, but the level of transmission security decreases the more error-prone the channel becomes. If the bit-error ratio of a channel is too high to achieve a required level of throughput using an ARQ method, and the code ratio of an FEC method is too small, then a combination of the two error-protection methods is applied.

This kind of combination in which an ARQ method is set up on an FEC system is called a hybrid ARQ/FEC method or an HARQ method. These methods combine the advantages of both techniques: the FEC system improves the residual bit error ratio of the channel; the ARQ technique eliminates any remaining errors. The principal structure of a hybrid ARQ/FEC system is presented in Figure 2.51.

The selection of the correct code allows the channel throughput, reduced because of the transmission of the redundancy bits of the FEC method, to be optimized despite the retransmission of disrupted packets.

2.8 Fundamentals of Random Access*

This section starts with a description of the random-access protocols that are essential for the initiation of contact between a mobile station and a radio network. First, the Slotted-ALOHA access protocol will be examined in detail. This will be followed by an introduction to the methods described in the literature for control of random access in Slotted-ALOHA systems. These are

*With the collaboration of Dietmar Petras and Martin Steppler

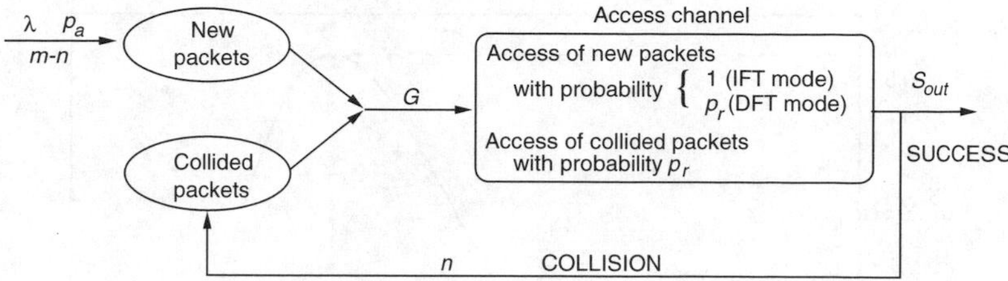

Figure 2.52: The Slotted-ALOHA access method (m and n are the number of not-collided and collided stations, respectively)

important for the transmission of information (and not only for signalling) in radio networks that transmit using asynchronous time-division multiplexing.

2.8.1 Slotted-ALOHA Access Methods

The access method used for initial access by a mobile station in a mobile radio network is based on a Slotted-ALOHA protocol. The protocol has been covered in depth in numerous publications (e.g., [8] and [17]). Some of the important features of Slotted-ALOHA are explained here.

2.8.1.1 Fundamentals of ALOHA

The ALOHA access protocol is used in communications networks in which a large number of uncoordinated users are competing for the same channels. The basic idea behind an ALOHA system is simply that users are allowed to send data any time they wish.

The protocol does not exclude the possibility of collisions, i.e., the fact that simultaneous or overlapping transmission by different stations can occur that cannot be decoded by the receiver. Collided packets must be resent. However, they cannot be resent until a randomly selected waiting time has passed; otherwise the same stations would collide again.

New packets are sent after they are created to ensure that there is as small a delay as possible with initial access (see Figure 2.52). If a collision occurs with other transmitted packets, the channel will be occupied but will have no throughput. Collided and successful packets together create a traffic load, which is measured by the arrival rate G (packets/slot). Two modes of transmission may be distinguished, IFT and DFT; see Section 2.8.3.2.

In contrast to *pure ALOHA*[1, 2], with the Slotted-ALOHA protocol transmission can only take place at the beginning of a *time slot* with a constant length. Therefore this is a time-discrete variation of the ALOHA protocol that offers double the amount of throughput S_{out}. If a situation with a large number of independent stations is assumed then the arrival process for new packets is a Poisson process (with an arrival rate λ). If collided packets are retransmitted after an independent random delay, the channel then experi-

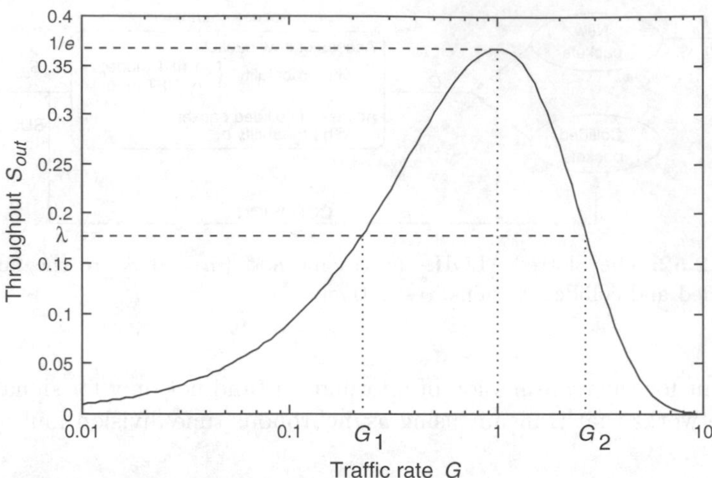

Figure 2.53: Throughput of Slotted-ALOHA

ences a Poisson arrival process at the rate G, and it can be shown [34] that the throughput at $G = 1$ reaches its maximum, which is $1/e = 0.368$. The probability that transmission will be successful is

$$S_{out} = Ge^{-G} \tag{2.53}$$

which is derived as follows and is presented graphically in Figure 2.53. Assuming an arrival rate $\lambda < e^{-1}$, the equilibrium point at G_1 initially becomes significant and the throughput is $S_{out} = \lambda$. However, the throughput at this point only constitutes an average value over a certain period of time. If the present traffic rate exceeds the value G_1 by a small amount, the throughput will be somewhat higher than λ. Consequently, data packets leave the system faster than they arrive, which causes the access rate to return to G_1. The system remains stable at point (G_1, λ) as long as the traffic data rate does not increase dramatically in the short term.

If the momentary arrival rate λ and consequently the traffic rate increase even further, e.g., to $G = 1$, then the system will become unstable (see the working point at G_2). But if the momentary arrival rate is somewhat lower than λ then the throughput S_{out} will also be lower than the arrival rate, and this will lead to an increase in the number of collided stations. When this happens, G continues to increase and the throughput S_{out} drops. This process continues: the throughput moves more and more towards 0. This is what makes the Slotted-ALOHA access protocol unstable.

It can also be shown [5] that the Slotted-ALOHA access method becomes unstable with fixed access probability per slot of $p < 1$ when the number of participating stations is very high. This instability can be avoided through the use of different approaches, as explained below.

2.8.1.2 Analysis of Slotted-ALOHA Access Methods

If the channel access rate G is normalized to the length of a time slot (G comprises new transmission attempts and retransmissions due to collisions), and a Poisson process is used to model all access, then the probability of a successful transmission attempt per slot can be described on the basis of Equation (2.53). The system is in a state of equilibrium when $S_{out} = \lambda$ (λ is also normalized to the slot duration). This formula offers no insight into the dynamic behaviour of a system, because feedback is produced owing to the number of collisions, which in turn alters the value of G. Generally there are too many unused slots if $G < 1$ and too many collisions if $G > 1$.

References [5, 10] describe a model for a Slotted-ALOHA system in which the stations directly waiting for the next transmission attempt are not accepting new data for transmission from their own application processes. The number m of stations is so large that the arrival process can be approximated as a Poisson process with rate λ. Furthermore, it is assumed that each station is carrying out a renewed transmission attempt per slot for its collided packet with fixed probability p_r. Consequently the number i of slots between a collision and a renewed attempt is geometrically distributed with probability

$$P(I = i) = p_r(1 - p_r)^{i-1} \tag{2.54}$$

If n indicates the number of collided stations at the beginning of a particular slot then each of the n stations, independently of one another, is transmitting with probability p_r. Each of the other $m - n$ stations will send its packet in this slot if it has arrived at the station during the previous slot. Because the arrival processes for the new packets of each station likewise are Poisson processes with mean rate λ/m, the probability that no data is arriving is $e^{-\lambda/m}$; consequently the probability that a station that has not yet collided will send its packet in the slot in question is $p_a = 1 - e^{-\lambda/m}$.

If $P_a(i, n)$ is the probability that i packets that have not yet collided are being sent in a specific slot, and $P_r(i, n)$ is the probability that i packets that have already collided are being sent, then

$$P_a(i, n) = \binom{m-n}{i}(1 - p_a)^{m-n-i}{p_a}^i \tag{2.55}$$

$$P_r(i, n) = \binom{n}{i}(1 - p_r)^{n-i}{p_r}^i \tag{2.56}$$

From this assumption, it is possible to produce a discrete-time Markov chain in which the state variable stands for the number of collided stations and the slot duration is the unit of time. In each unit of time the value of the state variable increases by the number of collided stations in the respective slot, but remains the same if a new packet is successfully transmitted or the slot is unused, and drops by one if a collided packet is successfully transmitted.

A transmission attempt is successful if

- there is no packet that has already collided and a newly arrived packet is sent in the slot in question;
- no packet has arrived and a collided packet is sent in the slot in question.

Consequently, the transition probability with a transition from state n to state $n+i$ can be expressed through the following equations [5, 10].

- If two or more packets arrive during a slot, then

$$P_{n,n+i} = P_a(i,n)\,, \qquad 2 \leq i \leq (m-n) \tag{2.57}$$

- With exactly one newly sent packet and at least one resent collided packet,

$$P_{n,n+1} = P_a(1,n)\,[1 - P_r(0,n)] \tag{2.58}$$

- With exactly one newly arrived packet and no attempt to resend packets that have already collided or with no newly arrived packet and no successful attempt to resend a packet that has already collided, the following applies:

$$P_{n,n} = P_a(1,n)P_r(0,n) + P_a(0,n)\,[1 - P_r(1,n)] \tag{2.59}$$

- With exactly one attempt to send a packet that has already collided,

$$P_{n,n-1} = P_a(0,n)P_r(1,n) \tag{2.60}$$

Let D_n be the expected difference in the number of collided stations for the next slot in state n from the updated number of collided stations. Therefore D_n in state n consists of the anticipated number of new transmitting stations, i.e., $(m-n)p_a$, minus the anticipated number of stations transmitting successfully together in a specific slot, which corresponds to the probability of a successful transmission attempt and is denoted by P_{succ}. Thus

$$D_n = (m-n)p_a - P_{succ} \tag{2.61}$$

with

$$P_{succ} = P_a(1,n)P_r(0,n) + P_a(0,n)P_r(1,n) \tag{2.62}$$

If G(n) is the number of transmission attempts in one slot in state n, then

$$G(n) = (m-n)p_a + np_r \tag{2.63}$$

From an analysis of the Markov chain, P_{succ} (on the condition that p_a and p_r are low) can be approximated (see [5]) as

$$P_{succ} \approx G(n)e^{-G(n)} \tag{2.64}$$

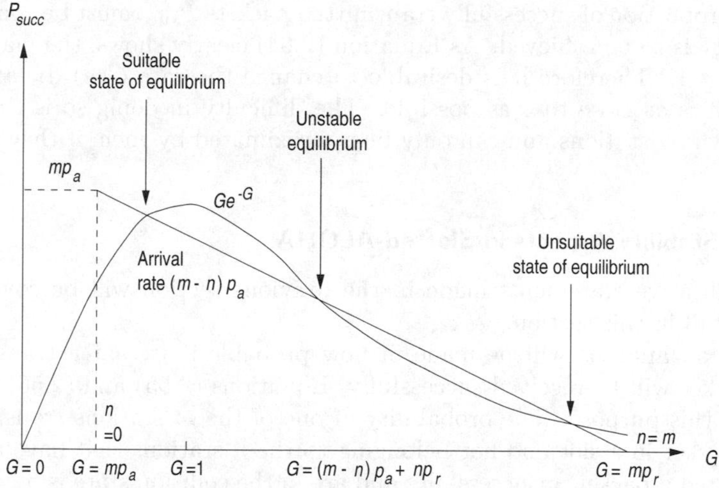

Figure 2.54: Stability in Slotted-ALOHA

with the probability of an empty slot being about $e^{-G(n)}$. This approximation therefore confirms the above assumption of G as the parameter for a Poisson distribution (see the result in Equation (2.53)).

The relationships that have just been explained are presented graphically in Figure 2.54 for $p_r < p_a$. In Equation (2.61) the deviation D_n is the distance between the curve that represents successfully transmitted data packets and the straight line that describes the arrival rate. Two stable states of equilibrium are evident.

Let us consider the effects of varying the parameter p_r. If p_r is increased, then the delay until the new transmission of collided packets is reduced. However, because of the linear dependence between n and the rate $G(n) = (m-n)p_a + np_r = mp_a + n(p_r - p_a)$, $G(n)$ also increases by n if p_r is increased and $p_r > p_a$. If the horizontal axis in Figure 2.54 is considered for a fixed n, this change to G corresponds to a shortening of the horizontal scale for G and therefore a horizontal compression of the curve Ge^{-G}. This means that there is a decrease in the number of collided packets that are needed in order to achieve an unstable equilibrium.

If, on the other hand, p_r is decreased, the delay will be higher with the renewed attempt to transmit (see Section 2.8.3), although a state of unstable equilibrium will not be reached as quickly. With a reduction of p_r, the graph of Ge^{-G} widens. This means that a stable equilibrium point is now only located in the area of the straight lines. This area has a large number of the m stations in a collided state, which means that no new data can be sent from these stations. This has a negative effect on the delay of the newly generated packets there.

The proportion of successfully transmitted packets P_{succ} must be controlled if stability is to be achieved. As Equation (2.64) clearly shows, the maximum is $G(n) = 1$. Therefore it is desirable to change the rate $G(n)$ dynamically so that it is as close to 1 as possible. The difficulty in doing so is that n is not known by stations and can only be approximated by them if they receive feedback.

2.8.1.3 Stability Aspects in Slotted-ALOHA

The qualitative statements made in the previous section will be covered in more detail in this section.

First, an analysis will be made of how probable it is that a transmitted data packet will be received successfully. Equations (2.55) and (2.56) will be used for this purpose. The probability of one of the m stations transmitting a new packet in a slot and not belonging to the n stations that have already transmitted a packet unsuccessfully and are in the collision state is $p_a = \lambda/m$. The stations undertaking another attempt to send their packets transmit with the probability p_r.

If the transmission result in the kth slot is indicated as e_k, then

$$e_k = \begin{cases} 0, & \text{if no station is transmitting in the } k\text{th slot} \\ 1, & \text{if exactly one station is transmitting in the } k\text{th slot} \\ c, & \text{if two or more stations are transmitting in the } k\text{th slot} \end{cases} \tag{2.65}$$

It is assumed that e_k in the kth slot is known and can be used to establish the transmission probability for the collided stations in the $(k+1)$th slot.

The throughput of n collided stations can be indicated by the probability that exactly one data packet will be sent in slot k. This is based on the probability that exactly one of the n previously collided stations is transmitting but no station is transmitting for the first time, and the probability that exactly one station is transmitting for the first time but none of the n collided stations is transmitting data packets. Therefore the probability of a successfully transmitted data packet is

$$\begin{aligned} P(e_k = 1 \mid n \geq 1) &= P_a(1,n)P_r(0,n) + P_a(0,n)P_r(1,n) \\ &= (m-n)\left(1-\frac{\lambda}{m}\right)^{m-n-1}\frac{\lambda}{m}(1-p_r)^n \\ &\quad + \left(1-\frac{\lambda}{m}\right)^{m-n} np_r(1-p_r)^{n-1} \end{aligned} \tag{2.66}$$

It is assumed here that there is at least one station in the system that has already collided. For a fixed arrival rate $\lambda = 0.2$ and a finite number of 25 stations, the probability of successful access can be shown graphically (see Figures 2.55 and 2.56). The probability of successful access can also be viewed as the momentary achievable throughput. From this diagram, it can be seen that for a certain number of collided stations (*backlog*) n the corresponding transmission probability is $p = p_r$, which, in the case being considered, possesses the greatest probability for successful access. So Figure 2.55 shows that with 10 collided stations the momentary throughput for $p = 0.1$ is approximately 0.38, whereas for $p = 0.25$ it is only about 0.17 and for $p = 0.5$ it is almost equal to 0.

Figure 2.56 shows a three-dimensional representation of expected throughput as a function of backlog n and the arrival rate λ for the different transmission probabilities $p = 0.5$, 0.25 and 0.1.

The relationships shown above are analyzed more precisely in [17] with respect to the stability of the access channel. The assumption there is that there are m users in the system and that non-collided stations are generating new packets with the probability of p_a. Consideration should be given to the (n, λ) level in which the straight line $\lambda = (m - n)p_a$ is indicated as a straight traffic line. For the fixed value p an equilibrium curve is described through the quantity of points for which the arrival data rate λ is exactly equal to the anticipated throughput rate $S_{out}(n, \lambda)$. An example of this curve is illustrated in Figure 2.57.

Within the shaded area in the figure, $S_{out}(n, \lambda)$ is greater than λ; outside this area, λ is greater than $S_{out}(n, \lambda)$. Three traffic lines corresponding to station numbers m, m' and m'' have been drawn. The arrows on the traffic lines indicate the direction (tendency) in which the number n develops. A traffic line can intersect with the equilibrium curve at several points. These points should be indicated as equilibrium points (n_e, λ_e). An equilibrium point is described as being *stable* if it can be regarded as a sink in respect to tendency n; or it can be regarded as *unstable* if the point considered has the characteristics of a source.

A stable equilibrium point is described as a working point if $n_e \leq n_{max}$; on the other hand, it is referred to as a saturation point if $n_e > n_{max}$. A traffic line is considered stable if it has exactly one stable equilibrium point; otherwise it is considered unstable. In a stable channel the point (n_e, λ_e) determines the throughput and access delay for a limited period of time. On the other hand, an unstable channel displays a *bistable* quality [8]; the throughput there is only achievable for a limited period of time before the working point moves in the direction of a saturation point. Line 3 in Figure 2.57 has a saturation point as its only stable equilibrium point. Therefore the traffic offered on the basis of m'' stations is too high with the given values for p_a and p.

Taking a specific traffic line, p_o can be considered as the optimal transmission probability (this will be explained in more detail in the following sections), where n is minimized in the working point and λ maximized. This value of p can cause the system to become unstable; optimal throughput is only achievable for a limited period of time. The following measures can be undertaken to stabilize the access channel: (1) selection of a smaller value for p or (2) reduction in the number m of stations allowed to transmit. The first option produces a higher value for n in the working point, which could result

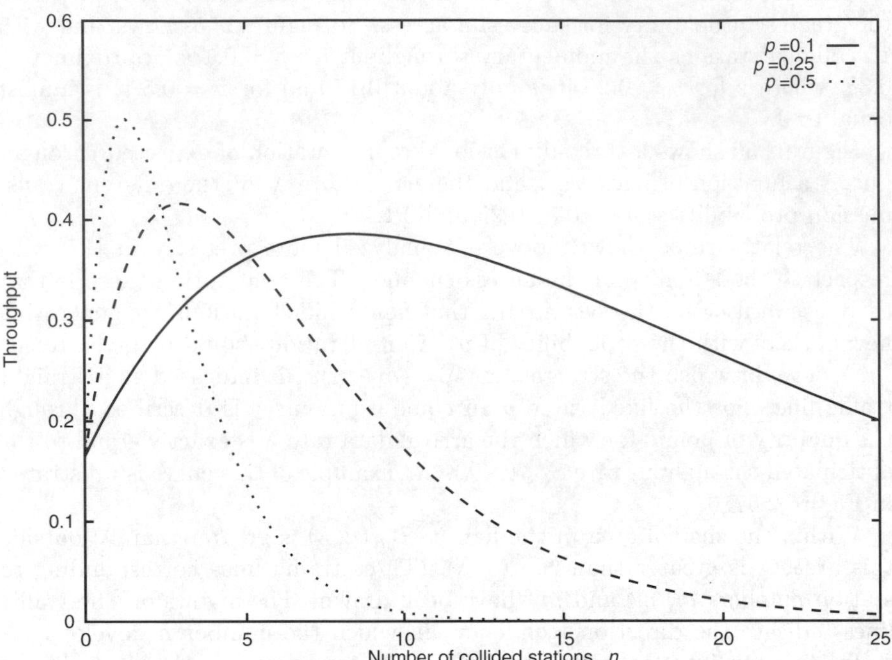

Figure 2.55: Throughput with $\lambda = 0.2; m + n = 25$

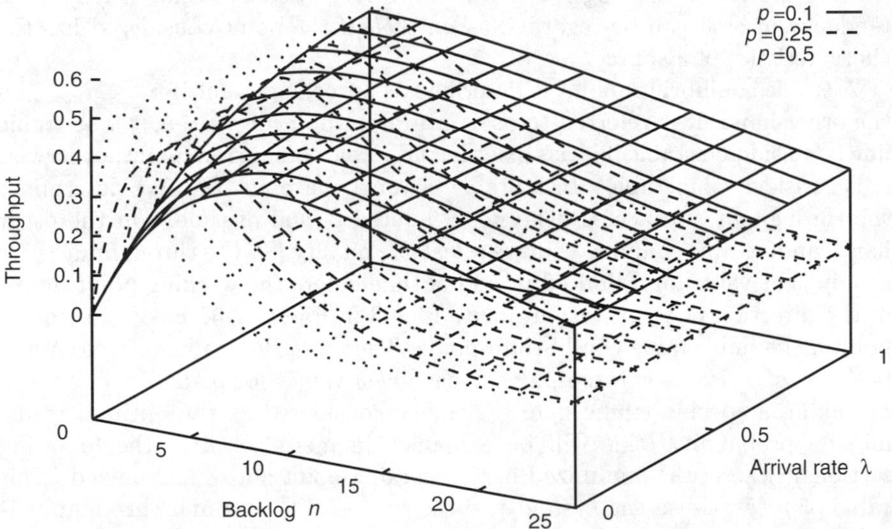

Figure 2.56: Representation of throughput equation (2.66)

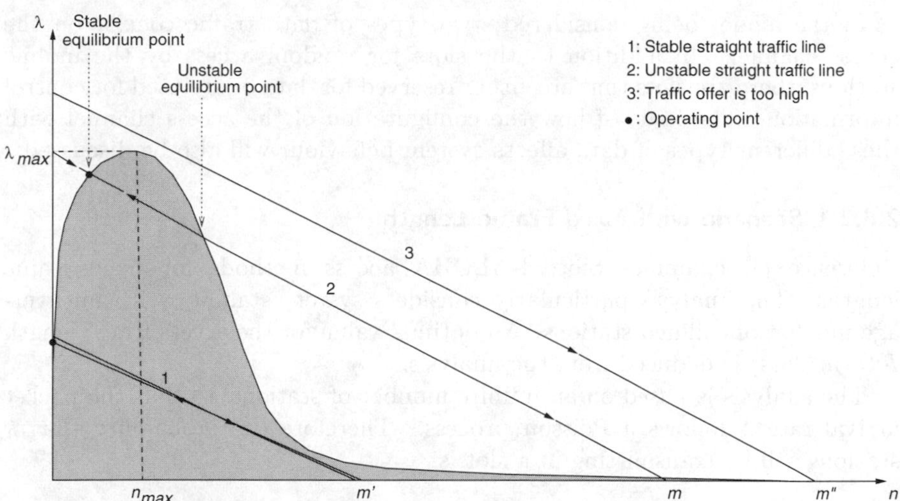

Figure 2.57: Stable and unstable data traffic

in making the access delay unacceptably high; the second option implies that $\lambda \ll \lambda_{max}$ (and thus $p_a \ll 1$) in the working point. However, this would mean a waste of channel capacity. One way to prevent the system from moving into a range of unstable equilibrium is to have it instruct the stations to discard all packets that are in a colliding state.

A better approach is to use the dynamic control algorithms discussed below.

2.8.2 Slotted-ALOHA with Random Access Frames

In most mobile radio systems with FDM/TDM channels the random access channel is not continuously available during frame length FL but only at certain times. The individual mobile stations are able to transmit spontaneously on the access channel over an interval of the length of FL slots. We shall look more closely at how access frames are structured with different user numbers and traffic rates. Consideration will also be given to the fact that slots used for reservations and signalling data are not available for random access.

With the access methods analysed here a base station sends information over access parameters, including frame length FL, to all stations at regular intervals. For its first access attempt a mobile station transmits in the next available slot. It then waits a certain period of time for confirmation from the base station. If the mobile station does not receive confirmation during this period of time (because of a collision or transmission error), it waits until the beginning of the next access frame and then randomly accesses one of the FL slots for a renewed attempt. The base station is able to control the value n of collided stations in such a way that the access channel is not overloaded because of too many collisions.

In the model being considered, two types of data traffic coexist on the access channel. In addition to the slots for random access by the mobile stations, there are slots that are either reserved for data or are used for control information. The issue of how the configuration of the access channel with these different types of data affects system behaviour will also be discussed.

2.8.2.1 Scenario with Fixed Frame Length

Reference [4] examines Slotted-ALOHA access methods for fixed frame lengths. The analysis particularly considers system stability with an average number of collided stations. An optimal value for the access frame length FL (in slots) is deduced from the analysis.

The analysis is based on an infinite number of stations, so that the packet arrival rate λ follows a Poisson process. Therefore the probability that k stations will be transmitting in a slot is

$$p_a(k) = \frac{\lambda^k}{k!} e^{-\lambda} \tag{2.67}$$

At the beginning of an access frame n collided stations from the previous frame attempt to transmit equally distributed in one of the FL slots of the current frame. The probability that a station will transmit in a particular slot is therefore $p = 1/(FL)$, and the probability that i of the n stations will transmit in a slot is derived from the binomial distribution:

$$p_{in} = \binom{n}{i} \left(\frac{1}{FL}\right)^i \left(1 - \frac{1}{FL}\right)^{n-i} \tag{2.68}$$

If the average number of collided stations per access frame is specified as N, then this value is made up of the three contributions N_1, N_2, N_3. The portion N_1 consists of the stations that were not successful in the previous access frame and collided again with the new frame. If i is the number of collided stations from the previous access frame that are transmitting again in the considered slot (and collide again) and k is the number of stations that want to transmit for the first time in the considered slot, then the number of stations colliding in this slot is equal to

$$n_{c1} = \sum_{k=0}^{\infty} (k+i)\, p_a(k) \qquad (i \geq 2) \tag{2.69}$$

The average number N_1 of stations that are colliding again is then

$$N_1 = FL \cdot \sum_{i=2}^{n} p_{in} \sum_{k=0}^{\infty} (k+i)\, p_a(k) \tag{2.70}$$

The contribution N_2 includes the collided stations in the previous frame as well as the stations that are transmitting their packets for the first time in the considered slot:

$$N_2 = FL \cdot p_{1n} \sum_{k=1}^{\infty} (k+1)\, p_a(k) \tag{2.71}$$

The contribution N_3 describes a case in which two or more stations are transmitting for the first time in a particular slot and are colliding. Therefore an access frame that is FL slots in length gives

$$N_3 = FL \cdot p_{0n} \sum_{k=2}^{\infty} k p_a(k) \tag{2.72}$$

The total number of collided stations per access frame is $N_1 + N_2 + N_3$, and according to [4] can be described as follows:

$$N = FL \cdot \left\{ \lambda - \left(1 - \frac{1}{FL}\right)^{n-1} e^{-\lambda} \left[\left(1 - \frac{1}{FL}\right) \lambda + \frac{n}{FL} \right] \right\} + n \tag{2.73}$$

The difference Δ of the number of collided stations from one access frame to the next one is

$$\begin{aligned} \Delta &= N - n \\ &= FL \cdot \left\{ \lambda - \left(1 - \frac{1}{FL}\right)^{n-1} e^{-\lambda} \left[\left(1 - \frac{1}{FL}\right) \lambda + \frac{n}{FL} \right] \right\} \end{aligned} \tag{2.74}$$

The minimum of Δ works out as

$$n_{min} = -\frac{\lambda\,(FL-1) \ln\left[1 - 1/(FL)\right] + 1}{\ln\left[1 - 1/(FL)\right]} \tag{2.75}$$

If the ln function in Equation (2.75) is expanded into a Taylor series (with the condition $FL \gg 1$), it follows that

$$n_{min} = FL \cdot (1 - \lambda) + \lambda \tag{2.76}$$

Δ for an arrival rate $\lambda = 0.2$ and frame lengths FL of 5, 10 and 15 slots is illustrated in Figure 2.58.

This shows that $\Delta(n)$ has two zero values. Furthermore, it is obvious that the first zero value (Z_1) is a stable equilibrium position. Reference [4] has examined system behaviour using different arrival rates and frame lengths as parameters. What emerged was that a larger value for FL also increases the area in which Δ is negative, and, as expected, this contributes towards the stabilization of a system. It was proved that a system is unstable with a fixed frame length—something that is already known from frame-less systems.

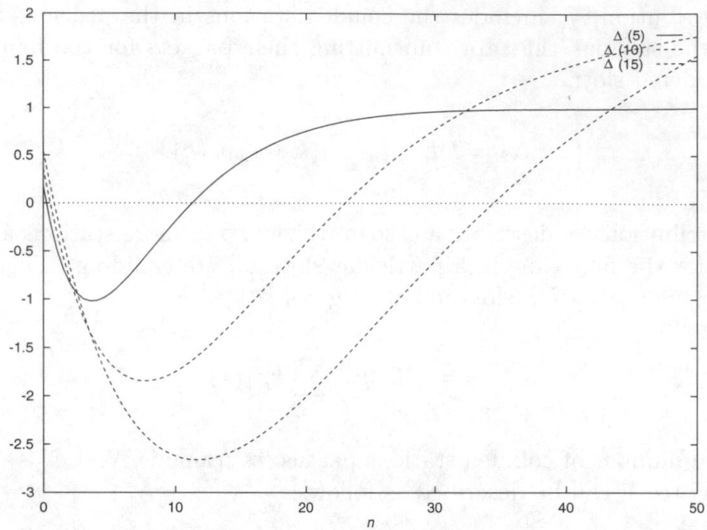

Figure 2.58: Difference $\Delta(FL)$ in the number of collided stations in successive frames when $\lambda = 0.2$

In addition to stability, another important parameter of the access method that is of interest is the average access time T_m, which in [4] is derived from the mean waiting time T_c between collisions of the same packet and the number of all packets n_t sent per frame:

$$T_m = \frac{T_c}{n_t} \tag{2.77}$$

Until now the implication has been that access frames follow each other seamlessly and have the duration T_f. As Figure 2.59 shows, the delay after

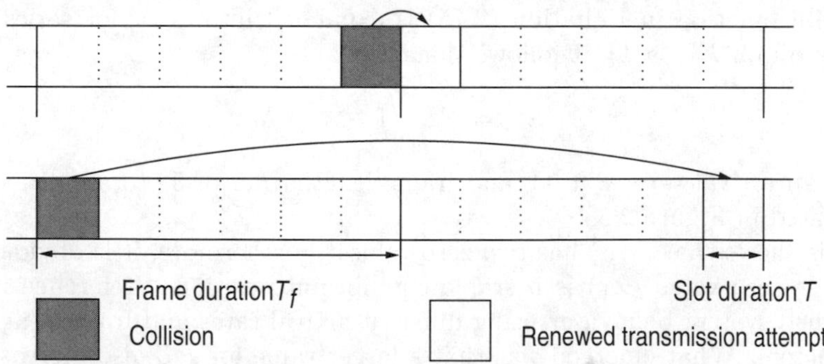

Figure 2.59: Distribution of the delay after a collision

a collided slot is evenly distributed over an interval between 0 and $2T_f$ and therefore $T_c = nT_f$. So Equation (2.77) is expressed as

$$T_m = \frac{\sum_{i=1}^{N_f} n_i T_f}{GN_f \cdot FL} \tag{2.78}$$

with n_i standing for the number of collided packets in the ith slot, N_f the number of preceding frames and G the average number of packets generated during a time slot of length T. The average number of collided packets during a frame is

$$N = \frac{1}{N_f}\sum_{i=1}^{N_f} n_i \tag{2.79}$$

and therefore it follows that T_m is

$$T_m = \frac{T}{G}N = \frac{N}{\lambda} \tag{2.80}$$

with λ denoting the number of packets generated per unit of time.

It is clear from Equation (2.80) that T_m is not directly dependent on FL, but is indirectly dependent on the parameter N, which according to Equation (2.73) is dependent on FL.

2.8.2.2 Scenario with Variable Frame Length

When frame length FL is varied based on the number of collided stations n, it should be noted that although an increase of FL reduces the number n, it also simultaneously increases the access time.

In the previous section we learned how to minimize the difference Δ with a specific arrival rate λ through selection of the frame length FL. The optimal frame length that follows from Equation (2.75) is

$$FL = \frac{n - \lambda}{1 - \lambda} \tag{2.81}$$

This frame length can be considered an adaptively gained optimal frame length for the momentary state of a system. However, the values for λ and n cannot be determined directly by the base station. Within the framework of the different algorithms that optimize throughput in Slotted-ALOHA using the above optimal frame length (or transmission probability), the next section will describe how these values can be determined using different estimation algorithms based on a sequence of empty, collided or successfully transmitted slots.

From Equations (2.81) and (2.74) the difference would be, with a variable frame length,

$$\Delta_v = \frac{n - \lambda}{1 - \lambda}\left[\lambda - \left(\frac{n-1}{n-\lambda}\right)^{n-1} e^{-\lambda}\right] \tag{2.82}$$

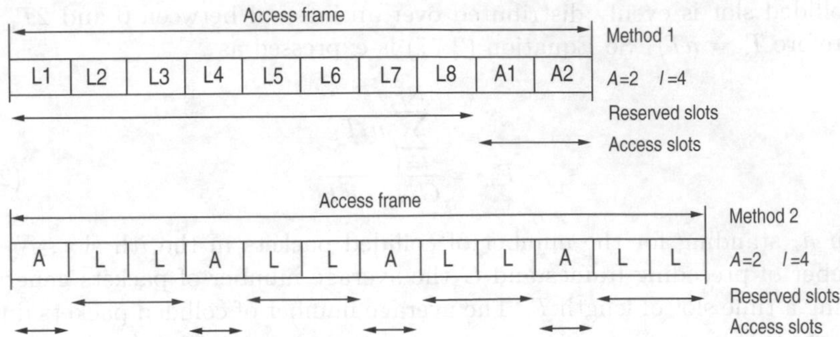

Figure 2.60: Different uses of an access frame

Δ_v is negative for $\lambda < 1/e$ and decreases with increasing n; thus the system is stable. With higher values of λ, however, Δ_v is positive and increases with n; when this is the case, the system is unstable.

Since the analysis assumes an unlimited number of users (to enable the rate λ to be used as a constant), and in practice the frame length FL cannot be varied between 0 and ∞, the result shown in Equation (2.82) should be used with caution.

2.8.2.3 Configuration of an Access Channel

If $FL = A$ is the number of slots available for random access, and lA $(l > 1)$ is the number of reserved slots in a frame, then the frame consists of $A+lA$ slots. The reserved slots can be positioned in different ways in the frame. Reference [4] explores frames in which reserved slots are grouped at the beginning of each frame (method 1) as well as a configuration in which the slots that are reserved and appropriate for access are arranged in alternating order (method 2). This situation is clarified in Figure 2.60.

The difference Δ can also be established analytically and can be determined as a function of n with set parameters λ and FL. The function $A = f(n)$ looks similiar to the one in Figure 2.58, although it is shown that the curve for method 2 already has its first zero positioned at lower values for n and also has a more distinct minimum. Therefore it proves to be more appropriate. It is also possible to visualize this situation by considering that no random access is permitted during the phases of reserved slots and that all stations receiving data during this time will access the first slot released after the reserved period. So, after a long reserved period, there is a backlog of access requests and consequently an increase in the number of collisions in this time slot.

It can be shown [4] that the optimal frame length with method 2 is

$$A = FL_{M2} = \frac{n - \lambda(l+1)}{1 - \lambda(l+1)} \tag{2.83}$$

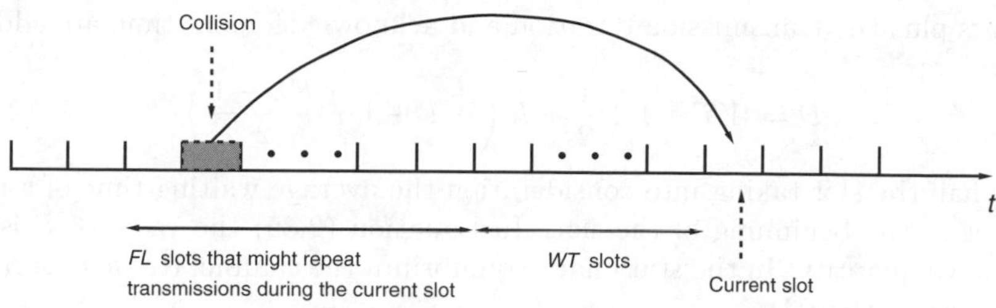

Figure 2.61: Repetition of a collided packet

and produces a minimum number of collided stations. With this method the system only operates in a stable state if the arrival rate is

$$\lambda < e^{-(l+1)} \tag{2.84}$$

2.8.3 Access Delay with Slotted-ALOHA

An important parameter of access algorithms is the access delay. It plays an important role with regard to time-critical services. In some cases it is more important to control the system so that the access delay remains within a particular time requirement than to maximize throughput. As was shown in the previous section, an algorithm that stabilizes the system to an optimum also minimizes the access delay at the same time.

Therefore the following discussion describes how access delay is dependent on the parameters waiting time WT and frame length FL. This is followed by a description of an algorithm used to determine the access delay from the calculated approximation for the arrival rate λ and the backlog of collided packets n.

2.8.3.1 Derivation of Access Delay

The access delay denotes the period of time from the first attempt to transmit to the completion of successful transmission including the duration of the transmission. A Slotted-ALOHA system with a finite number of stations is considered below. Data packets are produced at the total arrival rate λ. Stationary equilibrium is assumed. If a packet is not acknowledged after a fixed period of WT slots, it is resent k slots later, with k being randomly selected uniformly from the FL slots (see Figure 2.61).

The average delay time D of a packet is defined in [15] as the duration of the packet transmission (1 slot) plus the acknowledgement time WT when transmission has been successful after the first attempt. With R as the number of repeated transmissions, R times the average waiting time between two attempts plus the transmission duration and acknowledgement time are added. Therefore

$$D = WT + 1 + \frac{1}{2} + R\left(WT + 1 + \frac{FL - 1}{2}\right) \tag{2.85}$$

with half the slot taking into consideration the average waiting time of a new packet at the beginning of the slot. In Equation (2.85) the value of R is an unknown quantity. In the stationary equilibrium the channel traffic G is $R+1$ times larger than λ:

$$G = \lambda(1 + R) \tag{2.86}$$

R is established by defining p_n as the probability for the successful transmission of newly generated packets and p_r as the success probability of a previously collided packet. Thus the following applies for the probability $P\{i\}$ that exactly i repetitions are required until successful transmission has taken place:

$$P\{i\} = (1 - p_n)\, p_r\, (1 - p_r)^{i-1} \tag{2.87}$$

The average number of repetitions is then

$$R = \sum_{i=1}^{\infty} iP\{i\} = \frac{(1 - p_n)\, p_r}{[1 - (1 - p_r)]^2} = \frac{1 - p_n}{p_r} \tag{2.88}$$

Here p_n and p_r are unknown probabilities, which have been derived in [10]:

$$p_r = \frac{1}{1 - e^{-G}}\left(e^{-\frac{G}{FL}} - e^{-G}\right)\left(e^{-\frac{G}{FL}} + \frac{G}{FL}e^{-G}\right)^{FL-1} e^{-\lambda} \tag{2.89}$$

$$p_n = \left(e^{-\frac{G}{FL}} + \frac{G}{FL}e^{-G}\right)^{FL} e^{-\lambda} \tag{2.90}$$

The unknowns in Equations (2.85) and (2.86) are derived on this basis, and the delay can be deduced from the arrival rate and the momentary channel traffic in the following two equations:

$$D = WT + 1 + \frac{1}{2} + \frac{1 - p_n}{p_r}\left(WT + 1 + \frac{FL - 1}{2}\right) \tag{2.91}$$

$$\lambda = \frac{G}{1 + R} = \frac{Gp_r}{1 + p_r - p_n} \tag{2.92}$$

Equations (2.91) and (2.92) have been evaluated numerically in [15] for a finite number of stations, and are represented for different frame lengths in Figure 2.62.

2.8.3.2 Algorithms for Determining Access Delay

This section presents a method adapted from [10] that enables access delay to be calculated from the system values for arrival rate and backlog.

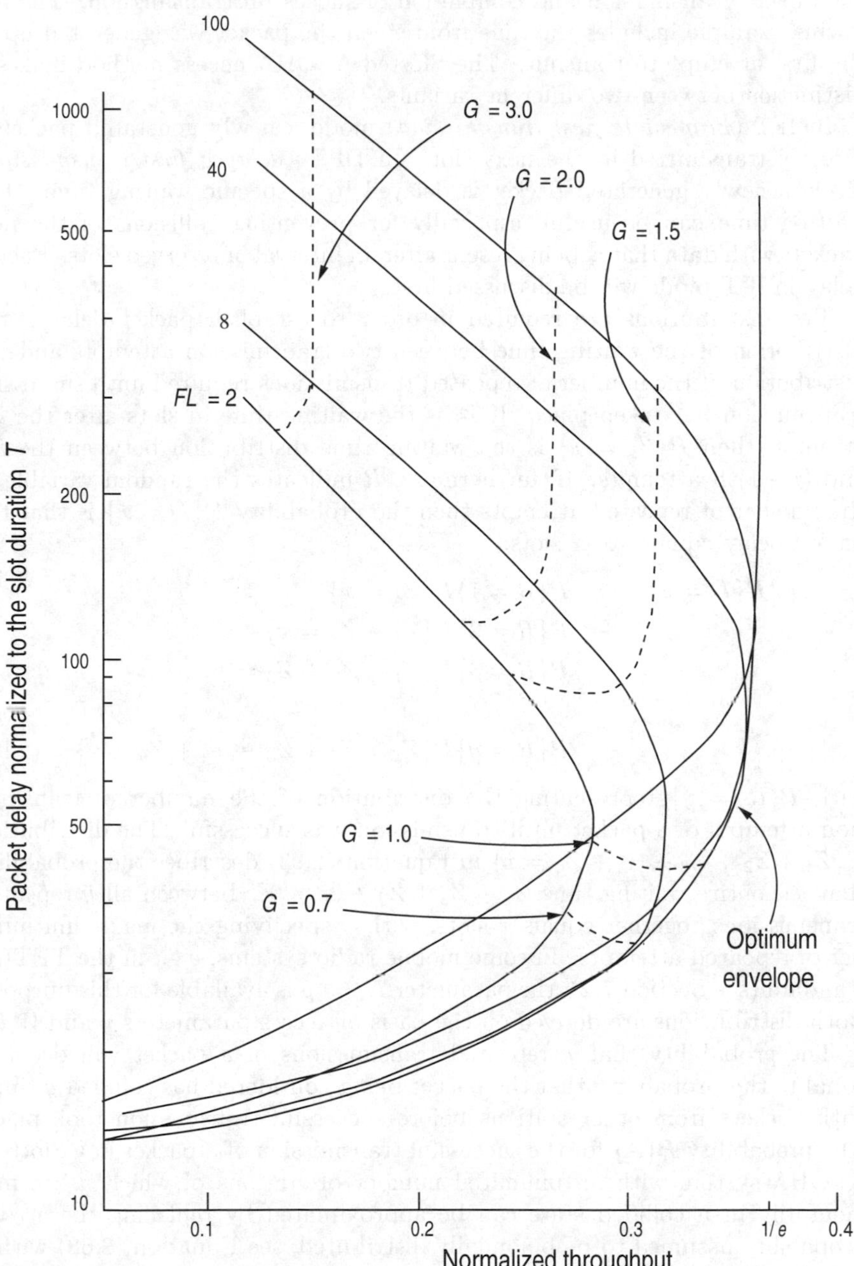

Figure 2.62: Packet delay versus throughput

Until now packet delay D has been defined as the time between the first attempt of transmission and completion of successful transmission. The following example includes the time from when the packet was generated up to the first attempt to transmit. The Slotted-ALOHA access method makes a distinction between two different variants.

In IFT (*immediate first transmission*) mode a newly generated packet is directly transmitted in the next slot. In DFT (*delayed first transmission*) mode a newly generated packet is delayed by a specific waiting time. The waiting time can be useful, especially for preventing collisions of the new packets with data that is being resent after an interval of reserved slots. Packet delay in IFT mode will be discussed first.

Two distributions are required in order to establish packet delays: the distribution of the waiting time between two transmission attempts and the distribution of the number of repeated transmissions required until successful transmission has taken place. If Z_i is the waiting time in slots after the ith attempt, then $P\{Z_i = x\}$ is the waiting time distribution between the ith and $(i+1)$th attempts. If, furthermore, R indicates the random variable of the number of repeated attempts then the probability $P\{D = x\}$ is that the packet delay equates to x slots:

$$\begin{aligned} P\{D=x\} &= P\{R=1\}P\{Z_1=x\} \\ &+ P\{R=2\}P\{Z_1+Z_2=x\} \\ &+ P\{R=3\}P\{Z_1+Z_2+Z_3=x\} \\ &+ \ldots \\ &+ P\{R=y\}P\{Z_1+Z_2+Z_3+\ldots+Z_y=x\} \end{aligned} \tag{2.93}$$

with $P\{R = y\}$ representing the distribution of the number of transmission attempts of a packet until transmission was successful. The distribution $P\{Z_1 + Z_2 + Z_3 + \ldots + Z_y = x\}$ in Equation (2.93) describes the probability that the entire waiting time $Z_1 + Z_2 + Z_3 + \ldots + Z_y$ between all y repeated transmissions together equals x slots, with y specifying the maximum number of repeated attempts. In some mobile radio systems, e.g., in the TETRA standard (see Section 7.2), the parameter $N_u = y$ is available for this purpose. Both distributions are derived on the basis of access parameters y and WT.

The probability that y repeated transmissions of a packet will occur is equal to the probability that the packet being considered has collided y times with packets from other stations before successful transmission took place. The probability $P\{A\}$ for the successful transmission of a packet in a Slotted-ALOHA system with an unlimited number of stations of which n are momentarily in a collided state can be approximated by replacing the arrival probability assumed to be binomially distributed; see Equation (2.68) with a Poisson distribution according to Equation (2.67). The probability of success is then

$$P\{A\} = \lambda e^{-\lambda}\left(1-\frac{1}{FL}\right)^n + e^{-\lambda} n \frac{1}{FL}\left(1-\frac{1}{FL}\right)^{n-1} \tag{2.94}$$

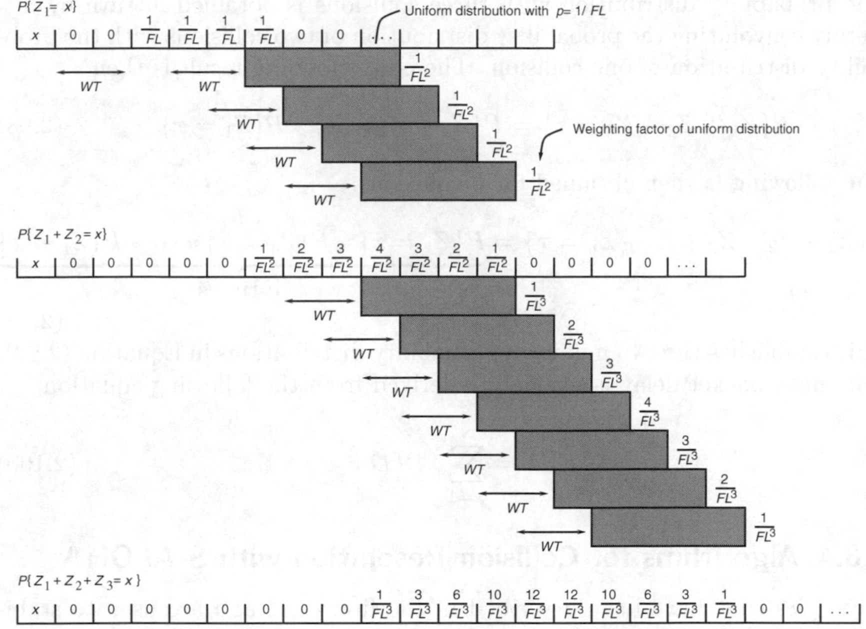

Figure 2.63: Waiting time distribution with $WT = 2$ and $FL = 4$

The probability that transmission will not be successful is therefore $P\{\bar{A}\} = 1 - P\{A\}$. This defines the distribution of the number of repeated attempts as

$$P\{R = y\} = P\{\bar{A}\}^y P\{A\} = (1 - P\{A\})^y P\{A\} \tag{2.95}$$

The final element that still needs to be determined is the waiting time distribution of a packet whereby the waiting time in y waiting intervals together equals x slots.

First, the probability that the time between a collision and a renewed attempt to transmit equals x slots is deduced on the basis of WT. This involves expressing the random distribution $P\{Z_1 = x\}$ independent of WT and FL as a discrete sequence of numbers. Thus

$$P\{Z_1 = x\} = \begin{cases} 0, & \text{for } 0 \le x \le WT \\ 1/(FL), & \text{for } WT < x \le WT + FL \\ 0, & \text{for } x > WT + FL \end{cases} \tag{2.96}$$

As the example in Figure 2.63 shows, the probability distribution with two collisions is produced from the discrete convolution of the probability distribution with one collision. Thus

$$P\{Z_1 + Z_2 = x\} = P\{Z_1 = x\} * P\{Z_1 = x\} \tag{2.97}$$

The probability distribution with three collisions is obtained in turn by discretely convoluting the probability distribution of two collisions with the probability distribution of one collision. The corresponding result is then

$$P\{Z_1 + Z_2 + Z_3 = x\} = P\{Z_1 + Z_2 = x\} * P\{Z_1 = x\} \tag{2.98}$$

The following is then obtained for y collisions:

$$P\{Z_1 + Z_2 + Z_3 + \ldots + Z_y = x\} = \underbrace{P\{Z_1 = x\} * P\{Z_1 = x\} * \cdots * P\{Z_1 = x\}}_{y \text{ times}} \tag{2.99}$$

This establishes the two unknown probability distributions in Equation (2.93). The mean packet delay is ultimately derived from the following equation:

$$E[D] = \sum_{x=1}^{\infty} xP\{D = x\} \tag{2.100}$$

2.8.4 Algorithms for Collision Resolution with S-ALOHA

Strategies that have been described in the literature and are used to stabilize Slotted-ALOHA access procedures while optimizing throughput are introduced below.

First we shall introduce some simple procedures that always involve evaluating the usage of the last slot, and on the basis of this evaluation we shall determine the new access probability for stations in a state ready to transmit. This is followed by a discussion of the more complicated algorithms that determine the transmission probability on the basis of a method for estimating the *backlog*, i.e., the number of collided packets.

The methods presented in this section can be used to control the access probability of stations centrally, thereby stabilizing a Slotted-ALOHA system, keeping access delays to a minimum and maximizing throughput.

2.8.4.1 Exponential-Backoff Algorithm

Backoff algorithms (for controlling renewed transmission of a collided packet) are very simple methods for achieving stability. They are used, for example, in standard IEEE 802.3-Ethernet-LANs. With this method the access channel is controlled in such a way that the access probability in each station is determined as $p_r = 2^{-i}$ on the basis of the number of collisions i of a packet, which means that access is distributed equally over the next 1 to 2^i slots. This algorithm extends the period of time in which a packet can be retransmitted by the number of collisions, thereby controlling the length of the access frame (which, however, ends here after successful access has taken place).

In the method suggested in [14], a station observes an access channel and determines whether the last slot has been accessed successfully, a collision has occurred or the slot has remained empty. If the number of stations transmitting in slot k is denoted by $z(k)$, then the result is $z(k) = 0$, $z(k) = 1$ or $z(k) \geq 2$ depending on whether the last slot was empty, transmission was successful or a collision occurred. With $p_r(k)$ as the transmission probability and I_A as the state function in slot k, where A is defined by $z(k)$,

$$p_r(k+1) = \min\left\{p_{max}\ ,\ p_r(k)\left(\frac{1}{q}I_{z(k)=0} + I_{z(k)=1} + qI_{z(k)\geq 2}\right)\right\} \qquad (2.101)$$

in which $0 < p_{max} \leq 1$, $0 < q \leq 1$, and I_A has the value 1 if A occurs; otherwise I_A equals zero.

The above algorithm implies that the station serving as an example considers the momentary value of p to be too low (high) because of an empty (collided) slot and therefore increases (decreases) the value of p. The value of p_{max} is 1, and the values $\frac{1}{4}$, $\frac{1}{2}$ and $\frac{1}{\sqrt{2}}$ are examined in [14] for the parameter q. Since this paper likewise assumes that newly generated packets are also being sent with the probability p, the approach used is the DFT mode of Slotted-ALOHA, and the throughput can immediately be determined using the fixed parameter p_r. If P_r in Equation (2.56) is taken for the value $i = 1$, since all stations are transmitting with the same probability, it follows that the throughput S_{DFT} of the Slotted-ALOHA protocol with fixed p_r is given by

$$S_{DFT}(m) = \binom{m}{1} p_r(1-p_r)^{m-1} \qquad (2.102)$$

For $m \geq 2$ the exponential backoff method can be described as a discrete Markov chain [14]. The system states $X_k(k = 0, 1, ...)$ in this description are selected so that $p_r = q^k$. If π_k is the state probability of the state X_k then it follows from the analysis of the Markov chain that the throughput of the exponential backoff algorithm is given by

$$S_{EB}(m) = \sum_{k=0}^{\infty} \binom{m}{1} p_r(1-p_r)^{m-1}\pi_k \qquad (2.103)$$

According to Equations (2.102) and (2.103) in [14], throughput is calculated as a function of n with different parameters p_r or q, with π_k being deduced from a recursion equation. The method with fixed p_r shows that with $m \to \infty$ the system becomes unstable and throughput is zero. In comparison, S_{EB} for a given parameter q converges with a specific positive value, even for very large values m. For $q = 0.5$ and with $m = 1000$ stations the result is $S_{EB} \simeq 0.311$; this corresponds to about 85 % of the maximum achievable throughput e^{-1} and in normal operations equates to a satisfactory result. The collision resolution method presented in [14] has therefore been proved to be stable.

2.8.4.2 The Pseudo-Bayesian Algorithm

This algorithm (see [28]) is based on a method that establishes the estimated value of a backlog in order to stabilize the Slotted-ALOHA access channel.

In contrast to the versions of Slotted-ALOHA access explored previously, with this method newly arriving packets are also sent with the probability p_r. Therefore the backlog n here is a combination of packets that have already collided and new ones. The access rate is consequently $G(n) = np_r$ and the probability of an access being successful is $np_r(1-p_r)^{n-1}$. However, this modification has no major effect on access delay if the parameter p_r is around 1 when traffic volume is low.

The task of the algorithm is to determine an estimated value $\hat{n}$ for the number of collided stations n at the beginning of each slot. Each packet will then be transmitted with probability

$$p_r(\hat{n}) = \min\{1, 1/\hat{n}\} \tag{2.104}$$

The min operation establishes an upper limit for the transmission probability and causes the access rate $G = np_r$ to become 1. For a given slot k a new estimated value for n is determined for the next slot $k+1$ based on the following rules.

If an empty or a collided slot is received, the estimated value for the backlog is calculated as

$$\hat{n}_{k+1} = \max\{\lambda, \hat{n}_k + \lambda - 1\} \tag{2.105}$$

with λ standing for the arrival rate normalized to the slot duration. The addition of λ to the previous backlog value applies to newly generated packets, whereas the max operation ensures that the estimated value is never lower than the contribution of the newly arrived packets. A successful transmission is considered by subtracting 1 from the previous estimated value of the backlog. A 1 is also subtracted for each empty slot, thereby causing a reduction to $\hat{n}$ if there are too many empty slots.

If a collision is detected, the new estimated value of the backlog is calculated as

$$\hat{n}_{k+1} = \hat{n}_k + \lambda + (e-2)^{-1} \tag{2.106}$$

Here again λ is added to the previous backlog value to take into account the newly arrived packets. The addition of $(e-2)^{-1}$ effects an increase to $\hat{n}$ if the number of collisions is too high. For high backlog values and an appropriate accurate estimated value for the backlog, the access rate $G(n)$ therefore becomes 1. If this rate is approximated using the Poisson distribution, then empty slots occur with the probability $1/e$ and collisions with the probability $1-2/e$. Consequently the anticipated change $\delta(\hat{n})$ for the estimated value $\hat{n}$ because of empty and collided slots can be stated as

$$\delta(\hat{n}) = \frac{-1}{e} + \left(\frac{e-2}{e}\right)\left(\frac{1}{e-2}\right) = 0 \tag{2.107}$$

If this algorithm is used in practice, there is the problem that λ will not be known at the outset and furthermore that it varies. Therefore the method must either estimate the rate λ from the average values of successfully transmitted packets (this will be covered in more detail later) or the algorithm will determine a fixed value. It has been proved (in [5] and elsewhere) that the system remains stable at $\lambda < 1/e$.

The following approximation is given in [28] to estimate the arrival rate λ:

$$\hat{\lambda}_{k+1} = 0.995\hat{\lambda}_k + 0.005 I_{z(k)=1} \tag{2.108}$$

2.8.4.3 Stabilization through MMSE Estimation

Another approach for stabilizing Slotted-ALOHA and reducing access delay times is offered by the MMSE (*minimum mean-squared error*) method, which estimates the number of collided stations. With this method transmission probability is determined dynamically on the basis of the estimated collision rate in order to maximize throughput. The technique presented in [35] is based on the same assumption as the one in the Slotted-ALOHA model that newly generated packets will be sent immediately in the next available slot. This is therefore the IFT mode of Slotted-ALOHA.

The control algorithm presented determines the transmission probability

$$p_r(k) = \min\left(\beta\ ,\ \frac{\alpha}{\hat{n}_k}\right) \tag{2.109}$$

with $\hat{n}_k$ as the MMSE estimation of the number of collided stations at the beginning of the kth time slot. The parameter β is to be selected so that $0 \leq p_r(k) \leq 1$. The study of the method in [35] was based on $\beta = 0.5$. The parameter α, which is supposed to minimize access delay times, will be covered in detail later. First, the asymptotic MMSE estimation of the number of collided stations in the $(k+1)$th slot is given:

$$\hat{n}_{k+1} = \hat{n}_k + \begin{cases} -\alpha/(\lambda+\alpha), & \text{if } I_{z(k)=1} \\ 0, & \text{if } I_{z(k)=0} \\ \lambda/[1-(\lambda+\alpha)/(e^{\lambda+\alpha}-1)], & \text{if } I_{z(k)\geq 2} \end{cases} \tag{2.110}$$

and

$$\hat{n}_{k+1} \leftarrow \max(0\ ,\ \hat{n}_{k+1}) \tag{2.111}$$

I_A is the state function again (see Equation (2.101)). Since $n_t \geq 0$ in a real system, Equation (2.111) guarantees that the estimated value $\hat{n}_t \geq 0$ will be maintained.

The access method is stable for arrival rates that satisfy the condition

$$\lambda \leq (\lambda+\alpha)e^{-(\lambda+\alpha)} \tag{2.112}$$

The maximum throughput (on the basis of a theorem mentioned in [35]) is achieved for this α, which allows both sides in Equation (2.112) to have about the same value. This applies to

$$\alpha = 1 - \lambda \tag{2.113}$$

which is synonymous with

$$\lambda_{max} = e^{-1} \tag{2.114}$$

The estimation algorithm is thereby simplified to

$$\hat{n}_{k+1} = \hat{n}_k + \begin{cases} -(1-\lambda), & \text{if } I_{z(k)=1} \\ 0, & \text{if } I_{z(k)=0} \\ 2.392\lambda, & \text{if } I_{z(k)\geq 2} \end{cases} \tag{2.115}$$

This recursive formula assumes knowledge of λ. The following algorithm for determining this arrival rate appears in [35]:

$$\hat{\lambda}_{k+1} = \frac{1}{k+1}\left[\hat{n}(k+1 \mid k,\, \hat{\lambda}_k) + \sum_{j=0}^{k} \epsilon_j + \lambda_0\right] \tag{2.116}$$

with

$$\epsilon_j = \begin{cases} 1, & \text{if } I_{z(k)=1} \\ 0, & \text{otherwise} \end{cases} \tag{2.117}$$

in which $\hat{n}(k+1 \mid k,\, \hat{\lambda}_k)$ represents the MMSE estimation at the time of the $(k+1)$th slot and takes into consideration all slots up to the kth slot and the rate $\hat{\lambda}_k$ as the value of the actual arrival data rate. The constant $\lambda_0 \in (0,\, e^{-1})$ is required in order to initialize the estimated algorithm. The parameter α given in Equation (2.110) can be adapted using $\hat{\lambda}_k$.

2.8.4.4 Stabilization Using Stochastic Approximation

Another technique used to optimize throughput and to keep it approximately constant is suggested in [11]. This algorithm does not take into account the number of collided stations, but instead presents a method, based on a traffic theoretic analysis, that each time produces a new transmission probability $p_r(k)$ at the point of the kth slot according to feedback on channel state. The technique determines $p_r(k)$ so that the throughput, which according to Equation (2.64) can be given as $S_{out}(k) \approx G(k)e^{-G(k)}$ and is not dependent on the momentary value of n, is always close to its optimal value; thus $G(k) = 1$.

The method is based on the transmission result in the last slot, and determines the new transmission probability depending on whether there is an empty slot, a successful transmission attempt or a collision:

$$p_r(k+1) = \min\left\{p_{max}\ ,\, p_r(k)\left(a_0 I_{z(k)=0} + a_1 I_{z(k)=1} + a_c I_{z(k)\geq 2}\right)^{\gamma}\right\} \tag{2.118}$$

with

$$(a_0, a_1, a_c) = \left(e^{\frac{1-2e^{-1}}{1-e^{-1}}}, 1, e^{\frac{-e^{-1}}{1-e^{-1}}}\right) \tag{2.119}$$

A concrete value for p_{max} is not given in [11]. However, an analysis is presented showing that when two stations are competing for a channel, throughput will be maximized if

$$p_{max} = \frac{1-\lambda}{2-\lambda} \tag{2.120}$$

p_{max} is dependent on the arrival rate, which has to be estimated and in some circumstances can vary. Therefore [9] gives a value for p_{max} that is independent of this rate. It uses the highest possible packet rate that keeps the system stable ($\lambda = 1/e$) in Equation (2.120) and therefore produces $p_{max} \approx 0.38$. For the parameters in Equation (2.119), $(a_0, a_1, a_c) \approx (1.52, 1, 0.56)$ applies. The behaviour of a system with the values of the parameters γ lying between 0.1 and 4 is explored in [11]. It emerges that the average throughput of a system decreases with γ. The average number of collided stations is relatively high, however, with values of γ in the area $\gamma \leq 0.1$ and an arrival rate of $\lambda = 0.32$. In [9] $\gamma = 0.3$ is regarded as optimal.

2.8.4.5 The CONTEST Algorithm

Different methods, including the RCP (*retransmission control procedure*), are suggested in [16] for stabilizing the Slotted-ALOHA protocol. With this procedure transmission probability is determined as p_o or p_c, depending on traffic volume. The value p_o is selected so that throughput is optimized and p_c is low enough to stabilize the system.

The CONTEST (*Control-Estimation*) algorithm is based on this procedure. The decision whether to use transmission probability p_o or p_c depends on whether the number of collided stations n exceeds a particular limit value $\hat{n}$. With this value of n the traffic rate of the access channel is

$$\hat{G}_o = \hat{n}p_o + (m-n)p_a \qquad (2.121)$$

$$\hat{G}_c = \hat{n}p_c + (m-n)p_a \qquad (2.122)$$

with m as the total number of stations and p_a as the probability of new packets being sent. On the assumption that the traffic volume on the access channel is Poisson-distributed, the following critical values can be given because of the probability of an empty slot: $\hat{f}_o = e^{-\hat{G}_o}$ and $\hat{f}_c = e^{-\hat{G}_c}$. Furthermore, it is also assumed that the transmission result in earlier slots is observed over an interval of length W slots. If $\bar{f}$ is also the portion of empty slots in interval W up to time slot k, then $\bar{f}$ approximates the probability of an empty slot in time slot $k+1$, on the assumption that there will be little change to the traffic rate during the period of time being considered. The following algorithm is given, in which $p(k)$ is the transmission probability to be determined for the new slot.

1. Step:
   - $k \leftarrow k+1$
   - $p(k) = p_o$
2. Step:
   - if $\bar{f} < \hat{f}_o$ GOTO step 4.
3. Step:
   - GOTO step 1.
4. Step:
   - $k \leftarrow k+1$
   - $p(k) = p_c$
5. Step:
   - if $\bar{f} < \hat{f}_o$ GOTO step 1.
6. Step:
   - GOTO step 4.

For p_o either a low transmission probability or the optimal one dependent on n and λ, derived in the next section, can be used. For p_c the value $p_c = K \cdot n$ with $K \geq 1$ is suitable for the satisfactory stabilization of a system.

Therefore the above algorithm can be expanded in all sorts of different ways. Similarly to a high-traffic-load situation, observation of the access channel can produce a limit for determining a low-traffic-load situation and provide an appropriate procedure for adapting the access parameters.

2.8.4.6 Adaptive Determination of Optimal Transmission Probability

The optimal frame length derived from Equation (2.83) corresponds to the optimal transmission probability $p_r = (1-\lambda)/(n-\lambda)$ stated in [7] and [23].

In [23] the mean value $\bar{n}$ is determined from the sequence of empty slots. It is assumed thereby that n as well as the arrival rate λ remain almost constant during an observed period of time of length x slots. The probability of an empty slot is then the product of the probability that new stations are not transmitting in slots being considered and the probability that none of the stations which has already collided is transmitting, and can be indicated as

$$P_0 \approx (1-p_r)^{\bar{n}} e^{-\lambda} \tag{2.123}$$

Furthermore, if an interval of x slots is given in which each one is empty with a probability P_0, and if e_0 is the number of empty slots in the period being observed, then e_0 is binomially distributed with mean value $E_0 = xP_0$. Then, from Equation (2.123), n becomes

$$n \approx \frac{\ln(E_0/x) + \lambda}{\ln(1-p_r)} \tag{2.124}$$

Reference [7] presents a different algorithm for deducing the value of n as well as the optimal transmission probability for collided stations p_r. Based on this algorithm, p_r is determined depending on n according to

$$p_r(\hat{n}) = \begin{cases} (1-\lambda)/(\hat{n}-\lambda), & \text{if } \hat{n} \geq 1 \\ 1, & \text{if } 0 \leq \hat{n} < 1 \end{cases} \tag{2.125}$$

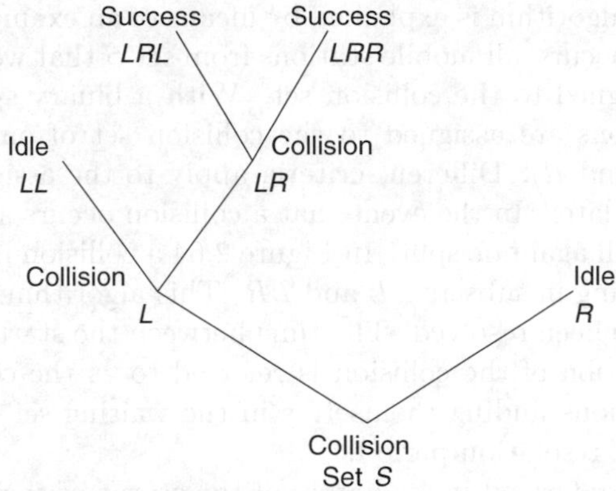

Figure 2.64: Example of a splitting algorithm

The new number of collided stations in each slot is determined using the following estimation algorithm based on the feedback channel state information:

$$\hat{n}_{k+1} = \hat{n} + U(f_k, \hat{n}_k) \tag{2.126}$$

with

$$U(f_k, \hat{n}) = \max(\hat{n}_{min} - \hat{n}, u_f), \qquad \text{with } \hat{n} \geq \hat{n}_{min} \text{ and } f = 0, 1, c \tag{2.127}$$

In the algorithm, $\hat{n}_{min}$ represents the lower limit; u_0, u_1 and u_c are the real constants. Reference [7] gives the parameters $u_0 = 2 - e \approx -0.7183$, $u_1 = 0$, $u_c = 1$ and $\hat{n}_{min} = 2$ for this algorithm when the arrival rate is $\lambda = 0.32$. A simulation experiment produces the lowest average access delay when $\lambda = 0.32$ with the parameters $(u_0, u_1, u_c) = (-0.4, -0.4, 0.9)$.

2.8.4.7 The Splitting Algorithm

The splitting algorithm is another method for resolving collisions. This entails assigning one of two possible sets to each station that wants to transmit using random access. All stations that are not participating in a current collision resolution are in a *waiting set*. If a new transmission request originates in a station, it takes its place in the waiting set. All the mobile stations that are participating in the current collision resolution phase are found in the *collision set*. At the beginning of a collision resolution phase, part of the waiting set is allocated to the previous empty collision set and therefore participates in the collision resolution. No other stations can be allocated subsequently to the collision set. Stations that have transmitted successfully are dropped from the collision set. This process is not completed until the collision set has been completely emptied. Then a new collision resolution phase begins with a new collision set.

The splitting algorithm is explained by means of an example in Figure 2.64. After a collision occurs, all mobile stations from set S that were involved in the collision are assigned to the collision set. With a binary splitting algorithm the mobile stations are assigned to the collision set of one of the two new subsets (sets L and R). Different criteria apply to the assignment and these will be discussed later. In the event that a collision occurs again in one of the two subsets, it will again be split. In Figure 2.64 a collision has occurred in set L, thereby resulting in subsets LL and LR. This algorithm is continued until all collisions have been resolved. The time between the start of a collision and the actual resolution of the collision is referred to as the collision resolution period. The stations finding themselves in the waiting set wait for the start of a new collision resolution period.

If stations are informed immediately of the status of their access attempt, this information can also be used in resolving the collision. These two improvements to the splitting algorithm help to increase throughput. These are introduced here.

There are different ways in which mobile stations can be allocated to the subsets:

- They are distributed evenly among the subsets.
- The distribution is according to arrival time.
- The mobile station calculates a priority for itself and uses it to determine the subset.
- Parts of the mobile station's identification number are used in the allocation.

Optimizing the splitting algorithm If information about the success of the collision resolution of the other sets is available, there are two possibilities for optimizing the splitting algorithm to increase throughput.

In Figure 2.64 a collision has occurred in subset L and no mobile station has transmitted in subset LL (`Idle`). Therefore a collision would inevitably occur with subset LR, so this subset can immediately be split into LRL and LRR. After a collision has occurred in subset L, i.e., at least two mobile stations have allocated themselves to this subset, the probability that a third station will find itself in subset R is small. It is therefore more advantageous to assign subset R to the waiting set again. These two improvements will increase throughput to 43 % (without optimization) and to 48.7 % (with optimization) instead of the maximum throughput of 36.8 % without splitting.

Bibliography

[1] N. Abramson. *The ALOHA system—Another alternative for computer communication.* In *Fall JOINT Computer Conference*, Vol. 37, pp. 281–285. AFIPS Conference Proceedings, 1970.

[2] N. Abramson. *Packet switching with satellites.* In *NCC*, Vol. 42, pp. 695–703, New York, AFIPS Conference Proceedings, 1973.

[3] A. Baier, W. Koch. *Potential of CDMA for 3rd generation mobile radio.* In *Mobile Radio Conference*, pp. 282–290, Nice, France, Nov. 1991.

[4] G. Benelli, G. F. Cau, A. Radelli. *A performance evaluation of Slotted Aloha multiple access algorithms with fixed and variable frames for radiomobile networks. IEEE Transactions on Vehicular Technology*, Vol. 43, pp. 181–193, 1994.

[5] D. Bertsekas, R. Gallager. *Data Networks, 2nd edn.* Prentice-Hall, Englewood Cliffs, New Jersey, 1992.

[6] G. K. Chan. *Effects of sectorization on the spectrum efficiency of cellular radio systems. IEEE Transactions on Vehicular Technology*, Vol. 41, pp. 217–225, 1992.

[7] L. P. Clare. *Control Procedures for Slotted-ALOHA systems that achieve stability.* In *ACM SIGCOMM'86 Symp.*, pp. 302–309, Aug. 1986.

[8] A. B. Corleial, M. E. Hellmann. *Bistable behaviour of Aloha-type systems. IEEE Transactions on Communications*, Vol. 23, pp. 404–409, 1975.

[9] G. A. Cunningham. *Delay versus throughput comparisons for stabilized Slotted ALOHA. IEEE Transactions on Communications*, Vol. 38, No. 11, pp. 1932–1934, 1990.

[10] J. Gotthard. *Paketdatennetze*, 1991.

[11] B. Hajek, T. van Loon. *Decentralized dynamic control of a multiaccess broadcast channel. IEEE Transactions on Automatic Control*, Vol. 27, pp. 559–569, 1982.

[12] M. Hata. *Empirical formula for propagation loss in land mobile radio services. IEEE Transactions on Vehicular Technology*, Vol. 29, pp. 317–325, 1980.

[13] ITU. Telecommunication Standardization Sector (ITU-T). *Information Technology—Open Systems Interconnection (OSI)—Basic Reference Model: The Basic Model.* ITU-T Recommendation X.200, International Telecommunication Union (ITU), Geneva, 1994.

[14] D. G. Jeong, W. S. Jeong. *Performance of an exponential backoff scheme for Slotted-Aloha protocol in local wireless environment. IEEE Transactions on Vehicular Technology*, Vol. 44, pp. 470–479, 1995.

[15] L. Kleinrock, S. Lam. *Packet-switching in a slotted satellite channel.* In *National Computer Conference AFIPS Conf. Proc*, Vol. 39, pp. 703–710. AFIPS Press, 1973.

[16] L. Kleinrock, S. Lam. *Packet switching in a multiple broadcast channel: dynamic control procedures. IEEE Transactions on Communications*, Vol. 23, pp. 891–904, 1975.

[17] L. Kleinrock, S. Lam. *Packet switching in a multiple broadcast channel: performance evaluation. IEEE Transactions on Communications*, Vol. 23, No. 4, pp. 410–423, 1975.

[18] L. D. Landau, E. M. Lifschitz. *Elektrodynamik der Kontinua*, Vol. VIII of *Lehrbuch der theoretischen Physik.* Akademie-Verlag, Berlin, 1985.

[19] M. Lebherz, W. Wiesbeck. *Beurteilung des Reflexions- und Schirmungsverhaltens von Baustoffen. Sonderdruck aus: Bauphysik*, Vol. 12, 1990.

[20] W. C. L. Lee. *Mobile Communication Design Fundamentals.* Howard & Sams, Indianapolis, USA, 1986.

[21] W. C. Y. Lee. *Mobile Communications Engineering.* McGraw-Hill, New York, 1982.

[22] S. Lin, D. J. Costello. *Error Control Coding—Fundamentals and Applications*, Vol. 1 of *Computer Applications in Electrical Engineering Series.* Prentice–Hall, Englewood Cliffs, New Jersey, 1983.

[23] N. B. Meisner, J. L. Segal, M. Y. Tanigawa. *An adaptive retransmission technique for use in a Slotted-ALOHA channel. IEEE Transactions on Communications*, Vol. 28, pp. 1776–1788, 1980.

[24] S. Mockford, A. M. D. Turkmani, J. D. Parsons. *Characterisation of Mobile Radio Signals in Rural Areas.* In *Seventh International Conference on Antennas and Propagation*, pp. 151–154, University of York, United Kingdom, IEE, April 1991.

[25] Y. Okumura, E. Ohmori, T. Kawano, K. Fukuda. *Field Strength and Its Variability in VHF and UHF Land-Mobile Service. Review of the Electrical Communications Laboratory*, Vol. 16, pp. 825–873, September 1968.

[26] N. Papadakis, A. G. Kanatas, A. Paliatsos, P. Constantinou. *Microcellular Propagation Measurements and Modelling at 1.8 GHz.* In *Wireless Networks—Catching the Mobile Future—Personal, Indoor and Mobile Radio Communications PIMRC '94*, pp. 15–19, The Hague, The Netherlands, September 18–22, 1994.

[27] J. D. Parsons. *The Mobile Radio Propagation Channel.* Pentech Press Publishers, London, 1992.

[28] R. L. Rivest. *Network control by Bayesian broadcast. IEEE Transactions on Information Theory*, Vol. 33, pp. 323–328, May 1987.

[29] A. A. M. Saleh, R. A. Valenzuela. *A statistical model for indoor multipath propagation. IEEE Journal on Selected Areas in Communications*, Vol. 5, No. 2, pp. 128–137, Feb. 1987.

[30] D. L. Schilling, (et al.). *Spread spectrum for commercial communications. IEEE Communications Magazine*, pp. 66–79, April 1991.

[31] H. D. Schotten, B. Liesenfeld, H. Elders-Boll. *Large families of code-sequences for direct-sequence CDMA systems. AEU International Journal of Electronics and Communications, Hirzel-Verlag, Stuttgart, Germany*, Vol. 49, pp. 143–150, 1995.

[32] B. Sklar. *Digital Communications: Fundamentals and Applications.* Prentice-Hall, Englewood Cliffs, New Jersey, 1988.

[33] G. L. Stüber. *Principles of Mobile Communications.* Kluwer Academic Publishers, Boston, Dordrecht, London, 1996.

[34] A. S. Tanenbaum. *Computer Networks.* Prentice-Hall, Upper Saddle River, New Jersey, 3rd edition, 1996.

[35] S. C. A. Thomopoulos. *A simple and versatile decentralized control for Slotted ALOHA, Reservation ALOHA and Local Area Networks. IEEE Transactions on Communications*, Vol. 36, pp. 662–674, 1988.

[36] J. Wiart. *Micro-cellular modelling when base station antenna is below roof tops.* In *Proceedings 44th Vehicular Technology Conference, Creating Tomorrow's Mobile Systems*, pp. 200–204, Stockholm, Sweden, IEEE, June 1994.

3
GSM System*

3.1 The GSM Recommendation

The early 1980s were marked by the development of a number of national and incompatible radio networks in Europe; see Table 1.2 and Figure 1.3. The seven different mobile radio networks made the prospect of the mobile telephone unattractive to many potential customers because of high tariffs and equipment costs.

For this reason, at its general meeting in Vienna in June 1982, the *Conférence Europeéne des Administrations des Postes et des Télécommunications* (CEPT) (see Appendix B.2.2) decided to develop and standardize a Pan-European cellular mobile radio network. The aim was for the new system to operate in the 900 MHz frequency band allocated to land mobile radio.

A working group, called *Groupe Spéciale Mobile* (GSM), was set up under the direction of CEPT. There were no guidelines on how the new mobile radio system was to transmit analogue or digital speech and data. The decision to develop a digital GSM network was not made until the development stage. But it was agreed from the beginning that the system being planned—called the GSM mobile radio system after the working group that developed it—should incorporate and consider new technology from the area of telecommunications, such as ITU-T *Common Channel Signalling System No. 7* (CCS7), ISDN and the *International Standards Organization* (ISO)/*Open Systems Interconnection* (OSI) reference model.

Six working groups and three supporting groups were formed to cope with the enormity of the standardization work. The tasks of the different GSM working groups are listed in Table 3.1.

The GSM objectives for its PLMN were to offer [1]:

- A broad offering of speech and data services.
- Compatibility with the wireline networks (ISDN, telephone networks, data networks) using standardized interfaces.
- Cross-border system access for all mobile phone users.
- Automatic roaming and handover.

***With the collaboration of Peter Decker and Christian Wietfeld**

Table 3.1: Tasks of the GSM working groups

GSM working groups	Tasks
Working Party 1	Definition of services and service quality
Working Party 2	Definition of access, modulation and coding procedures
Working Party 3	Definition of protocols for signalling between mobile stations, mobile functions and fixed communications networks
Working Party 4	Specification of data services
Working Party 5	Development of UMTS
Working Party 6	Specification of network management features
Speech Coder Experts Group (SCEG)	Definition of technique for digitization of speech at a low bit rate
Security Experts Group (SEG)	Responsibility for all aspects of security (access, coding, authentication)
Satellite Earth Systems (SES)	Support of GSM through satellite systems

- Highly efficient use of frequency spectrum.
- Support of different types of mobile terminal equipment (e.g., car, portable and hand-held telephones).
- Digital transmission of signalling as well as of user information.
- Supplier-independence.
- Low costs for infrastructure and terminal equipment.

The GSM group tested a number of prototypes for digital cellular radio systems, and in 1987 decided on a standard that combined the best characteristics of different systems. A timetable drawn up at the same time for the implementation of the plan gained the full support of the *European Union* (EU) (see Table 3.2).

By 1987, comprehensive guidelines for the new digital mobile radio system had already been established by the GSM group. By signing the *Memorandum of Understanding on the Introduction of the Pan-European Digital Mobile Communication Service* (MoU) on September 7th, 1987, the 13 participating countries confirmed their commitment to introducing mobile radio based on the recommendations of the GSM.

Later, in March 1989, the GSM working party was taken over by ETSI (see Appendix B.2.3), and since 1991 has been called the *Special Mobile Group* (SMG). Today the abbreviation GSM stands for *Global System for Mobile Communication*, thereby underlining its claim as a worldwide standard.

In the meantime all the European countries as well as a large number of other countries in the world have signed the GSM-MoU agreement and have

Table 3.2: Original timetable for introducing the GSM system

Date	Phase
February 1987	Invitation for tenders
Mid 1988	Letters of Intent
End 1988	Validation of interfaces
Mid 1990	System validation
March 1991	Start of equipment deliveries
June 1991	Operation of first base station
1993	Coverage to metropolitan areas and major roads
since 1995	Area-wide operation

Table 3.3: The series of GSM recommendations

Series	Content
00	Preamble
01	General aspects, terminology and service introduction phases of the GSM *Public Land Mobile Network* (PLMN)
02	Definition of telecommunications services, technical aspects concerning tariffs and international billing procedures
03	Definition of network functions such as traffic routing, handover, security issues relating to network access, network planning
04	Description and definition of protocols and interfaces between *Mobile Station* (MS) and *Base Station* (BS)
05	Radio path functions such as multiplexing, channel coding, synchronization and interleaving
06	Speech processing and speech coding functions
07	Adaptation of terminal equipment and transmission rates
08	Description of interface functions between *Base Station Subsystem* (BSS) and *Mobile-services Switching Centre* (MSC)
09	Definition of *Interworking Function* (IWF) between one or more GSM networks and different fixed networks
11	Equipment specifications and type approval guidelines
12	Operation and maintenance of a GSM network

developed or will be developing mobile radio systems in their countries based on the GSM recommendations (see Table 3.34).

The planned official start to the GSM system was delayed by one year. Only five countries were in a position to undertake test operations on 1st July 1991. The reason for the delay was the level of complexity of the digital network and its components, which is reflected in the voluminous specifications which today totals around 8000 pages. In 1990 alone another 500 GSM *change requests* were passed. The entire set of GSM recommendations is divided into 13 series, which cover different aspects of the GSM system, as shown in Table 3.3 (see also Appendix E).

The GSM recommendations contain detailed specifications for the radio interface which in part are borrowed from the concepts for the analogue national cellular standard and ITU-T Rec. X.25. However, large parts of the radio interface are specific to the GSM system. Some of the important features of GSM include:

Frequency band The frequency range between 935 and 960 MHz is used as the base station transmitting frequency (*downlink*) and the frequencies between 890 and 915 MHz are used as the base station receiving frequency (*uplink*). The carrier frequencies of the FDM radio channels have 200 kHz channel spacing in each band, thus providing 124 FDM channels. With TDM, eight communications channels (time slots) are supported per FDM channel.

Handover Handover from one base station to another is a mechanism that allows the connection quality of calls between users to be maintained, interference to be minimized and traffic distribution to be controlled. In addition, procedures are defined for the re-establishment of a connection if a handover fails.

Power control In the area over 30 dB the equipment of the mobile user and of the base station controls power in 2 dB steps in order to minimize interference.

Discontinuous transmission (DTX) GSM offers the option of discontinuous transmission of speech using voice activity detectors. With DTX, the mobile station's battery power is only used when speech or data is being transmitted, which minimizes interference and improves the utilization of frequency spectrum.

Synchronization Depending on the system, all frequencies and times are synchronized with a highly stable (0.005 ppm) reference, which can be coupled with a frequency normal.

The following features distinguish GSM from other European mobile radio systems:

- Europe-wide coverage
- Europe-wide standardization
- digital radio transmission
- extensive ISDN compatibility
- protection against eavesdropping
- support of data services

GSM is regarded as an important advance compared with predecessor systems and is considered to be representative of so-called 2nd-generation (2G) systems. Along with important technological advances—particularly the introduction of digital transmission—the standardization of the interfaces between subsystems in GSM has provided manufacturers and network operators with flexibility in their development work and configurations.

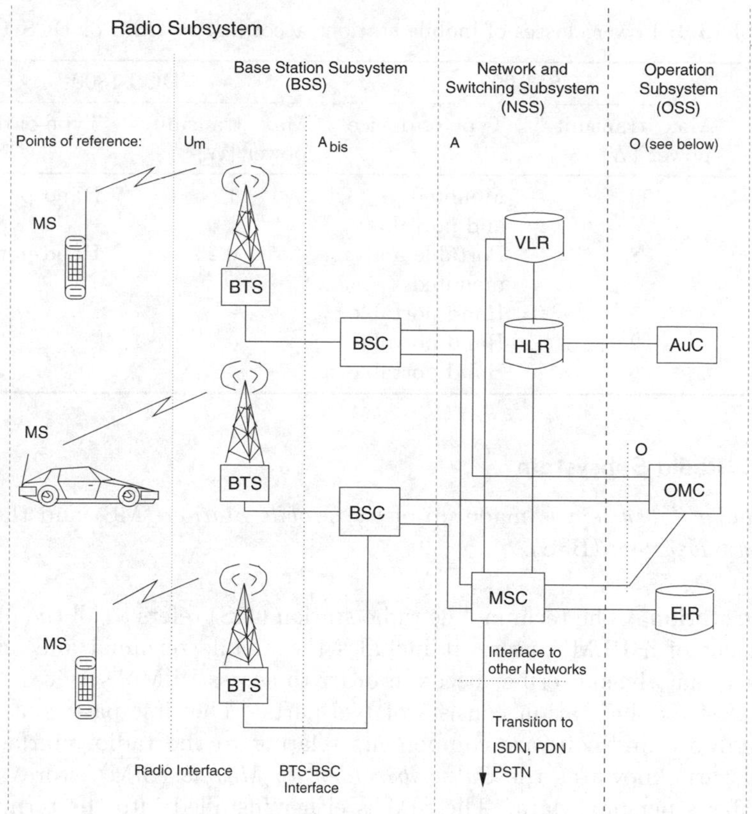

Figure 3.1: Functional architecture of the GSM mobile radio network

3.2 The Architecture of the GSM System

3.2.1 Functional Structure of the GSM System

In GSM specification 1.02 the GSM system is divided into the following subsystems [19]:

- *Received Signal Strength* (RSS)
- *Network and Switching Subsystem* (NSS) and
- *Operation Subsystem* (OSS).

These subsystems and their components are represented in the simplified version of the functional architecture in Figure 3.1.

Table 3.4: Power classes of mobile stations according to GSM or DCS 1800

	GSM 900		DCS 1800	
Class	Max. transmit. power [W]	Type of device	Max. transmit. power [W]	Type of device
1	20	Mounted and portable	1	Hand-portable
2	8	Portable and mounted	0.25	Hand-portable
3	5	Hand-portable	–	—
4	2	Hand-portable	–	—
5	0.8	Hand-portable	–	—

3.2.1.1 Radio Subsystem

The *Radio Subsystem* is made up of the *Mobile Station* (MS) and the *Base Station Subsystem* (BSS).

Mobile station The term mobile radio station (MS) refers to all the physical equipment of a PLMN user. It includes the mobile terminal and the user interface that the subscriber needs in order to access PLMN services.

A GSM mobile station consists of two parts. The first part contains all the hardware and software components relating to the radio interface; the second part, known as the *Subscriber Identity Module* (SIM), stores all the subscriber's personal data. The SIM is either installed into the terminal or provided as a *smart card*, which is either about the size of a credit card or much smaller and has the function of a key. Once it has been removed from a device, it can only be used for emergency calls, if the network so allows. A mobile subscriber can use the SIM to identify himself over any mobile station in the network, and accordingly a mobile phone can be personalized using the SIM. In addition, each mobile station has its mobile *Equipment Identity* (EI).

The following numbers and identities are assigned for the administration of each mobile station within a GSM network; see Figure 3.58:

- *International Mobile Subscriber Identity* (IMSI)
- *Temporary Mobile Subscriber Identity* (TMSI)
- *Mobile Station International ISDN Number* (MSISDN)
- *Mobile Station Roaming Number* (MSRN)

Mobile stations can be installed in automobiles or provided as portable/hand-portable devices and, according to GSM Rec. 2.06, are divided into five different classes depending on the allowable transmitter power; see Table 3.4.

These classifications also characterize different types of devices: mounted, portable and hand-portable devices. Equipment for the GSM-900 class 1 (8–20 W) has not yet been developed. Instead, portable and mounted equipment is typically found in class 2 (5–8 W). Hand-portable equipment mostly conforms with class 4 (0.8–2 W). Class 5 (up to 0.8 W) is also being planned for hand-portable equipment, but places a considerable strain on cellular radio signal supply. This is one of the reasons why it is more suitable for urban environments with small cells, but it is hardly being used anywhere yet. An MS can have facilities for both voice as well as data transmission.

In addition to the network-dependent radio and protocol functions that enable access to operation in the network, a mobile station outwardly has at least one other *interface* to the mobile subscriber (see Section 3.2.2). It is intended either for a human user (man–machine interface) or for coupling the *Terminal Adapter* (TA) of another terminal, such as a computer or a fax machine or a combination of the two. The GSM specifications leave the conversion and extent of the interface technique up to the manufacturer.

A user interface usually consists of the following components:

- Microphone.
- Speaker.
- LCD display field.
- Alphanumeric keyboard.
- So-called soft keys.

Soft keys are function keys used to switch a terminal to different operating states. They are not assigned a specific function, as is the case with hard keys, e.g., on a drinks dispenser. Consequently the user must be informed of the respective function before using the keys.

Soft keys are extremely useful with hand-held mobile phones. The subscriber can use his mobile device with one hand because of the soft key menu functions that are displayed on the mobile, without having to press key combinations at the same time, as is required with the hard key version of a control panel.

Unlike the conventional telephone, where the user is identified through the fixed network connection, radio connections form an anonymous network. Therefore subscriber identification is a prerequisite in a mobile radio network alone for operational reasons. The stored subscriber-related data in a SIM identifies the subscriber when he checks in, and his location area is derived from the serving base station—an automatic procedure when the terminal is used.

In older devices the SIM is installed into the equipment, but the new approach is to plug it in as a card; there are two versions of this:

- smart card, also called standard SIM card
- plug-in SIM card

The only difference between the two cards is their size. The standard SIM card is the size of a credit card based on standard ISO 7816, whereas the plug-in module is smaller in size and based on the GSM Rec. 02.17 [6]. In addition to their size, the cards are also used differently. Whereas the standard SIM card can be activitated simply by being inserted into the card slot provided in the mobile telephone, the smaller module slides into the equipment mounted on a cut-down card, which involves first removing the battery. The smaller plug-in SIM card has been successful with hand-held mobile telephones.

The subscriber-related data is stored in the non-volatile memory of the SIM. It can be changed statistically as well as temporarily. The permanent data includes the following elements [6]:

- SIM card type.
- IC card identification: serial number of the SIM; identifies card holder at the same time.
- SIM *service table*: list of additional services subscribed.
- *International Mobile Subscriber Identity* (IMSI).
- *Personal Identification Number* (PIN).
- *PIN Unblocking Key* (PUK).
- Authentification key K_i.

Before a SIM card is assigned to a subscriber, it is first initialized with this data, and only then can the subscriber use the card to check into the network. On the other hand, the dynamic data, which is permanently updated when the terminal is switched on, accelerates the checking-in process because relevant information is already stored centrally and there is no need for it to be requested from the network. This includes the following data items [6]:

- Location information: consists of a TMSI, a LAI, a periodically changed location updating timer, and update status.
- Ciphering key K_c for encoding, and its sequence number.
- BCCH information: list of carrier frequencies for cell selection during handover and call set up.
- List of blocked PLMNs.
- HPLMN search: period of time in which an MS roams the home network before it tries to check into another network.

Other optional data items can be found in [6]. All SIM data is copied in the memory of the MS only for the duration of the active operating state and then deleted. Manufacturers of mobile terminals have the option of additionally providing intermediate storage of less important data, such as short messages and the last-called telephone number. However, this data can only be called up if the equipment is turned on again with the same SIM card that was used for its previous deactivation [6].

PIN Except for emergency calls, mobile equipment can only be operated if the SIM card has first been activated. This is done by the subscriber punching in a PIN code, which can be between four and eight digits long, after switching on the equipment. When the SIM card is provided by the service provider, the PIN is generally preset with a four-digit number, which the subscriber can change as often as he likes. After the PIN has been correctly entered, the network responds and the mobile is automatically checked in.

A PIN can be but should not be disabled, because the subscriber will run the risk of potential thieves using the mobile free of charge until use of the card is suspended. Anyone who steals an activated mobile phone can only use the SIM card fraudulently until the first time the equipment is switched off or the battery runs out. If an incorrect PIN is inserted three times in a row, the card will be suspended. The subscriber then needs an unlocking key PUK.

Some cards are available with a second PIN to protect some of the numbers stored in the card. This specifically protects personal telephone numbers and names entered on the card from unauthorized access. The security mechanisms and maximum allowable code length of the PIN2 are identical to those of the PIN [6].

PUK A blocked SIM card can only be released through the use of a *PIN Unblocking Key* (PUK). The subscriber is allowed 10 attempts in which to enter the correct PUK code or else the card will be blocked permanently and can only be unblocked by the service provider. The PUK is an eight-digit permanent number that is divulged to the subscriber when he receives the card [6].

3.2.1.2 Base Station Subsystem (BSS)

The BSS comprises all the radio-related functions of the GSM network. Depending on the radio transmitting and receiving capabilities of the *Base Transceiver Station* (BTS), which because of limited transmitter power only supplies coverage to a specific geographical area within the network, radio cells are created in which the mobile subscriber is free to roam or communicate. The size of the individual cells depends on a number of parameters, including characteristics of radio wave propagation, local morphology, and expected user density in the region.

A BSS uses transceivers and the following hardware and software to enable it to connect a mobile subscriber to a number in the *Public Switched Telephony Network* (PSTN) and allow it to communicate:

- Signalling protocols for connection control.

- Speech codecs (coders/decoders) and data-rate adaptation (*Transcoding and Rate Adaption Unit*, TRAU) for network access.
- Digital signal transmission for coded data.

These functions already give an indication of some of the other important tasks of the BSS. Various interfaces have been specified between the BSS and GSM network elements and other networks for the exchange of information between subscribers and the GSM network or other networks; see Figure 3.1. The interface to the mobile subscriber is called the U_m-interface. It contains specific parameters for digital radio transmission, such as GMSK modulation, data rate, status of carrier frequencies in the 900 MHz band and channel grid. The BSS is connected to the GSM fixed network over the A-interface (familiar from ISDN) with *Mobile-services Switching Centres* (MSC), the NSS switching centres that provide the subscriber connectivity to each other and to the external network. The A-interface likewise contains specific digital transmission parameters, including *Pulse Code Modulation* (PCM), a 64 kbit/s data rate and a 4 kHz voice bandwidth.

Network availability and quality are established by the network *Operating and Maintenance Centre* (OMC) of the GSM operator over an O-interface, which provides direct access to BSS units.

The elements making up the BSS include:

- *Base Transceiver Station* (BTS)
- *Base Station Controller* (BSC).

BTS The BTS comprises the transmitting and receiving facilities, including antennas and all the signalling related to the radio interface. Depending on the type of antenna used, the BTS supplies one or several cells, so, for example, sectorized antennas can supply three cells arranged at 120° to each other (see Chapter 2.4).

In a standardized GSM structure the *Transcoding and Rate Adaption Unit* (TRAU) is part of the BTS. It contains GSM-specific speech coding and decoding as well as rate adaptation for data transmission.

BSC The BSC is responsible for the management of the radio interface through the BTS, namely for the reservation and the release of radio channels as well as handover management. Its other tasks include *paging* and transmitting connection-related signalling data adapted to the A-interface from/to the MSC.

A BSC generally manages several BTSs, and is linked to the NSS via an MSC.

3.2.1.3 Network and Switching Subsystem (NSS)

Switching and network-oriented functions are carried out in a *Network and Switching Subsystem* (NSS). It forms the gateway network between the radio network and the public partner networks (e.g., *Public Switched Telephony Network* (PSTN), *Integrated Services Digital Network* (ISDN), *Public Switched Data Network* (PSDN)). In their entirety not only are the elements of an NSS purely physical components but, more importantly, the switching subsystem provides a large number of functions that are the responsiblity of the manufacturer and network operator to implement appropriately.

The NSS components include the *Mobile-services Switching Centre* (MSC), the *Home Location Register* (HLR) and the *Visitor Location Register* (VLR).

Mobile Services Switching Centre (MSC) The MSC is a high-performance digital switching centre that carries out normal switching tasks and manages the network. Each MSC is usually allocated several base station controllers, and in the geographical area assigned to it carries out the switching between mobile radio users and other PLMNs and also forms the link between the mobile radio network and the wireline networks (PSTN, ISDN, PDN). The MSC is responsible for all the signalling required for setting up, terminating and maintaining connections, carried out in accordance with CCS7, and mobile radio functions such as *call rerouting* when there is strong interference, as part of a *handover* and the allocation and deallocation of radio channels.

Transmission functions for data services are supported through the use of specific *Interworking Functions* (IWF) that are integrated into each MSC. The respective communications channel functions are carried out by facilities called *Data Service Units* (DSU). The DSU contains functions such as rate adaptation, modem and codec of layer 1, and protocol functions of layer 2.

The other tasks of the MSC include the supplementary services familiar from ISDN, such as call forwarding, call barring, conference calling and call charging to the user called. The MSC can be envisaged as an ISDN switching centre that has been expanded to include the necessary mobility-related switching functions.

Home Location Register (HLR) All important information (quasi-permanent static data) relating to each mobile subscriber, including telephone number, MS identification number, equipment type, subscription basis and supplementary services, access priorities and authentication key, is stored in the database referred to as the home location register. Temporary (dynamic) subscriber data (e.g., current *Location Area* (LA) of the MS and *Mobile Station Roaming Number* (MSRN)) that are necessary for setting up a connection are also stored. When a mobile user leaves his momentary LA, the temporary data held in the HLR is immediately updated. The home location register usually falls under the responsibility of an MSC. Each mobile subscriber and his related data are registered in only one HLR in which all the billing and

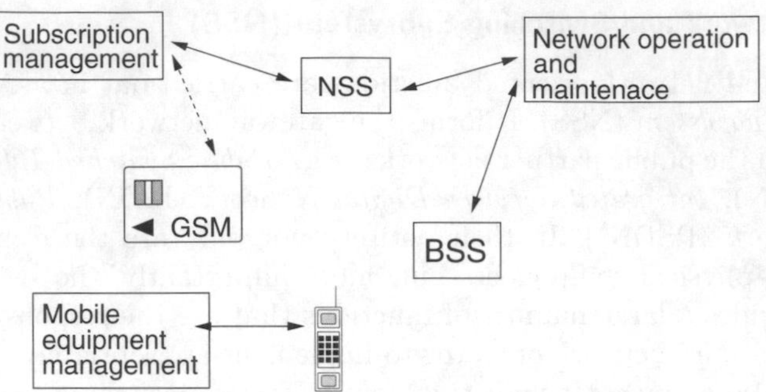

Figure 3.2: Structure of an OSS

administrative tasks are carried out. In many existent GSM networks there is only one HLR being implemented.

Visitor Location Register (VLR) The VLR is under the control of an MSC and is used to manage the subscribers who are currently roaming in the area under the control of the MSC or, more precisely, in one of possibly several LAs of the MSC. It stores information (e.g., authentication data, *International Mobile Subscriber Identity* (IMSI)), telephone number, agreed services) transmitted by the responsible HLR for the mobile users operating in the area under its control, thereby allowing the MSC to make a connection. The VLR also controls the allocation of roaming numbers (MSRN) to the mobile stations as well as of the TMSI. A special dialogue updates the VLR if a mobile user moves through several of the MSC's LAs. The same procedure applies when there is a change of MSC. The VLR avoids frequent interrogation of the HLR.

The functions *location area update* and *call set up* and the roles played by the HLR and the VLR in these functions are described in Sections 3.7 and 3.8.

3.2.1.4 Operation Subsystem (OSS)

The OSS in GSM comprises all the important functions for operation and maintenance. The user is only indirectly aware of these functions through his experience with a smoothly functioning mobile radio network.

The functions of an OSS are allocated to three areas of responsibility (see Figure 3.2):

- Subscription management.
- Network operation and maintenance.
- Mobile equipment management.

The following network elements are part of the OSS:

- *Operating and Maintenance Centre* (OMC).
- *Authentication Centre* (AuC).
- *Equipment Identity Register* (EIR).

Subscription management Subscription management is able to authenticate a GSM user from the personal data stored in the HLR (see Section 3.13.1) and provide him with the agreed services (*subscriber data management*). This data provides the network operator and the service provider with a call-charging basis.

Subscriber data management The subscriber data is stored and managed in the HLR; information relating to data security is in the AuC. The HLR can provide restricted access to elements from other networks, e.g., in order to allow service providers access to tariff and services data and to ensure the consistency of data stored in different locations. As has already been mentioned, the SIM card is a dynamically changeable data storage unit during the active operation of a mobile station.

Call charging Similarly to ISDN, the mobile radio user is charged for services used on the basis of so-called *call tickets*. These call tickets are used for billing irrespective of where a call is made in the network. The billing location can be the MSC in which the mobile subscriber is currently active or a *Gateway Mobile-services Switching Centre* (GMSC) where a communication is connected to an external network.

The HLR only stores call-related data. Call billing is handled by the responsible OSS subscriber management. At the same time tariff data is also transmitted between the MSCs or GMSCs and the HLR over the *Common Channel Signalling System No. 7* (CCS7).

Network operation and maintenance The control of network operation and maintenance tasks uses a separate switching network to connect operating personnel network elements. The network is based on the concept of *Telecommunications Management Network* (TMN) developed by the ITU-T. The TMN forms an integrated network with its own databases that offer the operator options for monitoring, control and intervention.

The TMN functions are divided into individual layers similar to the network element functions in the ISO/OSI reference model:

Business management Controls the interaction between network and services and provides information about other service and network developments.

Service management Used for the execution of all contractual aspects of a service between supplier and customer.

Network management Supports all network elements and helps to activate functions with similar elements of a network.

Network element management Facilitates access to individual network elements.

GSM uses standardized concepts for network management, thereby facilitating the integration of the network elements of different suppliers.

The TMN has links with defined interfaces to the network elements of the active network and to the workstation computers of operating personnel. OSS network elements that are connected to several BSS or *Maximum Segment Size* (MSS) units are referred to as OMCs. A radio OMC, for example, is responsible for several BSCs and their BTSs.

Mobile equipment management The management of mobile equipment by the OSS only concerns information about owner and equipment identity, whereas the MSS coordinates the movements of the equipment, including *roaming*, *handover* and *paging*. For example, an OSS can search for stolen or defective equipment using its own database, an EIR, for storing data about equipment and its ownership (some operators have not established the EIR).

Operation and Maintenance Centre (OMC) The OMC centrally monitors and controls the other network elements and guarantees the best possible service quality for a network. It relies on services of the network management and control functions allocated to the network elements by the hierarchical network management system (TMN). Operator commands are used for intervention into the network elements, while the network management is alerted of any unexpected occurrences in the network. The OMC is connected to all network elements over the standardized O-interface (an X.25-interface). The management functions of the OMC include administration of subscribers and equipment, billing, and generation of statistical data on the state and the capacity utilization of network elements.

Authentication Centre (AuC) The AuC contains all the information required to protect a subscriber's identity, and his mobile communication against eavesdropping, and his right to use the radio interface. Because the radio interface is generally susceptible to unauthorized access, special measures (e.g., authentication key assigned to each subscriber and coding of transmitted information) were undertaken in order to prevent the fraudulent use of GSM–PLMN connections. Authentication algorithms and encryption codes are stored in the AuC, and strict rules apply for access to this information (see Section 3.13).

Equipment Identity Register (EIR) The EIR is a central database in which subscriber and equipment numbers (*International Mobile Equipment Identity*,

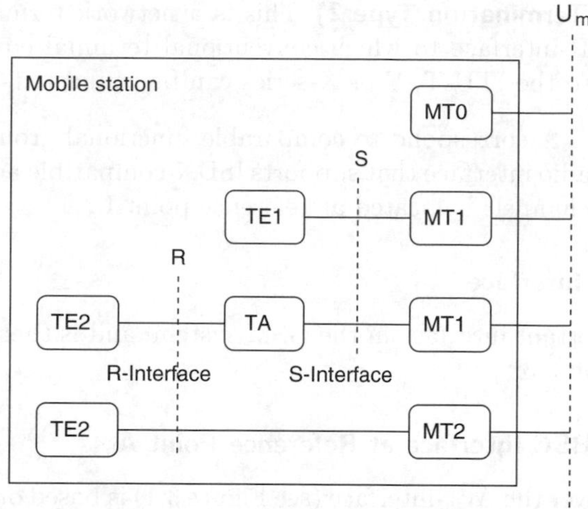

Figure 3.3: Mobile station network terminations with reference points R, S, U_m

IMEI) are stored, and is connected over an interface to the NSS network elements and the OSS. The database contains a *white*, a *black* and a *grey* list. The white list contains the IMEI list of valid mobile radio stations; the black list contains all the IMEIs of stolen or suspended mobile radio stations. The grey list includes a list of IMEIs for malfunctioning equipment that is not receiving any services.

3.2.2 Interfaces of the GSM System

3.2.2.1 User Interface of the Mobile Station

A GSM mobile station consists of the *Terminal Equipment* (TE) to which the subscriber has direct access, a *Terminal Adapter* (TA) (if required) and a part that contains the functions shared by all the services and referred to as *Mobile Termination* (MT) in the GSM specifications. The subscriber interface on the TE contains the network termination and the different equipment functions (see Figure 3.3).

The following mobile network terminations are used:

MT0 (Mobile Termination Type 0) A network termination for the transmission of speech and data integrating the terminal equipment, the terminal equipment functions and sometimes a TA.

MT1 (Mobile Termination Type 1) A network termination with an external ISDN S-interface to which an ISDN terminal (TE1) can be connected. Conventional terminal equipment (TE2) corresponding to the ITU-T, V or X-series can be connected to an MT1 through the use of an ISDN TA.

MT2 (Mobile Termination Type 2) This is a network termination with an external R-interface to which conventional terminal equipment corresponding to the ITU-T, V or X-series can be connected.

TE1, TE2 and TA correspond to comparable functional groups of the ISDN concept. The radio interface that supports ISDN-compatible access over traffic and signalling channels is located at reference point U_m.

3.2.2.2 Radio Interface

This is an important interface in the GSM system, and is therefore covered in detail in Section 3.3.

3.2.2.3 BTS–BSC Interface at Reference Point A_{bis}

Transmission over the A_{bis}-interface (see Figure 3.1) is based on ISDN *Primary Rate* (PR) and *Basic Rate* (BR) interfaces.

Because PLMN network operators frequently are not also the operators of the telecommunications networks, a submultiplex technique that transmits four 16 kbit/s channels over a 64 kbit/s BR channel was standardized to save on line costs.

3.2.2.4 BSS–MSC Interface at Reference Point A

Speech and data are transmitted digitally over the A-interface (see Figure 3.1), over PR (PCM-30) systems based on the ISDN standard (ITU-T-Series G.732). A PCM-30 system has 30 full-duplex channels at 64 kbit/s, with a transmission rate of 2.048 Mbit/s full-duplex. Two channels each with 64 kbit/s are required for synchronization and signalling (D_2-channel).

3.2.2.5 BSC/MSC–OMC Interface at Reference Point O

The O-interface is based on ITU-T recommendation X.25, which was specified for the attachment of data terminal equipment to packet-switched networks. Physically this interface can be implemented over a 64 kbit/s channel. The option exists to use interfaces of circuit-switched networks, e.g., V.24bis or X.21.

3.3 The Interface at Reference Point U_m

This radio interface (also called air interface) is located between the MS and the rest of the GSM network. Physically the information flow takes place between the MS and the BTS. But, viewed logically, the mobile stations are communicating with the *Base Station Controller* (BSC) and the *Mobile-services Switching Centre* (MSC). The gross transmission rate over the radio interface is 270.833 kbit/s.

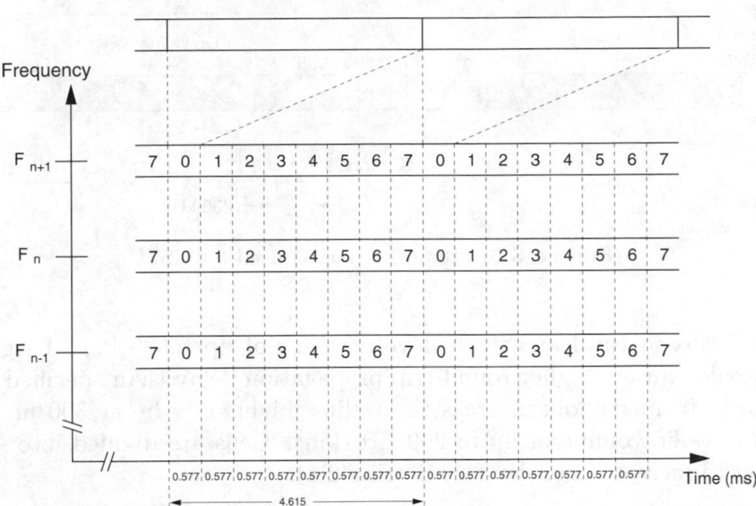

Figure 3.4: Realization of physical channels using FDM and TDM

3.3.1 Multiplex Structure

Along with speech coding and modulation, multiplexing is also very important. In the GSM recommendations a combination of *Frequency Division Multiplex* (FDM) and *Time Division Multiplex* (TDM) has been standardized, providing multiple access by mobile stations to these systems (FDMA, TDMA).

Figure 3.4 shows how a physical channel is produced through a combination of FDM and TDM (see channel 0 on frequency F_{n+1} and Sections 3.3.1.1 and 3.3.1.2).

GSM utilizes the cellular concept, already proven successful in analogue mobile radio networks, in which a geographical area is divided into planned radio cells (in the simplest case hexagons), with one BTS per cell with which the mobile stations can make contact. The radio cells, each having the exclusive use of specific FDM channels, are combined into groups (*clusters*). The same frequencies are only reused after a sufficiently long distance in neighbouring clusters (see Section 2.3).

The cell radius can vary according to user density. The likelihood that a mobile user will leave a cell during a call, thereby necessitating a *handover*, is less in large radio cells than in small cells. Small cells, on the other hand, make more efficient use of a frequency band because they operate with a lower transmitter power, the cluster is less spread out and consequently the available frequencies can be reused at smaller physical intervals. In practice, the size of cells is determined by traffic volume, the maximum transmitter power of the BTS of the frequencies allocated to a cell and morphological conditions.

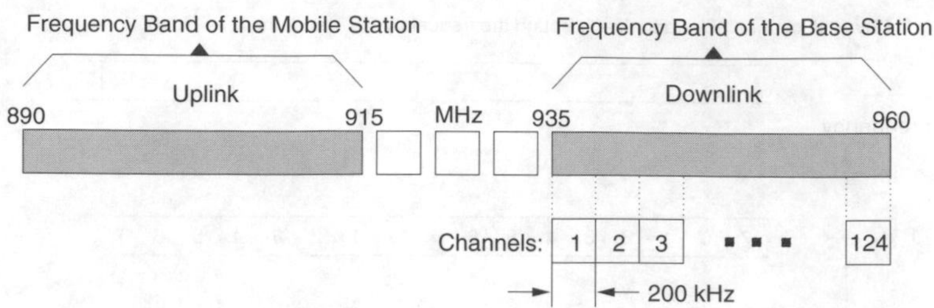

Figure 3.5: Frequency bands used by GSM

Thus cells in rural areas can have a radius of up to 35 km. Larger cell radii would cause a higher round-trip propagation delay than specified in the standard. In metropolitan areas the radius might only be at 300 m, which allows a traffic volume of up to 200 Erl./km^2. Cells are divided into sectors in order to increase capacity (see Section 2.4).

3.3.1.1 Frequency-Multiplexing Structure

One of the most important criteria in designing a radio interface was efficient utilization of the available frequency band. In Europe two 25 MHz wide frequency bands in the 900 MHz band were reserved for GSM. Transmission from the mobile unit to the base station (*uplink*) takes place in the 890 MHz to 915 MHz range; in the reverse direction (*downlink*) the 935–960 MHz frequency band is used in a *Frequency Division Duplex* (FDD) mode of operation. 15 MHz at the lower band limit and 1 MHz at the upper band limit will not be available until 2001. After current use is discontinued, an additional 10 MHz between 880 and 890 MHz and between 925 and 935 MHz will be available as a GSM extension band (see Appendix C). A duplex interval of 45 MHz exists between the transmit and receive frequencies.

The frequency bands are divided into 200 kHz bandwidth channels, therefore providing a total of 124 FDM channels each for transmitting and receiving operations (see Figure 3.5).

Each mobile station can occupy all 124 carrier frequency pairs, although according to the GSM specifications use of channels 1 and 124 should be avoided if possible. The respective 200 kHz bandwidth is kept as a guard band for the neighbouring systems in the frequency band. If the carrier frequencies on the uplink are denoted by F_u and those on the downlink as F_d then the GSM band can be defined as

$$F_u(n) = 890.2 \text{ MHz} + 0.2(n-1) \text{ MHz} \quad (1 \leq n \leq 124) \tag{3.1}$$

$$F_d(n) = 935.2 \text{ MHz} + 0.2(n-1) \text{ MHz} \quad (1 \leq n \leq 124) \tag{3.2}$$

and the extension band as

$$F_u(n) = 880.2 \text{ MHz} + 0.2(n-1) \text{ MHz} \quad (1 \leq n \leq 50) \tag{3.3}$$

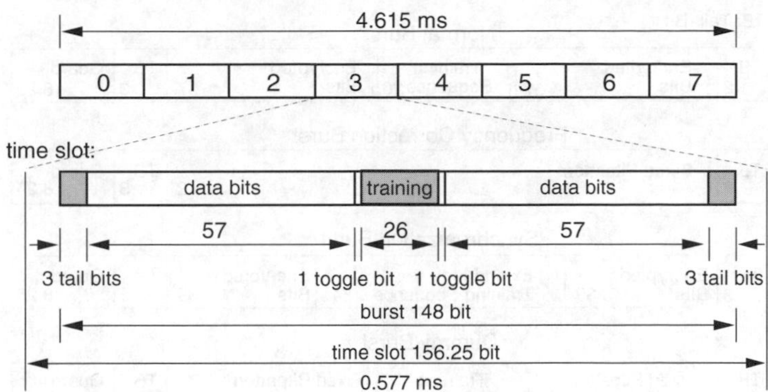

Figure 3.6: Structure of a TDMA frame

$$F_d(n) = 925.2 \text{ MHz} + 0.2(n-1) \text{ MHz} \quad (1 \leq n \leq 50) \tag{3.4}$$

3.3.1.2 Time-Multiplexing Structure

With the TDM method a carrier frequency is divided into eight physical TDM channels in which the time axis is divided into eight *periodic time slots* of 0.577 ms duration. Eight time slots are combined into a TDM frame of 4.615 ms duration (see Figure 3.6). Because these time channels are used in multiple access, the frame is referred to as *Time Division Multiple Access* (TDMA) frame in the GSM recommendations.

A physical channel is characterized by its carrier frequency and the time slot available to it, which recurs every 4.615 ms. Each time slot has a length corresponding to the duration of 156.25 bits or 0.577 ms (15/26 ms). This length is produced from the transmission rate of the modulation method (1625/6 kbit/s) and the number of bits to be transmitted in a slot. A slot is used by a *burst* with a length of 148 bits, which, corresponding to the guard time, is 8.25 bits shorter in duration than the slots to avoid overlapping with other bursts. Data is transmitted in bursts. If messages are longer than a burst, they are split up among several bursts and then transmitted.

Overall there are five types of bursts (see Figure 3.7 [16]) which differ from one another in function and content. The tail bits that occur in all bursts are defined as modulation bits and always have the same value as specified in the standard. The bursts are sent so that the bits with the lowest value are transmitted first.

Normal burst For transmitting messages in traffic and control channels.

Access burst Used for call set up. This burst is shorter than the others because it does not require the MS to be fully synchronous with the BTS.

Synchronization burst Sent by the base station and used for synchronization.

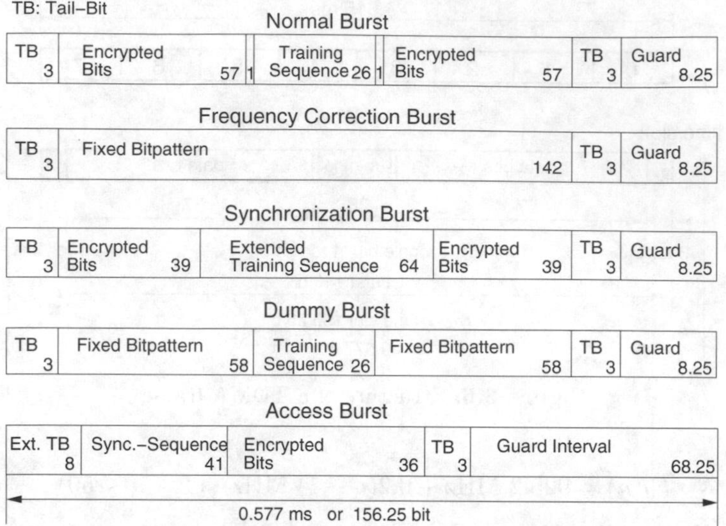

Figure 3.7: Bursts used in GSM

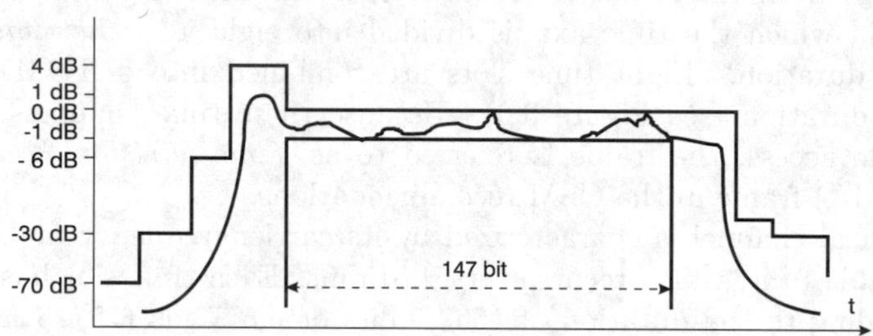

Figure 3.8: Envelope of the radio signal of a burst

Frequency correction burst Sent by the base station and used for frequency correction at the mobile station to prevent possible interference from adjacent frequencies.

Dummy burst Placed in an empty slot if no data is being sent.

The signalling characteristics of a burst over time are not allowed to exceed the area of a prescribed mask (see Figure 3.8). In the area of the tail bits and the guard space the signal can deviate considerably from the standard 0 dB. It is clear that neighbouring bursts only minimally overlap in the same TDMA frame.

The time-division multiplexing technique is applied to the uplink and to the downlink channel. So that the mobile stations do not have to transmit and receive at the same time, the TDMA frames from the uplink are transmitted with a delay of 3 time slots (see Figure 3.9). The parameter *Timing Advance*

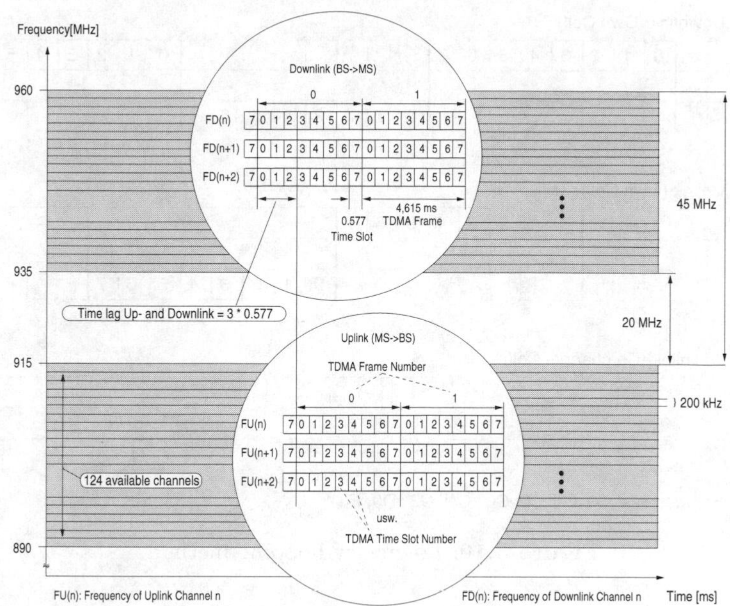

Figure 3.9: Time delay between uplink and downlink

(TA) is used by the BTS to compensate for the round-trip signal propagation delay BTS-MS-BTS. The value of the 6 bit of TA indicates to a receiving mobile how many bit durations (3.7 µs each) it must transmit its burst earlier than as derived from the received slot tact signal to reach synchronization with the slot tact defined by the BTS.

3.3.2 Frequency Hopping (FH)

Since multipath reception and co-channel interference can affect the quality of certain FDM channels, an optional method called *Frequency Hopping* (FH) is applied. With this method the frequency is changed after each transmitted frame of a channel (see Figure 3.10). The frequency change, which can last approximately 1 ms, takes place between the receiving or the transmitting time slots.

The sequence of frequencies in a hopping cycle through which a mobile station passes is calculated with an algorithm implemented in each MS. The advantage of this procedure is that all mobile subscribers are guaranteed transmission channels with nearly the same quality. During data transmission, interference from co-channels in the cycle is limited for each frequency to the duration of one burst only and can be eliminated through error handling; effective error-correction procedures are standardized for speech and data transmission.

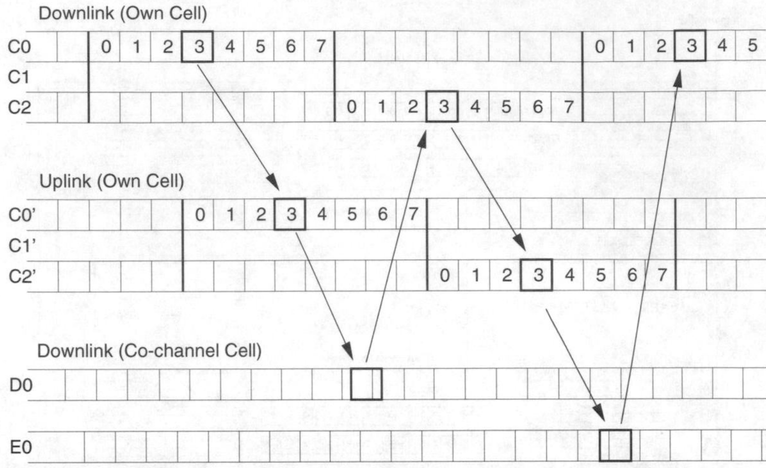

Figure 3.10: Frequency hopping method

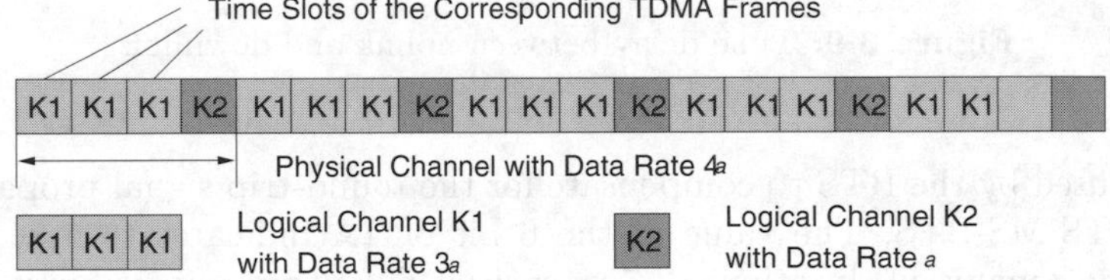

Figure 3.11: Relationship between logical and physical channels

3.3.3 Logical Channels

Logical channels are defined through the allocation of time slots of physical channels. Consequently the data of a logical channel is transmitted in the corresponding time slots of the physical channel. During this process, logical channels can occupy a part of the physical channel or even the entire channel. For instance, if a physical channel has a transmission rate of $4a$, then a logical channel K1 with a data rate of $3a$ and a second logical channel K2 with a data rate a can transmit on the same physical channel (see Figure 3.11).

The GSM recommendations define several logical channels for data transmission on the basis of this principle, dividing them into two main groups: traffic channels and control channels.

Table 3.5: Traffic channels in the GSM recommendation

Traffic channel	Abbreviation
Full-rate TCH for speech	TCH/FS
Half-rate TCH for speech	TCH/HS
9.6 kbit/s full-rate TCH for data	TCH/F9.6
4.8 kbit/s full-rate TCH for data	TCH/F4.8
4.8 kbit/s half-rate TCH for data	TCH/H4.8
$\leq$ 2.4 kbit/s full-rate TCH for data	TCH/F2.4
$\leq$ 2.4 kbit/s half-rate TCH for data	TCH/H2.4
Cell broadcast channel	CBCH

3.3.3.1 Traffic Channels

Traffic Channels (TCH) are logical channels over which user information are exchanged between mobile users during a connection. Speech and data are digitally transmitted on these channels using different coding methods.

Different transmission capacities are required depending on the type of service used (e.g., speech transmission, short-message service, data transfer, facsimile). A distinction is therefore made between the following traffic channels:

B_m-channel Transmission over a B_m-channel (*m=mobile*), which is also called a full-rate traffic channel (*full-rate TCH*), is carried out at a gross data rate of 22.8 kbit/s. Digitalized and coded speech only require 13 kbit/s for transmitting voice information. The remaining capacity in voice transmission is used for error correction. It is possible to transmit data at 12, 6 or 3.6 kbit/s over a B_m-channel.

L_m-channel The half-rate traffic channel (*half-rate TCH*) transmits at a gross rate of 11.4 kbit/s. The number of channels in GSM can be doubled in a given frequency band because of the speech codecs available for half-rate channels. Efficient speech coding algorithms were developed in 1995; they were introduced commercially in 1997/98. Half-rate TCHs allow data to be transmitted at bit rates of 6 or 3.6 kbit/s.

Table 3.5 lists the traffic channels specified in the GSM recommendation.

3.3.3.2 Control Channels

Control information is used for signalling and for system control and is not passed down to the subscribers. Typical signalling tasks include the signalling for establishing, maintaining and releasing traffic channels, for mobility management and access control to radio channels.

Control information is transmitted over so-called *Control Channels* (CCH), which, following ISDN, are also referred to as D_m-channels. The control channels offer the mobile stations a packet-oriented continuous signalling service

Table 3.6: Control channels in GSM

Direction	Group	Channel	Channel identification
MS ← BS	BCCH	BCCH	Broadcast Control Channel
MS ← BS		FCCH	Frequency Correction Channel
MS ← BS		SCH	Synchronization Channel
MS ← BS	CCCH	PCH	Paging Channel
MS → BS		RACH	Random Access Channel
MS ← BS		AGCH	Access Grant Channel
MS ↔ BS	DCCH	SDCCH	Stand-Alone Dedicated Control Channel
MS ↔ BS		SACCH	Slow Associated Control Channel
MS ↔ BS		FACCH	Fast Associated Control Channel

enabling them within the PLMN to receive messages from the base stations and to send messages to the base stations at any time.

Because the control and management of a mobile radio network is far more complex from the standpoint of signalling than a fixed network, three groups of control channels were defined in GSM:

- *Broadcast Control Channel* (BCCH),
- *Common Control Channel* (CCCH),
- *Dedicated Control Channel* (DCCH)).

Table 3.6 contains a list of all the control channels defined in the GSM recommendations, and in the directional column indicates the directions possible on each channel (uplink, downlink or both).

Broadcast Control Channel (BCCH) This channel is used to transmit information about the PLMN from the base station to the mobile stations in the radio cell through a point-to-multipoint connection. The kind of information conveyed over a BCCH includes identification of the network, availability of certain options such as *Frequency Hopping* and *Voice Activity Detection* (VAD) and identification of the frequencies being used by the base station and neighbouring base stations.

One of the subchannels of the BCCH is the *Frequency Correction Channel* (FCCH), used for transmitting a frequency correction burst to the mobile station for possibe correction of the transmitting frequency.

Another subchannel of the BCCH is the *Synchronization Channel* (SCH)), used for transmitting synchronization bursts to a mobile station to allow it to time-synchronize.

Messages transmitted over the BCCH and its subchannels are transmitted exclusively in simplex mode by the base station to the terminal equipment.

Common Control Channel (CCCH) This designation is an umbrella term for control channels that handle the communication between the network and the mobile phone. Included among the CCCH channels are:

Paging Channel (PCH) This channel exists only on the downlink, and is activated for the selective addressing of a called mobile terminal during a connect request from the network (incoming call).

Random Access Channel (RACH) This access channel only occurs on the uplink, and allows the mobile station, using an S-ALOHA access protocol, to request channel capacity from the base station to establish a connection.

Access Grant Channel (AGCH) The base station uses this logical channel to respond to a message received over the RACH from a mobile station. In accordance with the call set up mechanism selected by the network operator, the mobile station is allocated a *Stand-Alone Dedicated Control Channel* (SDCCH) or a TCH over the AGCH that only exists on the downlink; see Section 3.5.1.

Dedicated Control Channel (DCCH) This designation is an umbrella term for three bidirectional point-to-point control channels that are used to transmit signalling messages for call control at different bit rates. The three DCCH channels are:

Stand-Alone Dedicated Control Channel (SDCCH) This channel is always used when a traffic channel has not been assigned, and is allocated to a mobile station only as long as control information is being transmitted. The channel capacity available from an SDCCH is 782 bit/s, which is much lower than that of a TCH. Control information transmitted on the SDCCH includes registration, authentication, location area updating and data for call set up.

Slow Associated dedicated Control Channel (SACCH) This channel is always allocated parallel to a TCH or an SDCCH. It is used to transmit at a data rate of 383 bit/s system information from the network to the mobile station and measurement data on signal strength and receive quality from the MS to the network.

Fast Associated dedicated Control Channel (FACCH) This channel is set up in the short term only when a traffic channel exists and then it uses its time slots. As an example, an FACCH is set up for an impending handover and the necessary control data is transmitted over the FACCH. This channel can handle bit rates of 4600 bit/s or 9200 bit/s.

3.3.4 Hierarchy of Frame Structures

The division of the GSM frequency spectrum into 124 FDM channels and its TDMA channel structure, the tasks of the logical channels and the use of the different bursts have been explained in previous sections.

In GSM the TDMA frames, which contain eight time slots for transmitting bursts, are combined into *multiframes*. A distinction is made between multiframes of two different lengths:

- 26-frame multiframe
- 51-frame multiframe

The bursts of the TCHs and the SACCHs and FACCHs assigned to them are transmitted in 26-frame multiframes. Each traffic channel is allocated 1 of 8 (with half-rate transmission, 16) time slots of a TDMA frame. The associated 8 SACCHs are transmitted in the 12th TDMA frame. The last TDMA frame (25) of the 26-frame multiframe is only used if another 8 SACCHs are required for half-rate transmission. Time slots are stolen from the TCHs for the transmission of the FACCHs.

The data of the *Frame Control Channel* (FCCH), SCH, BCCH, RACH, AGCH, PCH, SDCCH, SACCH/C and CBCH channels is sent in 51-frame multiframes. It was generally specified that speech and data would be transmitted in 26-frame multiframes and signalling data (except for SACCH/T and FACCH) in 51-frame multiframes. However, there was some deviation from this rule when the packet-data service was introduced (see Section 4.1).

Fifty-one of the 26-frame multiframes and 26 of the 51-frame multiframes are combined into a *superframe*, and 2048 superframes produce a *hyperframe* (see Figure 3.12). It takes almost 3.5 hours to transmit a hyperframe.

Figure 3.13 shows the relationship between 51-frame and 26-frame multiframes. The BCCH frames in a 51-frame multiframe are decoded every 240 ms during the unused 26th frame of the 26-frame multiframe.

3.3.5 Combinations of Logical Channels

Logical channels are mapped onto a physical channel through a process of cyclically recurring multiframe patterns. The correct positioning timewise of the cycles is achieved through synchronization of BTS and MS. The BSS provides a reference counter in each cell that is used to number the time slots and serves as the mapping scheme for the multiframes. In GSM each time slot is given a number. This number along with the TDMA frame number allows each time slot to be clearly identified (see Figure 3.9). Because of a frame structure encompassing several levels, the counter for the TDMA frame numbers can go as high as 2 715 648. This high a number was necessary for cryptographic reasons, because the enciphering algorithm uses the TDMA frame number as an input parameter.

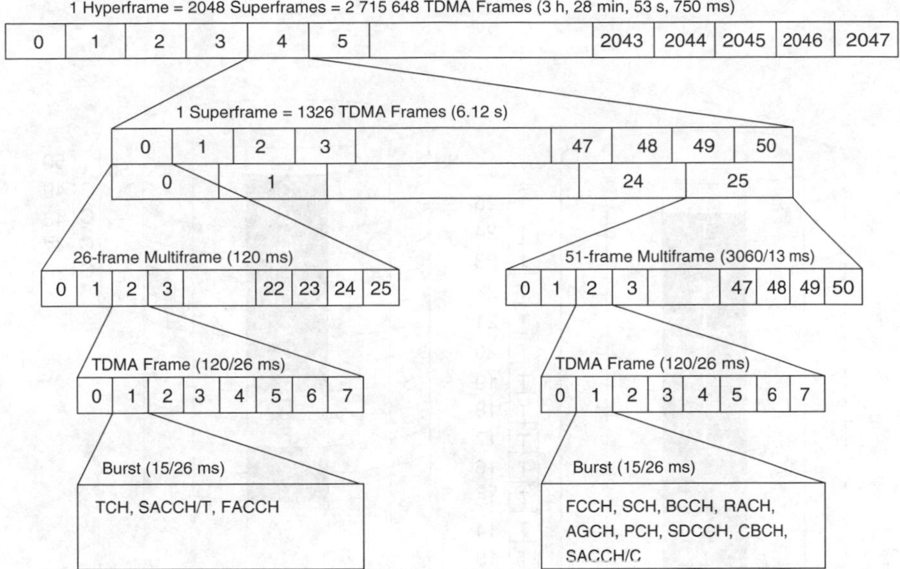

Figure 3.12: The structure of a transmission medium with TDMA frames, multiframes, superframes and hyperframes

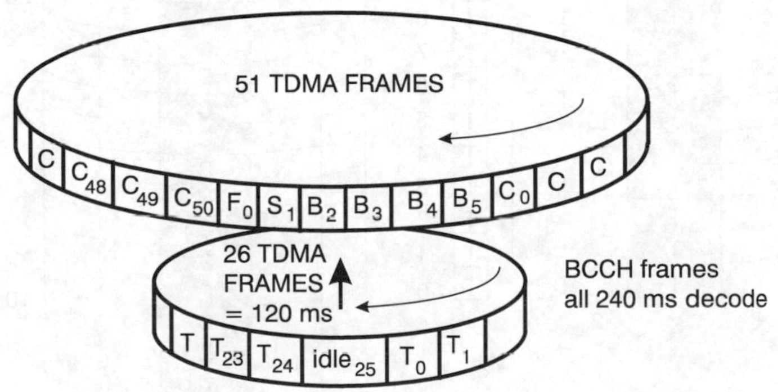

Figure 3.13: The relationship between 51-frame and 26-frame multiframes

Figure 3.14 shows how traffic and signalling channels are mapped onto a physical (frequency) channel. For each TDMA frame of the corresponding multiple frame, a slice of the 'cylinder' shown is provided that carries eight time slots (the TDMA frame) [33].

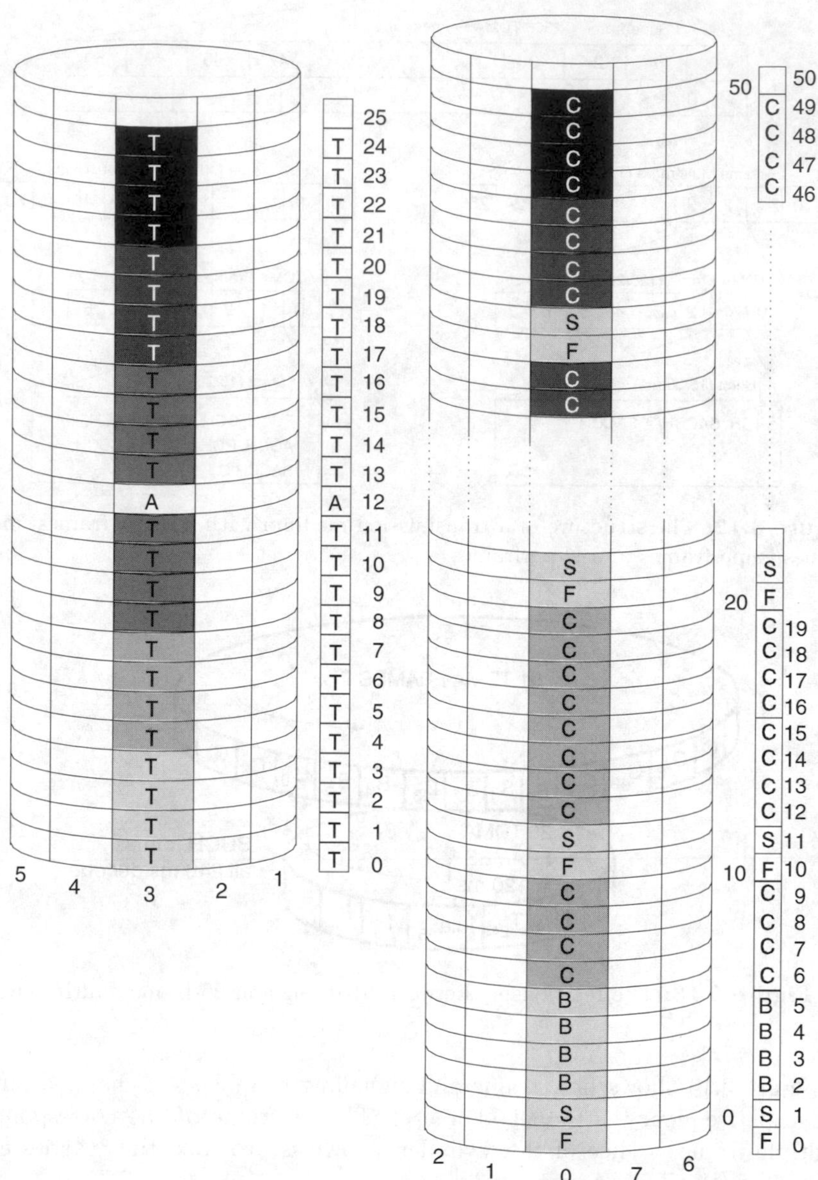

Figure 3.14: Traffic and control channel with 'wind up' time axis

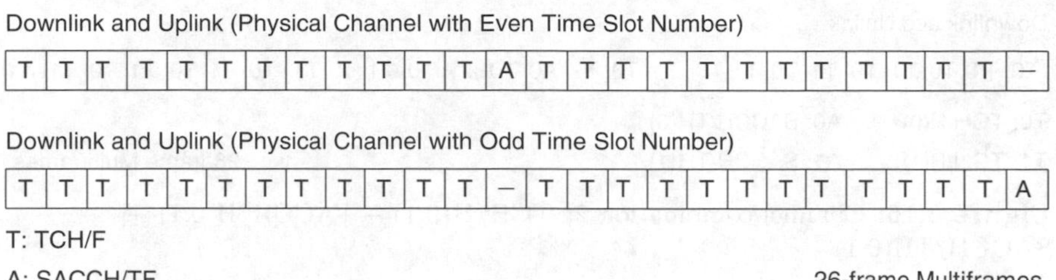

Figure 3.15: Channel combination 1: TCH/F + FACCH/F + SACCH/TF

3.3.5.1 Approved Channel Combinations

Logical channels are not simply distributed over the two multiframes, but follow a particular pattern. The following is a set of approved *Channel Combinations* (CC) [16], represented by their multiframes:

CC1:	TCH/F + FACCH/F + SACCH/TF
CC2:	TCH/H(0,1) + FACCH/H(0,1) + SACCH/TH(0,1)
CC3:	TCH/H(0) + FACCH/H(0) + SACCH/TH(0) +TCH/H(1)
CC4:	FCCH + SCH + BCCH + CCCH
CC5:	FCCH + SCH + BCCH + CCCH + SDCCH/4(0,1,2,3) + SACCH/C4(0,1,2,3)
CC6:	BCCH + CCCH
CC7:	SDCCH/8 + SACCH/8

A physical channel exactly contains one of these combinations. The group CCCH divides into the channels PCH, AGCH and RACH. The numbers in brackets after the logical channels give the numbers of the logical subchannels. This means that in the case of SDCCH/4 there are four SDCCHs on a physical channel, each of which is used by one mobile station.

The shared use of a physical channel by several logical channel types does not mean that all of these channels can be used at the same time. Although the mapping of combinations of logical channels onto a physical channel may mean that several logical channels can appear on one physical channel, they occur sequentially at intervals of at least one TDMA frame length. The sequential numbering of the TDMA frames ensures that the physical layers of both communicating partners use the currently appropriate logical channel for transmitting and receiving.

The multiframes in the approved channel combinations are shown in Figures 3.15–3.21. The mapped multiframe combinations are derived from tables found in Series 05.02 of the GSM standard [16]. The tables describe which time slots and which frequencies the respective logical channel is allowed to use.

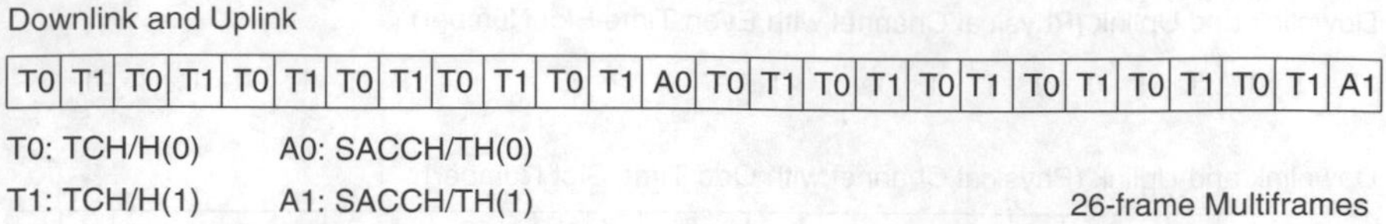

Downlink and Uplink

T0	T1	T0	T1	T0	T1	T0	T1	T0	T1	T0	T1	A0	T0	T1	T0	T1	T0	T1	T0	T1	T0	T1	T0	T1	A1

T0: TCH/H(0) A0: SACCH/TH(0)

T1: TCH/H(1) A1: SACCH/TH(1) 26-frame Multiframes

Figure 3.16: Channel combination 2: TCH/H(0,1) + FACCH/H(0,1) + SACCH/TH(0,1)

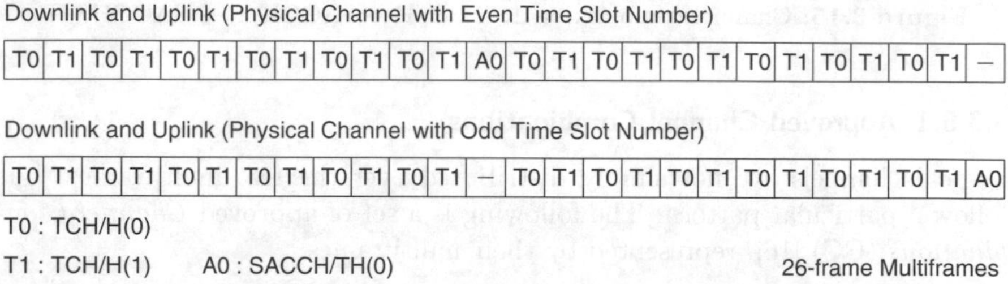

Downlink and Uplink (Physical Channel with Even Time Slot Number)

T0	T1	T0	T1	T0	T1	T0	T1	T0	T1	T0	T1	A0	T0	T1	T0	T1	T0	T1	T0	T1	T0	T1	T0	T1	–

Downlink and Uplink (Physical Channel with Odd Time Slot Number)

T0	T1	T0	T1	T0	T1	T0	T1	T0	T1	T0	T1	–	T0	T1	T0	T1	T0	T1	T0	T1	T0	T1	T0	T1	A0

T0 : TCH/H(0)

T1 : TCH/H(1) A0 : SACCH/TH(0) 26-frame Multiframes

Figure 3.17: Channel combination 3: TCH/H(0) + FACCH/H(0) + SACCH/TH(0) + TCH/H(1)

A cell in a GSM-PLMN has a set of $n+1$ frequency channels available for its use. These frequency channels are denoted by $C_0, C_1, \ldots, C_n$. The indices have no relationship to the frequency. The described channel combinations are not allowed to occupy just any random time slots or frequencies.

For example, the combination 4 shown in Figure 3.18 is only allowed to be transmitted in time slot 0 of the carrier frequency C_0. On the downlink the BTS must transmit in each time slot number 0 the logical channel of combination 4 to enable the mobile station to carry out performance measurements.

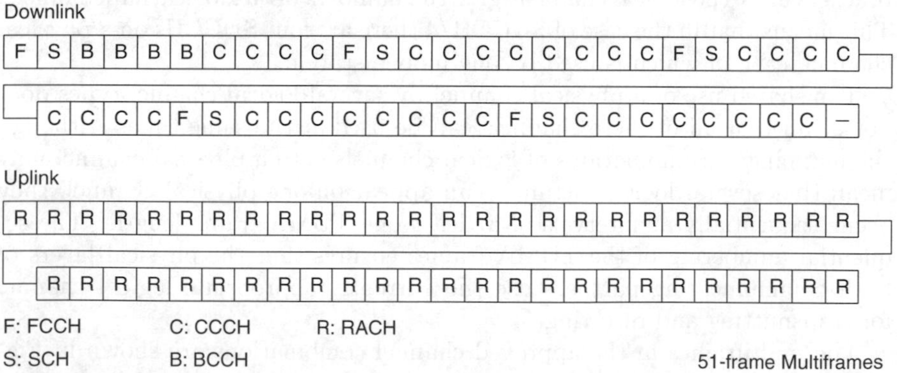

Downlink

F	S	B	B	B	B	C	C	C	C	F	S	C	C	C	C	C	C	C	C	F	S	C	C	C	C

C	C	C	C	F	S	C	C	C	C	C	C	C	C	F	S	C	C	C	C	C	C	C	C	–

Uplink

R	R	R	R	R	R	R	R	R	R	R	R	R	R	R	R	R	R	R	R	R	R	R	R	R	R

R	R	R	R	R	R	R	R	R	R	R	R	R	R	R	R	R	R	R	R	R	R	R	R	R

F: FCCH C: CCCH R: RACH

S: SCH B: BCCH 51-frame Multiframes

Figure 3.18: Channel combination 4: FCCH + SCH + BCCH + CCCH

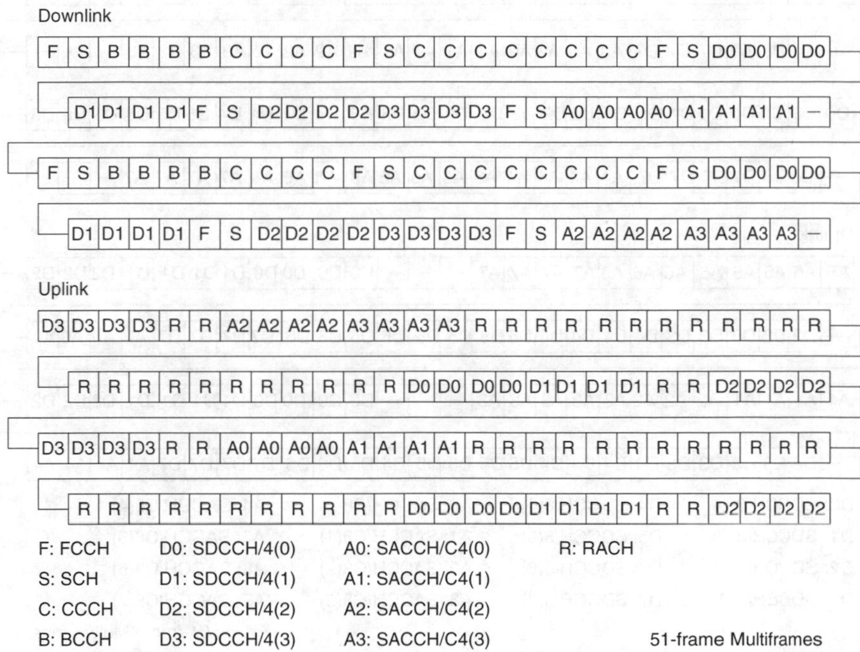

Figure 3.19: Channel combination 5: FCCH + SCH + BCCH + CCCH + SDCCH/4 (0, 1, 2, 3)

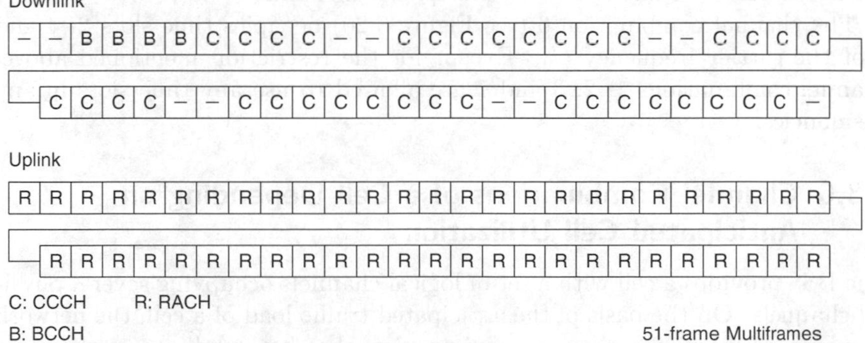

Figure 3.20: Channel combination 6: BCCH + CCCH

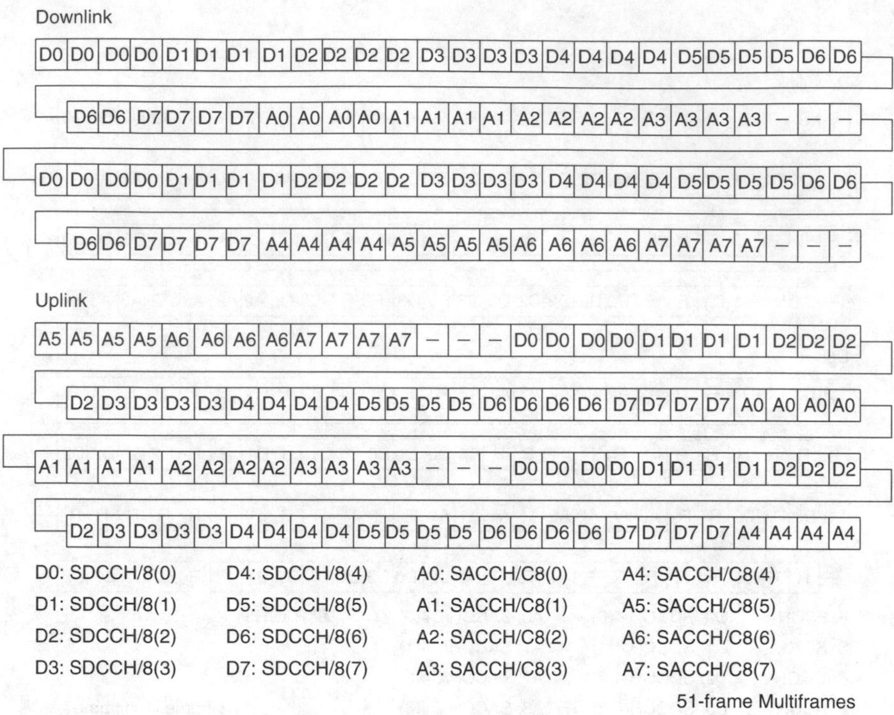

Figure 3.21: Channel combination 7: SDCCH/8 + SACCH/8

If no information is available for sending, a dummy burst is transmitted (see Section 3.3.1.1).

The combination 5 shown in Figure 3.19 is only allowed to occupy time slot 0 in frequency C_0. None of the other combinations, except for combinations 4 and 5, are allowed to use this physical channel.

The channel combination 6 (see Figure 3.20) occupies time slots 2, 4 and 6 of the carrier frequency C_0. Except for the restriction mentioned above, channel combinations 1, 2, 3 and 7 are allowed to use any time slots in any frequencies.

3.3.6 Channel Combinations of a Cell Depending on Anticipated Cell Utilization

The BSS provides a cell with a set of logical channels occupying several physical channels. On the basis of the anticipated traffic load of a cell, the network operator establishes a channel configuration that must adhere to the rules mentioned in the last section. Each individual *Transceiver* (TRX) can offer eight CCs in each time slot. The time slot is identified through a *Timeslot Number* (TN). Three common combinations are briefly presented here:

- Low-capacity cell with one TRX:
  - TN 0: FCCH + SCH + BCCH + CCCH + SDCCH/4(0,1,2,3) + SACCH/C4(0,1,2,3)
  - TN 1 to 7: TCH/F + FACCH/F + SACCH/TF
- Medium-capacity cell with four TRXs:
  - Once on TN 0: FCCH + SCH + BCCH + CCCH
  - Twice (on TN 2 and TN 4): SDCCH/8 + SACCH/8
  - 29 times: TCH/F + FACCH/F + SACCH/TF
- High-capacity cell with 12 TRXs:
  - Once on TN 0: FCCH + SCH + BCCH + CCCH
  - Once on TN 2: BCCH + CCCH
  - Once on TN 4: BCCH + CCCH
  - Once on TN 6: BCCH + CCCH
  - 5 times: SDCCH/8 + SACCH/8
  - 87 times: TCH/F + FACCH/F + SACCH/TF

After successful synchronization, the mobile station is informed through the system information of the BCCH which channel combinations on which physical channels are being offered to it by the BSS. Depending on its current operating state (idle state or dedicated state), it uses a particular subset from this offering of channels.

What is noticeable is that a BCCH always appears together with the logical channels SCH und FCCH and that it can always be found in time slot 0. This condition helps to facilitate synchronization when an MS makes its first contact with a BTS. Additional combinations 6 (see Figure 3.20) are added when traffic is expected to be heavy.

3.3.7 Layer 1: Physical Transmission

GSM layer 1 is essentially comparable to the physical transmission layer (*physical layer*) in the ISO/OSI reference model (see Section 2.5). Unlike the ISO/OSI model, an entity of layer 3, *Radio Resource Management* (RRM), directly accesses this lowest layer (see Figure 3.22).

This access allows channel allocations to be controlled and layer-3 information to be queried on the state of the *physical layer* and the radio connection, such as channel parameters, error ratios and received signal strength. This information is required by layer 3 in order for it to fulfil the typical cellular networks tasks such as *power control*, *handover* and *roaming*.

Layer 1 of a GSM implementation is responsible for the radio transmission of traffic and signalling information. Its main tasks include:

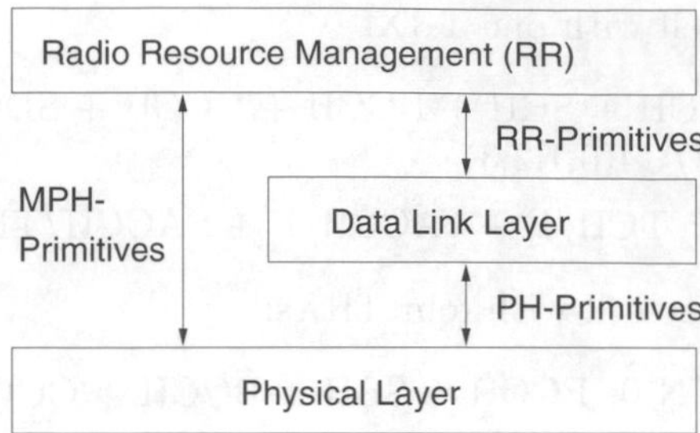

Figure 3.22: Interfaces of layer 1 (*physical layer*)

- Producing bursts, multiplexing the bursts in TDMA frames, and transmitting the frames on the available physical channels over dedicated control and traffic channels.
- Seeking and seizing a BCCH by the MS.
- Judging channel quality and received signal strength.
- Error detection and correction mechanisms (defective blocks are not forwarded to layer 2).
- Synchronization with frame transmission.
- Encoding of data stream.

The GSM system uses a digital 0.3 GMSK modulation format. GMSK stands for *Gaussian Minimum Shift Keying*, in which *minimum shift* means that there is no discontinuity between the phase of the carrier and the time. The modulation format 0.3 GMSK means that the modulated signal passes through a Gauss filter with the product of the 3 dB bandwidth B and the bit period T producing the value $BT = 0.3$.

3.3.7.1 Channel Coding and Interleaving

This section describes how GSM layer-2 messages are conveyed by means of bursts. The two dedicated signalling channels SDCCH and FACCH are especially important. In [14] they are referred to as the *main signalling link*. Table 3.8 presents a general overview of the coding procedures used for logical channels. The following information is provided for each type of channel:

- The number of bits per 456 bit-block of user information which are systematic redundancy and which are filler bits
- The coding rate of the convolutional code used
- The depth of bit interleaving

Table 3.8: Coding methods for logical channels

Channel type	Bit/Block Data+Parity+Tail	Convolutional coding rate	Bit/ Block	Interleaving depth
TCH/FS			456	8
Class I	$182 + 3 + 4$	1/2	(378)	
Class II	$78 + 0 + 0$	-	(78)	
TCH/F9.6	$4 \cdot 60 + 0 + 4$	244/456	456	19
TCH/F4.8	$60 + 0 + 4$	1/3	228	19
TCH/H4.8	$4 \cdot 60 + 0 + 4$	244/456	456	19
TCH/F2.4	$72 + 0 + 4$	1/6	456	8
TCH/H2.4	$72 + 0 + 4$	1/3	228	19
FACCHs	$184 + 40 + 4$	1/2	456	8
SDCCHs, SACCHs	$184 + 40 + 4$	1/2	456	4
BCCH, AGCH, PCH	$184 + 40 + 4$	1/2	456	4
RACH	$8 + 6 + 4$	1/2	36	1
SCH	$25 + 10 + 4$	1/2	78	1

Different channel coding methods are applied on the various channels for the different services (see Section 2.7.1).

A Fire code—a linear, binary block code derived from the cyclic block code family—is used with most signalling channels. This code has a generator polynomial that offers good error detection and/or error correction when errors occur in groups (*burst errors*):

$$(X^{23} + 1)(X^{17} + X^3 + 1)$$

Forty Fire code parity bits are added to the 184 data bits of a layer-2 *Protocol Data Unit* (PDU) (see Figures 3.23 and 3.24). This allows groups of errors of up to 11 bits to be detected and corrected. When layer 1 recognizes a defective block and can no longer correct the errors, it does not forward this block to layer 2.

All channels have *Forward Error Correction* (FEC) protection through convolutional coding. The respective code ratios are given in Table 3.8. A convolutional code with the ratio $r = 1/2$ and the influence length $k = 5$ (see Section 2.7.3.2) is used on the control channels.

Interleaving allows correlated errors caused by a short signal fading to appear as single errors at the decoder of the receiver (see Figure 3.25).

Two types of interleaving are in use, bit interleaving and block interleaving. The bit interleaving with a depth of n is based on an algorithm, where the 456 information bits of a convolutional coded layer-2 block (see Figure 3.23) are arranged into rows mod n into a table (see left-hand upper side of Figure 3.25).

The bits are read out for transmission from the table row by row (see right-hand upper side and lower side of Figure 3.25).

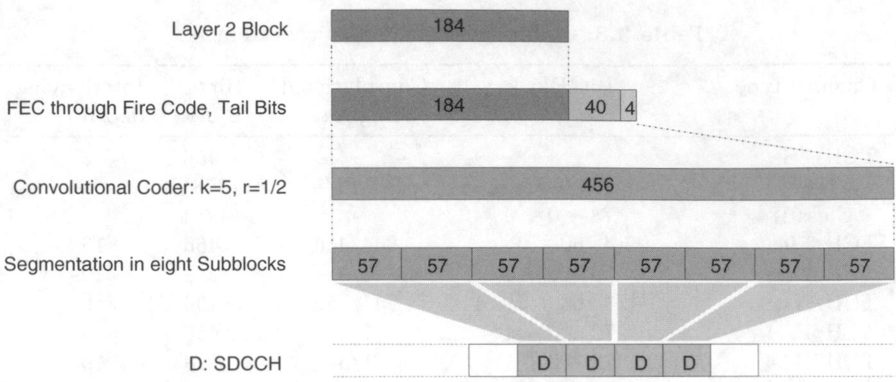

Figure 3.23: Coding and interleaving in an SDCCH

During this bit interleaving procedure, a block interleaving is introduced in addition whereby the first four rows of 57 times 8 bit (right-hand upper side of Figure 3.25), which each fill half a normal burst of a TCH, are inserted into bit positions in this burst with even bit numbers. The remaining four rows are transmitted on odd bit positions in the normal burst. Each of these rows corresponds to one subblock shown in Figure 3.23.

The de-interleaver at the receiver rearranges the interleaved bits into the original sequence. It is clear from the algorithm that a group of errors of maximum length $n - 1$ applied to the interleaved bit stream on the radio medium is resolved into single bit errors after de-interleaving.

The block interleaving procedure further contributes to spread originally neighbouring bits during the radio transmission to improve the resistance against the so-called error bursts resulting from signal fading.

Figure 3.26 (b) shows the error protection mechanisms for the *Random Access Channel* (RACH); see also Figure 3.32. Because a mobile station's call request should be recognized quickly by the base station and collisions can sometimes occur that must quickly be resolved, the 8-bit message of the accessing mobile station (code for type of access and random number along with check sum) is modified through the *Base Station Identification Code* (BSIC), then convolutionally coded to 36 bits and transmitted without interleaving.

On the FACCHs block diagonal interleaving occurs in eight half bursts; on all other control channels (with the exception of RACH) interleaving is in four whole bursts. This is illustrated on the basis of an SDCCH (see Figure 3.23) and an FACCH/F (see Figure 3.24).

For example, at a transmission rate of 9.2 kbit/s, every 20 ms 184 bits are transmitted on an FACCH. For the detection of non-correctable errors 40 parity check bits calculated according to the Fire code are added by the physical layer. All bits are then protected by a convolutional code of ratio $r = 1/2$. This provides sufficient error correction potential on the receiver side. The resulting 456 bits ($= 4 \cdot 114$) are then distributed over the four time

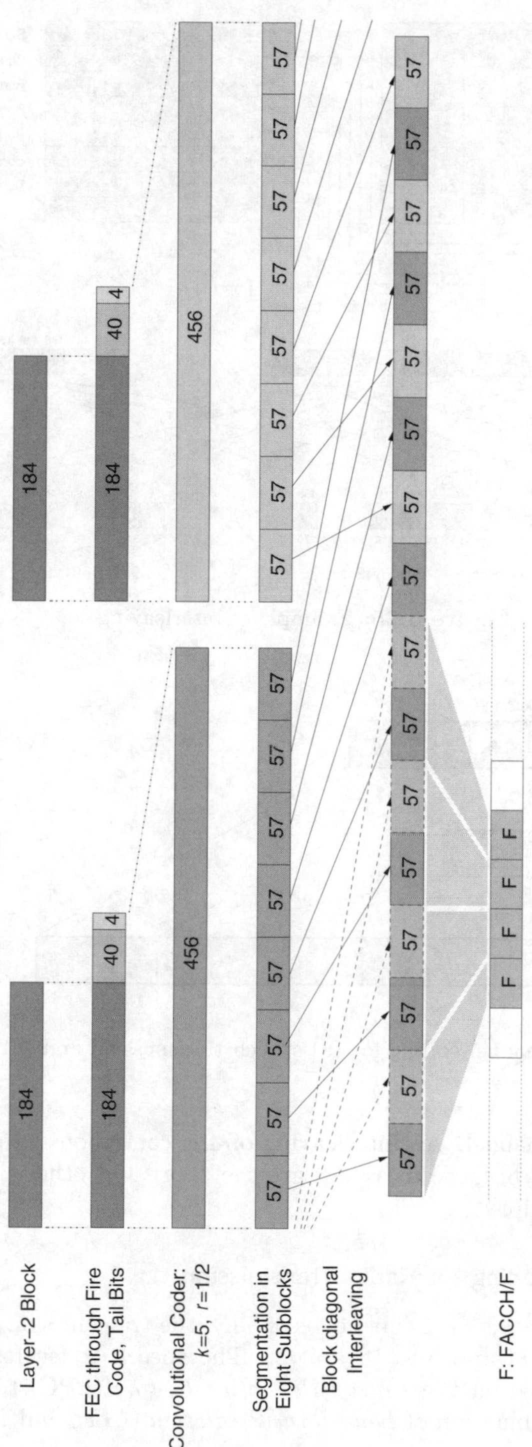

Figure 3.24: Coding and interleaving in an FACCH/F

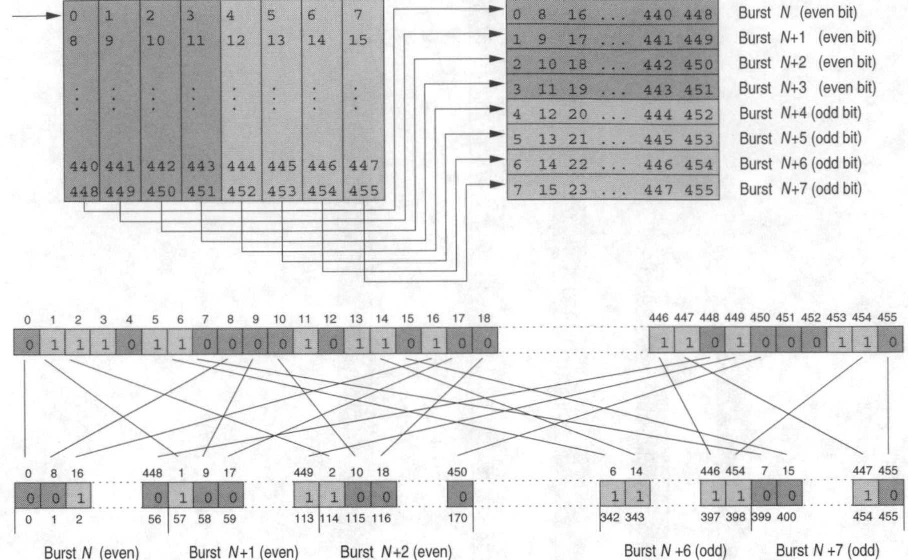

Figure 3.25: Example of interleaving

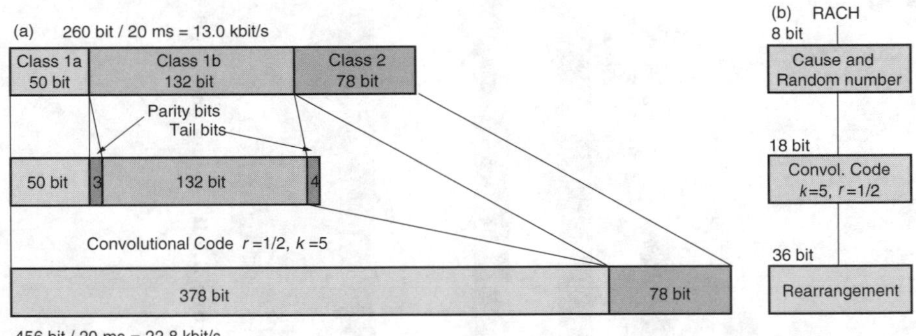

Figure 3.26: Channel coding for (a) speech transmission and (b) signalling

slots of a logical channel. An interleaving of 4 is continuously applied to the signalling channels because messages are too short and otherwise too many slots would be occupied.

3.3.7.2 Speech Coding for Radio Transmission

In line with GSM goals on frequency economy, the transmission rate was not to exceed a value of approx. 16 kbit/s. The speech codec RPE-LTP was selected. It is based on the *Linear Predictive Coding* (LPC) technique and incorporates a combination of *Long-term Prediction* (LTP) and *Regular Pulse*

Excitation (RPE). The codec has a net bit rate of 13.0 kbit/s. Together with the redundancy for error protection, 22.8 kbit/s (gross bit rate) are transmitted on the radio channel. Speech is produced in packets of 20 ms each from the speech codec. With consideration given to the channel characteristics and the possible bit error ratio of the speech codec parameters, the following three-stage channel coding method was selected (see Figure 3.26 (a)):

Ordering of the bits according to importance The bits of the speech codec parameters are arranged in descending order of importance, with the most important bit, No. 1, first and the least important bit, No. 260, last.

CRC for the most important 50 bits (Class 1a) These are extended by 3 parity bits as a *Cyclic Redundancy Check* (CRC). The CRC bits are inserted after bit No. 50, and the bits that follow are renumbered. The associated generator polynomial is $g(x) = x^3 + x + 1$ and is implemented through a cyclic shift register with appropriate taps. On the receiving side these CRC bits are used for error detection and can sometimes indicate one or more errors within the most important bits.

Convolutional coding of the most important 185 bits (Class 1a+1b) for error correction Four tail bits (0) are attached to the first 185 bits, and these 189 bits are then recoded to 378 bits using a convolutional coding with ratio r =1/2. Since this protects the first and last bits better than it does the bits in the middle of a sequence, the first 185 bits are rearranged further so that the most important bit ends up first, the second most important bit is last (185), the third most important bit is second, and so forth. The remaining 78 bits of the speech codec parameters (Class 2) are transmitted without protection so that altogether 456 bits are produced per speech coding frame and transmitted with the interleaving factor 8.

With interleaving (distribution of data over several time slots of a logical channel), transmission errors occurring in bursts are distributed more evenly over the speech data received, and transmission errors occur as single bit errors and therefore can be corrected.

Normal bursts (see Figure 3.7) with speech data are divided into two separate blocks, each with 57 bits and one control bit, which transport data of different (consecutive) speech coder frames so that eight sequential time slots contain the data from a speech coder frame of 20 ms duration and in addition the same amount of data from a second speech coder frame. Four consecutive time slots carry 456 user bits in 18.5 ms, namely the data output generated every 20 ms by the speech codec. The control bit indicates whether a block is carrying user data or data from the *Fast Associated dedicated Control Channel* (FACCH).

3.3.7.3 Half-Rate Speech Codec

To be able to provide a higher traffic capacity for speech in hot-spot areas without the need to allocate more frequency channels or to reduce the cell size, a half-rate speech coder has been standardized that requires a net bit-rate of 6.5 kbit/s only.

The speech quality has been claimed to be comparable to the full-rate coder, however, it does not behave well when music during wait for connection to an extension line of a branch exchange or when other non-speech tone-signals have to be transmitted. Some operators have distributed dual-mode (full- and half-rate alternating) mobile phones to their customers especially to those residing in large cities. In general, although a half-rate speech coder would allow for more than double the capacity of a cell for speech, most operators have decided to make only a limited or even no usage of half-rate coders and instead aim at introducing higher quality speech under full-rate speech.

3.3.7.4 Enhanced Full Rate (EFR) Speech Coder

The *Enhanced Full Rate* (EFR) speech coder is an *Algebraic Codebook Excited Linear Predictive* (ACELP) coder that has been introduced into products since 1998.

It produces 244 code bits for every 20 ms long speech frame, thereby corresponding to a bit rate of 12.2 kbit/s. These code bits are divided into three main categories upon their relative subjective importance. The most important 65 bits are protected by a half-rate convolutional encoder and an eight-bit cyclic redundancy checksum. If any of these bits are in error, the decoder will reject the speech frame as being in error. The following 117 bits are protected by the same half-rate convolutional code, whereas the remaining 78 bits are not offered any protection at all.

3.3.7.5 Discontinuous Voice Transmission

As an option, GSM permits the use of *Discontinuous Transmission* (DTX) for the transmission of voice signals, thereby reducing the amount of power handsets need in order to transmit and co-channel interference. Speech pauses are recognized by a function referred to as *Voice Activity Detection* (VAD) and transmission is halted until data is again available. The traffic channel remains assigned to the connection and does not produce interference to co-channels during transmission pauses.

To give the receiver the impression that a connection still exists during transmission pauses, the receiver side produces something called *comfort noise*, which is provided by the measurement results of the previous speech transmission phase and is activated by the transmission of updated measurement values from the transmitter during long speech pauses.

3.3.7.6 Adaptive Equalization of the Radio Signal

The 26 bits of the training sequence transmitted in the middle of each time slot are used by the receiver in order to adjust the parameters of its matched filter to the current channel characteristics and to be able to select the path with the strongest signal in multipath propagation of the radio waves in its equalizer.

In the GSM frequency band, shadowing may happen to the radio signals of mobile and base stations; this is partly compensated by the fact that radio waves are reflected by buildings, mountains, inshore waters, etc., and fill in shadowed areas. The radio path of reflected waves is longer than the direct path. With the bit rates selected for GSM this difference in path lengths can amount to several bit durations (3.7 μs). Equalization of superimposed multipath signals by the receiver can compensate for delay differences of up to 16 μs (3 km difference in path length).

The communications signal is recovered when an estimation of the impulse response of the radio channel is carried out per time slot based on the training sequence contained in the received burst, and the receive signal is passed through an appropriate parameterized matched filter. The training sequence with the known information it contains is also used for determining the current quality of a radio channel through counting the bit errors.

3.3.7.7 Coding for Signalling and Data Transmission

Because longer transmission times are acceptable, interleaving factors of up to 19 are used for bearer services in data transmission (see Figure 3.27).

3.3.7.8 Throughput and Delay

Figure 3.28 shows the transmission rates (throughput) and delay times of all the logical channels (see Table 3.12). The *delay* results from the respective periodic recurrence (*recurrence interval*) of the time channel plus a period corresponding to the selected interleaving depth (e.g., 14 ms with a large number of signalling channels).

3.3.7.9 Synchronization

If a mobile station wishes to make contact with the network through a base station, it must first detect the control channel (BCCH). The time slot and TDMA frame synchronization will then be established. The frequency of the mobile station's oscillator is automatically matched to the carrier frequency of the receive signal.

Recognition of the control channel (BCCH) The BCCH reserves one time slot (TN=0) per cell on a fixed carrier frequency. An important feature is that, on the frequency on which the BCCH is being sent, transmission is

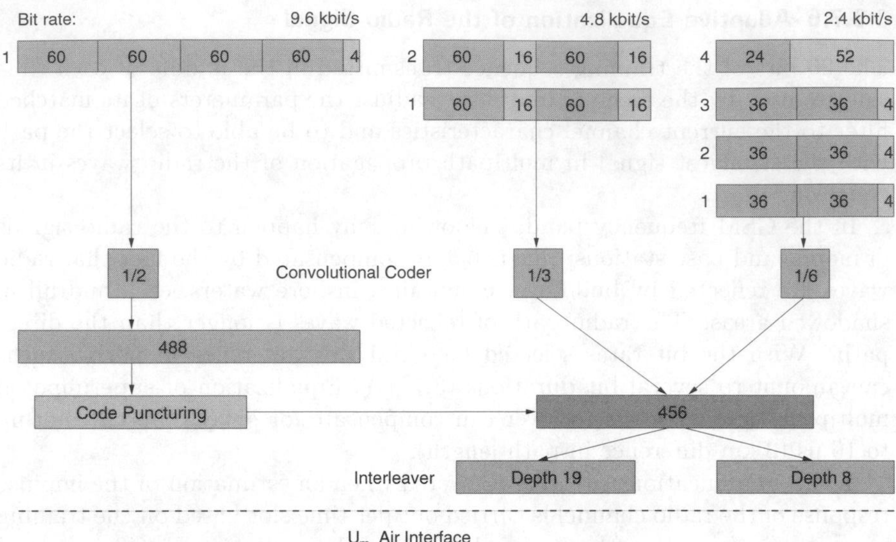

Figure 3.27: Forward-error correction of subscriber data with the different GSM bearer services

continuous with constant power (see also [18]). Therefore, the cluster size for the frequencies that carry the BCCH is typically much larger than for the other frequencies used in a cell.

There it is possible to transmit in each time slot with power attenuated to the individual connection. On other frequencies, in particular, the sender in unused time slots is switched off. Frequency hopping is not used on the BCCH because this method must first be set up by the system information, which is sent on the BCCH.

If a mobile station is switched on, it conducts measurements of the mean signal level on all the frequencies known to it. A preliminary selection is

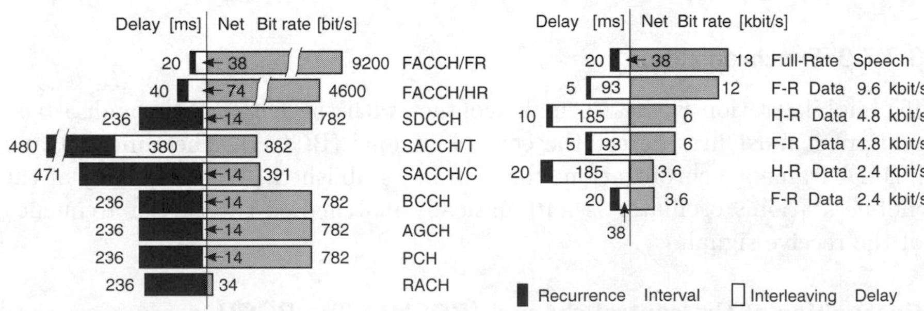

Figure 3.28: Logical channel throughput and delay times

Table 3.9: Time synchronization using counters

Counter	Abbreviation	Value range
Quarter-bit number	QN	0–624
Bit number	BN	0–156
Time slot number	TN	0–7
TDMA frame number	FN	0–2 715 647

Table 3.10: System parameters `BS_CCCH_SDCCH_COMB`, `BS_CC_CHANS` and `CCCH_CONF`

CCCH_CONF	BS_CC_CHANS	BS_CCCH_SDCCH_COMB
000	1	FALSE
001	1	TRUE
010	2	FALSE
100	3	FALSE
110	4	FALSE

made on the basis of the strongest mean receiving field strength. The BCCH can only be identified with satisfactory certainty if the *Frequency Correction Burst* (FCB) and the *Synchronization Burst* (SCB), which are transmitted on the same physical channel as the BCCH, are detected. The FCB is used for frequency adjustment, and the SCB is needed for establishing the time slot and TDMA frame synchronizations.

Time Synchronization Time synchronization means that the mobile station is intercepting the frequency channel at the appropriate time and is sending a burst at the appropriate moment. This requires that all logical channel combinations of all eight physical channels of a frequency as well as the uplink and the downlink be time-synchronized with respect to the communicating stations (BTS and MSs).

This synchronization is based on the frame hierarchy of the different frame types presented in Section 3.3.4. Each mobile station has a set of counters that must run synchronously with the reference counters of the BSS and are incremented in steps over time modulo the cycle duration of the respective counter. For each TDMA frame these counters produce a number that increases as time goes on. As long as a GSM-PLMN is in operation, the frame numbers continue to run; after a cycle of 2 715 648 the numbers start to repeat themselves. The counters for time synchronization are given in Table 3.9.

The *Quarter-bit Number* (QN) is derived from the training sequence. Because the SCH is always located on a physical channel of TN $= 0$, the mobile station sets the value for TN at 0 when it is receiving an SCB. The value of the *Frame Number* (FN) appears in coded form in the information bits of the SCB.

After the initial setting of these counters, the mobile station can use its own clock to update them. It increases the value of QN every 12/13 µs. The value of the *Bit Number* (BN) is produced from the integral part of QN/4. TN is incremented when QN changes from 624 to 0. FN goes up by one when TN increases from 7 to 0.

Recognition of the channel configuration of a cell If the BCCH is located and the MS's counter has been updated, the mobile station reads the system information in order to decode the channel structure available to this cell. The system information contains the parameter `CCCH_CONF` that establishes the number of physical channels (`BS_CC_CHANS`) supporting CCCHs. `CCCH_CONF` contains information on whether CCCHs and SDCCHs are combined (`BS_CCCH_SDCCH_COMB = TRUE`) or not combined (`BS_CCCH_SDCCH_COMB = FALSE`) on a physical channel (see Table 3.10).

If `CCCH_CONF` takes on the value of `000` then combination 4 (see Figure 3.18) is supported and there is no combination 6 (see Figure 3.20). The value `001` means that only combination 5 (see Figure 3.19) is used in the cell. As far as the remaining three values are concerned, combination 4 is again available (see Figure 3.18), and, depending on the utilization level of the cell, is expanded by one, two or three combination 6s (see Figure 3.20).

The number of physical channels carrying CCHs is at the most four. Each CCCH is allocated its own group of mobile stations (`CCCH_GROUP`), which, when in idle mode, listen in on the PCH; there is a maximum of four radio paging groups [16].

The mobile station is now in a position to respond to a call or to initiate its own call. The logical channels that it needs, i.e., the CCCHs and the BCCH, are known to it.

3.3.7.10 Synchronization of Base Stations and Mobile Stations

The TDMA method requires that the signals of all mobile stations using the same carrier frequency reach the base station within the time slots assigned. They are not allowed to overlap. Because of the propagation speed of the signal at approximately 300,000 km/s, the synchronization reference signal of the base station reaches the mobile stations closest to it sooner than those further away.

According to the GSM recommendation, a cell is allowed to be within a 35 km radius of the base station. The maximum loop duration from the base station to the edge of the cell and back is 0.23 ms for a 70 km stretch. If only the reference signal were being used for time synchronization, the time slots would have to contain the appropriate guard time; 40 % of the time slot would be wasted. In the GSM system the base station measures the loop duration and informs the mobile station by how much bit time before the reference signal it should be transmitting. This allows the guard time of each slot to be reduced to 30 μs (8.25 bits). This guard time is needed to allow the receiver to select the best out of a number of available radio paths of an incoming signal and still receive within the limits of a time slot.

3.3.8 GSM Layer 2: Data Link

In the ISO/OSI model this layer is responsible for the secure transmission of data over individual links between two directly connected network elements (nodes) as well as for the error handling of data packets (see Section 2.7.4). The GSM specifications in regard to the *data link layer* are oriented on the ISDN standards, such as ISDN/*Link Access Procedure on the D-channel* (LAPD) based on X.200/Q.920, X.25/LAPB and *High Level Data Link Control* (HDLC)/ISO 3309/4335. Some individual adaptations were made because, for example, unlike the *Link Access Procedure on the D-channel* (LAPD) protocol, no limiting flags are required. The synchronization is already provided by layer 1, and because of the existence of several logical channels a special data link control protocol had to be specified. Analogously to ISDN, this protocol is called LAPD_m, where m stands for mobile. It is used between MS and BS, whereas between BTS and BSC over the A_bis interface the LAPD protocol is used, and between BSC and MSC over the A-interface the *Message Transfer Part* (MTP) protocol familiar from CCS7 is employed (see Figure 3.29).

3.4 Signalling Protocols in the GSM Data Link Layer

The communication protocols in the GSM system in some cases deviate from the ISO/OSI reference model (ITU-T recommendations X.200 to X.219). Although the protocols used in GSM can be structured on the ISO/OSI model, other protocol functions come into play because of the special characteristics of a cellular radio network. Therefore certain wide-ranging tasks, such as the evaluation and allocation of required capacity in a radio path and management-related services, do not necessarily only affect one particular ISO/OSI layer.

Because of the physical qualities of the transmission medium radio and the characteristics of GSM, suitable communication protocols are required. The signalling between the network elements (BSS, MSC, HLR, VLR, AuC, EIR and OMC) of the GSM system is based on *Common Channel Signalling System No. 7* (CCS7), ITU-T Series Q.700-795. A *Mobile Application Part* (MAP) was developed specifically in order to accommodate the radio signalling in a GSM-PLMN. Figure 3.29 shows the architecture of the signalling protocols up to layer 3 and how they are distributed among the nodes of the GSM-PLMN.

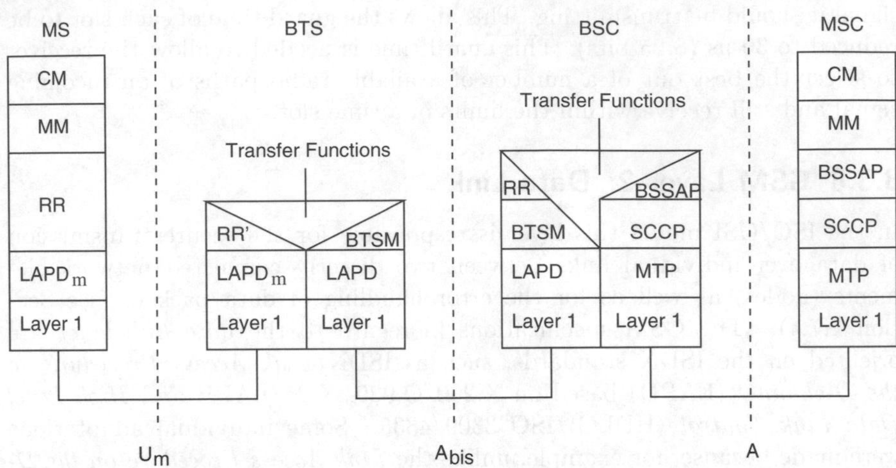

Figure 3.29: Architecture of the signalling protocols and distribution among GSM nodes

The MAP part is implemented in all the switching centres directly linked to the mobile network. It consists of several *Application Service Elements* (ASE), which are required for transactions involving registration and database inquiry and for determining a mobile subscriber's current location.

In addition to the MAP, the layered functions of CCS7 also include the following components [32]:

- The *Transaction Capability Application Part* (TCAP) consists of two sublayers: a transaction sublayer and a component sublayer. The transaction sublayer is responsible for controlling the transactions or dialogues in a point-to-point connection. Each message is given a counter that is used to identify all messages at the other end of a connection. The component sublayer controls the different operations involved, such as a response or a notification that an operation has been completed successfully (or unsuccessfully). The TCAP supports the MAP; the services it provides correspond to those of ISO/OSI layer 7.
- The *Intermediate Service Part* (ISP) corresponds to ISO/OSI layers 4 to 6, and for the time being is still empty because these layers have no significance in a message-switched environment.
- The *Signalling Connection Control Part* (SCCP) represents a part of layer 3, and is used to set up a point-to-point connection for the transmission of individual messages.
- The *Message Transfer Part* (MTP) corresponds to layers 1 and 2 and part of layer 3, and is used to transmit messages between two nodes connected over a subnet. The MTP is divided into three sublayers, with

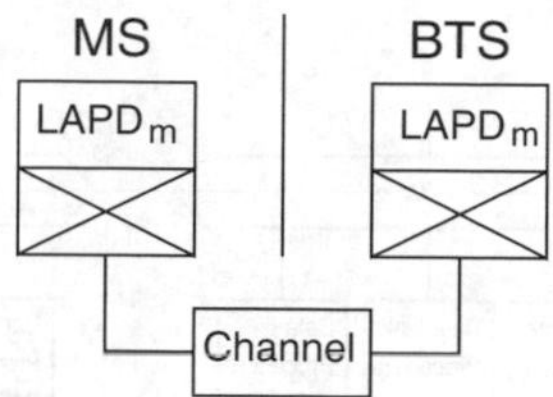

Figure 3.30: Position of signalling protocol $LAPD_m$

the two lower sublayers fulfilling protective functions for message transfer and the third having responsibility for operating and maintenance functions in the network.

Radio Resource Management (RRM) is supported to a limited degree in the base station and is implemented through the functions of *BTS Management* (BTSM). In the BSC these functions correspond to the *Base Station Subsystem Application Part* (BSSAP). As is shown, the entities of *Mobility Management* (MM) communicate directly between MS and MSC over an appropriate protocol. This establishes the service of *Call Management* (CM). The services CM, MM, BSSAP and SCCP fall under layer 3 (see Section 3.5).

The protocols of layers 1–3 defined in Series 04 and 08 of the GSM recommendations are required for signalling. The following sections will delve into the important characteristics of these three layers. Special emphasis will be given to two features of the $LAPD_m$ protocol that are dictated by channel structure (*change of dedicated channel*) and the air interface (*contention resolution*).

3.4.1 The $LAPD_m$ Protocol

This section presents an overview of the functions and services of $LAPD_m$. Two special procedures, *contention resolution* and *change of dedicated channel*, are described in detail. Extensive information on the service primitives and procedures used can be found in [12, 13].

The $LAPD_m$ protocol protects the data of the signalling protocols of the network layer and the short-message service on the radio channel. The participating partner entities are the MS and the BTS (or BSS) (see Figure 3.30). As is clear from the similarity in the names, the LAPD of ISDN was used as the basis for the development of $LAPD_m$.

3.4.1.1 Identification of Connection Endpoints

The data link layer provides its services to the network layer at a *Service Access Point* (SAP) (see Section 2.5). An SAP has one or more *Connection Endpoints* (CEP). Together an SAP and a CEP form the *Data Link Connection Identifier* (DLCI).

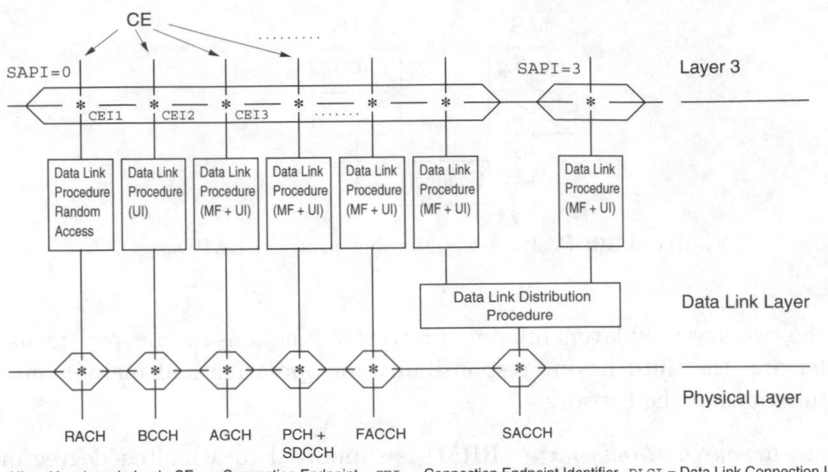

Figure 3.31: Allocation of layer-3 messages

The DLCI consists of:

Service access point identifier The *Service Access Point Identifier* (SAPI) appears binary-coded in the address field of each frame. The layer-3 entity determines which SAPI will be used by a layer-3 message according to the service being used.

Channel identification The SAPI determines which logical channel will be used for sending a message. The underlying channel structure with its existing time synchronization makes it possible for the channel identification to be managed locally in the terminal systems. It is transmitted as vertical control information between layer 3 and layer 2. This helps to keep the layer-2 protocol header short.

As can be seen in Figure 3.31, a message in layer 3 is clearly allocated to a logical channel. So far two different values for SAPI have been defined in the corresponding GSM recommendation 04.06 [13]:

- For the subprotocols *Call Control* (CC), *Mobility Management* (MM) and *Radio Resource Management* (RRM) of layer 3, SAPI = 0 is applied.
- With messages of the *Short Message Service* (SMS) an SAPI = 3 is used. Consideration is being given to providing additional SAPIs for this service in the future.

SAPI = 0 has a higher priority than SAPI = 3. Should other functions be added to the higher layers in the future, other SAPIs will be defined [12]. Coding possibilities are available for eight SAPI values.

3.4.1.2 Functions and Procedures

LAPD_m is used for transmitting signalling messages over the air interface between two layer-3 partner entities. LAPD_m procedures affect:

- several layer-3 entities
- several layer-1 entities
- signalling over the *Broadcast Control Channel* (BCCH)
- signalling over the *Paging Channel* (PCH)
- signalling over the *Access Grant Channel* (AGCH)
- signalling over *Dedicated Control Channels* (DCCHs)

There are no procedures in LAPD_m to support the messages of the RACH. These messages are transported without any involvement of layer 2 between layers 3 and 1 of a subsystem (MS or BSS).

LAPD_m includes the following functions for implementing the services to layer 3:

- Operation of several layer-2 connections on different logical channels. These connections are differentiated from one another through the DLCI.
- Recognition of different types of frames.
- Transparent transport of layer-3 messages over the layer-2 connection between the layer-3 partner entities.
- Sequential control of the sequence of information (I)-frames.
- Error-handling in regard to format of frames and functional sequence.
- Notification of layer 3 when non-resolvable errors occur.
- Flow control.
- Resolution of competing situations on a dedicated channel after successful random access on the RACH (see Section 3.4.1.4).
- Change of a dedicated channel, without termination of a connection (see Section 3.4.1.4).

3.4.1.3 Service Primitives of LAPD_m

The service primitives are formed as follows:

DL_XX_req/ind/conf

Data from the higher layers is contained as parameters in the primitives DL_Data_req/ind, DL_Unit_Data_req/ind. GSM recommendation 04.06 includes an overview of all the data link layer service primitives.

Table 3.11: Frame types in $LAPD_m$

Frame type	Function	Use
SABM	Set Asynchronous Balanced Mode	First frame necessary to enter confirmed mode
DISC	DISConnect	First frame necessary to exit confirmed mode
UA	Unnumbered Ack.	Confirmation of two preceding frames
DM	Disconnect Mode	Reply showing disconnected mode
UI	Unnumbered Information	Information frame in unconfirmed operation
I	Information	Information frame in confirmed operation
RR	Receive Ready	Proceed to transmit
RNR	Receive Not Ready	Stop transmission
REJ	REJect	Negative confirmation

3.4.1.4 Frame Types in $LAPD_m$

In addition to the frame types already presented in Section 2.7.4.1, there are other frame types that relate to the establishment and termination of a dedicated connection. These frame types are described to the extent that it is possible to understand the establishment and termination as well as the peculiarities of a dedicated connection. Details of the coding of PDUs and functional processes can be found in [12, 13]. Table 3.11 presents the different types of frames used in $LAPD_m$.

A frame can be either a command frame or a response frame. An Information frame (I-frame) is always a command frame. Supervisory frames (S-frames) can occur as either command frames or response frames. SABM, DISC and UI are command frames, DM and UA response frames. If a layer-2 entity receives a command frame with a pollbit P = 1 then it must transmit a response frame with a finalbit F = 1. A command frame with P = 0 can be responded to by a command frame P = 0 or by a response frame with finalbit F = 0.

Competition in setting up a dedicated connection As soon as an MS is allocated a dedicated channel, it attempts to use it to set up a layer-2 connection in acknowledged mode (see Section 3.4.1.5). To do so, the MS often sends an *Set Asynchronous Balance Mode* (SABM) to the BS for the SAPI 0.

In standard HDLC protocols (e.g., ISO 3309/4335/6159/6256) such as LAPD (see ITU-T Rec. I.441) an SABM frame is not allowed to transport information from a higher layer. In $LAPD_m$, however, it is essential that the SABM frame (and the *Unnumbered Acknowledge* (UA) frame) contains a signalling message for identification of the MS. This message is referred to as the *Initial Address Message* (IAM) [15].

When an MS requests channel capacity for a dedicated connection using random access, an 8-bit-long Channel_req message is sent to the network. The

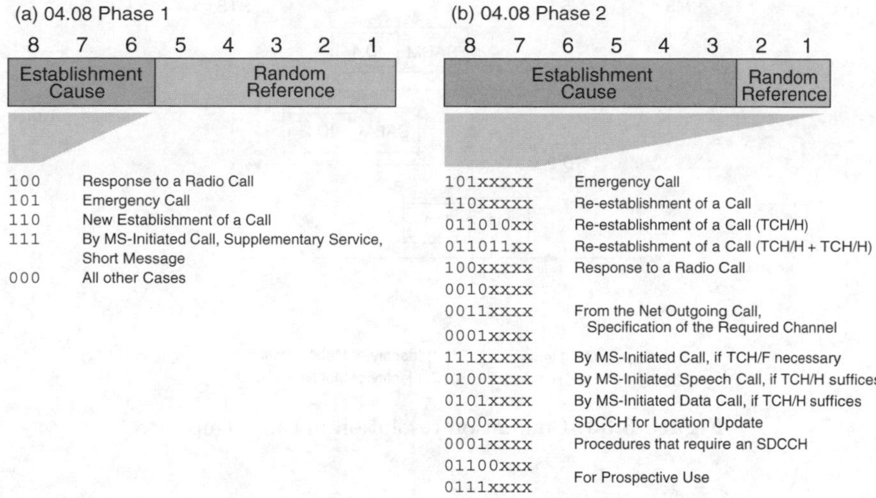

Figure 3.32: Content of Channel_req message

content of this message is presented in Figure 3.32. It shows that 2 or 5 bits are available for the representation of a random number. At this point in time this random number is the only means by which the mobile station can be identified by the BSS. In response the BSS sends an Immediate_Assignment message [15] that contains the Channel_req information element along with the allocated dedicated channel. This message is transmitted on the point-to-multipoint channel AGCH. The mobile station for which the message element Channel_req from its random access burst and the message element from the response of the BSS are one and the same thing seizes the allocated channel.

With the length of the random number as it is indicated here, it can no longer be considered random. In highly loaded cells a situation will inevitably arise in which two mobile stations will send a Channel_req with the same content at the same time. The BSS's allocation of a dedicated control channel will be picked up by both mobile stations. This means that both will seize the same dedicated channel. Until one of the MSs clearly identifies itself (providing IMSI, TMSI, IMEI), it is not certain that only one of the two mobile stations is seizing the channel. As a way of preventing these collisions LAPD_m offers a procedure called *contention resolution* (see Figure 3.33).

After the dedicated channel is activated, both mobile stations attempt to set up a bidirectional layer-2 connection in acknowledged mode. The identity (IMSI, TMSI, IMEI) of the MS is transmitted in the SABM frame. As a response frame to the SABM, the BSS must send back an UA response frame. The UA frame contains the identity of the first SABM frame received. Both mobile stations compare the identity contained in the UA frame with their own. If the identity matches, the respective MS changes to acknowledged mode (see Section 3.4.4). Messages from the higher layers can then be

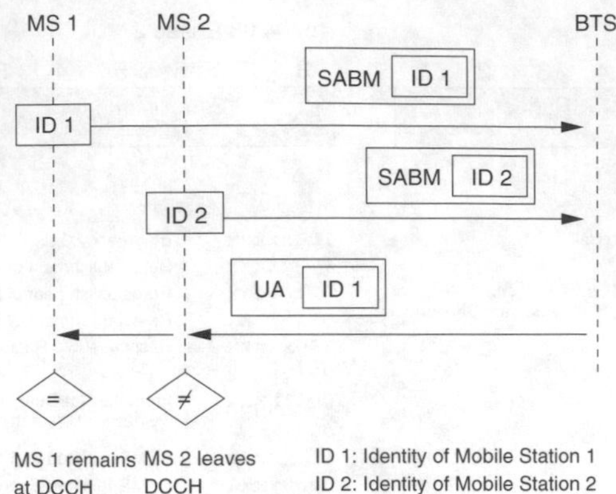

Figure 3.33: Contention resolution in call set up

transmitted bidirectionally and safely over the dedicated layer-2 connection. If the identities do not match the connection will be terminated.

Change of dedicated channel This procedure [13] only occurs on the signalling channels for SAPI = 0. A change of the dedicated channel is initiated by a BSS or an MSC. A dedicated channel is changed either at handover or during call set up if the first allocated dedicated channel was an SDCCH. In the first case the change is from one TCH to a new TCH in the same or in a different cell (intracell/intercell handover). In the second case the change is from an SDCCH to a TCH (nonOACSU/OACSU strategy); see Section 3.5.1.

If the request to change the channel is initiated by the BSS, the mobile station suspends the current dedicated channel (*suspension*). It stores the positions of the transmitting and receiving window of the I-frame that has not yet been acknowledged but should be transmitted next. The mobile station then sets up a connection on the new dedicated control channel and resumes with transmission (*resumption*).

3.4.1.5 Modes of Information Transmission

Basically two different types of $LAPD_m$ operation are specified in GSM recommendation 04.05 [12]:

- *Unacknowledged operation*
- *Multiple frame operation*

The point-to-multipoint channels BCCH, PCH and AGCH only permit unacknowledged operation. The channels for a point-to-point connection, SDCCH, FACCH and SACCH, support both types of operation.

Unacknowledged operation The information of layer 3 is transported in *Unacknowledged Information* (UI) frames. UI frames are not acknowledged. There are no control mechanisms such as timers or sequential numbering. Therefore no flow control or error correction takes place in this mode. This operating mode is available to all logical channels with the exception of the RACH.

Multiple frame operation The information of layer 3 is transmitted in I-frames. A send-and-wait logical link control protocol with window size 1 is used in this mode. The information frames are numbered sequentially modulo 8. Receipt of the I-frames must be acknowledged by the receiving side. This allows errors to be eliminated through the retransmission of the unacknowledged I-frames (error correction by repeated transmission). It also controls the flow of I-frames. If errors from layer 2 cannot be corrected, an error message is sent to layer 3.

3.4.1.6 Termination of a Layer-2 Connection

Multiple frame operation can be terminated in several different ways:

Normal release Layer 3 initiates a controlled termination of a layer-2 connection. The acknowledged connection between partner entities is terminated through an exchange of a *Disconnect* (DISC) command frame and a UA response frame.

Local end release No exchange of command and response frames occurs. This termination is initiated and controlled by layer 3.

Abnormal local end release Here too there is no exchange of command and response frames. The command for this call termination comes from layer 3.

If layer 3 seizes the initiative for the termination of a connection, it communicates the type of termination using the appropriate primitives of layer 2 [13]. No termination mechanism exists in mode *unacknowledged operation.*

3.4.2 Services of the Physical Layer

The services required by layer 1 are specified in [9]:

- A layer-1 connection for the transparent transmission of frames is provided. In contrast to the LAPD protocol, which distinguishes individual layer-2 frames from one another through bit patterns (*flags*), in this case a fixed block format is provided: the bursts.
- An indication is given whether a logical channel has been activated.
- Layer-2 frames are transmitted in the same sequence in which they arrived from layer 3.

- Synchronization to frame transmission.
- Use of channel coding procedures, as described in Section 3.3.7.1.
- Transmission (MS) and receipt (BTS) of random access bursts.

3.4.3 Influence of the Physical Layer on LAPD$_m$

The influence of the physical layer on LAPD$_m$ is as follows:

- The error protection function in layer 2 is achieved through forward error correction in layer 1.
- The physical layer controls the transmission of layer-2 frames.
- The value of the control timer T200 [13] is coordinated with the channel structure.

3.4.3.1 Error Protection Functions of Layer 1

The channel coding measures covered in Section 3.3.7.1 protect data from transmission errors. LAPD$_m$ does not add an additional CRC check sum to layer-3 messages. It depends on the Fire coder in layer 1. Consequently the error correction measures of LAPD$_m$ are limited to block sequence control mechanisms and repeat requests of information frames.

3.4.3.2 Time Control of Layer 2

The physical layer does not provide queueing for the transmission of layer-2 messages waiting to be sent. Messages are never allowed to be buffered between an entity of layer 2 and layer 1, because the layer-2 entity must transmit a response or command frame appropriate to the momentary situation.

The physical layer in GSM uses the following primitive in order to inform the layer-2 entity above it when the next layer-2 block can be transmitted, adding a parameter indicating the logical channel [12]:

```
Ready_To_Send(channel type)
```

On one hand, this primitive must appear early enough to allow sufficient time for the formation of the next layer-2 frame. On the other hand, the time in which the control timer T200 can run out must already have expired. Figure 3.34 shows the ideal timing.

3.4.3.3 Block Recurrence Time

The structure incorporating logical channels imposes transmitter bit timing on layer 2. This timing, called *block recurrence time*, depends on how frequently and at which intervals a logical channel will occur. The coder and interleaving techniques, as well as the size of the blocks arriving from layer 2, also have an effect. The block recurrence times do the following

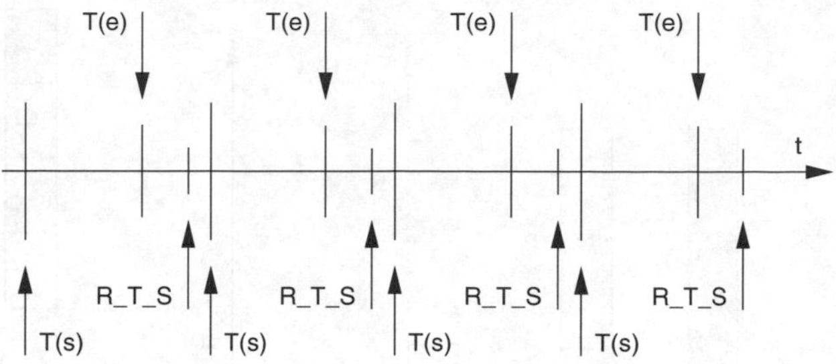

Figure 3.34: Time conditions with Ready_To_Send

- allow the calculation of the transmission rates of the logical channels of a physical channel, and
- allow the value of the control timer T200 to be optimized in the $LAPD_m$ protocol.

Table 3.12 presents an overview of the block recurrence times of the logical channels. The parameters n, p and r indicate the number of blocks in a block recurrence period and are specified by the network operator [15, 16].

3.4.3.4 Time-Critical Conditions

The $LAPD_m$ data link protocol must be in exact agreement with the time conditions dictated by the channel structure. The following points must be taken into account:

1. Consideration of the block recurrence times of the logical channels.
2. The shortest anticipated response time for an I-frame, i.e., attention must be given to the time synchronization of the uplink and the downlink.
3. Timer values for frame control must be set on the FACCH in such a way that not too many speech data bursts are displaced by slot-stealing.

Condition 2 reinforces the fact that blocks can be transmitted and received alternately on an SDCCH during a block recurrence period (see Figure 3.35). This explains why the control timer T200 that monitors the blocks on the SDCCH is set at such short intervals that it runs out before the next transmission [13] and shortly after the expected arrival of a response frame. The value of a T200 on an SDCCH is 220 ms. Therefore it is definitely clear which frame type should be sent next before the next block recurrence time.

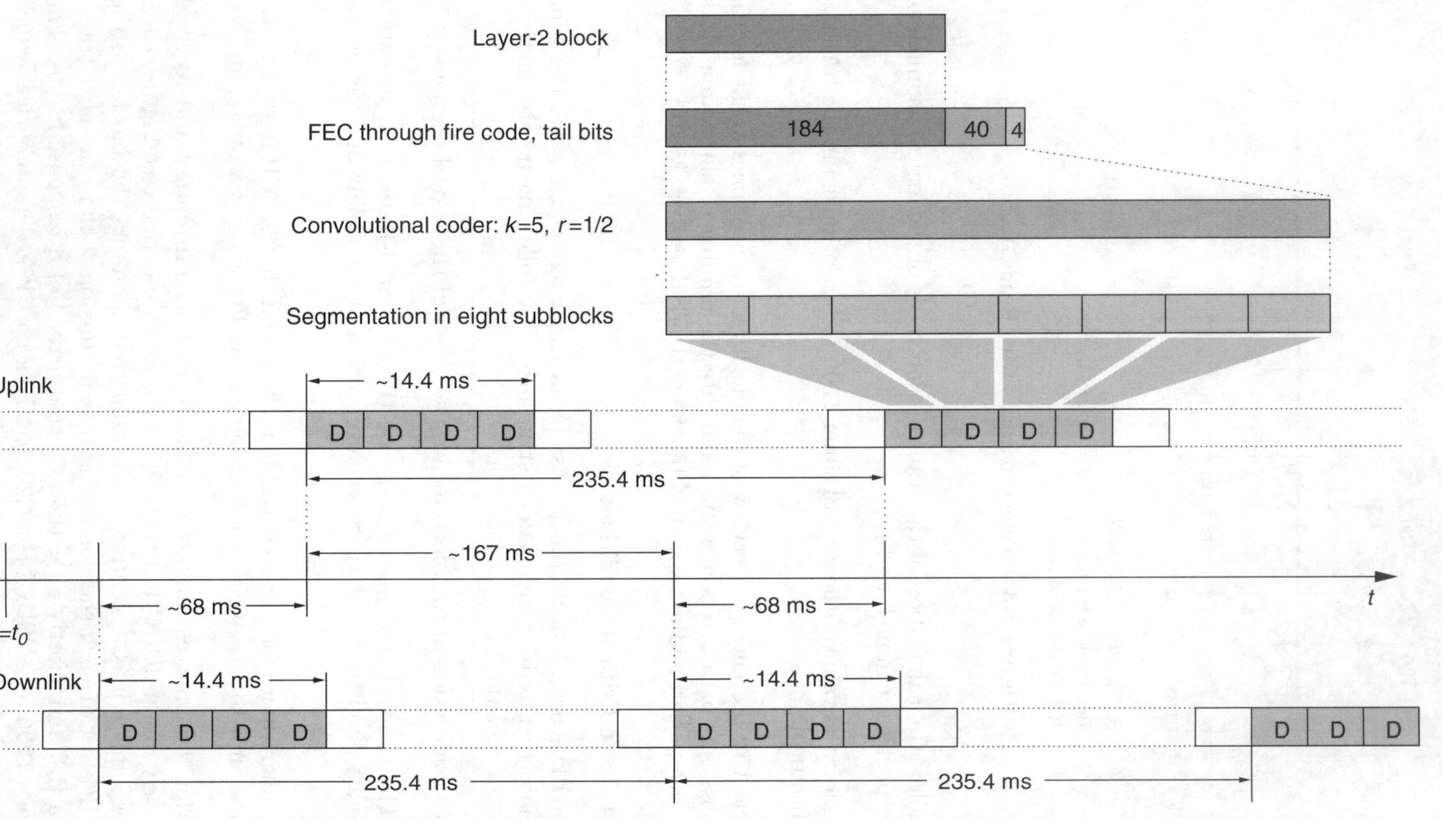

Figure 3.35: Time interleaving of uplink and downlink on an SDCCH

Table 3.12: Block recurrence times of the logical channels

Channel type	Net bit rate before layer 1 [kbit/s]	Block length [bit]	Block recurrence time [ms]
TCH/FS	13.0	182 + 78	20
TCH/HS			
TCH(9.6 kbit/s)	12.0	60	5
TCH(4.8 kbit/s)	6.0	60	10
TCH(≤2.4 kbit/s)	3.6	72	20
FACCH/F	9.2	184	20
FACCH/H	4.6	184	40
SDCCH	598/765 (≃ 0.782)	184	3060/13 (≃ 235)
SACCH/T	115/300 (≃ 0.383)	168 + 16	480
SACCH/C	299/765 (≃ 0.391)	168 + 16	6120/13 (≃ 471)
BCCH	598/765 (≃ 0.782)	184	3060/13 (≃ 235)
AGCH	$n \cdot 598/765$ (≃ 0.782)	184	3060/13 (≃ 235)
PCH	$p \cdot 598/765$ (≃ 0.782)	184	3060/13 (≃ 235)
RACH	$r \cdot 26/765$ (≃ 0.034)	8	3060/13 (≃ 235)

Condition 3 is important for the FACCH because the block recurrence time (see Table 3.12) is very short. An FACCH burst competes simultaneously with a burst carrying user data. In this case the value of the control timer T200 must be set large enough so that not too many user information data bursts are displaced or delayed in succession. The value of a T200 on an FACCH is 150 ms.

3.4.4 LAPD$_m$ Services

The services of the physical layer are provided by the layer-1 services. Based on the two operating modes presented in Section 3.4.1.5, there are two different kinds of information services. Both are able to coexist in the same connection in layer 2.

Layer 2 does not provide any services that support random access. The parameters for the primitives DL_Random_Access_req/ind/conf and PH_Random_Access_req/ind/conf are transmitted in transparent mode between layers 1 and 3. A differentiation is made between four different services:

Priorities Two different layer-2 connections can be operated at the same time. But messages with SAPI = 0 are given higher priority than those with SAPI = 3.

Segmentation and Reassembly (SAR) In acknowledged operation layer-3 messages waiting to be sent and not fitting into a layer-2 information frame are segmented. When they arrive at the receiving layer-2 partner entity, the segments are reassembled.

Unacknowledged information transfer There is no acknowledgement of I-frames in this mode. However, it is possible for the higher layers to use similar acknowledgement procedures as $LAPD_m$ in acknowledged mode.

The characteristics of this service can be summarized as follows:

- A layer-2 connection is provided between two layer-3 entities for non-acknowledged mode.
- A unique identification of layer-2 connection endpoints is possible, thereby enabling a layer-3 entity to specify its partner entity.
- The priority of layer-3 messages is considered.
- There is no assurance that a message has arrived.

Acknowledged information transfer This service utilizes the acknowledged mode of operation. It has the following features:

- A layer-2 connection between two layer-3 entities is provided.
- A unique identification is provided for layer-2 connection endpoints, thereby enabling a layer-3 entity to specify its partner entity.
- The priority of layer-3 messages is considered.
- It provides an assurance that the correct sequence of successive I-frames will be maintained.
- Layer 3 is notified when sequencing errors or unresolved errors occur.
- Flow control.
- Segmentation and reassembly of layer-3 messages.
- Provision of procedures for *suspension/resumption* that ensure that no information is lost when there is a change of the dedicated control channel.

3.5 The Network Layer in GSM

The layer-3 signalling protocols provide the functions for establishing point-to-point connections between two mobile subscribers in a GSM-PLMN or between a mobile subscriber and a user of another network. If a connection has been established, it must then be maintained, even when there is radio channel interference. If one of the two mobile users wishes to terminate the connection, layer 3 ensures that the connection is terminated properly.

The network layer is divided into three separate sublayers, each offering their own clearly delineated functions. The specifications for the subprotocols of the network layer are found in [14, 15]. The lowest sublayer, *Radio Resource* (RR), is based on the services of $LAPD_m$. The CC entity of the CM sublayer offers layer 4 the services of the network layer (see Figure 3.29). The individual sublayers carry out the following tasks:

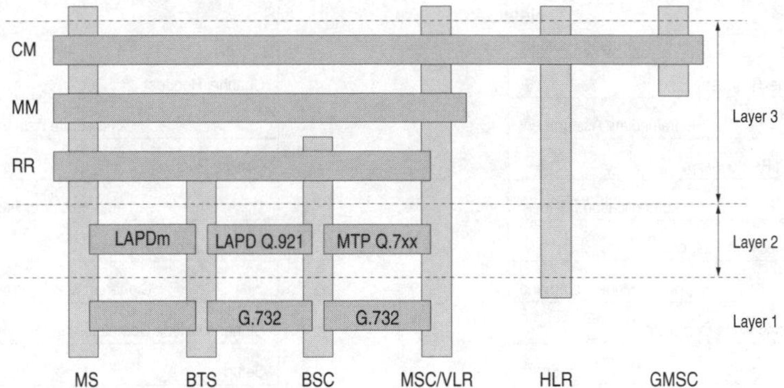

Figure 3.36: Network elements and related signalling protocols

Radio Resource Management (RRM) Set up, maintenance and termination of a dedicated radio channel connection.

Mobility Management (MM) Registration, authentication and allocation of new TMSI.

Call Control (CC, in the CM sublayer) Set up, maintenance and termination of circuit-switched calls.

The CM sublayer contains three independent entities: *Call Control* (CC), *Short Message Service* (SMS) and *Supplementary Services* (SS). The SS entity provides call-based and non-call-based services such as call diversion and billing. The SMS service permits short messages to be sent on the SDCCH and SACCH control channels. The call set up mechanisms will be discussed later so that the emphasis here can be on the CC entity. The protocols for SMS and SS are specified in Recs. 04.10, 04.11 and 04.12.

A complete overview of the network elements involved in the various signalling protocols and on the range covered by the related layers and sublayers is given in Figure 3.36.

3.5.1 Connection Establishment

A connection is established according to the message flow shown in the sequence chart of Figure 3.37. The following procedures have to be passed in chronological sequence:

1. Paging*.

2. Random access.

*With an outgoing call no paging by the BTS occurs, instead of the Paging_Response message a Service_Request message is sent containing the same information.

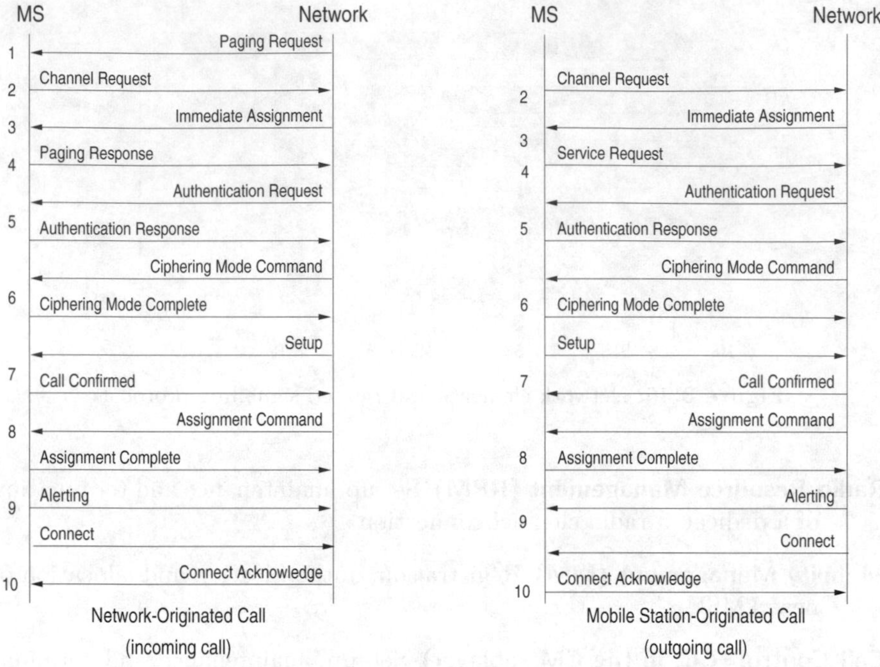

Figure 3.37: Connection establishment messages

3. Assignment of SDCCH.
4. Definition of type of service.
5. Authentication.
6. Ciphering.
7. Assignment of a TCH.
8. Alerting at called terminal.
9. Connect after the called terminal is responding.

To be able to make optimal use of the radio resources, three algorithms for connection establishment are contained in the GSM recommendations as alternatives:

- *Off Air Call Setup* (OACSU).
- *Non-Off Air Call Set Up* (Non-OACSU).
- *Very Early Assignment* (VEA).

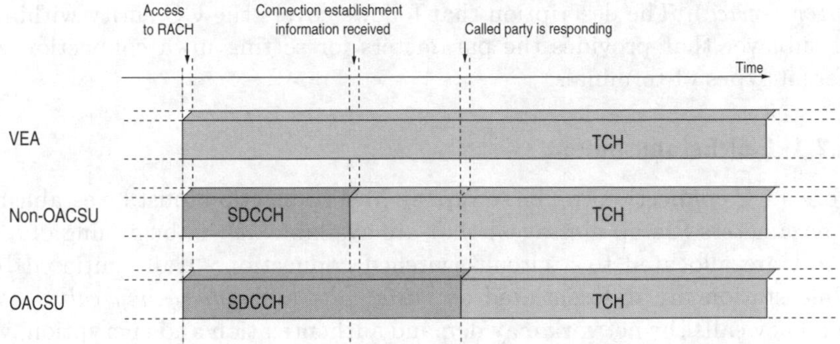

Figure 3.38: Alternate strategies for connection establishment

These alternatives differ in the time instant when a TCH is assigned to a call initiated by the mobile station (see Figure 3.38).

The VEA algorithm assigns the TCH for signalling directly after a random access message has been received at the BTS, and allows the fastest establishment of a connection, since the transmission rate of the TCH is much higher than that of the SDCCH. A disadvantage is the inefficient use of the TCH, since a percentage of outgoing calls is not answered by the called terminal and the TCH capacity is wasted then. Inefficient use of the TCH happens also when a location update is performed by using a TCH. According to the GSM recommendations, connection establishment in a cell must always apply the same algorithm. If the VEA algorithm is implemented, the resulting load to the TCHs from their use for signalling has to be taken into account when dimensioning the trunk of TCHs for a given traffic load of a cell.

With the OACSU algorithm the TCH is assigned at the latest possible time instance, namely when the called terminal does respond. With the Non-OACSU algorithm the TCH is assigned as soon as the BSS has received the complete information to establish a call from the calling MS.

3.5.2 Services of the CC Sublayer

Simultaneous calls can be managed on different MM connections. The tasks of a CC entity is to provide the transport layer with a point-to-point connection between two physical subsystems. This connection is characterized by parameters relating to individual terminal equipment. For example, facsimile transmission in GSM requires coding and encryption measures different from those for the transmission of speech. These parameters must be set separately for each individual connection.

The term *call* describes a point-to-point connection that transmits data between two terminals via a channel. The point-to-point connection is set up and provided by the CC sublayer at the initiation of layer 4. The connection between terminals uses individual terminal-related data link control protocols (*bearer service*). The description that follows covers the CC entity within the CM sublayer that provides the parameters for setting up a connection with different types of terminals.

3.5.2.1 Call Establishment

Before a CC connection can be set up, an MM connection must be established. All parameters (Setup messages) that are available at the beginning of a call request are allocated to a circuit-switched connection. Calls initiated by a mobile station are differentiated as *basic calls* and *emergency calls*. With emergency calls the network may demand authentication and encryption; with basic calls these procedures are executed in a normal sequence.

A connection is initiated through the dialling of a number. The called terminal accepts the call with Connect. The calling terminal then acknowledges the Connect message with Connect_ack. This completes the call set up. The primitive MNCC_Setup_Compl_ind is sent from layer 3 to layer 4 in the subsystem network when the connection has been fully established.

3.5.2.2 Call Clearing

This is the termination of a point-to-point connection.

3.5.2.3 Change of Call-Related Parameters with an Existing Connection

These procedures are carried out during the active state of a call:

User notification This procedure allows the network to inform the mobile station of any occurrence affecting the call. Conversely, the mobile station is able to notify the terminal of the other side of the connection. An example would be the separation of the terminal from the network connection.

Change of parameter arrangements of a call (Call rearrangement) A rearrangement of parameters is necessary when a terminal attached to a mobile station is changed.

DTMF protocol control *Dual-Tone Multiple Frequency* (DTMF) is a digital set of tones produced by the terminal user pressing a key (0, ... 9, A, ... D, *, #). Sequences of DTMF tones can be used to operate voice mailboxes and answering machines. The control tones can only be used during the operation of the teleservices voice, voice/data transmission and voice/facsimile transmission.

In the network the DTMF tone is transmitted on the traffic channels (not over the signalling network). Between the mobile station and the network this tone is transmitted on the FACCH. The reason for this difference in transmission is the speech codec, which would distort the DTMF tones so severely that it would not be possible to maintain the quality requirements of other networks.

Change of channel parameters during a call (In-call modification) This service provides users with a point-to-point connection for the transmission of control data for different services in succession during the same call. Each change of service involves a change in the channel parameters, i.e., a change of channel or a modification of the parameters of the channel. The channel parameters of teleservices that are changed when they are being used are:

- Alternating between voice and transparent or non-transparent data transmission.
- Change from voice to data transmission (transparent or non-transparent).
- Alternating between voice and facsimile transmission Group 3.

The bearer services for this kind of teleservice must be switchable during a call. Therefore this service can only be accessed by special (*multiple capability*) mobile stations.

3.5.2.4 Other Procedures

The following procedures cannot be allocated to any of the above groups:

- Inband tones and announcements.
- Status enquiry. A CC entity has the possibility of querying the accuracy of the current state of a partner entity in the event of a fault. If it is established that the partner entity is in an incompatible state, the call is terminated.
- Call reestablishment.

Calls cannot collide. When the network and a mobile station initiate a call at the same time, the calls can be kept separate from one another through their different *Transaction Identifier* (TI)s; see Section 3.5.6.

3.5.3 Services of the MM Sublayer

The functions that support the mobility of a mobile terminal are in the MM sublayer. The network is alerted by the mobile station when it is switched on or off or when it leaves a location area. This requires a location update. The MM sublayer contains the functions for controlling the authorization of a mobile user (authentication procedure) and supplying a new code.

There are three groups of MM procedures.

3.5.3.1 General MM Procedures

These procedures can always be initiated if an RR connection exists. The procedures in this group include the following.

- Initiated by the network:

  Allocation of new TMSI (TMSI reallocation) The use of a TMSI instead of an IMSI ensures that the identity of the mobile user is kept confidential. A TMSI is only adequate for identification in an area that is allocated to a VLR. Outside this area it must be combined with a *Location Area Identifier* (LAI). This procedure for the allocation of a new TMSI must be carried out when a mobile user moves into a different VLR area (see Section 3.7).

  The allocation of a new TMSI is usually carried out in conjunction with another procedure, for example during location updating or call set up.

  Authentication Authentication serves two purposes. On the one hand, it provides a mechanism for checking whether the identity of the mobile station is valid. This can be implemented on the basis of the IMSI or, if it is available, a TMSI. On the other hand, it provides the mobile station with a new code.

  Identification A mobile station receives a request from the network to send one of the identification parameters IMSI, TMSI or IMEI to the network.

- Initiated by the mobile station:

  Removal of IMSI (IMSI Detach) This procedure is used when a SIM is removed (detached) from a mobile station or a mobile station is switched off. The mobile station is then regarded by the network as unavailable.

3.5.3.2 Special MM Procedures

One of the following special procedures can only be initiated when no MM connection exists or when none of the other special procedures is active. Any general MM procedure except for *IMSI Detach* may be initiated during a special MM procedure.

Location updating (LU) This procedure updates the registration of the location of a mobile station. It can only be carried out if the mobile station contains a SIM. There is a reason for updating location when the LAI that the mobile station is receiving on the BCCH is different from the stored LAI. This occurs when a mobile station has moved into another location area or when it is switched on in an area whose LAI is not stored in the SIM. It therefore follows that only a mobile station can implement a location update. The procedure is the same as for cell set up, but without the steps 9 and 10 shown in Figure 3.37, since only MM signalling data are exchanged during *Location Updating* (LU). For details see Section 3.7.

Periodic updating Periodic updating only takes place when the appropriate parameter within system information is available on the BCCH.

IMSI Attach This is the opposite of the procedure *IMSI Detach*. After this procedure is carried out, a mobile station is identified as being available by the network.

All three procedures can be classed with the Location_Updating procedure in [15] (see Section 3.7). A special information element indicates which of the three procedures is being carried out.

3.5.3.3 Procedures for the Management of an MM Connection

These procedures are used for establishing, maintaining and terminating an MM connection between a mobile station and the network. The MM connection is for data exchange in the CC sublayer. More than one MM connection can be used at the same time. An MM connection can only be established if there is no MM procedure being run. There are three procedures:

- Establishing an MM connection, initiated by a mobile station. An MM connection may only be established if the location area has been successfully updated beforehand. Emergency calls are an exception.
- Establishing a connection, initiated by the network.
- Transmission of information about an MM connection. Each entity in the CM sublayer has its own MM connection. The messages of these entities are not given priority.

3.5.4 Services of the RR Sublayer

This sublayer contains procedures for the management of shared transmission resources. These include the physical channels and the layer-2 connections on the logical channels.

If an RR connection has been established, the RR sublayer provides the sublayers above it with a reliable RR connection. The duration of a point-to-point connection between two terminals normally exceeds the duration of a single RR connection. This means that there may be a change of dedicated channels during a connection, e.g., handover. In this context it is clearer to refer to an RR session during which an RR connection is suspended several times and re-established on another physical channel.

The dedicated channel is also changed during the set up of a connection if a non-OACSU strategy is pursued (see Section 3.4.1.4). In addition, the sublayer provides procedures for the transmission of messages on the BCCH and the CCCHs when no RR session exists. An example of this is automatic cell selection.

3.5.4.1 Idle Mode

When in idle mode, the mobile station is not allocated a dedicated channel but listens to the control channels BCCH and CCCH. In a subsystem a mobile station automatically changes cells when it moves beyond a cell boundary, and the higher sublayers are informed of this change. System information is passed on from the RR sublayer to the higher sublayers.

3.5.4.2 Set Up and Termination of an RR Connection

When an RR connection is established, it consists of an active physical bidirectional point-to-point connection with the appropriate layer-2 connection in acknowledged mode. This layer-2 connection is characterized as SAPI = 0. A mobile station is only able to maintain one RR connection. The MM sublayer gives the command for an RR connection to be established or terminated. A channel is activated when it is to be used for data.

3.5.4.3 RR Connection Established

In this operating state two dedicated channels are allocated to a mobile station. One of them is an SACCH, the other one is either an FACCH or a SDCCH. If an RR session has been set up, the RR sublayer provides the following services:

- Transmission of messages on all layer-2 connections. Layer-2 connections are characterized by their operating mode and the type of logical channel (see Figure 3.31).
- Establishment and termination of acknowledged mode in layer 2 connections with SAPI = 3 (SMS).
- Establishment and termination of acknowledged mode on SDCCH, FACCH or SACCH.
- Suspension and resumption of data transmission during a change of dedicated channel or handover.
- Automatic cell selection and handover in order to maintain an RR session.
- A switch to transmission mode on a physical channel. During call set up this includes a change from *signalling only* to *user data* according to the VEA method; see Section 3.5.1. This can involve a change of channel type and coding and decoding mechanisms and an activation of the encryption process.

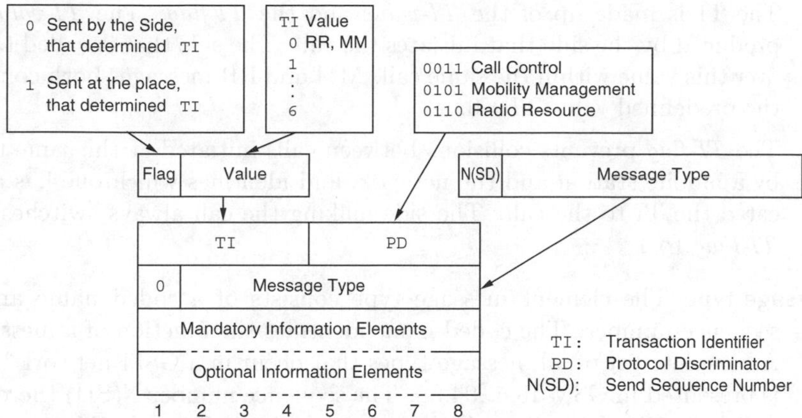

Figure 3.39: Establishment of a layer-3 message

- Sequential control of messages in the CC and MM sublayers (see Section 3.5.5).

The RR sublayer uses the services of layer 2 and also has direct access to the services of layer 1. The latter includes automatic tracking of the BCCH and indicating whether continuous transmission is taking place on an SACCH. The tracking is for monitoring a connection.

3.5.5 Format and Coding of a Layer-3 Message

A layer-3 PDU consists of up to 249 octets, which are combined into one block. Each PDU consists of the following elements (see Figure 3.39):

Protocol Discriminator (PD) The PD indicates the subprotocol that has produced the message. It is the first part of a message and occupies the first four bits of the first octet. Figure 3.39 describes how it is coded. There is also a 4-bit code for SMS messages and SS messages that are not part of a call.

Transaction Identifier (TI) The TI keeps transactions within a mobile station separate from one another. A transaction is a communication during which a particular service is used. Related to the CC protocol, this means that a TI is able to separate the signalling messages of different calls from one another. Current applications involving parallel transactions are *call waiting* and *call on hold*. Likewise a call on hold can be retrieved through *call retrieving*. In the future it will be possible to implement *multiple calls*. This will allow *conference calls* to be set up with participants being switched into a call or disconnected from it.

The TI corresponds to *call reference* of the ISDN layer-3 protocol. As shown in Figure 3.39, the TI forms the second part of every message.

The TI is made up of the *TI-value* and the *TI-flag*. The *TI-value* is produced by the side that initiates the call. The side that is called takes over this value within the same call. MM and RR messages both contain the predefined value TI = 0.

The *TI-flag* prevents collisions between calls initiated at the same time by a mobile station and the network, and identifies which one has allocated the TI to the call. The side making the call always switches the *TI-Flag* to 1.

Message type The element message type consists of a coded name and a sequence number. The coded name identifies the function of a message. A complete list of all message types that occur in a GSM network layer is presented in GSM Rec. 04.08. The sequence number N(SD) (here SD stands for *Send Duplicated*) is used to monitor MM and CM messages. It is incremented modulo 2.

MM and CM messages can be duplicated if there is a change of a dedicated control channel and the last layer-2 frame that contains an MM or CC message still has not been acknowledged by the layer-2 partner entity before the MS leaves the old channel.

Information elements: A message contains

- mandatory information
- optional information

Mandatory information elements are always transmitted before the optional elements. An optional information element must always be preceded by an *Information Element Identifier* (IEI). This does not apply to mandatory information, because the existence and sequence of the information elements are clearly identified by the PD and the message type. In this case the IEI is removed from the information element.

If the length of an information element varies, then its length will be marked by a *Length Indicator* (LI). In a mandatory information element the length is determined exactly by the PD and message type.

The information elements are divided into four types. The content of the information elements is abbreviated in Figure 3.40 with *Contents of Information Element* (CIE). An analysis of messages and error handling can be found in [15].

Type 1: Information element with half an octet of content.

Type 2: Information element without content.

Type 3: Fixed length information element containing at least 1 octet.

Type 4: Information element of a variable length.

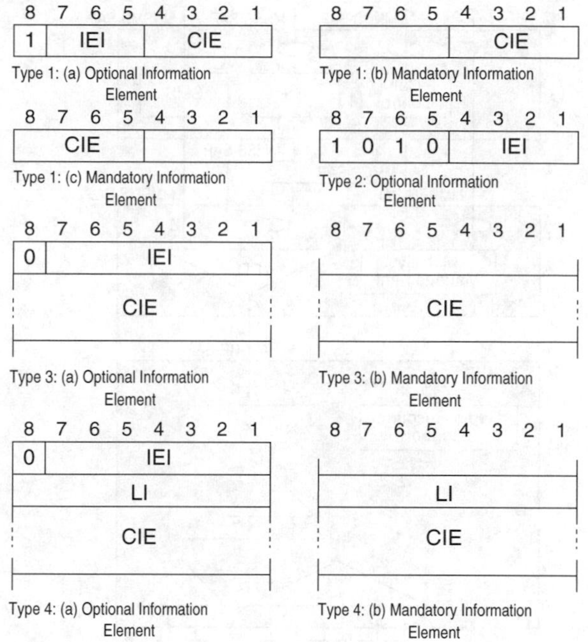

Figure 3.40: Optional and mandatory information element types 1 to 4

3.5.6 Routing of Layer-3 Messages

According to the directives of the OSI reference model, the header of an (N)-PDU is checked in an (N)-layer. Depending on the content of this header, the (N)-PDU is forwarded. If the content of the (N)-PDU is routed to the $(N+1)$-layer, the header is removed.

A method that does not comply with the ISO/OSI reference model is used for routing layer-3 messages to the correct sublayer. The advantage is that it reduces the number of octets per message.

This method is described based on the example of a mobile system subsystem. As is shown in Figure 3.41, the RR and the MM sublayers provide the routing and multiplexing functions. Messages arriving from an $LAPD_m$ entity are routed according to their PD.

Messages with a PD value the same as RR remain within the RR sublayer, whereas all other messages are routed to the MM sublayer. Depending on its PD and the TI, a message is sent from the MM sublayer to the correct CM entity or it remains in the MM entity.

The MM sublayer forwards messages from the CM entities and its own messages to the RR sublayer in the uplink direction. In the process it multiplexes the parallel transactions of a call. The RR sublayer distributes the messages according to their PD and the updated channel configuration to the correct *Service Access Point* (SAP) of layer 2.

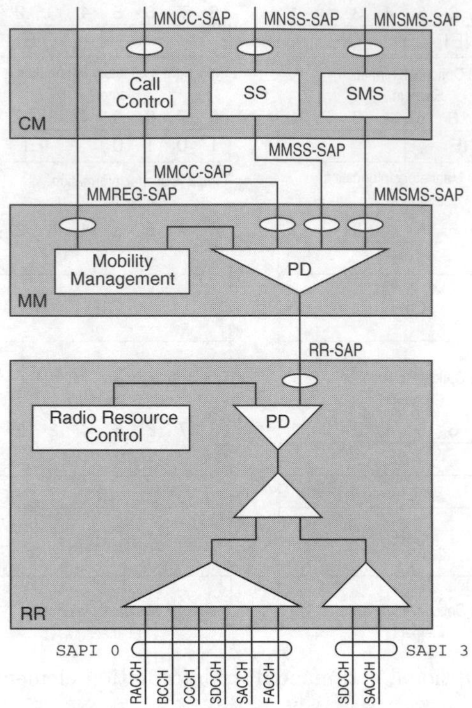

Figure 3.41: Routing of a layer-3 message

3.5.7 Primitives of the Sublayers

The following primitives together with their parameters are described in GSM Rec. 04.07. The respective information elements are specified in GSM Rec. 04.08.

Primitives for the RR sublayer The primitives for the RR sublayer are formed according to the following rule:

RR_XX_req/ind/conf

Primitives for the MM sublayer The primitives for the MM sublayer are formed according to the following rule:

MM_XX_req/ind/conf

Primitives for the CC sublayer The primitives for the CC sublayer are formed according to the following rule (MN = *Mobile Network*):

MN_XX_req/ind/conf

3.6 GSM Handover*

With new mobile radio networks an existing connection is automatically switched from one base station to another when the mobile subscriber moves to a different radio cell. GSM Rec. 05.08 even provides signalling functions for multiple *handover*. Handover always takes place when a mobile station leaves the area of a base station, and the MSC sometimes at the same time.

3.6.1 Handover Reasons

There can be many different reasons for a handover. Each mobile terminal attempts to use the radio channel that will provide the best connection quality, i.e., the best C/I (*carrier-to-interference ratio*). Co-channel interference is unavoidable because of multiple use of the same time and frequency channels due to existing cell layouts, and consequently quality can be poor (i.e., bit-error ratio high) despite a high signal level. The connection of a mobile terminal to the base stations can be the cause of interference to other mobile stations, even if it is a high-quality one. The interference can be minimized if the interfered station changes to a different radio channel. It is also possible for mobile users to have the same good receive quality from more than one cell. The service quality of the network can then be optimized if mobile users are equally distributed over the available cells.

It should be clarified that when co-channel interference occurs it is not the mobile stations that are causing direct interference to each other; instead the signals of the serving base station are superimposed on the mobile receiver (victim) by the signals of the co-channel base station and, in addition to the signals of the serviced MS, interference signals of one or more mobile stations from other cells of the system are occurring at the receiver of the serving base station, see Figure 3.42.

The possibility of being able to change cells during an active network connection without the connection being broken off is one of the most important functions of cellular networks in the support of mobile communications.

Section 3.6.3 deals with the principal handover variations and criteria. In addition, it provides a description of how, while actively connected to the GSM network, an MS produces a measurement report that is the basis for a handover decision by the network. Section 3.6.5 presents the handover criteria stipulated in GSM Rec. 05.08 [18] and an example of a suggested basic algorithm for a handover decision.

The GSM protocol for an intra-MSC handover is covered in Section 3.6.8. Section 3.6.9 explains the content of layer-3 handover messages.

***With the collaboration of Martin Junius and Markus Scheibenbogen**

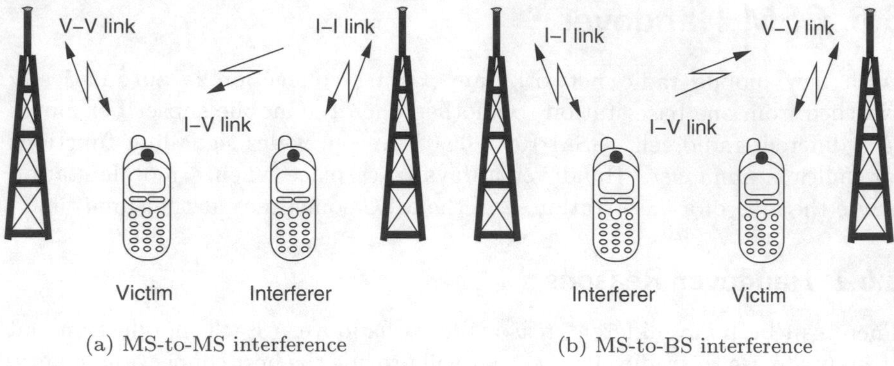

(a) MS-to-MS interference (b) MS-to-BS interference

Figure 3.42: Cochannel interference—interferers and victims

3.6.2 GSM Recommendations

The recommendations relevant to the handover process and power control are:

03.03: *Numbering, Addressing and Identification* Reference [7] includes a description of the *Base Station Identification Code* (BSIC), which is contained in the measurement report.

03.09: *Handover Procedures* Reference [8] describes the signalling for the handover process in the network.

04.08: *Mobile Radio Interface Layer 3 Specification* Reference [15] describes the layer-3 functions *Call Control*, *Mobility Management* and *Radio Resource Management*.

05.05: *Radio Transmission and Reception* Reference [17] includes a description of the power class of the MS and power levels.

05.08: *Radio Subsystem Link Control* Reference [18] describes handover, cell selection, power control, measurement reports and parameters, as well as a sample algorithm for handover and power control.

08.08: *Mobile Switching Centre to Base Station System Interface Layer-3 Specification* Reference [10] describes the *Base Station Subsystem Mobile Application Part* (BSSMAP) of signalling system CCS7 and message formats, including signalling for handover.

GSM offers *Mobile-assisted Handover* (MAHO), which means that handover decisions are made on the basis of measurement reports received from an MS on its radio field situation.

3.6.3 Handover Preparation

A handover is based on a connection being measured and evaluated continuously by the respective BS and MS. The decision algorithm mainly contributes to the spectral efficiency of the radio network and the service quality as seen by the mobile subscriber. A mobile station leaving the coverage area of a base station must receive coverage from a neighbouring base station in order to keep a connection intact. *Connection cut-off* or *call drop* are not acceptable to the mobile user during a conversation, and consequently it is very important to keep some quality of service level. Without automatic handover, the subscriber or the mobile station would be forced to establish a new connection. Handovers can also be initiated by the network when the traffic volume of a cell periodically reaches too high a level or when neighbouring cells are being underutilized.

3.6.3.1 Handover criteria

According to the different handover types described above, different criteria apply for the initiation of a handover. An effort is made to apply the measurement and evaluation of the receiving situation in the immediate past to the anticipated situation of the near future and on this basis to decide whether a handover seems appropriate. This extrapolation only offers a certain degree of probability of success, which depends on the criteria used and its evaluation.

Along with the signal level of the participating receivers, which can be used to estimate the path loss of a radio signal, the measured bit-error ratio is very important as an indication of the signal quality of the uplink and downlink. It is determined in the Viterbi decoder of the receiver through a comparison of certain received bit patterns with the known *training sequence* of each burst. In time-multiplexed systems time slots for transmitting bursts must be wider than the burst itself, so that it does not violate the time slot boundaries because of the *round-trip signal propagation delay* of the signal from the base station to the MS and back. In GSM the round-trip delay is measured and corrected by the base station, so that the distance between the mobile and base station is known and a handover can be initiated in time in case the MS leaves the supply area (cell) and another suitable base station is available. Otherwise the connection must be cut off. As explained later, the base station control decides the time and type of handover. The MS transmits its measurement results to the BTS once or twice per second, depending on signalling volume.

Information on the signal quality of the uplink and the downlink of alternative neighbouring cells is not available during a connection. Therefore during the handover decision potential target cells are evaluated on the basis of signal level measurements undertaken by the mobile station of the BCCHs of neighbouring cells, which work without power control but transmit continuously and consequently offer a constant signal level.

The GSM recommendation does not specify an algorithm for the handover decision or target cell selection. An example of a procedure that could be followed appears in the Appendix of GSM Rec. 05.08 (see Section 3.6.5). Network operators and manufacturers have the option of defining and using a handover algorithm based on the specified parameters.

The appropriate parameters are as follows:

- Permanent data, such as the transmitter power of:
  - the *Mobile Station*
  - the BTS of the supplying cell
  - the BTSs of neighbouring cells.
- Results of real-time measurements carried out by the MS:
  - downlink signal quality (gross bit-error ratio)
  - downlink receive level of the current channel
  - downlink receive levels of neighbouring base stations (BCCHs).
- Results of real-time measurements carried out by the BTS:
  - uplink signal quality (gross bit-error ratio)
  - uplink receive level of the current channel
  - uplink receive level of neighbouring cells.
- Traffic-oriented aspects (cell capacity, number of free channels, number of new connections waiting for a traffic channel in the BTS).

3.6.3.2 Measurement Report

Because of the complexity and importance of the parameters mentioned, a description of the *measurement report* is given below. The following explanations refer on the one hand to the information on measurement data sent by the MS to the network (*measurement reporting*), and on the other hand on the parameter measurements carried out in neighbouring cells by the MS (*neighbouring cells measurements*).

Measurement reporting So that the information required for a handover decision is available at all times, the MS measures system parameters on an almost continuous basis during a connection and sends this information to the network. This means that the MS is usually transmitting a measurement report to the network every 480 ms (block recurrence time of the SACCHs), but at a minimum once per second. It contains parameters that define not only the current network connection but also the radio conditions to neighbouring cells, which could possibly be used as target cells in a handover. The number of neighbouring cells monitored simultaneously by an MS depends on the network topology, and is stipulated by the network operator based on the cluster size. A maximum of six other cells are assumed in the measurement protocol.

An MS measures the receive level and quality of each downlink burst on the traffic channel allocated to its connection. Within a TCH multiframe (120 ms), measurement values for 24 TCH bursts and 1 SACCH burst are calculated; another time slot remains unused.

The interval between each individual measurement is 4.615 ms (duration of a TDMA frame), which corresponds to a route of approximately 4 cm at a speed of 50 km/h. Related to the duration of an SACCH block (480 ms $= 4 \cdot 120$ ms), this equates to 100 measurement values, the mean values of which are transmitted in the measurement report.

The measurement report is transmitted on the SACCH, which occurs every 120 ms and with an interleaving depth of 4 has a block recurrence time of 480 ms.

The SACCH only transmits control data, and is not dependent on the user data transmitted on a TCH. Discontinuous transmission on a TCH is only possible if control data is transmitted separately from user data so that the measurement report periodically expected by the network can be provided (see *Downlink DTX Flag* in Section 3.6.9.2). With a combined control and user data channel, discontinuous transmission would only be possible on a limited basis because of the measurement report, which is transmitted periodically.

Measurement of channels of neighbouring cells Measurement of the signal levels of neighbouring cells is more difficult because an MS must also establish which neighbouring cells it can even receive, and divide the possible measurement times among those cells capable of receiving.

Thanks to the delay of three slots introduced between *Uplink* (UL) and *Downlink* (DL) slots of a traffic channel (see Figure 3.9), an MS can operate in half-duplex mode, and not only alternate the transmission of bursts in the appropriate time slots of the UL and DL frequency channels of its own active connection but also find out the channel characteristics of neighbouring cells in the unused time slots with the same receiver.

The period of time between the transmission of measurement reports varies, and depends on channel type. For example, the following applies to generating a measurement report in an MS with TACH/Fs (TACH = TCH + SACCH, F = full-rate channel).

The possible measurement times lie between transmission of a burst on UL and reception of a burst on the DL of the traffic channel (see Figure 3.43). With a 26-frame multiframe (26 TDMA frames) of 120 ms duration, this equates to 24 time intervals of approximately 2.3 ms duration that are not suitable for the measurement of the channels of neighbouring cells. Furthermore, an interval of approximately 6.9 ms occurs because of an unused time slot in the 26-frame multiframe, and is used to seek out neighbouring cells capable of receiving.

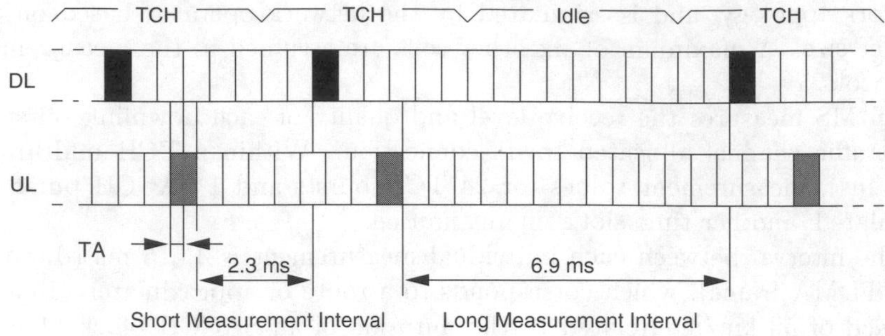

Figure 3.43: Measurement points for neighbouring cells

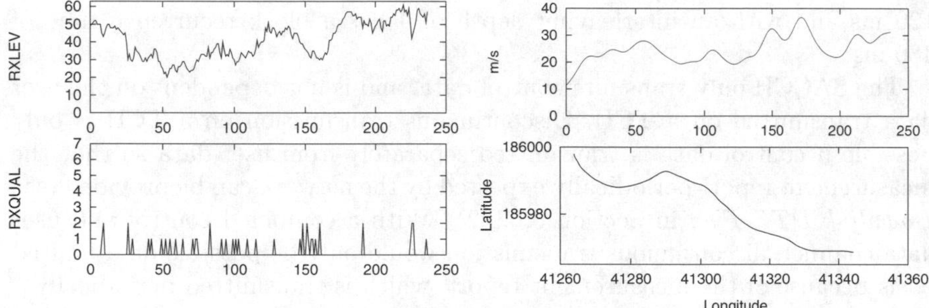

Figure 3.44: Measurement data: example 1

The long interval is also used by the MS to find a suitable neighbouring cell. The criterion used is the BCCH bearer of the neighbouring cell, which is operated with no power control and continuous transmission (no *discontinuous transmission*) and therefore offers a constant signal level.

The length of these intervals determines what technical equipment is required by an MS so that it is able to carry out measurements not only in long intervals but also in short ones and in some cases on frequencies it is not using itself. The frequency synthesizer of the MS must be capable of changing sending and receiving frequency in less than 1 ms in order to have around 300 µs available per measurement.

An SACCH block has 100 measurement values divided among the suitable neighbouring cells. The measurement report transmits the signal level mean value of the six most suitable neighbouring cells.

3.6.3.3 Real Measurement Data

Figures 3.44 and 3.45 show examples of mobile station measurement data recorded during two measurement trips in an existing network.

The downlink level (`RXLEV`) and the downlink quality (`RXQUAL`) can be seen in the two diagrams on the left of each figure; the speed of the MS (m/s) applied over the time (time intervals as multiples of the 480 ms SACCH frame duration) is shown in the diagram on the top right of each figure. The diagram on the bottom right of each figure shows the absolute position of the MS (longitude and latitude in arc seconds).

The two measurement data profiles clearly differ from one another. During the entire connection, measurement trip 1 was well supplied with a high signal level and in turn the quality of the connection was good. Measurement trip 2, on the other hand, shows a strong deterioration in signal level as the curve moves towards the end and this also reflects the very poor quality of the connection.

3.6.3.4 Analysis of Measurement Data

The measurement data clearly shows that the signal level characteristics can be divided into the following parts:

- Long-term mean value
- Short-term fading
- Shadowing

Figure 3.46 shows a superimposition of these components using simulation data. The short-term fading can no longer, however, be described here as Rayleigh or Rice fading but rather as an average calculated over 480 ms for a corresponding signal profile in GSM.

Figure 3.47 shows that the signal level profile is affected by shadowing and topographical circumstances.

3.6.4 Measurement Reports

In GSM *measurement reports* that are transmitted periodically from MS to BS on the SACCH assigned to each communication are available for each connection. The repetition duration of the SACCH produces a fixed time grid of 480 ms in which the measurement reports occur.

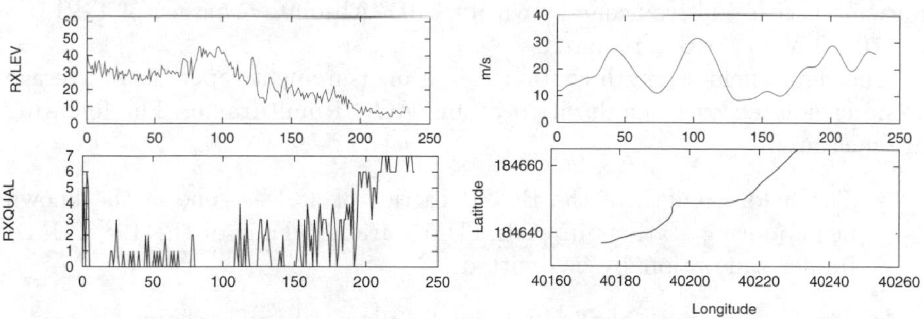

Figure 3.45: Measurement data: example 2

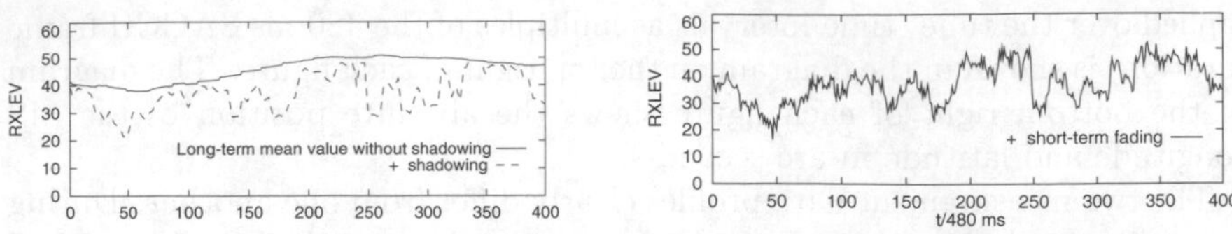

Figure 3.46: Long-term mean value, shadowing, short-term fading

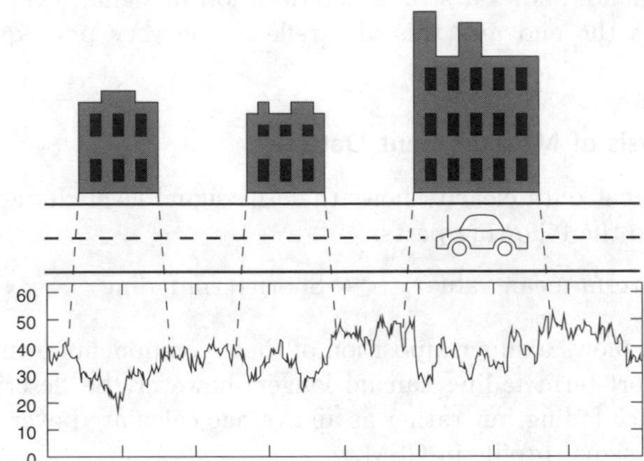

Figure 3.47: Level affected by topography

3.6.4.1 Measurements

During a connection, the MS and the base station measure signal field strength and signal quality and the base station also measures the absolute distance to the MS.

Signal field strength From -110 dBm to -48 dBm the signal field strength is measured with relative accuracy within 1 dB. Absolute accuracy of 4 dB (up to -70 dBm) or 6 dB is required.

The signal field strength contained in a measurement report is an average value calculated over the duration of an SACCH multiframe. The following are measured:

- The field strength of the BCCH carrier of at least one of the known neighbouring stations in each TDMA frame. Four of the 104 TDMA frames can optionally be omitted.
- The field strength of all bursts in the allocated traffic channel.

This averaged data is transmitted in 64 steps as `RXLEV` parameters (see Table 3.13). The signal field strength coded as `RXLEV` is indicated below as the receive level or simply as level.

Table 3.13: Signal field strength RXLEV and signal quality RXQUAL

dBm			RXLEV
	…	−110	0
−110	…	−109	1
−109	…	−108	2
−108	…	−107	3
⋮			⋮
−51	…	−50	60
−50	…	−49	61
−49	…	−48	62
−48	…		63

Bit error [%]			Average	RXQUAL
	…	0.2	0.14	0
0.2	…	0.4	0.28	1
0.4	…	0.8	0.57	2
0.8	…	1.6	1.13	3
1.6	…	3.2	2.26	4
3.2	…	6.4	4.53	5
6.4	…	12.8	9.05	6
12.8	…		18.10	7

Signal quality Signal quality is measured as the bit-error ratio before channel decoding, and is divided into eight RXQUAL levels of quality (see Table 3.13). The required measurement accuracy and the allocation probability to the correct level is required to be 75 % for (RXQUAL 1) and 95 % to levels (RXQUAL 5–7).

The signal quality contained in the measurement report is an average value calculated over the duration of an SACCH multiframe. This value is indicated below as the receive quality or simply as quality.

Distance The absolute distance between mobile station and base station can be determined from the *Timing Advance* (TA) for the MS. This value is produced from the *round-trip propagation delay* of the burst sent by the MS relative to the time prescribed by the time slot tact produced by the BS. The difference is measured by the base station. TA is calculated by the base station in multiples of the duration of a bit.

The distance of the MS is calculated from the timing advance as

$$d\mathrm{TA} = \frac{\mathrm{TA} \cdot c t_{\mathrm{bit}}}{2} = \frac{\mathrm{TA} \cdot 3 \cdot 10^{8}\ \mathrm{m/s} \cdot 3.69 \cdot 10^{-6}\ \mathrm{s}}{2} = \mathrm{TA} \cdot 554\ \mathrm{m}$$

c is the radio wave propagation speed, t_{bit} is the duration of a bit in GSM; the factor 1/2 is the result of the propagation time delay having to be calculated twice, namely as the round-trip propagation delay BTS–MS–BTS.

If a maximum error of ±0.5 bits is assumed for this measurement, then the overall accuracy in establishing the distance is approximately 1 km. Since microcells can have a radius of ≤1 km, this value is too inaccurate, and in practical terms is only useful as a means to detect that the MS has definitively left the planned cell area.

Measurement values supplied While a connection is taking place with a traffic channel, the MS supplies the following values:

Table 3.14: Measurement_Report message

Information element	Transmission direction	No. of octets
`Protocol Discriminator`	MS → network	2
`Transaction Identifier`	MS → network	2
`Message Type`	MS → network	2
`Measurement Results`	MS → network	16

- Receive level `RXLEV` of the traffic channel.
- Receive quality `RXQUAL` of the traffic channel.
- Receive level `RXLEV` of the BCCH bearers of up to six neighbouring cells, specifically those with the highest receive level and a known and valid *Base Station Identification Code* (BSIC).
- Frequency of the BCCH bearer of these neighbouring cells.
- BSIC of these neighbouring cells. The BSIC is not a unique identification of the neighbouring base stations but a so-called colour code used to differentiate base stations transmitting at the same frequency in adjacent clusters.

3.6.4.2 SACCH Transmission

The measurement reports are transmitted over the SACCH as the `Measurement Results` information elements of a Measurement_Report message. An SACCH block has a length of 184 bits and the block recurrence time is 480 ms. The measurements carried out by an MS during a period of 480 ms are available to the base station for a further 480 ms. The time delay between an individual measurement and the receipt of the measurement data at the base station is also approximately 0.5–1 s.

A Measurement_Report message is structured as shown in Table 3.14 and Figure 3.48 (also see [15]).

The components of a `Measurement Results` information element are as follows:

- `DTX`: indication whether an MS is using discontinuous transmission.
- `RXLEV` Serving Cell: receive level of the supplying base station; the `RXLEV` levels from 0...64 described above are coded with 6 bits.
- `RXQUAL FULL` or `RXQUAL SUB`: receive quality of serving base station measured over all time slots or a subset. The eight `RXQUAL` levels are coded with 3 bits.
- # `NCELL`: number of measured neighbouring stations.

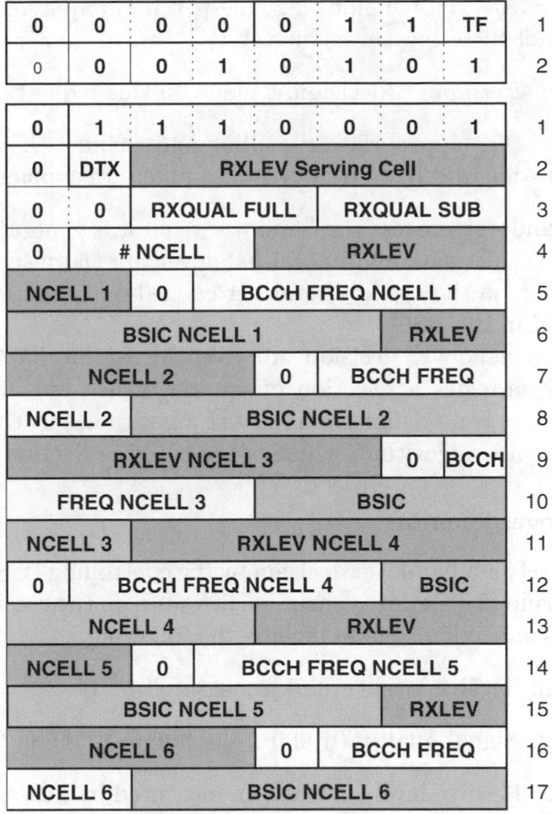

Figure 3.48: Components of `Measurement Results` information elements of a Measurement_Report message

- `RXLEV NCELL` n: receive level of the neighbouring stations.
- `BCCH FREQ NCELL` n: BCCH frequency of the neighbouring stations, coded as an index in the BCCH channel list.
- `BSIC NCELL` n: `BSIC` of the neighbouring station as colour code. This always consists of 3 bits with the colour code of the PLMN and the base station [7].

3.6.5 Handover Decision

The handover process in GSM is implemented in the following steps:

1. *Measurement values*: Measurement and transmission of radio data as described.

2. *Handover request*: A handover request, if it is required, is generated on the basis of these measurement values.

3. *Handover decision*: A decision is made on this request.

4. *Handover*: The appropriate signalling and channel change on both the mobile station and fixed network sides finally take place.

In a *normal* handover process the handover request is generated by the BSC, and the handover decision and actual handover are the responsibility of the MSC. Depending on the type of handover carried out, functions 3 and 4 can be implemented in the BSC.

Criteria for a handover decision are given in Section 3.6.3.1. GSM Rec. 05.08 [18] only provides a selection of criteria, which are described in Section 3.6.5.1. There is no definitive handover algorithm, but the recommendation includes a basic algorithm, which is presented in Section 3.6.6.

3.6.5.1 Handover Request

As the solid basis for a handover decision by the controlling BSS, each BTS and each mobile terminal generate measurement values on the receiving conditions of the current connection. These include the following:

- `RXLEV_UL`: Receive level (uplink), measured by BTS
- `RXQUAL_UL`:Signal quality (uplink), measured by BTS
- `RXLEV_DL`: Receive level (downlink), measured by MS
- `RXQUAL_DL`: Signal quality (downlink), measured by MS.

The receive levels can be measured directly. The values for quality are calculated from the occuring bit error ratios and coded into eight levels of quality. The downlink measurement data is transmitted by the MS over the SACCH. The MS also measures the bearer signal `RXLEV_NCELL`(n) of the broadcast channel BCCH of the neighbouring cells, and likewise notifies the BTS over the SACCH. On the basis of this data, the BSC can decide whether a handover must be initiated and which of the adjacent cells are in a position to provide further radio coverage to the MS. The advantages of this method are as follows:

- Most of the measurement effort is undertaken by the MS.
- Abnormal radio propagation, radio shadowing, etc., are taken into account because the connection quality of both sides (uplink and downlink) is evaluated.
- The receive level of neighbouring cells is already measured by the MS before the handover takes place.

Table 3.15: Handover_Required message

Information element	Transmission direction	No. of octets
`Message Type`	BSS → MSC	1
`Cause`	BSS → MSC	3
`Response Request`	BSS → MSC	1
`Cell Identifier List pref.`	BSS → MSC	$2n + 3 \dots 7n + 3$
`Current Radio Env.`	BSS → MSC	$15 \dots n$
`Environment of BS` n	BSS → MSC	$7 \dots n$

The following must be considered before a handover decision is made:

- The measurement results must be averaged over an appropriate period of time in order to avoid a superfluous handover because of short-term signal level fluctuations.
- There must be a check to determine whether the connection quality can be affected by a change in transmitting power.
- The best radio cell is selected to ensure that the MS can also continue to be supplied with coverage if required.

The algorithm proposed in the GSM recommendations works on a relatively simple principle: several measurement reports are generated in a BSC and compared with threshold values. A handover is implemented if the quality values exceed the established threshold or the level values fall short of it.

The handover requests generated in the BSC are normally routed to the MSC. The signalling for a handover request from BSC to MSC is carried out through a Handover_Required message, the structure of which is shown in Table 3.15 [10]. The reason for the handover, a request for a response, a list of target cells sorted in order of best values and the current radio measurement values are included.

Handover requests arriving from the BSS are processed in the MSC. The decision criteria include the availability of traffic channels in the target cell, the interference level in the target channel and other criteria relative to the network.

3.6.5.2 Handover Execution

A handover is initiated by an MSC, which first sends a Handover_Request message to request the necessary resources in the target cell (BSS 2) (see Figure 3.49). The target cell acknowledges with a Handover_Request_ack message. The MSC then sends a Handover_Command message to the previous cell (BSS 1), which in turn sends the same message to the MS. The MS then changes channel in the new target cell and sends a Handover_Complete message, which is also sent from the new cell (BSS 2) to the MSC. Finally the

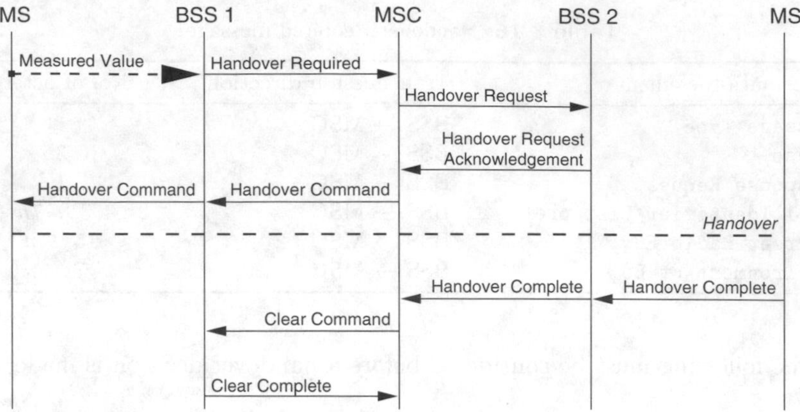

Figure 3.49: Handover (BSS 1 → BSS 2) executed with an MSC

MSC sends a Clear_Command message to the old cell, which acknowledges the message with Clear_Complete.

This procedure applies to a normal handover signalling case. During actual operation, several different situations can occur:

Intracell handover The channel for the connection is changed within the cell, e.g., if the channel has a high level of interference. The change can apply to another frequency of the same cell or to another time slot of the same frequency.

Intercell/Intra-BSC handover In this case there is a change in radio channel between two cells that are served by the same BSC.

Inter-BSC/Intra-MSC handover A connection is changed between two cells that are served by different BSCs but operate in the area of the same MSC.

Inter-MSC handover A connection is changed between two cells that are in different MSC areas.

In the first two situations the handover can be executed by the BSC if it supports internal handovers. If this is not the case, intracell handover must also be carried out by the MSC.

3.6.5.3 Power Control

A reduction in transmitter power results in less co-channel and adjacent-channel interference, and reduces the power needed by mobile stations—which is especially important for handheld mobile equipment. Conversely, an increase in transmitter power makes it possible to compensate for short-term supply gaps, e.g., signal level fluctuations due to shadowing, thereby possibly preventing the need for handover.

Power control functions are analogous to the handover process in the BSC. Decisions on changing transmitter power are always made in the BSC. In turn, the measurement reports are the basis for the decision on whether transmitter power should be increased or decreased.

Like the handover algorithm, the algorithm for this process is not standardized. The basic algorithm in [18] also specifies power control.

3.6.6 Sample Algorithm GSM 05.08

Appendix A of GSM Rec. 05.08 [18] includes the specifications for a basic algorithm for handover and power control that is used to decide whether handover is necessary on the basis of measurement data provided by the MS and the fixed network. Network operators and manufacturers of GSM components use algorithms that range from a lightly modified or partially abbreviated version of 05.08 to heavily modified versions with new parameters, criteria, and so forth. The basic principles in terms of functions, however, correspond to the 05.08 algorithm.

3.6.6.1 Preliminary Steps and Threshold Comparison

The averaging of values and comparisons are generated every 480 ms; this time reference is dictated by the SACCH.

Averaging and power-budget value All measurement values (level, quality, distance MS–BS) are first averaged in order to reduce the effect of short-term signal fading. The averaging is influenced by the following parameters:

`HREQAVE`: Number of measurement reports averaged,

`HREQT`: Number of averaged values in a Handover_Required message.

Averaged values are provided for downlink and uplink level, downlink and uplink quality, and level of neighbouring cells, as well as distance MS–BS. The `HREQT` values that in turn represent an average value of `HREQAVE` measurement values are also indicated.

In addition, the *power budget* is calculated from the averaged values of the neighbouring stations:

$$\begin{aligned}\texttt{PBGT}(n) &= [\min(\texttt{MS\_TXPWR\_MAX}, P) - \texttt{RXLEV\_DL} - \texttt{PWR\_C\_D}] \\ &\quad - [\min(\texttt{MS\_TXPWR\_MAX}(n), P) - \texttt{RXLEV\_NCELL}(n)]\end{aligned} \tag{3.5}$$

This value represents the difference between the path loss on the downlink and the path loss anticipated in the respective neighbouring cell. Thus the value is greater than zero for neighbouring cells with better radio coverage and less than zero for those with worse coverage.

Table 3.16: GSM 05.08: threshold values for power control

Power control			
Transmitter power	Value	P/N	Threshold
+ DL	RXLEV_DL	P1/N1	< L_RXLEV_DL_P
+ UL	RXLEV_UL	P1/N1	< L_RXLEV_UL_P
− DL	RXLEV_DL	P2/N2	> U_RXLEV_DL_P
− UL	RXLEV_UL	P2/N2	> U_RXLEV_UL_P
+ DL	RXQUAL_DL	P3/N3	> L_RXQUAL_DL_P
+ UL	RXQUAL_UL	P3/N3	> L_RXQUAL_UL_P
− DL	RXQUAL_DL	P4/N4	< U_RXQUAL_DL_P
− UL	RXQUAL_UL	P4/N4	< U_RXQUAL_UL_P

+	Increase transmitter power	DL	Downlink
−	Reduce transmitter power	UL	Uplink

MS_TXPWR_MAX is the maximum allowable transmitter power for an MS in a current cell or in a neighbouring cell, RXLEV_DL is the averaged downlink level, RXLEV_NCELL(n) is the averaged level for a corresponding neighbouring cell, and PWR_C_D is the difference between the maximum allowable transmitter power and current transmitter power of a BS.

Threshold value comparison for power control The averaged values are always used in the threshold value comparison. The parameters are the threshold values and the P from N criteria. It is always necessary for the Pn from Nn values to exceed or fall short of the threshold if the transmitter power is to be increased or reduced, respectively. Table 3.16 summarizes the different posssibilities.

Threshold value comparison for handover The appropriate threshold values and P from N criteria are also used for initiating handover requests (see Table 3.17).

Figure 3.50 presents an overview of the threshold values. Figure 3.51 shows the behaviour of PBGT criteria during handover between two cells, the HO_MARGIN value represents an hysteresis value.

3.6.6.2 BSS-Decision Algorithm

When a threshold value comparison establishes that handover is necessary, the BSC must send a Handover_Required message to the MSC. It contains a list of the neighbouring cells that have been evaluated, indicating the following conditions:

$$\texttt{RXLEV\_NCELL}(n) > \texttt{RXLEV\_MIN}(n) + \max(0, \texttt{MS\_TXPWR\_MAX}(n) - P) \quad (3.6)$$

Table 3.17: GSM 05.08: threshold values for handover

Handover			
Cause	Value	P/N	Threshold
RXLEV DL	RXLEV_DL	P5/N5	< L_RXLEV_DL_H
RXLEV UL	RXLEV_UL	P5/N5	< L_RXLEV_UL_H
RXQUAL DL	RXQUAL_DL	P6/N6	> L_RXQUAL_DL_H
RXQUAL UL	RXQUAL_UL	P6/N6	> L_RXQUAL_UL_H
Intracell DL	RXLEV_DL	P7/N7	> RXLEV_DL_IH
	RXQUAL_DL		> L_RXQUAL_DL_H
Intracell UL	RXLEV_UL	P7/N7	> RXLEV_UL_IH
	RXQUAL_UL		> L_RXQUAL_UL_H
DISTANCE	Distance	P8/N8	> MS_RANGE_MAX
PBGT	PBGT(n)		> 0
			> HO_MARGIN(n)

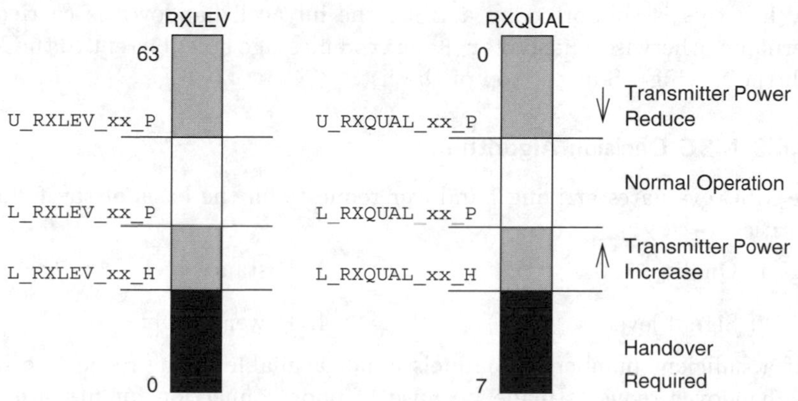

Figure 3.50: GSM 05.08 threshold values

$$\text{PBGT}(n) > 0 \tag{3.7}$$

These conditions must be met by the respective neighbouring cell so that it can be considered as the target cell for the handover. The list is sorted on the basis of the PBGT value; the first cell is the one with the highest PBGT(n) value.

If a handover is considered to be *imperative* (the reason being level, quality or distance) then the condition Eq. (3.7) need not be met, and the list also contains neighbouring cells with PBGT(n) < 0.

An intracell handover is a special case, because two criteria are evaluated: if the signal level is good but the quality is bad, this usually means that co-channel interference exists. This can be avoided by a change in frequency or time channel.

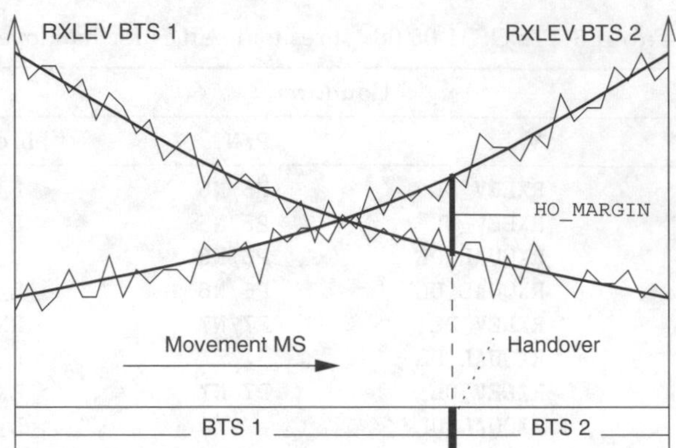

Figure 3.51: GSM 05.08 power budget handover

When this is supported by a BSC, the intracell handover is carried out internally; otherwise a `Handover_Required` message must be sent to the MSC, with the serving BS at the top of the list.

3.6.6.3 MSC Decision Algorithm

The MSC evaluates arriving handover requests on the basis of the following priorities:

1. Quality
2. Signal level
3. Distance
4. Power budget

If a sufficient number of channels is not available in the respective target cell, handover requests made because of poor connection quality are given priority.

There is also a provision for giving priority to individual cells specifically to distribute traffic load. Therefore in a hierarchical cell layout the superimposed *macrocell* is only selected as a target for handover if a free channel is not available in the microcells.

3.6.6.4 Modifications

In practice the GSM 05.08 algorithm is used with modifications in GSM networks. Modifications are appropriate with averaging techniques for measurement values, in the administration of parameters by cell and with decision criteria.

Averaging The `P` from `N` decisions, based on averaged measurement values, in effect present another averaging technique specifically for threshold values. The dependence of the parameters makes optimization difficult. An option exists for simplifying this by introducing separate averaging parameters for each handover cause (quality, level, distance, power budget).

Otherwise modified averaging methods can be employed that, for example, do not consider outliers.

Cell administration The parameters for handover can be administered on a cell basis and thereby adapted optimally to local conditions. There is also an option of specifically switching on/off individual types of handover requests per cell.

Decision criteria The BSC decision criteria can also be changed, e.g., it can be practical to allow negative values to be used for `PBGT`(n) and `HO_MARGIN`.

3.6.7 Problems in the GSM Handover Process

As has already been mentioned, handover in GSM is based on radio measurement data. This is also what causes the main problem, because radio propagation in an actual environment is unpredictable and highly irregular.

In particular, shadowing caused by obstacles can produce some undesirable effects. This is mostly manifested by the fact that too many handovers are taking place.

3.6.7.1 Pingpong Handover

One very undesirable effect that occurs relatively frequently is so-called *pingpong handover*. This is a handover to a neighbouring cell that returns to the original cell after a short time. The cause of a pingpong handover is the power budget criterion. A handover is mainly executed on the basis of this criterion in cells with good radio coverage and only minimal disruption due to interference. The parameter `HO_MARGIN` determines which level of hysteresis must be exceeded so that a change to a neighbouring cell takes place (see Figure 3.51). In normal operation a 5–10 dB `HO_MARGIN` is selected to prevent minor variations in signal level of different base stations from causing a handover.

Strong shadowing created by large obstacles can cause fading up to 30 dB. If such an obstacle is found in the line-of-sight of the serving base station but not of the neighbouring station then it is possible that a handover may be triggered through power budget although the MS continues to find itself in a well-supplied cell. As soon as the MS moves out of the shadowing, the level again becomes normal and a handover takes place to the original cell.

With medium to higher mobility of an MS, this results in a handover to a neighbouring station and back to the original base station within a short period of time ($<$10 s) (*pingpong handover*).

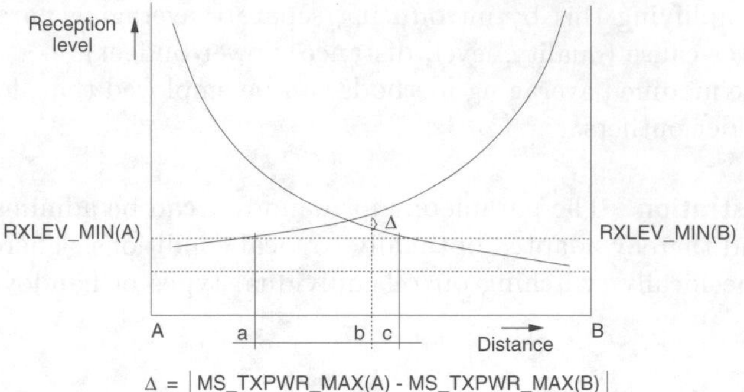

Figure 3.52: Recommended location for a handover during movement from base station A to base station B

This can be avoided if either the hysteresis value is increased so high that signal fading does not trigger a power budget handover or the averaging length is selected so that the time span in which an MS is in shadowing is bridged.

However, neither is appropriate in actual operation. A high hysteresis value would practically eliminate a power budget handover and delay a change in the boundary area of t wo cells far beyond the boundary. A high averaging length makes the handover process too sluggish and leads to a situation in which a really necessary handover sometimes is not carried out in time and the connection is then broken off.

This is why, with the 05.08 algorithm and the modified variants, pingpong handover cannot be avoided even by using appropriate parameter adjustments.

3.6.7.2 Number of Handovers

It was planned that the number of handovers per connection would be kept to a minimum in GSM because of the complexity of the signalling in the fixed network as well as from and to an MS. The traffic channel must be used for signalling in a handover between a mobile station and a base station, and this causes data loss and a corresponding forfeiture of voice quality.

It is therefore essential that unnecessary handovers be avoided. Control of transmitter power is optional for BTS and MS. Figure 3.52 shows a theoretical profile of the average receive power during the movement of an MS from BTS A to B. A handover is theoretically only possible between locations a and c, but it is recommended at position b because that is where the level of BTS B has first fallen below the allowable hysteresis value $\Delta =$ `HO_MARGIN`, in other words is sufficiently above that of BTS A.

Owing to propagation-related signal level fluctuations at the receiver, the curve characteristics shown only apply to the statistical average (see the actual curve characteristics in Figure 3.51). Therefore the location of a handover is randomly distributed in the area around b. When an MS has poor reception

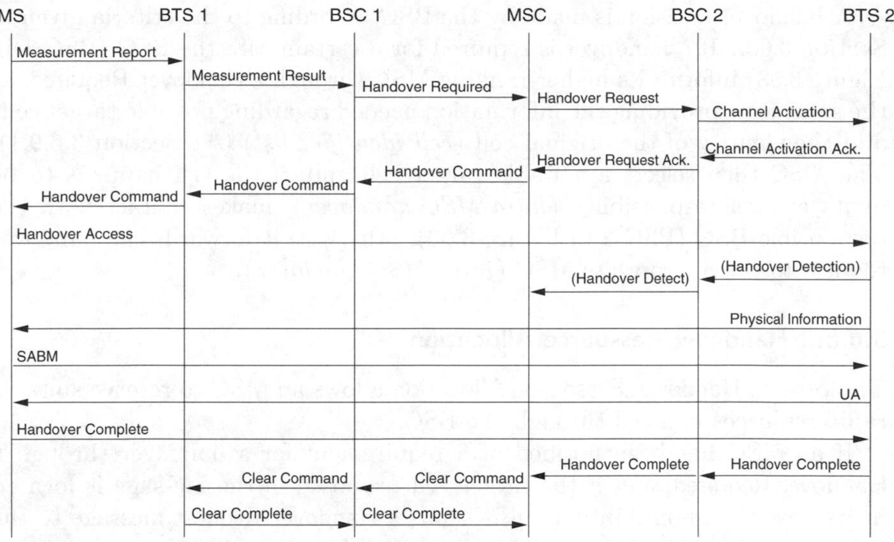

Figure 3.53: Intra-MSC handover in a GSM system

and a poor parameter selection exists for the handover algorithm, several handovers instead of only one may be necessary before the MS leaves the area where reception is critical and a stable allocation to a BTS has taken place.

3.6.8 Intra-MSC Handover

Figure 3.53 shows the message sequence chart of the protocol for a handover between two different BSCs within the same MSC area (intra-MSC or inter-BSC handover) based on GSM Rec. 04.08 [15], Rec. 08.08 [10] and Rec. 08.58 [11].

The handover protocol can be divided into three separate parts [10]:

- Indication of a necessary handover: Handover_Required_ind
- Allocation of radio resources: Handover_Resource_Allocation
- Handover execution: Handover_Execution

3.6.8.1 Handover Required Ind

With Handover_Required_ind a BSS requests a handover for one of its active (*dedicated mode*) MSs.

Based on the measurement protocol described in Section 3.6.3.1 and Section 3.6.3.2 and a supplementary measurement of the corresponding MS uplink signal parameters by the BTS, the network decides on the necessity of a handover.

A handover decision is made by the BSC according to the criteria given in Section 3.6.5. If a handover is required for a certain MS, the BSC (BSC 1 in Figure 3.53) informs its higher-ranking MSC through a Handover_Required, at the same time providing the information needed regarding possible target cells and the identity of the original cell (*cell identifier list*) (see Section 3.6.9.1). The MSC then selects a suitable target cell and, if the cell happens to be in its area of responsibilty (*intra-MSC handover*), makes contact with the responsible BSC (BSC 2 in Figure 3.53); otherwise it forwards the handover request to the appropriate MSC (*inter-MSC handover*).

3.6.8.2 Handover Ressource Allocation

The process Handover_Ressource_Allocation allows an MSC to reserve suitable radio resources in a cell through the BSC.

If an MSC has been notified of a requirement for a handover through a Handover_Required, and if the target cell proposed in the message is located in its area of responsibility, then it sends a Handover_Request message to the BSC responsible for this cell. This message informs the BSC of the situation of the target cell and the characteristics required of the channel that is to be allocated to the MS. If a pure data or speech channel is involved, it provides the channel type used between the MSC and the BSC.

The channel type required for the radio interface can be different from the current channel type [10], e.g., immediately after a connection has been established in which the MS was only able to obtain a half-rate channel instead of the full-rate channel required. On this occasion a direct change of channel (*directed retry*) takes place.

The network operator can assign a priority to a Handover_Request depending on the urgency of the handover execution. Details of the content of a Handover_Request message can be found in Section 3.6.9.2.

A BSC selects Channel_Activation to request that the BTS responsible for the target cell check its radio resources on a specific available channel and provide it with a handover reference number (`Handover Reference`) that identifies the MS when it uses Handover_Access to access the channel. This is the only way in which a channel can be allocated to an MS without the identity of the user being divulged.

In addition to the channel characteristics required and the handover reference number, the BTS is given the technical parameters relating to transmission, e.g., for the power control of the new channel. The content of the Channel_Activation message is described in Section 3.6.9.3, whereas *power control* is explained in more detail in Section 3.6.9.5.

If the BTS is able to reserve a suitable channel for the MS, it confirms this to the BSC through a Channel_Activation_ack. The content of this message (see Section 3.6.9.4) also includes the current `Frame Number` of the BTS, which the BSC uses to determine the time displacement factor with which the MS begins its normal transmission activity after it has successfully accessed the channel.

The MSC is then informed through Handover_Request_ack that a channel has been reserved. This completes the process Handover_Ressource_Allocation.

3.6.8.3 Handover Execution

For the MS the handover decision and initiation take place unnoticed in the background. As shown in Figure 3.53, it is first made aware of a handover when it receives the command Handover_Command.

The execution of a handover begins with a Handover_Request_ack message from the new BSC to the MSC that a requested channel has been reserved. It receives a data field `Layer 3 Information` with all the required information for the MS in order to register for the reserved channel with the new BTS. The MSC therefore forwards this data field unchanged through a Handover_Command to the original BTS, which sends it to the respective MS. The MS conveys the situation of the new cell only on the basis of the elements beacon frequency, contained in the data field, and BSIC, which provides a unique identification of the correct cell when several cells are using the same frequency for a BCCH or when there is a cell cluster with respect to the BCCH frequency; see [33]). The MS is not made aware of the total identity of the cell until the completion of handover over the SACCH of the new BTS.

Another important element in the data field is the handover reference number used by the new BTS to check whether the MS is authorized to receive the reserved channel. The MS notifies the BTS of the number using a Handover_Access message (see Section 3.6.9.7).

The Handover_Command indicates whether the MS should be executing an asynchronous or a synchronous handover. In both cases it is able to switch quickly to the frame and time slot tact of the new channel. An important difference in the procedure is the way in which the time advance parameter (TA) is determined.

With synchronous handover the MS itself calculates the parameter `TA` with which it begins radio transmission on the uplink. So that the BTS can specify the parameter for synchronization, the MS first sends some *access bursts* without time advance before it continues with normal transmission. This cancels out the need for a Physical_Information message, so that the MS goes straight to establishing a layer-2 connection to the network using an SABM frame.

With asynchronous handover the MS is dependent on the support of the network. In this case it sends *access bursts* so long until it receives a response from the network in the form of a Physical_Information message, the content of which allows it to calculate the time advance parameter (see Section 3.6.9.8). As with a synchronous handover, this is followed by the establishment of a layer-2 connection.

The timer `T3124` monitors and controls the process of an asynchronous handover in an MS between the first Handover_Access and the arrival of the message Physical_Information. If the timer has expired, the handover process is broken off and the MS returns to its original channel. Since the FACCH is

Table 3.18: Information elements of a Handover_Required message [10]

Information element	Transmission direction	No. of octets
Message Type	BSC → MSC	1
Cause	BSC → MSC	3–4
Response Request	BSC → MSC	1
Cell Identifier List (preferred)	BSC → MSC	$2n+3\ldots7n+3$

used to transmit the signalling data for a handover, the timer value according to GSM recommendation 04.08 [15] is 320 ms.

Immediately after receipt of the expected Handover_Access, the BTS informs its higher-ranking BSC through a Handover_Detection message that the reserved channel has been seized. With Handover_Detection the BSC thereupon informs the MSC, which then changes the connecting path to the MS through the network, sometimes not waiting for a Handover_Complete in order to reduce the switching times on the network side. After the MS has switched to normal transmitting (with time displacement), it establishes a layer-2 connection to the BTS using an SABM frame with an $LAPD_m$ protocol. The BTS acknowledges the frame with a UA frame, etc.

As soon as the layer-2 connection has been acknowledged by the network, the MS informs the BTS through Handover_Complete that the handover has been successfully executed. This message is forwarded by the BSC to the MSC, which consequently reroutes the network connecting path to the new BSC. It uses a Clear_Command to release the reserved network resources at the original BSC and corresponding BTS. Consequently the MS can now only use the new radio channel. The BSC sends an acknowledgement Clear_Complete to the MSC after the resources have been released.

3.6.9 Intra-MSC Handover Protocol

The content of layer-3 messages from the intra-MSC handover protocol based on GSM Rec. 04.08 [15], 08.08 [10] and 08.58 [11] is introduced below. This section does not provide a complete description, but focusses only on the important elements of layer-3 messages. $LAPD_m$ switching messages have already been discussed in Section 3.5.

3.6.9.1 Handover Required

If a BSC makes a positive handover decision for an MS that already has a dedicated channel, it indicates to the MSC with Handover_Required (see Table 3.18) that a handover is required for the MS concerned because of a particular reason (*cause element*).

`Message Type` This 8-bit (1-octet) long information element clearly identifies the type of message and is contained in all layer-3 messages. Section 3 of GSM Rec. 08.08 [10] particularly deals with message types used in the management of radio resources (*Radio Resource Management*, RRM).

`Cause` The `cause` element describes the reason for the occurrence of a particular message. It is based on a 3- to 4-octet long data field with the actual cause coded in the last or in the last two octets (see [10]).

The following are typical of the 40 possible causes of a handover:

- *Uplink quality*
- *Uplink strength*
- *Downlink quality*
- *Downlink strength*
- *Distance*
- *Better cell*
- *Response to MSC invocation*
- *O&M intervention*
- *Directed retry*

`Response Request` This 1-octet long information element indicates that a BSC is waiting for a Handover_Required_Reject in case the Handover_Required message has not resulted in a handover.

`Cell Identifier List` This list contains all the target cells being considered for a handover. It is produced on the basis of the detailed measurement protocol in the BSC when a positive handover decision has been made according to a specific network-oriented algorithm. Depending on the number of cells (indicated as n in Table 3.18) and the coding used for cell identity, it can have different data field sizes. The number of target cells possible is generally determined by the OSS or by the responsible network operator, but, according to the GSM recommendations, should be in the area between 1 and 16. The LAI or the *Cell Identity* (CI) code (5 or 2 octets/cell) is normally used for cell identification in a GSM network. However, according to [10], other codes can also be selected.

3.6.9.2 Handover Request

An MSC sends this message to inform the respective BSC that the relevant target cell for a handover is located in its area of responsibility (see Table 3.19). The MSC provides the identification of the target cell in which the BSC is to reserve appropriate radio resources.

`Channel Type` This information element contains all the information on channel characteristics that the BSC requires in order to select the correct channel. The information includes application (voice, data or signalling), channel type and speech codec or data rate.

Table 3.19: Information elements of a Handover_Request message [10]

Information element	Transmission direction	No. of octets
`Message Type`	MSC → BSC	1
`Channel Type`	MSC → BSC	5
`Encryption Information`	MSC → BSC	3–20
`Classmark Information 1`[a]	MSC → BSC	2
`Classmark Information 2`	MSC → BSC	5
`Cell Identifier`	MSC → BSC	5–10
`Priority`	MSC → BSC	3
`Circuit Identity Code`[b]	MSC → BSC	3
`Downlink DTX Flag`[c]	MSC → BSC	2
`Cell Identifier` (target cell)	MSC → BSC	3–10
`Interference Band to be Used`	MSC → BSC	2
`Cause`	MSC → BSC	3–4
`Classmark Information 3`[d]	MSC → BSC	3–14

[a]Only one of the two elements is required.
[b]Only incorporated if the `Channel Type` element indicates the transmission of either speech or data.
[c]Only added if there are speech TCHs. If the indicator is omitted then it has no effect on the current DTX mode of the BSC.
[d]The MSC forwards this element only if it receives it itself.

The following different channel types are appropriate [10]:

- Full-rate TCH,
- Half-rate TCH for speech or data,
- SDCCH.

There is also some flexibility in the selection of channel type. For instance, if a half-rate or a full-rate channel is not available, the other channel type is allowed as an option. Furthermore, the BSC can be instructed to change the channel that has initially been allocated to the desired channel type as soon as it becomes available again. An exact breakdown of the different possibilities and the associated coding can be found in [10].

Similarly to ISDN, full-rate channels are referred to as B_m-channels and half-rate channels as L_m-channels. The control channels are called D_m-channels [15].

The following rates are available for the transmission of data in a transparent bearer service: 9.6, 4.8, 2.4, 1.2, 0.6 kbit/s and 1200/75 bit/s (network → MS/MS → network).

The non-transparent services offer the following data rates [10]: 12 (B_m-channel) or 6 kbit/s (L_m- channel), which vary depending on available resources.

`Encryption Information` This information element includes the necessary ciphering information for control data transmission in order to set the target BTS to ciphering mode on the mobile side.

The basis for encryption is the ciphering key K_c mentioned in Section 3.13. `Encryption Information` notifies the BSC and the BTS of the ciphering key and the index of the coding algorithm used. In total there are up to seven different algorithms, however; it is also possible to select to transmit without ciphering. The information element consists of three fixed ciphering keys and one ciphering key up to 17 octets long.

`Classmark Information` Mobile stations differ from one another in a variety of different ways, for instance in their maximum transmitter power and in the bearer services they support. A network must be familiar with the characteristics of each active MS in order to avoid causing errors. Therefore the participating BSS network elements are informed of the characteristic features of the respective MS before each new connection is established.

The element `Classmark Information Type 2` is one of three types of data field, which (in contrast to the other field types) contains a complete list of characteristics relating to the mobile terminal and is made up of the following elements [15]:

`Revision Level` This is the revision number for the mobile terminal, which characterizes the current stage of development through 3 bits. Terminal equipment that technologically is classified as belonging to the first phase of GSM standardization is identified by the code `000` [33].

`RF Power Capability` This indicates the maximum transmitter power of a terminal. As shown in Table 3.4, the transmitter power of equipment is specified according to its classification.

The appropriate information is required for the dynamic control of the transmitter power (*power control*) and in preparation for handover.

`Encryption Algorithm` This element, a 2-bit long code word with the value `00` [15] for the identification of terminal equipment with A5 encryption, is used to indicate to the network whether a terminal is using an encryption algorithm and, if so, which one. Only one alternative, algorithm A5, is planned for GSM phase 1. According to the currently valid MoU, terminals with the A5 algorithm have been introduced in all the participating countries. As indicated in `Encryption Element`, A5 can be produced as one of seven variants by the equipment manufacturer.

`Frequency Capability` This element enables the network to support terminal equipment in different frequency bands. A GSM-900 terminal currently must have the capability of being able to receive at least all the carrier frequencies in the 2×25 MHz band. It will not be possible for all terminal equipment to support the expansion to a 35 MHz band (*Extended GSM* (EGSM) band). Consequently a BSC must be able to distinguish between different types of equipment and allocate an appropriate channel based on the frequency area supported.

This element is a 3-bit long code, which has the setting 000 to characterize terminal equipment that is able to operate with 2×25 MHz wide bands.

`Short-Message Capability` This `Classmark Type 2` element is coded by one bit and relates to *short messages*. It indicates whether mobile side support exists for this bearer service (1) or not (0) [15]. `Classmark Type 1` represents an abbreviated version of the data field in order to map certain layer-3 messages, such as Location_Updating_req, to a single layer-2 frame, and only contains the most important elements [15]:

- `Revision level`
- `Encryption algorithm`
- `RF Power capability`

`Cell Identifier` From the *cell identifier list* presented in Section 3.6.9.1, the MSC selects a suitable target cell for a handover and forwards the corresponding identity code for the cell (`Cell Identifier, Target`) to the respective BSC. The MSC also provides the identity of the original cell (`Cell Identifier, Serving`) in which a handover decision has been made for a particular MS. The coding of the cell identity corresponds to the form given in the `Cell Identifier List`.

`Priority` This information element indicates the importance of message types specified for processing. In addition to 14 different levels of priority, existing connections or impending calls are classified in order of priority levels for the allocation of radio channels. What is also allowable is that if a channel group is momentarily occupied, a Handover_Request will not immediately be rejected but placed in a queue to be processed as quickly as possible.

`Circuit Identity Code` This information element defines PCM multiplex mode and the corresponding time slot of the fixed network channel provided by the MSC.

`Downlink DTX Flag` *Discontinuous Transmission* (DTX) indicates an optional transmission mode for speech and non-transparent data services in which background noise (*comfort noise*) is inserted during periods when minimal information is being transmitted. Since DTX can be carried out independently of the transmitting direction, consideration must be given to two DTX modes: uplink and downlink DTX.

DTX affects the operating behaviour of the MS, the TRAU (see Section 3.2.1.2) and the BTS, which adapts its behaviour dynamically to the

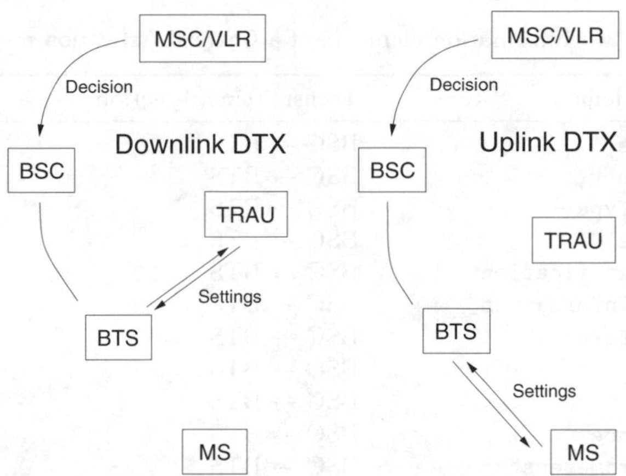

Figure 3.54: Process of discontinuous transmission

data transmission modes of the MS (UL) and of the TRAU or MSC/IWF (DL).

The MSC decides whether DTX will be applied, whereas the BSC controls the service and informs the BTS on a particular connection path whether DTX mode will be supported and, if so, which one. As shown in Figure 3.54, the BTS configures the participating radio resources TRAU (downlink-DTX) and MS (uplink-DTX).

Unlike downlink-DTX, which is associated with a particular connection, uplink-DTX is a cell parameter that is conveyed to an MS in *standby* over the BCCH and to an active MS over the SACCH, and therefore applies to all uplink connections. A downlink-DTX can only be changed in combination with a change in transmission mode (half-rate $\leftrightarrow$ full-rate channel, data rate adaptation) [33]. This kind of situation can arise with handovers, and is indicated to the target BSC by a `Downlink DTX Flag` during the allocation of a channel.

`Interference Band to be Used` With this information element a BSC considers the maximum allowable interference band and the lowest allowable interference level during channel allocation.

3.6.9.3 Channel Activation

A BTS receives this message from its higher-ranking BSC when it is reserving a channel. It is provided with the necessary channel information (see Table 3.20).

`Channel Number` A BSC uses a Channel_Activation message to notify the BTS of the channel or subchannel available for transmission. If a channel

Table 3.20: Information elements of a Channel_Activation message [11]

Information element	Transmission direction	No. of octets
`Message Type`	BSC → BTS	1
`Channel Number`	BSC → BTS	2
`Activation Type`	BSC → BTS	2
`Channel Mode`	BSC → BTS	6
`Channel Identification`	BSC → BTS	4–10
`Encryption Information`[a]	BSC → BTS	3–20
`Handover Reference`[b]	BSC → BTS	2
`BS Power`[c]	BSC → BTS	2
`MS Power`[c]	BSC → BTS	2
`Timing Advance`[c d]	BSC → BTS	2
`BS Power Parameters`[e]	BSC → BTS	2
`MS Power Parameters`[e]	BSC → BTS	2
`Physical Context`[f]	BSC → BTS	2

[a]Only available when user data is to be ciphered.
[b]Only part of a Channel Activation message when a handover is taking place.
[c]Only required when the elements `BS Power`, `MS Power` and/or `Timing Advance` occur in order to establish the initial transmitting power and the content of the first layer-1 message header.
[d]Required during a channel change within a cell.
[e]Only required with *Power Control*. The maximum allowable transmitting power is indicated in the `BS` or `MS Power` element.
[f]This is an optional element that contains additional channel information.

is free, the BTS acknowledges the BSC's message with a `Channel Number` element as part of the message Channel_Activation_ack (see Section 3.6.9.4). The following channel types and combinations are possible with the channel number [11]:

- B_m + SACCH (+ FACCH)
- L_m + SACCH (+ FACCH)
- SDCCH/4 + SACCH
- SDCCH/8 + SACCH
- Uplink CCCH (RACH)
- Downlink CCCH (PCH and AGCH)

`Activation Type` The information element `Activation Type` is used to clarify the purpose of a channel activation and reservation, namely that a BSC has one of the following intentions [11]:

- *Intracell handover* (channel change within the same cell).

- *Intercell handover* (channel change between different cells), asynchronous or synchronous.
- *Call set up* in conjunction with an *immediate assignment procedure* or a *normal assignment procedure.*

`Channel Mode` The information element `Channel Mode` allocates the transmission modes for both interfaces (U_m- and A_{bis}-interfaces) to a BTS. For the functionality of the TRAU this means that an uplink or downlink DTX may have to be established (see Section 3.6.9.2). Both transmitter directions are identified by a flag (1 bit), with the value `1` standing for DTX and the value `0` indicating that DTX is not allowed.

The `Channel Mode` element also indicates the purpose of use and the type of radio channel requested. Since this parameter area corresponds totally to the `Channel Type` element of a Handover_Request message, the reader is referred to Section 3.6.9.2 for an exact description.

`Channel Identification` The `Channel Identification` element provides a BTS with a complete description of the channel to be reserved by the `Channel Description` element. With frequency hopping, a list of the channels available for alternating use, called `Mobile Allocation` element, is added. Both elements are part of the `Layer 3 Information` data field, which is forwarded to the MSC via a Handover_Request_ack message. An explanation of this parameter therefore appears in Section 3.6.9.4.

`Handover Reference` This information element allows an MS to identify itself as the authorized candidate by means of Handover_Access when a channel is being allocated. It contains an access number in the area between 0 and 255 (1 octet) that is allocated to the reserved channel either until the authorized MS has requested it or the handover process has been broken off.

So that the MS as well as the BTS are notified separately of the reference number, the BSC passes this information on to the BTS and, once a channel reservation has been made, to the MSC.

`BS Power` The parameter `BS Power` is used to adapt the BTS transmitter power. It reduces the transmitter power based on the existing nominal value (20–30 dB, depending on the manufacturer) in increments of 2 dB. This covers a range between 0 and 30 dB [11]. With this parameter, the network operator can have a direct influence on the particular BTS coverage area. Additional information on the mechanism of *power control* is provided under `Power Command` in Section 3.6.9.5.

`MS Power` The parameter `MS Power` gives the transmitter power for the mobile side.

Table 3.21: Information element of a Channel_Activation_ack message [11]

Information element	Transmission direction	No. of octets
`Message Type`	BTS → BSC	1
`Channel Number`	BTS → BSC	2
`Frame Number`	BTS → BSC	3

`Timing Advance` The information element `Timing Advance` (TA) contains a time displacement factor used by the MS to compensate for signal propagation delays that originate because of the distance between MS and BTS on the uplink and downlink channel. Because the BTS requires fixed synchronization between both channels (distance of three burst periods = 1.73 ms), the MS, by timely transmission, ensures that the signal propagation delays are offset.

The coding of the time displacement factor is carried out in an octet, with the value being the dimension of the bit duration (48/13 µs ≈ 3.69 µs) [15].

`BS Power Parameters` The data field `BS Power Parameters` contains information for the BTS transmit/receive facility (TRX) for regulating its transmitter power.

`MS Power Parameters` The `MS Power Parameters` establishes data field recommended values and limits for *Power Control* of a mobile station controlled by the BTS, analogous to `BS Power Parameters`.

`Physical Context` The data field `Physical Context` contains information about the current transmit/receive process related to the current TRX. It should be passed from one TRX channel to the next, for example in a handover. The BSC has no influence on this data. The extent and content of this data field are left to the manufacturer.

3.6.9.4 Channel Activation Ack

Through Channel_Activation_ack, a BTS announces to a BSC that, as expected, it was able to reserve the desired channel (see Table 3.21).

`Frame Number` This information element informs the BSC of the current TDMA frame number and consequently the current time grid of the BTS. In an asynchronous handover situation the BSC requires this information in order to obtain the value of the `Starting Time` element in the `Layer 3 Information` data field (see Section 3.6.9.5), which the MS uses for channel access. Four numbers that, according to a formula given in [15], can constitute an area between 1 and 2 715 648 represent the frame number in the `Frame Number` as well as in the `Starting Time` elements. The resulting frame is produced from a modulo operation between the generated number and 42 432.

Table 3.22: Information elements of a Handover_Request_ack message [10]

Information element	Transmission direction	No. of octets
`Message Type`	BSC → MSC	1
`Layer 3 Information`	BSC → MSC	9–56
`Chosen Channel`[a]	BSC → MSC	2
`Chosen Encryption Algorithm`[b]	BSC → MSC	2

[a]Is at least included when the BSC has made a channel selection.
[b]Is at least included when the BSC has selected a coding algorithm.

3.6.9.5 Handover Request Ack

If a suitable channel can be found in the desired target cell, the responsible BSC informs the higher-ranking MSC that it and the target BTS are able to support the handover by issuing Handover_Request_ack (see Table 3.22). It also notifies the MSC of the identity of the reserved channel (`Channel Description`) to which the MS is to be rerouted. A complete description of the channels and channel allocation is given in the data field `Layer 3 Information`.

`Layer 3 Information` This data field contains all the information about the new channel, including the parameters required for a channel allocation (see Table 3.23).

`Cell Description` This information element describes the target cell. Since the MS has been presynchronized with the target cell through the measurement protocol, the details on BCCH frequency and the BSIC are sufficient for establishing the cell identity. Both elements have a code length of one octet.

`Channel Description` This information element provides a general description of a channel [10]:

- Channel type (B_m-TCH, L_m-TCH, SDCCH),
- Type of allocated control channel (SACCH),
- Time slot number,
- Indicator for support of dynamic frequency change (*frequency hopping*),
- Channel number (1–124).

`Power Command` Similarly to DTX described in Section 3.6.9.2, dynamic power control increases spectral efficiency and battery operating time. The transmitter power is reduced to the extent that the signal quality is

Table 3.23: Information elements of a `Layer 3 Information` data field [15]

Information element	Transmission direction	No. of octets
`Cell Description`	BSC → MSC	2
`Channel Description`	BSC → MSC	3
`Handover Reference`	BSC → MSC	1
`Power Command`	BSC → MSC	1
`Synchronization Indication`[a]	BSC → MSC	1
`Cell Channel Description`[b]	BSC → MSC	17
`Channel Mode`[c] `Channel Description`[d]	BSC → MSC	4
`Channel Mode 2`[e]	BSC → MSC	2
`Frequency Channel Sequence`[f]	BSC → MSC	10
`Mobile Allocation`[g]	BSC → MSC	2–10
`Starting Time`[h]	BSC → MSC	3

[a]If omitted means there is an asynchronous handover.
[b]Only occurs when the target cell supports *frequency hopping*.
[c]Included if there is a change in the transmission mode of the allocated channel.
[d]Only appears if the mobile side is utilizing two dedicated channels in which case it allocates the second channel.
[e]Included if there is a change in transmission mode in the optional `Channel Description` element.
[f]Is a combination of the allocation and channel descriptions intended for systems using *frequency hopping*. The data quantity is adapted so that a Handover_Command can be transmitted in a single signalling block. `Cell Channel Description` and `Mobile Allocation` are then omitted.
[g]Only necessary with *frequency hopping* and generally can be used with all channel types, but appears in conjunction with a `Cell Channel Description` element.
[h]Available when a change in carrier frequency occurs with a channel change.

still acceptable. Because co-channel interference depends on the transmitter power of the channels concerned, it can sometimes be reduced through power control.

In the GSM system the transmitter power in both directions of the radio route between BTS and MS (uplink and downlink) is controlled separately by the base station subsystem (BSS). Depending on the type of terminal equipment, the control range for the uplink channel is between 20 and 30 dB, and, depending on the manufacturer, for the downlink channel it is up to 30 dB. The transmitter power is controlled bidirectionally every 60 ms in 2 dB increments.

The BSS calculates the transmitter power required by the mobile side from the receive level measured by the BTS and the maximum possible transmitter power for a terminal, at the same time taking into account the signal quality measurement, for which the BTS is also responsible. For the downlink the BSS determines the required transmitter power for the BTS through an evaluation of the periodic detailed measurement report of the MS.

The use of power control in the BSC or in the BTS is up to the manufacturer. The associated BTS-BSC protocol is only specified through dummies.

For handover the BSC establishes the transmitter power for both the mobile station and the fixed network. But the terminal equipment power is usually predetermined by the cell, and is passed on to the target BTS in the `MS Power` element of the `Channel_Activation` message as well as to the MS through a `Power Command` in the `Layer 3 Information` data field. In the `Power Command` 5 bits are reserved for the coding of the signal power level, thereby enabling it to assume 32 different values [15].

`Synchronization Indication` This element indicates at handover whether or not the target cell is synchronized with the original cell. It is coded with one bit, and with unsynchronized cells it has the value `0` and otherwise the value `1` [15].

`Cell Channel Description` The element `Cell Channel Description` is a data field that provides a complete overview of all available and non-available channels in the new cell. Channels 1–124 are numbered in sequence and indicated with `1` if they belong to the cell and with `0` if they do not [15].

`Channel Mode` This element indicates the coding and decoding modes, and differentiates between the following transmission modes [15]:

- signalling only
- speech full-rate channel (B_m-channel)
- data channel
- speech half-rate channel (L_m-channel)

`Frequency Channel Sequence` This data element contains a channel sequence that should be used in frequency hopping. In total, 16 different channels are entered, and the one with the lowest channel number (1–124) is explicitly listed again. Two channel numbers always form an octet [15].

`Mobile Allocation` The data field `Mobile Allocation` contains all the channels listed in the `Cell Channel Description` that belong to the new cell (channel number is indicated as `1`), identifying all the channels that can be used in the frequency hopping of the MS with `1` and all others with `0`. It contains a standard form of eight channels per octet. If the number of overall channels available to a cell is less than a multiple of eight then the remaining positions in the third octet are stuffed with `0` [15].

`Starting Time` This informs an MS with which TDMA frame it can make a connection. The calculation of the frame number is implemented in the same way as the `Frame Number` element described in Section 3.6.9.4.

Table 3.24: Information elements of a Handover_Command message [10]

Information element	Transmission direction	No. of octets
`Message Type`	MSC → BSC	1
	BSC → BTS	
	BTS → MS	
`Layer 3 Information`	MSC → BSC	9–56
	BSC → BTS	
	BTS → MS	

Table 3.25: Information elements of a Handover_Access message [15]

Information element	Transmission direction	No. of octets
`Handover Reference`	MS → BTS	1

Table 3.26: Information elements of a Physical_Information message [15]

Information element	Transmission direction	No. of octets
`Message Type`	BTS → MS	1
`Timing Advance`	BTS → MS	1

The MS can then be instructed to wait to transmit for a maximum of up to 42 431 frame durations (around 195.8 s) [15].

`Chosen Channel` This informs an MSC which of the following channel types has been allocated to the mobile side [10]:

- SDCCH
- L_m-TCH
- B_m-TCH

`Chosen Encryption Algorithm` This information element informs an MSC whether the BSC supports ciphering and which of the seven different A5 types is being used [10].

3.6.9.6 Handover Command

After a layer-3 Handover_Request_ack message has been received by an MSC to acknowledge a channel query, the MSC forms a Handover_Command from it that essentially only includes the BSC data field `Layer 3 Information` (see Table 3.24). Via BSC and BTS, this message reaches the relevant MS, which receives all the information it requires about the channel characteristics and access parameters through the data field mentioned.

Table 3.27: Information elements of a Handover_Complete message [15]

Information element	Transmission direction	No. of octets
`Message Type`	MS → BTS	1
`RR Cause`	MS → BTS	1

3.6.9.7 Handover Access

With Handover_Access an MS sends a request to the target BTS to allocate the reserved channel to it. This layer-3 message only contains the `Handover Reference` element (1 octet) known from the `Layer 3 Information` data field, which is initially sufficient for the identity control of the MS in order to allocate a channel (see Table 3.25). A more comprehensive authenticity test of the MS and the user takes place after handover.

3.6.9.8 Physical Information

A Physical_Information message is directed by the network to an MS to instruct the mobile side to stop sending a Handover_Access message and to activate the reserved channel for the MS (see Table 3.26).

3.6.9.9 Handover Complete

Once a connection has been established between MS and BTS layer 2 through the exchange of an SABM and a UA frame, the MS notifies the network with Handover_Complete that it has completed the handover successfully (see Table 3.27). This message is sent back over the network entities to the MSC that originated the Handover_Command [10].

`RR Cause` An MS uses this information element to notify the network of the reason why it has left the original channel. A number of different reasons and the related coding are mentioned in [15, Section 10 and Appendix F].

3.6.9.10 Clear Command

If a handover has been terminated by the mobile side with Handover_Complete then the MSC has the task of also completing the handover process on the network side. To do so, it sends the BSC responsible for the original cell a Clear_Command in order to initiate the release of the allocated radio resources there (see Table 3.28). The BSC then forwards this message to the relevant BTS.

`Layer 3 Header Information` This element informs a BSS of the content of a layer-3 message header that is to be used for sending user data over the

Table 3.28: Information elements of a Clear_Command message [10, 15]

Information element	Transmission direction	No. of octets
`Message Type`	MSC → BSC BSC → BTS	1
`Layer 3 Header Information`[a]	MSC → BSC BSC → BTS	4
`Cause`	MSC → BSC BSC → BTS	3–4

[a]Serves no particular purpose and is an optional element. However, it should be sent all the same in order to facilitate the functional process between network elements unless the MSC is certain that the BSC can manage without it. This will no longer constitute part of this message in any future GSM recommendation.

Table 3.29: Information elements of a Clear_Complete message [10]

Information element	Transmission direction	No. of octets
`Message Type`	BTS → BSC BSC → MSC	1

radio interface. At this point in time, however, it is of no significance in the handover process.

`Cause` The `Cause` element has already been briefly described in Section 3.6.9.1. The following are some typical reasons for releasing a channel [10]:

- *Call Control*
- *O & M Intervention*
- *Equipment Failure*
- *Handover Successful*
- *Protocol Error between BSC and MSC*

3.6.9.11 Clear Complete

As soon as the BTS has released the channel allocated to the MS that is being supplied by another cell after handover is completed on the mobile side, it immediately reports the process to its higher-ranking BSC with a Clear_Complete (see Table 3.29).

The BSC then forwards this message to the MSC, thereby also concluding the network part of the handover protocol.

3.6.10 Inter-MSC Handover

In an inter-MSC handover the channel is switched to a radio cell managed by another MSC. There are two different scenarios:

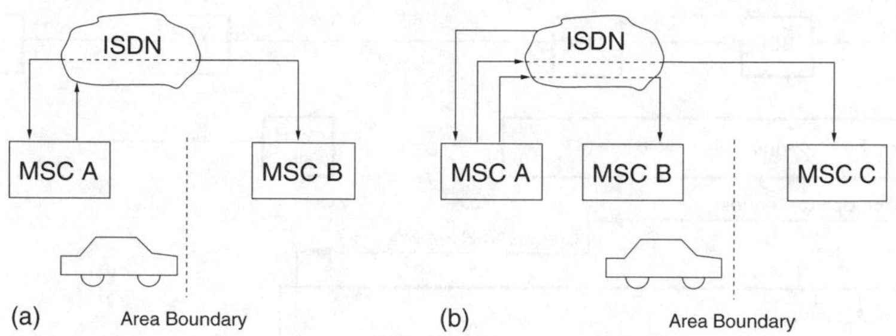

Figure 3.55: (a) First area change (*basic handover*). (b) subsequent area change (*subsequent handover*)

- In a *basic handover*, handover first takes place from one MSC service area to another (see Figure 3.55a). In the process the connection from MSC A is routed to MSC B over the wireline ISDN network.
- A *subsequent handover* occurs when at least one *basic handover* has taken place (see Figure 3.55b). Also in this case MSC A continues to be responsible for controlling the connection. Here too the connection is routed by MSC A over the fixed ISDN network to MSC C.

3.7 Location Update

When a mobile user leaves a location area, the HLR and the VLR must be updated immediately to ensure that the user can be reached no matter which location he is currently moving in. The mobile station continuously evaluates the receive quality of all BSs capable of receiving, and logically attaches itself to the one with the strongest signal. If in the process it has allocated itself to a base station that is located outside the previously valid location area, it initiates a *location update*, thereby informing the network of its new location. Two different situations can arise:

- Change within the same VLR area.
- Change of VLR area.

If there is a change in VLR area at the same time the location area changes, and the sequence of the process is as is shown in Figure 3.56. Because no entry for the mobile station exists in the new VLR, its data must be transmitted from the HLR and deleted in the old VLR. If a mobile station requests a location update, a four-character long temporary mobile subscriber identity (TMSI) as well as the old location area identification (LAI) usually also exists. This makes it possible to determine which was the former VLR and consequently also the international mobile subscriber identity (IMSI) of the MS.

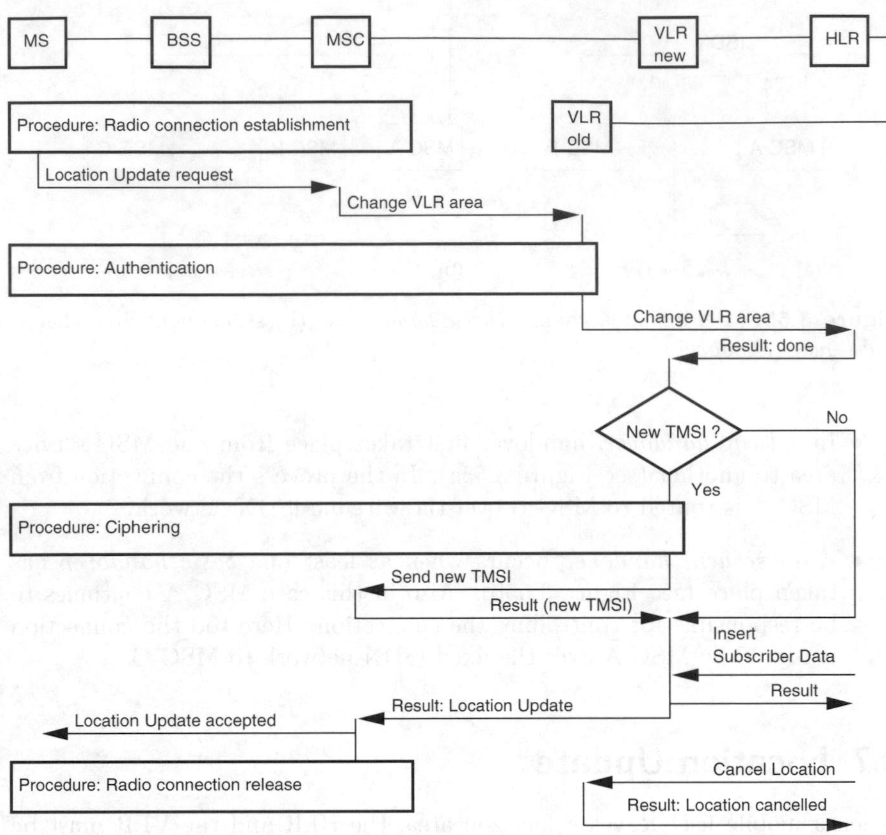

Figure 3.56: Sequence of location update procedure

Alternatively the mobile station can also immediately be requested to transmit its IMSI. After the user has been authenticated, the new VLR informs the HLR about the location change by sending an Update_Location message. The HLR subsequently sends the necessary user data to the new VLR using Insert_Subscriber_Data.

Cancel_Location then deletes the subscriber's entry in the former VLR. Through Location_Updating_Accepted the new VLR confirms to the mobile station that the transfer has taken place.

3.7.1 Roaming Support

Roaming is a service provided by a mobile radio network allowing a mobile user freedom of movement while still being within reach of coverage. It allows subscribers to make calls and to receive calls, and independently to make use of several services at the same time. There is a difference between roaming within

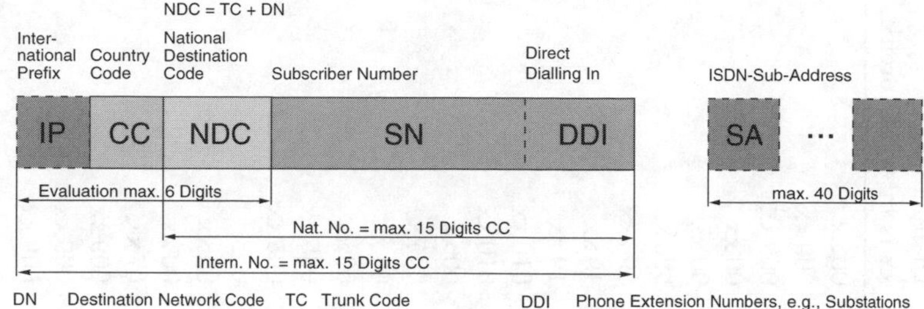

Figure 3.57: Numbering structure according to ITU-T Rec. E.164

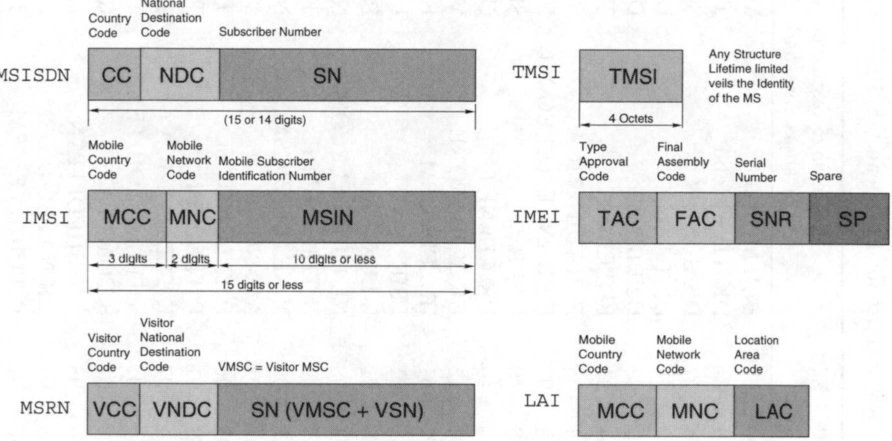

Figure 3.58: Structure of GSM numbers

the same network (across location areas), national roaming (between different networks—this is not yet supported, however) and international roaming.

3.7.2 Numbering Plan for Roaming

The numbering in GSM follows the rules of ITU-T Rec. E.164 for ISDN (see Figure 3.57). In addition to an MSISDN telephone number, a mobile subscriber has another number—an IMSI with a maximum length of 15 digits that provides him with unique identification. The numbers are structured as follows (ITU-T Rec. E.212); see Figure 3.58:

- A *Mobile Station International ISDN Number* (MSISDN) consists of the home *Country Code* (CC), the *National Destination Code* (NDC)—identifies the home PLMN and is therefore the HLR address—and the *Subscriber Number* (SN) in the HLR; thus MSISDN=CC+NDC+SN.

Table 3.30: Country and network codes for European GSM networks

Country	Network operator	MCC	MNC	Message displayed or short message		Dialling code xx is internat. prefix
Belgium	RTT Belgacom	206	01	BEL MOB-3	MOB-3	00 xx
Denmark	Tele Danmark Mobil	238	01	DK TDK-Mobil	TD MOB	009 xx
Denmark	Dansk Mobil Telefon	238	02	DK SONOFON	SONO	009 xx
Germany	Deutsche Telekom AG	262	01	D1-Telekom	D1	00 xx
Germany	Mannesmann Mobilfunk	262	02	D2 PRIVAT	D2	00 xx
Finland	Telecom Finland	244	91	FI TELE FIN	TELE	990 xx
Finland	Oy Radiolinja AB	244	05	FI RADIOLIJNA	RL	990 xx
France	France Telecom	208	01	F FRANCE TELECOM	FT	19 xx
France	SFR	208	10	F SFR	SFR	19 xx
Great Britain	Cellnet	234	10	UK CELNET	CLNET	010 xx
Great Britain	Vodafone	234	15	UK VODAFONE	VODA	010 xx
Ireland	Eircell	272	01	IRL EIR-GSM	E-GSM	00 xx
Italy	SIP	222	01	I SIP	I SIP	00 xx
Luxemburg	P&T Luxembourg	270	01	L LUXGSM	P&T L	00 xx
Netherlands	PTT Telecom	204	08	NL PTT TELECOM	NL PTT	09 xx
Norway	Tele Mobil Norway (Nor. Telecom)	242	01	N TELE-MOBIL	TELE	095 xx
Norway	NetCom GSM A/S	242	02	N NETCOM GSM	N COM	095 xx
Austria	PTV Austria	232	01	A E-Netz	MN-E	0 xx
Portugal	Telecom. Moveis Nacionals	268	06	P TELEMOVEL	TMN	00 xx
Portugal	Telecel	268	01	P TELECEL	TLCL	00 xx
Spain	Telefonica	214	07	E TELEFONICA	TLFCA	07 xx
Sweden	Televerket Radio (Sw. Telecom)	240	01	S TELIA MOBITEL	TELIA	009 xx
Sweden	Comvik GSM AB	240	07	S COMVIQ	IQ	009 xx
Sweden	AB Nordic Tel	240	08	S EURIPOLITAN	EURO	009 xx
Switzerland	Swiss PTT Telecom	228	01	CH NATEL D GSM	NAT D	00 xx
Turkey	PTT Turkey	286	01	R TURKIYE GSM	TR GSM	00 xx

- The *International Mobile Subscriber Identity* (IMSI) consists of the home *Mobile Country Code* (MCC), the code for the mobile network, i.e., the HLR address *Mobile Network Code* (MNC), as well as the *Mobile Subscriber Identity* (MSIN) in the HLR. Table 3.30 lists the MCC and MNC codes for the current European GSM networks.

In addition, a temporary address, the *Mobile Station Roaming Number* (MSRN), provides the link to the mobile subscriber's current location. An MSRN is linked to location, is temporarily allocated by a VLR (and stored in the HLR) at the request of the MSC, and is required to make a connection from the fixed network to a mobile station over the GMSC. It consists of the country code of the visited mobile network (*Visited Country Code*, VCC), the local code (area that the user is currently visiting (*Visited NDC*, VNDC), identification of the *Visited MSC* (VMSC) and the *Visited Subscriber Number* (VSN) allocated by the VLR. The MSRN protects the identity and the location of the mobile user from being discovered through eavesdropping on the signalling traffic.

If a mobile station wishes to send a signal in the network, it uses a local temporary number for identification, a so-called *Temporary Mobile Subscriber Identity* (TMSI), to conceal the IMSI. The TMSI is allocated and coded and transmitted together with an *Location Area Identifier* (LAI) from the current *Visitor Location Register* (VLR) to the mobile station. Therefore a TMSI only has local significance and is only valid within the area administered by a VLR. A TMSI is periodically changed to ensure the confidentiality of the mobile subscriber's transmitted information. The LAI in the first two elements is constructed in the same way as the IMSI; the third element contains the *Location Area Code* (LAC).

Roaming has a constant high requirement for signalling resources in a mobile radio network, even if a user is not communicating. Therefore the current location and access rights of a mobile user must constantly be updated and, if required, be specified. Because roaming not only necessitates a change in location area in the home PLMN but also between international networks, several networks are sometimes affected.

3.8 Connection Set Up

Similarly to ISDN, a service-integrated GSM mobile radio network allows different telecommunications services to be provided on the same subscriber line. Therefore, in addition to a destination address, the terminal placing a call must also specify details on the desired transmission characteristics of a connection and the conditions of compatibility for the terminal called. The called unit can then select the terminal equipment suitable for the desired service. In contrast to ISDN, there are additional requirements concerning the specification of the mobile subscriber's location address with a mobile-terminated call and the selection of special interworking functions between the

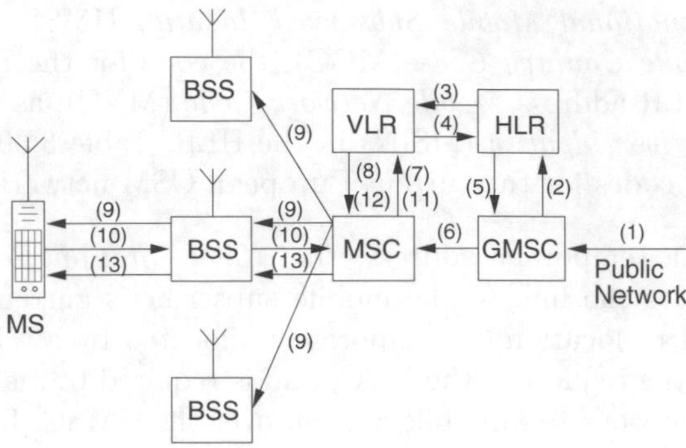

Figure 3.59: Mobile-terminated call

protocols used at the radio interface of the mobile network and the protocols of the fixed network, e.g., ISDN.

3.8.1 Mobile-Terminated Call

If a caller from the fixed network wants to reach a mobile subscriber, he must dial his ISDN number (MSISDN). If the switching centre of the fixed network recognizes that the number is one within a PLMN, the call with an *Initial Address Message* (IAM) is routed from the public network to the next *Gateway Mobile-services Switching Centre* (GMSC) (1) (see Figure 3.59).

From the code NDC contained in the IAM message, the GMSC establishes the mobile subscriber's home location register (2). The HLR checks that the number exists and, if the services desired were also included in the IAM message, that the mobile subscriber is authorized to use the services. The HLR also asks the VLR to provide a mobile station roaming number MSRN for the pending connection request (3).

After receipt of the MSRN (4), which the HLR uses to determine which MSC is responsible for the current location area, it communicates the number to the GMSC (5), which then establishes a connection to the responsible MSC (6). The MSC instructs the VLR to forward the availability status of the mobile station (7). If the MS is available, the VLR notifies the MSC (8), which then initiates a paging call to the subscriber in all the radio cells of the MS's location area (9). If the mobile subscriber answers (10), then, after all security procedures have been completed (11), the MSC is instructed by the VLR (12) to establish the connection in the radio network (13).

3.8.2 Mobile-Originated Call

A mobile subscriber in a GSM-PLMN can request a connection at any time. A mobile-originated call begins with an operation that initiates access to the network and informs the VLR of the type of request and the transmission

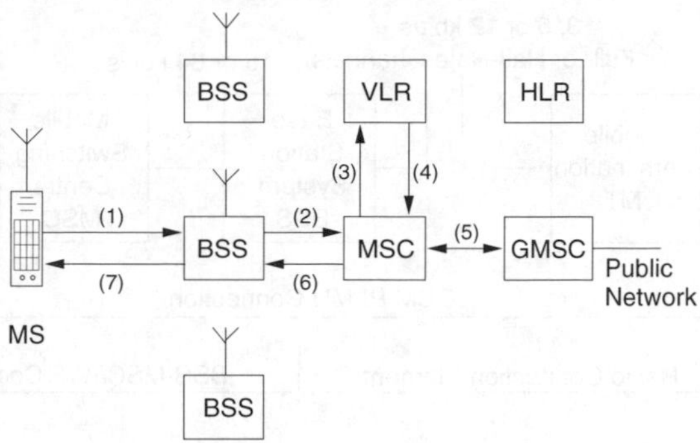

Figure 3.60: Mobile-originated call

ciphering key currently stored in the mobile station along with the subscriber identification. To avoid an unnecessary occupancy of the radio channel, the desired call number is indicated in a display on the terminal equipment so that in case of input error the user may correct the number before pressing the calling key.

The steps that take place during a connection request are illustrated in Figure 3.60. After completion of the required security procedures and the resulting call request (1), the mobile station making the call forwards a RIL3_CC_Setup message to the MSC containing the telephone number of the called subscriber as well as requirements in respect of quality of service of the network path and compatibility of the subscriber unit being called (2).

To meet the desired requirements, the MSC must check the mobile subscriber's authorization by sending a query to the VLR (3 + 4) as well as reviewing the resources available to it (IWF, free line to fixed network, etc.) (5). If the MSC has been successful, it allocates the resources to the connection, selects the necessary transcoding function (IWF), and, if a connection has been established successfully, sends the mobile station a RIL3_CC_Alerting message (6 + 7), which is converted into an alert signal in the mobile station.

3.9 Data Transmission and Rate-Adaptation Functions

Figure 3.61 illustrates the principles of a GSM-PLMN connection. Different adaptation functions must be carried out to ensure that a connection has been established correctly. In respect of the physical interface, the so-called S-interface (ITU-T: I.430-series; see Figure 3.3) is to be supported. Mobile stations are connected via two 64 kbit/s B-channels as traffic channels and a D-channel as the control channel. However, the whole range of ISDN services cannot be offered because of the restricted bandwidth of the radio channels.

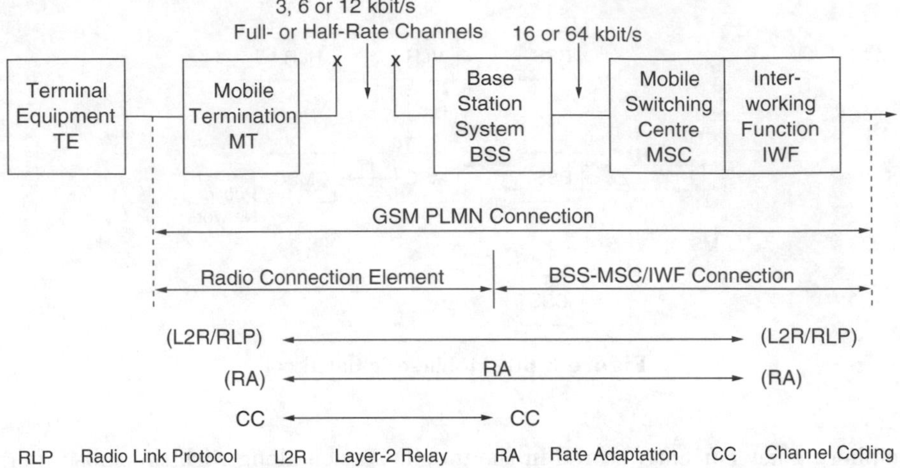

Figure 3.61: Connection for data transmission in PLMN

Conventional R-interfaces are also supported. The data interfaces are supported according to ITU-T X.21, X.21bis, X.25 and V.24.

3.9.1 Rate Adaptation to Traffic Channel Performance

Rate Adaptation (RA) is defined according to ITU-T X.30/V.110. Modifications were required because the radio interface only supports data rates lower than 64 kbit/s. ITU-T describes a three-level process with the functions:

RA 0 This converts an asynchronous input stream into a synchronous output stream of $2^n \cdot 300$ bit/s.

RA 1 This function is dependent upon synchronous input data as well as status information on the subscriber interface, control information and synchronization patterns. Rate adaptation is then carried out through multiplexing on an 8 or 16 kbit/s output stream in the form of 80 bit-blocks. In ISDN the rates 8/16 kbit/s are called intermediate rates.

RA 1' This modification of RA 1 produces the GSM intermediate rates of 12, 6 and 3.6 kbit/s. Intermediate rates are produced through the elimination of the synchronization patterns and control information. The latter is explicitly conveyed over the L_m-control channel (also called the D_m-channel). With the 3.6 kbit/s output rate even some redundant data bits are removed.

RA 2 Through so-called bit stuffing, this function forms a datastream of 64 kbit/s from the blocks received. This third level is only used in the base station.

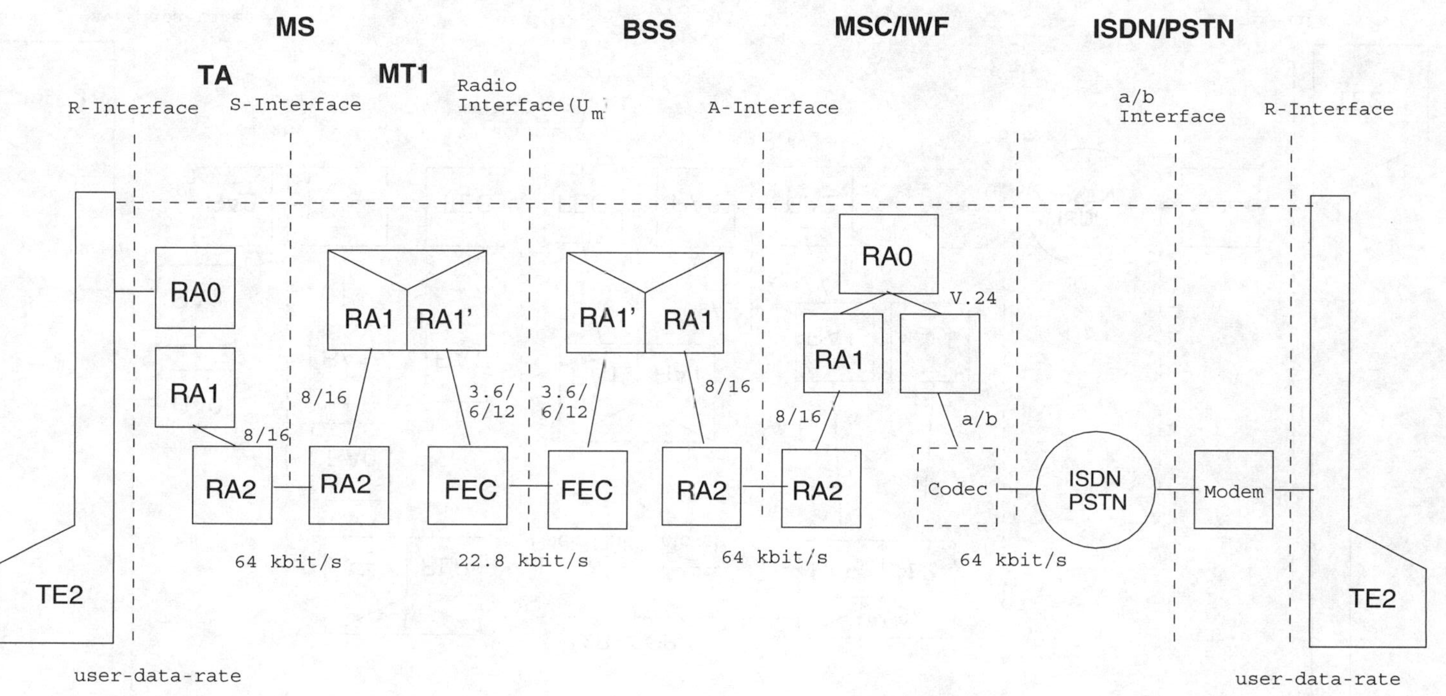

Figure 3.62: Example of asynchronous transparent data transmission

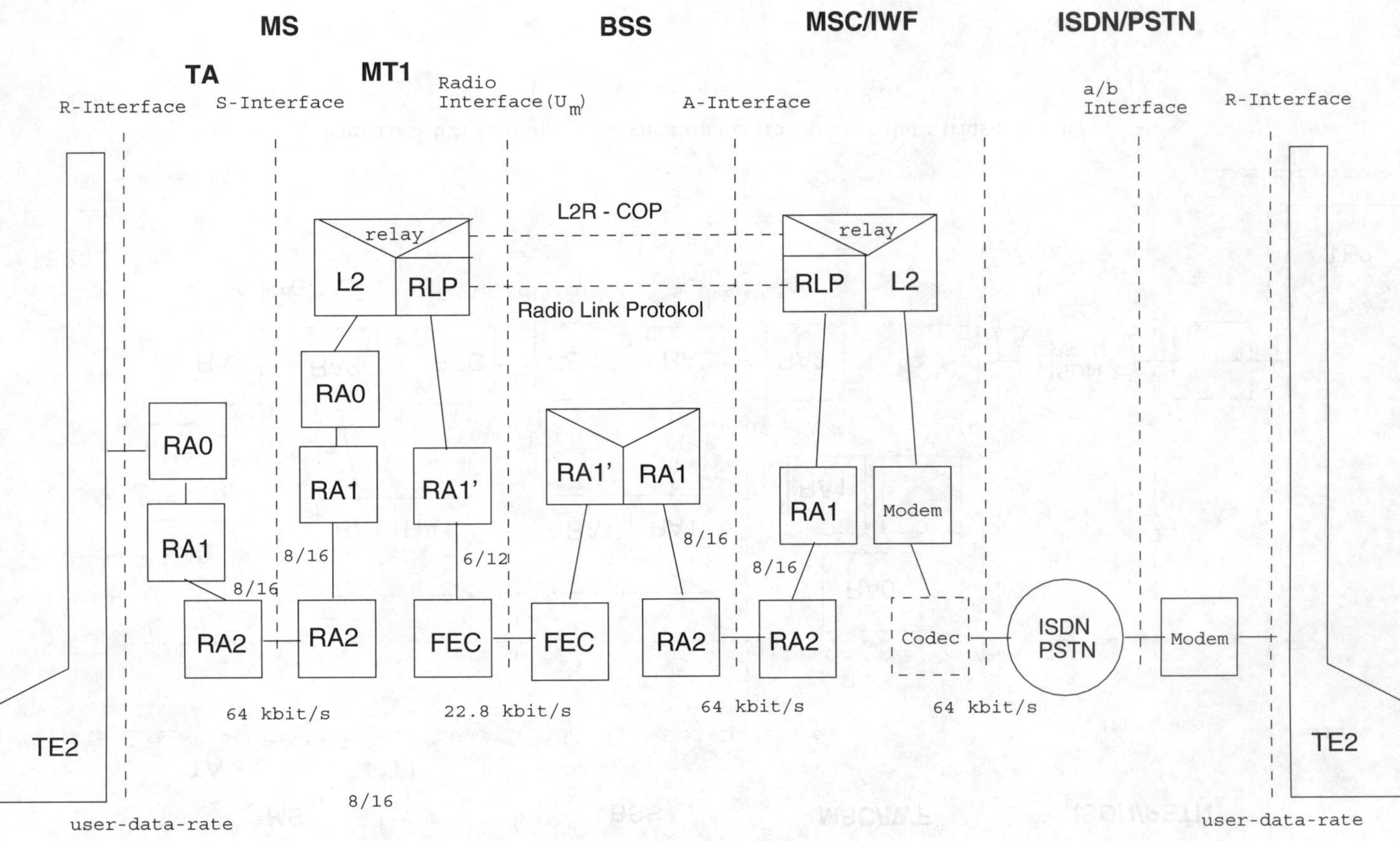

Figure 3.63: Example of asynchronous non-transparent data transmission (see Section 3.9.4)

Figure 3.62 shows the application of rate adaptation with transparent data transmission and the R-interface; Figure 3.63 shows a case of non-transparent data transmission with the R-interface. The S-interface to which a mobile station with an ISDN interface can be directly connected also appears in both illustrations.

3.9.2 Rate Adaptation in the Connection BTS/Transcoder to an MSC or MSC/IWF

Because an MSC can only switch 64 kbit/s channels, the GSM intermediate rates must be increased to 64 kbit/s through function RA 2, which is usually allocated to the speech-transcoder function (TRAU). The functions for (non-)transparent services are almost identical. Bits that in transparent transmission indicate the transmission rate are used in transparent transmission to indicate the alignment of the *Radio Link Protocol* (RLP) blocks and to control DTX, if applicable.

3.9.3 Layer-2 Relay Function and Radio Link Protocol

A traffic channel transmits with error correction (FEC) (see Section 3.3.7.1). The RLP function (see Section 3.9.4) offers an *Automatic Repeat Request* (ARQ) protocol that extends from the mobile station to the network *Interworking Functions* (IWF) in the MSC in order to detect errors that were corrected/not corrected by the FEC procedure and to eliminate them by repeating the transmission (see Figure 3.63). The *Layer-2 Relay* (L2R) function converts the layer-2 protocol of the terminal equipment into a *Connection-Oriented L2R Protocol* (COP) that uses transmission protected by an RLP. RLP and L2R functions are only available in the mobile radio network termination and in the network gateways, because only MSCs with IWFs are the main interworking points (GMSC) so that problems of protocol synchronization can be kept to a minimum.

3.9.4 Radio-Link Protocol (RLP)

A bearer service with error correction (FEC) is sufficient for transparent telecommunications services in a GSM network. Some applications, however, require a higher level of transmission quality that needs an ARQ protocol, which means that a bearer service with error detection is being used. For the non-transparent service a layered full-duplex protocol, called the *Radio Link Protocol* (RLP), is used, which corresponds to layer 2 of the ISO/OSI reference model.

3.9.4.1 Characteristics of the Radio Link Protocol

The *Radio Link Protocol* supports the transmission of user information up to 9.6 kbit/s without any noticeable delay. According to GSM Rec. 04.22,

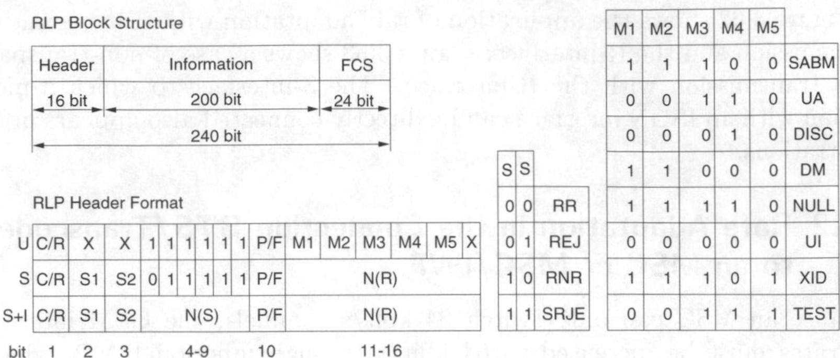

Figure 3.64: Protocol data units of RLP

throughput can typically be maintained in 90 % of a cell area for 90 % of the time, with a maximum delay of 120 ms per direction. The FEC method of the bearer service used offers a *Bit Error Ratio* (BER) of approximately 10^{-3} at 9.6 kbit/s and approximately 10^{-7} at 4.8 kbit/s with a full-rate traffic channel. With the ARQ mechanism a BER of approximately 10^{-7} at 9.6 kbit/s is possible. The RLP was further developed from the HDLC protocol (ISO IS 4335) because radio network characteristics differ from those of wireline connections for which HDLC was intended. Figure 3.64 shows the three formats of the RLP protocol data units and the control commands used. U- and S-blocks do not carry information. The main differences with HDLC are:

- No synchronization patterns (flags) are used.
- Blocks are numbered modulo 64 instead of modulo 8 or 128.
- The check sum (*Frame Check Sequence*, FCS) has been extended from 2 to 3 bytes.
- RLP permits simultaneous transmission of user and status information (see S+I-block).
- There are no addresses, because of exclusive use of only one connection at a time.
- Commands/responses are differentiated by a C/R-bit, the P/F-bit (`poll/final`) is set accordingly.
- An X-bit (E, *evolutionary*) is reserved (e.g., for point-to-multipoint connections).

3.9.4.2 Functioning of the Radio Link Protocol

Stations can establish, cancel or terminate a connection at any time, which is called *Asynchronous Balanced Mode* (ABM) *Procedure.* RLP supports point-

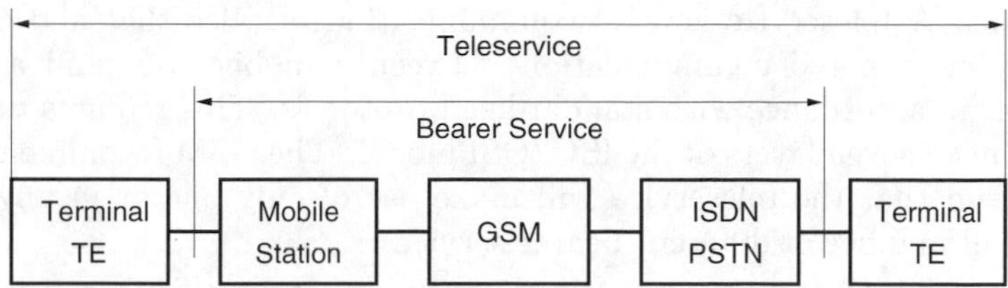

Figure 3.65: Relationship between bearer services and teleservices in a point-to-point connection in GSM

to-point duplex connections. The RLP entity can be found either in *Asynchronous Disconnected Mode* (ADM) or in ABM in which data transmission is taking place. U-blocks are unnumbered control blocks for establishing or terminating a communication link. Their function is coded in the "MMMMM" field. S-blocks (*supervisory*) are control blocks with sequence numbers; they acknowledge data blocks that have been received correctly and initiate the retransmission of data blocks detected to have errors or temporarily suspend the transmission of data. Their function is coded in the "SS" field. The definitions of U- and S-blocks with RLP are closely similar to those of HDLC. In RLP information blocks, called I-blocks in HDLC, also carry control information. They are therefore called S+I-blocks. The N(R)-fields (*Receive Sequence Number*) in S- and S+I-blocks report on receiving status. S+I-blocks also have an N(S)-field (*Send Sequence Number*) for numbering transmitted blocks modulo 64.

3.10 Services in the GSM Mobile Radio Network

An important element in the conception of the GSM was the definition of the services standardized in the network (see Figure 3.65).

The GSM network enables the integration of different voice and data services and also offers gateway functions for interconnection to other telecommunications networks for voice and data transmission. In addition to the voice service, the GSM recommendations are proposing a phased introduction of data and teleservices.

The telecommunications services offered by GSM are divided into three main categories:

Bearer Services Bearer services are all telecommunications services that allow mobile subscribers to transmit signals in bit-transparent mode between the mobile subscriber and a user of any partner network. They only offer pure transport services as defined for the lower three layers of the ISO/OSI reference model, namely connection-oriented channel and packet-switched data transmission.

Teleservice A teleservice is a telecommunications service that allows applications-oriented communication between a mobile user and a second user in accordance with standardized protocols. This requires protocols from all seven layers of the ISO/OSI model. The GSM recommendations intend that the teleservice will make use of only one or in any case a small number of different bearer services.

Supplementary services The supplementary services are further facilities that are offered to subscribers in addition to the services mentioned above. These are not independent services, and are always offered in conjunction with the teleservices or the bearer services.

Support for Value-Added Services Value-added services like *Tariff Area Indication*, *E-mail Access*, *Electronic Banking* or *Travel Information* are not standardized in the GSM specifications. However, in the GSM specifications two mechanisms are foreseen to facilitate the provision of operator-specific value-added services.

Figure 3.65 shows a point-to-point connection in GSM and the relationship between bearer services and teleservices.

3.10.1 Service Introduction Phases

Not all the services specified in the standard can and will be available to subscribers at the same time the network is operational. Therefore GSM Rec. 01.06 presents a concept for a phased introduction of the different services according to their importance and the anticipated market situation. The services are classified accordingly as *essential* (E) or *additional* (A):

E-service must be provided by all mobile communications networks (PLMN).

A-service can be offered by the network operator.

E-services are relegated to one of three introduction phases (see Table 3.31), with the index providing the values (1,2,3) to indicate which services must be offered first. Thus E1-services were already available at the same time when most GSM networks were implemented, whereas the subsequently ranked services, with consideration of upward compatibility, were introduced at a later time. On the one hand, this approach takes into account the requirements of network operators who want to provide their services in line with prevailing market conditions, and, on the other hand, it addresses the necessity for harmonization in the telecommunications area.

3.10.2 Bearer Services

GSM supports different variants of data transmission, thus functioning as a *bearer* for data transmission. The bearer services allow GSM to be used for non-voice services up to 9600 bit/s. Because voice transmission is expected

Table 3.31: The introduction phases for services in the GSM network

Phase	Introduction in Europe	Services
E1	1991	Telephone services as well as some supplementary services
E2	1994	Telephone services and a limited number of data and supplementary services
E3	1996	Telephone services, advanced data and supplementary services

to represent the main traffic component and it is projected that data services will equate to less than 10 % of the overall transmission volume, only some of the bearer services were classified as *essential*.

In GSM networks a radio path might be interrupted briefly owing to radio shadowing. Because of the redundancy available in natural speech, this is less serious for voice transmission than it is for data transmission. Appropriate measures such as error-recovery protocols are necessary for the data services to eliminate the effects of interference.

The bearer services support transparent and non-transparent, synchronous or asynchronous data services. In a non-transparent service the time transparency is violated, i.e., data experiences variable delays and throughput is dependent on the current signal quality seen by an MS.

The transparent bearer service This is one in which only functions of ISO/OSI layer 1 are involved and none of the higher layers.

A user is allocated a traffic channel without the support of ARQ protocols but with a defined transmission rate. Because a transparent bearer service uses a channel coding method with *Forward Error Correction* (FEC), data is transmitted with constant throughput and constant delay. The use of error-correction methods with different code ratios produces the following possible data rates:

- With a full-rate channel: 9.6 / 4.8 / 2.4 kbit/s.
- With a half-rate channel: 4.8 / 2.4 kbit/s.

The quality of a connection (measured on the basis of its bit-error ratio) depends on the quality of the radio channel.

The non-transparent bearer service This uses protocols whose functions correspond to layer 2 of the ISO/OSI reference model in order to protect communication through error detection and repeated transmission and to optimize throughput by data flow control.

A non-transparent service is based on a transparent bearer service and also uses the *Radio Link Protocol* (RLP). In addition to the *reject* command, where transmission is repeated once errors have occurred of all the received frames starting with the first frame number indicated in the reject command, it also allows the *selective-reject* mechanism in which only the disrupted frame needs to be repeated (see Section 3.9.4 and Figure 3.64).

Unlike HDLC, control commands can also be transmitted in information frames. The maximum window size of 61 is adapted to the delay caused by bit interleaving.

This hybrid ARQ protocol results in a residual bit-error probability of $< 10^{-7}$, with throughput and transmission delay times subject to strong fluctuations due to radio field conditions. With a non-transparent data service the error-correction method FEC corresponds to that with a data rate of 9.6 kbit/s, although equipment can also transmit at a lower bit rate.

The following are some of the services available as bearer services:

- Data circuit-switched, duplex-asynchronous (E2/A)
- Data circuit-switched, duplex-synchronous (A)
- Circuit-switched access to *Packet Assembly Disassembly* (PAD) of a packet-switched data network (Datex-P) (E2/A) (see ITU-T Recs. X.25, X.28, X.29)
- Synchronous duplex access to a packet-switched data network (E3/A)
- Alternate speech/unrestricted digital
- Speech followed by data
- 12 kbit/s unrestricted digital.

3.10.2.1 Data Circuit-switched, Duplex-Asynchronous

In cooperation with ISDN or the telephone network, which throughout Europe offers a full-duplex data transmission rate of 1200 bit/s, GSM is planning to provide bearer services with a transmission rate of between 1200 bit/s and 9600 bit/s. In the beginning two services at 300 bit/s and 1200 bit/s were supported by GSM through an appropriate gateway into the telephone network (PSTN). Because of the anticipated high demand and the relatively simple implementation, these bearer services were classified as E2.

3.10.2.2 Data Circuit-switched, Duplex-Synchronous

In the analogue telephone network (PSTN) and in the circuit-switched data network four synchronous data transmission rates of 1200, 2400, 4800 and 9600 bit/s are used. Correspondingly, four PLMN bearer services have been standardized for GSM.

Table 3.32: Bearer services in the GSM network

Bearer service	Transmission rate	Mode	Partner network
Data cda	300–9600 bit/s	T/NT	PSTN/ISDN
Data cds	1200–9600 bit/s	T	PSTN/ISDN/CSPDN
PAD cda	300–9600 bit/s	T/NT	PSPDN
Data cda	2400–9600 bit/s	T/NT	PSPDN

cda	circuit data asynchronous	CSPDN	Circuit-Switched Data Network
cds	circuit data synchronous	PSPDN	Packet-Switched Data Network

3.10.2.3 Circuit-switched Access to PAD

Access from data terminal equipment over the public telephone network to the data packet-switched service (*Packet Assembly Disassembly*, PAD) has been regarded as an attractive and easily implementable option and given the classification E2. Two bearer services with transmission rates incoming/outgoing of 300/300 bit/s and 1200/1200 bit/s have been standardized for this purpose. A service with 1200/75 bit/s (videotex) has also been specified, and although it is important in the fixed and mobile networks of some countries, it has been classified as A.

3.10.2.4 Data Packet-Switched, Duplex-Synchronous

Packet-switched data network services with transmission rates of 2400, 4800 and 9600 bit/s provide the only option for transmitting data with high rates and quality of service to a CEPT-wide area. Consequently PLMN bearer services that use ARQ protocols with flow control were defined. Although important, these bearer services have been classified as E3 because of the high implementation cost. At the request of individual operators, a 1200 bit/s service (classified as A) has also been specified.

Table 3.32 presents a summary of the bearer services described, with T standing for transparent and NT for non-transparent transmission mode.

3.10.3 Teleservices

Two important characteristics of the teleservices are system-optimized coded voice transmission and data transmission in which the use of special adapters ensures compatibility with terminal equipment in ISDN or over a modem to the PSTN (see Figure 3.3).

The following teleservices are standardized in the GSM Rec. 02.03 [5]. The parantheses indicate whether a service is classified as *essential* (E1, E2, E3) or *additional* (A) or is marked for *Further Study* (FS).

- Telephone service (E1)
- Emergency call service (E1)

- Three videotex access profiles (A)
- Telefax Group 3 (E2)
- Access to electronic mail (A)
- Three short-message services:
  - *Short-Message Mobile-Terminated Point-to-Point* (E3)
  - *Short-Message Mobile-Originated Point-to-Point* (A)
  - *Short-Message Mobile Cell Broadcast* (FS).

3.10.3.1 Telephone Services

The telephone service is the most important service offered, and therefore its introduction is classified as E1. Telephone services are provided with a number of additional features such as call diversion, call barring and closed user groups. The transmission of voice is exclusively carried out digitally and has been optimized for telephony, which as anticipated is the prevailing form of communication. The most important aspect of this optimization is the economic use of the frequency spectrum. Because the speech codecs transmit the analogue data signals processed by a data modem in a distorted form, they are replaced by special modems during data transmission over the communications channel of the air interface.

3.10.3.2 Emergency Call Services

The emergency call service is available with the introduction of GSM (E1) to enable voice communication to be established with the regional emergency centre in the area responsible for the mobile station on the basis of a standard access procedure or by dialling a national emergency number.

3.10.3.3 Short-Message Services

A *Short Message Service* (SMS) for point-to-point connections allows messages up to 160 bytes in length to be sent between a GSM mobile station and an *SMS centre* (see Figure 3.66) [3]. Longer messages are broken down into parts 160 bytes in length, albeit without the assurance that the sequence will be maintained.

The following features typify the *Short Message Service*:

- The transmission of messages between the relevant GSM network entities is carried out over signalling channels (SACCH and SDCCH).
- The SMS is a store-and-forward service that is implemented through the *SMS centre*.

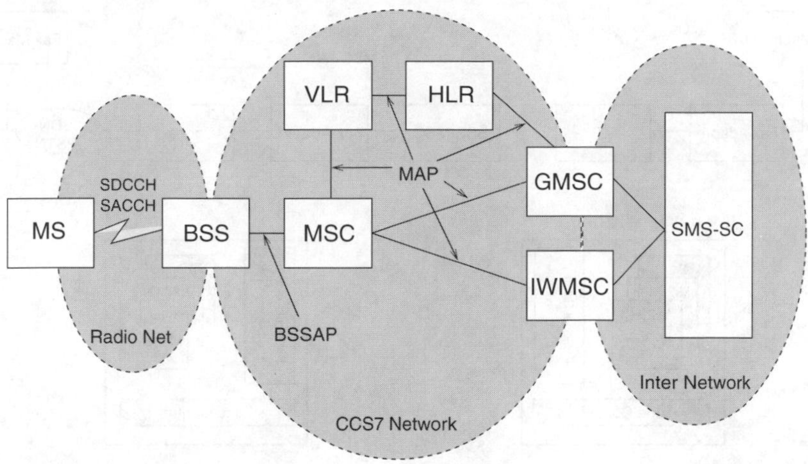

Figure 3.66: Implementation of the short-message services in GSM

Standardized interfaces, such as for the packet-switched public data network or CCS7, are usually required for access to different service providers. The transmission of short messages requires the same functions as normal data and voice communication (e.g., setting up a signalling channel, authentication) and includes an acknowledgement of the MS.

Because short delays (some seconds) can be tolerated with SMS, and short messages can also be transmitted in parallel to an existing connection, there is a better utilization of the signalling channels. Mobile terminated messages are indicated on the display field or, if the receiver is temporarily not available, stored and then resent within a predetermined period of time. Mobile originated messages can either be stored ones or messages that have been entered through the keyboard of a mobile station or an external terminal.

The SMS point-to-multipoint service is also called *Cell Broadcast* (CB). With this service messages are transmitted to all mobile stations that are located in a particular region. This allows messages to be sent to a narrow local area of reference (so-called location aware services).

3.10.3.4 Videotex Access Services

The access of mobile users to the databases of interactive videotex services was considered an appealing service and was therefore standardized. Since all three communications protocols for videotex available in the CEPT countries are not compatible, three access services were specified. These allow an appropriately equipped mobile station to use the videotex service in its own country or one it is visiting.

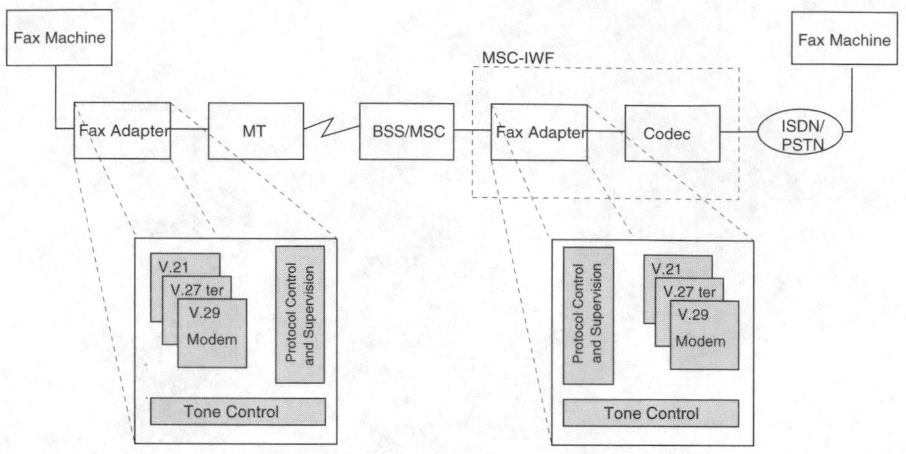

Figure 3.67: Fax adapter function and interworking in GSM

3.10.3.5 Facsimile Service

This data service provides for the use of standard Group 3 facsimile machines, which transmit digitally in the analogue telephone network and are available today as compatible equipment in all CEPT countries [4]. The GSM specifications relate essentially to ITU-T Recs. T.30 and T.4, with facsimile signalling described in T.30 and image transmission between facsimile units described in T.4. The adaptation of facsimile equipment to the public mobile communications network is executed through a fax adapter both on the mobile side and on the side of the interworking function (see Figure 3.67).

The signalling and message data received by the modem in the fax adapter of the originating GSM terminal is transmitted to the adapter of the partner facsimile unit (in the MSC/interworking unit) over the GSM traffic channel and sent from there by modem over the fixed network to the corresponding facsimile unit. Different modem procedures are required for the signalling and transmission of messages.

The main tasks of a fax adapter involve monitoring and processing protocol T.30 and, in non-transparent mode, the T.4 protocol as well to ensure that the activities of the facsimile units are reacted to appropriately. Communication between fax adapters is under the control of the respective protocols, which are service-dependent.

The transparent facsimile service The transparent facsimile service was the first one introduced by all network operators. The coded data and associated signalling are transmitted via a transparent bearer service.

A full-duplex traffic channel was selected in order to ensure that synchronization would be fast and secure, although the T.30 protocol provides for half-duplex transmission only. During the signalling phase, in which the data rate is 300 bit/s, additional redundancy can be achieved through retransmission of the bits. During data transmission, document lines can be more affected by strong interference caused by conditions in the radio field than is normally the case in a fixed network.

Aside from the signalling phases, the duration of a fax transmission is the same as in a fixed network. When transmission quality in a transparent service is poor, a facsimile unit reduces the transmission rate. This reduction is registered by the T.30 control and leads to a change in channel operating mode on the radio interface (*Channel Mode Modify Procedure*). The procedure involves selecting a bearer service with a lower transmission rate whereby a change of channel coder is requested in the mobile station as well as in the base station system. In some cases the bit-rate adaptation must also be changed in the IWF.

The non-transparent facsimile service The non-transparent facsimile service requires an adapter to adapt a terminal to the bearer service provided by GSM. The adapter has modem functionality for the attached facsimile unit, and controls and manipulates the facsimile protocol by storing data, transcoding and adapting the facsimile rate to the radio channel quality.

In contrast to the transparent facsimile service, transmission between fax adapters is in half-duplex mode corresponding to the T.30 protocol. For improved transmission quality on the traffic channel, the RLP, which transmits in full-duplex mode and also protects data transfer over the radio interface and between fax adapters, is used in non-transparent mode. Consequently, with a non-transparent facsimile service it is possible to send at a transmission rate of up to 9.6 kbit/s. Use of the RLP also means, however, that the retransmission of transmitted frames that have been received with errors takes extra time. This results in the need for intermediate storage measures in the transmission chain, which must be compatible with T.30 or T.4.

Fluctuation in radio channel quality, which causes lower transmission rates and requires documents to be stored in the sending facsimile adapter, was considered a problem in the standardization. Therefore in 1994 the service was specified as optional (A). As soon as a timer in the receiving fax unit expires, a transmission is broken off. Simulation tests carried out by the author [2] have shown that these terminations do not occur in normal illuminated cells, because a handover is usually initiated if the receiving quality is poor (7 dB $> C/I >$ 5 dB) (see Section 3.6). Therefore the non-transparent facsimile service represents an interesting alternative to the transparent service, since it provides a higher quality of service with a much lower bit-error ratio.

3.10.3.6 Access to Electronic Mail (E-Mail)

A PLMN does not have *electronic mail* (e-mail) but offers an access service to systems in fixed networks that exist in many CEPT countries based on ITU-T series X.400.

3.10.4 Supplementary Services

Supplementary services are offered to subscribers who use the teleservices and bearer services described above, expanding or supporting these services as follows:

Subscriber identification This function group specifies services that allow for or restrict the identification of the other mobile user. It also provides a possibility for registering unwanted calls, even if the caller does not want to reveal his identity.

Call rerouting There is a difference between unconditional and conditional rerouting of call requests. With unconditional call rerouting, a connection is automatically switched to the line of another user, whereas conditional call rerouting is only possible if, for example, the line is occupied, the subscriber does not respond or the radio network is overloaded.

Call forwarding In contrast to call rerouting, this function supports the forwarding of a connection that already exists.

Call holding: This feature allows a subscriber to maintain an existing connection and at the same time temporarily establish another connection.

Conference call This extends the number of users participating in a connection from two to more. The basis for a successive structuring of the conference call is the supplementary service *call holding.*

Closed user group This function supports the forming of logical subnetworks within the entire GSM. Communication is only possible between the registered subscribers of a subnetwork or of a subscriber group. This service can be used, for example, for setting up a company-specific GSM subnetwork to which only the employees of the company have access.

Call restriction This function group makes it possible to impose either a total restriction or a partial restriction on calls. The restriction can apply to incoming and/or outgoing calls. Examples include:

- Barring outgoing calls (possibly international ones).
- Barring incoming calls, for instance if the user is outside his own home network.

3.10.5 Support for Value-added Services*

Owing to the restrictive standardization of the GSM network, operators have limited opportunity to differentiate their service portfolio from their competitor's offering. In many cases the tariffs are the only means of distinguishing

*With the collaboration of Eckhardt Geulen

one GSM operator from another. Value-added services are, however, beyond the boundaries of the GSM standard and offer a way to increase the attractiveness of an operator for a specific group of users. Such services are not treated in the GSM specifications, but GSM provides dedicated mechanisms to facilitate the provision of value-added services. As it is up to the respective network operator to provide that type of a service, the term "operator-specific" services is used as well.

3.10.5.1 Unstructured Supplementary Service Data (USSD)

The MAP-process *Unstructured Supplementary Service Data* (USSD) allows for the communication between a GSM user and a network-based application. In the current phase 2 of USSD both network- and user-initiated USSD dialogues are specified. This communication is transparent to both the MS and the intermediate network nodes. Transparency to the MS means that the messages sent from the network application are straightforwardly displayed on the handset, and similarly the messages typed in by the user are sent to the network without further processing in the mobile station.

USSD Application The value-added services making use of the USSD mechanism can be referred to as the *USSD Application*. It is a sequence of input/output strings exchanged between the network and the user, possibly triggered by certain events. The USSD application is fully stored in the network, since the handset functions only as a simple man–machine interface. This explains why control over USSD dialogues is always with the network-based applications, including in the case of user-initiated USSD dialogues.

USSD applications can either be implemented in the MSC/VLR or in the HLR. Thus it is possible that in the roaming case USSD sequences from a user are either interpreted locally in the VLR or forwarded to the HLR in the home PLMN. Which of the two methods applies is specified by the string entered by the user and the capabilities of the visited network.

Structure of a USSD Sequence Dedicated key stroke sequences (DTMF signals) are defined for all standard operations like call set up and supplementary service control. Any of these sequences are translated by the mobile station into functional signalling, which means that predefined messages are sent over the air interface in order to invoke the desired funtion. In principle, all key sequences that cannot be interpreted in the mobile station are packed into an USSD message and sent to the NSS. However, to have a certain amount of additional functional signalling available for future use, a format for USSD messages has been defined (see Figure 3.68).

The *start delimiter* consists of one, two or three star or hash symbols. The *service code* consists of two symbols, and first of all identifies the location of the USSD application (visited or home PLMN) and secondly specifies which application has to be started upon reception of the USSD string. The field

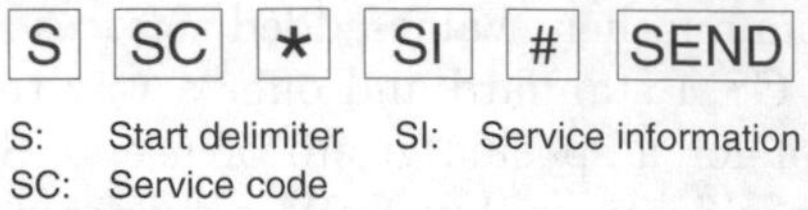

S | SC | * | SI | # | SEND

S: Start delimiter SI: Service information
SC: Service code

Figure 3.68: Structure of a USSD sequence

service information can carry service-specific information. Its length can vary between zero and a number of symbols limited by the maximum USSD string length.

An example of a value-added service offered by means of the USSD mechanism is the e-mail service allowing the user to read electronic mails on the mobile station. The USSD sequence entered by the user to activate the service could be, for example, `**13*2#`, where `13` would specify the e-mail service (located in the home PLMN) and `2` could mean to only transmit sender and subject field of the messages received in the mail server. Upon reception of this USSD sequence, the network would send e-mail messages directly to the display of the user's mobile station.

USSD Dialogues The USSD process is dialogue-oriented. Instead of simply sending unrelated messages back and forth, a context is established to monitor proper exchange of messages in an USSD dialogue. The dialogue is in any case controlled from the network side, including in the case of a user-initiated dialogue. In that case the control is taken over by the network as soon as the first message is received in the node hosting the USSD application. This implies that the mobile station cannot send subsequent requests to one application without being explicitly requested to do so.

The network can send two different types of messages: the reception of a notification is not acknowledged by the mobile station, whereas a request must be acknowledged.

Performance of USSD For a USSD dialogue it does not matter whether or not a traffic connection is established. In both cases USSD messages can be send over the air interface—only the way in which this is done differs. In the case of an established connection the *Fast Associated dedicated Control Channel* (FACCH) is used, whereas in the other case USSD messages are sent over the *Stand-Alone Dedicated Control Channel* (SDCCH). A fact to be considered for massive use of the USSD mechanism in the case of established connections is that the FACCH is realized by so-called pre-emptive dynamic multiplexing. This means that part of the *Traffic Channel* (TCH) is cut out to form the FACCH. Thus the quality of the traffic is lowered.

The performance of USSD is different for the two cases described above. The theoretical throughput for USSD traffic over the FACCH is no more than 1200 bit/s. For the SDCCH the value is 670 bit/s. Owing to interleaving and other mechanisms, there are strong dependencies on the amount of data to be transmitted as well as on other parameters. But in any case the figures show that the use of USSD is limited to notification-type services and that USSD cannot be used as a serious bearer for data transport.

3.10.5.2 Customized Applications for Mobile Network Enhanced Logic

The concept of the *Intelligent Network* (IN) allows easy deployment of new services in telecommunication networks. In an IN service control logic is stored in a central place (*Service Control Point*, SCP) and can be used to control many instances of a service in different *Service Switching Functions* (SSF). The standard IN functions are today mostly intended for fixed networks. As more and more operators run both fixed and mobile networks, a requirement has arisen to re-use parts of fixed network IN installations in mobile networks. At the same time, the arguments for the introduction of fixed network IN are valid for mobile networks as well. Finally the operator's wish to provide operator-specific value-added services as described above drove the standardization for a feature called *Customized Applications for Mobile network Enhanced Logic* (CAMEL).

Introducing IN into GSM The GSM system has been designed without consideration of the IN standard. CAMEL can be seen as a way to introduce IN into GSM. The network model for the introduction of CAMEL is depicted in Figure 3.69.

The CAMEL feature allows the separation of service logic from the MSC. Therefore a *Service Switching Function* (SSF) is added to the MSC, and a so-called *CAMEL Service Environment* (CSE) has been introduced within the GSM network. This CSE contains a *Service Control Function* (SCF) similar to a traditional IN solution. With the help of CSEs, *Operator-Specific Servicess* (O-SS) can be offered to customers.

Mobile Originating Calls Subscribers to the CAMEL feature will be marked with a *CAMEL Subscription Information* (CSI). If an active originating CSI is found in the VLR during the call set up of a mobile station, the *Visited Service Switching Function* (VSSF) sends an `InitialDetectionPoint` message to the *GSM Service Control Function* (GSMSCF) and the VMSC suspends the call processing. According to GSM Specification 03.78 [25], the `InitialDetectionPoint` must always contain

- the service key
- called and calling party numbers
- calling party's category
- location number
- bearer capability

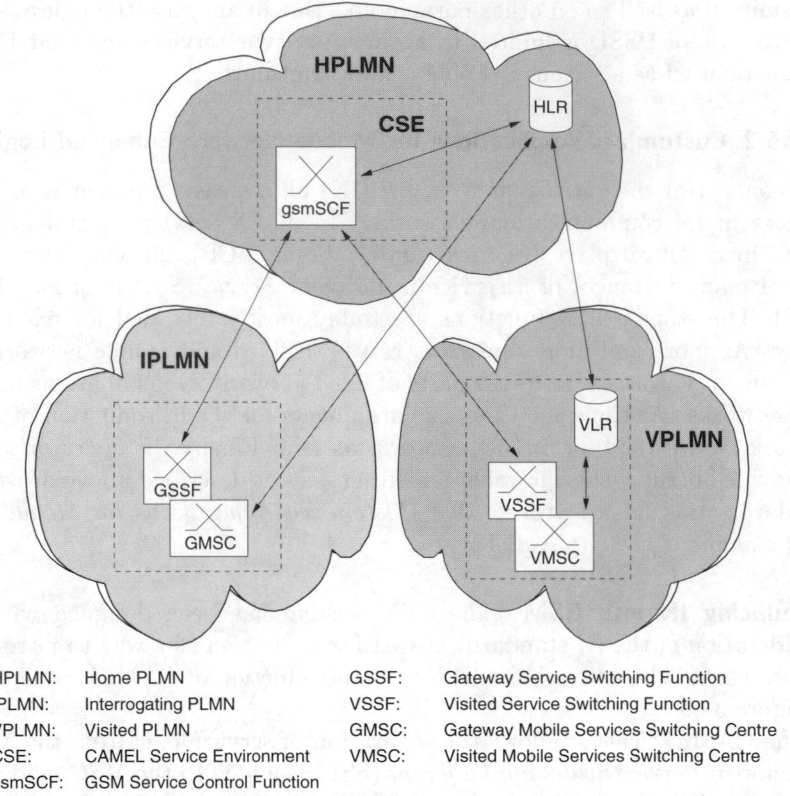

Figure 3.69: The principle of introducing CAMEL into GSM

- event type *Basic Call State Model* (BCSM)
- location information
- the *International Mobile Subscriber Identity* (IMSI).

After the service logic has been processed, CAMEL-specific handling is initiated from the GSMSCF.

The signalling procedure in the case of mobile-originating calls is illustrated in Figure 3.70. A Third Party Provider identified by the number n (TPP n) might then be involved.

Mobile Terminating Calls In the case of a mobile-terminating call, the GMSC in the interrogating PLMN identifies the HLR of the called party with the help of the *Mobile Station International ISDN Number* (MSISDN). Then the GMSC sends a `RoutingInformationRequest` to the HLR. The HLR checks the CSI of the called party and sends the information stored in the subscriber record back to the GMSC. Now, the GMSC acts according to CSI. If

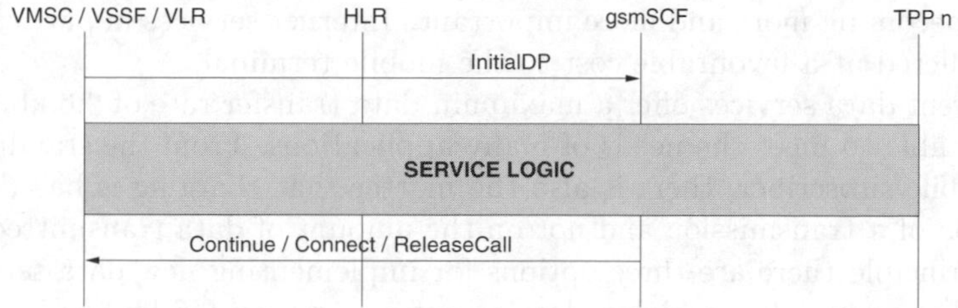

Figure 3.70: Signalling procedure for mobile-originating calls

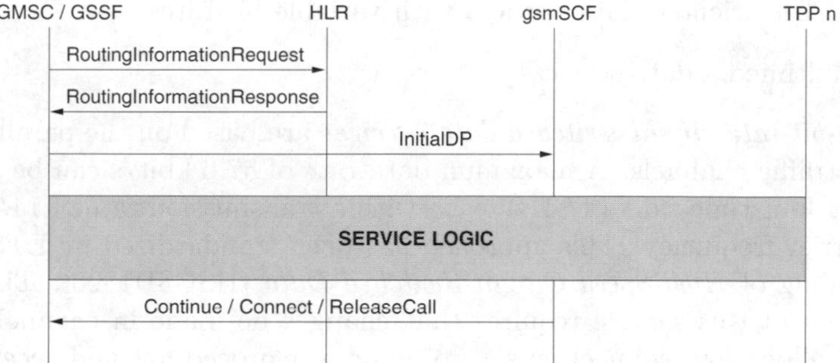

Figure 3.71: Signalling procedure for mobile-terminating calls

the terminating CSI is active, the call processing is suspended as soon as the trigger criteria of a *Detection Point* (DP) of the CAMEL service logic is fulfilled. An `InitialDP` message is sent to the CSE and the service logic execution is started. Thereafter, CAMEL specific handling is initiated. Figure 3.71 shows the signalling procedure for mobile-terminating calls.

The new mobile data services SAT, MExE and WAP are described in Section 4.6.

3.11 Advanced Voice and Data Services in GSM*

The data services introduced by GSM in mid-1994 are based on circuit-switched transmission. As with voice transmission, each user is provided with an exclusive connection over a TCH. Since data sources often have fluctuating traffic volumes, circuit switching results in an inefficient utilization of the radio channels.

Owing to the continuing strong growth projected in the number of mobile subscribers, frequency economy as well as flexibility in the use of radio chan-

*With the collaboration of Götz Brasche

nels is becoming more and more important. Internet services in particular are to be offered at a favourable cost to the mobile terminal.

Current data services offer a maximum data transfer rate of 9.6 kbit/s and are not able to meet the needs of many applications. From the standpoint of the mobile subscriber, there is also the matter that charging is based on the duration of a transmission and not on the amount of data transmitted.

In principle there are three options for implementing new data services in GSM that would offer a higher data transfer rate than 9.6 kbit/s:

- High-bit-rate circuit-switched data services.
- Packet-oriented data services with variable bit rates.
- Multimedia data services.

High-bit-rate circuit-switched data services are based on the parallel use of several traffic channels. A maximum data rate of 57.6 kbit/s can be achieved by using four time slots (TCHs), each with a transmission rate of 14.4 kbit/s, of a carrier frequency. This approach has been standardized by ETSI under the heading of *High-Speed Circuit-Switched Data* (HSCSD) [20]. The implementation of this service requires that changes be made in channel coding, channel allocation, connection set up, handover procedures and access to the fixed network (*interworking*). This function became commercially available at the end of 1998.

In contrast to the HSCSD service, a *packet-oriented data service* concept not only provides a high data rate but also offers flexibility in the use of channel capacity for applications with variable bit rates because of the possibility to multiplex several connections at the same time on the same traffic channel or on several traffic channels used in parallel. A maximum data rate of 115 kbit/s is aimed at. Because GSM transmits using circuit-switching, significant modifications would be required in order to integrate a packet-switched service. This technique became commercially available in 2000 and will prepare for new data services carried over UMTS; see Section 6.1.

The *multimedia data service* is a circuit-switched service, and will extend the GSM services to that expected from 3rd-generation systems. Under the acronym *Enhanced Data Rates for GSM Evolution* (EDGE), a technique totally compatible with the GSM channel spacing of 200 kHz has been standardized at ETSI/SMG in 1998. A modified air interface with especially a new modulation method applied to the 200 kHz channel as a whole is able to reach 384 kbit/s per carrier. EDGE was initiated by Ericsson, and will be applicable to all the existent GSM frequencies such as 900, 1800 and 1900 MHz. More over, EDGE under the name of IMT-SC is a member of 3rd generation systems. The related products and services are projected to be available by 2001.

In addition to new data services, future applications will also require new voice services (e.g., for group communication), which until now have only been offered in trunked radio systems like the TETRA system, see Section 7.2.

Since 1994, at the instigation of the International Railway Federation (*Union Internationale des Chemins de fer*, UIC) under the heading *Advanced Speech Call Items* (ASCI), group and broadcast services with fast call set up and priority control in the list of GSM Phase 2+ working points are being run and processed.

3.11.1 ASCI—Advanced GSM Speech Call Items

Currently, in Europe the national railways are using many different non-compatible radio systems. National railcars need separate communication systems for national and for international rail services. Therefore the development of a standard *European Train Control System* (ETCS) is underway under the overall control of the UIC. Although the GSM radio channel is only specified for a maximum terminal speed of 250 km/h, field studies have shown that no noticable restrictions occur even at speeds of 300 km/h.

The new communications platform has been called GSM-R(ailway) and will be operated directly below the GSM extension band.

Group and broadcast radio calling with fast connection set up must be integrated into GSM in order to meet the requirements of UIC train radio control. Both services are being enhanced with a priority control mechanism adopted from ISDN. The ETSI standard designs consequently comprise specifications for:

- *enhanced Multi-Level Precedence and Pre-Emption* (eMLPP) [21, 26].
- *Voice Group Call Service* (VGCS) [23, 28].
- *Voice Broadcast Service* (VBS) [22, 27].

3.11.1.1 Voice Broadcast Service

The *Voice Broadcast Service* (VBS) allows users of the mobile radio and fixed networks to send messages to several so-called *listeners.* Figure 3.72 provides a clarification of the logical service concept.

If a broadcast call is initiated by a mobile station (see Figure 3.73), the identity of the appropriate cell with the requested group identity is routed to the *Group Call Register* (GCR) of the applicable MSC.

With a call initiated from the fixed network the appropriate user and requested group identity is transmitted to the GCR. The GCR then sends back a list of cells in which the call is to be broadcast according to the make-up of the group and the location. The responsible MSC routes the list to the appropriate MSCs. The MSCs instruct the appropriate BSCs to set up a broadcast channel in each of the relevant cells and to send out a call *notification* on a newly specified signalling channel. In contrast to the conventional GSM use of voice, this process is not described as *paging* because the mobile stations are not explicitly being addressed and are not responding to the call notification. In the corresponding GSM Rec. 08.58 [24] these GSM control

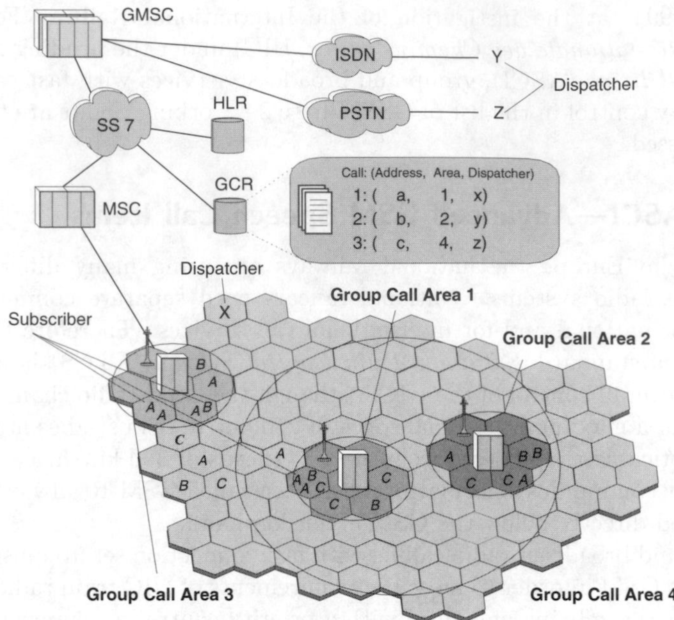

Figure 3.72: Logical concept for expanded voice services

channels have been expanded to include the *Notification Common Control Channel* (NCCH).

A notification is sent out at periodic intervals until the call has been completed. Mobile stations that receive a call notification switch over to the broadcast channel indicated and listen in on the appropriate downlink. The call initiator remains on his dedicated channel during the course of the call, and completes the call after having transmitted his message.

Accordingly, a broadcast call can be established and managed in the same way as a GSM point-to-point connection, except for the additional signalling required for the rerouting. This means that no additional handover procedures are required during an impending cell change.

This does not apply to the mobile stations participating in the broadcast call, although a so-called *idle mode cell reselection* algorithm can be used for them.

The way to prevent a mobile station from moving into a cell in which the broadcast call is not being transmitted is by not including the cell in the list. Signalling requirements are minimized through the fact that the mobile stations are only notified of the frequencies of the signalling channels of the surrounding cells, i.e., the frequencies on which the call notification is being sent. The mobile stations must then listen in on the corresponding control channels in order to determine the actual broadcast channel.

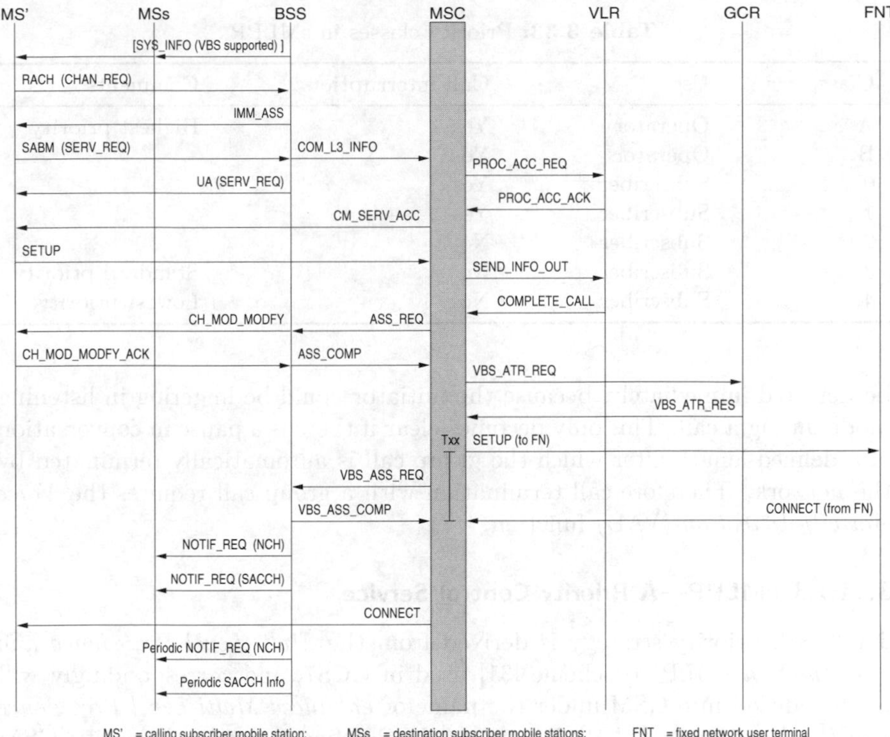

Figure 3.73: Call set up in the voice broadcast service

3.11.1.2 Voice Group Call Service

The *Voice Group Call Service* (VGCS) supported by ASCI provides a service that allows the fixed network or the mobile stations to set up a group call channel on which the group members can listen in or also transmit. After the call indicator has sent his message, he releases the channel and transfers over to *listener mode.*

As with VBS, the addressee group of VGCS is divided into the following:

- Mobile stations that are members of the group and are located in a predefined geographical area.
- A fixed group of fixed network stations.

As soon as all the group call participants finish talking, any one of the users can request to be allocated the channel. If the channel is allocated, the user is allowed to transmit until he releases the channel and in turn changes to listening mode.

A broadcast call is normally explicitly ended by the initiator. A connection between initiator and network that is disrupted owing to interference cannot

Table 3.33: Priority classes in eMLPP

Class	Use	Call interruption	Comments
A	Operator	Yes	Highest priority
B	Operator	Yes	
0	Subscriber	Yes	
1	Subscriber	Yes	
2	Subscriber	No	
3	Subscriber	No	Standard priority
4	Subscriber	No	Lowest priority

be detected immediately, because the initiator could be lingering in listening mode during a call. This only becomes clear if there is a pause in conversation of a defined length after which the group call is automatically terminated by the network. Therefore call termination with a group call requires the *Voice Activity Detection* (VAD) function.

3.11.1.3 eMLPP—A Priority Control Service

The ASCI priority strategy is derived from the *Multi-Level Precedence and Pre-Emption* (MLPP)) scheme [31] used in CCS7, and correspondingly will be introduced into GSM under the name of *enhanced Multi-Level Precedence and Pre-Emption* (eMLPP). Whereas MLPP defines a five-level priority, GSM is planning seven priority classes (see Table 3.33). The MLPP priorities 0–4 will correspond to the eMLPP priorities 0–4.

In addition to these five classes, two further classes, A and B, have been specified for the GSM system (see Table 3.33). They are exclusively being reserved for processes within the network, e.g., for the configuration of group and broadcast calls VGCS and VBS described above.

Calls with priority A or B can only be made locally—in other words only within the coverage area of an MSC. If this kind of priority is used globally, in other words a GSM call is routed over an ISDN network, the priority classes A and B are changed to priority 0.

The maximum priority allocated to a mobile user, and reflected in the monthly basic tariff, is negotiated with the service provider when the contract is signed and is stored in the SIM card.

3.11.2 HSCSD—The High-Speed Circuit-Switched Data Service

The *High-Speed Circuit-Switched Data* (HSCSD) service simultaneously allocates several full-rate traffic channels (TCH/F9.6) to a mobile station within a 200 kHz frequency channel for the duration of a transmission.

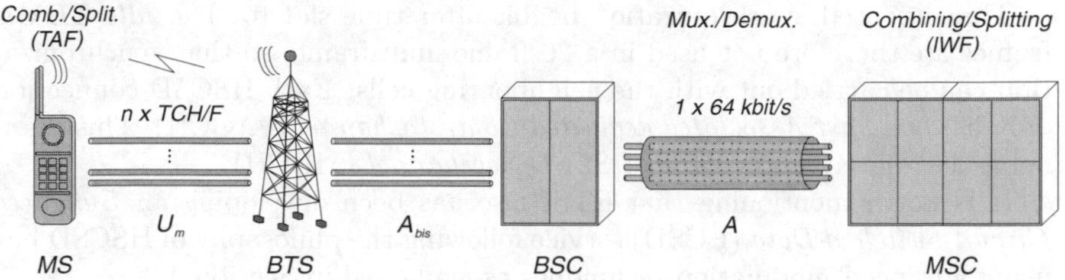

Figure 3.74: Architecture of an HSCSD

With the parallel use of all eight time slots, and depending on the bearer service used, data rates of up to 76.8 kbit/s are achievable on the basis of TCH/F9.6 coding in conformance with GSM (see Table 3.8). (Costly transmitting and receiving facilities are required in mobile stations if more than four channels are used at the same time. The number in the standard is currently restricted to four channels reaching 57.6 kbit/s with different coding.)

3.11.2.1 The Logical Architecture

In the current GSM the functions needed for data transmission are mainly embedded in the *Terminal Adaptation Function* (TAF) of the mobile station and in the *Interworking Function* (IWF) of the MSC. In principle this functional division has been retained in the HSCSD service (see Figure 3.74). Essentially only a *splitting/recombining* function needs to be added to the above components of the MS or MSC when several traffic channels are used simultaneously.

Logically only *one* connection exists between MS and MSC. The segmentation and reassembly are therefore based on the consecutive numbering of the individual data frames.

The time slots used on the radio interface at the A_{bis} reference point between BTS and BSC are mapped transparently. The HSCSD channels are then multiplexed onto a 64 kbit/s connection at the A reference point (or E reference point between two MSCs).

3.11.2.2 The Radio Interface

At the radio interface an HSCSD connection can consist of up to eight traffic channels (TCH), called *Multi-Slot Assignment* (MSA)). All the channels in an HSCSD connection use the same frequency hopping procedures and the same training sequence. For security reasons, however, separate enciphering is being applied for each channel. The channel coding, interleaving and rate adaptation of the current traffic channels has been retained in order to keep the implementation costs to a minimum. Each subchannel is allocated an SACCH. This provides individual transmitter power control and improves the interference level.

Time-oriented synchronization begins after time slot 0. The *idle* TDMA frames are therefore not used in a 26-frame multiframe, so that synchronization can be carried out with the neighbouring cells. Each HSCSD connection only has one *Fast Associated dedicated Control Channel* (FACCH). This channel is designated as the *Main HSCSD Subchannel* (MHCH).

It is worth mentioning that ETSI also has been developing an *Enhanced Circuit-Switched Data* (ECSD) service following the philosophy of HSCSD but using advanced modulation techniques as explained in Section 4.2.

3.11.2.3 Bearer Services

In accordance with the current GSM bearer services, transparent and non-transparent bearer services are supported in the HSCSD service (see Figure 3.65). The transparent service guarantees a constant data rate, even with fluctuating quality of service. The number of allocated channels can be increased if the quality of service falls below a threshold value. The non-transparent bearer service, in contrast, guarantees that quality of service will remain constant with fluctuating throughput.

Transparent bearer service The transparent bearer service uses the ITU-T X.30/V.110 protocol [30], which provides a three-level rate adaptation to the user interface (R- and S-interface) (see Figure 3.62).

The time displacement of the individual HSCSD channels between TAF and IWF does not have a considerable effect on frame length. In practice, a data frame that is being sent in one of the channels cannot overtake a frame being sent in another channel. Nevertheless there must be a safeguard to recognize that the frames are being transmitted in the correct sequence.

In GSM the status of the V.24 interface used for data transmission at the network gateway is transmitted in the status bits `SA`, `SB` and `X`. Because more than one subchannel can be used for a logical connection in the HSCSD service, the redundant status bits can be used for the numbering of the channels without causing a reduction in the repetition rate of the status bit per connection. An extra bit can be used for a modulo 2 numbering within the subchannels in order to prevent problems caused by the time displacement of the subchannels. Rate adaptation for data rates that are not multiples of 9.6 kbit/s can be effected through a corresponding number of filler bits in the last four V.110 frames.

Non-transparent bearer service Like the corresponding bearer service in GSM, the non-transparent HSCSD bearer service is based on the RLP protocol (see Section 3.9.3).

According to the service concept, an RLP entity will manage all the subchannels (see Figure 3.75), because, with the numbering only based on one transmitting and receiving window, any number of RLP frames can be sent on the subchannels.

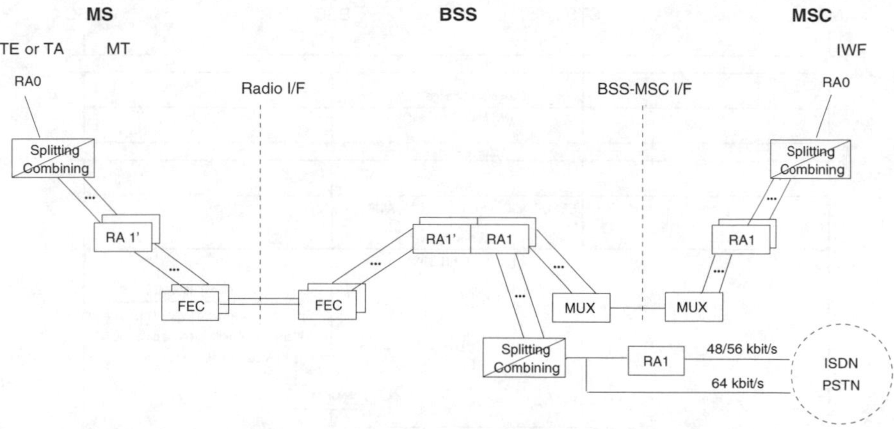

Figure 3.75: The non-transparent HSCSD bearer service

A maximum window size of 61 is allowed for a conventional RLP. A total of 6 bits is sufficient for numbering each transmitting or receiving window. In a non-transparent HSCSD service a maximum of 8 traffic channels can be allocated to an RLP entity. The addressing space must be increased correspondingly for the management of the transmitting or receiving window. An additional 6 bits are required in the RLP header for the management of the transmitting/receiving window; this reduces the data rate by 3 % (Since it is not necessary for each RLP frame to use the status byte of the *Layer-2 Relay* function, L2R, the maximum data rate of a subchannel of a 9.6 kbit/s subchannel can be maintained.)

Pseudo-asymmetrical transmission A pseudo-asymmetrical transmission mode has been defined for the non-transparent bearer service. This allows mobile stations with half-duplex transmitting and receiving facilities to receive in multi-slot mode but to transmit in single-slot mode. This will result in a higher data rate—at least on the downlink. A mobile station will decide which of the allocated lower channels is to be used in the uplink direction and which are to remain free. As in DTX mode, the frames are regarded by the network as not being used and are rejected by the IWF. Although not specified in the standard, with additional signalling the time slots that are not being used can be made available to other services.

3.11.2.4 Signalling

There is a greater chance of blocking with the HSCSD service because parallel traffic channels are being occupied at the same time.

In the current GSM a bearer service exists for each data rate. Depending on the configuration, different variable data rates are possible with the HSCSD

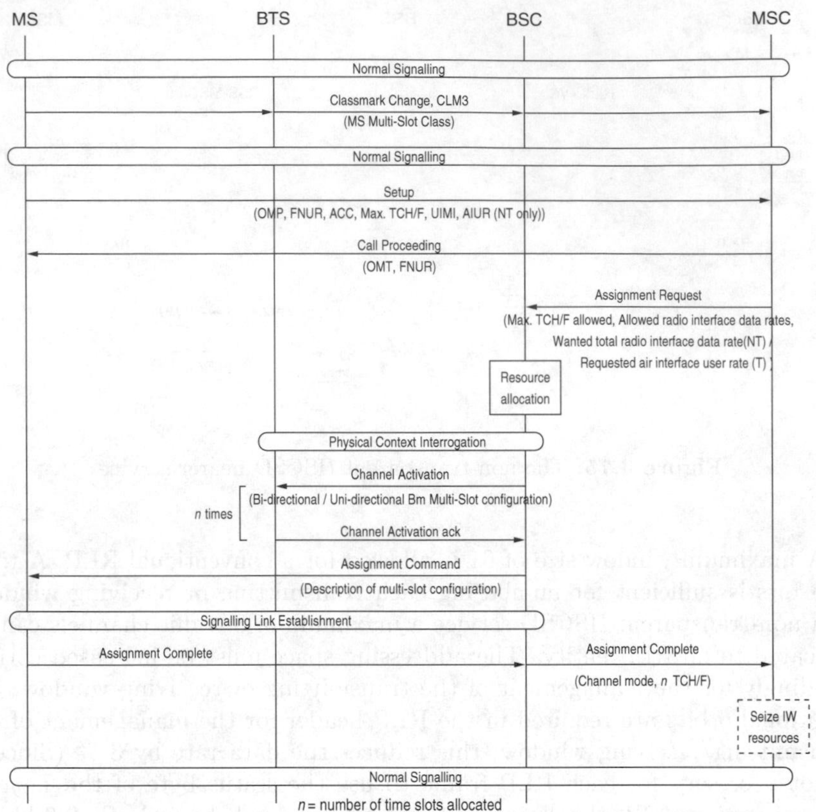

Figure 3.76: Call set up procedure in HSCSD

service. It is not useful to define another bearer service for each possible data rate. Therefore the data rate for the HSCSD service is merely considered a quality of service parameter. This *Flexible Bearer Services* (FBS) concept provides for a desirable and required data rate (acfDNC or *Required Number of Channels* (RNC)). The desired data rate represents the maximum and the required data rate the minimum data rate necessary to maintain the quality of service required by an application.

Depending on radio channel availability, a mobile station can be allocated any data rate between a desired and a required data rate. This concept promises a low blocking rate, particularly with handovers.

If a target cell does not have a sufficient number of free channels available during a handover, the FBS concept can be applied to maintain the connection as long as the target cell has enough channels for the required data rate. The number of blockings can be reduced when there is an overload, since channels can be released up to the minimum data rate. The two parameters are transmitted in the Set Up message when a connection is established. The so-called *Bearer Capability Element* (BCE) is expanded accordingly. When transmission is carried out with a fixed data rate, the parameters for desired and maximum data rate have the same value.

In addition to these two new parameters, the characteristics of the mobile station in terms of its *Multi-Slot Assignment* (MSA) capability and the channel coding supported (*class mark*) must be communicated in the BCE.

Figure 3.76 illustrates the call set up procedure. After receipt of a Set Up message, the MSC sends a modified Assignment_req message to the BSC. This message contains a list of channels at the A-interface in the `Message Content` element. The parameters *Desired Number of Channels* (DNC) and *Radio Network Controller* (RNC) are added to the `Channel Type` element. Before this message can be transmitted, the MSC reserves the required capacity. If the requested number of radio channels cannot be allocated, a Channel_Update procedure is used between MSC and BSC in order to adapt the resources to the A-interface.

A separate Channel_Activation message is required for all lower channels before the selected channel configuration as well as information on channel coding can be forwarded in an Assignment_Command message to the MS. In this message the time slot numbering parameter indicates the first of the consecutively allocated time slots. The `Channel Description` element accepts the time slots.

3.11.3 GPRS—The General Packet Radio Service

Within the framework of the continuing development of GSM Phase 2+, ETSI has also developed a packet-oriented service concept for the transfer of data. Standardization of the new service *General Packet Radio Service* (GPRS) was completed in 1998 (see Appendix E, Table E.2 for an overview of the GPRS standards). According to its importance, GPRS is described in a separate section, see Section 4.1.

3.12 Interworking Function (IWF)

Interworking functions allow GSM-supported services to be connected to those of the fixed networks and vice versa (see Figure 3.61).

3.12.1 Gateway to the Public Switched Telephone Network

The *Public Switched Telephony Network* (PSTN) in the CEPT countries has developed in a variety of different ways over a period of many years. As a result, there is a strong reliance on the respective PSTN for the interworking functions needed. The GSM recommendations for the PLMN side of the IWFs are very detailed, whereas only general requirements have been formulated for the PSTN side. The problem is that a switching centre is unable to specify the service desired for calls coming from PSTNs. Therefore a technique is required to differentiate between the types of connection to enable the mobile radio network to select the correct IWFs. Two methods have been selected (GSM Rec. 09.07):

1. Allocation of separate numbers (MSISDN) for each service (optional) to each mobile user, with each number representing a particular bearer service or teleservice, e.g., voice, facsimile group 3, transparent synchronous data transfer at 9.6 kbit/s, and each service requiring special IWFs. The associated information is stored in the HLR under the MSISDN together with the MSRN, is established when the mobile user checks-in, and can only be changed through administrative channels.

2. The called mobile user conveys the request for a specific service in his *call confirmation message*. This method is obligatory for all networks and mobile stations because it is also applied to ISDN-terminated calls. The mobile user here is assigned only one MSISDN for all services. Based on GSM 03.04, all calls are routed by the querying switching centre. The mobile user is then sent a *call set up message* but without any bearer service information included. The information relating to the service desired is sent by the mobile user in his call acknowledgement. Only then does the MSC/IWF select the appropriate resources and establish a connection.

3.12.2 Gateway to ISDN

Three bearer services are available for the interworking between *Integrated Services Digital Network* (ISDN) and GSM-PLMN:

Circuit-switched transmission (3.1 kHz audio) The same solutions being applied to PSTN gateways are being used for this service. The main difference is in the available signalling capabilities of the fixed network, which can allow the selection of IWFs and indication of connection type and teleservice. However, this possibility depends on how this service is currently implemented in the fixed network.

Circuit-switched unrestricted digital bearer service The IWF permits access to circuit-switched data transmission from and to ISDN in accordance with the ISDN standard rate adaptation based on ITU-T X.30 and V.110. This means that a connection can be established with any ISDN subscriber with an asynchronous or synchronous terminal, with transmission rates up to 9.6 kbit/s. Conventional terminals with a V.24 or an X.21 interface or ISDN terminals with an S-interface can be used (see Figure 3.3). Only a minimum number of IWFs are needed for transparent services, because 64 kbit/s are already offered by the MSC. Specifications are still needed for terminal-to-terminal synchronization and status bit filtering. Non-transparent services require additional IWFs.

The RLP must be properly terminated and the standard ISDN block structure restored.

Packet-switched transmission Because in ISDN access to the packet services takes place over the S-interface either over a B-channel (64 kbit/s) or a D-channel (16 kbit/s), but GSM only has B_m- or L_m-channels, the data rates do not fit. Rate adaptation will also be carried out in this area.

3.12.3 Gateway to the Public Switched Packet Data Network

There are two different bearer service for the interworking between the GSM network and the *Public Switched Packet Data Network* (PSPDN):

1. Access to the PAD service, duplex, asynchronous with transmission rates of 300–9600 bit/s, with two possibilities for implementation:

   (a) *Basic PAD access* is carried out over an asynchronous bearer service, e.g., a PSTN, to an existing PAD unit in the wired network. Existing IWFs to the PSTN can be used for data services, and additional functions such as billing are not required. The disadvantage is that users are committed to the options and transmission rates of the wired network (in some countries only 1.2 kbit/s). The mobile user must be registered by the PSPDN of each country visited.

   (b) With *dedicated PAD access* the GSM-PLMN has its own PAD unit with the service quality parameters specified by the network operator, e.g., a maximum data rate of 9.6 kbit/s. No registration with a PSPDN is required for the mobile subscriber. The disadvantage is cost because of the need to implement additional functions, particularly for billing.

2. Access to a PSPDN with the rates 2.4/4.8/9.6 kbit/s, duplex, synchronous. There are two possibilities:

   (a) Access over the X.32 interface and the PSTN as the transit network. There is a problem because of incompatible X.32 implementations in Europe. Access to a PSPDN is through dedicated *access units.* A PSPDN regards them as packet-switching and communicates with them over an X.75 protocol. This procedure requires close cooperation with the PSPDN operator.

   (b) Access via ISDN using the X.31 protocol. The IWFs GSM/ISDN can then be used.

3.12.4 Gateway to the Public Switched Data Network

There is a gateway between the *Public Switched Data Network* (PSDN) and GSM; most of the procedures used are those already available for PSTN and ISDN.

3.12.5 Interworking Functions for Teleservices

GSM-SMG intentions to support the terminals existing in PSTNs have caused considerable problems. The two-wired telephone interface (a/b) of this equipment must be adapted to the digital mobile radio interface; support of the S- and R-interfaces is also required (see Section 3.10.3).

- Access to videotex: PSTN gateways or PAD access can be used minimizing the need for IWFs to videotex. Roaming mobile users face problems because of the different videotex profiles in different countries (see GSM 03.43).
- Access to facsimile, group 3: This has turned out to be a particularly complex service gateway. There are two different scenarios:
  (a) Support of existing terminals using a two-wire or four-wire analogue interface. In this case the equipment termination must contain a modem function for converting speech band data signals into data streams, thereby also increasing the complexity of the mobile network termination.
  (b) Use of terminals with a data interface, e.g., V.24. This would offer the advantage that existing PSTN gateways for data transmission could be used. The disadvantage is that these terminals are not yet available in the market. The *Personal Computer* (PC) interface *Personal Computer Memory Card Interface Association* (PCMCIA) is currently the preferred one in the market.

3.13 Security Aspects

Open access to the network over the radio interface leads to the danger that communication between mobile subscribers could be open to eavesdropping by third parties or that network resources could be subject to unauthorized use at the cost of registered subscribers. As a precautionary measure, access to and use of the mobile communication network are protected by security procedures that are essentially based on proof of identity of the mobile subscriber through authentication, enciphering of message transmission (including the signalling data) as well as anonymization of the identity of the mobile subscriber through a time change of the user identification (TMSI). GSM digital message transmission provides for the use of cryptographic procedures.

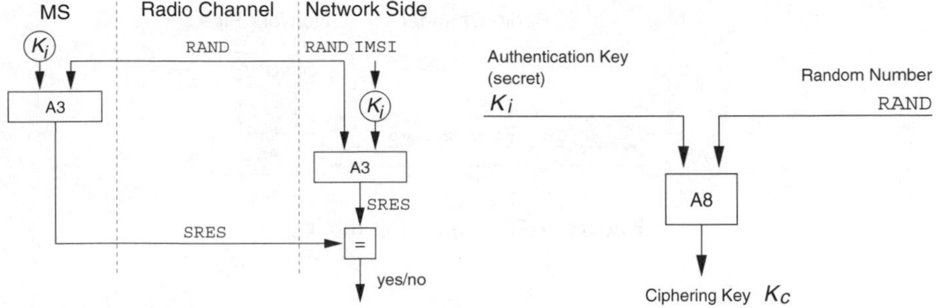

Figure 3.77: Authentication of a mobile subscriber

Figure 3.78: Generation of a ciphering key

With GSM the storage of secret communication data is carried out with the mobile user in a non-manipulable module (SIM), which can be in the form of either a *plug-in* module or an *Integrated Circuit* (IC) card and is also protected against unauthorized use by a *Personal Identification Number* (PIN). Identification of a mobile subscriber to GSM is only possible with a valid SIM. The module gives the user the advantage of being able to have unrestricted access to any agreed services using any other mobile radio unit and to be charged for them. The data stored in the SIM includes the subscriber ciphering key K_i as well as information that is necessary for using the system flexibly and efficiently (see Section 3.2.1).

3.13.1 Authentication

The identity of a mobile subscriber is proved to the network through the process of authentication.

A subscriber joining a mobile radio network is allocated an international unique subscriber identity (*International Mobile Subscriber Identity*, IMSI), a secret authentication key K_i and a secret algorithm A3, which are stored in the SIM. If the subscriber is to identify himself, he is provided with a random number RAND from the network, which is used by the mobile station to calculate the authenticator $\texttt{SRES} = K_i(\texttt{RAND})$ with K_i and A3 (see Figure 3.77). It sends the result to the VLR, where it is checked against the internal value SRES for agreement. If the result is positive, the mobile subscriber is authorized to proceed; otherwise all existing transactions are immediately terminated.

3.13.2 Confidentiality of User and Signalling Data

All messages with subscriber-related information are transmitted in protected mode. A ciphering key K_c is generated from a random number RAND with the authentication key K_i and A8 algorithm (see Figure 3.78).

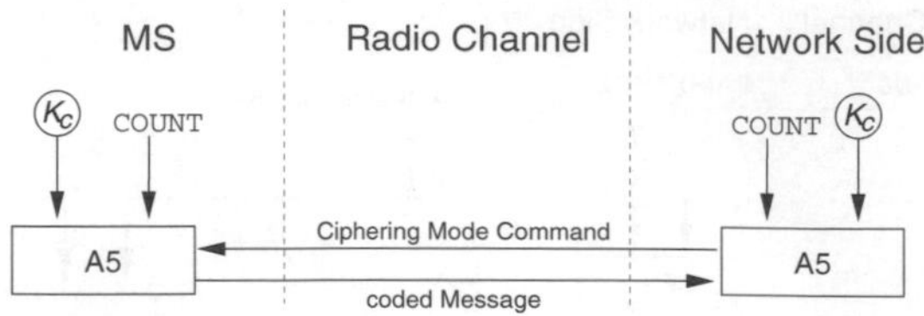

Figure 3.79: Ciphering mode

K_c is not transmitted over the radio path, but is stored in the mobile station and recalculated with each authentication procedure. Synchronization between the mobile station and the network is ensured whereby the ciphering key is allocated a key number (`Count`), which is also stored in the mobile station and supplied to the network each time a new message is transmitted. If the network recognizes that data requiring protection is being transmitted, it initiates a ciphering procedure whereby it sends a Ciphering_Mode_Command message to the mobile station. The mobile station in turn encodes or decodes on the basis of a stream cipher procedure using the A5 algorithm and the key K_c (see Figure 3.79).

3.13.3 Confidentiality of Subscriber Identity

The *Temporary Mobile Subscriber Identity* (TMSI) is changed periodically to guarantee the confidentiality of the information transmitted by a mobile user. This prevents a mobile radio connection from being allocated to a certain mobile subscriber as a result of eavesdropping. The TMSI is always assigned temporarily by the current VLR and is transmitted coded to the mobile station. A TMSI is changed no later than when there is a change in the location area controlled by a VLR. However, it can also be changed even if there is no change in VLR. Figure 3.80 shows a diagram of the procedure location area update from the standpoint of security.

3.13.4 The Transport of Security-Related Information between MSC, HLR and VLR

The *Authentication Centre* (AuC) is set up separately with special protection or integrated into the HLR, and guarantees the security of the network. It carries out the following tasks:

- Generation of key K_i as well as its allocation to IMSI.
- Generation of records `RAND`/`SRES`/K_c per IMSI for transfer to the HLR.

For updating the location area a VLR requires security-related information, which it obtains as follows:

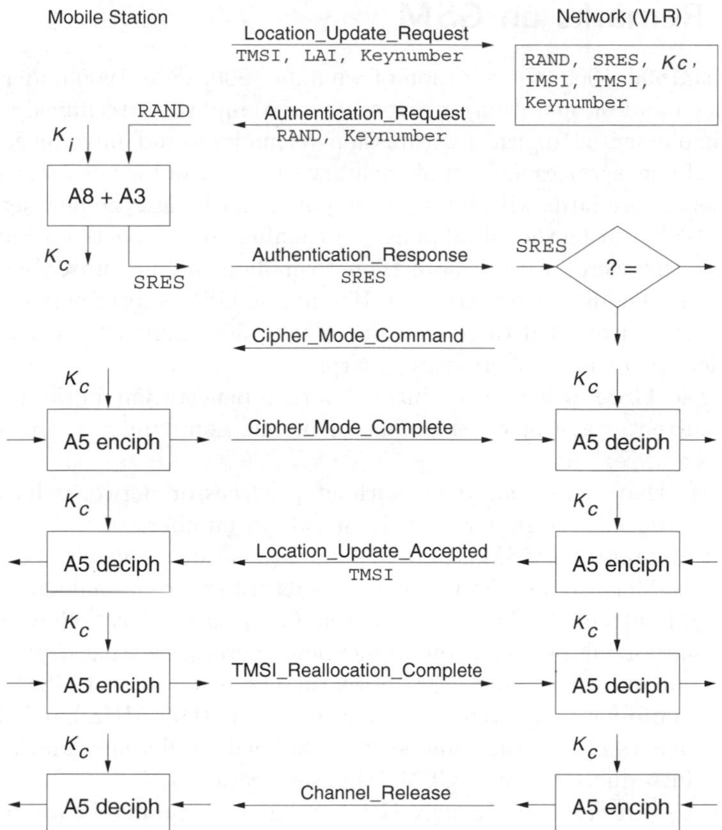

Figure 3.80: Location area update from the standpoint of security

- If a mobile station identifies itself using an IMSI, the VLR sends a request to the HLR for five records RAND/SRES/K_c, which are allocated to this IMSI.
- If a mobile station identifies itself using a TMSI and momentary location area identification LAI, the new VLR requests the IMSI as well as the records RAND/SRES/K_c from the old VLR that have not been used up.

If a mobile station roams in the area of a VLR for a longer period of time, after several authentications the VLR requires new RAND/SRES/K_c records. It receives them by sending a request to the HLR of the respective mobile user. Authentication itself takes place in the VLR whereby it sends the RAND to the MSC. The MSC establishes the authenticator SRES and transmits it to the VLR, which compares the value of the authenticator with that received from the mobile station. If the authentication is successful, the VLR allocates a TMSI to the IMSI. The ciphering key and the TMSI are sent to the MSC.

3.14 Remarks on GSM

GSM 900/1800/1900 (the addition of a number 900/1800/1900 underlines the frequency band of operation) incorporates two important technical advances: it has implemented digital transmission technology and introduced a large number of new services and supplementary services, and it has replaced existing national standards with international ones. Although the new services offered by GSM during the initial phase were mainly attractive to business users, by 1998 there were already more than 45 million mobile subscribers in Europe, by 1st August 1998 there were 100 million GSM subscribers worldwide, and by the end of 2000 there were more than 300 million GSM subscribers worldwide with the numbers rising sharply.

The role GSM will play in future market penetration initially depends on the current stage of development of mobile communication in different countries and regions.

Markets that were opened up without predecessor networks have shown that GSM subscribers grew relatively quickly in number.

In existing markets GSM is able to eliminate national capacity bottlenecks, making it sufficiently attractive to entice subscribers from analogue networks to the GSM system. In developed markets GSM particularly offers the possibility of supporting personal communication, thereby creating new communications opportunities. Table 3.34 shows the extent of current GSM coverage (without claiming completeness). In Europe the 1800 MHz band has been opened since 1992 and the same system, but with different radio front-end, has been introduced there as GSM 1800, see Section 3.15.

In the *United States of America* (USA) many regional mobile radio network operators have decided to operate GSM networks—albeit in the 1900 MHz frequencies, which means that special radio front ends are required for the licenced bands. These systems are called GSM 1900 systems. Operators of digital mobile radio systems with a different radio interface use the services and protocols of GSM in network and switching subsystems (see Figure 3.1).

Since the introduction of dual-band GSM phones it makes sense to list GSM 900 and GSM 1800/1900 operators together. 1900 MHz is used in the *United States* (US), Canada and Japan, the rest of the world uses 900 MHz (usually first two operators) and 1800 MHz (usually third and fourth operator).

Table 3.34: GSM coverage at 900, 1800 and 1900 MHz (as of September 2000)

Country	Operator Name	Network Code
Albania	AMC	276 01
Algeria	AMN	603 01
Amer. Samoa	Blue Sky	544 11
Andorra	STA-Mobiland	213 03
Antigua	Apua PCS	
Armenia	Armentel	283 01

Continued

Table 3.34: GSM coverage (as of 09/2000) (continued)

Country	Operator Name	Network Code
Australia	Optus	505 02
	Telecom/Telstra	505 01
	Vodafone	505 03
	One.Tel	505 08
Austria	Mobilkom Austria	232 01
	max.mobil.	232 03
	Connect Austria	232 05
	Telering	
Azerbaijan	Azercell	400 01
	JV Bakcell	400 02
Bahrain	Batelco	426 01
Bangladesh	Grameen Phone Ltd	470 01
	TM International	470 02
	Sheba Telecom	
Belgium	Proximus	206 01
	Mobistar	206 10
	KPN Orange	206 20
Benin	Libercom	616 01
	Spacetel	
	Telecel	
Bolivia	Nuevatel	
Bosnia	Eronet	
	Mobilna	218 05
	PTT Bosnia	218 19
Botswana	Mascom Wireless	652 01
	Vista Cellular	652 02
Brunei	DSTCom	528 11
Bulgaria	MobilTel AD	284 01
Burkina Faso	OnaTel	613 01
Burundi	Safaris	
	Spacetel	
	Telecel	
Cambodia	CamGSM	456 01
	Cambodia Samart	456 02
	Cambodia Shinawatra	
Cameroon	PTT Cameroon Cellnet	624 01
	Mobilis	624 02
Cape Verde	Cabo Verde Telecom	625 01
Canada	Microcell	302 37
Chile	Entel Telefonia	
China	China Telecom	460 00
	China Unicom	460 01
Congo	African Telecoms	
	Congolaise Wireless	
Croatia	HR Cronet	219 01
	Vipnet	219 10
Cyprus	CYTA	280 01
Czech Rep.	Eurotel Praha	230 02
	Radio Mobil	230 01
	OSKAR	230 02
Denmark	Sonofon	238 02
	Tele Danmark Mobil	238 01

Continued

Table 3.34: GSM coverage (as of 09/2000) (continued)

Country	Operator Name	Network Code
	Mobilix	238 30
	Telia	238 20
Egypt	MobiNil	602 01
	Click GSM	602 02
Estonia	EMT	248 01
	Radiolinja Eesti	248 02
	Q GSM	248 03
Ethiopia	ETA	636 01
Faroe Isl.	Faroese Telecom	
Fiji	Vodafone	542 01
Finland	Radiolinja	244 05
	Sonera	244 91
	Alands Mobiltelefon	244 05
	Telia	244 03
	Finnet	244 09
	Lännen Puhelin	244 09
	Helsingin Puhelin	244 09
France	France Telecom	208 01
	SFR	208 10
	Bouygues Telekom	208 20
Fr. Polynesia	Tikiphone	547 20
Fr. W. Indies	Ameris	340 01
Georgia	Superphone	
	Geocell	282 01
	Magticom	282 02
Germany	D1, DeTeMobil	262 01
	D2, Vodaphone	262 02
	E-Plus Mobilfunk	262 03
	Viag Interkom	262 07
Ghana	Franci Walker Ltd	
	ScanCom	620 01
Gibraltar	GibTel	266 01
Greece	Panafon	202 05
	STET	202 10
	Cosmote	202 01
Greenland	Tele Greenland	290 01
Guam	Guam Wireless	
Guinea	Telecel	611 ??
	Mobilis	611 01
	Lagui	611 02
Hong Kong	HK Hutchison	454 04
	SmarTone	454 06
	Telecom CSL	454 00
	New World PCS	454 10
	Sunday	454 16
	Peoples Telephone	454 12
Hungary	Pannon GSM	216 01
	Westel 900	216 30
	Primatel/Vodaphone	216 70
Iceland	Post & Simi	274 01
	TAL	274 02
India	Airtel	404 10

Continued

Table 3.34: GSM coverage (as of 09/2000) (continued)

Country	Operator Name	Network Code
	Essar	404 11
	Maxtouch	404 20
	BPL Mobile	404 21
	Command	404 30
	Spice Cell	404 31
	Spice - Karnataka	404 44
	Skycell	404 40
	RPG MAA	404 41
	Spice Punjab	404 14
	Sterling Cellular	404 11
	BPL Maharshtra	404 27
	Koshika	
	Bharti Telenet	
	Cellular Comms	
	TATA	404 07
	Escotel Haryana	404 12
	Escotel Kerala	404 19
	Escotel UP	404 56
	JTM Andhra Pradesh	404 49
	JTM Karnataka	404 45
	Evergrowth Telecom	
	Aircell Digilink	404 15
	Hexacom India	404 70
	Reliance Telecom	
	Fascel Limited	404 05
	AT&T Guajarat	404 24
	AT&T Goa	404 22
	BPL Kerala	404 46
	BPL Tamil Nadu	404 43
	Aircell	404 42
Indonesia	TELKOMSEL	510 10
	PT Satelit Palapa	510 01
	Excelcom	510 11
	PT Indosat	
Iraq	Iraq Telecom	418 ??
Iran	T.C.I.	432 11
	Kish Free Zone	
Ireland	Eircell	272 01
	Digifone	272 02
	Meteor	272 03
Israel	Partner Communications	425 01
Italy	Omnitel	222 10
	Telecom Italia Mobile	222 01
	Wind	222 88
	Blu	222 98
Ivory Coast	Ivoiris	612 03
	Comstar	612 01
	Telecel	612 05
Jordan	JMTS	416 01
	Mobilecom	
Kazakhstan	K-Cel	401 02
	K-Mobile	401 01

Continued

Table 3.34: GSM coverage (as of 09/2000) (continued)

Country	Operator Name	Network Code
Kenya	Kencell	639 03
	Safaricom	639 02
Kuwait	MTCNet	419 02
	Wataniya Telecom	
Kyrgyz Rep.	Bitel Ltd	437 01
La Reunion	SRR	647 10
Laos	Lao Shinawatra	457 01
Latvia	LMT	247 01
	BALTCOM GSM	247 02
Lebanon	Libancell	415 03
	Cellis	415 01
Lesotho	Vodacom	651 01
Liechtenstein	Natel-D	228 01
Lithuania	Omnitel	246 01
	Bite GSM	246 02
Luxembourg	P&T LUXGSM	270 01
	Tango	270 77
Lybia	Orbit	
	El Madar	
Macao	CTM	455 01
Macedonia	PTT Makedonija	294 01
Madagascar	Sacel	646 03
	Madacom	646 01
	SMM	646 02
Malawi	TNL	650 01
Malaysia	Celcom	502 19
	Maxis	502 12
	TM Touch	502 13
	Time	502 17
	Digi Telecom	502 16
Malta	Telecell	278 01
Marocco	O.N.P.T.	604 01
Mauritius	Cellplus	617 01
Moldova	Voxtel	259 01
	Moldcell	259 02
Monaco	Itineris	208 01
	SFR	208 10
	Office des Telephones	
Mongolia	MobiCom	
Montenegro	Pro Monte	220 02
Mozambique	Telecom de Mocambique	634 01
	T.D.M. GSM1800	
Namibia	MTC	649 01
Netherlands	PTT Netherlands	204 08
	Libertel	204 04
	Telfort Holding NV	204 12
	Ben	204 16
	Dutchtone	204 20
New Caledonia	Mobilis	546 01
New Zealand	Vodaphone NZ	530 01
Nigeria	United Net	
	Africell	

Continued

Table 3.34: GSM coverage (as of 09/2000) (continued)

Country	Operator Name	Network Code
	Celia	621 03
	Comm Invest	
	Integrated Mobile	
	Nigerian Mobile	
	Reliance	
Norway	NetCom	242 02
	TeleNor Mobil	242 01
Oman	General Telecoms	422 02
Palestinia	Jawwal	425 05
Pakistan	Mobilink	410 01
	Pak Telecom	
Papua	Cellnet	310 01
Philippines	Globe Telecom	515 02
	Islacom	515 01
	Smart	515 03
Poland	Plus GSM	260 01
	ERA GSM	260 02
	IDEA Centertel	260 03
Portugal	Telecel	268 01
	TMN	268 06
	Optimus	268 03
Qatar	Q-Net	427 01
Romania	MobiFon	226 01
	MobilRom	226 10
	Cosmorom	
Russia	Mobile Tele... Moscow	250 01
	United Telecom Moscow	
	NW GSM, St. Petersburg	250 02
	Dontelekom	250 10
	JSCKuban	250 13
	BM Telecom	250 07
	Beeline	250 99
	Extel	250 28
	Far Eastern Cell	250 12
	Baykal West	
	Ermak RMS	250 17
	Gorizont	
	SCS Rus	250 05
	NCC	250 03
	OAO	
	Primtelefone	
	Uraltel	250 39
	North Caucasian	250 44
	XXI	250 93
	Udmurtia	
	Smarts	250 07
Rwanda	Rwandacell	635 10
San Marino	Omnitel	222 10
	Telecom Italia Mobile	222 01
	Wind	222 88
	Blu	222 98
Saudi Arabia	Al Jawal	420 01

Continued

Table 3.34: GSM coverage (as of 09/2000) (continued)

Country	Operator Name	Network Code
	EAE	420 07
Senegal	Sonatel	608 01
	Sentel	608 02
Seychelles	SEZ SEYCEL	633 01
	Airtel	633 10
Sierra Leone	Celltel	
Singapore	Singapore Telecom	525 01
	MobileOne	525 03
	Sin Tel 1800	525 02
	Starhub	525 05
Slovak Rep.	Eurotel	231 02
	Globtel	231 01
Slovenia	Mobitel	293 41
	SI.Mobil	293 40
Somalia	Barakaat	637 01
South Africa	MTN	655 10
	Vodacom	655 01
Sri Lanka	MTN Networks Pvt Ltd	413 02
Spain	Airtel	214 01
	Telefonica Spain	214 07
	Amena	214 03
Sudan	Mobitel	634 01
Swaziland		
Sweden	Comviq	240 07
	Europolitan	240 08
	Telia Mobile	240 01
Switzerland	Swisscom 900	228 01
	Swisscom 1800	228 01
	diAx	228 02
	Orange	
Syria	SYR MOBILE	417 09
Taiwan	LDTA	466 92
	Mobitai	466 93
	TransAsia	466 99
	TWN	466 97
	Tuntex	466 06
	KG Telecom	466 88
	FarEasTone	466 01
	Chunghwa	466 11
Tanzania	Tritel	640 01
Thailand	TH AIS GSM	520 01
	Total Access Comms	520 18
	WCS	520 10
	Hello	520 23
Tunisia	Tunisian PTT	605 02
Turkey	Telsim	286 02
	Turkcell	286 01
UAE	UAE ETISALAT-G1	424 01
	UAE ETISALAT-G2	424 02
Uganda	Celtel Cellular	641 01
	MTN	641 10
UK	Cellnet	234 10

Continued

Table 3.34: GSM coverage (as of 09/2000) (continued)

Country	Operator Name	Network Code
	Vodafone	234 15
	Jersey Telecom	234 50
	Guernsey Telecom	234 55
	Manx Telecom	234 58
	One2One	234 30
	Orange	234 33
Ukraine	Mobile comms	255 01
	Golden Telecom	255 05
	Radio Systems	255 02
	Kyivstar JSC	255 03
USA	Bell South	310 15
	Sprint Spectrum	310 02
	Voice Stream	310 26
	Aerial Comms.	310 31
	Omnipoint	310 16
	Powertel	310 27
	Wireless 2000	310 11
	Pacific Bell	
	Airadigm	
Uzbekistan	Daewoo GSM	434 04
	Coscom	434 05
	Buztel	434 01
Vatican City	Omnitel	222 10
	Telecom Italia Mobile	222 01
	Wind	222 88
	Blu	222 08
Venezuela	Infonet	734 01
	Digitel	734 02
Vietnam	MTSC	452 01
	DGPT	452 02
Yugoslavia	Mobtel BK-PTT	220 01
	Pro Monte	220 02
	Telekom Serbia	220 03
Zaire	African Telecom Net	
Zambia	Zamcel	
	Telecel	
Zimbabwe	NET*ONE	648 01
	Telecel	648 03
	Econet	648 04

3.15 ETSI/DCS 1800 Digital Mobile Radio Network

The ETSI/DCS 1800 (*Digital Cellular System at 1800 MHz*) mobile radio network is regarded as an extension as well as a competitor to the GSM 900 network. The *Digital Cellular System* (DCS) 1800 standard developed by ETSI is based on the GSM recommendations but applied to the higher frequency range at 1800 MHz. This is the reason why nowadays DCS 1800 is termed

GSM 1800. The frequency ranges assigned in Europe are 1710–1785 MHz (uplink) and 1805–1880 MHz (downlink) (see Appendix C). A total of 374 carrier frequencies are therefore available to the system.

The entire technical infrastructure of the DCS 1800 system is directed at the mass market for personal mobile communication—a so-called *Personal Communication System* (PCS) planned to be comparable to the system licensed for the *Federal Communications Commission* (FCC)—the frequency regulator in the USA—frequency band at 1900 MHz, with:

- Low communication costs similar to those of the fixed network.
- Very high network capacity.
- Lightweight, compact radio telephone.
- Radio coverage also in buildings.

The DCS 1800 system meets the following requirements of PCS systems:

- High traffic density of 500 Erl./km^2.
- Lower power consumption of 250 mW up to 2 W.
- Use of GSM half-rate codec.
- Cost-effective implementation of the network.

The DCS 1800 system follows the GSM recommendations, except for the frequency-specific specifications, which is evident from the common architectural components (see Figure 3.81) with network elements which for the most part have already been shown in Figure 3.1: *Base Transceiver Station* (BTS), *Base Station Controller* (BSC), *Mobile-services Switching Centre* (MSC), *Operating and Maintenance Centre* (OMC), *Service Creation and Accounting Centre* (SCAC), *Home Location Register* (HLR), *Visitor Location Register* (VLR) und *Equipment Identity Register* (EIR).

The deviations of DCS 1800 from the GSM 900 specifications were published in 11 Delta recommendations (see Appendix E). The changes affect the radio interface definition, which differs owing to the low transmitter power required (250 mW up to 2 W in DCS 1800 compared with 5 up to 10 W in GSM). Because of the low transmitter power and the transmission attenuation, which on average is 10 dB higher at 1800 MHz than at 900 MHz, the cell radii in DCS 1800 are smaller than in GSM 900. With DCS 1800, cell radii of a maximum of 1 km are in use in urban areas, whereas in open country radio cells of up to 8 km (macrocells) are supported.

There will be two important different cell sizes in urban areas: microcells with a radius of over 150 m and picocells for the support of mobile radio communication inside buildings. Smaller cells have the advantage of higher traffic capacity, thus providing better support of personal communication. However, DCS 1800 must plan for considerably more cells than are required

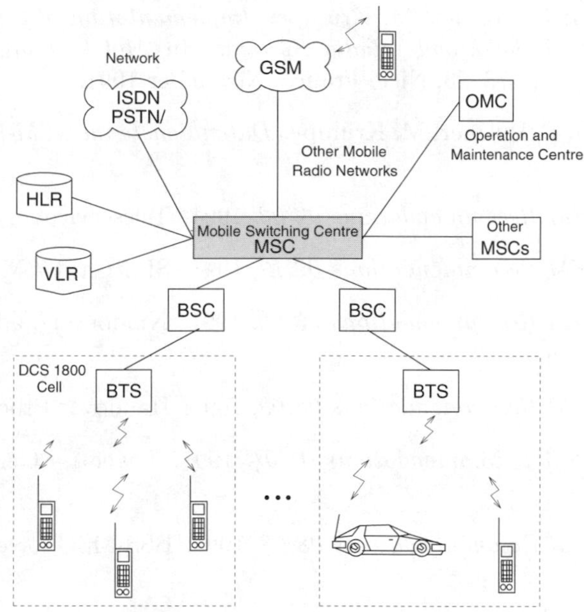

Figure 3.81: The architecture of the DCS 1800 system

in GSM, resulting in a higher investment in the development of the network infrastructure.

So that there is no great deviation between the GSM 900 and the DCS 1800 specifications, it was agreed that during the second phase of DCS 1800 standardization ETSI would combine the different documentations into a joint range of specifications. Other aspects to be examined in the second phase cover local routing and roaming between GSM 900 and DCS 1800, extension of the DCS 1800 service to include coin-operated telephones, and interworking with Inmarsat and satellite services [29].

Great Britain was considered to be the trailblazer in the introduction of the DCS 1800 system, which is called a *Personal Communication Network* (PCN) there and was put into operation in early 1993.

Bibliography

[1] J. C. Arnbak. *The European (R)evolution of Wireless Digital Networks. IEEE Communications Magazine*, Vol. 31, pp. 74–82, September 1993.

[2] P. Decker, B. Walke. *Performance Analysis of Fax Transmission on Non-Transparent GSM Data Service.* In *Mobile and Personal Communication*, pp. C340.03–C340.17, Brighton, UK, IEEE, December 1993.

[3] I. Dittrich, P. Holzner, M. Krumpe. *Implementation of the GSM-Data-Service into the Mobile Radio System.* In *Mobile Radio Conference (MRC'91)*, pp. 73–83, Nice, France, November 1991.

[4] I. Dittrich, P. Holzner, M. Krumpe. *Datendienste im GSM-Mobilfunksystem*, 1992.

[5] ETSI. *GSM Recommendations 02.03*, 1991. Teleservices.

[6] ETSI. *GSM Recommendations 02.17*, 1991. SIM Card / Module.

[7] ETSI. *GSM Recommendations 03.03*, 1991. Numbering, addressing and identification.

[8] ETSI. *GSM Recommendations 03.09*, 1991. Handover Procedures.

[9] ETSI. *GSM Recommendations 04.04*, 1991. Layer 1 - General requirements.

[10] ETSI. *GSM Recommendations 08.08*, 1991. BSS-MSC Layer 3 specification.

[11] ETSI. *GSM Recommendations 08.58*, 1991. BSC-BTS Layer 3 Specification.

[12] ETSI. *GSM Recommendations 04.05*, 1993. Data Link Layer - General aspects.

[13] ETSI. *GSM Recommendations 04.06*, 1993. MS-BSS Interface, Data link layer specification.

[14] ETSI. *GSM Recommendations 04.07*, 1993. Mobile radio interface, Layer 3 - General aspects.

[15] ETSI. *GSM Recommendations 04.08*, 1993. Mobile radio interface, Layer 3 specification.

[16] ETSI. *GSM Recommendations 05.02*, 1993. Multiplexing and multiple access on the radio path.

[17] ETSI. *GSM Recommendations 05.05*, 1993. Radio Transmission and Reception.

[18] ETSI. *GSM Recommendations 05.08*, 1993. Radio Sub-System Link Control.

[19] ETSI. TC-SMG. *European Digital Cellular Telecommunications System (Phase 2+), General Description of a GSM Public Land Mobile Network (PLMN), (GSM 01.02).* Draft Technical Specification 5.0.0, European Telecommunications Standards Institute, ETSI Secretariat, 06921 Sophia Antipolis Cedex, France, March 1996.

[20] ETSI. TC-SMG. *European Digital Cellular Telecommunications System (Phase 2+) (GSM 03.34) High Speed Circuit Switched Data: Stage 2 Service Description.* Draft Technical Specification 1.1.0, ETSI, ETSI Secretariat, 06921 Sophia Antipolis Cedex, France, November 1996.

[21] ETSI. TC-SMG 1. *European Digital Cellular Telecommunications System (Phase 2+); enhanced Multi-Level Precedence and Pre-emption service (eMLPP): Stage 1 (GSM 02.67).* Technical Specification 5.0.1, European Telecommunications Standards Institute, ETSI Secretariat, 06921 Sophia Antipolis Cedex, France, July 1996.

[22] ETSI. TC-SMG 1. *European Digital Cellular Telecommunications System (Phase 2+); Voice Broadcast Service (VBS): Stage 1 (GSM 02.68).* Technical Specification 5.1.0, European Telecommunications Standards Institute, ETSI Secretariat, 06921 Sophia Antipolis Cedex, France, March 1996.

[23] ETSI. TC-SMG 1. *European Digital Cellular Telecommunications System (Phase 2+); Voice Group Call Service (VGCS): Stage 1 (GSM 02.68).* Technical Specification 5.1.0, European Telecommunications Standards Institute, ETSI Secretariat, 06921 Sophia Antipolis Cedex, France, March 1996.

[24] ETSI. TC-SMG 3. *European Digital Cellular Telecommunications System (Phase 2+); Base Station Controller—Base Transceiver Station (BSC-BTS) interface; Layer 3 Specification (GSM 08.58).* Technical Specification 5.2.0, ETSI, ETSI Secretariat, 06921 Sophia Antipolis Cedex, France, July 1996.

[25] ETSI. TC-SMG 3. *European Digital Cellular Telecommunications System (Phase 2+); Customized Applications for Mobile Network Enhanced Logic (CAMEL), (GSM 03.78).* Technical Specification 0.8.0, European Telecommunications Standards Institute, ETSI Secretariat, 06921 Sophia Antipolis Cedex, France, July 1996.

[26] ETSI. TC-SMG 3. *European Digital Cellular Telecommunications System (Phase 2+); enhanced Multi-Level Precedence and Pre-emption service (eMLPP): Stage 2 (GSM 03.67).* Technical Specification 5.0.0, ETSI, ETSI Secretariat, 06921 Sophia Antipolis Cedex, France, February 1996.

[27] ETSI. TC-SMG 3. *European Digital Cellular Telecommunications System (Phase 2+); Voice Broadcast Service (VBS): Stage 2 (GSM 03.68).* Technical Specification 5.1.0, European Telecommunications Standards Institute, ETSI Secretariat, 06921 Sophia Antipolis Cedex, France, July 1996.

[28] ETSI. TC-SMG 3. *European Digital Cellular Telecommunications System (Phase 2+); Voice Group Call Service (VGCS): Stage 2 (GSM 03.68).*

Technical Specification 5.1.0, European Telecommunications Standards Institute, ETSI Secretariat, 06921 Sophia Antipolis Cedex, France, May 1996.

[29] A. D. Hadden. *Development of the DCS 1800 Standard.* In *Mobile Radio Conference (MRC'91)*, pp. 11–15, Nice, France, November 1991.

[30] ITU. ITU-T Study Group. *ITU-T Recommendation V.110: Support of Data Terminal Equipments with V-serives Type Interfaces by an Integrated Services Digital Network.* Technical report, International Telecommunication Union - Telecommunication Standardization Sector, Geneva, Switzerland, March 1992.

[31] ITU. ITU-T Study Group XI (Switching and Signalling). *ITU-T Recommendation Q.735: Multi-Level Precedence and Preemption (MLPP).* Technical report, International Telecommunication Union - Telecommunication Standardization Sector, Geneva, Switzerland, 1993.

[32] U. Janssen, B. Nilsen. *The Mobile Application Part for GSM Phase 2.* In *Mobile Radio Conference (MRC'91)*, pp. 65–72, Nice, France, November 1991.

[33] M. Mouly, M.-B. Pautet. *The GSM System for Mobile Communications.* M. Mouly and Marie-B. Pautet, 49, Rue Louise Bruneau, F-91129 Palaiseau, France, 1992.

4
Enhanced GSM Services*

In the framework of the evolution of the *Global System for Mobile Communication* (GSM) towards *Third-Generation* (3G) mobile communication systems known as *Universal Mobile Telecommunication System* (UMTS) new standards are presently integrated into the existing mobile radio networks. The driving force for this development is the predicted user demand for mobile data services that will offer mobile multimedia applications and mobile Internet access.

After *High-Speed Circuit-Switched Data* (HSCSD) has been introduced in some countries in 1999, the *General Packet Radio Service* (GPRS) will be available in 2001 in Europe and many countries worldwide. With these new services mobile multimedia applications with net bit rates of up to 117 kbit/s will be offered and established on the market. To realize mobile real-time applications as the next step the *European Telecommunications Standards Institute* (ETSI) has been developing the *Enhanced Data Rates for GSM Evolution* (EDGE) standard, which offers a net bit rate of up to 384 kbit/s by means of modified modulation, coding and medium access techniques. The packet-oriented part is the *Enhanced General Packet Radio Service* (EGPRS). The circuit-switched part is the *Enhanced Circuit-Switched Data* (ECSD) that extends the capabilities of HSCSD by introducing the EDGE philosophy mentioned. In the following the packet data service GPRS and the EDGE packet-switched services are introduced followed by a performance evaluation of GPRS and EDGE networks.

4.1 GPRS—The General Packet Radio Service

The main intention of integrating the GPRS into the GSM is to increase the number of connections per bearer by utilizing the given physical channels more efficiently than the existing Phase 2 service.

GPRS has been standardized by the ETSI as part of the GSM Phase 2+ development. It represents the first implementation of packet switching within GSM, which is essentially a circuit-switched technology.

Packet-switching means that GPRS radio resources are used only when users are actually sending or receiving data. Rather than dedicating a radio

*With the collaboration of Götz Brasche and Peter Stuckmann

channel to a mobile data user for a fixed period of time, the available radio resource can be shared between several users. The actual number of users supported depends on the application being used and how much data is being transferred. Through multiplexing of several logical connections to one or more GSM physical channels, GPRS reaches a flexible use of channel capacity for applications with variable bit rates.

GPRS is extremely efficient in its use of scarce spectrum resources and enables GSM operators to introduce a wide range of value-added services for market differentiation. It is ideal for 'bursty' type data applications such as email, Internet access or *Wireless Application Protocol* (WAP) based applications. It brings IP capability to the GSM network for the first time and enables connection to a wide range of public and private data networks using industry standard data protocols such as TCP/IP and X.25.

The standardization of the GPRS was widely concluded in 1997. However, regularly details were discussed in the boards of standardization and the GPRS standard was modified in 1998. After this, in the framework of the standardization of EDGE the GPRS standard has advanced to the EGPRS standard. First proposals for a packed-oriented data service in GSM have been published in 1991 [49, 4, 5, 48, 50].

4.1.1 Service Properties

An important aspect for the introduction of the new service is the acceptance expected from the mobile subscribers. The acceptance depends on the *Quality of Service* (QoS) parameters and especially on the pricing of the service providers. Economically priced services can only be offered, if development costs are kept to the minimum. Therefore, it was the premise to change the GSM system components as little as possible and to develop the new service in consideration of the limits of the existent tele- and bearer services.

GPRS integrates a packet-based air interface in the existing circuit-switched GSM network. The GSM infrastructure had not to be replaced. Only a couple of new network elements have been added (see Section 4.1.4).

The GPRS specification does not provide an upper limit for the amount of data, which can be transmitted per access. Since the transmission of long packets in packet-oriented systems reduces the multiplexing gain, GPRS is most suitable for

- the frequent, regular transmission of short data packets up to 500 bytes, and
- the irregular transmission of short to medium-sized data packets up to a few kbytes.

The basic approach to integrate the packet data service into the GSM standard represents the reservation and the logical subdivision of certain GSM channels. The number of channels allocated for GPRS can be dynamically adapted to the workload situation in the respective radio cell.

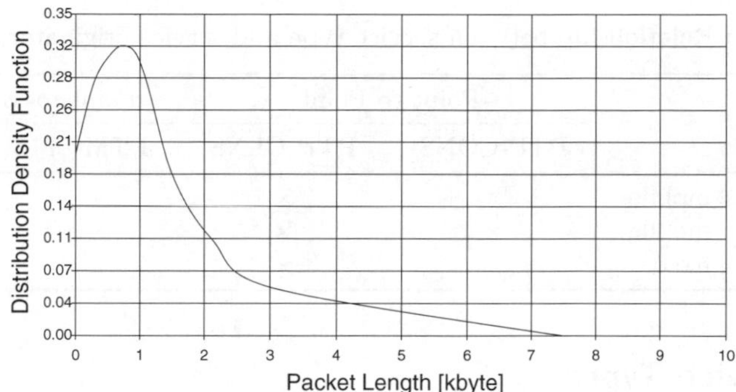

Figure 4.1: The distribution density function of the FUNET load model

The following traffic models were regarded for performance analysis of GPRS protocols in the framework of the protocol specification in the standardization process. They represent the three vertical markets prognosed for GPRS by ETSI at the beginning of the 1990s:

- Traffic reports.
- Fleet management.
- Goods/supply logistics.

Three different load models for identifying the typical applications have been defined by ETSI.

FUNET This traffic model is based on a statistical evaluation of the distribution of the length of electronic messages (e-mail) at the *Finish University and Research Network* (FUNET), which is approximated by a Cauchy distribution density function with a maximum message length of 10 kbytes and a mean value of 100 bytes (see Figure 4.1):

$$\text{Cauchy}(0.8.1) = f(x) = \frac{1}{\pi[1 + (x - 0.8)^2]} \tag{4.1}$$

Mobitex This traffic model is based on a statistical evaluation of the fleet management application in the Ericsson Mobitex System. Data packets with a uniform distribution of the length of 30 ± 15 bytes are sent on the uplink. The packet length on the downlink is at 115 ± 57 bytes.

Railway This model describes the anticipated distribution of packet lengths in train control applications through a curtailed negative exponential distribution with an average packet length of 256 bytes and a maximum length of 1000 bytes:

$$F(x) = 1 - e^{-x/256} \tag{4.2}$$

Table 4.1: Relationship between service type and service originator/receiver

Packet flow	Point-to-Point		Point-to-Multipoint	
	PTP-CONS	PTP-CLNS	PTM-M	PTM-G
fixed → mobile	×	×	×	×
mobile → mobile	×	×	×	×
mobile → fixed	×	×	–	×

4.1.2 Service Types

A PLMN provider is responsible for the data transfer between the *Service Access Point* (SAP) in the fixed network and the SAP in the MS. Two service categories are specified:

Point-To-Point (PTP) Using the PTP service single packets can be transmitted between two subscribers. The PTP service is offered both in a connection-oriented mode *Connection Oriented Network Service* (CONS) and in a connectionless mode *Connectionless Network Service* (CLNS). Applications using the PTP service can be classified regarding their communication properties into:

- **non-interactive** There is no dependence between the single data packets.
- **interactive or dialogue-oriented** There is a logical relationship between the service subscribers for a particular period. This period can range from several seconds to several hours.

Point-to-Multipoint (PTM) This service supports the transmission of data packets between a service user and a specified group inside a certain geographical region. It is divided into:

- **Multicast call (PTM-M)** *PTM Multi-cast* (PTM-M) comprises calls, which are broadcast in the area specified by the call initiator. Either all subscribers in this region or the members of a group are addressed.
- **Group call (PTM-G)** The messages are addressed explicitly to a specified group and are sent in all regions where group members are located.

Table 4.1 shows that services except the multicast service can be originated both from the mobile radio network (*Mobile Originated Call*, MO-C) and from the fixed network (*Mobile Terminated Call*, MT-C).

4.1.3 Parallel Use of Services

During a GPRS session circuit-switched services (speech as well as data) may still be initiated and used. Similarly, it is possible to send and receive GPRS data while carrying out a telephone call. Parallel use of these services is provided for PTP as well as PTM services, however, it causes fluctuating transfer rates according to traffic load and affects the QoS.

Furthermore, it is possible to additionally utilize the *Short Message Service* (SMS), for *Mobile Originated Calls* (MO-Cs) as well as *Mobile Terminated Calls* (MT-Cs). An SMS message may, however, be transmitted with delay or a lower data rate, according to the current load situation. An SMS cell broadcast message may be sent in parallel to GPRS services but is not allowed to be used simultaneously to a circuit-switched connection.

Three equipment and subscriber classes have been proposed to offer the user a packet-data service with various performance characteristics.

Subscriber class A Simultaneous usage of all services according service profile with unchanging speech quality.

Subscriber class B Restricted simultaneous usage of services with reduced speech quality.

Subscriber class C No simultaneous usage of services. For subscriber class C it is possible to receive short messages at any time, however.

4.1.4 Logical Architecture and Functional Structure

The existing GSM network does not provide sufficient functionality to realize a packet data service. Enabling GPRS on a GSM network requires the addition of components, which provide the packet-switched service (see Figure 4.2). Hence, the GSM network is extended by two additional nodes:

Gateway GPRS Support Node (GGSN): The GGSN serves as the interface towards external *Public Data Network* (PDN) or other *Public Land Mobile Network* (PLMN). Here, switching functions are fulfilled, e.g., the evaluation of the *Packet Data Protocol* (PDP) addresses and the routing to mobile subscribers via the SGSN.

Serving GPRS Support Node (SGSN): The SGSN represents the GPRS switching centre in analogy to the MSC. Packet data addresses are evaluated and mapped onto the IMSI. The SGSN is responsible for the routing inside the packet radio network and for mobility and resource management. Furthermore, it provides authentication and encryption for the GPRS subscribers.

For the communication between the SGSN and the GGSN within one PLMN, the Intra PLMN *IP Version 6* (IPv6) or *IP Version 4* (IPv4) backbone is used. SGSN and GGSN encapsulate (decapsulate resp.) the packets

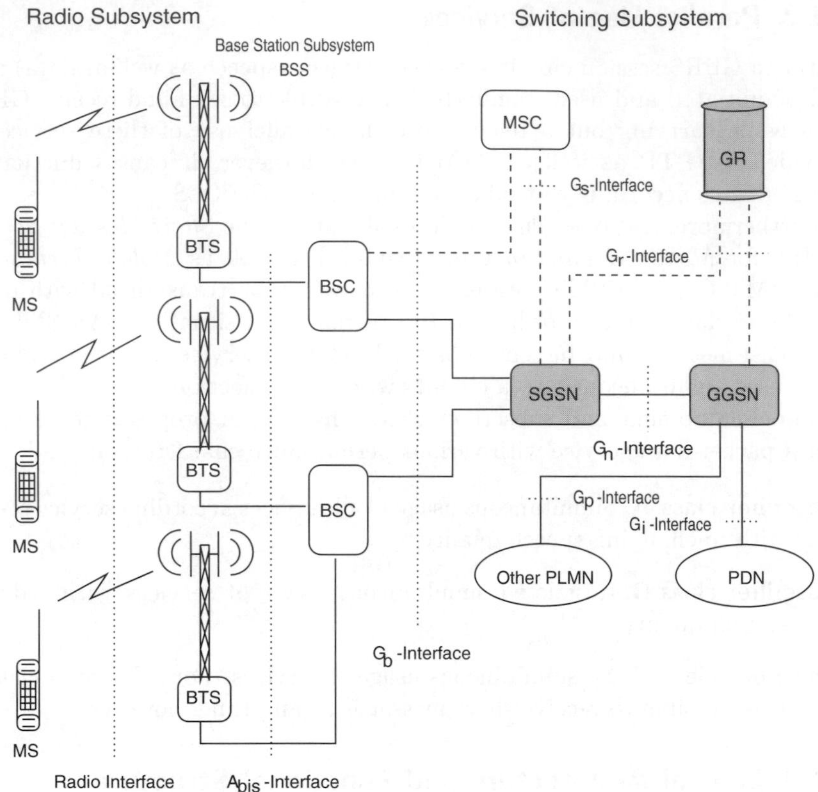

Figure 4.2: GPRS logical architecture

using a special protocol called the *GPRS Tunnelling Protocol* (GTP) that operates on top of standard *Transport Control Protocol* (TCP)/IP protocols. The SGSN and GGSN functions may also be combined into one physical node.

With the extension of the existing GSM network by GPRS interfaces and reference points had to be redefined. The defined interfaces are displayed in Figure 4.3. The dotted lines allude to signalling traffic between the corresponding elements. Solid lines mean that user data can be transmitted additionally at these reference points.

In GPRS the proper data traffic is handled by the SGSN. The MSC serves only for signalling.

4.1.5 Protocol Architecture

The GPRS protocol architecture follows the ISO/OSI reference model. The proposed protocol stack is shown in Figure 4.4. For GPRS, an interworking with TCP, IP, *Connectionless Network Protocol* (CLNP) and X.25 based networks is foreseen. Between SGSN and GGSN the encapsulated data pack-

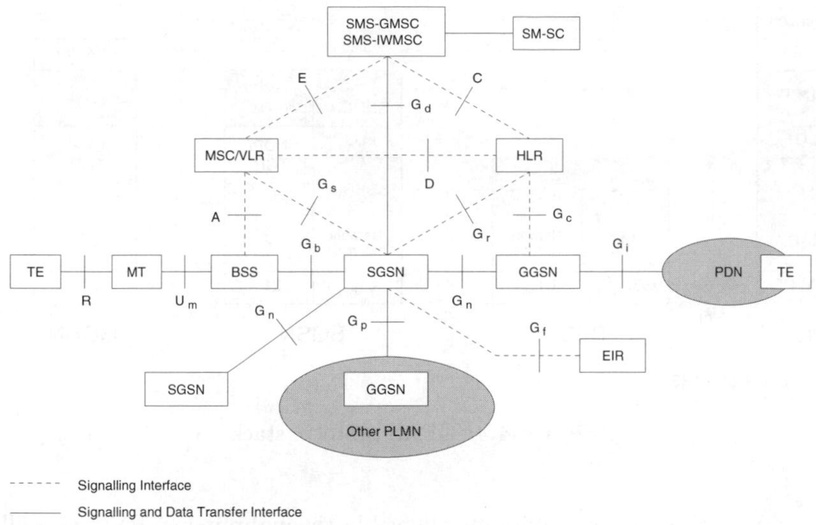

Figure 4.3: GPRS interfaces and reference points

ets with the signalling information will be transmitted with the help of the GTP, which realizes the IWF between the GSM protocol and specific network protocols.

4.1.5.1 Base Station Subsystem GPRS Protocol (BSSGP)

On the G_b interface, the *Base Station Subsystem GPRS Protocol* (BSSGP) [20] provides a connection-less link between BSS and SGSN. This protocol's main task is the management of flow control for the downlink transfer of *Logical Link Control* (LLC) PDUs. There is no flow control performed in uplink direction. The SGSN is expected to accept all information that is sent to it—the buffers and link capacity have to be dimensioned according to this premise to avoid loss of uplink data.

Flow Control between SGSN and BSS Each cell in the coverage area of one BSS is virtually connected to the SGSN by a *BSSGP Virtual Connection* (BVC), identified by a *BSSGP Virtual Connection Identifier* (BVCI). The BSS maintains one queue for each BVCI that contains the LLC PDUs for that particular cell. There are several ways to further split these queues, e.g., into one queue per MS, or into one queue that serves all LLC PDUs belonging to a certain delay or precedence class. The queues are filled with downlink LLC PDUs and emptied by the BSS forwarding these PDUs to the MSs addressed. The SGSN is regularly informed by the BSS about maximum queue size available for each BVC and MS queue at a given time, and also about the rate at which the available queue size increases. Thus, the SGSN is

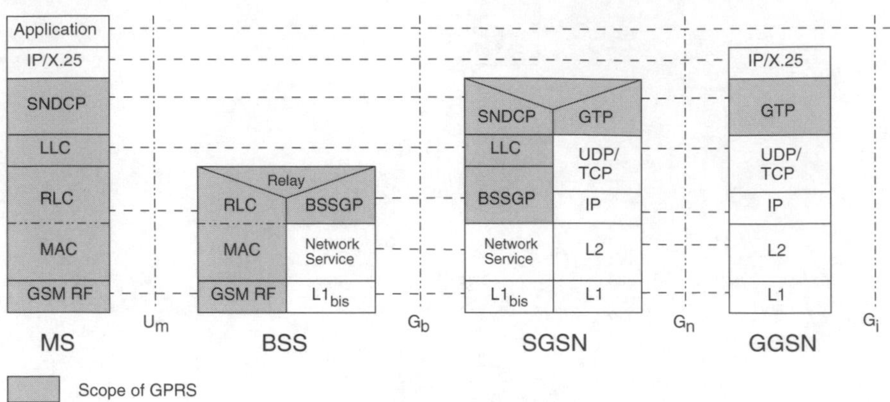

Figure 4.4: GPRS protocol stack

enabled to estimate the maximum allowable throughput per BVC as well as per MS belonging to the BVC.

Within the SGSN there are queues provided, similar to those in the BSS. Here the downlink LLC PDUs are queued and scheduled as long as the maximum allowable throughput per BVC and per MS is not exceeded. By this mechanism, it is ensured that both the MS and the BSS are capable of handling the incoming traffic.

BSS Context The SGSN can provide a BSS with information related to ongoing user data transmission. The information related to one MS is stored in a BSS context. The BSS may contain BSS contexts for several MSs. A BSS context contains a number of BSS *Packet Flow Contexts* (PFCs). Each BSS PFC is identified by a *Packet Flow Identifier* (PFI) assigned by the SGSN. A BSS PFC is shared by one or more activated *Packet Data Protocol* (PDP) contexts with identical or similar negotiated QoS profiles. The data transmission related to PDP contexts that share the same BSS PFC comprise one packet flow.

Three packet flows are predefined, and identified by three reserved PFI values. The BSS shall not negotiate BSS PFCs for these pre-defined packet flows with the SGSN. One pre-defined packet flow is used for best-effort service, one is used for SMS, and one is used for signalling. The SGSN can assign the best-effort or SMS PFI to any PDP context. In the SMS case, the BSS shall handle the packet flow for the PDP context with the same QoS that it handles SMS with.

The combined BSS QoS profile for the PDP contexts that share the same packet flow is called the *Aggregate BSS QoS Profile* (ABQP). The ABQP is considered to be a single parameter with multiple data transfer attributes as defined in Section 4.1.12.1. It defines the QoS that must be provided by the BSS for a given packet flow between the MS and the SGSN, i.e., for the U_m

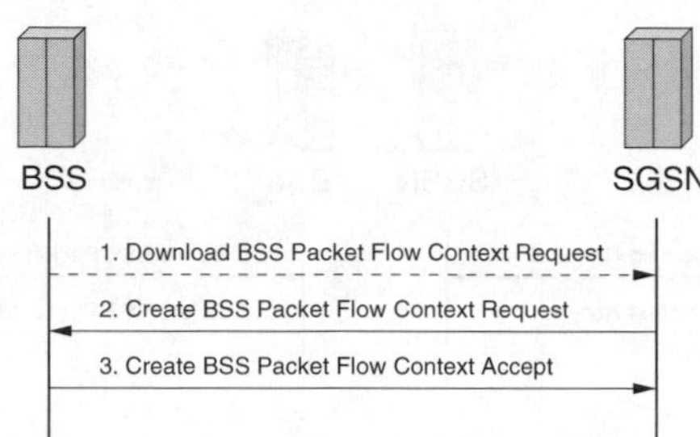

Figure 4.5: BSS PFC creation procedure

and G_b interfaces combined. The ABQP is negotiated between the SGSN and the BSS.

A BSS packet flow timer indicates the maximum time that the BSS may store the BSS PFC. The BSS packet flow timer shall not exceed the value of the `READY` timer for this MS. The BSS packet flow timer is started when the BSS PFC is stored in the BSS and when an LLC frame is received from the MS. When the BSS packet flow timer expires the BSS shall delete the BSS PFC.

When a PDP context is activated, modified, or deactivated (see Section 4.1.11.2), the SGSN may create, modify, or delete BSS PFCs.

BSS Packet Flow Context Creation On receiving a request to transmit an uplink or downlink LLC PDU for which no BSS PFC exists in the BSS, the BSS may request the download of the BSS PFC from the SGSN.

The SGSN may at any time request the creation of a BSS PFC, e.g., due to the activation of a PDP context.

The BSS PFC creation procedure is illustrated in Figure 4.5. If the BSS receives a request to transfer an uplink or downlink user data LLC PDU for which it currently does not have a BSS PFC, it sends a `Download BSS Packet Flow Context Request` message to the SGSN. Until the BSS receives the BSS PFC, the BSS will handle uplink and downlink transfers according to a default ABQP. For uplink transfers, the default profile is specific to the radio priority level.

On reception of the `Download BSS Packet Flow Context Request`, or in PDP context activation or modification procedures, the SGSN sends a `Create BSS Packet Flow Context Request` message to the associated BSS.

The BSS may restrict the requested ABQP given its capabilities and the current load. It then creates a BSS PFC, inserts the parameters in its BSS context, and returns a `Create BSS Packet Flow Context Accept` message to the SGSN. The BSS uses the negotiated ABQP when allocating radio resources and other resources such as buffer capacity.

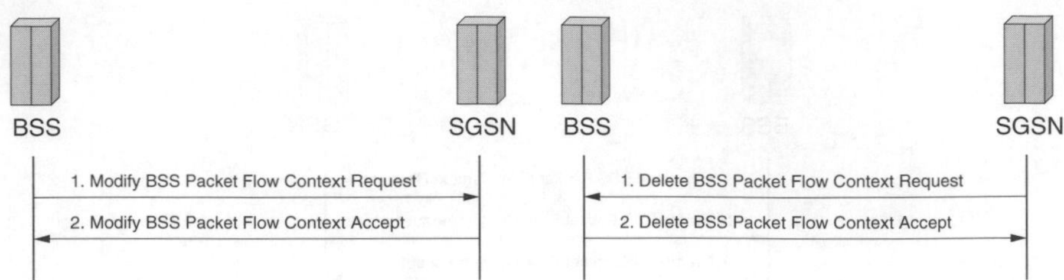

Figure 4.6: BSS-initiated BSS PFC modification procedure

Figure 4.7: SGSN-initiated BSS PFC deletion procedure

BSS Packet Flow Context Modification If the SGSN wishes to modify the contents of any existing BSS PFC, e.g., due to the activation, modification, or deactivation of a PDP context, it uses the BSS PFC creation procedure. The BSS will instead of creating a BSS PFC overwrite the existing parameters with the modified parameters.

The BSS itself can at any time request modification of the contents of an existing BSS PFC, e.g., due to a change in the resource availability at the BSS. Figure 4.6 illustrates the according procedure. The SGSN may again restrict the requested ABQP.

BSS Packet Flow Context Deletion The BSS can, owing to e.g., memory restrictions, at any time delete a BSS PFC without notifying the SGSN.

The SGSN may request the deletion of a BSS PFC with the SGSN-initiated BSS PFC deletion procedure, as illustrated in Figure 4.7.

4.1.5.2 Sub-Network Dependent Convergence Protocol (SNDCP)

The main functions of SNDCP [26] are:

- Multiplexing of several PDPs.
- Compression and decompression of user data and control information (e.g., TCP/IP header compression).
- Segmentation and reassembly of network PDUs.

Any (possibly compressed) N-PDU with SNDCP header is segmented by SNDCP if it is longer than the maximum payload size for an LLC frame. In the SNDCP protocol, an acknowledged and an unacknowledged mode of data transfer are defined. For data transfer in acknowledged mode, an SNDCP entity buffers an N-PDU until successful transmission of all SN-PDUs carrying a segment of the N-PDU. For unacknowledged data transfer mode, the SNDCP deletes an N-PDU immediately after it has been delivered to the LLC layer. Moreover, for unacknowledged mode, the SNDCP inserts a sequence number in all PDUs to allow PDUs to be reassembled in sequence at the receiver.

4.1.5.3 Logical Link Control (LLC)

The LLC layer [23] is responsible for the transportation of the data packets between the network layer of MS and SGSN. Both point-to-point and point-to-multipoint communication is supported. The essential functions are flow control and error correction (with ARQ and FEC mechanism).

The LLC layer includes functions for:

- Provision of one or more logical connections.
- Sequence control.
- Error detection and correction.
- Flow control and ciphering.

The protocol for the LLC sublayer was based on well-known data link protocols like the *Link Access Procedure on the D-channel* ($LAPD_m$) used in GSM and the *High Level Data Link Control* (HDLC) protocol standardized by ISO for *Open Systems Interconnection* (OSI).

The key modifications can be summarized as follows:

Variable frame length The GPRS protocol architecture with transmission protected at the radio interface by the RLC/MAC protocol allows a variable frame length at the LLC level. Therefore frame delimiters and bit stuffing are not necessary but an additional field is required in the frame header for specification of the frame length.

Variable address types New address types of different lengths are being introduced with the packet data service. The address field must therefore have a variable length, which is controlled by an expansion bit in the address field. Each *DLC Connection Identifier* (DLCC-ID) consists of a *Service Access Point Identifier* (SAPI) and a *Terminal Endpoint Identifier* (TEI), which can contain addresses such as the *Temporary Logical Link Identifier* (TLLI), TMSI, IMSI and *International Mobile GPRS Identity* (IMGI). The address types are allocated during the registration process and are controlled by the management entities.

Prioritized SAPIs The priority classes in GPRS are considered in the introduction of new SAPIs with priorities. Corresponding to four radio priority levels, at least four SAPIs are required. In accordance with the priority scheme based on eMLPP, it is conceivable that other SAPIs will be reserved for signalling or emergency calls. With the introduction of a priority system, it will also be necessary for flow control and error correction to be controlled on a priority basis.

Enhanced operation mode The protocol supports duplex communication for the acknowledged point-to-point service operating in the ABM. Since LAPD only supports PTP communication, the protocol for the PTM service is enhanced. Duplex communication for multipoint is not possible, so the operation mode being introduced for acknowledged data transfer is the *Asynchronous Unbalanced Mode* (AUM). Non-acknowledged data transfer is provided through the exchange of unnumbered information frames.

4.1.5.4 Radio Link Control (RLC) and Medium Access Control (MAC)

The idea of packet-switched services is to multiplex several users onto one single physical channel in order to use its capacity in a more efficient way. Furthermore, one single MS can use more than one *Packet Data Channel* (PDCH) simultaneously to increase the data rate. The maximum number of PDCHs that can be used in parallel is determined by the multislot capabilities of the MS, which can range from multislot capabilty 1 (one PDCH) up to multislot capability 8 (all eight PDCHs).

The Data Link Layer at the mobile U_m interface is divided in two sublayers: the *Radio Link Control* (RLC) layer and the *Medium Access Control* (MAC) layer, see Figure 4.4. The RLC sublayer provides services, which warrant a safe logical connection between MS and BSS, while the MAC sublayer controls the access to the physical medium, the radio link. The general objective of the packet service is to reserve dedicated physical GSM channels for the GPRS service and to divide these into logical channels.

RLC layer The RLC functions in GPRS provide an interface towards the LLC layer, especially the segmentation and reassembly of LLC PDUs into RLC data blocks depending on the used *Coding Scheme* (CS) (see Table 4.2). *Backward Error Correction* (BEC) is used to enable the selective retransmission of uncorrectable PDUs. The *Block Check Sequence* (BCS) for error detection in GPRS is provided already by the Physical Link Layer. There are two possible modes provided by the RLC/MAC layer: the acknowledged mode, used for reliable data transmission, and the unacknowledged mode, mainly used for real-time services, such as video or voice, where time delay is most critical, whereas bit errors are less important unless they top the accepted range. For RLC acknowledged mode, a selective ARQ protocol is used between the MS and the BS, which provides retransmission of erroneous RLC data blocks. As soon as a complete LLC frame is successfully transferred across the RLC layer, it is forwarded to the LLC layer.

MAC layer The GPRS MAC layer is responsible for providing efficient multiplexing of data and control signalling on the uplink and the downlink. The multiplexing on the downlink is controlled by the downlink scheduler, which has knowledge of the active MSs in the system and of the downlink traffic. Therefore, an efficient multiplexing on the PDCHs can easily be made. On the uplink, the multiplexing is controlled by channel reservation to individual MS. This is done by resource requests, which are sent by the MS to the network,

Table 4.2: Coding scheme parameters

Coding		Number of bits						
scheme	rate	USF (pre-coded)	Payload	BCS	Tail	Coded	Punct.	Data rate (kbit/s)
CS-1	1/2	3(3)	181	40	4	456	0	9.05
CS-2	≈ 2/3	3(6)	268	16	4	588	132	13.4
CS-3	≈ 3/4	3(6)	312	16	4	676	220	15.6
CS-4	1	3(12)	428	16	–	456	0	21.4

which then has to schedule the uplink PDCHs. Additionally, a contention resolution is needed when several MSs attempt simultaneously to access the control channel for resource requests. Collision detection and recovery are included when using the Slotted Aloha reservation protocol for recovery (see Section 2.8), which means that the MS has to wait a random amount of time before a new access attempt is done.

The MAC procedures include the provision of *Temporary Block Flow* (TBF), which allows a point-to-point transfer of data in one cell between the BSS and the MS. As a physical connection between two radio resource entities, the TBF is used to transport the RLC data blocks on the PDCHs. The TBF is maintained only for the duration of the data transfer and is released as soon as all data is sent (RLC send queue is cleared). In respect to *Voice over IP* (VoIP), the TBF is running only during talkspurts and is released when the sender buffer is cleared. A *Transport Format Identifier* (TFI) is assigned for the TBF by the network to associate the MS with the current TBF.

An *Uplink State Flag* (USF) is used by the network to control the multiplexing of different mobile stations on uplink PDCHs. It is included in the header of each RLC/MAC PDU on a downlink PDCH. The USF indicates which station is allowed to transmit during the next uplink radio block period. The MS, which has the ID indicated in the USF field, is allowed to transmit an RLC block in uplink direction on the same PDCH, on which it has received the radio block with the corresponding USF. Note that the USF field has only 3 bit which means that up to 7 MS can operate on one PDCH. One USF value is used to indicate the *Physical Random Access Channel* (PRACH) on the next uplink block period. Assuming a higher multislot capability, the MS occupies USFs on several PDCHs, which limits the total number of active MSs on one frequency.

Physical layer (PL) The *Physical Layer* (PL) at the air interface is divided into the *Physical Link Layer* (PLL) and the *Radio Frequency* (RF) layer. The specification of these sublayers is done in the GSM series 05 [19, 28]. While in the RF sublayer mainly modulation and demodulation are carried out, the PLL sublayer provides services for data transmission over the radio interface. The PLL sublayer is responsible for the FEC coding, allowing the detection

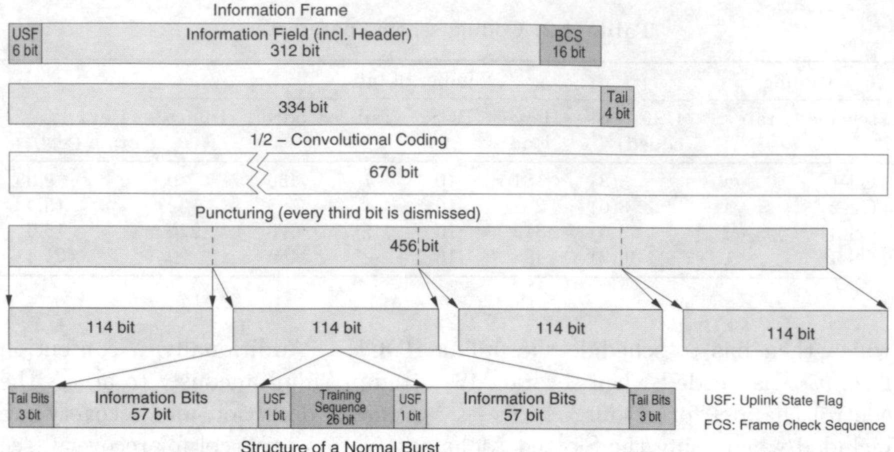

Figure 4.8: Coding of an information frame according to scheme 3

and correction of erroneous transmitted code words and the indication of uncorrectable code words. It is also responsible for the interleaving of one radio block over four bursts in consecutive TDMA frames and provides procedures for synchronization (*Timing Advance*, TA).

Channel coding and error correction The RLC/MAC protocol uses the punctured convolutional code of the GSM 1/2-convolutional coding. The decoding is carried out with the Viterbi *soft decision* decoder familiar from GSM.

The scheme used for the GSM SDCCH involving a convolutional code with a code ratio 1/2 and a 40-bit Fire code is used as standard for transmitting signalling information. For user information further code ratios are used, the redundancy of which adapts dynamically to the current signal quality (see Table 4.2).

The decoding of the USF is simplified whereby for coding schemes 2–4 a 12-bit long code word, which is not punctured, is generated for the USF. With schemes 2 and 3 the USF is precoded with 6 bit before the frame is convolutionally coded, with the first 12 bit not being punctured. When coding scheme 1 is used, the entire frame is coded and the USF must be decoded as part of the information data. Figure 4.8 shows an example of the coding of an information frame for coding scheme 3.

Both *stealing bits* of a GSM normal burst are used to indicate the coding scheme to be followed. Since four consecutive bursts accompany an RLC/MAC frame, the scheme can be protected against transmission error by an 8-bit block code with a Hamming distance 5.

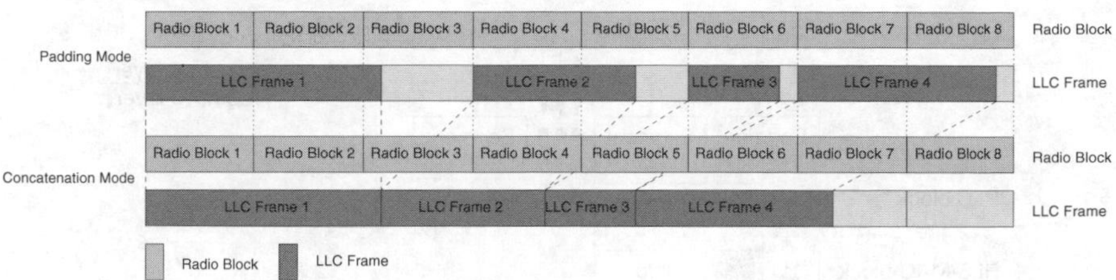

Figure 4.9: Segmentation of LLC Frames

Segmentation modes The segmentation of LLC frames can be performed in two different modes: padding mode or concatenation mode (see Figure 4.9).

Padding mode Using padding mode, only one LLC frame can be transported in one RLC data block, and if the length of the LLC frame does not fit into an integer number of RLC data block frames, the last RLC data block has to be filled up with padding bits. Therefore, transmission capacity is wasted.

Concatenation mode The concatenation mode is introduced corresponding to recommendation GSM 04.60. It provides the ability to transport several LLC frames within one single RLC data block. If the length of an LLC frame does not fit into an integer number of RLC data blocks, the free space in the last RLC data block can be used for the transmission of the next LLC frames.

If a consecutive LLC frame for instance is smaller than the remaining space in the actual RLC data block, it is possible to transmit the LLC frame as a whole in the RLC data block. The newly calculated free space can be used for the transmission of the beginning of the following LLC frame, and so on. Assuming a stream of LLC frames, the concatenation mode segments the transmission data into RLC data blocks without paying attention to the beginning or ending of the LLC frames.

Data flow Figure 4.10 comprises the packet data flow from LLC to PL. LLC frames are segmented, and provided with RLC and MAC header data, thus forming RLC/MAC blocks. Those are handed over to PL where a BCS and tail bits—necessary for the convolutional coder—are added. After the convolutional coding, the whole block is punctured and split into four radio blocks, which are then interleaved over four radio bursts in consecutive TDMA frames.

4.1.6 Radio Interface

The radio interface U_m is located between MS and BSS. The GSM recommendations combine a FDM and a TDM scheme to the radio resources.

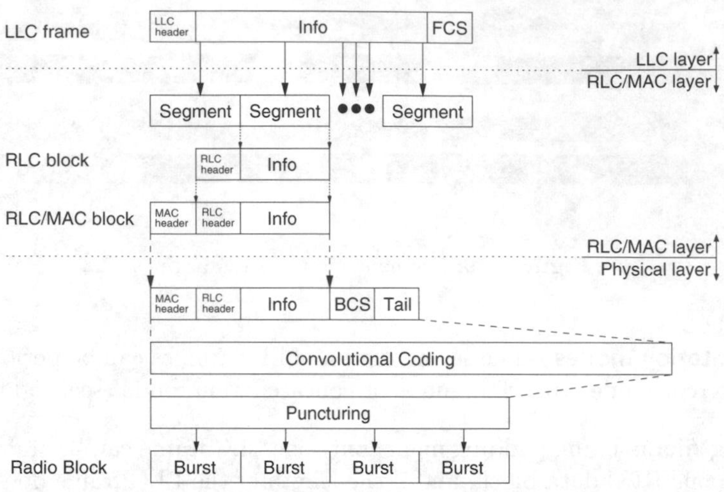

Figure 4.10: Packet data flow from LLC to PL

Frequency Division Multiplex (FDM) The spectrum is divided into several frequency bands with guard channel in between to prevent interference. Additionally, the uplink and the downlink are divided, reserving the frequency range 890–915 MHz for the uplink transfer and the range 935–960 MHz for the downlink direction.

Time Division Multiplex (TDM) Periodically, the frequency channel is allocated alternately to different logical channels. The time is divided into 8 *Timeslots* (TSs), recurring periodically. 8 TSs form a TDMA frame with a duration of 4.615 ms. Thus one slot takes 0.577 ms and is used by a burst of 148 bit, which is 8.25 bit shorter in duration to realize a guard time between the bursts in order to avoid overlapping with other bursts (see Figure 4.11).

There are four types of bursts differing in functionality and content:

Normal burst used for traffic and control channels.

Access burst call set up.

Synchronization burst to synchronize the sender and receiver.

Dummy burst indicates an empty slot.

The communication over the GPRS radio interface comprises functions of the RLC/MAC layer and the physical layer.

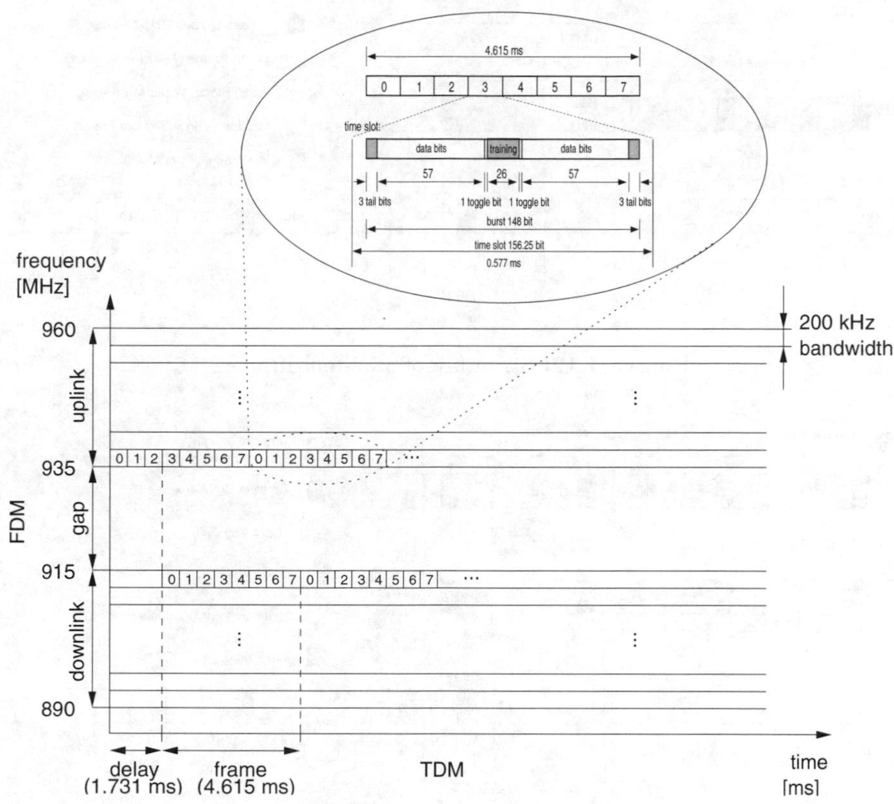

Figure 4.11: TDMA frame structure and normal burst

4.1.6.1 Channel concept

The basic approach to integrate the packet data service into the GSM standard represents the reservation and the logical subdivision of certain GSM channels for GPRS. The physical channels dedicated to packet data traffic are called *Packet Data Channels* (PDCHs).

Similar to GSM, logical channels PDCH are mapped onto physical channels using a cyclically recurring multiframe structure. Two 26-multiframes are assembled to one GPRS 52-multiframe. A 52-multiframe represents one physical GPRS channel consisting of 12 radio blocks and one `Idle` block, each block comprising four radio bursts distributed on TSs with the same TN in consecutive TDMA frames (see Figure 4.12, Figure 4.13 and Figure 4.14).

Since one TDMA frame represents eight physical channels to be shared by GSM and GPRS, there have to be strategies on how to distribute these resources, whether fixed or on-demand (see Figure 4.15). [24] specifies principles concerning this task. The specific details of implementation are left to manufacturers and network operators.

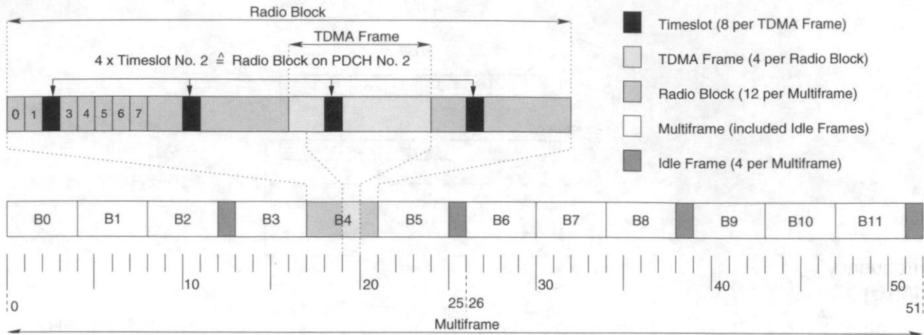

Figure 4.12: Structure of a 52-multiframe

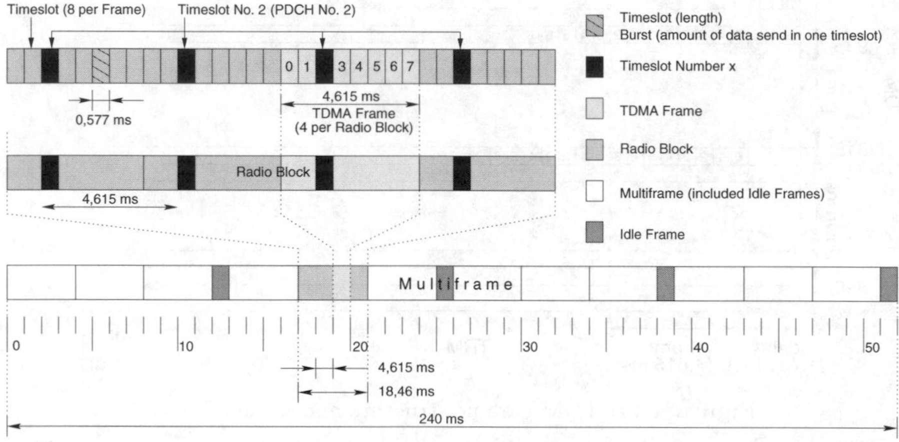

Figure 4.13: Time duration of a timeslot, TDMA frame, radio block and 52-frame

The master-slave concept One or more *Packet Data Channels* (PDCHs) are run as *Master Packet Data Channels* (MPDCHs) (see Figure 4.14), and provide *Packet Common Control Channels* (PCCCHs) that carry the necessary control and signalling information to initialize packet data transmission. These PCCCHs have to be run if signalling information is not transmitted via existing *Common Control Channels* (CCCHs).

Furthermore PDCHs operate as *slaves* and are utilized for transmission of user data over the *Packet Data Traffic Channels* (PDTCHs) and associated control over the *Packet Associated Control Channel* (PACCH).

The capacity-on-demand concept GPRS does not require fixed allocated PDCHs. Capacity assignment for packet data transmission can be done according to actual demand. The decision about the number of fixed and on-demand PDCHs rests with the radio network operator.

TDMA Frame Number

TDMA Slot Number													
0	B0	B1	B2	B3	B4	B5	B6	B7	B8	B9	B10	B11	Master
1	B0	B1	B2	B3	B4	B5	B6	B7	B8	B9	B10	B11	Slave Channels
2	B0	B1	B2	B3	B4	B5	B6	B7	B8	B9	B10	B11	
3	B0	B1	B2	B3	B4	B5	B6	B7	B8	B9	B10	B11	
4	B0	B1	B2	B3	B4	B5	B6	B7	B8	B9	B10	B11	
5	B0	B1	B2	B3	B4	B5	B6	B7	B8	B9	B10	B11	
6	B0	B1	B2	B3	B4	B5	B6	B7	B8	B9	B10	B11	
7	B0	B1	B2	B3	B4	B5	B6	B7	B8	B9	B10	B11	

Frame number scale: 0, 5, 10, 15, 20, 25, 30, 35, 40, 45, 50

Radio Blocks of the Master Channel — Idle Frame

Figure 4.14: GPRS channel structure

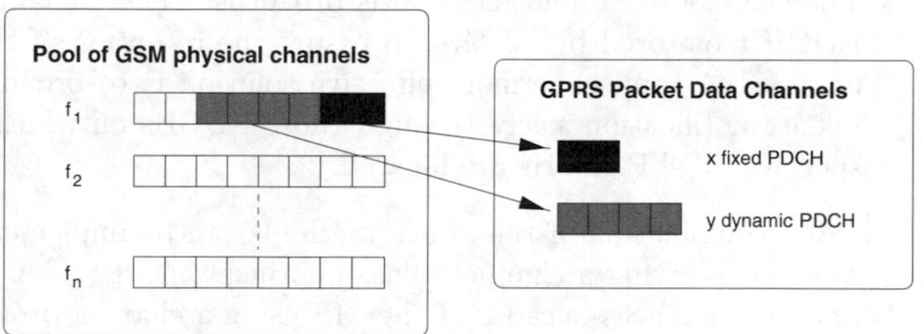

Figure 4.15: Assignment of GSM physical channels for GPRS

There are several mechanisms to increase or diminish the number of actually assigned PDCHs on a capacity-on-demand base.

Load monitoring The PDCH utilization is supervised by a load monitor instance that should be implemented as part of the MAC functionality. For GSM services, the common channel allocation instance inside the BSC is used.

Dynamic allocation of PDCHs Unused channels, whether by GSM or GPRS, may be allocated as PDCHs to increase the GPRS QoS. PDCHs are released in order to meet the obligations of higher priority services.

Release of PDCHs Fast release of PDCHs is an important criterion for a pool of radio resources to be dynamically shared by circuit as well as packet switched services. To achieve this goal, there are the following possibilities:

- Channel release is delayed until there are no more allocations on the PDCH in question.
- Each user having allocated the PDCH to be released has to be notified, e.g. by `Packet Resource Reassignment` messages from the network side. These messages have to be sent via PACCHs to each single MS. The PACCHs may be located on different PDCHs.

Table 4.3: GPRS logical channels

Group	Channel	Name	Direction	Function
PCCCH	PRACH	Packet Random Access Channel	UL	random access
	PPCH	Packet Paging Channel	DL	paging
	PAGCH	Packet Access Grant Channel	DL	access grant
	PNCH	Packet Notification Channel	DL	multicast
PBCCH	PBCCH	Packet Broadcast Control Channel	DL	broadcast
PTCH	PDTCH	Packet Data Traffic Channel	UL/DL	data
	PACCH	Packet Associated Control Channel	UL/DL	assoc. control

- The channel release notification is broadcast. There has to be one PDCH monitored by all MSs to assure the reception of this notification. A somewhat more effective solution is to broadcast the release notification solely through those PDCHs on which the respective MSs' PACCHs are located.

In practice, a combination of the above methods can be implemented. In case of an MS not receiving a channel release notification there will be temporary channel interferences because of the MS using a channel provided for other services. However, the MS will terminate the RLC link as soon as it receives an inappropriate response on the downlink, and initiate a retransmission on a different PDCH.

4.1.6.2 Logical channels

The packet data logical channels are mapped onto the physical channels dedi

Table 4.3 lists the GPRS logical channels and their functions. A detailed description for each channel is presented below.

Packet Common Control Channel (PCCCH) The PCCCH comprises logical channels for common control signalling used for packet data:

Physical Random Access Channel (PRACH) This channel is used by the MSs in the uplink direction to initiate uplink transfer, e.g., for sending data or paging responses. In other words the channel is used by the MS to initiate packet transfers or respond to paging messages. On this channel MSs transmit access bursts with long guard times. On receiving access bursts, the BSS assigns a *Timing Advance* (TA) to each terminal.

Packet Paging Channel (PPCH) This channel is used in the downlink direction to page an MS prior to downlink packet transfer. The PPCH uses paging groups in order to allow usage of *Discontinuous Reception* (DRX) mode. PPCH can be used for paging packet data services.

Packet Access Grant Channel (PAGCH) This channel is used in the establishment phase of the packet transfer to send resource assignments in the downlink direction to an MS prior to packet transfer. Additionally, resource assignment for a downlink packet transfer can be sent on PACCH if the MS is currently involved in a packet transfer.

Note: The PAGCH can only be mapped onto the Master Channel in the downlink direction (see Figure 4.50).

Packet Notification Channel (PNCH) This channel is used in the downlink direction to send a PTM-M notification to a group of MSs prior to a PTM-M packet transfer. The notification has the form of a resource assignment for the packet transfer. DRX mode will be provided for monitoring the PNCH. Furthermore, a `PTM-M new message` indicator may optionally be sent on all individual paging channels to inform MSs interested in PTM-M when they need to listen to the PNCH.

Packet Broadcast Control Channel (PBCCH) The PBCCH broadcasts packet data specific system information. If the PBCCH is not allocated, the BCCH transmits the system information to all GPRS terminals in a radio cell.

Packet Traffic Channel (PTCH) The PTCH is allocated to carry user data and associated control information.

Packet Data Traffic Channel (PDTCH) This is a channel allocated for data transfer. It is temporarily dedicated to one MS or in the PTM case to a group of MSs. In multislot operation, one MS may use multiple PDTCHs, e.g., more than one PDTCH, simultaneously for individual packet transfer.

Packet Associated Control Channel (PACCH) This channel is used to convey signalling information related to a given MS, such as ACK. The PACCH also carries resource assignment and reassignment messages comprising the assignment of capacity for PDTCHs and for further occurrences of PACCH. One PACCH is associated with one or several PDTCHs concurrently assigned to one MS.

4.1.7 Packet Transfer

Both the network and the MS can initiate the establishment of a *Temporary Block Flow* (TBF) on the PCCCH allocated in the cell. The access is carried out on the PCCCH in either one or two phases, whereas two-phase access is used if the requested RLC mode is unacknowledged mode to ensure a safe establishment. A short access procedure is used if the amount of data to send fits in 8 or less than 8 RLC/MAC blocks using *Circuit-Switched* (CS)-1.

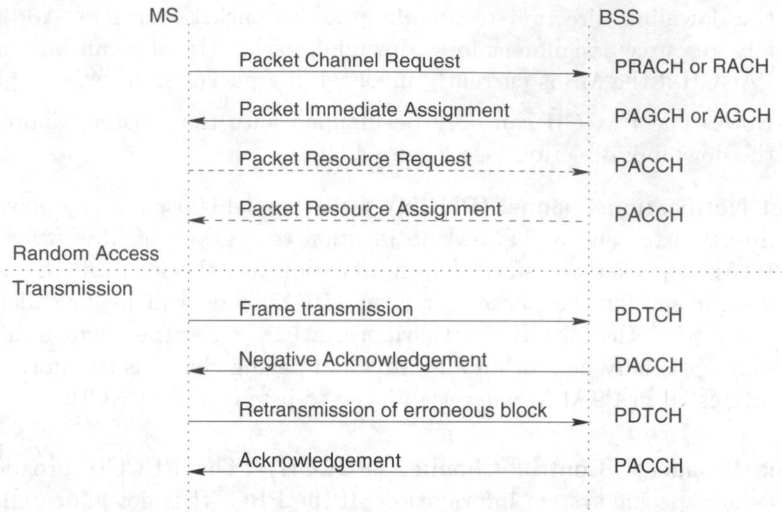

Figure 4.16: Uplink TBF establishment and data transmission

4.1.7.1 Uplink (initiated by the mobile station)

The MS enters the packet access procedure by sending a `Packet Channel Request` message on the PRACH and entering the packet transfer mode (see Figure 4.16). This `Packet Channel Request` message contains parameters required to indicate the mobile station's request of radio resources and the type of access needed. Access persistence control on PRACH can be steered either by the network (network steered method) or by the mobile station (mobile station steered method) to avoid collision failure.

In one phase access, the network responds to the `Packet Channel Request` with the `Packet Immediate Assignment` (`Packet Uplink Assignment`), reserving the resources on PDCHs for uplink transfer of a number of radio blocks.

In two phase access, the network responds to the `Packet Channel Request` with `Packet Immediate Assignment` (`Packet Uplink Assignment`), which reserves the uplink resources for transmitting the `Packet Resource Request`. The `Packet Resource Request` message carries the complete description of the requested resources for the uplink transfer. Thereafter, the network responds with a `Packet Resource Assignment` message (`Packet Uplink Assignment`), reserving resources for the uplink transfer.

If there is no response to the `Packet Channel Request` within a predefined time period (`T3168`), the MS retries after a random back-off time. On receipt of a `Packet Channel Request`, the network sends a `Packet Uplink Assignment` message on the same PCCCH on which the network has received the `Packet Channel Request` message.

Packet data traffic is bursty in nature. Sometimes, the BSS will receive more channel requests than it can serve within a certain time limit. To avoid the repeat of `Packet Channel Requests` the sender is notified with a `Packet Queueing Notification` that its message is correctly received and will be handled later.

Efficient and flexible utilization of the available spectrum for packet data traffic with one or more PDCHs in a cell can be obtained using a multislot channel reservation scheme. Blocks from one MS can be sent on different PDCHs simultaneously, thus reducing the packet delay for transmission across the air interface. The bandwidth may be varied by allocating one to eight time slots in each TDMA frame depending on the number of available PDCHs, the multislot capabilities of the MS, and the current system load.

As mentioned earlier, the master–slave channel concept requires mechanisms for efficient utilization of PDCH uplink(s). Therefore, the USF is used on PDCHs. The 3-bit USF, sent at the beginning of each radio block on the downlink, points to the next uplink radio block. It enables the coding of eight different USF states which are used to multiplex the uplink traffic. The channel reservation command includes the list of allocated PDCHs and the corresponding USF state per channel. To an MS, the USF marks the part of the channel it can use for transmission. An MS monitors the USF and, according to the USF value, identifies PDCHs assigned to it and starts transmission. This allows efficient multiplexing of blocks from a number of MSs onto a single PDCH. Additionally, the channel reservation command can be sent to the MS even before the total number of requested PDCHs is free. Thus, the status flags not only result in a highly dynamic reservation but also allow interruption of transmission because of a pending message or high-priority messages. One USF value is used to denote PRACH (`USF=free`). The other USF values (`USF=R1/R2/···/R7`) are used to reserve the uplink for different MS.

4.1.7.2 Downlink (initiated by the network)

A BSS initiates a packet transfer by sending a `Packet Paging Request` on the PPCH on the downlink in order to ascertain the location of a MS (if it is not already known). The MS responds to the page by initiating a procedure for page response very similar to the packet access procedure described earlier. If the location is known, the paging procedure is followed by the `Packet Resource Assignment` for downlink frame transfer containing the list of PDCHs to be used. The last procedure is also performed before downlink transfer, if the location of the MS is already known.

Since an identifier (e.g., TFI) is included in each radio block, it is possible to multiplex radio blocks destined for different MSs on the same downlink PDCH. It is also possible to interrupt a data transmission to one MS if a higher-priority data or pending control message is to be sent to some other MS. Furthermore, if more than one PDCH is available for downlink traffic, and

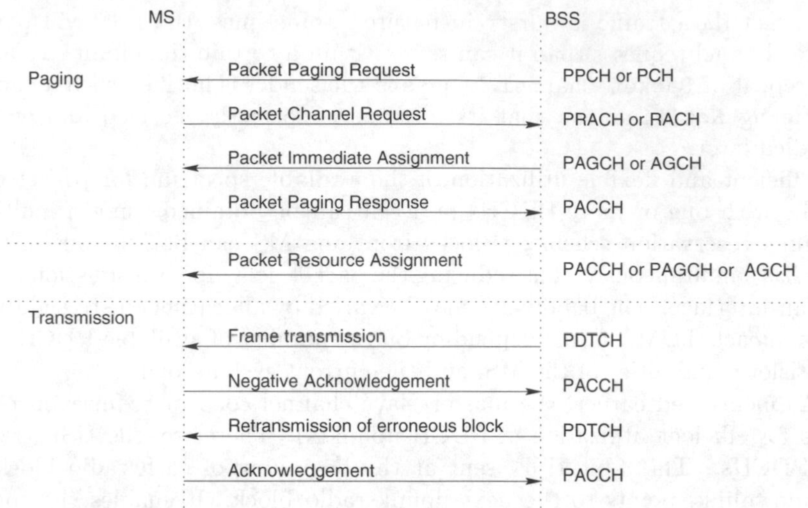

Figure 4.17: Downlink TBF establishment and data transmisson

provided the MS is capable of monitoring multiple PDCHs, blocks belonging to the same frame can be transferred on different PDCHs in parallel.

4.1.8 MAC Procedures in Packet Transfer Mode (Dynamic Allocation)

In the next two sections, the MAC procedures during up- and downlink RLC data block transfer are described.

4.1.8.1 Uplink

When the MS receives the complete uplink assignment, it begins to monitor the assigned PDCHs for the USF value. If there is already a TBF running, the MS waits for the moment of the TBF starting time, which is specified in the `Packet Uplink Assignment` message. Then the MS starts to use the new parameters. Otherwise, if there is no TBF running, the MS begins to monitor the PDCH for USFs as soon as the starting time expires.

It is possible to set the `RLC-DATA-BLOCKS-GRANTED` information element in the `Packet Uplink Assignment` message to allow the MS to send only a specified number within the TBF.

When transmitting RLC blocks on the uplink, the first three RLC data blocks of the uplink TBF contain a TLLI field in the RLC data block header. Each time the MS detects an assigned USF value on an assigned PDCH, one RLC block will be transmitted on the same PDCH in the next block period. When an RLC block is transmitted to the network, the MS starts the timer `T3180`. When the assigned USF value is detected, the MS resets `T3180`. On expiry of timer T3180, the MS performs the abnormal release with random-access procedure.

On the network side, the counter N3101 will be reset if a valid RLC block arrives, but it will be incremented for each radio block, for which no data is received. If the counter reaches its maximum value, the network starts timer T3169. After its expiry, the network may reuse the USF and TFI for another TBF.

4.1.8.2 Downlink

After the reception of the Packet Downlink Assignment message, the MS starts the timer T3190 to define when to stop waiting for the valid data from the network side following the initial Packet Downlink Assignment.

If there is no downlink TBF in progress and the received message contains a TBF starting time information element, the MS remains on the PCCCH until the TDMA frame number, indicated by the TBF starting time, arrives. Now the MS begins to receive on the assigned downlink PDCHs.

If there is already a downlink TBF in progress, the MS continues as before until the indicated TDMA frame number occurs, at which the MS begins to use the new assigned downlink TBF parameters.

Receiving a valid RLC data block without the *Feedback Information* (FBI) bit set to 1, the MS resets and restarts timer T3190 to define when to stop waiting for valid data from the network. Otherwise (FBI=1) the timer T3190 will be stopped. In this case the TBF release procedure is triggered and no retransmissions are made. On expiry of T3190 (TBF release procedure failed), the TBF will be terminated with abnormal release.

4.1.9 RLC Modes of Operation

The RLC ARQ protocol supports two modes of operation: *RLC acknowledged mode* and *RLC unacknowledged mode.*

4.1.9.1 Acknowledged mode

In this mode, the transmitting side numbers the RLC data blocks with the *Block Sequence Number* (BSN) to supervise the acknowledged transmission of the RLC blocks. With Packet Uplink/Downlink ACK/NAK messages, the receiving side can request retransmissions of RLC blocks if needed.

Since the radio resource management is located in the BS, the transmission of the acknowledgement messages is controlled by the BSS, since the BSS manages the resource. The network obtains acknowledgements for downlink transmission by *polling* the MS. The MS sends the ACK/NAK message in the reserved radio block which is allocated in the polling process. Packet Uplink ACK/NAK messages can be transmitted directly. In case of a negative acknowledgement, only those blocks listed as erroneous are retransmitted. In the case that the TBF ends before the LLC frame is transmitted completely, the missing part has to be transmitted during the new TBF.

If the transmit window is stalled, the MS indictates this by setting the *Stall Indicator* (SI) in all subsequent uplink RLC blocks until the condition disappears. Additionally, the timer `T3182` will be set and stopped when it receives a `Packet Uplink ACK/NAK` message. This timer is used to decide when to stop waiting for temporary `Packet Uplink ACK/NAK` during a stalled condition or after the last RLC data block has been sent for the current send window or for the entire TBF.

Release of the uplink TBF The mobile station sends the *Countdown Value* (CV) in each uplink RLC data block to indicate to the network the absolute BSN of the last RLC data block that will be sent in the uplink TBF. The network is then able to calculate the number of remaining RLC data blocks for the current uplink TBF. The final RLC data block transmitted in the TBF has CV set to the value 0.

If the last RLC block is sent with the CV value set to 0 and if there are no elements in the `V(B)` array (indicating the acknowledgement status of previous RLC data blocks) which are unacknowledged, the timer `T3182` has to be started. This timer will be stopped on the reception of a `Packet Uplink ACK/NAK` message. If this `Packet Uplink ACK/NAK` message orders more retransmissions, the MS has to wait until the uplink resources are reallocated before transmitting the requested RLC data blocks. The network sets the Final Ack Indicator when all data has been received properly and causes the MS to release the TBF. The network also includes a valid *Relative Reserved Block Period* (RRBP) field in the RLC block header and waits for the `Packet Channel Request`, which is specified by the RRBP. This RRBP value specifies a single uplink block in which the MS transmits a `Packet Channel Request` message or an associated control message (e.g., *Acknowledge* (ACK)) to the network.

Release of the downlink TBF The release of the downlink TBF is initiated by the BS by setting the FBI to the value 1 and sending the RLC block with a valid RRBP field. Additionally, the timer `T3191` is started to define when the current assignment is invalid on the mobile side so that the TFI can be reused.

The network is now waiting for the `Packet Downlink ACK/NAK` message, which indicates whether some retransmissions are required or not. If there are some, the network begins to retransmit them accordingly to the ARQ protocol. If no retransmissions are necessary, the TBF is released.

4.1.9.2 Unacknowledged mode

This mode comprises no retransmissions. The BSN is now used to reassemble the RLC blocks. The `Packet ACK/NAK` messages are sent to convey the necessary control signalling, as for instance the monitoring of the channel quality. The MS transmits *Window Size* (WS) RLC data blocks without receiving a `Packet ACK/NAK` message. Then it starts the Timer `T3182` to wait for a `Packet Uplink ACK/NAK` message. On a receipt of this message, the timer will be stopped. On expiry of the timer, an abnormal release with cell reselection will be performed.

Release of the uplink TBF The release of the uplink TBF is initiated by the MS, when it sends the final RLC block with CV set to 0. The network on the other side starts timer `T3169` and sends a `Packet Uplink ACK/NAK` message with the Final Ack Indicator set to 1. The MS then answers with a `Packet Channel Request` and the TBF release is completed.

Release of the downlink TBF Initiated by the network, the release of the downlink TBF is indicated with the enabled FBI field in the header of a downlink RLC block and with a valid RRBP field. Timer `T3191` is also started in the network. Receiving this RLC block, the MS transmits the `Packet Channel Request` message in the uplink block indicated by the RRBP, starts timer `T3192` and continues to monitor all assigned PDCHs. When receiving a subsequent RLC block with a valid RRBP and the FBI set to 1, the MS retransmits the `Packet Downlink ACK/NAK`. On receipt of the `Packet Channel Request` message, the network considers the TBF properly terminated. On expiry of `T3191`, the network terminates the TBF and releases all resources. On expiry of timer `T3192`, the MS stops monitoring the assigned downlink PDCH, sets the timer `T3194` and monitors the AGCH and PCH, or the PAGCH and PPCH until its expiry. When the timer expires, the MS begins to monitor only the assigned paging channel.

Abnormal release In an abnormal release case, the sending side begins to establish a new RLC connection with a new TFI. At the beginning of the new TBF, the old RLC blocks, which could not be sent within the old TBF, will be transmitted.

4.1.10 Routing and Mobility Management

Besides the realization of an effective radio resource management, central design criteria is the realization of mobility management and routing.

4.1.10.1 Routing for Point-to-Point Communication

When an MS initiates a transmission, the SGSN decapsulates the incoming packets, extracts the address information and forwards them to the corresponding GGSN, which initiates the routing to the PDN. In the respective PDN the network specific routing procedures are used to forward the packets to the peer entity (see Figure 4.18).

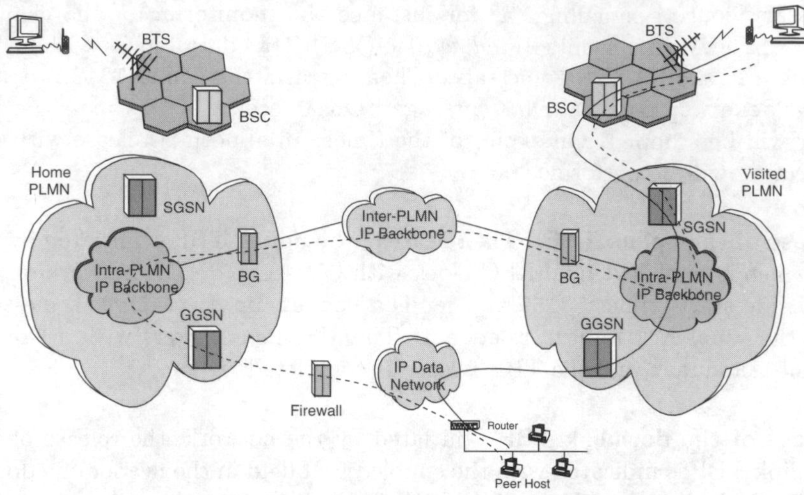

Figure 4.18: Simplified example of GPRS routing

Packets sent from the peer entity are routed through the PDN to the GGSN. The GGSN extracts the routing context associated to this destination address and gets the tunnel information from the corresponding SGSN. The packet is encapsulated and forwarded to the MS.

4.1.10.2 PDP context activation and mobility management

The previous description illustrates that there are two different packet encapsulation types used. Between the *GPRS Support Nodes* (GSNs) the packets are encapsulated by a GPRS specific tunnel protocol GTP. That gives the possibility to use *Packet Data Protocols* (PDPs) even if they are not supported by all SGSNs. The encapsulation between MS and SGSN is made to decouple the *Data Link Layer* (DLL) from the *Network Layer* (NL).

Before being able to send data an MS has to log into the GPRS, e.g., when the MS is switched on. In this *attachment procedure* between MS and SGSN a logical connection context is set up. As the result the TLLI is assigned and returned. After a successful login, routing contexts for several PDPs can be negotiated with the SGSN (see Table 4.4). As in GSM a *Cyphering Key Sequence Number* (CKSN) defines the encryption of the user data. The *GPRS Register* (GR) contains the information, if the MS has access to the respective PDP. If the access is permitted, the GGSN updates its routing context.

During a GPRS session these agreements are updated regularly. Based on the state diagram shown in Figure 4.19 the location management is done. While the MS informs the SGSN in the `Ready` state about every cell change, the location update in `Standby` is only performed when a *Routing Area* (RA) is changed.

Table 4.4: GPRS context

Parameter	Location of storage
Status of MS (active, idle, standby)	MS + SGSN
Authentication (ciphering key)	MS + SGSN
Data compression (y/n)	MS + SGSN
Routing data (TLLI, RA, Cell ID, PDCH)	MS + SGSN
Mobile station identity (IMSI, IMGI)	SGSN
Gateway GSN address (IP address)	SGSN
Charges acquisition parameters (number of bytes)	SGSN

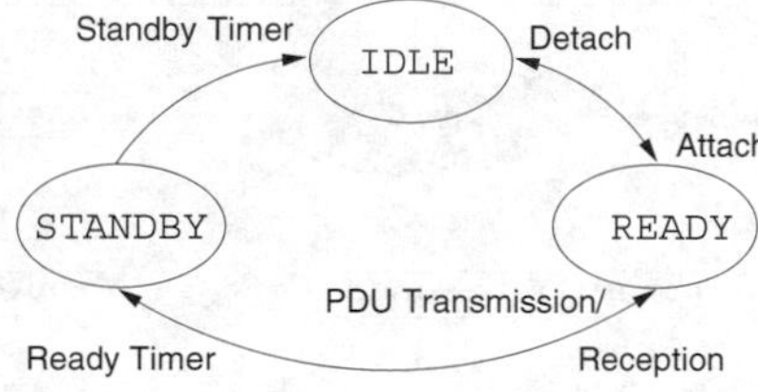

Figure 4.19: State model of mobility management

4.1.11 Session Management

The main function of the *Session Management* (SM) [15, 27] is to support PDP context handling of the MS. The SM comprises procedures for

- identified PDP context activation, deactivation and modification
- anonymous PDP context activation and deactivation.

Anonymous access does not need a *Global Mobility Management* (GMM) context to be established, and there is no *Packet Temporary Mobile Subscriber Identity* (P-TMSI) assigned to the MS. In the following, only identified access will be regarded.

QoS profile negotiation between MS and the network goes hand in hand with the activation or modification of PDP contexts. Therefore, SM procedures play an important role in the GPRS mechanisms for QoS support. The states and procedures of SM both on MS and network side will thus be explained in the following paragraphs.

4.1.11.1 Session Management States

This section deals with the SM states possible for both MS and the network. SM states are described for one SM entity, each SM entity being associated with one PDP context [16].

Figure 4.20 provides an overview of the possible states of an SM entity in the MS. There are five SM states possible:

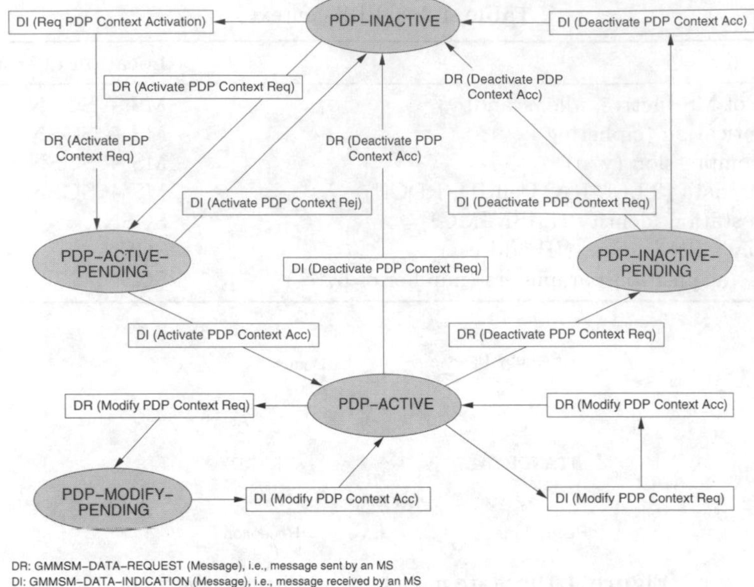

Figure 4.20: SM states in the MS

PDP-INACTIVE There is no PDP context existing.

PDP-ACTIVE-PENDING A PDP context activation was requested by the MS.

PDP-INACTIVE-PENDING The MS has requested the deactivation of its PDP contexts.

PDP-ACTIVE This state indicates that the PDP context is active.

PDP-MODIFY-PENDING This state applies when modification of the PDP context was requested by the MS.

`PDP-MODIFY-PENDING` is only valid for GPRS Release 99-capable MSs. Release 97/98 does not provide the possibility of PDP context modification initiated by MSs.

Next, the possible states of an SM entity on the network side are described. As illustrated in Figure 4.21, there are five SM states on the network side:

PDP-INACTIVE This state indicates that the PDP context is not active.

PDP-ACTIVE-PENDING A PDP context activation was initiated by the network.

PDP-INACTIVE-PENDING The network has requested the deactivation of the PDP context.

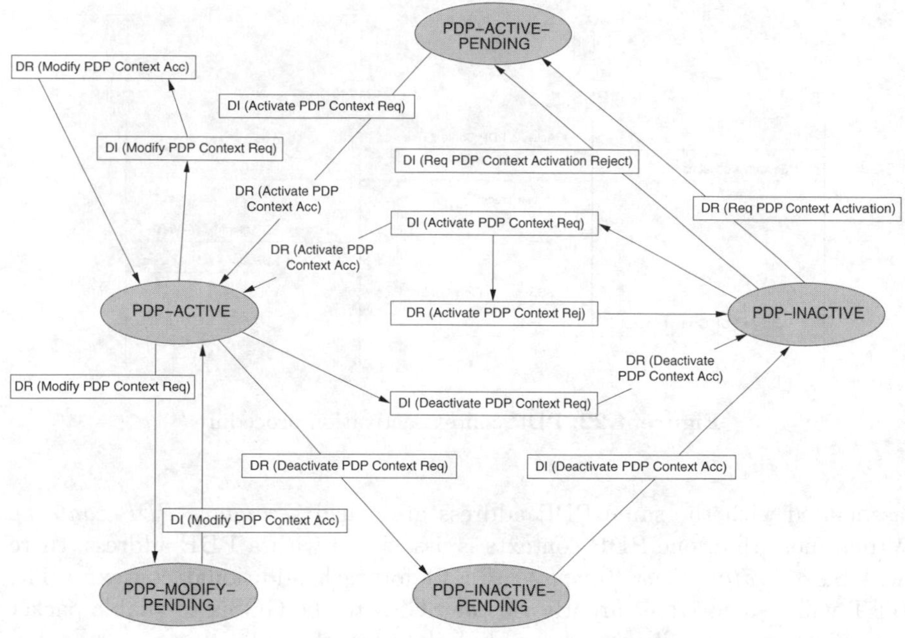

Figure 4.21: SM states on the network side

PDP-ACTIVE This state indicates that the PDP context is active.

PDP-MODIFY-PENDING A modification of the PDP context has been requested by the network.

Reception of a `Modify PDP Context Request` message and the according state transition are not possible for GPRS Release 97/98.

4.1.11.2 Session Management Procedures

In the following subsections PDP context activation, modification, and deactivation procedures will be explained in brief.

The purpose of the PDP context activation procedure is to establish a PDP context between the MS and the network for a specific QoS on a specific *Network Layer Service Access Point Identifier* (N-SAPI). The PDP context activation may be initiated by the MS or the initiation may be requested by the network.

In GPRS Release 97/98, there is only one PDP context possible per MS. In GPRS Release 99, each PDP address may be described by one or more PDP contexts in the MS or the network. The first PDP context activated for a PDP address is called the *primary PDP context*, whereas all additional contexts

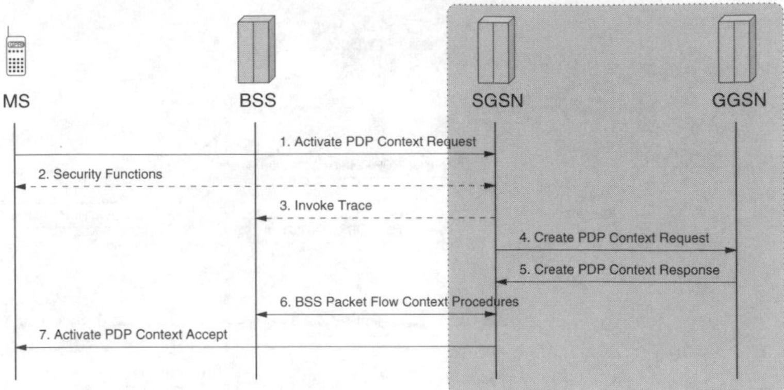

Figure 4.22: PDP context activation procedure

associated with the same PDP address are called *secondary PDP contexts*. When more than one PDP contexts is associated with a PDP address, there will be a *Traffic Flow Template* (TFT) for each additional context. The TFT will be sent transparently via the SGSN to the GGSN to enable packet classification and policing for downlink data transfer [14].

Figure 4.22 outlines the message sequence for a successful MS-initiated PDP context activation procedure. Upon receiving an `Activate PDP Context Request` message the SGSN will initiate procedures to set up PDP contexts. This includes, first of all, subscription checking. Security functions may as well be executed. If BSS trace is activated, an `Invoke Trace` message is sent to the BSS.

Furthermore, it is checked whether the SGSN load is below a predefined threshold and if the related radio cell is able to support the requested QoS. The current cell status is based on flow control information from the BSS sent over BSSGP (see Section 4.1.5.1).

If a GGSN is involved in a traffic flow between an MS and a serving host there is a `Create PDP Context Request` sent to the GGSN. The negotiated QoS profile from the SGSN must also meet possible restrictions due to the GGSN's load and capabilities. The GGSN then creates a new entry in its PDP context table to be able to route PDP PDUs between the SGSN and the external PDN. If there is already a primary PDP context active for the MS, the secondary PDP contexts are provided with a TFT associated with the relevant PDP address. The TFT contains attributes that specify an IP header filter that is used to map data packets received from the interconnected external PDN onto the newly activated PDP context. If these GGSN procedures have been successful, the GGSN sends a `Create PDP Context Response` back to the SGSN, containing the possibly modified negotiated QoS profile.

If the requested QoS has been accepted by both SGSN and GGSN, it is returned to the MS as the negotiated QoS profile by an `Activate PDP Con-`

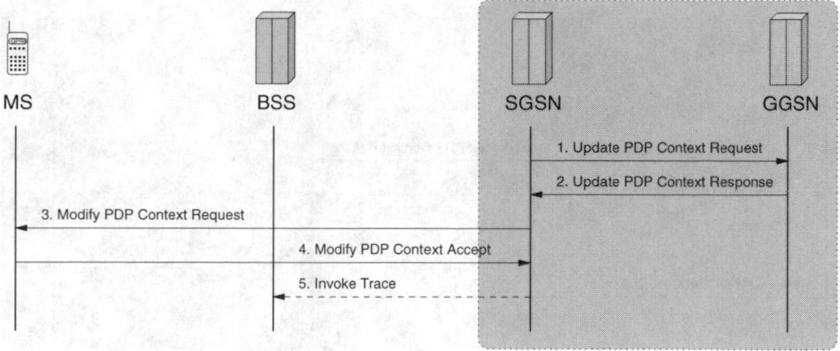

Figure 4.23: SGSN-initiated PDP context modification procedure

text Accept message, and to the BSS in charge during BSS PFC procedures (see Section 4.1.5.1).

Otherwise, if the cell is congested or the SGSN or GGSN load thresholds are exceeded, the PDP Context Activation Request is either rejected with a PDP Context Activation Reject message, or the SGSN or GGSN attempt to negotiate a QoS profile lower than requested. If the BSS cannot support the requested QoS the SGSN suggests best effort, which the MS either accepts, or it deactivates the PDP context.

While in GPRS Release 97/98 only the SGSN may initiate this procedure, PDP context modification may be initiated by MS, SGSN, or GGSN in a Release 99 conforming network. It takes place, e.g., in case of inter-SGSN RA updates, when the new SGSN is unable to maintain the QoS negotiated, or due to changed load conditions in the radio cell.

Figure 4.23 shows an SGSN-initiated PDP context modification procedure. The GGSN is informed about the context to modify by an Update PDP Context Request message, which it answers with an Update PDP Context Response. The MS concerned either accepts the following Modify PDP Context Request, or initiates a PDP context deactivation procedure.

Deactivation of a PDP context may as well be initiated by MS, SGSN, or GGSN.

Figure 4.24 shows the message sequence for an MS-initiated PDP context deactivation procedure. As on PDP context activation, security functions may first be conducted when an SGSN receives a Deactivate PDP Context Request. Thereafter, a Delete PDP Context Request is sent to the GGSN, which is answered by a Delete PDP Context Response. When both SGSN and GGSN have deleted the according PDP context, the MS is informed of completion by a Delete PDP Context Accept message.

The MS may include the *Tear Down Indicator Information Element* (IE) in the Deactivate PDP Context Request in order to indicate that all active PDP contexts sharing the same PDP address as the PDP context associated with this specific TI will be deactivated. If the tear down indicator IE is

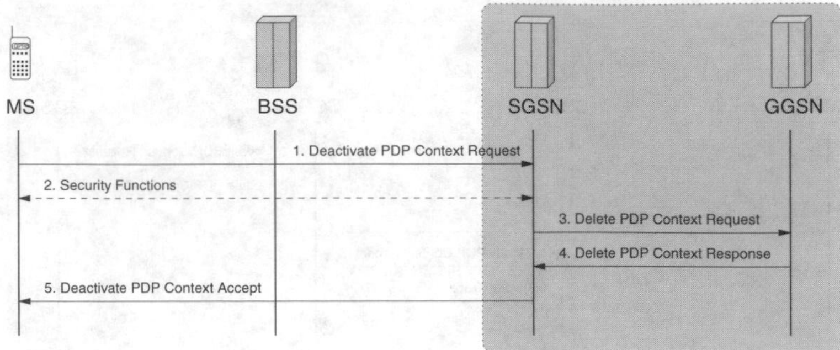

Figure 4.24: PDP context deactivation procedure

not included, only the PDP context associated with this specific TI is to be deactivated.

4.1.11.3 PDP Context Information Element

All PDP context information is stored within an appropriate *Information Element* (IE) [21]; as shown in Figure 4.25—the fields corresponding to QoS Subscribed (QoS Sub), QoS Requested (QoS Req), and QoS Negotiated (QoS Neg), as well as the PDP Address, are emphasized. They play an important role in the implementation of the mechanisms to support QoS.

All QoS-related information, whether subscribed, requested, or negotiated, is included in a QoS profile. The according IE is described in Section 4.1.12.2.

4.1.12 Quality of Service

As seen in Section 4.1.4, Section 4.1.5 and Section 4.1.11, there are several logical entities and protocols involved in GPRS QoS management.

PDP contexts as well as QoS profiles are negotiated between MS and SGSN. The BSS is provided with a PFC containing the *Aggregate BSS QoS Profile* (ABQP) and is responsible for resource allocation on a TBF base and scheduling of packet data traffic with respect to the according QoS profiles negotiated. Moreover, it has to regularly inform the SGSN about the current load conditions in the radio cell. The tasks of the GGSN comprise mapping PDP addresses as well as classification of incoming traffic from external networks on behalf of downlink TFTs. The GR, finally, holds the QoS-related subscriber information and delivers it, on demand, to the GSNs.

In Figure 4.26, part of a GPRS session is schematically outlined, depicting the instances involved, messages exchanged, and parameters negotiated for PDP context, PFC and TFT installation and renegotiation.

From a time-scale point of view, the mechanisms for QoS management within the GPRS might be regarded as a three-stage model (see Figure 4.27). On PDP context activation, the QoS parameters are negotiated. As long

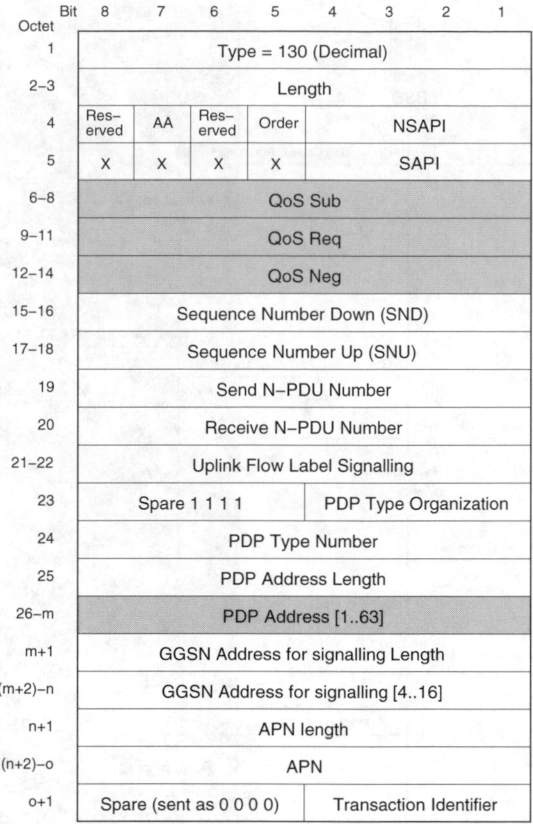

Figure 4.25: PDP context information element

as the PDP context remains active, these parameters should be guaranteed, unless there is a QoS renegotiation. The QoS profile is considered both for each TBF and for each radio block period. At TBF set up, radio resources, like the number of PDCHs, the associated USFs and TFIs, are assigned according to the negotiated QoS parameters. During the TBF, radio blocks are scheduled at the BSS in competition with other existent TBFs in the radio cell. This scheduling function has to be done considering the QoS profiles of the PDP contexts associated with the TBFs.

A QoS profile can be considered as a single parameter value that is defined by a unique combination of attributes. There are numerous QoS profiles available based on permutations of these different attributes, but each mobile network operator must choose to support only a limited subset, reflecting their planned range of GPRS subscriptions. In the following subsections, the QoS attributes defined in GPRS Release 97/98, as well as the changes made for Release 99, will be explained.

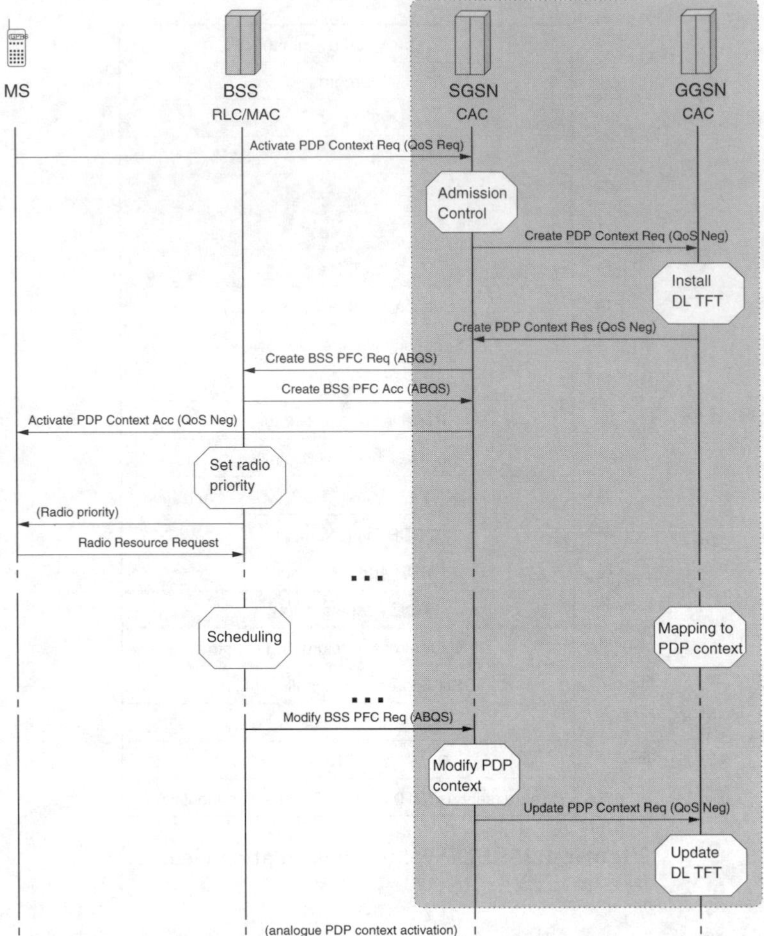

Figure 4.26: QoS negotiation and renegotiation procedures (example)

4.1.12.1 QoS Attributes According to GPRS Release 97/98

A QoS profile defines the QoS within the range of the following service classes [13, 25].

Precedence classes Under normal circumstances the network should try to meet all profiles' QoS agreements. The precedence specifies the relative importance to keep the conditions even under critical circumstances, e.g., momentarily high network load. The various precedence classes are presented in Table 4.5.

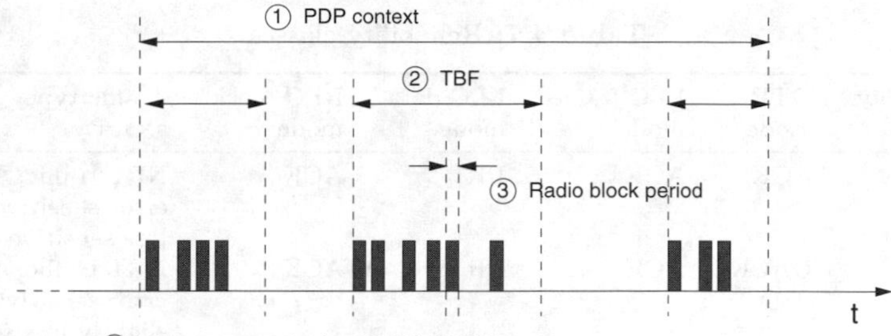

Figure 4.27: Three-stage model of QoS management

Table 4.5: Precedence classes

Precedence class	Identifier	To be served
1	High priority	preferably before classes 2 and 3
2	Normal priority	preferably before class 3
3	Low priority	without preference

Delay classes A packet's delay is defined by the time needed for transmission from one GPRS SAP to another. Delays outside the system, e.g., in transit networks, are not considered. [13] determines four delay classes (see Table 4.6). The network operator has to provide for convenient resources on the air interface to be able to serve the number of participants with a certain delay class expected within each cell. Although there is no need for all delay classes to be available, at least *best effort* has to be offered.

Reliability classes Data services generally require a low residual BER. Erroneous data is usually useless, while incorrectly received speech only leads to a worse perception. Reliability of data transmission is defined within the scope of the following cases:

Table 4.6: Delay classes

Delay class	128 byte packet		1 024 byte packet	
	Mean delay [s]	95 % [s]	Mean delay [s]	95 % [s]
1 (predictive)	0.5	1.5	2	7
2 (predictive)	5	25	15	75
3 (predictive)	50	250	75	375
4 (best effort)	unspecified			

Table 4.7: Reliability classes

Reliability classes	GTP mode	LLC frame mode	LLC data mode	RLC block mode	Traffic type security
1	ACK	ACK	PR	ACK	NRT traffic, error sensitive, loss sensitive
2	UACK	ACK	PR	ACK	NRT traffic, error sensitive, slightly loss sensitive
3	UACK	UACK	UPR	ACK	NRT traffic, error sensitive, not loss sensitive
4	UACK	UACK	UPR	UACK	RT traffic, error sensitive, not loss sensitive
5	UACK	UACK	UPR	UACK	RT traffic not error sensitive, not loss sensitive
(U)ACK	(Un)acknowledged			NRT	Non-Realtime
PR/UPR	Protected/Unprotected			RT	Realtime

- probability of loss of data
- probability of out-of-sequence data delivery
- probability of multiple delivery of data and
- probability of erroneous data.

The reliability classes specify the requirements for each layer's services. The combination of different modes of operation of the GPRS specific protocols GTP, LLC, and RLC, explained in Section 4.1.5, support the reliability requirements of various applications, e.g., *Real-Time* (RT) or *Non Real-Time* (NRT). The reliability classes are summarized in Table 4.7.

Peak throughput classes User data throughput is specified within the scope of a set of throughput classes that characterize the expected bandwidth for a requested PDP context. It is defined by the choice of peak and mean throughput class. Peak throughput is measured in byte/s at the reference points G_i and R (see Figure 4.3). Peak throughput specifies the maximum rate, at which data is transmitted within a certain PDP context. There is no guarantee given that this data rate is actually achieved at any time during transmission. Rather, this depends on the resources available and the capabilities of the MS. The operator may limit the user data rate to the peak data rate agreed on, even if there is capacity left for disposal. The peak throughput classes are presented in Table 4.8.

Table 4.8: Peak throughput classes

Peak throughput class	Peak throughput	
	[byte/s]	[kbit/s]
1	up to 1 000	8
2	up to 2 000	16
3	up to 4 000	32
4	up to 8 000	64
5	up to 16 000	128
6	up to 32 000	256
7	up to 64 000	512
8	up to 128 000	1 024
9	up to 256 000	2 048

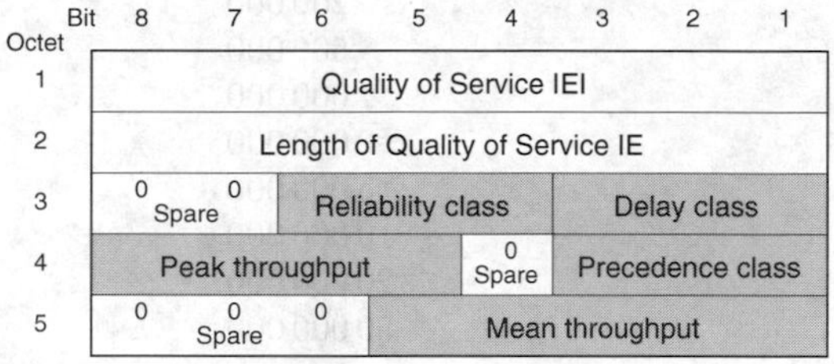

Octet \ Bit	8	7	6	5	4	3	2	1
1	Quality of Service IEI							
2	Length of Quality of Service IE							
3	0 Spare	0	Reliability class			Delay class		
4	Peak throughput				0 Spare	Precedence class		
5	0	0 Spare	0	Mean throughput				

Figure 4.28: QoS profile information element

Mean throughput classes indexMean throughput classes Like peak throughput, mean throughput is also measured in byte/s at the reference points G_i and R. It specifies the average rate data transmitted within the time remaining for a certain PDP context. The operator may limit the user data rate to the mean data rate negotiated, even if excessive capacity is available. If *best effort* has been agreed on as the throughput class, throughput is made available to a MS whenever there are resources needed and at disposal. Table 4.9 summarizes the classes of mean throughput.

4.1.12.2 QoS Profile Information Element

All QoS-related information to be exchanged between MS and SGSN is stored in a QoS profile. The according IE is shown in Figure 4.28. It consists of an IEI, a length field, five fields that contain the values of the service classes (see Section 4.1.12.1), and three fields filled with spare bits [27]. IEI and length fields are not part of a PDP context IE (see Section 4.1.11.3).

4.1.12.3 QoS in GPRS Release 99

It is evident from the above description that the QoS architecture defined in GPRS Release 97/98 shows some major weaknesses (see also [34, 46]):

Table 4.9: Mean throughput classes

Mean throughput class	Mean throughput	
	[byte/h]	≈ [bit/s]
1	100	0.22
2	200	0.44
3	500	1.11
4	1 000	2.2
5	2 000	4.4
6	5 000	11.1
7	10 000	22
8	20 000	44
9	50 000	111
10	100 000	220
11	200 000	440
12	500 000	1 110
13	1 000 000	2 200
14	2 000 000	4 400
15	5 000 000	11 100
16	10 000 000	22 000
17	20 000 000	44 000
18	50 000 000	111 000
31	Best effort	

1. The BSS is not aware of the negotiated QoS profile. This restricts the ability of the BSS to perform scheduling and resource management on the radio interface.
2. Neither MS nor GGSN can influence the QoS profile, even if they detect congestion in external networks as well as changing radio conditions or varying application requirements.
3. It is only possible to have one QoS profile for every PDP context that is associated with a specific service access. Thus, there can only be one QoS profile utilized by all applications for one PDP address.

Table 4.10: End-user performance expectations for selected services belonging to different traffic classes [18]

Traffic class	Medium	Application	Data rate (kbit/s)	One-way delay
Conversational	Audio	Telephony	4–25	< 150 ms
	Data	Telnet	< 8	< 250 ms
Streaming	Audio	Streaming (HQ)	32–128	< 10 s
	Video	One-way	32–384	< 10 s
	Data	FTP	—	< 10 s
Interactive	Audio	Voice messaging	4–13	< 1 s
	Data	Web-browsing	—	< 4 s/page

GPRS Release 99 defines several further QoS parameters with finer grained properties to meet requirements on different levels of service for applications [17], some of which are yet still under discussion:

- maximum bitrate
- guaranteed bitrate
- delivery order
- maximum SDU size
- SDU format information
- SDU error ratio
- residual bit error ratio
- delivery of erroneous SDUs
- transfer delay
- traffic handling priority
- allocation/retention priority
- source statistics descriptor ('speech'/'unknown')

Additionally, there are four distinct *traffic classes* introduced, with different parameters specifying their QoS requirements (see Table 4.10):

- *conversational*
- *streaming*
- *interactive*
- *background*

Delay-sensitive services belonging to the *conversational* class, e.g., do need absolute guarantees in terms of *guaranteed bitrate* and *transfer delay* attributes, while for *background* traffic none other than bit integrity is necessary.

In GPRS Release 97/98, the BSS cannot use QoS profile information to schedule resources on a continuous data flow, neither on the downlink, since there is no mechanism provided to download the QoS profile from the SGSN, nor on the uplink, because there is no QoS information available from the MS. With the introduction of Release 99, the BSS is not only provided with QoS profiles on a PFC base, but also with the ability to modify the QoS profile associated with a data flow in case of changing load conditions. Likewise, MS and GGSN may initiate QoS profile renegotiation, either because of changing application requirements, or due to congestion or a change in radio link quality.

Release 99 also solves the Release 97/98 problem of having only one PDP context installed per PDP address; thus all different applications running on top of this PDP address having to share one single QoS profile, independent of their specific requirements on, e.g., delay or reliability. GPRS Release 99 provides the possibility to install multiple PDP contexts per PDP address. Each PDP context is uniquely associated with a TFT which identifies the traffic flow. This makes it possible to assign different QoS profiles to simultaneous traffic flows, so that each application may receive the appropriate QoS requirement.

GPRS Release 99 conform architectural components will not be available before the year 2002. For interworking purposes with GPRS Release 97/98 conform network equipment, mapping rules between the different releases are defined [17]. These mapping rules will be used for application-level QoS profile definitions, *Admission Control* (AC) rules, and application-based scheduling.

4.2 Enhanced Data Rates for GSM Evolution (EDGE)

Third-Generation (3G) mobile radio networks like the *Universal Mobile Telecommunication System* (UMTS) will provide data services with higher data rates, e.g., UMTS is aiming to realize net bit rates of up to 128 kbit/s with wide coverage, up to 384 kbit/s in hotspots, and up to 2 Mbit/s in indoor scenarios. A stepwise way in the direction of 3G, however, will already be performed by extension and the further development of existing cellular systems. The advantage of such an evolution process is the faster availability of such services, since the infrastructure of existing 2G systems can be applied. Furthermore, there is the opportunity to prepare the customers for new, so-called 3G services. EDGE is a further development of the GSM data services HSCSD and GPRS and is suitable for circuit- and packet-switched services. The circuit-oriented part is the *Enhanced Circuit-Switched Data* (ECSD). The packet-oriented part is the *Enhanced General Packet Radio Service* (EGPRS). Applying modified modulation and coding schemes EDGE reaches very high raw bit rates of up to 69 kbit/s per GSM physical channel. If a user utilizes all 8 time slots in parallel, the theoretical maximum raw bit rate rises to 554 kbit/s. The maximum net bit rate achievable rises to about 384 kbit/s [31]. EDGE was introduced to the ETSI for the first time in 1997 for the evolution of GSM. After a successful feasibility study of the ETSI the standardization process for EDGE was initiated. Although EDGE was introduced for the evolution of GSM, this concept can be applied to increase the data rate in other systems.

The integration of EDGE into GSM networks will be done gradually in order to minimize investments. So in the beginning EDGE transceivers will be able to process ordinary GSM signals and replace the older GSM transceivers. As result, GSM and EDGE signals will coexist in the same frequency band.

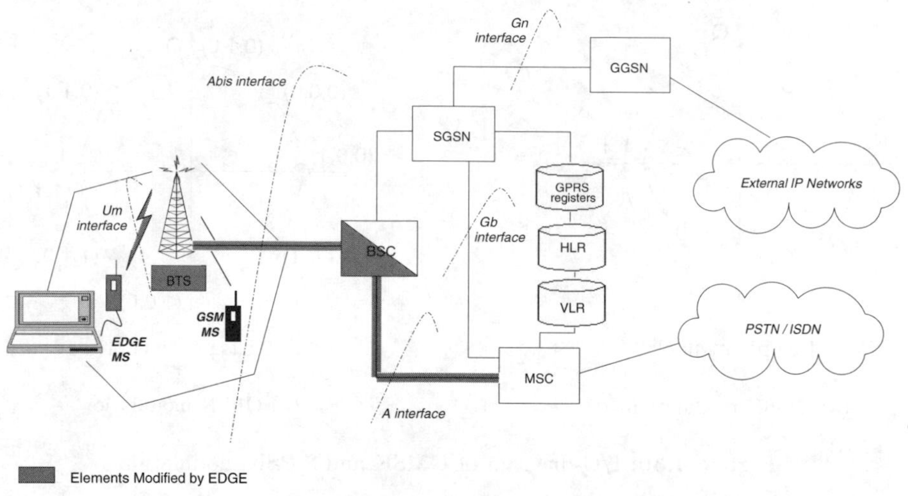

Figure 4.29: EDGE architecture

Additionally it is desired to perform minimal changes in cell and frequency planning.

Since the network architecture of the GSM will remain similar for EDGE, the modifications at the air interface are depicted. The components that are modified for the integration of EDGE are shown in Figure 4.29 as shaded blocks.

The major changes in the GSM standard are made in the radio interface. To support higher data rates, a modulation scheme called *8-Phase-Shift-Keying* (8-PSK) is introduced which will not replace the current GMSK but coexist with it. With 8-PSK it is possible to provide a higher data rate, which is necessary to support bandwidth extensive data applications.

The modifications mostly concern the RLC/MAC layer and the physical layer. Since these protocols are placed in the MS and the BTS, both have to be modified. In reality, the changes that have to be made comprise a new EDGE transceiver unit and software upgrades to the BSC and BS, which then can handle standard GSM or GPRS traffic and will automatically switch to EDGE mode when needed.

The core of EDGE is the Link Quality Control mechanism, which allows link adaptation as to a changing radio link quality. Additionally, soft information is stored during retransmissions to enable *Incremental Redundancy* (IR). The RLC/MAC protocol structure and retransmission mechanism proposed for EGPRS is based on the GPRS standard. An LLC frame is divided into a number of RLC blocks. The duration of an RLC block is 20 ms, whereas its information content ranges from 176 to 1184 bit, depending on the *Modulation and Coding Scheme* (MCS) is used. For error detection, the RLC blocks

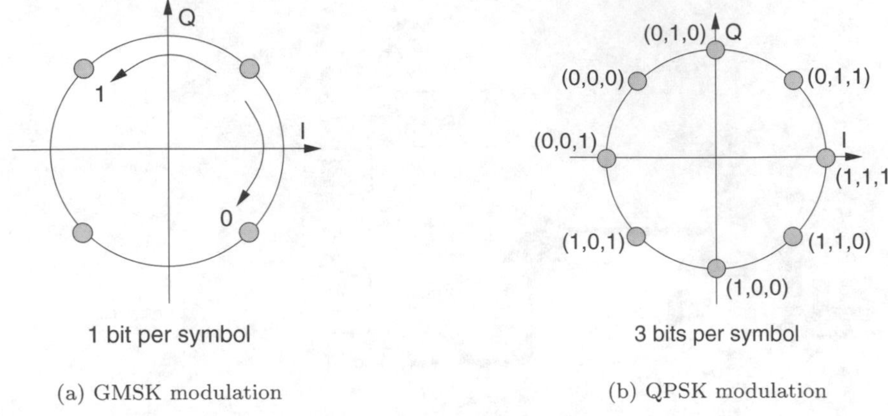

(a) GMSK modulation

(b) QPSK modulation

Figure 4.30: I/Q-diagram of GMSK and 8-PSK modulation

contain a CRC field. Upon reception of an RLC block, the receiver checks the CRC and determines if retransmission of the RLC block is necessary.

To give a more detailed introduction, in the following the modulation scheme 8-PSK used for EDGE is described and compared with GMSK currently used for GSM. Additionally the new MCSs are presented. Then the main characteristics of the hybrid ARQ mechanisms are given in Section 4.2.3, comprising LA and IR.

4.2.1 8-PSK Modulation versus GMSK Modulation

The modulation scheme that is used in GSM is called *Gaussian Minimum Shift Keying* (GMSK). In GPRS the same modulation type will be used and high bit rates are achieved through multislot operation. To allow higher bit rates within EGPRS, the *8-Phase-Shift-Keying* (8-PSK) modulation scheme is introduced in addition to GMSK [32, 39]. The differences between GMSK and 8-PSK is illustrated in an I/Q-diagram as shown in Figure 4.30.

Within GMSK transmitting a 0-bit or a 1-bit is represented by incrementing the phase with $\pm 1/2\,\pi$. Every symbol that is transmitted represents one bit, which means, that symbol rate and bit rate are equal. The symbol rate of 271 kbit/s is divided over 8 time slots. After burst formatting a data rate of 22.8 kbit/s per time slot results. Via different channel coding schemes this results in different user rates.

Within 8-PSK three consecutive bits are mapped onto one symbol in the I/Q-plane. With the same symbol rate as in GMSK of 271 kbit/s, bit rates of up to 813 kbit/s can be achieved. With interleaving, ciphering and burst formatting, this results in a gross data rate of 69.2 kbit/s per time slot. 8-PSK modulation still has the GSM spectrum mask and leaves the burst duration unchanged which offers the possibility to use both modulation schemes next

Table 4.11: EGPRS modulation and coding schemes and coding parameter

MCS	CR	HCR	Data (bit) per RB	Fam.	BCS	Tail	Data rate (kbits)
9 (8-PSK)	1.0	0.36	2×592	A	2×12	2×6	59.2
8 (8-PSK)	0.92	0.36	2×544	A	2×12	2×6	54.4
7 (8-PSK)	0.76	0.36	2×448	B	2×12	2×6	44.8
6 (8-PSK)	0.49	1/3	592 *(544+48)*	A	12	6	29.6 (27.2)
5 (8-PSK)	0.37	1/3	448	B	12	6	22.4
4 (GMSK)	1.0	0.53	352	C	12	6	17.6
3 (GMSK)	0.80	0.53	296 *(272+24)*	A	12	6	14.8 (13.6)
2 (GMSK)	0.66	0.53	224	B	12	6	11.2
1 (GMSK)	0.53	0.53	176	C	12	6	8.8

to each other. However, 8-PSK modulated data is less robust if the channel conditions are bad. The advantage of the new modulation scheme is support for higher data rates under good channel conditions, and to reuse at the same time the channel structure of the GPRS system. Since the BERs are significantly depending on the channel conditions, *Link Quality Control* (LQC) is of major importance for a high throughput in EGPRS.

Link Adaptation (LA) as part of LQC gives the possibility to change to another modulation scheme or channel coding if the channel quality changes. Through the adaptation to the radio channel conditions it is possible to offer the highest data rates in good propagation conditions close to the site of the BSs, whereas under lower quality conditions a more robust coding scheme is selected.

4.2.2 Modulation and Coding Schemes

The additional use of 8-PSK modulation enables the possibility of the introduction of new *Modulation and Coding Schemes* (MCSs). The possible MSCs are depicted in Table 4.11. MCS-1 to MCS-4 use GMSK as modulation scheme and equal the coding schemes used in GPRS, providing data rates up to 17.6 kbit/s. MCS-5 to MCS-9 use the previously described 8-PSK modulation scheme enhancing the data rate up to 59.2 kbit/s per TS. The MCS are classified in three different MCS families consisting of different basic payload sizes. Figure 4.31 gives an overview of how an RLC block is created depending on the MCS family. For instance, MCS-9 comprises four basic units of 37 octets each, whereas MCS-6 combines two basic units and MCS-3 only one. Through transmitting a different number of payload units within a 20-ms-block, different code rates are achieved, resulting in bit rates of 8.8 kbit/s up to 59.2 kbit/s per TS. According to the link quality, an initial MCS is selected for each RLC block. For retransmissions, it is only allowed to use the same MCS or to switch to another MCS of the same MCS family.

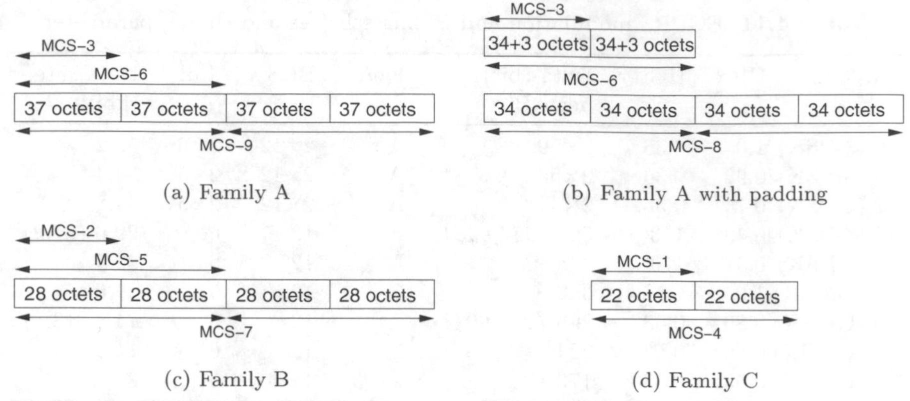

Figure 4.31: Basic block sizes and MCS families

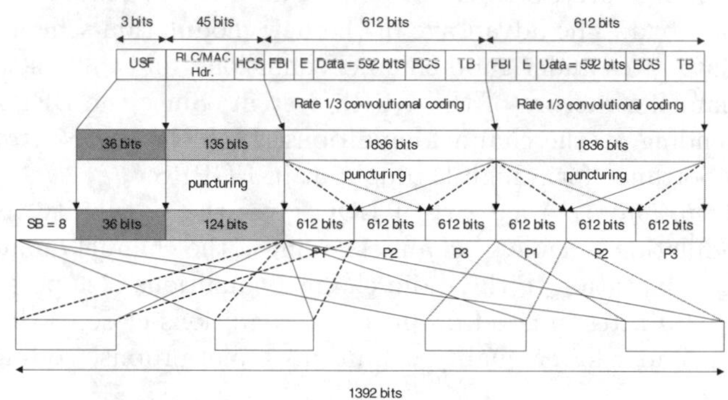

Figure 4.32: Coding and puncturing for MCS-9; uncoded 8-PSK, two RLC blocks

The following figures show the coding and puncturing used in Family A (MCS-9: Figure 4.32, MCS-6: Figure 4.33 and MCS-3: Figure 4.34). They differ in the amount of redundancy used for error correction, which determines accordingly the payload size.

Note that using MCS-7, MCS-8 and MCS-9, two RLC PDUs are transmitted within 20 ms. Both PDUs are interleaved over only two bursts, whereas their common header is interleaved over all four. Using a lower MCS, the RLC PDU (data and header) is interleaved over all four bursts. Figure 4.32 shows a MCS-9 RLC/MAC block consisting of two RLC PDUs of 592 bit (74 octets) payload and separate header containing the FBI field, extension field and the BSN (not pictured) to identify the RLC PDU within a TBF. The two RLC PDUs do not have to be in sequence; for example the first PDU can be a new RLC PDU and the second a retransmission of a previously sent RLC PDU. If there are no further RLC blocks to send, no retransmissions requested and no pending retransmissions foreseen, the second payload will be padded.

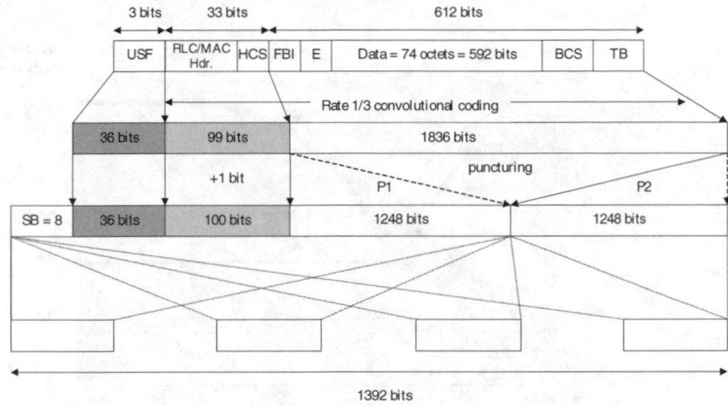

Figure 4.33: Coding and puncturing for MCS-6; rate 0.49 8-PSK, one RLC block

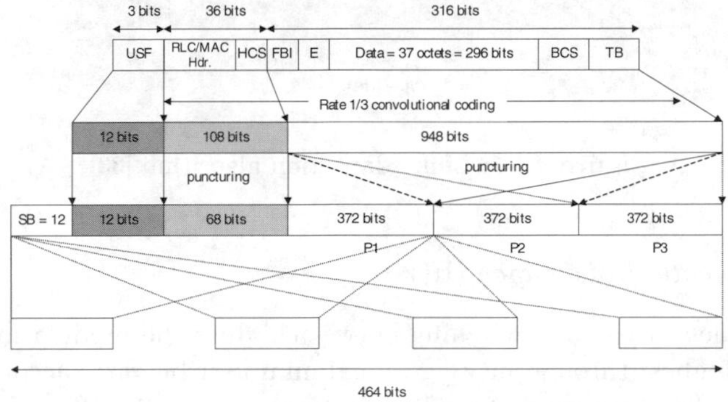

Figure 4.34: Coding and puncturing for MCS-3; rate 0.80 GMSK, one RLC block

Initially, the RLC data block will be encoded with Puncturing Scheme PS-1, whereas PS-2 and PS-3 are used for retransmissions, then enabling the use of IR by using the different punctured information in PS-1, PS-2 and PS-3. The EDGE standard foresees a link adaptation algorithm for the dynamic selection of MCSs to adapt to the quality of the radio link. This includes measurement and reporting the quality of the downlink and means ordering new modulation schemes (see Figure 4.35).

4.2.3 Link Quality Control

Link Quality Control (LQC) is a term used for techniques to adapt the channel coding of the radio link to the varying channel quality. Different modulation and coding schemes are optimal during different situations, depending on the link quality. The link quality control used for EDGE is performed through the techniques of

- *Link Adaptation* (LA)

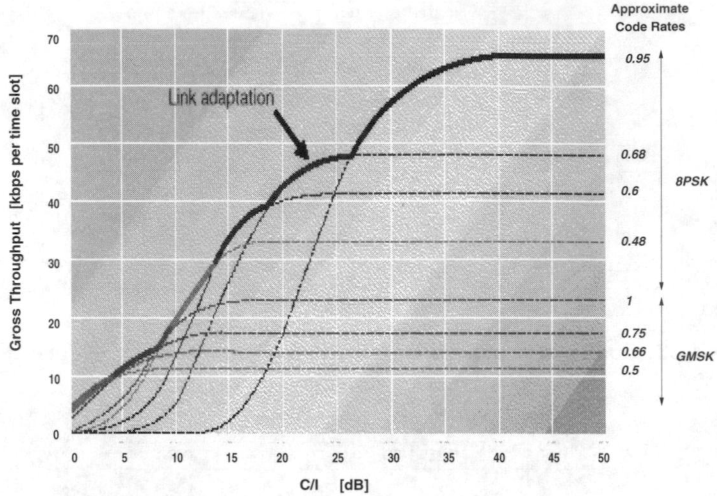

Figure 4.35: Link adaptation algorithm [33]

- *Incremental Redundancy* (IR)

LA provides a dynamic switching between coding and modulation schemes, so that the highest throughput (e.g., maximum user bit rate) according to the time-varying link quality (e.g. C/I) is achieved. With IR, information is first sent with very little coding. If decoding is successful, this will yield a very high user bit rate or throughput. However, if decoding is unsuccessful, then additional coded bits are sent until decoding is successful. The more coding bits that have to be sent, the less the resulting bit rate and the higher the delay. As a result of the link quality control, a low radio link quality will not cause a dropped call, but only give a reduced bit rate for the user. Furthermore, a tight frequency plan (with a small cluster size, say, 4) can be introduced while still providing high data rates for packet data services.

4.2.3.1 Link Adaptation (LA)

An LA scheme is proposed to maximize the user bit rate on the link through estimation of the time varying link quality in EGPRS and respectively adapt the most appropriate modulation and coding scheme. Through measurements of the link quality, the predictive algorithm estimates the performance of the currently used scheme and decides whether another MCS would perform better than the currently used one. The algorithm chooses the MCS with the best performance for the current measured radio link quality and therefore for the expected link quality during the next bursts (it is assumed that the link quality can vary with time and location). During active TBF the most appropriate MCS is chosen on the basis of the channel quality measurements concerning the used MCS family. Due to compatibility reasons it is not allowed to switch the chosen MCS family during retransmissions. Otherwise too much padding will decrease efficiency.

In order to perform the most efficient LA, measurements of the quality of the radio channel are the basis of the choice of the MCS. The operation of link quality control relies on measurements of the link quality, depending also on C/I values, frequency errors, time dispersion, interleaving gain due to frequency hopping, velocities and bit error probability.

4.2.3.2 Incremental Redundancy (IR)

Operating in RLC acknowledged mode, retransmissions are possible, when requested, or when capacity is available for pending retransmissions (RLC blocks not yet acknowledged). An ARQ scheme ensures an error-free data transmission by retransmitting erroneous blocks. When an error is detected in a block, all information about the block is usually discarded. The basic idea of IR is soft combining, i.e., not to forget what was already received in the erroneous transmission. Instead, the soft bit information from the erroneous blocks is saved and combined with the information received with the next (erroneous) retransmission. That way the *Block Error Probability* (BLEP) is reduced.

With IR mode PS-1 is used at first and if decoding is unsuccessful, PS-2 will be used etc. The IR functionality saves soft information of PS-1 and performs a joint decoding with PS-2 after the received retransmission with PS-2. A *Synchronization Downlink Burst* (SB) is used to signal which code rate and which puncturing scheme is used. Information about link quality control is given in [9] and [11]. Through measurements of the link quality every 60 ms (inter-arrival time of the measurement report), the most suitable MCS family is chosen, based on the mean and variance of the C/I. For the following explanation, MCS family A is assumed to handle examples in an easier way. All assumptions apply also for the other families. For retransmissions three possible scenarios are drawn. The first one assumes that the link quality remains stable during the retransmissions. Then the RLC data blocks can be retransmitted with the original used MCS. If the link quality worsens, an MCS downgrading becomes necessary. Otherwise if the link quality improves, an MCS upgrading is possible. In the following the upgrading and downgrading actions are described according to the assumptions made for the EGPRS simulator (see Section 4.3.10).

Upgrading When performing an MCS upgrade, retransmissions can be sent with their old MCS, allowing the use of IR by using different puncturing schemes (PS-1, PS-2, PS-3). That way, the *Block Error Ratio* (BLER) is reduced, by using the information of the different puncturing schemes. Otherwise no IR is possible.

Downgrading If the channel quality worsens, an MCS downgrading is necessary to provide a low BLEP through a higher protection of the bits. Thus, the retransmissions are carried out using a lower MCS, rejecting the possibility of the use of IR because of the incompatible puncturing of the different MCSs. For instance, when MCS-6 is chosen, when previous RLC blocks were sent in MCS-9, no coding problem occurs as MCS-9 comprises two individual RLC blocks consisting of two basic blocks each. MCS-6 also comprises two basic blocks, therefore only the coding, puncturing and interleaving will change and not the number of RLC blocks distinguished by separate BSN. Retransmission of a previous MCS-9 RLC block results in a retransmission of two separate RLC blocks in MCS-6.

Assuming MCS-6 and downgrading to MCS-3, the payload data has to be split and the header duplicated, creating two MCS-3 RLC blocks with the same BSN. Both MCS-3 blocks are retransmitted consecutively with the split bit set to 1 for the first and 2 for the second MCS-3 block to allow an accurate assembly at receiver side requiring the error-free arrival of both the first and second split block. Performing downgrading, all consecutive RLC blocks will be sent using the lower MCS, remaining in this MCS even if a changed channel quality allows upgrading. This applies only for retransmissions, whereas the newly segmented LLC frame will be sent with the actual MCS, taking advantage of the better channel quality and the transmission capacity of the higher MCS.

Note that it is not possible to acknowledge the split blocks separately, hence the receiver has to wait until both split blocks have arrived before they enter the receiver window. The RLC block, consisting of two split blocks is acknowledged positively if both split blocks have arrived error-free. Otherwise both split blocks have to be requested by using the BSN of the former MCS-6 PDU.

4.3 Traffic Performance of GPRS and EGPRS

4.3.1 Simulation Environment

The (E)GPRS Simulator GPRSIM is a pure software solution based on the programming language C++. Up to now in the (E)GPRS architecture models of *Mobile Station* (MS), *Base Station* (BS), and *Serving GPRS Support Node* (SGSN) are implemented. The simulator offers interfaces to be upgraded by additional modules.

For implementation of the simulation model in C++ the *Communication Networks Class Library* (CNCL) [3] is used that is a predecessor to the *SDL Performance Evaluation Tool Class Library* (SPEETCL) [40, 1]. This allows an object-oriented structure of programs and is especially applicable for event-driven simulations. The complex protocols like LLC, RLC/MAC, the Internet traffic load generators and TCP/IP are specified formally with the *Specification and Description Language* (SDL) and are translated to C++ by means of the Code Generator SDL2CNCL [36] and are finally integrated into the simulator.

Different from usual approaches to building a simulator, where abstractions of functions and protocols are being implemented, the approach of the GPRSIM is based on the detailed implementation of the standardized protocols. This enables a realistic study of the behaviour of EGPRS and GPRS. In fact, the real protocol stacks of (E)GPRS are used during system simulation and statistically analysed under a well-defined traffic load. The event control is performed by event handlers that are activated by arriving events and that send events to other event handlers after processing. The scheduling is done by a scheduler, which determines the order of processing of the events. Each event has a priority and a defined processing time. The simulation time advances in discrete steps, when all events, the processing time of which coincides with the current simulation time, are processed. Through this contemporaneity in the simulation process the simultaneous reaction of, e.g., several mobile stations to an event is represented realistically.

4.3.2 Structure of the GPRS Simulator

The software architecture of the GPRSIM and the information flow between the modules is shown in Figure 4.36. The simulator comprises the modules MS, BS, SGSN, the transmission links, the load generator, session control modules, a graphical user interface for presentation and a module for statistical evaluation. Multiple instances of MS and BS can be generated and studied in a multicellular environment.

The formal stucture of MS, BS and SGSN is similar. They contain the implementations of the respective protocol stacks. The transmission links are represented by error models. While the G_b Interface is regarded as ideal, block errors on the radio interface U_m can be simulated based on look-up-tables, which map the actual *Carrier-to-Interference* (C/I) to a *Block Error Probability* (BLEP).

The generator comprises:

- Internet applications (*World Wide Web* (WWW), E-Mail, *File Transfer Protocol* (FTP), WAP, telematics, real-time and streaming applications)
- the packet-traffic models FUNET, Mobitex and Railway (see Section 4.1.1) proposed by ETSI for performance analysis
- a circuit-switched generator to model voice traffic coexisting to the packet oriented data traffic.

The module *Channel Management* supervises the physical GSM channels available in the respective cell and allocates channels for the GPRS resource

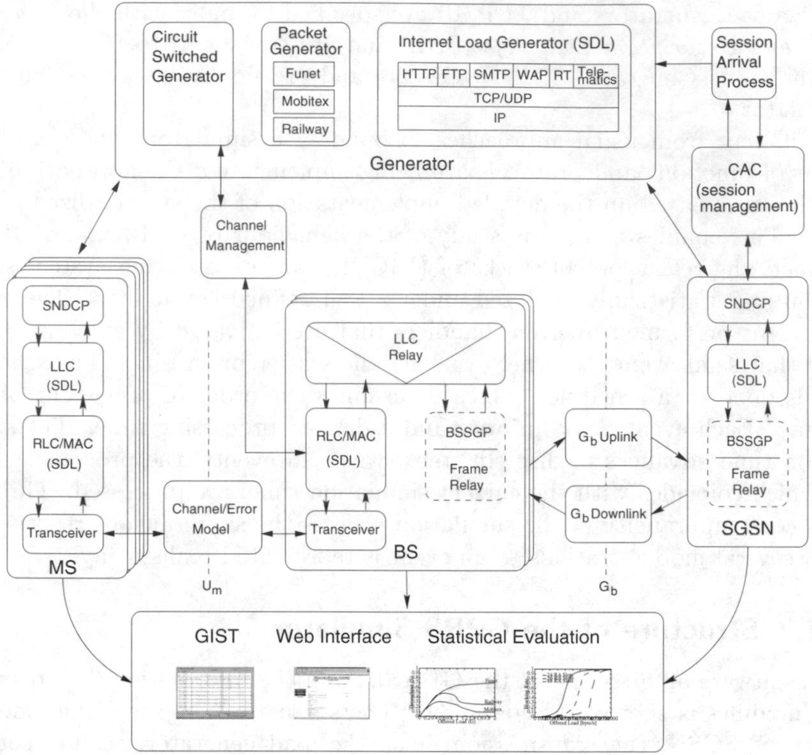

Figure 4.36: Structure of the (E)GPRS simulator

management entity, if the channels are not used by circuit-switched services or if they are not allocated as dedicated PDCHs for (E)GPRS.

The layers BSSGP and frame relay are not represented in the EGPRSIM because the focus was set on the radio interface. The respective classes do not provide any functionality and simply forward the service data units to the peer entity.

The output of the simulator comprises a graphical presentation of the protocol cycles and the statistical evaluation results of the performance measures.

In the following sections the different modules are presented with their functionality and interactions.

4.3.3 Packet Traffic Generators

The GPRSIM comprises an Internet load generator based on the Internet traffic models for the applications *Hypertext Transfer Protocol* (HTTP), FTP, and *Simple Mail Transfer Protocol* (SMTP). The transport and network protocols TCP and IP are also implemented.

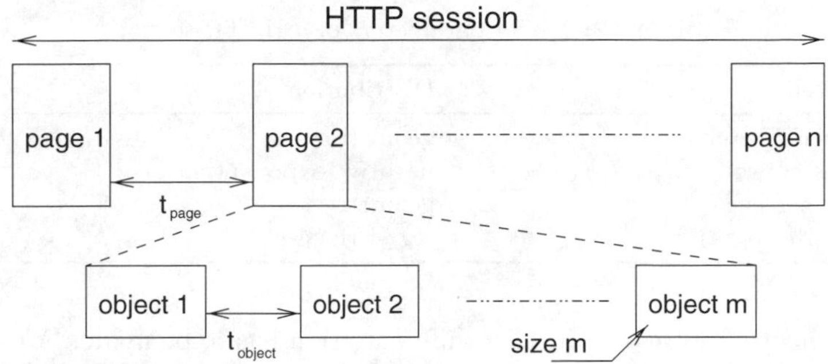

Figure 4.37: Sequence chart and important parameters for an HTTP session

The conception of the Internet load generator is the specification of a system *client*, which behaves like the user running an Internet session, and a system *server*, which represents the peer entity for the Internet service. These systems for modelling Internet traffic were specified in SDL, translated to C++ code with the SDL2CNCL [36] code generator and communicate with the other GPRSIM modules via the SDL process environment.

4.3.3.1 Application models

The following three subsections describe the applications WWW, E-Mail and FTP generating protocol specific traffic load. Related documentation can be found in [2], [30] and [37] while the main part of the later implementation is based on the work of [2] and [47]. The parameters of these models coming from US traffic measurements have been updated by parameters given by ETSI/*3rd Generation Partnership Project* (3GPP) prognoses for the behaviour of mobile Internet users [12]. Traffic models for WAP, telematics and real-time usage are available as well.

HTTP model To describe the load model's parameters, Figure 4.37 depicts the sequence chart for a complete session on a WWW Browser. HTTP sessions consist of requests for a number n of pages. This number is assumed to follow a geometric distribution with a mean value of $\overline{n} = 5$.

Since not only the number of pages describes the behaviour of an HTTP traffic source, the delay between two pages must also be defined. In the following definition the time t_{page} between two pages is defined as the delay between complete reception of the previous page until start of transmission of the next one (see Figure 4.37). This value is strongly depending on the user's behaviour to surf around the Web. The measurements reported in [2] indicate a behaviour for t_{page} following a negative exponential distribution with a mean value of $\overline{t_{page}} = 33$ s. ETSI proposes $\overline{t_{page}} = 12$ s [12].

Table 4.12: Model parameters of a HTTP session

Parameter	Distribution	Mean
Pages per session	geometric	5.0
Intervals between pages (s)	negative exponential	12.0
Objects per page	geometric	2.5
Object size (byte)	$\log_2$-Erlang-k	3700

Referring to Figure 4.37 further differentiation has to be made. A page can be divided into a couple of *objects*. Three parameters need to be defined to generate these objects. First of all the number m of generated objects has to be specified. [47] proposes a mean value of $\overline{m} = 2.5$ objects per page following a geometric distribution.

The delay t_{object} between two objects is given by a negative exponential distribution. In [2] the value is measured as $t_{object} = 0.5$ s. In contrast to the delay between subsequent pages, this value is user-independent but dependent on the server hardware since the server sends the related objects automatically. In this framework the parameter of the response time from the packet-data network is set to zero because only the performance of the GPRS network is regarded.

Third, the size of each object has to be defined using an approximation. The amount of data for each object transferred is in reality not only defined by the object size but also by additional header information. This additional data will be ignored since it represents only a minor fragment of the total amount. The work in [47] refers to measurements [2] and proposes a $\log_2$-Erlang-k-distribution with a mean of $\overline{m} = 3700$ byte. The value k is determined to $k = 24$ referencing the model of [47]. Table 4.12 gives an overview of the parameters describing the HTTP model.

FTP Model The FTP model represents an unidirectional data source. It describes an object transmission from an FTP server to an FTP client. The parameterization is based on extensive *Wide Area Network* (WAN) measurements documented in [37]. Over 95 % of the measured FTP data connections are performed by a `GET` command. Therefore it is sufficient to regard data transfer from server to client. Offered traffic originated by FTP control connections is not considered in the following model. Parameters describing an FTP session are:

- The total amount of data per session.
- The size of each transferred object (ftp data connection).
- The interval between two object transmissions.

The amount of data per session characterizes the duration of a session. In [37] a $\log_2$-normal-distribution is proposed to describe the total amount

Table 4.13: Model parameters of an FTP session

Parameter	Distribution	Mean	Variance
Total bulk of data (byte)	$\log_2$-normal	32768	10000
Bulk of data per object (byte)	$\log_2$-normal	3000	1000
Interval between connections (s)	$\log_{10}$-normal	4	2.55

Table 4.14: Model parameters of an SMTP session

Bulk of data (byte)	Distribution	Mean	Variance
E-Mail size	$\log_2$-normal	10000	1000
Base quota	constant	300	—

of data per session. As parameters the mean of $\overline{x}_{arith.} = 32,768$ and the standard deviation of $\sigma_{x,arith.} = 10,000$ are suggested. For the amount of data per object—per FTP data connection—a $\log_2$-normal-distribution is proposed there likewise. This distribution is characterized by the mean of $\overline{x}_{arith.} = 3000$ and a standard deviation of $\sigma_{x,arith.} = 1000$. The given number of objects includes both the file transfer itself and, e.g., the listing of a directory. The interval between two object transmissions can be modelled according to [38] by a *log_{10}-normal-distribution* with mean $\overline{x} = 4$ and variance $\sigma_x^2 = 2.55$. These parameters are summarized in Table 4.13.

SMTP model The SMTP model describes the load resulting from the transfer of messages downloaded from a mail server by an electronic mail user. The bidirectional phase at the beginning of an SMTP session is not represented by this model. It only describes the phase of the message transfer from SMTP server to SMTP client.

The only parameter is the amount of data per SMTP session (E-Mail). Measurement results from [37] can be characterized by a *log_2-normal distribution* and a fixed added quota of 300 byte. The parameters of this distribution are shown in Table 4.14.

4.3.3.2 System client

The system client represents the client for Internet traffic. Four different processes together represent the block `Internetclient` inside the system (see Figure 4.38). While three of them determine the overall system behaviour, the fourth manages administrative purposes.

- Process `LoadGenClient` requests objects from the Internet server according to the traffic model. Two parameters are generated by the `TCP_client`:

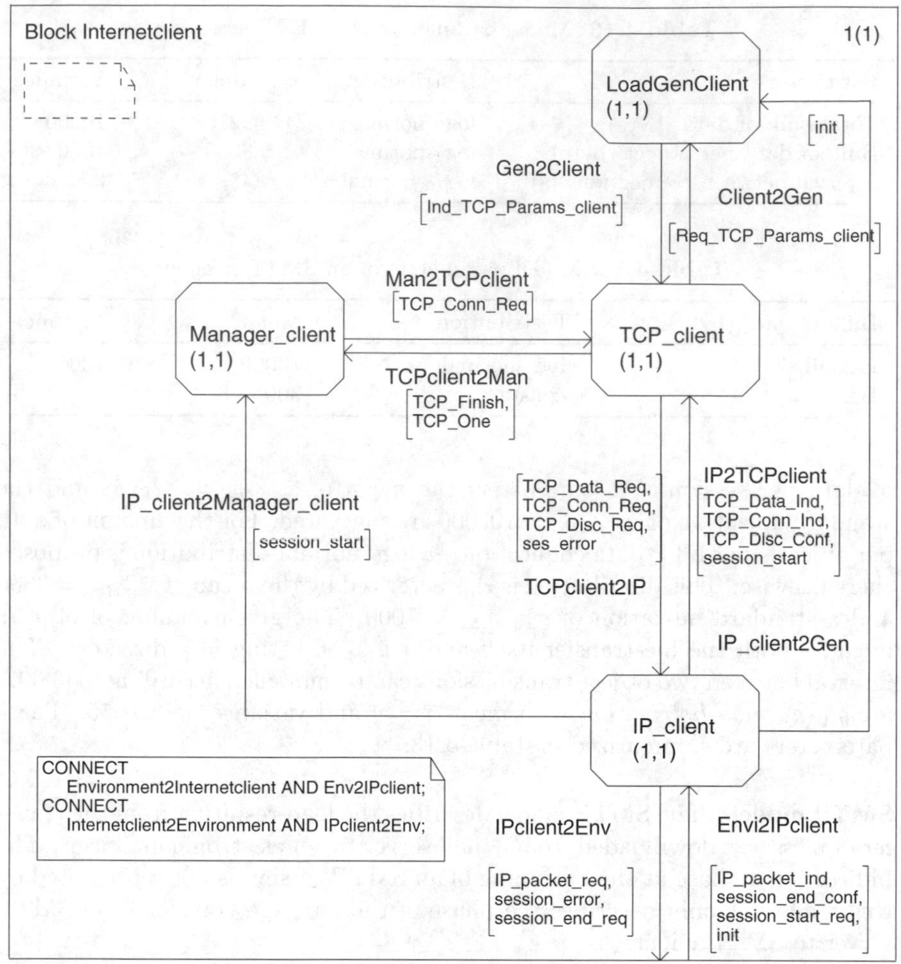

Figure 4.38: Graphical representation of the block `Internetclient` in SDL

– the number of Web pages or files to request,
– the delay between each page/file.

While three applications are supported (HTTP, FTP and SMTP), additionally the used protocol is indicated.

- At session start the `TCP_client` requests the transmission parameters from the `LoadGenClient` and receives information about the protocol to be used, the number of Web-pages or files to request from the server and the delay between the requests. Following the TCP protocol specification, the TCP segments are sent to the process `IP_client`. When a TCP connection ends, a message is sent to the `Manager` process that initiates the session end.

- The process `IP_client` receives the segments generated from the `TCP_client` and encapsulates them together with the IP header. After that the IP packet is sent to the environment. The packets received from the environment are directly handed out to the process `TCP_client`.

4.3.3.3 System server

The system server models the server of the Internet session. It contains the block `Internetserver`, which is similar in structure to the block `Internet-client`.

- The process `LoadGenClient` is replaced by the process `LoadGenSrv` corresponding to the different functionality of load generation. It has to generate random values for the size of requested objects, files or messages. The distinction of the different application protocols is done by the protocol information received by the `TCP_Srv` on TCP connection set up.
- The process `TCP_Srv` receives the load parameters generated by the `LoadGenSrv`. After the transmission of the TCP PDUs as segments the `TCP_Srv` starts the connection release procedure.
- The process `IP_Srv` works the same as the process `IP_client`.

The structure of the block `Internetserver` is shown by Figure 4.39.

Besides the Internet load generator a packet generator based on the models FUNET (electronic mail), *Mobitex* (fleet management) and *Railway* (railway application) as proposed by ETSI (see Section 4.1) are implemented.

4.3.4 Circuit-switched Traffic Generator

The *Circuit-Switched* (CS) traffic generator generates events with an interarrival time determined by a negative-exponential distribution. These events correspond to calls initiated in a cell. When one event arrives, directly the next event is generated to realize the arrival process behaviour. A channel assign request is sent to the channel management module. If there is a traffic channel available, the request is confirmed and the call duration, also negative-exponentially distributed, is determined. After the call duration an event is sent to the channel management module to release the assigned channel. The traffic value in Erlang is given by the two configurable mean values of the call interarrival and the call duration times.

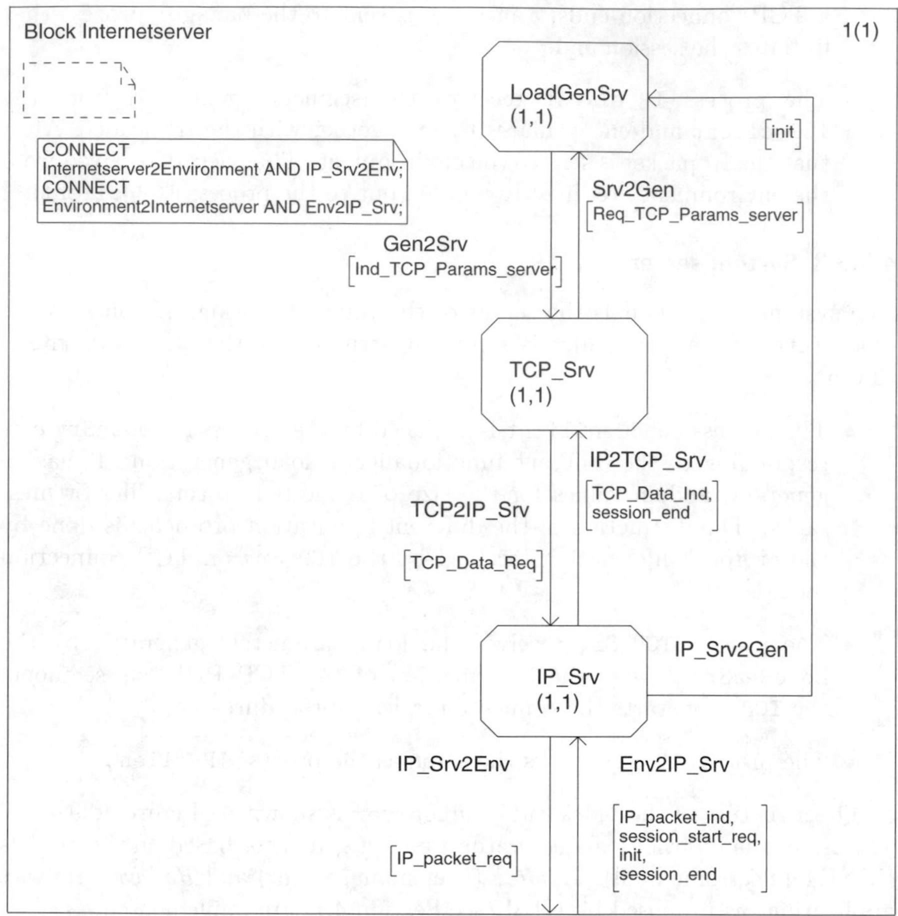

Figure 4.39: Graphical representation of the block `Internetserver` in SDL

4.3.5 Channel Management

The channel management module is the dynamic channel allocation instance controlling the pool of GSM physical channels for (E)GPRS and GSM described in GSM 03.64. The *Traffic Channel* (TCH) can be used as both CS channel and *Packet Data Channel* (PDCH). It communicates with the CS generator and the *Radio Resource Management* (RRM) instance, which is realized in the RLC/MAC layer. The state of this module is characterized by the number of TCHs per radio cell, the maximum number of fixed PDCHs, the maximum number of on-demand PDCHs and the number of CS connections presently set up. With these numbers the channel management module determines the number of PDCHs currently available for GPRS and EGPRS. CS connections are prioritized with preemption, i.e., a new CS request allocates

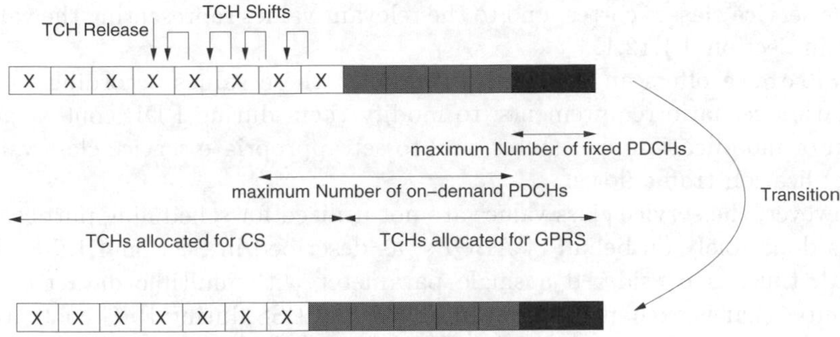

Figure 4.40: TCH and PDCH allocation table

a PDCH used so far by GPRS or EGPRS immediately, if no other TCH is free. All TCHs that are not used by CS connections are available for GPRS, if the number of PDCHs does not exceed the sum of maximum number of fixed and on-demand PDCHs. It is assumed that TCHs used for CS connections are consecutive beginning with channels not lying on the frequency used for GPRS. If a CS connection is released, the TCHs are shifted in sequence so that the TCHs are in sequence again after release (see Figure 4.40). In real systems this is done by only one TCH change. This mechanism is known as the repacking algorithm.

Transitions, i.e., when the number of PDCHs available for GPRS changes, are immediately indicated to the RRM of RLC/MAC using update messages.

4.3.6 Quality of Service Management

In this section, a description of the concepts developed to model the *Quality of Service* (QoS) management functions presented in Section 4.1.11 and Section 4.1.12 is provided, as well as a brief description of the relevant implementation [46].

4.3.6.1 QoS Profile and Aggregate BSS QoS Profile

First of all, a data structure is needed for the data defining the various aspects of QoS within the GPRS system regarded.

The class `QoSProfile` contains all parameters necessary to specify a mobile user's subscribed QoS, a mapping function of the application type onto an appropriate traffic class (see Section 4.1.12.3), and the traffic classes' definition in terms of the following service classes (see Section 4.1.12.1):

- precedence class
- delay class
- reliability class
- peak throughput class
- mean throughput class

The service classes correspond to the relevant values representing the values listed in Section 4.1.12.1.

`QoSProfile` offers an interface to initialize these values according to the simulation scenario requirements, to modify them during PDP context activation or modification procedures, and to set appropriate service class values for application traffic flows.

However, the service class values are not utilized for scheduling purposes—this is done solely on behalf of ABQPs, as described in Section 4.1.5.1. The ABQP, thus, is considered a single parameter with multiple data transfer attributes that is exchanged between SGSN and BSS during PFC-context related procedures. For implementation purposes, it is not necessary to combine subscription and traffic class—which will define the ABQP—into one parameter. They are instead encapsulated in a C++ `struct`, as shown in the following source code.

```
typedef struct aggregateBSSQoSProfileType : public CNObject
{
  int               tlli;
  subscriptionType  subscription;
  trafficClassType  trafficClass;
};
```

For QoS differentiation both application classes (conversational, streaming, interactive and background) and subscriber classes (Premium, Standard and Best-Effort) are considered.

4.3.6.2 PDP Context

Since communication paths between the GPRSIM modules involved in PDP-context related signalling are recreated only logically (see Section 4.3.6.4), there are very few fields of the PDP context IE necessary to be modelled within the scope of the simulator. Hence, the class `PDPContext` simply contains some constructors, QoS Sub, QoS Req, QoS Neg fields, and the PDP address.

```
class PDPContext : public CNObject
{
  public:      //***** Constructors *****/
    PDPContext();
    PDPContext(int);
    PDPContext(const PDPContext&);
    PDPContext(CNParam *) {}

  public:      //***** Public interface *****/
    int         PDPAddress;

    QoSProfile* QoSSub;
    QoSProfile* QoSReq;
    QoSProfile* QoSNeg;

    PDPContext& operator=(const PDPContext&);

  private:     //***** Internal private members *****/

 // {{{ Member functions required by CNCL...
};
```

4.3.6.3 GPRS Register (GR)

The class `GR`, representing the *GPRS Register* (GR) within the HLR, is responsible for the storage of subscriber related QoS profile data. It provides an array of pointers to objects of type `QoSProfile`. This array's size is equivalent to the number of MSs for the respective simulation scenario.

The tasks of `GR` comprise the initialization of the array of QoS profiles according to the relation of *premium* to *standard* to *best-effort* subscribers specified for the respective scenario, and the delivery of subscriber QoS profile data on request from an SGSN handling a PDP context activation. Moreover, the class `GR` has to provide an interface to the generator to allow traffic load generation in conformance with the subscriber QoS profiles valid for the simulation scenario under examination.

4.3.6.4 Connection Admission Control (CAC)

As outlined in Section 4.1.12 the *Connection Admission Control* (CAC) function plays a decisive role in QoS-oriented traffic management in a GPRS environment. Within the scope of the GPRSIM, it has to fulfil various tasks, from *Admission Control* (AC) and QoS profile negotiation to *Resource Request* (RRq) status monitoring. Within the context of the GPRSIM the class `CAC` represents the QoS-management related functions of both the SGSN and the GGSN. Communication paths from MS via BSS to the *Channel Access Control* (CAC) module within the GSN are recreated only logically and do not utilize any physical signalling resources of the system.

As already shown in Figure 4.26, the module `CAC` is responsible for handling PDP context activation and deactivation requests, keeping track of traffic load and radio resource utilization status within the radio cell on behalf of BSSGP flow control information from the BSSs, and performing PDP context renegotiation. Thus, there are interfaces necessary for communication with the modules GR, generator, BS transceiver, and the RLC/MAC modules via the SDL process environment.

The message flow between the module CAC and the other GPRSIM modules involved in the AC process is summarized in Figure 4.41.

The module CAC must provide methods and objects to store, update, and evaluate RR utilization data as well as traffic-flow-related QoS profile data (*Traffic Flow Template*, TFT). Further, it has to perform *Admission Control* on the base of this data, and react to changing resource utilization in a radio cell by engaging PDP context renegotiation procedures.

To perform *Admission Control*, there is an AC policy to be implemented in a way that all active traffic flows can be served according to the QoS profiles negotiated. Preferably, there is only a limited number of privileged connections allowed simultaneously if the QoS guarantees given by subscription do not run the risk of being violated.

Furthermore, any *Standard* traffic should receive the resources necessary to meet its QoS requirements. Thus, it might be preferable to reject a PDP

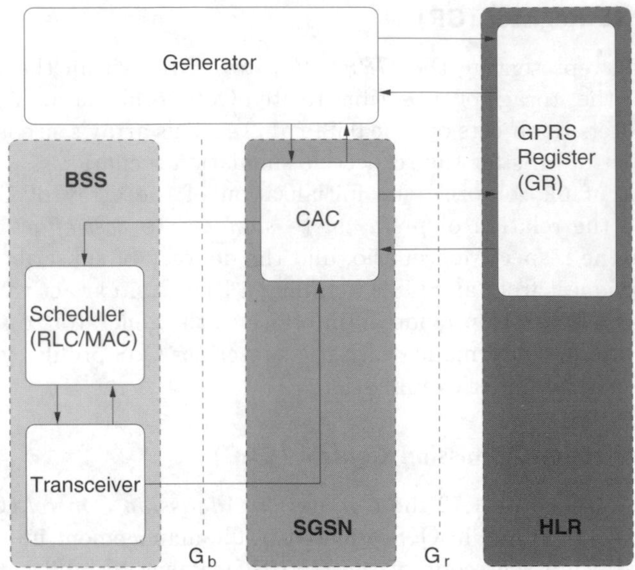

Figure 4.41: Traffic flow between the GPRSIM modules involved in QoS management

context activation request rather than to endanger the quality of all sessions. On the other hand, it might be advantageous to pre-empt background or even interactive traffic flows to allow for an additional *Conversational* traffic flow to be admitted.

Therefore, PDP context activation requests are differentiated according to their subscribed QoS profiles and their traffic classes. The CAC function first examines whether there are additional PDP contexts allowed for the requested QoS profile. If this proves to be true, in a second step it is decided on the basis of the radio cell resource utilization whether to grant access or not. For this purpose, it is necessary for the CAC to keep track of the radio cell resource utilization status. The BSS channel management module thus has to regularly transmit according data.

To avoid a total withdrawal of resources from the *Standard* traffic classes with lower QoS requirements, e.g., other than *Conversational*, there is a share reserved for this kind of traffic from the pool of radio resources in the cell. In general, all resources are open to traffic of any kind. In times of high load, however, traffic flows with more demanding QoS requirements are allowed to pre-empt flows belonging to applications with lower QoS requirements, but only up to a certain limit (see Figure 4.42), where P and I represent the appropriate limits.

If this limit is reached, the requested QoS is not granted, but rather degraded to the next-lower-prioritized class. Within the framework of this example, no deactivation of the PDP context by an MS, because of not receiving

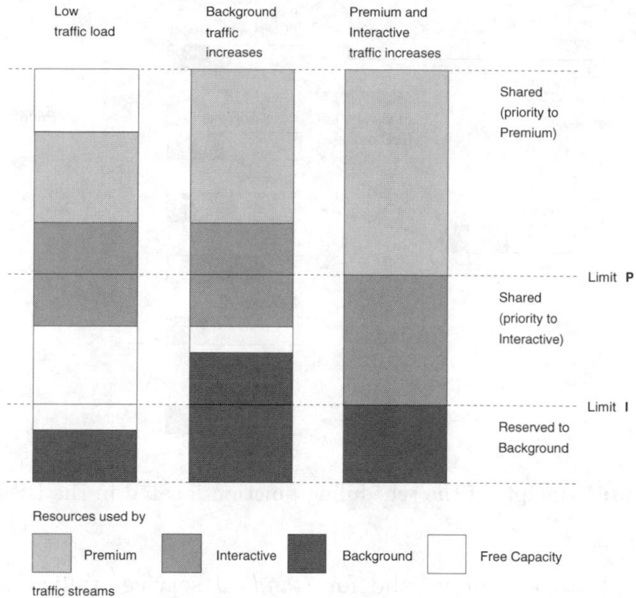

Figure 4.42: Admission control policy (example)

the QoS requested, is initiated. The service requests that have to be downgraded are nevertheless counted for evaluation of QoS guarantees that had to be violated.

4.3.6.5 Scheduler

On behalf of the QoS profile negotiated the BS MAC layer has to perform scheduling of the radio blocks. For evaluation purposes, only the downlink direction is regarded within this framework, since the traffic models used within the generator module of the GPRSIM produce a considerably higher amount of data traffic for this direction, compared with the uplink. Thus, downlink direction may be regarded the bottleneck in respect of resource utilization. Traffic-class based uplink scheduling is nevertheless performed on USF assignment.

For future implementation of further traffic generator modules, e.g., for *Conversational* applications like VoIP, or *Streaming* applications like *video on demand*, it is necessary to also regard the uplink QoS.

The scheduling mechanism implemented for both downlink and uplink direction follows a three-stage principle (see Figure 4.43). First, incoming radio blocks are distributed into one of three queues according to the QoS subscription associated with the respective traffic flow. It is differentiated between *Premium* (*Gold Card*) service, *standard* service and *best-effort* service.

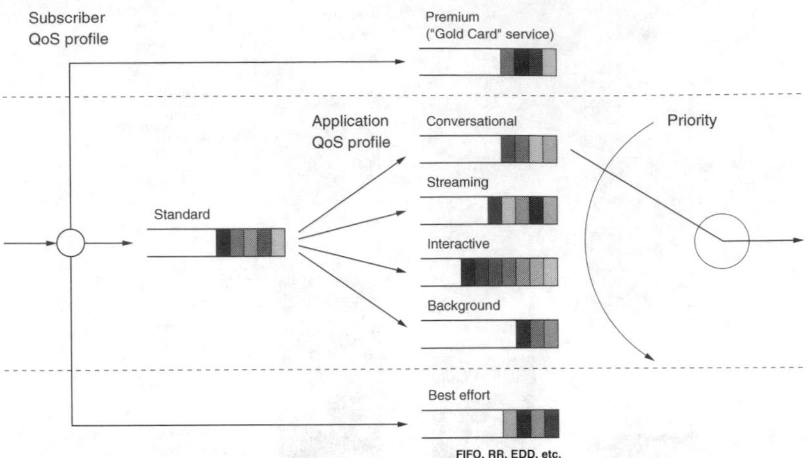

Figure 4.43: Principle of the scheduling function located in the BS MAC layer

The second stage is only valid for *standard* service traffic. Regarding a packet's application QoS profile, the appropriate traffic class queue is chosen from *Conversational*, *Streaming*, *Interactive*, or *Background*. Best-effort traffic from the first stage is put into a fifth queue. Within the traffic class queues, packets are scheduled according to their TLLI and a RR algorithm. Other algorithms than RR may also be implemented on this stage, e.g., *First Come First Serve* (FCFS) or *Earliest Due Date* (EDD).

The third stage is built by a simple priority mechanism, serving the traffic class queues in order from highest priority *(premium)* to lowest priority *(best-effort)*.

All three stages are part of the GPRSIM RLC/MAC layer module and specified in SDL. Communication with the module CAC is performed via the SDL Process Environment.

4.3.7 The SNDCP Layer

The *Sub-Network Dependent Convergence Protocol* (SNDCP) class has two main functions. First TCP/IP header compression is performed. Then the IP packets, coming from the IP layer are adapted to the LLC layer. As the *Maximum Transfer Unit* (MTU) in the LLC layer is 1520 byte, bigger IP packets have to be segmented and reassembled by SNDCP. In addition SNDCP has a queueing function in the GPRSIM. With the function `put_ip_packet` the given parameter `ip_packet` is segmented into SNDCP PDUs and queued as LLC *Interface Data Unit* (IDU). When the LLC layer is ready to send new data, an LLC IDU can be dequeued.

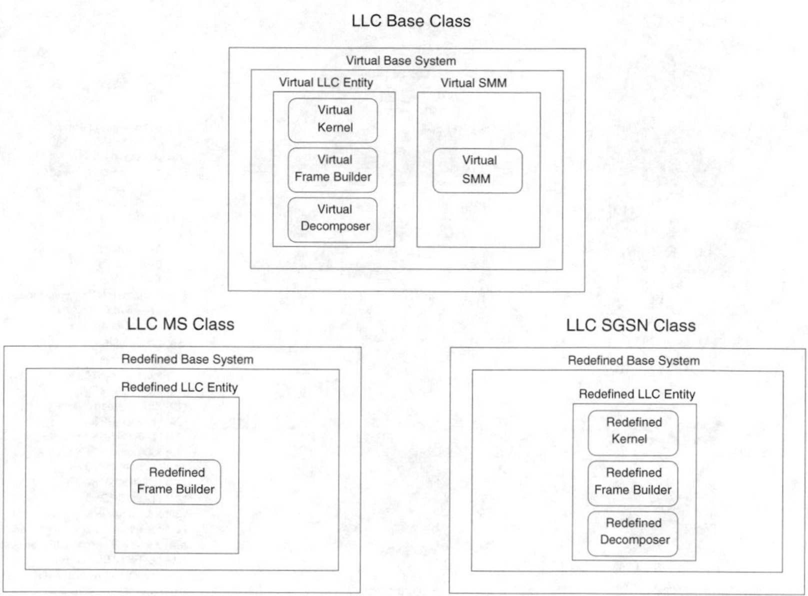

Figure 4.44: Structure of the LLC Layer as an SDL92 System

4.3.8 The LLC Layer

The LLC protocol is also specified in SDL (see Figure 4.44). Since the simulator is realized in C++ with the utilization of the CNCL, the SDL specification is translated to C++ code with the code generator SDL2CNCL [36]. The generated code has been integrated into the GPRSIM.

The class `SDLProcess_SystemManagement` represents the interface with the network layer. It prepares the signals received from the SDL Process environment for the LLC layer and sends the according signals to the Environment.

The classes `SDLSystem_LLC_MS` and `SDLSystem_LLC_SGSN` represent the LLC entities of the MS and the SGSN. They contain several processes that organize and control the transfer procedures. The class `SDLProcess_Kernel` is the kernel of the LLC layer. Here received frames and service primitives are processed and the related reactions are performed. In the kernel it is decided, which frames can be sent and which services are offered to the network layer. In the course of a transfer the kernel enters different well-defined states, which determine the reaction to received signals. The states are presented in Figure 4.45. To guarantee a clear presentation, sender and receiver are presented separately. Transitions between states are caused by specific events that are pictured by arrows.

The kernel informs the class `SDLProcess_FrameBuilder`, which frames have to be sent. This class composes the frames accordant to the frame type

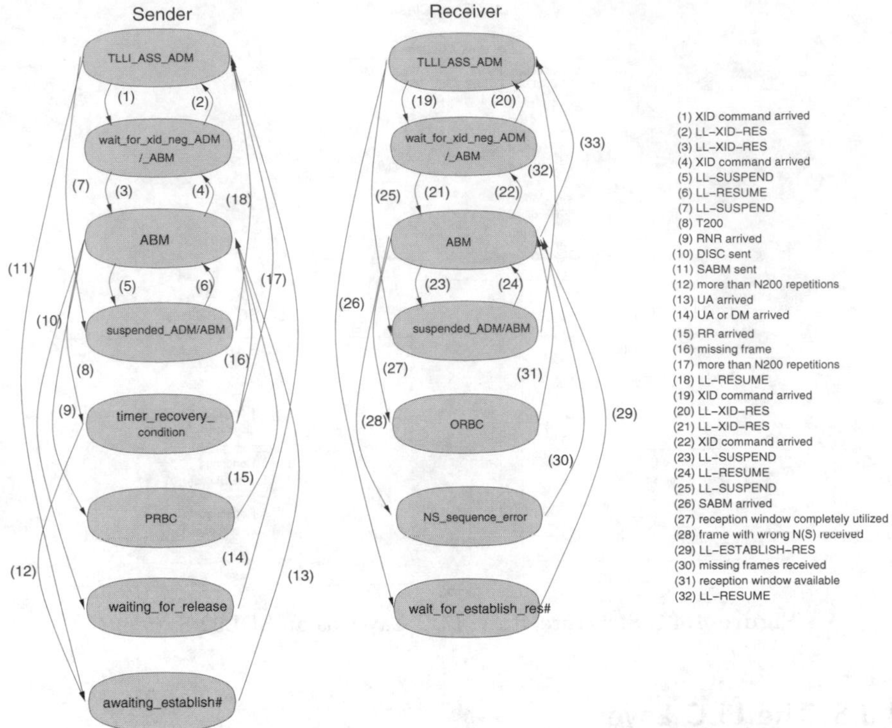

Figure 4.45: State machine of the process LLC entity

or command and transfers them with an RLC/MAC- or `BSSGP-DATA-REQ` to the underlying protocol instances.

The class `SDLProcess_Decomposer` receives an LLC PDU by an `RLC/MAC-` or `BSSGP-DATA-IND`. After evaluation of the control field the data of the information field is forwarded to the LLC kernel.

In the following paragraph the states and events of Figure 4.45 are described.

TLLI_ASS_ADM In this state the LLC entity waits for the reception of a frame or a LLC service primitive. The transfer of *Unnumbered Information* (`UI`) frames or the exchange of Exchange Identifier (`XID`) parameters is not possible in this state.

awaiting_establish: An `LL-ESTABLISH-REQ` primitive from the network layer has arrived. The LLC layer has sent a `Set Asynchronous Balanced Mode` (`SABM`) message and waits for the acknowledgement (`Unnumbered Acknowledge` (`UA`)). After that it moves into the `Asynchronous Balanced Mode` (`ABM`) state.

wait_for_establish_res: A SABM was received and the network layer was informed by a LL-ESTABLISH-IND. The LLC entity waits for an LL-ESTABLISH-RES from the network layer. After reception a UA response is sent and the new state is ABM.

ABM: In this state the acknowledged data transfer is possible. The LLC entity waits for an LL-DATA-REQ from the network layer or the reception of an Information (I)-frame from the peer entity.

awaiting_release: The LLC entity has received an LL-RELEASE-REQ from the network layer. After sending the Disconnect (DISC) message it waits for a UA or a Disconnect Mode (DM) response from the peer entity.

wait_for_xid_neg_ADM: After receipt of an XID command in Asynchronous Disconnected Mode (ADM) and the forwarding to the network layer the entity waits for the negotiated parameters from the network layer. After that the parameters are transmitted to the peer entity.

wait_for_xid_neg_ABM: After receipt of an XID command in ABM state and the forwarding to the network layer the entity waits for the negotiated parameters from the network layer. After that the parameters are transmitted to the peer entity.

suspended_ABM: The LLC entity is temporary suspended and waits for an LL-RESUME-REQ from the network layer. After that it goes to ABM state.

suspended_ADM: The LLC entity is temporary suspended and waits for an LL-RESUME-REQ from the network layer. After that it goes to ADM state.

PRBC: In Peer_Receiver_Busy_Condition, the peer entity is not capable to receive further I-frames.

ORBC: In Own_Receiver_Busy_Condition, the peer entity is not capable to receive further I-frames, but can transmit I-frames.

timer_recovery_condition: Timer T200 has expired and the LLC entity calls for certain frames or commands from the peer entity.

NS_sequence_error: An I-frame with a wrong sequence number was received. The LLC entity calls for the missing frames and goes back to ABM after reception.

4.3.9 The RLC/MAC Layer

This protocol implementation models the RLC/MAC Standard GSM 04.60, Release 99, Version 8.2.0 [22]. The full functionality of RLC and MAC is presented. The architecture of this layer in MS and BS is similar, but the functionality of the processes is different. In the following the structure and the functions of the BS MAC layer are described. After that the differences in the MS MAC implementation are outlined.

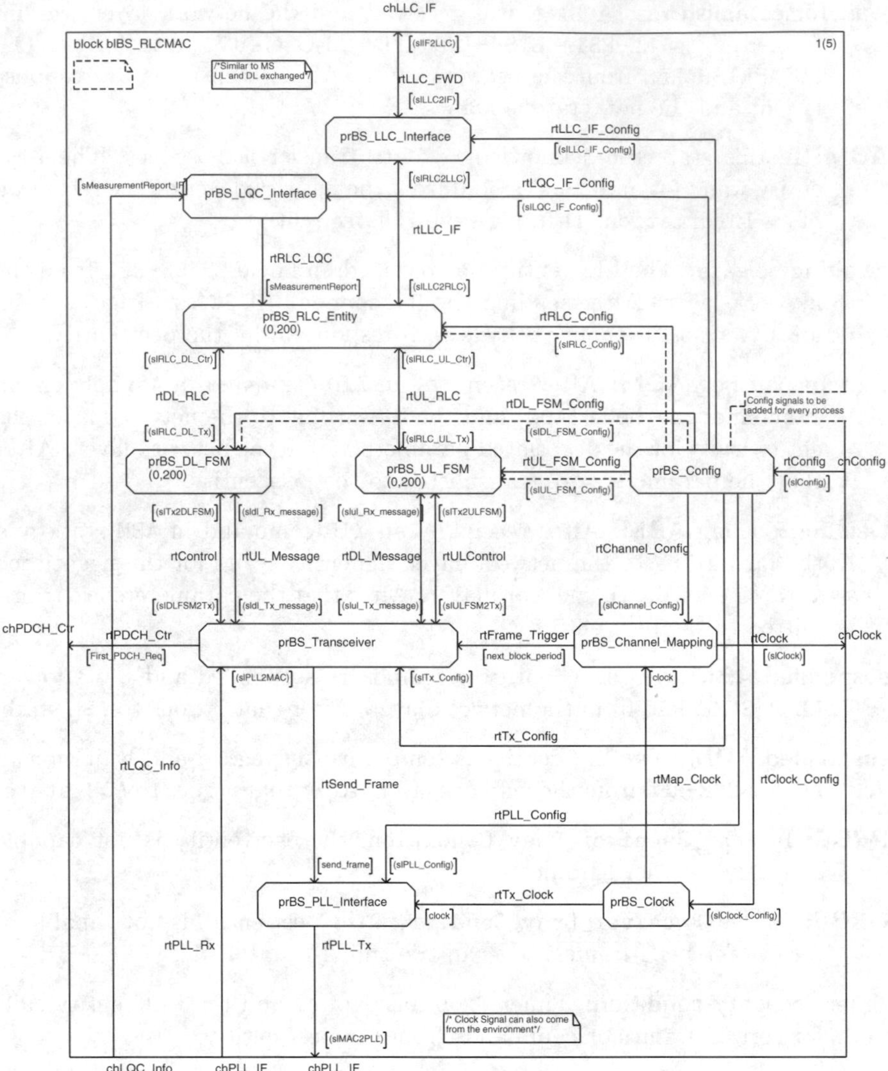

Figure 4.46: RLC/MAC base station architecture

4.3.9.1 RLC/MAC in the Base Station

In Figure 4.46 the structure of the RLC/MAC layer in the base station and the interaction between the functional processes and the environment are shown.

After the start of a simulation run the process `prBS_Config` obtains the configuration parameters from the environment and creates and initializes the instances `prBS_RLC_entity`, `prBS_DL_FSM` and `prBS_UL_FSM` for every mobile station associated with this base station. The other processes exist only once in the system. The data flow in the system is defined as follows. In the transmit direction LLC PDUs are requested from the higher layer always if the LLC queue is not empty and if there is enough space for new data to transmit in the send buffer of the RLC entity.

If necessary a new downlink TBF initiation is requested to the downlink *Finite State Maschine* (FSM). The control blocks, which are part of the TBF establishment process are given to the `prBS_Transceiver` and are transmitted as soon as radio resources are available. As soon as a TBF is established RLC blocks can be transmitted if downlink resources are available. These resources are managed by the BS transceiver, which requests new RLC blocks by asking the related RLC entity for new blocks to transmit. The order of RLC entities that are requested is defined by the scheduling function.

Complete coded radio blocks for each block period are given to the `prBS_PLL_Interface`. This process forwards one block after the other to the environment at the correct time regarding the multiframe structure. In the receive direction radio blocks are received by the BS transceiver from the environment and are forwarded to the related uplink or downlink FSM. If these radio blocks contain RLC blocks, i.e., data or acknowledgement blocks, they are forwarded by an FSM to the RLC entity. RLC data blocks are saved in the receive window and reassembled into LLC PDUs, which are finally given to the LLC layer.

In the following the internal functionality of the processes is described.

prBS_Config The Configuration process gets protocol settings and scenario parameters from the environment. Processing these values the process creates the requested number of process instances for each associated MS in the cell. After that all processes are initialized with configuration parameters. Through this mechanism the processes can be configured from outside the system.

prBS_LLC_Interface The LLC interface ensures the communication between the RLC entity of the logical link and the related LLC queue. In the transmit direction signals from an LLC instance are mapped onto the RLC entity associated to this logical link. In the receive direction reassembled LLC PDUs as well as other interworking signals from the RLC entity are forwarded to the associated LLC entity or queue.

prBS_LQC_Interface The LQC interface forwards measurement reports about the link quality to the related RLC entity.

prBS_RLC_Entity The RLC entity provides the transmission service for LLC PDUs between MS and BS. An RLC connection involves two peer entities. It uses the radio resources offered by the MAC function. A selective repeat ARQ protocol is used to ensure high reliablity. Each RLC endpoint has a receive window to receive the RLC data blocks and a transmit window to transmit

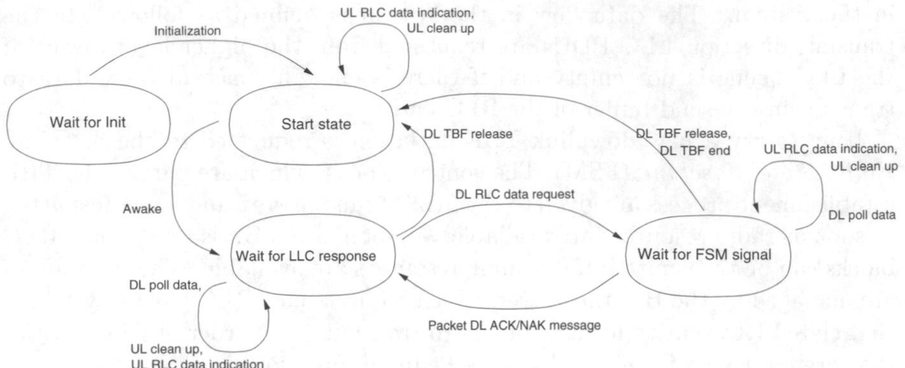

Figure 4.47: State machine of the process RLC entity at the base station

the RLC data blocks. For GPRS the size of this windows is 64. This means that maximum 64 data blocks can be transmitted until the oldest block in the transmit window must be acknowledged to allow transmissions of further blocks. For EGPRS the WS depends on the number of allocated PDCH. The values ranges from 64 to 1024.

Measurement reports in the form of a proposed optimal *Modulation and Coding Scheme* (MCS) (EGPRS), *Coding Scheme* (CS) (GPRS) resp., are received every 60 ms. The respective state machine is shown in Figure 4.47. After initialization the process is in state `Start`.

Transmit Procedures The LLC layer informs the RLC entity about the aim of sending LLC PDUs. After the TBF establishment of the MAC FSM, the RLC entity asks the LLC layer for a number of data octets whose value depends on the available free places in the send buffer and the used channel coding scheme. The CS used for each transmission is either a fixed default value or chosen based on the current channel quality (LA). After the requisition of data, the RLC entity waits for LLC response.

The LLC entity delivers a maximum number of PDUs according to the requested octets that are segmented into RLC data units. If the contents of an LLC PDU do not fill an integer number of RLC data units, the beginning of the next LLC PDU is placed within the final RLC data unit of the actual LLC PDU. No padding is necessary unless the last LLC PDU in the TBF does not fill an integer number of RLC data units.

The RLC header is generated and the RLC data blocks are stored in a send buffer, where it waits for transmission. For GPRS, the send buffer can contain a maximum number of 128 (*Sequence Number Space* (SNS)) RLC data blocks. For EGPRS the SNS range is 2048. Now the RLC entity waits for FSM signals.

When resources for data transmission are available, the MAC function polls for data and is served directly with blocks from the send buffer. The moment an RLC data block is given to the MAC the corresponding element in the acknowledge state array `V(B)` is set to `PENDING_ACK`. The elements in `V(B)` indicate the acknowledgement status (`NAK`, `PENDING_ACK`, `ACK`, `INVALID`) of the related RLC data blocks.

The data blocks are taken out of the send buffer in the following order: First, RLC data blocks whose corresponding elements in `V(B)` have the value `NAK` are delivered to the MAC, starting with the oldest one. If no data block with a `NAK` element exists and the transmit window is not stalled, the RLC data block with `BSN = V(S)` is selected. The send state variable `V(S)` denotes the block sequence number of the next in-sequence RLC data block to be transmitted. If the transmit window is stalled or there is no new block to send, the oldest RLC data blocks whose value in `V(B)` is `PENDING_ACK` are sent again.

For each RLC data block sent a timer is started. Only when it is expired is a negative acknowledgement for the corresponding RLC data block accepted. The reason is that directly after an RLC data block is retransmitted (e.g., `PENDING_ACK`—reason see above) a Packet ACK/*No Acknowledge* (NAK) message with an negative acknowledgement of the previous transmission can arrive. Then it is not reasonable to set the corresponding element in `V(B)` to `NAK` because this results in a direct retransmission of the same RLC data block.

The transfer of Packet ACK/NAK messages is managed by the RLC entity of the base station. In the downlink direction an acknowledgement message is demanded by the poll bit in the RLC header of the downlink data blocks. An ACK/NAK message is sent in the following cases:

- The receiver has received a configurable number of radio blocks.
- The *Stall Indicator* (SI) in the received data block indicates that the transmit window is stalled.
- All blocks within the TBF are received and the next ACK/NAK message is the last within the TBF.

After the receipt of a `PaDowACK/NAK` message the values of `V(B)` are updated from the values received from the peer entity in the *Received Block Bitmap* (RBB) of the acknowledgement message. The transmit window is moved and the LLC layer is asked for new data octets. If the LLC layer has no more LLC PDUs to transmit and all previous PDU are successfully transmitted, the RLC entity initiates a regular TBF end and changes into the `Start` state.

An abnormal release during an active TBF can be performed by the MAC or by the RLC. The RLC entity initiates a TBF release, if no `PaDowACK/NAK` message has arrived in a configurable time after it has sent the last RLC data block in a downlink TBF.

An abnormal release which is performed by the MAC is indicated to the RLC. The transceiver is initiated and goes into the `Start` state where it has to wait for a new request from the LLC layer. If the LLC layer has no more

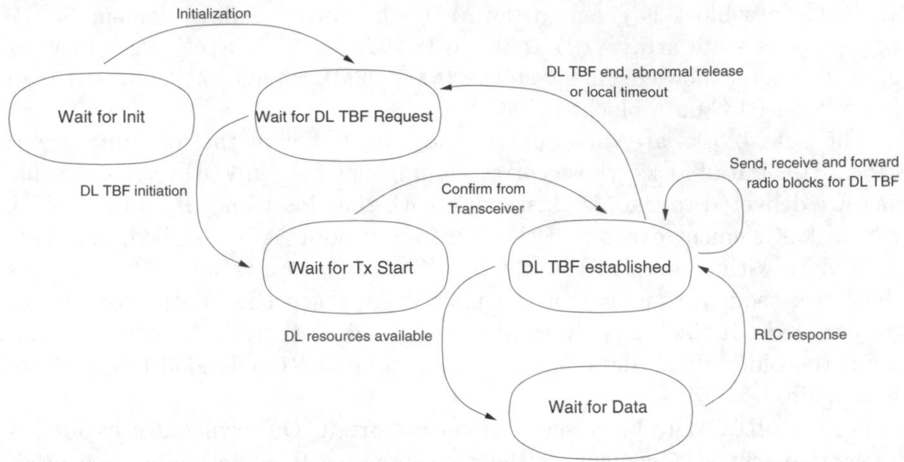

Figure 4.48: State machine of the process downlink FSM at the base station

LLC PDUs to transmit and all previous PDUs are successfully transmitted, the RLC entity initiates an regular TBF end and changes into the `Start` state.

The GPRSIM operates with open-ended TBFs. The last RLC data block in a downlink TBF is indicated by setting the FBI. The logical connections between the RLC entities are different from the TBFs provided by the MAC. Delayed termination of the TBF context is provided. Before delayed termination is done the TBF context of the MAC layer can be used to transmit data after a short inter-packet-train gap.

Receiver Procedures In the `Start` state uplink data can be received at any time. RLC data blocks are collected in the receive buffer until all RLC data blocks comprising an LLC PDU have been received. As soon as an LLC frame is successfully transmitted across the RLC Layer, it is reassembled and forwarded to the LLC Layer. For each RLC data block, which is received, the corresponding element in the receive state array `V(N)` is set to `RECEIVED`. The elements in `V(N)` indicate the receive status (`RECEIVED`, `INVALID`) of the related RLC data blocks.

When an acknowledgement message has to be sent, a Packet ACK/NAK message is composed. It contains a *SubSlot Number* (SSN) and an RBB according to the standard (GSM 04.60 [22]).

In the case of an abnormal release of the uplink TBF, the receive buffer is initialized and the receiver is ready for processing a new Uplink TBF.

prBS_DL_FSM The downlink FSM manages the state sequence and behaviour of the downlink TBF. It realizes functions to receive RLC/MAC control and data blocks. The state machine is shown in Figure 4.48.

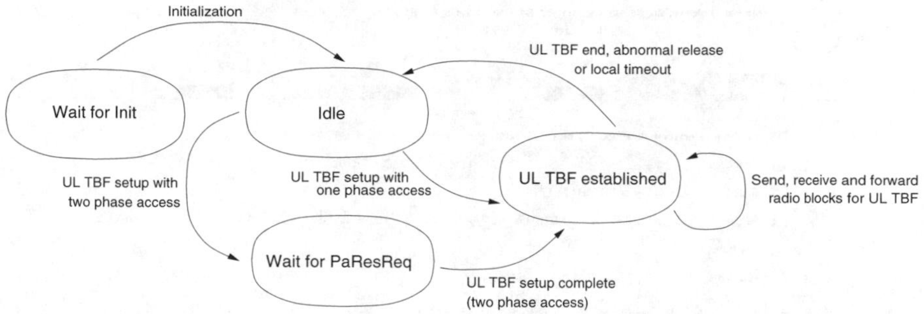

Figure 4.49: State machine of the process uplink FSM at the base station

After initialization the process waits for a downlink TBF request by the RLC entity. This is corresponding to an idle state where no TBF exists. When an initiation arrives, a `PaDowAss` message is coded and sent to the transceiver. After that the process waits for the confirmation of the transceiver process implying that the downlink TBF set up has been performed. The process goes into the state `DL TBF established` after this confirmation. In this state the transmission and reception of downlink associated RLC/MAC blocks are possible. If the transceiver has allocated radio resources for a logical link the related downlink FSM is notified. On this polling signal an RLC data block from the related RLC entity is demanded and after the RLC response the data block is forwarded to the transceiver process or the transceiver is notified that there is no data to transmit. For every state it is ensured that a timer resets the process into the `idle` state, if the process stays a certain time in a wait state without receipt of a response signal. If a regular TBF end is indicated by the RLC entity, the TBF context is released after a delay time in order to leave the TBF available to new LLC data which has to be transmitted after a short inter-packet-train gap.

prBS_UL_FSM Similar to the process downlink FSM the uplink FSM manages the state sequence and behaviour of the uplink TBF. It manages the receipt of RLC/MAC control and data blocks. The state machine is shown in Figure 4.49.

After initialization the process moves into the `idle` state and waits for an uplink TBF request in form of a `PaChaReq` message sent by the peer entity and received and forwarded by the transceiver process. After decoding this control block the process continues with the procedure of one-phase or two-phase access. Performing one-phase access the process answers with a `PaUplAss` message, which is coded and given to the transceiver. After that the uplink TBF is established. Performing two-phase access first a `PaUplAss` is coded and given to the transceiver and the process waits for a `PaResReq` message from the peer entity. After reception of this message the process sends a `PaResAss` message to the transceiver. Then the uplink TBF is established.

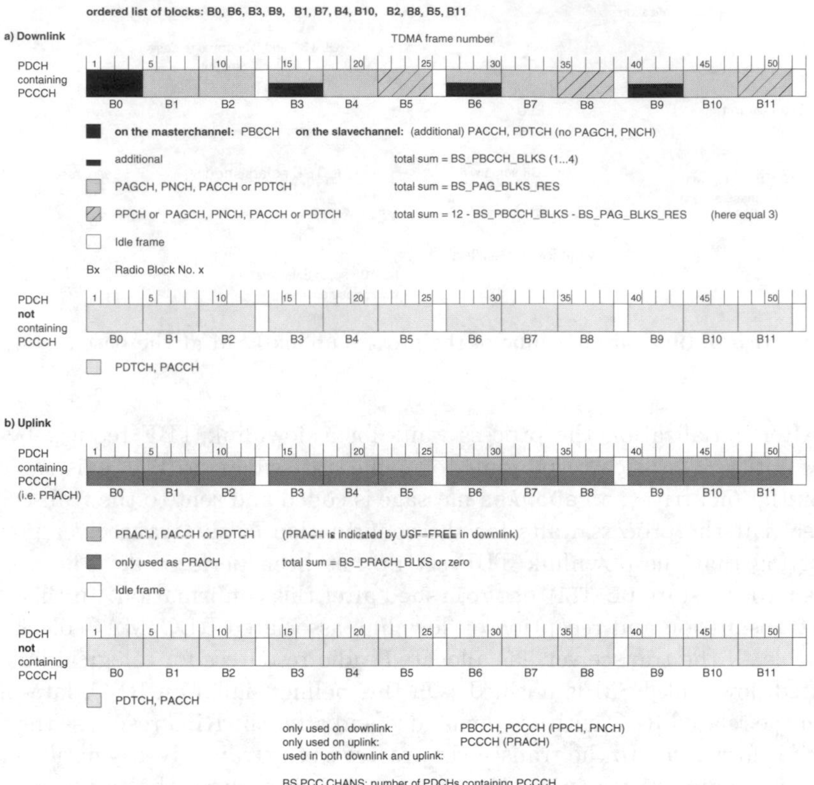

Figure 4.50: Channel map for uplink and downlink

In this state RLC/MAC blocks can be sent, received and forwarded to the upper and lower processes. Similar to the downlink FSM errors are resolved by resetting the process into the idle state, if a timeout occurs. If a regular TBF end is indicated by the RLC entity, the TBF context is released after a delay time in order to let the TBF open for new LLC data which has to be transmitted after a short inter-packet-train gap.

prBS_Channel_Mapping This process recreates the multiframe structure of (E)GPRS. It receives the clock signal which corresponds to the beginning of each timeslot. For each timeslot this process decides which logical channel is represented by the current timeslot for the uplink and the downlink. This is easily performed by a channel map which maps a position in the multiframe structure on a logical channel identifier. The channel map is shown in Figure 4.50. The optional parameters such as the number of PCCCHs and PACCHs within the PDCHs are given during the initialization phase.

Additionally the process sends a trigger signal to the transceiver in order to indicate a new block period and indicate the channel parameters for this block period such as the number of PDCHs available and the type of logical channel of each block in the block period. Other trigger signals indicate to the environment whenever a new slot, a new TDMA frame and a new 52 multiframe begins.

prBS_Transceiver The principal data structures in the transceiver are

- A resource database, which stores the current information about the radio resources available and allocated (e.g., TFI, USF).
- A TBF information table, which stores the states and parameters of the uplink and downlink TBFs of each logical link.
- Message queues for common and associated control messages to transmit.
- Buffers for received control blocks, which have to be processed and forwarded.
- Queues containing `PaChaReq` messages, which wait for resource allocation.
- TBF history queues, which contain scheduling information.

As part of the process the receiver receives radio blocks (Rx blocks) and processes the header information of these blocks. It extracts the logical link that is addressed, updates the data structures depending on the message type and its parameters and stores the blocks in the receive buffer. Received blocks are forwarded later to the uplink or downlink FSM of the related logical link. If a `PaChaReq` message arrives, resources have to be allocated to an uplink TBF regarding the multislot class of the MS. If no resources are available, a queuing notification is stored in the common control queue for transmission.

Control messages requested for transmission from the uplink or downlink FSMs (Tx blocks) are stored in the common or associated control queue. If a `PaDowAss` message is requested for transmission by the uplink FSM for a downlink TBF establishment, downlink resources have to be allocated regarding the multislot class of the MS and coded in the `PaDowAss` message, which is stored in the common control queue.

Always when a new block period is indicated by the channel mapping process, a number of radio blocks for the downlink is scheduled depending on the number of available PDCHs. For the uplink an uplink TBF is scheduled for each uplink PDCH available. The related USF field of each downlink radio block is set to the USF of the scheduled uplink TBF. The scheduling function works as follows. For each PDCH that is available in the current block period the channel type is regarded. According to this channel type common control or associated control messages are taken from the message

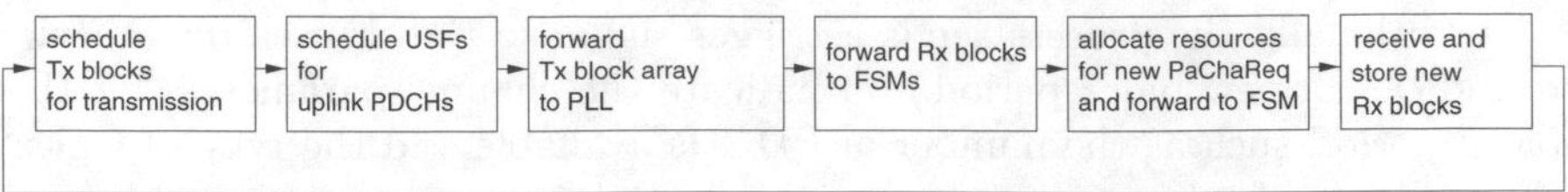

Figure 4.51: Task sequence in a block period at the base station

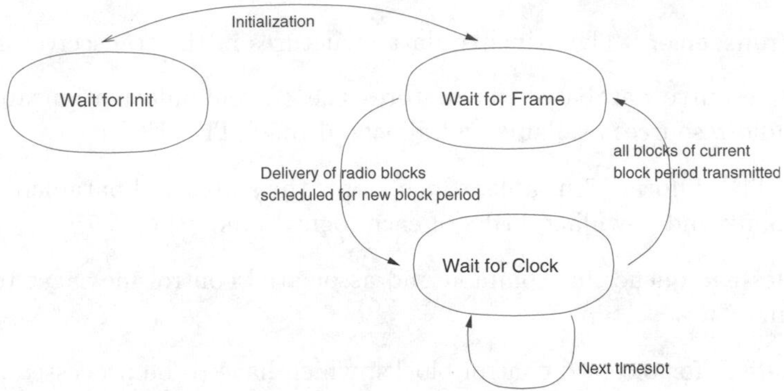

Figure 4.52: State machine of the process PLL interface

queues. If no control messages are scheduled, data blocks are demanded over the downlink FSM from the RLC entity. The sequence of the logical links that are polled for data blocks to transmit is determined by a round robin and priority scheduling strategy. The USF for each radio block is determined by another scheduling strategy for the uplink TBFs which are active. This is also done by a round robin and priority scheduling strategy. After a block for each PDCH is scheduled the blocks are given to the PLL interface. Then the radio blocks, which have been received in the block period, are forwarded to the corresponding uplink or downlink FSMs. The task sequence for one block period is shown in Figure 4.51.

prBS_Clock This process triggers the channel mapping process and the PLL interface each time a new timeslot period begins.

prBS_PLL_Interface This process is responsible for the regular transmission of radio blocks over the radio channel. It is triggered by the clock process and transmits a new radio block if the related first burst period begins. When all blocks of the current block period are transmitted, the process waits for the next scheduled radio blocks which will be transmitted in the next block period. The state machine of the process is shown in Figure 4.52.

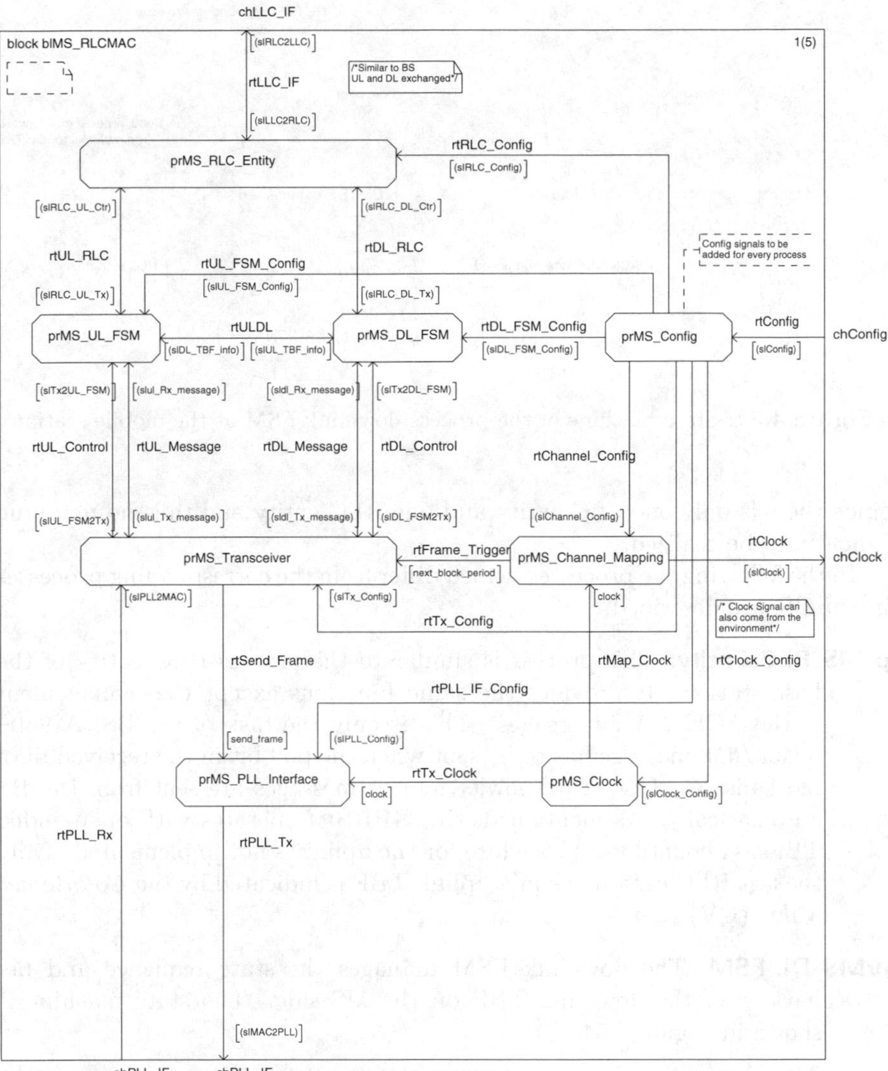

Figure 4.53: RLC/MAC mobile station architecture

4.3.9.2 RLC/MAC in the Mobile Station

In Figure 4.53 the structure of the RLC/MAC layer in the base station is shown.

The processes `prMS_PLL_Interface` and `prMS_Clock` are identical. Since there is only one logical link, one uplink TBF and one downlink TBF associated with the mobile station, each process exists only once in the system initialized by the process `prMS_Config`. The LLC interface is not necessary

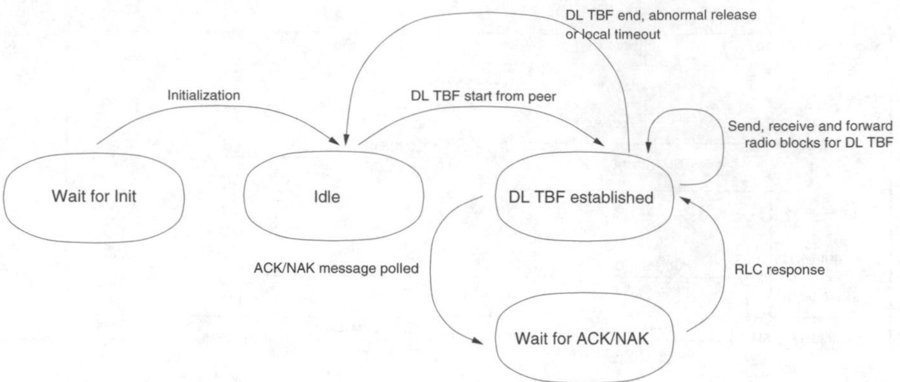

Figure 4.54: State machine of the process downlink FSM at the mobile station

since there is only one LLC entity and one RLC entity and thus no mapping function to be realized.

In the following the processes, which differ from the corresponding processes in the BS, are described.

prMS_RLC_Entity: This process is similar to the process RLC entity of the base station. It provides the same functions except the management of the ACK/NAK messages. This is only the task of the BS. A `PaDowACK/NAK` message has to be sent when the poll bit in the received data blocks is set. Uplink acknowledgement messages are sent from the BS automatically. As mentioned, the GPRSIM operates with open-ended TBFs. A countdown procedure for the uplink is not implemented. Only the last RLC data block in a uplink TBF is indicated by the *Countdown Value* (CV) zero.

prMS_DL_FSM: The downlink FSM manages the state sequence and behaviour of the downlink TBF on the MS side. The state machine is shown in Figure 4.54.

After initialization the process waits for a `PaDowAss` message from the peer entity forwarded by the MS transceiver process. This is corresponding to an idle state where no TBF exists. When this message arrives the TBF is established. In this state RLC/MAC blocks can be received and forwarded. If a Packet ACK/NAK message is polled by the peer entity, the RLC entity is asked for such a message and the process moves into the wait state. After the response from RLC it goes back to the former state. For every state it is ensured that a timer resets the process into the idle state, if the process stays a certain time in a wait state without reception of a response signal. If a regular TBF end is indicated by the RLC entity, the TBF context is released after a delay in order to let the

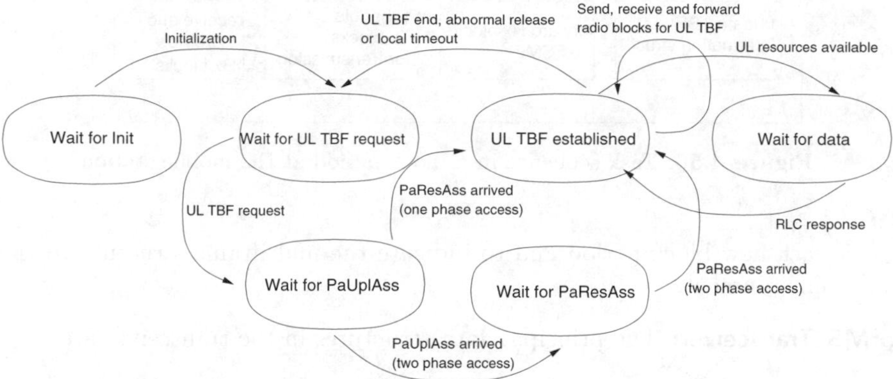

Figure 4.55: State machine of the process uplink FSM at the mobile station

TBF open for new LLC data which has to be transmitted after a short inter-packet-train gap.

prMS_UL_FSM: Similar to the process downlink FSM the uplink FSM manages the state sequence and behaviour of the uplink TBF at the mobile station. The state machine is shown in Figure 4.55.

After initialization the process waits for an uplink TBF initiation from the RLC entity. A `PaChaReq` message is coded and given to the transceiver process for transmission. The access mode, which can be either one-phase or two-phase access, is part of this block and is dependent on a configuration parameter. In one-phase access the process waits for a `PaUplAss` (`PaImmAss`) Assignment message and goes into the TBF Established state, when it arrives. In two-phase access the process waits for a `PaUplAss` message. When it arrives, a `PaResReq` message is sent to the transceiver and the process waits for a `PaUplAss` (`PaResAss`) message from the peer. After it is received the process goes into the state TBF Established and is ready for sending, receiving and forwarding of RLC/MAC blocks. If a demand is made from the transceiver to transmit data, the process asks the RLC entity for a data block and the process waits for the RLC response. After this response a data block is forwarded to the transceiver or the transceiver is informed that there is no data to transmit. Similar to the downlink FSM, errors are resolved by resetting the process into the Idle state, if a timeout occurs. If a regular TBF end is indicated by the RLC entity, the TBF context is released after a delay time in order to let the TBF open for new LLC data which has to be transmitted after a short inter-train gap.

prMS_Channel_Mapping: This process models the multiframe structure of (E)GPRS for the mobile station. Since logical channels are indicated by the USF, this process has the sole task to trigger the transceiver

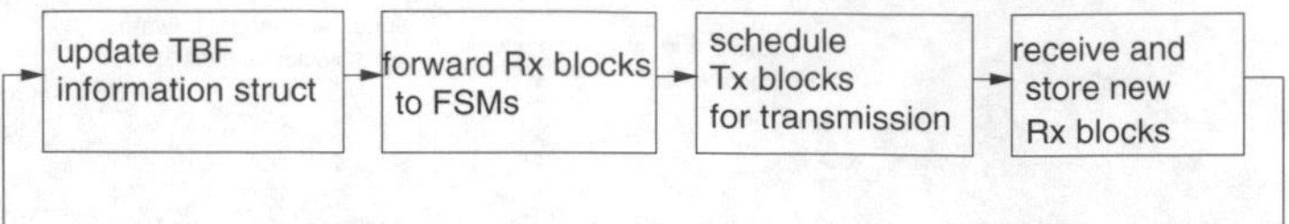

Figure 4.56: Task sequence in a block period at the mobile station

each new block period and to indicate the multiframe structure to the environment.

prMS_Transceiver: The principal data structures in the transceiver are

- A TBF information table, which stores the information of this logical link.
- An USF array, which indicates the USF indicated by the peer entity in the last block period.
- Buffers for received blocks, which have to be processed and forwarded.

The receiving procedure is similar to the transceiver at the base station. Radio blocks are received and the header information of these blocks is regarded. It extracts the TBF that is addressed. If the station is addressed by this TBF identifier, the process updates the data structures depending on the message type and its parameters and stores the blocks in the receive buffer. Additionally the USF for the related PDCH is stored in the USF array. Stored blocks are forwarded later to the uplink or downlink FSM. If a resource assignment message arrives, resource allocation information has to be stored in the TBF information structure. If the station is polled for a `PaResReq` or a `PaACK/NAK` message the blocks are requested from the higher processes and directly scheduled for transmission at the correct timeslot. The transmission of RLC/MAC blocks works as follows. When a new block period is indicated by the channel mapping process, the PDCHs that are allocated for transmission are regarded. If a poll command was indicated by the peer, a `PaResAss` or a `PaACK/NAK` message is demanded from the higher layers. If the poll bit was not set and the USF is set on the value `FREE` then a `PaChaReq` message may be sent. If the indicated USF has another value, its own USF entry for the regarded PDCH is compared with this value. If it is equal, the higher processes are requested for data to transmit. At the end of the block period the RLC/MAC blocks that were decided to transmit in the related slots are given to the PLL interface. The task sequence for one block period is shown in Figure 4.56.

4.3.10 Transceiver

The transceiver modules have the task to forward the radio blocks, which are given from the RLC/MAC layer, to the peer entity or to forward the radio blocks, which are received from the peer entity, to the RLC/MAC layer. An error flag, which is part of the radio block data structure is set, if a collision of packet channel request blocks of several MS is detected or if the channel model generates a block error. Additionally the delay, which corresponds to the duration of the block transmission on the channel interface, is realized here.

4.3.10.1 Channel Models

The channel models define the state of the radio channel and the resulting block errors for uplink and downlink radio blocks. The GPRSIM provides three channel models.

Channel model 0 (error-free channel) This channel model assumes an error-free channel. It can be used for GPRS and EGPRS simulations. Only when a collision of packet channel request blocks of several MS is detected, are the corresponding blocks rejected by setting the error flag.

Channel model 1 (GPRS channel model) The channel model 1 is a simple model for pure GPRS simulations. Mobile stations not supporting EGPRS may be active. Input of the error model is a fixed C/I ratio and a fixed *Coding Scheme* (CS) which can be selected before simulation start. Control block errors are not regarded. The look-up function giving a *Block Error Probability* (BLEP) for every C/I ratio is taken from link level simulations. A block error occurs if one or more bits in a coding block (1 block = 4 bursts) is erroneous after decoding. Perfect error detection is assumed for the simulations. The TU3 (Typical Urban) channel model in GSM 05.05 [29] is used in all link level simulations reported later. Frequency hopping is not used. The channel fades according to 3 km/h on a 900 MHz channel during the burst. The results are obtained without antenna diversity.

In Figure 4.57 the BLER versus C/I that formed the basis for the look-up function used in the channel model 1 is shown. When radio blocks are exchanged between the transceiver modules, the channel model determines, whether the arrived block is correctly received at the receiver or not.

Channel model 2 (EGPRS channel model) This channel model is usable for both EGPRS and GPRS. Figure 4.58 shows a block diagram of the model.

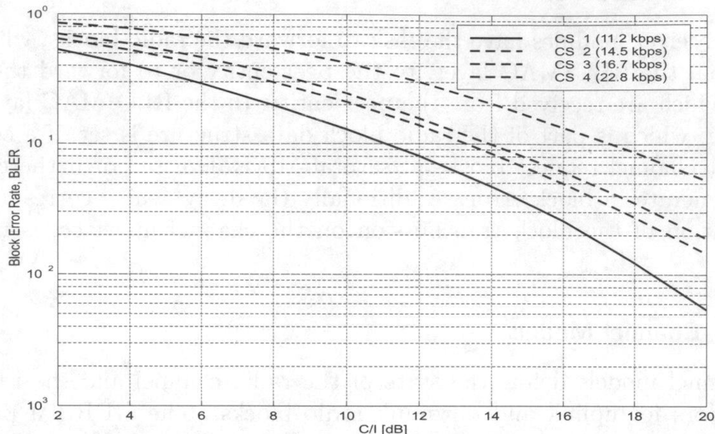

Figure 4.57: BLER over C/I reference function used for channel model 1 (GPRS)

4.3.10.2 Mobility model

The *mobility model* determines the actual distance between MS and BS within every time step of the simulation. The distance is used by the *radio propagation model*, together with statistical behaviour assumptions of the channel and the transmission activities of MS and BS, to calculate the signal-to-interference ratio on the receiving side. Within the *air interface error model* the BLEP is derived and the radio block is set either to error-free or defective.

The implemented mobility model determines the distance between BTS and MS in every time step (a radio block period) of the simulation. During start-up time a randomly located start point on the cell border (determined by a start angle α_{start}) and a given velocity of movement are assigned to every MS.

When the simulation starts, the MS starts moving with the assigned velocity on a horizontal line through the circular cell with radius R. If a mobile station reaches the cell boundary, it starts on the cell boundary and moves again on a straight line through the cell.

Figure 4.59 illustrates how a single MS moves through the cell, reaches the cell boundary and starts again moving from a random location on the cell boundary. This model is a rather simple model and does not take into account the fact that mobile stations could change their direction or their velocity. But for the basic effects of LQC, examined within the example studies presented later, it is sufficient. In Figure 4.60 the actual distance between one MS and the BTS is shown over the simulation time where the MS has a velocity of 100 km/h.

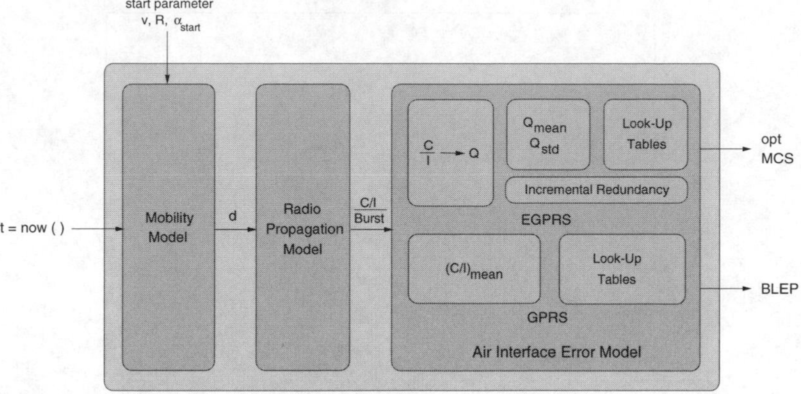

Figure 4.58: Block diagram of channel model 2

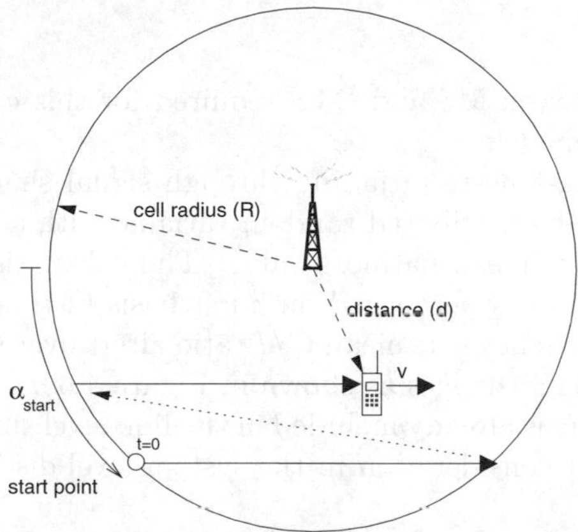

Figure 4.59: Mobility model

4.3.10.3 Radio Propagation Model

The influence of radio propagation is estimated in two steps: first, the C/I ratio is calculated using:

$$\frac{C}{I}[\text{dB}] = 10 \cdot log\left(\frac{d^{-\gamma}}{6 \cdot D^{-\gamma}}\right) \text{with} D = R \cdot \sqrt{3 \cdot C} \tag{4.3}$$

where d is the distance between MS and BS, R the cell radius, C the cluster size and D the co-channel distance and γ the path-loss factor.

Since the long-term fading mean-value changes relatively slow with the distance between MS and BTS, the distance-dependent calculation of the pathloss-related C/I value is carried out only once per radio block. The

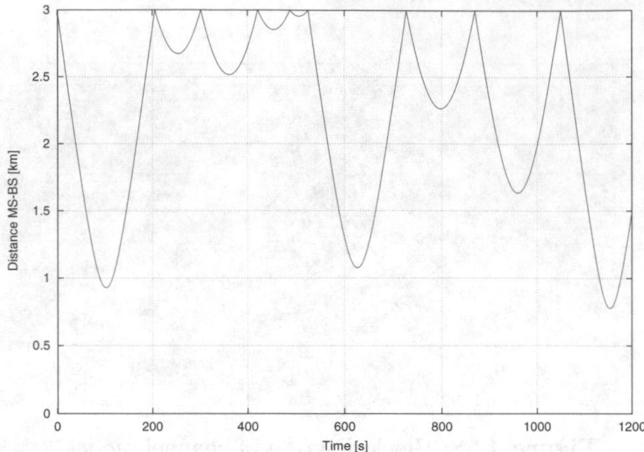

Figure 4.60: Distance between MS and BS using the mobility model

actual distance between MS and BTS required for this calculation is gained from the mobility model.

In a second step, long-term fading through signal shadowing is modelled by adding a normally-distributed random variable with a mean of zero and a given variance to the mean pathloss value. The calculation of the impact of shadow fading on C/I is performed on burst basis (4 times per radio block). Figure 4.61 illustrates how the mean C/I ratio alters over time, if the distance between MS and BTS changes as shown in Figure 4.60.

Short-term fading is already included in the link level simulations and needs therefore no further consideration in the system-level discussed here.

4.3.10.4 Air Interface Error Models

Within the air interface error model the decision is made if a received data or control block is either error-free or defective. To decide this a set of mapping curves is used. These curves are gained from link level simulations and allow the mapping of C/I values to the corresponding block error probabilities for every radio block [31, 63, 64].

EGPRS air interface error model: The value of the C/I ratio on burst basis is mapped onto a generic burst quality measure Q which is directly related to the BLEP. The mapping from the C/I ratio to Q is done according to the curve as given in Figure 4.62.

In a second step the values Q_{mean} and Q_{std} are calculated for every data block. The number of bursts per data block depend on the selected MCS. If the MS operates with MCS-1–MCS-6 every data block is interleaved over 4 bursts. For MCS-7, MCS-8, and MCS-9 one data block is interleaved over 2 bursts. If IR is used, the Q values of the previ-

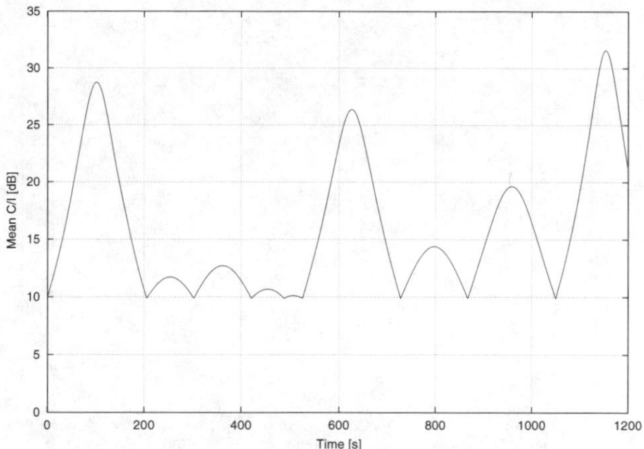

Figure 4.61: Mean C/I Ratio over Time, using the mobility model

ous transmissions are considered. The Q_{mean} and Q_{std} values are then mapped to the BLEP, using appropriate mapping curves, which are obtained from link level simulations [11]. For each MCS one mapping table is implemented in the simulator. As an example, two of these curves are shown in Figure 4.63.

GPRS air interface error model: For GPRS transmissions the values of the C/I ratio on burst basis are used to calculate an average C/I. Since in GPRS no IR is possible, no standard deviation has to be taken into account. The average C/I is mapped directly onto a BLEP. The look-up functions of the BLEP and probability are derived from effects on the physical layer, such as fast fading, frequency hopping, channel coding and equalization.

In Figure 4.64 the BLER versus C/I, which is taken from link level simulations and is used for GPRS in the channel model 2, is shown.

Beside the BLEP value the air interface model determines the optimal MCS for the actual channel quality.

4.3.10.5 Channel model parameters

The parameters for the (E)GPRS channel model 2 are listed in Table 4.15.

4.3.11 Input-Output Behaviour

4.3.11.1 Input/Initialization

The initialization is performed comprising the following steps:

- Read the simulation parameters from the file *.sim_defaults*.

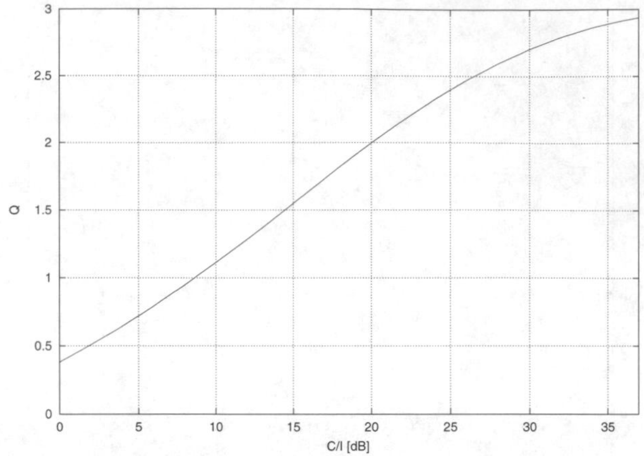

Figure 4.62: C/I to Q mapping used for channel model 2 (EGPRS)

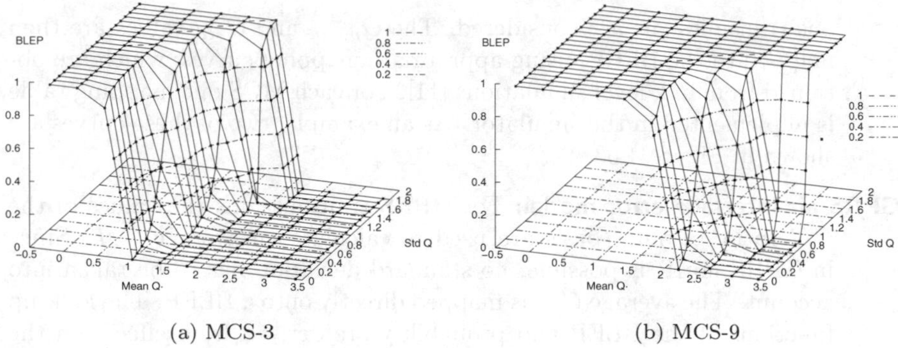

(a) MCS-3 (b) MCS-9

Figure 4.63: Mapping of Q_{mean} and Q_{std} to BLEP

- Read the Look-Up-Tables.
- Create the several class instances.
- Send the *Start Events* to the generator.
- Create and initialize the SDL systems.
- Start the scheduler.

The parameter file `.sim_defaults` is read at the start of the simulation, so simulation parameters can be changed without recompiling the program.

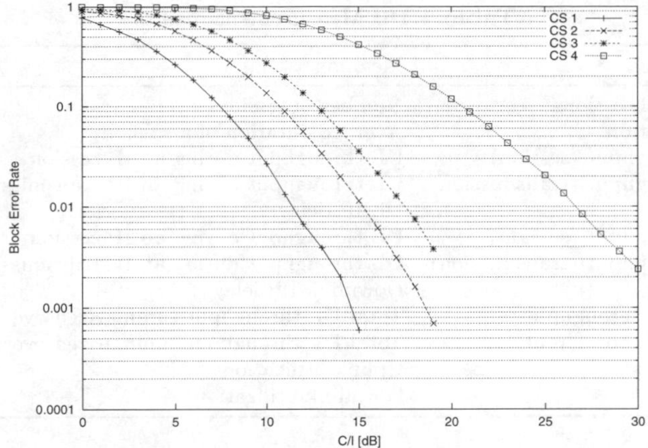

Figure 4.64: BLEP over C/I reference function used for channel model 2 (GPRS)

Table 4.15: EGPRS channel model parameters

Parameters	Default values
Velocity of the MS	100 km/h
Cell radius	3 km
C/I Variance (Shadowing)	7
Path loss parameter γ	3.7

4.3.11.2 Output/Statistics

After a simulation run, statistics are produced in a results file. The parameters are given in Table 4.16, while the evaluation of further measures to obtain further system aspects is also possible.

4.3.11.3 Graphical presentation

To make the protocols' functionality transparent the simulation flow can be presented with a graphical user interface. The tool *Graphical Interactive Simulation Result Tool* (GIST) [1] is utilized for this. GIST is an independent process called by the main program. It offers the simulation an interface to exchange data with this tool. This data is recorded comparable to a video recorder and is also available after the simulation has terminated. Navigation over the simulation time can be done by using the buttons used in a video recorder. They provide the controls of simulation sequences with a configurable speed.

The GIST graphic (see Figure 4.65) shows a 52-multiframe. For every timeslot the bursts on uplink and downlink and the USF are pictured. Also the LLC protocol sequence is represented by LLC message sequence charts.

Table 4.16: Measurement results

Parameter	Meaning
Conventional sessions	Number of sessions
Session duration	Average duration of a session
UL IP throughput (session)	UL throughput during an IP session
UL IP throughput (transmission)	UL throughput during an IP transmission
UL IP delay	UL IP delay
DL IP throughput (session)	DL throughput during an IP session
DL IP throughput (transmission)	DL throughput during an IP transmission
DL IP delay	Downlink IP delay
Total UL IP throughput	Total UL throughput summarized over all channels
Total DL IP throughput	Total DL throughput summarized over all channels
UL utilization	Uplink utilization
DL utilization	Downlink utilization

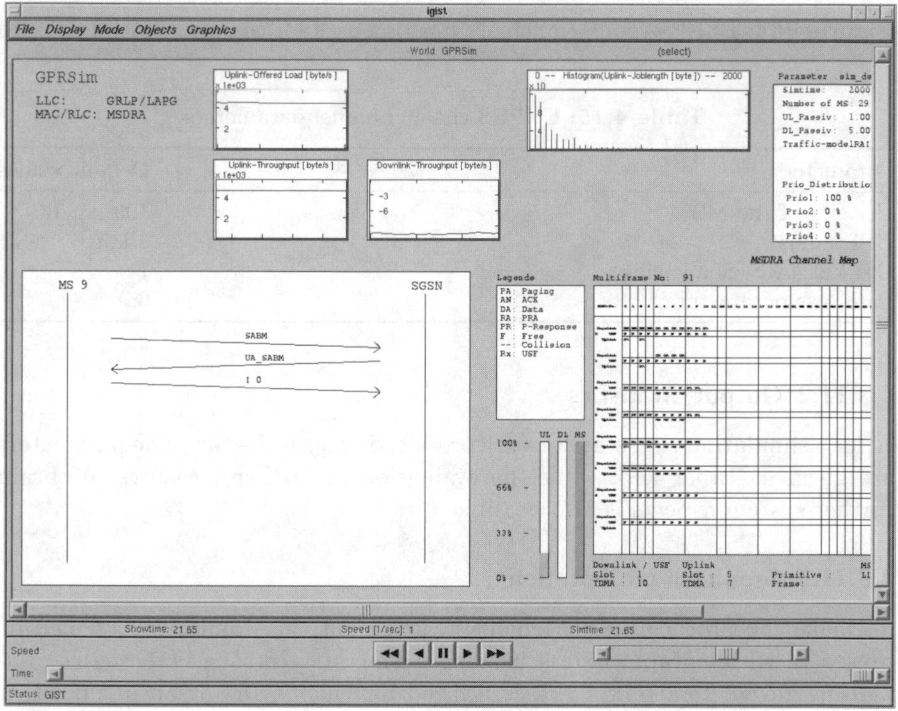

Figure 4.65: Graphical GIST presentation of the MAC and LLC transfer

4.3.12 Web Interface

The web interface was created to simplify the simulator handling and to enable start and evaluation of simulation runs over the Internet.

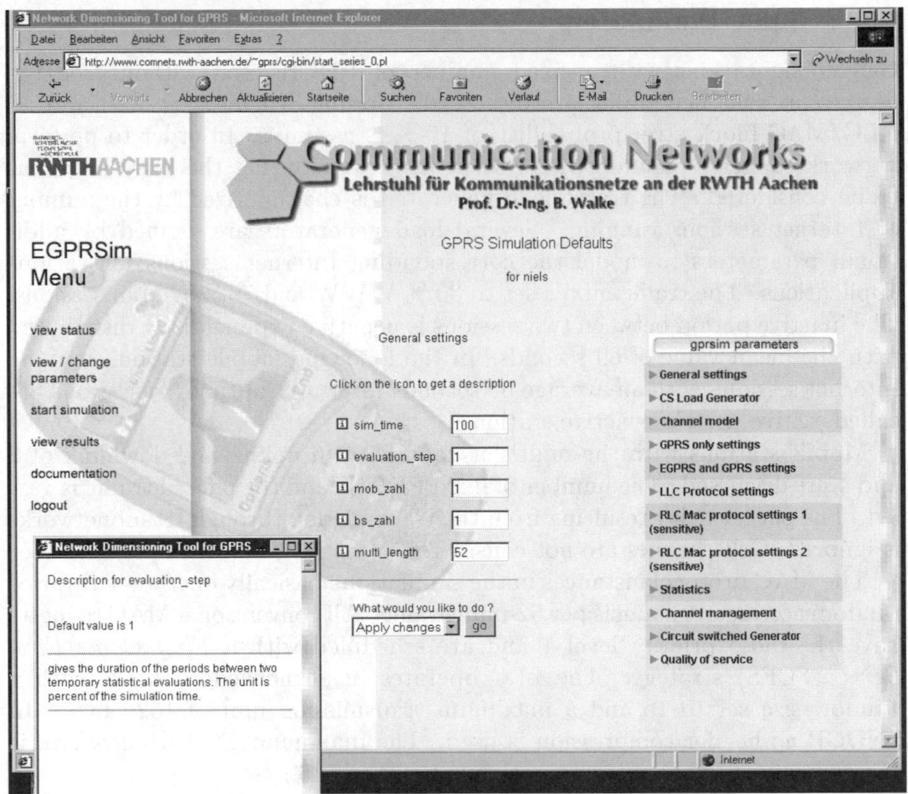

Figure 4.66: The GPRSIM web interface

After authentication, the user arrives at the start page of the simulator, where he can decide between the menu points simulation parameters and simulation progress. At the menu simulation parameters the user can change a certain number of parameters and start a new simulation after confirmation (see Figure 4.66). It is also possible to get a survey on the existing parameters. To avoid conflicts between several users and to guarantee discretion between users from different companies, for each user a new domain is created. At the menu simulation progress the user can supervise the progress of a current simulation run and read the temporary evaluation files that are created in specified intervals and at the end of the simulation.

4.4 Simulation Results for GPRS

In the following the performance measures from both the user and the system sides for several GPRS scenarios are presented [45, 44].

4.4.1 Input Parameters

Cell configurations with a fixed number of PDCHs and without coexisting circuit switched traffic are discussed. Coding-Scheme 2 is used and an RLC/MAC block error probability of 10 % is assumed. In order to perform network capacity planning the system behaviour during the busy hour has to be considered. The traffic load generated is characterized by the number of Internet sessions running. Several load generators are defined by additional parameters to model the corresponding Internet sessions for various applications. The traffic mix is set to 30 % WWW and 70 % E-Mail sessions. The inactive period between two sessions is negative-exponentially distributed with the mean value of 60 seconds. In the following mobile stations running Internet sessions with an average 60 seconds inter-arrival times of sessions are called 'active users' or 'active stations'.

Mobile stations with the multislot capability in uplink and downlink of 2 and 4 are discussed. The number of fixed PDCHs and the offered traffic is varied. The packet delay resulting from the transmission through IP subnetworks is ignored. Cell changes are not considered.

The MAC protocol instances in the simulations basically operate with three random access subchannels per 52-multiframe. All conventional MAC requests have the radio priority level 1 and are scheduled with a *First Come First Serve* (FCFS) strategy. The LLC operates in acknowledged mode with a window size set to 16 and a maximum transmission unit of 1520 byte. In SNDCP no header compression is used. The maximum TCP window size is 16 segments with the maximum segment size of 536 byte.

4.4.2 Performance and Capacity Measures

The following measures describe the GPRS performance from the user's point of view.

Mean IP throughput per user For this, the downlink IP packet throughput is measured during transmission periods, e. g., the download period of a single object of an *Hypertext Markup Language* (HTML) page. This is the important parameter for the QoS from a user's point of view. The statistical evaluation of this measure is done by counting the amount of IP bytes transmitted in each TDMA frame period, if a packet train is running. Thus, the throughput in inactive periods is not evaluated. Each value divided by the TDMA frame duration represents a value in the evaluation sequence. At the end of the simulation the mean throughput is calculated from this evaluation sequence.

Mean IP packet delay The end-to-end delay of IP packets is evaluated by means of time stamps given to the packets, when the IP layer performs an SNDCP data request for transmission. When the packet arrives at the receiver, the difference of the current time and the time stamp value is calculated. This value is one entry of the evaluation sequence.

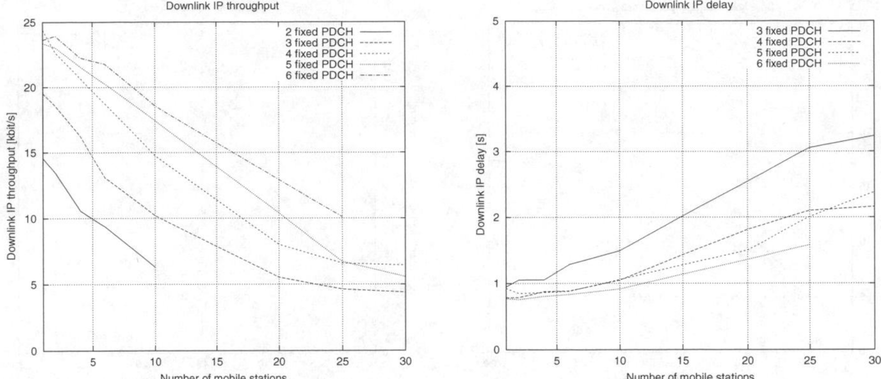

Figure 4.67: Performance measures for different numbers of fixed PDCHs

The following values describe the GPRS performance from the system's point of view.

Mean throughput in the cell The mean IP throughput in the cell, also called system throughput, is calculated from the total IP data transmitted on all PDCHs and for all users during the whole simulation, divided by the simulation duration. Since the loss of IP packets over subnetworks is not modelled, this parameter equals the offered IP traffic in the radio cell.

PDCH utilization The PDCH utilization is the ratio of the number of MAC blocks utilized for MAC data and control blocks to the sum of data, control and idle blocks. Thus existing capacity reserves in the scenario under consideration may be seen from this measure.

4.4.3 Fixed PDCH Scenarios

Figure 4.67 and Figure 4.68 show the performance and system measures over the number of mobile stations with the multislot capability of 4 uplink and downlink slots in the cell for a different number of PDCHs available for GPRS. The maximum downlink IP throughput per user during a transmission is about 25 kbit/s and decreases with the number of mobile stations. The delay increases from about 1 second to about 3 seconds per IP packet, when 25 stations are active. With 20 to 25 mobile stations being active the system comes into saturation, which can be seen both in the functions for the PDCH utilization and the cell throughput (see Figure 4.68).

The following graphs are based on the same simulations, but show performance parameters plotted over the offered IP traffic (see Figure 4.69 and Figure 4.70).

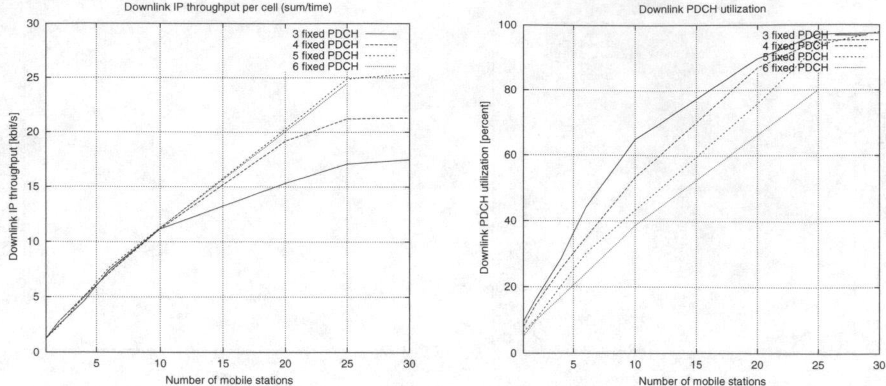

Figure 4.68: System measures for different numbers of fixed PDCHs

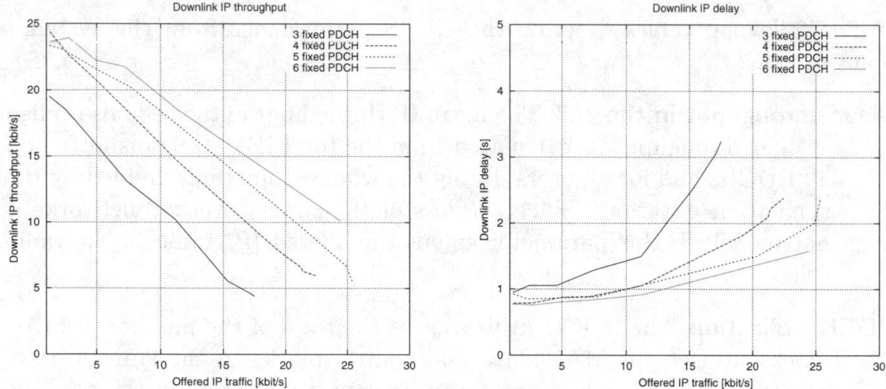

Figure 4.69: Performance measures over the offered IP traffic

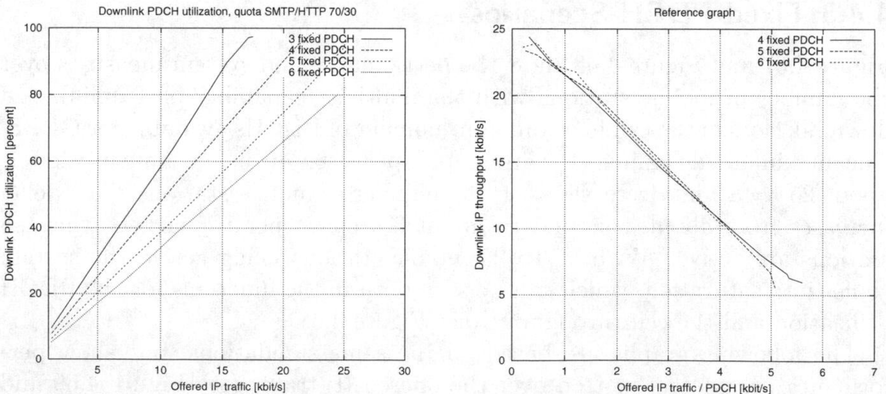

Figure 4.70: Cell throughput over the offered IP traffic and reference graph

The downlink IP throughput per user during transmission decreases almost linearly with an increasing offered IP traffic. For 3 fixed PDCHs the IP packet delay reaches a pole with an offered IP traffic of around 20 kbit/s. The poles for configurations with more than 3 PDCHs are not reached in the simulations, since the capacity limit is not reached for up to 25 stations studied.

Lengthening the graphs for the downlink IP throughput per user until the intersection with the x-axis and dividing through the number of fixed PDCHs leads to a reference graph (see Figure 4.70 right-hand side). In the reference graph, the maximum portable traffic per PDCH is 7 kbit/s. This graph can directly be used for capacity planning and network dimensioning. The aim is to find the necessary number of channels to guarantee a certain QoS. It is also possible to calculate the QoS, e.g. the downlink IP throughput per user, based on the available number of channels.

Capacity planning can be done following the next five steps:

1. Definition of the desired QoS.
2. Estimation of the number of users per cell.
3. Calculation of the user-specific offered traffic and the offered traffic per cell.
4. Determination of the acceptable traffic per PDCH with the desired QoS from the reference graph.
5. Calculation of the needed number of PDCHs:

$$\text{PDCH} = \frac{\text{Estimated offered traffic}}{\text{Acceptable offered traffic per PDCH}}$$

.

In the simulator the ascertained offered traffic per user is 540 kbyte/h.

4.4.4 Estimating the Number of Channels Needed

To visualize the dimensioning procedure introduced above, the following example is given. In this case, the QoS is given and the number of needed channels is calculated. The calculation of the offered traffic is based on the given scenario.

Definition of the desired QoS: In this example the mean DL IP throughput per user during transmission of 12.5 kbit/s is the QoS limit that is desired.

Estimation of the number of users per cell: 10

Calculation of the offered traffic per user: 540 kbyte/h = 4320 kbit/h = 1.2 kbit/s

Calculation of the total offered traffic: 43200 kbit/h = 12 kbit/s

Determination of the carriable traffic per PDCH: (from the respective reference graph)

→ acceptable traffic per PDCH = 3.5 $^{kbit}/_{s}$/PDCH
→ needed PDCHs: 3.4

For dimensioning it is necessary to round off the calculated number of PDCHs.

4.4.5 Estimating the Downlink IP Throughput Per User

Calculation of the downlink IP throughput per user during transmission for a certain number of PDCHs and an estimated number of users is the reversion of the previous approach. Let the number of given PDCHs be 4 and the estimated number of users per cell 10.

For the regarded load scenario (70 % E-Mail and 30 % WWW), the offered traffic per user is about 540 kbyte per hour, resulting in a total offered traffic for 10 users of 12 $^{\text{kbit}}/_{\text{s}}$:

- offered traffic/user = 540 kbyte/h = 4320 kbit/h = 1.2 $^{kbit}/_{s}$
- total offered traffic = 43200 kbit/h = 12 $^{kbit}/_{s}$
- carried traffic per PDCH → carried traffic = 3 $^{kbit}/_{s}$/PDCH
- quality of service → throughput per user = 12 $^{kbit}/_{s}$

4.4.6 Effects of the Multislot Capability

The simulation results shown in Figure 4.71 and Figure 4.72 visualize the difference between the multislot capabilities of 2 and 4 in uplink and downlink direction. The scenario is modelled according to the previous section with 4 fixed PDCHs.

The maximum downlink IP throughput in the simulation with multislot capability 2 is about 60 % of the value reached in the simulation with multislot capability 4. The same performance is reached for the different capabilities in situations with higher traffic load, since the PDCH utilization allows no higher throughput.

The results in Figure 4.73 and 4.74 are based on the same simulations but show the performance parameters over the offered IP traffic.

As depicted in the previous graphs, the downlink IP throughput is equal for the different multislot capabilities at 15 $^{\text{kbit}}/_{\text{s}}$, for the scenarios considered. The IP delay has a pole for an offered IP traffic of 25 $^{\text{kbit}}/_{\text{s}}$ not depending on the selected multislot class.

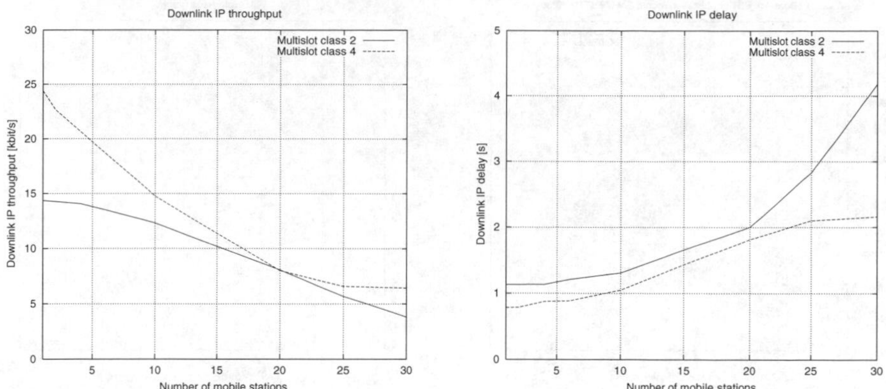

Figure 4.71: Performance measures for different multislot capabilities

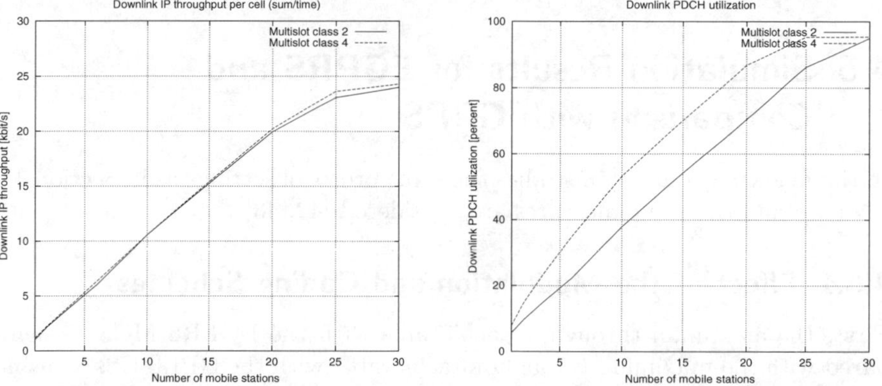

Figure 4.72: System measures for different multislot capabilities

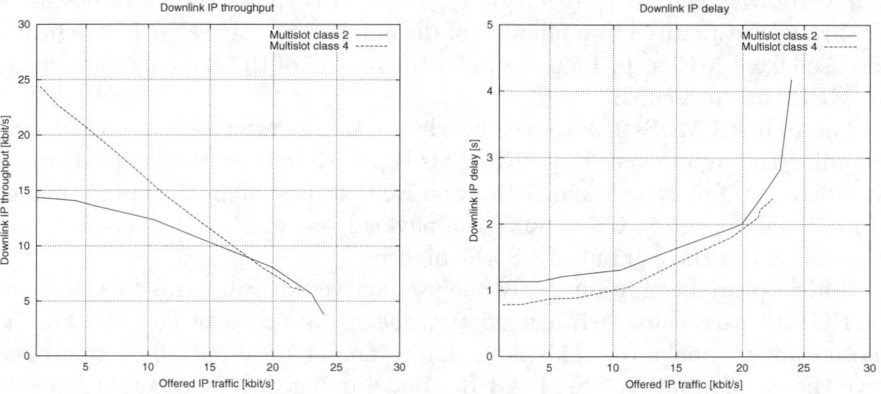

Figure 4.73: Performance measures for different multislot capabilities over offered IP traffic

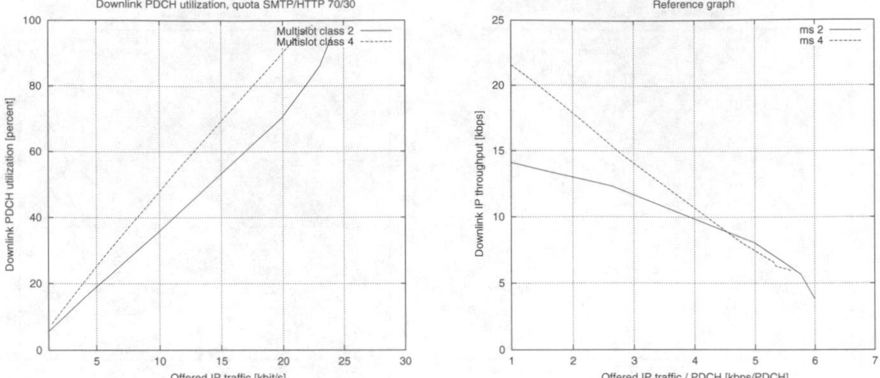

Figure 4.74: System measures for different multislot capabilities over offered IP traffic

4.5 Simulation Results for EGPRS and Comparison with GPRS

In the following simulation studies the same protocol settings as in Section 4.4 are used and the same measures are considered [42, 43].

4.5.1 Effect of the Modulation and Coding Schemes

First, the maximum throughput achievable with the EGPRS MCSs is compared with the maximum throughput achievable with the GPRS CSs focusing on Internet applications WWW and E-Mail. Additionally the IP delay and PDCH utilization for the different channel codings are examined. An optimal, error-free channel is assumed. Figure 4.75(a) shows the mean downlink IP throughput per user as a function of the number of active EGPRS users and the used fixed MCSs. In Figure 4.75(b) the results of the same measurements for GPRS are presented.

The EGPRS MCSs 1-3 achieve nearly the same performance as the corresponding coding schemes in GPRS. This is plausible since the payload size of the corresponding schemes in GPRS and EGPRS is similar. The performance of MCS-4 differs from CS-4. Since the payload size of MCS-4 is smaller than with CS-4, the throughput of CS-4 is higher.

Under optimal conditions (only one user active, MCS-9, error-free channel) an EGPRS user can reach a maximum mean data rate of 66 kbit/s for the chosen application mix. This is an improvement of about 50 % compared with the throughput for CS-4 in GPRS, but starting with 10 users the benefits from EGPRS MCSs becomes more significant, here the gain rises over 100 %. Figure 4.76 shows the IP packet delay for the different channel codings in

EGPRS and GPRS. A comparison shows a clear gain through the EGPRS MCSs.

Next, the system measures are shown. Figure 4.77 compares the PDCH utilization for EGPRS and GPRS, Figure 4.78 compares the mean IP throughput in the cell.

The diagrams in Figure 4.79(a)–(f) show the performance and system measures over the total offered IP traffic. Here the typical curve for a delay is indicated by the pole for MCS 1 and CS 1 and 20 kbit/s offered IP traffic.

The graphical presentation in Figure 4.78 shows that with ten active MSs the mean utilization of the PDCHs is nearly 100 % for all GPRS CSs. The two highest EGPRS MCS (MCS 7-9) at 10 MSs leave some reserves for further users. As expected, the maximum values of the mean cell throughput (Figure 4.79) are measured for the highest MCS of GPRS and EGPRS (CS-4 for GPRS and MCS-9 for EGPRS). The value for MCS-9 lies between 30 % for 10 mobiles up to *ca.* 50 % for 20 MSs higher than for CS-4.

The results presented within this section do not indicate the maximum throughput, since the measured values depend on the application mix and the configuration of e.g. the WWW sessions.

4.5.2 The Effects of LQC on the Block Error Probability

It is expected that LA and IR reduce the error probability of transferred data blocks. Figure 4.80(a)–(d) shows the measured BLEP as a function of the number of MSs and various combinations of the LQC mechanisms. The measurements were carried out for a cluster size of 3 and 7. Since the mean channel quality within a cell with cluster size 3 is lower than within a cell with cluster size 7 the positive effect of IR is more significant. Also the differences of LA and init LA regarding the BLEP becomes more significant. For cluster size 3 two velocities, 6 and 100 km/h, were examined.

The performance values such as throughput and delay cannot be directly derived from the BLEP values. If e.g. all data blocks are coded with MCS-1 the error rate is very low, but since the payload of MCS-1 coded blocks is very small, only slight data rates can be reached.

4.5.3 Performance Evaluation Regarding Different LQC Mechanism

Now the performance and system measures, especially the IP throughput performance, are discussed considering the EGPRS LQC functions LA and IR compared to fixed MCSs.

First, a scenario with cluster size 3, cell radius of 300 m and an MS velocity of 6 km/h is regarded. Simulations have shown that fixed MCS-5 reaches the best performance for this scenario compared to other fixed MCSs. Additionally, the performance is considered for LA without IR and LA with IR. The latter represents full LQC capability.

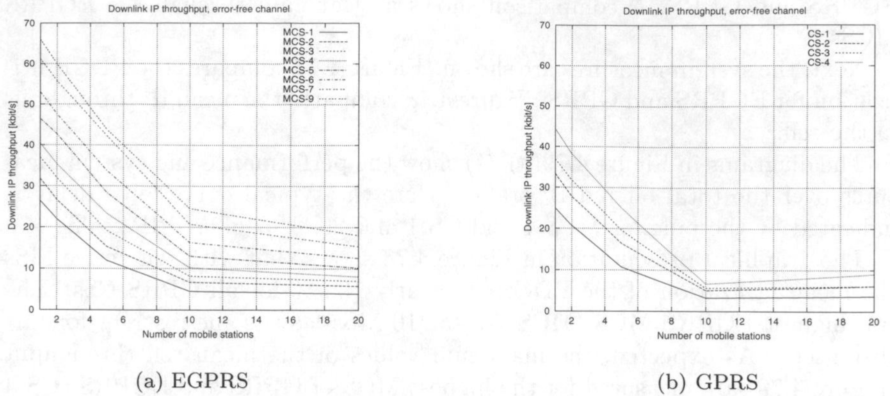

(a) EGPRS (b) GPRS

Figure 4.75: Mean throughput per user for EGPRS and GPRS (error-free channel)

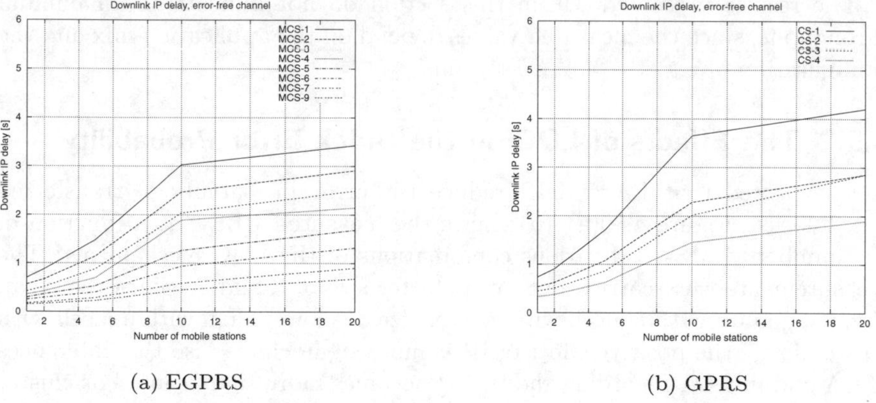

(a) EGPRS (b) GPRS

Figure 4.76: Mean IP packet delay for EGPRS and GPRS (error-free channel)

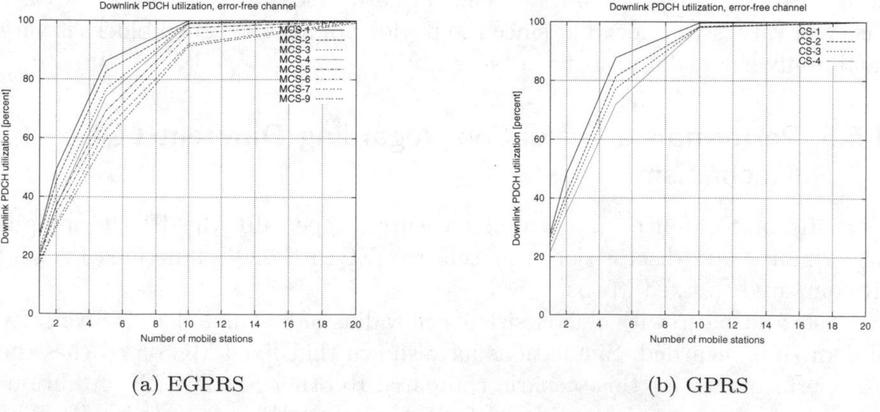

(a) EGPRS (b) GPRS

Figure 4.77: Mean PDCH utilization for EGPRS and GPRS (error-free channel)

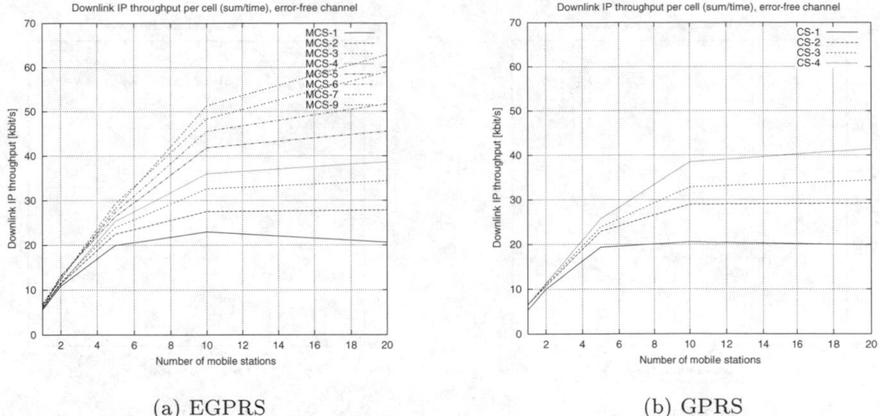

(a) EGPRS (b) GPRS

Figure 4.78: Mean IP throughput in the cell for EGPRS and GPRS (error-free channel)

Figure 4.81(a) shows that the throughput performance with LA is increased by up to 20 % compared to the fixed MCS-5. If additionally IR is used, the performance gain is not significant to LA without IR. In Figure 4.81(b) the mean downlink IP packet delay is presented. Compared to the operation with the well-chosen fixed MCS-5, IR is able to decrease the delay by up to 10 %. LA achieves a delay performance gain of even 20 to 25 %, while the additional use of IR does not change the delay performance significantly. The LQC functions do not have a major effect on the system throughput per cell and the PDCH utilization compared to the well-chosen fixed MCS-5 (see Figure 4.81(c) and Figure 4.81(d)). In the range of 10 to 20 MS the cell throughput ranges between 37 and 41 kbit/s. In Figure 4.81(e) and Figure 4.81(f) the performance measures throughput and delay are shown over the offered IP traffic.

Now the throughput and delay performance is discussed regarding the LQC functions in comparison to a less robust fixed MCS chosen. If LA is not used and a less robust MCS is chosen, e.g. MCS-6, the throughput performance can be increased with IR. Figure 4.82(a) shows a gain of up to 30 %. If the MCS is chosen even more aggressively, e.g. MCS-9 (see Figure 4.82(c)), the performance cannot be fully recovered compared to fixed MCS-5.

Finally, the throughput performance considering different cluster and cell sizes is considered. Figure 4.82(e) shows that the mean downlink IP throughput per user during transmission periods reaches up to 45 kbit/s in the scenario with cluster size 7 and 3000 m cells. While the performance gain in this scenario compared to the scenario regarded above ranges between 15 % and 30 % in low load situations, an increase of up to 45 % is reached in high load situations. The throughput does not fall below 14 kbit/s for the scenario with cluster size 7 even with 20 MS that are active in the radio cell.

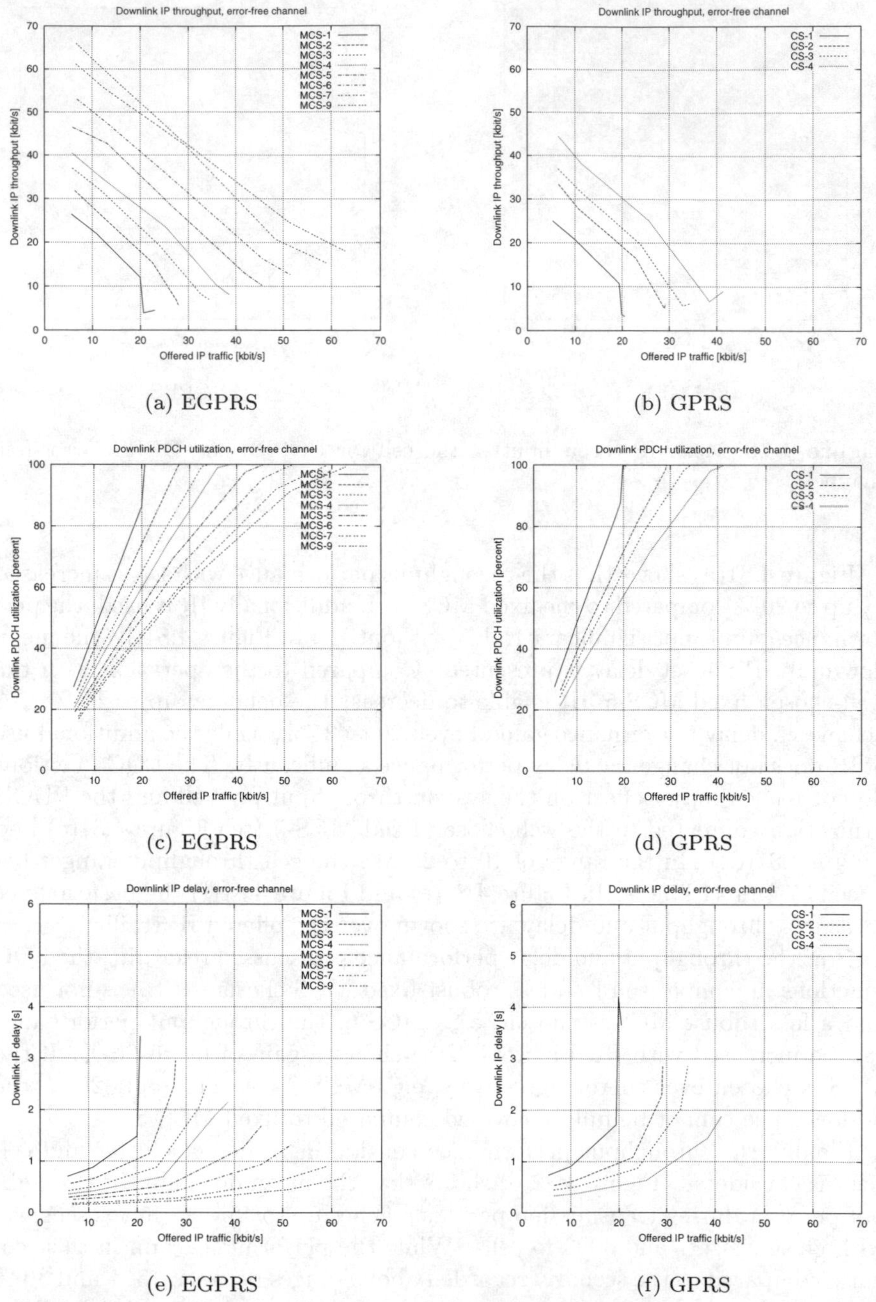

(a) EGPRS (b) GPRS

(c) EGPRS (d) GPRS

(e) EGPRS (f) GPRS

Figure 4.79: Mean IP throughput, PDCH utilization and IP packet delay versus offered traffic for EGPRS and GPRS (error-free channel)

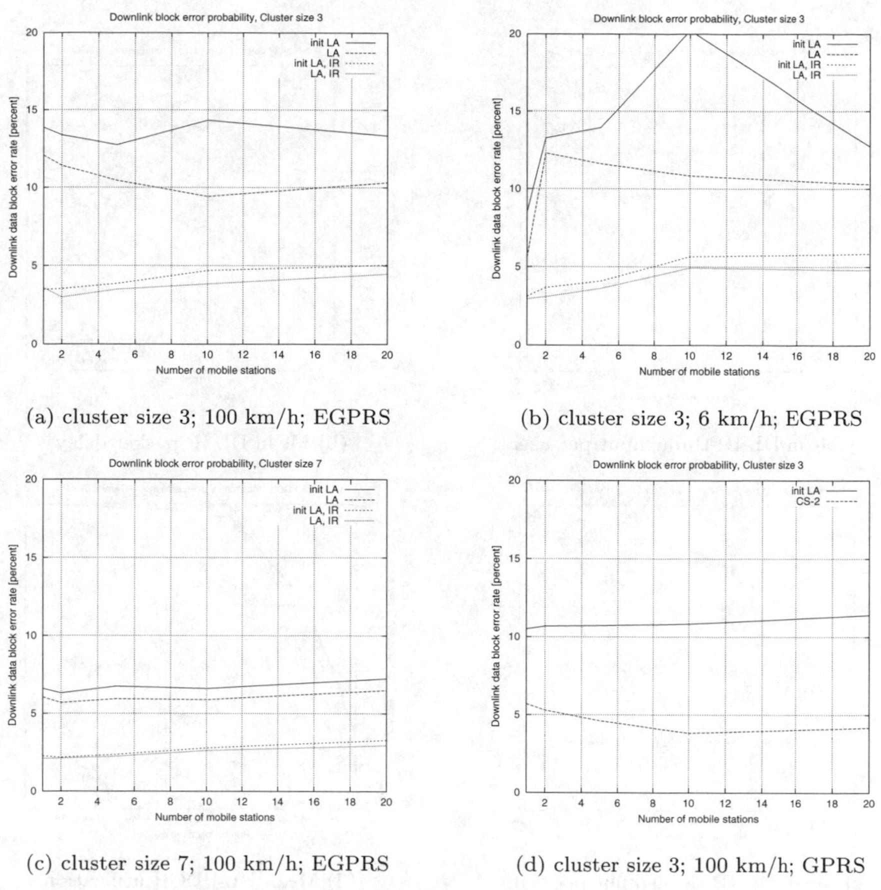

(a) cluster size 3; 100 km/h; EGPRS

(b) cluster size 3; 6 km/h; EGPRS

(c) cluster size 7; 100 km/h; EGPRS

(d) cluster size 3; 100 km/h; GPRS

Figure 4.80: Downlink data block error probabilities (BLEP)

4.6 Mobile Data Services Using GSM as a Bearer Service

This section explains the basic concepts of the service platforms *SIM Application Toolkit* (SAT), *Mobile Station Execution Environment* (MExE) and *Wireless Application Protocol* (WAP) using GSM as a bearer service. Owing to the fact that WAP is the accepted and the first available standard for future mobile data applications, the different parts of this standard are examined more closely. As the common usage of the WAP architecture is requesting *Wireless Markup Language* (WML) decks for displaying contents with a microbrowser on WAP capable devices, the characteristics of WML are also stated. An example WML page can be found at the end of this section.

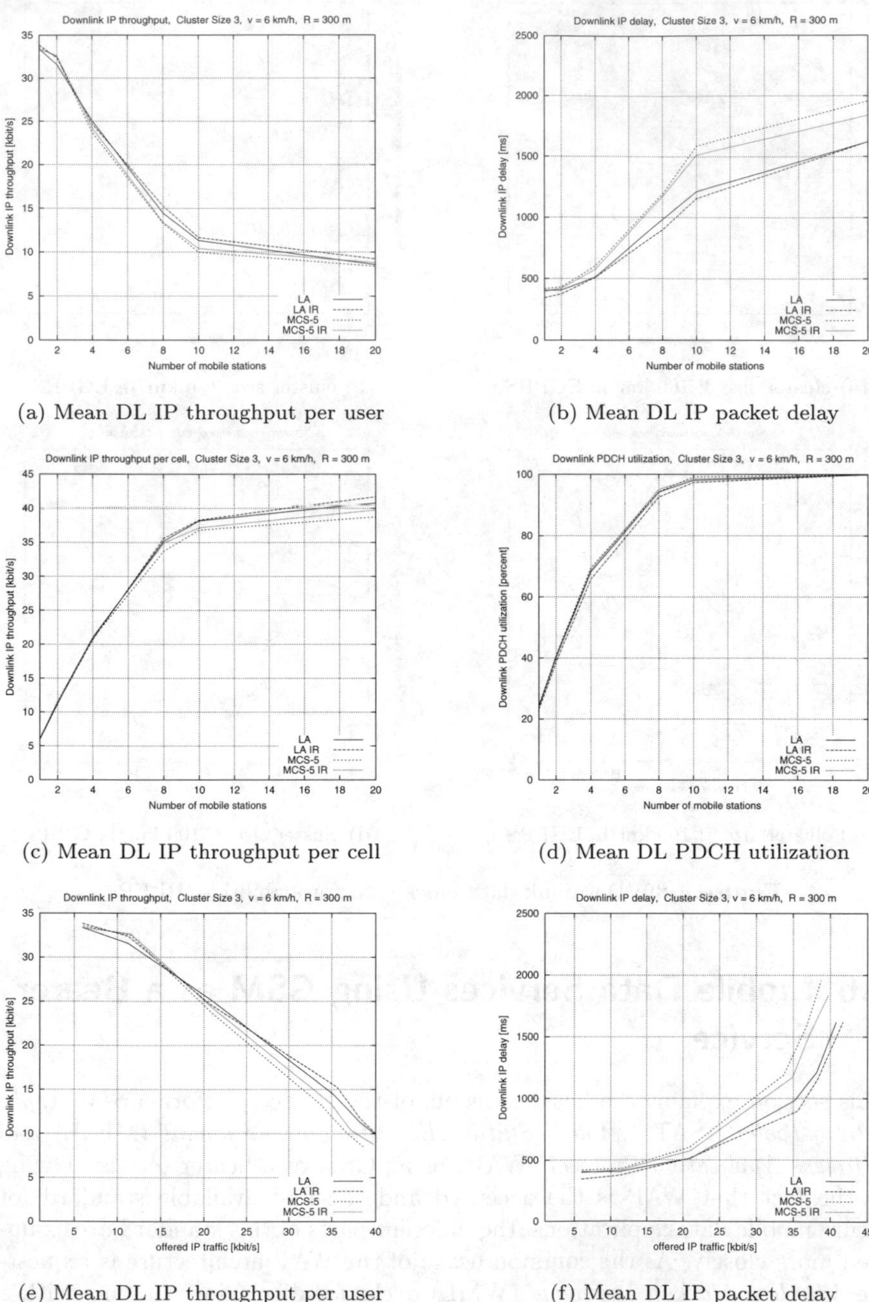

(a) Mean DL IP throughput per user

(b) Mean DL IP packet delay

(c) Mean DL IP throughput per cell

(d) Mean DL PDCH utilization

(e) Mean DL IP throughput per user

(f) Mean DL IP packet delay

Figure 4.81: Performance and system measures for different LQC capabilities

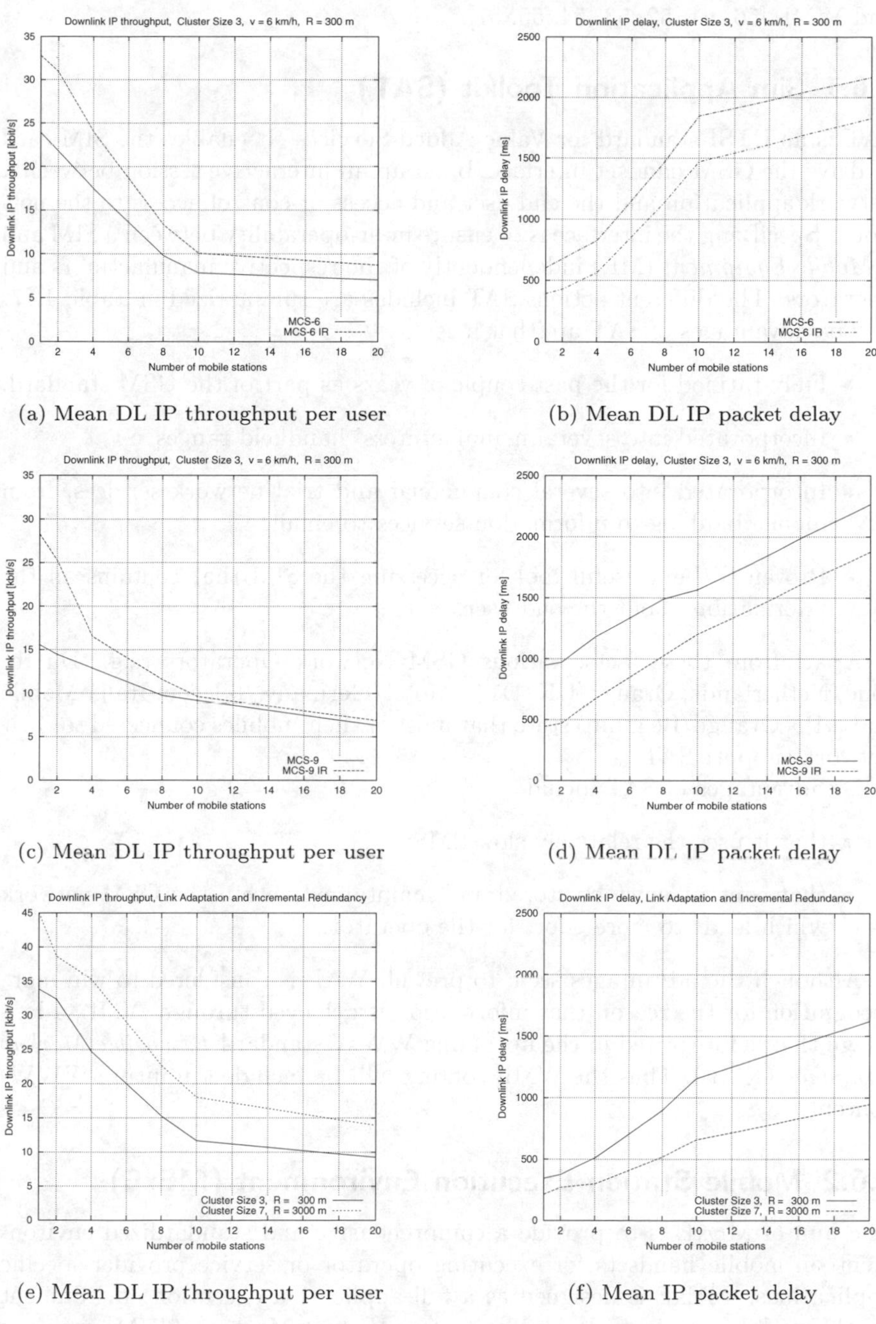

(a) Mean DL IP throughput per user

(b) Mean DL IP packet delay

(c) Mean DL IP throughput per user

(d) Mean DL IP packet delay

(e) Mean DL IP throughput per user

(f) Mean IP packet delay

Figure 4.82: Performance measures for different fixed MSCs and IR

For further information on SAT see [7, 8], on MExE see [10] and on WAP and WML [56, 51, 52, 53, 54, 55, 57].

4.6.1 Sim Application Toolkit (SAT)

SAT is an ETSI standard for Value Added Services. It enables the SIM card to drive the GSM handset interface, build up an interactive session between a network application and the end user and access or control access to the network. Specifying the interface is to ensure inter-operability between a SIM and a *Mobile Equipment* (ME) independently of the respective manufacturers and operators. The different actions SAT includes are summarized in Table 4.17.

The advantages of SAT are that it is:

- Fully ratified for the past couple of years as part of the GSM standard.
- Incorporated into several manufacturers' handheld ranges.
- Incorporated into several commercial and trial network services, from mobile banking to information services to email.
- Proven to be a useful tool for accessing the SIM that contains all the information about the end user.

Apart from these facts, various GSM Network Operators (e.g., Dutchtone/Netherlands, Orange/UK, D1 T-Mobil/Germany, Telecom Italia Mobile and KPN Orange/Belgium) state that most of their mobiles connected to their network support SAT.

Disadvantages of SAT include:

- that it uses the relatively slow SMS;
- that content must be stored and maintained within the GSM network which leads to more effort for the operator.

Although the advantages seem to prevail, WAP is considered to win more recognition for the reason that information is achieved through WML, a language that is integrated in the upcoming WWW standard *Extensible Markup Language* (XML). Thus the WML-content will be included in future WWW-content.

4.6.2 Mobile Station Execution Environment (MExE)

The aim of MExE is to provide a comprehensive and standardized environment on mobile handsets for executing operator or service provider-specific applications. MExE is designed as a full application execution environment on the mobile handset. It builds a *Java Virtual Machine* (JVM) into the user's mobile phone. MExE shares several similarities with WAP, in that

Table 4.17: Sim Application Toolkit Features

Command	Action
Select Item	The SIM sends a sub-menu with a list of items to the handset and returns the item selected by the user
Display Text	The SIM sends this command to exchange information with the user
Get Inkey	The SIM sends this command to ask the user a single character
Get Input	The SIM sends this command to ask the user a character string
Play Tone	The SIM sends this command to instruct the handset to play a predefined audio tone
Set Up Call	The SIM sends this command to request an automatic call to a given number
Send Short Message	The SIM instructs the handset to send a short message (i.e., with request for a certain information)
Send SS	The SIM sends a Supplementary Service control (i.e., to request call forwarding or call barring)
Send USSD	A *Unspecified Service Signalling Data* (USSD) is a data message that uses the same channel as SS and is faster than SMS
Provide Local Information	The SIM asks for information about the current location of the mobile (country code, mobile network code, location area and cell identity value)
Cell Broadcast download	This is used to push data or applications to the SIM
SMS PP Download	Allows download point-to-point (PP) SMS messages with data or applications
Call Control	The SIM handles all calls originated from the handset with the result of authorization, refusal or authorization with changes (i.e., shortcut numbers)
Mobile Originated Short Message Control	The SIM handles all SMS originated from the handset (compare Call Control)
Set Up Menu	The SIM sends a new or removes a menu item displayed by the handset
Menu Selection	Tells the SIM which menu item has been selected at the handset
More Time	Allows the SAT task in the SIM more time for processing
Event Download	Tells the SIM that one of the following events happened: mobile terminated call, call connected, call disconnected, user activity and idle screen available (to display messages)
Set Up Event List	The SIM give the handset a list of events it wants to know about if they happen (compare Event Download)
Timer Management/Timer Expiration	Allows SIM to handle timers in the mobile equipment
Power On Card/Power Off Card	Toggles power of a second installed SIM
Get Reader Status	Is there a second card installed?
Perform Card APDU	Sends commands to a second installed card
Poll Interval/Polling Off	Sets the polling of the handset to check for an installed SIM
Refresh	The SIM tells the handset that changes to the files have taken place

both protocols have been designed to work with a range of GSM mobile network services from SMS to GPRS and later with UMTS. Whereas WAP includes support for text, graphics and scripting, MExE allows full application programming. Because programming and running Java applications require significant processing resources of the mobile client, MExE is primarily aimed at the next generation of powerful mobile handset. On the other hand, MExE terminals can be regular phones, because MExE incorporates a capability indication method called classmarks. MExE classmarks define the MExE-related services that a particular terminal supports. There will be classmarks that match and those that exceed WAP functionality. The MExE mobile client can inform the MExE server of its classmark and therefore its capabilities.

MExE will be available in the year 2003 later than WAP because the processing power to run the Java applications is not currently available, but under development in mobile terminals. MExE and WAP are not necessarily competitors, but may rather coexist. For further information on MExE see [10].

4.6.3 Wireless Application Protocol (WAP)

The *Wireless Application Protocol* (WAP) is a result of the WAP Forum's efforts to promote industry-wide specifications for technology useful in developing applications and services that operate over wireless communication networks. WAP specifies an application framework and network protocols for wireless devices such as mobile phones, pagers, and Personal Digital Assistants (PDAs), normally referred to as 'Thin-Client' devices [62]. The specifications extend and leverage mobile networking technologies (such as digital data networking standards) and Internet technologies (such as XML, URLs, scripting, and various content formats). The effort is aimed at enabling operators, manufacturers, and content developers to meet the challenges in building advanced differentiated services and implementations in a fast and flexible manner [41].

A *Thin-Client* device can range from low-end, display limited cellular telephones to full functionality, high-speed lap-top computers.

A client is termed *thin* based on any of the following attributes:

- Display and data entry constraints.
- Memory or *Central Processing Unit* (CPU) processing constraints.
- Availability of transport bearer services.

4.6.3.1 History of WAP

Prior to the formation of the WAP forum there were several vendors working on the *Thin-Client* problem:

- Nokia with *Narrow Band Sockets* (NBS) and the *Tagged Text Markup Language* (TTML), aimed at bringing HTML content to the handset. It is already supported in the terminal 8010i.

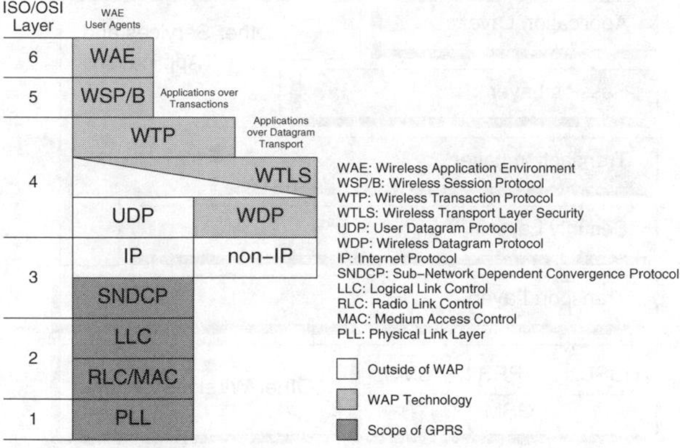

Figure 4.83: WAP architecture and applications access

- Ericsson with the *Intelligent Terminal Transfer Protocol* (ITTP), aimed at providing enhanced call control.
- Unwired Planet (nowadays `phone.com`) with the *Handheld Device Markup Language* (HDML), aimed at providing a web-like environment for the development of network-based applications and services.

This diversity was threatening to fragment the wireless data service providers.

With industry pressure, these three companies, along with Motorola formed the WAP Forum on June 26, 1997. Each company entered with the aim of minimizing the impact on current work in progress. For political reasons carriers were excluded from the meetings (though they joined the WAP Forum soon, after its founding). The goal of the forum was to develop a protocol which comprised the best of all three proposals, which led to the creation of WAP.

4.6.3.2 WAP Architecture

The WAP Architecture Specification is intended to present the system and protocol architectures essential to achieving the objectives of the WAP Forum. The WAP Architecture Specification acts as the starting point for understanding the WAP technologies and resulting specifications. As such, it provides an overview of the different technologies and refers to the appropriate specifications for further details. Figure 4.84 shows an overview of the WAP infrastructure while Figure 4.83 shows how applications can access the WAP stack; alternatively, using the whole stack, simply using Transaction services or even using only the Transport Protocols like *Wireless Datagram Protocol* (WDP) or *User Datagram Protocol* (UDP).

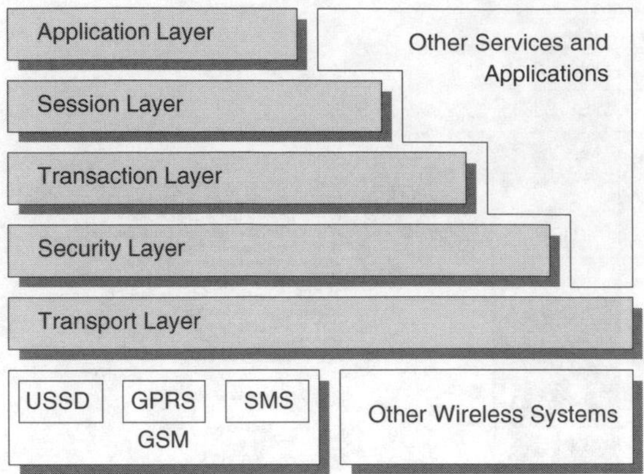

Figure 4.84: WAP protocol stack

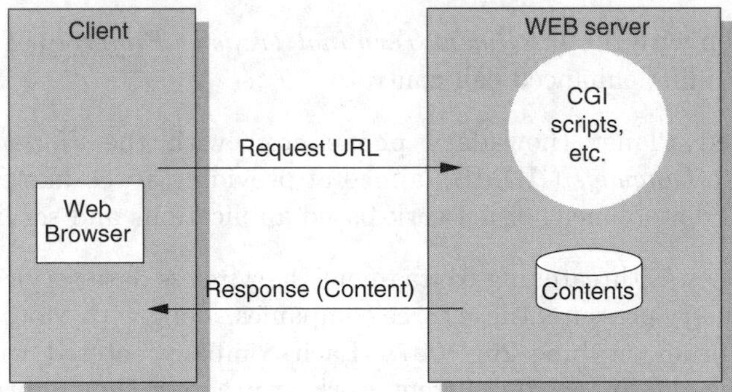

Figure 4.85: WWW programming model

4.6.3.3 Wireless Application Environment (WAE)

The WWW Model The Internet's World Wide Web provides a very flexible and powerful logical model. Applications present content to a client in a set of standard data formats that are browsed by client-side user agents known as Web browsers (or simply browsers). Typically, a user agent sends requests for one or more named data objects (or content) to an origin server. As shown in Figure 4.85 an origin server responds with the requested data expressed in one of the standard formats known to the user agent, e.g., HTML.

The WWW standards include all the mechanisms necessary to build a general-purpose environment:

- All resources on the WWW are named with Internet-standard *Unified Resource Locator* (URL).
- All classes of data on the WWW are given a specific type allowing the user agent to correctly distinguish and present them appropriately.

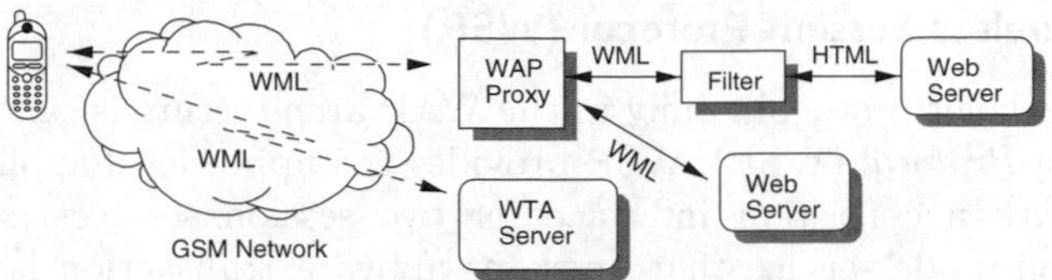

Figure 4.86: WAP architecture

Furthermore, the WWW defines a variety of standard content formats supported by most browser user agents. These include the *Hypertext Markup Language* (HTML), the JavaScript scripting language, and a large number of other formats (e.g., bitmap image formats).

- The WWW also defines a set of standard networking protocols allowing any browser to communicate with any origin server. One of the most commonly used protocols on the WWW today is the *Hypertext Transfer Protocol* (HTTP).

The WWW infrastructure and model have allowed users to easily reach a large number of third-party content and applications. It has allowed authors to easily deliver content and services to a large community of clients using various user agents, e.g., Netscape Navigator and Microsoft Internet Explorer.

The WAE Model *Wireless Application Environment* (WAE) adopts a model that closely follows the WWW model. All content is specified in formats that are similar to the standard Internet formats. Content is transported using standard protocols in the WWW domain and an optimized HTTP-like protocol in the wireless domain. WAE has borrowed from WWW standards, including authoring and publishing methods wherever possible. The WAE architecture allows all content and services to be hosted on standard Web origin servers that can incorporate proven technologies (e.g., CGI). All content is located using WWW standard URLs. WAE enhances some of the WWW standards in ways that reflect the device and network characteristics. WAE extensions are added to support mobile network services such as call control and messaging. Careful attention is paid to the memory and CPU processing constraints that are found in mobile terminals. Support for low bandwidth and high-latency networks is included in the architecture as well. WAE assumes the existence of gateway functionality responsible for encoding and decoding data transferred from and to the mobile client. The purpose of encoding content delivered to the client is to minimize the size of data sent to the client over-the-air as well as to minimize the computational power required by the client to process that data. The gateway functionality can be added to origin servers or placed in dedicated gateways as illustrated in Figure 4.86.

4.6.3.4 Wireless Session Protocol (WSP)

The Session layer protocol family in the WAP architecture is called the *Wireless Session Protocol* (WSP). WSP provides the upper-level application layer of WAP with a consistent interface for two session services. The first is a connection-mode service that operates above a transaction layer protocol *Wireless Transaction Protocol* (WTP), and the second is a connectionless service that operates above a secure or nonsecure datagram transport service. The WSPs currently offer services most suited for browsing applications. WSP provides HTTP 1.1 functionality and incorporates new features such as long-lived sessions, a common facility for data push, capability negotiation, and session suspend/resume. The protocols in the WSP family are optimized for low-bandwidth bearer networks with relatively long latency.

4.6.3.5 WSP Architectural Overview

WSP is designed to function on the transaction and datagram services. Security is assumed to be an optional layer above the transport layer. The security layer preserves the transport service interfaces. The transaction, session, or application management entities are assumed to provide the additional support that is required to establish security contexts and secure connections. This support is not provided by the WSP protocols directly. In this regard, the security layer is modular. WSP itself does not require a security layer; however, applications that use WSP may require it.

WSP provides a means for organized exchange of content between cooperating client/server applications. Specifically, it provides the applications means:

- to establish a reliable session from client to server and release that session in an orderly manner
- to agree on a common level of protocol functionality using capability negotiation
- to exchange content between client and server using compact encoding
- to suspend and resume the session.

WSP actually defines two protocols: One provides connection-mode session services over a transaction service, and another provides nonconfirmed, connectionless services over a datagram transport service. The connectionless service is most suitable when applications do not need reliable delivery of data and do not care about confirmation. It can be used without actually having to establish a session. In addition to the general features, WSP offers means to do the following:

- provide HTTP/1.1 functionality
- extensible request-reply methods

- provide composite objects
- provide content-type negotiation
- exchange client and server session headers
- interrupt transactions in process
- push content from server to client in an unsynchronized manner
- negotiate support for multiple, simultaneous asynchronous transactions.

Basic Functionality The core of the WSP design is a binary form of HTTP. Consequently, the requests sent to a server and responses going to a client may include both headers (meta-information) and data. All the methods defined by HTTP/1.1 are supported. In addition, capability negotiation can be used to agree on a set of extended request methods so that full compatibility to HTTP/1.1 applications can be retained. WSP provides typed data transfer for the application layer. The HTTP/1.1 content headers are used to define content type, character set encoding, languages, and so forth, in an extensible manner. However, compact binary encodings are defined for the well-known headers to reduce protocol overhead. WSP also specifies a compact composite data format that provides content headers for each component within the composite data object.

This is a semantically equivalent binary form of the *Multipurpose Internet Mail Extensions* (MIME) *multipart/mixed* format used by HTTP/1.1. WSP itself does not interpret the header information in requests and replies. As part of the session creation process, request and reply headers that remain constant over the life of the session can be exchanged between service users in the client and the server. These may include acceptable content types, character sets, languages, device capabilities, and other static parameters. WSP will pass through client and server session headers as well as request and response headers without additions or removals. The life-cycle of a WSP session is not tied to the underlying transport. A session can be suspended while the session is idle to free up network resources or save battery. A lightweight session re-establishment protocol allows the session to be resumed without the overhead of full-blown session establishment. A session may be resumed over a different bearer network. Figure 4.87 shows a successful WSP session establishment.

4.6.3.6 Wireless Transaction Protocol (WTP)

A transaction protocol is defined to provide the services necessary for interactive browsing (request/response) applications. During a browsing session, the client requests information from a server, and the server responds with information. The request/response cycle is referred to as a transaction in

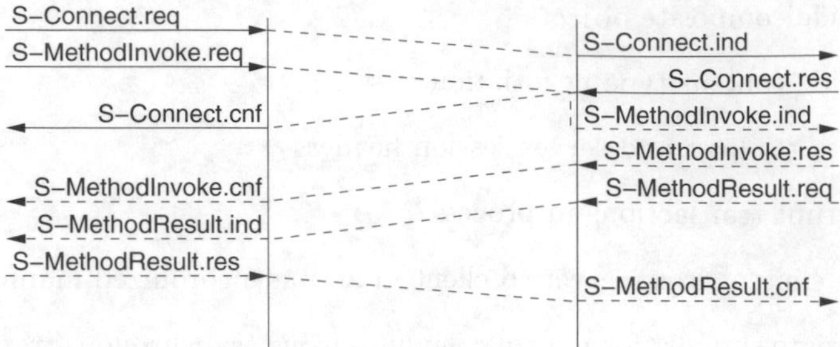

Figure 4.87: WSP succesfull session establishment

Table 4.18: Utilization of WTP by WSP

WSP Facility	WTP Transaction Classes
Session Management	Class 0 and Class 2
Method Invocation Class 2	Class 2
Session Resume	Class 0 and Class 2
Push	Class 0
Confirmed Push	Class 1

the following. The objective of the protocol is to reliably deliver the transaction while balancing the amount of reliability required for the application with the cost of delivering the reliability. WTP runs on top a datagram service and optionally a security service. WTP has been defined as a lightweight transaction-oriented protocol that is suitable for implementation in thin clients (mobile stations) and operates efficiently over wireless datagram networks. The benefits of using WTP include:

- Improved reliability over datagram services. WTP relieves the upper layer of retransmissions and acknowledgements, which are necessary if datagram services are used.
- Improved efficiency over connection-oriented services. WTP has no explicit connection set up or teardown phases.
- WTP is message oriented and designed for services oriented toward transactions such as browsing.

Transaction Class 0 Class 0 is an unreliable datagram service. It can be used by WSP [61], for example, to make an unreliable *push* within a session using the same socket association. This class is intended to augment the transaction service with the capability for an application using WTP to occasionally send a datagram within the same context of an existing session using WTP.

Transaction Class 1 Class 1 transactions provide a reliable datagram service. It can be used by applications that require a *reliable push* service. The basic behaviour for class 1 transactions is as follows: One `invoke` message is sent from the Initiator to the Responder. The `invoke` message is acknowledged by the Responder. The Responder maintains state information for some time after the acknowledgement has been sent to handle possible retransmissions of the acknowledgement if it gets lost and/or the Initiator retransmits the invoke message. At the Initiator, the transaction ends when the acknowledgement has been received. The transaction can be aborted at any time. If the User acknowledgement function is enabled, the WTP user at the Responder confirms the invoke message before the acknowledgement is sent to the Initiator.

Transaction Class 2 Class 2 transactions provide the basic invoke/response transaction service. One WSP session MAY consist of several transactions of this type. The basic behaviour for class 2 transactions is as follows: one invoke message is sent from the Initiator to the Responder. The Responder replies with exactly one result message that implicitly acknowledges the invoke message.

4.6.3.7 Wireless Transport Layer Security (WTLS)

Wireless Transport Layer Security (WTLS) is a security protocol based on the industry-standard *Transport Layer Security* (TLS) protocol, whose best-known implementation is *Secure Socket Layer* (SSL). WTLS is intended for use with the WAP transport protocols and has been optimized for low-bandwidth bearer networks with relatively long latency. The primary goal of the WTLS layer is to provide privacy, data integrity, and authentication between two communicating applications. WTLS provides functionality similar to TLS 1.0 and incorporates new features such as datagram support, optimized handshake, and dynamic key refreshing.

The WTLS layer operates above the transport protocol layer. The WTLS layer is modular and it depends on the required security level of the given application whether it is used or not. WTLS provides the upper-level layer of WAP with a secure transport service interface that preserves the transport service interface below it. In addition, WTLS provides an interface for managing (e.g., creating and terminating) secure connections.

The WTLS protocol is composed of three layers:

- The WTLS handshake protocol manages secure connections, provides client and server authentication and is used to exchange key material.
- The WTLS record layer provides privacy and data integrity.
- The alert layer is used to report error conditions to each other and to handle the close alert.

WTLS is designed to function on connection-oriented and/or datagram transport protocols. Security is assumed to be an optional layer above the transport layer. The security layer preserves the transport service interfaces. The session or application management entities are assumed to provide additional support required to manage (e.g., initiate and terminate) secure connections.

4.6.3.8 WAP Identity Module (WIM)

WAP security functionality includes the WTLS introduced in the previous section and application level security, accessible using the Wireless Markup Language Script [60].

For optimum security, some parts of the security functionality need to be performed by a tamper-resistant device, so that an attacker cannot retrieve sensitive data. Such data is especially the permanent private keys used in the WTLS handshake with client authentication, and for making application level electronic signatures (such as confirming an application layer transaction).

In WTLS, the master keys, protecting secure sessions, are relatively long lasting which could be several days. This is in order to avoid frequent full handshakes which are relatively heavy both computationally and due to large data transfer. Master keys are used to derive MAC keys and message encryption keys which are used to secure a limited number of messages, depending on usage of WTLS. The *WAP Identity Module* (WIM) is used in performing WTLS and application layer security functions, and especially to store and process information needed for user identification and authentication. The functionality presented here is based on the requirement that sensitive data, especially keys, can be stored in the WIM, and all operations where these keys are involved can be performed in the WIM.

An example of a physical implementation of a WIM is a smart card. In the phone, it can be the SIM card or an external smart card. The way which a phone and a smart card interact is specified as a command-response protocol, using *Application Protocol Data Units* (APDU) specific to this application. This specification is based on ISO 7816 series of standards on smart cards and the related GSM specifications [7].

4.6.3.9 Wireless Datagram Protocol (WDP)

The WDP protocol operates above the data capable bearer services supported by multiple network types. WDP offers a consistent service to the upper-layer protocols (security, transaction, and session) of WAP and communicates transparently over one of the available bearer services. The services offered by WDP include application addressing by port numbers, optional segmentation and reassembly, and optional error detection. The services allow for applications to operate transparently over different available bearer services. The model of protocol architecture for the WTP is given in Figure 4.88.

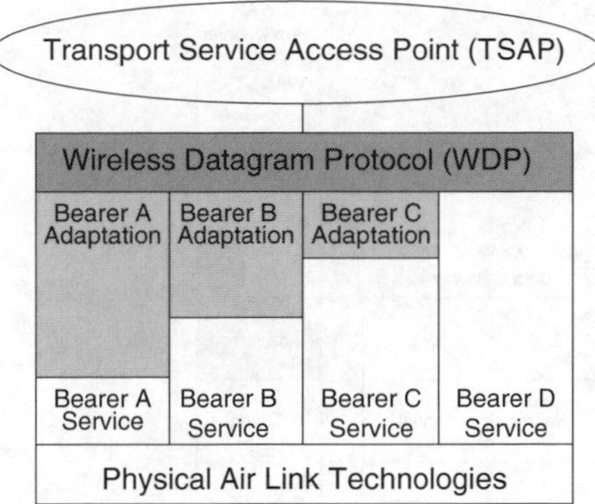

Figure 4.88: Wireless Datagram Protocol architecture

The varying heights of each of the bearer services shown in Figure 4.88 illustrate the difference in functions provided by the bearers and thus the difference in WDP necessary to operate over those bearers to maintain the same service offering at the *Transport Service Access Point* (TSAP) is accomplished by a bearer adaptation. WDP can be mapped onto different bearers, with different characteristics. In order to optimize the protocol with respect to memory usage and radio transmission efficiency, the protocol performance over each bearer may vary. However, the WDP service and service primitives will remain the same, providing a consistent interface to the higher layers.

Figure 4.89 shows a general model of the WAP protocol architecture and how WDP fits into that architecture. In Figure 4.89 the shaded areas are the layers of protocol to which the WDP specification is specifically applicable. On the mobile, the WDP protocol consists of the common WDP elements shown by the layer labelled WDP. The adaptation layer is the WDP layer which maps the WDP functions directly onto a specific bearer. The adaptation layer is different for each bearer and deals with the specific capabilities and characteristics of that bearer service. The bearer layer is the bearer service such as GSM SMS, USSD, IS-136 R-Data, or CDMA packet data. At the gateway the adaptation layer terminates and passes the WDP packets on to a WAP proxy/server via a tunnelling protocol, which is the interface between the gateway that supports the bearer service and the WAP proxy/server. For example, if the bearer were GSM SMS, the gateway would be a GSM *SMS Centre* (SMSC) (see Section 3.10.3.3) and would support a specific protocol (the tunnelling protocol) to interface the SMSC to other servers. The subnetwork is any common networking technology that can be used to connect two communicating devices; examples are wide area networks based on TCP/IP or X.25, or LANs operating TCP/IP over Ethernet. The WAP proxy/server may offer application content or may act as a gateway between the wireless WTP protocol suites and the wired Internet.

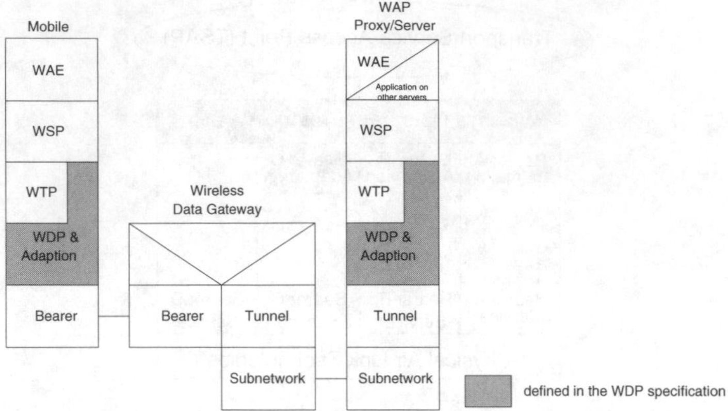

Figure 4.89: General WDP architecture

The Figures 4.90, 4.91, 4.92 and 4.93, respectively, show how WDP operates over the different bearer services.

WAP requires a full duplex datagram service from the bearer network. Unfortunately, GSM USSD does not provide such a service.

The USSD dialogue provides a two-way-alternate interactive service to the user. This means that only the entity (mobile phone or network node) with the turn may send and its correspondent is permitted only to receive. In order to be able to use the USSD dialogue as a full duplex service a special protocol has to be specified that deals with the management of the dialogue. The protocol designed to hide the complexity of the USSD dialogue is the *USSD Dialogue Control Protocol* (UDCP). UDCP is mapped directly onto the USSD protocol and is located in the mobile and the end node in the GSM network. The end points of UDCP are identical to the end points of the USSD dialogue. For further information about UDCP please refer to [59].

4.6.3.10 Wireless Markup Language (WML)

WML is a markup language based on XML [65] and is intended for use in specifying content and user interface for narrowband devices, including cellular phones and pagers. WML is designed with the constraints of small narrowband devices in mind. These constraints include:

- small display and limited user input facilities
- narrowband network connection
- limited memory and computational resources.

WML includes four major functional areas:

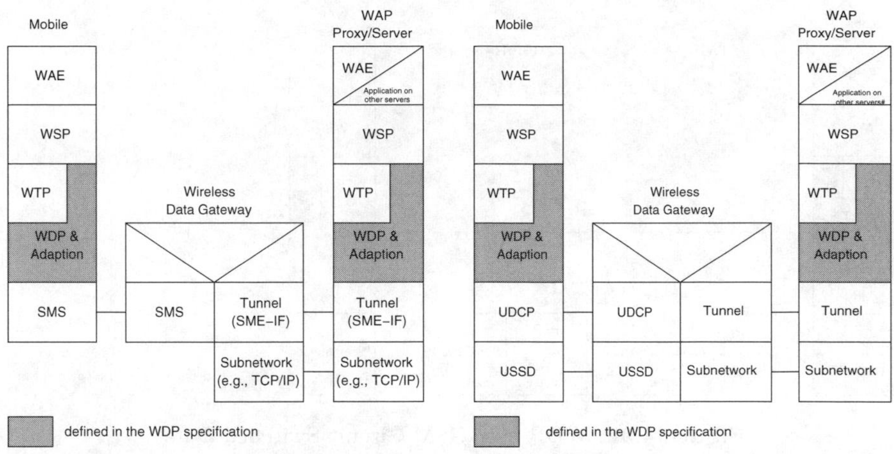

Figure 4.90: WDP over SMS

Figure 4.91: WDP over USSD

Text presentation and layout WML includes text and image support, including a variety of formatting and layout commands. For example, emboldened text may be specified.

Deck/card organizational metaphor All information in WML is organized into a collection of cards and decks. Cards specify one or more units of user interaction (e.g., a choice menu, a screen of text, or a text entry field). Logically, a user navigates through a series of WML cards, reviews the contents of each, enters requested information, makes choices, and moves on to another card. Cards are grouped together into decks. A WML deck is similar to an HTML page in that it is identified by a URL [35] and is the unit of content transmission.

Intercard navigation and linking WML includes support for explicitly managing the navigation between cards and decks. WML also includes provisions for event handling in the device, which may be used for navigational purposes or to execute scripts. WML also supports anchored links, similar to those found in [66].

String parameterization and state management All WML decks can be parameterized using a state model. Variables can be used in the place of strings and are substituted at runtime. This parameterization allows for more efficient use of network resources.

The official WML specification is developed and maintained by the WAP Forum. This specification defines the syntax, variables, and elements used in a valid WML file. The actual WML 1.1 *Document Type Definition* (DTD) is available for those familiar with XML at `http://www.wapforum.org/DTD/wml_1.1.xml`. A valid WML document must correspond to this DTD or it cannot be processed.

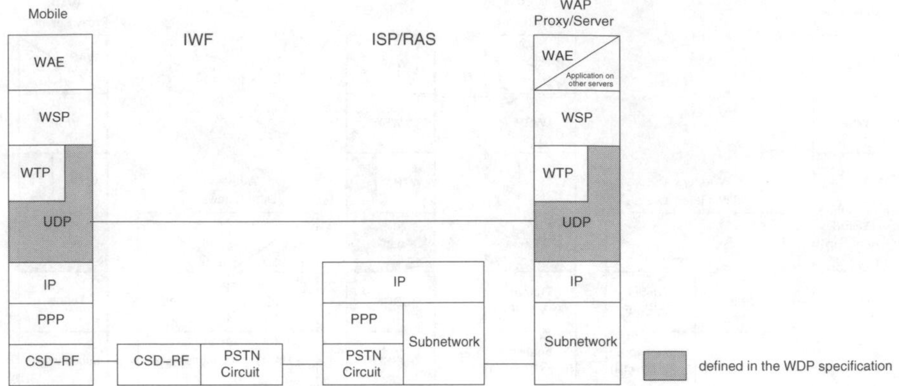

Figure 4.92: WDP over GSM Circuit Switched Data

WML files may be stored as static files on a web-server, but for enhanced services, most probably the files are generated dynamically (*on the fly*) using *Common Gateway Interface* (CGI) scripts, Microsoft's *Active Server Page(s)* (ASP) or *PHP Hypertext Protocol* (PHP).

The example Source Code 4.1 on page 404 of a simple WML deck is loaded with one class 2 WTP transaction. Please see Figure 4.94 on how it could be displayed.

In the beginning of the Sourcecode the proper identification as a XML document, using the type WML is done. Afterwards, the single cards of the deck are declared.

On loading, it would first display the text "Hello World" with a link named "Second Card" and a link "Third Card", the caption is "First Card". If the user chooses the link to card number two, no further transmission is needed, i.e., all data was transmitted during the first transaction. On the card with the caption "Second Card" the only option is to go back to the first card.

On the third card, captioned "Third Card", the user can select one of the optionally links to

- the deck `help.wml`, named "Help" (would be loaded with a new transaction)
- the deck `example.wml`, named "Reload Example" (assumed, that this deck is named `example.wml` the example would be retransmitted)
- go to the first card, named "Go to first card " (no further transaction needed, as this is the present deck)

They are shown if the user selects the box (in Figure 4.94 indicated by the dotted lines).

Below the selection, **Boldface**, *Italic* and <u>Underlined</u> would be printed. On the bottom of card number three the picture `test.wbmp` would be displayed,

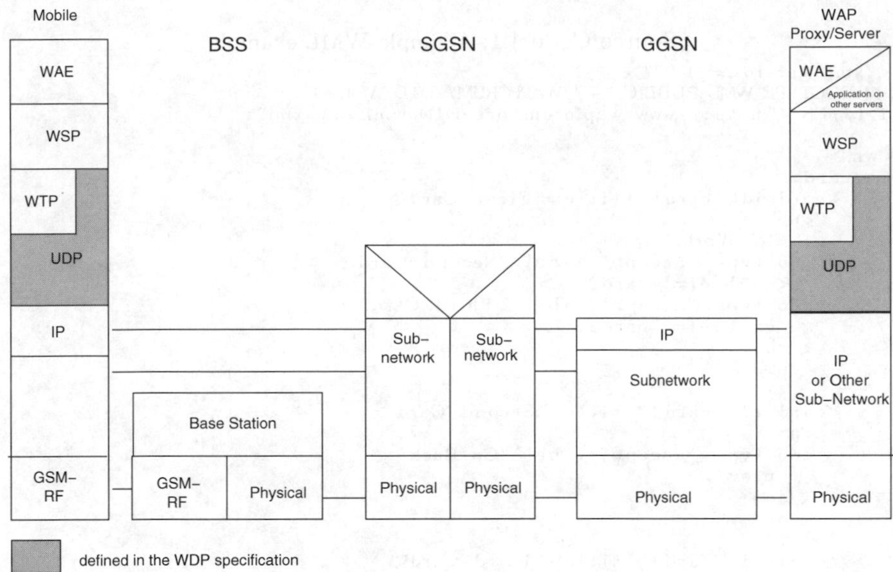

Figure 4.93: WDP over GPRS

having a height of 30 pixels, a width of 50 pixels, a horizontal border of 5 pixels and a vertical border of 3 pixels. The picture would be separately loaded in a transaction with displaying "Description of Picture", if any error occurs for this action (i.e., not loadable, not displayable by device).

This deck example gives only a fraction of the possibilities of WML, described more in [52]. It has to be noted, that most possibilities concern the linking, the displaying of text and processing of variables.

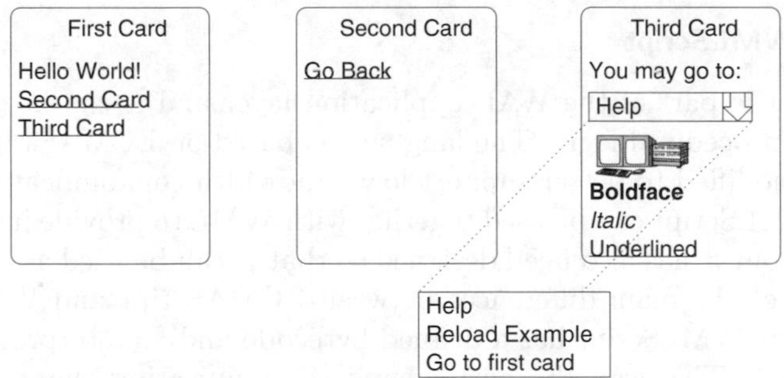

Figure 4.94: View of the example WML page

Source Code 4.1: A simple WML example

```
<?xml version="1.0"?>
<! DOCTYPE WML PUBLIC "-//WAPFORUM//DTD WML
1.1//EN" "http://www.wapforum.org/DTD/wml_1.1.xml">

<wml>
  <card>
    <card id="card1" title="First Card">
      <p>
      Hello World!
      <do type="accept" label "Second Card">
        <go href="#card2" />
      <do type="accept" label "Third Card">
        <go href="#card3" />
      </do>
      </p>

    <card id="card2" title "Second Card">
      <p>
      <do type="accept" label "Go Back">
        <prev/>
      </do>
      </p>

    <card id="card3" title "Third Card">
      <p>
      You may go to:
      <select>
        <option onpick="/help.wml">Help</option>
        <option onpick="/example.wml">Reload Example</option>
        <option onpick="#card1">Go to first card</option>
      </select>
      <b>Boldface</b>
      <i>Italic</i>
      <u>Underlined</u>
      <img src="test.wbmp" alt="Description of Picture"
      vspace="3" hspace="5" align="bottom" height="30" width="50">
      </p>

  </card>
</wml>
```

4.6.3.11 WMLScript

WMLScript is part of the WAP application layer and it can be used to add client side procedural logic. The language is based on ECMAScript [6] but it has been modified to better support low bandwidth communication and thin clients. WMLScript can be used together with WML to provide intelligence to the clients but it has also been designed so that it can be used as a standalone tool. One of the main differences between ECMAScript and WMLScript is the fact that WMLScript has a defined bytecode and an interpreter reference architecture. This way the narrowband communication channels available today can be optimally utilized and the memory requirements for the client kept to the minimum. Many of the advanced features of the ECMAScript language have been dropped to make the language smaller, easier to compile into bytecode and easier to learn. For example, WMLScript is a procedural language and it supports locally installed standard libraries.

WMLScript Libraries WMLScript supports the usage of libraries. Libraries are named collections of functions that belong logically together. These functions can be called by using a dot ('.') separator with the library name and the function name with parameters: An example of a library function call:

```
function dummy(str)
  {
      var i = String.elementAt(str,3," ");
  };
```

Standard Libraries Many of the advanced features of the JavaScript language have been removed to make the language easier to compile into byte-code and easier to learn. Library support has been added to the WMLScript to replace some of the functionality that was removed from ECMAScript to make the WMLScript more efficient. This feature provides access to built-in functionality and a means of future expansion without unnecessary overheads. There is a set of libraries defined to provide access to core functionality of a WAP client. This means that all libraries, except Float, are present in the client's scripting environment. Float library is optional and only supported by clients that can support floating-point arithmetic operations.

Parameters An open issue regarding the call of an WMScript method payment method is the question of restriction for the parameter list passed to the function. At present there are no indications in the WAP specifications that the number of passed arguments should be restricted. Independently of the condition if you call an internal or external function, the following characteristics apply to WMLScript methods calls:

- all parameters of a function are passed by value
- the number of the called function's parameters must match the number of the parameters declared for the function
- parameters are handled in a function like local variables
- a function always returns a value.

The number of arguments in a function call must match the number of arguments accepted by the function.

Passing of Function Arguments Arguments must be present in the operand stack in the same order as they are presented in a WMLScript function declaration at the time of a WMLScript or library function call. Thus, the first argument is pushed into the operand stack first, the second argument is pushed next, etc. The instruction executing the call must take the arguments from the operand stack and use them to initialize the appropriate function variables.

4.6.3.12 WAP Push

The WAP Push framework [58] introduces a means within the WAP effort to transmit information to a device without a previous user action. In the normal client/server model, a client requests a service or information from a server, which then responds in transmitting information to the client. This is known as *pull* technology: the client *pulls* information from the server (see Figure 4.85). The WWW is a typical example of pull technology, where a user enters a URL (the request) which is sent to a server, and the server answers by sending a web-page (the response) to the user. In contrast to this, there is the *push* technology, which is also based on the client/server model, but where there is no explicit request from the client before the server transmits its content. Another way to say this is that whereas *pull* transactions of information are always initiated by the client, *push* transactions are server-initiated.

Since the push initiator, i.e., a WWW server in the Internet, shares no protocol with the WAP client in the WAP domain, it cannot contact it without an intermediary. In order to perform a push operation, the push initiator contacts the *Push Proxy Gateway* (PPG) from the Internet side, delivering content for the destination client using Internet protocols. The PPG does what is necessary to forward the pushed content to the WAP domain, and the content is then transmitted over the air in the mobile network to the destination client. In addition to providing some simple proxy gateway services, the PPG may be capable of notifying the Push Initiator of the final outcome of the push operation, and it may wait for the client to accept or reject the content. It may also provide the Push Initiator with client capability lookup services, letting a Push Initiator select the optimal strategy of this particular content for this particular client. The Internet-side PPG access protocol is called the *Push Access Protocol* (PAP), while the WAP-side (*Over-The-Air* (OTA)) protocol is called the Push-over-the-Air protocol. The OTA protocol is based on WSP services (see Section 4.6.3.4), while PAP uses XML messages that may be tunnelled through various well-known Internet protocols, for example HTTP.

Bibliography

[1] Aixcom GmbH. `http://www.aixcom.com/Produkte_e.php`.

[2] Martin F. Arlitt, Carey L. Williamson. *A Synthetic Workload Model for Internet Mosaic Traffic*. In *Proceedings of the 1995 Summer Computer Simulation Conference*, pp. 24–26, Ottawa, Canada, July 1995.

[3] Communication Networks,Aachen University of Technology, anonymous ftp to ftp.dfv.rwth-aachen.de:/pub/CNCL. *CNCL: a C++ library for event driven simulation, statistical evaluation and random number generators and distributions*, 1992/1993.

[4] P. Decker. *A Packet Radio Protocol Proposed for the GSM Mobile Radio Network.* In *Workshop on Mobile Multimedia Communication (MoMuC-1)*, Tokyo, IEEE, December 1993.

[5] P. Decker. *Entwurf und Leistungsbewertung hybrider Fehlersicherungsprotokolle für paketorientierte Sprach- und Datendienste in GSM-Mobilfunksystemen.* Dissertation, RWTH Aachen, Lehrstuhl für Kommunikationsnetze, 1995.

[6] European Association for Standardizing Information and Communication Systems (ECMA). *ECMAScript Language Specification.* `http://www.ecma.ch/ecma1/STAND/ECMA-262.HTM`, 1999.

[7] ETSI. *GSM 11.11, V6.2.0, Digital cellular telecommunications system (Phase2+); Specification of the Subscriber Identity Module—Mobile Equipment (SIM-ME) interface.* `http://www.etsi.org/`, 1997.

[8] ETSI. *GSM 11.14, V7.1.0, Digital cellular telecommunications system (Phase2+); Specifica tion of the SIM Application Toolkit for the Subscriber Identity Module - Mobile Equipment (SIM-ME) i nterface.* `http://www.etsi.org`, 1998.

[9] ETSI. *Tdoc SMG2 EDGE 095/98.Link Quality Control Proposal for EGPRS rev 1.* Technical report, European Telecommunications Standards Institute, Paris, France, December 1998.

[10] ETSI. *GSM 02.57, V8.0.0, Digital cellular telecommunications system (Phase2+); Mobile Station Execution Environment (MExE); Service description.* `http://www.etsi.org`, 1999.

[11] ETSI. *Tdoc SMG2 EDGE 038/99. EDGE: Link Quality Control Aspects for Mobile Testing.* Technical report, European Telecommunications Standards Institute, Toulouse, France, March 1999.

[12] ETSI 3GPP. *Selection Procedures for the Choice of Radio Transmission Technologies of the Universal Mobile Telecommunication System UMTS (UMTS 30.03, 3G TR 101 112).* Technical report, European Telecommunications Standards Institute, Sophia Antipolis, France, April 1998.

[13] ETSI 3GPP. *Digital cellular telecommunications system (Phase 2+) (GSM); Universal Mobile Telecommunications System (UMTS); General Packet Radio Service (GPRS); Service Description; Stage 1 (3G TS 22.060 version 3.2.0 Release 1999).* Technical Specification ETSI TS 122 060, European Telecommunications Standards Institute, Sophia Antipolis, France, January 2000.

[14] ETSI 3GPP. *Digital cellular telecommunications system (Phase 2+) (GSM); Universal Mobile Telecommunications System (UMTS); General Packet Radio Service (GPRS); Service Description; Stage 2 (3G*

TS 23.060 version 3.2.1 Release 1999). Technical Specification ETSI TS 123 060, European Telecommunications Standards Institute, Sophia Antipolis, France, January 2000.

[15] ETSI 3GPP. *Digital cellular telecommunications system (Phase 2+) (GSM); Universal Mobile Telecommunications System (UMTS); Mobile Radio Interface layer 3 specification, Core Network protocols—Stage 3 (3G TS 24.008 version 3.2.1 Release 1999).* Technical Specification ETSI TS 124 008, European Telecommunications Standards Institute, Sophia Antipolis, France, January 2000.

[16] ETSI 3GPP. *Digital cellular telecommunications system (Phase 2+) (GSM); Universal Mobile Telecommunications System (UMTS); Mobile radio interface signalling layer 3; General aspects (3GPP TS 24.007 version 3.5.0 Release 1999).* Technical Specification ETSI TS 124 007, European Telecommunications Standards Institute, Sophia Antipolis, France, September 2000.

[17] ETSI 3GPP. *Universal Mobile Telecommunications System (UMTS); QoS Concept and Architecture (3G TS 23.107 version 3.1.0 Release 1999).* Technical Specification ETSI TS 123 107, European Telecommunications Standards Institute, Sophia Antipolis, France, January 2000.

[18] ETSI 3GPP. *Universal Mobile Telecommunications System (UMTS); Service aspects; Services and Service Capabilities (3G TS 22.105 version 3.8.0 Release 1999).* Technical Specification ETSI TS 122 105, European Telecommunications Standards Institute, Sophia Antipolis, France, March 2000.

[19] ETSI TC-SMG. *Digital cellular telecommunications system (Phase 2+); Channel coding (GSM 05.03 version 8.4.0 Release 1999).* Draft European Standard ETSI EN 300 909, European Telecommunications Standards Institute, Sophia Antipolis, France, May 2000.

[20] ETSI TC-SMG. *Digital cellular telecommunications system (Phase 2+); General Packet Radio Service (GPRS); Base Station System (BSS) – Serving GPRS Support Node (SGSN); BSS GPRS Protocol (BSSGP) (GSM 08.18 version 8.3.0 Release 1998).* Technical Specification ETSI TS 101 343, European Telecommunications Standards Institute, Sophia Antipolis, France, May 2000.

[21] ETSI TC-SMG. *Digital cellular telecommunications system (Phase 2+); General Packet Radio Service (GPRS); GPRS Tunnelling Protocol (GTP) across the Gn and Gp Interface (GSM 09.60 version 7.4.0 Release 1998).* Draft European Standard ETSI EN 301 347, European Telecommunications Standards Institute, Sophia Antipolis, France, July 2000.

[22] ETSI TC-SMG. *Digital cellular telecommunications system (Phase 2+); General Packet Radio Service (GPRS); Mobile Station (MS) – Base Station System (BSS) interface; Radio Link Control / Medium Access Control (RLC/MAC) protocol (GSM 04.60 version 8.4.0 Release 1999).* Draft European Standard ETSI EN 301 349, European Telecommunications Standards Institute, Sophia Antipolis, France, May 2000.

[23] ETSI TC-SMG. *Digital cellular telecommunications system (Phase 2+); General Packet Radio Service (GPRS); Mobile Station (MS) – Serving GPRS Support Node (SGSN) Logical Link Control (LLC) layer specification (GSM 04.64 version 8.3.0 Release 1999).* Technical Specification ETSI TS 101 351, European Telecommunications Standards Institute, Sophia Antipolis, France, March 2000.

[24] ETSI TC-SMG. *Digital cellular telecommunications system (Phase 2+); General Packet Radio Service (GPRS); Overall description of the GPRS radio interface; Stage 2 (GSM 03.64 version 8.3.0 Release 1999).* Technical Specification ETSI TS 101 350, European Telecommunications Standards Institute, Sophia Antipolis, France, April 2000.

[25] ETSI TC-SMG. *Digital cellular telecommunications system (Phase 2+); General Packet Radio Service (GPRS); Service description; Stage 2 (GSM 03.60 version 7.4.0 Release 1998).* Draft European Standard ETSI EN 301 344, European Telecommunications Standards Institute, Sophia Antipolis, France, April 2000.

[26] ETSI TC-SMG. *Digital cellular telecommunications system (Phase 2+); General Packet Radio Service (GPRS); Subnetwork Dependent Convergence Protocol (SNDCP) (GSM 04.65 version 8.0.0 Release 99).* Draft European Standard ETSI TS 101 297, European Telecommunications Standards Institute, Sophia Antipolis, France, 2000.

[27] ETSI TC-SMG. *Digital cellular telecommunications system (Phase 2+); Mobile radio interface layer 3 specification (GSM 04.08 version 7.7.1 Release 1998).* European Standard ETSI EN 300 940, European Telecommunications Standards Institute, Sophia Antipolis, France, October 2000.

[28] ETSI TC-SMG. *Digital cellular telecommunications system (Phase 2+); Physical layer on the radio path; General description (GSM 05.01 version 8.3.0 Release 1999).* Technical Specification ETSI TS 100 573, European Telecommunications Standards Institute, Sophia Antipolis, France, April 2000.

[29] ETSI TC-SMG. *Digital cellular telecommunications system (Phase 2+); Radio transmission and reception (GSM 05.05 version 8.4.0 Release 1999).* Draft European Standard ETSI EN 300 910, European Telecommunications Standards Institute, Sophia Antipolis, France, May 2000.

[30] Victor S. Frost, Benjamin Melamed. *Traffic Modeling for Telecommunications Networks.* IEEE Communications Magazine, March 1994. pp. 70–81.

[31] Anders Furuskär, Mikael Höök, Stefan Jäverbring, Hakan Olofsson, Johan Sköld. *Capacity Evaluation of the EDGE Concept for Enhanced Data Rates in GSM and TDMA/136.* In *Proc. of Vehicular Technology Conference*, USA, IEEE, February 1999.

[32] Anders Furuskï, Sara Mazur, Frank Müller, Hakan Olofsson. *EDGE: Enhanced Data Rates for GSM and TDMA/136 Evolution. IEEE Personal Communications*, pp. 56–65, June 1999.

[33] Anders Furuskär, Häkan Olofsson. *Aspects of Introducing EDGE in Existing GSM Networks. Ericsson Review*, No. 1, pp. 28–37, 1999.

[34] H. Gudding. *Capacity Analysis of GPRS.* Master thesis, NTNU Norwegian University of Science and Technology, Trondheim, Norway, March 2000.

[35] Internet Engineering Task Force (IETF). *Uniform Resource Locators (URL).* `http://www.ietf.org/rfc/rfc1738.txt`, 1994.

[36] W. Olzem. *Integration of Protocols Specified in SDL into an Event-driven Simulation Environment.* Technical report, RWTH Aachen, Lehrstuhl für Kommunikationsnetze, Kopernikusstr. 16, D-52074 Aachen, Germany, January 1995.

[37] Vern Paxson. *Empirically-Derived Analytic Models of Wide-Area TCP Connections. IEEE/ACM Transactions on Networking*, Vol. 2, No. 4, pp. 316–336, August 1994. `ftp://ftp.ee.lbl.gov/papers/WAN-TCP-models.ps.Z`.

[38] Vern Paxson, Sally Floyd. *Wide-Area Traffic: The Failure of Poisson Modeling. IEEE/ACM Transactions on Networking*, Vol. 3, No. 3, pp. 226–244, July 1995. `ftp://ftp.ee.lbl.gov/papers/WAN-poisson.ps.Z`.

[39] N. Sollenberger, et al. *The Evolution of IS-136 TDMA for Third-Generation Wireless Services. IEEE Personal Communications*, Vol. 6, 1999.

[40] Martin Steppler. *Performance Analysis of Communication Systems Formally Specified in SDL.* In *Proceedings of the First International Workshop on Simulation and Performance '98 (WOSP '98)*, pp. 49–62, Santa Fe, New Mexico, USA, 12th–16th October 1998.

[41] P. Stuckmann, H. Finck, T. Bahls. *A WAP Traffic Model and its Appliance for the Performance Analysis of WAP over GPRS.* In *Proc. of*

the IEEE International Conference on Third Generation Wireless and Beyond (3Gwireless '01), San Francisco, USA, June 2001.

[42] Peter Stuckmann, Jörg Franke. *The Capacity and Performance Gain Reachable with Link Quality Control in EGPRS Networks.* In *Proc. of the IEEE International Conference on Third Generation Wireless and Beyond (3Gwireless '01)*, San Francisco, USA, June 2001.

[43] Peter Stuckmann, Jörg Franke. *Performance Characteristics of the Enhanced General Packet Radio Service for the Mobile Internet Access.* In *Proc. of the Second International Conference on 3G Mobile Communication Technologies*, London, UK, March 2001.

[44] Peter Stuckmann, Frank Müller. *GPRS Radio Network Capacity and Quality of Service using Fixed and On-Demand Channel Allocation Techniques.* In *Proc. Vehicular Technology Conference (VTC spring 2000)*, Tokyo, Japan, May 2000.

[45] Peter Stuckmann, Frank Müller. *GPRS Radio Network Capacity Considering Coexisting Circuit Switched Traffic Sources.* In *European Conference on Wireless Technologies 2000*, Paris, France, October 2000.

[46] Peter Stuckmann, Frank Müller. *Quality of Service Management in GPRS Networks.* In *Proc. of the IEEE International Conference on Networking (ICN '01)*, Colmar, France, July 2001.

[47] Peter Stuckmann, Peter Seidenberg. *Quality of Service of Internet Applications over GPRS.* In *Proc. of the European Wireless'99*, pp. 251–255, Munich, Germany, Oct. 1999. ISBN 3-8007-2490-1.

[48] B. Walke, W. Mende, P. Decker, R. Crumbach. *The Performance of CELLPAC: A Packet Radio Protocol Proposed for the GSM Mobile Radio Network.* In *Mobile Radio Conference*, pp. 57–63, Nice, France, November 1991.

[49] B. Walke, W. Mende, G. Hatziliadis. *CELLPAC, a packet radio protocol applied to the cellular GSM mobile radio network.* In *Proc. 41. Vehicular Technol. Conf.*, pp. 408–413, St. Louis, Miss., USA, IEEE, May 1991.

[50] Bernhard Walke, Götz Brasche. *Concepts, services and protocols of the new GSM Phase 2+ General Packet Radio Service. IEEE Communications Magazine*, Vol. 35, 1997.

[51] WAP Forum. *Wireless Application Protocol—Wireless Datagram Specification.* `http://www.wapforum.org`, 1999.

[52] WAP Forum. *Wireless Application Protocol—Wireless Markup Language Specification.* `http://www.wapforum.org`, 1999.

[53] WAP Forum. *Wireless Application Protocol—Wireless Session Protocol Specification.* http://www.wapforum.org, 1999.

[54] WAP Forum. *Wireless Application Protocol - Wireless Transaction Protocol Specification.* http://www.wapforum.org, 1999.

[55] WAP Forum. *Wireless Application Protocol—Wireless Transport Layer Security Specification.* http://www.wapforum.org, 1999.

[56] WAP Forum. *Wireless Application Protocol–Wireless Application Environment Overview.* http://www.wapforum.org, 1999.

[57] WAP Forum. *Wireless Application Protocol Architecture Specification.* http://www.wapforum.org, 1999.

[58] WAP Forum. *Wireless Application Protocol: WAP 1.2. Push Architectural Overview.* http://www.wapforum.org/what/technical.htm, 1999.

[59] WAP Forum. *Wireless Application Protocol: WAP over GSM USSD Specification.* http://www.wapforum.org/what/technical.htm, 1999.

[60] WAP Forum. *Wireless Application Protocol: Wireless Markup Language Script Specification.* http://www.wapforum.org/what/technical.htm, 1999.

[61] WAP Forum. *Wireless Application Protocol: Wireless Session Protocol Specification.* http://www.wapforum.org/what/technical.htm, 1999.

[62] Wireless Communications Alliance (WCA). *Overview of WAP.* http://wca.org/sbc/, 1999.

[63] Jeroen Wigard, Preben Mogensen. *A Simple Mapping from C/I to FER and BER for a GSM Type of Air-Interface.* In *Vehicular Technology Conference (VTC)*, 1996.

[64] Jeroen Wigard, Thomas Toftegård Nielesen, Per Henrik Michaelsen, Preben Mogensen. *BER and FER of Control and Traffic Channels for a GSM Type of Air-Interface.* In *Vehicular Technology Conference (VTC)*, 1998.

[65] World Wide Web Consortium. *Extensible Markup Language (XML) 1.0.* http://www.w3.org/TR/1998/REC-xml-19980210, 1998.

[66] World Wide Web Consortium. *HTML 4.01 Specification.* http://www.w3.org/TR/html401/, 1999.

5
Other Public Mobile Radio Systems

5.1 Airline Telephone Network for Public Air–Ground Communication

In 1993 ETSI RES 5 submitted a standard for the *Terrestrial Flight Telephone System* (TFTS), specifying the radio interface and the interfaces to public telecommunications networks. At the same time the *European Airlines Electronic Committee* (EAEC) specified the airline equipment and interfaces to cabin facilities. Commercial operations began in 1994. In July 1994, after inviting international tenders, the Ministry of Post and Telecommunications granted a licence to DeTeMobil for the operation of TFTS. DeTeMobil was to supply radio coverage to all airspace up to an altitude of 4500 m. The service was available by 1996.

Thirteen network operators in Europe have signed an MoU for the introduction of TFTS and an agreement on cooperation with the major European airlines in order to resolve related commercial, organizational, technical and operational issues [1, 2].

5.1.1 TFTS Cellular Network

TFTS is a cellular system that uses direct radio links to ground stations (GS) that are connected to the fixed network to provide public communication services for air passengers (see Figure 5.1).

There are three types of ground stations differentiated by area covered (cell) and related transmitter power:

- *En-route* (ER) GS for altitudes from 13 to 4.5 km, with cell radii up to 240 km
- *Intermediate* (I) GS for altitudes below 4.5 km, with cell radii up to 45 km
- *Airport* (AP) GS, with cell radii of 5 km

Handover between areas is part of the system. According to WARC'92, two 5 MHz wide bands have been specified for operation of the TFTS:

- 1670–1675 MHz for uplink (*ground-to-air*)

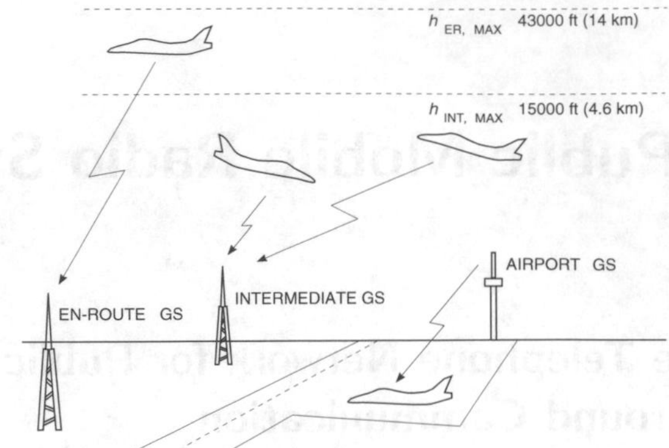

Figure 5.1: Coverage areas and ground stations

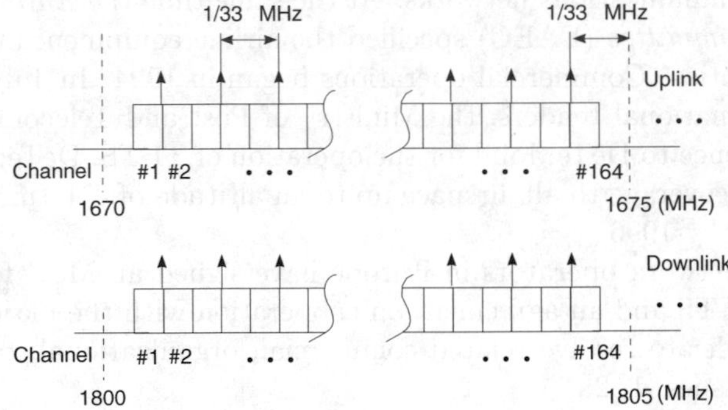

Figure 5.2: TFTS channel map

- 1800–1805 MHz for downlink

The system offers automatic dialled connections to PSTN/ISDN without any limitation on target subscribers, with the same quality of service as customary in PLMNs.

In addition to speech, data services such as facsimile, data transfer at 4.8 kbit/s and DTMF signalling are supported. Calls from the ground to an aircraft are only allowed to be made for operational purposes or for paging.

The user is billed directly by (credit) card for services used.

5.1.2 Frequency and Time-Multiplexing Channels

Each 5 MHz band is divided into 164 FDM channels (each 30.45 kHz wide); see Figure 5.2.

Each FDM channel transmits at 44.2 kbit/s gross. On the uplink this capacity is divided into 17 time channels based on the TDM method, and on

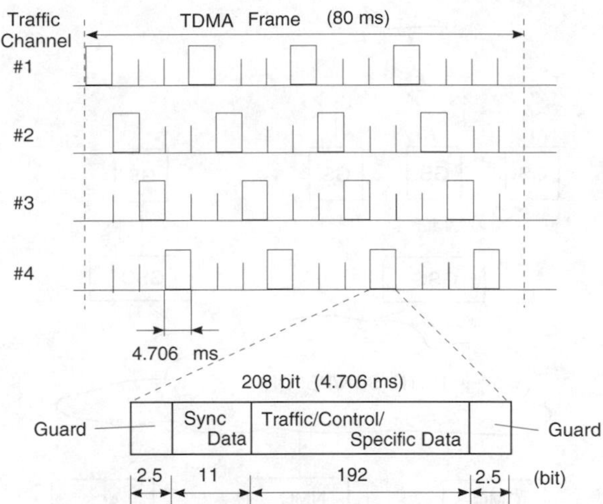

Figure 5.3: Frames and time slots

the downlink into 17 time channels based on the TDMA method. Each FDM channel carries four voice channels.

According to Figure 5.3, 17 time slots are combined into a frame of 80 ms duration, and 20 frames form a superframe of the duration of 1.6 s. Each time slot contains 208 bits and has a duration of 4.706 ms.

5.1.3 Voice and Data Transmission

Voice signals are digitally coded into blocks of 192 bits and transmitted at 9.6 kbit/s in time slots. A 9.6 kbit/s voice channel occupies 4 of the 17 time slots of an FDM channel; the 17th slot is used for network control.

As soon as voice codecs are available for a 4.8 kbit/s transmission rate, the number of voice channels will be doubled. Data services at 4.8 kbit/s require 2 time slots per frame, therefore each FDM channel carries 8 data channels.

5.1.4 Functional Characteristics

Each aircraft has transmitting and receiving facilities (*transceiver*), which can be tuned selectively to one of the different FDM channels. Four communications can be carried out on the same FDM channel at the same time. Ground stations can transmit to different aircrafts simultaneously (on different time channels) on each one of their FDM channels.

Signals are transmitted digitally with linear $\pi/4$-DQPSK (*Differential Quadrature Phase Shift Keying*) modulation, and require a simple non-coherent receiver.

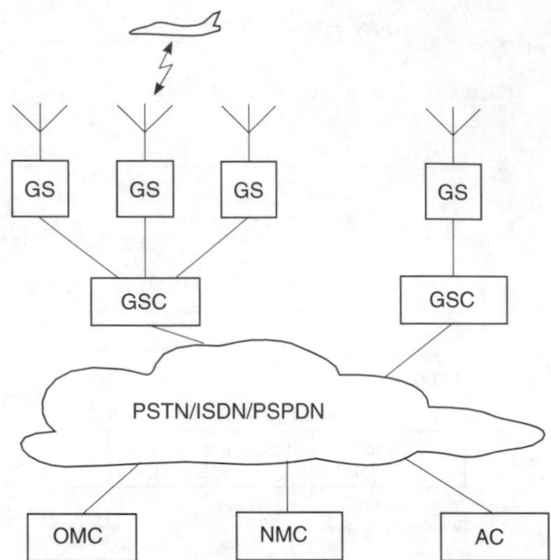

Figure 5.4: Architecture of a complete TFTS network

Handover can be initiated by the mobile or the ground station, and is controlled by signal quality, distance and flight state. The particular ground station selected as a target is the one towards which the mobile station is moving.

The distance between mobile and ground stations is estimated on the basis of signal propagation delay time. This information also determines the network synchronization for the ground stations capable of receiving. Ground stations are linked to the fixed network through *Ground Switching Centres* (GSC) (see Figure 5.4).

The GSC has responsibility for all the ground stations linked to it, and its tasks include mobility management, connection establishment to mobile subscribers, handover control and dynamic frequency management. The TFTS fixed network additionally contains three management components, namely:

- Operations and maintenance centre (OMC)
- Network management centre (NMC)
- Administration centre for billing (AC)

The MoU group produced a coordinated introduction plan for the TFTS ground network to enable the system to be introduced throughout Europe. This effort required cooperation between telecommunications network operators and airlines.

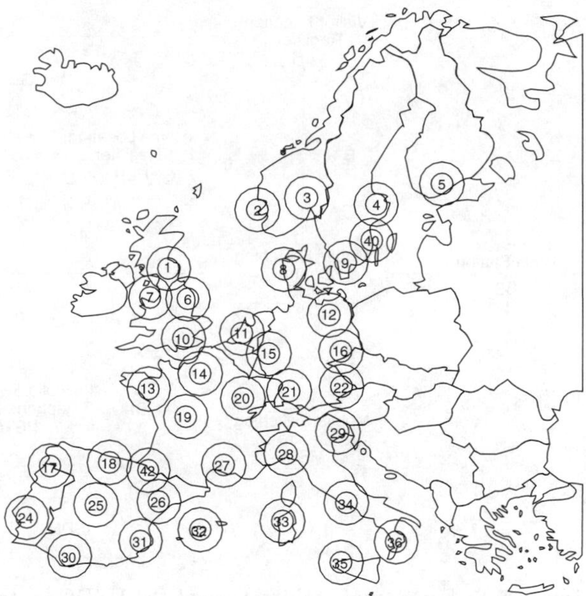

Figure 5.5: Cellular coverage through en-route ground stations in Europe

5.1.5 Ground Stations and Frequency Plan

En-route ground stations are spaced approximately 380 km apart according to a hexagonal grid, with a nominal range of approximately 240 km, which cannot be exceeded for signal propagation reasons (see Figure 5.5).

Cochannel ground stations are planned at a distance of 760 km, and neighbouring channel cells at a distance of at least 600 km. Cell planning is more difficult compared with terrestrial cellular networks because of the need to incorporate flying altitudes.

5.2 The US Digital Cellular System (USDC)

During the 1980s there was an impressive increase in the number of subscribers to the public cellular mobile radio network in the USA. Because approval for the installation of new base stations and antennas is expensive and difficult to obtain in larger cities, only a portion of this increased need for capacity could be accommodated through a reduction in cell sizes. A permanent solution turned out to be the development of a digital system capable of coping with increased capacity without the need for new base stations.

In March 1988 the *Telecommunication Industries Association* (TIA) set up the *TR-45.3* subcommittee to develop the standard for a cellular digital system. This digital system, the *American Digital Cellular System* (ADC),

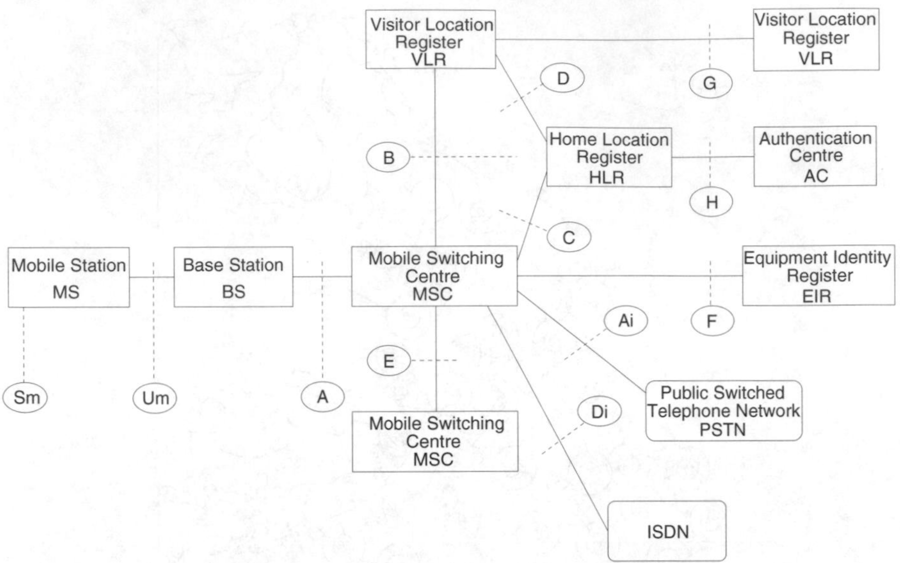

Figure 5.6: Functional architecture of the USDC system

was to support and be compatible with the existing analogue mobile radio network, the *American Mobile Phone System* (AMPS); see [3]. The digital system operates in the frequency range of the analogue AMPS system at the same time, which allows individual channels to change over gradually to digital technology. A characteristic of this system is that terminal equipment can be used for analogue as well as for digital operation (dual-mode). In addition to increased capacity, the ADC standard enables the introduction of new services, such as authentication, a data service and a short-message service, which were not supported by AMPS.

In 1990 the digital standard was accepted by industry as Interim Standard 54 (IS-54). The North American digital system with the architecture illustrated in Figure 5.6 is now called US Digital Cellular (USDC). In addition, a number of standards have been accepted by FCC for the *Personal Communication System* PCS 1900 market, e.g., IS-134.

5.2.1 Technical Data on the USDC System

The USDC system uses the 824–849 MHz frequency band for transmission between mobile station and base station (uplink), and in the reverse direction (downlink) the 869–894 MHz band. The duplex separation between the transmit and receive frequency is therefore 45 MHz. The frequency bands are divided into FDM channels with a 30 kHz bandwidth, thereby providing 832 frequency carriers.

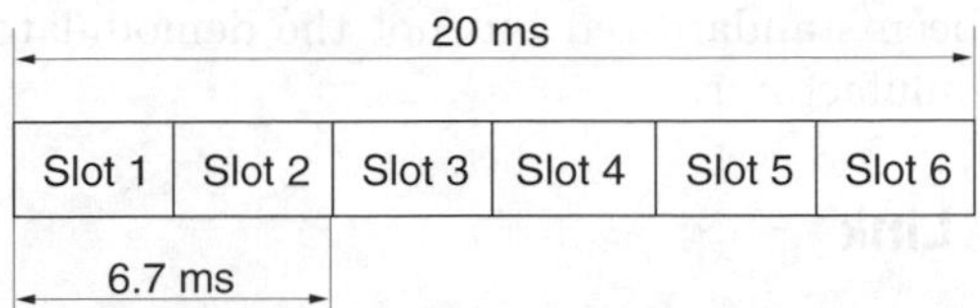

Figure 5.7: Structure of a TDMA frame in the USDC system with half-rate channels

The modulation technique used is $\pi/4$-DQPSK (*Differential Quadrature Phase Shift Keying*), a four-level scheme that, although it produces higher spectral efficiency than GMSK, places a heavy demand on the linearity of the output amplifier. In addition, for optimal detection at the receiver input, filters with a transmission function capable of describing the root of the Nyquist transmission function are required, and this is something that inexpensive filters can only approximate. In contrast to GMSK, $\pi/4$-DQPSK contains different amplitude components. The eight different phase states in $\pi/4$-DQPSK modulation are all in one circle, but the four allowed phase transitions from one phase to another do not run in the circle. This means that not only the phase but also the amplitude is covered in the specifications for modulation.

Like the GSM system, the USDC system operates in time-division multiplexing (TDM) and multiple access (TDMA) mode, albeit with three voice channels being transmitted over one carrier. The length of the TDMA frame is 20 ms and is divided into three time slots each of 6.7 ms duration. The modulation data rate per FDM channel (3 time slots, 30 kHz) is 48.6 kbit/s. After the development and introduction of a half-rate codec, a TDMA frame will contain six time slots (see Figure 5.7) [5].

The USDC system uses a VSELP speech codec (*Vector Sum Excited Linear Prediction*) which, compared with GSM, results in lower source rates. With a full-rate codec, voice coding together with error-protection coding produces an overall transmission rate of 13 kbit/s, whereas the total rate on the SACCH is 0.6 kbit/s.

5.3 CDMA Cellular Radio According to US-TIA/IS-95

TIA Interim Standard 95 was developed by QUALCOMM. Unlike IS-54, which guarantees compatibility of a digital system with analogue, the IS-95 standard defines a CDMA transmission system. It includes the lowest three levels of the OSI reference model. The transmission system of the LEO system Globalstar will be based on the IS-95 standard with modifications (see Section 16.3.3). The physical layer is described below. However, only the modulators have been standardized but not the demodulators; these can be specified by the manufacturer.

5.3.1 Forward-Link

Forward-link uses coherent QPSK modulation in which transmitter and receiver must be phase-synchronized for demodulation. Walsh sequences are used for channel separation (see Section 2.6.4). A short PN sequence is used for each in-phase and quadrature-phase for the spreading. A long PN sequence individually assigned to the user is used for the traffic channel. Demodulation is carried out through a pilot tone that is also transmitted.

5.3.1.1 Modulator

Figure 5.8 shows the modulator for the forward link. A number of physical channels are available for establishing a connection. The first thing that must be carried out when a mobile station is switched on is synchronization. Phase synchronization and frame synchronization are achieved through the transmission of a pilot tone. The network synchronization is then carried out over the synchronization channel. This involves transmitting the paging channel data rate and power control information. Data for channel allocation is sent over the paging channel. Information is transmitted over the traffic channel.

Pilot channel The all-one Walsh sequence W_0 is combined with a short code and transmitted to the modulator. With a value set of $(0, 1)$ the two codes are added modulo 2, or with a bipolar $(-1, 1)$ approach they are multiplied.

The Walsh sequences are the lines of the Hadamard matrix, and are formed according to the following recursion:

$$H_1 = 0 \quad \text{and} \quad H_{2N} = \begin{pmatrix} H_N & H_N \\ H_N & \bar{H}_N \end{pmatrix} \tag{5.1}$$

in which N must be a power of two and $\bar{H}_N$ is the negation of H_N. The next two matrices are formed in the same way:

$$H_2 = \begin{pmatrix} 0 & 0 \\ 0 & 1 \end{pmatrix} \quad \text{and} \quad H_4 = \begin{pmatrix} 0 & 0 & 0 & 0 \\ 0 & 1 & 0 & 1 \\ 0 & 0 & 1 & 1 \\ 0 & 1 & 1 & 0 \end{pmatrix} \tag{5.2}$$

All Walsh sequences of the same matrix are orthogonal to each other. The IS-95 standard uses $2^6 = 64$ Walsh sequences. The Globalstar system will probably use $2^7 = 128$ sequences.

In IS-95 the short code is formed with two irreducible polynomials (the polynomials 121 641 and 117 071 are primitive. Note that because code sequences can be produced with a polynomial and its reciprocal polynomial,

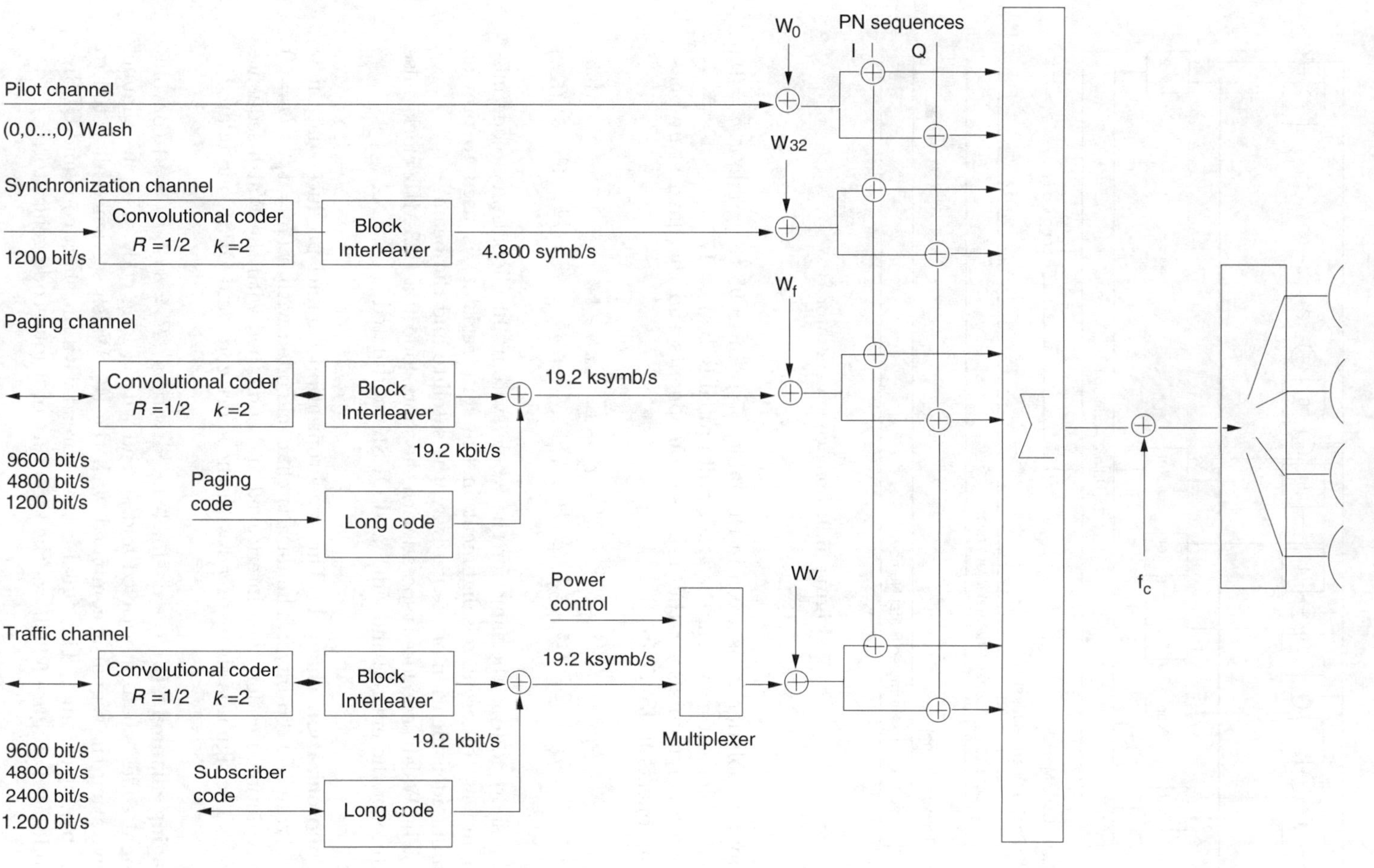

Figure 5.8: Modulator for forward-link

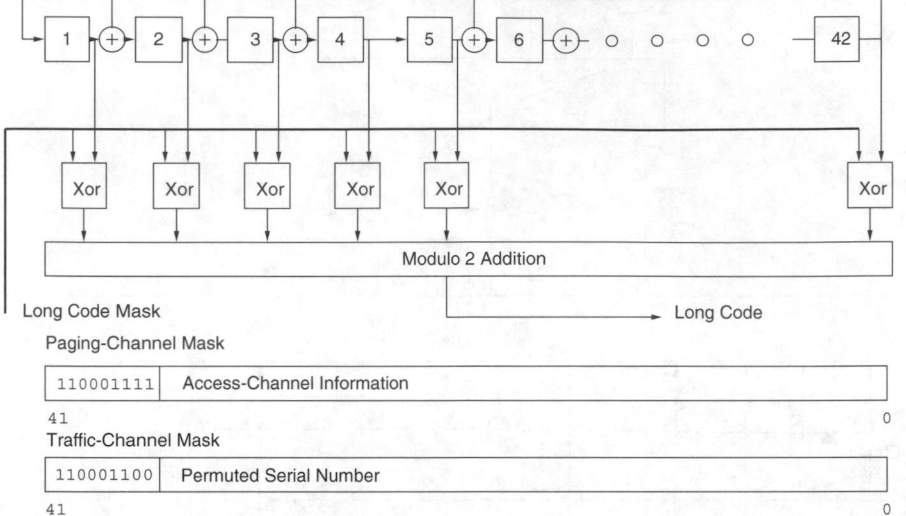

Figure 5.9: Long code generator

only one polynomial is given in the tables [7]). In IS-95 the grade is $n = 15$; in the Globalstar system the grade will probably be $n = 17$.

The polynomials for the in-phase components and the quadrature-phase components in IS-95 are

$$P_I = x^{15} + x^{13} + x^9 + x^8 + x^7 + x^5 + 1 \tag{5.3}$$

$$P_Q = x^{15} + x^{12} + x^{11} + x^{10} + x^6 + x^5 + x^4 + x^3 + 1 \tag{5.4}$$

The short code is the same for the whole system. In Globalstar a code misalignment (different misalignment in the shift register) is used to provide unique identification of the gateway, the satellite and the beam.

The Walsh sequence is spread with the short code at a 1.23 MHz clock-pulse rate over the entire bandwidth and QPSK-modulated.

Synchronization channel The synchronization channel produces data flow at a rate of 1200 bit/s. The data is channel-coded with a ($R = 1/2,\ K = 9$) convolutional coder, then interleaved and combined with the Walsh sequence W_{32}. The signal is then spread with the short code and QPSK-modulated.

Paging channel Data is channel-coded with a ($R = 1/2, K = 9$) convolutional coder, then interleaved and spread with a long code. For the channel separation the signal is combined with the W_p Walsh sequence allocated to the paging channel. The signal is then spread with the short code and QPSK-modulated. Figure 5.9 shows the structure of a long-code generator.

This involves setting up a shift register with 42 delay elements, with the outputs linked by a 42-bit long mask. The outputs are added modulo 2 and generate the long code.

Traffic channel The vocoder (standardized in accordance with IS-96), which is capable of producing different data rates as required, delivers the data to the channel coder and the interleaver. Each user has a personal secret key number which forms part of the long code mask for the traffic channel. The long code is linked to the output of the interleaver. On this basis, power control data and traffic channel data are alternatively spread using a user Walsh sequence W_u. The data flow is combined with the short code and QPSK-modulated.

5.3.1.2 Power Control on the Forward-Link

In IS-95 power control is carried out in a closed loop on the forward-link. This requires a periodic reduction in the transmitted power of the base station. The reduction continues until the user notices an increase in the frame error ratio. The user then sends a command for the power to be increased. The measurement increments of power control are relatively small and in the area of 0.5 dB. The dynamics covers an area of ±6 dB. The power changes occur every 20 ms.

5.3.2 Return-Link

Non-coherent orthogonal 64-correlated Walsh modulation is used in the return-link in IS-95. This modulation can be interpreted as FSK modulation with the Walsh sequences corresponding to different frequencies.

The long code is used here for the channel separation. In forward-link Walsh sequences are used for channel separation, whereas here the Walsh sequences are used for modulation. The data is spread with the short code and transmitted using QPSK.

5.3.2.1 Modulator

There are two physical channels: an access channel and a traffic channel, differentiated only by the long code mask. Figure 5.10 shows the modulator for the return-link of the traffic channel.

Access channel A base station receives access requests on the access channel. First a preamble of three frames of 96 zeros per frame is transmitted. Then the user's long code is transmitted. The data rate is always 4.8 kbit/s. Another eight bits containing only zeros are added after each net data frame. The data is channel-coded with a ($R = 1/3$, $K = 9$) convolutional coder, scrambled by the interleaver and modulated orthogonally. The long-code generator ($n = 41$), which is combined with a paging mask, and the short-code generator

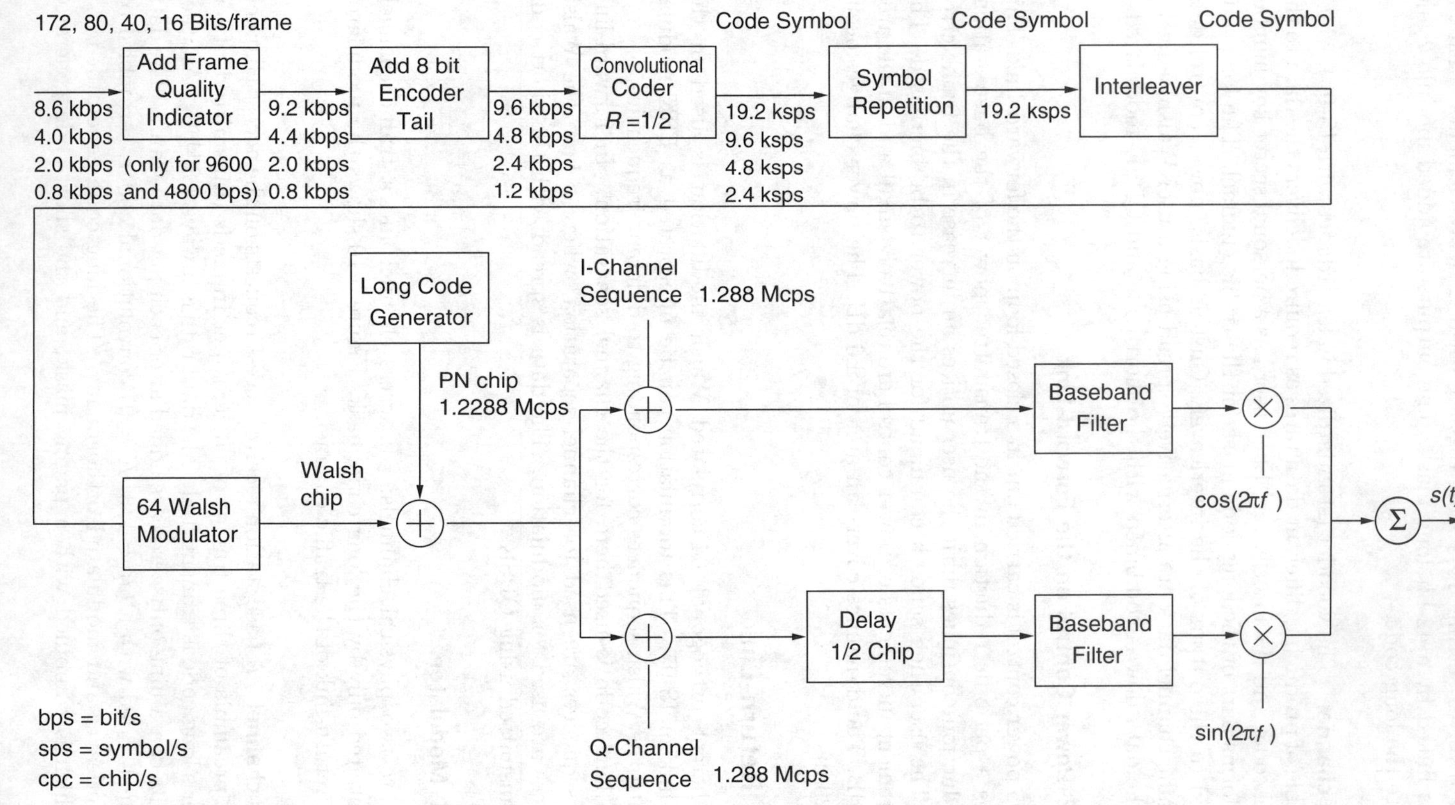

Figure 5.10: Modulator for the return-link

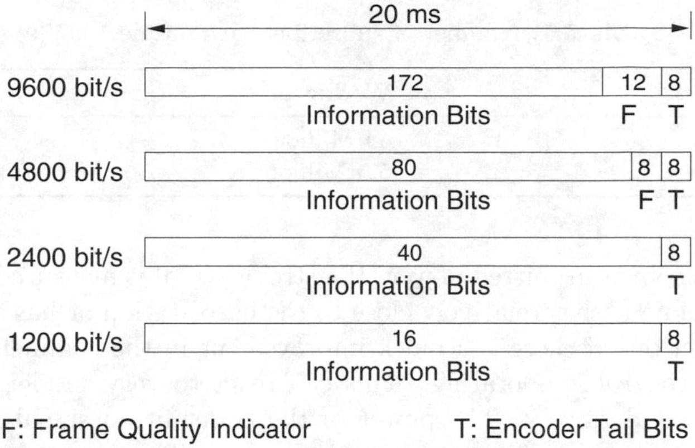

20 ms

9600 bit/s 172 Information Bits 12 F 8 T

4800 bit/s 80 Information Bits 8 F 8 T

2400 bit/s 40 Information Bits 8 T

1200 bit/s 16 Information Bits 8 T

F: Frame Quality Indicator T: Encoder Tail Bits

Figure 5.11: Frame structure for different data rates of the IS-95 vocoder

($n = 15$) are used in the spreading. With a bandwidth of 1.2388 MHz the signal is digitally band-pass filtered. At the same time it is sampled at four times the data rate.

Traffic channel Figure 5.11 illustrates the frame structure for the different data rates of the vocoder. The data rate can be altered dynamically in order to adjust to the data volume. For example, with voice transmission the data rate is reduced to 1200 bit/s during intervals when there is no speech.

A CRC check sum is generated at transmission of 9600 bit/s or 4800 bit/s. It has two functions:

1. To protect the net data.
2. To provide support in establishing the data rate at the receiver.

The CRC polynomials are given in [8]. The data is channel-coded with a ($R = 1/3$, $K = 9$) convolutional coder, interleaved und modulated orthogonally. The user-specific long-code generator ($n = 41$) and the short-code generator ($n = 15$) are used in the spreading.

5.3.2.2 Power Control on the Return-Link

The power is controlled through an interplay between two mechanisms:

- Power control with *open loop*
- Power control with *closed loop*

Each mobile station attempts to estimate free-space attenuation. With IS-95 a tone is sent from the base station on the pilot channel. The power of the pilot tone is measured by the mobile station, and is also used by it to

Table 5.1: Number of subscribers as of June 1997 [9]

	Worldwide	USA
GSM	44 Million	0.5 Million
IS-95 CDMA	5.5 Million	0.6 Million

estimate its own transmitted power. If a strong signal is measured, this means that the mobile user is relatively close to the base station or has an unusually good connection. If there is a quick improvement in the channel state, *open-loop power control* is operating. This can react to very sudden changes (in the Microsecond range). The power of the transmitted signal is controlled analogously, corresponding to the receive power. The power gradation is in the area of 85 dB. However, only the mean value of the transmitted power can be computed. Despite the speedy reaction to power changes, it is not possible to compensate for *Rayleigh fading* with open loop. The reason is that forward-link and return-link are in different frequency bands, and Rayleigh fading occurrences of the two connections are statistically independent of one another. A closed loop is used to suppress the fast Rayleigh fading.

With closed-loop power control the input power is measured at the base station and compared with the desired level of power. Should there be a discrepancy between the two, the mobile station is instructed to carry out a discrete change in transmitter power. The power changes are in the area of 0.5 dB and are transmitted every 1.25 ms. Field tests have shown that this time is sufficient to combat multipath fading.

5.3.3 Experiences Gained with IS-95 CDMA Systems

Seen from today, IS-95 is a narrowband CDMA system when compared with the forthcoming UMTS; see Chapter 6.

Since mid-1992, systems according to TIA/IS-95 standard developed by QUALCOMM have been introduced in the USA by many cellular operators, i.e., just half a year later than GSM in Europe. IS-95 systems have also been established in South Korea and Hong Kong. Seen from June 1997, GSM and IS-95 had both reached a quite different acceptance in terms of the number of subscribers; see Table 5.1 [9].

Until 1997 it had been claimed by QUALCOMM that, owing to the much larger radio distance that could be covered with CDMA signals, the coverage of a given area would only need less than half of the base stations compared with GSM to provide the same quality of service. This has not been proven in the field; e.g., investigations in the number of base stations deployed to serve the same area by different systems (and operators) have resulted in the statistics shown in Table 5.2 [9].

One reason for this result is that radio engineering for a given traffic load is much more difficult with CDMA systems compared with FD/TDMA/FH

Table 5.2: Number of base stations in Tampa/Florida [9]

Operator	Aerial GSM	AT&T D-AMPS	GTE CDMA	Prime Co CDMA
Number of base stations	16	21	23	19

systems (GSM, D-AMPS). With CDMA systems the cells are shrinking with increased traffic load (so-called cell breathing). Even after six years of experience, optimization of CDMA systems in the field, which combines power control, frequency re-use, and error protection (besides others), is still considered a very complex task requiring a lot of time with measurements and adustments to be performed in the running system.

There is no seamless handover available across frequency bands with IS-95 systems, and a number of value-added services attracting many GSM subscribers are still not offered [9]. It has been found that the GSM infrastructure cost is only one-third of that of IS-95.

In spite of this experience, CDMA has been selected as the transmission technique for the radio interface of the UMTS; see Chapter 6. There the situation is somewhat different: much more spectrum is available for UMTS, allowing higher spreading factors, e.g., for speech transmission, and a lot of experience will be available from IS-95 systems when the UMTS is introduced around the year 2003.

Anyway, it appears worth repeating what has been stated in [9] as the result of an extensive investigation of the reasons for deciding to operate IS-95 CDMA systems:

"CDMA is a religion... You get: I believe..."

5.4 The Personal Digital Cellular System (PDC) of Japan

A study of the public mobile radio market in Japan shows a high concentration of users in metropolitan areas such as Tokyo (50%) and Osaka (22%). To eliminate the bottlenecks caused by a lack of frequencies in these areas, the Ministry of Post and Telecommunications (MPT) made the decision in April 1989 to develop a digital mobile radio standard. Compared with existing analogue systems, this digital mobile radio system was to be more cost-effective and offer a higher level of capacity and security, as well as new services.

The new system, formerly called *Japanese Digital Cellular* (JDC) but now referred to as the *Personal Digital Cellular* (PDC) system, was specified by the *Research & Development Center for Radio Systems* (RCR). In addition, there is the *Personal Handyphone System* (PHS), which is a PCS system for the cordless mass-market; see Chapter 12.

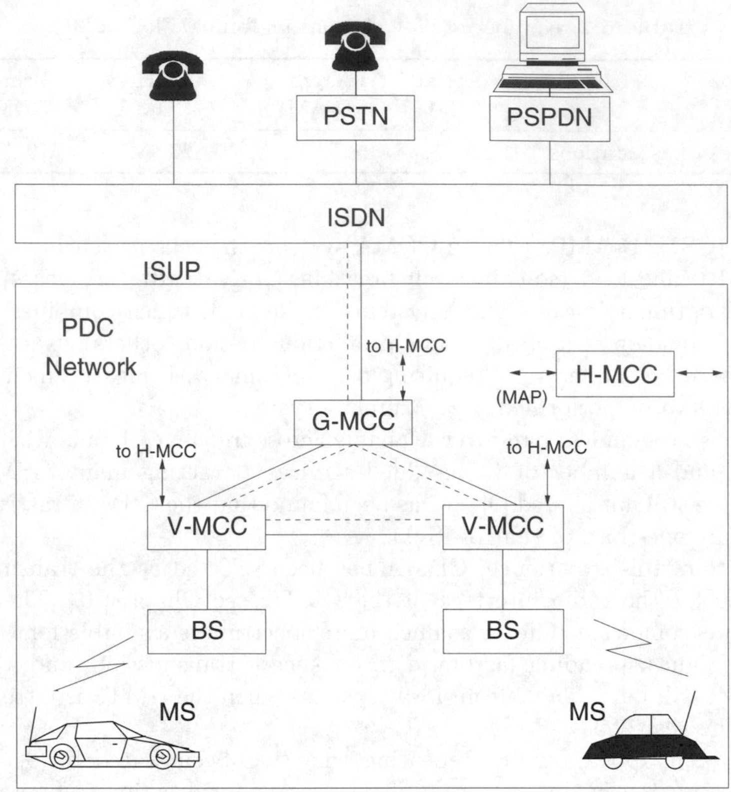

Figure 5.12: The functional architecture of the PDC system

As can be seen in the system architecture shown in Figure 5.12, the *Mobile Communication Control Centers* (MCC) are divided into *Gate-MCC*, *Visit-MCC* and *Home-MCC*, with only the G-MCCs being connected to the fixed network to save on the cost of infrastructure [4].

5.4.1 Technical Data on the PDC System

The technical parameters of the PDC standard are similar to those of the American USDC system, albeit with some important differences.

In Japan the digital system does not directly replace the existing analogue system, because the frequency bands for PDC are above and below those of the analogue system. In contrast to the USDC and GSM systems, mobile stations transmit on a higher frequency (940–960 MHz, uplink) than the base station (810–830 MHz, downlink). Furthermore, additional frequencies were provided in the 1500 MHz frequency band for the PDC system. The duplex separation is 130 MHz in the 800 MHz band and 48 MHz in the 1500 MHz band. The frequency bands themselves are divided into 25 kHz channels.

Table 5.3: Comparison of the technical parameters in the GSM, USDC and PDC systems (assuming GSM full-rate traffic channels)

Parameter	GSM	USDC (ADC)	PDC (JDC)
Frequency range [MHz]			
MS-BS	890–915	824–849	940–960
BS-MS	935–960	869–894	810–830
MS-BS (partial)			1477–1513
BS-MS (partial)			1429–1465
Access method	TDMA	TDMA	TDMA
Duplex method	FDD	FDD	FDD
Duplex separation [MHz]	45	45	130 48
Channel grid [kHz]	200	30	25
No. of channels	8 (16)	3 (6)	3 (6)
Frequency carrier	124	832	800
No. of traffic channels	$124 \cdot 8$	$832 \cdot 3$	$800 \cdot 3$
Modulation	GMSK	$\pi/4$-DQPSK	$\pi/4$-DQPSK
Speech codec	RPE-LTP	VSELP	VSELP
TCH transmission rate [kbit/s]	22.8	13	11.2
Data rate [kbit/s]	270.8	48.6	42
Min. C/I [dB]	> 9	> 12	> 13
Max. speed [km/h]	250	100	100
User data rate [kbit/s]	9.6	4.8	4.8
User cap. [Erl./km^2/MHz]	$[1.1 \ldots 1.6]$	$[1.64 \ldots 2.00]$	$[2 \ldots 3.63]$

Similarly to the American USDC system, the PDC system uses four-level $\pi/4$-DQPSK modulation [10].

With the use of a full-rate codec, three voice channels are transmitted over one carrier using time-division multiplexing and multiple access (TDMA); once the half-rate codec is introduced, it will be six. The TDMA frame has a duration of 20 ms, and each of the three time slots comprises 280 bits. Since two bits per symbol are transmitted with the $\pi/4$-DQPSK-modulation, the total transmission rate for the three time slots is 42 kbit/s.

Because a VSELP codec is used, the transmission rate (speech coding together with error-protection coding) amounts to a total of 11.2 kbit/s. The signalling data is transmitted over the SACCH at a rate of 0.75 kbit/s.

5.5 Comparison of Some Second-Generation Cellular Systems

Table 5.3 provides a comparison of the technical parameters for the radio interfaces of the digital public mobile radio systems (GSM, USDC and PDC).

Studies show that the user capacity in the PDC system is almost double that of the GSM system and that the USDC system is about 1.5 times that of the GSM system [6]. The high spectral efficiency in the PDC and USDC systems is essentially attributed to the different modulation techniques and different speech codecs used in these systems.

The advantages of GSM over USDC and PDC are higher user data rates and a lower minimal carrier-to-interference ratio. $\pi/4$-DQPSK transmitters produce considerable broadband noise. Because of an efficient speech codec, the channel and source coding, and interleaving, the GSM system achieves a higher level of voice quality than the other two systems. Comparisons of the spectral efficiency of mobile radio systems are not appropriate, mainly because the systems are designed for different qualities of service yet this parameter is usually ignored in any comparative analysis.

Unlike the GSM system, the PDC system only uses antenna diversity and no equalizers. In multipath propagation antenna diversity can be more advantageous than an equalizer, which has a high power consumption because of the complexity of the calculations.

The GSM recommendations have been successfully accepted in Europe as well as internationally. What was important for the success of GSM was European union, open standardization and early availability of the system.

Bibliography

[1] E. Berrutto, et al. *Terrestrial flight telephone system for aeronautical public correspondence: Overview and handover performance.* In *Digital Mobile Radio Conference DMR IV*, pp. 221–228, Nice, France, Nov. 1991.

[2] G. D'Aria, et al. *Terrestrial flight telephone system: Integration issues for a pan-European network.* In *Digital Mobile Radio Conference DMR V*, pp. 123–130, Helsinki, Finland, Dec. 1992.

[3] D. J. Goodman. *Wireless Personal Communications Systems.* Addison-Wesley, Reading, Massachusetts, 1997.

[4] K. Kinoshita, M. Kuramoto, N. Nakajima. *Developments of a TDMA digital cellular system based on Japanese standard.* In *41st IEEE Vehicular Technology Conference*, pp. 642–647, St Louis, May 1991.

[5] G. Larsson, B. Gudmundson, K. Raith. *Receiver performance for the North American Digital Cellular System.* In *41st IEEE Vehicular Technology Conference*, pp. 1–6, St Louis, May 1991.

[6] R. W. Lorenz. *Digitaler Mobilfunk (Systemvergleich). Der Fernmeldeingenieur*, No. 1/2, 1993.

[7] W. W. Peterson, E. J. Weldon Jr. *Error Control Coding*, Vol. 1. MIT Press, Cambridge, Massachusetts, 2nd edition, 1972.

[8] TIA. *TIA/EIA IS-95 INTERIM Standard*, July 1993.

[9] S. Titch. *Blind Faith. Telephony*, pp. 24–50, Sep. 1997.

[10] K. Tsujimura. *Digital cellular in Japan.* In *Mobile Radio Conference (MRC'91)*, pp. 105–107, Nice, France, Nov. 1991.

6
Third-Generation Cellular: UMTS*

Dramatic developments have been taking place in the mobile radio area all over the world during the last couple of decades. Mobile communications is one of the fastest growing markets in the telecommunications area. According to projections, there will be a linear increase in the number of subscribers to the major GSM networks operated in Europe by the end of the decade.

The political environment in Europe is the main reason for the rapid development. Without a free exchange of information, the concept of an internal market striving for a free flow of goods between EU states would be inconceivable. This was the line of thinking behind the liberalization and deregulation of the telecommunications industry, which promoted and accelerated competition and opened up the markets.

Another reason for this rapid development is the advances being made in the microelectronics, microprocessor and transmission technology areas. These advances are enabling the use of ever smaller terminal equipment, with computing power previously only possible with mainframes, and with low power consumption—factors that have improved customer acceptance.

In Europe the development of uniform standards, the introduction of European-wide radio systems and the participation of industry in the standardization process through the establishment of ETSI have further contributed to the widespread success of mobile communications.

The chronological development of different kinds of mobile radio networks which conform to different user needs is presented in Figure 1.2 [47].

The systems that fall into the category of first-generation mobile communications systems in which mobility is only ensured within a specific network area are the different analogue cellular systems (e.g., C-Netz, NMT) that are being closed in these days, cordless systems (CT1/CT2) and various national paging systems.

The second generation includes the digital systems such as GSM 900/1800/1900, USDC (IS-136), PDC, IS-95 and ERMES, which underwent further development and were expanded or were first introduced during the first half of the 1990s.

Along with these public cellular systems that provide PSTN/ISDN services at a mobile terminal, there are other systems that fall into the second-

*With the collaboration of Arndt Kadelka, Matthias Lott, Peter Seidenberg and Marc Peter Althoff**

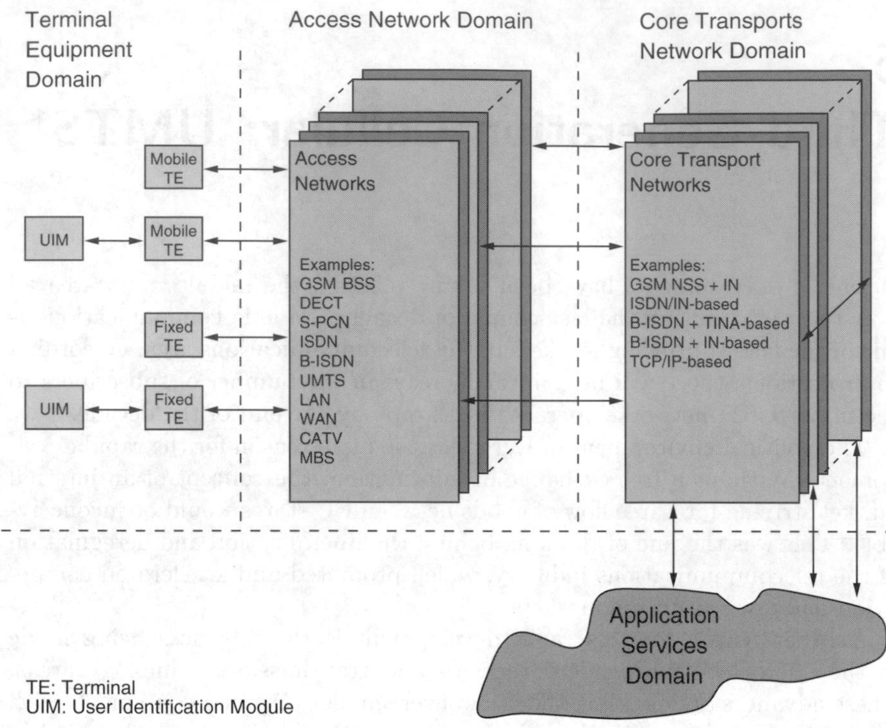

Figure 6.1: Global multimedia mobility architecture

generation or the transitional category between the second and third generations, and cater specifically to mobile or moving applications. These include trunked radio (ETSI/TETRA, see Section 7.2), cordless communications (ETSI/DECT, see Chapter 10, and the *Personal Handyphone System*, PHS, see Chapter 12), local broadband communications (ETSI/HIPERLAN 2, see Section 14.2, IEEE 802.11, see Section 14.3), mobile personal satellite radio (IRIDIUM, Globalstar, see Chapter 16) and other systems integrating aspects of these systems.

Third-generation mobile radio systems, which use intelligent networks to incorporate public mobile radio services that previously were operated separately, have already been developed recently. Under the designation *Global Multimedia Mobility* (GMM), ETSI has developed an architecture that defines mobile radio networks as the access networks to an integral transport platform that is based on broadband (B)-ISDN and provides mobility-supported value-added services (see Figure 6.1). It is planned that the third generation mobile communications networks (UMTS and FPLMTS or IMT-2000), which aim to support the services of the terrestrial broadband ISDN and the Internet, will lead to a universal worldwide public mobile radio system that is expected to be operational by the year 2003.

The main characteristics of third-generation mobile radio systems are [4]:

- Support of all features currently being offered by different radio systems.
- Support of new services with high quality of service, the same as in the fixed network.
- High capacity, which will support high market penetration.
- High spectral efficiency.
- Lightweight, small (pocket-sized) and inexpensive handheld equipment for mobile telephone use.
- High security, comparable to that of the fixed network.

High demands are being placed on the third-generation systems, e.g.:

- Services (voice and data, teleservices, bearer services, supplementary services).
- Different bit rates (low bit rates for voice; data rates up to 2 Mbit/s).
- Variable bit rates and packet-oriented services.
- Use of different sized cells (macro, micro, pico) for *indoor* and *outdoor* applications, with seamless handover between indoor and outdoor base stations.
- Operation in non-synchronous base station subsystems.
- Advanced mobility characteristics (UPT, roaming, handover, etc.).
- Flexible frequency management.
- Flexible management of radio resources.

6.1 UMTS (Universal Mobile Telecommunications System)

In Europe work has been carried out on the development of a third-generation mobile radio system called UMTS (*Universal Mobile Telecommunications System*) in the EU programmes RACE (1989–1994) (*Research and Development in Advanced Communications Technologies in Europe*) and ACTS (1995–1998) (*Advanced Communication Technologies and Services*) in cooperation with ETSI. Work on UMTS has also been done in COST (*European Cooperation in the Field of Scientific and Technical Research*) projects [23].

Initially, the technical subcommittee (STC) SMG 5 at ETSI has been given the responsibility for producing the UMTS standard. From 1999 on a Third

Generation Partnership Project (3GPP) has been formed to coordinate the efforts of ETSI, *Association of Industries and Businesses* (ARIB) and *Telecommunication Industry Association* (TIA)—the standardization institutes of Europe, Japan and the United States, respectively—under the umbrella of the *International Telecommunication Union* (ITU) for the definition of a worldwide standard on the basis of UMTS/FPLMTS/IMT-2000. Currently (2001), the 3GPP has split up into two groups namely UTRAN working on the development of the combined European and Asian views of 3rd Generation Systems and cdma2000 working on the US vision of a 3G-System. There is also the UMTS Forum, comprising the European signatories to the *UMTS—Memorandum of Understanding of the Introduction of UMTS* defined in 1996.

The main tasks of 3GPP are:

- Study and definition of services, system architecture, the air interface and the network interfaces for 3G systems.
- Generation of basic technical documentation.
- Coordination of ETSI, ARIB, T1P1 (of TIA).
- Cooperation with national research programmes.

The aim of the UMTS concept is to provide users with a handheld terminal that will cover all areas of application—at home, in the office, en route by car, in a train, in an aircraft and as a pedestrian. UMTS will therefore offer a common air interface that will cover all fields of application and have the flexibility to integrate worldwide the different mobile communications systems available today, such as mobile telephone and telepoint, trunked radio, data radio, and satellite radio systems, into one system, Figure 6.2.

What will play an important role in UMTS is the concept of intelligent networks (IN) that will provide call charging and mobility management for the localization and routing of calls across networks operated by different service providers and operators. UMTS is the first system to offer mobile users roaming during an existing connection, with handover between networks with different applications and different operators [21].

UMTS offers transmission capacity comparable to ISDN for services such as video telephony and wideband connections, and will support the service concept *Universal Personal Telecommunication* (UPT) [6] (see Chapter 17). With UMTS it is possible to transmit voice, text, data and images over one connection, and subscribers will own a personal telephone number allowing them to be reached anytime, anywhere in the world.

The first series of standards for UMTS has been completed in March 1999. The projection is that UMTS, which according to Figure 6.4 will use the frequency band between 1.885 and 2.2 GHz, will be introduced around 2003.

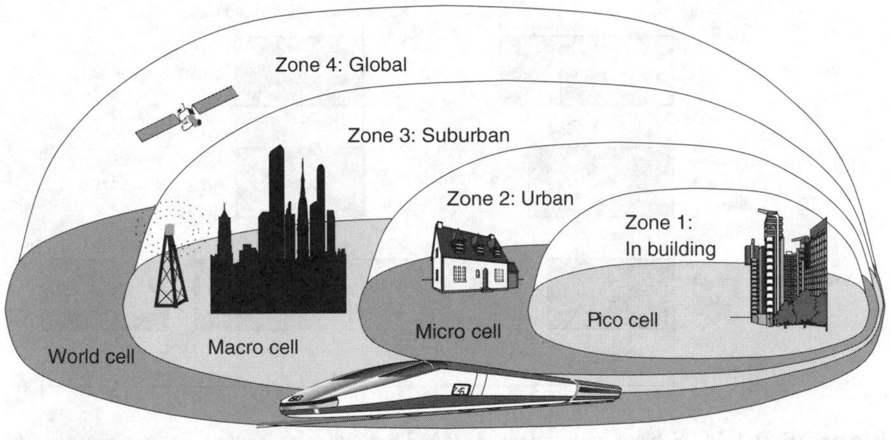

Figure 6.2: Worldwide integration of different mobile communication systems into UMTS

6.1.1 Spectrum Assigned to UMTS

The UMTS Forum has a preference for the frequencies indicated in Figure 6.3, staggered timewise as shown, and is promoting the *refarming* of bands previously used for other purposes (see Appendix D) and working towards including asymmetrical bands along with the symmetrical ones. The planned frequency allocations to IMT-2000/UMTS are shown in Figure 6.4 (see also Figure 18.6).

There, the current allocations of the *Radiocommunication Standardization Sector of ITU* (ITU-R) to IMT-2000/UMTS according to WARC 1992 can be seen for Europe, Japan and the USA. What is remarkable from Figure 18.6 is that the USA according to the licensing of 2nd generation systems (*Personal Communication System*, PCS) in the 1850–2200 MHz band will not have bands in common for 3GPP systems with Europe and Japan.

The assignments of extension bands according to WRC 1998 is shown in Figure 6.4. It is remarkable there

- that the spectrum being allocated to 2nd generation systems, e. g. GSM and DECT in Europe is foreseen to be refarmed as an extension band for 3GPP.
- extension bands for 3GPP will be available in the different regions on a different time scale according to the current allocations, e. g., in Europe the spectrum 2520–2670 MHz will be available from 2008 on but the spectrum occupied by GSM systems will be available only after the year 2015.

Compared to the request for the spectrum of the UMTS Forum shown in Figure 6.3 the ITU-R spectrum shown in Figure 6.4 does not appear to be

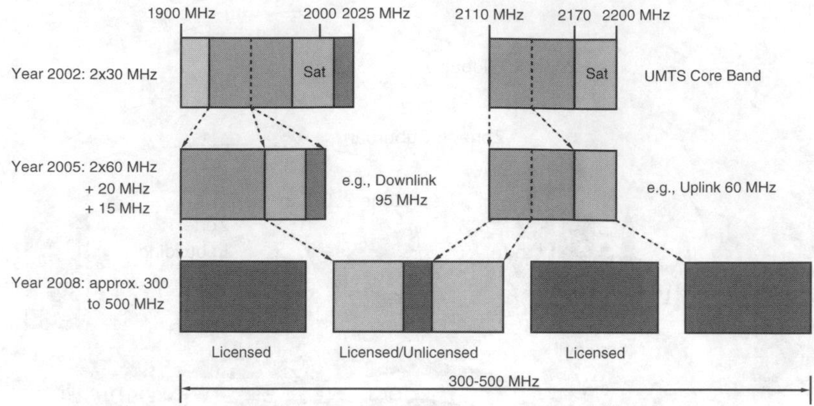

Figure 6.3: UMTS frequency spectra, UMTS Forum's perception of timetable for development

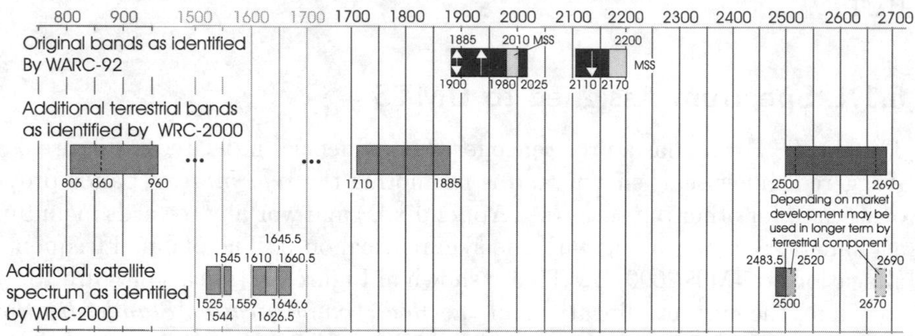

Figure 6.4: IMT-2000 extension bands

sufficient to cover the spectrum needs of 3G. Various initiatives are under way to prepare for more spectrum for 3G systems, e. g.:

- cooperation of cellular and broadcast operators to combine their systems to make available a hybrid digital mobile radio network based on, say, *Digital Video Broadcasting Terrestrial* (DVB-T) for high datarate downlink transmission and UMTS for symmetrical traffic support. The *Information Society Technology* (IST) project *Dynamic Radio for IP-Services in Vehicular Environments* (DRIVE) [28] is following the idea to demonstrate the feasibility of such a service offered for multiband terminals with DVB-T reception in the current broadcasters' spectrum (460–860 MHz) and UMTS transmission/reception either in the spectrum shown in Figure 6.4 or in spectrum below 860 MHz refarmed to be used for cellular.

- cofarming of spectrum assigned to the defence community to enable a use of a band either for cellular radio or for defence related purposes, under well-defined conditions, where a cellular operator would have to pay a leasing fee for using the defence-owned band to the defence-representing organization. Apparently flexible multiband terminals would be needed to make possible this mode of spectrum usage [48].

6.1.2 UMTS and Other 3G Standards

Originally it was planned to specify one air interface only for the IMT-2000 systems able to cover all the different services and applications aimed at. From the decision made in January 1998 (see Section 6.3), it was clear that at least two air interfaces will be specified for UMTS—one based on paired bands with *Frequency Division Duplex* (FDD) transmission, and another air interface operating in a single band with *Time Division Duplex* (TDD) transmission. Both standards will use DS-CDMA for radio transmission and channel access, and are addressed as FD-CDMA and TD-CDMA systems respectively. Besides UMTS, a further developed GSM system named EDGE (see Section 3.11) has been accepted as a member of the IMT-2000 3G standards by ITU-R that is capable of providing wideband services compatible with the GSM 2+ phase of standardization. From Section 6.14 it can be seen that EDGE and *Wideband Code Division Multiple Access* (W-CDMA) are similar in traffic performance for the usage scenarios studied there.

The main driving force for UMTS at present comes from manufacturers aiming to introduce new products into the market and operators aiming to get under the label UMTS access to more bandwidth for voice services only. Mobile data was still only a few percent of business in 1998. The European Commission has issued guidelines for the licensing of UMTS bands to operators demanding that 50 % of the services offered should be data services for multimedia applications.

The demand for more bandwidth could of course easily be covered by assigning UMTS frequency bands to be used by GSM networks, and does not need the introduction of a completely new air interface. This in fact has happened by accepting EDGE as a 3G system by 3GPP.

6.2 FPLMTS (Future Public Land Mobile Telephone System); IMT-2000 (International Mobile Communications at 2000 MHz)

In 1985 the CCIR (see Annex B.1.2) set up a working group, the Task Group 8/1 (previously IWP 8/13), for the purpose of specifiying all the requirements and system parameters for a *Future Public Land Mobile System* (FPLMTS). The following requirements for an FPLMTS were drawn up by the working group [7, 16, 22]:

- Small, lightweight handheld equipment.
- Worldwide use of terminal equipment, i.e., uniform frequencies worldwide.
- Integration of different mobile radio systems and international roaming.
- Integration into the fixed telephone networks (ISDN compatibility).
- Integration of mobile satellite radio.
- Use of terminal equipment on land, in the air and at sea.

As with the UMTS, the aim with the FPLMTS is to integrate all existing services (mobile telephony, cordless telephony, paging, trunked radio, etc.) into one service. Many of the aspects of FPLMTS are the same as those of UMTS; however, since the ITU activities are globally based, there are some differences between the two systems. For example, FPLMTS defines several air interfaces for dealing with the different requirements of densely populated areas (e.g., in Europe) versus sparsely populated areas (third world countries) [33]:

- R1: radio interface between *Mobile Station* (MS) and *Base Station* (BS)
- R2: radio interface between *Personal Station* (PS) and *Personal Base Station*
- R3: radio interface between *Satellite Base Station* and *Mobile Earth Station* (MES)
- R4: additional air interface for paging FPLMTS terminals

The plan is to use FPLMTS as a temporary or permanent substitute for fixed networks in developing countries and in rural areas where it is not economically feasible to set up fixed networks.

At W(A)RC 1992 a spectrum of 230 MHz in the frequency bands 1885–2025 MHz and 2110–2200 MHz was allocated to the FPLMTS system worldwide (see Figure 6.4). These frequency bands were not exclusively reserved for FPLMTS, and can also be used in other systems. So in Europe, for example, the lower part of the allocated frequency band is occupied by GSM 1800 and the DECT system.

The earliest date being envisaged for the operation of FPLMTS is sometime between 2001 (W-CDMA in Japan) and 2002 (UMTS in Europe). Since about 1995, FPLMTS has often been referred to as IMT-2000, but both designations refer to the same system operating around 2000 MHz.

6.3 3GPP Standardized Systems

Although the original aim of ITU-R was to prepare with the FPLMTS initiative the introduction of one air-interface standard for 3G systems worldwide, it appeared during the process of definition of the details that this would be impossible owing to the economic interests of the participating actors. Japan and Europe teamed to bring out an air-interface standard based on *Wideband Code Division Multiple Access* (W-CDMA), now called *UMTS Terrestrial Radio Access* (UTRA), and ended up during the phase of ETSI involvement (until 1998) with two versions of *UMTS Terrestrial Radio Access* (UTRA), namely (see Table 6.1):

UTRA-FDD a W-CDMA system with symmetrical channels of the same capacity for up- and downlink, separated in the frequency domain (*Frequency Division Duplex*, FDD)

UTRA-TDD a wideband system using TDMA channels each spread with a number of codes to become CDMA channels, able to support asymmetric communication channels for up- and downlink transmission, separated in the time domain (*Time Division Duplex*, TDD)

The USA developed cdma2000, an evolution from IS-95b combining a number of IS-95 narrowband CDMA channels for up- and downlink transmission separated by FDD. To harmonize the physical channel characteristics of UTRA and cdma2000, the 3GPP has been established that resulted in a number of commonalities of both systems, e.g., the chiprate of 3.84 Mcps, frame length and modulation scheme. Meanwhile 3GPP has split its support into two directions: UTRAN and cdma2000.

It became evident that it would be very difficult to achieve identical specifications to ensure global equipment compatibility. Therefore initiatives were started to create a single forum for standardization of a common UTRA specification. The *3rd Generation Partnership Project* (3GPP) was set up in 1998 with this aim in mind (see Section B.1.6). It involves the following partners: TTC/ARIB for Japan, ETSI for Europe, TTA for Korea, T1P1 for USA, and, (in 1999) CWTS for China [30, 5].

At about the same time, the North America cellular market formed the 3GPP2 group to work on the rival cdma2000 radio technology, and the *Universal Wireless Communications Consortium* (UWCC) was enlarged to cover the *Universal Wireless Communications* (UWC) 136 or IMT-SC (*Single Carrier*) technology. These two industrial groups rely on the *American National Standards Institute* (ANSI) 411 mobility protocols defined in the *Telecommunication Industry Association* (TIA) TR45.2 committee.

The two partnership projects, along with UWCC and the ETSI project working on the 2 Mbit/s DECT system, have all worked via their host standards bodies to complete the ITU framework for IMT-2000 radio technologies, as shown in Figure 6.5 and Table 6.1.

Table 6.1: Members of the IMT-2000 terrestrial radio interface family

Parameter	UTRA FDD	UTRA TDD	TD-SCDMA	cdma2000	UWC-136 (EDGE)	DECT
Official name	IMT-DS	IMT-TC	IMT-TC	IMT-MC	IMT-SC	IMT-FT
Carrier spacing	5 MHz	5 MHz	1.6 MHz	$N \cdot 1.25$ MHz	30/200/1600 kHz	1.7 MHz
Chiprate	3.84 Mcps	3.84 Mcps	1.28 Mcps	$N \cdot 1.28$ Mcps	24.3/270.833/2600 kSymbols/s	700 kSymbols/s
Duplex scheme	FDD	TDD	TDD	FDD	TDD/FDD	TDD
Frame length [ms]	10	10	10	10...80	40/4.615/4.615	10
Modulation	QPSK	QPSK	QPSK	QPSK	π/4-DQPSK, GMSK, 8-PSK, QAM	—
Number of timeslots	—	15	7	—	8 (GSM)	10
Evolution from	—	—	—	IS-95b	GSM, IS-136	DECT

DS: direct sequence
TC: time code
MC: multi carrier
SC: single carrier
FT: frequency time
TD-SCDMA: time division synchronous CDMA

FDD: frequency division duplex
TDD: time division duplex
EDGE: enhanced datarate for GSM evolution
DECT: digital enhanced cordless telecommunications
UTRA: UMTS terrestrial radio access

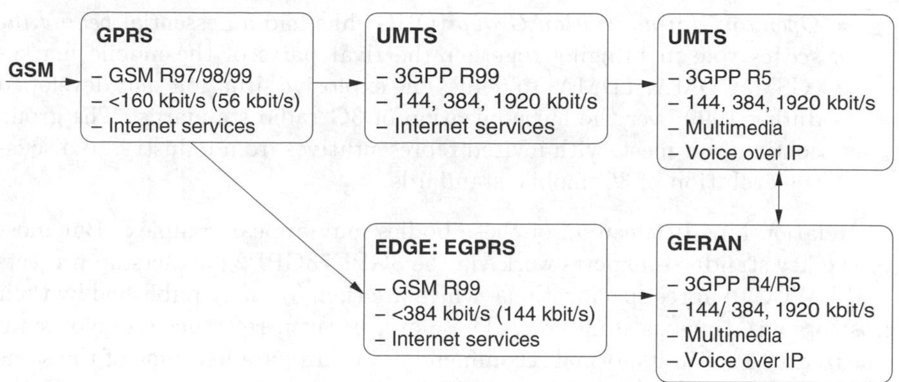

Figure 6.5: 2G to 3G Migration

While 3GPP, 3GPP2, UWCC and ETSI EP DECT are the lead bodies for the different versions of the 3G mobile standards, they are by no means the only groups working in this field. There are a number of other important organizations [30, 5]:

- ITU has a number of Study Groups working on IMT-2000 (the official generic term for 3G mobile). Within ITU-T, the lead body is the new *Special Study Group* (SSG) IMT-2000, while the lead in ITU-R is now assigned to Working Party WP8F, which replaces the former TG8/13 groups.
- *Mobile Wireless Internet Forum* (MWIF), which has the mission of *driving the acceptance and adoption of a single mobile wireless and Internet architecture that is independent of the access technology.* The forum's main aim is to look for synergy between the GSM/UMTS and CDMA/cdma2000 markets.
- *3G mobile Internet* (3G.IP), which has the role of *actively promoting a common IP-based wireless system for third generation mobile communication technology to ensure rapid standards development and take-up by operator, vendors and application developers.* The main target is to look for synergy between the GSM/UMTS and TDMA/UWC136 markets, and thereby encourage the common use of a packet data backbone based on the GPRS.
- *Internet Engineering Task Force* (IETF) is becoming more and more involved in mobile standards issues as IP technology is introduced into mobile networks. Key working groups for 3G mobile are MOBELEIP (for mobility), *Session Initiation Protocol* (SIP) (for IP-based call control) and SIGTRAN (for signalling transmission).

- *Operator Harmonization Group* (OHG) has had an essential *behind the scenes* role in bringing together the rival parts of the mobile market (GSM, CDMA, TDMA) to defuse the explosive situation that developed during 1999 over the harmonization of 3G radio standards. The group continues to meet, with invited representatives from industry, to discuss the evolution of 3G mobile standards.

Relationships between all of these bodies may appear complex. But most of the key standards experts work via the 3GPP/3GPP2 partnership projects and UWCC to agree specifications, which are then formally published by their host organizations as standards. These are, in turn, referenced and/or summarized by ITU in its formal recommendations. In parallel, some of the same experts are also involved in the requirements-setting forums, such as HOH, MWIF and 3G.IP, where they can discuss inter-system issues and harmonization. As Table 6.1 shows, besides UTRA and cdma200, there are three more IMT-2000 family members standardized that also count as 3G systems:

TD-SCDMA the synchronous modification of UTRA-TDD with a bandwidth per carrier of 1.6 MHz, different number of time-slots per 10-ms frame that has its support from CWTS of the People's Republic of China.

EDGE a further development of the GSM system, where advanced modulation schemes are used to substantially improve the data rate for asymmetrical services with the need to have a higher datarate on the downlink.

DECT especially aimed at indoor and close-to-buildings services.

EDGE and DECT, both systems standards resulted from European supporters, namely Ericsson and Siemens, respectively. It is clear from these alternatives of IMT-2000 systems for 3G solutions that no unique standard can be expected worldwide; instead dual-mode or even multi-mode terminals will be needed always to be connected independent of the home-used standard of a customer. In addition, 3G terminals for a long while will need a 2G air-interface implemented in addition, typically GSM, to be able to connect independent of the roll-out status of 3G systems.

6.4 Services for UMTS and IMT 2000

ETSI has published a preliminary list of services [9] that are to be supported by UMTS and are based on the ITU-R/CCIR recommendations for FPLMTS and the specifications of various European research projects. These UMTS-supported services are described below.

6.4.1 Carrier Services

UMTS should be able to support ISDN as well as broadband ISDN bearer services. The following services are to be integrated [9]:

- Circuit-switched services:
  - Transparent 64, 2·64, 384, 1536 and 1920 kbit/s with user data rates of 8, 16 and 32 kbit/s
  - Voice transmission
  - 3.1, 5 and 7 kHz audio transmission
  - Alternative voice or transparent data transmission with user data rates of 8, 16, 32 and 64 kbit/s
- Packet-switched services:
  - Virtual calls and permanent virtual channels
  - Connectionless ISDN
  - User signalling

Wideband ISDN services with a transmission rate of 2 Mbit/s (so-called *wideband services*) are also to be offered by UMTS to mobile users. According to ITU-T, these services will be classified as interactive or distribution services.

Interactive services fall into the category of conversational services, message services or interrogation services. Conversational services are implemented through end-to-end connections, which can be either symmetrical bidirectional, asymmetrical bidirectional or unidirectional. Message services offer communication between users that is not time transparent. Interrogation services are used for the inquiry and receipt of centrally stored data.

With distribution services information can be transmitted continuously from one central location to any number of users, with the users unable to influence the start or the end of a transmission. Another distribution service offers users the possibility of influencing the start of the information transmission.

Asynchronous Transfer Mode (ATM) was specified by ETSI as the transmission technology in the fixed (core) networks. In order to derive requirements for the radio interface from the bearer services being supported, ETSI, in accordance with the functional descriptions of B-ISDN and the *ATM Adaptation Layer* (AAL) (see Section 13.2.4), divided the bearer services into four classes [10]. These four classes of bearer services differ from each other in their time responses, bit rates and types of connection. Maximum bit ratio, maximum bit-error probability and maximum delay time are specified within each class of bearer service for the different communication scenarios.

At present it appears to be clear that the core network of 3G systems will be packet based using advanced Internet technology able to differentiate service classes according to the user's requirements specified in the IPv4 and IPv6 datagram headers, respectively (see Section 6.6).

6.4.2 Teleservices

The teleservices to be supported by UMTS are divided into three classes [9]:

1. Teleservices that already exist in the fixed network in accordance with ITU-T/CCITT recommendations of the E, F and I series:
   - Telephony:
     - Voice
     - Inband facsimile (telefax groups 2 and 3)
     - Inband data transmission (using modem)
   - Teleconferencing:
     - Multiparty, added value services
     - Group calls
     - Acknowleged group calls
     - Multiple calls
2. UMTS teleservices and applications, e.g.:
   - Audio and video transmission
   - Paging
   - Broadcast services
   - Database inquiries
   - Data transmission
   - Directory services (e.g., telephone book)
   - Mobility services (e.g., navigation or localization)
   - Electronic mail
   - Emergency calls
   - Emergency call broadcasts
   - Short-message services:
     - Initiated by user
     - Terminated by user
     - Voice messages
     - Facsimile
     - Electronic mail
   - Teleaction services (e.g., remote control)
   - Teleshopping
   - Video monitoring
   - Voice messages
3. The services with the largest need for bandwidth are *Multimedia* (MM) and *Interactive Multimedia* (IMM), such as data, graphics, images, audio and video, and combinations thereof. With UMTS it should be possible to use more than one of these media at the same time. Multimedia allows the transmission of more than one type of information, e.g., video and audio information. No further specifications exist yet for this service [9].

6.4.3 Supplementary Services

In the standardization of supplementary services a differentiation has principally been made between traditional non-interactive PSTN/ISDN services and personalized interactive supplementary services. The service provider has the option of making these services accessible to user groups or to individual users. The following classes of supplementary services have been proposed in accordance with the GSM and ISDN standards:

Number identification, e.g., abbreviated dialling, protection against undesirable calls, calling party identification

Call offering, e.g., call forwarding

Call termination, e.g., call holding

Multiparty communication, e.g., conference call

Group communication, e.g., communication in closed user groups

Billing, e.g., credit balance

Additional information, e.g., user-to-user signalling

Call rejection, e.g., blocking all incoming calls

A list of different service attributes is available in [9].

6.4.4 Value-Added Services

Personal mobility Using a smart card, subscribers are able to transfer their telephone numbers to any terminal.

Virtual home environment (VHE) and service portability This allows the users to set up their own personalized service portfolios and use them in any other network. VHE emulates those services that are not actually offered in the visited network (see Section 6.13), so that users notice nothing differently from their own home network environments. Moreover, this is how the preliminary UMTS services are provided.

Bandwidth-on-Demand This offers an efficient use of resources for services that have heavily varying requirements for transmission bandwidth, such as short-message services and video. Furthermore it allows users the independent option of selecting between a higher bandwidth for a maximum quality of service or a lower bandwidth for more favourable costs.

6.4.5 Service Parameters

A service is characterized by different service parameters, some of the most important being:

- Net bit rate
- Usage level
- Symmetry of a service
- Coding factor
- Maximum bit-error ratio based on channel decoding
- Maximum delay allowed in data transmission

The net bit rate is the number of bits that have to be transmitted on average within a certain period of time.

The delay parameter describes how long a waiting time is allowed in the transmission of these bits. For example, a voice service requires a small delay whereas a packet-based service has minimal requirements for the delay times of individual packets. However, data transfer requires a considerably lower bit-error ratio than a voice service, because the redundancy of the voice codec can be fully utilized. A higher channel coding factor is needed for achieving a lower bit-error ratio in order to protect data during transmission over a radio link.

The usage level parameter describes how often a connection is being used to transmit data. For example, the usage level of a voice service is less than 0.5 because a user is generally either listening or speaking and the active user needs some time to breathe during speaking.

A service is also defined by its symmetry. This value determines which bandwidth is required for a connection in one or the other direction. The voice service is an example of a symmetrical service, because in channel-oriented systems the same bandwidth is used for both transmission directions. Internet browsing (e.g., world wide web, WWW) is a typical example of an asymmetrical service, because it requires considerably less bandwidth for requesting than for receiving data. Table 6.2 lists the characteristics of some of the services.

6.4.6 Service-Specific Traffic Load

The effective service bandwidth can be calculated from the data of the service parameters net bit rate, symmetry and coding factor [14]. The service bandwidth describes the bandwidth used to provide a particular service.

The traffic generated by the use of a service is calculated by taking the average duration of this usage and the frequency of usage. The effective call duration T_{eff}, which is calculated on the basis of the usage level N and the average call duration T_{call}, is produced as

$$T_{eff} = NT_{call} \tag{6.1}$$

For the systems being planned, the frequency of usage of a service can only be estimated. It is measured in BHCA (*busy hour call attempts*), and indicates the average frequency of the usage of a service by a user during the peak traffic hour.

If it is known which portion of the overall usage of services is an individual service then it is possible to calculate the effective bandwidth needed by a user. The share of the service in the overall usage is then indicated with the penetration D. The penetration varies with different operating environments (see Section 6.5). The traffic produced by a user utilizing a service is calculated in *Equivalent Telephony Erlang* (ETE) [14]:

$$\frac{\text{ETE}}{user} = T_{eff} \times \text{BHCA} \times D \times \frac{service\,bandwidth}{telephony\,bandwidth} \tag{6.2}$$

Table 6.2: Quality of service parameters

Service	Call duration	Data rate [kbit/s]	Residual bit-error ratio	Delay [ms]
Telephony				
– Voice	2 min	8–32	10^{-4}	40
– Teleconferencing	1 h	32–128	10^{-4}	40
Video telephony	2 min	64–384	10^{-7}	40–90
Video conferencing	1 h	384–768	10^{-7}	90
Message services				
– SMS and paging	cl	1.2–9.6 (1.2–2.4 type)	10^{-6}	100
– Voice mail	2 min	8–32	10^{-4}	90
– Facsimile mail	1 min	32–64	10^{-6}	90
– Video mail	tbd	64	10^{-7}	90
– e-mail	cl	1.2–64	10^{-6}	100
Distribution services	tbd	1.2–9.6 (2.4 type)	10^{-6}	100
Database use	tbd	2.4–768	10^{-6}	200+
Teleshopping	tbd	2.4–768	10^{-6}	90
Electronic mail	tbd	2.4–2000	10^{-6}	200
Message dist.	cl	2.4–2000	10^{-6}	300
Tele-action services	tbd	1.2–64	10^{-6}	100–200

tbd: to be defined cl: connectionless

An ETE therefore corresponds to an Erlang of voice service with a transmission bandwidth of 16 kbit/s. This equation was used to produce an example of the traffic load for voice telephony, video telephony and the facsimile service.

The throughput represents the speed at which the user data is transmitted. This data quantity is increased by a constant factor through the coding used for error detection and correction. Finally, consideration must be given to the form of symmetry. For example, with the telephony services data is transmitted in both directions, whereas with the facsimile service it is mainly in one direction.

6.4.6.1 Voice Telephony

Voice telephony is a symmetrical service with a usage level of 0.5 or less. The net bit rate of the voice codec is 16 kbit/s. Since the requirements for bit error ratio are low, a coding factor of 1.75 is sufficient. For the average call duration 120 s is assumed. This equates to an effective service bandwidth of 56 kbit/s and an effective call duration of 60 s (see Table 6.3).

The estimated values for penetration D and for the frequency of calls during a busy hour produce the ETE/user values shown in Table 6.4 for the voice service in different communications environments (see Section 6.5).

Table 6.3: Service bandwidth and effective call duration of some services

	Voice telephony	Video telephony	Fax service
Throughput [kbit/s]	16	64	64
Channel coding factor	1.75	3	3
Symmetry	2	2	1.1
⇒ Service bandw. [kbit/s]	56	384	211.2
Usage level	0.5	1	1
Call duration (average) [s]	120	120	156
⇒ Eff. call duration [s]	60	120	156

Table 6.4: Calculation of traffic load for voice telephony

Operating environment	D	BHCA per user	ETEs per user
Business use indoors	0.5	1.0	$8.33 \cdot 10^{-3}$
Residential area	0.3	0.13	$6.50 \cdot 10^{-4}$
City centre, in vehicle	0.4	0.5	$3.33 \cdot 10^{-3}$
City centre, pedestrian outdoors	0.4	0.5	$3.33 \cdot 10^{-3}$
Aircraft	0.4	0.5	$3.33 \cdot 10^{-3}$
Local high bit rate	0.5	1.0	$8.33 \cdot 10^{-3}$

Table 6.5: Calculation of traffic load for video telephony

Operating environment	D	BHCA per user	ETEs per user
Business use indoors	0.13	1.0	$2.97 \cdot 10^{-2}$
Residential area	0.08	0.13	$2.67 \cdot 10^{-3}$
City centre, in vehicle	0.04	0.5	$4.57 \cdot 10^{-3}$
City centre, pedestrian outdoors	0.04	0.5	$4.57 \cdot 10^{-3}$
Aircraft	0.04	0.5	$4.75 \cdot 10^{-3}$
Local high bit rate	0.13	1.0	$2.97 \cdot 10^{-2}$

6.4.6.2 Video Telephony

Video telephony is a symmetric service and has a usage level of one, i.e., transmission is always in both directions of a connection. The effective service bandwidth for the video telephony service is 384 kbit/s and the effective call duration 2 min (see Table 6.3).

The estimated values for the penetration D and for the BHCA produce the ETE/user values in Table 6.5 for different communications environments.

6.4.6.3 Facsimile

A throughput of 64 kbit/s is assumed for the facsimile service, which corresponds to the transmission rate of the facsimile service currently being offered by ISDN. The facsimile service is an asymmetrical service with a coding factor

Table 6.6: Calculation of traffic load for facsimile services

Operating environment	D	BHCA per user	ETEs per user
Business use indoors	0.3	0.06	$2.94 \cdot 10^{-3}$
Residential area	0.15	0.03	$7.35 \cdot 10^{-4}$
City centre, in vehicle	0.1	0.002	$3.27 \cdot 10^{-5}$
City centre, pedestrian outdoors	0.1	0.002	$3.27 \cdot 10^{-5}$
Aircraft	0.15	0.002	$4.90 \cdot 10^{-6}$
Local high bit rate	0.3	0.06	$2.94 \cdot 10^{-3}$

Table 6.7: Resulting total traffic loads

Operating environment	ETE/user Voice+ video+fax	Total ETE/user	User density (per km^2)	Total traffic density [ETE/km^2]
Business use indoors	$4.10 \cdot 10^{-2}$	$4.92 \cdot 10^{-2}$	180000	$8.85 \cdot 10^{3}$
Residential area	$3.76 \cdot 10^{-3}$	$4.52 \cdot 10^{-3}$	380	1.72
City centre car telephone	$7.94 \cdot 10^{-3}$	$9.52 \cdot 10^{-3}$	2050	$1.95 \cdot 10^{1}$
City centre pedestrian	$7.94 \cdot 10^{-3}$	$9.52 \cdot 10^{-3}$	730	6.95
Aircraft	$7.91 \cdot 10^{-3}$	$9.49 \cdot 10^{-3}$	0.24	$2.28 \cdot 10^{-3}$
Local high bit rate	$4.10 \cdot 10^{-2}$	$4.92 \cdot 10^{-2}$	108000	$8.85 \cdot 10^{3}$

of 3. This produces an effective service bandwidth of 211.2 kbit/s. The effective call duration is 156 s (see Table 6.3).

The estimated values for penetration D and for BHCA produce the ETE/user values in Table 6.6 for different communications environments.

6.4.6.4 Resultant Overall Traffic Loads

The procedures presented in the sections above can be used to calculate the traffic generated by a user in each of the services listed. Table 6.7 gives the total traffic generated by a user in the different communications environments (see Section 6.5). If a specific user density is assumed for each communications environment [9] then it is always possible to arrive at a value for the traffic load.

This traffic load describes in ETEs the traffic originating from an area. The requirement for frequency spectrum can be calculated if an assumption is made on the efficiency of the radio interface (see Section 6.5).

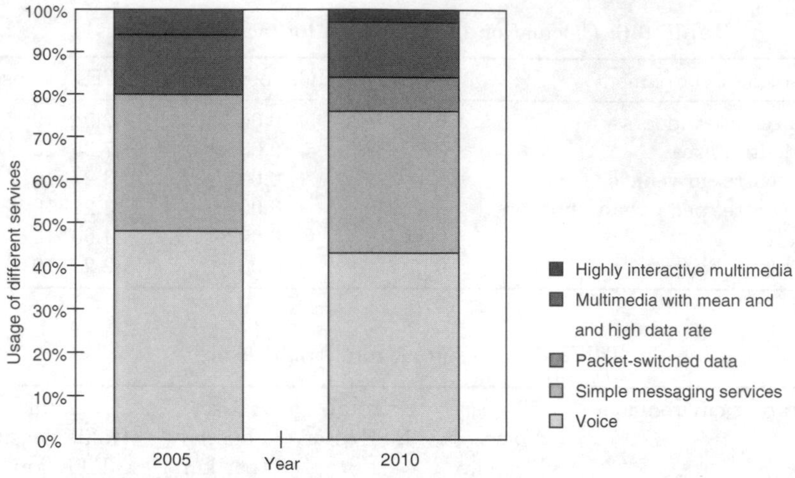

Figure 6.6: Anticipated service spectrum for UMTS

Table 6.8: User density in the year 2010

Environment	User density [per km^2]
City (indoors)	180 000
Suburbs (indoors or outdoors)	7 200
City, pedestrian	108 000
City, auto	2 780
Rural areas (total)	36

6.5 Frequency Spectrum Needed for UMTS

This section presents the UMTS Forum assessments on the frequency spectrum required for UMTS [14]. They are based on estimates of market penetration, future user density, service characteristics and characteristics of the radio interface.

In determining the bandwidth needs the UMTS Forum makes its assumptions based on the breakdown of different categories of service shown in Figure 6.6. In addition, assumptions are made on anticipated user numbers in relationship to the communications environment. The figures for the year 2010 are given in Table 6.8.

The service characteristics compiled in Table 6.9 are also taken into account, with a 16 kbit/s voice codec assumed. The voice service is a symmetrical service with the same transmission rates on the uplink and the downlink. Simple message services are those services that are similar to the *Short Message Service* (SMS) in GSM. The asymmetric MM services represent typical Internet services (WWW using the http protocol), whereas the interactive

Table 6.9: Overview of service characteristics

Service	Net rate [kbit/s]	Coding factor	Symmetry	Eff. call duration [s]	Service bandwidth [kbit/s]
High interactive MM	128	2	1/1	144	256/256
High data rate MM	2000	2	0.005/1	53	20/4000
Med. data rate MM	384	2	0.0026/1	14	20/768
Packet-sw. data	14	3	1/1	156	43/43
Simple mess. serv.	14	2	1/1	30	28/28
Voice	16	1.75	1/1	60	28/28

Table 6.10: Number of calls in a city during busy hour

	2005			2010		
Service	Business	Ind.	Outd.	Business	Ind.	Outd.
High interact. MM	0.12	0.06	0.004	0.24	0.12	0.008
High data rate MM	0.12	0.06	0.004	0.12	0.06	0.004
Med. data rate MM	0.12	0.06	0.004	0.12	0.06	0.004
Packet-sw. data	0.06	0.03	0.002	0.06	0.03	0.002
Simple mess. serv.	0.06	0.03	0.002	0.06	0.03	0.002
Voice	1	0.06	0.06	1	0.85	0.85

Table 6.11: Bandwidth requirements for the years 2005 and 2010 [MHz]

Year	2005	2010
High interactive multimedia	22	82
Multimedia with medium and high data rates	113	241
Packet-switched data	12	9
Simple message services	2	2
Voice	220	220
Overall	339	554
Overall (with guard bands)	406	582

multimedia service represents a symmetrical connection such as is required for video conferencing.

Together with the ratio of the average number of active users to the overall number, measured during the busy hour (see Table 6.10), the bandwidth requirements for UMTS can be calculated from the information supplied in the service characteristics and user density (see Table 6.11).

The projected bandwidth requirements for each service for the years 2005 and 2010 are presented in Figure 6.7.

The maximum requirement for bandwidth projected for the year 2010 is 554 MHz for traffic bands and 28 MHz for guard bands. The basic standards for UMTS were completed in March 1999, refinements were finished

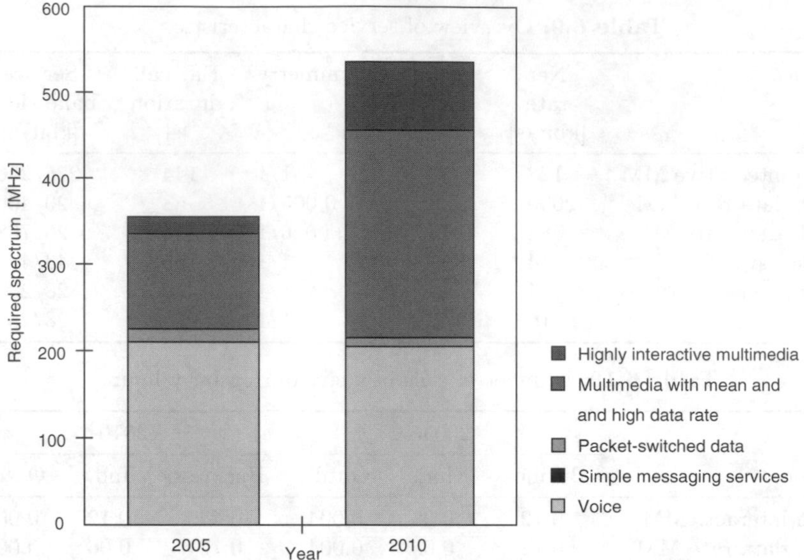

Figure 6.7: Anticipated requirements for frequency spectrum for UMTS services

in June 2000, and UMTS itself is expected to be introduced in about 2002 in Europe. The UMTS Forum has a preference for the frequencies given in Figure 6.3, staggered timewise as shown, with bands previously used for other purposes set aside for *refarming*, and is aiming to have asymmetrical as well as symmetrical bands.

The need for frequency spectrum in individual countries can vary depending on population density and economic development. The UMTS Forum for the introduction phase had asked for allocation of a so-called *core* band. Since UMTS is one 3G member system within the IMT 2000 family, either part or all of the core band is to be available for IMT 2000 worldwide. The 1900–1980 MHz and 2010–2015 MHz as well as the 2110–2170 MHz bands are being provided for terrestrial applications. The 1980–2010 MHz and 2170–2200 MHz bands are to be used for satellite-supported applications.

Estimations by the UMTS Forum and ITU-R for the years 2005 and 2010 about the expected asymmetry of services are presented in Figure 18.7. It becomes clear from that figure that the downlink in a radio cell will need about twice the capacity of the uplink. For the introduction of UMTS a symmetrical spectrum allocation is being used but the UMTS extension bands (see Figure 6.3) will add more downlink capacity to the system in operation. How to do that is currently under discussion.

In the year 2000 in most of the European countries UMTS spectrum has been licensed, namely 2×60 MHz for UTRA-FDD and 30 MHz for UTRA-TDD. The larger countries typically have licensed 6 operators, each 2×60 plus 5 MHz unpaired spectrum. Since auctions have been organized to identify

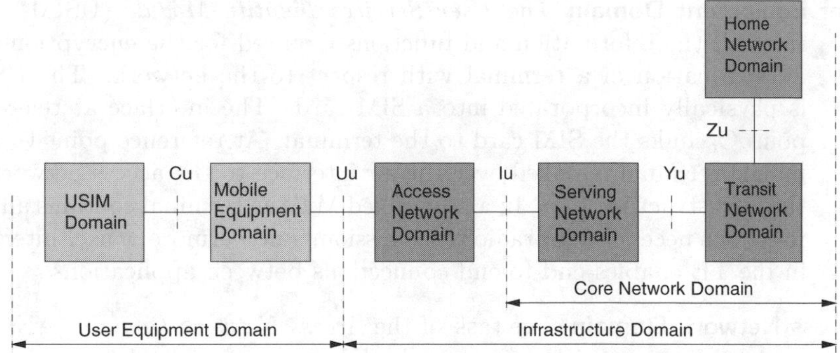

Figure 6.8: Breakdown of UMTS into domains and corresponding interfaces

those operators having greatest ability to quickly build up the new technology in large areas very high licence fees have resulted in some countries, e. g., up to 16,000 US $ per kHz spectrum per year.

This high licence fee is paired by a similar amount of money for the operator to roll out the new system and another cost will result from subsidizing the user terminals. Alltogether it is expected that services using up a remarkable capacity of a cell will be quite expensive to use and that operators will have to look for services where they need not transmit much data but where the user or some third party is willing to pay a lot for the bits transmitted compared to a voice service; SMS of GSM and i-mode of NTT DoCoMo of Japan are examples of this class of services. Voice services are not well-suited to pay back the money spent in advance by the operators owing to the relative high bitrate needed and the limited capacity of about only 50 simultaneous calls per 5 MHz carrier per cell possible.

This fact highly encourages operators to introduce packet-based voice services to reduce the capacity needs of these services. From these arguments it appears that the predictions of the UMTS Forum of 1998 of the service usage in UMTS for the year 2005 might be dramatically erroneous; instead packet-data might be the main cash cow.

6.6 Architecture of UMTS

6.6.1 UMTS Domain Concept

The main structure of UMTS is based on the architecture of *Global Multimedia Mobility* (GMM), as illustrated in Figure 6.1. The breakdown of UMTS into different *domains* is therefore based more on a physical separation of the different domains than on a functional one. Figure 6.8 presents the architecture of UMTS with the different domains and corresponding interfaces.

Three domains are distinguished:

User Equipment Domain The *User Services Identity Module* (USIM) contains all the information and functions required for the encryption and authentication of a terminal with respect to the network. The USIM is physically incorporated into a SIM card. The interface at reference point C_u links the SIM card to the terminal. At reference point U_u the mobile terminal is linked over the air interface to the access network in the infrastructure area. In a part called MT the terminal contains all the functions necessary for radio transmission. Furthermore, a user interface in the TE enables end-to-end connections between applications.

Access Network Domain The task of the *Access Network Domain* (AND) is to provide users with access to the UMTS network and to implement a connection to the transport network. The AND is realized either through an UTRAN or a GSM-BSS.

Core Network Domain In UMTS the *Core Network Domain* (CND) is an integral platform that can consist of different transport networks—such as PDN (e.g., Internet), GSM, N-ISDN or B-ISDN—that are linked together over network gateways (*Interworking Unit*, IWU). The CND is further divided into the subdomains *Serving Network Domain* (SND), *Home Network Domain* (HND) and *Transit Network Domain* (TND). The SND is linked to the access network at reference point I_u and contains all the local functions for the user that track the movement of the user within the network. The SND also has the task of switching channel and packet-switched connections.

The HND incorporates the functions in the transport network that relate to a fixed location which is separate from a user's permanent location. These specifically include the functions required for managing important information relating to a user or for the provision of location-specific services that are not offered by the SND. The HND contains the key functions required for the provision of services by a *service provider*.

The CND also contains the TND that implements the interface to other networks.

6.6.2 Releases of the UMTS Standard

The different versions of the UMTS Standard are bunched together in so-called *releases*. The first complete series is called *Release99* or just R99 [18]. The standard consists of the document series 23 to 35 where R99 is covered by documents with versions *3.x.y*.

After R99 it was planned to release R00 but the concept of naming the releases after years was abandoned in September 2000 since it could be foreseen that the versions could not be finished in time. Instead, R00 was split into two releases called R4 and R5. The term R00 was abandoned. Furthermore, in mid-2001 a renaming of R99 to R3 has been decided.

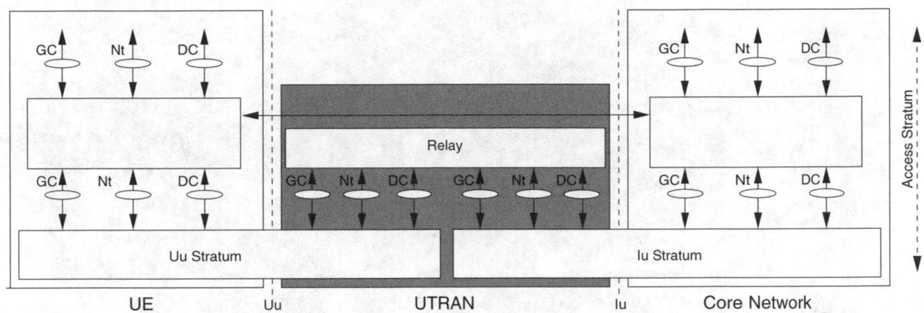

Figure 6.9: The AS in the UMTS Terrestrial Radio Access Network

Documents belonging to R4 will have version numbers *4.x.y* whereas R5 will have *5.x.y*.

Every new release will incorporate new features to enhance the capabilities for terminals and networks built to that release's specifications. The first UMTS equipment available will be built according to R99. Once a release is finished, further modifications to that release have to be done via so-called *change requests*.

6.7 The Access Stratum (AS)

The two protocol stacks at reference points U_u and I_u in conjunction with the *relay* function of the UTRAN form the *Access Stratum* (AS). Figure 6.9 shows how UTRAN and the AS are incorporated into the UMTS architecture.

The AS is responsible for carrying out transparent transmission of information between *Core Network Domain* (CND) and *User Equipment Domain* (UED). It offers services over the following service access points:

General Control (GC): This service access point provides unacknowledged distribution services that transmit non-user-specific information to terminals within a specific geographical area.

Notification (Nt): The Nt service access point offers distribution services for the unacknowledged transmission of user-specific information. Paging and notification services are provided over the Nt.

Dedicated Control (DC): The DC service access point enables call set up and termination and the transmission of usage data. The call set up service also makes it possible for messages to be transmitted at the same time as call set up is taking place. A quality of service can be assigned to the different transmission services.

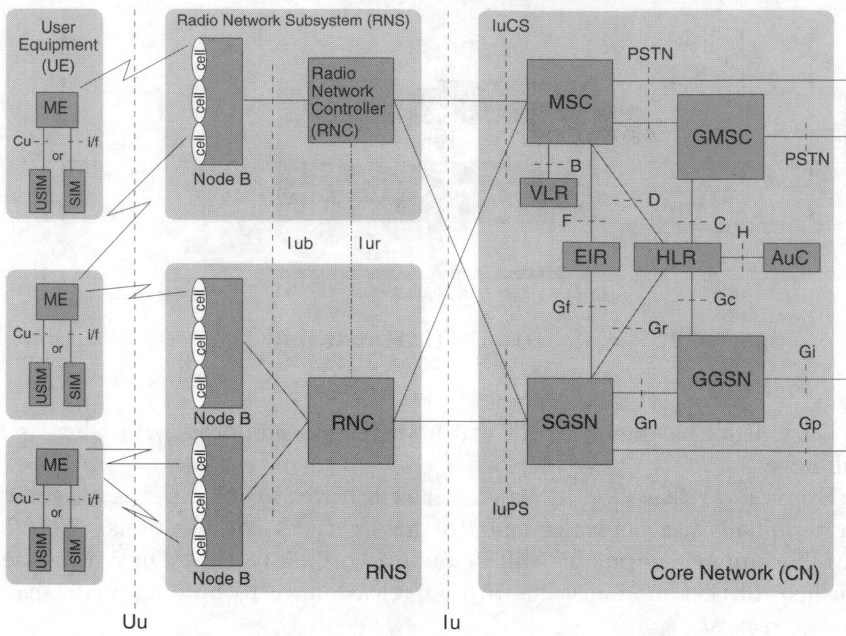

Figure 6.10: Architecture of the access stratum

6.7.1 The Core Network

The *Core Network* (CN) is logically divided into a *Circuit Switched Domain* (CSD) for channel-switched services and a *Packet Switched Domain* (PSD) for packet-switched services. Consequently, the I_u interface between CN and UTRAN is logically broken down into a I_uCS and a I_uPS. The CSD contains all the functional units of the CN necessary for providing a channel-switched service, including signalling. A feature of the channel-switched service is that resources are reserved during connection set up and not released again until connection termination. In the PSD packets can be transmitted independently of one another without the need to reserve resources of the network per user beforehand.

The functional structure of the core network is shown in Figure 6.10. The individual functional elements are part of either the PSD or the CSD or are used jointly by both. The MSC, the GMSC and the VLR are part of the CSD. The SGSN and the GGSN belong to the PSD. The functional elements of the core network essentially correspond to those of GSM in the CSD and those of the GPRS in the PSD.

6.7.2 UMTS Terrestrial Radio Access Network

UMTS Terrestrial Radio Access Network (UTRAN) is a network that contains the functions and the protocols used for data transmission over the terrestrial radio access network and is embedded within the AND. These include the protocols at reference point U_u between the MT and the AND and the protocols at reference point I_u between the AND and the transport network. The UTRAN network forms the interface between these two reference points and therefore contains entities of the U_u- and I_u protocol stacks.

6.7.3 Functional Structure of UTRAN

The *UMTS Terrestrial Radio Access Network* (UTRAN) consists of a number of *Radio Network Subsystems* (RNSs) each of which is connected over the I_u interface to the transport network CN (see Figure 6.10). An RNS consists of a RNC and one or more *nodes B*. *Node B* is a logical unit responsible for radio transmission in one or more cells and communicates with the RNC over reference point I_{ub}. A *node B* on the user side (*User Equipment*, UE) only contains the physical layer of the U_u protocol stack.

An RNC in the *Serving Radio Network Subsystem* (SRNS) is allocated to each connection between an UE and the UTRAN. If it is necessary for additional radio resources to be provided to a connection from another RNC (for example, for a soft handover), then the affiliated RNS is referred to as a *Drift Radio Network Subsystem* (DRNS). An RNS is classified as an SRNS or a DRNS, depending on the connection. A logical connection exists between the SRNS and the DRNS at reference point I_{ur}, which is mapped physically onto the I_u interface. An end-to-end connection between the UE and the CN exists only over the I_u interface in the SRNS.

6.7.4 Functions of UTRAN

Admission Control: The purpose of admission control is to prevent overloading in a radio network. Based on available interference and load measurements, it decides which radio resources should be reserved in the system. These include, for example, access control for new connections, the reconfiguration of existing connections and the reservation of resources for macrodiversity and handover. The *admission control* function is embedded in the SRNS.

Congestion control: If a radio network becomes overloaded even though no new connections are being accepted, then congestion control has the task of returning the system to a stable state in a way that is largely unnoticeable to the users.

System information broadcasting: The task of this function is to provide mobile stations with all the information from the access layer and from higher layers they need for operation.

Encryption on the radio channel: UTRAN offers an encryption function to protect against the undesired decoding of signals on the radio interface. The encryption uses a cyphering algorithm with a session-specific key. Air interface encryption is a function of UTRAN as well as of UE.

Handover: The handover function implements mobility management at the air interface. The handover function is also used to ensure compliance with the quality of service requested by a CN. The handover function can be controlled either by the network or from the UE. Therefore, it can be embedded in both the SRNS and the UE. The handover function in an UTRAN network also allows connections to be switched to other networks or from other networks to the UTRAN (for example, a GSM/UMTS handover).

SRNS relocation: If a mobile station is using the radio resources of an SRNS and a DRNS, the role of the respective RNC can change during the course of the connection due to the mobility of the mobile station. Since end-to-end connections between the UE and a CN always only exist over the I_u reference point on the SRNS, the UTRAN must also change the I_u interface to the CN if there is a change in the SRNS. The SRNS initiates the change of SRNS through functions in the RNC and the CN.

Configuration of radio network: This function is used in the configuration of the radio cells and the general transport channels (*Broadcast Channel* (BCH), *Random Access Channel* (RACH), *Forward Access Channel* (FACH) and *Paging Channel* (PCH) as well as in the activation and deactivation of radio resources in the cells.

Radio channel measurements: The following parameters are measured for the purposes of evaluating radio channel quality:

- The receive signal level of serving and neighbouring cells.
- The measured or estimated bit error rates on connections to serving and neighbouring cells.
- Propagation conditions, categorized into different types (for example, for fast and slow fading).
- The estimated distance to the serving BS.
- Doppler shift.
- Current level of synchronization.
- Received interference power.
- The total power received by each UE on the downlink.

Because CDMA transmission technology is used at the radio interface, interference signals can be differentiated from wanted signals (which is not the case with GSM, for example). In CDMA systems the different physical channels are separated by codes and each code corresponds to a certain power. The measurement function is embedded both in the UE and in the UTRAN.

Macro diversity: Macro diversity is a function that allows data streams to be duplicated and sent simultaneously over various physical channels in different cells to a terminal. Vice versa, macro diversity also enables the data stream sent by a mobile station to more than one base station to be received and merged again. The data stream can be retrieved both in the *Serving Radio Network Controller* (SRNC) and the *Drift Radio Network Controller* (DRNC), or even in node B. The function of macro diversity is only used in FDD mode and is found in UTRAN.

Radio bearer control: This function is used in the reconfiguration of radio bearer services and in the provision or release of radio bearer services for call set up and termination as well as for handover.

Radio resource management: The allocation and release of radio resources are functions of the RNC. This function is required, for example, when resources for macro diversity or for improvement to the quality of the bearer services are needed.

In TDD mode the switching point between the uplink and the downlink can be varied to enable the transport of asymmetrical traffic loads. This can produce interference between all base and mobile stations in a system, which is only preventable through dynamic channel allocation. Interference in a system can also be controlled through so-called fast dynamic channel allocation that controls time slot capacity and therefore the number of simultaneously activated codes. Therefore, this function is also related to the function for access control. *Dynamic Channel Allocation* (DCA) is not mandatory. TDD networks with *Fixed Channel Allocation* (FCA) will have to use the same switching point in all cells.

Data transmission over the radio interface: An UTRAN network transmits user and control data over the radio interface. This requires the mapping of the bearer services onto the radio interface. This functionality includes:

- Segmentation and reassembly of messages
- Acknowledged and unacknowledged transmission

Power control: A control of transmitter power reduces interference and maintains the prescribed connection quality. There are two types of power control in an UTRAN: one with two interleaved control loops and one with an open control loop. Open control loops are used on both the uplink and the downlink to determine transmitter power (for example, for random access). BSs are mobile controlled, whereas the mobile stations either are base station controlled or use system parameters sent by the base stations to determine the appropriate transmission power. Power control with interleaved control loops requires a signalling connection between the UE and the UTRAN.

Power control in FDD mode: Power control on the uplink consists of an outer and an inner control loop. The outer closed control loop is used for the long-term monitoring of connection quality and on the basis of the measurements on the transport channels determines the desired value of the inner control loop in terms of signal quality, i.e. C/I ratio. This value in turn controls the transmitter power on the dedicated physical channels. In FDD mode the inner control loop is closed. Measurements on the *Dedicated Physical Control Channel* (DPCCH) on the uplink are used as the actual values (see Section 6.9.3.1). Power control commands are transmitted over the DPCCH of the downlink. The outer control function on the uplink is embedded in the SRNC of the UTRAN; the inner control in node B and the UE.

An outer control loop also determines the desired value of the inner control loop on the downlink. This desired value is established based on measurements of UE on the transport channels. The SRNC uses this measurement data to determine the control range for the outer control loop. Measurements on the DPCCH of the downlink are used as the actual values for the inner control. Power control commands are transmitted over the DPCCH of the uplink.

Power control in TDD mode: In principle, the same power control function exists in TDD mode as in FDD mode. However, the inner control loop of the uplink is not closed in TDD mode. Transmission quality is calculated on the basis of measurements on the downlink in the UE. The SRNC notifies the UE of the desired value in terms of C/I ratio.

Channel coding: Systematic redundancy is added to data streams for the protection of data transmission. The type and the rate of coding can vary for the different logical channels and the different bearer services.

Random access: This function is used to detect and handle initial instances of random access by UEs. Because a Slotted ALOHA protocol is used for random access, this function is also responsible for the resolution of collisions.

6.8 The Radio Interface at Reference Point U_u

The protocol stack at reference point U_u is divided into the physical layer, data link layer and the network layer. The data link layer breaks down into the *Medium Access Control* (MAC), *Radio Link Control* (RLC), *Packet Data*

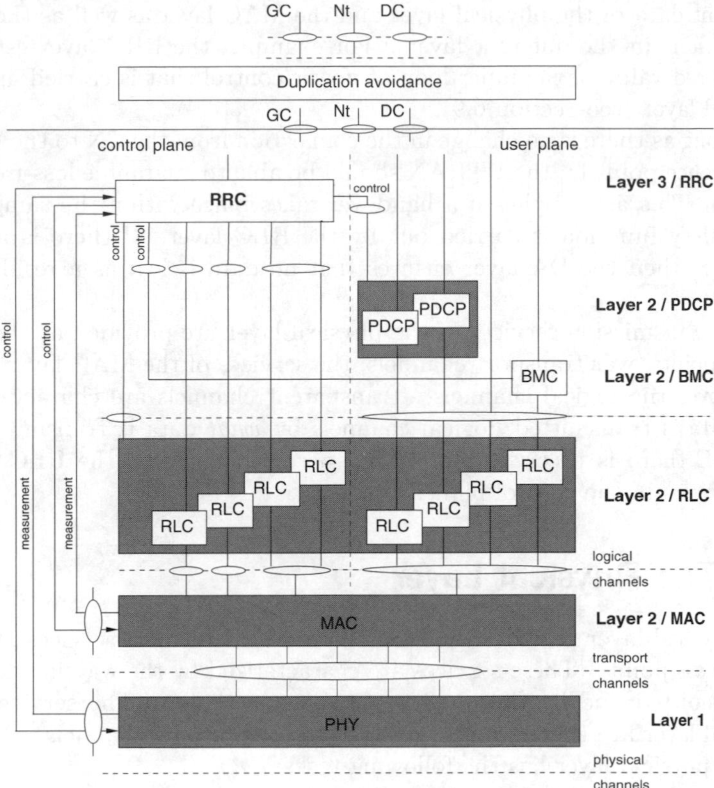

Figure 6.11: The protocol stack at reference point U_u

Convergence Protocol (PDCP) and *Broadcast/Multicast Control* (BMC) sublayers (see Figure 6.11). The network layer consists of the two parts *Radio Resource Control* (RRC) and *Duplication Avoidance* (DA), with only the RRC layer also ending in the UTRAN. The corresponding entity of DA is not part of the UTRAN and instead is located in the *Core Network*. In layer 3 and in the RLC layer a distinction is made between the user plane and the control plane, with the *Packet Data Convergence Protocol* (PDCP) and the *Broadcast/Multicast Control* (BMC) layers solely being part of the user plane. Ellipses between the layers symbolize the locations of the service access points for communication with the respective partner entity.

The RRC layer manages and controls the use of radio resources and therefore has links over control service access points to all other layers. This enables the RRC to control the configuration of the layers. Consequently, these control service access points are not used in the communication between partner entities but solely between layers of the same protocol stack and the RRC. The connections between RRC and lower layers enable the receipt of measurement data of the physical layer and the MAC layer as well as the control of functions in the different layers. For example, the RRC layer establishes the desired value of the inner loop of power control that is carried out in the physical layer (see Section 6.9).

As long as there is no change in the connection from the CN to the UTRAN at reference point I_u, the UTRAN should be able to guarantee loss-free transmission. This also applies if a handover takes place within the same SRNS. This safety function is carried out by the RLC layer. If there is a change of SRNS, then the DA layer ensures that no data is lost as a result of this process.

The transmission services of the physical layer are provided at the service access points over transport channels; the services of the MAC layer are provided over the logical channels. Transparent channels are characterized by *how* data is transmitted, logical channels by *what* data is transmitted. For each UE there is precisely one service access point from the RLC and the MAC layers in the control plane.

6.9 The Physical Layer

The physical layer provides the MAC layer with transport services at its service access points. These services are characterized by the way in which data is transmitted and by the quality of the transmission. The services of the physical layer are therefore also referred to as transport channels.

The physical layer has the following tasks:

- Error protection and detection for the transport channels using *Forward Error Correction* (FEC).
- Measurement of transmission characteristics of the radio channel and provision of the measured parameters, such as BER, transmitter power and C/I ratio, to the RRC layer.
- Duplication and merging of data streams in the case of macro diversity and soft handover.
- Weighting and merging of different physical channels.
- Multiplexing of transport channels and demultiplexing of the *Coded Composite Transport Channel* (CCTrCH).
- Matching of transmission rate.
- Mapping of the CCTrCH onto physical channels.
- Spreading and modulation of physical channels.
- RF-Signal processing.

- Frequency, chip, bit, time slot and frame synchronization.
- Power control.

The physical layer maps the transport channels onto physical channels. Each transport channel is allocated a transport format or a set of transport formats that establish the type of mapping. In particular, a transport format stipulates the channel coding, the interleaving and the bit rate (see Section 6.9.2.1).

6.9.1 Multiple Access

Information transmission over the radio interface in a UMTS is based on a combination of the multiple access techniques TDMA, *Frequency Division Multiple Access* (FDMA) and CDMA. What is characteristic of the CDMA technique is that a narrowband radio signal is transmitted in a broad frequency spectrum with the narrowband signal mapped to a broadband signal through a unique user code. This is referred to as code spreading (see Section 2.6.3).

Each user in a radio communications system is always assigned a unique user code. This unique code is used to spread the signal spectrum being transmitted into multiples of the original bandwidth. The senders then transmit these signals simultaneously and in the same frequency band.

The receiver has to know the user code of the sender and searches for the broadband signal according to the chip pattern of the sender's code sequence. Using an auto correlation function the receiver is able to synchronize with the code channel of the transmitter (called acquisition) and contract the signal back to its original bandwidth. The signals of the other transmitters the codes of which are uncorrelated with the selected code sequence are not transformed back to the original bandwidth and consequently only contribute to the noise level of the received signal.

With a certain number of code channels on the same frequency channel the *Signal to Noise Ratio* (SNR) can fall below the value required for reception by the correlator. The CDM method thereby also restricts the number of users that can use the same channel.

In the receiver the usage signal must be regenerated from the noise-like receive signal through correlation with the code sequence of the transmitter. According to [27], the optimal maximum-likelihood receiver for a CDMA system with synchronous transmission in K code channels consists of a number of K-matched filters, followed by a correlator that calculates the 2^K information sequences possible. The information sequence most likely to be transmitted is then selected. In the case of asynchronous transmission, the correlation requirement increases to include the calculation of $2^{N \cdot K}$ correlation metrics. N then stands for the number of information bits per code sequence. In practice, this kind of receiver is too complex. Therefore, less complicated receiving methods are used although they are less than optimal in terms of error probability and system capacity.

Single detection and joint detection are two basic methods for providing simpler types of receivers. In the case of single detection the receive signal is channelled to a single decoder where non-relevant $K - 1$ code sequences and additive noise act as a disturbance variables.

Joint detection is orientated towards the optimal receiver; however, the correlation and the decision are simplified to the extent that the complexity is no longer exponential in K but rises linearly according to the number of code channels [8].

Interference cancellation techniques offer a halfway option between the two methods of single and joint detection. With this technique the overall signal received is used in decoding a channel. The transmit signal is then reconstructed through spreading from the estimated bit sequence and copied from a similarly delayed version of the overall signal. The result is a signal with reduced interference that is used to estimate the bit sequence for the next channel. This technique is easier to implement than joint detection and more efficient than single detection [17].

Some of the remaining interference is due to the number of code channels that are occupied simultaneously. Therefore, depending on the system, the quality of service can drop with the number of active users increasing.

Either TDD or FDD can be used as the duplex method on the radio interface. Since the duplex method has a considerable influence on the multiple access technique as well as on the parameters and functions of the physical layer, a distinction is made between TDD mode and FDD mode. FDD and TDD mode were formerly called W-CDMA and *Time Division—Code Division Multiple Access* (TD-CDMA), respectively.

6.9.1.1 FDD Mode

In FDD mode multiple access is based on a combination of CDMA and FDMA. Individual user signals in the same transmission direction are kept distinguishable from one another through different spreading codes or different transmission frequencies (see Figure 6.12). The frequency spacing between two FDMA channels is a nominal 5 MHz. Within the frequency spectrum of an operator this spacing can be reduced in 200 kHz steps to 4.4 MHz in order to increase the frequency spacing to neighbouring bands.

For most physical channels an FDMA channel is divided time-wise into frames with the length of 10 ms, each frame consisting of 15 time slots. This organization is not designed to separate user signals but to allow the implementation of periodic functions, e.g., power control. Variable transmission rates can be achieved either through a change of the spreading factor or through multicode transmission.

6.9.1.2 TDD Mode

In TDD multiple access is carried out through a combination of CDMA, FDMA and TDMA (see Figure 6.13). A TDMA frame is 10 ms in length

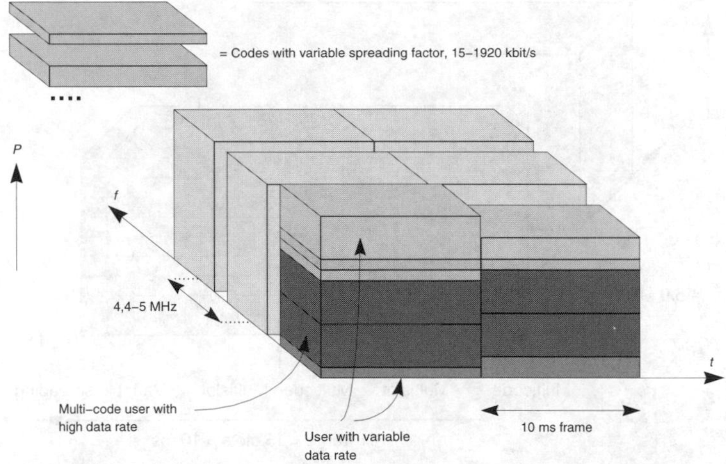

Figure 6.12: Multiple access in FDD mode

and divided into 15 time slots each of which can have up to 16 CDMA channels. The individual time slots can be allocated independently of one another either to the uplink or to the downlink. This produces an optimal distribution of radio resources when traffic in a system is heavily asymmetric. At least one slot has to be assigned for the uplink and downlink each. Different cells using the same frequency cannot use different timeslot UL/DL allocations as long as no DCA is applied.

As is the case in FDD mode, the distance between two FDMA channels is between 4.4 MHz und 5 MHz. In TDD variable transmission rates are possible either through multicode or through multislot transmission or both. However, only two code channels can be used simultaneously on the uplink by a UE.

The TDD mode in many aspects is comparable to the organization of the DECT system (see Section 10.4.1.2) that does not have any code spreading of TDMA channels but otherwise is very similar in design. DECT uses TDD duplexing with a variable switching point in the 10 ms frame and multislot assignment to a connection, both being part of the standard for data transmission. A consequence of this is that the *Dynamic Channel Selection* (DCS) function in DECT (see Section 10.5) similarly has to be performed in the TDD mode where it is part of the RLC function (see Section 6.10.2).

Table 6.12 provides a summary of the key parameters of the UTRAN radio interface. The parameters will be explained in detail in the sections that follow.

6.9.2 Transport Channels

Transport channels are the services provided by the physical layer to the MAC layer. They are devided into the dedicated and the common channels. Ded-

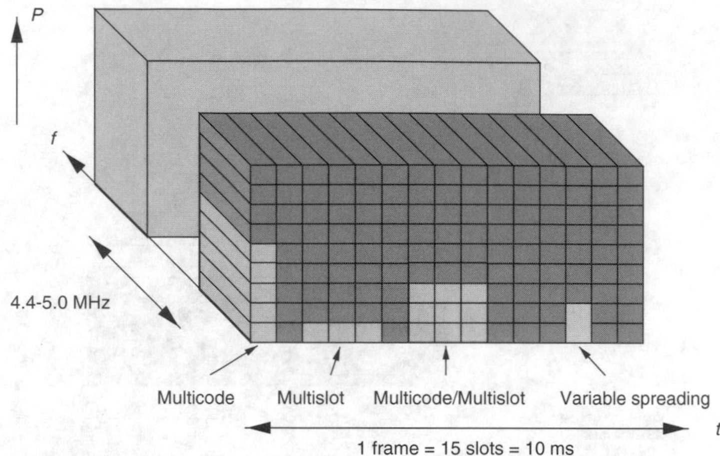

Figure 6.13: Multiple access in TDD mode

Table 6.12: Characteristic parameters of the UTRAN radio interface

Parameter	FDD mode	TDD mode
Multiple access methods	CDMA/FDMA	CDMA/FDMA/TDMA
Duplex methods	FDD	TDD
FDMA channel spacing (MHz)	4.4–5	4.4–5
TDMA frame length (ms)	10	10
Slots/frames UL	15	1–14 (of 15)
Slots/frames DL	15	1–14 (of 15)
Transmission rate (MChips/s, gross)	3.84	3.84
Modulation	QPSK	QPSK
Spreading factor UL	4–256	1–16
Spreading factor DL	4–512	1 or 16

icated transport channels carry data of one UE, common transport channels carry data used by different or more than one UEs.

There is one dedicated transport channel:

Dedicated Channel (DCH) A DCH is a bidirectional channel that is available exclusively to a specific UE. The transmission rate of a DCH can be changed every 10 ms. A DCH uses fast power control and can be available either in an entire cell or, through the use of directed antennas, only in parts of a cell.

The following common transport channels are also available:

Random Access Channel (RACH) A RACH is used to transfer relatively small amounts of data on the uplink and is always available in an entire cell. A RACH is not a collision-free channel because all UE compete for transmission capacity. It is used for random access and for transmitting non-time-critical control or usage data.

Common Packet Channel (CPCH) A CPCH is a transport channel on the uplink in FDD mode. It is used to transmit packet data. Similar to the RACH, several UEs of a cell compete for the transmission capacity of a CPCH. The channel can be available either in an entire cell or, if directional antennas are used, only in parts of a cell. A dedicated channel on the downlink is always allocated to a CPCH to support fast power control on the CPCH.

Forward Access Channel (FACH) A FACH is a shared transport channel on the uplink used to transmit relatively small quantities of data. Slow power control with an open loop is applied to a FACH. The transmission rate of a FACH can be changed at 10 ms intervals. Directional antennas can be used with a FACH.

Downlink Shared Channel (DSCH) A DSCH is a transport channel on the downlink. Its transmission capacity is divided up among several UE. A DCH or a *DSCH Control Channel* (DSCCH) is always allocated to the DSCH.

DSCH Control Channel (DSCCH) A DSCCH is a channel on the downlink and is allocated to a DSCH. It is used for signalling the reservation of resources on the DSCH.

Uplink Shared Channel (USCH) A USCH is exclusively available in TDD mode. It is an uplink channel and is used by more than one UE.

Broadcast Channel (BCH) A *Broadcast Channel* (BCH) is used to broadcast system information in a cell. A BCH therefore exists on the downlink and has a fixed transmission rate.

Synchronization Channel (SCH) The SCH transmits information for synchronization on the downlink. The SCH exists only as a transport channel in TDD mode. In FDD mode each physical channel provides its own synchronization. Like a BCH, a SCH is transmitted in an entire cell and has a low, fixed transmission rate.

Paging Channel (PCH) A PCH is a transport channel on the downlink; it is used to broadcast messages to certain UE. Paging and notification functions are implemented over the PCH. A PCH is transmitted in an entire cell.

6.9.2.1 Data Transmission over Transport Channels

One or more transport blocks are periodically transmitted over the transport channels of the physical layer. A set of these transport blocks is called a

Transport Block Set (TBS). Each transport block contains one or more protocol data units of the MAC layer.

What characterizes a TBS is that all transport blocks of a set are the same size and that the same error protection is applied to all transport blocks for transmission over the same transport channel. Each TBS is described by a so-called *Transport Format* (TF). The number, the size and the period of the transport blocks in a set and the error protection on the transport channel are all specified in the TF. The TF is described by a semi-static and a dynamic part. The semi-static part specifies error protection and transmission delay. It is specified by the following parameters:

- The type of error protection:
  - Coding scheme: the choice is between turbo codes, convolutional codes or no coding
  - Coding rate
  - Parameters for static adaptation of the transmission rate through puncturing
  - For the uplink a limit for the puncture rate
- Length of checksum
- *Transmission Time Interval* (TTI). This describes the period in which transport blocks or a TBS is to be transmitted over the transport channel in FDD mode. In TDD mode this period is already determined through the time slot and frame structure. Therefore, the TTI parameter does not apply in the semi-static part of the transport format in TDD mode but can be determined in the dynamic part.

The dynamic part of the transport format contains details about the following:

- Size of the transport block in bits
- Size of the TBS in bits. Since all transport blocks of a TBS are the same size, the size of a set is an integer multiple of the size of a transport block.
- In TDD mode the TTI can be indicated as an alternative for non-realtime services.

Since the transmission capacity of the physical layer is limited, only a finite number of possible sizes of TBS is available for each assignment of error protection. The transport formats, which all have the same semi-static part and are valid for the same transport channel, form a *Transport Format Set* (TFS).

Within the physical layer several transport channels that use the same coding, interleaving and puncturing can be multiplexed to a CCTrCH. This

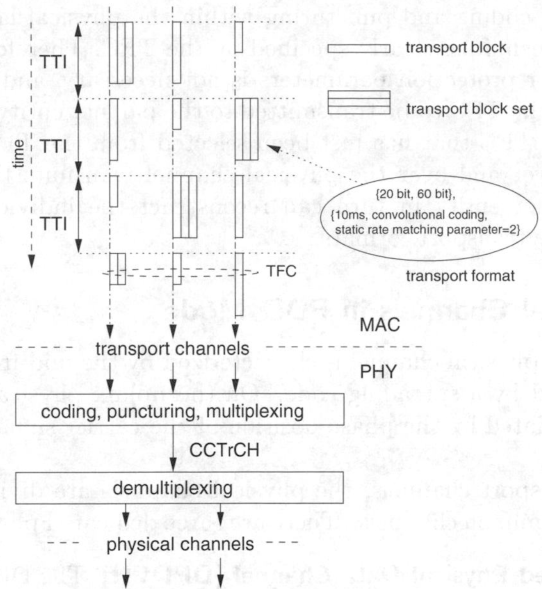

Figure 6.14: Data transmission over transport channels

CCTrCH is an internal channel of the physical layer and is mapped onto one or more physical channels. Since the transmission capacity of CCTrCHs is also limited, it is not possible for just any transport formats to be used on the different transport channels at the same time. The only combinations of transport formats of different transport channels that are valid are those that allow multiplexing to a CCTrCH. This combination is called *Transport Format Combination* (TFC). The set of all possible combinations is called *Transport Format Combination Set* (TFCS). Figure 6.14 shows how data is transmitted over transport channels and the physical layer.

Within a TFS individual transport formats can be referenced over an index called a TFI. The same also applies to a TFCS. The index, which clearly identifies a TFCS, is called a *Transport Format Combination Identifier* (TFCI).

The TFCS is prescribed to the MAC layer by the RRC layer. The MAC layer can freely select a TFC from the TFCS. Since all the transport formats of a TFCS have the same semi-static part, the MAC layer only has control over the dynamic part, i.e., over how much transmission capacity of a transport channel will be utilized. The MAC layer informs the physical layer of the selection of a transport format by transmitting the appropriate TFI.

The physical layer must ensure that the prescribed QoS is provided for all transport channels. The RRC layer determines the configuration of the physical layer, thereby also ensuring compliance with the requirements of the semi-static parts of the transport formats.

Multiplexing, coding and puncturing within the physical layer take place according to a technique clearly specified by the TFC. Therefore, an explicit exchange of error protection parameters is not necessary, and it is sufficient for the index of the TFC to be transmitted to the partner entity. The physical layer selects the TFC that has just been selected from the TFIs transmitted by the MAC layer and over the physical channel transmits the TFCI from which the partner entity in turn can reconstruct the individual TFIs and consequently the transport formats.

6.9.3 Physical Channels in FDD Mode

In FDD mode a physical channel is characterized by the mid-frequency of the radio carrier and by a spreading code. On the uplink physical channels are further differentiated by the phase position of the carrier signal (0 or π) (see Section 6.9.4.3).

Like the transport channels, the physical channels are divided into dedicated and into common channels. There are three dedicated physical channels:

Uplink Dedicated Physical Data Channel (DPDCH) The DPDCH only exists for the uplink and is used to transmit user data from the data link layer or the layers above it. A layer 1 connection consists of several DPCCHs or none at all.

Dedicated Physical Control Channel (DPCCH) A DPCCH is a physical channel that controls data transmission between partner entities of the physical layer for the uplink. It is only used for the transmission of information from the physical layer. Exactly one DPCCH exists for each layer 1 connection.

Dedicated Physical Channel (DPCH) The DPCH is only available for the downlink and carries out the tasks handled by the DPDCH and the DPCCH for the uplink. This involves multiplexing the information of the DPDCH and the DPCCH to the DPCH.

The following physical channels are also specified:

Synchronization Channel (SCH) The SCH is a channel for the downlink and is used in cell search and in the synchronization of mobile stations. It is divided into two subchannels *Primary Synchronization Channel* (P-SCH) and *Secondary Synchronization Channel* (S-SCH).

Common Control Physical Channel (CCPCH) Distribution services on the uplink are implemented over the CCPCH. A CCPCH is divided into the two subchannels *Primary Common Control Physical Channel* (P-CCPCH) and *Secondary Common Control Physical Channel* (S-CCPCH). The P-CCPCH is used to transmit information of the BCH; the S-CCPCH transmits information of the FACH and the PCH.

Common Pilot Channel (CPICH) The CPICH supports macro diversity on the downlink. It is used to transmit the same predefined code sequence in different cells. The CPICH is divided into a *Primary Common Pilot Channel* (P-CPICH) and a *Secondary Common Pilot Channel* (S-CPICH) that are differentiated by their physical features, such as spreading code and availability in a cell.

Physical Random Access Channel (PRACH) The PRACH is a channel that supports the RACH and is used for random access and for the transmission of small quantities of data.

Physical Common Packet Channel (PCPCH) The PCPCH transmits the packet data of the CPCH using a CSMA/CD technique.

Paging Indication Channel (PICH) The *Page Indicator* (PI) for the implementation of paging on the downlink is transmitted over a PICH. A PICH is always allocated to an S-CCPCH over which the PCH is transmitted.

Acquisition Indication Channel (AICH) The AICH is a physical channel of the downlink. It is used to signal the success of random access on a PRACH or a PCPCH.

AP-AICH The *Access Preamble—Acquisition Indication Channel* (AP-AICH) carries the access preamble acquisition indicators of the CPCH.

Physical Downlink Shared Channel (PDSCH): The PDSCH transmits data over the DSCH on the downlink. A DPCH is always allocated to a PDSCH. Several UEs share this channel. Its data is multiplexed to different codes.

CD/CA-ICH The *Collision Detection/Channel Assignment Indicator Channel* (CD/CA-ICH) carries the collision detection information for the CPCH. Additionally, a channel assignment can be transmitted via this channel.

CPCH Status Indicator Channel (CSICH) The CSICH carries CPCH status information.

The physical channels that are directly used for data transmission are mapped onto a uniform frame structure. Figure 6.15 illustrates the burst structure for the different physical channels. Each 15 bursts are combined to form one 10 ms long frame; 72 frames produce one *superframe* of 720 ms duration.

The SCH, AICH, AP-AICH, PICH, CPCH, CD/CA-ICH, CSICH and PRACH channels use their own slot and frame structures.

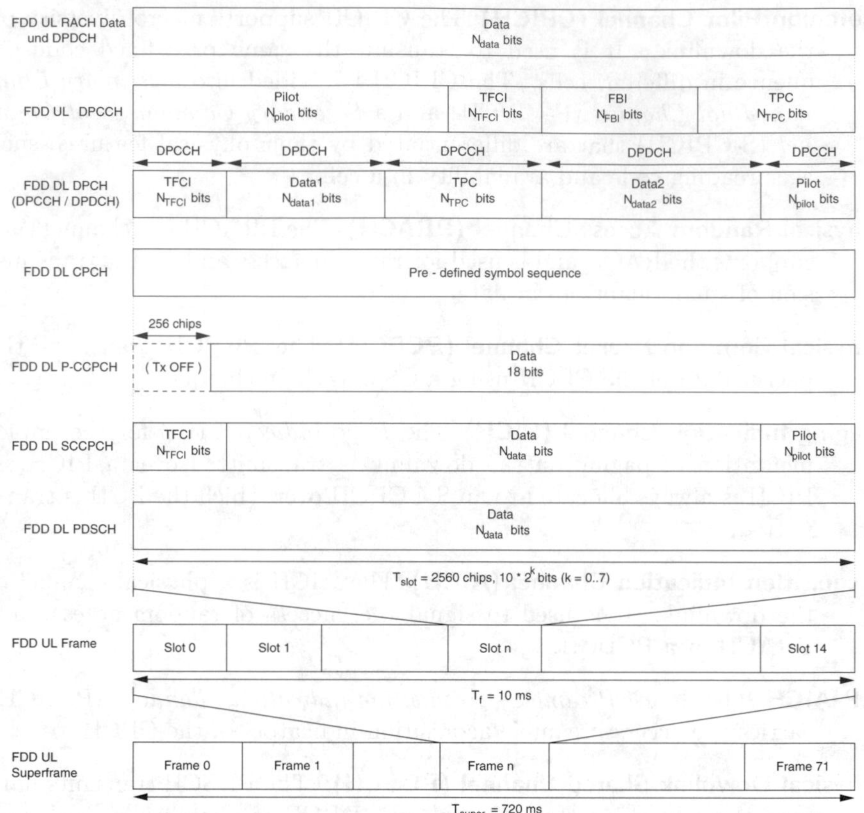

Figure 6.15: Different burst types in FDD mode

6.9.3.1 Dedicated Physical Channels

Each layer 1 connection on the uplink has exactly one DPCCH and a number of DPDCHs. Several parallel DPDCHs that are part of the same connection and consequently are allocated to the same DPCCH can be used for multicode transmission. The DPDCH is transmitted over the inphase or quadrature branch of the modulator, the DPCCH always only on the quadrature branch. Each burst consists of 2560 chips. The number of bits that are transmitted in a DPDCH burst depends on the respective spreading factor and is between 4 and 256. The spreading factor can be calculated as follows using the parameter k:

$$F_S = \frac{256}{2^k}$$

Table 6.13 presents an overview of the valid spreading factors and the resulting transmission rates for a DPDCH burst.

Table 6.13: Different configurations of a DPDCH burst

Burst format (k)	Bit rate (kbit/s)	Spreading factor	Bits/frame	Bits/slot	N_{data}
0	15	256	150	10	10
1	30	128	300	20	20
2	60	64	600	40	40
3	120	32	1200	80	80
4	240	16	2400	160	160
5	480	8	4800	320	320
6	960	4	9600	640	640

Table 6.14: Different configurations of a DPCCH burst

Burst format	N_{Pilot}	N_{TFCI}	N_{FBI}	N_{TPC}
0	6	2	0	2
1	8	0	0	2
2	5	2	1	2
3	7	0	1	2
4	6	0	2	2
5	5	2	2	1

A DPCCH is a physical channel with a fixed spreading factor of 256 chips per bit, i.e., with a constant transmission rate of 15 kbit/s. A DPCCH burst consists of four fields with variable sizes (see Table 6.14). Due to the constant spreading factor, the DPCCH has a constant transmission capacity of 10 bits per burst or 150 bits per frame.

The receiver uses the pilot bits for estimation of the channel impulse response. The TFCI field is optional and, if available, informs the receiver of the current transport formats of the transport channels that are multiplexed to a DPDCH. A TFCI is valid for the duration of an entire frame. An FBI is required for functions that need feedback from UE to the UTRAN on the layer 1 plane. These particularly include the functions for transmitter macro-diversity and for site selection diversity. The last field of a DPCCH burst is used for *Transmitter Power Control* (TPC) bits. Irrespective of whether one or two bits are available for this field, information about an increase or a reduction in transmitter power is the only information transmitted in this field. DPCCH and DPDCH channels generally have different spreading factors.

A dedicated physical channel, the DPCH, exists only on the downlink. Since, in contrast to the uplink, the individual physical channels are separated by the carrier phase, layer 1 and layer 2 information in time-multiplex is transmitted over the DPCH (see Figure 6.15). A DPCH burst consists of 2560 complex-valued chips. The number of bits per burst is determined by the spreading factor F_S.

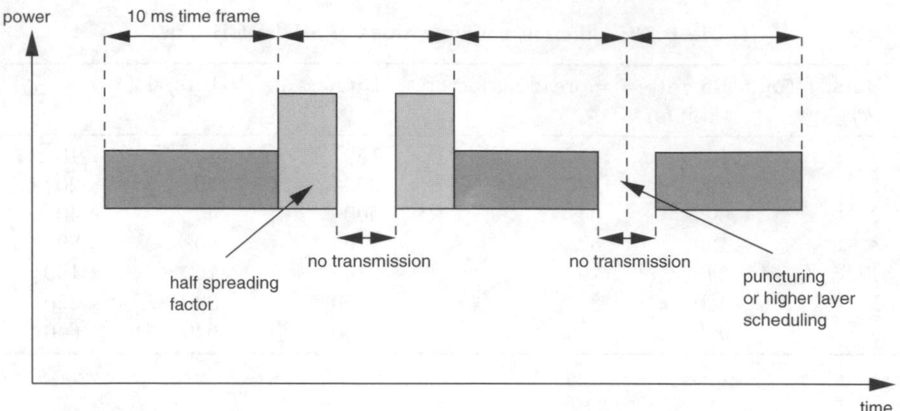

Figure 6.16: UTRA FDD compressed mode transmission

The two data fields of a DPCH burst carry information from higher layers. Table 6.15 lists the 17 different configurations of a DPCH burst. Since two bits are transmitted per symbol on the downlink, the symbol rate always corresponds to half the bit rate. DTX information can be transmitted instead of the TFCI in bursts 12 to 16.

The type of DPCH burst to be used is negotiated at the time of connection set up and can be renegotiated during the connection. As is the case on the uplink, the TFCI field is optional. The TFCI describes the parameters, in particular the transmission rates of the DCHs that are transmitted over the DPCH. If the transmission rate of an individual DPCH is not sufficient for carrying a CCTrCH, several DPCHs can be operated simultaneously. With this multicode transmission, layer 1 information is only transmitted on one of the different parallel DPCHs. The corresponding fields in the other bursts remain empty and are not transmitted, i.e., the transmitter power of the respective DPCHs during this time is zero.

6.9.3.2 Compressed Mode

To realize a handover to a different frequency channel or to a different system like GSM the mobile station has to be able to receive and decode signals at different frequency carriers. Without an additional receiver in the mobile station, the downlink transmission has to be interrupted to allow the mobile station the measurement of other base station's pilot signals. To avoid a loss of information during the interruption period the downlink transmission pause is realized using the so-called *compressed mode*.

As depicted in Figure 6.16 there are two possibilities of clearing a space within the downlink transmission without information loss. Either the spreading factor can be halved before and after the pause or the transmission rate is adjusted by puncturing and higher layer scheduling. The transmission gap

Table 6.15: Different configurations of a DPCH burst

	Bit rate	F_S	Bits/frame			Bits/	DPDCH		DPCCH		
	[kbit/s]		DPDCH	DPCCH	total	Slot	N_{Data1}	N_{Data2}	N_{TFCI}	N_{TPC}	N_{Pilot}
0	15	512	60	90	150	10	0	4	2	0	4
1	15	512	30	120	150	10	0	2	2	2	4
2	30	256	240	60	300	20	2	14	2	0	2
3	30	256	210	90	300	20	2	12	2	2	2
4	30	256	210	90	300	20	2	12	2	0	4
5	30	256	180	120	300	20	2	10	2	2	4
6	30	256	150	150	300	20	2	8	0	2	8
7	30	256	120	180	300	20	2	6	2	2	8
8	60	128	510	90	600	40	6	28	2	0	4
9	60	128	480	120	600	40	6	26	2	2	4
10	60	128	450	150	600	40	6	24	2	0	8
11	60	128	420	180	600	40	6	22	2	2	8
12	120	64	900	300	1200	80	12	48	4	8	8
13	240	32	2100	300	2400	160	28	112	4	8	8
14	480	16	4320	280	4800	320	56	232	8	8	16
15	960	8	9120	480	9600	640	120	488	8	8	16
16	1920	4	18720	480	19200	1280	248	1000	8	8	16

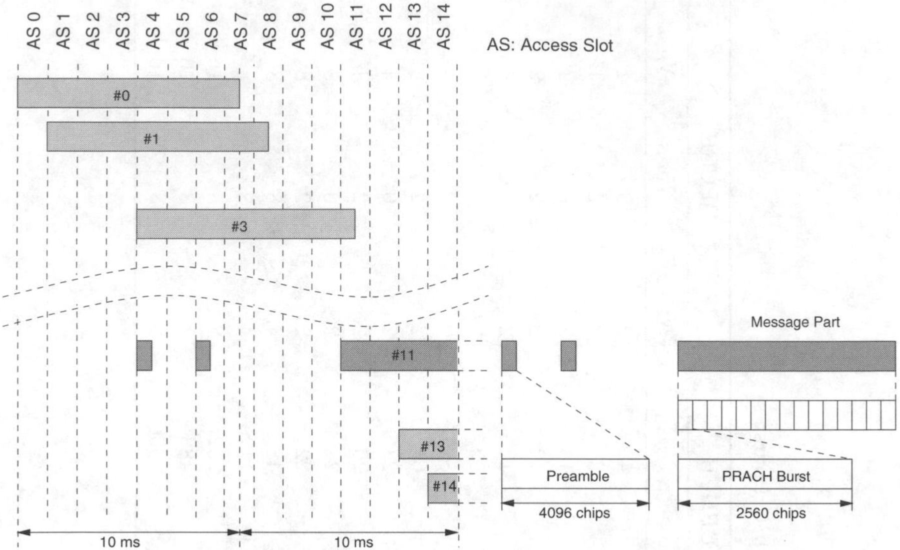

Figure 6.17: Frame structure of the random physical access channel

can have a maximum duration of 7 timeslots and can be located within one timeframe or symmetrically at the end of one frame and at the beginning of the following frame (see Figure 6.16).

6.9.3.3 The Physical Random Access Channel

Random access in UTRA-FDD is based on a slotted-ALOHA protocol with fast acknowledgement. Random access by a UE is possible at periodically recurring fixed times, called *Access Slot* (AS). A total of 15 such ASs exist during a 20 ms period, i.e., two time frames (see Figure 6.17). The time period between two successive access slots corresponds to the duration of 5120 chips. An *Access Channel* (ACS) is divided into 12 RACH subchannels. An access time is clearly established by the subchannel and the *System Frame Number* (SFN). Mobile stations are notified of which groups are available in a cell over the BCH. This provides the possibility of allocating different RACH subchannels to different classes of service so that competition with other classes of service is not necessary.

Random access is divided into a competition phase and a data transmission phase. During the competition phase each UE uses a chip sequence called a *preamble* to successfully compete against other UEs. Sixteen different preambles are available and are obtained through the 256-fold spreading of a so-called signature. This signature is a *Hadamard* code with the length 16. Therefore, up to 16 UEs can access each AS collision-free.

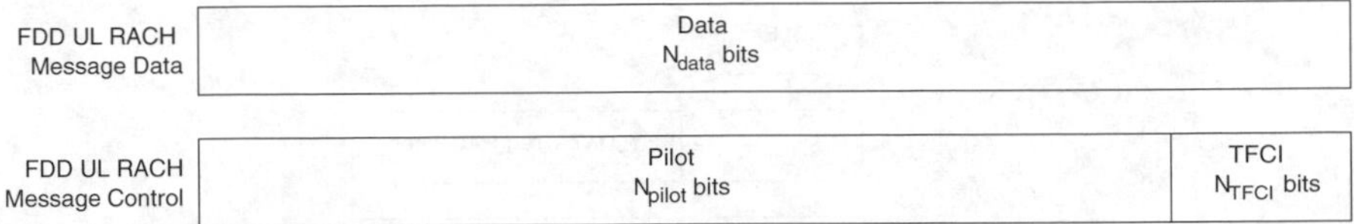

Figure 6.18: Data and signalling burst of a PRACH

Table 6.16: Fields of a PRACH burst

Burst format (k)	Bit rate [kbit/s]	Spreading factor	Bits per frame	Bits per slot	N_{Data}	N_{Pilot}	N_{TFCI}
Data 0	15	256	150	10	10	—	—
Data 1	30	128	300	20	20	—	—
Data 2	60	64	600	40	40	—	—
Data 3	120	32	1200	80	80	—	—
Control 0	15	256	150	10	—	8	2

After its success during the competition phase derived from a positive ACK received on the AICH, the UE transmits a 10 or 20 ms long random access message called a *message part.* This message consists of 15 or 30 PRACH bursts for each 2560 chips. Up to 15 messages can be transmitted simultaneously during one random access frame (see Figure 6.17). The message part is used to transit both layer 2 and layer 1 information. The same as with the DPDCH and the DPCCH, modulation separates this information into different carrier phases. Figure 6.18 shows the structure of the data and control parts of a PRACH burst.

The number of bits transmitted in the data part is dependent on the spreading factor. Table 6.16 lists the parameters of a PRACH burst. The code spreading of the data part corresponds to that of the DPDCH. The pilot bits in the control part are for estimation of the channel impulse response and enable coherent reception of the message part. The transport format of a random access message is transmitted in the TFCI field of a PRACH burst.

A UE must receive the following information about the BCH before it can carry out random access:

- The spreading and scrambling codes available in the cell
- The signatures available for generating the preamble
- The spreading factors allowed for the message part
- The transmitter power of the CCPCH
- The time parameter of the AICH
- The difference in transmitter power for preamble and message parts

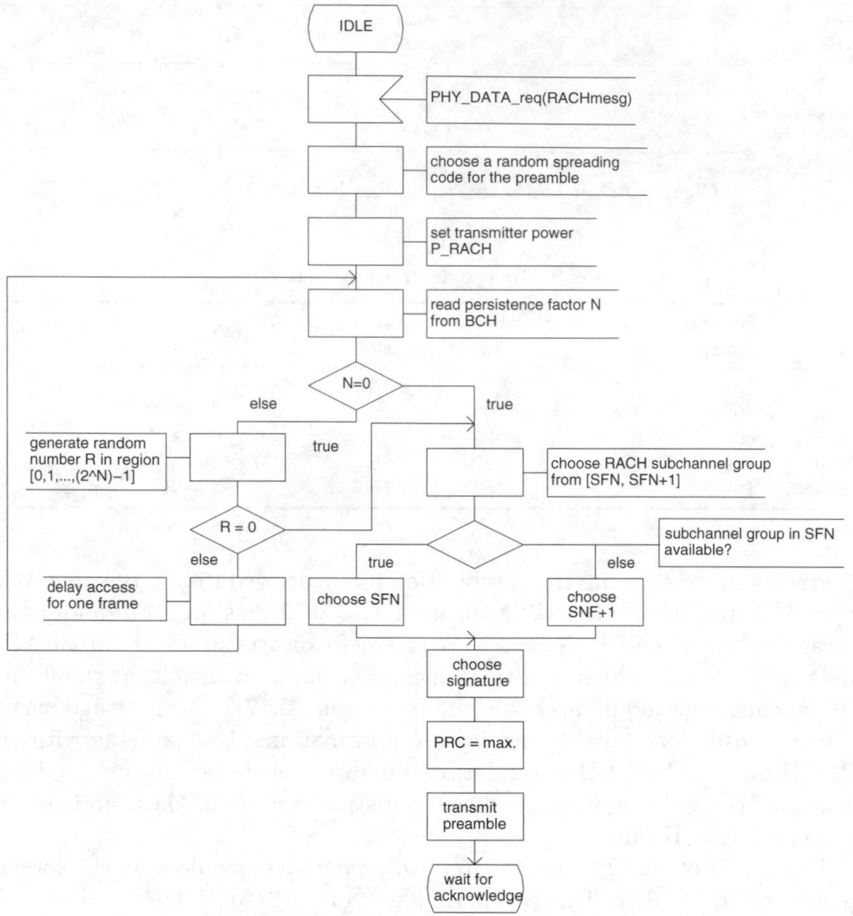

Figure 6.19: Protocol for random access (1)

- The control steps ΔP_0 and ΔP_1 for controlling the transmitter power when there is a negative acknowledgement or no acknowledgement after random access of an UE.

The protocol for transmission of a random access message is shown in Figures 6.19 and 6.20. The UE first randomly selects a valid spreading code for the preamble. A dynamic persistent algorithm is used to prevent the access channel from being blocked when there is an overload. The persistence factor N is stipulated by the UTRAN and made known over the BCH. If the UE concludes that it can access the channel, it randomly selects one of the subchannels corresponding to the *Access Service Class* (ASC). The subchannel identifies the available access slots in the current access frame. If no slots are available in the first part of the access frame, the UE selects

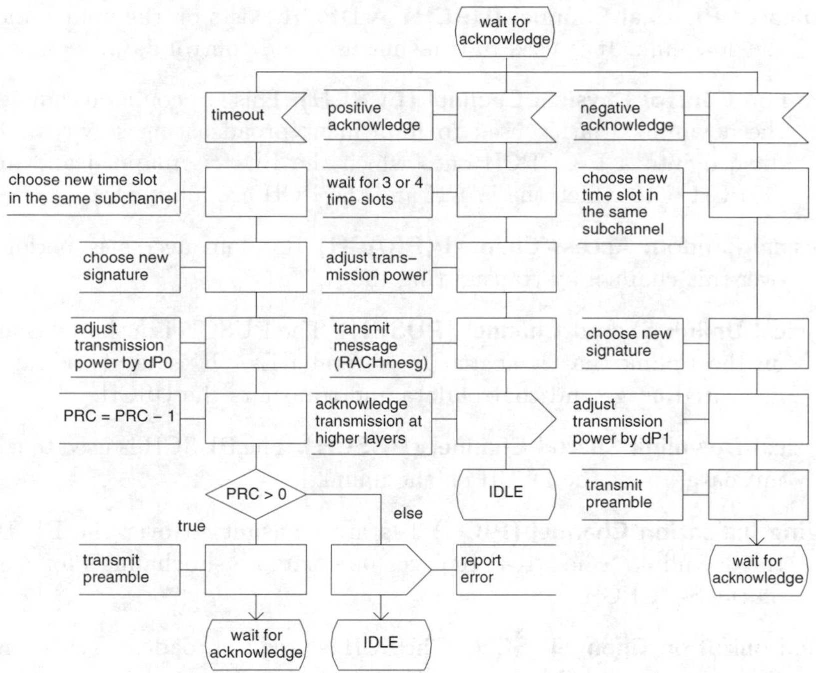

Figure 6.20: Protocol for random access (2)

an access slot from the second part. It then transmits the preamble and awaits an acknowledgement over the AICH. If the UE receives a negative acknowledgement or none at all, it selects a new signature and a new access slot and retransmits the preamble, albeit with the transmitter power changed by ΔP_0 or ΔP_1. ΔP_0 is generally positive, i.e., it increases the transmitter power. This mechanism implements a type of slow power control with closed loop that should ensure that a minimum of interference power is produced by the random access. If a certain number of access attempts in which no acknowledgement is received is exceeded, then this fact is signalled to the higher layer and no further attempts are undertaken. In the case of positive acknowledgements, random access messages are transmitted with a delay of 3 or 4 time slots and the higher layer is notified of the successful access.

6.9.4 Physical Channels in TDD Mode

In TDD mode a physical channel is determined by spreading code, time slot and frequency channel. A physical channel can occupy a time slot either on a regular basis in each time frame or only in a subset of all frames. There are six common physical channels in addition to one dedicated physical channel:

Dedicated Physical Channel (DPCH) A DPCH exists on the uplink and on the downlink. It is used to transmit user and control data.

Common Control Physical Channel (CCPCH) This is a common channel on the downlink that is used to implement broadcasting services. It is divided into a P-CCPCH onto which the BCH is mapped and an S-CCPCH onto which the FACH and the PCH are mapped.

Physical Random Access Channel (PRACH) Random access is performed over this channel by transmitting the RACH.

Physical Uplink Shared Channel (PUSCH) The PUSCH is a shared channel on the uplink that is shared by several UEs. It is used to transmit dedicated usage and control data and transmits the USCH.

Physical Downlink Shared Channel (PDSCH) The PDSCH is used to transmit data about the DSCH on the uplink.

Paging Indication Channel (PICH) PIs are transmitted over the PICH on the downlink. The PICH replaces one or more subchannels for paging on the S-CCPCH.

Synchronization Channel (SCH) The SCH is used to broadcast synchronization information within one cell.

Figure 6.21 presents the different types of bursts that exist for the DPCH. Each burst consists of two data fields between which is found a data sequence for channel estimation, called a *midamble.* A burst is 96 chips or 192 chips shorter than a time slot to enable the compensation of propagation delays of different user signals. There are principally three different types of bursts, differing from one another in the length of the individual fields and labelled as type 1, type 2 or type 3. At 512 chips the *midamble* of a type 1 burst is double the length of that of a type 2 burst. This enables channel calculation of up to 16 different user signals. Therefore, the type 1 burst is the preferred burst on the uplink but this does not reduce its usability on the downlink.

Because of its shorter midamble, the type 2 burst can only be used on the uplink if not more than four users are transmitting simultaneously in the time slot being considered. The burst type 3 has the same number of training sequences as the burst type 1 but it provides a longer guard period. This burst type can be used in situations where the synchronization between base and mobile station is insufficient for shorter guard periods. This can be the case for the initial access or the access to a new cell after handover. Therefore the burst type 3 is only used in the uplink. Table 6.17 lists the lengths of the individual fields of the burst types 1 and 2 based on the spreading factors.

Only a spreading factor 16 or 1 can be used on the downlink. In the latter case, this means that spreading does not occur. Therefore, on the downlink variable transmission rates can only be implemented within a time slot if multicode transmission takes place. A UE can transmit a maximum of two

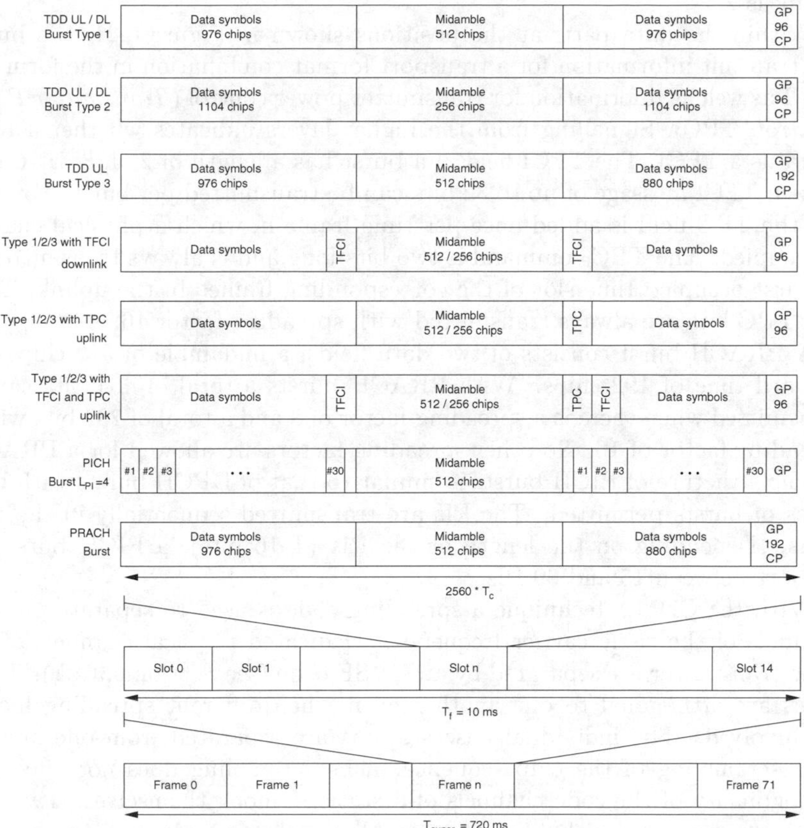

Figure 6.21: Different types of bursts in TDD mode

Table 6.17: Characteristics of DPCH bursts type 1 and 2

	Bits per data field		Bit rate (1 slot)		Bit rate (1 slot/frame)	
Spreading factor	Type 1 [bit]	Type 2 [bit]	Type 1 [kbit/s]	Type 2 [kbit/s]	Type 1 [kbit/s]	Type 2 [kbit/s]
1	1952	2208	5856	6624	390.4	441.6
2	976	1104	2928	3312	195.2	220.8
4	488	552	1464	1656	97.6	110.4
8	244	276	732	828	48.8	55.2
16	122	138	366	414	24.4	27.6

bursts simultaneously on the uplink, the minimum spreading factor on the uplink is 2.

Within the data part, at the positions shown in Figure 6.21, the bursts can transmit information for a transport format combination in the form of a TFCI as well as information for transmitter power control (*Transmitter Power Control*, TPC). Signalling from the higher layers indicates whether a burst contains a TFCI. The TFCI field of a burst has a length of 2, 4, 8 or 16 bits. Thus a TFCI message of up to 32 bits can be transmitted per burst. For each UE the TPC field is added once per time frame in which a physical channel is occupied. The TPC command is two bits long and is always transmitted in the first occupied time slot of the corresponding frame. In the uplink, TFCI and TPC bits are always transmitted with spreading factor 16.

A PRACH burst consists of two data fields, a midamble of 512 chips and a guard time of 192 chips. With PRACH bursts a total of 464 bits can be transmitted when there is a spreading factor of 8 and a total of 232 bits with a spreading factor of 16. No other spreading factors are allowed for a PRACH.

The structure of PICH bursts is similar to that of DPCH bursts with both types of bursts permitted. The PIs are transmitted sequentially in the data fields. Depending on the length of the PIs (4–16 bits), a PICH burst can support between 15 and 69 PIs.

With the CDMA technique a spreading code is used to separate physical channels of the same carrier frequency. Dedicated physical channels of the same transmitter are separated by an OVSF code (see Section 6.9.4.1). These codes are orthogonal to one another even when different spreading factors are involved. The individual base stations are separated from one another by a scrambling of the chip sequence. The scrambling does not affect the orthogonality of the code channels of a station among themselves. Two chip sequences generated with different scrambling codes are only quasi-orthogonal. The scrambled chip sequence is transmitted using QPSK modulation. in TDD, all UEs in one cell use the same scrambling code. Therefore, in TDD the codes of the UEs of one cell are orthogonal if they are synchronized on a per-chip basis. Since this precision is not easily achievable, especially in macrocellular environments, the base station has to use joint detection to take into account the time shifts of the different uplink signals.

6.9.4.1 Orthogonal Codes with Variable Spreading Factors

Orthogonal codes with variable spreading factors (*Orthogonal Variable Spreading Factor*, OVSF) are used to separate dedicated physical channels that can have different transmission rates or spreading factors. Each bit of a channel with the lowest bit rate R_{min} is spread with a spreading code of the length $N = 2^n$. Consequently, the bit duration of a doubled bit rate of $2R_{min}$ is half as long as the bit duration of the channel with the lowest bit rate. The spreading code length is then $N/2 = 2^{n-1}$. As a generalization, a bit rate of

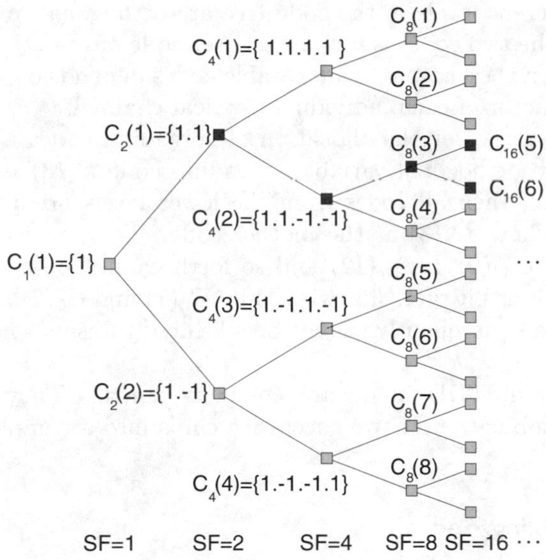

Figure 6.22: Tree structure of an OVSF code

$2^k R_{min}$ produces a code length of 2^{n-k} for $k = 1 \ldots n$. OVSF codes are Walsh sequences that are generated from the single-element unit code $C_1(1) = [1]$.

Equation (6.3) describes the recursive generation of the OVSF code by a quadratic generator matrix of the dimension $[N \times N]$. C_N identifies the set of N spreading codes with the length of N chips.

$$C_N = \begin{bmatrix} C_N(1) \\ C_N(2) \\ C_N(3) \\ C_N(4) \\ \vdots \\ C_N(N-1) \\ C_N(N) \end{bmatrix} = \begin{bmatrix} C_{N/2}(1)C_{N/2}(1) \\ C_{N/2}(1)\overline{C_{N/2}(1)} \\ C_{N/2}(2)C_{N/2}(2) \\ C_{N/2}(2)\overline{C_{N/2}(2)} \\ \vdots \\ C_{N/2}(N/2)C_{N/2}(N/2) \\ C_{N/2}(N/2)\overline{C_{N/2}(N/2)} \end{bmatrix} \tag{6.3}$$

C_N is a column vector with $N = 2^n$ elements produced from $C_{N/2}$. $\overline{C_{N/2}}$ is the binary complement to $C_{N/2}$. The quantity of the codes produced this way can be presented in a tree structure (see Figure 6.22), because each code is generated with a spreading factor N from a code with the spreading factor $N/2$. A quantity of 2^k spreading codes with a length of 2^k chips is available in each k plane. OVSF codes up to a length of 256 or 512 chips exist in FDD mode UL or DL respectively.

Since a code in the code tree is generated through the duplication of a code in the lower level, codes of different levels are only orthogonal to one another if the shorter one is not found again in the longer one. This means that two codes from different levels in the code tree are orthogonal to one another as long as one of the two codes is not the mother code of the other one. Because of this restriction the number of allocatable codes depends on the bit rate and the spreading factor of each individual physical channel.

A sample scenario for the allocation of spreading codes is shown in Figure 6.22. If the orthogonal variable spreading code $C_8(3)$ is allocated to a physical channel, then all codes from the lower levels originating from this spreading code have $C_8(3)$ as the mother code. This means that the codes $C_{16}(5)$, $C_{16}(6)$, $C_{32}(9) \ldots C_{32}(12)$ and so forth cannot be allocated to users who require a lower bit rate. Likewise codes $C_2(1)$ and $C_4(2)$ are not orthogonal to $C_8(3)$ and consequently cannot be allocated to users who need a higher bit rate.

OVSF codes in TDD mode are complex-valued. They are generated through the combination of two successive chips into a complex-valued symbol.

6.9.4.2 Scrambling Codes

In FDD mode, scrambling codes with either long or short periods can be used on the uplink. Irrespective of the period, 2^{24} each complex-valued short and long codes are available. Long scrambling codes are generated from a real-valued Gold code with a length of $2^{25}-1$ chips through the linking of the Gold code with a copy of the same code that has been shifted by $2^{24}+16$ chips.

Short scrambling codes have a length of 256 complex chips and are generated through a similar procedure as the long scrambling codes from a four-digit code sequence of the length 2^8-1.

Long, complex-valued 38,400-chip codes are used on the uplink to scramble spread physical channels. The scrambling codes are generated from a real-valued Gold code with a length of $2^{18}-1$ chips. The real part of the scrambling code corresponds to the Gold code, the imaginary part corresponds to the code shifted by 2^{17} chips.

TDD mode has 128 different real-valued scrambling codes of the length 16.

6.9.4.3 Spreading and Modulation on the Uplink in FDD Mode

Figure 6.23 shows the spreading and modulation of the DPCH, divided into one or up to 6 DPDCH and a DPCCH on the uplink. Each physical channel is spread with its own OVSF code and each code sequence is then weighted with an amplitude factor. This amplitude factor can have 16 different discrete values and is used to adapt the signal power of the DPDCH and the DPCCH to each other in the case of different spreading factors. The spread and weighted channels are then multiplexed to the inphase and quadrature components of a complex-valued baseband signal. The DPCCH is always mapped onto the quadrature component, the DPDCHs alternately on the inphase and on the quadrature components. The resulting complex-valued chip sequence is then scrambled with a complex-valued long or short scrambling code and finally

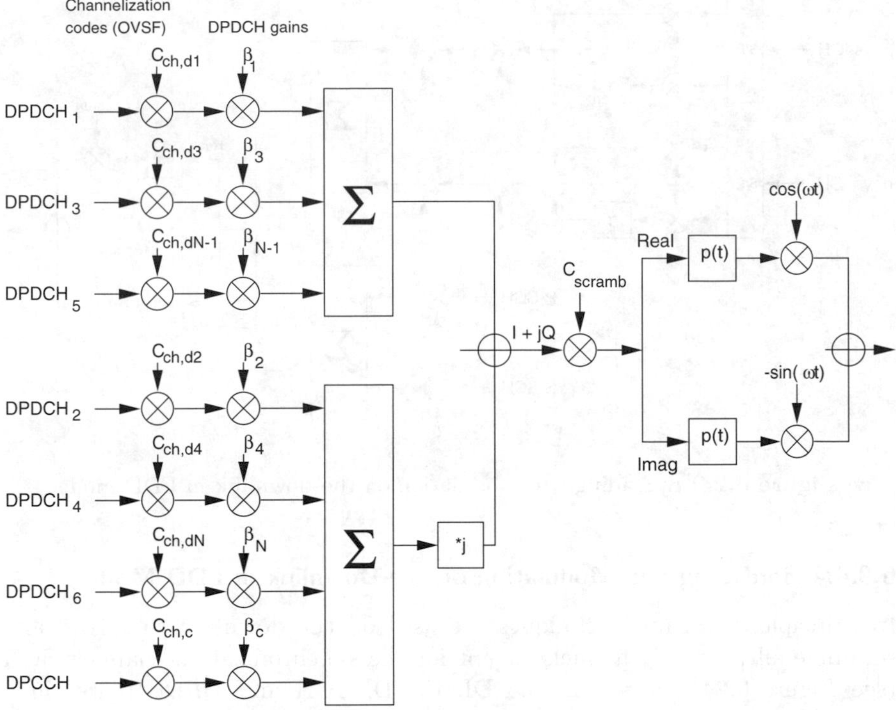

Figure 6.23: Transmission of DPDCH and DPCCH on the uplink in FDD mode

QPSK-modulated and transmitted. A *Root Raised-Cosine* filter with a *roll-off* factor of 0.22 is used as the pulse-forming filter. The modulation rate is 3.84 Mchips/s.

Exactly one DPCCH and up to six DPDCHs can be transmitted simultaneously, with the DPCCH always using the spreading code $C_{256}(1)$. If more than one DPDCH is transmitted, all DPDCHs have the same spreading factor. Since all DPCCH and DPDCH channels are multiplexed either to the quadrature or the inphase branch of modulation, this technique allows each physical channel considered individually to be transmitted with *Binary Phase Shift Keying* (BPSK) instead of *Quaternary Phase Shift Keying* (QPSK) modulation.

A PRACH and a PCPCH are spread and modulated with the same techniques used for a DPCH and a DPDCH. The message part of a PRACH and a PCPCH is mapped onto the inphase branch and the control part onto the quadrature branch. Both paths are spread and weighted and then the complex-valued aggregate signal is scrambled with a complex-valued code. The same scrambling codes are not used simultaneously with the DPCH, the PRACH or the PCPCH.

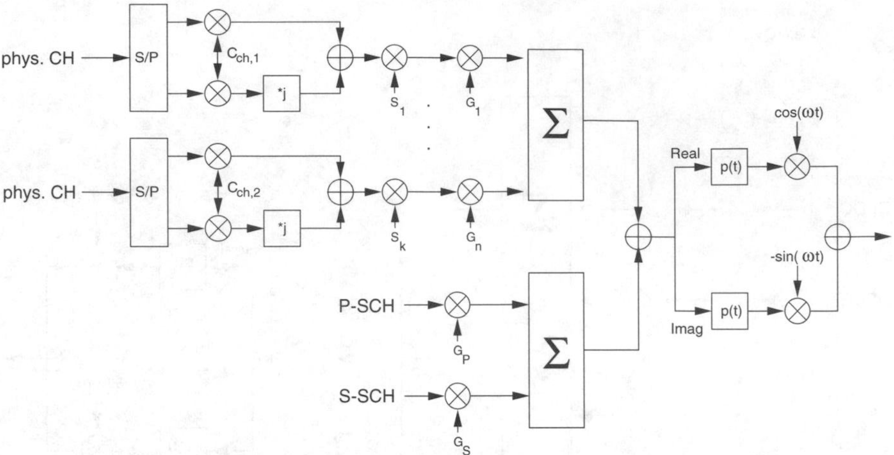

Figure 6.24: Spreading and modulation on the downlink in FDD mode

6.9.4.4 Spreading and Modulation on the Downlink in FDD Mode

In principle, the same techniques are used on the downlink to spread and scramble all physical channels except for the synchronization channels SCH (see Figure 6.24). Since in the DL the DPCCH and DPDCH are time-multiplexed, QPSK modulation is applied to the serial-parallel converted data stream. The SCH bits with an even index are multiplexed to the inphase branch and bits with an odd index to the quadrature branch. The two resulting bit streams are always spread with a real-value OVSF code. The spreading is implemented through a multiplication of the bipolar code times the bipolar representation of the bit sequence. During this process DTX indicators are mapped onto the value zero and therefore transmitted with or without the value zero.

In the next step each complex-valued symbol that consists of one bit each from the inphase and quadrature phases is scrambled with a complex-valued code. The resulting complex-valued chip sequences of all physical channels are each weighted with a factor and then added up. The chip sequences of the P-SCH and the S-SCH are already complex-valued and are added to the aggregate sequence of the remaining channels. The resulting signal is then transmitted with QPSK modulation at a chip rate of 3.84 Mchips/s. This means that in contrast to the uplink two-valued symbols are always transmitted for each channel.

6.9.4.5 Spreading and Modulation in TDD Mode

Figure 6.25 illustrates how the data parts of a TDD burst are spread, scrambled and modulated. Every two successive bits are combined to form a

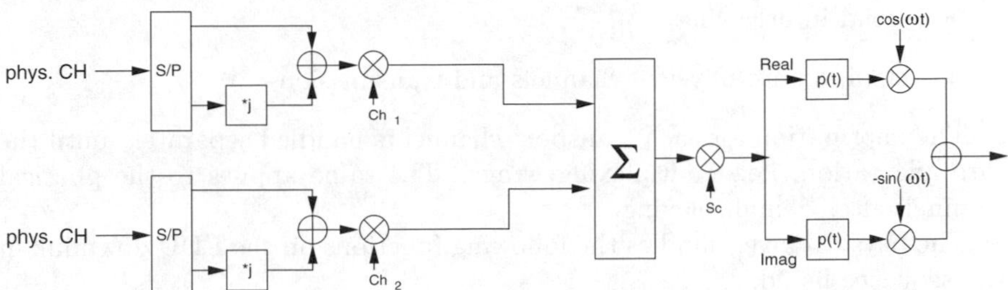

Figure 6.25: Spreading and modulation on the uplink in TDD mode

complex-valued symbol. This is followed by spreading with a complex-valued OVSF code (see Section 6.9.4.1).

All physical channels that have been spread this way are added, and the aggregate signal is scrambled with the user-specific real-valued scrambling code. The resulting chip sequence is transmitted with QPSK modulation at a chip rate of 3.84 MChip/s.

Spreading and modulation on the downlink are the same as on the uplink. However, up to 16 DPCHs can be transmitted simultaneously on the downlink but only 2 on the uplink.

6.9.5 Channel Coding, Multiplexing and Interleaving

Transport block sets with the period of a TTI are transmitted by the MAC layer to the physical layer. On the FDD uplink and in the TDD mode the following functions of the physical layer in the sequence listed ensure that transmitted information is protected and that this information is mapped onto the physical channels:

- Addition of a checksum (*Cyclic Redundancy Check*, CRC) to each transport block
- Concatenation of transport blocks and segmentation into code blocks
- Channel coding
- Frame-length adaptation
- First interleaving
- Frame-length segmentation
- Rate matching
- Multiplexing of transport channels to CCTrCH
- Demultiplexing of CCTrCH to physical channels

- Second interleaving
- Mapping onto physical channels and transmission

The information for each transport channel is handled separately until the rate adaptation, i.e., multiplexing, stage. The same applies to the physical channels after demultiplexing.

The physical layer handles the following functions on the FDD downlink in the sequence listed:

- Addition of checksum CRC to each transport block
- Linking of transport blocks and segmentation into code blocks
- Channel coding
- Rate matching
- First insertion of DTX indicators
- First interleaving
- Frame-length segmentation
- Multiplexing of transport channels to CCTrCH
- Second insertion of DTX indicators
- Demultiplexing of CCTrCH to physical channels
- Second interleaving
- Mapping onto physical channels and transmission

Here too information from the transport channels is handled separately until the multiplexing stage and for the physical channels until after the demultiplexing. The functions of the uplink are described in the following sections, followed by the functions on the downlink that differ.

6.9.5.1 Attachment of a Checksum

For error detection a checksum can be attached to each transport block in a transport block set. This checksum is 24, 16, 12, 8 or, if no error detection is required, 0 bits long. Parity bits are generated systematically through the use of the following generator polynomials over GF(2):

$$\begin{aligned}
g_{24} &= X^{24} + X^{23} + X^6 + X^5 + X + 1 \\
g_{16} &= X^{16} + X^{12} + X^5 + 1 \\
g_{12} &= X^{12} + X^{11} + X^3 + X^2 + X + 1 \\
g_8 &= X^8 + X^7 + X^4 + X^3 + X + 1
\end{aligned}$$

Table 6.18: Channel coding techniques for different transport channels

Transport channel	Coding	Coding rate	Block length according to coding
BCH	Convolutional coding	1/2	$2n + 16$
PCH	Convolutional coding	1/2	$2n + 16$
RACH	Convolutional coding	1/2	$2n + 16$
CPCH, DCH	Convolutional coding	1/2	$2n + 16$
	Convolutional coding	1/3	$3n + 24$
DSCH, FACH	Turbo coding	1/3	$3n + 12$
	No coding	1	n

6.9.5.2 Linking of Transport Blocks and Segmentation into Code Blocks

The different channel coding schemes possible require blocks with restricted lengths as input. All transport blocks of a TBS are therefore ordered sequentially before channel coding takes place, and this data stream is divided into same-sized blocks depending on the coding technique selected. The maximum length of these blocks is as follows:

- for convolutional coding 504 bits
- for turbo coding 5114 bits
- there is no restriction for uncoded transmission

If a bit stream within a TTI is not divisible into same-sized blocks with this maximum length, zeros are added until an even division is possible.

6.9.5.3 Channel Coding

The transport channel type determines which channel coding scheme can be used. In principle, convolutional coding with an influence length of 9 bits and coding rates of 1/2 or 1/3 or turbo coding with a coding rate of 1/3 are possible. In addition to these schemes, information can also be transmitted over transport channels uncoded. Table 6.18 lists the different channel coding techniques available for different transport channels.

6.9.5.4 Frame Length Adaptation and Segmentation (Uplink)

If the TTI of a transport channel is the multiple of a frame length (10 ms), then the data stream being transmitted within a TTI is evenly divided between several successive time frames (frame length division). Since the transport format for the TFS being transmitted is constant for the duration of a TTI, this division must be effected in such a way that the blocks transmitted in the individual time frames are of the same size and entire time frames can always be filled. The size of the individual blocks conforms with the transport format and the parameters of rate adaptation on the uplink (see Section 6.9.5.6). These parameters are established by higher layers. If a division of the data set of a TTI into equal-sized blocks is not possible, then bits can be added so long until the number of bits is an integer multiple of the desired block length (frame length adaptation).

6.9.5.5 First Interleaving

Before its division into individual time frames, the length-adapted data stream of a transport channel is subject to block interleaving. The bits to be interleaved are written by line into a matrix that has as many columns as the number of time frames that fit into a TTI, e.g., two columns for one TTI of 20 ms. The columns of the matrix are then permutated among themselves according to a prescribed sequence and finally the resulting matrix is read out by column.

6.9.5.6 Rate Matching

When physical channels are occupied on the uplink, they must also be used. This means that the bit rate of the CCTrCH always has to be the same as the sum of the transmission rates of the occupied channels at any given time. Since the data set that is transmitted in a TTI over a transport channel can vary, the rate adaptation function ensures that the data stream multiplexed from different transport channels has a constant rate. It does so through bit repetition or through a puncturing of the bit stream. Rate adaptation is controlled by higher layers. Systematic bits of turbo-coded data are not punctured. If none of the transport channels of a CCTrCH transmits any data during an interval, no adaptation takes place, i.e., the DPDCH remains empty.

6.9.5.7 Multiplexing of Transport Channels to CCTrCH

The CCTrCH is created from the serially-multiplexed data streams of different transport channels within a TTI. The number of transport channels mapped onto a CCTrCH is determined by the capabilities of the UE. The same applies to the size of the TBS of a transport channel.

There can only be one CCTrCH at a time for each UE on the uplink; several parallel CCTrCHs are permitted for a UE on the downlink. The transmitter power of all CCTrCHs transmitted simultaneously over physical channels is controlled by the same fast power control with closed loop. It is also possible for different CCTrCHs to require different signal-interference ratios.

There are dedicated and common CCTrCHs, depending on which transport channels are being multiplexed to a CCTrCH. Dedicated transport channels can only be mapped to dedicated CCTrCHs, common transport channels only to common ones. Table 6.19 lists some possible mapping of transport channels to physical channels in FDD mode (see also Figure 6.28).

Table 6.19: Mapping of transport channels to physical channels

	Transport channel	Physical channel
UL	RACH	PRACH
UL	CPCH	PCPCH with a control part and one or more message parts
UL	one or more DCH multiplexed to the same CCTrCH	one DPCCH and one or more DPDCH
DL	BCH	P-CCPCH
DL	one or more FACH + PCH	one S-CCPCH for each FACH-PCH pair
DL	one or more DCH multiplexed to the same CCTrCH	one or more DPCH, control part only in one DPCH
DL	DSCH and one or more DCH multiplexed to the same CCTrCH	PDSCH and one or more DPCH, control part only in one DPCH
DL	FACH and one or more DCH multiplexed to the same CCTrCH	S-CCPCH and one or more DPCH, control part only in one DPCH
DL	DSCH and FACH and one or more DCH multiplexed to the same CCTrCH	S-CCPCH and PDSCH and one or more DPCH, control part only in one DPCH
DL	various DCH multiplexed to various CCTrCH	various DPCH, control part only in one DPCH per CCTrCH

6.9.5.8 Demultiplexing CCTrCH on Physical Channels

Several transport channels can be multiplexed to a CCTrCH (see Figure 6.14) to enable a maximum utilization of available radio resources. Depending on the transmission rate of the CCTrCH, this channel is split up into one or more physical channels with the same spreading factor even though physical channels cannot transmit data from different CCTrCH. Therefore, the necessary signal-to-interference ratio should be the same for all bits of a CCTrCH. The parameters for error protection and multiplexing for each CCTrCH are specified by a TFC which can be changed every 10 ms. The communicating entity is either notified of this TFC or finds out about it independently.

The CCTrCH is divided up into different physical channels as a result of the distribution of the data stream to the same physical channels. In each case one contiguous block of bits of the CCTrCH is allocated to a channel.

6.9.5.9 Second Interleaving

The second interleaving is implemented according to the same principles as the first one and is carried out separately for each physical channel. It involves the interleaving together of information of the same time frame. Data waiting in a time frame for transmission over a physical channel is written line-by-line into a 30-column matrix which, if necessary, is padded out with bits. The matrix is read out column-by-column up to the filler bits according to a fixed permutation of the columns.

6.9.5.10 Mapping to Physical Channels and Transmission

After the second interleaving, the bursts that are ultimately to be modulated and transmitted after spreading and scrambling are generated from the data streams (see Section 6.9.4.3).

The functions of the physical layer are clarified in Figure 6.26. The figure shows how a dedicated transport channel with a rate of 12.2 kbit/s on the uplink transmits 244 bits every 20 ms. Another DCH, one with a rate of 2.4 kbit/s, is multiplexed to the CCTrCH. This CCTrCH is mapped onto a DPDCH with a spreading factor of 64 (see burst format 2 in Table 6.13). A 16-bit long checksum is added to the 244-bit block. Together with the eight bits required for the initialization of the convolutional coding, this produces a block 804 bits long after channel coding. After interleaving this block can be divided without extension into two 402-bit parts that are adapted to the time frame length. The transmission rate is adapted to the CCTrCH through a repetition of the bits. This process produces two blocks per 20 ms, each with 495 bits, which together with blocks from the second DCH are multiplexed to a CCTrCH. Since only one DPDCH with burst format 2 is required for transmission of this CCTrCH, there is no demultiplexing on the different physical channels. After the second interleaving the DPDCH bursts are generated from the data stream and modulated and transmitted after spreading and scrambling.

6.9.5.11 Rate Adaptation on the Downlink

If the data provided deviates from a set maximum value during the TTI, transmission is interrupted until the maximum bit quantity to be transmitted in an TTI of this transport channel has been reached. Rate adaptation on the downlink ensures through puncturing or repetition that the number of bits is an integer multiple of the block length required for transmission on a physical channel.

6.9.5.12 Insertion of DTX Indicators on the Downlink

It is possible for a transport channel on the downlink to be allocated a fixed position within a CCTrCH. In this case the position of the transport channel is known within the time frame, and the same maximum quantity of bits is reserved for this transport channel in each time frame. If the data transferred within a TTI deviates from the maximum value during voice transmission, the block is padded after coding and adaptation with DTX indicators that are not transmitted. This fixed positioning of a transport channel makes it easy for the receiving entity to recognize the transport format if no TFCI is transmitted.

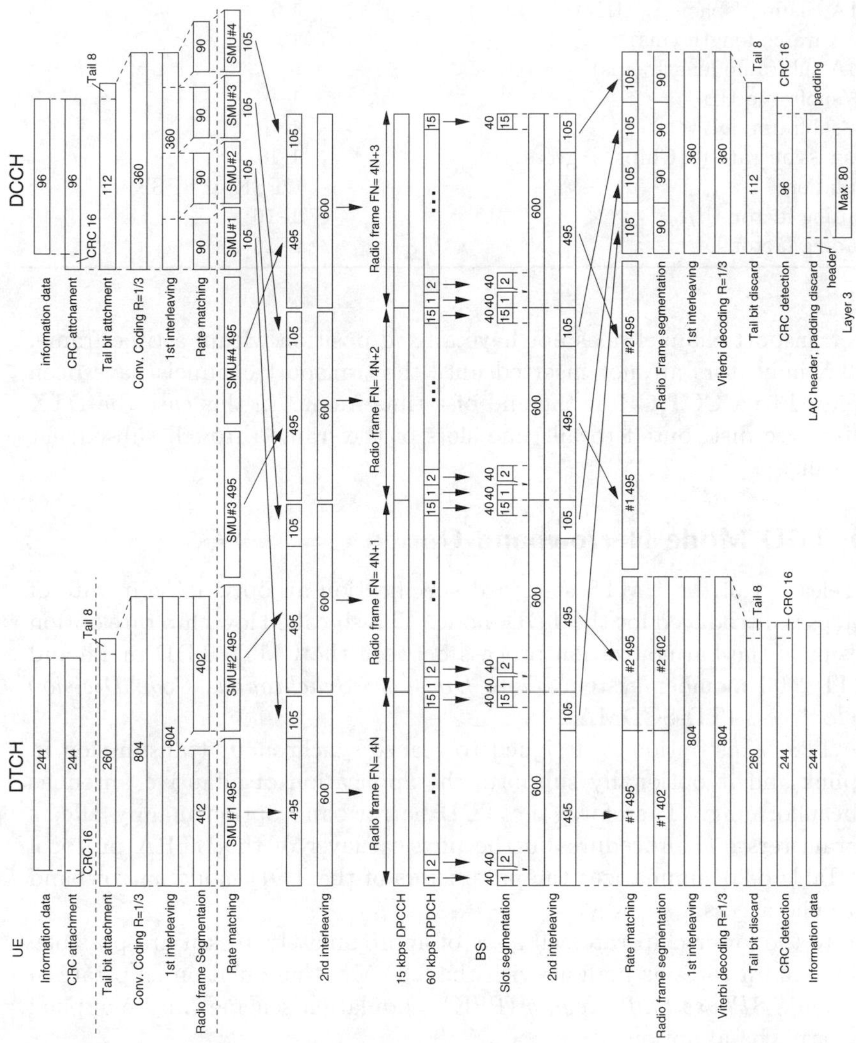

Figure 6.26: Transmission of a dedicated channel with 12.2 $^{\mathrm{kbit}}/_{\mathrm{s}}$ on the uplink

Table 6.20: Characteristic parameters of the UTRA TDD mode narrowband option radio access

Parameter	
Multiple access methods	CDMA/FDMA/TDMA
Duplex methods	TDD
FDMA channel spacing (MHz)	1.6
TDMA frame length (ms)	10
TDMA subframe length (ms)	5
Slots/subframe UL	1–6
Slots/subframe DL	1–6
Transmission rate (MChips/s, gross)	1.28
Modulation	QPSK or 8PSK
Spreading factor UL	1–16
Spreading factor DL	1–16

If a transport channel does not have a fixed position within a time frame, the DTX indicators are not inserted until the transport channels have been multiplexed to a CCTrCH at the end of a time frame. In this case the DTX indicators are distributed to all time slots of the frame through subsequent interleaving.

6.9.6 TDD Mode Narrowband Option

With release 4 of the UMTS standard specification an optional chip rate of 1.28 Mcps is introduced for the TDD mode. This so-called low chip rate option is a result of the harmonization process between the UMTS TDD mode and the IMT-2000 member system *Time Division—Synchronous Code Division Multiple Access* (TD-SCDMA).

The 1.28 Mcps option is designed to realize synchronous transmission in the uplink and it optionally supports the application of adaptive antennas with beamforming. Therefore, the TDD narrowband option mainly affects the parameters and procedures in the physical layer of the UTRA protocol stack. Table 6.20 summarizes the parameters of the TDD mode narrowband option radio access.

Due to the lower chip rate cell sizes of approximately 10 km are possible. To achieve data rates as high as with the 3.84 Mcps modulation chip rate, a QPSK and a 8*Phase Shift Keying* (PSK) modulation scheme can be applied in the narrowband option.

For the low chip rate option a TDMA frame is divided into two subframes of 5 ms duration each. Since the structure of each subframe is the same this allows a fast update of power control, uplink synchronization and adaptive antenna beamforming with a period of 5 ms.

Figure 6.27 shows the TDMA frame structure for the TDD low chip rate option. Each subframe has 7 traffic time slots of 864 chips duration each.

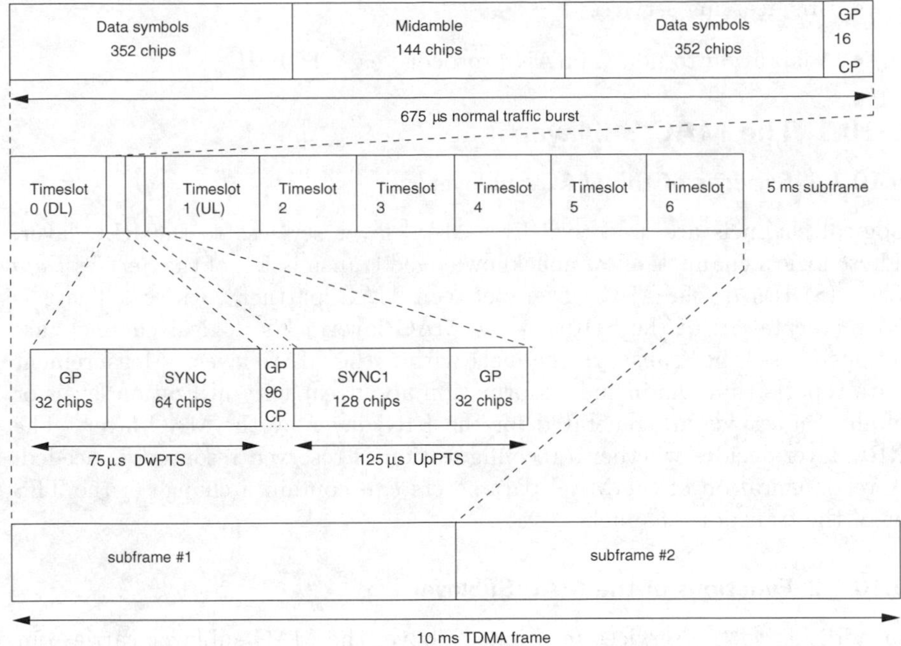

Figure 6.27: TDMA frame structure in the TDD narrowband mode

Between traffic time slot 0 which is always used for the downlink and time slot 1 which is alway used for the uplink, a *Downlink Pilot Channel* (DwPCH) and *Downlink Pilot Channel* (UpPCH) is inserted. Each subframe has two switching points, one between the *Downlink Pilot Time Slot* (DwPTS) and the *Downlink Pilot Time Slot* (UpPTS) and one after timeslot 1.

There is one normal traffic burst type as shown in Figure 6.27. As in wideband TDD, a burst consists of two data fields separated by a midamble and tailed by a guard period. Bits for transmitter power control, TFCI or additional uplink synchronization bits are included in the data fields of the bursts in the case they are needed.

The DwPCH and UpPCH bursts are transmitted in each subframe. These bursts are used for downlink and uplink synchronization, respectively.

6.10 The Data Link Layer

The data link layer consists of a total of four sublayers, with the BMC and the PDCP sublayers always only being part of the user plane (see Section 6.8). The data link layer essentially provides the following services:

- Data transmission with different qualities of service
- Layer 2 connection set up and termination

- Broadcasting services
- Adaptation to non-UTRAN protocols, e.g., TCP/IP

6.10.1 The MAC Sublayer

6.10.1.1 Services of the MAC Sublayer

Logical channels are used to deliver MAC layer services to the RLC layer. These logical channels allow unacknowledged transmission of the *Service Data Unit* (SDU)s of the MAC layer between MAC partner entities. There is no segmentation of the SDUs in the MAC layer. The logical channels are mapped onto the transport channels within the MAC layer. Measurement data reports that can include information about capacity utilization levels or quality of service are generated for the RRC layer by the MAC layer. The RRC layer decides whether a reconfiguration of reserved resources is needed. A reconfiguration of the MAC parameters can contain a change of the TFS or of the transport channel.

6.10.1.2 Functions of the MAC Sublayer

In addition to the services mentioned above, the MAC sublayer carries out the functions described below:

Selection of suitable transport format To ensure that the transport channels are used effectively, the MAC layer separately selects from the valid transport format set a transport format for each transport channel depending on the current transmission rate. The TFCS is supplied by the RRC layer (see Section 6.9.2.1).

Management of priorities In its selection of a suitable transport format combination (TFC) the MAC layer can take into account different priorities between the individual transport channels through an allocation of the transmission rate. These priorities either are service-specific or can be specified by the RLC layer in order to prevent a buffer overflow. In the case of very bursty traffic, such as packet data transmission, the MAC layer can control priorities by selecting the shared transport channel for individual packets.

Management of common transport channels The common transport channels require identification of the UE. This function is handled by the MAC layer which inserts the *Radio Network Temporary Identifier* (RNTI) into the MAC-PDU or removes it, as the case may be.

Multiplexing logical channels to transport channels: When logical channels are mapped onto transport channels, different services can be multiplexed to a transport channel. Conversely, transport channels are demultiplexed to logical channels.

Changing the type of transport channel On the instructions of the RRC layer, the MAC layer transmits either over a shared or a dedicated transport channel.

Encryption This function of the MAC layer is only used in combination with the transparent transmission mode of the RLC layer (see Section 6.10.2).

Random access Random access is based on the protocol described in Section 6.9.3.3. Each *Access Service Class* (ASC) is assigned at least one *backoff* parameter along with a set of access time slots and signatures. The MAC layer notifies the physical layer of the parameters possible for the transmission of a MAC-PDU over the RACH or PRACH.

6.10.1.3 Logical Channels

Logical channels are the MAC layer services for the RLC layer. Logical channels that are used to transmit information of the control plane are combined with the group of *control channels*. All logical channels that transmit information of the user plane belong to the group of *traffic channels*. Logical channels are defined by the type of information transmitted over them.

There are six different logical control channels:

Synchronization Control Channel (SCCH) The SCCH is used in TDD mode to broadcast information about the configuration of the BCCH.

Broadcast Control Channel (BCCH) The BCCH broadcasts system information on the downlink.

Paging Control Channel (PCCH) The PCCH is used for paging on the downlink.

Dedicated Control Channel (DCCH) The DCCH is a bidirectional channel that transmits dedicated control information between UE and network. This channel is established during the set up of RRC connections.

Common Control Channel (CCCH) The CCCH is a bidirectional logical channel that handles the exchange of control information between UE and UTRAN. It is used when no RRC connections exist between network and UE or when the UE is already using a common transport channel and is changing over to a new cell (*cell reselection*).

Shared Channel Control Channel (SHCCH) Information for the control of common channels used jointly by different UEs is transmitted over a SHCCH. This channel is bidirectional and exists in TDD mode only.

Two different logical traffic channels are also available:

Dedicated Traffic Channel (DTCH): A DTCH can exist either on the uplink or on the downlink. It is a point-to-point channel and always belongs to a certain UE. It is used to transit data of the user plane.

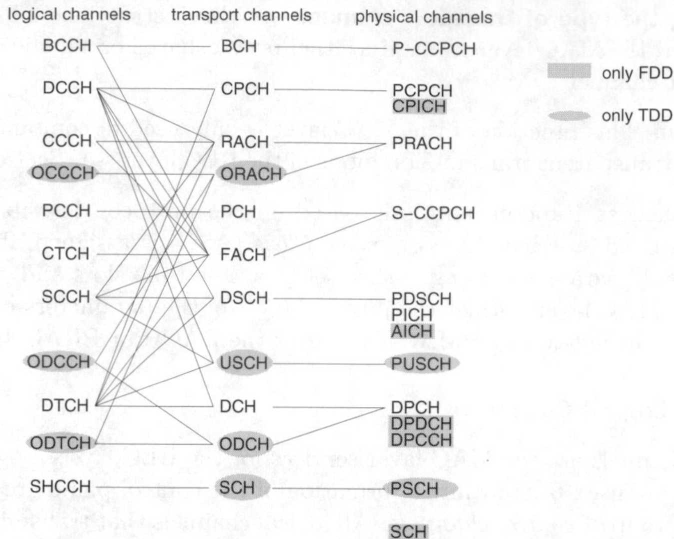

Figure 6.28: Mapping of logical channels, transport channels and physical channels to one other

Common Traffic Channel (CTCH): The CTCH broadcasts data of the user plane to all or a group of UEs. It is a unidirectional point-to-multipoint channel on the downlink.

Figure 6.28 presents an overview of the mapping of logical channels onto transport channels and from these channels to physical channels, as well as the other way around.

Table 6.21 lists the mappings possible of logical channels onto transport channels in both transmission directions.

Table 6.21: Mapping of logical channels onto transport channels

	BCCH	PCCH	DCCH	CCCH	SHCCH	CTCH	DTCH
BCH	DL						
PCH		DL					
CPCH			UL				UL
RACH			UL	UL	UL		UL
FACH	DL		DL	DL	DL	DL	DL
USCH			UL		UL		UL
DSCH			DL		DL		DL
DCH			UL/DL				UL/DL

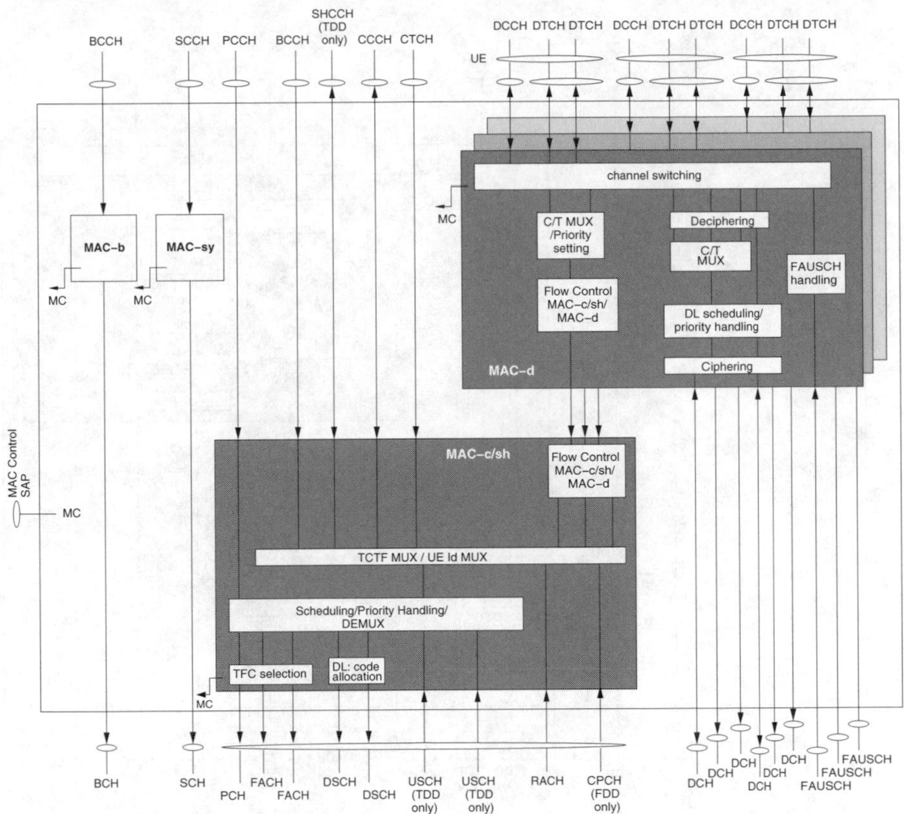

Figure 6.29: Structure of the MAC layer in UTRAN

6.10.1.4 MAC Entities

Figures 6.29 and 6.30 illustrate how the MAC layer is structured in UTRAN and in the UE. There are four different MAC entities: MAC-b, MAC-c/sh, MAC-d and MAC-sy. The latter exists only in TDD mode. Each MAC entity has a *MAC-control-Service Access Point* over which the RRC layer can control the parameters of the individual entities.

The MAC-b entity forms the connection between BCCH and BCH. One instance of this entity exists in each UE and for each cell. The MAC-sy entity manages the synchronization channels in TDD mode. It is a cell-specific channel and one instance of this entity exists in each UE and for each cell.

Dedicated logical channels and dedicated transport channels are linked together over the MAC-d entity. Exactly one instance of this entity exists in the UE and exactly one instance per UE exists in a UTRAN. After input from the RRC layer the MAC-d entity selects the transport channel suitable for transmission of the MAC-SDU. If this involves a common transport channel

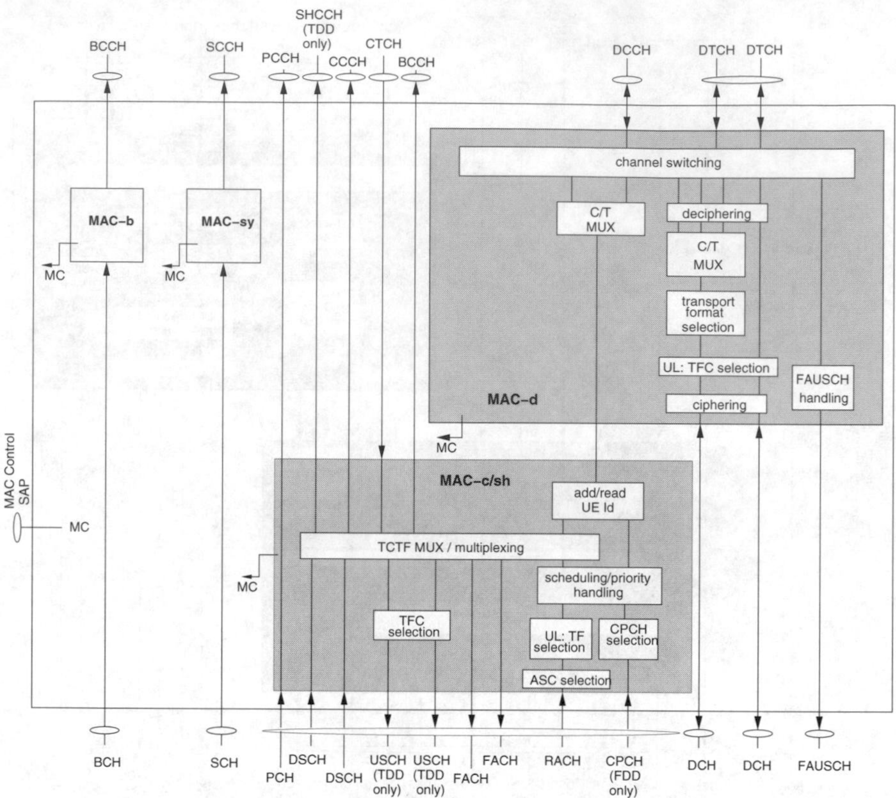

Figure 6.30: Structure of the MAC layer in UE

or if various dedicated logical channels are multiplexed to a dedicated transport channel, then the selected channel is referenced in a C/T field in the MAC-PDU. A dedicated logical channel can be mapped simultaneously onto a DCH and, through the MAC-c/sh entity, onto a DSCH.

With regard to the MAC-c/sh entity, flow control in the MAC-d entity ensures that there is a minimal delay of layer 2 signalling over dedicated logical channels and minimizes the number of discarded or repeated transmissions when the FACH or the DSCH is overloaded.

The TFC for establishing priorities among the transport channels in the MAC-d entity is selected on the uplink (see Section 6.10.1.2). Along with this allocation of transmission bandwidth, prioritization is also dealt through control of the transmission sequence of the MAC-PDUs on the various transport channels.

Non-dedicated logical control channels are mapped onto non-dedicated transport channels over the MAC-c/sh entity. The mapping of dedicated logical channels (DCCH, DTCH) onto non-dedicated transport channels takes

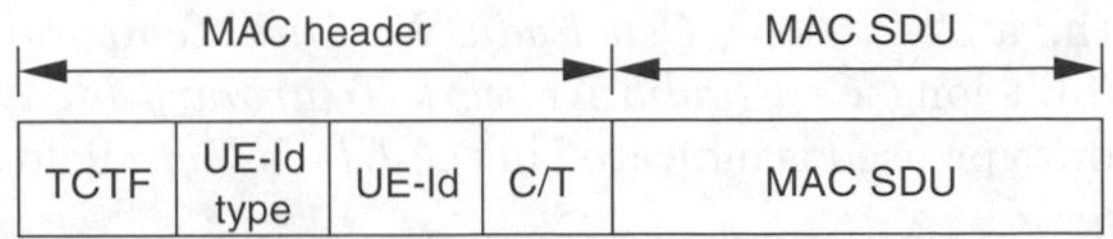

Figure 6.31: The structure of a MAC-PDU

place over the MAC-d entity. Exactly one instance of the MAC-c/sh entity exists for the UE and for each cell.

The logical channels BCCH, CCCH, CTCH and SHCCH and the MAC-SDUs from the MAC-d entity are mapped onto the transport channels RACH and FACH within the MAC-c/sh entity. A *Target Channel Type Field* (TCTF) in the header of the MAC-PDU identifies which logical channels the MAC-SDU belongs to. The TCTF is added to or removed from the MAC-c/sh entity, as the case may be. If a dedicated logical channel is transmitted over the MAC-d and the MAC-c/sh, a field referencing the respective UE is added to the header of the MAC-PDU.

The MAC-c/sh entity controls the priorities between logical channels and clearance over the transport channels for all transport channels that are shared by different logical channels. On the uplink this function is for the RACH and the CPCH; on the downlink for the PCH, the FACH and the DSCH.

The parameters for random access (ASC and the *backoff* parameters) are specified within the MAC-c/sh entity and communicated to the physical layer.

6.10.1.5 MAC-Data-PDU

Figure 6.31 shows how a MAC-PDU is structured. Depending on the logical channels, one or more data fields can be added to a MAC-SDU. The length of an SDU always corresponds to the length of an RLC-PDU, because, as mentioned above, there is no segmentation in the MAC layer.

A TCTF is two or eight bits long in FDD mode, one to three bits long in TDD mode. The length is determined by the transport channel (USCH, FACH or RACH). The TCTF is omitted if allocation of the logical channel and the transport channel is explicit. This is the case if

- The BCCH is mapped onto the BCH.
- The DTCH or the DCCH is the only logical channel.

The TCTF is only used in the MAC-c/sh entity.

If a transport channel transmits various logical channels, a C/T field identifies each logical channel. This field is 4 bits long and indicates which type of logical channel is involved, especially in the case of transmission over a RACH or a FACH. The C/T field is only used in the MAC-d entity. It can only be omitted if no logical channels are multiplexed to transport channels.

The various UE for the shared transport channels are referenced over the *UE-Id* field. There are two different types of identification numbers, each of a different length: a 32-bit long *Cell Radio Network Temporary Identity* (U-RNTI) and a 16-bit long *Cell Radio Network Temporary Identity* (C-RNTI). The identification type used is indicated in the *UE-Id-Type* field and stipulated by the RRC layer.

6.10.2 The RLC Sublayer

6.10.2.1 Services of the RLC Sublayer

The RLC layer provides the following services of the layer above it:

Set Up and termination of RLC connections Exactly one RLC connection exists per bearer service.

Transparent data transmission With transparent data transmission no protocol information is added in the RLC layer to data that is being transmitted. However, the PDU of the layer above can be segmented into several parts and reassembled in the partner entity. This process is implemented according to a scheme known to both communicating partners. Additionally, the entity for transparent transmission contains both a send and a receive buffer.

Unacknowledged data transmission This service transmits PDUs without guaranteeing their receipt. However, in this transmission mode the RLC layer can detect whether errors have occurred. If necessary, the PDUs of the higher layers are segmented so that a RLC-PDU may also contain parts of different PDUs of the layer above. The RLC-PDUs or the PDU parts are numbered consecutively to enable the detection of errors. The partner entity only forwards those PDUs to the higher layer that have been received in the correct sequence. Otherwise the entire PDU is discarded. There is no retransmission in this case.

Acknowledged data transmission With acknowledged data transmission the RLC layer guarantees that the PDU being transmitted is received free of errors by the partner entity. The transmitting entity is signalled if errors occur in the transmission and then retransmits the defective PDU. If a PDU is repeatedly transmitted with errors, the higher layer is notified and the PDU discarded. This service ensures that PDUs are transmitted in the same sequence in which they are sent by the higher layer. For reasons of efficiency there is no need for sequential transmission if the higher layer is able to process PDUs that are received in the wrong sequence.

Setting the quality of service The parameters for the protocol for acknowledged data transmission can be set by the RRC layer. This enables different levels of quality of service to be achieved for the acknowledged data transmission service.

Table 6.22: Mapping of RLC transmission services onto logical channels

Service	SCCH	BCCH	PCCH	CCCH	SHCCH	DCCH	DTCH	CTCH
transp.	DL	DL	DL	UL	UL/DL		UL/DL	
unack.				DL	DL	UL/DL	UL/DL	DL
ack.						UL/DL	UL/DL	

Reporting of non-resolvable transmission errors If error protection with the mechanisms of the RLC layer cannot be provided in acknowledged transmission mode, the higher layer is notified accordingly.

Transparent, acknowledged and non-acknowledged data transmission services are mapped onto different logical channels. The mappings that are possible are presented in Table 6.22.

6.10.2.2 Functions of the RLC Sublayer

To provide transport services, the RLC sublayer requires the following functions:

Segmentation and reassembly With this function the large PDUs of the higher layer are split up into small *Payload Unit* (PU)s and then reassembled again in the communicating entity. The size of the RLC-PDU can be adapted to the valid TFS of the MAC layer during transmission.

Linking of RLC-SDUs When RLC-SDUs are segmented, the length of the SDU should not be an integer multiple of the length of the PU. The last segment of an SDU can therefore be linked to the first segment of the next SDU and combined into a PU.

Padding of PDUs If a PDU cannot be padded through linking to other PDUs, then it is filled with bits to produce the prescribed size.

Error detection When SDUs are segmented, the individual PUs are numbered consecutively. Therefore, the partner entity can determine whether the transmission is complete when it reassembles the SDUs. If part of an SDU is defective, the entire SDU is discarded.

Error correction In acknowledged mode of data transmission error correction is implemented through the retransmission of defective PDUs. Various protocols are available to deal with this, such as *Selective Repeat*, *Go-Back-N* and *Send-and-Wait ARQ*.

Sequential transmission of RLC-SDUs This function allows the RLC layer to ensure that RLC-SDUs are transmitted to the higher layer connected to the partner entity in the same sequence as they were received from the local higher layer. If this function is not used, the SDUs are sent in the sequence they had after error correction.

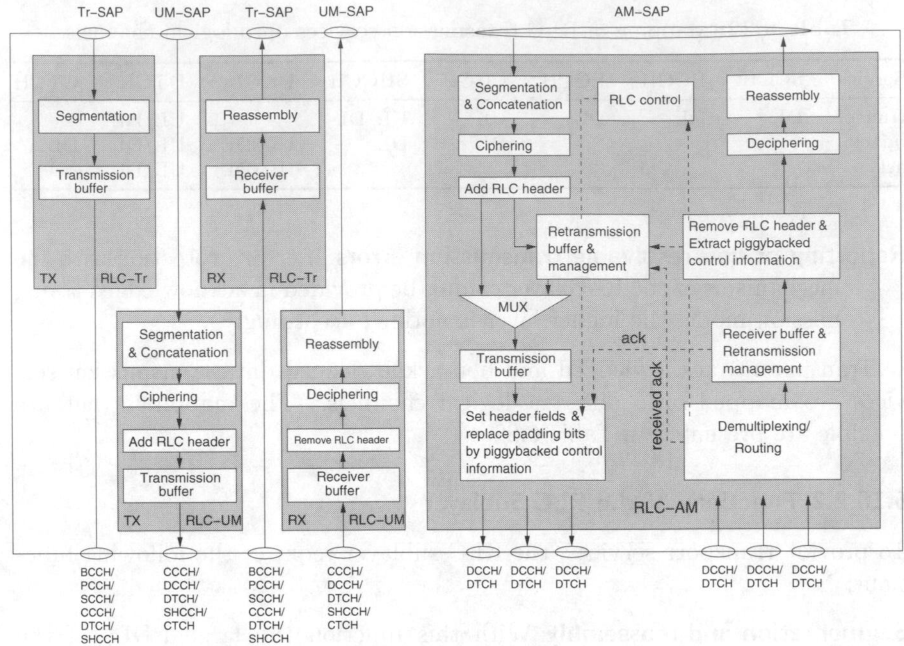

Figure 6.32: Entities of the RLC layer

Detection of multiple transmission Multiple transmissions of SDUs are detected and discarded to ensure that only one version of an SDU is sent to the partner entity.

Flow control This function allows the receiving partner entity to control the transmission rate of the transmitting partner entity.

Encryption Data encryption is only used within the RLC layer for acknowledged and non-acknowledged transmission services.

Suspension and resumption of data transmission The transmission of data can be suspended and then resumed over a different logical channel. The RLC connection remains intact.

6.10.2.3 RLC Entities

Figure 6.32 shows the entities of the RLC layer. Transparent data transmission is provided by an instance of the *TX-RLC-Tr* entity on the transmitting side and an instance of the *RX-RLC-Tr* entity on the receiving side. RLC-SDUs are only segmented if necessary. One of the logical channels indicated in Table 6.22 is used for transmission. Transparent data transmission over a CCCH is only possible on the uplink.

The *RLC-UM* entities provide unacknowledged transmission services. Because detection of missing RLC-PDUs is possible, the consecutive numbering of a RLC-PDU ensures that only those RLC-SDUs that were received without errors are forwarded to the higher layer. Only one type of PDU is always used for transparent and non-acknowledged transmission services.

In acknowledged transmission mode the *RLC-AM* entity receives the SDUs from the higher layer. These SDUs are segmented into equal-sized PUs. The length of a PU is semi-static and after a RLC connection set up can only be changed by the RRC layer. One PU can contain more than one SDU. The position of the individual SDUs within the PU is indicated by length indicators that are placed at the beginning of the PU. The PUs are sent to a buffer for retransmission and over a multiplex unit to the transmission buffer. The multiplexer determines which PDUs are transmitted and when. Before transmission, the header of the RLC-PDU is completed, e.g., with the setting of a polling bit. If needed, status information can be added instead of the filler bits. This information is then transmitted (*piggybacked*) in its PDU. This PDU has a variable size and is not stored in a buffer when it is retransmitted. Different logical channels can be used for the transmission of data-PDUs and status-PDUs. The data- and status-PDUs are separated in the partner entity. The data-PDUs are broken down again into individual PUs and, if necessary, into status information. The PUs are temporarily stored in the receive buffer until they can be assembled into a complete SDU. If some PUs have not been received or have only been received with errors, the buffer unit sends a negative acknowledgement to the transmitting entity. If all PUs are received correctly, the headers of the PDUs are removed and the SDUs reassembled.

6.10.3 Data Flow Through the Data Link Layer

Data can be transmitted with or without the addition of protocol information in either sublayers of the data link layer. The four transmission modes resulting from the combination of transmission services are presented in Figure 6.33. The services of the different layers are only differentiated by whether protocol information has been added.

Segmentation or reassembly can always take place within the RLC layer. The RLC-PDU is mapped by the MAC layer onto a TBS. The size of the RLC-PDU does not necessarily have to correspond to the size of the TBS. The transmission mode used to implement an RRC connection depends on the logical channel and ultimately also on the transport channel over which the service is being provided.

The following example uses a scenario in which data transmission in the user plane takes place over a DTCH.

DTCH over FACH or RACH The mapping of a DTCH onto a FACH or a RACH indicates use of a non-transparent mode in the RLC layer. In this case, the MAC-PDU contains both a TCTF and a *C/T* field as well

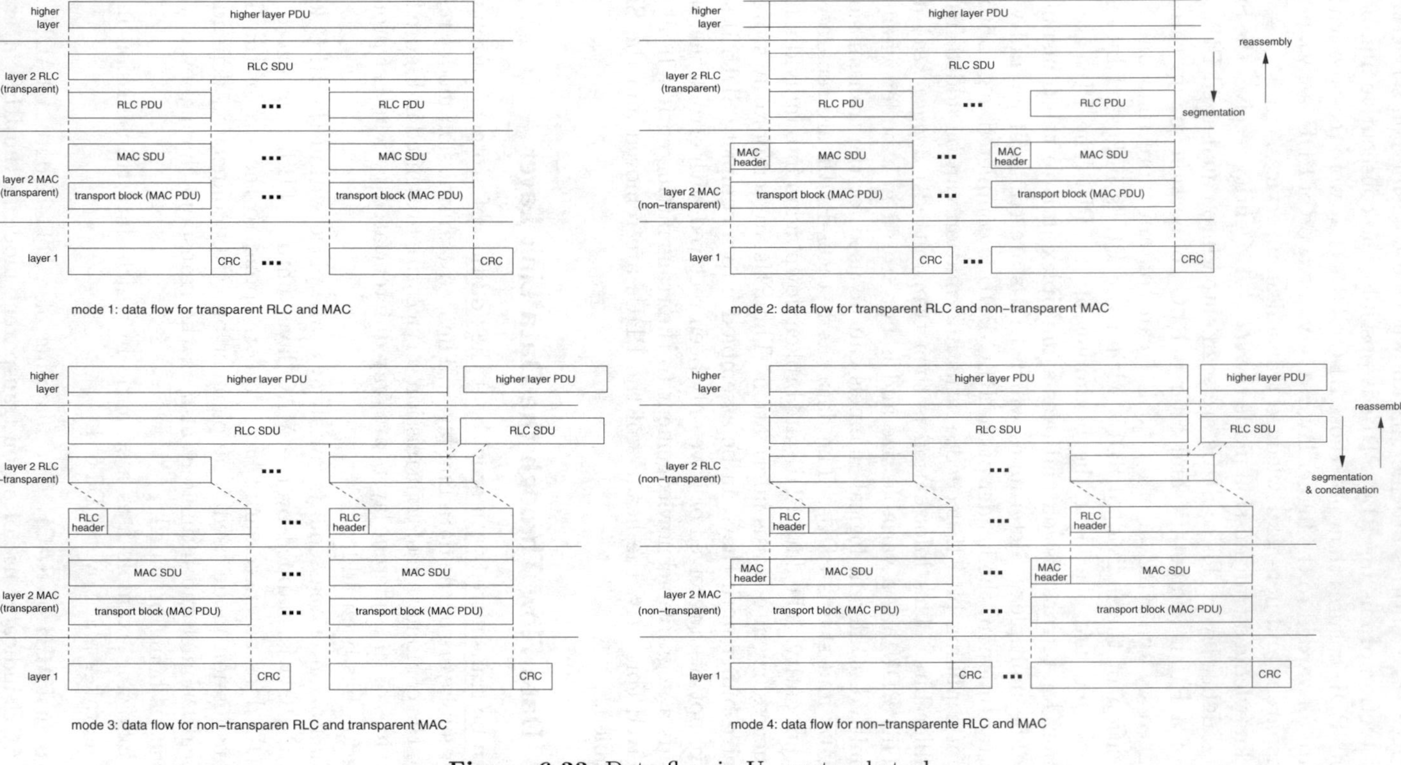

Figure 6.33: Data flow in U_u protocol stack

as identification of the UE (see Section 6.10.1.5). This corresponds to the fourth mode in Figure 6.33.

DTCH over DSCH In principle, the same applies here as in the case above, except that MAC protocol information can be omitted in TDD mode. If this is the case, the third mode is used.

DTCH over USCH The USCH also implies use of a non-transparent transmission service in RLC. The MAC-PDU only has to transmit protocol information if more than one DCCH and DTCH is being multiplexed to the USCH. In this case, mode three or four is possible.

DTCH over DCH In RLC a DTCH can be mapped onto a DCH either in transparent or in non-transparent transmission. The need for MAC protocol information depends on whether more than one DTCH is being multiplexed in the MAC layer. In principle, therefore, any of the modes of the physical layer can be used for the transmission of user data over a DTCH or a DCH.

DTCH over CPCH The same applies here as in the mapping of a DTCH onto a FACH or a RACH.

6.10.4 The BMC Sublayer

The task of the BMC layer is to transmit messages to all or some of the UE in a cell. The BMC sublayer only exists in the user plane and is transparent for all services except the broadcasting services *(broadcast/multicast)*. In a UTRAN this layer consists of one instance of the BMC entity per cell. BMC-PDUs are transmitted over a CTCH via the unacknowledged transmission service of the RLC.

6.10.5 The PDCP Sublayer

The PDCP sublayer is an adaptation layer for the transparent transmission of the network layer-PDUs of other protocols, such as IP over the U_u interface.

6.10.5.1 Functions of the PDCP Sublayer

In addition to providing algorithms for the adaptation of outside network protocols, the PDCP layer carries out the following functions:

Header Compression of NL PDUs *Header-compression* algorithms can be used to increase data transmission efficiency. These algorithms are always specifically for a particular network layer protocol or a combination of protocols, e.g., for TCP/IP. The algorithm and corresponding parameters are negotiated for each PDCP entity by the RRC layer and communicated over the *Control Service Access Point* (C-SAP). The signalling between the communication entities carrying out compression

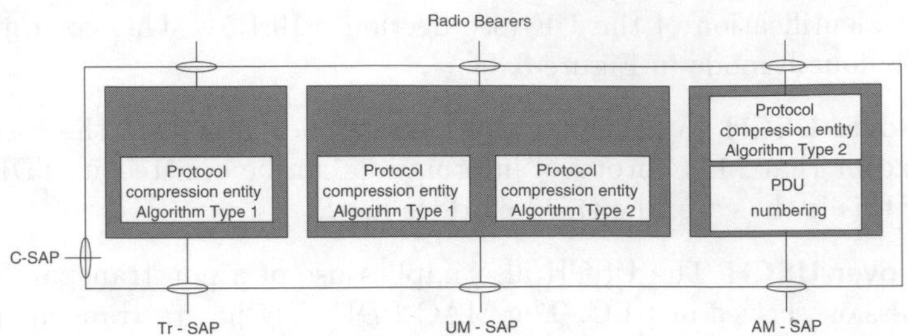

Figure 6.34: The structure of the PDCP sublayer

and decompression is within the user plane. Each PDCP entity can use any number of compression algorithms. If a number of different algorithms are available for selection in an entity, then the algorithm that is valid for the current PDCP-SDU is referenced through an entry in the header of the PDCP-PDU. For this purpose the communicating entities keep a list of available algorithms. This list can be changed through a reconfiguration of the PDCP entity by the RRC layer.

Data transmission For data transmission the PDCP layer uses the acknowledged, the non-acknowledged as well as the transparent transmission services of the RLC layer. In the case of acknowledged transmission, the PDCP entity stores the PDUs until acknowledgement has been received by the RLC layer. Otherwise a PDU is destroyed as soon as it has been transferred to the RLC layer.

Buffering of PDCP-SDUs PDCP-PDUs can be numbered and buffered to prevent packets from being lost in the case of a reallocation of SRNC. Should there be a change of SRNC, it will then be known which PDUs have already been received successfully. Stored PDUs are destroyed after receipt of an acknowledgement by the RLC layer.

6.10.5.2 Entities of the PDCP Sublayer

The structure of the PDCP sublayer is presented in Figure 6.34. The C-SAP of the RRC layer is used in the configuration of the entities, particularly the compression techniques. The bearer services are provided over the service access points C-SAP of the PDCP layer. Each PDCP entity has exactly one PDCP-SAP and can implement any number of compression techniques. In the case of acknowledged transmission over the AM-SAP of the RLC layer, the PDCP entity has the additional task of carrying out the numbering and buffering functions.

6.11 The Network Layer

6.11.1 Functions of the Network Layer

Broadcasting messages A network usually sends system information periodically to all UEs. It is the task of the RRC layer to ensure that these messages are ordered correctly and sent at the right time. Broadcast messages at the access layer are typically specific to a cell, whereas the messages of a CN can also apply to larger areas.

Paging Paging messages are always directed to one UE and are broadcast by the RRC layer of a UTRAN upon request by higher layers.

RRC connection set up and termination between UE and UTRAN RRC connections are initiated by UE. Connection acceptance control in the communicating entity provides the appropriate radio resources. If a RRC connection is broken off, the UE can request that the connection is reestablished. If the connection is ended, the RRC layer releases the radio resources.

Set up, termination and configuration of bearer services If data transmission services are requested from higher layers, the RRC layer sets up bearer services in the user plane. Several different services can be set up simultaneously for one UE, with the lower layers notified of the valid parameters for each respective service over the C-SAPs.

Allocation, release and configuration of radio resources The RRC layer is responsible for managing the radio resources for services in the control and user planes. In addition to having control over connection acceptance, it is also responsible for the distribution of resources to different bearer services of the same RRC connection. The RRC layer can allocate or reconfigure radio resources for an existing RRC connection separately for the uplink and the downlink. Another function controls the reservation of radio resources for the DCH of different UEs on the uplink. This allocation is effected through fast signalling over a broadcasting channel.

Monitoring QoS The RRC layer guarantees the quality of bearer services. This requires that sufficient radio resources are available to the different services.

Mobility functions for RRC connections The mobility functions particularly include *handover* and *location update*. Mobility functions usually necessitate a measurement and evaluation of radio parameters such as connection quality.

Transmission and evaluation of measurement data reports A UE performs measurements on instruction of the RRC layer in the UTRAN. The

RRC layer of the UTRAN notifies the UE of which parameters are to be measured, when they should be measured and how they should be transmitted. Another task of the RRC layer is the transmission of the measurement data from UE to UTRAN.

Routing of PDUs of higher layers On the UE side this function ensures that PDUs are forwarded to the right entities of higher layers; in the UTRAN it provides for the addressing of the right entity of the *Radio Access Network Application Part* (RANAP).

Power control A number of different power control techniques are used in the UTRAN. The power control function of the RRC layer represents the external loop of the interleaved power control and consequently determines the desired value for the inner loop.

Slow dynamic channel allocation in TDD mode In TDD mode in macrocellular networks a cluster size larger than one is sometimes necessary. Consequently, the channel allocation function adapts the radio resources used in a cell, particularly the time slots, to the respective interference situation. Measurements taken over a longer period of time are used as the basis for the decision on the preferred use of time slots.

Cell selection in idle mode The selection of an allocated cell takes place in the RRC layer of the UE when the UE is in idle mode. This allocation is based on measurements of supply quality in the lower layers.

Integrity assurance of RRC messages If necessary, the RRC layer can add a *Message Authentication Code Identifier* (MAC-I) to the beginning of its messages in order to guarantee data integrity.

Encryption control The RRC layer controls the use of encryption methods with data transmission over radio networks.

Timing advance in TDD mode *Timing advance* parameters are controlled in the RRC layer in TDD mode.

6.11.2 Structure of the Network Layer

The structure of the RRC layer on the network side is presented in Figure 6.35. Control information and measurement values are exchanged between the RRC layer and the respective layers below it over C-SAPs. The services of the RRC layer are provided over the service access points *General Control* (GC), *Notification* (Nt) and *Dedicated Control* (DC). On the UE side the *Routing Functional Entity* (RFE) forwards messages to different UE mobility control and connection control entities; on the UTRAN side it forwards messages to different CNDs.

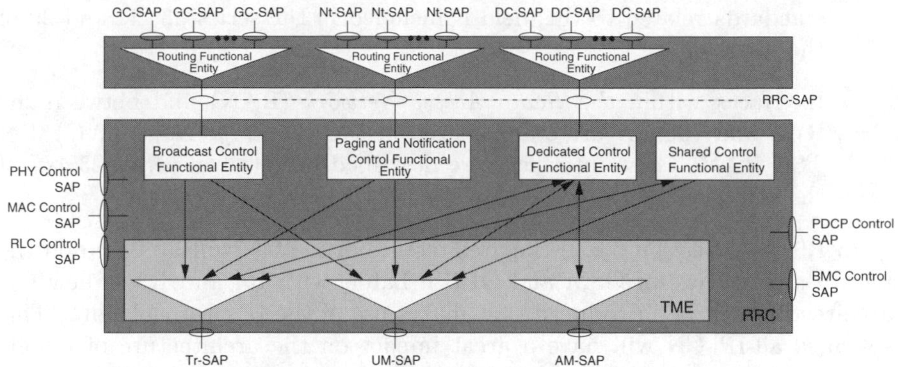

Figure 6.35: The structure of the RRC layer in UTRAN

All broadcasting services provided over a GC-SAP are implemented in the *Broadcast Control Functional Entity* (BCFE). The BCFE can only use the non-acknowledged or transparent transmission services of the RLC layer.

The *Paging and Notification Functional Entity* (PNFE) contains all functions for supplying broadcasting and notification services provided over the Nt-SAP. Like the BCFE, the PNFE can only use the transparent or the non-acknowledged RLC service.

UE-specific functions are located in the *Dedicated Control Functional Entity* (DCFE). The RRC services of this entity are provided over the DC-SAP. The DCFE uses all transmission services of the RLC layer. In TDD mode the DCFE is also supported by functions of the *Shared Control Functional Entity* (SCFE) that control the reservation of the PDSCH and the PUSCH.

The different entities access the SAPs of the RLC layer over the *Transfer Mode Entity* (TME). Since the protocols for GC, Nt and DC SAPs can physically terminate in different units, such as Node-B or SRNC, the various services can also be mapped onto different RLC-SAPs of the same type. Therefore, the SAPs of the RLC layer shown in Figure 6.35 only indicate the type of SAP and not the SAP itself.

In principle, the structure of the RRC layer in UE is the same as in an UTRAN, except that unidirectional connections in the TME are in the opposite direction in the UE (as shown in Figure 6.35).

6.12 UMTS Fixed Network Architecture

The architecture of the UMTS fixed network infrastructure is standardized by different *Technical Specification Groups* (TSGs) of the 3GPP:

- The nodes and interfaces within the CN are standardized in the TSG CN. This includes the interfaces to external networks and the *Open Service Architecture* (OSA) (see also Section 6.13.3.2). Most of the standards related to this field is included in the series 23, 24 and 29 of the standard.

- Interfaces within the *Radio Access Network* (RAN) and between the RAN and the CN are standardized in the *Working Group* (WG) 3 of TSG RAN. These interfaces are described in detail in series 25.4xx of the standard.

In R99 (see Section 6.6.2) the architecture of the CN will be based on the known architecture of GSM and GPRS. Later, with R4 and R5, a new CN architecture will be introduced that makes use of the IP protocol suite. This so-called all-IP CN will have a great impact on the architecture of mobile radio networks and their services and is described in Section 6.12.2.1.

6.12.1 Architectural Overview of R99

6.12.1.1 Nodes and Interfaces

Figure 6.10 shows an overview over the R99 fixed network nodes including the interfaces between them.

Most of the nodes in the CN known from GSM/GPRS are kept in R99 [35]. The MSC implements the interface between the RAN and connected fixed networks. For clarity reasons it is sometimes called *Third-Generation MSC* (3GMSC). It is responsible for the *Circuit Switched Domain* (CSD) and connects to the RAN via the Iu-CS interface. The other nodes like HLR, EIR and VLR are described in Section 3.2.1.3.

The packet-switched domain is based on a modified GPRS network architecture (see Section 4.1). The use of the *GPRS Tunnelling Protocol (User Plane)* (GTP-U) has been slightly modified and now also involves the RAN. This is described in more detail in Section 6.12.1.2.

6.12.1.2 The Iu-Interface between RAN and CN

The interface between the CN and the RAN is called the Iu-interface. Each RNC has no more than one interface towards the circuit-switched domain and one towards the packet-switched domain. Since it is desirable from the network operator's point of view to have a multi-vendor option when setting up a network, the Iu-interface is open. This means that RNCs and CN elements of different manufacturers interoperate if designed according to the standard.

Iu-Interface between RNC and MSC—Circuit-Switched This interface is also called Iu-CS. Since the UTRAN may consist of one or more RNCs there may be more than one Iu-CS interface towards the CN.

The structure of the protocol stack at the Iu-CS interface is depicted in Figure 6.36. In R99, the common transport medium for both user and control plane is ATM. Atop this reside three different protocol suites.

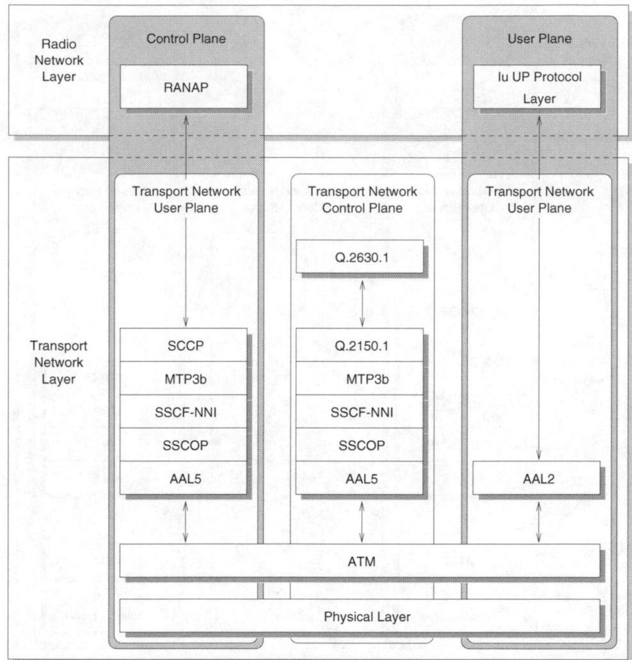

Figure 6.36: Protocol stack at the Iu-interface in the CSD

The protocol stack on the left is the control plane for the communication between MSC and RNC via the *Radio Access Network Application Part* (RANAP). It uses SCCP of the *Common Channel Signalling System No. 7* (CCS7) as signalling bearer [41]. SCCP provides a reliable transfer of control plane signalling messages.

MTP3b provides routing functionality and also load sharing. Below MTP3b, ATM protocols are used. For the control plane, ATM AAL5 with variable bit rates is used to segment the data into ATM cells. The cell headers are built according to the ATM *Network-To-Network Interface* (NNI) [43].

The protocol stack in the middle is used for control plane signalling within the transport network control plane. It is also based on CCS7.

In the user plane data is transmitted directly on AAL2. AAL2 provides efficient use of bandwidth combined with low packetization delay. It is mainly used for speech traffic in telecommunication systems [25]. No further protocols are used in between.

Iu-Interface between RNC and SGSN—Packet-Switched Figure 6.37 shows the Iu-PS-interface between the RNC and the CN in the *Packet Switched Domain* (PSD). This interface connects the RNC to a SGSN. Again, RANAP is used for the control plane of the radio network layer. RANAP

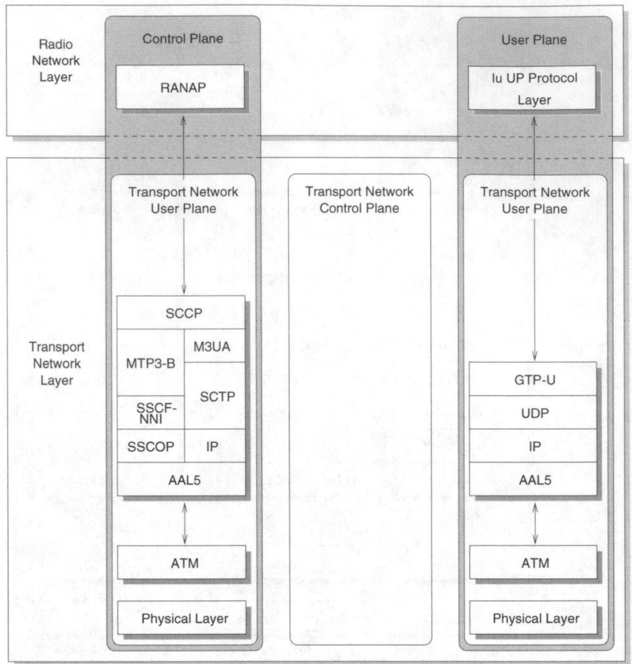

Figure 6.37: Protocol stack at the Iu-interface in the PS domain

either uses the same protocol stack used for the CSD or IP. In this case, the SCCP interfaces with the *MTP3 User Adaptation Layer* (M3UA) protocol, which adopts SCCP to the underlying *Simple Control Transport Protocol* (SCTP), which in turn runs over IP. SCTP is a transmission protocol similar to TCP, that was designed by the IETF to transport small chunks of streaming data between IP hosts. It supports multiple signalling connections via the same SCTP connections. The transmission itself is again based on ATM AAL5.

In the user plane, the protocol stack looks totally different: in contrast to GPRS, the GTP-U protocol reaches into the RAN to the RNC. GTP-U uses UDP/IP as a transport bearer. As for the signalling connection, ATM AAL5 is used.

The Iu-Signalling Protocol RANAP *Radio Access Network Application Part* (RANAP) is an application running atop SCCP, which provides all the control plane signalling functionality specific to UMTS. It is a protocol common to the CSD and the PSD.

The standard [44] names the following basic functions of RANAP:

Relocating serving RNC This function is used to switch the SRNC functionality for a given *Radio Access Bearer* (RAB) between different physical RNCs. The procedure is called *SRNC relocation*.

RAB management This function is responsible for setting up, modifying and releasing RABs. In case of RAB set up, a queueing functionality is provided, if the RNC cannot fulfil a set up request immediatly. The RNC is also able to request the release of a RAB.

Release of all Iu connection resources An Iu connection is a signalling connection between UTRAN and CN set up for exchange of signalling information for a single UE. This function is used to explicitly release all resources related to one Iu connection. This may happen, if a transaction between UE and CN is complete or if a UE has performed a SRNS relocation and is now handled by a new Iu connection. Again, also the RNC is able to request the release of all Iu connection resources.

SRNS context forwarding function In case the UE accesses the CN (PSD) via a new RNC, a new SRNS is used. In this case the CN can request the SRNS context from the old RNC and transfer it to the new RNC. This context contains the next valid sequence number for the GTP-U and the PDCP for both uplink and downlink. This makes it possible to continue the packet stream via the new SRNS at exactly the right place.

Identity management Information elements can be exchanged between CN and UTRAN to inform the RNC of the permanent *Non Access Stratum* (NAS) UE identity (i.e. IMSI). This ID can then be used for identifying the UE.

Transport of NAS information between UE and CN If there is no Iu connection established for a given UE, this connection is set up. Via that link *Non Access Stratum* (NAS) information elements can be exchanged between the UE and the CN. These include mobility management, circuit-switched call control and support for setting up packet-switched services.

Overload control at the Iu-interface This function allows adjusting the load in the Iu-interface by removing traffic from the Iu-interface. Both UTRAN and CN may be source of the overload control information element. In case of UTRAN, two reasons are possible: either processor overload or overload in the capability to send signalling messages to the UE. For the CN, only processor overload is applicable.

Resetting the Iu This function is used for bringing the Iu-interface back to a defined state in case of an failure. The reset can be triggered both by the UTRAN and the CN. All RAB affected will be released. This information element overrides all other procedures.

Paging a UE This function is used by the CN to trigger the broadcast of a paging message by the UTRAN towards a UE. Included in the paging message is the paging area in which a given UE shall be paged for. If no region is given, the RNC will try to locate the UE in the whole RNC area. Furthermore, the RNC is able to detect if a signalling connection to the UE exists in the other domain. In that case the paging request will be rerouted via that connection instead of using a paging channel on the air interface.

UE trace The CN can activate a trace of all UE activity by the UTRAN. Within the control message a OMC can be specified to which the trace shall be directed. If during an active trace a SRNS relocation takes place, the CN must re-enable the UE trace at the new SRNC.

Security mode control in the UTRAN This function is used to set the security keys (ciphering and integrity protection) to the UTRAN, and setting the operation mode for security functions. Since encryption of data is located in the RLC sublayer for acknowledged and unacknowledged mode and in the MAC layer for transparent mode and these layers are found in the UTRAN, parameters related to these features are configured with this message. In case a UE has established connections to both domains, user data encryption shall be handled according to the requirements of each domain.

Location management The CN controls, how and when the UTRAN informs the CN about the position of a UE under control of the RNC. The location information can be sent directly or only, when a service area boundary is crossed by a UE. The location information can be coded as a service area identifier or as geographical coordinates.

Data volume reporting function The CN can request the UTRAN to report the amount of data, that could not be transported via a specific RAB. This command is only applicable in the PSD, where volume-based billing in the CN will count only the data volume that was successfully transported to the UE [40].

Error reporting Each function can report failures during execution to all involved nodes. RANAP also includes a generic error reporting mechanism to report generic errors, e.g., due to syntax errors in a received message.

The physical connection between the CN and the RNC can be based on several different standards named in [42]. This ranges from 1.5 Mbit/s T1 lines and 2 Mbit/s E1 lines up to *Synchronous Transfer Mode* (STM)-1 and STM-4 lines with 155 Mbit/s and 622 Mbit/s respectively [32]. STM-4 can be transmitted only via optical fibre, the other standards can also use an electrical interface.

In a PLMN a RNC can serve up to several hundred Node Bs. In these cases STM lines can be expected, whereas in small pico-scenarios the smaller links will be used.

GPRS Tunnelling Protocol (User Plane) In the *Packet-Switched* (PS) domain, the user data is transported via the *GPRS Tunnelling Protocol (User Plane)* (GTP-U). This protocol is used in UMTS to encapsulate and transport *Transport Layer PDUs* (T-PDUs) through the CN and into the UTRAN. It provides a tunnelling functionality to these PDUs and delivers them at the endpoint of the tunnel. Usually, the PDUs transported will also belong to the TCP/IP protocol suite, but GTP-U can also transport other protocols.

The protocol architecture is described in detail in [46]. More details on GTP-U can be found in Section 6.12.1.5.

6.12.1.3 The Iub Interface between RNC and Node B

The interface between the Node B and the RNC is called Iub. This interface has to fulfil the following basic tasks:

- Transport of user data
- Signalling on handling the user data
- Configuration and *Operating and Maintenance* (OAM)

To do this, the Iu_b interface provides the following functions:

Management of Iub transport resources The RNC sets up transport resources (ATM AAL2 connections) to transport user data from the RNC to the Node B and vice versa.

Logical OAM Some resources are physically implemented in the Node B but belong to the RNC logically (i.e., a channel). In this case, transportation of OAM-message elements is necessary.

System information management The RNC schedules system information messages, that will be transmitted in the cell periodically. It is the task of the Node B to fulfil the schedule. If required, the Node B has to update the system information autonomously according to the guidelines given by the RNC.

Traffic management of common and shared channels The configuration of the common and shared channels at the air interface is controlled by the RNC. As an example, the RNC initiates paging messages transmitted via the paging channels and configures, which UEs are paged for. The RNC also provides scheduling for data transmission over the shared channels.

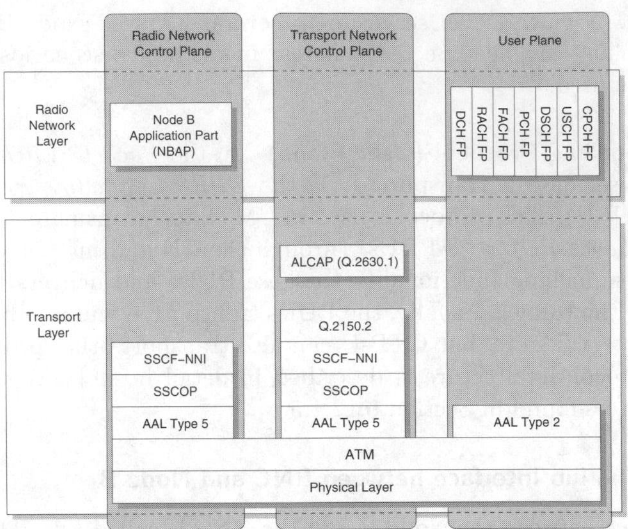

Figure 6.38: Protocol stack at the Iub interface between RNC and Node B

Traffic management of dedicated channels The RNC controls the operation and configuration of the different dedicated channels. This includes Combining and Splitting of data streams from different Node Bs, handover decision, allocation of physical resources like codes, uplink and downlink outer loop power control, admission control and power and interference reporting.

Synchronization The Iub-interface provides mechanisms to control the timing and synchronization of the Node B.

Implementation-specific OAM It will be possible to transport implementation specific OAM configuration mechanisms via the Iub interface without interfering with the normal operation of the interface.

These functions are realized with the *Node B Application Part* (NBAP) protocol. Figure 6.38 shows the protocol architecture at the Iub-interface. Details on the protocol can be found in [45].

At the physical level, the same interfaces as at the Iu-interface can be used. Additionally, multiples of 64 kbit/s timeslots on a fractional E1 link can be used to connect small Node Bs to an RNC.

6.12.1.4 The Iur Interface between different RNCs

Different RNCs can be interconnected via the Iur-interface. This interface was not available in 2G networks and is used to guarantee continous operation of a radio link even when a UE moves between areas controlled by different RNCs.

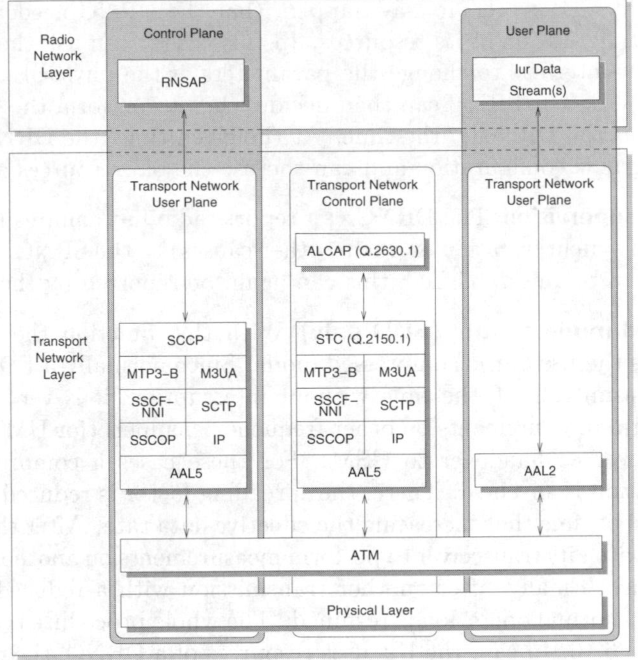

Figure 6.39: Protocol stack at the Iur-interface between different RNCs

In this case it is either possible to perform a hard handover between different RNCs. This means setting up a new Iu-connection and redirecting all traffic via the new RNS. As an alternative, it is possible to make use of the Iur interface and use resources controlled by both RNCs. In this case, the old RNC is still serving the Iu-interface and is therefore called SRNC. The new RNC also attributing resources to this specific connection is called DRNC.

The management of resources at the DRNC is performed with the *Radio Network Subsystem Application Part* (RNSAP) protocol. The protocol stack at the Iur-interface is depicted in Figure 6.39.

All the functionality is implemented in a control protocol called RNSAP. The RNSAP provides the following functions:

Management of the radio link in the DRNS With this function the SRNC is able to control the dedicated resources allocated to this specific radio link at the DRNC. This includes setting up a new link, adding or removing additional radio links to a given UE in the DRNS and reconfiguring already established links. On the other hand, the DRNC may inform the SRNC that it needs to free the resources of a UE controlled by the SRNC. This is called *Radio Link Pre-emption.*

Physical channel reconfiguration Since the DRNC allocates resources for UEs controlled by another RNC, it no longer has full access to these resources. However, it may happen, that the DRNC needs to perform a reconfiguration of its resources. In this case it can ask the SRNC via the Iur-interface to change the parameters of the physical channel of a given UE. The SRNC can then decide when to perform the change and informs the DRNC of the time. At the given time, the DRNC switches to the new configuration and can the use the old resources again.

Radio link supervision The DRNC can report radio link failures (i.e., due to loss of synchronization or due to other causes) to the SRNC. If the radio link can be re-established, this can again be reported by the DRNC.

Compressed mode control (FDD only) With this function the SRNC can toggle the use of the compressed mode. Since normally a FDD terminal is transmitting all the time without interruption, it is very difficult to perform measurements on other frequency channels (for UMTS internal handover or handover to GSM). For these cases a compressed mode operation is specified. There, the spreading factor is reduced for a short period of time thus increasing the effective data rate. After that, the UE can switch its transceiver to perform measurements on another frequency channel. Finally, after another transmission with a reduced spreading factor, normal operation is resumed. The whole procedure is depicted in Figure 6.16. In case the UE uses resources of a DRNC, the compressed mode operation is controlled by the SRNC via RNSAP.

Measurements on dedicated resources The SRNC can request the DRNC to perform measurements on dedicated resources. The DRNC can report on the *Signal to Interference Ratio* (SIR), the transmitted code power value, the received code power value (TDD only), the received timing deviation (TDD only), the round trip time (FDD only) and some others. The SRNC can use this information in its radio resource control algorithms.

Downlink power drifting correction (FDD only) Power Control is very important to FDD operation. In case a UE uses resources from different Node Bs, the UE transmits just a single power control command (increase or decrease power). This command is interpreted by all involved Node Bs. As long as the command is received correctly, both Node Bs are controlled in the same way. If one Node B misinterprets the command there is a gap between the two Node Bs. This gap may increase over time and is called *Power Drifting*. RNSAP provides a mechanism to close this gap again even when the involved Node Bs belong to different RNC areas.

CCCH signalling transfer With this function, a SRNC can transmit information to a UE via the CCHs at cells under control of the DRNC.

Paging The SRNC can trigger a paging of a UE via the paging channels in cells under control of the DRNC.

Common Transport Channel Resources Management A SRNC can also allocate common transport channel resources at the DRNC for user data transmission via RACH, FACH or CPCH (FDD only).

SRNS relocation execution When a UE moves further into the area controlled by the DRNC, it will become more efficient to switch the controlling RNS. This includes switching the Iu-connection from one RNC (the old SRNC) to the DRNC, which will in turn become the controlling RNC. This procedure is called SRNS relocation. With this function a prepared switch of controlling RNC is executed. After that, radio resource control will take place at the new RNC.

Downlink power timeslot correction (TDD only) The RRC entity of the given connection in the SRNC receives the experienced downlink interference level from the UE. This information can be relayed to the DRNC via the Iur interface. The DRNC can then apply an individual correction factor to the downlink transmit power for each timelsot.

Error handling Finally, RNSAP provides mechanisms for error handling.

The Iur-interface is an open interface, so it will be possible to connect RNCs from different vendors via this interface. For transport, the same media and protocols as at the Iu-interface can be used.

6.12.1.5 Protocol Stacks within the Core Network

CS domain The transportation of user data in the CSD is based on the same principles as in GSM Phase 2+. As a transport medium, ATM AAL2 will be used.

PS domain In the packet switched domain, the architecture is similar to GPRS. As mentioned above, the IP protocol stack is used not only in the CN, but extends into the UTRAN. Figure 6.40 shows the protocol stack in the PSD.

The basic principle of user data transmission in the PSD is as follows [37]:

- Once a UE activates a PDP context in the GGSN, incoming IP traffic for this UE is encapsulated into the GTP-U protocol and transported via UDP and IP towards the SGSN.
- The SGSN determines, which *Receiver Address* (RA) the UE is in and forwards the packet via another GTP-U link to the appropriate RNC.
- In the RNC, the encapsulated IP packet is taken out of the envelope and transported via the PDCP across the Uu interface towards the UE.

As a consequence, the encapsulated user IP traffic does not mix with PLMN-internal IP traffic at the Gn-interface but tunnels through the PLMN.

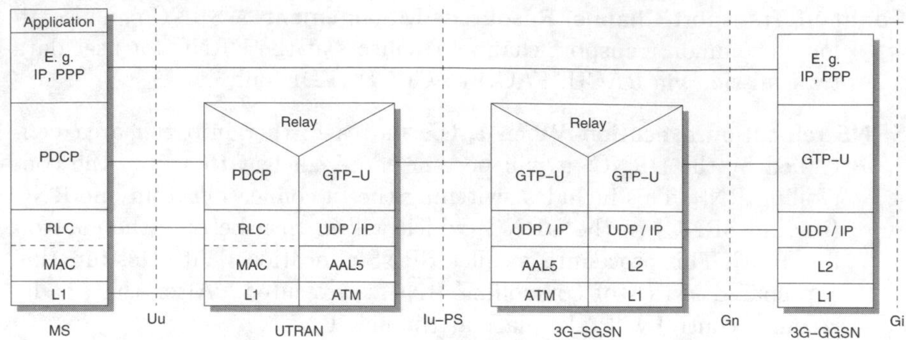

Figure 6.40: Protocol stack in the PS CN domain

GTP-U is not limited to transportation of IP traffic. It can also transport various other protocols, but IP is the main application. For R99, IPv4 will be used for transportation of GTP-PDUs. Later, IPv6 shall be supported [37].

The signalling between GGSN and SGSN takes place via the *GPRS Tunnelling Protocol (Control Plane)* (GTP-C) protocol. This protocol is used for PDP context management, routing management and the setting up and release of user protocol tunnels. Signalling between RNC and SGSN is done with RANAP.

A very detailed description of the GTP-protocol suite can be found in [46].

6.12.2 Advancements with R4 and R5

Several features have not been introduced into R99 but are planned for R4, R5 or even beyond. Examples from the workplan for R4 plans are:

- Further work on QoS support via the ATM AAL2 links within the RAN
- Low Chiprate option. Based on the Chinese contribution to the IMT-2000 family, UMTS R4 will introduce a modified air interface for the TDD mode called *Low Chiprate TDD option.* See Section 6.9.6 for details.
- Improvements within the RAN. This includes the introduction of *Robust Header Compression* (RHC) for use in the PDCP and synchronization between different Node Bs in TDD.
- The *Open Service Architecture* (OSA) will be used to define further commercial applications and interfaces.
- Support of *Multimedia Messaging Service* (MMS) as a successor to the SMS known from GSM.
- Work on the security features and over-the-air programming issues of the MExE, a software runtime environment within the terminal.

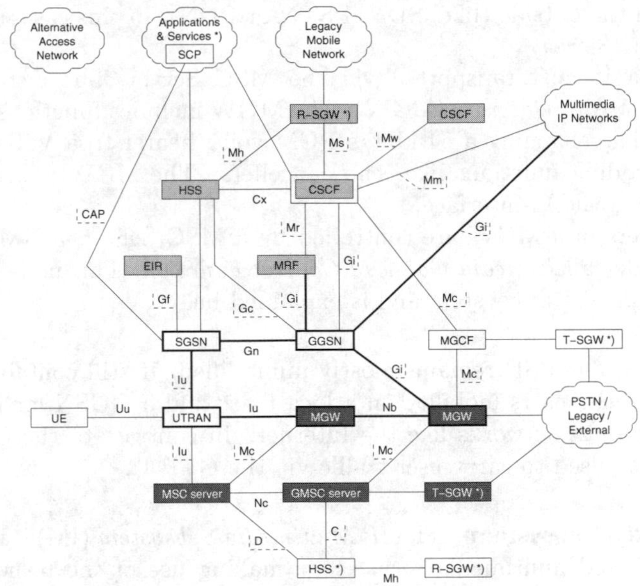

Figure 6.41: Architecture of a R5 all-IP UMTS Core Network

- Extensions to the *SIM Application Toolkit* (SAT) interface in the USIM.

.

This list is far from being complete and the feature list changes all the time. In the 3GPP project plan more details can be found [1].

6.12.2.1 All-IP Core Network

With R5, a complete redesign of the CN architecture is planned [36]. Most of the signalling will be based on IP. Several new nodes are introduced and new protocols will enable easy interworking with the Internet world.

Figure 6.41 depicts the architecture of an all-IP-based CN. The bold lines show the interfaces carrying user traffic whereas the dashed lines show the signalling connections. The network can be split into three distinct parts, the *Circuit Switched Domain* (CSD), the *Packet Switched Domain* (PSD) and the *IP Multimedia Subsystem* (IM). It should be noted that some of the logical nodes mentioned in the following can be integrated into a single physical node.

CS Domain In contrast to the known integrated architecture of GSM and UMTS R99, the signalling has been separated from the switching of the user traffic and put into a different logical node. Signalling, especially call control and mobility management towards the radio interface, is handled by a node called *MSC Server*. At the edge of the PLMN, an MSC Server may also include *Interworking Functions* (IWFs) to convert protocols used in the PLMN to the protocols on the outside (like ISDN, PSTN or PDN). In this case, the node is called *GMSC Server*.

User data is not transported via the MSC Server, but instead passes through the *Media Gateway* (MGW). The MGW includes functions for media conversion, i.e. changing a 64 kbit/s PCM voice bearer to a VoIP bearer or other transcoding functions like echo cancellers. The MGW is connected to the UTRA via the Iu-interface.

The bearers of a MGW are controlled by a MSC Server, a GMSC Server or a new node called *Media Gateway Control Function*. This node belongs to the IP Multimedia Subsystem and is explained below.

PS Domain The PSD remains mostly unmodified. It still contains a SGSN for control of the users mobility on a local base and a GGSN for interfacing other packet data networks like the Internet. In contrast to the CSD, these nodes are also used to carry user traffic via the GTP-U.

IP Multimedia Subsystem The *IP Multimedia Subsystem* (IM) is introduced to offer IP-based multimedia services by making use of the bearer services offered by the PSD [39]. While all the user traffic is routed through the PSD, the IM is used to perform all the necessary database queries and to do all the signalling related to the new services.

Located in the IM are most of the new nodes: most important is the so-called *Call State Control Function* (CSCF). This unit has several functions:

Incoming Call Gateway (ICGW) The CSCF acts as an entry point for the signalling of incoming calls. It performs the necessary database queries for location management and routes the call to its destination.

Call Control Functions (CCF) Since the CSCF controls the flow of user data through the network, it needs functionality for call control signalling. In the CCF functions like call forwarding, *Virtual Private Network* (VPN) services etc. will be realized.

Serving Profile Database (SPD) To perform its tasks, the CSCF interacts with a new integrated database called *Home Subscriber Server* (HSS). This HSS contains user- and service-related information and replaces the HLR.

Address Handling (AH) It may be necessary to do translation of addresses, i.e., for call rerouting or call forwarding. This task is done by the AH function in the CSCF.

During a service set up, several CSCFs may be involved in the handling of the call. Three distinct roles of the CSCF can be identified:

Proxy-CSCF (P-CSCF) The P-CSCF is the CSCF handling all communication with a given UE in the network. All service control related signalling

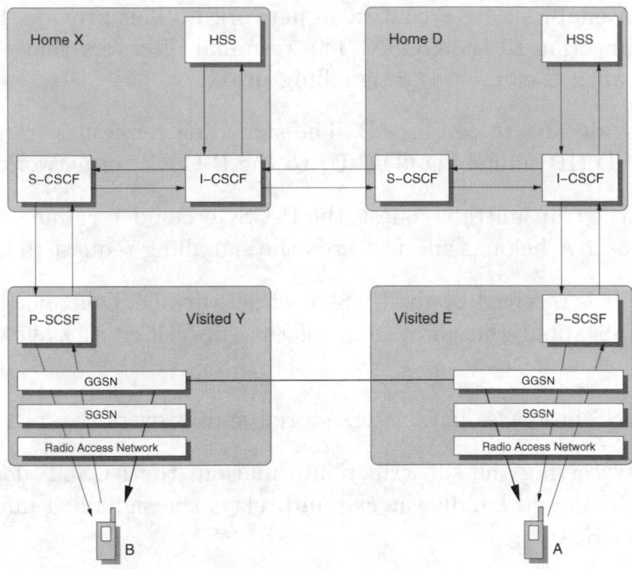

Figure 6.42: Interaction between different CSCFs for a call set up

to this UE passes through this CSCF. Since all the signalling it does is invisible to the UE, it acts like a proxy for the UE.

Interrogating-CSCF (I-CSCF) The first CSCF in a PLMN that is involved in the handling of a call is called I-CSCF. It performs the necessary database queries making use of its above mentioned functions and decides how to proceed with the call.

Serving-CSCF (S-CSCF) The S-CSCF finally is responsible for providing the service.

Between the CSCF-nodes, a protocol standardized by the IETF called *Session Initiation Protocol* (SIP) is used. This protocol is used in the Internet for setting up conferencing, telephony and multimedia calls between two or more participants (`www.sipforum.com`). There are already some SIP-enabled products available. The use of SIP is also planned for the signalling between the UE and the P-CSCF.

The following example explains how different CSCFs interact in a mobile to mobile telephone call using the IM [29].

User A is subscriber to services of the network D. He is currently roaming in the area served by network E. This user wants to start a voice call to user B, who has a subscription at network X. User B currently roams in the area of network Y. The following signalling will occur during the set up of the call:

1. User A enables a PDP context in network E. This provides him with an IP-connection to the GGSN. The terminal discovers that the P-CSCF in network E can act as a signalling proxy.
2. User A decides to call user B. The signalling request is transported via the PSD (IP tunnel via GTP-U) to the P-CSCF in network E.
3. From the transmitted request, the P-CSCF can determine, to which network user A belongs and forwards the signalling request to the network.
4. There it is received by the I-CSCF of network D. This node interrogates the HSS about the parameters of the subscriber, the allowed services etc.
5. The HSS has these parameters stored and answers the query.
6. After receiving the subscriber information, the I-CSCF decides which S-CSCF should handle the call and relays the signalling information to this S-CSCF.
7. The S-CSCF determines that the call should be routed to network X. This means setting up a signalling connection to the I-CSCF of network X.
8. In network X, the same procedure as in network D takes place again. The I-CSCF queries the HSS of network X on the subscriber information and location of user B.
9. The HSS answers the query and provides the I-CSCF with the appropriate information.
10. The I-CSCF now determines which S-CSCF should handle the call and forwards all necessary information.
11. In the S-CSCF the call is handled and a connection is set up to P-CSCF in network Y, where user B is currently attached to.
12. This P-CSCF does all the signalling with user B. Once the connection is established, the information is relayed back via S-CSCF (network X), S-CSCF (network D), I-CSCF (network D) and the P-CSCF of network E to user A. This answer message contains the current IP-adress of user B attached to network Y.
13. User A now transmits his speech data as VoIP packets directly to this IP address. The data tunnels through the PSD of network E, passes through any intermediate IP networks and is received by the GGSN of network Y. There it is encapsulated in a GTP-U-tunnel and transmitted via a SGSN to user B.

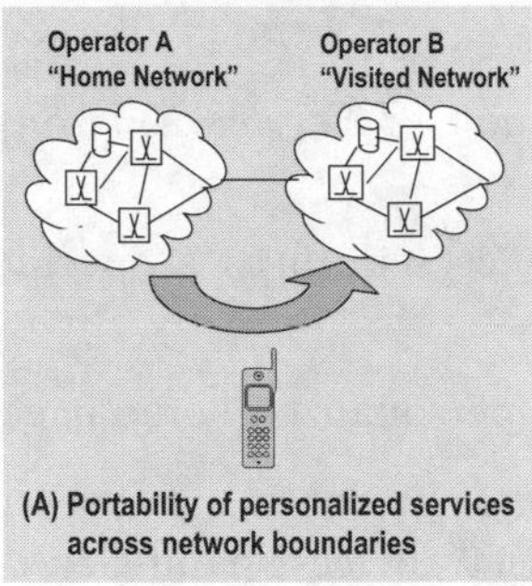

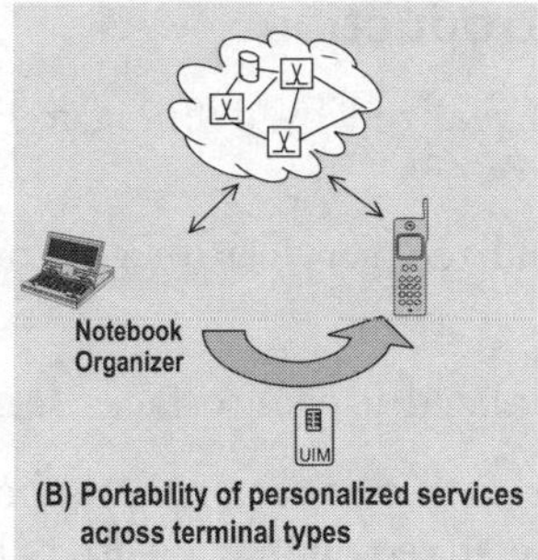

Figure 6.43: Scope of the Virtual Home Environment

In 3GPP, it has already been decided to use SIP both for network-to-network signalling and also for network-to-user signalling. The details of the protocols have not been decided upon yet so it will take some time until an all-IP based CN will be commercially available. Most of the components of a R99 CN cannot be used in such an architecture.

The main driver for the use of SIP and IP in the CN is that this approach makes it very easy to introduce new services of any kind. Furthermore, since the call handling passes through the home network of a user, specific services a user has subscribed to are available to him even if he is roaming abroad. Therefore each network operator can deploy his own services and these are available to his users everywhere, independent of the services available in the visited network. With the all-IP CN architecture, it will be easy to introduce a *Virtual Home Environment* (VHE).

6.13 Virtual Home Environment*

Features to support roaming from the home network to visited networks have been introduced in GSM based on standardized tele-services (voice calls, data calls, SMS), see also Section 3.10. To allow for more service differentiation between service providers, UMTS services will be built on standardized service capabilities rather than fully specified end-user services [34]. Service capabilities are building blocks which enable service providers to design end-user services in a flexible and personalized way. As a result of the increased flexibility, roaming to different serving networks (UMTS terminology for visited networks) has become more complex. The *Virtual Home Environment* (VHE) concept addresses therefore the portability of personalized services across different serving networks as well as different terminal and bearer types (see Figure 6.43):

***With the collaboration of Christian Wietfeld**

6.13.1 Introduction

The VHE is realized by defining a *Personalized Service Environment* (PSE) which consists of:

- personalized services (e. g., conference set up to a specified group of users),
- personalized user interfaces (e. g., a personalized home page),
- consistent set of services from a user's perspective irrespective of access (mobile, fixed, home, visited, handset devices, palm-size device, etc.), especially in the case of global roaming.

The VHE will allow the user to configure his own personal profile and *Man-Machine Interface* (MMI) for his terminal(s). This will provide the mechanism to set up calls, invoke value added services, etc., often even when not directly supported by the serving network. This personal profile should always work in the same way, as far as possible, whether on a fixed network, cordless, cellular or satellite, public or private, whether at home or roaming. Although VHE is being fully introduced only with UMTS, several relevant technologies have been introduced already in GSM networks.

The VHE is accomplished by either preserving, transferring, emulating or mapping of service elements from the home network (service provider) into the visited network.

Examples of VHE mechanisms include:

- Home-to-visited core network interaction:
  - relaying of service control to visited network
  - direct service control of visited network by home network (e.g., CAMEL)
  - shadowing of service control from home to visited network
  - adaptation to differing capabilities of home and visited network
- Download of messages, data and programs 'over the air' from the home network to the terminal/ *UMTS Subscriber Identity Module* (UIM) (SIM Toolkit, MExE)
- Transparent mobile terminal/UIM-to-home network (service provider) interaction based on mobile execution environments (client-server interaction, Java-like, WAP-like approaches)
- Support of user-specific MMIs
- UIM-to-visited network interaction: service profile/service logic transportability with UIM support (home service logic is carried to the visited network via the UIM).

To provide a better insight in the various technologies and tools to accomplish the VHE, the next section introduces a generalized VHE [26].

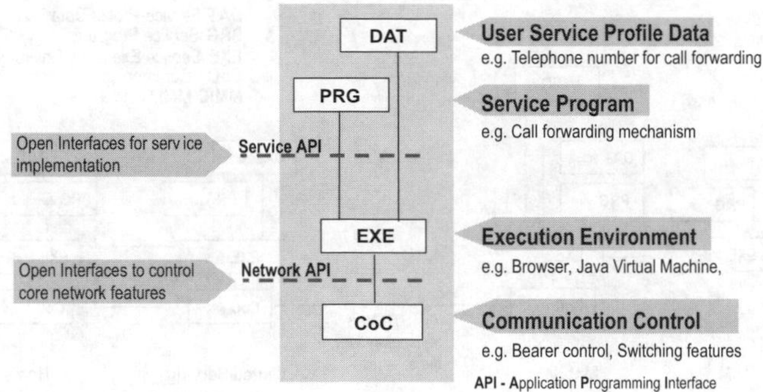

Figure 6.44: Core functions involved in the execution of VHE services

6.13.2 VHE Architecture Model

As core element of the architecture, the following *functional components* are introduced (see Figure 6.44):

- The Service Program (PRG) describes the behaviour of a service and its corresponding service elements by means of (standardized) commands. The behaviour described by the PRG may be standardized, network- or even user-specific.
- The Service Execution Environment (EXE) provides a (standardized) platform to execute a service program and provides access to the communication resources. The service execution environment is accessed via (standardized) Application Programmers Interfaces (API), e.g., Java-based. The execution environment also protects the communication control from unauthorized access.
- The Service Profile/Data (DAT) provides user- or network-specific input data to run a service program.
- The Communication Control (CoC) handles actual communication (i.e., allocates bearers, handling of SMS, controls switching functions such as the set up of conference calls, etc.).

Open *Application Programmers Interface* (API)s are able to define services as well as to access the communication control functionality. The Service APIs will be typically accessible for any application/service programmer (e.g., in case of WAP the WML (see Section 4.6.3.10) has been introduced), whereas network APIs typically are restricted to inter-network communication and access by third party service providers. These functional elements can be replicated in all relevant domains of the UMTS network architecture (see Figure 6.45).

The corresponding *network components* are:

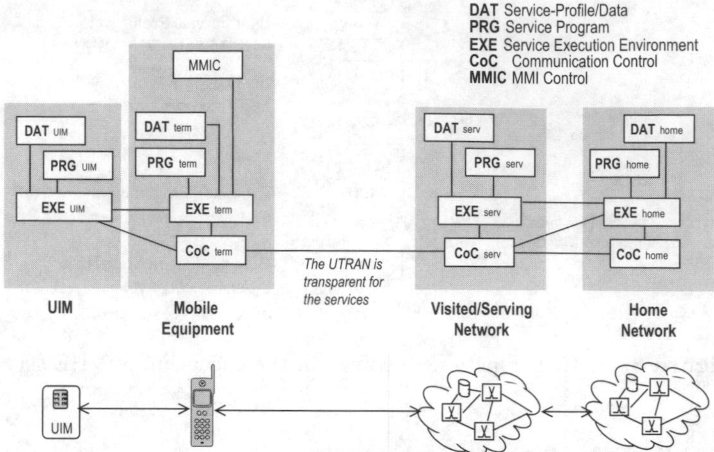

Figure 6.45: VHE architecture model

- The *Mobile Equipment* (ME) which provides CoC, EXE, DAT, PRG. For the mobile terminal, the MMI Control *MMI Control* (MMIC), which provides network-/user-specific control of MMI (triggered by Execution Environment), is additionally introduced to cater for the interaction between the user and the underlying terminal capabilities.

- The UIM, which may provide user-specific and often also home network (service provider) specific DAT and PRG as well as an EXE.

- The *Home Network* (also sometimes denoted as service provider network), which holds CoC, DAT, PRG as well as EXE. The home network in particular stores the *user profiles*, which contain *user interface* information (e. g., menu settings, terminal settings, language settings, etc.) and *service profile* information (e. g., list of supported services and associated service data such as redirection numbers, redirection conditions, call screening lists, etc.).

- The *Visited Network* (in the UMTS environment also denoted as serving network), which holds—similar to the home network—CoC, EXE, PRG, DAT.

A key characteristic of the VHE architecture model is that service data (in particular user profiles) and service programs may be stored in a distributed way in the UMTS network (e. g., service provider, serving network, terminal, UIM). The data and program codes may be transferred in a flexible way in the network (either 'downloaded' or 'pushed') as required by the service provider and/or user. Secure administration mechanisms (e. g., authentication, validity checks, update procedures, location, etc.) of the transferred programs and data have to be employed to ensure network integrity and user privacy.

6.13.3 Overview of Concrete VHE Realization Options

The following scenarios for the realization of VHE can be differentiated by identifying, where the actual service execution *(service control)* is located:

- Service Execution in the Home Network (Service Provider)
- Service Execution in the UIM
- Service Execution in the Mobile Equipment
- Service Execution in the Serving Network
- or any combination of the above options

The following sections will outline how these options can be realized by existing and future GSM (Phase 2+) and UMTS functionalities, e. g., CAMEL, SIM Application Toolkit, MExE.

6.13.3.1 Service Execution in the Home Network (Service Provider)

In this scenario the service will be controlled by the home network by directly accessing call control functionality in the serving network. This principle is used by CAMEL [13] where the control of the services is located in the CSE of the home network. The serving network is directly controlled through the *GSM Service Switching Function* (GSMSSF). This scenario can be used, e. g., for the support of operator-specific services, cross-phase compatibility (support of UMTS services by non-UMTS networks, e.g., 2nd generation systems), support of UMTS for simple terminals, support of (user-specific) supplementary services.

The service control as specified in CAMEL can be described in terms of the architecture model in the following way (see Figure 6.46, (A)): The execution environment of the home network directly interacts with the communication control in the serving network. The corresponding interface is either standardized (e.g., as described by the CAMEL specifications) or bilaterally agreed between home and serving network. No service program and service data needs to be transferred between home and serving network.

6.13.3.2 Service Execution in the Serving Network

In this scenario the service will be controlled by the serving network. To transfer the service logic and/or service-specific/user-specific data to the serving network different possibilities exist, which are described in more detail in the following sub-sections:

- using ITU CS 2 mechanism as defined as part of the IMT-2000 network architecture [19, 20]
- making use of OSA functionalities [38]

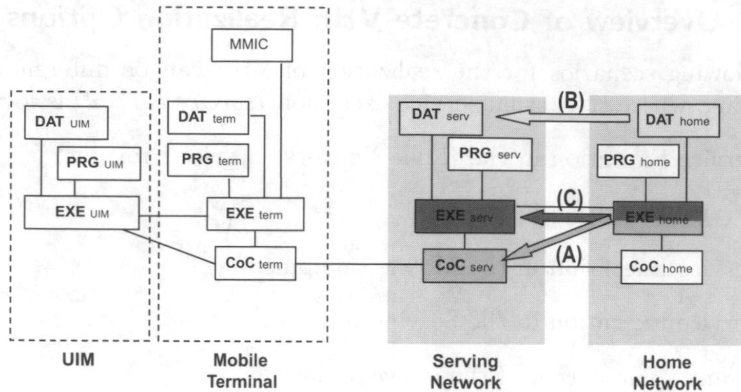

Figure 6.46: Support of personalized services through CAMEL (A), IN (B) and OSA (C)

- downloading from the home network (service provider) to the serving network
- downloading from the UIM to the serving network

Using Intelligent Network (IN) mechanism The service control according to IN principles (as defined in Capability Set 2 of the IN ITU-T standards, see [20] and Chapter 17) may imply the download of user-specific service profile data from the home network to the serving network (see Figure 6.46 (B)). The execution environment of the serving network fully controls the access to the communication control. No program code is exchanged, as the behaviour and input parameters of the services are either standardized or bilaterally agreed between home and serving network.

Using the Open Service Architecture (OSA) interfaces The above described IN mechanisms have their origins in a fixed network environment. They are considered to be not flexible and open enough to cover all requirements of future mobile network environments. To provide a more flexible interface to control services in a serving network, the *Open Service Architecture* (OSA) concept has been introduced in UMTS [38]. The service capabilities of a network can be accessed through a so-called *Service Capability Server* (SCS). The SCS (often also denoted as OSA gateway) provides a comprehensive set of APIs to access a network's service capabilities by a so-called OSA client. Example service capabilities include: call control, access to a user's location, access to terminal capabilities and message transfer functions. Although the OSA concept was primarily designed to allow third party application developers to provide network-oriented services (such as setting up of phone conferences, delivering messages, etc.), these mechanisms are very well

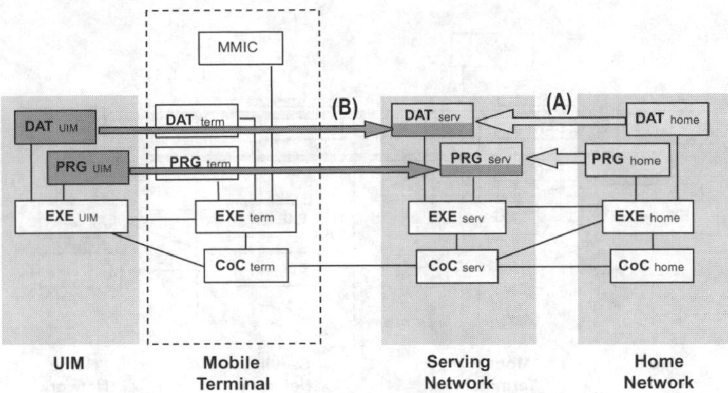

Figure 6.47: Support of personalized services by download of data and service programs to the serving network (later phases of UMTS)

suited to support the realization of the VHE requirements. As an OSA gateway will also provide IP-based interfaces, this approach is especially suited for IP-based environments.

In terms of the VHE architecture model, the SCS or OSA gateway serves as an Execution Environment, which provides a standardized access to the Communication Control as well as to User Service Profile Data (e. g., the CAMEL Service Environment or the HLR functionality) of the visited network (see (C) in Figure 6.46). The OSA client can be associated with the home network (as a Service Program running on the Execution Environment of the Home Network).

Downloading from the home network (service provider) to the serving network Another approach is the download of complete service programs (e. g., Java code) and associated service/user profile data (see Figure 6.47 (A)). The execution environment in the serving network uses the downloaded programs and data to interact with the communication control. Due to concerns regarding security and network integrity, this mechanism is foreseen for later stages of UMTS.

Downloading from the UIM to the serving network As an extension of the approach described above, the service/user profile data and even service programs may be stored also in the UIM and downloaded from the UIM to the serving network (see Figure 6.47 (B)). By doing so, the user may modify the service data and program according to his current needs without needing interaction with his home network. Due to the above mentioned security and network integrity concerns, this mechanism is not foreseen for early phases of UMTS.

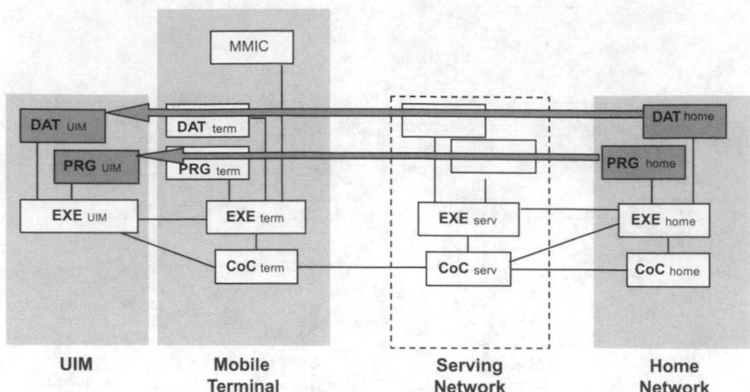

Figure 6.48: Mapping of the SIM Application Toolkit (SAT) onto the VHE architecture model

6.13.3.3 Service Execution in the UIM

In this scenario the service will be controlled by the UIM that will use functionality of the ME to execute the service, e. g., for the personalization of the MMI by adding new menu entries. This principle is used, e. g., by the *SIM Application Toolkit* (SAT) [11]. The SAT is a capability of the UIM to store service data and programs as well as to provide an execution environment that interacts with the mobile terminal (see Figure 6.48). As the SAT is linked to the security features of the UIM, the supported services include specifically banking and payment transactions.

6.13.3.4 Service Execution in the ME

In this scenario the service will be controlled by the ME that interacts as client with a specific server in the network. The functionality described by the *Mobile Station Execution Environment* (MExE) specifications (see Section 4.6.2) is based on this approach [12]. The MExE concept covers both WAP (see Section 4.6.3) [15] as well as Java-enabled clients. MExE clients are classified according to the following so-called classmarks: MExE class mark 1: WAP client; MExE class mark 2: WAP and Java client; MExE class mark 3: Java client (including a WAP browser as Java application).

WAP-enabled clients are equipped with a browser that supports a specific mobile-optimized mark-up language (WML) as well as a scripting language (WMLscript, see Section 4.6.3.11). Both are derived from Internet standards (HTML resp. Javascript). The WAP browser provides a universal user interface, which may run on any device type. By controlling services through the WAP browser, services may be easily ported from one device type to another. Java-enabled clients host a mobile-optimized Java Virtual Machine to allow for the execution of service programs.

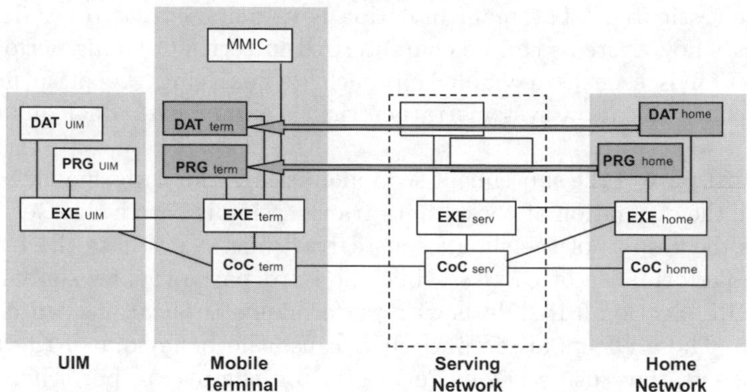

Figure 6.49: Mapping of MExE on VHE architecture model

The case of MExE can be mapped to the architecture model in the following way (see Figure 6.49): the execution environment in the terminal uses service programs and user-specific data provided by the ME or the UIM to interact with the communication control and MMI control. The service program and data will be typically downloaded from a server in the home or even serving network. Apart from the provisioning of the basic bearer services (e. g., a standard data call), the serving network is fully transparent for the service execution.

6.13.4 Summary

As described in the previous paragraphs, the *Virtual Home Environment* requirements can be met by making use of various technologies, ranging from *SIM Application Toolkit*, WAP, CAMEL to OSA and Java-enabled clients. The increased flexibility in UMTS networks will lead to new personalized services and increased potential for differentiation between mobile operators. As a result of the implementation of VHE mechanisms, the user can switch seamlessly between terminals and serving networks, without changing his Personalized Service Environment.

6.14 Performance of UMTS and Comparison with EGPRS*

This section presents simulation results and provides a comparison of the UMTS and EGPRS systems when they are used to access the Internet. Packet data is moving more and more into the focus and therefore the WWW service

*With the collaboration of Michael Meyer and Peter Stuckmann

is here investigated. The main intention is to demonstrate to students and enterprises how systems can be characterized in terms of traffic performance. This section is seen as a viable approach to analysing the performance of mobile communication systems rather than a study with final quantitative results.

The first part of the subsequent section describes the performance measures on which the evaluation of packet data transmission is based. The second part presents the results of the simulations carried out to evaluate the behaviour and the performance of UMTS, while the third part provides similar results for EGPRS. Section 6.14.4 deals with performance implications caused by the Internet. These effects are important because the behaviour of the Internet Protocol suite has also a significant impact on the overall performance. For example, IP packet losses might occur in the Internet, which cause certain protocol reactions leading to performance influencing interactions. Finally, a comparison of UMTS and EGPRS referring to the simulated packet data traffic is made in the last part of this section.

All the presented results are valid only under the assumptions made. No attempt is made to cover all possible parameter settings of the systems nor to include the newest changes in the standards (Release 99).

6.14.1 Performance Measures for Evaluation of Wireless Packet Data Systems

To evaluate packet data communication systems on an end-to-end basis, performance measures are needed which allow a performance comparison of different systems. This section will introduce the measures which are used within this study to evaluate and compare the UMTS and the EGPRS systems.

6.14.1.1 Packet Bit Rate

Contrary to Section 4.4.2 where the IP throughput per user is regarded as a performance measure, here a measure called *Packet Bit Rate* (PBR) is applied.

The packet delay is a widely used measure for the evaluation of packet transmission. However, the transmission delay is influenced both by the packet size and the instantaneous system state itself. Thus it is difficult to interpret resulting delays. In order to evaluate the system performance when the packet size is not constant, a normalized measure is required. The PBR is suited to this requirement. The PBR is defined as the packet size divided by the time it takes to transmit the packet from one peer to the other and is measured in kbit/s. Typically it is applied to application layer data, i.e., files or Web-objects, which may correspond to many IP packets. The advantage of looking into the PBR at the application layer is its relevance to characterize the user perceived performance. Conceptually, all overhead introduced by the lower layers of the protocol stack are included. Such overheads may come from protocol header fields, segmentation, signalling or retransmissions. Since the user usually does not care which part of the network is introducing delays, the PBR is a very natural way to describe what the user perceives.

6.14.1.2 System Throughput

A second important measure applied in this evaluation is the system throughput. The system throughput gives the total throughput of the system measured in kbit/s. If the offered load increases, the throughput saturates and the system becomes congested. This allows the determination the system capacity and offers a measure to evaluate how the system performs overall. In the presentation of the simulation results the system throughput is used in order to describe the behaviour of UMTS and EGPRS for different simulated scenarios.

6.14.1.3 Spectral Efficiency

To compare different communication systems, the system throughput as presented above is not suitable, since various systems may use different bandwidths. Therefore a measure is needed, which normalizes the system throughput for a given bandwidth. To satisfy this requirement the *spectral efficiency* will be used as defined in Equation (2.31):

$$\textit{Spectral Efficiency}\left[\frac{\mathrm{bit/s}}{\mathrm{MHz}\cdot\mathrm{km}^2}\right] = \frac{\textit{Traffic}}{\textit{Bandwidth}\cdot\textit{Area}}. \tag{6.4}$$

It puts the system throughput into a relationship with the allocated bandwidth. As the allocated bandwidth per cell is different for the simulated systems, this measure is used to compare the system performance of EGPRS and UMTS.

6.14.2 Performance of WWW Traffic over WCDMA (UMTS, FDD mode)

This section analyses the system performance of UMTS systems carrying WWW traffic-based on results of the performed simulation studies. Most measurements are based on an end-to-end perspective in terms of throughput and PBR under various load conditions. Due to the asymmetry of WWW traffic, the performance was measured in downlink direction. The uplink carries only the requests of the WWW client and also signalling traffic and is therefore of minor interest. The UMTS system behaviour has been investigated for various WWW traffic load situations. The applied simulation parameters are listed in Table 6.23. Please note that just a single cell is simulated. Therefore, the impact of softhandover is not considered in this study. In addition, it was assumed that there is always a sufficient number of orthogonal codes available to support all users. In cases when too many users transmit and introduce too much interference, those users that are using the highest power

Table 6.23: WWW simulation parameters for UMTS

Parameter	Values
Number of MS	10, 30, 50, 100, 150, 200, 250
MS max. data rate (kbit/s)	64, 128, 256
E_b/N_0-target (dB)	1.5 ⇒ BLEP ≈ 10 %
Cell size (km)	3
Number of 5 MHz carriers	1
Maximum TCP transmit window (bytes)	16384
TCP max. segment size (bytes)	512
Mean Internet packet loss ratio (%)	1

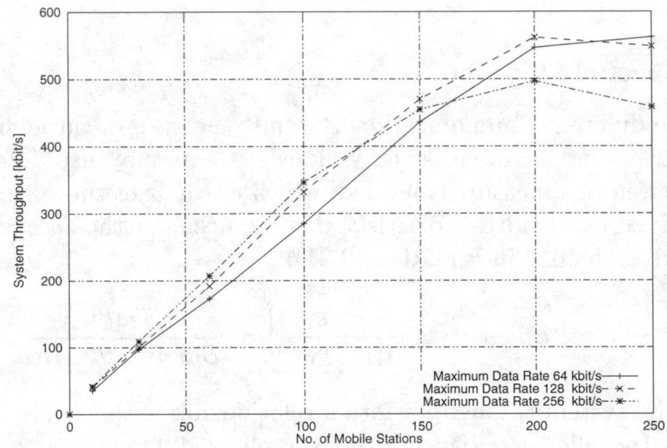

Figure 6.50: System throughput for WWW-traffic in UMTS

are downgraded to lower data rates. These are typically the users at the cell border. Such an environment is denoted as interference limited.

6.14.2.1 WWW System Throughput Performance in UMTS

Figure 6.50 shows the UMTS system throughput on application layer for a varied number of WWW users and for different maximum allowed data rates per MS. The maximum information data rate on the MAC layer per MS is limited to 64 kbit/s, 128 kbit/s or 256 kbit/s and a system throughput curve for each of these cases is plotted.

The results exhibit the typical behaviour of systems, where multiple user share a common resource, here the power budget of a cell. The traffic load increases when the number of MS is increased. At a certain point the system is saturated and becomes congested. The maximum system throughput varies with the maximum allowed data rate per UE, as can be seen from Figure 6.50.

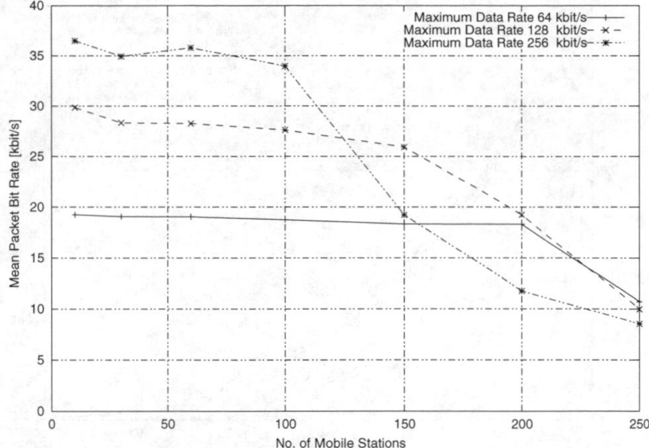

Figure 6.51: Mean packet bit rate in UMTS

The results show that in the low load case, the highest system throughput can be achieved, if the mobile stations are allowed to transmit with high data rates. With 60 MS in the system, the system throughput is 172.9 kbit/s, 191.6 kbit/s and 206.9 kbit/s for a maximum data rate per UE of 64 kbit/s, 128 kbit/s and 256 kbit/s, respectively. With an increasing system load this order changes and the highest system throughput is reached for the 64 kbit/s and 128 kbit/s scenarios at a level of 562 kbit/s.

System throughput considerations as presented above, permit determination of the system capacity for the applied simulation parameters. However, it makes no assertion about perceived data rates from the user's point of view. To show how the system performs from this viewpoint, the next section discusses the UMTS system performance in terms of the PBR.

6.14.2.2 WWW Packet Bit Rate Performance in UMTS

Figure 6.51 depicts the mean PBR measured for application layer packets versus the number of MS. Corresponding to the different maximum data rates on MAC layer of 64 kbit/s, 128 kbit/s and 256 kbit/s, three curves are plotted.

The behaviour in relation to the mean PBR is analogous to the system throughput behaviour. In low load cases the highest PBR is reached in the scenario where UEs are allowed to transmit data with 256 kbit/s. In this case the mean PBR is on a nearly constant level of approximately 35 kbit/s if less than 100 MS are in the system. For more than 100 MS the mean PBR decreases and reaches a level of 8.5 kbit/s for 250 MS. In the scenario where UEs transmit data with a maximum data rate of 128 kbit/s, the mean PBR is 29.8 kbit/s for 10 MS. It decreases slowly to a level of 25.9 kbit/s for 150 MS and then faster to 10 kbit/s when 250 MS are in the system. In the case of a maximum bit rate of 64 kbit/s per UE the mean PBR is approximately 18 kbit/s

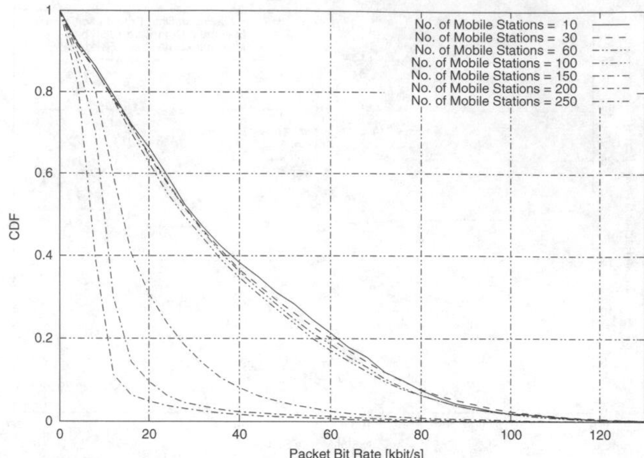

Figure 6.52: Complementary cumulative distribution function (CDF) of PBR and maximum data rate per MS with max. data rate of 256 kbit/s

until the load corresponds of 200 MS. From this level it decreases to a mean PBR of 11 kbit/s in case of 250 MS.

In order to better understand this behaviour PBR distributions have been produced. The obtained PBR curves for various load situations show the following behaviour:

- Even in low load situations, the PBR does not reach the maximum available bit rate. In the 256 kbit/s scenario (see Figure 6.52) the highest observed PBR is 150 kbit/s. Less than 10 % of all application packets exceed a PBR of 80 kbit/s. For the 64 kbit/s scenario no packet exceeds a PBR of 37 kbit/s and in the 128 kbit/s scenario 80 kbit/s is the upper limit reachable.

- Up to a certain number of MS the PBR distributions vary only slightly for each scenario and all PBR curves are close together. This indicates that the different users are well decoupled until the interference becomes the limiting factor.

- In high load situations the PBR decreases rapidly.

In summary, the simulation results show that only a small fraction of application packets gain high PBRs and the maximum observed PBRs are far below the offered bit rate. To explain this behaviour, the next section deals with protocol overhead considerations. Furthermore, in Section 6.14.4 simulations carried out with bulk data traffic are presented in order to further explain this system behaviour.

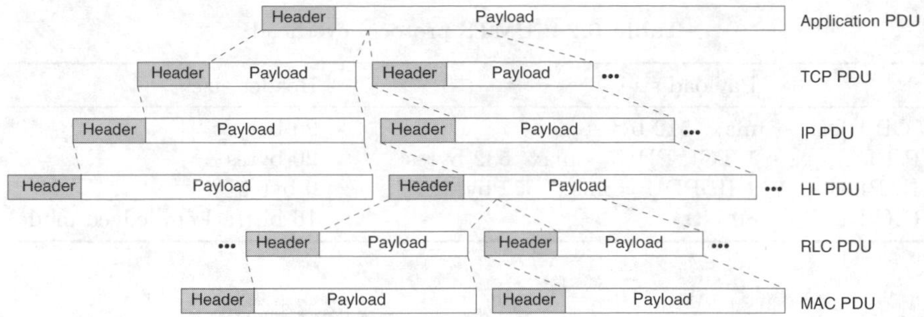

Figure 6.53: UMTS Protocol Data Units

6.14.2.3 UMTS Protocol Overhead Considerations

In order to explain the differences between measured PBRs and the maximum offered bit rates, some overhead considerations are made in this section. To transmit user data a certain amount of overhead has to be added. Overhead results from the use of various protocols, which add a certain amount of header overhead. Furthermore, in order to transfer an SDU from a higher layer, the protocol may segment it into several pieces, each of which is given a header and sent as a separate PDU. Figure 6.53 shows the PDUs and headers of the various protocol layers in the case of IP traffic over UMTS. Besides the header overhead, the overhead owing to signalling and the overhead owing to required retransmissions because of lost or erroneously received packets decrease the overall system capacity.

In Table 6.24 the corresponding header and payload sizes which are used in the system simulator are listed. The overhead due to TCP/IP headers is approximately 8 % if a maximum segment size of 512 bytes is assumed (meanwhile, a technique called TCP-header compression over the air-interface has been introduced that showed a significant improvement). In addition, not all segments carry the maximum payload, which results in an even larger relative overhead. Nevertheless, the UMTS standard foresees the application of TCP/IP header compression algorithms, which compress the header to a few bytes. For the reason of simplicity these algorithms were not applied in the presented simulations.

In the considered example, the overhead due to RLC headers is 16 bits for 80 bits payload, thus 20 %. It has to be mentioned that the standard allows larger RLC blocks. But increasing the block size increases the block error rate for a given bit error rate. Thus, the trade-off between header-overhead and tolerable block error rate should be considered.

Sources of signalling overhead are the control mechanisms of the various protocol layers in the system. TCP adds a certain amount of overhead resulting from the TCP connection establishment and release procedures. The applied HTTP 1.0 web model foresees that a TCP connection is set up for

Table 6.24: UMTS protocol overhead

	Payload size	Header size
TCP PDU	max. 512 bytes	20 bytes
IP PDU	1 TCP PDU ⇒ max. 532 bytes	20 bytes
HL PDU	1 IP PDU ⇒ max. 552 bytes	0 bytes
RLC PDU	80 bit	16 bit (acknowledged mode)

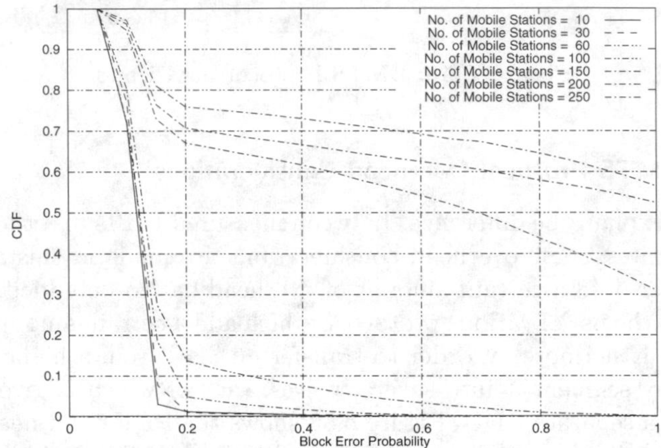

Figure 6.54: BLER in UMTS with max. data rate of 256 kbit/s

each WWW object requested. Therefore the static overhead owing to TCP signalling is dependent on the WWW packet size. Small application packets lead to an increasing impact on the overhead due to TCP signalling. An average packet size of 6000 bytes as used in the simulations implies that on average a TCP connection has to be established and released for every 12 TCP packets.

The amount of overhead due to retransmissions depends on the block error rate. Figure 6.54 shows that in low load situations a BLER between 5 % and 15 % is achieved. Fast power control is responsible to maintain this value. If the load increases, the interference and thus the BLER also increase. This happens when the upper transmit power limit is reached and therefore an increase of transmit power is not possible to compensate the perceived interference. If this is the case the E_b/N_0 target can no longer be maintained which results in a higher BLER. Since the RLC acknowledged mode is used, all erroneously received RLC PDUs have to be retransmitted. Obviously, an increasing BLER decreases the possible throughput.

The different kinds of overhead sum up to an aggregated overhead of 39 % if the system is not congested. This leads to a maximal possible throughput of 39 kbit/s, 78 kbit/s and 156 kbit/s in the 64, 128 and 256 kbit/s scenarios. This explains the maximum PBRs as presented in Section 6.14.2.2. However, only a very small portion of packets reach these maximum possible PBR. This was illustrated for the 256 kbit/s scenario in Figure 6.54.

6.14.2.4 Conclusions for WWW-traffic over WCDMA

Before the main conclusions of this section are summarized, it will again be noted that the main purpose is to provide an overview of how the system performs and which effects occur. Quantitative values should be seen just as an example, because the parameter space is by far too large to be fully investigated here and some system characteristics were ignored to simplify the simulations.

The simulation results presented so far have shown that UMTS behaves well for an increased application load. The aggregated system throughput is increasing linearly up to a certain knee point. After this point is reached the throughput remains more or less stable. The user-perceived performance is significantly less compared to the data rate offered by UMTS. This is owing to protocol overhead resulting from headers, retransmissions and signalling. The results have indicated that the users are well decoupled until congestion and too much interference occurs.

6.14.3 Simulations with WWW Traffic for EGPRS

To be able to compare the results for *Wideband CDMA* (WCDMA), the same traffic scenarios were set up for EGPRS with 8 PDCHs. For EGPRS, the simulation studies were carried out with various cluster sizes, different *Multislot Capability* (MSC) and under various traffic loads. The MSC gives the maximum number of time slots a mobile station is allowed to use simultaneously. On the uplink MSC 2 is used for all simulations whereas on the downlink MSC 2 and MSC 4 are analysed. This reflects the asymmetric traffic pattern of WWW. The main simulation parameters are listed in Table 6.25.

6.14.3.1 WWW System Throughput Performance in EGPRS

Figure 6.55 shows the system throughput for WWW traffic in EGPRS over the number of MS. The six curves represent three different cluster sizes with two sets of different MSCs. Obviously, the impact of the MSC is small. The system throughput for an assigned MSC of 4 is slightly higher. This is valid for all considered cluster sizes. The maximum throughput is reached for cluster size 12 and MSC 4 with 161 kbit/s. The system throughput decreases with smaller cluster sizes. This is due to the fact that the interfering cells are closer and therefore the interference increases. In the scenarios with cluster sizes of 9 and 12 the system saturates when a load of 60 MS is reached and with cluster size 4 the system saturates with 45 MS. Thus, the cluster size is an important design parameter for EGPRS.

Table 6.25: WWW simulations parameters for EGPRS

Parameter	Values
Number of MS	10, 30, 45, 60, 75
Cluster size	4, 9, 12
Cell size (km)	3
Number of PDCHs at 200 kHz	8
Multislot capability downlink	2, 4
Multislot capability uplink	2
Incremental redundancy	applied
Link adaptation	applied
Maximum TCP send window (bytes)	16384
TCP segment size (bytes)	512
Mean Internet packet loss rate (%)	1
Simulation duration (s)	1800

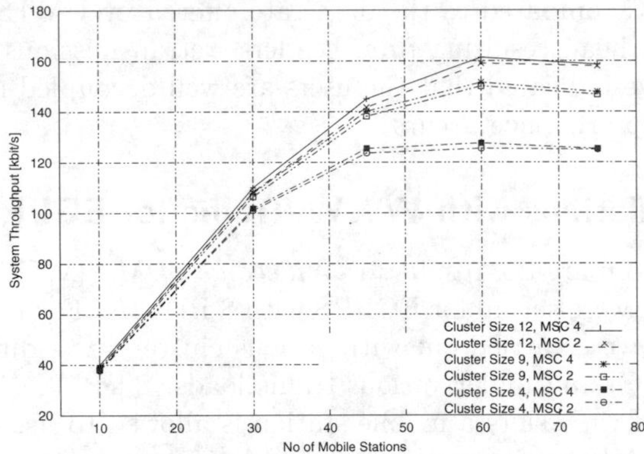

Figure 6.55: EGPRS system throughput and different multislot-capabilities in DL with WWW traffic

To show under which channel conditions the system throughput was reached, the actual C/I ratio was measured. Figure 6.56 depicts the distribution of the observed C/I ratio, for the three different cluster sizes. As expected, it can be seen that channel conditions are on average better for larger cluster sizes. Since link adaptation is used, the MCS is selected dependent on the actual channel condition of a simulated link.

The portions of used MCSs during the simulations are shown in Figure 6.57. In the case of cluster size 12, almost 80 % of the time MCS 9 is used, and even with cluster size 4, the MCS is never below MCS 3. This distribution of the used MCSs indicates that the channel conditions are on average very good, which results from the path loss exponent of 3.7 used for the simulations. Knowing the obtained C/I ratio and MCS distributions, the system throughput seems to be relatively small. Using MCS 9 a maximum data rate of

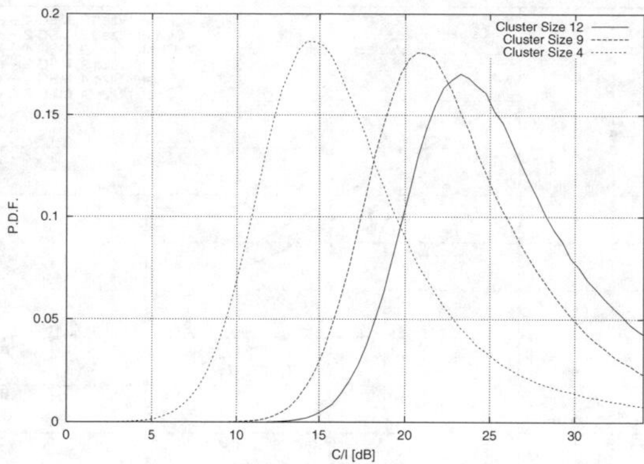

Figure 6.56: Distribution of C/I for different cluster sizes

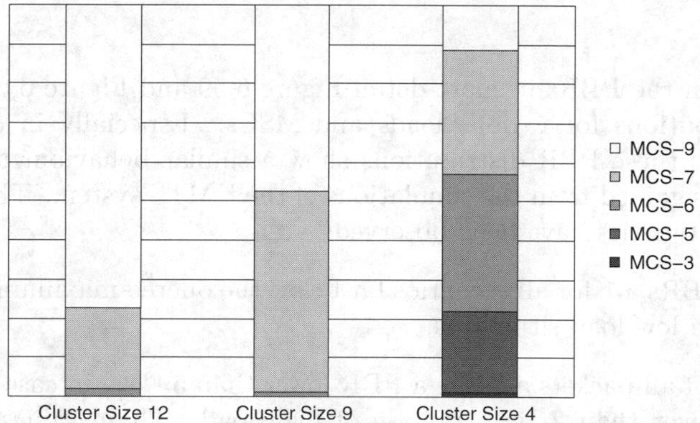

Figure 6.57: Modulation and coding scheme used for different cluster sizes

59.2 kbit/s per packet data channel is possible on RLC/MAC level. If 8 PDCHs are available, this results in a theoretical capacity limit of 473.6 kbit/s. As it was shown for UMTS, this limit is further reduced by protocol and signalling overhead. Section 6.14.3.3 deals in more detail with protocol and signalling overhead in the EGPRS system, but first the user perceived performance is described.

6.14.3.2 WWW Packet Bit Rate Performance for EGPRS

Figure 6.58 shows the mean PBR obtained from the WWW traffic simulations in EGPRS over the number of MS.

As expected, the mean PBR is high in case of few MS and decreases with increasing load. An interesting effect is that for low load, the chosen MSCs influences the performance, while for higher loads this influence disappears.

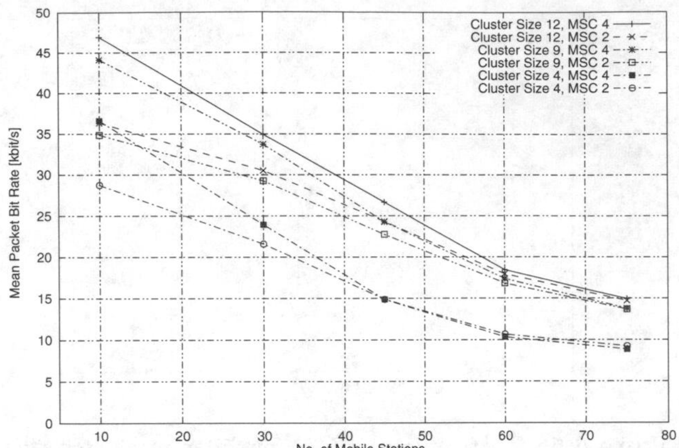

Figure 6.58: Mean PBR for different cluster sizes and MCSs

To explain the PBRs in more detail Figure 6.59 and Figure 6.60 show the PBR distributions for various loads and MSCs. Especially in case of low system loads these PBR distributions show a similar behaviour as the PBR distributions gained from the simulations of the UMTS system. The following main characteristics have been observed:

- The PBRs are for all scenarios far below the offered maximum data rate, even in low load situations.
- 10 % of all packets achieve a PBR lower than 5 kbit/s in case of low load and lower than 2 kbit/s in case of high load. These values are nearly independent of cluster size and MSC.
- For a given MSC the PBRs for large cluster sizes are higher than those achieved with small cluster sizes. This behaviour seems to be reasonable, since on average higher MCSs are used in case of larger cluster sizes.
- The PBRs between scenarios with MSC 4 and MSC 2 are only different for a low system load. With increasing load, the distinction disappears. However, even in low load situations the PBRs achieved by a system with MSC 4 is less than twice as high as those PBRs achieved with MSC 2.
- With increasing load the PBRs decrease continually.

Compared to the simulation results with WWW traffic over UMTS as presented in Section 6.14.2.2 the main difference is the behaviour related to an increasing system load. In UMTS the PBR curves for a given maximum allowed data rate are close together up to a certain load. Then the PBRs decrease dramatically. The PBRs of the EGPRS decrease more continuously with a

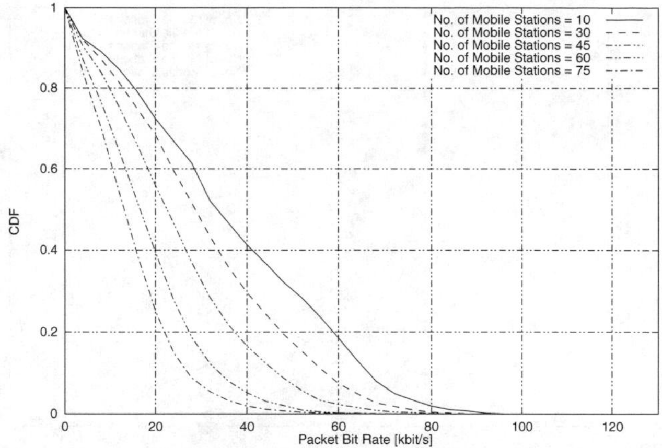

Figure 6.59: Complementary cumulative distribution function of PBR, cluster size 12, MCS 2

Table 6.26: EGPRS Protocol Overhead

PDU Type	Payload size (bytes)	Header size (bytes)
TCP	max 512	20
IP	1 TCP PDU ⇒ max 532	20
SNDCP	1 IP PDU ⇒ max 552	10
LLC	1 SNDCP PDU ⇒ max 566	4
RLC/MAC	depends on MCS	

growing number of MS. This is due to the radio block scheduling applied in EGPRS. For the UMTS study it was assumed that a dedicated channel has been assigned for each user, while in EGPRS the users have to share their assigned channels. Therefore, in EGPRS they influence each other already for low loads.

6.14.3.3 EGPRS Protocol Overhead Considerations

In order to explain the relatively low performance compared to the possible throughput on RLC/MAC level, some theoretical considerations about the system overhead are outlined. As already stated in Section 6.14.2.3, overhead results from protocol headers, from signalling and from retransmissions due to block errors on the air interface, if an ARQ mechanism is applied. Figure 6.61 illustrates the overhead added to the user data in form of headers within the various protocol layers.

The overhead generated in terms of RLC/MAC headers is not considered, since the possible throughput is given on RLC/MAC level. In Table 6.26, the header sizes, added within the simulations are listed.

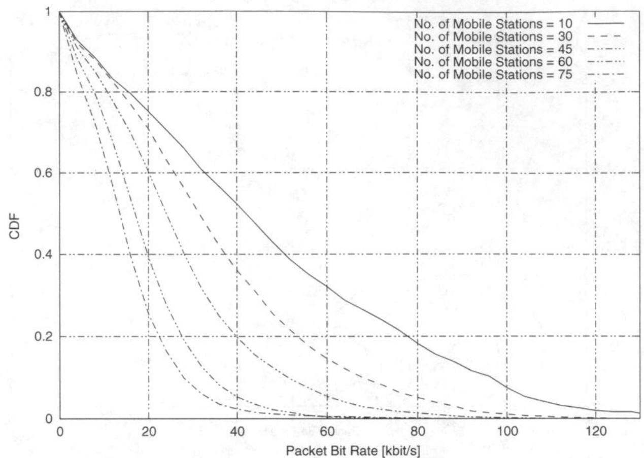

Figure 6.60: Complementary cumulative distribution function of PBR, cluster size 12, MCS 4

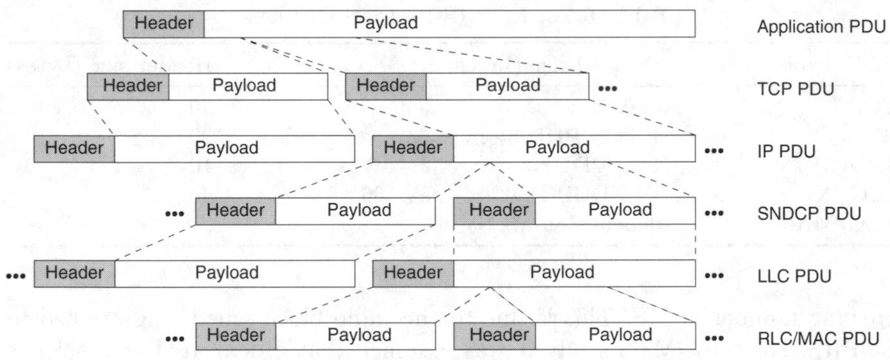

Figure 6.61: EGPRS PDUs

Considering the basic unit of one TCP PDU the overhead due to TCP-, IP-, SNDCP- and LLC-headers is 10.5 %. This fraction of overhead increases, if the TCP segment is smaller than the maximum TCP segment size. As stated in Section 6.14.2.3 this is not negligible for WWW traffic, caused by the large number of small packets.

In EGPRS, traffic in uplink direction requires some downlink signalling due to the TBF establishment procedure. In the optimal case one packet uplink assignment message as well as one packet uplink ACK/NAK message can be assumed. In this context it is important to keep in mind that overhead due to signalling on RLC/MAC layer is not only the actual signalling message. In addition, small signalling packets prohibit the transmission of large user data chunks, because they block the radio resources. For example, a control message of 20 bytes requires one radio block. The same radio block could

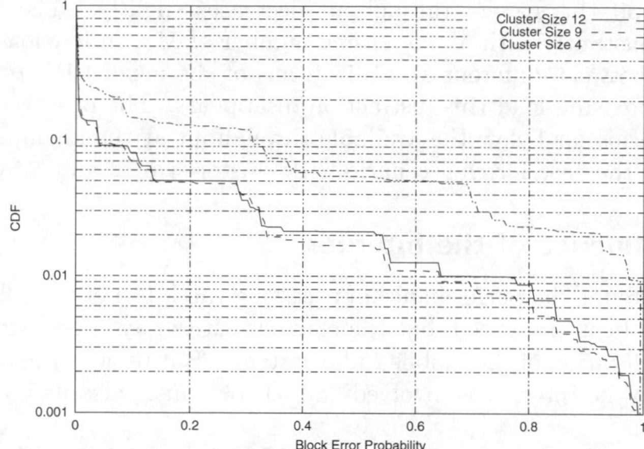

Figure 6.62: CDF of BLER in EGPRS for different cluster sizes

deliver 148 bytes assuming MCS 9. As already explained in Section 6.14.2.3, in the case of WWW traffic over TCP a certain amount of overhead results from the TCP connection establishment and release procedures. In addition, the TCP acknowledgements in UL-direction lower the downlink capacity due to the TBF establishment and release procedure as described above.

The overhead resulting from retransmissions depends on the BLER. Figure 6.62 shows the measured BLER in EGPRS for the three different cluster sizes. In contrast to the UMTS system, for a TDMA system like EGPRS the BLER is less dependent on the system load within the considered cell. Therefore, no strong dependency between interference and system load exists. The results show, that the distribution of the BLER is equal for cluster size 9 and 12 and slightly higher for cluster size 4. In the scenarios with cluster sizes 9 and 12, for 92 % of all transmitted data blocks the BLER is less than 10 %.

The total overhead in EGPRS due to headers, signalling and retransmissions leads to a significant decrease in system throughput, especially if high MCSs are used. This is caused by the large portion of TCP connection establishments and releases due to small packet sizes and the large amount of TBF establishments and releases in UL-direction due to TCP acknowledgement traffic.

6.14.3.4 Conclusions for WWW Traffic over EGPRS

The simulation results for WWW traffic over EGPRS using 8 PDCHs on a 200 kHz FDM channel have shown that the maximum possible throughput for the different scenarios is between 125 kbit/s and 161 kbit/s, which is reached when 60 users are in the system. With more than 60 users the system becomes congested and the system throughput decreases slightly. The mean PBR decreases constantly with an increasing number of MS. In low load cases, the distinction between different MSCs in terms of the mean PBR is significant. With an increasing load this distinction disappears. The observed PBRs are for all scenarios far below the available throughput at the air interface. This is especially the case for the scenario with cluster size 12 and MSC 4.

6.14.4 Influence of the Internet

So far, results for the performance of accessing the Internet with EGPRS and UMTS have been presented, but the characteristics have been explained just under consideration of the mobile radio systems. But besides the mobile radio network also the Internet is involved and has obviously also its impact on the performance.

The Internet can be characterized by a black box consisting of an unknown number of nodes and routes between them. It provides an unreliable transport of IP packets, which means that packet losses are natural. It is up to the network peers to ensure transport reliability, if required by the application. This is the case for an application like WWW, because only complete Web-pages are acceptable. The same is true for file transfers.

The transport layer protocol (*Transport Control Protocol*, TCP) [31] is the most commonly used protocol in the Internet, if data reliability needs to be achieved. TCP is too complex to be described here in full detail, but some features are essential for the understanding of the influence of TCP on the end-to-end performance of wireless networks. TCP is a connection-oriented protocol comprising an ARQ functionality to ensure reliability and in-sequence delivery. Since packet losses in the Internet are mostly due to congestion and buffer overflow in a network node, TCP includes also functionality for congestion control. Basically, it uses IP packet losses as congestion signal and adapts its send rate by reducing the send window size. Thus the send window is an adaptive parameter of TCP. Based on the assumption that congestion causes packet losses, the functionality for congestion control and error recovery are intertwined in TCP [49].

There are two different methods to detect packet losses. One is based on a timer which is started for a send packet. If the timer expires before an acknowledgement has been received, the packet is retransmitted. The other method is denoted as fast retransmission and is based on so-called duplicate acknowledgements. TCP works with cumulative acknowledgements and if the third acknowledgement in a sequence is not acknowledging new data, it is assumed that the corresponding packet is lost and a retransmission is performed. Both events, timeout and reception of the third duplicate acknowledgement, trigger a window adaptation. After a timeout the window is reduced to one TCP segment, while during a fast retransmission the window is reduced effectively to one half of the previous value. Thus, a timeout slows down the

Table 6.27: Bulk data simulation parameters for EGPRS

Parameter	Values
Number of bulk processes	1
Packet size (kbyte)	0.1, 1, 10, 50, 100, 200, 400
Cluster size	12
Cell size (km)	3
Distance MS—BS (km)	0.5 $\Rightarrow$ MCS-9
Multislot capability downlink	4
Multislot capability uplink	2
Maximum TCP send window (byte)	16384
TCP segment size (byte)	512
Mean Internet packet loss rate (%)	1, 0
Simulation duration (s)	50

connection tremendously, while a fast retransmission slows down, but keeps a certain speed.

The increase of the TCP window is controlled by two algorithms, which are applied in different phases of the connection. The slow start algorithm is applied after connection set up and after TCP time-outs. The second algorithm, congestion avoidance, is used after a fast retransmission. During slow start the window increases exponentially, while during congestion avoidance the window is enlarged linearly with an increase of one segment size per round trip time.

An important means to characterize a TCP connection is the so-called bandwidth-delay product. It is defined as the product of underlying data rate and the round trip time of the connection. For example a UMTS link with 64 kbit/s and a round trip time of 400 ms would have a bandwidth delay product of 3.2 kbyte. This means that 3.2 kbyte of data are necessary to fully utilize the link. As long as the TCP window size is smaller than the bandwidth delay product, the link is not efficiently used. In this case TCP is limiting the performance.

In the following results will be presented that will explain the most important effects. A bulk data source has been used in order to eliminate effects from randomly generated file sizes. The results apply for the EGPRS system but equivalent effects do occur for UMTS. The parameters corresponding to the simulations presented are listed in Table 6.27.

Figure 6.63 shows the PBRs for the various data packet sizes (i.e., file sizes) without IP packet loss. It can be seen that the PBR is dependent on the packet size. When transmitting packets of 100 byte on an end-to-end basis, it is impossible to obtain a PBR higher than 8 kbit/s. With packet sizes of 1000 and 10000 byte, the PBRs do not exceed the values of 34 and 134 kbit/s respectively. The PBRs for bulk data packet sizes of 50000, 100000, and 200000 byte are all at around 185 kbit/s which seems to be the upper

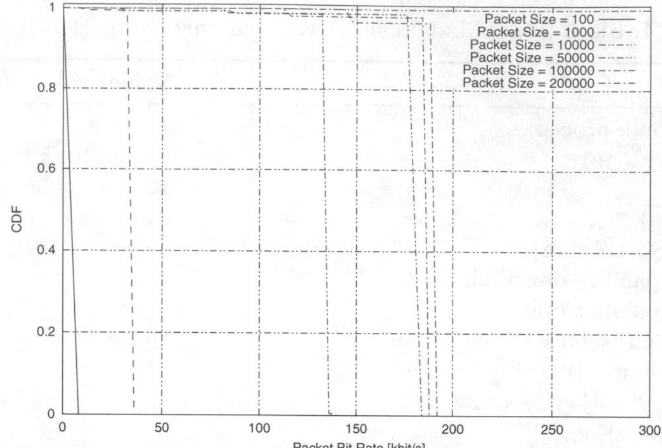

Figure 6.63: CDF of PBR for mean Internet loss rate 0 %

limit for this scenario. Since MSC 4 is used throughout the simulation, the maximum possible throughput on the RLC/MAC layer is given as 236.8 kbit/s. Reduced by the overhead the value of 185 kbit/s seems to be plausible. That means that in the case of large application packet sizes, the users perceive only 80 % of the offered data rate.

In the case of small file sizes like 10000 byte, the slow start effect of TCP has also a significant impact on the performance. Here, the TCP window size is smaller than the bandwidth-delay product of the link (approximately 12 kbyte), which means the provided link cannot be fully utilized. If the file size is only 100 byte, there is only one TCP segment to send. The transport through the Internet, the set up of the radio bearer, and the transmission over the air interface contribute to the delay around 100 ms, while the actual air interface transmission time is only 20 ms.

The next effect, which will be explained, occurs in the case of packet losses in the Internet. Packet losses might occur also in the core network of the mobile radio network or at the air interface, but here it is assumed that these parts do not introduce packet losses. If a packet is detected as lost, TCP has to perform a retransmission and adapts also the send rate by reducing the window.

Figure 6.64 depicts the system behaviour if the mean Internet loss rate is set to 1 %. All other parameters have been kept unchanged compared to case a). The results show, that for packet sizes of 100 and 1000 bytes, the behaviour of the curves is nearly unchanged, but for larger packets the PBR decreases dramatically. Independent of the packet size, none of the transmitted application packets exceeds a PBR larger than 150 kbit/s. A closer look into the TCP congestion avoidance algorithm shows that with 1 % IP packet loss rate the TCP transmit window remains sometimes significantly below the bandwidth-delay product. Packet losses occur too frequently to

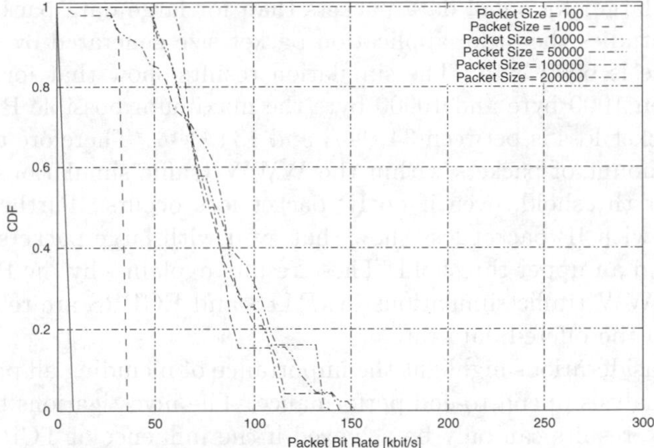

Figure 6.64: CDF of PBR for mean Internet loss rate 1 %

allow TCP's congestion avoidance algorithm to enlarge the window enough that the next detected packet loss does not lead to a send window below the bandwidth delay product. This effect is well known as the bias of TCP against large round-trip times and appears for all links where large round trip times lead to a large bandwidth-delay product. Consequently, the TCP send window is the limiting factor for the communication link in case b). The bottleneck has moved from the wireless access link to the Internet. This becomes obvious if one considers that attaching an access link with even increased data rate would not improve the performance.

It is also important to note that not negligible delays of cellular access links contribute to a large round trip time and large bandwidth-delay products. Thus, it needs to be emphasized here that, besides the data rate of the access link, the overall delay is a critical parameter for the TCP performance.

Our findings correspond to the results described in [24]. There it is stated that the average throughput of a TCP connection for the download of a file with infinite size is given by

$$\text{Throughput} = \frac{\text{MSS}}{\text{RTT}} \cdot \frac{C}{\sqrt{p}}, \tag{6.5}$$

where MSS is the *Maximum Segment Size*, RTT is the *Round Trip Time* and p denotes the probability of packet loss. The derivation of the formula assumes that the downloaded file has infinite size and the environment is network limited. C is a constant factor that considers the influence of the packet loss distribution and is set to $C = 1.31$ for random packet losses. Using the simulation parameters in Equation (6.5), a throughput of 117 kbit/s for EGPRS for case b) is attained.

Our simulation results demonstrate two effects. First, for small packets it is impossible to reach a high PBR. This is due to the fact that the relative overhead is larger for small data packets than for large data packets. In our simulation studies the mean application packet size generated by the WWW traffic source is 6000 byte. The simulation results show that for packets of sizes between 1000 byte and 10000 byte the maximum possible PBR in case of no IP packet loss is between 34 kbit/s and 134 kbit/s. Therefore the PBR of a certain amount of packets within the WWW traffic simulations is limited by an upper threshold, even if no IP packet loss occurs. Furthermore, the simulations with IP packet loss show that even with large packets the PBRs do not exceed an upper threshold. These results explain why the PBR gained from the WWW traffic simulations in UMTS and EGPRS are relatively low compared to the offered data rate.

These considerations highlight the importance of including all protocol layers in the analysis of end-to-end performance. The investigations have shown that realistic results can only be achieved if the influence of TCP and Internet packet losses are considered as well as detailed radio link layer aspects. Neglecting these effects would have led to overoptimistic results.

Consequently, as well as optimizing the wireless access TCP should also be configured in an optimal way. Although a complete description is beyond the scope of this section, the main recommendations:

- Equation (6.5) indicates that the TCP maximum segment size should be as large as possible. Path MTU discovery [31] should be used to ensure this.
- The TCP socket buffer size should be two times larger than the bandwidth-delay product of the considered path to avoid link under-utilization.
- An initial window size of 2 or 3 segments should be used [2, 3] to improve the performance during slow-start.

These recommendations help to improve the end-to-end performance.

6.14.5 Comparison of UMTS and EGPRS

The comparison based on the collected simulation results is first done by viewing the simulation results as presented in the previous sections in terms of the main system characteristics. Since different carrier bandwidths are allocated to both systems, this comparison is not based on absolute capacity and PBR results, but in terms of the observed system behaviour under various loads. In order to compare the system capacities of UMTS and EGPRS, the second part of this section will present a comparison by applying the spectral efficiency measure approach as defined in Section 6.14.1.3. Note that this comparison is based on the gained simulation results which are only valid for a specific set of parameters. The variation of other parameters (e.g., cell size or path loss exponent) may lead to different results.

6.14.5.1 Comparison in Terms of the System Behaviour and Main Characteristics

The simulation results as presented in Section 6.14.2 and 6.14.3 show that the system performance and the behaviour of the two systems are comparable for the end-to-end delivery of WWW traffic. For both systems the maximum end-to-end PBRs reached are far below the offered data rates and only a small portion of application packets achieve these high PBRs. This applies especially for the scenarios with high offered data rates.

Further simulations, carried out with bulk data, have shown that this behaviour is caused by a relatively large amount of small WWW packets and a decreased maximum TCP send window size caused by the TCP congestion avoidance mechanism in case of IP packet losses. Both systems, UMTS and EGPRS show a similar behaviour with respect to these effects.

However, the simulation results also point out some different characteristics of the two systems. In EGPRS the system throughput does not achieve as high values as expected, caused by the loss of capacity due to signalling as explained in Section 6.14.3.3. A high amount of signalling traffic in EGPRS occurs, caused by TBF establishment and release procedures which are required for the transfer of TCP ACKs in UL-direction and TCP connection establishment and release procedures. Compared to this behaviour observed in the EGPRS system, the UMTS system is more flexible. If a RAB is established, the transmission of TCP ACKs in UL-direction does not utilize DL-resources. Furthermore, the loss of capacity due to signalling in UMTS is proportional to the amount of signalling data.

The simulated system behaviour in terms of the PBR distribution characteristics show that in UMTS the PBR is nearly independent of the offered load until congestion occurs. In the case of congestion the PBR decreases rapidly. In EGPRS the achieved PBRs decrease more continually with an increasing system load.

6.14.5.2 Comparison in Terms of the Spectral Efficiency

Finally, the spectral efficiency for both systems will be considered because EGPRS and UMTS use significantly different carrier bandwidths. However, the spectral efficiency gives no insight into the performance from the user's point of view. To compare both systems from this perspective, it is here proposed to investigate the mean PBR as a function of the spectral efficiency. Figure 6.65 shows how the systems perform in terms of the mean PBR over the spectral efficiency. It allows a comparison between the UMTS 256 kbit/s scenario with the EGPRS MSC 4 scenario. A good system performance is expressed by the combination of a high PBR and a high spectral efficiency. Since the figure is derived from the simulations presented in Section 6.14.2 and 6.14.3 the various points describe the system performance for the different load situations in terms of MS. Thus, for EGPRS the given points correspond to a load of 10, 30, 45, 60 and 75 MS, where a small load leads to a high PBR

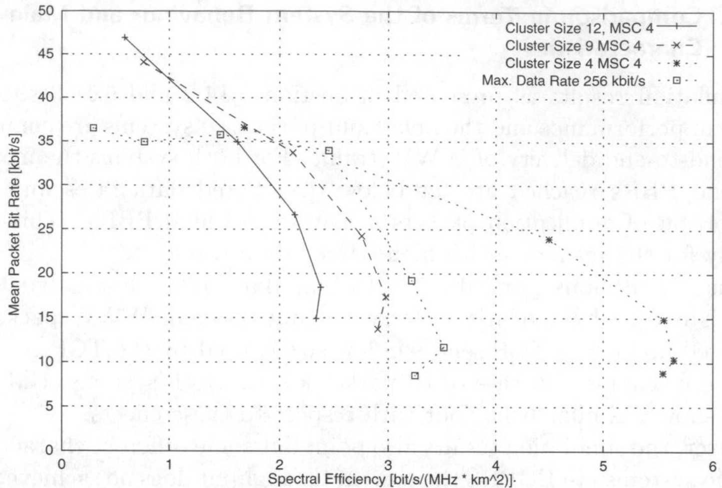

Figure 6.65: Mean PBR over spectral efficiency for 256 kbit/s in UMTS and MSC 4 in EGPRS

and a low spectral efficiency. For UMTS the loads of 10, 30, 60, 100, 150, 200 and 250 MS are considered.

Figure 6.65 depicts the relationship between spectral efficiency and the mean PBR reached by the systems under various loads. According to this figure the EGPRS system with cluster size 12 shows a good performance in low load situations since it reaches a higher mean PBR for a given spectral efficiency compared to the UMTS system. With increasing system load, the performance situation changes and for a spectral efficiency above 1.6 bit/(s · MHz · km^2) the UMTS system achieves higher mean PBR/spectral efficiency combinations. This is owing to the fact that the mean PBR of the EGPRS system decreases faster with increasing load as for the UMTS system. This behaviour corresponds to the observations that have been made in connection with the mean PBR behaviour of the UMTS and EGPRS systems in Section 6.14.2.2 and Section 6.14.3.2. A comparison between the EGPRS system with cluster size 9 and the UMTS system leads to similar results. The comparison between the EGPRS system with cluster size 4 and the UMTS system shows that the UMTS system offers only for a load at around 60 MS and more a slightly higher mean PBR/spectral efficiency combination, whereas the EGPRS system with cluster size 4 performs for all other load situations better than UMTS.

These results show that the EGPRS system performs well in terms of high PBRs when a large cluster size is selected and the system load is low, whereas the EGPRS system with a low cluster size performs well in the area of low PBRs and in high load situations. This is owing to the fact that the EGPRS cluster size 4 scenario reaches the highest spectral efficiency. The UMTS system shows a more constant behaviour over the system load compared to all scenarios of the EGPRS system. This constant behaviour applies until the system becomes congested.

Finally, it should again be pointed out that this section was intended to provide the skeleton for a thorough system performance analysis from an application layer perspective. We attempted to explain the approach taken, to highlight interesting effects and to formulate some general conclusions. However, moving standards, complex parameter sets, various applications and an infinite amount of scenarios make a general final statement impossible whether EGPRS or UMTS is superior.

Bibliography

[1] 3GPP. *3GPP Project Website.* `http://www.3gpp.org`, 2001.

[2] M. Allman, S. Floyd, C. Partridge. *Increasing TCP's Initial Window.* Network Working Group Request for Comments (RFC): 2414, September 1998.

[3] M. Allman, V. Paxson, W. Stevens. *TCP Congestion Control.* Network Working Group Request for Comments (RFC): 2581, April 1999.

[4] A. Baier, H. Panzer. *Multi-rate DS-CDMA radio interface for third-generation cellular systems.* In *7th IEE European Conference on Mobile Personal Communications*, p. 255, The Brighton Centre, UK, Dec. 1993.

[5] Stephan M. Blust. *Wireless Standards Development—A New Paradigm. IEEE Vehicular Technology Society News*, pp. 4–12, 11 2000.

[6] W. Broek, A. Lensink. *A UMTS architecture based on IN and B-ISDN developments.* In *7th IEE European Conference on Mobile Personal Communications*, pp. 243–249, The Brighton Centre, UK, Dec. 1993.

[7] M. Callendar. *Standards for global personal communications services.* In *Mobile Radio Conference (MRC'91)*, pp. 229–234, Nice, France, Nov. 1991.

[8] D. S. Chen, S. Roy. *An adaptive multiuser receiver for CDMA systems. IEEE Journal on Selected Areas in Communications*, Vol. 12, No. 5, pp. 808–816, 1994.

[9] ETSI. *Framework for services to be supported by the Universal Mobile Telecommunications System (UMTS).* Draft UMTS DTR/SMG-050201, ETSI, July 1995.

[10] ETSI. *Overall requirements on the radio interface(s) of the Universal Mobile Telecommunications System (UMTS).* Draft UMTS ETR 04-01, ETSI, Sept. 1996. Ref. DTR/SMG-050401.

[11] ETSI. *GSM 11.14, V7.1.0, Digital cellular telecommunications system (Phase2+); Specification of the SIM Application Toolkit for the Subscriber Identity Module - Mobile Equipment (SIM-ME) interface.* `http://www.etsi.org`, 1998.

[12] ETSI. *GSM 02.57, V8.0.0, Digital cellular telecommunications system (Phase2+); Mobile Station Execution Environment (MExE); Service description.* `http://www.etsi.org`, 1999.

[13] ETSI. *GSM 02.78, V8.0.0, Digital cellular telecommunications system (Phase2+); Customized Applications for Mobile network enhanced Logic (CAMEL), Service definition—stage 1.* `http://www.etsi.org`, 1999.

[14] UMTS Forum. *Spectrum for IMT 2000.* Technical Report, UMTS Forum, Oct. 1997.

[15] Wireless Application Protocol Forum. *Wireless Application Protocol - Wireless Transaction Protocol Specification.* `http://www.wapforum.org`, 1999.

[16] R. E. Fudge. *FPLMTS.* In *7th IEE European Conference on Mobile Personal Communications*, The Brighton Centre, UK, Dec. 1993.

[17] Harri Homa, Antti Toskala, editors. *WCDMA for UMTS—Radio Access for Third Generation Mobile Communications.* Wiley, Chichester, 1 edition, 2000.

[18] J. Huber, D. Weiler, H. Brand. *UMTS, the Mobile Multimedia Vision for IMT-2000, A Focus on Standardization.* In *IEEE Communications Magazine*, pp. 129–136, Sep. 2000.

[19] ITU-T. *Recommendation Q.1701: Framework for IMT-2000 networks.*

[20] ITU-T. *Recommendation Q.1702: Network Functional Model for IMT-2000.*

[21] A. C. Kerkhof, E. Spaans. *Accounting in UMTS.* In *7th IEE European Conference on Mobile Personal Communications*, p. 221, The Brighton Centre, UK, Dec. 1993.

[22] E. Lycksell. *Network architecture for FPLMTS.* In *5th Nordic Seminar on Digital Mobile Radio Communications (DMR V)*, pp. 203–212, Helsinki, Finland, Dec. 1992.

[23] A. Maloberti, P. P. Giusto. *Activities on third generation mobile systems in COST and ETSI.* In *Mobile Radio Conference (MRC'91)*, pp. 235–242, Nice, France, Nov. 1991.

[24] M. Mathis, J. Semke, J. Mahdavi, T. Ott. *The Macroscopic Behaviour of the TCP Congestion Avoidance Algorithm. ACM Computer Communications Review*, July 1997.

[25] Juan Noguera. *ATM Adaptation Layer 2.* `http://www.computer.org/students/looking/spring1998/noguera/aal2.html`, 1998.

[26] E. Postmann, C. Wietfeld, R. Becher. *Proposed architectural model for the support of VHE (Virtual Home Environment), ETSI SMG 3 (TDoc SMG 3 98S105).* Technical report, March 1998.

[27] J. G. Proakis. *Digital Communications.* McGraw-Hill, New York, 3rd edition, 1995.

[28] IST-DRIVE Project. *Dynamic Radio for IP-Services in Vehicular Environments.* `http://www.ist-drive.org`, 2000.

[29] Adam Roach. *SIP in 3GPP.* `http://www.sipforum.com`, 2000.

[30] P. Sehier, J-M. Gabriagues, A. Urie. *Standardization of 3G mobile systems. Alcatel Telecommunications Review*, No. 1, pp. 11–18, 2001.

[31] W.R. Stevens. *TCP/IP Illustrated*, Vol. 1. Addison-Wesley, Reading, Massachusetts, 1994.

[32] A.S. Tanenbaum. *Computer Networks.* Prentice-Hall, Inc., Englewood Cliffs, New Jersey, 3. edition, 1996.

[33] W. Tuttlebee, editor. *Cordless Telecommunications in Europe.* Springer, Berlin, 1990.

[34] UMTS/3GPP. *TS 22.121, Universal Mobile Telecommunication System (UMTS); The Virtual Home Environment.* Technical report.

[35] UMTS/3GPP. *TS 23.002 Universal Mobile Telecommunication System (UMTS); Network Architecture (Release 99).* Technical report.

[36] UMTS/3GPP. *TS 23.002 V5.2.0 TSG Services and Systems Aspects; Network Architecture (Release 5).* Technical report.

[37] UMTS/3GPP. *TS 23.060 V3.6.0 General Packet Radio Service (GPRS); GPRS Service Description Stage 2.* Technical report.

[38] UMTS/3GPP. *TS 23.127 Universal Mobile Telecommunication System (UMTS); Virtual Home Environment/Open Service Architecture.* Technical report.

[39] UMTS/3GPP. *TS 23.228 V5.0.0 TSG Services and Systems Aspects; IP Multimedia (IM) Subsystem - Stage 2 (Release 5).* Technical report.

[40] UMTS/3GPP. *TS 25.401, Universal Mobile Telecommunication System (UMTS); UTRAN Overall Description.* Technical report.

[41] UMTS/3GPP. *TS 25.410 V3.3.0 Universal Mobile Telecommunication System (UMTS); UTRAN Iu Interface: General Aspects and Principles.* Technical report.

[42] UMTS/3GPP. *TS 25.411 V3.3.0 Universal Mobile Telecommunication System (UMTS); UTRAN Iu Interface Layer 1.* Technical report.

[43] UMTS/3GPP. *TS 25.412 V3.6.0 Universal Mobile Telecommunication System (UMTS); UTRAN Iu Interface Signalling Transport.* Technical report.

[44] UMTS/3GPP. *TS 25.413 V3.5.0 Universal Mobile Telecommunication System (UMTS); UTRAN Iu Interface RANAP Signalling.* Technical report.

[45] UMTS/3GPP. *TS 25.433, Universal Mobile Telecommunication System (UMTS); UTRAN Iub Interface NBAP Signalling.* Technical report.

[46] UMTS/3GPP. *TS 29.060 V3.7.0 General Packet Radio Service (GPRS); GPRS Tunneling Protocol (GTP) across the Gn and Gp Interface.* Technical report.

[47] B. Walke. *Mobile data communications in Germany—A survey.* In *Proceedings of 6th International Symposium on Personal and Indoor Mobile Radio Communications*, pp. 799–804, The Hague, Netherlands, Sept. 1994.

[48] B. Walke. *Spectrum Issues for Next Generation Cellular.* In *Visions of the Wireless World*, Brussels, Belgium, Proc. of Workshop of the IST project 'Wireless Strategic Initiative', 12 December 2000.

[49] G. Xylomenos, G.C. Polyzos, P. Mähönen, M. Saaranen. *TCP Performance Issues over Wireless Links. IEEE Commun. Magazine*, pp. 52–58, 4 2001.

7 Trunked Mobile Radio for Group Communications

In addition to the public radio telephone service and the paging service, there are other radio services that are not accessible by the public. These radio systems, called the *non-public land mobile radio network*, have access to frequencies that cannot be used by the public but only by specific users or groups of users.

Probably the best known non-public mobile radio service is analogue *Private Trunked Mobile Radio* (PTMR), which has been used for many years by large firms such as airlines, taxi and transport companies, the railways, and ports, as well as by government departments and organizations responsible for public safety. What is characteristic of previous PTMR systems is that they have one radio channel that is used exclusively by all the mobile terminals of a specific user group. An analysis of conventional commercial radio systems reveals a number of weaknesses that affect both the customer and the operator:

- Because of too many PTMR users, the fixed allocation of radio channels in congested areas leads to a frequency overload.
- Radio supply areas are too small.
- There is the possibility of eavesdropping by unauthorized persons.
- There is no link to the public telephone networks.
- There is limited support of voice and data transmission.

Frequency overload was the main reason for considering new radio systems and infrastructures. This led to the introduction of *trunked mobile radio systems* as the successors to analogue PTMR.

Although it is not possible for trunked mobile radio systems to expand the frequency spectrum available, they are able to improve the quality of service both for the end user and for the network operator through the optimization of frequency utilization and increased channel use. Advances in trunked mobile radio technology have resulted not only in providing user groups with one channel as in PTMR but also in making a trunk of channels available jointly to a large number of users. A channel is allocated to the user by the system only when required, and then immediately withdrawn after use. Whereas in PTMR a user would have to wait until a channel allocated to his user group was free, a trunked mobile radio user can start speaking as soon as any one of the channels in the channel group is free. In trunked mobile radio, traffic volume is divided evenly over all the available radio channels, with the trunking of the channels achieving a trunking gain, i.e., the loss probability p_v becomes less and less as the number of channels in a group increases and each channel is constantly utilized (see Appendix A.2). The traffic capacity [in Erl./(MHz · km^2)], increases with the *trunk group size*.

In addition to frequency economy, trunked mobile radio systems offer other advantages:

- Low installation cost compared with separate radio control centres.
- Radio supply areas corresponding to the economic areas of activity.
- Higher range.
- No undesirable eavesdropping by others.
- Increased availability because of allocation of channels according to need.
- Optional access to the public telephone network.
- Expanded services because of selective calling, variable group calling and priority calls.
- Improvement in quality of service in voice and data transmission.
- Orderly call queuing operation.

It is worth noting that trunking is not an unique feature of the class of systems described in this chapter, instead, all the cellular systems described in this book use trunking of frequency channels. In fact, the name trunking is misleading. What is really unique with this class of systems is that they efficiently support group communication and offer a number of related services not available in other cellular systems to allow groups to be spread over a number of cells during operation.

Typically for these systems is also that the requirements on the availability of radio connectivity are much higher than with public cellular, e.g., 95 % throughout the country including radio coverage within buildings. Since the traffic generated from the users per area element is quite small, the cell layout is not oriented towards reaching high traffic capacity but towards good coverage.

7.1 The MPT 1327 Trunked Mobile Radio System

The pacesetter in standardized trunked mobile radio systems was Great Britain, where the *Ministry of Post and Telecommunication* (MPT) developed the trunked mobile radio standard MPT 1327/1343, which is also used in Europe as the technical standard for the first generation of (analogue) trunked mobile radio networks. In the USA the *Telecommunication Industry Association* (TIA) had developed the *American Public Safety Communications Officials* (APCO) standard that is quite similar in functionality to MPT 1327.

Following are some of the services offered in an MPT 1327 trunked mobile radio network:

- A *normal call* can be either an individual or a group call.
- A *priority call* can be either an individual or a group call.
- The mobile telephones called do not respond when they receive a *recorded announcement*.
- A *conventional central station call* in which a radio unit wishing to make a call is not immediately allocated a channel but is required to wait until the central station sets up the call at a convenient time.
- A *conference call* in which additional users can participate in a set up call.
- An *emergency call*, which can be either a voice or a data call placed by an individual or by a group.
- A *data call* can take place between different signalling systems, and is either an individual call or a group call that is transmitted either as a normal or a priority call.
- *Call forwarding or call diversion* to another user or group is possible.
- *Status messages* can be interchanged between different radio units or between radio units and the system, whereby there are 30 different special-purpose messages available.
- *Radio telegrams* are up to 184 bits long and can be interchanged between the radio units or between the radio units and the system.
- A *short telephone call* permits access to a private branch exchange and to the public telephone network.

In a trunked radio network a distinction is made between two different types of radio channel: the *control channel* and the *traffic channel*. All switching-related organizational functions between the system controller and the mobile radio devices are carried out over the control channel through the exchange of data. The main tasks of the control channel include:

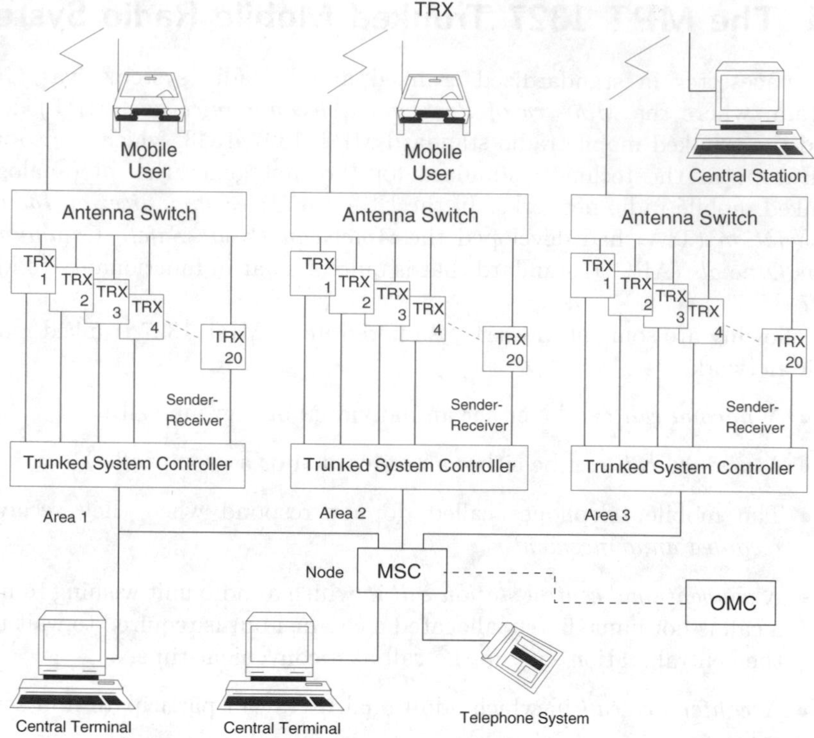

Figure 7.1: Principle structure of a trunked mobile radio network

- Notification of call requests
- Establishment and termination of calls
- Allocation of communications channels to mobile stations

Trunked radio systems can be operated as local systems with only one base station or as area-wide (cellular) systems with cell sizes from 3 km up to 25 km, e.g., in metropolitan areas with a 6 km diameter of the cells.

The basic structure of a cellular MPT 1327 trunked radio network consists of several cells, each with a radio base station (*transceiver*, TRX), a *Trunked System Controller* (TSC) and a central node, the *Master System Controller* (MSC), which also implements the gateway to the public telephone network or to the branch exchange networks (see Figure 7.1).

The TSC controls a radio cell and manages the traffic channels and their allocation to the mobile stations when a call is made. Since roaming is allowed in a multiple-cell trunked radio network, the TSC also maintains a home and visitor location register of all the subscribers allocated to the radio cell or who are temporarily operating in the cell. If a call is in progress during a cell change, it is not taken over by the new cell but is broken off; handover is not supported. *Operating and Maintenance Centres* (OMCs), which monitor the system, carry out statistical evaluations and record charges, are coupled to the MSC.

In addition to the MPT 1327 standard described, which defines the signalling protocols between the TSC and the mobile devices, the following standards are also of importance:

- MPT 1343 specifies the operations of the terminal equipment and defines the functions for system control and access to the traffic channel.
- MPT 1347 specifies the functions of the fixed network of the system as well as directives on the allocation of identity numbers.
- MPT 1352 describes the procedures for checking the conformity of the network elements of different manufacturers.

Trunked radio networks can operate in any frequency band suitable for mobile communications. In Europe trunked radio networks operate in the 80–900 MHz range.

One example is the Chekker network operated by Deutsche Telekom AG in Germany in accordance with the MPT 1327 standard in the 410–418 MHz (uplink) and 420–428 MHz (downlink) frequency bands. Up to 20 radio channels, each with a 12.5 kHz bandwidth, are available per cell. One channel can normally service 70–80 users. The maximum transmitter power per base station is 15 W.

Messages are transmitted digitally on the control channel, whereas with the MPT standard user information is transmitted on the traffic channels in analogue. Mobile stations use the control channel in half-duplex mode, whereas the base station transmits on this channel in duplex mode. The necessary signalling data is exchanged on the allocated traffic channel during a user connection.

Phase Shift Keying (PSK) modulation has been selected for speech modulation. *Fast-Frequency Shift Keying* (FFSK) modulation is used for data. The transmission rate for signalling data is 1.2 kbit/s; the data transmission rate possible is 2.4 kbit/s.

Systems with a small number of channels can employ a technique allowed by the MPT protocol in which the control channel can be used as a communications channel if the need arises.

Mobile stations in a trunked radio system access the control channel in accordance with a *random access* method, called the *Slotted-ALOHA* (S-ALOHA) protocol.

In a trunked radio network a call is set up through a series of steps. All checked-in radio units follow the sequence of operations on the control channel in *standby* mode. When a call request is made, indicated by a keystroke on the mobile terminal to the central station, the central station checks the availability of the subscriber terminal being called and informs the respective subscriber, in some cases through a paging signal over the control channel. If the subscriber called answers, a free traffic channel is automatically assigned to the respective parties. The maximum call duration in the Chekker service is 60 s (billing for the service is on a monthly fixed basis). When a call has been completed, the terminals switch back to the control channel. In the event that all the radio channels are occupied, an automatic queuing buffer system ensures that radio channels are allocated on an orderly basis, depending on waiting time or priority.

Trunked radio networks are divided into two categories based on user type:

- *Public networks*, which are operated by an operating company, whose users include small to medium-sized firms (e.g., towing services, haulage firms, other services).
- *Private networks*, which are operated by large groups, such as port authorities, automobile manufacturers, airline companies and the police.

7.2 Second-Generation Trunked Mobile Radio Systems: The TETRA Standard*

Despite the introduction of GSM throughout Europe, it is expected that the subscriber numbers for trunked mobile radio systems will continue to grow steadily, possibly reaching around five million by the year 2003.

None of the first-generation trunked radio systems currently on the market offers satisfactory voice and data services or is technically capable of dealing with the anticipated number of subscribers. In an effort to harmonize the trunked radio market in Europe, and taking all these factors into account, ETSI decided in 1988 to produce a standard for a digital, Pan-European trunked radio network. The first working title for this system, which was developed by the Technical Subcommittee RES 06, was *Mobile Digital Trunked Radio System* (MDTRS). However, in late 1991 the new name *Terrestrial Trunked Radio* (TETRA) was introduced for MDTRS.

Two families of standards have been produced for TETRA (see Table 7.1):

- *Voice plus Data* (V+D) Standard
- Data only (Packet Data Optimized Standard, PDO)

The TETRA V+D standard is envisaged as the successor to existing first-generation trunked radio networks, whereas the PDO standard defines a second-generation packet radio system. Both standards use the same physical transmission technology and largely the same transmit/receive equipment.

***With the collaboration of Martin Steppler and Peter Sievering**

European-wide standardization is forcing the issue of interoperability, i.e., manufacturer-independency of terminal equipment in the TETRA network, as well as interworking between different TETRA networks and the fixed networks. Local and regional voice and radio data applications are being replaced by a European trunked radio system that covers all voice and data services and satifies current requirements for bit rates and transmission delay. The main application areas for TETRA are fleet management, telemetry, service companies and for communication within government departments and organizations responsible for security.

Network operators, legislators, manufacturers and users were included in the standardization process to ensure that the ETSI-TETRA standard would have the chance of widespread implementation in the European market. The first TETRA products were available in late 1996. In 1997 the system was able to offer individual and group calls and data and other services described in detail in Section 7.2.2.

The PDO version of the standard has not found a technical implementation so far and therefore is not described here. For further details see [28].

The V+D version has found support by many manufactures and is currently in use in Europe as both, public systems, e. g., Dolphin in UK and Germany, and private systems (airports, transport etc.). Systems based on TETRA V+D have been established from 2000 on in many European countries for the public safety forces (police, fire brigades, customs, etc.) as the common platforms. The standard is expected to reach its break-through in 2002 in Europe with the decision of major countries in favour of it. It is worth mentioning that another digital trunked radio system, TETRAPOL, is competing with TETRA V+D and has found already acceptance in a number of European countries in public safety forces applications. Since TETRAPOL is not an ETSI standard it is not described here, see [26, 24, 27, 25] for reference.

Table 7.1: The series of the TETRA standard

Series	Content
01	General network description
02	Definition and description of air interface
03	Definition of interworking function
04	Description of air interface protocols
05	Description of user interface
06	Description of fixed network stations
07	Security aspects
08	Description of management services
09	Description of performance characteristics
10	Supplementary services—Level 1
11	Supplementary services—Level 2
12	Supplementary services—Level 3

Table 7.2: Technical data on TETRA V+D in Europe

Frequencies	Uplink: 380–390 MHz, Downlink: 390–400 MHz Uplink: 410–420 MHz, Downlink: 420–430 MHz Uplink: 450–460 MHz, Downlink: 460–470 MHz Uplink: 870–888 MHz, Downlink: 915–933 MHz
Channel grid	25 kHz
Modulation	$\pi/4$-DQPSK
Bit rate	36 kbit/s gross; 19.2 kbit/s net (in 25 kHz channel)
Channels/carriers	4 TDMA voice or data channels in 25 kHz
Access methods	TDMA with S-ALOHA on the random access channel (reservation offered with packet data)
Frame structure	14.17 ms/slot; 4 slot/frame; 18 frame/multiframe; 60 multiframe/hyperframe; slot length: 510 bit
Neighbouring channel protection	−60 dBc
Connection set up	<300 ms circuit-switched; <2 s connection-oriented
Transmission delay of a 100 byte reference packet	<500 ms in a connection-oriented service, <3–10 s in a connectionless service, depending on transmission priority

7.2.1 Technical Data on the TETRA Trunked Radio System

The trunked radio system TETRA can be used as a local or a multicell network. Since the transmitter power of the terminal equipment is 1 W, 3 W or 10 W, the maximum cell radius in rural areas is limited to 25 km. Several frequencies in the ranges between 380 MHz and 470 MHz and 870 MHz up to 933 MHz have been allocated to the frequency bands for the uplink and the downlink (see Table 7.2). The possibility of using the 1.8 GHz band is being examined.

The TETRA system uses $\pi/4$-DQPSK modulation and offers a gross bit rate of 36 kbit/s in a single 25 kHz channel. With an average quality of service guaranteed by the channel coding, the net bit rate is at 19.2 kbit/s. Without channel coding it is possible to achieve a maximum net bit rate of 28.8 kbit/s (see Table 7.3). With V+D four TDMA voice or data channels are available per carrier.

Slotted-ALOHA (with reservation in data transmission) is used as the access procedure with V+D. The frame structure consists of four 510-bit time slots per frame, 18 frames per multiframe and 60 multiframes per hyperframe, the latter representing the largest time unit and taking approximately one minute (see Section 7.2.4.2). An exact profile of the channel coding for V+D is given in Section 7.2.4.3.

The time required to establish a connection should not exceed 300 ms for a circuit-switched call or 2 s for a connection-oriented transmission of packet data (*Connection Oriented Network Service*, CONS). The transmission delay of a 100-byte reference packet in connection-oriented transmission with V+D should be a maximum of 500 ms; with connectionless transmission, depending on the respective transaction priority, it should be a maximum of 3 s, 5 s or 10 s.

7.2.2 Services of the TETRA Trunked Radio System

The TETRA V+D system provides packet data services, circuit-switched data, and voice services. The packet-oriented services differentiate between the following types of connections:

- Connection-oriented packet data transmission in accordance with ISO 8208 *Connection Oriented Network Service* (CONS) and services based on ITU-T recommendation X.25.
- Connectionless packet data transmission in accordance with ISO 8473 *Connectionless Network Service* (CLNS) for acknowledged point-to-point services and/or TETRA-specific acknowledged point-to-point and non-acknowledged point-to-multipoint services.

Circuit-switched voice can be transmitted unprotected over bearer services or (preferably) protected over teleservices (see Table 7.3). The teleservices for voice transmission offer five different types of connection:

Individual call Point-to-point connection between calling and called subscribers.

Group call Point-to-multipoint connection between calling subscriber and a group called through a common group number. The call is set up quickly because no confirmation is required. The communication takes place in half-duplex mode through the activation of a *push-to-talk* switch.

Direct call (Direct Mode, DM) Point-to-point connection between two mobile devices with no use of the infrastructure. A mobile station establishes a connection with another mobile station without the services of a base station, maintains the connection and takes over all the functions needed for local communication normally handled by the base station. Frequency ranges not normally used in the network are used for this purpose. At least one of the stations must have a connection to a base station on another channel [2]. For example, a connection can be established between two subscribers one of which is not operating in the supply area of the base station.

Acknowledged group call Point-to-multipoint connection between a calling subscriber and a group called through a shared group number in which the presence of the group members is confirmed to the calling subscriber through an acknowledgement. If one of the group members is not present or is on another call, the TETRA infrastructure informs the calling

Table 7.3: Bearer services and teleservices for V+D in the TETRA standard

Bearer services	7.2–28.8 kbit/s circuit-switched, unprotected speech or data 4.8–19.2 kbit/s circuit-switched, minimally protected data 2.4–9.6 kbit/s circuit-switched, highly protected data Connection-oriented packet transmission (point-to-point) Connectionless packet transmission in standard format (point-to-point) Connectionless packet transmission in special format (point-to-point, multipoint, broadcast)
Tele-services	4.8 kbit/s speech Coded speech

subscriber. If not enough members in the group can be reached, the caller can decide whether to discontinue or to maintain the call. An option is to have the group members who initially were not available to switch into the conversation later.

Broadcast call Point-to-multipoint connection in which the subscriber group dialled through a broadcast number can only hear the calling subscriber.

The bearer services and teleservices provided in the standard for the protocol stack V+D are listed in Table 7.3.

The TETRA system supports the following data and text services:

- Group calls
- Status messages
- Data messages
- Emergency call messages
- Electronic mail (e-mail)
- Facsimile and videotex

In addition, different supplementary services are offered, e.g.:

- Indirect access to PSTN, ISDN and PBX over a *gateway*.
- *List Search Calls* (LSC) in which subscribers or groups are called in the order of sequence of the entries in a list.
- *Include calls* in which by dialling the respective number an additional subscriber can be included in an existing telephone call.
- Call forwarding and call diversion.
- Call barring of incoming or outgoing calls (BIC/BOC).
- *Call Authorized by Dispatcher* (CAD), in which a request is made to place a particular kind of call.

- *Call Report* (CR), allowing the telephone number of a calling subscriber to be recorded so that the called subscriber can return the call later.
- *Calling number/connected line identification* (CLIP, COLP). This function can be prevented using *calling/connected line identification restriction* (CLIR). *Talking party identification* is also possible.
- *Call Waiting* (CW), indicating to the subscriber who has called while the talking party was on another call.
- *Call holding, connect-to-waiting*, allowing a subscriber to interrupt his current call to take another call and then to continue with the original call.
- *Short-number Addressing* (SNA), allowing a user to make a call using an abbreviated calling number. The TETRA infrastructure converts the abbreviated number to the subscriber number.
- Priority calls.
- Priority calls with interruption.
- Access priority.
- *Advice of Charge* (AoC) is a service that indicates the cost of a call to a subscriber either during or after a call.
- Discrete eavesdropping of a conversation by an authorized person.
- *Ambience Listening* (AL), allowing the user of a mobile device to be restricted to emergency calls only.
- Dynamic group number allocation.
- *Transfer of Control* (TC), permitting the initiator of a multipoint connection to transfer control of a call to another person involved in the call.
- *Access Slot* (AS), allowing an authorized user to select the cell for setting up a call or a subscriber currently being served to determine the cell.
- *Late Entry* (LE) is an invitation to potential subscribers in a multipoint connection to be switched into an existing call.

Series 10, 11 and 12 of the V+D standard contain a complete compilation of the supplementary services, along with detailed definitions and descriptions [11, 12, 13].

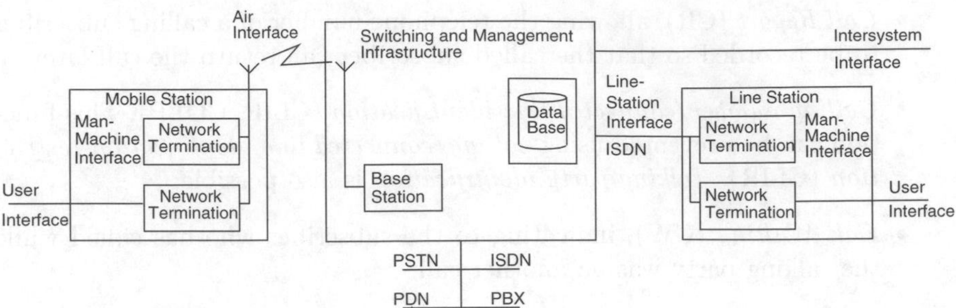

Figure 7.2: The architecture of the TETRA system

7.2.3 Architecture of the TETRA Standard

7.2.3.1 Functional Structure of the TETRA System

With few differences, the TETRA system is structured like the GSM one (see Figure 7.2). It has the following three subsystems:

- Mobile station
- Line station
- Switching and management infrastructure

Mobile station (MS) The mobile station (MS) comprises all the subscriber's physical equipment: the radio telephone and the interface that the subscriber uses to access the services.

As in GSM, a mobile station consists of two parts: the telephone device, which contains all the necessary hardware and software for the radio interface, and a *Subscriber Identity Module* (SIM), which contains all the subscriber-related information. The SIM can either be in the form of a smart card the size of a cheque card or permanently mounted in the device. The first version has the advantage that it allows a quick change in ownership of the mobile station. The third option is to key in a *login code* to convey the subscriber-specific information. In this case too the mobile unit is not restricted to one particular user.

In addition to the subscriber's identification, each mobile device has a *TETRA Equipment Identity* (TEI) that is specific to the device. This number is input by the operator; only the operator is able to bar the use of the device or release it for use. This means that a stolen device can be disabled immediately and unauthorized access is virtually impossible.

The following numbers and identities are allocated to ensure that a mobile station can be uniquely addressed and managed:

- *TETRA Subscriber Identity* (TSI)
- *Short Subscriber Identity* (SSI)
- *TETRA Management Identity* (TMI)
- *Mobile Network Identity* (MNI)
- *Network Layer SAP Addresses* (NSAP)

The TSI consists of three parts: a *Mobile Country Code* (MCC) that contains the country identification, a *Mobile Network Code* (MNC) that identifies the respective TETRA network, and a *Short Subscriber Identity* (SSI) that identifies the subscriber. When a connection is to be established within the home network, only the SSI is used as the address. This reduces the volume of signalling data.

The TMI is used for the management functions of the network layer. The NSAP is employed for the addressing of external, i.e., non-TETRA networks, and is optional. For instance, it can be used to establish a connection to ISDN.

Similarly to GSM, mobile telephones can be installed in vehicles or used as portable/hand-portable devices. All the standard services listed in Section 7.2.2 can be accessed by a mobile station. Supplementary services are offered by the network operator or must be contracted at the same time as the mobile phone contract so that the user can access them.

Line station (LS) In principle a line station is structured in the same way as the mobile station, but with the switching and management infrastructure connected over ISDN. For example, the owner of a company that operates a fleet of lorries would use a line station as the central station for his network. A line station offers the same functions and services as a mobile station.

Switching and management infrastructure (SwMI) The *Switching and Management Infrastructure* (SwMI) forms the local control unit of the TETRA system. It contains base stations that establish and maintain communication between mobile stations and line stations over ISDN. It carries out the required control tasks, allocates channels and switches calls. It carries out authentication checks and supports the relevant databases such as the *Home Data Base* (HDB), which contains the telephone numbers, the equipment numbers and the subscribed basic and supplementary services for each individual subscriber in the home network, as well as the *Visited Data Base* (VDB), which contains information on visitors to the network copied from their HDB. It also handles call charging.

7.2.3.2 Interfaces of the TETRA System

Subscriber interface of the mobile station A TETRA mobile station is called a *Mobile Termination* (MT). Its functions cover radio channel resource

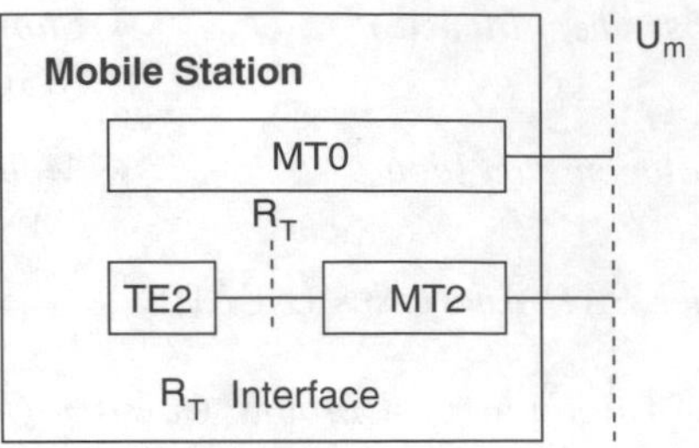

Figure 7.3: Mobile station network interfaces at reference points R_T and U_m

and mobility management, speech and data coding/decoding, transmission security as well as data flow control. The following versions are used:

- MT0 (*Mobil Termination Type 0*) contains the named functions with the support of non-standardized terminal interfaces that contain the terminal equipment functions (see Figure 7.3).
- MT2 (*Mobil Termination Type 2*) also supports the named functions, and has an R_T interface for terminal equipment based on the TETRA standard (see Figure 7.3).

The terminal equipment (TE2) is directly accessible by the subscriber, and corresponds to comparable function groups of the GSM and ISDN concept. The interface that supports access using traffic and signalling channels is situated at reference point U_m.

Subscriber interface of the line station The line station has a *Network Termination Functional Group* (NT) because it is linked to the fixed network over a *Transmission Line* (TL) interface via an ISDN line (see Figure 7.4). The NT is planned in two versions, and contains the functions defined in ITU-T Rec. I.411 [16]:

- NT1 supports the functions of the NT functional group, special functions of the TL interface and, possibly, the NT2 interface.
- NT2 likewise supports the functions of one or possibly several NTs and an interface to the functional groups TE1 and TETRA TA.

The TE functional group is available in two versions, with TE1 containing an ISDN interface and TE2 a TETRA-specific interface. Both versions support the *man–machine* and eventually also a TETRA TA or an ISDN interface. The *Terminal Adapting* (TA) functional group is responsible for data-rate adaptation as well as for flow control. In the current version it has interfaces to a TE2 and to an NT2 functional group.

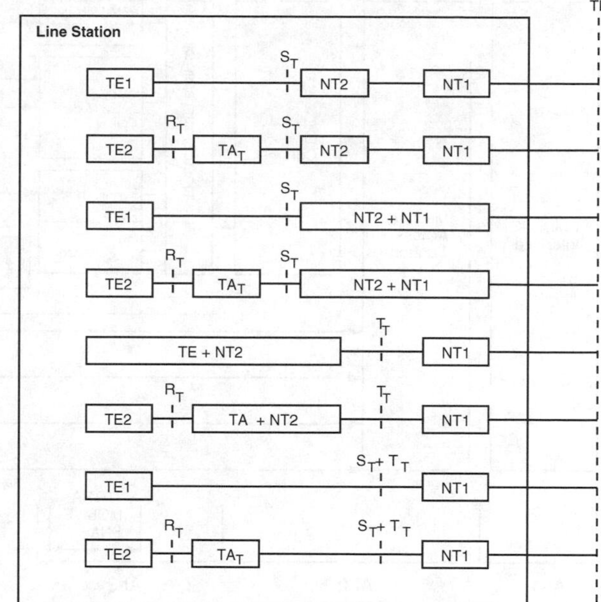

R_T TETRA R Reference Point
S_T TETRA S Reference Point
T_T TETRA T Reference Point
TL : Transmission Line
TE : Terminal Equipment
TE1: TE presenting an ISDN Interface
TE2: TE presenting a TETRA Interface
NT1/2 : Network Termination
TA_T: TETRA Terminal Adapting Functions

Figure 7.4: Line station network links

The radio interface The radio interface at reference point U_m is discussed in Section 7.2.4.2.

7.2.4 The Voice+Data Protocol Stack

This section will cover the Voice+Data protocol stack in general. This will be followed by detailed discussion of the radio interface, the physical layer and the link layer, with emphasis on functions, service elements, data structures and states.

7.2.4.1 Structure of the Voice+Data Protocol Stack

The protocol stack comprises three layers (*Air Interface*, AI) (see Figure 7.5): the physical layer; the data link layer, which is divided into *Medium Access Control* (MAC) and *Logical Link Control* (LLC); and the *Network Layer* (N), which is divided into several sublayers and offers management services to base and mobile stations. The MAC layer is based on two protocol stacks: the *user plane*, which is responsible for information transport, and the *control plane* for signalling.

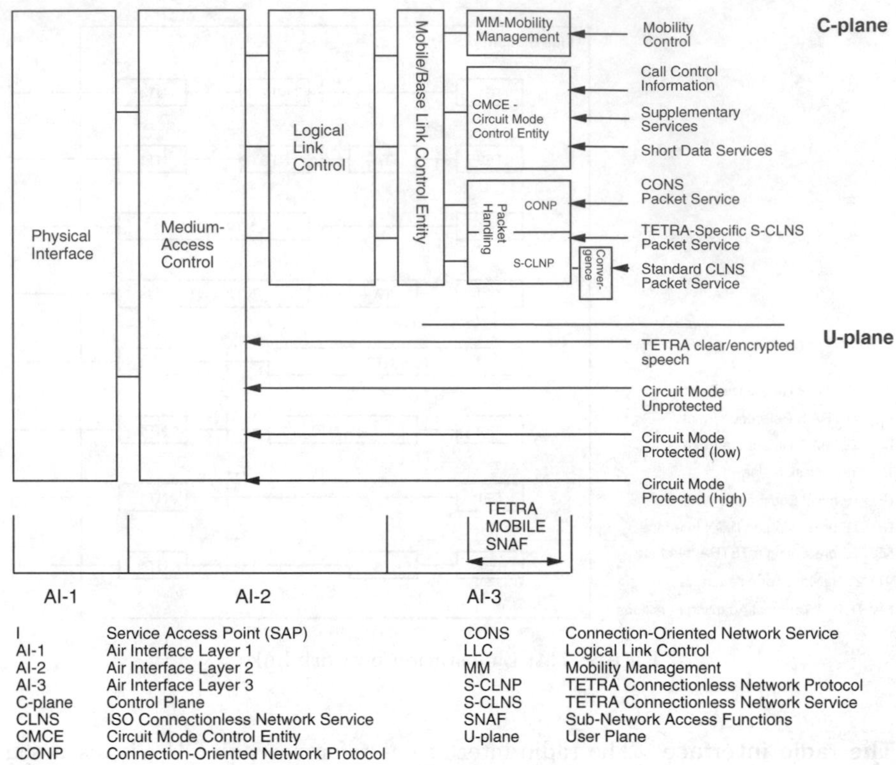

Figure 7.5: Architecture of the Voice+Data protocol stack

7.2.4.2 The Radio Interface at Reference Point U_m

The radio interface U_m is between the mobile station and the switching and management infrastructure (see Figure 7.2).

The multiplexing structures Similarly to GSM, the TETRA standard uses a combination of *Frequency Division Multiplex* (FDM) and *Time Division Multiplex* (TDM) with multiple access (TDMA) of the mobile stations (see Figure 7.6). These procedures, along with speech coding and modulation, play an important role in the standard.

The TETRA system uses a cellular concept in which a supply area is divided into cells and has an SwMI installed in the centre. A mobile station is able to measure the receive level of the FDM channel allocated to it. If the level falls below a certain limit, a *cell-reselect* procedure is triggered that reroutes the call or data transmission to another cell with a momentary interruption of a minimum of 300 ms. Different versions of the call-reselect procedure are being planned in the standard, with a basic version being obligatory for all mobile stations. The versions based on it are optional and will facilitate a

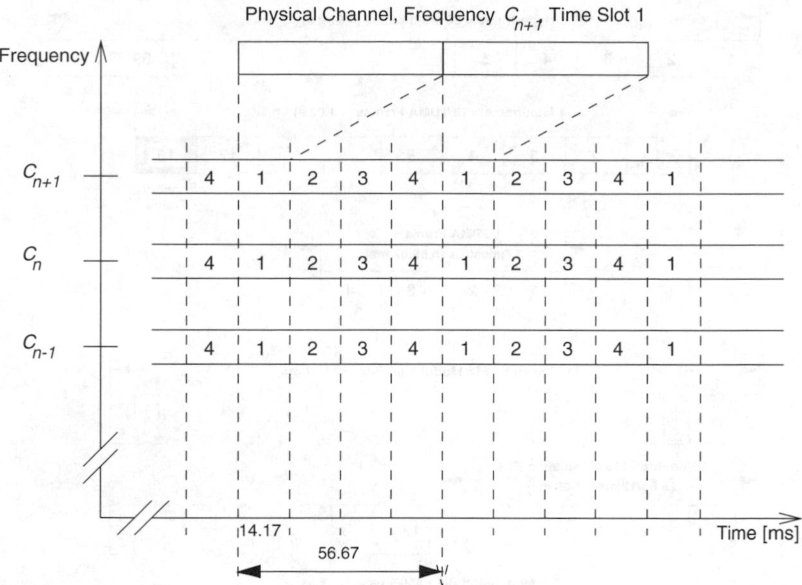

Figure 7.6: Implementation of physical channels using FDM and TDM

speedier change of cell. It is not mandatory to have a change of FDM channel implemented. The TETRA system does not have any real handover functions, because the users of mobile stations (e.g., taxi fleets) normally do not have a large radius of action.

Frequency-Division Multiplexing Structure Several frequency bands that have been allocated to the TETRA network in Europe do not completely comply with the specifications of the TETRA standard [9] and use additional frequencies above 870 MHz or 915 MHz (see Table 7.2). The frequency bands for both uplink and downlink are the same width. The carrier frequency separation is 25 kHz, and each uplink and downlink band is divided into N carrier frequencies. A G kHz wide band has been added to each boundary of the band in order to prevent interference from outside the band. Thus N and G constitute the total bandwidth. The following formulas can then be used to calculate the carrier frequencies. For the uplink

$$F_{up}(c) = F_{up,min} + 0.001G + 0.025(c - 0.5) \text{ MHz}, \quad c = 1, \ldots, N$$

and for the corresponding downlink

$$F_{dw}(c) = F_{up}(c) + D \text{ MHz}, \quad c = 1, \ldots, N$$

D stands for the constant duplex separation between the uplink and the downlink carrier frequencies. $F_{up,min}$ is the cutoff frequency on the lower boundary of the respective frequency band.

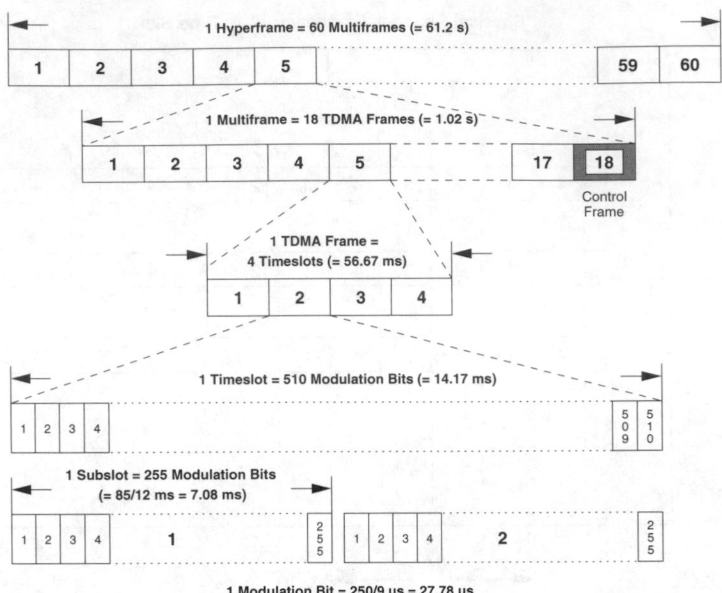

Figure 7.7: TDMA structure of the Voice+Data system

Time-Division Multiplexing Structure As can be seen in Figure 7.6, the time axis is split into four time slots, each 14.17 ms in duration corresponding to 510 bits through the use of the TDM technique on each carrier frequency. A periodic time slot produces a physical TDM channel onto which a logical channel is mapped. The logical channel is characterized by its carrier frequency and the time slot recurring at 56.67 ms intervals.

Figure 7.7 shows the TDMA structure for the Voice+Data system. It consists of hyper-, multi- and TDMA frames as well as the time slots and subslots which only occur in uplink traffic. A subslot (half a time slot) consists of 255 bits and a duration of 7.08 ms. Four time slots are combined to form a TDMA frame, and are numbered consecutively here from one to four (*Timeslot Number*, TN). A TDMA frame is 56.67 ms in length. Eighteen cyclically numbered frames (*Frame Number*, FN) are combined into a *multiframe*, which is 1.02 s in length.

The 18th frame of a multiframe is reserved for the signalling channels, and is called a *control frame*. If required, other frames can also be reserved for signalling purposes. A *hyperframe* consists of 60 multiframes, and represents the largest structure possible. It has the length of 61.2 s. In contrast to the downlink, the entire frame structure on the uplink is staggered by two time slots to enable a mobile station to transmit and receive at the same time. The frame synchronization of the mobile station is adaptively dependent on the signal propagation delay.

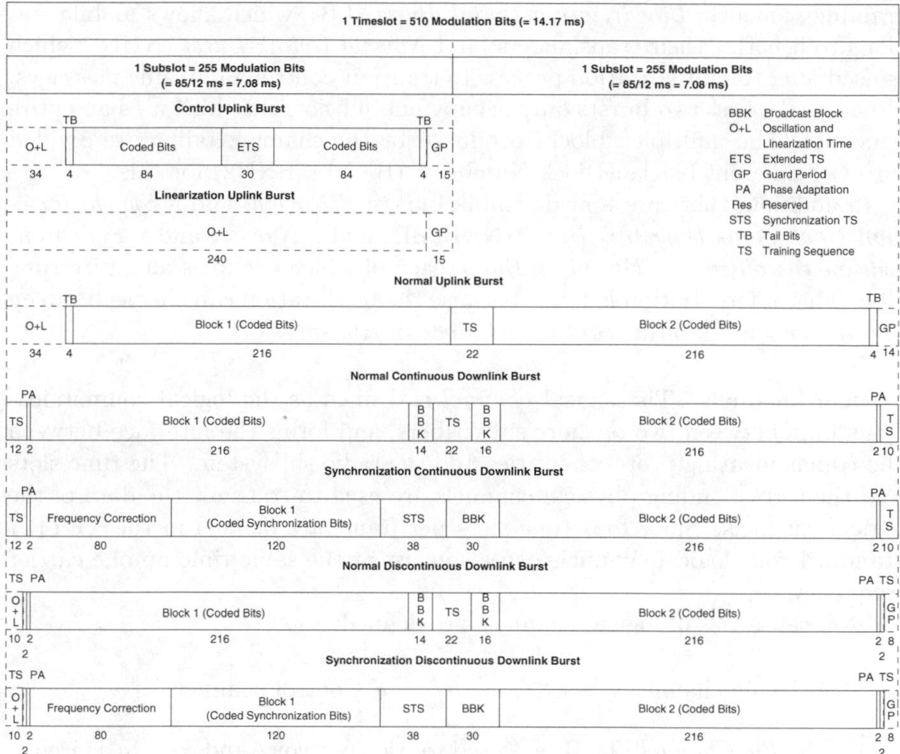

Figure 7.8: V+D downlink and uplink bursts

A burst is a set of data bits modulated on a carrier frequency (see Figure 7.8). In V+D it represents the physical content of a time slot, in other words the physical channel information. There are three different types of channels:

- A control channel, *Control Physical Channel* (CP), which exclusively transmits the control channel data.
- A traffic channel, *Traffic Physical Channel* (TP), onto which the logical voice and data channels are mapped.
- An *Unallocated Physical Channel* (UP), which is not allocated to any mobile station and is used for sending broadcast and dummy messages.

An existing physical channel always uses the same time slot in successive TDMA frames.

Figure 7.8 (see [7]) presents an overview of the burst structure, which is clarified further below. Three different uplink bursts are defined in the standard: *Control Uplink Burst* (CB), recognized on the basis of its expanded training sequence; *Linearization Uplink Burst* (LB), which allows mobile stations to linearize their transmitters; and *Normal Uplink Burst* (NUB), which is used after the initialization process to transmit control and traffic messages. Whereas the first two bursts only occupy one subslot, the third uses an entire time slot. The multiplex blocks produced by the channel coding are divided into two different blocks (Block Number 1 (BKN1), BKN2) for this.

In addition, there are four downlink bursts: a *Normal* and a *Synchronization Continuous Downlink Burst* (NDB, SB) and a *Normal* and a *Synchronization Discontinuous Downlink Burst*, each of which occupies an entire time slot. This differentiation is made because the base station can choose between *continuous mode* and *discontinuous mode* of transmission.

Logical Channels The logical *channel* is defined as the logical communications path between two or more subscribers, and forms the interface between the communicating protocol entities and the radio subsystem. The time slots and the corresponding physical channels are used to transmit the data of the logical channels. Since four time slots per frame are defined in the TETRA standard, four logical channels can also exist at the same time on one carrier frequency.

Two categories of logical channels are defined:

- Traffic channels
- Control channels

The *Traffic Channel* (TCH) are used to transmit voice and data in a circuit-switched connection. The *Control Channel* (CCH) transmit the signalling information and data packets. The following logical channels are defined:

TCH There are four different traffic channels that are suitable for speech and data services: the TCH/S (S for *speech*) for transmitting speech; and the TCH/7.2, TCH/4.8 and TCH/2.4 channels for transmitting data. As their labels imply, they offer net bit rates of 7.2 kbit/s, 4.8 kbit/s and 2.4 kbit/s, respectively. The difference in bit rates is due to the differing requirements of the error protection methods used (see Section 7.2.4.4). The speech codec provides a data rate of 4.8 kbit/s, and therefore uses the TCH/4.8.

CCH Five different control channels are available for the transmission of signalling and packet data information.

BCCH The *Broadcast Control Channel* is a unidirectional downlink channel that is used by all mobile stations in general. There are two categories of BCCH: the *Broadcast Network Channel* (BNCH) provides network information, and the *Broadcast Synchronization Channel* (BSCH) supplies information for time and cryptographic synchronization.

LCH The *Linearization Channel* is used by the mobile and base stations for linearizing their transmitters. Here too there are two types: the

Common Linearization Channel (CLCH) for the uplink of the mobile stations, and the *Base Station Linearization Channel* (BLCH) for the downlink.

SCH The *Signalling Channel* is shared by all the mobile stations, but is able to contain information for only one or for one specific group of mobile stations. The TETRA system functions require that there is at least one SCH per base station. There are three categories of SCH, which depend on the length of a message. A bidirectional *Full-Size Signalling Downlink Burst* (SCH/F) always occupies a complete time slot, and the unidirectional *Half-Size Uplink/Downlink Signalling Channels* (SCH/HU, SCH/HD) always occupy half a time slot, in other words, a subslot.

AACH The *Access Assignment Channel* is sent on all downlink slots in the broadcast block and contains details about the allocation of the next uplink and downlink slots on the corresponding radio channel. It is transmitted in each *Broadcast Block* (BBK) of a downlink burst (see Section 7.2.2).

STCH The *Stealing Channel* is a bidirectional channel that is associated with a TCH. It *steals* some of the capacity of the TCH for the purpose of transmitting control information. In half-duplex mode the STCH is a unidirectional channel and is rectified with the respective TCH. The STCH is used with a high priority in signalling, e.g., during the cell change procedure (*cell reselect*).

Mapping logical channels onto physical channels The logical channels are mapped onto physical channels in the lower part of the MAC layer. Table 7.4 shows how the mapping of the logical to the physical channels is defined.

Some of the important aspects of the table are explained below. The BCCH and the CLCH are mapped onto the control frame (18th frame of a multiframe) of a physical control or traffic channel through functions of the time slot and the multiframe number. The following algorithms are used:

$$\begin{aligned} \text{Downlink:} \quad & \text{BNCH, if } \texttt{FN} = 18 \text{ and } (\texttt{MN} + \texttt{TN}) \bmod 4 = 1 \\ & \text{BSCH, if } \texttt{FN} = 18 \text{ and } (\texttt{MN} + \texttt{TN}) \bmod 4 = 3 \\ \text{Uplink:} \quad & \text{CLCH, if } \texttt{FN} = 18 \text{ and } (\texttt{MN} + \texttt{TN}) \bmod 4 = 3 \end{aligned}$$

Furthermore, the base station is able to map the CLCH onto the uplink subslot 1 and the BLCH to the downlink block 2 of a *Control Physical Channel* (CP). It accomplishes this on a slot-to-slot basis and indicates it on the AACH. Mobile stations can linearize their transmitters to any CP when there is a CLCH if in so doing they are not violating any of the rules on mapping and the linearization process is longer than one multiframe. The BLCH is also mapped onto a downlink block 2 if an SCH/HD or a BSCH is mapped onto

Table 7.4: Mapping logical channels to physical channels

Logical channel	Direction	Burst type	SSN/BKN	Physical channel	FN	TN
BSCH	DL	SB	BKN1	CP, TP	18	$4-(\mathtt{MN}+1) \bmod 4^a$
				UP	1…18	1…4
BNCH	DL	NDB, SB	BKN2	CP, TP	18	$4-(\mathtt{MN}+3) \bmod 4^a$
				CP, UP	1…18	1…4
AACH	DL	NDB, SB	BBK	CP, TP, UP	1…18	1…4
BLCH	DL	NDB, SB	BKN2	CP, UP	1…18	1…4
				TP	18	
CLCH	UL	LB	SSN1	CP, TP	18	$4-(\mathtt{MN}+1) \bmod 4^a$
				CP, UP	1…18	1…4
SCH/F	DL	NDB	BKN1, BKN2	CP	1…18	1…4
	UL	NUB				
SCH/HD	DL	NDB, SB	BKN1, BKN2	CP, UP	1…18	1…4
				TP	18	
SCH/HU	UL	CB	SSN1, SSN2	CP	1…18	1…4
					18	
TCH	DL	NDB	BKN1, BKN2	TP	1…17	1…4
	UL	NUB				
STCH	DL	NDB	BKN1, BKN2	TP	1…17	1…4
	UL	NUB				

[a] Mapping to respective time slot is stipulated

the first block. However, there cannot be more than one BLCH per four multiframes on one carrier.

If a CLCH is not mapped onto the first subslot, an SCH/F or two SCH/HUs can be mapped onto the uplink. Otherwise only subslot 2 can be used for an SCH/HD. An SCH/F or two SCH/HUs can be mapped onto the downlink if a block 2 is not being used for a BNCH.

The base station indicates on the AACH which logical channel type will be used on the next uplink time slot. This indication is only one frame in length and valid for one physical channel. The logical channel number on the downlink is determined by the type of *Training Sequence* (TS) used. When several downlink traffic channels are used by a connection, the uplink and downlink SCHs are mapped onto the control frame (`FN` 18) and to the lowest time slot number. If several uplink traffic channels are used by a connection then the uplink and the downlink SCHs are also mapped onto the control frame, but to the highest time slot number.

Logical traffic channels (TCH) are mapped onto frames 1–17 of the physical traffic channels (TP), in other words onto block 1 and block 2. An STCH can be mapped onto all frames available for traffic, and always first steals the first block of a time slot. This is indicated by a special training sequence.

7.2.4.3 The Physical Layer in the TETRA Standard

The physical layer forms the physical radio interface of the TETRA system. It generates the bursts, which consist of a range of symbols and are transmitted and received.

It is responsible for the following functions, which are explained in detail below:

- Radio-oriented:
  - Modulation and demodulation,
  - Transmitter and receiver management,
  - Radio frequency characteristics,
  - Fine tuning of radio parameters.
- Bit and symbol-oriented: symbol synchronization.
- Burst formation:
  - Receiving and transmitting of data from and to MAC layer,
  - Slot-flag coding/decoding,
  - Scrambling and deciphering.

The modulation method is $\pi/4$-*Differential Quaternary Phase Shift Keying* (DQPSK) at a modulation (gross) bit rate of 36 kbit/s.

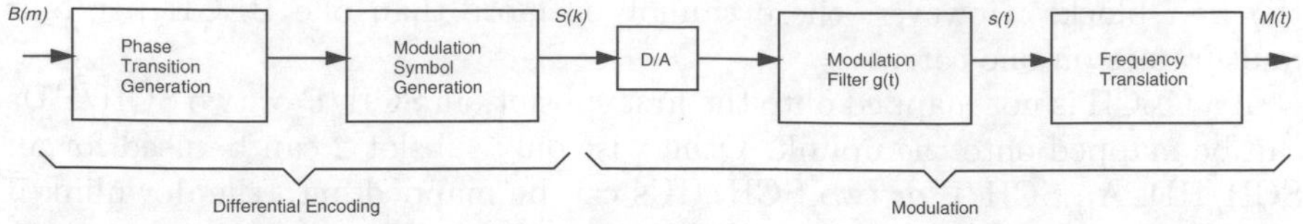

Figure 7.9: Block diagram of the modulation process

The modulation process is illustrated in Figure 7.9. The sequence $B(m)$ of the modulation bits to be transmitted is represented by a differential coding in a sequence of modulation symbols $S(k)$ based on the following rule:

$$\begin{aligned} S(k) &= S(k-1)e^{jD\Phi(k)} \\ S(0) &= 1 \end{aligned} \tag{7.1}$$

As shown below, the phase translation $D\Phi(k)$ is dependent on $B(m)$:

$B(2k-1)$	$B(2k)$	$D\Phi(k)$
1	1	$-3\pi/4$
0	1	$+3\pi/4$
0	0	$+\pi/4$
1	0	$-\pi/4$

It follows from this definition that $S(k)$ can assume eight different values. The modulated signal $M(t)$ on the carrier frequency f_c is

$$M(t) = \text{Re}\{s(t)e^{(j\cdot 2\pi f_c t + \Phi_0)}\} \tag{7.2}$$

with Φ_0 representing the phase offset and $s(t)$ the complex envelope of the modulated signal:

$$s(t) = \sum_{k=1}^{K} S(k)g(t - t_k) \tag{7.3}$$

Here $g(t)$ stands for the inverse Fourier transform of the *square-root raised-cosine* spectrum and K is the maximum number of symbols.

The management functions of the transmitter and the receiver are responsible for the choice of frequency bands and the transmit power. The output power is controlled so that it is increased in very short ramps until the full strength of transmitter power is reached in order to minimize interference in the system. This requires an adherence to a number of limit values, which is guaranteed by the management function. Optimization in this area will result in a reduction in the bit-error ratio. Detailed explanations of this function can be found in [9, 10].

A fine tuning of the radio parameters results in a frequency correction using a special *frequency correction sequence*, which is localized in a synchronization burst within the BSCH. This ensures that there is only minimal deviation from the carrier frequency. It also requires a high level of accuracy in the oscillator of the mobile station. The *power control* function also ensures that a mobile station is always using the adjusted transmitter power. This is controlled by the MAC layer.

Similarly to frequency correction, symbol synchronization is achieved through a specific training sequence, which is longer for an initial synchronization than it is for existing connections. The synchronization information is contained in bursts. This enables the physical layer to recognize clearly the boundaries of a burst. As a way of ensuring that a mobile station remains synchronized for a longer period of time, after completion of the synchronization process timers are started that indicate when the next frame, multiframe or hyperframe will begin or end. This requires a high level of accuracy in the time control of the mobile station.

This process allows a mobile station to detect the beginning and the end of a burst, and to use this knowledge to transmit data in order to map the MAC-PDUs (*Protocol Data Unit*) to the bursts as well as to place its own specific information correctly. During reception this process is reversed. The information specific to the physical layer is removed and the MAC-PDU is retrieved from the burst and forwarded to the MAC layer. The different types of bursts are explained in detail in Section 7.2.4.2 and [9].

The slot-flag appears in two versions and indicates through the bursts contained in the training sequences whether an entire slot (`SF=0`) or half a slot (`SF=1`) is being occupied by the signalling data. Its other tasks are *scrambling* and *descrambling*. Information on the scrambling code is passed on to the receiving station in a non-coded MAC-PDU by a socalled *colour code* characterizing a base station. The scrambling is part of the channel coding and is described in Section 7.2.4.4.

7.2.4.4 The Data Link Layer in the TETRA Standard

The architecture of the *data link layer* of the TETRA standard V+D is presented in Figure 7.10. It is organized into the sublayers MAC and LLC, with the MAC layer divided into a *lower* and an *upper* MAC at the *TETRA MAC Virtual SAP* (TMV-SAP) (*service access point*). On the upper boundary of the data link layer three service access points of the *Link Control Entity* (LCE) (see Figure 7.5) offer their services to the network layer: TLA (*TETRA-LLC-A*), TLB and TLC. The TLA-SAP provides services for the bidirectional transfer of addressed signalling and data messages. The TLB-SAP provides non-addressed data transfer in which broadcast messages with system information are sent unidirectionally from the base station to the mobile station. The TLC-SAP is only available on the mobile station side, and is used for control and management information. The LLC and MAC sublayers correspond over the SAPs TMA (*TETRA-MAC-A*), TMB and TMC, which have the same functions as the corresponding SAPs of the LLC sublayer. The MAC sublayer contains another SAP (TMD) for the user plane, which

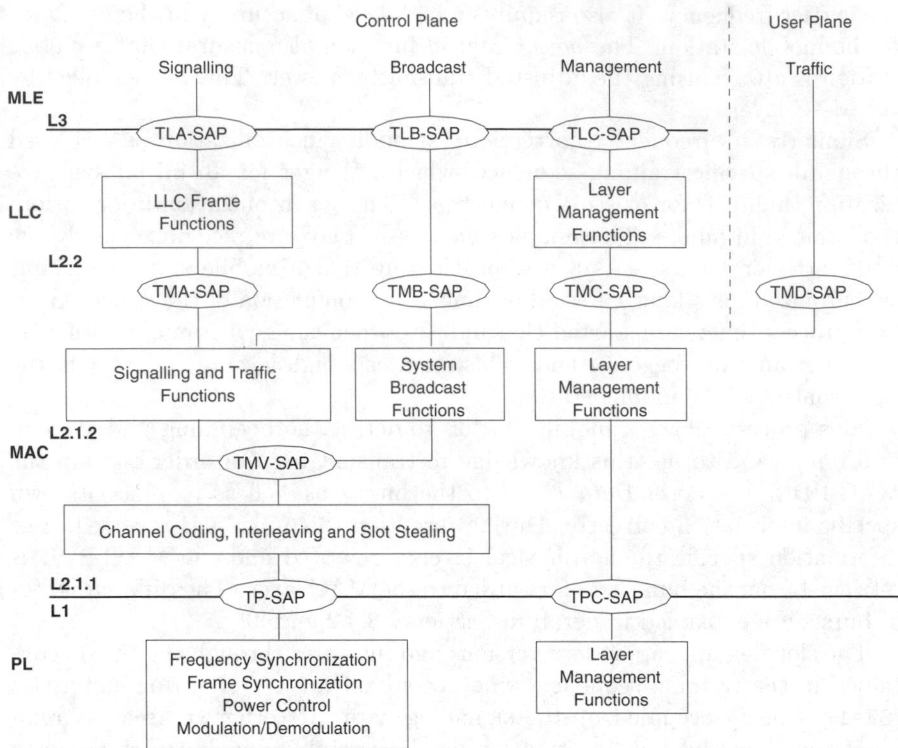

Figure 7.10: Architecture of the Voice+Data data link layer

transports information over a circuit-switched connection. The lower and upper MAC sublayers communicate over a virtual TMV-SAP, which provides services for actual radio transmission, such as channel coding, interleaving and slot stealing. At the interface to the physical layer the TPC-SAP (analogously to TLC- and TMC-SAP) offers access to the local layer management. Through the TP-SAP, the lower MAC layer corresponds to the physical layer.

Medium access control As shown in Figure 7.10, the functions of the *Medium Access Control* (MAC) sublayer mainly consist of channel coding, channel access control and radio resource management, which, depending on transmission mode, offer their services at the three SAPs—TMA, TMB and TMC—for signalling and packet data mode or at the TMD-SAP for user traffic mode. After the set up of a circuit-switched voice or data transmission connection for which the traffic mode has been defined, signalling messages can be transmitted with the help of a slot-stealing mechanism.

A base station is allowed to transmit continuously or discontinuously. In the latter case the respective base station interrupts its transmission if there is no further information to be transmitted or if it is sharing the same radio channel

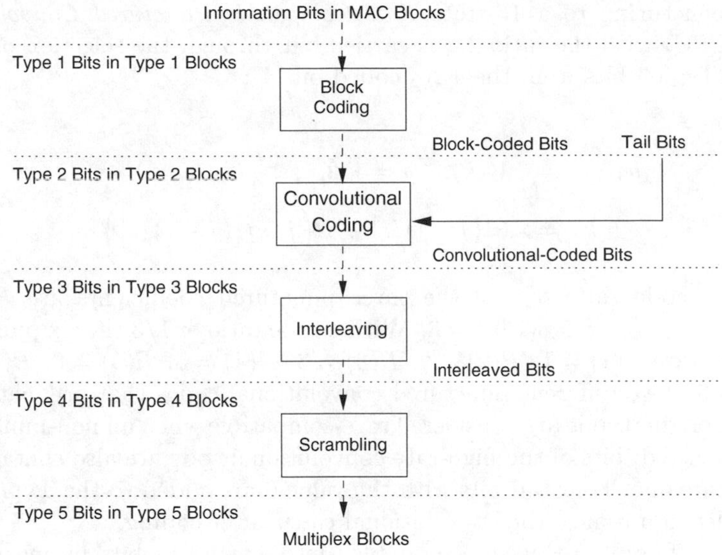

Figure 7.11: TETRA channel coding

for signalling purposes with other base stations that also find themselves in *timesharing mode*. The functions, data structures, service elements and states of the MAC layer are discussed below.

Channel coding Channel coding in the Voice+Data standard is carried out as shown in Figure 7.11. Information bits contained in MAC blocks, called Type 1 bits, are coded by a (K_2,K_1) block coder. Except for the AACH, in which Type 2 bits are produced by means of an abbreviated (30,14) Reed-Muller code [8, 17], at this initial stage of channel coding a systematic cyclical block code [18] is produced from K_1 Type 1 bits $b_1(1), b_1(2), \ldots, b_1(K_1)$, and $K_2 = K_1 + 16$ Type 2 bits $b_2(1)$,$b_2(2)$,...,$b_2(K_2)$.

The code words f are sent together with Type 1 information bits b_1, thereby producing the Type 2 bits b_2 as

$$b_2(k) = \begin{cases} f(k-1) & \text{for} \quad k = 1, 2, \ldots, 16 \\ b_1(k-16) & \text{for} \quad k = 17, 18, \ldots, K_2 = K_1 + 16 \end{cases} \tag{7.4}$$

Since $K_2 \in \{76, 108, 140, 284\} \neq 2^m - 1$ for $m \in \mathbb{N}$, the block code used here is, according to [17, 18], neither a Hamming code nor a *Broadcast Channel* (BCH) code.

At the next stage of channel coding the K_2 Type 2 bits b_2 are coded by a (4,1,5)-rate-compatible, punctured convolutional coder with the code ratio 1/3 or 2/3 to K_3 Type 3 bits b_3 with $K_3 = \frac{t}{2} K_2, t \in \{3, 6\}$. First the output V is computed by a 16-state convolutional coder with the ratio 1/4 from the Type 2 bits $b_2(k)$.

The puncturing to a 16-state *Rate-Compatible Punctured Convolutional code* (RCPC) with the ratio $2/t$ is carried out through the selection of $K_3 = (t/2)K_2$ Type 3 bits from the 4 K_2 coded bits V:

$$\begin{aligned} b_3(j) &= V(k), \quad j = 1, 2, \ldots, \frac{t}{2}K_2, \\ k &= 8\lfloor (j-1)/t \rfloor + P\Big(j - t\lfloor (j-1)/t \rfloor\Big) \end{aligned} \tag{7.5}$$

With a code ratio of $2/3$ the three punctured coefficients are $P(1) = 1,\ P(2) = 2,\ P(3) = 5$, with $t = 3$. With a code ratio of $1/3$ the six punctured coefficients are $P(1) = 1,\ P(2) = 2,\ P(3) = 3,\ P(4) = 5,\ P(5) = 6,\ P(7) = 7$, with $t = 6$. Two different punctured convolutional codes that originate from the same original code are considered rate-compatible when all non-punctured, i.e., not deleted, bits of the high-rate convolutional code are also contained in the low-rate one. Four tail bits with the value 0 are added to the Type 2 bits b_2 in order to initialize the convolutional coder after coding.

In block Type 3 b_3 bits are scrambled into Type 4 b_4 bits by means of a (K, a) block interleaver where $K = K_3$, e.g., $a = 101$ (as shown in Figure 7.12), in order to prevent bursty-type transmission errors:

$$\begin{aligned} b_4(k) &= b_3(i), \quad i = 1, 2, \ldots, K = K_3 = K_4, \\ k &= 1 + \big((a \cdot i) \bmod K\big). \end{aligned} \tag{7.6}$$

For 432-bit long Type 3 blocks an alternative of bit interleaving over N blocks in a sequence of Type 4 blocks is being planned in two stages. In the first stage $b_3'(m,k)$ interleaved bits are generated from Type 3 b_3 bits as k bits of the m block:

$$b_3'(m,k) = b_3\big(m - j,\ j + (i \cdot N)\big), \qquad \begin{aligned} k &= 1, 2, \ldots, 432, \\ m &= 1, 2, \ldots, N, \end{aligned} \tag{7.7}$$

$$\begin{aligned} j &= \big\lfloor (k-1)/(432/N) \big\rfloor, \\ i &= (k-1) \bmod (432/N) \end{aligned}$$

The $B_3'(m)$ blocks thus created are interleaved into Type 4 $B_4(m)$ blocks:

$$b_4(m,i) = b_3'(m,k), \quad \text{with} \quad i = 1 + (103k) \bmod 432. \tag{7.8}$$

At the last stage K_4 Type 4 b_4 bits are coded into K_5 Type 5 b_5 bits through the use of a linear feedback register in which a coding sequence p, which is formed from a generator polynomial and an initialization sequence for P, is added to b_4 modulo 2:

$$b_5(k) = b_4(k) + p(k), \quad k = 1, 2, \ldots, K_5 = K_4. \tag{7.9}$$

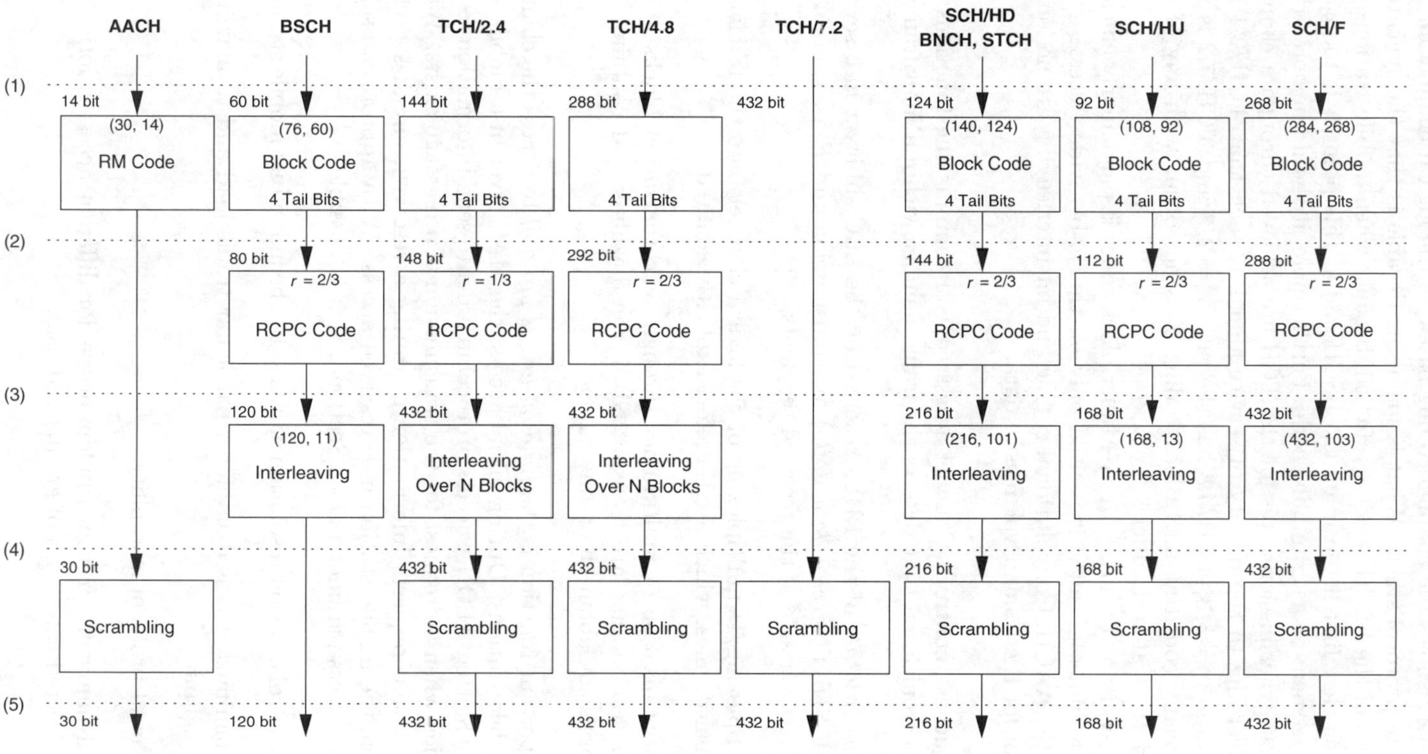

Figure 7.12: Structure of V+D channel coding

Figure 7.12 shows the channel coding based on the logical channel for the Voice+Data protocol stack. What is mainly noticeable is that the synchronization data of the BSCH is not scrambled, because when mobile stations are switched on they first listen in on the BSCH and at this point it is too early for any agreement on a scrambling code. Furthermore, it should be noted that the different transmission rates for the TCH are achieved through a channel coding according to the quality of service aimed at such that TCH/2.4 has the code ratio $r = 1/3$, TCH/4.8 has the ratio $r = 2/3$ and TCH/7.2 is not convolutionally coded at all. Block coding does not occur with any of the variants of the traffic channel.

Since, according to [17], a Reed–Muller block coder is particularly efficient for small volumes of data, it is used only on the 14-bit wide Access_Assign PDUs of the AACH. Convolutional coding and bit interleaving are no longer carried out for these small volumes of data.

Channel access control Channel access control contains the functions for frame synchronization that detects the frame number within a multiframe:

- *Fragmentation* of the PDUs received by the LLC sublayer into several MAC-SDUs (*service data units*) for transmission and *Reassociation* on the receiver side of the received segments.
- Multiplexing/Demultiplexing of the logical channels and the formation of multiframes, which are therefore explicitly counted.
- Synchronization of multiframes through synchronization blocks of the base station that are transmitted on the downlink and contain, e.g., information about the frame number.

An important function is the *random-access protocol* to access the channel by the mobile stations. During initial access, the MAC layer of a mobile station uses a Slotted-ALOHA access protocol in order to send information to the base station when no request for the transmission of a message exists. When the base station requests information or during a reserved, circuit-switched connection, the mobile station uses reserved access. With appropriate selection of the access parameters (see Section 2.8) it is possible:

- To control collision resolution of access by the individual mobile stations.
- To minimize access delay for a particular traffic load and to maximize throughput.
- To avoid protocol instability.
- To dynamically prevent random access for different access priorities or for selected categories of groups and users.

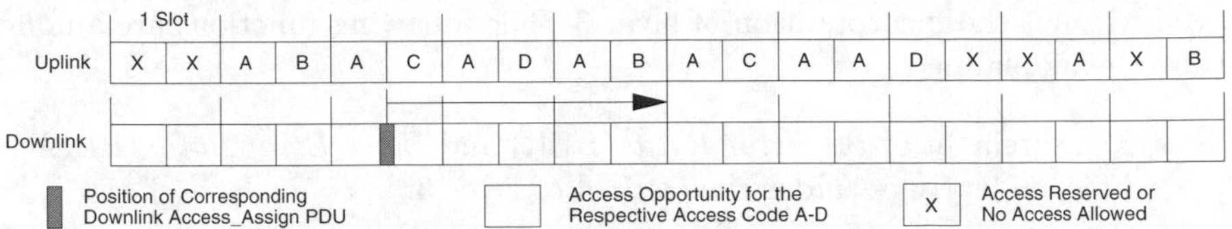

Figure 7.13: Access procedures of the V+D random-access protocol

- But at the same time, and independently of the *Grade of Service* (GoS), to offer access to different categories of groups and users.

A base station offers different mobile stations the possibility of random access through the use of four *access codes*, which are referred to as A, B, C and D and should be interpreted as priorities. The access code represents a combination of user services and is permanently allocated to mobile stations by the network operator. It is not necessary for every code to be accessible by all mobile stations. A mobile station is only allowed access if the conditions for access to the subslot of a particular code have been met. Base stations inform mobile stations that they have access by the means of two different PDUs. An Access_Define PDU (see Section 7.2.4.4) is sent in the intervals specified by the network operator, and provides information about the access code and the priority allowed for sending as well as the time period and number for a repeated access. An Access_Assign PDU that is contained in each downlink time slot is used to indicate the access rights of mobile stations to the uplink time slots, and a *traffic usage marker* indicating permission to transmit and receive in the current uplink and downlink time slot. The information of both PDUs is stored until receipt of an update.

Example of uplink subslots characterized by different access codes or a reservation and indicated in an Access_Assign-PDU are shown in Figure 7.13. A mobile station wishing to send a message first compares the transmitted conditions for access with those that it is allowed. If access is allowed, it uses the MAC_Access-PDU that contains the message to be transmitted and sets its access timer. If the message is fragmented, it makes a reservation request for several time slots in the MAC_Access-PDU sent to the base station. The base station responds to the MAC_Access-PDU with a MAC_Resource-PDU, in which it acknowledges the successful receipt of the message. The mobile station is also notified whether time slots have been reserved, and if so, which ones. These are represented by crosses in Figure 7.13. If a mobile station still has not received a response to the Mac_Access-PDU when the access timer has expired, it repeats the random access. The random-access protocol is explained in detail in [5].

Radio resource management Some of these functions are only available in mobile or base stations, others in both. They should be available at all times, even without the incorporation of layer 3. The following functions are among those being planned:

- Measurement of *Bit Error Ratio* (BER) and *Block Error Ratio* (BLER), independently or under the control of layer 3.
- Path-loss calculation through monitoring of current and neighbouring cells. Calculation of path-loss parameters based on information from the current cell as well as periodic *scanning* other cells. Determination of path-loss parameters of neighbouring cells from the scanned signal level.
- Address management for individual, group and broadcast calls. Two addresses can be used: a copy of the *individual* or *group subscriber identity*, or an *event label* that can be used by the MAC layer to reduce the amount of signalling required during existing connections.
- Management of *power control* in layer 1.
- Set up of a connection through the use of the details on frequency, time slot and *colour code* of the *Mobile Link Entity* (MLE).
- Storage of control data and speech frames until time of transmission.

Service elements of TMA, TMB, TMC and TMD service access points (SAP) A TMA-SAP is used by both LLC connection types—basic link and advanced link (see Section 7.2.4.4)—for the transmission of signalling and packet data information (see Figure 7.10). Five *Service Primitive* (SP) are available for this purpose.

The LLC sublayer uses the TMA_Unitdata_req-SP to ask the MAC sublayer to send LLC frames, which at this interface are called TM-SDUs (SDU=service data unit).

The TMA_Unitdata_ind-SP is used by the MAC sublayer in order to forward received messages.

A TMA_Cancel_req-SP can be used to clear a TMA_Unitdata_req that was sent by the LLC sublayer.

With a TMA_Release_ind-SP an LLC *advanced link* connection can be interrupted (see Section 7.2.4.4) if the MAC sublayer releases a channel and all *advanced links* on this channel are quietly interrupted.

In the TMA_Report_ind-SP the MAC sublayer presents information to the higher layers on the sending of Request procedures. During the initialization process, this SP indicates whether a connection has been established successfully and the partner entity has acknowledged receipt of the message. This SP is also used to notify the higher layers of errors, such as failed random access, which occur during these processes.

The TMB-SAP carries out the transfer of non-addressed broadcast messages that contain information on network and system organization. In this

Table 7.5: Mapping of V+D-MAC-PDUs to logical channels

MAC-PDU	Logical channels	Type of information
Access_Assign	AACH	MAC internal information
Access_Define	SCH/HD, SCH/F, STCH	
MAC_Access	SCH/HU	TMA-SAP information
MAC_End_HU	SCH/HU	
MAC_Data	SCH/F, STCH	
MAC_Resource	SCH/HD, SCH/F, STCH	
MAC_Frag	SCH/HD, SCH/F	
MAC_End	SCH/HD, SCH/F, STCH	
MAC_Traffic	TCH	TMD-SAP information
MAC_U_Signal	STCH	
Sync	BSCH	TMB-SAP information
Sysinfo	BNCH	

case no special TMB-SAP-related functions exist in the LLC sublayer, and it is necessary for Requests at the TLB-SAP to be directly mapped into TMB-SAP Requests and TMB-SAP Indications into TLC-SAP Indications. This is why the same service primitives (explained further in Section 7.2.4.4) also exist at both these SAPs.

The information for local layer management is transmitted in the same way over the TMC-SAP. There are no TMC-SAP-relevant functions in the LLC sublayer. Therefore Requests and Responses at the TLC-SAP have to be mapped to Requests and Responses at the TMC-SAP, and vice versa Indications and Confirms at the TLC-SAP to Indications and Confirms at the TMC-SAP.

In a circuit-switched voice or data connection the voice or data frame is forwarded to the MAC sublayer through the TMD-SAP, which forms the interface between the TETRA speech codec and the MAC sublayer. Voice frames are permitted to contain coded or non-coded speech; the information they contain is irrelevant to the MAC sublayer. Individual time slots or subslots can be *stolen* for end-to-end signalling between users of circuit-switched services. Depending on whether a time slot contains voice, data or signalling information, or whether an entire or half a time slot is stolen, appropriate adapted channel coding is carried out. The TMD_Unitdata-SPs Request and Indication are used to transfer the data or voice frame that has been received or is to be transmitted. The TMD_Report_ind-SP informs the MAC service user of the status of the current transmission.

Data structures Twelve different PDU types are defined in the MAC sublayer (see Table 7.5).

Access_Assign This 14-bit (net) long PDU is contained in the *Broadcast Block* (BBK) (see Figure 7.8) of each downlink time slot, and carries information on the current time slot and access rights to the uplink time slot, which is delayed by two time slots and is associated with this downlink time slot. The content varies depending on whether the PDU is being sent in one of the frames from 1 to 17 or in frame 18.

Access_Define The random code specifically valid for random access, the ALOHA parameter and the access procedure are stipulated by the BS in this downlink-PDU, which occupies either half a time slot or an entire time slot and uses the logical SCH/HD or SCH/F channel. At the user plane (see Figure 7.10) the STCH channel is also occupied.

MAC_Access This PDU contains LLC data and is sent on the uplink and in half subslots for random access. A TM-SDU is sometimes fragmented and divided among several MAC_Access- and MAC_Frag-PDUs or MAC_End_(HU)-PDUs [4]. Filler bits are added when a message is shorter than the number of user data bits available.

Depending on whether a 24-bit wide layer-2 address (*Short Subscriber Identity*, SSI) is used for the initial access or a 10-bit wide *Event Label* (EL) allocated by a BS to reduce the addressing overhead for a transaction, either 56 bits or 70 bits of net user data can be transported in this PDU. The beginning and the end of a fragmented TM-SDU are indicated in the MAC_Access-PDU and in the MAC_End_(HU)-PDU. The MAC_Access-PDU is mapped to the SCH/HU channel.

MAC_End_HU This PDU is only sent on the uplink, and a fragmentation indicates that it contains the last fragment. It always occupies only half a time slot.

MAC_Data This is for the transfer of LLC data in full time slots on the uplink or, with a *stolen* channel, in the first and, if required, in the second subslot. In the second subslot the MAC_Data-PDU can be used to send other data of the control plane. This PDU can only be used in reserved transmission. With a MAC_Access-PDU or a MAC_Data-PDU, a base station obtains uplink reservation requests from the length data contained in the *length indication or capacity request*. A fragmentation of the message is sometimes also indicated. Depending on the type of addressing—(SSI or EL)—231 bits or 245 bits of user data net can be sent.

MAC_Resource This PDU is used by a base station to transmit information of the control plane on the downlink. It is also able to confirm that a random access has been successful and in some cases to allocate capacity on the uplink. Depending on the addressing and whether a half or a full time slot is used, the maximum length of user data is 95 bits or 239 bits. A MAC_Data-PDU or a MAC_Resource-PDU can be used to indicate in the PCI that the second subslot is being stolen and contains a MAC_U_Signal-PDU.

MAC_Frag This PDU is used to send further fragments of fragmented messages of the control plane from the LLC sublayer on the downlink. The header attached to the PDU is 4 bits long, which means that 264 user information bits can be sent.

MAC_Traffic Voice and data frames of circuit-switched connections are transmitted with this PDU. Except for the 2 bits used for the identification of the PDU type, this PDU exclusively contains user data.

MAC_U_Signal End-to-end signalling at the user level is carried out through this PDU from the TMD-SAP of the sender to the TMD-SAP of the partner entity. It is always mapped onto the STCH. Similarly to the MAC_Traffic-PDU, this PDU also exclusively contains user data except for the first three bits. An indication whether one or two subslots are being stolen is given in the third bit. Even if the second subslot is occupied, it can contain LLC data.

MAC_End This PDU is only sent on the downlink, and occupies either a whole time slot on the SCH/F channel or half of one on the SCH/HD channel. It is sent with the last fragment of fragmented LLC signalling data.

Sync Per broadcast, a base station sends non-addressed Sync- and Sysinfo-PDUs on the logical BSCH and BNCH channels to the mobile stations, which decode all of the information contained in both PDUs.

A Sync-PDU establishes the initialization code (*colour code*) responsible for the channel coding, and contains the current multiframe, frame and time slot numbers (see Section 7.2.4.2). It also notifies mobile stations whether it is transmitting continuously or discontinuously and whether it is in *Timesharing mode* with other base stations.

Sysinfo A mobile station computes the path loss for the current cell using the parameters sent by the base station per broadcast in this PDU. The PDU notifies the mobile station of the carrier frequency and the frequency band of the main control channel and provides details about the cell cluster and power control algorithm used.

Table 7.5 lists the logical channels over which the MAC-PDUs are forwarded to the lower MAC sublayer for channel coding, and which type of information is contained in these PDUs. The data of the STCH is either transmitted over the SCH/HU or the SCH/HD. The allocation of a whole time slot (SCH/F) to the STCH is also possible.

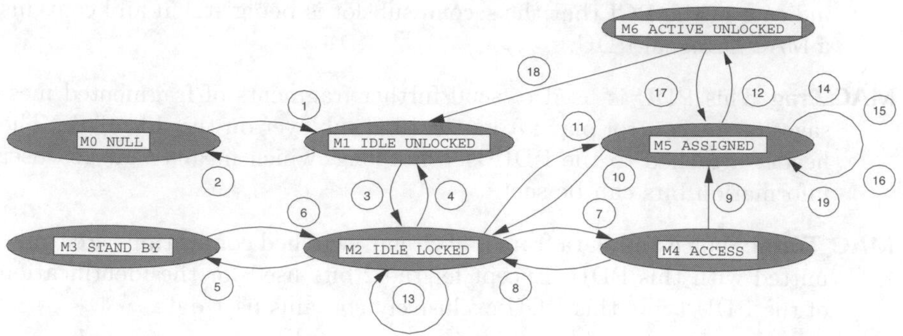

Figure 7.14: State diagram of a V+D mobile station in the MAC sublayer

State Diagram and Transitions of a Mobile Station The MAC protocol of a mobile station comprises seven states (see Figure 7.14 and Table 7.6), with M0 NULL corresponding to the switched-off state. When a mobile station is switched on (1), it is in the state M1 IDLE UNLOCKED. After it is switched off (2), which can take place in any state, it returns to the M0 NULL state. Only after an MS has received the Sync and Sysinfo PDUs from the BS per broadcast and a cell has reached a sufficient level of quality, does it select this cell and inform the MLE accordingly.

It then moves to the state M2 IDLE LOCKED (3). In the event that the MAC sublayer determines a deterioration in signal quality and notifies the MLE, it will then be instructed by the MLE to find a new cell with sufficient quality, and will return (4) to the state M1 IDLE UNLOCKED.

Depending on the actions of the user, after a specific period of inactivity the MS is operated in power-saving mode and moves to M3 STAND-BY state (5).

After the timer, which can be set to a time interval of between 1 and 432 frames (57 ms to 25.9 ms) based on the user's discretion, expires, there is again an exit from power-saving mode and the MS returns (6) to the state M2 IDLE LOCKED.

In this state an MS only receives. It monitors the SCH for the paging messages sent to it by the BS and for general system information. The MS also prepares for future transmitting actions whereby it analyses Access_Define-PDUs received over the SCH and observes the AACH. If there are no paging messages being received or data being sent, the MS (13) scans the signal strength of the neighbouring cells and attempts to decode all the network information of these cells.

If the MS receives a paging message or packet data from the BS or carries out a random access, it finds itself (7) in active M4 ACCESS state and, in the case of a circuit-switched connection, waits until it is allocated a *Traffic Channel* (TCH). When it is in this state, all the packet data is transmitted if the connection is not a circuit-switched one. If the timer runs out or the

Table 7.6: States of a V+D mobile station in the MAC sublayer

1	The MS is switched on.
2	The MS is switched off. This changeover can be carried out from any state.
3	A cell with sufficient quality is found, the MLE layer informed and the cell selected.
4	The MAC layer has informed the MLE layer of the deterioration of signal quality and is instructed to return to search mode.
5	The MS switches to power-saving mode.
6	The MS leaves power-saving mode.
7	The MS carries out a random access or is paged. In both cases there is a wait in the `ACCESS` state until the allocation of a traffic channel (circuit-switched connection).
8	Time-out or end of packet-data transmission.
9	Receipt of the channel allocation.
10	End of a transmission (transmission trunking) or of a conversation (message trunking).
11	Direct channel allocation (group call or call diversion—transmission trunking).
12	Signal loss during a call. MS begins to restore the connection after a waiting period.
13	Search of neighbouring cells during idle state.
14	Search of neighbouring cells during assigned state.
15	Slot-stealing procedures.
16	Voice or data transmission with message trunking.
17	Return after brief signal loss.
18	Continuation of call reestablishment procedure.
19	Handover (fast call reestablishment) to a neighbouring cell.

transmission of packet data has been completed, the MS returns (8) to the `M3 IDLE LOCKED` state.

Once a BS has exclusively allocated a channel to an MS or to a group of MSs, there is a transition (9) to the `M5 ASSIGNED` state. The speech and data frames of a circuit-switched connection are transmitted in this state. The end-to-end signalling between users takes place through the use of slot-stealing mechanisms (15).

Similarly to state M2, an MS regularly measures (14) the signal strength of the neighbouring cells during a call and tries to decode all the network information of these cells. If after the receipt of this information the MS prefers a different cell, it requests a handover (19), which activates the fast-call-reestablishment procedure.

At the end of a transmission (*transmission trunking*) or a call (*message trunking*), the MS moves (10) to the state `M2 IDLE LOCKED`. With a group call or the forwarding of a call with transmission trunking, there is a direct channel allocation and a return (11) to the `M5 ASSIGNED` state. Since the channel is

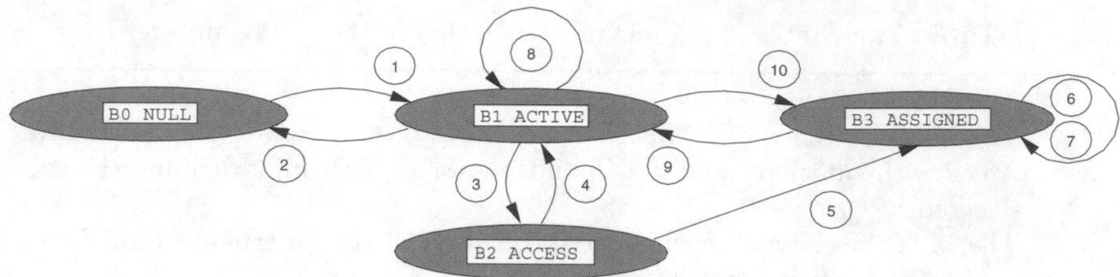

Figure 7.15: State diagram of the V+D base station in the MAC sublayer

permanently occupied in a message trunking situation, transmitting requests are processed in this state (16).

In V+D transmission trunking is the method in which a *Traffic Channel* (TCH) is only occupied if there is actual voice activity, and then released again immediately. With message trunking the traffic channel is permanently occupied for the entire duration of a call, which, compared with transmission trunking, results in a lower protocol overhead but also lower throughput.

If a mobile station loses contact with the base station (12), it begins to reestablish the connection after a short waiting period while in `M6 ACTIVE UNLOCKED` state. After successful restoration of the connection, it returns (17) from this short signal break to the state `M5 ASSIGNED`. If this is not the case, it moves (18) to the state `M1 IDLE UNLOCKED` and continues the establishment of the connection in this position.

State Diagram and Transitions of the Base Station Four states are defined for the V+D base station in the MAC sublayer (see Figure 7.15 and Table 7.7), with the BS in state `B0 NULL` before the switching-on process. The transition to state `B1 ACTIVE` takes place after the BS has been switched on (1). When it is switched off (2), the BS returns to the original state, no matter which state it is currently occupying.

Since the BS produces a new instance of the MAC entity for each connection, it can assume a number of states B1 to B3 at the same time.

Synchronization and system information (Sync- and Sysinfo-PDUs) are sent to inactive MSs in state `B1 ACTIVE`. In addition, the BS transmits access definitions (Access_Define-PDUs) and access assignments (Access_Assign-PDUs) to enable MSs to prepare themselves for random access in the future. However, data transfer traffic does not take place in this state.

After a BS has successfully recognized (3) and received a random-access request from an MS, it moves into state `B2 ACCESS`. In the event that the BS wants to establish a connection with the MS, the latter is informed (paging message) and a transition (3) to B2 takes place. Packet data is transmitted between BS and MS in this state. Upon completion of this transmission or the running out of a timer, the BS returns (4) to the state `B1 ACTIVE`.

Table 7.7: State transitions of the V+D base station in the MAC sublayer

1	The BS is switched on.
2	The BS is switched off.
3	Successful reception of a random access or paging message by an MS.
4	Time-out or end of packet-data transmission.
5	Traffic channel (TCH) is allocated.
6	Slot-stealing in traffic mode.
7	Sending permission to transmit (message trunking).
8	Sending permission to transmit (transmission trunking).
9	End of transmission (transmission trunking) or a call (message trunking).
10	Direct allocation of a traffic channel (TCH) for a group call or an existing connection (transmission trunking).

If an MS has requested the allocation of a traffic channel during its random access, the BS acknowledges this and moves (5) into state `B3 ASSIGNED`. In this state voice and data frames are exchanged and transmission capacity for end-to-end signalling between users can be stolen (slot-stealing, 6). In the case of message trunking the connection is permanent and transmission requests are acknowledged in this state (7).

Upon completion of a call (message trunking) or a transmission (transmission trunking), the BS returns (9) to state `B1 ACTIVE`, where it acknowledges further requests for transmission (8) in the case of transmission trunking and directly allocates a traffic channel for an already-existing connection (9). A direct channel allocation is also made in the case of group calls.

Logical Link Control

Overview and functions of the LLC sublayer The *Logical Link Control* (LLC) for TETRA provides two types of connection at the TLA-SAP: a *Basic Link* (BL) and an *Advanced Link* (AL) (see Figure 7.10). Acknowledged or non-acknowledged data transmission is possible with both types of connection. A BL constitutes a bidirectional, connectionless path between one or more MSs and a BS, and is immediately available after the synchronization of MS and BS. In contrast, an AL is defined either as a bidirectional, connection-oriented path between an MS and a BS or as a unidirectional path from a BS to several MSs. An AL offers a better *Quality of Service* (QoS) than a BL and always requires that a connection be established. Compared with a BL, which only provides for the use of expanded error protection (*Frame Check Sequence*, FCS) to minimize the number of undetected faulty messages, an AL offers a more favourable quality of service by providing flow control, segmentation of layer-2 service data units, a window mechanism that allows the transmission of more than one SDU without having to wait for an acknowledgement of the previous SDU, and the possibility to select different throughputs. FCS is always used with an AL.

The LLC sublayer of an MS supports simultaneously up to four ALs, each differentiated by its own *Connection Endpoint* (CEP) Identifier. Each of these CEP identifiers is associated with the time slots used by the MAC sublayer. For each AL and each circuit-switched service there is a BL that can use the respective number of reserved time slots of the corresponding ALs or line-switched services.

The transmission mode of an LLC connection is not contingent on the other LLC connections of an N-connection (*Network*, N), e.g., in point-to-multipoint mode a sending MS can use an AL to the BS from one MS to several MSs, and conversely the BS can use unacknowledged BLs for the receiving MS.

Service elements of the TLA, TLB and TLC-SAP The TLA-SAP is used for addressed data transmission and for the control of data transmission procedures in layer 2. Each independent entity of a service is represented by a separate connection endpoint at the TLA-SAP. Up to four AL entities can be maintained by an MS at the same time. The information flow is possible from an MS to a BS and in the reverse as well as within an MS or a BS.

The service primitives listed in Table 7.8 are available at the TLA-SAP and are partially used by both connection types BL and AL or by only one of them. An acknowledged connection is set up and maintained through the four primitives of the TL_Connect-SP (*Service Primitive*, SP). The procedure is initiated with Request and indicated with Indication by the partner entity, which in turn sends back a Response using a Confirm to provide information about the connection set up to the initiating unit. The TL_Data-SP, which is defined separately for BL and AL, is available for the acknowledged transmission. For a BL there is also a Response-SP, which can be used to respond to a message with an acknowledgement or a new message and itself is not acknowledged. Non-acknowledged message transmission is carried out by a TL_Unitdata-SP. However, the Confirm-SP listed in Table 7.8 is optional and only indicates that transmission has been successful. The TL_Disconnect-SP is used to terminate an AL that has been initiated with Request and completed with Confirm, i.e., a confirmation from the communicating partner. A TL_Release-SP can be used to end an abnormally interrupted connection locally in the MS or the BS. A TL_Report_ind-SP is used internally for MLE information on the status of processes it has terminated.

The service elements at the TLB- and TLC-SAPs are summarized in Table 7.9. Using the TL_Sync- and TL_Sysinfo_req-SPs at the TLB-SAP of a BS, the MLE sublayer requests that the BS send the MS the synchronization and system information for the cell selection of the MS. The TL_Sync- and TL_Sysinfo_ind-SPs at the TLB-SAP of the MS are used in the MS to forward the data received from the LLC sublayer to the MLE sublayer.

The service elements of the TLC-SAP are responsible for monitoring a connection. Thus the TL_Configure-SP is used to configure layer 2 on the

Table 7.8: Service elements of the V+D TLA-SAP

Service element	Description	Link type
TL_Cancel_req	Request to cancel transmission before it has even started	BL, AL
TL_Connect_req/conf TL_Connect_ind/resp	Sending/receiving parameters required for setting up or maintaining a connection	AL
TL_Data_req/conf TL_Data_ind/resp	Sending and receiving data with acknowledgement	BL
TL_Data_req/conf TL_Data_ind	Sending and receiving data with acknowledgement	AL
TL_Disconnect_req/conf TL_Disconnect_ind	Termination of a connection between two corresponding AL entities	AL
TL_Release_req TL_Release_ind	Internal termination of a connection without informing the communicating partner	BL, AL
TL_Report_ind	Message to MLE about the status of an operation terminated by the MLE	BL, AL
TL_Unitdata_req/conf TL_Unitdata_ind	Sending and receiving data without acknowledgement	BL, AL

basis of the selected cell parameters and the state of the respective MS. The TL_Measurement_ind-SP reports to the higher layers on the quality of a connection based on measurement results and the parameters of the neighbouring cells determined through *monitoring* or *scanning* (see Section 7.2.4.4). Through the TL_Monitor_List-SP, the monitoring process is applied to all the channels indicated in a list and the TL_Monitor-SP indicates the measurement results for a channel. The TL_Report-SP is used in the same way as at a TLA-SAP. The TL_Scan-SP starts the scanning process of a channel and the TL_Scan_Report-SP reports on the results. With Request and Confirm the TL_Select-SP is used to carry out a request from the MAC sublayer to the MLE sublayer to select a channel or to change channel, and through Indication and Response it is used in the notification of the MLE by the MAC sublayer of the channel change requested by the BS.

All the service primitives of the LLC and MAC sublayers and their interrelationships are listed in Table 7.10. Detailed explanations are available in [3, 6].

Table 7.9: Service elements of V+D TLB- and TLC-SAPs

Service element	Description	SAP
TL_Sync_req/ind	Broadcast of synchronization parameters for cell selection	TLB
TL_Sysinfo_req/ind	Broadcast of system information for cell selection	
TL_Configure_req/conf	Configuration of layer 2 on basis of cell parameters provided	TLC
TL_Measurement_ind	Notification of measurement results on the quality of a connection	
TL_Monitor_ind	Notification of measurement results on the path loss of these channels	
TL_Monitor_List_req	Request to ascertain path losses of a specific number of channels	
TL_Report_ind	Report to MLE on the status of an action terminated by the MLE	
TL_Scan_req/conf	Scanning of a particular channel	
TL_Scan_Report_ind	Regular reporting on the signal quality of a channel and the current cell parameters after a channel has been scanned	
TL_Select_req/conf TL_Select_ind/resp	Selection of a specific channel for radio transmission	

Data structures The *Protocol Data Unit* (PDU) of the LLC sublayer are called LLC frames and are defined according to the type of connection used: AL or BL.

Basic link With a BL the service data units of the *Service Primitive* (SP) received by the MLE are segmented not in the LLC sublayer but in the MAC sublayer, supplied with an LLC header and optionally with a check sum (FCS). Whereas an acknowledged service is only planned for point-to-point communication, the non-acknowledged service can also be used in a point-to-multipoint communication. The following PDUs have been specified for the BL:

BL_Adata This frame is used for connectionless transmission to confirm receipt of a frame and to send SDUs that arrive from the MLE and should be acknowledged. Another BL_Adata-PDU can only be sent again after the last PDU to be transmitted has been acknowledged through a BL_Adata-PDU or a BL_Ack-PDU.

BL_Data This frame is used for acknowledged, connectionless transmission, and contains the MLE's SDU which is to be transmitted. As with a BL_Adata-PDU, a new BL_Data frame cannot be sent until the last one to be sent has been acknowledged.

Table 7.10: Relationship between V+D LLC and MAC service elements

LLC-SAP	LLC service element	MAC service element	MAC-SAP
TLA	TL_Connect_req/resp	TMA_Unitdata_req	TMA
	TL_Data_req/resp		
	TL_Disconnect_req		
	TL_Unitdata_req		
	TL_Connect_ind/conf	TMA_Unitdata_ind	
	TL_Data_ind/conf		
	TL_Disconnect_ind/conf		
	TL_Unitdata_ind		
	TL_Cancel_req	TMA_Cancel_req	
	TL_Release_req/ind	none	
	TL_Report_ind	TMA_Report_ind	
		TMA_Unitdata_ind	
TLB	TL_Sync_req/ind	TMB_Sync_req/ind	TMB
	TL_Sysinfo_req/ind	TMB_Sysinfo_req/ind	
TLC	TL_Configure_req/conf	TMC_Configure_req/conf	TMC
	TL_Measurement_ind	TMC_Measurement_ind	
	TL_Monitor_ind	TMC_Monitor_ind	
	TL_Monitor_List_req	TMC_Monitor_List_req	
	TL_Report_ind	TMC_Report_ind	
	TL_Scan_req/conf	TMC_Scan_req/conf	
	TL_Scan_Report_ind	TMC-Scan-Report_ind	
	TL_Select_req/conf/ind	TMC-Select_req/conf/ind	
		TMD-Report_ind	TMD
		TMD-Unitdata_req/ind	

BL_Udata This frame is used for non-acknowledged, connectionless transmission to send the data of the MLE. In this case several frames can be sent one after another because no acknowledgement is required.

BL_Ack This acknowledges the receipt of individual BL_Data-PDUs. If there is a message from the MLE, the BL_Ack frame can also contain layer-3 information.

Advanced link A layer-2 SDU can be longer with transmission over an AL than over a BL because the LLC sublayer offers AL segmentation as a service parameter. In contrast to the fragmentation in the MAC sublayer used with a BL, in an LLC segmentation each segment is assigned a number to enable the receiver to recognize the segment sequence. This numbering also allows a receiver to request a segment that has been lost, thereby eliminating the need for the complete message to be repeated again. The services of the BL are also available. The following PDUs have been defined for the AL:

AL_Set-Up The prerequisite for each AL is the establishment of a connection. This frame is used in setting up an AL for non-acknowledged as well as acknowledged transmission, and contains the respective QoS parameters. The called party confirms that a connection has been established and, in the case of acknowledged transmission, its acceptance of the QoS, also by means of an AL_Set-Up frame. If it cannot fulfil the QoS parameters, it sends its own QoS parameters, which must then be acknowledged by the calling party using a new AL_Set-Up.

AL_Data This frame is used for the acknowledged transmission of layer-3 information. It contains the respective segment of the TL-SDU, and all the segments except for the last one are transmitted with this PDU. The window process of an AL permits a number of TL-SDUs to be sent before confirmation of receipt of a complete TL-SDU has been received. However, the window is not used in flow control.

AL_Data-AR (*Acknowledge Request*, AR) This frame has the same significance as the AL_Data frame. In addition, the other side is requested to send back an acknowledgement immediately.

AL_Final This frame indicates to the partner entity that the segment contained in this frame is the last one of the TL-SDU being transmitted.

AL_Final_AR Similarly to AL_Final and AL_Data-AR, this frame contains the last segment being transmitted, and also requests an acknowledgement.

AL_Udata After a connection has been established with unacknowledged transmission, this PDU is used in the unidirectional exchange of information. Each segment of a TL-SDU, except for the last segment of the complete message, is sent with this PDU. The segments are numbered so that the receiver notices if a segment has been lost.

AL_Ufinal This frame transmits the last segment of a non-acknowledged transmission to the receiver. Thus it is only used as the conclusion of one or more AL_Udata.

AL_Ack If an acknowledgement has been accepted or the window has closed, this acknowledges all the Tl-SDU segments that have been received without errors. An AL_Ack frame therefore contains a request for selectively non-acknowledged segments to be resent.

AL_RNR (*Receiver Not Ready*, RNR) The receiver can send this frame at any point in time to control the flow. This frame replaces an AL_Ack frame when a receiver at times is not able to receive new PDUs. However, the segments indicated as faulty can continue to be sent. New frames cannot be sent until an AL_Ack frame has been received or after the timer, which is started with the receipt of the AL_RNR frame, runs out.

AL_Disc An AL_Disc frame is used to terminate an AL in acknowledged or non-acknowledged transmission. In the case of acknowledged transmission the receiver also reponds with an AL_Disc frame.

State diagrams The states in the LLC sublayer are only explained superficially in the standard for a *basic link* and with inconsistencies in the one for an *advanced link*, making it difficult to provide reliable data with any assurance that it is correct.

Basic link For a basic link there are only two states for acknowledged data transmission in the LLC sublayer that can be entered after synchronization has taken place between MS and BS. The LLC sublayer is otherwise in `IDLE` state. The `TX-READY` state is one in which messages can be sent through a TL_Data_req-SP and any operations during transmission can be indicated through TL_Report_ind. Confirmation by the partner side is provided through a TL_Data_resp-SP.

The state `RX-READY` is entered by the receiving station; here too the TL_Report_ind-SP reports on the proceedings during the receiving process. The incoming message is indicated with a TL_Data_ind-SP and acknowledged with a TL_Data_conf-SP.

The standard does not include states for non-acknowledged data transfer.

Advanced link Figure 7.16 provides an overview of the states of the LLC sublayer with an advanced link, with explanations following.

With the TL_Connect_req-SP it is possible to change from any state into the `WAIT-OUT-CONNECTED` state. An AL_Set-Up-PDU is generated that notifies the partner of the request for a connection. The proceedings of a connection establishment are indicated by a TL_Report_ind-SP. If the connection does not materialize, which is conveyed by different AL_Set-Up-PDUs and by the AL_Disc-PDU, there is a transition to state `IDLE`. If the connection is established successfully, the LLC sublayer moves to the `CONNECTED` state through a TL_Connect_conf-SP.

If an LLC sublayer in `IDLE` state receives a request to set up a connection from a TL_Connect_ind-SP, it changes to state `WAIT-IN-CONNECT`. If a connection is set up successfully, it moves from here to the state `CONNECTED` with a TL_Connect_resp-SP. If the connection is to be set up with other QoS parameters, a TL_Connect_resp-SP is generated with this information, a changeover is made to the `WAIT-OUT-CONNECTED` state, and the process is continued as described above. If a connection does not materialize, the TL_Disconnected_req-SP is used to move to state `WAIT-DISCONNECTED`. If access to the channel fails, the LLC sublayer reports this with a TL_Report_ind-SP and changes to the `IDLE` state.

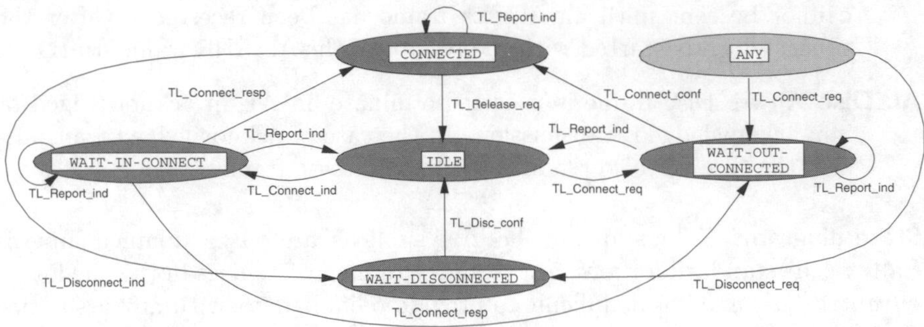

Figure 7.16: State diagram of an advanced link in the V+D LLC sublayer

If the sending LLC sublayer receives a TL_Disconnect_req-SP, a change is made from every state, except `IDLE`, to the `WAIT-DISCONNECTED` state through the dispatch of an AL_Disc-PDU. When a connection in `CONNECTED` state has been abnormally interrupted (indicated by a TL_Report_ind-SP), i.e., not initiated by the LLC sublayer, it moves into the `IDLE` state by means of an internal TL_Release_req-SP. The state `CONNECTED` in turn consists of a number of states; detailed information on this is given in [3, 6].

- `AL-TX-READY`: An MS is ready for the acknowledged transmission operation.
- `AL-RX-READY`: An MS is ready for the acknowledged receiving operation.
- `AL-UNACK-READY`: An MS is ready to receive/send non-acknowledged data.

7.2.5 Direct Mode

The *Direct Mode* (DM) is a mode of operation that supports the direct communication between mobile stations using frequencies which are outside the control of a TETRA network. Table 7.11 lists the available services in DM operation. The half-duplex communication is limited to frequencies that have been reserved for DM operation, but may be monitored by the TETRA network.

7.2.5.1 Reference models for the Direct Mode

Reference models describe the interfaces between the different types of terminal and the network elements that are involved (see Figure 7.17). Five types of reference models are differentiated in TETRA DM:

Direct Mode MS: This basic reference model for DM operation describes the simple point-to-point or point-to-multipoint communication between

Table 7.11: Bearer services and teleservices for the Direct Mode in the TETRA standard

Bearer services	7.2 kbit/s circuit-switched, unprotected data
	4.8 kbit/s circuit-switched, protected (low) data ($N^\star = 1$)
	4.8 kbit/s circuit-switched, protected (low) data ($N^\star = 4$)
	4.8 kbit/s circuit-switched, protected (low) data ($N^\star = 8$)
	2.4 kbit/s circuit-switched, protected (high) data ($N^\star = 1$)
	2.4 kbit/s circuit-switched, protected (high) data ($N^\star = 4$)
	2.4 kbit/s circuit-switched, protected (high) data ($N^\star = 8$)
Tele-services	Individual Call
	Group Call

⋆ N=Interleaving depths

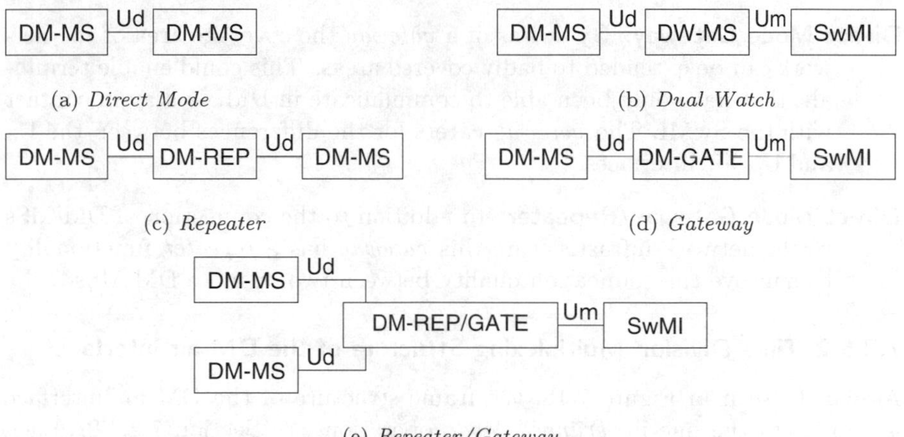

Figure 7.17: Reference models for TETRA *Direct Mode* operation

two MS using the DM air interface U_d. The MS that initiates a call set up in DM provides the synchronization reference and is defined as *master*. Accordingly, every synchronizing MS is defined as *slave*.

Dual Watch MS: This reference model applies to a MS that is capable of dual watch operation. The MS can be in one of the following states:

1. Idle in DM and *Trunked Mode* (TM) operation and monitoring a selected DM channel as well as the trunked mode signalling channel.
2. Communicating with another DM-MS using the U_d interface with monitoring the signalling channel of the network infrastructure over the air interface U_m at the same time.

3. Communicating with the network infrastructure in TM mode operation over the air interface U_m with monitoring a selected DM channel at the same time.

A communication over both air interfaces at the same time is not possible. Normal MSs can either be in DM or TM operation, without being able to monitor the other air interface simultaneously.

Direct Mode Repeater: The DM *repeater* receives the signals of one DM-MS on an uplink time-slot and sends them to another MS in DM operation on a downlink time-slot. By means of a *repeater* larger distances between MSs without coverage can be achieved. The *repeater* can use different frequencies for uplink and downlink to achieve more flexibility concerning the frequency usage.

Direct Mode Gateway: By means of a *gateway* the coverage area of the network can be extended to badly covered areas. This could enable terminals, that have only been able to communicate in DM, to stay in contact with the SwMI. The *gateway* caters for the differences between the U_d and U_m air interfaces.

Direct Mode Gateway/Repeater: In addition to the connection of DM MSs to the network infrastructure this *gateway* has a *repeater* functionality to improve the connection quality between two or more DM MSs.

7.2.5.2 Time-Division Multiplexing Structure of the DM air interface

As can be seen in Figure 7.18, the frame structure of the DM air interface is similar to the one in *Trunked Mode* operation, see Section 7.2.4.2. Each DM radio channel is divided into multiframes each 1.02 s in duration. Each multiframe contains 18 frames, each 56.67 ms long. The 18th frame of a multiframe is reserved for the *control frame.* Each frame consists of four time slots of 14.167 ms length, which transport the data bursts. Because of the half-duplex communication in DM, data are transmitted on uplink and downlink on only one channel of 25 kHz bandwidth. A time-division duplexing function allowing to realize both transmit directions in two different slots of the same carrier is also defined.

7.2.6 Performance Evaluation of TETRA

This section presents simulation results and provides a performance measure of the TETRA V+D air interface (reference point U_m). Results regarding the maximum number of users which can be served by TETRA systems are provided in [21]. For a performance evaluation of the TETRA PDO random access protocol please refer to [22]. Starting with an outline of the ETSI scenarios, we then introduce the TETRA system simulator TETRIS and conclude with a presentation of the traffic performance results.

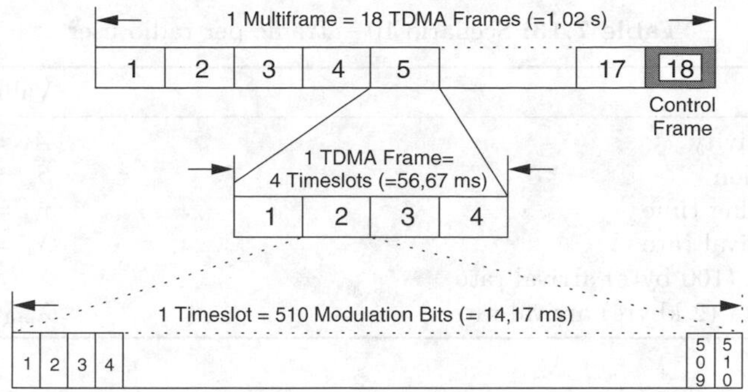

Figure 7.18: Frame structure of the DM air interface

Table 7.12: Scenario 10 – General Parameters

Parameter	Value
Type of area	BU
Covered area	50 km^2
Subscriber density	50 km^{-2}
Subscriber distribution	Gaussian
Class of terminals	80% portable, 20% vehicle
Velocity	3–50 km/h
Grade of Service	5%

7.2.6.1 ETSI Scenarios

The TETRA designer's guide [14] describes ten scenarios for the comparison of TETRA systems. For each scenario detailed specifications have been laid down concerning speech activity and offered data traffic of the mobile end user. Furthermore, the channel model to be used, the size of the scenario area, the number and type of the mobile stations, mobile or hand radio terminal, and their maximum velocity have been defined. Due to the fact, that scenario 10 defines the highest amount of offered load per terminal, this scenario has been chosen for the performance analysis in this section. Scenario 10 describes the parameters of a public or private network for airlines ground services, airport security, fire brigades and so on. Table 7.12 depicts the general parameters defined by scenario 10.

As can be seen from Table 7.13, the total traffic load per mobile terminal is:

$$\lambda = \lambda_s + \lambda_{sd} + \lambda_{md} = 24.1 \text{ h}^{-1} \tag{7.10}$$

The speech arrival rate λ_s has been calculated using $\lambda_s = A_s/\bar{\beta}_s$. The mean waiting time is defined as the duration between the dialling of a subscriber or group number and the successful completion of the call set up.

Table 7.13: Scenario 10 – Traffic per radio user

Parameter	Value
Speech activity	$A_s = 20$ mE
Call duration	$\bar{\beta_s} = 20$ s
Mean waiting time	$\bar{\tau_w} = 4$ s
Speech arrival rate	$\lambda_s = 3.6\ \mathrm{h}^{-1}$
Short data (100 byte) arrival rate	$\lambda_{sd} = 20\ \mathrm{h}^{-1}$
Middle data (2 kbyte) arrival rate	$\lambda_{md} = 0.5\ \mathrm{h}^{-1}$

Sixty percent of the voice calls are assumed to be group calls, the mean group size is 20 mobile terminals. We assume that the 2500 radio users are distributed over four TETRA cells. Configurations with 400, 500, 600, 700, and 800 users per radio cell are evaluated, taking into account a non-uniform distribution of the radio users over the four cells.

TETRA systems allow queueing of set up calls and the Erlang C formula (see Appendix A.1.2) is applicable for trunking capacity estimation under the call queuing strategy. Due to the mean number of 625 radio users per cell the total mean offered speech traffic is $A = 12.5$ E. To reach a call blocking probability of 5 % at least 20 traffic channels and 1 control channel are required, i.e. at least 6 carrier frequencies, accordingly 150 kHz spectrum per radio cell, are required.

7.2.6.2 Simulation Concept

For the traffic performance evaluation of the TETRA protocol stack the protocols of the air interface at reference point U_M have been implemented because they are the key elements.

The structure of the simulator TETRIS is depicted in Figure 7.19. It is similar in structure as that described for the analysis of the GPRS and EDGE systems, see Section 4.3.1. The protocol stack of the TETRA V+D system has been specified with the help of *Formal Description Techniques* (FDT) to guarantee not only syntactically and semantically unambiguous formal descriptions of the communication protocols but also interoperable and compatible implementations of these protocols independent of their implementation [20]. The SDL is the most widely used FDT in the area of telecommunications [23]. With the help of the C++ code generator SDL2SPEETCL, which converts SDL phrase representation to C++ source code, the mobile and base station protocol stacks have been embedded in the C++ simulation environment (see Section 14.2.8). The C++ implementations are based on the SPEETCL[1]. The SPEETCL provides generic C++ classes as well as a simulation library with strengths in random number generation, statistics evaluation, and event driven simulation control.

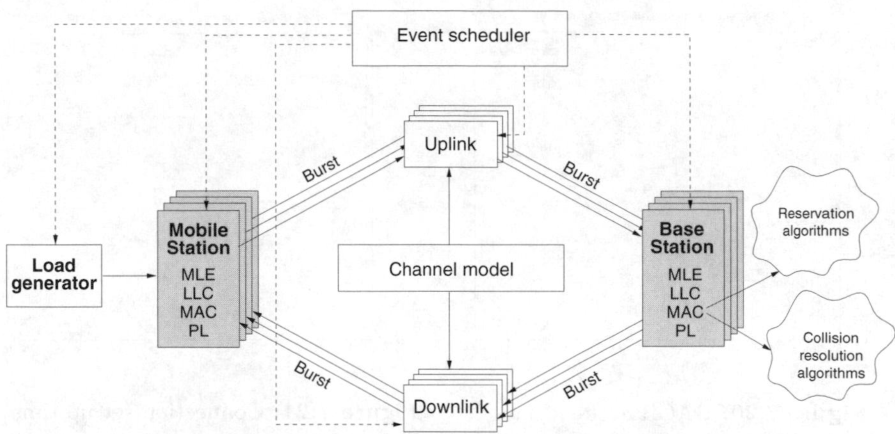

Figure 7.19: Structure of the simulation environment

The core of the simulator is the simulation control, which creates mobile and base stations and assigns the traffic generators to create specific traffic loads to the individual mobile stations. Depending on the scenario (see Section 7.2.6.1) the traffic generators are controlled to offer a certain traffic load. A traffic load is defined by inter-arrival times and the size of the data units. The SPEETCL contains traffic load generators for applications like speech, video, HTTP, FTP, SMTP and *Wireless Application Protocol* (WAP).

Mobile and base stations communicate via uplink and downlink channels, created by the simulation control, by exchanging bursts. With the help of error pattern files transmission errors can be introduced on the uplink or downlink dependent on the actual C/I value at the respective receiver. The pattern files have been generated taking into account a channel propagation model, the characteristics of the TETRA physical layer, and the receiver characteristics.

Mobility and the Mobility Management protocol are not taken into account because of the low terminal speed in comparison to the size of a TETRA cell. For further studies on the TETRIS simulation tool please refer to [19].

7.2.6.3 Performance Measurements

Figure 7.20 shows the complementary distribution function of the RACH access delay for the traffic load parameters of scenario 10. The RACH is used to activate point-to-point connections as well as group communications.

The access delay is defined as the duration between the creation of a connection set up request and the reception of the acknowledgement of a successful connection set up sent by the BS. The access delay is influenced by the structure of the logical channels and the collision resolution algorithm.

If the TETRA BS cannot assign capacity to an MS within a preset time after a successful access on the RACH, the MS accesses the RACH again to

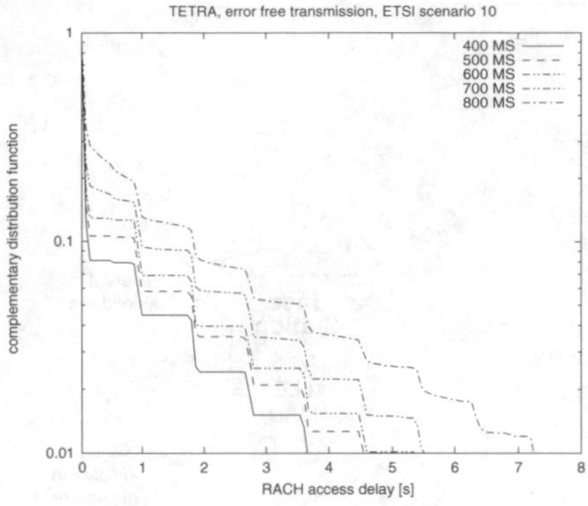

Figure 7.20: RACH access delay

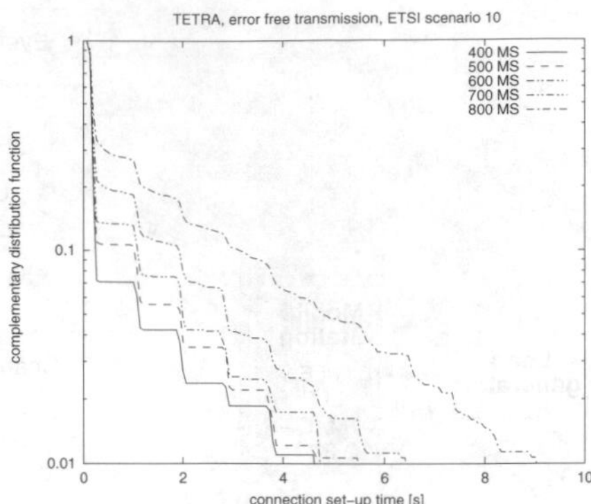

Figure 7.21: Connection set up time

repeat the capacity request. In the simulation results presented, this time-out is set to 18 TDMA frames. Thus, the access delay, as shown in Figure 7.20, also takes into account the retransmission on the RACH due to this time-outs.

With a higher traffic load offered, the probability for a repetition of the capacity request is increased because of the limited number of traffic channels. In case of 400 MSs this probability is about 8 % and with 800 MSs it becomes more than 20 %.

The connection set up time, as shown in Figure 7.21, includes both, the access delay and the waiting time for the assignment of a traffic channel. The measurements include point-to-point as well as group communications.

Because of the low collision probability the RACH access delay for the first access is small and the connection set up time is mainly influenced by the waiting time for a free traffic channel. In all the traffic load settings studied, the blocking probability is smaller than 5 %.

7.2.7 The Future Development of TETRA

In the year 2000 ETSI agreed to develop extensions to the TETRA standard to keep TETRA competitive to the third generation mobile radio networks. The following extensions and characteristics are planned for TETRA Release 2:

- Higher bit-rate packet data services for the support of multimedia and data applications;
- Selection and standardization of an advanced speech codec to enable the simple transition between TETRA networks and other networks of the third generation without transcoding and to reach an optimal voice quality under the use of the latest codec technology;

- Further improvements of the air interface to increase the spectral efficiency, the capacity of the network, quality of service and terminal quality (life span of the battery, size and costs);
- Development or adoption of standards with the goal to reach improved inter-working and roaming between TETRA and other mobile radio networks such as GSM, GPRS, UMTS and other networks of the third generation that are based on the IP;
- Advancement of the TETRA SIM to the USIM of UMTS, to offer TETRA specific services whilst supporting inter-working with the public mobile radio networks like GSM, GPRS and UMTS;
- Extension of the range of TETRA up to approx. 120–200 km for applications like airborne public safety, maritime, rural telephony, pipeline monitoring and railways;
- Development of new ETSI deliverables for the support of user- and market-induced requirements.

7.2.8 Digital Advanced Wireless Services

The ETSI working group 4 (WG4) of the TETRA project has worked so far on the *Packet Data Optimized* (PDO) standard [28]. Since 1998 the major task of WG4 consists of the development of the *Digital Advanced Wireless Services* (DAWS) standard. DAWS is a follow-up standard to the packet-oriented information transfer of TETRA PDO in highly mobile environments. As the North American TIA has started a similar work in the *American Public Safety Communications Officials* (APCO) 34 project, ETSI and TIA have decided to further their cooperation under the project *Mobility for Emergency and Safety Applications* (MESA).

The sole groups of target users of DAWS are Public Safety and Law Enforcement Authorities. DAWS is the mobile counterpart to BRAN/HIPERLAN, as can be seen from Figure 7.22 [15]. A growing demand for broadband mobile services within tele-medicine, firefighting and mobile robotics is expected. This requires the application of mobile carriers capable of processing up to 155 Mbit/s.

The specific requirement of the target users is to keep the capability gap regarding public available telecommunications technology and the technology used by Public Safety Agencies as small as possible.

Figure 7.23 shows a sample DAWS network structure [15]. A DAWS network can be directly connected to the Internet to provide high-speed wide-area Internet access. It can be installed throughout a corporate campus, within a residence or as an ad hoc network to provide full, seamless mobility throughout all areas covered with DAWS networks.

Currently no particular technology is proposed. Detailed technology assessments will have to wait for the final definition of service scenarios and

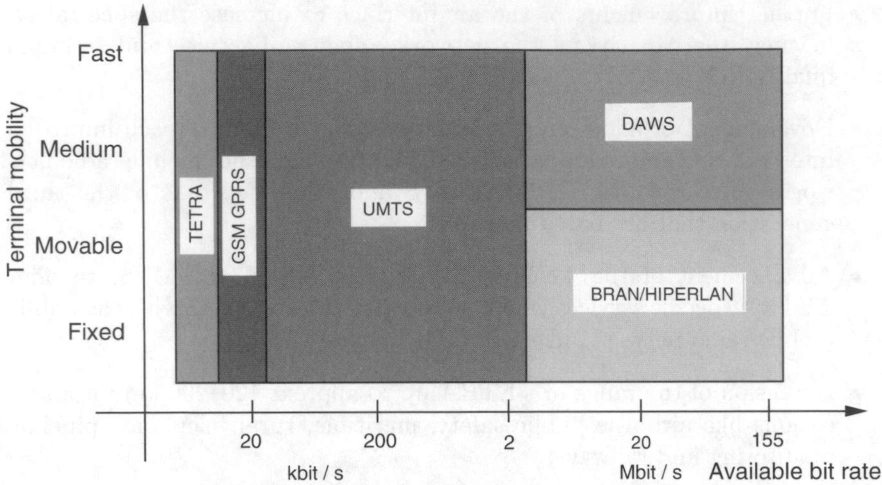

Figure 7.22: DAWS and other ETSI standards

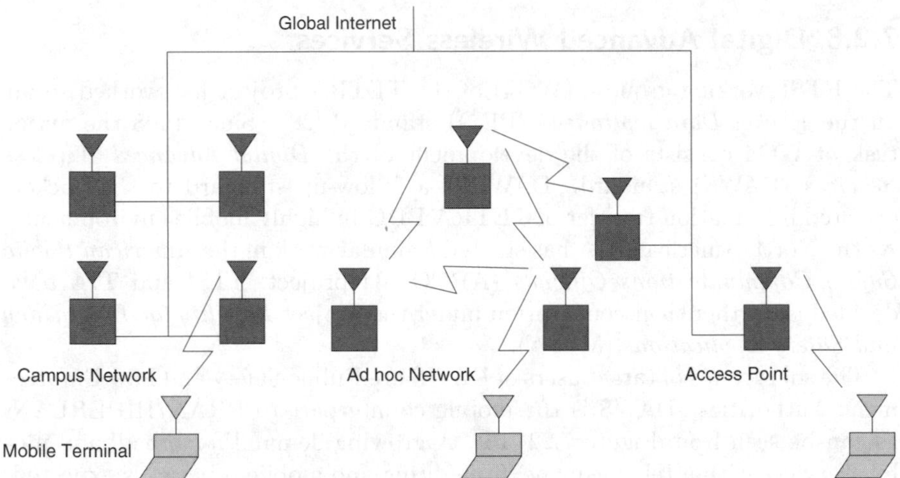

Figure 7.23: A sample DAWS network

applications. The assignment of spectrum is another key issue for the success of this project, which needs to be available by the time the first live test beds are set up in 2003.

Bibliography

[1] AixCom GmbH. http://www.aixcom.com.

[2] G. Cayla. *TETRA the new digital professional mobile radio.* In *5th Nordic Seminar on Digital Mobile Radio Communications (DMR V)*, pp. 113–118, Helsinki, Finland, Dec. 1992.

[3] ETSI. *TETRA 04.11: V+D Layer 2 Service Description.* Working document, European Telecommunications Standards Institute, Sophia Antipolis, France, Nov. 1994.

[4] ETSI. *TETRA 04.12: V+D Layer 2 PDU description.* Working Document v1.2.6, European Telecommunications Standards Institute, Sophia Antipolis, France, March 1994.

[5] ETSI. *TETRA 04.14: V+D Layer 2 MAC Protocol.* Working Document, European Telecommunications Standards Institute, Sophia Antipolis, France, Nov. 1994.

[6] ETSI. *TETRA 04.15: V+D Layer 2 LLC Protocol.* Working Document, European Telecommunications Standards Institute, Sophia Antipolis, France, Nov. 1994.

[7] ETSI. *TETRA 05.02: Channel Multiplexing for V+D.* Working Document, European Telecommunications Standards Institute, Sophia Antipolis, France, Nov. 1994.

[8] ETSI. *TETRA 05.03: Channel Coding.* Working Document, European Telecommunications Standards Institute, Sophia Antipolis, France, Nov. 1994.

[9] ETSI. *TETRA 05.05: Radio Transmission and Reception.* Working Document, European Telecommunications Standards Institute, Sophia Antipolis, France, Nov. 1994.

[10] ETSI. *TETRA 05.08: Radio Sub-System Link Control for V+D.* Working Document, European Telecommunications Standards Institute, Sophia Antipolis, France, Nov. 1994.

[11] ETSI. (RES06). *TETRA Voice + Data, Part 10.* European Telecommunication Standard, European Telecommunications Standards Institute, Sophia Antipolis, France, Nov. 1994.

[12] ETSI. (RES06). *TETRA Voice + Data, Part 11.* European Telecommunication Standard, European Telecommunications Standards Institute, Sophia Antipolis, France, Nov. 1994.

[13] ETSI. (RES06). *TETRA Voice + Data, Part 12.* European Telecommunication Standard, European Telecommunications Standards Institute, Sophia Antipolis, France, November 1994.

[14] ETSI. (RES06.1). *Scenarios for Comparison of Technical Proposals for MDTRS.* Working document, European Telecommunications Standards Institute, Sophia Antipolis, France, June 1991.

[15] ETSI. (WG4). *TETRA DAWS.* Technical Requirements, European Telecommunications Standards Institute, Sophia Antipolis, France, April 1999.

[16] ITU. *I.411 ISDN User-Network Interfaces—Reference configurations.* ITU-TS, Geneva, 1988.

[17] S. Lin, D. J. Costello. *Error Control Coding—Fundamentals and Applications*, Vol. 1 of *Computer Applications in Electrical Engineering Series.* Prentice-Hall, Englewood Cliffs, New Jersey, 1983.

[18] B. Sklar. *Digital Communications: Fundamentals and Applications.* Prentice-Hall, Englewood Cliffs, New Jersey, 1988.

[19] Martin Steppler. *TETRIS—A simulation tool for TETRA systems.* In *Mobile Kommunikation*, Vol. 135 of *ITG-Fachbericht*, pp. 403–410, Berlin, Offenbach, Informationstechnische Gesellschaft (ITG) im Verein Deutscher Elektrotechniker (VDE), VDE-Verlag GmbH, September 1995.

[20] Martin Steppler. *Performance Analysis of Communication Systems Formally Specified in SDL.* In *Proceedings of The First International Workshop on Simulation and Performance '98 (WOSP '98)*, pp. 49–62, Santa Fe, New Mexico, USA, 12th–16th October 1998.

[21] Martin Steppler. *Maximum Number of Users Which Can Be Served by TETRA Systems.* In Bernhard Walke, editor, *Tagungsband der European Wireless '99*, Vol. 157 of *Reihe ITG-Fachbericht*, pp. 315–320, München, VDE-Verlag, 6th–8th October 1999.

[22] Martin Steppler, Michael Büter. *Collision Resolution for TETRA Systems.* In Bernhard Walke, editor, *Proceedings of 2. EPMCC '97*, Vol. 145 of *ITG-Fachbericht*, pp. 559–565. Informationstechnische Gesellschaft (ITG) im Verein Deutscher Elektrotechniker (VDE), VDE-Verlag GmbH, 30th September–2nd October 1997. ISBN 3-8007-2307-7.

[23] Martin Steppler, Matthias Lott. *SPEET – SDL Performance Evaluation Tool.* In Ana Cavalli, Amardeo Sarma, editors, *Proceedings of the 8th SDL Forum '97*, pp. 53–67, Evry, France, Elsevier Science Publishers, 23rd–26th September 1997. ISBN 0-444-82816-8.

[24] TETRAPOL FORUM. *TETRAPOL Specification: General Network Design: Voice and Data Services in Network and Direct Mode.* PAS 0001-1-2, TETRAPOL FORUM, Bois d'Arcy, France, March 1998.

[25] TETRAPOL FORUM. *TETRAPOL Specification: Radio Air Interface.* PAS 0001-2, TETRAPOL FORUM, Bois d'Arcy, France, July 1998.

[26] TETRAPOL FORUM. *TETRAPOL Technical Report (TR): Guide to TETRAPOL features: System Technical Report.* PAS TTR 0001-1-1, TETRAPOL FORUM, Bois d'Arcy, France, January 1998.

[27] TETRAPOL FORUM. *TETRAPOL Specification: General Network Design: General Mechanisms.* PAS 0001-1-3, TETRAPOL FORUM, Bois d'Arcy, France, January 1999.

[28] B. Walke. *TETRA Packet Data Optimized (PDO) Standard*, 2001. This document is available via `http://www.comnets.rwth-aachen.de/MRN/PDO.pdf`.

8
Paging Systems

It is often important to be able to reach certain people very quickly. The conventional telephone network is not always optimal, because a line may be engaged or the person being called might not be available. Mobile telephone systems offer a high degree of reachability of a mobile subscriber, but MSs are not always switched on and are expensive to use.

Paging systems fill a particular gap (see Table 8.1). They allow unidirectional transmission of information in the form of a tone or a numeric or alphanumeric message to the person being contacted, whose location area is not known (see Table 8.2). A terminal is required that is constantly ready to receive but cannot transmit, and consequently is small, lightweight and inexpensive.

A paging message is automatically initiated when a telephone user dials the paging service and, using a telephone keyboard, an Internet terminal or a PC, conveys the pager number of the called person or a short message for it to the computer that responds (see Figure 8.1).

One characteristic of the paging system is that the person sending the message can never be certain that it has been received. There can also be a considerable gap between the time a message has been sent and when it is received by the addressee, and during peak load times this can amount to 10 minutes.

Table 8.1: Public operation of paging systems in Europe

Since 1974	European Paging Service (Eurosignal)
Since 1989	Cityruf
Since 1990	Euromessage
Since 1996	ERMES

Table 8.2: Call types of different paging systems

Call type	Eurosignal	Cityruf	ERMES
Tone only [No. of paging no.]	Up to 4	Up to 4	Up to 8
Numeric [No. of num. char.]	—	Up to 15	20–16 000
Alphanumeric [No. of char.]	—	Up to 80	400–9000
Transparent data transm. [max. length]	—	—	4000 bits

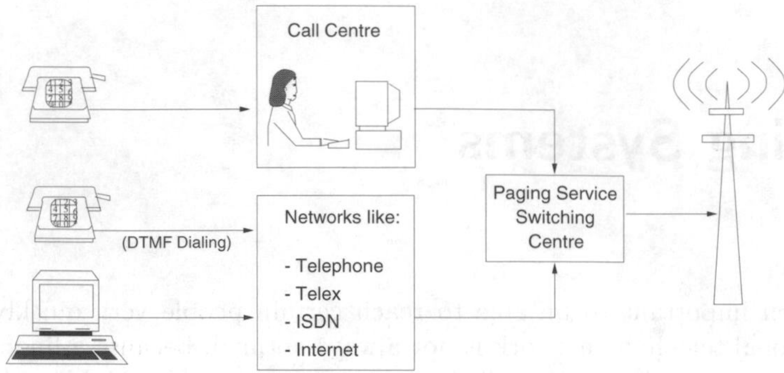

Figure 8.1: Principle of paging

Strengths and Weaknesses of Paging Systems

Strengths

- Inexpensive alerting and information service.
- Small receivers.
- The user of a paging system can be reached anywhere if the receiver is carried on his/her person.
- No additional antenna required (this applies only partially in the case of Eurosignal).
- Discrete communication of messages (messages are signalled through vibration, messages can be read on display).

Weaknesses

- The sender of a paging message receives no confirmation that the message has been received.
- Falsified paging messages and pages by malicious callers can cause the user of a pager to take inappropriate action.

Types of Calls

Individual calls The paging message is transmitted to the defined paging areas (one or more) or in the dialled paging area.

Collective calls A collective call number consists of n individual telephone numbers (receivers with different addresses). The receivers within a *paging area* are called in *consecutive* order.

Group calls Several receivers with the same address are called *simultaneously* in one or several paging areas.

Target calls By dialling additional numbers, the caller specifies the paging area in which the paging message should be transmitted.

It is anticipated that the number of users of pagers will increase in the future, despite the availability of services such as the *short-message service* (SMS) offered by GSM that are regarded as competition. A pager is an ideal extension to the mobile telephone because it can often be reached in places outside the radio coverage range of mobile telephone systems.

The miniaturization of terminals, including waterproof wristwatches that incorporate a radio pager receiver, offers the potential for a mass market for paging services that will extend beyond commercial applications to private and leisure use.

8.1 Paging Service 'Cityruf'

Paging systems that operate in different frequency bands (150–170 MHz or 440–470 MHz) in accordance with POCSAG-Code (the Post Office Code Standardization Advisory Group pager system was developed as long ago as 1981) (*CCIR Radio-Paging Code No.1*) are now being used in many parts of Europe. Subscribers can be reached by practically all communications networks over the POCSAG radio calling service (see Table 8.3). A typical example of the POCSAG paging service is Cityruf, which was introduced in Germany in March 1989.

A radio paging system based on POCSAG (see Figure 8.1) consists of:

- A paging service switching centre that forms the link between the various communications services (telephone, data, etc.) and the paging network, in which incoming data is prepared for processing in the paging computer.
- Paging transmitters with transmitter power up to 100 W.
- A paging concentrator, which is used to switch on a paging transmitter at the paging service switching centre and is allocated a paging measurement receiver for the automatic propagation control of the modulation paths.
- A paging receiver.

As the name implies, the Cityruf network is not envisaged to be a wide-area service. It is a regional paging service for cities, and its coverage area is divided into internetworked paging areas. About 50 paging areas with a maximum diameter of 70 km have been implemented in the final configuration. However, subscribers to the Cityruf service are not only booked in locally or regionally but in several paging areas and, in fact, throughout Germany. The transmitter systems of Cityruf are designed in such a way that they can guarantee reception within buildings without the need for additional antennas. The maximum number of subscribers that the system is capable of addressing is two million. Fifteen paging messages can be sent in one second.

Users of Cityruf have a choice between different calling classes based on different monthly charges:

Calling class 0 (tone only) Input is over the telephone. Tone-only devices issue up to four optically and acoustically different signals.

Calling class 1 (numeric) A maximum of 15 digits or special characters can be entered directly over the telephone using a supplementary device or with a dual-tone multiple-frequency (DTMF) signal generator, and then appear in the display of the terminal.

Calling class 2 (alphanumeric) Text messages up to 80 characters long can be entered, for example over the Internet or using an acoustic coupler device over the PSTN. Several messages can be transmitted in succession.

The following types of calls are available for the three calling classes:

Individual calls A radio call is broadcast in the registered paging areas or to the selected paging area.

Group calls A number of receivers are addressed over a group number at the same time.

Collective calls Up to 20 individual paging numbers are assembled in a list and then dialled up automatically in succession.

Target calls The receiver of a call is assigned a special number. To reach this person, a caller first dials the paging number and then the suffix number of the radio paging area in which the call is to be transmitted.

In addition to terminals and accessories for Cityruf, radio paging network operators offer the supplementary service Inforuf, which allows the user to receive information such as stock market updates, business news and weather reports. To access this information, users require a special Inforuf receiver that receives the Inforuf signals along with the Cityruf signals and is capable of storing 80 000 characters and reading them out on its 80-character display. The current information providers include Reuters, Telerate and pooled information services.

Cityruf is transmitted on the following frequencies:

- 465.970 MHz
- 466.075 MHz
- 466.230 MHz

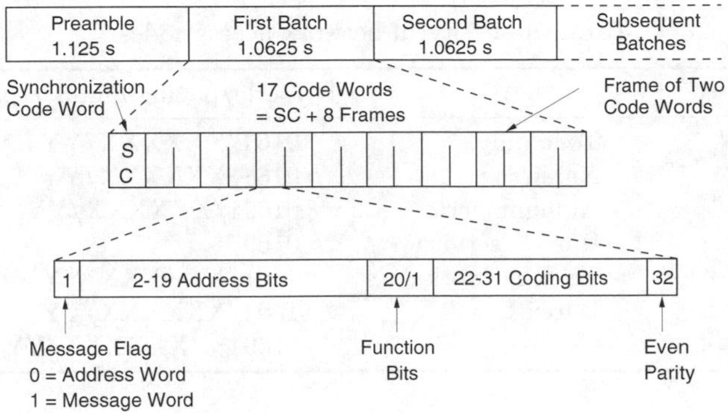

Figure 8.2: Message coding and block format with POCSAG

The transmission rate is 512 bit/s or 1200 bit/s. The digital signals are NRZ (*Non-Return to Zero*) coded and modulated through the use of differential frequency-shift keying (DFSK). The transmitters emit *bursts* of data blocks with code words (see Figure 8.2) [2].

Each burst begins with a 1.125 s long preamble, which is followed by a number of data blocks of 1.0625 s duration. The preamble enables synchronization to take place on the signal pulse on the receiver side, and is a prerequisite for error-free message coding. The data blocks consist of 17 code words, the first of which is used in the synchronization. The remaining 16 code words, each with 32 bits, contain an initial bit that indicates whether the code word is an address or a message, 20 address or message bits, 10 bits for error detection and correction, and a parity bit. The 16 code words form eight frames, each of which contains two code words. Each terminal is only addressed in a specific frame. The message for a receiver can be of any length, and is sent in the form of message words based on the address of the person receiving the message.

Paging areas can be divided into several radio coverage areas, with all transmitters within an area transmitting in a common frequency. The three frequencies are not used simultaneously in a transmitting zone, but are staggered and cyclical. Transmission never takes place simultaneously on the same frequency in the adjacent transmitting zones, so that a three-site cluster occurs. The time (time slot) during which transmission can take place on a frequency can be adapted to traffic volume. The advantages of this procedure are that adjacent zones can be decoupled through radio engineering and a receiver only needs to be operational when its frequency is being sent. If no messages are being sent on the frequency of the receiver or if it recognizes that a radio call is not directed at it, the receiver remains in battery-saving idle state.

Cityruf receivers are small in size, include storage options for characters which are received and use minimal power only. Table 8.3 lists the service codes required for network access.

Table 8.3: Cityruf network access codes

Access over		Code/paging number/paging area
PSTN/ISDN	Tone only	0164/XXXXXX/YY
	Numeric	0168/XXXXXX/YY
	Alphanumeric	01691/XXXXXX/YY
	Bureau service	016951
IDN	Telex network	1691/XXXXXX/YY
	Datex-L	1692/XXXXXX/YY
T-Online		*1691#/XXXXXX/YY

Development of Service

1989 Start of service to the public.

1990 Internetworking of Cityruf with the systems ALPHAPAGE in France, TELEDRIN in Italy, EUROPAGE in Great Britain.

1991 Introduction of the *Inforuf* service for closed user groups, introduction of alphanumeric messages with multifrequency dialling using the access code 0168, internetworking of Cityruf with radio paging system in Switzerland.

1992 Automatic dialling through telephone answering machines, alphanumeric access through access code 01691 at higher transmission rates (up to 24 kbit/s).

1994 Remote control and telemonitoring with Cityruf, input of numeric messages using voice.

8.2 Euromessage

In addition to Cityruf, which is a national service, there is also a European-wide radio paging service based on POCSAG called Euromessage (*European Messaging*). This is an extension of Cityruf that was set up in March 1990 to include an international service through the internetworking of the national radio paging services in Germany (Cityruf), France (Alphapage), Italy (Teledrin) and Great Britain (Europage).

Subscribers to this service wishing to be reached whilst abroad must notify the service operator with details about the period of time they will be away and in which paging area they can be located. All messages that arrive for a subscriber during this period are then rerouted to the appropriate paging area.

Euromessage is considered to be an intermediate solution until the availability of the Pan-European standardized radio paging service ERMES.

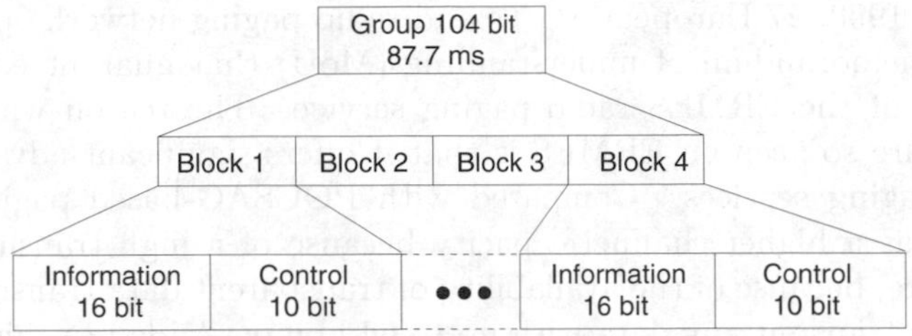

Figure 8.3: The RDS block format

8.3 RDS Paging System

RDS (*Radio Data System*), which was specified by the European Broadcast Union (EBU) and passed in 1984, is used for transmitting supplementary information over VHF radio broadcast transmitters such as:

- Transmitter recognition
- Programme information
- Alternative transmitter frequencies
- Traffic information
- Paging

The supplementary information consists of digital data, which is combined into groups of 104 bits each (see Figure 8.3). Each of these groups has a code to indicate which type of supplementary information it contains. Groups that contain tuning and switching information are transmitted more frequently than other groups.

The countries that use RDS for transmitting radio paging services are Sweden since 1978, France since 1987 and Ireland. Other countries, including Spain and Norway, are planning to introduce the service. Germany introduced RDS in 1988.

The RDS paging service offers the advantages that investment costs are low and subscribers can be reached anywhere because of countrywide VHF radio broadcast coverage with the existing transmitter network and shared use of VHF frequencies. Alphanumeric messages cannot be transmitted in the RDS system.

8.4 ERMES

ERMES (*European Radio Messaging System*) is a European-wide paging service that was developed as the result of a resolution adopted by the CEPT countries in 1986. The standardization work begun under the responsibility of CEPT was continued by ETSI in 1989 and completed in 1992.

By late 1990, 27 European PTTs and radio paging network operators had signed a memorandum of understanding (MoU) that guaranteed the implementation of the ERMES radio paging service. The reason why European countries are so keen on ERMES is that it offers significant advantages over existing paging services. Compared with POCSAG-based paging systems, ERMES has a higher channel capacity because of a high transmission rate; furthermore, because of the availability of transparent data transmission, it is possible to transmit any data with up to 64 kbit/s. Added to this is the possibility of international roaming and, in the case of several operators, national roaming.

ERMES represents the first standardized paging system in Europe that operates in the same frequency band throughout Europe and guarantees the accessibility of its subscribers throughout Europe.

Because of its 6250 bit/s bit rate, the capacity of an ERMES channel is four times higher than a 1200 bit/s POCSAG channel. Between 300 000 and 400 000 subscribers can be served per channel, which equates to a system capacity of around six million subscribers. ERMES pagers are being designed to require extremely little battery power and are smaller than equivalent POCSAG models.

8.4.1 The Services of the ERMES Paging System

ERMES offers its subscribers a number of basic services as well as supplementary services. The basic services that each operator is required to provide include [3]:

- Tone-only calling in which ERMES supports up to eight different tone signals per *Radio Identity Code* (`RIC`). This means that a tone receiver with a `RIC` can produce eight different alarm signals.
- Numeric radio paging in which the receiver has a display with at least 20 digits and also supports the tone-only function.
- Alphanumeric paging with a minimum of 400 characters and a receiver that is also suitable for tone-only calls and numeric pages.
- Transparent data transmission at 64 kbit/s, which can also be used for process control, telemetry and alarm activation.
- Roaming: the network recognizes the paging area in which a subscriber is located.

ERMES is planning the following supplementary services that can be offered by an operator as an option:

- Standard text allows an alphanumeric page to be sent through the input of DTMF codes over the telephone. Each code is linked to a standard text, such as "Meet in one hour".

- Group call (see Section 8.1).
- Collective call (see Section 8.1).
- Call forwarding to another receiver.
- Storage of incoming messages, which the subscriber can then request to be sent at a later time.
- Numbering of messages and automatic retransmission of the number of the last message. The receiver will recognize from the sequence number whether certain messages have been received.
- Repetition of call when requested.
- Temporary call barring.
- Transmission of calls with different categories of priority.
- Target call in which the caller determines the paging area in which a call should be broadcast.
- Closed user group.
- Display of the category of urgency of a message and acceptance of a message based on its urgency.
- The person calling can indicate the time when a message should be sent.
- Encryption of messages.

8.4.2 ERMES Network Architecture

The system structure for ERMES is presented in Figure 8.4 [4]. The *Paging Network Controller* (PNC) processes the input received over the telephone, data network and from other networks, with consideration of the services agreed with a subscriber and stored in the network controller. With ERMES, messages for a tone or numeric pager can be conveyed over DTMF (*Dual-Tone Multiple Frequency*) signals. Paging text messages can be input over any conventional data network (ISDN, PSPDN, CSPDN, etc.) using standardized UPC protocols. Furthermore, the broadcast of a radio page can be initiated through arrangement with the operator. The network controller is therefore responsible for the following tasks:

- Provide a user with access to ERMES over the fixed network.
- Control and administer the database containing subscriber data.
- Provide the possibility of roaming through its link to other ERMES networks over an interface in accordance with CCITT Rec. X.200.

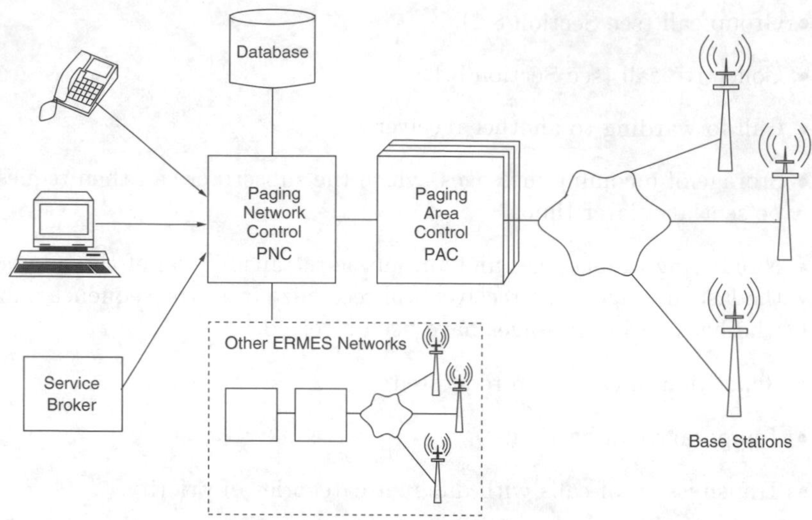

Figure 8.4: ERMES system structure

- Control radio transmission in the service areas.

Up to 64 *Paging Area Controllers* (PAC) can be connected to a network controller (see Figure 8.4). An area controller organizes the broadcast of a call and, in accordance with the agreement with the subscriber, activates one or more base stations in order to ensure that coverage is provided to the paging area concerned. Depending on the priority of incoming pages, the PAC forwards them to the base stations. The base stations (BS) transmit with up to 100 W and can reach receivers within a radius of up to 15 km.

8.4.3 Technical Parameters of the ERMES Paging System

Sixteen channels with a bandwidth of 25 kHz in the 169.4125–169.8125 MHz frequency band have been allocated to ERMES in Europe. With the use of the 4-PAM/FM modulation procedure the data transmission rate is 6.25 kbit/s or 3.125 kBaud/s.

An abbreviated cyclical code (30,18) derived from BCH (31,21) is used as an error-correction code. This code is supported by interleaving with a depth of nine.

Network operators and users have been assigned identities to ensure that messages in the ERMES system are routed correctly [1]. An *operator identity* `OPID` is 13 bits long (see Figure 8.5).

The zone and country codes are derived from CCITT Rec. E.212. The 3-bit long operator code supports eight different operators per country. Should further operator codes be necessary, additional country codes will be granted to differentiate between them.

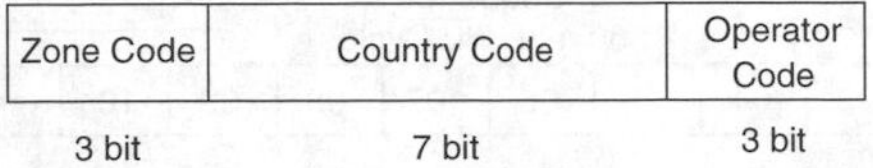

Figure 8.5: The structure of an OPID

Zone Code	Country Code	Operator Code	Initial Address	Batch Type
3 bit	7 bit	3 bit	18 bit	4 bit

Figure 8.6: The structure of an RIC

The structure of the user identity (*Radio Identity Code*, RIC) consists of 35 bits and is shown in Figure 8.6. The RIC contains the 13-bit long OPID and a further 22 bits for the initial address and the batch type, which together form a so-called local address. A receiver is able to have several user identities stored.

Depending on its description, an ERMES network can be set up as a:

- *Time Division Network* (TDN): Transmission in neighbouring or overlapping paging areas is at different time intervals.
- *Frequency Division Network* (FDN): Transmission in neighbouring or overlapping paging areas is on different frequencies.

A combination of the two types of operation is also possible. The channel and block structures of the transmission protocol are presented in Figure 8.7.

The periodic sequence forms the primary structure of 60-minute duration. It consists of 60 *cycles*, each of which lasts one minute and is used in the coordination between different networks. Because the receivers are designed to save on battery power, they can only listen in on one or a few cycles. Each cycle contains five subsequences, each 12 s in duration. In TDN the transmission for a paging area depends on the traffic volume and takes place in at least one of these subsequences. Each subsequence is divided into 16 *batches*, with each of the first 15 batches containing 154 words (30 bit/word) and the 16th having a length of 190 words. Through the last 4 bits in the RIC, the receiver is allocated one of the 16 possible types of batches and can be addressed in the batch type specified for it. A batch is divided into four parts:

Synchronization Partition The preamble word (PW) and the synchronization word (SW) enable the receiver to carry out the symbol or code word synchronization for the batch.

System Information Partion This part consists of:

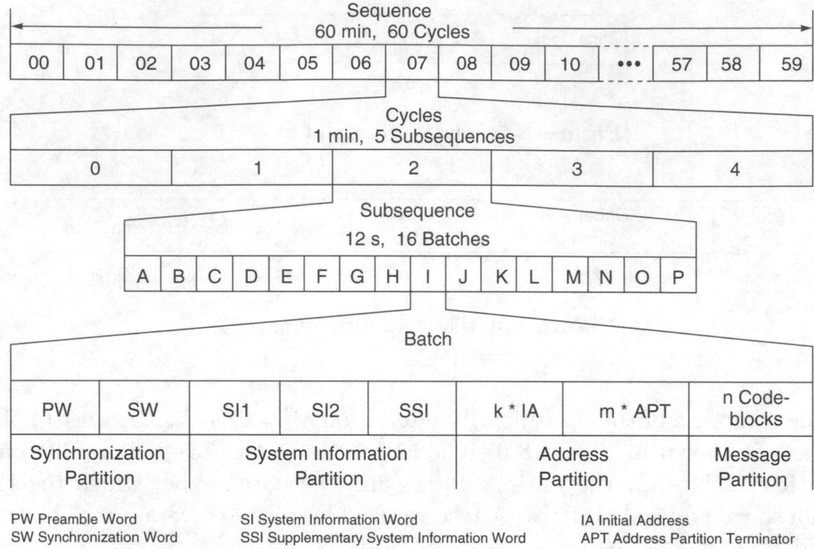

Figure 8.7: Message coding and block format in ERMES

- Two system information words (SI1, SI2), of which SI1 notifies the receiver of the operator identity (ID) and the paging area (PA) number. SI2 gives the position of the batch in the sequence and the channel of the transmitter.
- An additional system information word that contains either the area code, the time and the date or only the date.

Address Partition This partition contains a number of initial addresses (IA) and at least one *Address Partition Terminator* (APT). The initial addresses are transmitted in descending order. If the initial address received by a receiver is smaller than its own then it cannot receive any further radio pages in this batch.

Message Partition This part contains the local address of a receiver, the message number, the type of message and the message itself.

Bibliography

[1] G. Edbom. *The Concept for World Wide Radio Paging.* In *41st IEEE Vehicular Technology Conference*, pp. 840–847, St Louis, May 1991.

[2] R. J. Horrocks, R. W. A. Scarr. *Future Trends in Telecommunications.* Wiley, Chichester, 1993.

[3] A. Örtqvist. *ERMES's Role in Europe.* In *5th Nordic Seminar on Digital Mobile Radio Communications (DMR V)*, p. 119, Helsinki, Finland, Dec. 1992.

[4] B. Walke. *Mobile data communications in Germany—A survey.* In *Proceedings of 6th International Symposium on Personal and Indoors Mobile Radio Communications*, pp. 799–802, The Hague, Netherlands, Sept. 1994.

9
Cordless Telephone Systems

Cordless telephone systems are mobile radio systems that, although their transmission range is not suitable for wide-area coverage, are attractive to many users because of the services offered and their low cost.

Cordless telephony (CT) can be used within the area of a subscriber line, and provides a radio link within a radius of approximately 50 m inside a building or 300 m outdoors between the fixed terminal and a mobile part. This requires the fixed terminal, as the base station, and the mobile part each to be equipped with transmitting and receiving capabilities.

Along with their suitability for home use (a base station, a mobile part), CT systems lend themselves to other uses:

- Microcellular cordless private branch exchanges, e.g., for use in offices and industrial plants.
- Public cellular systems with local coverage and as regional or nationwide networks for public cordless coin-operated telephones from which calls can be made from appropriate portable terminals without use of previously conventional telephone cells (Telepoint service).
- Wireless access technology for stationary users of telecommunications networks.

The first cordless telephones in Europe came from the USA and the Far East. This CT0-designated equipment, which was not authorized in the European countries, used analogue transmission technology at 1.6 or 4.7 MHz with eight channels (each with 25 kHz), and because of the lack of security provisions was particularly susceptible to eavesdropping (see Table 9.1). Furthermore, the equipment could be used (fraudulently) as a telephone handset to access other base stations and their lines.

The disadvantages of CT0 equipment led to the development of the CT1 standard for cordless telephony in 1983 under the aegis of CEPT. Two bands with a 2 MHz bandwidth and 45 MHz duplex spacing at 900 MHz were stipulated as the frequency bands for these systems, which incorporated analogue transmission technology (see Table 9.1). The frequencies from 914 to 916 MHz were designated for the link between the mobile and the fixed parts, and the frequencies from 959 to 961 MHz for the transmission from the fixed part to the cordless telephone. In accordance with the CT1 standard, FDMA technology is used to access the total 40 channels available, each with a 25 kHz

Table 9.1: The main parameters of cordless analogue telephone systems

System	CT0	CT1	CT1+
Signal transmission	Analogue	Analogue	Analogue
Frequency band [MHz]	1.6/4.7	914–915	885–887
		959–961	930–932
No. of channels	8.0	40	80
Bandwidth [MHz]	0.4	2	4
Channel spacing [kHz]		25	12.5
Access procedure	FDMA	FDMA	FDMA
Duplex procedure	FDD	FDD	FDD
Channel allocation	Fixed	Dynamic	Dynamic
Cellular networks	No	Limited	Limited
Max. transmitter power [mW]		10	10
Range [m]	< 1000	< 300	< 300
Handover	No	No	No
Capacity [Erl./km²]	1.0	200	200

bandwidth. These channels are allocated dynamically; there is a fixed allocation of frequencies to individual devices or local areas.

It soon emerged that the number of channels available with the CT1 standard was inadequate for metropolitan areas, and this led to the decision by different countries, including Germany, to introduce a system with 80 channels in accordance with the CT1+ standard in 1989. With CT1+ the mobile part transmits to the fixed part in the 885–887 MHz frequency range, whereas the frequencies between 930 and 932 MHz are used for the link between the fixed part and the mobile device (see Table 9.1). An organizational channel is being planned for CT1+.

Unauthorized access to the base station has been almost completely eliminated with the CT1 and CT1+ standards because of a number of codes that provide a clear identification between the mobile and fixed parts; however, these systems are not fully protected from eavesdropping.

9.1 CT2/CAI and Telepoint

In contrast to other European countries, in the mid-1980s Great Britain started using a version of cordless telephony incorporating the T-standard from the USA. Because these systems did not provide adequate capacity (eight channels) and there was concern about a foreseeable frequency collision with GSM, Great Britain was not keen to continue with the CT1-standard, and a digital standard for cordless telephony was developed at the initiative of the network operator *British Telecom.*

With the aim of making cordless telephony attractive to a broad range of customers, it was planned that CT2 terminals should also be used for Telepoint applications. The Telepoint concept allows users with the appropriate equipment to set up a connection to the public telephone network over a Telepoint base station within a radius of up to 300 m of a public and highly frequented area (pedestrian area, railway station, airport, shopping centre, etc.), but with no facilities for receiving calls [1].

The CT2 standard is frequently supplemented with the abbreviation CAI (*Common Air Interface*), which denotes a radio interface between the fixed part and the mobile part developed by the Department of Trade and Industry (DTI) in Great Britain with the participation of British industry. The interface was required to enable existing non-compatible terminal equipment to access the Telepoint service.

The CT2/CAI standard was designed with enough flexibility so that the same handset can be used to make calls at an office, at home and to the public network. CAI basically allows incoming as well as outgoing calls. The restriction that exists with Telepoint, i.e., that only outgoing calls are possible, is only due to the terms of the licence and not to any technological limitations.

CT2/CAI has now been adopted as an ETSI Interim-Standard.

9.2 Technical Parameters of CT2/CAI

Technically, CT2 is based on the CT1 standard; however, digital transmission has been introduced (see Table 9.2). The system has more than 40 frequency channels, each with a 100 kHz bandwidth, in the frequency range between 864

Table 9.2: The main parameters of cordless digital telephone systems

System	CT2/CAI	DECT
Signal transmission	Digital	Digital
Frequency band [MHz]	864–868	1880–1900
No. of channels	40	120
Bandwidth [MHz]	4	20
Channel separation [kHz]	100	1.728
Access procedure	FDMA	FDMA/TDMA
Duplex procedure	TDD	TDD
Channel allocation	Dynamic	Dynamic
Voice channels/carriers	1	12
Coding	ADPCM 32 kbit/s	ADPCM 32 kbit/s
Modulation data rate	72 kbit/s	1.152 Mbit/s
Modulation	Two-level GFSK	GFSK
Cellular networks	Limited	Yes
Max. transmitter power [mW]	10	250
Range	< 300	< 300
Handover	No	Yes
Capacity [Erl./km²]	250	10 000

and 868 MHz, which, unlike CT1, does not collide with other standardized mobile radio services in Europe.

Two-level GFSK (*Gaussian Frequency Shift Keying*) is used for modulation. The nominal bit rate per channel is 72 kbit/s, which means that data transmission is also possible, albeit at a relatively low net rate.

CT2/CAI is the first ever mobile radio system in which the base station as well as the mobile terminal transmit on the same radio channel. Yet the two transmission directions are not separated by different frequencies; instead the pingpong technique is employed, in which the transmission direction is changed every millisecond on the same frequency. With this procedure, called *Time-Division Duplex* (TDD), the expensive filters required with *Frequency-Division Duplexing* (FDD) for switching transmission direction are replaced by simple switches.

The functions of the protocols of layers 1 to 3 specified for the air interface in the CAI standard correspond to those of the ISO/OSI reference model:

Layer 1 The specifications apply to the physical transmission system, including modulation procedure, frame structure, synchronization, time behaviour and bit rate, along with channel selection and link control.

Layer 2 The tasks of this layer include error detection, error correction, message acknowledgement, connection control and identification.

Layer 3 The functions and message elements of this layer are used in the signalling for connection control, similarly to the ISDN-D-channel protocol. It is responsible for recognizing the type of messages, for setting up and terminating calls and for maintaining connections.

Three logical subchannels are defined by the standard:

- The B-channel, in which speech and, with the use of a modem, data can be transmitted.
- The D-channel, for signalling.
- The SYN-channel, which transmits information on bit and burst synchronization.

Depending on the application, these subchannels are allocated different channel capacities in four different socalled multiplex frames (see Figure 9.1) [2].

Multiplex 1.2 This defines a 66-bit long burst of duration 1 ms that contains 64 B-channel bits and 2 D-channel bits. Since this frame is used with an existing connection, there is no need for an SYN channel. If the synchronism is lost when this operating mode is used, the connection must be reinitialized. Because this burst is sent every 2 ms, the data rate in the B-channel is 32 kbit/s and that in the D-channel 1 kbit/s.

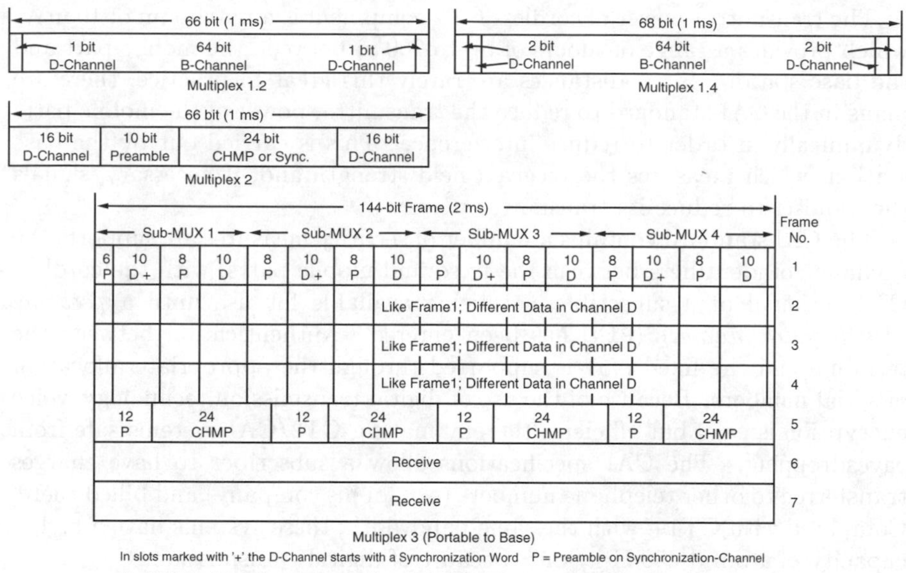

Figure 9.1: The structures of the multiplex frames with CAI

Multiplex 1.4 This is set up similarly to the Multiplex 1.2 burst, although a total of 68 bits are transmitted in 1 ms, of which four are D-channel bits. Therefore the data rate on the D-channel is 2 kbit/s. This multiplex frame is used with an existing connection if the base station and the mobile device have indicated during call setup that they are able to support the 68-bit long burst.

Multiplex 2 This is a 66-bit long burst consisting of 32 D-channel bits and 34 SYN-channel bits, and is used during normal call setup and to restore an interrupted connection. A total of 16 kbit/s are available to the D-channel and 17 kbit/s for the SYN-channel. The SYN-field consists of a 10-bit long preamble, followed by one of three different 24-bit long synchronization patterns—a so-called *channel marker for portable.*

Multiplex 3 This defines a 10 ms long burst that is used to set up and restore a connection that has been initiated by a handset. Here 10 ms are divided into five 2 ms long frames, each of which is subdivided into four identical *subframes.* The first four frames contain 20 D-channel bits and 16 preamble bits in each subframe, whereas the fifth frame only contains SYN-channel bits.

Since the B-channel is used for transmitting speech, the data rate is 32 kbit/s. Speech is coded through *Adaptive Differential Pulse Code Modulation* (AD-PCM).

The transmitter power of cordless CT2 equipment is a maximum of 10 mW, which provides a range outdoors of up to 200 m between the mobile part and the base station. Since distances are rarely this great in practice, there are plans in the CAI standard to reduce the transmitter power of the mobile parts dynamically in order to reduce interference. This is carried out by the base station, which measures the received field strength and, if necessary, signals the handset to reduce its transmitter power.

The CAI standard contains a number of security measures. It supports the exchange of identifiers between the fixed and mobile parts, with the cordless CT2 terminal of a subscriber not being available for use until a *personal identification number* (PIN) has been entered. Authentication between the terminal and the fixed part is supported through the appropriate allocation of serial numbers. Because of the use of digital transmission technology, voice encryption is easy but efficient, thereby making CT2/CAI systems safe from eavesdropping. The CAI specifications allow a subscriber to have charges transferred to other telephone numbers (e.g., of his company) and billed there. Compared with CT1+ with the same bandwidth, these systems have a higher capacity of around 250 Erl./km^2 because of digital transmission.

Bibliography

[1] M. P. Clark. *Networks and Telecommunications: Design and Operation.* Teubner, 1991.

[2] W. Tuttlebee, editor. *Cordless Telecommunications in Europe.* Springer, Berlin, 1990.

10
DECT*

Besides cellular mobile radio networks that are primarily envisaged for use outdoors, systems that have been specifically designed for use in buildings are also important. In recent years cordless telephones with a range of a few hundred metres have become increasingly popular in private households. There is (along with CT2/CAI) a digital alternative to these analogue devices that offers better voice quality and a greater security against eavesdropping, as well as other advantages: the DECT system.

The abbreviation DECT originally stood for *Digital European Cordless Telecommunications*, but to underline its claim of being a worldwide standard for cordless telephony DECT today stands for *Digital Enhanced Cordless Telecommunications*. This standard was specified by the *European Telecommunications Standards Institute* (ETSI) in 1992. A DECT network is a microcellular digital mobile radio network for high user density and primarily for use inside buildings. However, an outdoor application is also possible.

DECT systems allow complete cordless private branch exchanges to be set up in office buildings. Calls can be made over the normal subscriber line as well as between mobile stations through the DECT base station subsystem. When a staff member leaves his office, he usually can no longer be reached for incoming calls over a wireline telephone connection, although he may only be located in another part of the building. But if he uses a DECT terminal, he can continue to be reached under his normal telephone number wherever he may be anywhere on the premises. To enable users to continue to receive calls after they have left the DECT coverage area, ETSI specified an interface between GSM and DECT (see Section 10.15).

The DECT standard permits the transmission of voice and data signals. Consequently cordless data networks can also be set up on a DECT basis. The use of ISDN services (*Integrated Services Digital Network*) is also possible. Users are able to move freely within different cells without risking an interruption to their calls. The handover process switches calls from one radio cell to the next without interrupting the call.

In outdoor areas the maximum distance between base and mobile station is approximately 300 m; in buildings, depending on the location, it is up to 50 m. Larger distances to the base station can be bridged through the installation of appropriate base stations using the relay concept (see also Section 10.12).

---

*With the collaboration of Christian Plenge and Markus Scheibenbogen

The first time DECT systems were presented to the public was at the CeBIT exhibition in Hanover in March 1993.

Since then, the costs for a DECT mobile have been decreased to be comparable to those of analogue cordless telephones.

10.1 Possible Applications of DECT Systems

The size of a DECT system determines how it is installed. The labels *large* and *small* are relative and relate to the number of mobile stations to be operated within the DECT coverage area. A DECT system is capable of automatically localizing up to 1000 subscribers in one *location area* (LA). With a larger number of users additional location areas are required that fall under the administration of the DECT system (see Section 10.1.2).

10.1.1 DECT Fixed Networks

DECT fixed networks are normally designed as dedicated private branch exchanges.

If the DECT systems are installed by a network operator, each customer-specific DECT system can be allocated its own location area. All DECT location areas are then interconnected over a backbone ring and administered centrally by a DECT system, as illustrated in Figure 10.8. Each customer has his own *DECT fixed station* (DFS) with his own identification of the location area.

Despite the efficient location administration of DECT systems, it was planned that each customer would be allocated his own location area to ensure that the channel capacity of other customers would not be affected by the calling activity of his own mobile stations.

Private home base stations These offer a possibility of using the DECT system in small private households (see Figure 10.1). The home base station supplies the entire area of a house and can support one or several mobile stations that are supplied by the base station. The home station consists of a *DECT fixed system* (DFS), which controls the system, a simple *database* (DB) for user administration and a *fixed part* (FP), which provides the radio supply to the mobile stations. An *interworking unit* (IWU) is provided for connection to an external network [18].

Wireless private branch exchanges In terms of installation, central systems with a DFS to which several FPs are connected are suitable for large private households or small companies (see Figure 10.2). Fixed terminals can also be connected. The system is administered by the *DECT fixed system*. The correct location area of the mobile stations is stored in a database (DB), and an IWU enables a connection to the external networks.

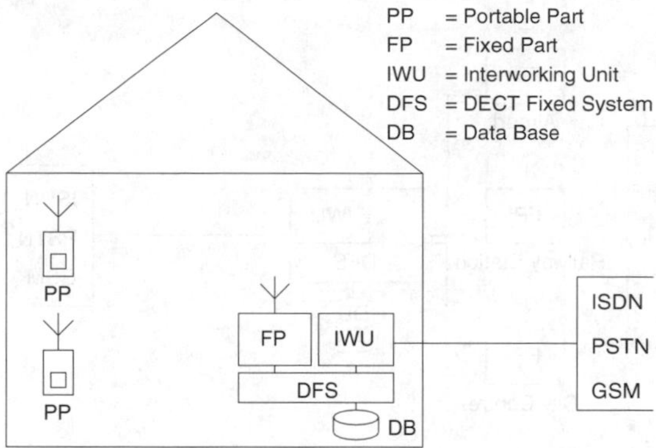

Figure 10.1: Private home base station

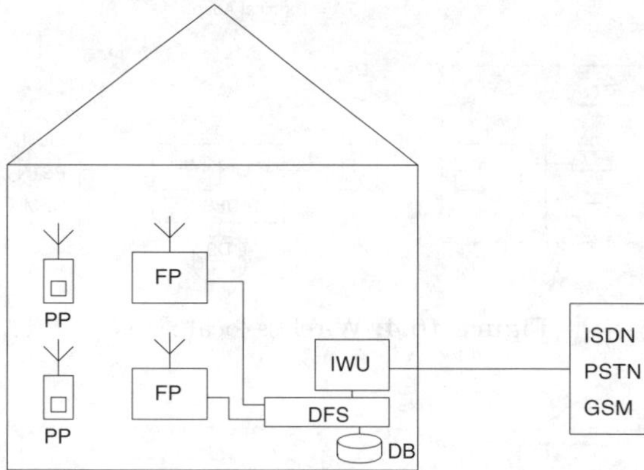

Figure 10.2: Wireless private branch exchanges

Public Telepoint systems These systems provide DECT mobile stations access to the public telephone network over "other" FPs at public sites (see Figure 10.3). Possible application areas are public facilities with high user density such as airports, train stations and city centres. These private and public DECT systems consist of a number of FPs that are administered by a *DECT fixed system* (DFS). Access to the public telephone network can be provided by a network interface. (With DECT, PPs in principle can also be reached in the area of a Telepoint. The restriction to outgoing calls is dictated by the respective licensing agreement.)

Wireless local rings This is a ring to which a *terminal adaptor* (TA) and all the terminals are linked (see Figure 10.4). The terminal adaptor provides a connection over the radio interface to one of the public FPs in the area. The terminals (e.g., telephone, facsimile) are linked to the terminal

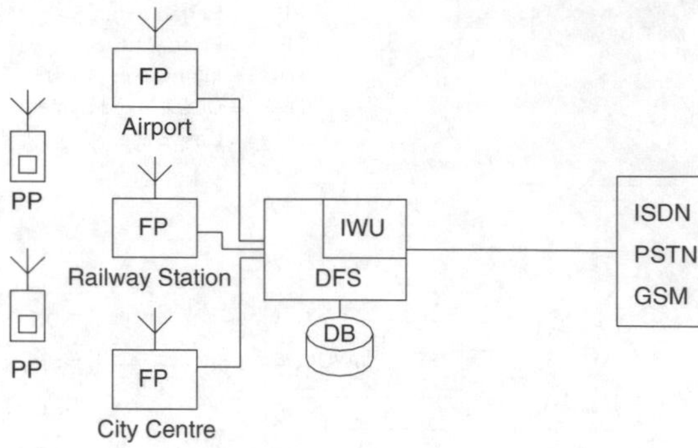

Figure 10.3: Public Telepoint system

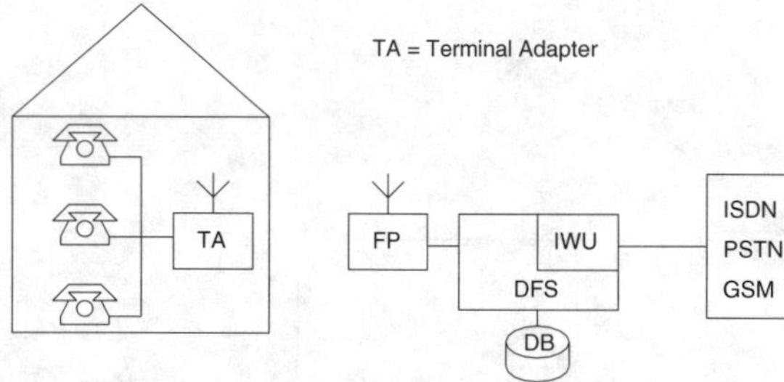

Figure 10.4: Wireless local rings

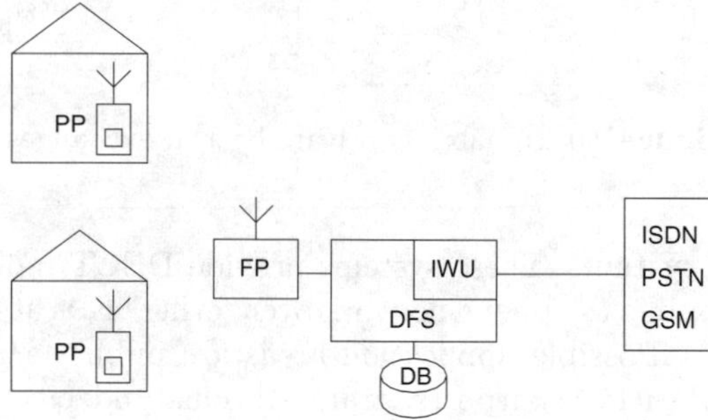

Figure 10.5: Neighbourhood Telepoint

adaptor over conventional lines. If necessary, the terminal adaptor establishes a radio connection to the next FP. The FP is administered in the same way as with the public Telepoint service.

Neighbourhood Telepoint This is a combination of the public Telepoint service and a private home base station (see Figure 10.5). Private households do not have their own FP; instead one FP supplies a number of

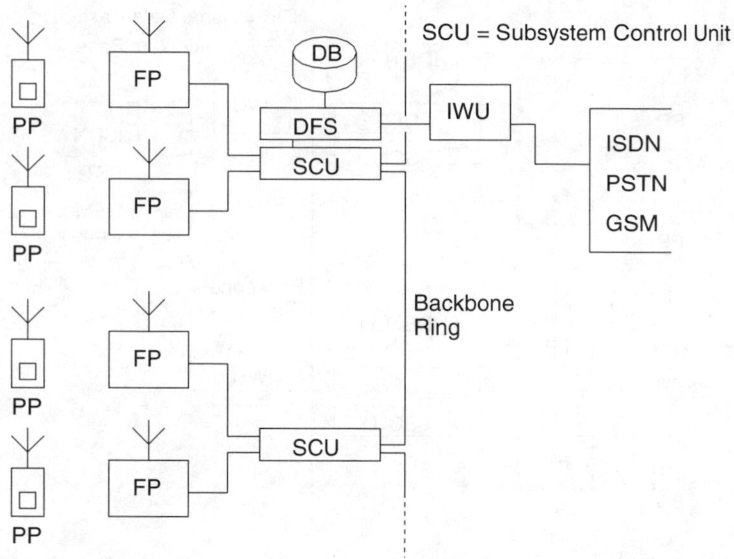

Figure 10.6: Private branch exchanges with ring and central DFS

different private households on the basis of the Telepoint principle. The FPs are installed so that several households will always be supplied by one FP. A DECT fixed system connected over a network interface with an interworking unit (IWU) to the public telephone network manages the FPs.

Because of the way they are designed in terms of size and subscriber capacity, the following systems are also suitable for use by companies with scattered operating locations. One or more nodes can be provided for DECT at each company location. Each implementation of the system must be flexible and tailored to the needs of the customer in each individual case.

Private branch exchanges with ring and central DFS These private branch exchanges consist of a backbone ring with a number of switching nodes (see Figure 10.6). Each switching node consists of a *Subsystem Control Unit* (SCU) and several *base stations*, i.e., FPs. One of the nodes contains the DFS which is responsible for all the nodes connected to the backbone ring and controls them from a central function. The DFS also contains the *home database* (HDB) and the *interworking unit* (IWU), which constitutes the external connection. All signalling data transmitted in this system must be transmitted over several nodes (SCUs) as well as the backbone ring. Transmission channels of the network that are no longer available for transporting user data are used for this purpose. A disadvantage is that the entire network suffers if the central system breaks down. Some relief is provided by the decentral systems,

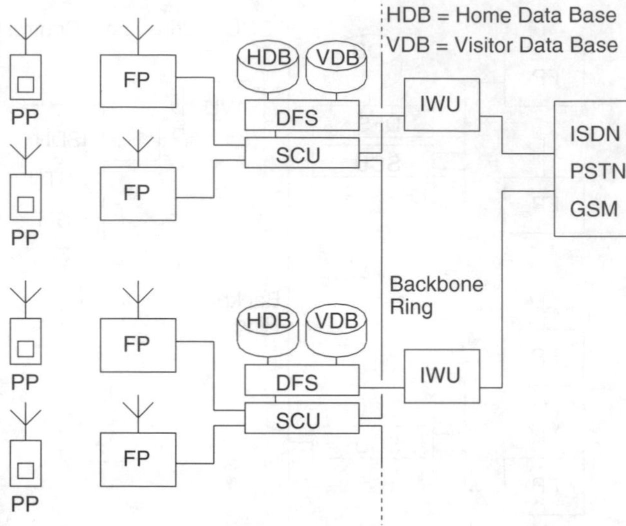

Figure 10.7: Private branch exchanges with ring and distributed DFS

which ensure that there is a reduction in signalling traffic and that only the respective mobile stations are affected when the subsystem breaks down [18].

Private branch exchanges with ring and decentral DFS These are set up similarly to the systems with a central DFS (see Figure 10.7) and consist of a backbone ring that interconnects a number of nodes. A control unit is located at each of these nodes and several fixed stations are connected to a node.

This system operates as a decentral structure because each control unit is controlled by its own DECT fixed system. Furthermore, each DECT fixed system has an interworking unit (IWU) that connects it to the external networks. The data is also stored at decentral locations. Subscriber administration is carried out in a *home database* (HDB), from which important data for operations can be retrieved over the network.

To reduce the signalling required when a mobile subscriber is roaming the network, certain information about each subscriber who moves out of the coverage area of his home database is written to the *visitor database* (VDB) of the new coverage area. The network provides access to the data in the visitor database, and the backbone ring abdicates responsibility for the signalling. If a DFS suffers a breakdown, only the subscribers moving in the respective coverage area will be affected [18].

Private branch exchanges with direct connection of FP These branch exchanges have the same decentral structure as the decentral private

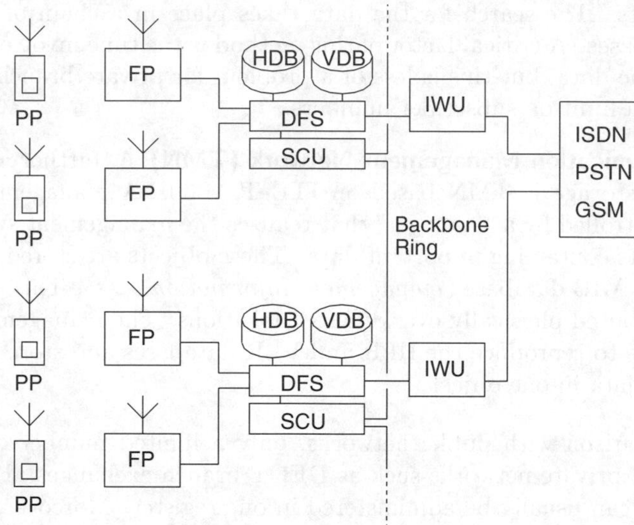

Figure 10.8: Private branch exchanges with direct connection of FP

branch exchanges (see Figure 10.8). The difference between them is that the fixed stations of these private branch exchanges are directly connected to the respective DFS, which has the advantage that there is a reduction in the signalling traffic within a node and consequently a higher capacity for user data [18].

10.1.2 Data Storage

Databases can be set up in different ways. Three different types are presented below:

SS 7-MAP The SS 7-MAP (*Common Channel Signalling System Number 7, Mobile Application Part*) is a method used by public cellular networks such as GSM900 and GSM1800. User data is stored in two registers: a *home location register* (HLR) and a *visitor location register* (VLR). Incoming calls are routed over the HLR to the VLR and from there to the mobile terminal. Outgoing calls only use the VLR. Access of the HLR is not required for these calls.

X.500 Another method is the storage of data based on the standardized register service ITU-T X.500. The database is distributed physically over different locations but logically centralized. The data therefore gives the impression of being stored in a *Directory Information Base* (DIB). All data is organized as objects in a hierarchical *Directory Information Tree* (DIT), in which each branch can be filed in a physically separate *Directory System Agent*. A *Directory User Agent* accesses the individual objects. The search for the data takes place in a chain of interlinked databases. A critical factor of this method is the time involved in accessing the data, but this is less of a problem for private branch exchanges with a limit on subscriber numbers.

Telecommunication Management Network (TMN) A further method for data storage is TMN based on ITU-T M.30. A management system is controlled by a "manager" that realizes the management system using objects containing important data. These objects are stored in a hierarchical MIB database (*management information base*). This MIB can be distributed physically over different locations. The management system is able to reproduce the HLR and VLR databases and store all location area data in one object.

In comparison with public networks, only a limited number of users are managed in private networks such as DECT branch exchanges. The data for these users can usually be administered in one register. Moreover, it is easier to change objects in the frequently changing organizational structure of office environments than in conventional databases. This makes the HLR/VLR concept less suitable for private branch exchanges than the other alternatives. A comparison between X.500 and TMN shows that X.500 is more flexible because it is not required to copy the HLR/VLR principle.

10.2 The DECT Reference System

The DECT Reference System describes the defined logical and physical components of the DECT system, the interfaces between the different units and the connection points to other networks. The global logical structuring of a local DECT network is explained below. This is followed by examples of different physical implementations.

10.2.1 Logical Grouping of DECT systems

The logical groups of the DECT network are organized according to their functionalities with the intervening interfaces D1, D2, D3 and D4, which, however, do not describe how they are physically implemented (see Figure 10.9).

10.2.1.1 Global Network

The *global network* supports the national telecommunication services. It carries out address conversion, routing and relaying between the individual connected socalled local networks. The global network is usually a national (sometimes an international) network. Examples of local networks include:

- *Public Switched Telephone Network* (PSTN)

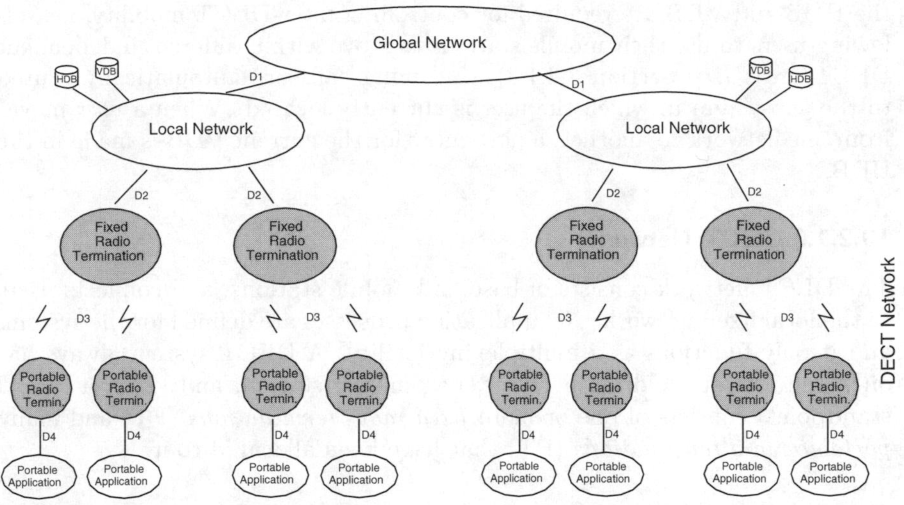

Figure 10.9: DECT Reference System: Logical grouping

- *Integrated Services Digital Network* (ISDN)
- *Packet Switched Public Data Network* (PSPDN)
- *Public Land Mobile Network* (PLMN)

10.2.1.2 Local Networks

Each local network provides a local telecommunications service. Depending on the actual installation, this can vary from a simple multiplexer to a highly-developed complex network. If the subordinate *DECT fixed radio termination* (FT) does not have a switching function then the local network must take over this function. Yet it should be noted that there can be a difference between logical definition and physical implementation, e.g., it is possible for several networks with their functions to be combined into one unit.

A local network converts the global identification numbers (e.g., ISDN numbers) to the DECT-specific `IPUI` (*International Portable User Identity*) and `TPUI` (*Temporary Portable User Identity*). The following networks are often found under the local network:

- analogue or digital *Private Automatic Branch Exchange* (PABX)
- *Integrated Services Private Branch Exchange* (ISPBX)
- IEEE 802 LAN: local area network based on IEEE 802

All the typical network functions must be embedded outside the DECT system and occur either in a local or in a global network. Similarly to GSM, the HDB and VDB are required for controlling inter-DECT mobility, i.e., allowing users to use their mobile stations to move within different independent DECT areas (see Section 3.2.1.3). Incoming calls are automatically routed to the subsystem in which the user is currently located. When a user moves from one network to another, a new entry for the current VDB is made in the HDB.

10.2.1.3 DECT Network

The DECT network consists of base and mobile stations, and connects users to the local fixed network. No application processes are defined for the system, and it only functions as a multiplexing facility. A DECT system always has only one network address per user, i.e., mobile station, and (from a logical standpoint) consists of one or more *fixed radio terminations* (FT) and many *portable radio terminations* (PT) that have been allocated to it.

Fixed Radio Termination The FT is a logical grouping of all the functions and procedures on the fixed network side of the DECT air interface. It is responsible for:

- layer-3 protocol processes in the C-(*control*) layer (except for mobility)
- layer-2 protocol processes in the U-(*user*) layer
- layer-2 switching (routing and relaying) in the respective DECT network

Except for handover and multicell management, the FT contains no switching functions. Although it can manage a large number of call entities, it cannot establish a direct connection between two users. This can only be carried out outside the logically delineated area of the FT in the local network.

Portable Radio Termination and Portable Application These two parts constitute the logical groups on the mobile side of the DECT network. Whereas the *portable radio termination* with all its protocol elements for OSI layers 1, 2 and 3 is defined in the standard, it is up to the manufacturer of the equipment to define the acceptable application. Therefore it is not standardized.

10.2.2 Physical Grouping of DECT Systems

Whereas the logical structure of the DECT network is clearly defined, the physical grouping can assume different forms. It is adapted to each customer's requirements, and therefore can be conceptualized as a single base station with up to 12 simultaneously communicating mobile stations if it is equipped with one transceiver or as an independent switching centre for office buildings. The logical interfaces D1,...,D4 are thereby partially integrated into a common physical unit and consequently can no longer be clearly apportioned (see Figure 10.10).

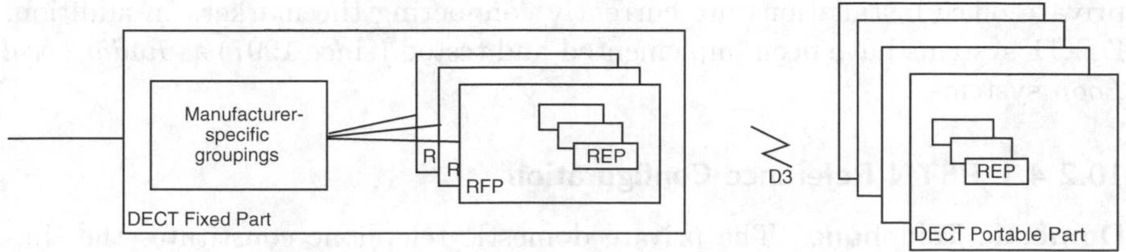

Figure 10.10: DECT Reference System: physical grouping

10.2.2.1 DECT Base Station (Fixed Part)

Physically a DECT system can be split into two parts: a DECT *fixed part* (FP) on the fixed side and a DECT *portable part* (PP) on the mobile side. The fixed part on the wireline side can contain one or more logical groups of the *fixed radio termination* with common control.

An FP can be divided into two physical subgroups:

- *Radio Fixed Part* (RFP): responsible for only one cell in the network
- *Radio End Point* (REP): corresponds to a transceiver unit in the RFP

10.2.2.2 DECT Mobile Device (Portable Part)

The two logical groups *portable radio termination* and *portable application* are physically combined into one *portable part* (PP), typically a handset. A PP normally only has one *radio endpoint.*

10.2.3 DECT Authentication Module (DAM)

A mobile station can be used by different users. Each user must identify himself before being granted access to the DECT network. This function is handled by a *DECT Authentication Module* (DAM), which contains information for the identification (*International Portable User Identity*, `IPUI`) and authentication (*Authentication Key*, `K`) of the user and can be inserted into the PP. The DAM contains all the required encryption procedures.

10.2.4 Specific DECT Configurations

The DECT system description [3] lists several typical DECT configurations. Different physical implementations are required, depending on the higher-ranking network:

• PSTN • ISDN • X.25 • IEEE 802 LAN • GSM

Examples of possible installations are given below. Several suppliers are already marketing private household systems at reasonable prices. Complex private office installations are currently conquering the market. In addition, DECT systems have been implemented and tested (since 1997) as *Radio Local Loop* systems.

10.2.4.1 PSTN Reference Configuration

Domestic Telephone The private domestic telephone constitutes the simplest DECT configuration. Here the network is connected to a PSTN over a *subscriber interface*, just like a plain old telephone (POT) (see Figure 10.11). The functional characteristics resemble those of the earlier CT 1 and CT 2 generations of cordless domestic telephones (see Chapter 9). There is no provision for a local network.

PBX Similarly to the domestic telephone, in a simple implementation of a DECT-PBX, individual FPs are connected to the switching unit of the PSTN. Technically it would then be very complicated to allow a changeover from one FP to another FP during a call (handover).

In Figure 10.12 the *fixed part* contains several *radio fixed parts*, each of which always serves one cell, thereby giving the system its cellular character. A mobile station establishes a connection to the strongest RFP. If the user is roaming in the area of a neighbouring cell, the FP is then in a position of executing an internal layer-2 handover. The protocols for cell control are controlled by the common control (CC) function, which physically can be integrated into the FP or into the PBX.

Radio Local Loop A DECT system can also be incorporated as a local access network into the PSTN (see Figure 10.13). In this case the radio connection is transparent to the user. The user's wireline telephone is linked to a *Cordless Terminal Adapter* (CTA), which carries out the radio transmission to the RFP. In the radio in the local loop (RLL) area, suppliers of fixed networks are currently trying to avoid the high cost of local network cabling by using DECT-RLL systems for the so-called *last mile*. A comparable configuration using ISDN as the local fixed network and DECT as the RLL system is shown in Figure 10.14.

10.2.4.2 GSM Reference Configuration

Aside from the X.25 reference configuration, which is not covered here, the connection between the two mobile systems GSM and DECT should be mentioned. This development offers users the possibility of coupling a locally based DECT system with the national GSM mobile radio system (see Figure 10.15). From the standpoint of GSM, the portable application, PT, FT and possibly also a local network form a mobile user unit. The D1 reference point is the R reference point in the GSM standard (see Figure 3.3). Detailed coverage of the integration of DECT and GSM systems is provided in Section 10.15.

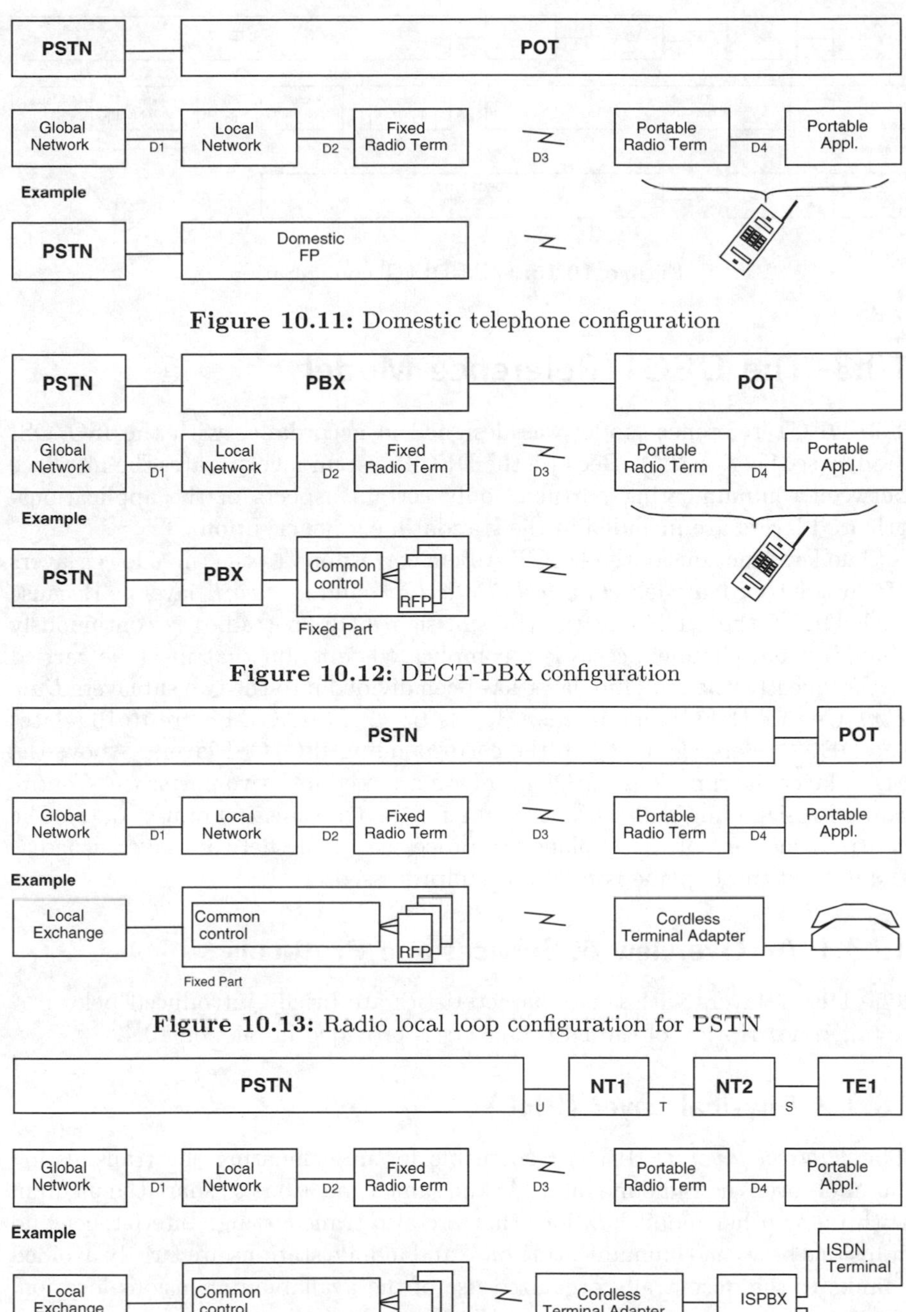

Figure 10.11: Domestic telephone configuration

Figure 10.12: DECT-PBX configuration

Figure 10.13: Radio local loop configuration for PSTN

Figure 10.14: Radio local loop configuration for ISDN

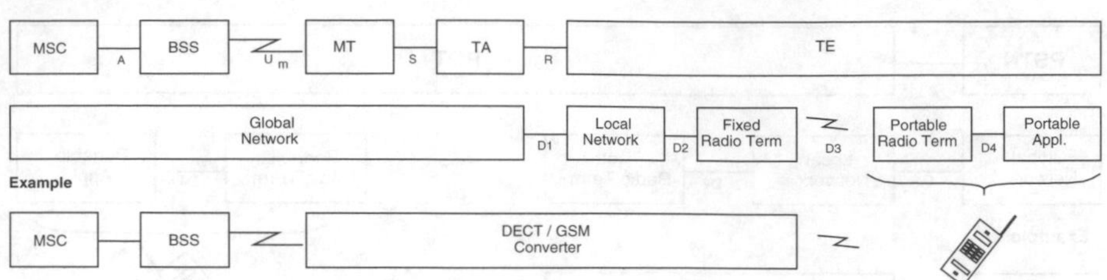

Figure 10.15: GSM-DECT configuration

10.3 The DECT Reference Model

The DECT reference model was designed in accordance with the ISO/OSI model (see Section 2.5). Because the DECT system incorporates the interface between communicating partners, only certain aspects of the applications-oriented layers are included in the standard, e.g., encryption.

The key functions of the DECT system correspond to the three lower layers of the ISO/OSI model: *Physical*, *Data Link* and *Network* layers. Because with DECT the quality of the transmission medium (radio) is continuously changing and channel access is a complicated function that must be carried out frequently, the data link layer has been divided into the two sublayers *Data Link Control* (DLC) and *Medium Access Control* (MAC). Figure 10.16 relates the DECT reference model to the corresponding ISO/OSI layers. Above the MAC layer the functions of the layers are grouped into two parts: the *Control plane* for signalling and the *User plane* for the transmission of user data. The control functions of the C-plane are processed in the network layer, whereas the data of the U-plane is passed on unprocessed.

10.3.1 An Overview of Services and Protocols

The DECT layers with their characteristics are briefly introduced below. A detailed description of the DECT layers is provided in Section 10.4.

10.3.2 Physical Layer (PHL)

The *physical layer* (PHL) is responsible for implementing the transmission channels over the radio medium. At the same time it has to share the medium with many other mobile stations that are also transmitting. Interference and collisions between communicating base and mobile stations are largely avoided thanks to the decentrally organized use of the available dimensions location, time and frequency (see Figure 10.17). Each dimension contains several possibilities for seizing a channel for interference-free transmission.

The TDMA (*Time-Division Multiple-Access*) method is the one used for the time dimension. Each station sets up its channel in any available time slot

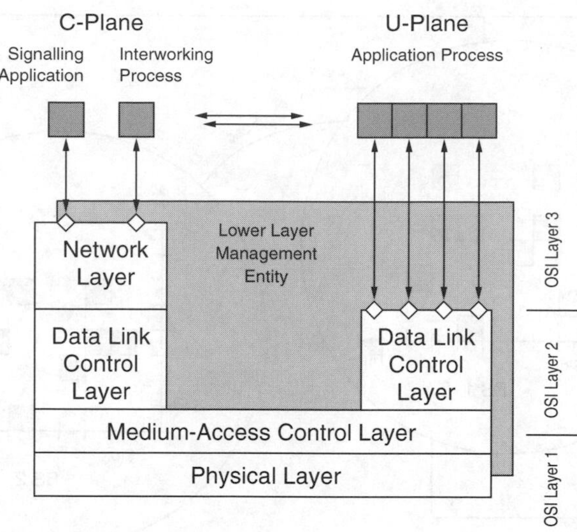

Figure 10.16: DECT reference model

Table 10.1: Physical data on the DECT system

Frequency band	1880–1900 MHz
No. of carrier frequencies	10
Carrier spacing	1.728 MHz
Max. transmitter power	250 mW
Carrier multiplexing	TDMA
Basic duplex method	TDD
Frame length	10 ms
No. of slots (frame)	24
Modulation	GFSK with $BT = 0.5$/GMSK
Modulated overall bit rate	1152 kbit/s
Net data rate	32 kbit/s data (B-field) unprotected
for standard connections	25.6 kbit/s data (B-field) protected
	6.4 kbit/s signalling (A-field)

and is then able to transmit in that time slot at a constant bit rate. Because of the TDD (*Time-Division Duplexing*) method the uplink and downlink of this channel are in slot pairs on the same frequency. Consequently a duplex transmission always occupies two time slots, which are separated from another by a fixed interval.

The FDMA (*Frequency-Division Multiple-Access*) method with 10 different frequencies is used for the frequency dimension. This means that each station is able to select and occupy a slot pair on any frequency for its transmission.

Owing to the attenuation of the propagation of the radio waves, frequencies and time slots can be reused at appropriate distances. These resources are

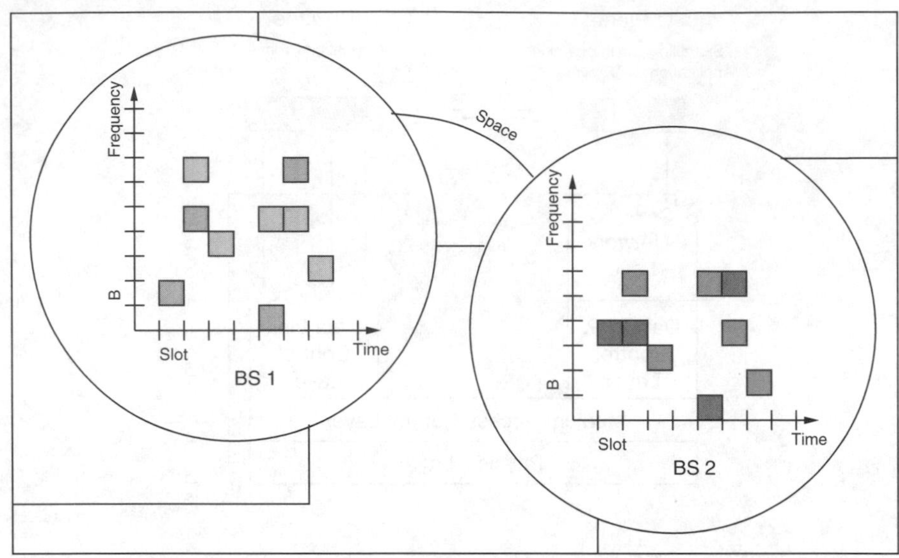

Figure 10.17: Three-dimensional use of spectrum

occupied based on a dynamic procedure for channel selection that takes into account the local traffic load on the system.

Key technical data for the DECT system is listed in Table 10.1 (see also Table 9.2). With only 12 duplex voice channels in 1.73 MHz, i.e., 144 kHz/channel pair, DECT is very generous in its use of spectrum (GSM only requires 50 kHz/channel pair). However, because of its dynamic channel selection, the resulting small random reuse distances and the microcellular coverage, DECT provides a much higher capacity (Erl./km^2), producing almost unbelievably high values (10 kErl./km^2), especially in high-rise buildings because of the reuse distances at every two levels over the same base area.

10.3.3 Medium-Access Control (MAC) Layer

The MAC layer (*Medium-Access Control (MAC) Layer*) (see Figure 10.16) is responsible for setting up, operating and releasing channels (*Bearer*) for the higher layers. The different data fields of the MAC protocol are protected by cyclical codes, which are used for error detection at the receiver. The MAC layer ensures that service-specific control data is added to each time slot.

The MAC layer comprises three groups of services:

BMC The *Broadcast Message Control Service* is offered on at least one physical channel in each cell, even if there is no user transmitting. This channel is called a *beacon channel*, and produces a continuous connectionless point-to-multipoint connection on the downlink in which a base station transmits its system-related data. This identifies the base station to the mobile unit. By evaluating the received signal, the terminal can determine the current channel quality at the same time.

CMC The *Connectionless Message Control Service* supports a connectionless point-to-point or point-to-multipoint service that can be operated bidirectionally between a base station and a mobile user.

MBC The *Multi-Bearer Control Service* offers a connection-oriented point-to-point service. An entity transmitting in one or both directions is able to support several bearers, thereby achieving a higher net data rate.

Each of these three services incorporates its own independent *service access point* (SAP), which links it to the next-highest layer and can combine several logical channels together.

10.3.4 Data Link Layer

The protocol stack directly above the MAC layer divides into two parallel parts. Similarly to the MAC layer, extensive error protection in the C-plane of the data link layer improves the reliability of the data transmission. Along with a point-to-point service, the C-plane offers a broadcast service to the network layer above it. The U-plane is responsible for processing user data on the radio section, with the spectrum of services ranging from the transmission of unprotected data with minimal delays, e.g., speech, to the provision of protected services with variable delays for data transmission. The required data rate of an existing connection can be changed at any time.

10.3.5 Network Layer

The network layer sets up, manages and terminates connections between users and the network. The U-plane in DECT has no role in the network layer and forwards all data unprocessed in a vertical direction. The C-plane carries out the signalling and is responsible for controlling data exchange. Five protocols based on the *link control entity* are provided for this purpose. Along with the call and connection entities, there is a *mobility management* service that carries out all the tasks necessary to support the mobility of mobile stations. In addition to the data for location area management, messages for authentication and encryption data are also transmitted.

10.3.6 Management of the Lower Layers

The management of layers 1–3 (*Lower Layer Management Entity* (LLME) (see Figure 10.16) contains procedures that affect several protocol layers. Processes such as generating, maintaining and releasing physical channels (*Bearers*) are initiated and controlled from this entity. Furthermore, the LLME selects from the physical channels which are available and evaluates the quality of the received signal.

10.4 Detailed Description of Services and Protocols

10.4.1 Physical Layer (PHL)

A physical channel is the physical transmission path between two radio units. Radio transmission places a great demand on transceivers to ensure that there is good reception of the usage signal. The receiver sensitivity for the required bit error ratio (BER) of 0.001 is −83 dBm (60 dBμV/m) [6]. This was increased to −86 dBm for applications run by public services [11]. In DECT a normal telephone connection requires two independent channels between the device end points on the transmission medium radio.

10.4.1.1 FDMA and Modulation Techniques

Because of FDMA access, the DECT system has the possibility of selecting from a number of frequencies in its channel selection.

It operates in the 1880–1900 MHz frequency band. Within this band 10 carrier frequencies are defined, the mid-frequency f_c of which can be calculated as follows:

$$f_c = f_0 - c \cdot 1728 \text{ kHz}, \quad \text{with } c = 0, 1, \ldots, 9 \text{ and } f_0 = 1897.344 \text{ MHz} \quad (10.1)$$

In an active state the deviation from the mid-frequency should be a maximum of ±50 kHz.

The modulation techniques used are either *Gaussian Frequency Shift Keying* (GFSK) with a bandwidth–time product $BT = 0.5$ or *Gaussian Minimum Shift Keying* (GMSK).

If a transmit signal is formed from two orthogonal bandpass signals with different mid-frequencies, this is referred to as *Frequency Shift Keying* (FSK). A Gaussian filter acting as a low-pass filter (GFSK) removes the high-frequency parts of the signal in order to keep the band of the signal's spectrum as small as possible. If the modulation index is 0.5 and the possibility of a coherent demodulation of the radio signal exists, this shift procedure is referred to as minimum-shift keying (MSK). With GMSK the Gaussian filter also comes into play. For cost reasons, receivers in DECT systems are usually not built with coherent demodulation/modulation [24].

The transmission of a binary `1` in a DECT system produces a frequency increase of $\triangle f$ = 288 kHz to f_c + 288 kHz. For the transmission of a `0` the frequency is decreased by $\triangle f$ to f_c − 288 kHz.

The standard does not provide for any equalizers. With a bit duration of 0.9 μs, waves that reach the receiver with a delay due to multipath propagation (see Figure 2.7) cause signal dispersion (see Section 2.1.7), which with a 300 m alternative path length already corresponds to the symbol duration and makes reception impossible even when sufficient signalling power exists. The literature recommends using 16 of the 32 synchronization bits in the S-field in

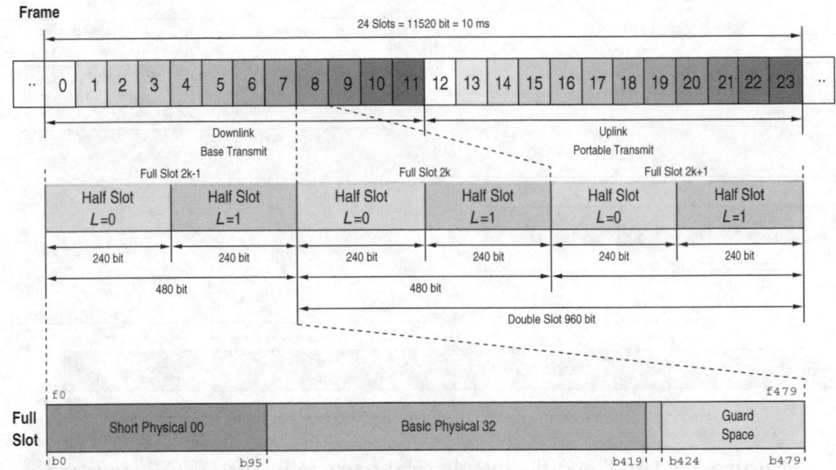

Figure 10.18: Time-multiplex elements of the physical layer

accordance with Figure 10.19 to estimate the impulse response of a channel in the receiver, which equates to a good (but not one conforming to the standard) implementation of an equalizer [19].

10.4.1.2 TDMA Technique

Each station receives a protected, periodically recurring portion of the overall transmission rate of a frequency. According to Figure 10.18, the frame and time slot structure of the DECT system are explained in the physical layer.

The transmission capacity of each frequency is divided into 10 ms long periodically recurring frames, each of which has a length corresponding to the duration of 11.520 bits. This produces a gross frame transmission rate of 1152 kbit/s. A frame comprises 24 time slots, which are used as either full slots, double slots or half slots (see Figure 10.18).

In a normal *basic connection* the first 12 time slots are used for transmitting data from the base station to a mobile station (*downlink*), whereas the second part of the 24 slots is reserved for the direction from the mobile station to the base station (*uplink*). Since a duplex connection requires both an uplink and a downlink connection, the DECT system uses a technique called *Time-Division Duplexing* (TDD). If the base station occupies slot k in order to transmit to the mobile unit, slot $k + 12$ is reserved for the mobile station to enable it to send data to the base station. This rigid system of allocation is abandoned when it comes to more complex *advanced connections*, where a generous use of time slots is allowed in each transmission direction.

Each of the 24 time slots has a length of 480 bits (416 µs) and can be used in accordance with the slot type (full, half, double). Different *physical packets* are based on this structure. Each physical packet contains a synchronization

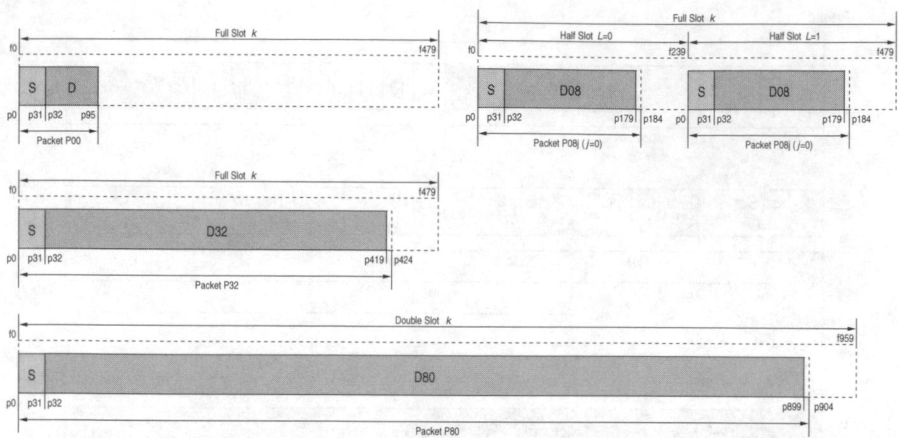

Figure 10.19: The different physical packets in the DECT standard

field S and a data field D. A physical packet is shorter than a time slot by one *guard period* to prevent an overlapping of packets from neighbouring time slots.

A P00 (*short physical packet*) is used for short connectionless transmissions in the beacon channel or for short messages. It comprises only 96 of the 480 bits and therefore provides a particularly large guard time in a slot. This is especially important at the start of a connection phase when a mobile station is not yet completely synchronized with the network and could cause interference to neighbouring time slots when it transmits a full slot.

A P08j message only requires a half slot for transmission, and therefore because of the decreased transmission rate ends up with double the number of available channels per frame; see Figure 10.19.

Because full slots are usually the ones preferred when a slot is requested, the packet P32 will be described here. The first 32 bits form the synchronization field S, which is used for clock and packet synchronization in the radio network. It consists of a 16-bit preamble, followed by a 16-bit long packet synchronization word. In the S-field the response from the mobile station contains the inverted sequence of the bit of the synchronization field sent by the base station.

The S-field is followed by the user data field D, which is 388 bits long. Some of the content of this field is evaluated by the MAC layer, and therefore will be covered in detail later (see Section 10.4.2.5).

The possibility of transmitting a so-called Z-field is offered by the D-field. This 4-bit long word contains a copy of the last 4 bits of the data field, which is also referred to as an X-field. A comparison of these two areas allows the receiver to establish whether a transmission is being disrupted because of errors in the synchronization of neighbouring DECT systems. These disruptions are referred to as *sliding collisions* within the system. A measurement of these

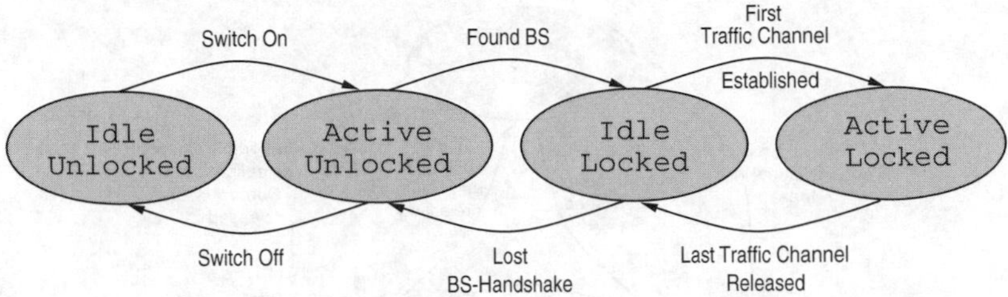

Figure 10.20: Operating states of a mobile station

disruptions results in the early detection of interference, and can be used as the criterion for an optimal handover decision. Packet P80 is the one with the highest data rate, and requires a double slot. User data rates up to 80 kbit/s can be achieved with this packet.

10.4.2 Medium-Access Control (MAC) Layer

10.4.2.1 Operating States of a Mobile Station

With reference to the MAC layer [7], a mobile station can find itself in one of the four states shown in Figure 10.20.

- `Active Locked` The synchronized mobile station has at least one connection to one or more base stations.
- `Idle Locked` The mobile station is synchronized with at least one base station. It currently does not have a connection, but is in a position to receive requests for connections.
- `Active Unlocked` The mobile station is not synchronized with any base station, and therefore cannot receive any requests for connections. It attempts to find a suitable base station for synchronization to enable it to move into the `Idle Locked` state.
- `Idle Unlocked` The mobile station is not synchronized with any base station, and is not in a position to detect base stations that are suitable.

If a terminal is switched off, it is in the `Idle Unlocked` state. When it is switched on, a mobile station tries to transit to the `Idle Locked` state. It begins to search for a suitable base station with which to synchronize. If it is successful, a transition to the `Idle Locked` state is made. In this state a mobile station is able to receive or transmit requests for connection. As soon as the first traffic channel is set up, it changes over to `Active Locked` state. If the last channel is cleared after the termination of a connection, the mobile station returns to the `Idle Locked` state. If it loses its synchronization with the base station, it must change to the `Active Unlocked` state and seek a new suitable base station.

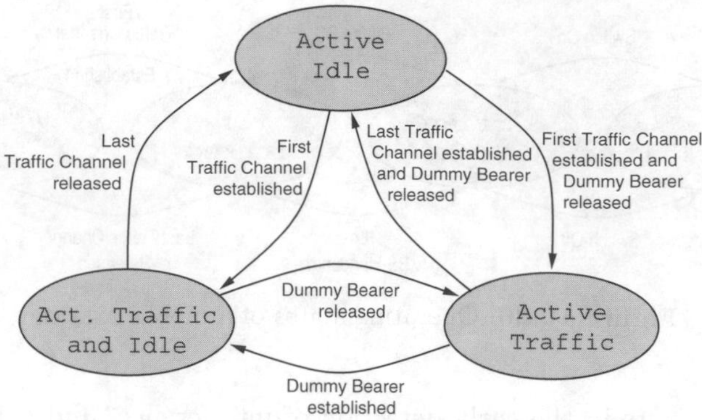

Figure 10.21: States of a base station

10.4.2.2 Operating States of a Base Station

A base station can find itself in one of four states. The `Inactive` state in which a base station is switched off is not shown in Figure 10.21.

- `Inactive` The base station is switched off and can neither receive nor transmit messages.
- `Active Idle` The base station is not operating a traffic channel, and therefore radiates a dummy bearer that the mobile receivers are able to detect when observing the physical channels.
- `Active Traffic` The base station operates at least one traffic channel. The dummy bearer is no longer supported.
- `Active Traffic and Idle` The base station maintains a dummy bearer in addition to at least one traffic channel (*traffic bearer*).

In the basic state `Active Locked`, the base station transmits a dummy bearer in order to enable mobile terminals to synchronize with its frame and slot clock. If a traffic bearer is established, the base station moves into `Active Traffic` state. The dummy bearer can then be eliminated. The transition on the opposite side takes place after the last traffic channel has been released. If a dummy channel is required when the traffic bearer is transmitted, the base station can move into `Active Traffic and Idle` state. Again the dummy bearer can be retained when the first traffic channel is set up. A transition is then made from `Active Idle` to `Active Traffic and Idle`.

10.4.2.3 Cell and Cluster Functions

The MAC layer functions have already been covered in the overview in Section 10.3.3. This layer is used to set up and manage the traffic channels

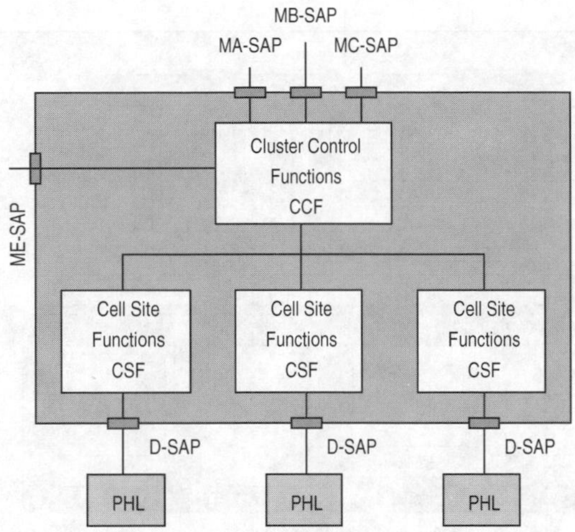

Figure 10.22: Classification of MAC services

(bearers) requested from the LLME and, when requested, to release them. The control information, which is introduced through different service access points in the MAC layer, is multiplexed and added to the actual user data in each time slot.

The different services of the MAC layer are divided into two groups (see Figure 10.22). The *cluster control functions* in the upper area are linked to the data link layer through the three service access points MA, MB and MC. The cell site functions in the lower part coordinate the transition to the physical layer. The two groups offer the following individual functions:

Cluster Control Functions (CCF) These functions control a cluster of cells. Each logical cluster of cells contains only one CCF, which controls all the cell site functions (CSF). Three different independent services are available within this cluster:

Broadcast Message Control (BMC) This function only exists once in each CCF, and distributes the cluster broadcast information to the respective cell functions. The BMS supports a number of connectionless point-to-multipoint services, which are directed from a base station to the mobile station. The BMC works with any type of traffic channel. An important service is the paging of mobile stations.

Connectionless Message Control (CMC) All information that concerns the connectionless service is usually controlled by one CMC in each CCF. In addition to transmitting information from the control level of the DLC layer, the CMC also processes user data from the so-called U-plane. The services can be operated in both directions.

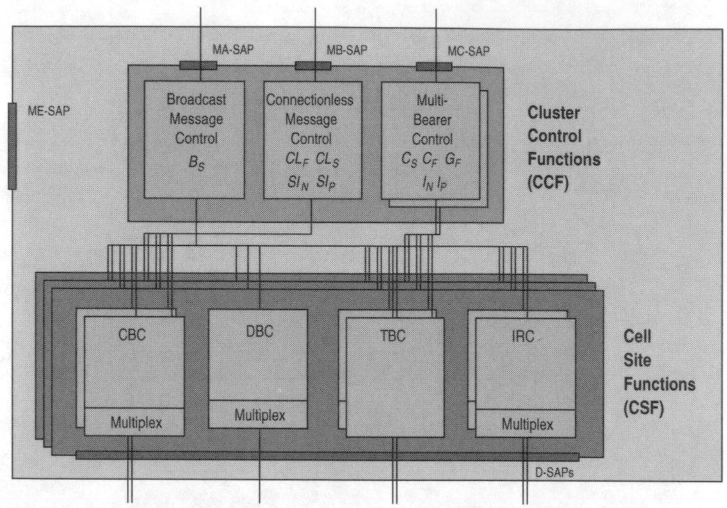

Figure 10.23: Overview of services and channels in the MAC layer

Multi-Bearer Control (MBC) This service comprises the management of all data that is exchanged directly between two corresponding MAC layers. An MBC capable of organizing several traffic channels exists for each connection-oriented point-to-point connection.

Cell Site Functions (CSF) These functions appear below the CCF services in the MAC layer, and stand in for the respective cell. Therefore several CSFs are controlled by each CCF. The following cell-oriented services are differentiated (see Figure 10.23):

Connectionless Bearer Control (CBC) Each connectionless bearer within the CSF is controlled by its own CBC.

Dummy Bearer Control (DBC) Each CSF has a maximum of two dummy bearers to provide a beacon function for the synchronization of the mobile stations in the event that no user connection exists in the cell.

Traffic Bearer Control (TBC) An MBC must request a TBC for a duplex connection.

Idle Receiver Control (IRC) This service controls the receiver in a cell when no connection is being maintained to a user; a cell can have more than one receiver, and consequently a related number of IRC services.

10.4.2.4 Service Access Points

The cluster and cell-oriented functions include several service access points (SAP), which can be used by the MAC layer entities to communicate with the next-highest OSI layer (*data link control layer*) and the next-lowest OSI layer (*physical layer*) (see Figure 10.22). The following SAPs are available between the CCF and the DLC layers:

- MA-SAP
- MB-SAP
- MC-SAP

Each cell-specific service has its own D-SAP for access to the physical layer. The MAC layer has a separate SAP, the ME-SAP, to support the lower layer management functions. This SAP is not formally specified, and therefore has no logical channels.

MA-SAP Information from the *Broadcast Message Control Service* is transmitted to the DLC layer over this access point. Data from the B_S channel as well as control data that controls the data flow on the B_S channel is transmitted. This logical *Higher-Layer Broadcast Channel* supports a connectionless simplex broadcast service from the base station to the mobile station.

MB-SAP This access point links *Connectionless Message Control* with the DLC layer, and contains four logical channels:

- CL_F
- CL_S
- SI_N
- SI_P

The control channels CL_F and CL_S support a connectionless duplex service between the base station and the mobile terminal. A continuous service exists from the base station to the mobile station but not in the opposite direction. The quantity of data permitted on the CL_S channel is 40 bits, which corresponds to the segment length of this channel. The fast CL_F channel has a permitted segment length of 64 bits, and the quantity of data can amount to four times the segment length.

The information channels SI_N/SI_P offer an unprotected/protected simplex service from the base station to the mobile station.

MC-SAP The multi-bearer control unit is linked to the DLC layer over the MC-SAP. Five logical channels are available for the transmission of information on data flow control and establishing, maintaining and releasing MAC connections:

- C_S
- C_F
- G_F
- I_N
- I_P

The control channels C_S and C_F offer two independent connection-oriented duplex services. The maximum throughput for the connection of a slow C_S with a segment length of 40 bits is 2 kbit/s. With a data segment length of 64 bits, the fast C_F control channel achieves a throughput of 25.6 kbit/s with full slots. The information of both control channels is supplied with a CRC checksum (*Cyclic Redundancy Check*) that provides error detection and correction through retransmission with an ARQ technique.

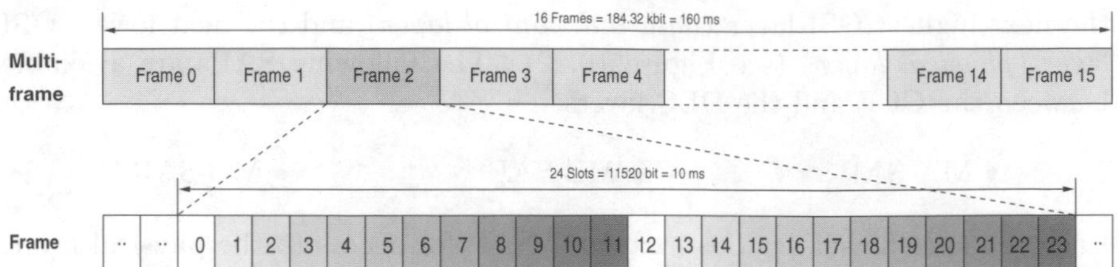

Figure 10.24: DECT multiframe organization

One of the two information channels I_P or I_N is used with each transmission to ensure that the higher layers are offered an independent connection-oriented duplex service. The I_N channel is for voice transmission, with the MAC layer supplying 4 bits (X-field) of error detection protection for the information. I_P channels are planned for the transmission of data with error-detection or error-correction coding.

The G_F channel is a connection-oriented simplex service with a data segment length of 56 bits and is used by the U-plane of the DLC layer. The MAC layer provides error detection for this channel.

10.4.2.5 MAC Multiplex Functions

Multiframe structure In the MAC layer the TDM frame structure implemented from the physical layer (see Section 10.4.1) is superimposed logically with a multiframe structure. A multiframe is composed of 16 individual frames (see Figure 10.24), and normally begins with the first half of the frame 0, used by the base station. The end of a multiframe is used by the mobile terminal in the last half of frame 15. The number of the current multiframe is transmitted in at least each eighth multiframe, and is used for encryption purposes. The frame number is recognized by frame 8 because this is where the logical channel Q_T is transmitted from the base station to the mobile station (see Figure 10.26).

D-field Bit mapping, the compilation of the D-field for subsequent transmission to the physical layer, is based on fixed rules. As already described in Section 10.4.1.2, the length of a D-field depends on the type of transmission requested. If a P32 transmission is requested, the length of the D-field is 388 bits. A different D-structure is used for P00, P08j and P80 operations.

The D-frame of a P32 packet comprises two parts, as shown in Figure 10.25. The A-field is 64 bits long and is used for the continuous transmission of control information. The B-field is available for actual user data, and in full slot operation is 324 bits in size.

The individual parts of the A and B-fields are explained below:

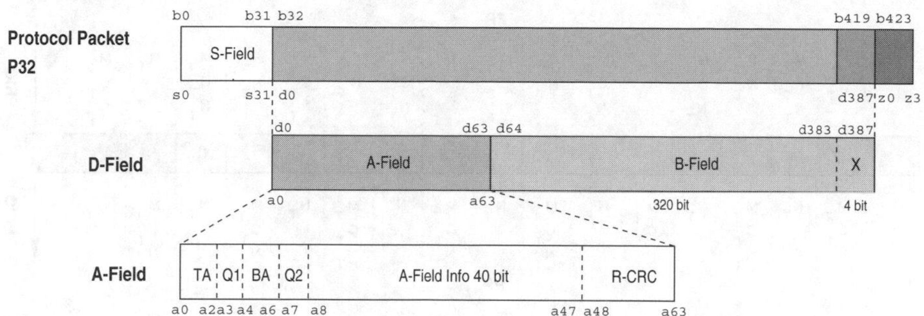

Figure 10.25: DECT time-multiplexing: D32-field

A-field The control field contains three sections (see Figure 10.25). The A-field information of 40 bits is joined to the header of the A-field, which has a length of 8 bits. An R-CRC field with 16 bits for protection of the control data forms the connection.

- **TA (a0...a2)** The 3-bit long TA field at the beginning of the header indicates the type of A-field information (a8...a47). There are five different logical channels, one of which always transmits data in the A-field. Here a distinction is made between the internal MAC channels N_T, Q_T, M_T and P_T and the C_S channel (for the control data of the higher layers).
- **Q1 (a3)** In connection-oriented transmission the Q1 bit in conjunction with duplex bearers carries out the quality control of a channel and serves as the handover criterion. In services which contain error correction it can be used for flow control.
- **BA (a4...a6)** This part describes the qualities of the B-field. In addition to normal protected and unprotected information transmission (*U-type*, I_P or I_N), expanded signalling (*E-type*, C_F or C_L) is also possible in certain cases.
- **Q2 (a7)** Like the Q1 bit, the Q2 bit is designated for the quality control of a connection. A combination of Q1 and Q2 constitutes a handover criterion.
- **A-field info (a8...a47)** Internal MAC messages can be transmitted within the 40-bit long tail field. This involves sending out different control information in consecutive time slots using an *E-type* multiplexer.
- **R-CRC (a48...a63)** The MAC layer protects the logical channels of the A-field with a *cyclic redundancy check* (CRC). This involves calculating and transmitting 16 redundancy bits, which with up to
  - 5 independent errors

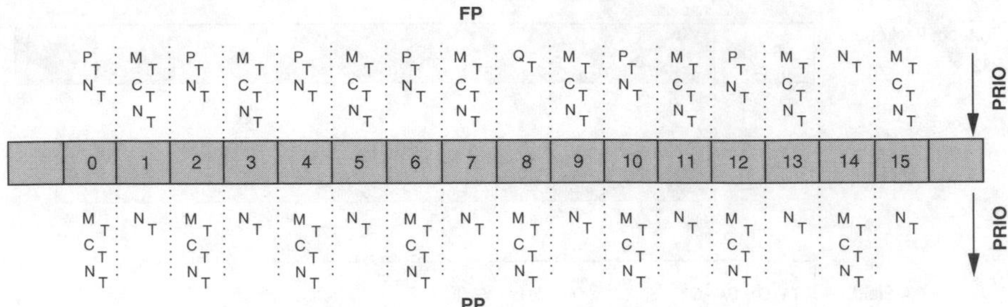

Figure 10.26: Priority of control information in a multiframe cycle

- bursty errors up to a length of 16
- error patterns with an odd number of errors

can be detected in the A-field [23]. An error-correction mechanism is not provided in the MAC layer.

The internal control data of the MAC layer is contained in the A-field information. One of the following options can be selected during multiplexing (see Figure 10.26):

N_T **Identities Information** Using its *Primary Access Rights Identifier* (`PARI`) and its *Radio Fixed Part Number* (`RPN`), the MAC layer of the base station generates its own *Radio Fixed Part Identity* (`RFPI`), which is transmitted to the terminal.

Q_T **System Information and Multiframe Marker** This channel, used once in each multiframe, is only transmitted by the base station, and therefore is used for indirect synchronization on the multiframe cycle. Information on the technical structure of the base station as well as on the current connection is provided.

P_T **Paging Information** This channel is the only one that a mobile station can even receive when it is in the `Idle Locked` state (see Figure 10.20), and contains the broadcast service (B_S) of a base station. In addition to the identification number of the base station, the terminal receives important MAC layer information. For example, blind slot data that evaluates the quality of the current bearer is transmitted and suggestions for other qualitatively good channels are passed on. In response to a Handover_Request, the handover areas approved by the base station are transmitted back.

M_T **MAC Control Information** Administrative tasks such as connection setup, maintenance and termination and handover requests are carried out on this channel. A Wait command is sent if there is a delay in setup. A reciprocal exchange of the channel list can accelerate the selection of the best channel for transmission.

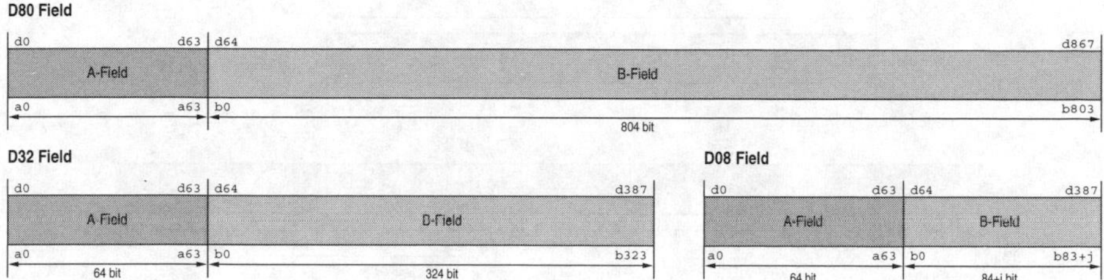

Figure 10.27: D-fields for the different physical packets

C_T **Control Information Higher Layers** Either C_L or CS_L information from the higher layers is transmitted on this channel. In other words, this is not an internal MAC channel.

Multiplexing of control channels during transmission is carried out according to a specific plan. Different control information for the A-field, which is transmitted on the basis of a priority list, is provided in each multiframe (see Figure 10.26). The base station always sends its system information (Q_T) in the eighth frame, whereas for the mobile station each uneven frame is reserved for the N_T channel. If a control channel with the highest priority attached to it remains unused in a frame, control information with a lower priority can be sent.

B-field With physical packet P32 the 324-bit long B-field, for which a protected and an unprotected format exist, follows the control field (A-field) in the D-field (see also Figure 10.27). In an unprotected transmission of data, as in the transmission of voice, 320 effective usage bits from the B-field are required for a data rate of 32 kbit/s. The remaining 4 bits (X-field) are used for error protection by the means of CRC. In a protected B-field format the 4 bits are kept for error protection, whereas the 320 effective data bits are divided into four blocks. The usage bits within each block are reduced to 64 so that an R-CRC check sequence can be formed with the remaining 16 bits. The net data rate reduces to 25.6 kbit/s (see Figure 10.28).

10.4.2.6 MAC Services

The different MAC services produced from the combination of different physical packets and the option for protected or unprotected transmission are listed below. These can be carried out as symmetrical or as asymmetrical services. With asymmetrical services the rigid separation between uplink and downlink is eliminated. Thus the service from a base station to a mobile station can use a slot in the first part of a frame as well as the corresponding slot in the second part of the frame for the downlink. It is different with services that offer a

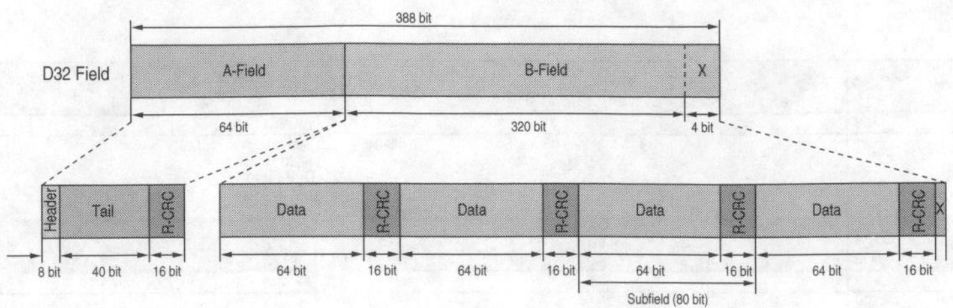

Figure 10.28: Elements of the A and B-fields

guaranteed throughput or guaranteed bit-error ratio, which then results in a variable throughput. In this case the MAC layer uses an ARQ protocol. The services of the MAC layers are listed individually in Tables 10.2 and 10.3.

10.4.2.7 Types of Bearers

MAC bearers are elements that are created by the *cell site functions* (CSF) and always correspond to a service entity in the physical layer. The following is a list of the different bearer types:

Simplex Bearer This type of bearer is used to produce a physical channel in one direction. A differentiation is made between long and short simplex bearers. Whereas the short bearer only contains an A-field, the long one also transmits a B-field next to the A-field. For example, *dummy bearer control* (DBC) (see Section 10.4.2.3) reserves a simplex bearer sent by the base station for the transmission of broadcast information.

Duplex Bearer A pair of simplex bearers transmitting in the opposite direction on the physical channels is called a duplex bearer. The two bearers use the same frequency and are time-deferred by half a frame (slot pair). A *traffic bearer controller* (TBC) uses a duplex bearer, e.g., for a traffic channel between the base station and the mobile station.

Double Simplex Bearer A pair of long simplex bearers transmitting in the same direction on two physical channels is called a double simplex bearer. Similarly to a duplex bearer, the two bearers should operate in one slot pair on the same frequency. This type of bearer only occurs in multibearer connections, and is also used for asymmetrical transmission.

Double Duplex Bearer A double duplex bearer consists of a pair of duplex bearers that is part of a shared MAC connection. Each of these duplex bearers is produced by a TBC. The two TBCs are controlled by an MBC.

Table 10.2: Partial list of symmetrical MAC services

S_T	I-channel cap. [kbit/s]	B-field multipl. scheme	Error detection	Error correction	Delay [ms]
2d	$k \cdot 80$	(U 80a, E 80)	—		15
2f	$k \cdot 32$	(U 32a, E 32)	—	—	15
2d	$8 + j/10$	(U 08a, E 08)	—	—	15
3d	$k \cdot 64.0$	(U 80b, E 80)	×	—	15
3f	$k \cdot 25.6$	(U 32b, E 32)	×	—	15
3d	6.4	(U 08b, E 08)	×	—	15
4d	$\leq k \cdot 64$	(U 80b, E 80)	×	×	Var.
4f	$\leq k \cdot 25.6$	(U 32b, E 32)	×	×	Var.
4d	$\leq$ 6.4	(U 08b, E 08)	×	×	Var.

S_T = Service type: xd = type x double slot, xf = type x full slot, xh = type x half slot

Table 10.3: Partial list of asymmetrical MAC services

S_T	I-channel cap. [kbit/s]	B-field multipl. scheme	Error detection	Error correction
5d	$k \cdot 80$	(U 80a, E 80)	—	—
5f	$k \cdot 32$	(U 32a, E 32)	—	—
6d	$k \cdot 64.0$	(U 80b, E 80)	×	—
	$m \cdot 64.0$	(U 80b, E 80)	×	—
6f	$k \cdot 25.6$	(U 32b, E 32)	×	—
	$m \cdot 25.6$	(U 32b, E 32)	×	—
7d	$\leq k \cdot 64.0$	(U 80b, E 80)	×	×
	$\leq m \cdot 64.0$	(U 80b, E 80)	×	×
7f	$\leq k \cdot 25.6$	(U 32b, E 32)	×	×
	$\leq m \cdot 25.6$	(U 32b, E 32)	×	×

10.4.2.8 Types of Connections

Connection-oriented entities exist in addition to the connectionless services that include broadcast services. Each *multibearer control* unit in the MAC layer is responsible for the maintenance of a connection. It controls one or more *traffic bearer control* entities, which are used in the administration of the bearers. A differentiation is made between *advanced connections* and *basic connections*:

Basic Connections A basic connection does not have a common connection number that is known to the base station and to the mobile station. Consequently only one basic connection can exist between a base station and a mobile terminal. It consists of a single duplex bearer. It is possible for two basic connections that serve the same DLC connection to exist together for a short time when a switch is being made in physical channel (handover).

Advanced Connections Because, unlike basic connections, advanced connections use a common number, several connections can exist between two stations. The individual bearers are assigned a logical number to allow a differentiation within a connection in the MAC layer.

Physical Connections Physical connections are not supported by MAC services (see Tables 10.2 and 10.3). They are used for non-standardized data transmission (see Section 10.4.3.8).

When a connection is being made, all the bearers requested from the DLC must be set up within 3 s (T200: *Connection Setup Timer*) or else the connection is considered aborted. Additional bearers can be set up or released if necessary during an ongoing connection in order to increase or reduce the overall transmission capacity.

Advanced connections can be either symmetrical or asymmetrical. With symmetrical connections the same services and number of bearers are used in both directions.

10.4.2.9 Connection Setup

Connection-oriented procedures within the MAC layer use two point-to-point connections: *connection* and *bearer*. Only the pure connection is apparent to the data link layer. The bearers that each connection uses for a transmission are managed within the MAC layer and are not apparent to the higher layers. Connection setup is usually initiated by a mobile terminal (*portable-initiated*). If there is a call from the fixed network, the mobile station must first be paged so that it is aware of it and can initiate the setup procedure.

Connection Setup The sequence of connection setup can be explained by a description of the MAC layer primitives (see Figure 10.29). The higher layers (DLC layer) initiate connection setup (MAC-Connect_Request) in the MAC layer. In addition to a MAC connection endpoint identifier (MCEI) that applies as the reference address for all subsequent primitives, this primitive contains a parameter that specifies the requested service. If the MAC layer is not able to provide the requested services, it sends a Disconnect_request to the DLC layer.

The instance of an MBC entity (*multibearer control*) produced by the request receives permission from the *lower-layer management entity* (LLME) to set up a connection between the base station and the mobile station. Depending on the service requested, either a basic or an advanced connection is set up.

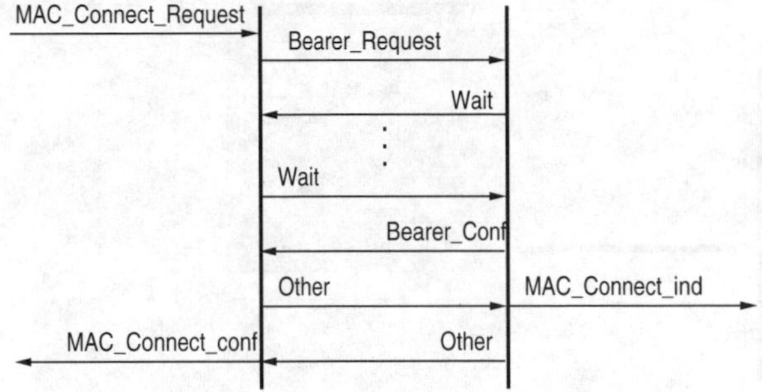

Figure 10.29: Connection setup of a basic connection

Bearer Setup If an MBC was established through connection setup in the MAC layer of the terminal, an attempt is made to set up the bearers required for the service. First the mobile station must be in the state `Idle Locked`, in other words, it must be synchronized with at least one base station in the cluster that will allow it to set up its connections. The MBC produces new instances of *traffic bearer control* (TBC) entities, which take over the required bearer setups. A bearer setup is based on the following process (see Figure 10.29):

1. The mobile station sends a Bearer_Request for a selected channel to a known base station. It uses the *first transmission* code within the 40 bits of the A-field.

2. The base station receives the request without any errors, and establishes a new TBC in its MAC layer.

3. The TBC in the base station requests the address of a supporting instance of an MBC entity from the lower-layer management entity.

4. If the base station is not yet ready to transmit a confirmation of the setup to the mobile station, it sends a Wait command. The mobile terminal receives this command and likewise responds with Wait.

5. After completion of the protocol process on the fixed side, the base station sends a confirmation (Bearer_Conf).

6. The mobile station receives the confirmation and directly sends Other in the next frame.

7. The base station receives the Other message, and likewise responds immediately with this primitive.

8. The mobile station receives the Other message, and the TBC informs its MBC that the setup has been successful (Bearer_Established_ind).

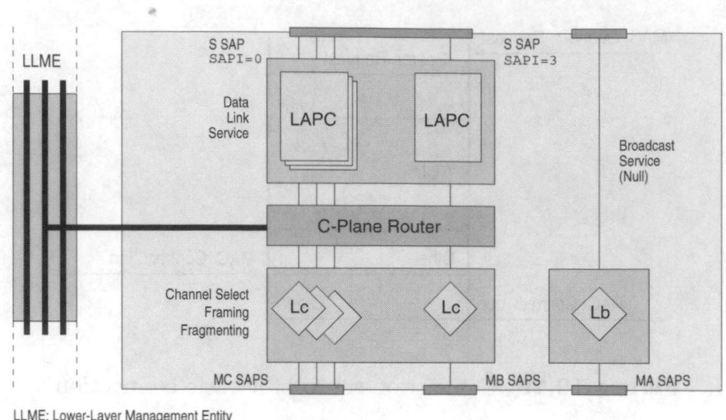

Figure 10.30: Control plane of the DLC layer

If an error occurs at any time during the connection setup, the setup attempt is discontinued and the setup procedure is then repeated. A maximum of 10 setup attempts (N200) not exceeding the connection setup timer (T200 = 3 s) is permitted.

10.4.3 Data Link Control Layer (DLC)

The DLC layer [8] is divided into a U-plane and a C-plane (see Figure 10.16). The C-plane constitutes the control layer for all internal DECT protocol processes relating to signalling (see Figure 10.30).

It provides two independent services:

- Data link services (LAPC+Lc)
- Broadcast services (Lb)

The U-plane controls the transmission of user data and offers the following services, which are accessible over its own service access points (see Figure 10.31):

- LU1 *TRansparent UnProtected service* (TRUP)
- LU2 *Frame RELay service* (FREL)
- LU3 *Frame SWItching service* (FSWI)
- LU4 *Forward Error Correction service* (FEC)
- LU5 *Basic RATe adaption service* (BRAT)
- LU6 *Secondary RATe adaption service* (SRAT)
- LU7 *64 kbit/s data bearer service*
- LU8-15 reserved for future services

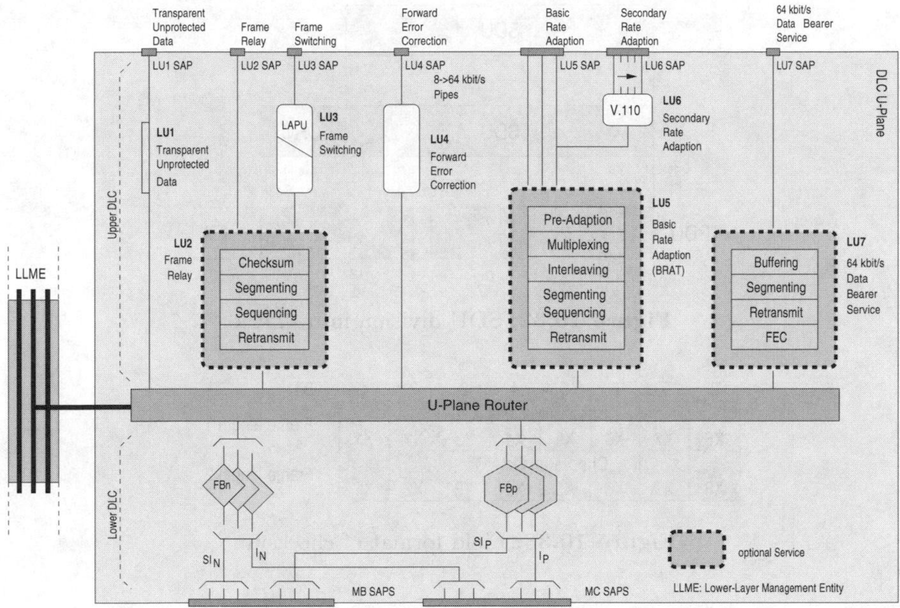

Figure 10.31: User plane of the DLC layer

- LU16 *ESCape* (ESC) for services other than standardized services

The specifications for services LU3 and LU4 have not yet been finalized. Subordinate frame structures that allow direct mapping of the MAC services are available for each of the LUx services (see Tables 10.2 and 10.3).

Each LUx service can occur in different classes of transmission:

Class 0 No LUx transmission repeat and no sequential control in the DLC layer. This means that error messages from the MAC layer are forwarded to the higher layers.

Class 1 No LUx transmission repeat; the DLC layer distributes the DLC frames in the correct sequence.

Class 2 Variable throughput with LUx transmission repeat.

Class 3 Fixed throughput with LUx transmission repeat.

The LUx services are described below.

10.4.3.1 Transparent Unprotected Service LU1

The LU1 service can only occur as a *Class 0* service, and is the most basic of the services. Although envisaged for voice connections, it can also be used for data services. No error protection is provided. FU1 is the subordinate frame service.

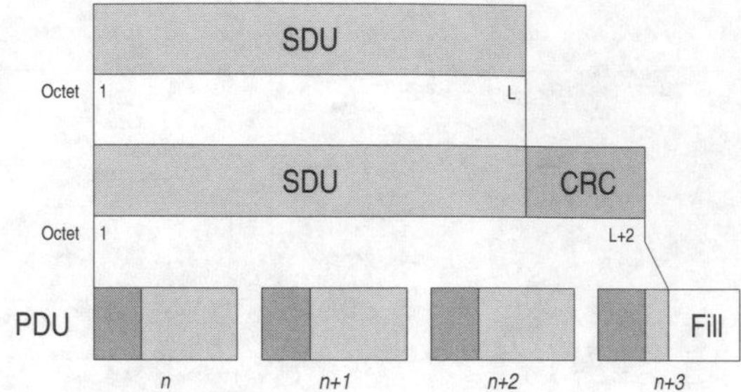

Figure 10.32: SDU division into PDUs

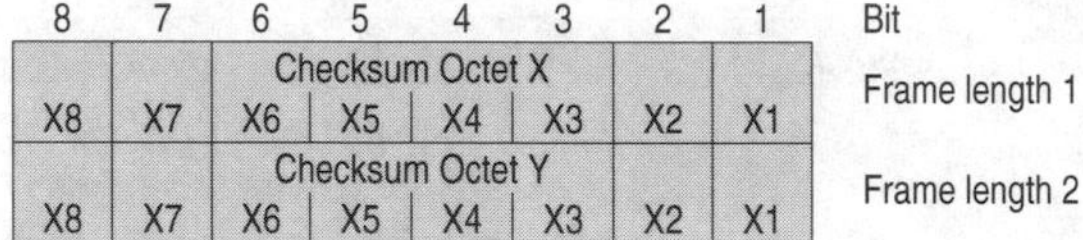

Figure 10.33: Field format of checksum

10.4.3.2 Frame Relay Service LU2

The LU2 service is a protected service for the transmission of data blocks (*frame relay*) that is accessed through the LU2-SAP. It operates in a generic field of user data, which is transferred in and out of the DLC U-plane as service data units (SDU).

LU2 supports the reliable transport of the SDUs and also respects the SDU boundaries. Three basic procedures are supported:

1. Addition of a checksum to each service data unit (SDU).
2. Distribution of the resultant data-SDU+checksum to data fields of protocol data units (PDU).
3. Peer-to-peer transmission of these PDUs.

A distinction is made between external and internal blocks (frames) (see Figure 10.32):

- SDUs are external blocks.
- PDUs are internal blocks.

Checksum procedures The 16-bit long checksum supports error detection for an entire SDU frame (see Figure 10.33).

A defective SDU is not retransmitted. The user can designate external error-handling protocols.

10.4.3.3 Frame Switching Service LU3

This service should be mentioned here, but the standardization for it has not been completed.

10.4.3.4 Service with Forward Error Correction LU4

The specifications for this service have also not yet been completed. It will include forward error correction and possibly also an ARQ entity.

10.4.3.5 Data Rate Adaptation Service LU5

The service LU5 is being introduced to support synchronous data streams with fixed data rates (64, 32, 16 and 8 kbit/s). A differentiation is made between protected and unprotected transmission. The protected service offers considerably more reliable transmission.

The difference lies mainly in the use of the different MAC services: I_P services are used for the protected service and I_N services for the unprotected one. This produces differences in the combination of MAC services for the data adaptation service, which justifies the separate treatment of protected and unprotected services. In general, the principle of these services can be described as follows. The service supports up to three independent data channels. There is only a limited combination of different possible data rates. Rate adaptation is carried out for the individual data connections. These data connections are then multiplexed onto a channel. An option is offered here for interleaving the data. The channel is segmented into individual frames that are provided with control information and transmitted over the radio channel. The entire process is performed in reverse on the receiver side so that up to three independent data channels are again available at the service access point.

10.4.3.6 Supplementary Data Rate Adaptation Service LU6

The LU6 service can only operate in combination with LU5. It offers data adaptation for terminals that comply with the V-series. This service converts data rates in accordance with CCITT Recommendation V.110. The rate adaptations executed in the LU5 service are presented in Table 10.4.

10.4.3.7 64 kbit/s Data Service LU7

This service was specifically developed to support the 64 kbit/s data service of ISDN. Since ISDN fixed networks have a lower bit-error ratio than DECT radio connections (with ISDN a BER $\leq 10^{-6}$ is expected; with DECT a handover is initiated at a BER $= 10^{-3}$), the aim of the protocol is the reduction of bit-error ratio. This is accomplished through a combination of forward error correction (FEC) with an RS code and ARQ procedures. An increased data rate and buffer storage are necessary in order to maintain the ISDN data

Table 10.4: Synchronous and asynchronous rate adaptation in the LU6 service

Input data	Output data rate [kbit/s]						
	Synchronous				Asynchronous		
rate [kbit/s]	8	16	32	64	8	16	32
0.05					×		
0.075					×		
0.11					×		
0.15					×		
0.2					×		
0.3					×		
0.6	×				×		
1.2	×				×		
2.4	×				×		
3.6					×		
4.8	×				×		
7.2		×				×	
9.6		×				×	
12			×				×
14.4			×				×
19.2			×				×
48				×			
56				×			

rate of 64 kbit/s. Realistically it is possible to transmit with a net data rate of 64 kbit/s or 72 kbit/s; the buffer storage produces a delay of 80 ms (see Figure 10.34).

The transmit buffer contains the last eight transmitted blocks. The receive buffer stores the arriving (maximum eight) blocks, thereby producing the time delay of 80 ms. If a defective block is received, it is rerequested and for eight transmit/receive cycles there is a possibility to receive the block again. A system of counters ensures that the block is written at the correct place in the receive buffer. A temporary change to the increased transmission rate of 72 kbit/s compensates for the time lost through the repeated transmission.

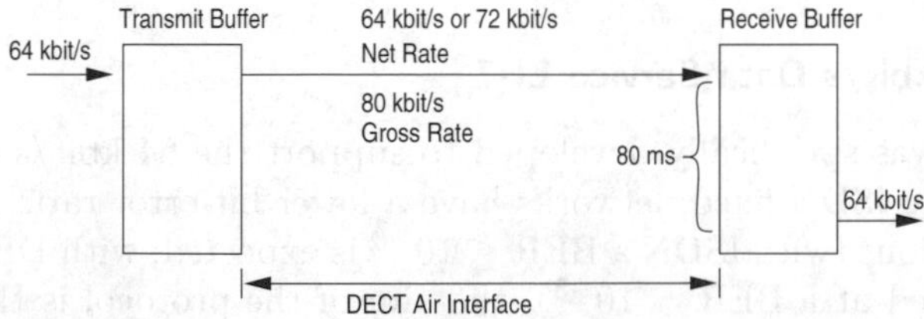

Figure 10.34: Implementation of ISDN data rate through error handling

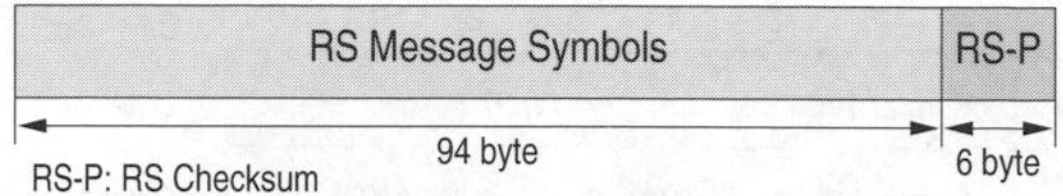

Figure 10.35: Content of a double slot B-field from the standpoint of BS coding

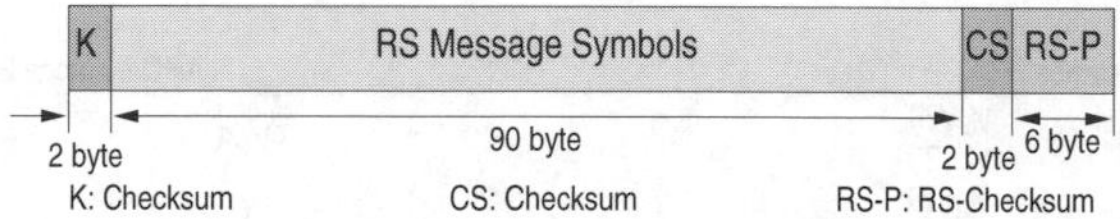

Figure 10.36: Content of an RS code word

Details of the algorithms as they are specified in the standard are discussed below.

The data fields The B-field of a double slot (see Figure 10.35) uses a (100,94) Reed–Solomon code with 94 bytes for the ARQ mechanism and 6 bytes for the checksum.

The ARQ mechanism divides the 94 message symbols into three groups (control, information and checksum, CS); see Figure 10.36.

The checksum field contains a checksum which the ARQ mechanism used to determine whether forward error correction was successful. The information field contains 90 bytes of user data when there is a 72 kbit/s transmission format; with a 64-kbit/s transmission format the last 10 bytes are filled with zeros (see Figures 10.37 and 10.38).

The control field contains the counters and format control parameters important for the ARQ mechanism (see Figure 10.39).

The send and receive buffer storage and status variables $V(O)$, $V(R)$, $V(S)$, $V(A)$ and $V_i(T)$ can be seen in Figure 10.40. These regulate the ARQ process by indicating which frame has been sent, received or acknowledged and from which storage position in the send storage read should take place or at which storage position in the receive storage write should take place. The param-

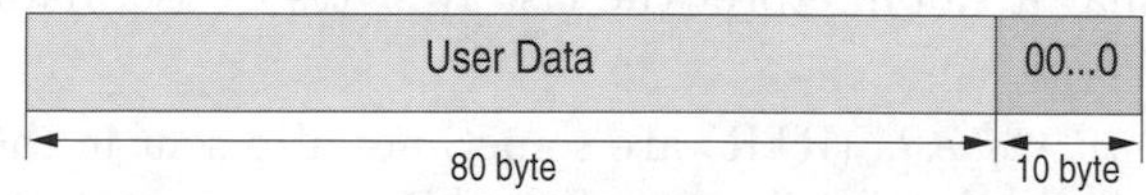

Figure 10.37: ARQ information field for 64 kbit/s user data rate

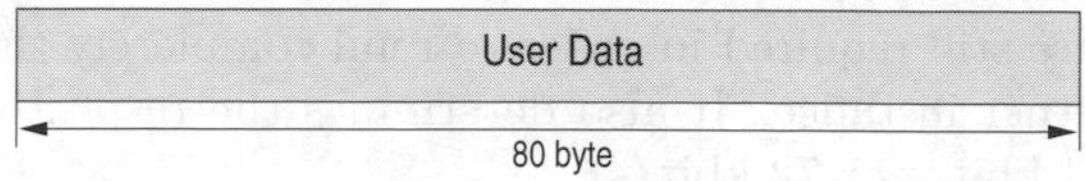

Figure 10.38: ARQ information field for 72 kbit/s user data rate

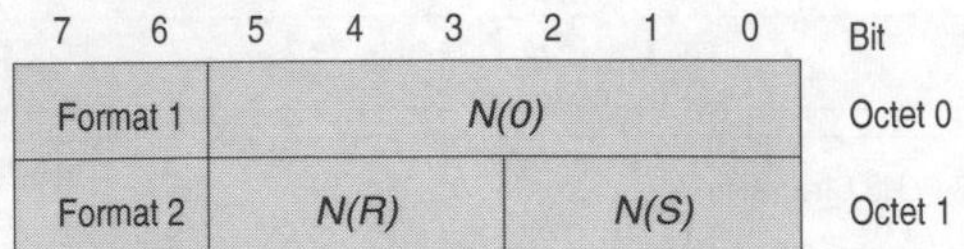

Figure 10.39: Content of ARQ control field

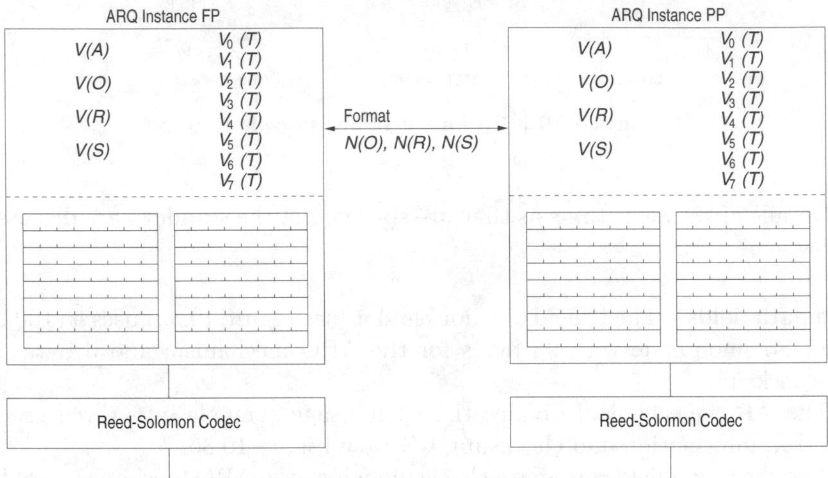

Figure 10.40: Components of the ARQ mechanism

eters $N(O)$, $N(R)$, $N(S)$ transmitted in the ARQ control field and formats derived from the status variables are used in the communication between ARQ entities.

ARQ Control Fields

Format field The transmission format is coded with a total of 4 bits that are divided over two 2-bit fields (format 1, format 2); see Figure 10.39. The receiver must be informed of which transmission format is being used so that it can interpret the last 10 bytes as user data or zeros (see Figure 10.37).

ReTransmit requests (RTR) are sometimes also sent in these fields. Table 10.5 shows the significance of the bits.

The Offset variable V(O) indicates how many additional bytes (in units of 10 byte) are still required in order to refill completely the receive buffer of the partner instance. It also determines the respective transmission format (64 kbit/s or 72 kbit/s).

When a frame is retransmitted, $V(O)$ is increased by 8 (corresponding to the deficit of 80 bytes in the occupation of the receive buffer to be

Table 10.5: Significance of bit values in ARQ format fields

	Format 1		Format 2		
Bit	8	7	8	7	Significance
	0	0	0	0	Format 64 kbit/s
	0	0	0	1	Format 64 kbit/s, RTR
	0	1	0	0	Format 72 kbit/s
	0	1	0	1	Format 72 kbit/s, RTR

compensated for). Retransmission is only permitted as long as $V(O) \leq 48$. If $V(O)$ is greater then retransmission is not possible because the data requested is no longer available in its entirety in the transmit buffer and, furthermore, would no longer have a chance of arriving in time at the partner entity. Instead, the next frame waiting to be sent in the transmit buffer is transmitted.

If $V(O) = 0$ then transmission is at the 64 kbit/s format and $V(O)$ is not altered. If $V(O) > 0$, then the format of 72 kbit/s is used. Because of the additional 10 bytes of user data also being transmitted, $V(O)$ can be reduced by 1 if transmission is successful, which corresponds to the reduction by 1/8 of the missing $8 \cdot 10$ bytes in the receive buffer of the partner entity.

The time variables V_n(T) indicate the position in the transmit buffer of the entity and in the receive buffer of the partner entity for the last eight transmitted blocks. When the transmission is repeated, the value $V_n(T)$ of the requested block is transmitted as $N(O)$ to inform the partner entity of the position in the receive buffer where it must write the block.

The offset number N(O) gives the position in the receive buffer in which the block is to be written. *N(O)* here corresponds to the distance of the last bit from the start of storage in units of 10 bytes. Figure 10.41 shows examples of two different *N(O)s*.

The send status variable V(S) is the sequence number for each new initial transmission of a block. It is increased modulo 8 by 1 and therefore can assume values between 0 and 7. $V(S)$ is the send sequence number $N(S)$ of the next block waiting to be transmitted.

The acknowledgement status variable V(A) identifies each block being acknowledged by the partner entity. Here $V(A) - 1$ is the send sequence number $N(S)$ of the last block to be acknowledged by the partner entity.

The send sequence number N(S) is the sequence number of the transmitted block. For an initial transmission it is set at $V(S)$, and for a repeated transmission it is set at $V(A)$. $N(S)$ is transmitted as a binary-coded number (3 bits).

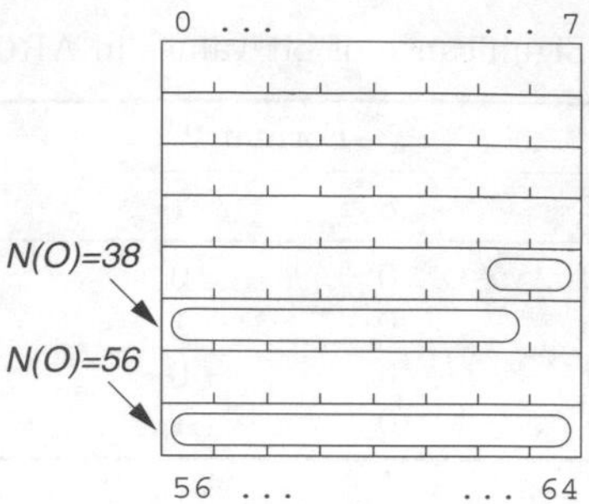

Figure 10.41: Relationship between $N(O)$ and the position of a block to be written in storage

The receive status variable V(R) contains the send sequence number $N(R)$ of the next block that is expected. If an error-free block with $N(S) = V(R)$ is received then $V(R)$ is increased by 1, and by an additional 1 for each following error-free block in the receive buffer.

The receive sequence number N(R) indicates to the receiver the send sequence number of the next expected block. $N(R)$ is set at $V(R)$ before a block is sent, and consequently acknowledges all frames including $N(R) - 1$ to the partner entity.

The ARQ information field contains 80 or 90 bytes of user data, depending on the transmission format (64 kbit/s or 72 kbit/s); with the format 64 kbit/s the last 10 bytes are zero.

The ARQ checksum field CS (16 bits) is calculated as follows (see Figure 10.36):

$$\text{checksum}(x) = \text{EK}(a(x) + b(x))$$

with EK( ) representing the ones complement (bit inversion) and

$$\begin{aligned} a(x) &= x^{736}(x^{15} + x^{14} + x^{13} + x^{12} + x^{11} + x^{10} + x^{9} + x^{8} + x^{7} + x^{6} \\ &\quad + x^{5} + x^{4} + x^{3} + x^{2} + x + 1) \text{mod}(x^{16} + x^{12} + x^{5} + 1) \end{aligned}$$

as well as

$$b(x) = x^{16}(ci(x))\text{mod}(x^{16} + x^{12} + x^{5} + 1) \tag{10.2}$$

Here $ci(x)$ is the polynomial representation in the $GF(2)$ of the combined ARQ control and information field.

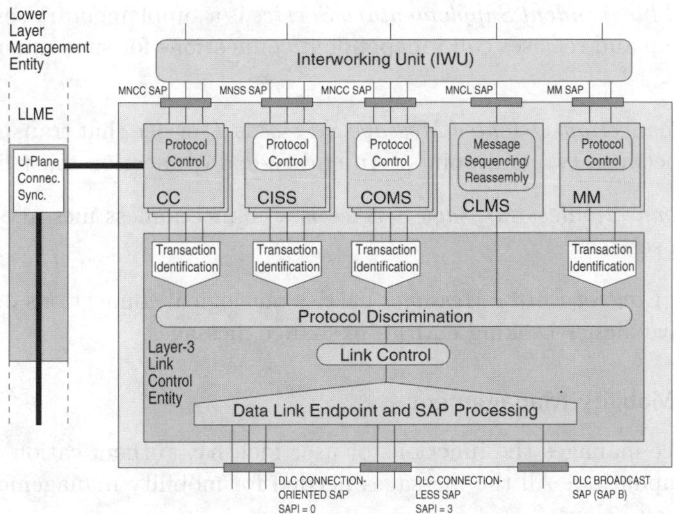

Figure 10.42: Services and protocols of the C-plane, layer 3 in DECT

10.4.3.8 ESCAPE Service LU16

This service permits the development of a separate user plane in which use of the different terminals is only possible if an additional agreement exists outside the standard. The LU16 service can use each subordinate DLC service and any transmission category. However, a protected MAC service must be selected. The transparent LU1 service is available for unprotected services. No service may be implemented that is already covered by a standard service.

10.4.4 Network Layer

The network layer (NWL) (see Figure 10.16) controls the connections between the mobile part (*Portable Radio Termination*, PT) and the network (*Fixed Radio Termination*, FT); see Figure 10.9. The U-plane in DECT has no tasks in the network layer, and forwards all data unprocessed in the vertical direction. The C-plane carries out signalling, and is responsible for controlling data exchange. Five protocols with corresponding tasks and based on the *link control entity* (LCE) are provided for this purpose (see Figure 10.42).

MM *Mobility Management* is responsible for all tasks required for the mobility of mobile stations. It processes authentication messages and encryption data relating to the location area.

CC *Call Control* is the entity used for establishing, managing and releasing connections.

CISS *Call Independent Supplementary Service* is a supplementary service that sets up and releases call-independent connections for supplementary services.

COMS *Connection-Oriented Message Service* is a service that transmits packets between two end points. An end-to-end connection must first exist.

CLMS *Connectionless Message Service* is a connectionless message-switched data service.

LCE *Link Control Entity Messages* carries out logical connections over which all five higher-ranking entities exchange messages.

10.4.4.1 Mobility Management

This service manages the functions of user identity, authentication and location area updating. All the messages needed for mobility management (MM) are discussed below.

Identity Messages

- Temporary_Identity_Assign
- Temporary_Identity_Assign_ack
- Temporary_Identity_Assign_Reject
- Identity_Request
- Identity_Reply

These messages can be used to query the FT about specific identification data of the PT and to assign a *Temporary Portable User Identity* (TPUI). The identification is always initiated by the FT. A Temporary_Identity_Assign message is sent by the FT to the PT to agree on a TPUI with the PT. After successful execution, the PT acknowledges the assignment with Temporary_Identity_Assign_ack. If an error occurs, the PT sends a Temporary_Identity_Assign_Reject. The FT uses Identity_Request to request an identity from the PT, to which the PT responds with an Identity_Reply.

Authentication Messages

- Authentication_Request
- Authentication_Reject
- Authentication_Reply

Authentication is carried out in two directions to guarantee to the network and to the PT that the PT or the network (as the case may be) has authorization. A new encryption parameter can be provided at the same time (see Sections 10.10 and 3.3). The procedure is initiated with an Authentication_Request message. The calculated reply is returned with Authentication_Reply. Should the authentication not succeed or another error occur, the partner entity is informed through Authentication_Reject, and is able to take the appropriate action.

Location Messages

- Locate_Request
- Locate_Accept
- Locate_Reject
- Detach

These messages inform the network of the current location area of a PT. This enables an incoming call to be forwarded quickly to the respective PT. The mobile station sends a Locate_Request message to the FT in order to establish a connection or to request a handover to another base station. The FT responds to a successful connection or handover with Locate_Accept, and to an error with Locate_Reject. If a mobile station is switched off, it informs the network with a Detach message.

Access Rights Messages

- Access_Rights_Request
- Access_Rights_Terminate_Request
- Access_Rights_Accept
- Access_Rights_Terminate_Accept
- Access_Rights_Reject
- Access_Rights_Terminate_Reject

With these messages the *International Portable User Identity* (IPUI) and the *Portable Access Rights Key* (PARK) can be stored or cleared again in the PT. There are the four PARK classes A to D. An Access_Rights_Request is sent by the PT to the FT to start the procedure. The FT responds with Access_Rights_Accept and the corresponding parameters, or with Access_Rights_Reject if there has been an error and the requested parameters cannot be transmitted. The Access_Rights_Terminate messages are the reverse of the messages described above.

Key Allocation Messages

- Key_Allocate

This message is used to replace an authentication key (AC) with a user authentication key (UAC).

Parameter Retrieval Messages

- MM_Info_Suggest
- MM_Info_Request
- MM_Info_Accept
- MM_Info_Reject

These messages supply information, e.g., for external handovers, which is transmitted with the help of existing connections. Using MM_Info_Suggest, the FT can propose to a PT that it carry out a handover or location update. The PT uses MM_Info_Request to request information from the FT. The FT supplies this information sending an MM_Info_Accept message. If the information requested by the PT cannot be sent, the FT informs the PT accordingly with MM_Info_Reject.

Ciphering Messages

- Cipher_Suggest
- Cipher_Request
- Cipher_Reject

Ciphering messages are used to transmit the ciphering parameters and to initiate or terminate the ciphering. The PT can make this recommendation to the FT with Cipher_Suggest. This then takes place through a Cipher_Request from the FT to the PT. In the event that the PT is unable to carry out the ciphering, it responds accordingly with Cipher_Reject.

10.4.4.2 Call Control (CC)

Call control is an instance of an entity responsible for call handling. It is used to set up, maintain and release calls.

Call Establishment Messages

- CC_Setup
- CC_Setup_ack
- CC_Info
- CC_Call_Proc
- CC_Alerting
- CC_Notify
- CC_Connect
- CC_Connect_ack

CC_Setup commences the call setup of call control, and can be initiated by both sides (FT and PT). Some of the setup parameters are sent at the same time; the other data is transmitted with CC_Info. A CC_Setup_ack message is optional and is only sent in the direction of a mobile station. CC_Call_Proc indicates to the mobile station that the call is being processed in the network. A CC_Alerting message indicates to the network that the mobile station is calling or to the mobile station that the user terminal being called is being paged. CC_Connect signals the connection to the partner layer of the U-plane and CC_Connect_ack acknowledges it. CC_Notify is a message that can be sent during an existing connection in order to convey a message (such as a Hold) to the called party.

Call Information Phase Messages

- CC_Info (see above)
- CC_Service_Change
- CC_Service_Accept
- CC_Service_Reject
- IWU_Info

A CC_Info message can be used to establish an end-to-end connection (more precisely from a PT to an IWU) for the exchange of information. With CC_Service_Change the communication parameters can be changed during the setup phase or during the active phase. A CC_Service_Accept message constitutes the corresponding acknowledgement and a CC_Service_Reject message a rejection of the parameter changes.

IWU_Information is used to transport `IWU Packet Elements` or `IWU to IWU Elements` when corresponding data is to be transmitted but cannot be attached to any other message.

Call-Related Supplementary Services

- Facility
- Hold
- Hold_ack
- Hold_Reject
- Retrieve
- Retrieve_ack
- Retrieve_Reject

Facility is used to request or acknowledge a supplementary service. The desired service is supplied within the messages with the appropriate parameters.

An MS can use Hold to transfer an existing connection to waiting status. The MSC acknowledges with Hold_ack. This waiting status can be rejected by the MSC with Hold_Reject. The Retrieve messages are the reverse of these commands. Retrieve is used by the MS to reactivate a call in waiting status. The MSC replies with Retrieve_ack or with Retrieve_Reject.

Call Release Messages

- CC_Release
- CC_Info
- CC_Release_com

CC_Release releases all U-plane and C-plane calls in the network layer. A CC_Release message is transferred to the link control entity (LCE), which then decides which procedure should be followed for releasing a call. CC_Release_com is the acknowledgement for the release of the call.

10.4.4.3 Call Independent Supplementary Services (CISS)

CISS Establishment Messages

- CISS_Register

CISS Information Phase Messages

- Facility

CISS Release Messages

- CISS_Release_com

Supplementary services consist of call-independent supplementary services (CISS) and call-related supplementary services (CRSS). CISS connections are call-independent. CISS_Register is used to set up new calls without end-to-end connections. CISS_Release_com indicates that the sender of a message is releasing the connection and also instructing the receiver to release the connection.

Facility is used to request or acknowledge a supplementary service. The desired service is provided within the message with the necessary parameters.

10.4.4.4 Connection-Oriented Message Service (COMS)

COMS Establishment Messages

- COMS_Setup
- COMS_Notify
- COMS_Connect

COMS Information Phase Messages

- COMS_Info
- COMS_ack

COMS Release Messages

- COMS_Release
- COMS_Release_com

COMS offers a point-to-point connection that can be used for transmitting packets. This connection can be established at any time using COMS_Setup. A successful connection is acknowledged with COMS_Connect. The connection phase uses a COMS_Info message to transmit the information of a COMS connection. A COMS_ack message acknowledges that one or more COMS_Info messages have been correctly received. The COMS connection is released through a COMS_Release command and acknowledged through COMS_Release_com.

10.4.4.5 Connectionless Message Services (CLMS)

Call Information Phase Messages

- CLMS_Variable
- CLMS_Fixed

CLMS can be used to transmit data when an end-to-end connection does not exist. The FT uses CLMS_Fixed and CLMS_Variable to transmit applications-related data to the PTs. In contrast to all the others, which have the S-format (see Section 10.4.4.8), a CLMS_Fixed message is in the B-format.

10.4.4.6 Link Control Entity Messages (LCE)

LCE Establishment Messages

- LCE_Request_Page
- LCE_Page_Response
- LCE_Page_Reject

The LCE entity in the physical layer is below the five entities (MM, CC, CISS, COMS, CLMS) above it, and is used to set up connections to the partner layer with LCE_Request_Page. The positive response to a connection setup request is LCE_Page_Response; the negative one is LCE_Page_Reject. Like CLMS_Fixed, LCE_Request_Page is a B-format message.

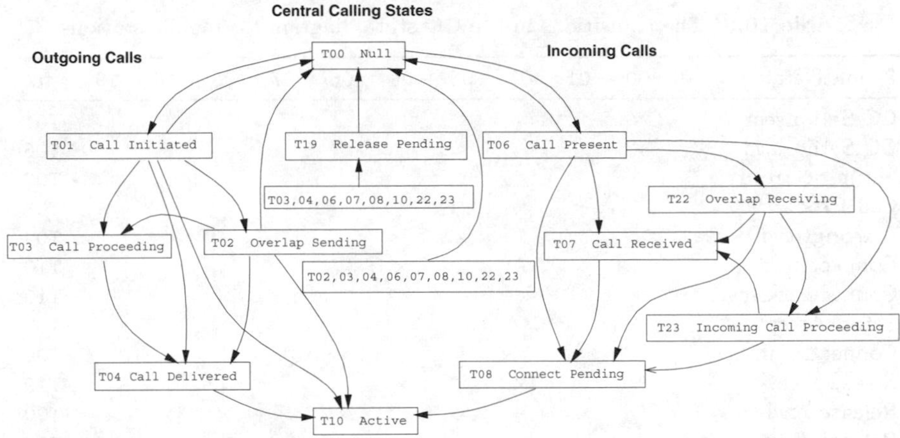

Figure 10.43: State diagram for a mobile station

10.4.4.7 State Diagrams for Connection Control

Along with the operating states of the mobile and base station (see Figures 10.20 and 10.21), each protocol can be represented by a state automaton. All the possible states and transitions of a system are listed in a corresponding state diagram. New states are achieved whereby an existing state is transferred through transition into another state. Associations exist between the states and transitions in a way that some states cannot be directly transferred into any other state.

These states can be generated for each protocol, and are of particular interest here for call control because it too uses the protocols below it and establishes complete connections. It is possible for more than one connection to exist between a mobile station and the network, but in this context a simple connection is considered only. Figure 10.43 shows the states of a base station (FT) and Figure 10.44 the states of a mobile station (PT).

State diagram for a mobile station (PT) Table 10.6 presents a summary of the most important messages and their effect on the state diagram (see Figure 10.43).

The states are divided into four groups.

Central calling states (PT)

- `T-00 Zero` No call exists.
- `T-19 Release Pending` The mobile station has sent a Release message to the base station but has not yet received a reply.
- `T-10 Active` A call (outgoing or incoming) exists and a connection has been established.

Table 10.6: The transitions in the CC state diagram for mobile stations

From T-state	00	01	02	03	04	06	07	08	10	19	to
CC_Setup_sent	×										T01
CC_Setup_rcvd	×										T06
Setup_ack_rcvd		×									T02
Call_Proc._rcvd		×	×								T03
Alerting_rcvd		×	×	×							T04
Connect_rcvd		×	×	×	×						T10
Connect_ack_rcvd								×			T10
Setup_Accept						×					T07
Connect_sent						×	×				T08
Release_sent			×	×	×			×	×		T19
Release_rcvd				×	×	×	×	×	×		T00
Release_Compl._rcvd		×	×							×	T00
Release_Compl._sent						×					T00

States for outgoing call (PT)

- `T-01 Call Initiated` The mobile station has sent a Setup message to the base station in order to initiate a call.
- `T-02 Overlap Sending` Another outgoing connection is being established via *overlap sending*.
- `T-03 Call Proceeding` The base station has acknowledged receipt of the Setup message.
- `T-04 Call Delivered` The mobile station receives an Alerting message informing it that the terminal it is calling is being paged.

States for incoming calls (PT)

- `T-06 Call Present` The mobile station has received a Setup message but has not yet responded to it.
- `T-07 Call Received` The mobile station indicates to the base station through Alerting that it is ringing.
- `T-08 Connect Pending` The user of the mobile station has taken the call, but the mobile station is waiting for an acknowledgement that the U-plane connection is being established.

Optional calling states (PT)

- `T-22 Overlap Receiving` Another incoming call is being set up in an overlapped state.

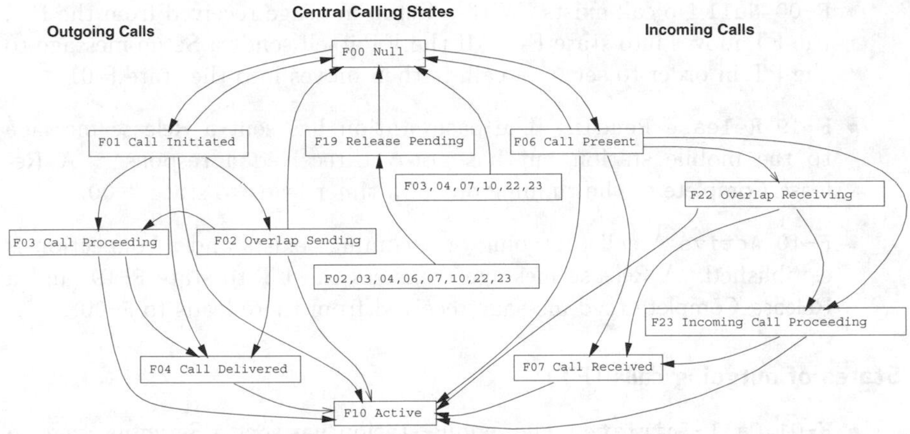

Figure 10.44: State diagram for a base station

Table 10.7: The transitions of a CC state diagram for base stations

From F-state	00	01	02	03	04	06	07	10	19	Final state
Setup_sent	×									F06
Setup_rcvd	×									F01
Setup_ack_sent		×								F02
Call_Proceeding_sent		×	×							F03
Alerting_sent		×	×	×						F04
Alerting_rcvd						×				F07
Connect_sent		×	×	×	×					F10
Connect_rcvd						×	×			F10
Release_sent						×	×	×		F19
Release_rcvd		×	×	×	×	×	×	×		F00
Release_Complete_rcvd						×			×	F00
Release_Complete_sent		×	×							F00

- `T-23 Incoming Call Proceeding` The mobile station acknowledges receipt of the setup information from the base station.

State diagram for a base station (FT) Table 10.7 presents a summary of the most important messages and their effect on the state diagram (see Figure 10.44).

As with mobile stations, the states are divided into four groups.

Central calling states (FT)

- `F-00 Null` No call exists. With a Setup message received from the PT, the FT moves into state F-06. If the FT itself sends a Setup message to the PT in order to set up a call, it then moves into the state F-01.
- `F-19 Release Pending` The base station has sent a Release message to the mobile station but has not yet received a response. A Release_Complete or the running out of a timer leads to state `F-00`.
- `F-10 Active` A call (outgoing or incoming) exists and a connection is established. A Release message transfers the FT to state `F-19`, and a Release_Complete_rcvd message received from there leads to `F-00`.

States of outgoing calls (FT)

- `F-01 Call-Initiated` The mobile station has sent a Setup message to the base station in order to establish a connection. The base station has not yet responded to it.
- `F-02 Overlap Sending` An outgoing call is being set up overlapping another.
- `F-03 Call Proceeding` The base station acknowledges that the Setup message from the mobile station has been received.
- `F-04 Call Delivered` The base station notifies the mobile station that the user terminal called is ringing.

States of incoming calls (FT)

- `F-06 Call Present` The base station has sent a Setup message, but the mobile station has not yet responded to it.
- `F-07 Call Received` The base station receives a message that the mobile station is ringing but the incoming call from the user is not yet being picked up.

Optional calling states (FT)

- `F-22 Overlap Receiving` An incoming call is set up overlapping with others.
- `F-23 Incoming Call Proceeding` The base station receives an acknowledgement that the setup information from the mobile station has been received.

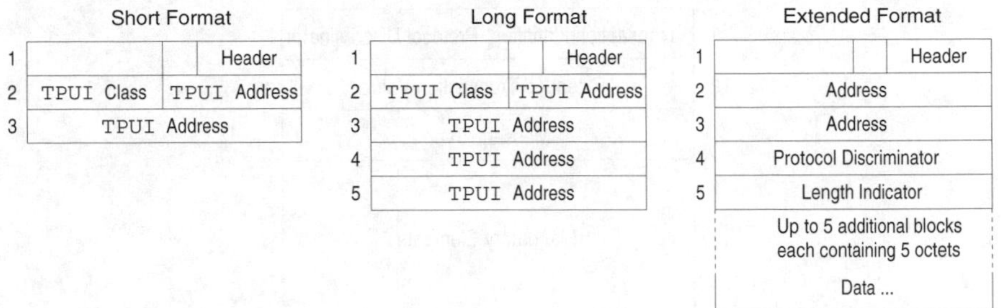

Figure 10.45: The structure of B-format messages

10.4.4.8 Setup of Signalling Messages

A message is a block that consists of a number of bits and contains related data. Each eight bits are combined to form an octet organized from an MSB (*most significant bit*) to an LSB (*least significant bit*), and the bits are numbered from 8 (MSB) to 1 (LSB). Octets form the elements from which a message is created, with the structure the same for all messages and a differentiation made in DECT between a B-format and an S-format.

B-format messages These may only be sent by the LCE or the CLMS protocol.

- LCE_Request_Page
- CLMS_Fixed

All messages have a fixed length to make it easier for them to be processed at the lower levels (MAC level). Different formats are available for this (see Figure 10.45):

- Short format (3 octets)
- Long format (5 octets)
- Extended format (5, 10, 15, 20, 25 or 30 octets)

LCE_Request_Page uses either the short or the long format. In both cases they are structured in the same way.

Bits 1 to 3 in the first octet contain the header. Bit 4 is a flag for the `TPUI` address. In the long format the length of the `TPUI` address is indicated in bit 4. Bits 5 to 8 have no significance. Depending on the information in the first octet, the `IPUI` address begins in the second octet.

CLMS_Fixed uses either the long or the extended format. The extended format has a header consisting of five octets. In addition to the target address, the length of the data which belongs to this message is provided. This data is combined into blocks, each consisting of five octets.

S-format messages Each S-format message consists of the following parts (see Figure 10.46):

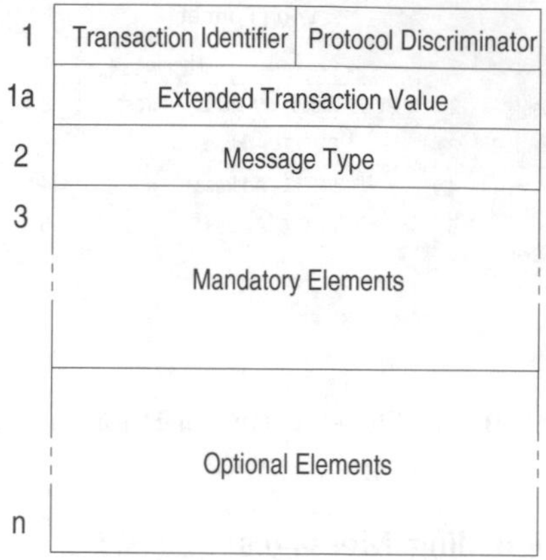

Figure 10.46: Structure of S-format messages

- Transaction identifier
- Protocol discriminator
- Message type
- Mandatory elements
- Optional elements

The first three elements are mandatory for each message, with the two elements protocol discriminator and transaction identifier contained in the first octet, each with four bits, and the message type in the second octet. The last two elements are specifically allocated to individual message types, and can be of different lengths.

The protocol discriminator occupies bits 1 to 4, and describes the type of connection (MM, CC, CISS, COMS, CLMS, LCE) to which the message belongs.

The transaction identifier indicates in bit 8 the direction in which the message is being transmitted. A 0 signifies the direction from the calling to the called user; a 1 the other direction. A transmission value is assigned to bits 5 to 7. If it is the value 1 1 1, the following octet is treated as an expanded transmission value (octet 1a).

The message type is identified in the subsequent octet: bits 1 to 7. This is where the code is for all MM, CC, CISS, COMS, CLMS and LCE messages; bit 8 is always 0.

The octets 3 to n can contain mandatory and optional elements. Some elements have a firmly fixed length (*Fixed Length Information Elements*) and often a variable length in the second octet (*Variable Length Information Elements*).

Locate_Request is a typical example of an MM message (see Figure 10.47). A typical CC message is CC_Setup (see Figure 10.48).

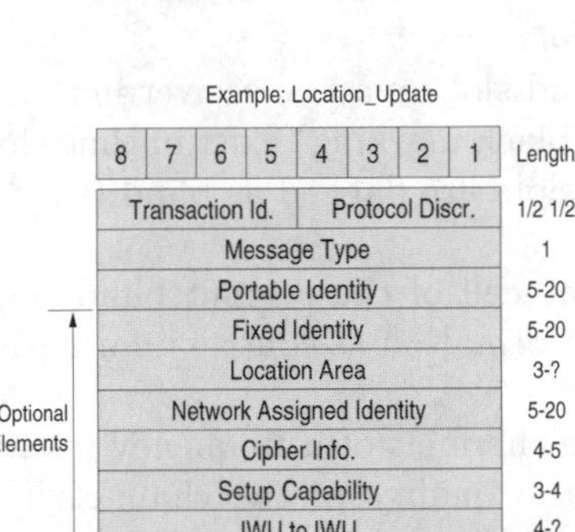

Figure 10.47: MM message structure based on the example of Locate_Update

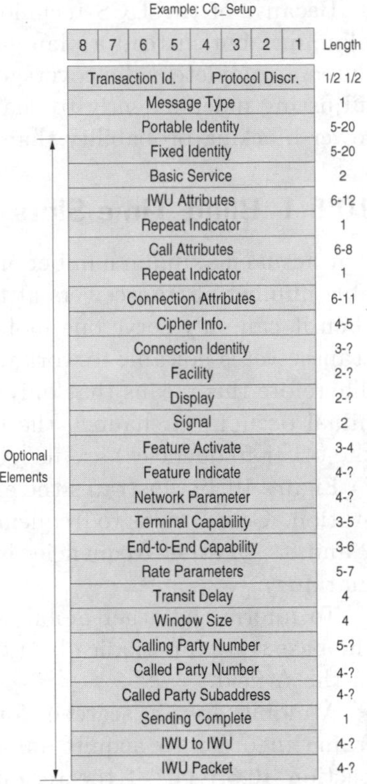

Figure 10.48: CC message structure based on the example CC_Setup

10.5 Dynamic Channel Selection

The DECT system is designed to cope with high voice and data traffic loads. Since the load on the cells is usually distributed unevenly and the peak loads are extremely variable timewise and geographically, the process of *Dynamic Channel Selection* (DCS) is applied. This means that basically the entire frequency spectrum with all (120) channels is available in each cell, and a mobile station is able to select a suitable channel for itself.

In cellular mobile radio systems channel assignments are made in accordance with a fixed plan in which certain frequencies are allocated to the base stations on the basis of anticipated traffic volumes. This technique, which is referred to as *Fixed Channel Allocation* (FCA), requires very careful cell planning because it is difficult to make any changes once the installation has been implemented. The system makes it difficult to absorb short-term dynamic load changes in individual cells.

Because of the DCS technique, DECT systems do not require frequency planning, but instead a planning of the base station locations, just to guarantee a complete radio coverage of the service area. The system is capable of adjusting independently to changes in traffic load, and therefore can offer a lower blocking probability than FCA networks (see Section 10.13).

10.5.1 Blind Time Slots

The actual maximum number of channels in a DECT cell depends on the available number of transceivers at the base station. If it has only one transceiver then it can only serve one mobile station within each time slot, because the transceiver is not able to work with two different frequencies at the same time. Therefore this means that only 12 duplex channels can be operated. If a terminal occupies a channel, the remaining channels in the same time slot (on the other frequencies) are flagged as *blind slots.*

Figure 10.49 illustrates the effects of a blind slot on a transceiver in a base station. Connections to frequencies 5 and 1 (black marking) exist in time slots 2 and 8. All other frequencies in these time slots are flagged as blind (shaded marking).

To inform the other mobile stations in the cell of the current blind slots, the base station periodically transmits this information as control information in the A-field.

A mobile station searches for the possible channels of a frequency in each frame in order to acquire information on the quality of the channels (see Section 10.5.1.1). If the terminal is occupying a channel for transmission, it can no longer monitor the channels of other frequencies in this time slot. Because of the more cost-effective and simpler hardware used, it is usually not in a position to check the quality of the time slots that appear before and after the selected traffic channel. Likewise these blocked channels are no longer available for use by the mobile station (see Figure 10.50).

10.5.1.1 Channel Selection by Mobile Stations

The base station and the mobile station must select a physical channel before the initial transmission of information by a bearer. First the mobile station undertakes a measurement of the signal levels of the individual base stations and assigns itself to the one with the strongest level. The mobile station makes up to three attempts to set up a connection before it moves to the base station with the next-highest signal level.

The signal level (*Radio Signal Strength Indicator*, RSSI) determines the suitability of a channel. Relative to the reference level of 1 mW, it indicates the measured signalling power on the channel. Depending on the type of bearer requested, a number of different RSSI measurements are relevant to the decision.

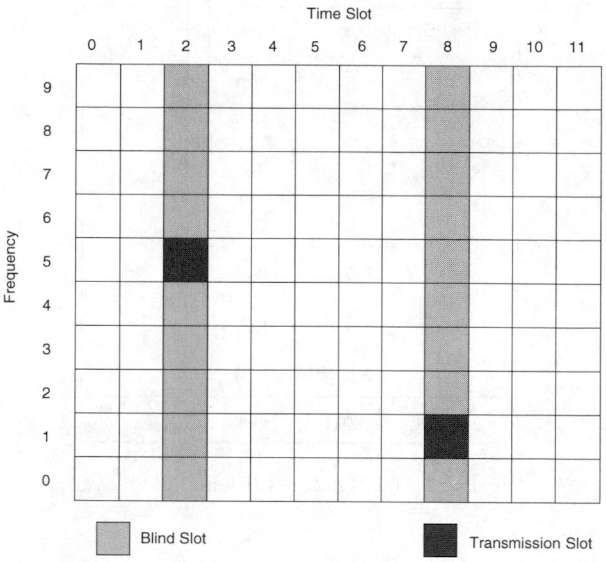

Figure 10.49: Blind slots in a base station with a transceiver operating two connections

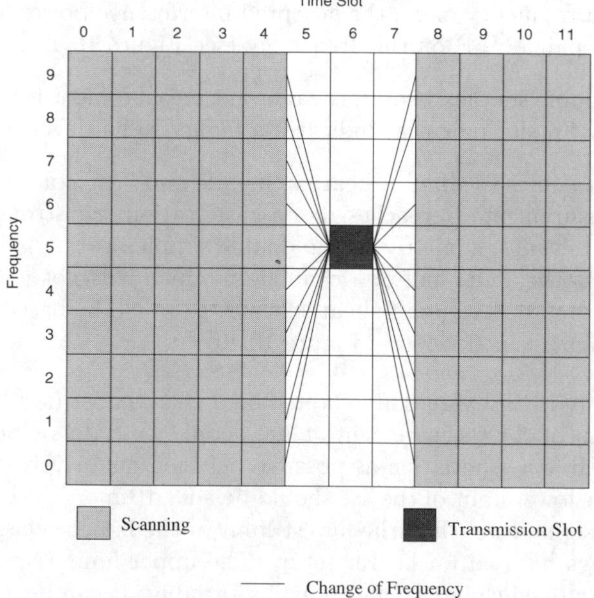

Figure 10.50: Channels not usable because of time required for PT to switch from communication to channel measurement

RSSI	Δ RSSI	Band	Comment
> max dBm	∞	busy	Busy, don't try
	≤ 6 dB	$b(n)$	Possible Candidates
⋮	⋮	⋮	
	≤ 6 dB	$b(4)$	
	≤ 6 dB	$b(3)$	
	≤ 6 dB	$b(2)$	
	≤ 6 dB	$b(1)$	
< min dB	∞	quiet	Quiet, always allowed

Figure 10.51: Channel list in MAC layer

- For a duplex bearer, the receiving channel in a terminal, in other words the downlink, is critical for the evaluation of a slot pair. As an example, using the RSSI measurement the mobile station establishes, e.g., in slot 2 with a frequency of f_x, the accepted interference power level for using the slot pair (2/14) on this frequency (see Figure 10.49).

- For a double simplex bearer, the relevant measurement is the one in that section of a slot pair that indicates a higher signal level.

- When it comes to simplex bearers, mobile and base stations have different measurement approaches. For a base station, the stronger of the two slots is relevant, similarly to the double simplex case. A mobile station evaluates the TDD half of a channel in which it is not transmitting its simplex bearer. An uplink bearer in slot 14 would be based on a positive measurement in slot 2 (see Figure 10.49).

The measured RSSI values are entered on a channel list (see Figure 10.51). The resolution of the measurement values should be 6 dB or better in order to achieve as fine a graduation as possible between qualitively good and bad channels. The lower limit of the list should be −93 dBm or lower. All channels with a lower signal level than the lowest limit of the list are classified as *quiet* and can always be used for bearer setup. The upper limit (*busy*) defines the area of the occupied channels, and remains variable. It can be adapted to the respective environment in which the DECT system is being operated and is normally at −33 dBm. Under no circumstances can the channels above this level be used in setup.

The channel list shown in Figure 10.51 has a granularity of 6 dB between the quiet and busy areas to allow channels to be entered appropriately in one of n rows. The following procedure is followed for the selection of a channel from one of the rows $b(1)$, $b(2) \ldots b(n)$.

When the quietest channel is to be selected, the *quiet* area must be checked to ensure that a sufficient number of channels is available. If this is the case, then no channel is selected from the area b directly above it. On the other hand, if no *quiet* entry is available, a switch is made to the $b(1)$ area. If a channel is not available in this area either, then the $b(2)$ area is checked. In general, a transition is made from $b(x)$ to $b(x+1)$. If no channel can be found, the channel selection process is terminated and the base station is identified as *busy*.

Channel selection must be completed within two seconds (`T210` = 2 s). A maximum of 10 attempts (`N202` = 10) can be carried out during this time. If more than one bearer is being set up for a connection, up to five times as many more attempts are allowed.

Channels other than those with strong signals can also be entered as *busy* on an expanded channel list. One reason for this restriction could be that a transceiver unit is not able to support all frequencies. Blind slots can also be included in the busy area to prevent the selection of these channels.

10.5.1.2 Power Control

The standard provides the option of two levels of power control, with 250 mW as the maximum permitted power.

10.5.2 Channel Selection and the Near/Far Effect

It is easy to understand that with the DCS technique under high load in an area with many base stations the service areas of the base stations are breathing. The reason for this is that under high load the near/far effect dominates the decision as to which terminal will be able to open a new connection, with preference given to the terminals close to a base station. More distant terminals might even be pushed out of the system when a terminal near to a base station decides to set up a new connection; see also Section 5.3.3.

When a connection is first being established, the mobile station (*portable part*, PP) attaches itself to the base station with the highest signal level. A condition for the success of the channel setup is an adequate C/I in one of the 120 channels. A mobile station that is located in the close proximity of its base station will receive a higher signal level than one that is located further away. Consequently, the likelihood of a successful connection setup is greater with the mobile station that is closer to the base station than with the one further away. This is referred to as the near/far effect, which is clarified further in Figure 10.52 [22].

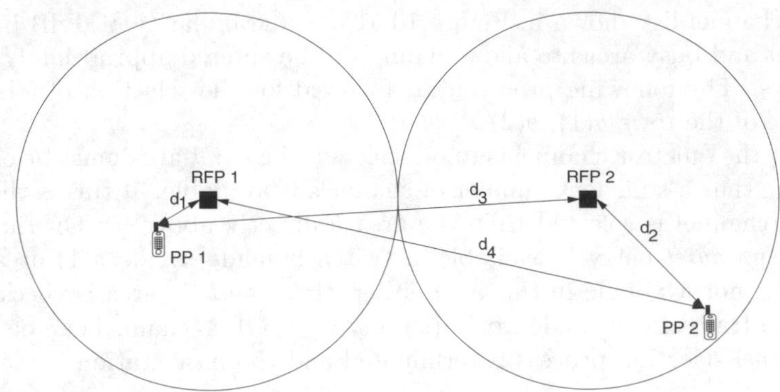

Figure 10.52: Channel displacement on the basis of the near/far effect

If a connection has been successfully established from the nearby mobile station (PP1) to its base station (*radio fixed part*, RFP1), it is possible that another mobile station (PP2) located further away and occupying the same channel to a different base station (RFP2) will be displaced from the channel due to high interference.

When two connections exist on one channel, the carrier-to-interference ratio C/I at the location of the nearest mobile station PP1 is

$$\left(\frac{C}{I}\right)_{PP_1} = \frac{{d_3}^{\gamma}}{{d_1}^{\gamma}}$$

with γ representing the attenuation coefficient of the radio propagation. At the location of mobile station PP2 that is located further away the following applies:

$$\left(\frac{C}{I}\right)_{PP_2} = \frac{{d_4}^{\gamma}}{{d_2}^{\gamma}}$$

If $d_1 \ll d_2$ and $d_3 \simeq d_4$ then it follows that:

$$\left(\frac{C}{I}\right)_{PP_1} > \left(\frac{C}{I}\right)_{PP_2} \tag{10.3}$$

In a DECT system a specific carrier-to-interference ratio C/I of, e.g., x dB, is required in order to set up and operate a connection on a channel. In the case considered it is possible that the nearer mobile station PP1 is measuring a carrier-to-interference ratio of more than x dB during its connection setup, and consequently will be successful in setting up a connection on this channel. This means that the carrier-to-interference ratio for the more remote mobile station (PP2) will will then fall below the required value of x dB, i.e., this mobile station will be displaced from its channel. As a result, it is even possible that the PP2 will be cut off from its connection if it is not able to find an alternative channel.

10.6 Speech Coding Using ADPCM

The 300–3400 Hz frequency range is usually used for the transmission of voice signals over telephone lines. Therefore the voice base frequency, which is between 120 and 160 Hz for men and between 220 and 330 Hz for women and children, is not transmitted. The hearing range for a person is between 16 Hz and 16 kHz. The highest perceptible frequency decreases from 20 kHz for a child to 10 kHz as one ages. The covered dynamic area is 130 dB. Despite the band limitations of a voice signal in the 300–3400 Hz frequency range, the level of syllable intelligibility is 91 %. Phrase intelligibility is even at 99 %.

In DECT systems *Adaptive Differential Pulse Code Modulation* (ADPCM) is employed in the digital signal coding of voice signals [10]. The assumption is that the analogue signal $x(t)$ has already been converted into the digital scanning sequence $\{x(t = k\,T)\}$. An analogue/digital conversion of the voice signal is required in order to produce this scanning sequence. First a band limitation of the signal to $f \leq f_g$ is carried out using an analogue filter.

The frequency f_g then becomes the cutoff frequency of the filter. This is followed by sampling with a frequency $f_A = 1/T \geq 2f_g$.

The objective of ADPCM is to reduce the number of binary decisions required in the quantization without reducing the voice quality. This can be done because of the correlation of the successive sampling values. There is a relatively small change in the signal $x(k)$ from sampling value to sampling value. Successive values $x(k)$ are dependent on one another. With this dependency of the signal values being exploited, the prediction is that the point in time $k = k_0$ is predicted from n past sample values.

ADPCM was standardized as ITU-T Rec. G.721 in 1984. The coding is based on sampling with $f_A = 8$ kHz and a quantization of the residual signal with $w = 4$ bits into 2^w levels. This produces the data rate $f_A w = 8\text{ kHz} \cdot 4 = 32$ kbit/s.

The voice quality is only negligibly worse than with logarithmically companded *Pulse Code Modulation* (PCM). Companding is the combination of a compressor and an expander — two characteristics used to produce an uneven quantization.

With PCM a distinction is made between an A-characteristic (used in Europe), which is approximated on the basis of 13 segments, and a µ-characteristic (used in the USA and in Japan). The compressor characteristics enable an irregular quantization to maintain a constant relative quantization error. This involves quantizing low signal values more finely than higher ones.

The quantizer adaptation is halted for stationary signals, thereby making the system also suitable for modem signals.

The achievable voice quality with 32 kbit/s ADPCM is comparable to the quality of 64 kbit/s-PCM (A-characteristic, µ-characteristic).

10.7 Handover

A handover takes place when a connection is switched to another base station because of poor quality, or a channel change within a cell is effected by the same base station. Because of the small cell sizes in microcellular systems it is assumed that roaming mobile stations will initiate a large number of handovers. Because a handover involves a great deal of signalling and the occupation of additional channels, there is an interest in minimizing the number of needed handovers.

The *Radio Resource Management* (RR) of a DECT system provides a decentral mobile-controlled handover (MCHO) algorithm that decides whether and when handover is necessary. *Seamless handover* (SH) in which the mobile station does not leave its current channel until a new one has already been set up is usually possible. In contrast to a *non-seamless handover*, the user usually does not notice that a channel or cell change has taken place.

There are two different physical forms of handover (see Figure 10.53):

Intracell Handover This occurs when the time and/or frequency channel within a cell changes. When this happens, the selected base station is retained.

Intercell Handover If the original base station is released when a channel change takes place and a new channel is set up by a new base station then this is referred to as an intercell handover. When this happens, a new frequency and/or slot selection can be made in the new cell.

In addition to the differentiation between two types of handover because of physical characteristics, it is also possible to define two logical steps in a channel change:

Internal Handover If the handover takes place within a closed DECT system (*fixed part*, see Section 10.2.2), this is an internal channel change or handover. In the protocol, this handover technically occurs either in the MAC layer or in the DLC layer.

External Handover A changeover between two independent DECT systems (*Fixed Part*) is referred to as an external handover. This is a high-level handover that is executed in the mobility management of the network layer. A short-term loss of service (*non-seamless handover*) can occur with this handover.

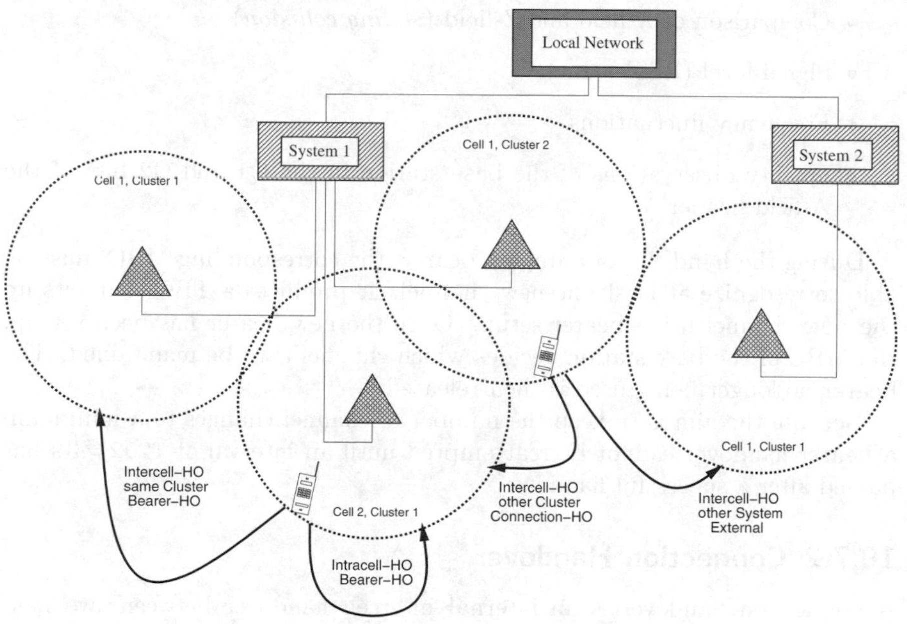

Figure 10.53: Physical and logical DECT handovers

10.7.1 Bearer Handover

A bearer handover is an internal handover, and requires fewer protocol efforts than any other handover process. It takes place within the MAC layer, and only affects certain slots. Along with an intracell handover, an intercell handover can also take place at the MAC level, as long as the t wo cells involved are within one cluster (see Figure 10.53).

While the process is taking place, the current bearer is retained until the new one has been completely set up. There are briefly t wo bearers at the same time jointly supporting the same connection. Bearer handovers are initiated on the basis of quality tests carried out by the MAC layer of the mobile device, taking the following into account:

- A-field: 16-bit R-CRC
- B-field (unprotected): 4-bit X-CRC
- Synchronization impulse
- Information for connection identity
- B-field (protected): four 16-bit R-CRCs in the field parts

- Comparison of X-field and Z-field (*sliding collision*)
- Signal level (RSSI value)
- Frequency fluctuations
- Quality observations of the base station in the Q1 and Q2 bits of the A-field header

During the handover of a duplex bearer, the corresponding MBC must be able to recognize at least one new channel. It produces a TBC that sets up the new channel using bearer setup. Once the new bearer has been set up, the MBC of the base station decides which channel is to be maintained. The bearer no longer being used is then released.

Because the aim is to keep the number of channel changes to a minimum, a bearer handover cannot be reattempted until an interval of T202 = 3 s has passed after a successful handover.

10.7.2 Connection Handover

A connection handover is an internal intercell handover between two base stations that belong to different clusters. It is executed in the DLC layer on the basis of its protocol. The service for the network layer is maintained during this process.

A connection handover occurs for one of the following reasons:

- Involuntary handover due to loss of service in the MAC layer
- Voluntary handover because of the poor quality of a connection

The LLME triggers a connection handover in which a new MAC connection is established. Similarly to a bearer handover, two MAC connections are operated during a voluntary handover process in order to guarantee a *seamless handover*. A seamless change of channel is not possible in an involuntary handover, because the old connection no longer exists at the start of the handover. The original type of connection (basic, advanced) is retained during the change in connection.

10.7.3 External Handover

The change from one DECT system to another is controlled by the mobility management of the network layer. This high-level handover from one network node to another should only be carried out in special cases, because of the amount of signalling required. A common management entity that coordinates the handover process must exist above the network layer for the execution of the change. During the transition, important processes such as encryption must be stopped and reactivated in the new system. There is no guarantee of a seamless handover, and it is possible that a connection will be broken off.

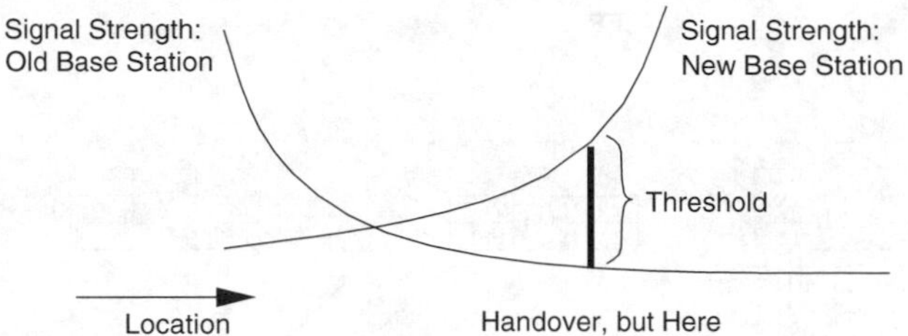

Figure 10.54: Hysteresis curve in an RSSI handover

10.7.4 Handover Criteria

Loss of channel quality in a DECT system is usually attributable to one of t wo reasons. One is that a user's mobility can cause a reduction in the signal level of the base station when the terminal moves out of the supply area. Another reason is that the somewhat random reuse distance of a channel results in more or less heavy interference. A handover is initiated early in order to counter these t wo effects. Both situations are described in detail below.

10.7.4.1 RSSI Handover

There can be several reasons for the decrease in a base station's signal level in the location where a mobile station is operating. The main reason is the change of location caused by the mobility of the terminal. If a mobile station begins to roam too far away from its immediate base station or moves into radio shadowing, there is the chance of a sharp drop in signal level described by the *Radio Signal Strength Indicator* (RSSI). A sudden loss of *line-of-sight* (LOS) to the base station can cause a loss of 15–30 dB.

If a mobile station roams beyond the logical cell boundary into a neighbouring cell, the signal strength of its own base station decreases more and more, while that of the neighbouring station increases. A change in cell is desirable in this case in order to improve the quality of the connection. The measured RSSI values are transmitted over a time window in order to filter out the effects of short-term signal loss. An intercell handover should not be initiated until the signal level averaged for the neighbouring station is higher by a certain threshold value than that for the terminal's own base station (see Figure 10.54), in order to prevent frequent intercell handovers on the cell boundaries without a significant improvement to the receive level.

10.7.4.2 C/I Handover

If, in spite of a high signal level, the quality of a channel deteriorates owing to interference from other radio connections then a handover must be carried out within the same cell or to a more favourable cell. This kind of disturbance is caused by co-channel interferers which are occupying the same channel in

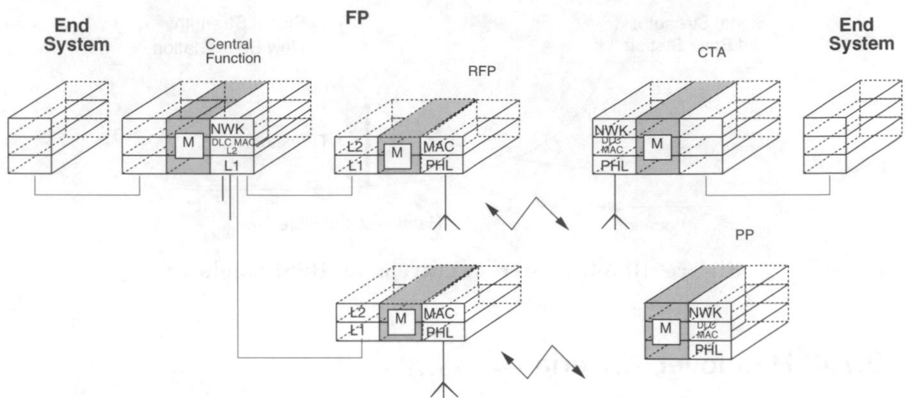

Figure 10.55: Protocol architecture for multicell systems (NWK= Network)

other cells. The parameters used for characterizing this event are the *Carrier to Interference* (C/I) ratio and the *Bit-Error Ratio* (BER). Limit values that take into account the technology used and the requirements for connection quality are defined for the C/I ratio and/or bit-error ratio.

A cyclical code for forming check bits for error detection (CRC) is used for the measurement of connection quality in the DECT system. With the 32 kbit/s ADPCM voice coding used, the maximum bit-error ratio allowed is 0.001. For GFSK modulation with $BT = 0.5$ this corresponds to a minimum C/I ratio of 11 dB, which does not yet take into account field strength fluctuations (fading effects) originating in the radio channel [24]. If shading caused by obstacles in the path between sender and receiver are also considered in addition to the multipath propagation, the result is a required threshold value of 31 dB for the C/I ratio.

10.8 Protocol Stacks for Multicell Systems

A DECT fixed part in a multicell network consists of explicitly defined subgroups that can be transferred to the radio fixed parts. The result is that some of the protocol operations are only working in the RFP, thereby not burdening the whole system. However, many of the logical functions must more or less be centralized into one unit, thereby forming the gateway to a local network through the use of an internetworking unit. Since most multicell networks only possess one cluster, the layer 2 (MAC layer) of the DECT system can physically be divided into t wo parts; see Figure 10.23. The cell site functions (e.g., TBC) are allocated to each RFP, whereas the cluster control functions (e.g., MBC) remain in the central network element. The internal protocols L1 and L2, as indicated in Figure 10.55, can be specified and configured for their use by the operators themselves.

Table 10.8: CC service primitives of the interworking unit

CC service primitive	Request	Confirm	Indication	Response
Setup	×		×	
Setup_Acknowledge	×		×	
Reject	×		×	
Call_Proceeding	×			
Alert	×		×	
Connect	×	×	×	
Release	×	×	×	
Facility	×		×	
Info	×		×	
Modify	×	×	×	
Notify	×		×	

A DECT system requires a control unit, called the *DECT Central System* (DCS), which controls and links together all the base stations (FP). All the data necessary for authentication and encryption must also be filed in this unit. The DCS must also provide access to external networks through the interconnection of an *interworking unit* (IWU). The fixed network side is configured differently by each of the individual manufacturers of DECT systems.

10.9 The DECT Network Gateway Unit

In order to transport information between DECT and other systems, the DECT system uses an interworking unit at the interface to other networks (see Figure 10.42). The interworking unit works at two parallel planes. All signalling data and protocols such as *call control* and *mobility management* are converted in the signalling plane (*C-plane*). The user data for an existing connection is transmitted over the user data plane (*U-plane*). This data must be converted to conform with the requirements of other networks.

10.9.1 Signalling Data

DECT [14] provides service primitives for the interworking unit at layer 3 that relate to the protocols *mobility management* (MM) and *call control* (CC). These service primitives are used to convert the DECT protocols into the protocols of external networks. The services are made available to the higher layers of the interworking unit by DECT layer 3.

The service primitives listed in Tables 10.8 and 10.9 consist of elements for the actual transport of information. They correspond to the elements of protocol messages in the DECT standard [14], which are described in the next section.

Table 10.9: MM service primitives of the interworking unit

MM service primitive	Request	Confirm	Indication	Response
Identity	×	×	×	×
Identity_Assign	×	×	×	×
Authenticate	×	×	×	×
Locate	×	×	×	×
Detach	×		×	
Access_Rights	×	×	×	×
Access_Terminate	×	×	×	×
Key_Allocate	×		×	
Info	×	×	×	×
Cipher	×	×	×	×

10.9.2 User Data

User data can consist of coded ADPCM speech. The data must be converted if the connected external network requires the user data in another form. The DECT user data is provided in the interworking unit at a uniform PCM-interface point. The user data is provided in PCM-coded form at this reference point, and can be accepted by external networks.

10.10 Security in DECT

Mobile telecommunications networks must be protected against unauthorized access and misuse. Criteria relating to system security were therefore established for DECT [15]:

- User identification
- Unauthorized use of a mobile station
- Base station identification
- Unauthorized use of a base station
- Illegal eavesdropping of user or signalling data

The security measures described below relate to these criteria.

10.10.1 User Identification

At the beginning of a call a user is requested by the mobile station (PP) to input the *Personal Identity Number* (PIN). This number is checked locally by the mobile station. Each subscriber has an *International Portable User Identity* (IPUI), which is clearly defined in an authorized area (*Portable Access Rights Key*, PARK). Together with the PARK, the IPUI provides a unique identification for the mobile station [9]. Every `IPUI` can be used with several different `PARKs`, with the costs calculated over the network authorized by the `PARKs`. A mobile station also requires an `IPUI` and one or more `PARKs` in order to initiate a call.

10.10.2 Portable Access Rights Key (PARK)

There are four categories of `PARKs`, depending on the size of the DECT system:

`PARK Category A` is the access right to small single- or multicell systems in private households.

`PARK Category B` is the access right to more complex private branch exchanges and local area networks (LANs).

`PARK Category C` corresponds to the public version of categories A and B.

`PARK Category D` is designated for public use when a DECT system is connected directly to GSM.

10.10.3 IPUI

An IPUI contains the *Portable User Type* (`PUT`) and the *Portable User Number* (`PUN`). A `PUT` is 4 bits long; the length of a `PUN` depends on the type of `IPUI`. For example, seven different types of `IPUI` for use in different applications are defined in DECT (see Figure 10.56).

`IPUI Type N` is the simplest type of `IPUI`. It consists of a `PUT` and a 36-bit long `PUN` (see Figure 10.56) and is used in conjunction with `PARK Class A` in private households.

`IPUI Type S` is provided for DECT connection to the ISDN because the 60-bit long `PUN` is coded like a PSTN or an ISDN number. It is used together with `PARK Category A`.

`IPUI Type O` is a local unique number in a private branch exchange (PABX or LAN) based on `PARK Category B`. The `PUN` can be specified by the operator of the branch exchange so that it is a locally unique number corresponding to the PSTN number. The `PUN` is 60 bits in length.

`IPUI Type T` is used for expanded branch exchanges. The 60-bit long `PUN` consists of a 16-bit long *Equipment Installer's Code* (`EIC`) and the additional number (44 bits). Large companies with a network of subsidiaries have the possibility of assigning each location with its own `EIC`. This enables each mobile station within the subsidiaries to be used just as it would be in the home subsidiary.

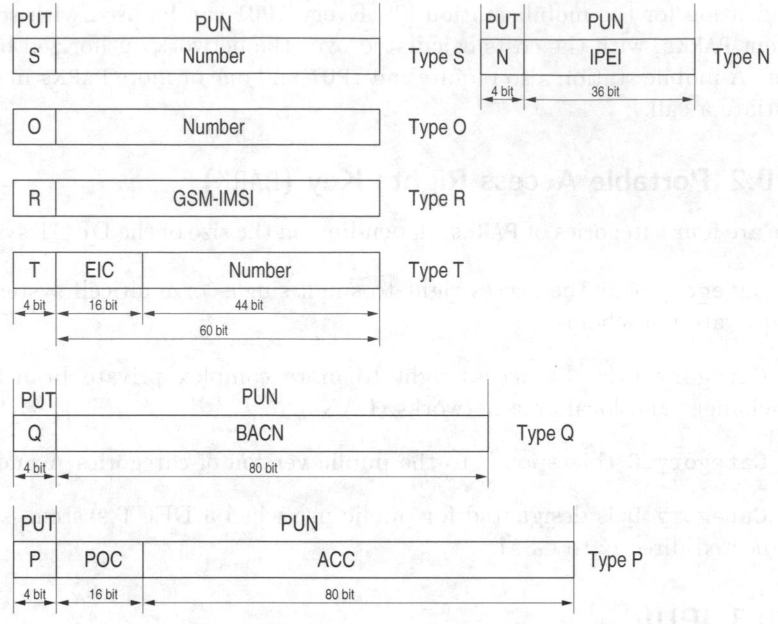

Figure 10.56: The structure of an IPUI

`IPUI Type P` is available for public Telepoint use or radio local-loop applications. The 96-bit `PUN` consists of a *Public Operator Code* (`POC`, 16 bits) and a *Telepoint Account Number* (`ACC`, 80 bits), which are used in the calculation of call charges. The operator is coded in the `POC`. The uniqueness of the `IPUI` is only guaranteed through the `POC` and the `ACC`.

`IPUI Type Q` corresponds to Telepoint use, similarly to `Type P`, but is calculated using an 80-bit long BCD-coded *bank account number* (`BACN`).

`IPUI Type R` is provided for public access to GSM via a DECT mobile station. The identity is unique, and allows for DECT and GSM charges to be calculated together. The 60-bit long `PUN` corresponds to the GSM network BCD-coded `IMSI`, which can be up to 15 digits long [2]. This `Type R` is only available to operators of GSM networks and is used in conjunction with the `PARK Category D`.

10.10.4 TPUI

In local applications a temporary telephone number (*Temporary Portable User Identity*, `TPUI`) agreed between FP and PP is used for contacting a mobile station. This protects the `IPUI`, which in this case is not transmitted, from eavesdropping. The `TPUI` consists of 20 bits and is either the last part of an `IPUI` or an assigned `TPUI` (see Figure 10.57).

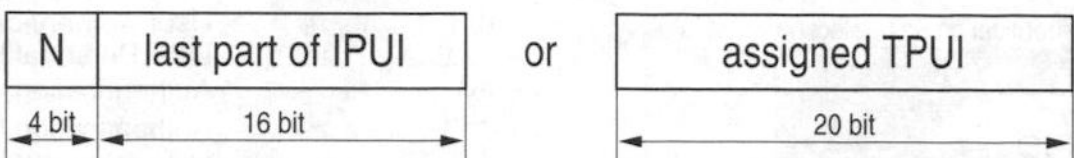

Figure 10.57: The structure of a TPUI

10.10.5 Authentication of a Mobile Station

Before a connection is set up, a mobile station is checked to ensure that the equipment is not recognized as stolen and that the user has authorized access to the network. This check is undertaken on the basis of an *Authentication Key* (`K`), which is stored in the mobile station (DAM) and in the base station (see Figure 10.58). Through the use of algorithm A11, a temporary key (*Session Authentication Key*, `KS`) is calculated from `K` and a value transmitted by the FP (`RS`). A random value (`RAND F`) selected by the FP is transmitted to the PP where on the basis of the KS and algorithm A12 an *encryption result* (`RES 1`) is calculated. This result is sent back as proof of identification of the `KS` to the FP, where the result is compared with that calculated in the FP. If there is any disparity between the two, the connection is immediately terminated. This identification check protects a mobile station from using an unauthorized or manipulated base station.

The advantage of a temporary key is that there is no need for the PP to transmit its secret authentication key when it is visiting a different coverage area. `K` is calculated from an authentication code (`AC`), user authentication key (`UAK`), or a combination of `UAK` and *user personal identity.*

10.10.6 Authentication of a Base Station

The process is the same as that described above for a mobile station, except that the roles of the PP and FP are reversed. The PP calculates a *Reverse Authentication Key* (`KS'`) from the value `RS` received from the FP and the authentication key `K` using algorithm A21. A random value (`RAND P`) sent from the PP to the FP is calculated to produce the result `RES 2` using A22 and `KS'` and then transmitted back to the PP (see Figure 10.58).

10.10.7 Equivalent Authentication Between Mobile and Base Stations

There are three different methods for carrying out this authentication: the direct method (see above) and two indirect methods.

The indirect methods rest on the premise that ciphering of the data is not possible unless the ciphering parameters are known. With the first method the PP is authenticated as described above. Using the data transmitted, the PP calculates a *Ciphering Key* (`CK`) and uses it to encipher all data. If the FP does not have the correct authentication key `K`, it will calculate the ciphering key

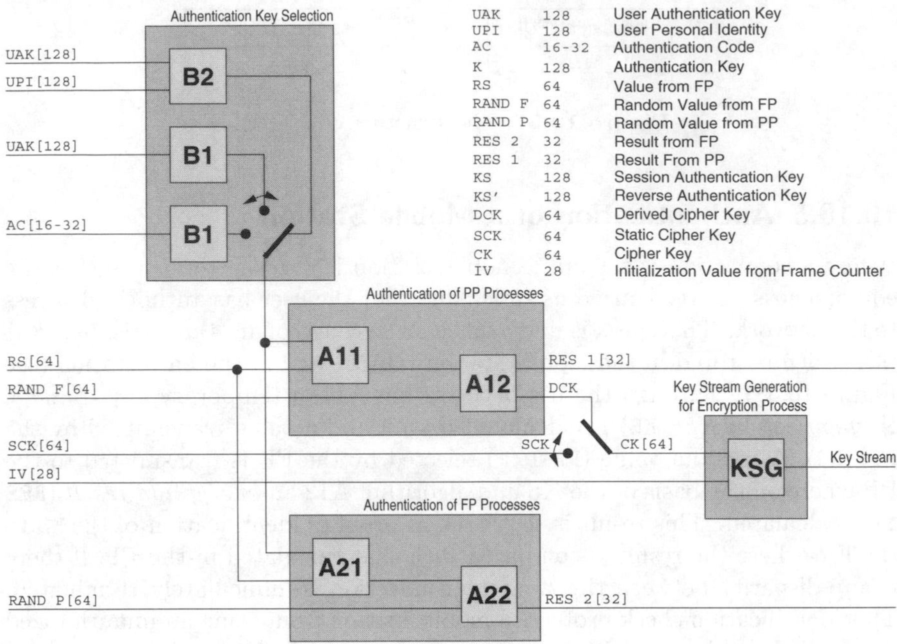

Figure 10.58: Authentication and ciphering

CK incorrectly, and therefore not understand the data sent by the PP—which will cause the connection to be terminated. The authentication is completed when the FP understands and is able to respond to the ciphered data.

With the second indirect method, a static ciphering key that is used by all PPs and FPs is agreed [15].

10.10.8 Ciphering of User and/or Signalling Data

When data is transmitted on the radio interface, it is sent in ciphered form to protect it from unauthorized eavesdropping. For this purpose a ciphering parameter (*Derived Cipher Key*, DCK) is calculated during the authentication. A *Static Cipher Key* (SCK) can also be used in the ciphering. A key stream that can encipher the data on the radio channel is calculated in a *Key Stream Generator* (KSG) (see Figure 10.58). The parameters and algorithms for security are listed in the top part of Figure 10.58.

10.11 ISDN Services

In discussions on the suitability of the DECT system as a public access network *(Radio in the Local Loop, RLL)* a stipulation was made that ISDN services would have to be available to subscribers. Appropriate solutions were specified

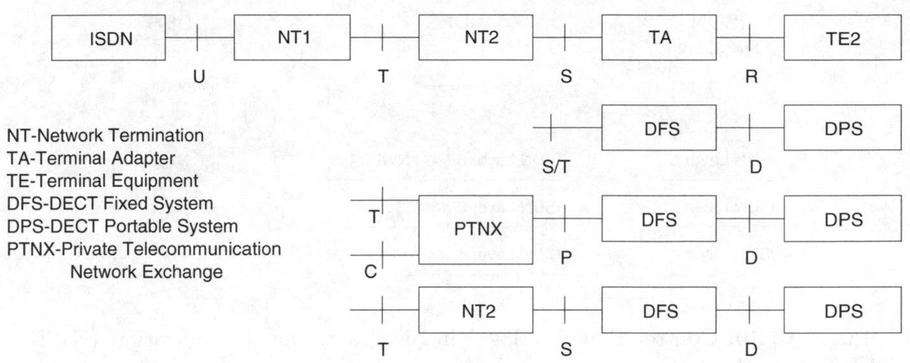

Figure 10.59: Reference configuration of an end-system (first line: ISDN reference model)

at ETSI/RES 3, and these included taking advantage of the possibility of using parallel DECT bearers for the same connection.

10.11.1 End System and Intermediate System

If a user wants to connect a DECT terminal to the ISDN, a conversion unit (*Interworking Unit*, IWU) must ensure that the protocols are converted into one another (see Figure 10.14).

The IWU converts the functions of one system into the functions of the other system. One application is a DECT terminal, which is connected to an ISDN basic connection and allows the use of certain ISDN services. This type of constellation is referred to as an *end system* [13].

Another application of the interworking unit is in linking ISDN terminals to an ISDN over a DECT radio interface, with the IWU on the network side undertaking the conversion ISDN↔DECT and the IWU on the user side the conversion DECT↔ISDN. This involves transmitting transparent ISDN channels over radio. This constellation is referred to as an *intermediate system* [17].

10.11.1.1 Reference Configuration of an End System

A reference configuration describes a functional system constellation on the basis of the reference points of the DECT or ISDN system (see Figure 10.59). The significance of these interfaces is explained in [21].

An end-system configuration exists when the *DECT fixed system* (DFS) and the *DECT portable system* (DPS) together assume the role of an ISDN terminal.

Figure 10.60 presents the IWU protocol stack for the C-plane. It shows that the lower three layers of the OSI reference model are converted, whereas the right side corresponds to the protocol stack in Figure 10.16.

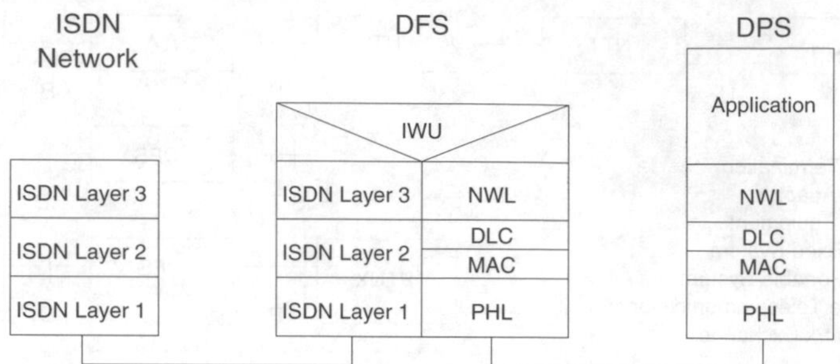

Figure 10.60: Control Plane of IWU in an end-system configuration (NWL = DECT network layer)

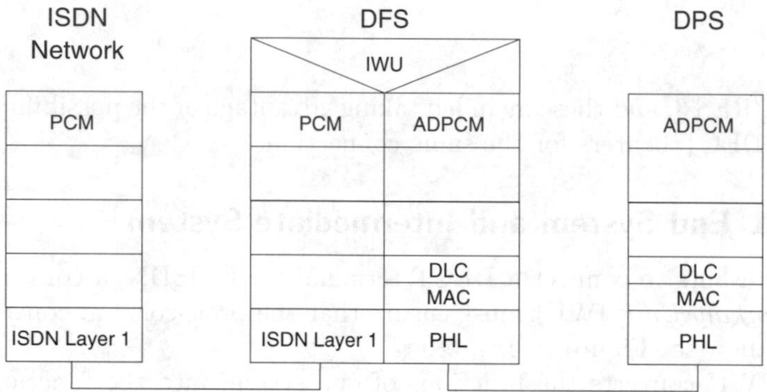

Figure 10.61: User plane of IWU in an end-system configuration using the voice service as an example

Figure 10.61 illustrates the protocol stack for the IWU for the U-plane in the voice service. It shows that layer 1 of ISDN is converted to the physical DECT layer and that the DLC and MAC layers are also contained in the IWU on the DECT side. Furthermore, it is clear that the ISDN speech coding PCM must be converted to the DECT coding ADPCM above layer 3.

Figure 10.62 presents the reference model specified for the 64 kbit/s data service. It shows that on the DECT side of the IWU the data link layer (DLC) carries out the backup of the 64 kbit/s-data service described in Section 10.4.3.7.

10.11.1.2 Intermediate System Configuration

For comparison purposes, the reference model for the C-plane of an intermediate system is illustrated in Figure 10.63. In contrast to the C-plane in an end system (see Figure 10.60), it can be seen that the layer-3 messages in ISDN are

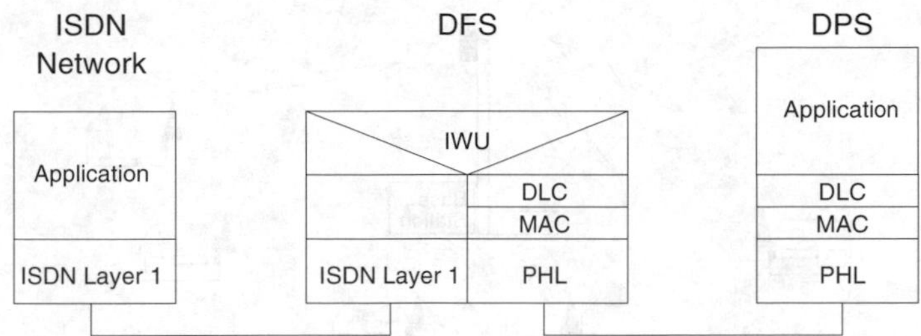

Figure 10.62: User plane of IWU in an end-system configuration using the 64 kbit/s data service as an example

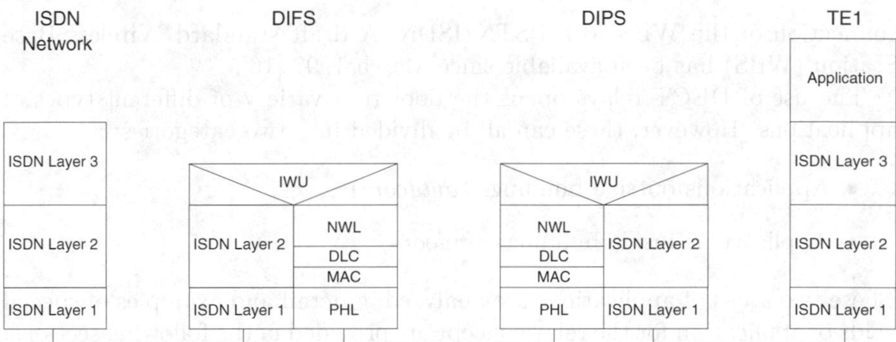

Figure 10.63: Control plane of IWU in an intermediate system configuration

forwarded from the intermediate system to the partner entity without being mapped to the DECT functions in the IWU.

To provide a 64 kbit/s data service, a DECT station must be able to transmit 640 bits — which corresponds to 80 bytes — of user data per sending cycle, because a new DECT frame is transmitted every 10 ms.

However, because a slot only contains 480 bits, additional channel capacity must be created through a bundling of slots. Two consecutive slots are used for this purpose. This solution is referred to as a double slot and is used in the standard [13] for the 64 kbit/s data service.

For comparison purposes, Figure 10.35 shows the distribution of a double slot (packet P80) in the different fields. The distribution of the D-field that is also shown is explained in detail in Section 10.4.3.7.

10.12 DECT Relays

A DECT relay is a wireless DECT base station (WBS) that provides coverage over radio to partial areas in DECT networks without requiring a direct

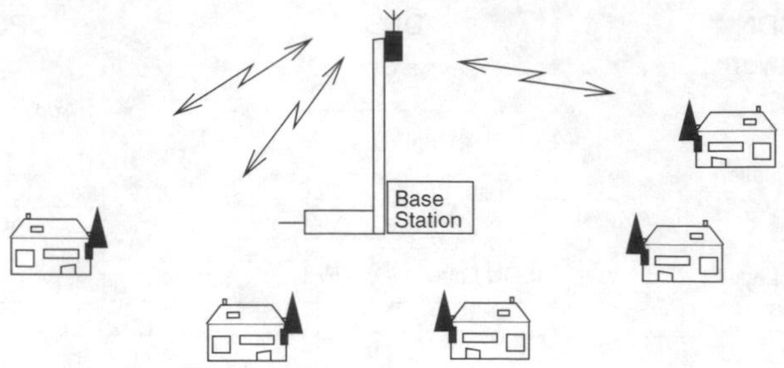

Figure 10.64: Radio local loop (RLL) system

connection of the WBS to a PSTN/ISDN. A draft standard Wireless Base Station (WBS) has been available since March 1997 [16].

The use of DECT relays opens the door to a variety of different types of applications. However, these can all be divided into two categories:

- Applications outside buildings (*outdoors*)
- Applications inside buildings (*indoors*)

These two areas of applications are analyzed in detail and examples of special fields of application for the relay concept are provided in the following sections.

10.12.1 Outdoor Applications

As an extension to DECT base stations (FT), operators use DECT relays in wireless local networks (RLL) to replace the previously wirelined *last mile* from the local exchange, concentrators or cabling distribution panels up to the home location of the customer with a DECT radio path (see Figure 10.64). The system therefore leaves the indoor area for which it had originally been designed, and takes over network functions outdoors. The elimination of customers' wireline connections first provides the flexibility needed to establish a coverage area quickly, and secondly offers a cost-effective option through the use of relays. The savings in cabling costs in the local network in which each household is connected to the local exchange through a two-wire line makes the wireless local network especially attractive.

Whereas these measures in the distribution network of a network operator can equate to lower charges for the customer, the next step attempts to make these operator routes available for public access. This means that a mobile DECT handset would not only be used for private use to communicate over DECT systems installed in homes but could also access the public network with its relays and base stations. Consequently, the Telepoint concept, which has already been tested but not proven successful in Germany, would be reintroduced in a different version over the RLL system.

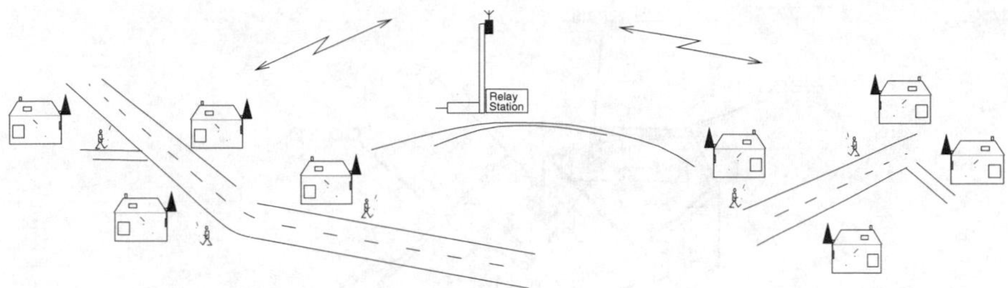

Figure 10.65: Radio local loop (RLL) with public access

This would create a local or regional wide-area network that could guarantee comprehensive secure mobile coverage to subscribers. An example of this network is illustrated in Figure 10.65. Mobile subscribers can be reached anywhere in the whole scheme, and are able to access the fixed network not only directly through the base station but also over the stationary DECT end systems if they are implemented as relays. The *Generic Access Profile* (GAP) [5] approved by ETSI in 1996 and the *Public Access Profile* (PAP) [11] lay down a manufacturer-independent access interface for DECT systems ensuring that public access to other systems is also possible from every PP.

10.12.2 Indoor Applications

Similar ground constellations as outdoors are also found indoors. Relays can be added to existing DECT systems in order to improve the coverage to certain areas. Also, when consideration is being given to installing a new DECT branch exchange, the savings on cabling a house can be an important factor in favour of DECT. Relay stations particularly lend themselves for use by companies with scattered sites and assembly plants, as well as places like construction sites where a work location is used for a limited period of time (see Figure 10.66). Except for access to power supply sources, relays eliminate the need for any cabled connections, and sometimes even solar power supplies can be used.

In contrast to RLL, in indoor use relays are considered integral components of the system. As with RLL systems with public access, the capacity of an indoor network is determined by the mutual interference of the cells among themselves. Similarly to outdoor situations, indoor relay systems, with a given number of send/receive facilities, are not restricted by the number of channels available (*trunk-limited*). However, the multiple occupancy of the radio medium through the use of relays causes a deterioration in the C/I values on the cell boundaries. This creates a capacity loss that cannot be offset even by an increase in the number of transceivers.

The strength of interference in both indoor and outdoor applications is similarly dependent on the spacing selected for the relays and base stations,

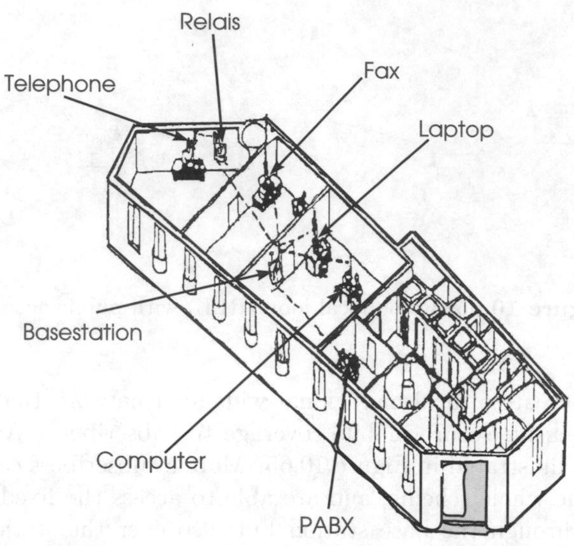

Figure 10.66: Indoor relay system

the transmitter power and the antennas used. It is therefore possible to make a direct comparison of the results from indoor and outdoor studies if the dimensions of the different scenarios are taken into account.

10.12.3 Relay Concept

Relay systems offer the following advantages:

Savings on cabling This is especially attractive to users of DECT indoor systems. Existing cabling is often not sufficient for linking an adequate number of base stations to the fixed network.

Wider radio coverage The radio coverage for a situation can be improved considerably through the use of relays. Areas that later turn out to have insufficient supply (shadowing) can be integrated into an overall system through the use of relays.

DECT implementation Relays allow users to connect to base stations and relay stations using conventional DECT-compatible terminals. Subsequent expansions to a system using relays therefore do not affect the mobile stations.

Retention of frequency and transmitter power Relays in a DECT implementation use the frequency channels authorized for the DECT system.

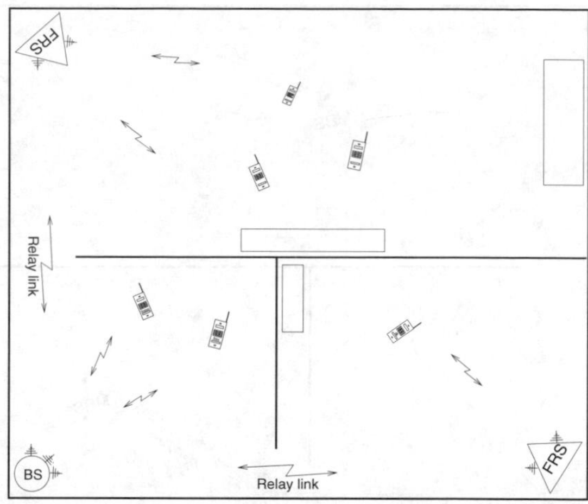

Figure 10.67: Indoor application with fixed relay stations (FRS) and RFP (= BS)

The 250 mW maximum transmitter power specified also applies to the relay systems, and therefore no additional approval is required for the operation of relays.

Mobility of relays for changing locations of use Because relays are not restricted to a particular network, the operator is able to adjust the system manually to cope with long-term load fluctuations in selected areas.

These advantages carry different weight in every network installation. Depending on the particular application, the indoor and outdoor areas can overlap to the extent that a system is referred to as a combined one.

In principle, all concepts can be divided into two categories.

10.12.3.1 Stationary Relay Stations

As the name implies, these are fixed relay stations (see Figure 10.67). Although they can be moved manually, they usually remain fixed to one location.

The DECT radio interface to the user is maintained, and no modifications are required to the hardware and software in the user's terminal. Therefore the user is not even aware that a *Fixed Relay Station* (FRS) is being used. The mechanism for forwarding a call from a relay to a base station (BS) that is linked to a network is a matter for the supplier of the system, and can be controlled and changed within the network. A high transmission quality will be achieved if an optimal location is selected and complex antenna systems for the link FRS↔BS are used.

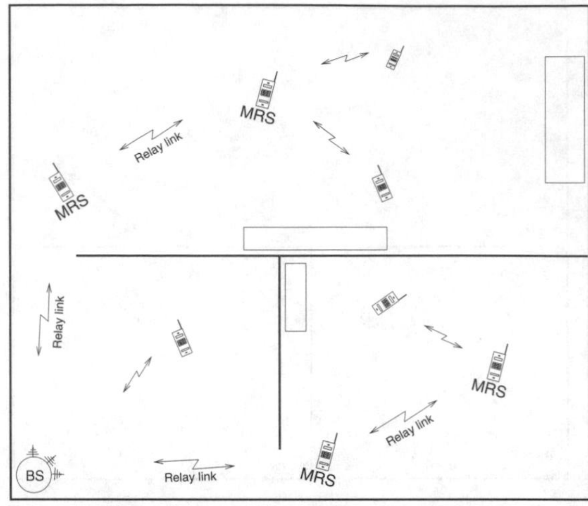

Figure 10.68: Indoor application with mobile relay stations (MRS)

10.12.3.2 Mobile Relay Stations

Mobile relay stations (MRS) are not in fixed positions like the FRS, but are movable. One possible implementation of a mobile relay could be the use of appropriately modified handsets (see Figure 10.68). Because each mobile terminal is able to offer relay operations along with the traffic channels it uses itself, this type of operation provides all the mobiles in a system with a considerably improved quality of service compared with fixed relay stations, albeit clearly placing higher demands on the terminals, particularly on battery use.

The quality of a connection depends on the mobility of the relays and the terminal, accompanied by all the associated risks, including the possibility of a connection being terminated. The issues that must be taken into account are the amount of routing required and the time delays that can occur because of the lack of control over the growing number of *hops* with mobile relay systems. A comparable concept for multihop communication is found with ETSI/HIPERLAN/2 (see Section 14.1.1.2).

The following discussion focusses exclusively on fixed relay concepts. An important advantage of the FRS variation is that the location can be determined precisely in advance and that there is a finite definable number of occurring sequential radio hops of connections.

10.12.4 Setting up a Relay Station

A DECT system that uses relays places a multiple burden on the radio medium. The mobile stations that are directly linked to the system in the

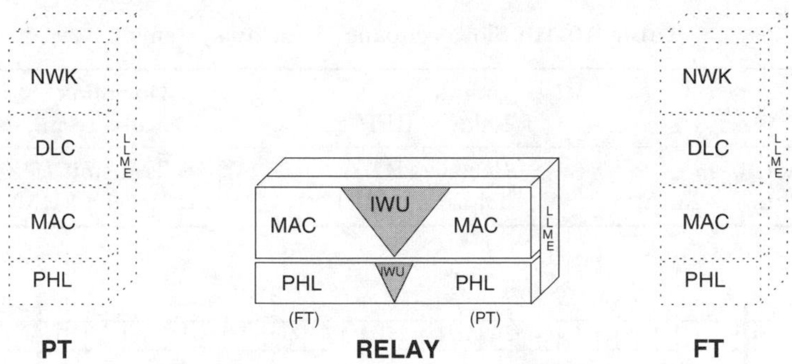

Figure 10.69: Reference model for relays

other RFPs must share the available channels with the relays in the immediate vicinity.

Relay systems are characterized as follows:

A relay acts the same as an RFP for each mobile station It offers the PP the same services and is therefore completely interchangeable with an RFP.

Unrestricted configurability A relay can be adapted to prevailing local conditions. In addition to a specification of the frequencies and the time slots allowed, independent *Dynamic Channel Selection* (DCS) from the base stations for access to the fixed network is also possible.

Coverage to undersupplied areas Optimal coverage is possible because of the flexibility allowed in specifying the location of the relays. It is recommended that the corresponding signal strengths selected should be higher than average to help eliminate interference to channels between the relays and the base stations.

Interference due to relays Because they change incoming radio channels to outgoing ones, relay stations cause a higher level of interference and a reduction in the capacity of a total system compared with wireline fixed stations (RFP). With unimpeded free-space propagation in outdoor situations, the quality of service in neighbouring cells can drop considerably. This loss can be reduced through the installation of antenna systems with directional beams in relays and RFPs.

10.12.4.1 Basic Relay Structure

A DECT relay station (*wireless relay station*, WRS) contains the key functions of the *radio fixed part* (RFP) and of the *portable part* (PP). But there is no provision for things like network gateways, user interfaces and speech coding of the PP and the RFP. To a base station the relay appears as a mobile station

Table 10.10: Slot occupancy in relay systems

	Uplink Relay – RFP	Downlink Mob. term. – Relay
Relay slots 0–11	Receive (RX)	Transmit (TX)
Relay slots 12–23	Transmit (TX)	Receive (RX)

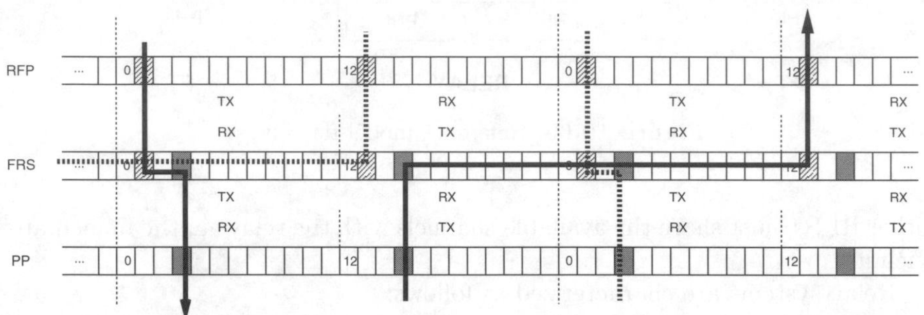

Figure 10.70: Example of slot occupancy with use of a relay station

with up to 12 traffic channels. A mobile handset perceives the relay as an RFP. The relay functions required are responsible for conversion of the PP ↔ relay connection to the allocated relay ↔ RFP connection (see Figure 10.69), where a relay is provided between the portable and the fixed radio termination (PT and FT).

If the relay operates as a pure repeater station then the time slots within the physical layer are interlinked through the relay functions in the physical layer. This transparent service allows data to be transmitted within the same half frame. If the relaying is carried out in the MAC layer, each connection in the relay is assigned its own separate MAC layer entity. This allows it to evaluate independently the quality of the channels and to execute a bearer setup or a handover using dynamic channel selection.

Each slot in a relay (FRS) can be used as either a send or a receive slot (see Figure 10.70). The allocation of slot areas in a relay is presented in Table 10.10.

This splitting-up of the transmitting and receiving areas makes it possible to retain the standard uplink and downlink separation in the RFP and in the mobile station. Consequently, no modifications are required to the terminals. This type of WRS is called a *cordless radio fixed part* (CRFP) [4]. A *repeater part* (REP) is a technically more complicated solution.

Figure 10.70 shows that a connection exists between the RFP and the relay (FRS) in slot pair 1/13. The corresponding downlink path between the relay and the mobile terminal (PP) is occupying slot pair 3/15. This means that an information exchange (black line in Figure 10.70) between the base station to

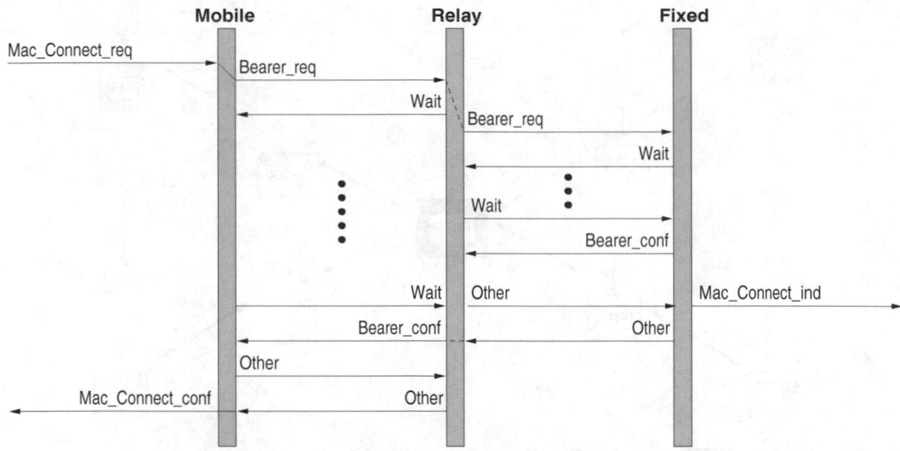

Figure 10.71: Connection setup using a relay station

the mobile station and back no longer only requires a half frame, but is taking up one-and-a-half frame lengths. When several relay stages are used, this time delay can cause unacceptable transmission delays (and resultant echo interference) for a connection.

Switching from transmitting to receiving mode between two slots requires high-speed transceiver hardware in the relay. If this kind of hopping is not possible, it is necessary for some of the slots to be set as *blind* so that the transceiver unit can be switched from sending to receiving and vice versa (see Figure 10.49). However, this reduces the slot capacity in the relay.

10.12.4.2 Connection Setup and Handover

When a mobile user is being connected to a network-coupled base transceiver over a relay system, several connections must be set up in sequence. As usual, the terminal first sends a request to what it knows to be the strongest base station, which in this case is considered as a relay (see Figure 10.29). If the bearer request has been received correctly, the base station responds with a standard WAIT command, to indicate that the request for a connection has been accepted. The next step is for the relay to establish a connection to the base station coupled to the network. The connection cycle for the mobile station can continue as soon as this connection has been implemented successfully (see Figure 10.71).

If a setup error occurs between the relay station and the RFP, another attempt is made to set up a connection between the two, with no reinitiation of the WAIT connection to the mobile station required. However, if the connection timer or the counter in one of the systems (the mobile unit, the relay or the RFP) runs out, the connection setup will be considered to have failed and all the existing hops will be terminated.

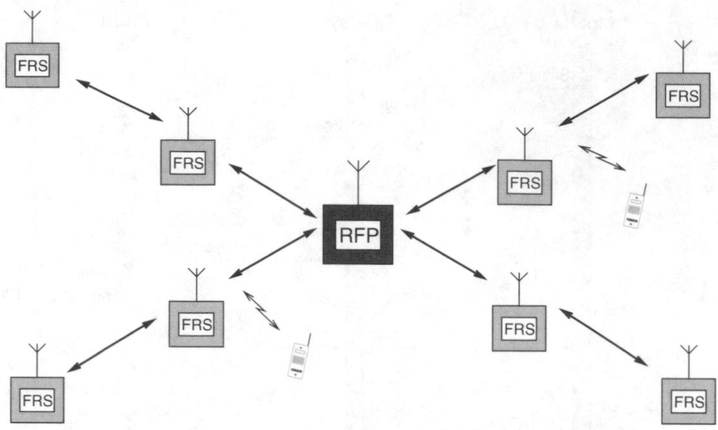

Figure 10.72: Star-shaped multirelay structure

A change of channel during an actual connection (*handover*) can occur for the traffic channel between the relay and the PP as well as between the RFP and the relay. If an intracell handover is required for a mobile station because of co-channel interference then a new connection is only established between the PP and the relay. The corresponding relay ↔ RFP link is not affected by this process.

If a terminal initiates an intercell handover — in other words a change to a different base station — then not only the PP ↔ relay connection but also the corresponding relay ↔ RFP connection will be terminated once a completely new transmission path has been established over a qualitatively more favourable base station. The relay is normally responsible for initiating handovers on the uplink path (relay ↔ RFP). In the same way as a PP, it measures the appropriate quality parameters and decides when an intracell or an intercell handover should be executed. This change of channel is not noticed by a downlink connection.

10.12.4.3 Multirelay Structures

Until now the discussion has centred on the installation of relays between mobile terminals and wireline base transceiver stations. Figure 10.72 illustrates the simple layout of a multirelay structure. Here a number of relays at different levels are arranged around a fixed base station that has responsibility not only for its own internal cellular traffic but also for the traffic between the different relay levels. This star-shaped structure incorporates the following advantages and disadvantages:

- The star structure allows a particularly easy hierarchical classification of the relay stations. Depending on their level in the hierarchy, the FRSs have special features such as transceiver equipment, antenna configurations and routing characteristics.

- The classification according to level makes it easier for mobile stations to seek out the base station with the best performance characteristics. The mobile station should keep the routing in the relay network to a minimum.
- A potential disadvantage of the star-shaped structure is the heavy network use of the RFP. Because the base station supports all the traffic in the relay arrangement, it requires a large number of transceivers. Considering the relatively high level of interference in the central area, this is something that should be avoided.
- Network security is limited because there is only one gateway over the base station to the fixed network for the entire system. The system can collapse if parts of this important station become inoperative.

Other more complex physical arrangements or logical structures such as ring-shaped and bus-type layouts and hexagonal arrangements are possible in addition to the star-shaped one.

A system's range can be increased through the use of relays on the edge of an existing network. However, since the relay must be located within the transmitting area of a higher-level base station, the system only expands by half the diameter of the relay cell. Figure 10.73 (a) shows that a large part of the relay field overlaps with that of the base station when omnidirectional antennas are used. The actual increase in the system's range is smaller than half the cell radius. As can be seen in Figure 10.73 (b), the range of radio coverage can be improved considerably through the use of directional antennas for the link between the RFP and the relay, and omnidirectional antennas for supplying the respective PPs. In this scenario the receiving radius of the relay station is utilized almost completely to provide coverage to one of its own cells. The directional beam can be realized in the DECT frequency band.

The use of sequential relay systems increases the signal propagation time. Whether these factors have any effect on the functioning of a system or whether there should be an upper limit on how many relay systems in sequence are used can be determined based on the following considerations:

1. A delayed transmission of information can particularly cause timers of the signalling protocols to expire if acknowledgements for transmitted messages take a too long time to arrive. This necessitates limiting the number of relay levels.
2. The handover algorithms for multilevel relay systems are very complex. Although this does not affect intracell handovers, it makes the intercell handover process clearly more complex. Figure 10.74 illustrates the

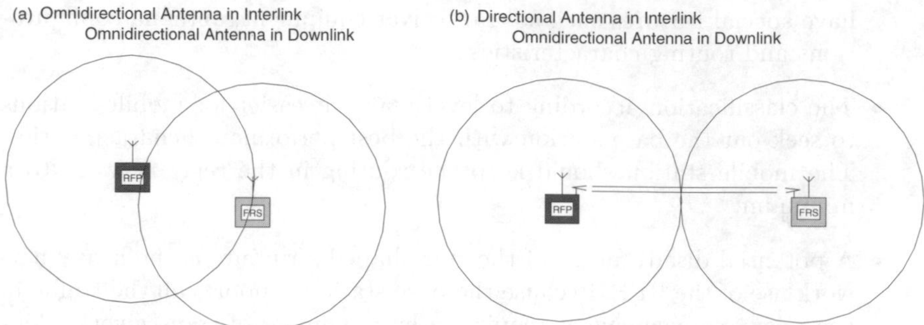

Figure 10.73: Range of a relay system with omnidirectional and directional antennas in the interlink of RFP and FRS

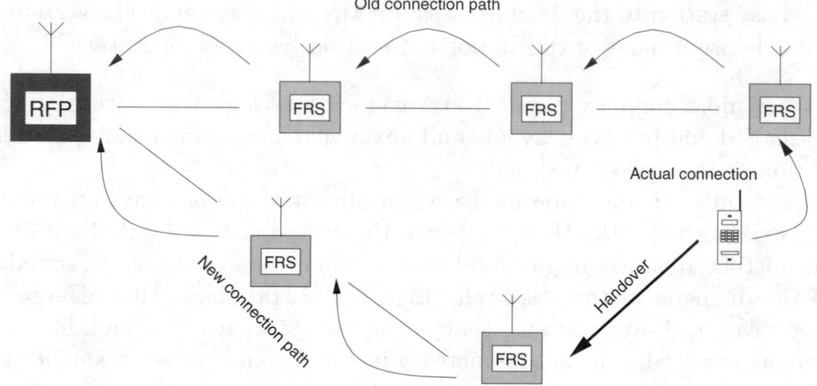

Figure 10.74: Handover process for relay systems

handover process between two relay systems. It shows that all higher-level connections must be reestablished and operated in parallel for a short time before the original connection can be terminated.

3. Synchronization of the individual relays is a problem in many relay systems. Whereas an impulse for synchronization over the appropriate hardware port normally guarantees the same cycle for all neighbouring base stations (RFP) connected to the same local network, this is not possible with relay systems coupled over radio. The only solution is for the system to synchronize itself following its own timing using the signal received from the higher level. This can lead to a shift in the synchronization, *sliding collisions* — in other words to a blocking of individual slots by non-synchronous stations.

4. Relays increase the channel load of a system. Interference in the network increases considerably with each relay level. Not even directional

antennas can always reduce this interference. Relays are therefore especially suitable for improving coverage in situations involving a small or medium amount of traffic.

The relay in Figure 10.70 only has one transceiver unit; two slot pairs are required for each connection: one pair as the downlink from the mobile terminal to the relay station, and one pair as the uplink from the relay to the next RFP. This reduces the amount of traffic that the entire system is able to support. The relay can only accept 6 instead of 12 independent bearers per RFP. The installation of a second receiver in the relay helps to solve the problem. The uplink side (RFP ↔ relay) and the downlink side (relay ↔ mobile) are then physically separated on two transmit/receive facilities.

10.12.5 Parameters for the Performance Evaluation of DECT Systems

A specification of appropriate performance characteristics is required before the quality of a concept can be described.

System load of a configuration When an RFP is replaced by a relay or a radio local loop application is installed, there is an increase in interference in the total system according to the number of relay systems and fixed base stations, and the capacity falls. An analysis of the individual average traffic loads per base station or relay produces information on performance requirements, equipment and suitable locations for RFPs and relay stations.

Change to signal propagation times The delay times in a system are increased with the addition of each relay level. It is necessary to analyse various time parameters, e.g., for setting up a connection.

Number of blocked setup attempts and terminated handovers These two parameters can be used to define the *grade of service* (GoS) [12] of a system, which is an important measure of a network's quality of service:

$$\text{GoS} = \frac{\text{number of blocked calls} + 10 \cdot \text{number of interrupted calls}}{\text{total number of calls}} \tag{10.4}$$

Therefore a lower grade of service means that the system has a high quality of service. As the GoS value increases, so too does the share of aborted connection setup attempts and interrupted connections.

Handover behaviour With an uneven distribution of system load within a network, it can be expected that the handover activity in the area of the relay locations will increase. It is there that interference caused by the double burden of uplink and downlink traffic is the greatest. The aim should be to strive for a relatively even distribution of the locations used for changing channels.

Radio Signal Strength Indicator (RSSI) and Carrier-to-Interference (C/I) ratio at terminal The actual quality of the supply to the user can be determined by the C/I ratio calculated by the mobile station on the basis of the RSSI level of the base stations at the PP location.

10.13 Traffic Performance of DECT Systems

Most of the following also applies to the *Personal Handyphone System* (PHS).

There is a familiar model from traffic theory that can be used to calculate the traffic performance of DECT systems. It is derived from a large number of independent traffic sources (PPs) and a known number of identical radio channels accessible by each station. According to Appendix A.2.2, an $M/M/n-s$ model can be used for modelling, and allows the calculation of loss probability p_l (see Equation A.13). At $s = 0$ waiting positions in the control buffer the duration of a connection can be generally distributed, and is only calculated on the basis of its mean value. The requests for connections by the PPs are modelled on the basis of a Poisson arrival process at the rate λ (calls per time unit): the product of the average duration of a call and the arrival rate is called offered traffic ρ and describes the average channel utilization rate minus the lost traffic rate (see Equation A.4).

From the probability p_l in Figure A.7, it can be inferred that with n available channels a connection will not be possible, because the channels of the base station concerned are temporarily occupied. The loss probability depends on the average utilization rate of the channels ρ of the base station. For example, with $n = 20$ channels the loss probability $p_l = 0.01$ would occur if each channel were used to 62 % capacity.

Simulation studies confirm that in some situations it is possible to make a reliable prediction of DECT traffic capacity (i.e., allowable percentage of utilization) with a given grade of service (see Equation 10.4) using the Erlang model shown in Appendix A.2.2. The requirement is that the arrangement should consist either of geographically isolated base stations (without neighbouring cells) or of stationary user terminals (as is the case with RLL systems) that can only be assigned to a specific base station. With a general base station layout, the correct assumption in terms of appropriate cluster size in the scenario considered is the one that determines the average number of available channels per cell.

10.13.1 Equipment and Interference-Limited Capacity

If the channels switched by the base station are not sufficient for dealing with a particular traffic load, in other words all 12 channels of the only transceiver of an RFP are occupied, and if the interference conditions would allow use of a 13th channel, then this is referred to as an equipment-limited capacity restriction of the base station. This capacity restriction can occur with geographically isolated or poorly equipped base stations in the cellular layout.

If the base stations are well equipped with transceivers then capacity is restricted by co-channel interference. Interference-limited capacity occurs when the hardware of the base stations would permit additional channels to be operated but other channels are not able to achieve the C/I value required for connection setup and therefore are not usable.

Owing to dynamic channel selection in interference-limited DECT systems, an average reuse distance between co-channel cells is created that depends on the path loss or morpho structure (see Section 2.3) and is very difficult to determine through measurement. It is known from simulation tests [22] of large-scale urban scenarios that average reuse distances are set at $p_l = 0.01$, which corresponds to a cluster size of about five.

10.13.2 Estimating the Capacity of DECT Systems

When occupied, each of the total of 120 available channel pairs produces an interference field that does not allow it to be reused locally until adequate propagation attenuation has taken place. The path-loss models to be used depend on each individual situation, with the result that cluster sizes of 7 and below are achievable if the RFPs are not installed within the range of visibility. (In cellular systems a cluster describes the number of neighbouring cells in which each frequency channel may only occur once. These clusters are differentiated from the group of RFPs/FRSs that are connected to the fixed network at the same access point and are also referred to as clusters [20].)

For simplification purposes, p_l=0.01 is equated with a grade of service (GoS) of GOS=1% in the following. The aim would be to seek a value of GOS = 0.5 % in order to catch up with the grade of service of wireline local networks.

If one assumes that there is only enough radio coverage by RFPs over an area that the supply areas of neighbouring cells barely overlap (or not at all), and that each RFP is equipped with 10 transceivers (thereby guaranteeing that each frequency/time channel on each RFP is available), then, depending on the cluster size assumed or the resulting average number of available channels per RFP, the following capacity utilization ρ is produced per channel with the given value p_l, which can be derived from Figure A.7 (see Table 10.11). The traffic capacity (Erl./RFP), i.e., the number of connections allowed at the same time, can be established from ρ and the number of channels n per RFP, depending on the cluster.

Instead of being equipped with 10 transceivers, RFPs are clearly adequately equipped with, for example, only one or two transceivers per RFP if the supply zones of the neighbouring RFPs overlap sufficiently so that a PP is able to reach approximately 17 to 24 time channels (of its own and all the neighbouring RFPs together), as is the case for the 7-cell to 5-cell clusters in Table 10.11).

If the radio coverage by RFPs is assumed so that the supply areas of the neighbouring cells barely overlap (or not at all), as is the case with fixed

Table 10.11: Traffic capacity per RFP depending on cluster size with 10 transceivers/RFP

Clusters	Channels/ RFP	$\rho(p_l = 1\%)$	Erl./ RFP	$\rho\,(p_l = 0.5\,\%)$	Erl./ RFP
4	30	0.68	20.4	0.63	18.9
5[a]	24	0.64	15.36	0.60	14.4
6[a]	20	0.62	12.4	0.56	11.2
7	17	0.58	9.86	0.53	9.01
9	13	0.525	6.8	0.46	6.0
12	10	0.455	4.55	0.39	3.9
15	8	0.39	3.1	0.32	2.5

Cluster sizes 5 and 6 are to be seen as mean values, which are not possible in cellular systems but can occur in DECT systems.

Table 10.12: Traffic capacity per RFP with one transceiver per RFP

Channels/RFP	ρ/N $(p_l = 1\%)$	Erl./RFP	ρ/N $(p_l = 0.5\,\%)$	Erl./RFP
12	0.5	6.0	0.44	5.3

radio local loop systems (RLL), and each RFP is equipped with only one transceiver, then the results are those shown in Table 10.12, which can easily be understood from Figure A.7.

According to Table 10.12, the traffic capacity per RFP represents a lower capacity limit (with a given loss probability).

If the radio coverage areas of the RFPs equipped with one transceiver overlap then, depending on the situation, there is a corresponding increase in the number of possible channels per PP, and the traffic capacity per RFP achieves the values shown in Table 10.11. Overlapping helps to achieve a load distribution between sites with little or high traffic, and is therefore sensible; it also produces savings in the hardware needed for transceivers.

The cluster size appropriate for a given scenario depends heavily on the path loss (model): with morpho structures with little shadowing, 7-cell clusters are likely for modelling local channel reuse; with strong shadowing caused by buildings, channel reuse would be expected to be smaller, corresponding to a cluster size down to five.

If the RFPs/FRSs are very close together (e.g., <50 m because of the high amount of traffic), the probability of line-of-sight links between RFPs increases dramatically, and the cluster size needed increases to values of 20 or higher (with a correspondingly lower number of available channels/RFP and consequently less traffic capacity per RFP). Under these conditions, the DECT system is pushed to the capacity limits due to interference.

From experience, the traffic capacity per RFP for the physical arrangement of RFPs in multilevel buildings is approximately 20 – 30 % less than indicated in Table 10.11 owing to the occurrence of three-dimensional interference.

Reference [22] presents calculation methods that have been validated through simulation experiments for interference-limited DECT systems; these methods use the C/I ratio and the path-loss factor as parameters, and examine the uplink and the downlink separately. A complete theory covering the calculation of capacity of systems with dynamic channel allocation does not yet exist.

10.14 Traffic Capacity of DECT RLL Systems with Several Competing Operators

There is a discussion in Europe about awarding licences to several competing DECT-RLL operators in the same coverage area in the DECT 1880–1900 MHz frequency band. In addition to the cases mentioned above, consideration must here be given to how much the different RLL operators would influence each other.

From a global traffic-theoretical standpoint, the explanations given in Section 10.13.2 continue to apply. Independently of the RFP/FRS density of the operators, the same overall capacity of the DECT systems, as calculated in Tables 10.11 and 10.12, would be available. This capacity has to be shared if there is more than one operator in the same coverage area.

It was pointed out in Section 10.13.2 that

- the transmitter power used per RFP/FRS and subsequent overlapping of the neighbouring cells of an operator, and
- the number of transceivers per base station

have a profound influence on the number of channels (with trunking gain) available to a PP. The achievable quality of service of each operator can be affected through the same measures. Simulation results show that when there are competing operators, each operator is forced to equip his RFPs with more transceivers than would be necessary for a single operator to take full advantage of the system capacity and to achieve the same quality of service [22]. Having a large number of transceivers also allows an operator to gain an advantage (on demand or continuously) at a cost to the others. The list that follows is certainly incomplete.

10.14.1 Using a Higher Density of Base Stations

In high traffic situations that (owing to the equipment installed and base station density) exceed the locally available system capacity of DECT, the operator who benefits at a cost to the other competitors is the one who installs his RFP/FRS stations closely together and hence close to the PPs of his subscribers (see Section 10.5.2). A PP that is situated close to its operator's base station is generally more apt to succeed in making a connection than a PP located further away. Depending on its signal strength, through causing interference it can even displace other PPs that are already connected but are operating relatively far away from their RFP/FRS (and therefore are probably served by competitors with smaller station density) from their channels, with the consequence that it will not be possible for them to find an alternative channel, and their connections will be interrupted.

The operator with higher base station density can have a noticeable affect on the quality of service of his competitors locally. This observation corresponds to the internationally accepted view that a frequency etiquette in the form of *spectrum sharing rules* is necessary when different radio systems are being licensed in the same radio spectrum in order to ensure that the spectrum capacity is divided evenly among the interfering systems.

10.14.2 Use of More than One Transceiver per Base Station

In Section 10.13.2 it was shown that the use of several transceivers per RFP/FRS increases the number of channels available for connection setup or handover of a PP, and consequently has a favourable effect on the loss probability p_l.

For operators with low station density the installation of a sufficient number of transceivers per RFP/FRS compensates for the disadvantages created by the near/far effect if the capacity limits of the system have not yet been reached locally.

10.14.3 Channel Reservation

The effect of RFPs transmitting, irrespective of station density, at maximum transmit power in order to achieve a greater overlapping of their coverage areas is the same as it is when RFPs are equipped with many transceivers per RFP.

According to the DECT standard, PPs generally only attempt to set up connections or handovers on those channels for which the required carrier-to-interference ratio (e.g., $C/I = 11$ dB plus fading reserve) has been measured. The framework of the standard allows for technical measures to give the impression that the channels are being occupied (with the aim of reservation) to prevent the competitor's PPs from accessing them.

The (load-dependent/dynamic) use of channels (e.g., as beacon channels) with the aim of reserving these channels can distort the competition between operators.

According to a wise old saying that can be applied to the use of radio systems with multiple access, "anyone who is too considerate or careful will never find a channel". Accordingly, an operator could specify that a lower carrier-to-interference ratio than the usual 20 dB is adequate for call setup for his system when the other operator's PP is currently attempting to use a channel (in order to displace the competitor's PP from the channel through deliberate interference). The competitor could then grab the freed-up channel for his own use. The DECT system is based on a system of cooperation between PPs: the law of the jungle rules with all its consequences when the competitors' PPs are not cooperative.

10.14.4 Problems Anticipated through Mutual Interaction

When operating in the same band, competing RLL/PCS operators should be required to ensure that their PPs and base stations are cooperative and fair with one another. It is difficult to define this concept, and almost impossible to delineate between what is fair and what is unfair. Moreover, the behaviour of a DECT system is controlled dynamically through parameters over the beacon channel; in other words, it cannot be monitored externally.

On this basis, it can be assumed that competition in the operation of public DECT systems will produce problems because of the coexistence of the systems at the same location. This also applies to competition in the operation of private and public systems — however, only at the location where they overlap.

It is possible to prevent the interference anticipated from competing public systems if only one public network operator is allowed per region or the operators (at least at the moment) use different frequency bands.

It is possible that the competitors in the DECT band in the future will be general licensed private and office systems and operators licensed to run public services. If this is the case, the following could apply:

1. Private systems and office systems will experience noticeable interference from RLL systems at so-called *hot spots*. Private operators have a right to confidentiality protection because public operators were licensed in *their* bands later.

2. In some cases public operators will adversely affect the services of private operators more than they should (see the comments in Section 10.14.5).

3. The quality of service of public systems will be adversely affected by private systems in so-called *hot spots*.

4. Competing operators of public systems in the same frequency band will take technical steps in an attempt to gain advantages from the standpoint of capacity and quality of service.

   It is known from a field study carried out in Sweden that users typically judged the quality of service of RLL systems to be lower than that of the copper wire lines of the local network [1]. Therefore each operator has a high motivation (within cost limitations) to compete with as high a quality of service as possible.

5. Because of the constraints of licensed systems for public use and the generally approved systems for private use being operated in the same frequency band, the DECT system is pushed to its technical limits more quickly than would be the case if these conditions did not exist. Therefore there is a danger with DECT systems that they appear to be less efficient than would be the case if normal conditions applied, that they are losing their reputation as systems with international applicability, and that they are suffering in the competition with other concepts.

10.14.5 Separating Competing Operators in the Spectrum

There are different solutions for alleviating or avoiding the problems mentioned:

1. Restricting public operators to a part of the DECT band. This then provides private operators with the confidentiality protection expected. The frequencies reserved for their use should be measured so that they are sufficient for normal use (e.g., three carriers). This proposal reduces the trunking gain of public operators.

2. Licensing of only one public DECT network operator on the condition that he must allow access to at least two competitive service providers per regional area to his network. The result of this proposal could be that some regions only have one service provider. The DECT licensee is in a similar situation as, e.g., an incumbent local loop operator awarding use of its local network to a third party.

3. Licensing in the DECT band, but partially separated from competitive public operators because of a restriction to a subset of available carrier frequencies, with some shared frequencies (e.g., allocation of 7 instead of 10 carriers per operator, with 3 of them for the exclusive use of each operator). The exclusively allocated frequencies are shared with the private systems.

   This proposal noticeably reduces trunking gain, because only 84 channels instead of 120 are available per operator, and, according to the calculation in Table 10.11, these are to be divided among clusters. However, there is a part of the band (36 channels) that each operator does not have to share with the competitors.

   With proposal 1 each of two operators would only have exclusive use of three carrier frequencies and shared use of one (48 channels together). This would produce a sharp drop in trunking gain.

4. Licensing in the DECT band but partial separation from competing public operators through the availability of additional DECT extension bands (adjacent to the DECT band).

   According to this proposal:

- The extension band will be shared by both operators (e.g., $x = 10$ MHz).
- Each operator will have exclusive use of part of the band (e.g., $x = 5$ MHz).

5. Licensing public operators in a DECT extension band (FPLMTS band) with a bandwidth of approximately 20 – 30 MHz and with two options:
   - Both operators are competitors in the same band.
   - Each operator is allocated exclusive use of part of the band.

   This solution corresponds to the licensing arrangements expected for third-generation cellular systems (FPLMTS). The advantages and disadvantages of both solutions have already been covered.
6. Licensing in the DECT band in accordance with one of the models (1, 2, 3) listed above for the introductory phase (in which traffic loads are still light), with the option of providing an extension band later for shared or (partially) separate use by competitive public operators, as the need arises.

Of all the options mentioned, under current conditions the last solution appears to be practical and the most viable.

10.15 Integration of DECT Systems into GSM900/1800

This section presents concepts for linking DECT systems to the GSM network. Thereby, an example discussion and outline of the solutions are given for the problems arising when integrating two different systems in a way that they can both be used from a so-called *dual-mode mobile station*. Examples of systems integration not covered here that can be similarily solved are integration of

- GSM and mobile satellite service
- GSM and UMTS
- Any second-generation mobile radio systems.

The DECT fixed network side is not specified in the respective standards, but instead is manufacturer-dependent implemented.

DECT and GSM protocols are designed to establish, to maintain and to disconnect connections (*call control*) and to support mobile station mobility (*mobility management*).

An *interworking unit* (IWU) that establishes connections to external systems is provided in all DECT systems. Manufacturers are therefore required to provide a suitable protocol, corresponding to the BSSAP protocol in the GSM network, between the *DECT fixed system* (DFS) and the IWU in order to translate the DECT system *mobility management* (MM) and *call control* (CC) protocols into those of GSM and vice versa (see Section 3.5). An interworking function is used in the conversion process.

A separation between the DECT and GSM functions is important for this purpose, to differentiate between which functions are to be carried out independently by the DECT fixed network side and which functions should be provided by GSM user administration. The IWU must provide all the data required by the GSM network in a way that enables the GSM network to communicate with the IWU as well as with a BSS. As a result, the IWU on the GSM side must simulate a BSS with all its protocols and timers, and interconvert the MM and CC protocols in the DECT system and the corresponding GSM protocols. To do so, the IWU needs the appropriate information, which is supplied by the DECT system. The IWU on the DECT side must be designed so that it is able to convert all the GSM network protocols into the corresponding DECT protocols. The same applies on the GSM side.

The details on integration of DECT into GSM were contained in the first edition of this book and are now available from `http://www.comnets.rwth-aachen.de/mrn`.

Bibliography

[1] G. Ekberg, S. Fleron, A. S. Jolde. *Accomplished field trial using DECT in the local loop*. In *Proceedings of 45th IEEE Vehicular Technology Conference*, Vol. 1, pp. 764–768, Chicago, USA, IEEE, July 1995.

[2] ETSI. *GSM Recommendations 03.03*, 1991. Numbering, addressing and identification.

[3] ETSI. *Radio Equipment and Systems (RES); Digital European Cordless Telecommunications, System Description Document.* Technical Report ETR 056, European Telecommunications Standards Institute, 1993.

[4] ETSI. *Radio Equipment and Systems (RES); Digital European Cordless Telecommunications; Radio in the Local Loop (RLL).* Technical Report ETR 139, European Telecommunications Standards Institute, Nov. 1994.

[5] ETSI. *Radio Equipment and Systems (RES); Digital European Cordless Telecommunications, Generic Access Profile (GAP).* Standard ETS 300 444, European Telecommunications Standards Institute, Dec. 1995.

[6] ETSI. *Radio Equipment and Systems (RES); Digital Enhanced Cordless Telecommunications, Part 2: Physical Layer (PHL).* Standard ETS 300 175-2, European Telecommunications Standards Institute, Sept. 1996.

[7] ETSI. *Radio Equipment and Systems (RES); Digital Enhanced Cordless Telecommunications, Part 3: Medium Access Control (MAC) Layer.* Standard ETS 300 175-3, European Telecommunications Standards Institute, Sept. 1996.

[8] ETSI. *Radio Equipment and Systems (RES); Digital Enhanced Cordless Telecommunications, Part 4: Data Link Control (DLC) Layer.* Standard ETS 300 175-4, European Telecommunications Standards Institute, Sept. 1996.

[9] ETSI. *Radio Equipment and Systems (RES); Digital Enhanced Cordless Telecommunications, Part 6: Identities and Addressing.* Standard ETS 300 175-6, European Telecommunications Standards Institute, Sept. 1996.

[10] ETSI. *Radio Equipment and Systems (RES); Digital Enhanced Cordless Telecommunications, Part 8: Speech Coding and Transmission.* Standard ETS 300 175-8, European Telecommunications Standards Institute, Sept. 1996.

[11] ETSI. *Radio Equipment and Systems (RES); Digital Enhanced Cordless Telecommunications, Part 9: Public Access Profile (PAP).* Standard ETS 300 175-9, European Telecommunications Standards Institute, Sept. 1996.

[12] ETSI. *Radio Equipment and Systems (RES); Digital Enhanced Cordless Telecommunications, traffic capacity and system requirements for multi-system and multi-service DECT applications co-existing in a common frequency band.* Technical Report ETR 310, European Telecommunications Standards Institute, Aug. 1996.

[13] ETSI. *Radio Equipment and Systems (RES); Digital Enhanced Cordless Telecommunictions (DECT) and Integrated Services Digital Network (ISDN) Interworking for end system configuration. Part 1: Interworking specification.* Standard ETS 300 434-1, European Telecommunications Standards Institute, April 1996.

[14] ETSI. *Radio Equipment and Systems (RES); Digital Enhanced Cordless Telecommunications, Part 5: Network Layer.* Standard ETS 300 175-5, European Telecommunications Standards Institute, Dec. 1997.

[15] ETSI. *Radio Equipment and Systems (RES); Digital Enhanced Cordless Telecommunications, Part 7: Security Features.* Technical Report ETS 300 175-7, European Telecommunications Standards Institute, September 1997.

[16] ETSI. *Radio Equipment and Systems (RES); Wireless Base Station, Digital European Cordless Telecommunications.* Standard ETS 300 700, European Telecommunications Standards Institute, March 1998.

[17] ETSI. *Radio Equipment and Systems (RES); Digital Enhanced Cordless Telecommunications (DECT); Integrated Services Digital Network (ISDN); DECT/ISDN interworking for intermediate system configuration.* Standard ETS 300 822, European Telecommunications Standards Institute, April 1999.

[18] O. Freitag, A. Krantzik. *Strategies for the implementation of DECT systems in ISPBX networks. TELENORMA Bosch Telecom*, pp. 617–622, 1991.

[19] J. Fuhl, G. Schultes, W. Kozek. *Adaptive equalization for DECT systems operating in low time/dispersive channels.* In *44th IEEE Vehicular Technology Conference*, pp. 714–718, Stockholm, 1994.

[20] A. Henriksson. *Multiapplication scenarios in DECT.* Working Document, submitted to ETSI Technical Committee Radio Equipment and Systems TC-RES-3R, June 1995.

[21] ITU. *I.411 ISDN User–Network Interfaces — Reference Configurations.* ITU-TS, Geneva, 1988.

[22] C. Plenge. *Leistungsbewertung öffentlicher DECT-Systeme*, 1997. ISBN 3-86073-389-3.

[23] B. Sklar. *Digital Communications: Fundamentals and Applications.* Prentice-Hall, Englewood Cliffs, New Jersey, 1988.

[24] W. Tuttlebee, editor. *Cordless Telecommunications in Europe.* Springer, Berlin, 1990.

11
Wireless Local Loop Systems

Wireline communications networks are expensive and time-consuming to configure. Growing demand for telecommunications services and the imminent liberalization of the telecommunications market in Europe have greatly increased the interest in radio-based communication (*Wireless Local Loop*, WLL), also referred to as *Fixed Wireless Access* (FWA) or *Radio in the Local Loop* (RLL), see [4, 2].

Future competitors of telecommunications monopoly service providers, such as power suppliers who already have their own wide-area networks, will usually lack access networks to the end users. Message transfer over existing power supply lines is currently being attempted using CDMA transmission, and allows rates of several Mbit/s. The CEN 50 065 standard, which enables the use of carrier frequencies of up to 140 kHz, is available for this purpose. In 1997 Northern Telecom conducted its first field tests in this area in Great Britain. If successful, the power supply network could be used in the future as a user access network for voice and data services, playing a major role in promoting competition and price reductions in telecommunications services.

Meeting the following requirements, wireless technology is very effective at bridging the "last mile" of the path between the fixed network access of private network operators (*Point of Presence*, POP) and the user:

- Fast and economic network configuration
- Economic network operations
- Flexible and expandable network structure
- Possible added-value through (restricted) mobility

For the user, a connection over a WLL system replaces the wireline fixed network connection and should guarantee the same quality of service characteristics—bit-error ratio, delay time and blocking probability—as the public analogue telephone network (*Public Switched Telephone Network*, PSTN) or even the *Integrated Services Digital Network*, ISDN.

Figure 11.1 shows the different levels of mobility for a WLL user. In addition to connections without mobility in which the customer uses a normal telephone connected through a cable to an antenna in the wall of the house, systems with restricted mobility also exist, allowing a user to roam in a restricted area, e.g., in a radio cell defined by an internal base station. Systems

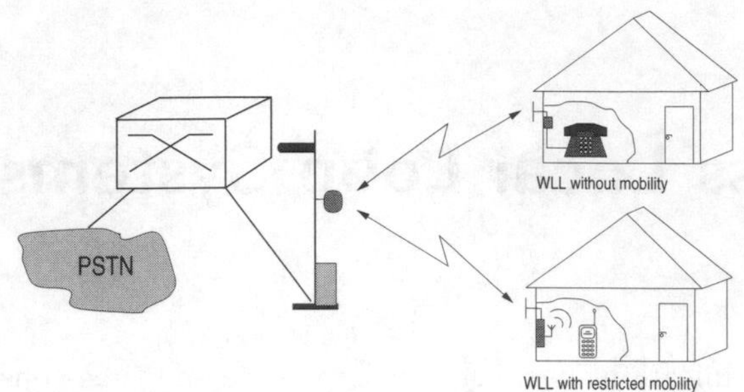

Figure 11.1: WLL without and with restricted mobility

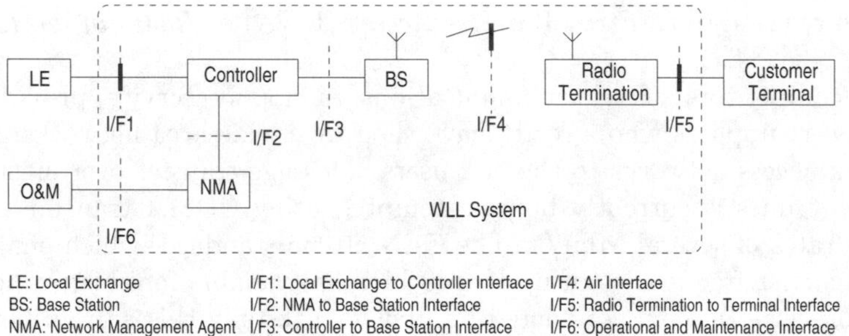

Figure 11.2: ETSI reference model for WLL systems

in which users enjoy the same kind of mobility as in cellular mobile radio networks (e.g. GSM) are called *Personal Communications Systems* (PCS); see Section 3.15.

Because of the great interest in WLL technology for the above reasons, the ETSI working party *Radio Equipment and Systems 3* (ETSI RES 3), founded in January 1993, published a report [1] in November 1994 addressing the marketing aspects, different WLL technologies as well as scenarios and characteristics that are of interest from the standpoint of the dimensioning of WLL systems, such as radio range and capacity aspects. The reference model presented in Figure 11.2 originates from this report.

The reference model shows the interfaces of a WLL system and its elements. In addition to antennas, the base station (BS) has facilities for measuring and controlling radio connections with the radio termination of the customer terminal. The controller connects the base station to the local exchange (LE) and controls the base station.

11.1 Technologies for WLL Systems

The following technologies are suitable for WLL systems [3]:

- Analogue cellular mobile radio
- GSM/DCS1800 derivations
- CDMA systems based on US-TIA standard IS-95
- Digital cordless radio networks:
  - DECT
  - *Personal Handyphone System*, PHS (Japan)
  - US *Personal Access Communication System* (PACS)
- Digital *point-to-multipoint* (PMP) radio relay systems

Directional antennas with, e.g., 5–12 dB gain are normally provided for stationary residential lines to enable a low bit-error ratio on the radio channel to be achieved even when base stations have a large coverage radius (typically 2–5 km). However, discussions are also taking place about systems in which the stationary customer line can get by with an indoor antenna, which then has to be small and unobtrusive. The base station then of course has to operate with smaller supply radii.

The suitability of the systems mentioned above depends heavily on the types of user—in other words on the services that are to be offered by the WLL system. A differentiation is made between the following types of user:

- Residential users who require an analogue (PSTN) connection or an ISDN basic rate interface (BRI).
- Small-business customers who operate a small branch exchange for which they require a number of ISDN BRIs (each with $n \cdot 144$ kbit/s) but for whom an ISDN primary rate interface (PRI) would be impractical and too expensive.
- Large business customers who operate large branch exchanges and require a gateway to data networks such as X.21 and X.25 and frame relay along with normal telephone applications. Their connection capacity needs are on the order of one or more ISDN PRIs ($n \cdot 2048$ kbit/s, $n = 1, 2, \ldots$).

11.1.1 Cellular Mobile Radio Networks

The Ericsson RAS 1000 System, which is based on the NMT standard (*Nordic Mobile Telephony*), was used by Deutsche Telekom AG (DTAG) in 1993 to connect around 13 000 customers in Potsdam, Germany. However, analogue cellular systems are no longer competitive compared with the other systems presented here because of the high cost per connection and the disadvantages compared with digital networks.

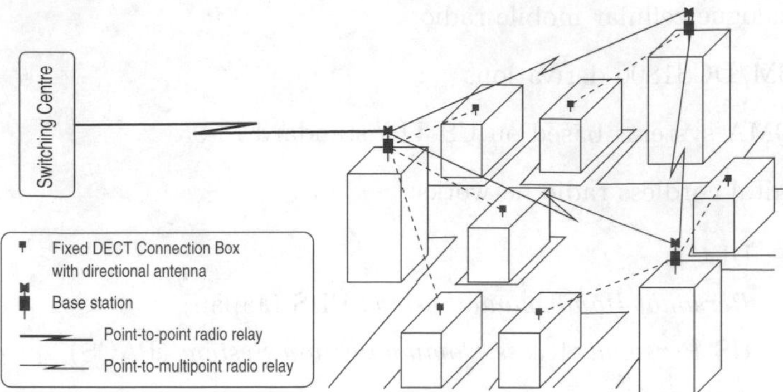

Figure 11.3: Customer coverage by over the roofs supply

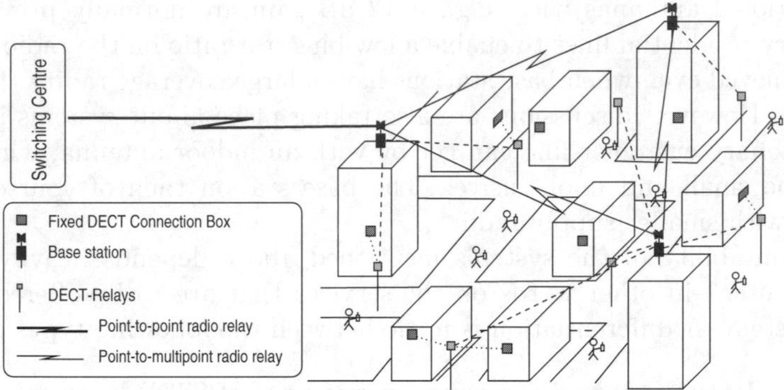

Figure 11.4: Customer coverage by below the roofs supply

Experiments have also been made with digital mobile radio networks. Studies have been conducted with systems such as GSM900/1800 to determine the feasibility of the possibility shown in the lower part of Figure 11.1. This would allow customers to communicate cost-effectively indoors using their home base stations and to use the same terminal to access the corresponding cellular network. However, mobile radio networks only offer narrowband channels and therefore cannot achieve the quality of service of a PSTN or ISDN/BRI access for wireline users. Nevertheless, all users who just want to place a call have access to the wireless connection to the mobile radio networks.

11.1.2 Digital Cordless Radio Networks

Some of the new network operators are pinning their hopes on WLL systems based on the DECT standard or on CT2 or PHS. At around 300–400 US$, the cost per connection is considerably less than with GSM-based solutions. DECT is able to support coverage *over the roofs* as well as *below the roofs* (see Figures 11.3 and 11.4). *Over the roofs* coverage would not be able to reach

Table 11.1: Frequency bands for WLL systems in Europe

Frequency band [GHz]			Applications	Channel grid [MHz]
24.549	...	26.061	PMP radio relay	3.5, 7, 14
17			Radio relay for Hiperlan/4	
3.41	...	3.60	PMP radio relay	3.5
2.5	...	2.67	PMP radio relay	
1.88	...	1.9	DECT-WLL	1.7

ditches, but with low-gain receiving antennas could penetrate through house walls. *Below the roofs* systems operate with base stations (*Radio Fixed Parts*, RFP) that are installed under the roof, thereby illuminating the ditches and enabling full mobility of the PCS system both indoors and outdoors. The traffic performance of cordless systems for the WLL application has been studied in [2].

11.1.3 Digital PMP Systems

Point-to-multipoint radio relay systems offer attractive connection options particularly for business customers, because they can provide the customer with channels with high transmission rates. The frequencies for PMP-WLL applications are listed in Table 11.1.

WLL systems based on PMP technology are offered by a large number of suppliers, and represent an interesting connection option in urban and rural areas because of their ability to bridge up to 20 km in borderline cases. Because the transmission capacity needs of stationary customers connected over PMP systems vary, there exist PMP radio relay systems that can dynamically change their (switched) channels in terms of capacity and allocate them to the areas where they are most urgently needed (see Figure 11.5).

PMP systems are frequently also implemented as multihop systems, as shown in Figure 11.6. Several radio links (*Line-of-Sight Radio*, LOS) belonging to point-to-point or point-to-multipoint (PMP) systems are arranged sequentially in order to bridge the path between the fixed network access (*Point of Presence*, POP) and the customer's connection. Another interesting option offered by a number of manufacturers is a combination of PMP systems with DECT systems connected to a PMP end terminal for the connection of residential customers.

11.2 Different WLL Scenarios

The competitive scenarios examined here are based on standard scenarios from ETSI RES-3 [1]. The following scenarios were defined:

1. An existing operator, new area to be developed.

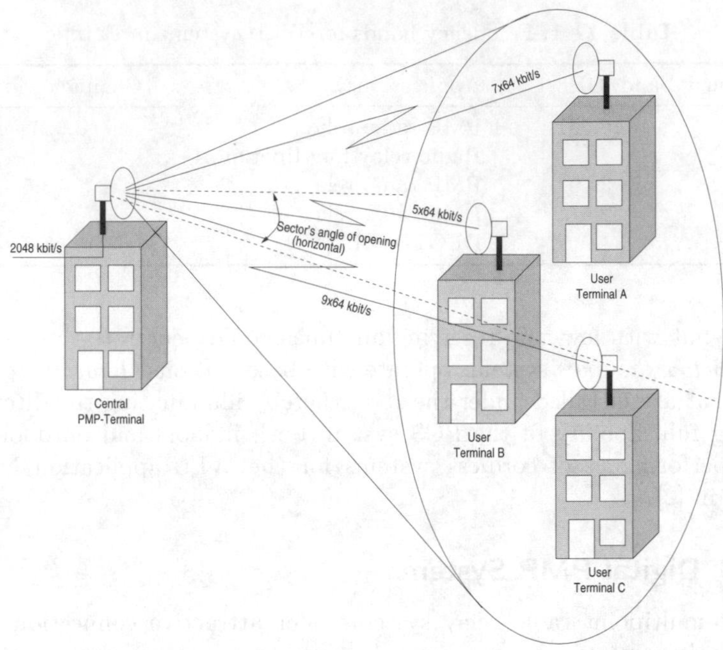

Figure 11.5: PMP radio system with a fixed or dynamic channel allocation controlled by the central terminal

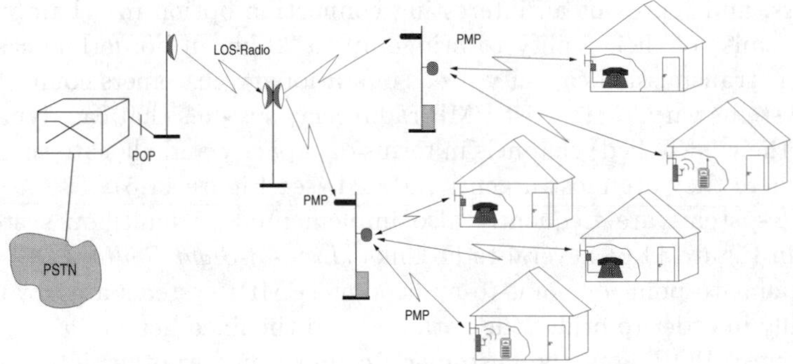

Figure 11.6: Multihop connection-oriented PMP system

2. Replacement of copper lines with WLL.

3. Reaching the capacity limits of an existing fixed network.

4. New operator in competition with existing operator.

ETSI scenario 4 is the one being considered as a model for capacity studies (see Figure 11.7). The assumption is that a provider of telecommunications

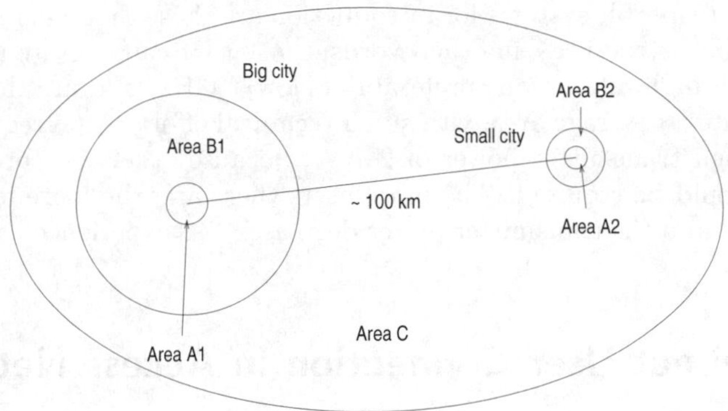

Figure 11.7: ETSI scenario 4

Table 11.2: Definition of areas of ETSI scenario

	Big city	Small city
Number of connections	500 000	50 000
Density of inner area	2000/km^2	1000/km^2
Density of outer area	500/km^2	500/km^2
Radius of inner area	4.5 km	2 km
Radius of outer area	16 km	5 km
Average traffic	70 mErl./user	70 mErl./user
Penetration	1 % increase of users p. a. during the first 10 years	

services is in competition with the national telephone company, and perhaps with other operators. Of the three areas defined in the model scenario, only areas A and B will be considered (see the assumptions in Table 11.2). It is assumed that it would not be viable to supply coverage to the sparsely populated area C.

On the assumption of a certain penetration of the respective area, e.g., connection of x % of the users of the WLL system, and with the knowledge of the capacity (in Erl./km^2) of the WLL system under discussion, the calculations in the model show that it is possible to implement voice telephony with any of the WLL technologies mentioned above. However, the initial investment required in developing a connection area is high, and operation of the service especially during the early phases is not economical because of the time involved in gradually enticing customers away from the fixed network operator. In the long run, all WLL technologies would appear to be cost-effective and promising. If the legal conditions allow a competitor direct access to individual subscribers via the wired local loop network by the so-called unbundled access, and if the costs for this use are low (as is the case in Germany at 2.3 Pfg./min), then this will have a considerable effect on the competitiveness of radio-based WLL systems, because they might appear to be too costly.

The most suitable systems for introduction are those that can manage without much infrastructure—in other words, those that can operate in frequency ranges with high diffraction (preferably below 1 GHz) and are therefore able to illuminate a coverage area with small technical effort. A prerequisite is sufficiently high transmitter power of 2–8 W, because otherwise too many base stations would be required. The more users there are, the more base stations are needed, and the transmitter power decreases—as experienced with cellular systems.

11.3 Direct User Connection in Access Network

In deregulated markets the onus is typically on the previous monopoly owner of the wireline local loop network (*incumbent*) to provide competitors with unbundled user access over their fixed networks. In so doing, the incumbent must allow the competitor, against an appropriate reimbursement of costs, to have direct access to the two-wire lines in the local loop. Non-unbundled access, on the other hand, means that competitors must use (and pay for) the incumbent's multiplex systems every time they want to reach one of their subscribers.

Owing to the changeover from analogue to digital technology (ISDN), old-established telecom companies (*incumbents*) actually have sufficient space in their main distribution frames to allow them to provide direct access within the corresponding rooms. It is not currently clear to what extent new network operators will be taking advantage of direct user access.

Bibliography

[1] ETSI (RES). *Radio in the Local Loop.*, Nov. 1994.

[2] Peter Stavroulakis. *Wireless Local Loops—Theory and Application.* John Wiley & Sons, Chichester, 2001.

[3] B. Walke, et al. *Technische Realisierbarkeit öffentlicher DECT-Anwendungen im Frequenzband 1880–1900 MHz*, Aug. 1995. Studie im Auftrag des Bundesministers für Post und Telekommunikation.

[4] W. Webb. *Introduction to Wireless Local Loop.* Artech House, Boston, 1998.

12
Personal Handyphone System (PHS)*

12.1 Development of the Personal Handyphone System in Japan

At the end of the 1980s, two cordless telephone systems, CT2 and DECT, had entered the second step of their development process. At that time, Japan had not yet developed any comparable technology; therefore the creation of a new Japanese cordless standard, which has become known as the Personal Handyphone System (PHS), was started in 1989. Disadvantages of conventional cellular telephone systems, such as high costs of infrastructure and cell-planning resulting in high communication fees, had motivated the development of a less expensive system.

The review by the *Telecommunications Technology Council*, a consulting organization for the Japanese Ministry of Posts and Telecommunications (MPT), and the technical study by the *Research and Development Centre for Radio Systems* (RCR) of PHS started at the beginning of 1991 in Japan. The PHS air interface was then standardized through the publication of the RCR STD-28 [1], Version 1, in December 1993. Various field trials in the Sapporo area in October 1993, and in the Tokyo area in April 1994, were then conducted to prove the feasibility of PHS for various demands and services, respectively.

The PHS service was commercially launched in Japan by three operator groups—(NTT (Nippon Telegraph and Telephone) Personal Group, DDI (Daini Denden Inc.) Pocket Telephone Group and the Astel Group)—in July 1995. The technology gained unprecedent popularity, with the number of subscribers reaching the six million mark by the beginning of 1997—just two years after its introduction. As of March 1998, the number of subscribers rised to about 6.7 million [7]. Thereafter the numbers were regressive, so by the end of May 2000 PHS only served 5.78 million customers. The operator groups found that the reason for the drop in numbers was not lack of interest, but drop of customers with weak payment morality. Comparison between the customer numbers of PHS and the Japanese cellular network implies migration of cus-

*With the collaboration of Matthias Siebert

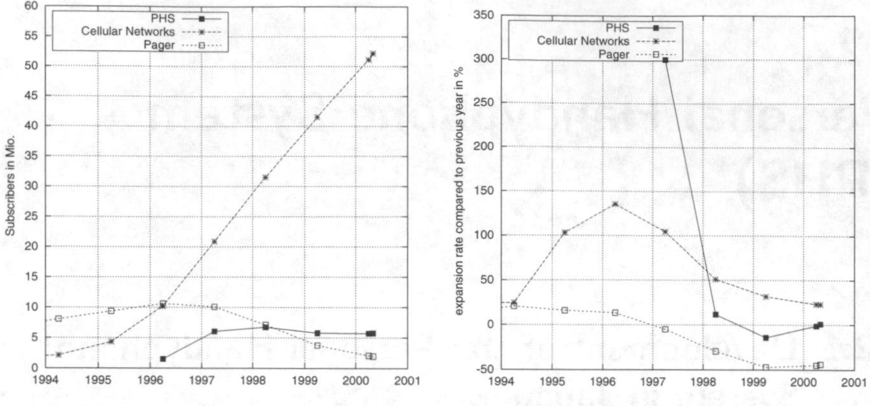

(a) Subscribers of different mobile radio systems in Japan

(b) Percentage of yearly expansion rate of mobile radio systems in Japan

Figure 12.1: Evolution of different mobile systems in Japan

tomers to be more likely (see Figure 12.1). Most of PHS's acceptance is based on its concept, being a cost-efficient but rather restricted alternative to the existing mobile network. Since mobile network prices dropped, as in Europe, PHS has lost significant market shares. Nevertheless other Asian countries, including Thailand and Hong Kong, and some South American countries, encouraged by the initial success of PHS, announced plans for the introduction of PHS-technology. An overview of the worldwide usage of PHS is given in Figure 12.2.

12.2 System Overview

The aims of PHS span those of cordless and cellular systems, encompassing the idea of a low-cost wireless handset that can be used in both indoor and outdoor environments to access fixed network supported services. Similar to cordless systems (see Chapter 9), PHS enables access to private communication systems for greater flexibility in the office environment or at home. Additionally it provides a platform for public network access by subscribers moving with pedestrian speed. A micro-cellular structure has been adopted; thus the radio transmit power can be much smaller than that of existing cellular telephone systems. As described for DECT, by means of dynamic channel allocation (DCA), cell station engineering is simplified and a multi-operator environment in the same service area is facilitated. However, some restrictions concerning the fixed control frequencies still require a certain amount of frequency planning in advance.

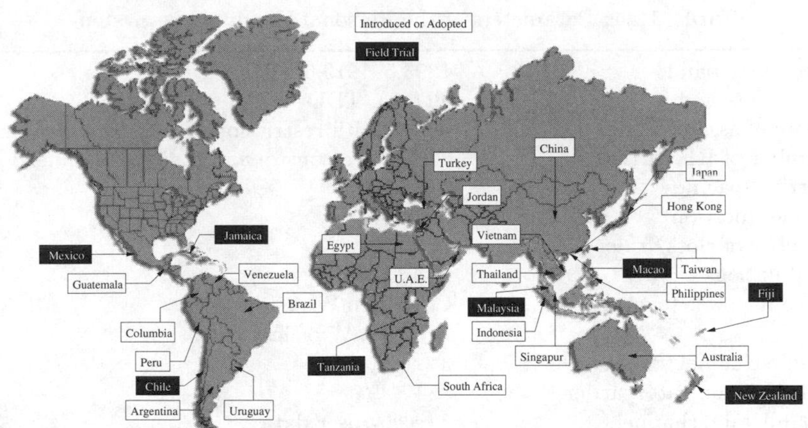

Figure 12.2: Worldwide deployment of PHS

Various services are supported. Apart from telephone services, the PHS air interface supports voiceband data and facsimile communications at rates up to 9.6 kbit/s. Also, there is the capability of data transmission at rates of 32/64 kbit/s. According to [9] the upper limit of data transmission has been raised to 128 kbit/s, using four channels simultaneously. An overview of the parameters of the Personal Handyphone System is given in Table 12.1.

Similar to CT2 (see Section 9.1), PHS defines only two network elements: A portable part (mobile terminal) and a fixed part (base station). The nomenclature for a base station is *cell station* (CS) and a PHS terminal is referred to as a *personal station*(PS).

12.2.1 Personal Station (PS)

Similarly to DECT (see Chapter 10), personal stations transmit with an average (peak) RF power of 10 mW (80 mW). As a special feature, comparable to the TETRA system (see Section 7.2), PHS also specifies direct radio communication between a pair of terminals. The direct mode of communication does not need the intervention of cell stations (CSs). As shown in Figure 12.6, ten frequency carriers are designated for such a use. However, since it is only a supplementary service that should not restrict the ordinary use of these frequencies, the direct communication must end within a time limit of three minutes.

As PHS is a system for private and public use, PSs support two modes of operation, namely public and private operation modes. The public operation mode enables the PS to access the public PHS service areas. The private operation mode enables a PS to access private systems like a wireless PBX or the home digital cordless system. The same PS can be used at home and

Table 12.1: Parameters of the Personal Handyphone System

Frequency band	1893.5–1919.6 MHz
Access method	TDMA/TDD
Channel assign. method	DCA (with restrictions for control channels)
Number of RF carriers	87 (incl. 6 control and 4 guard channels)
Carrier spacing	300 kHz
Frame duration	5 ms
Number of slots/frame	8
Modulation	$\pi/4$ DQPSK
Output power (average)	CS: 500 mW or less
	PS: 10 mW or less
Traffic channels/transceiver	3 (resp. 4)
Transmission rate/carrier	384 kbit/s
Net bit rate/channel	32 kbit/s user data
	6.8 or 12.4 kbit/s signalling information

in the office by selecting a relevant private mode within the pre-registered private systems.

Terminal mobility allows the PS a smooth transition between public and private PHS services by selecting the appropriate mode on the PS. This selection can also be done automatically (*Automatic Selection Mode*). When the PS is within the range of both services, it will operate in a predetermined mode. Optionally, an automatic dual-mode operation is possible, where the PS can be paged from both the public system and the private system when it is within the range of both.

12.2.2 Cell Station (CS)

A cell station consists of CS equipment and antennas. These two parts together create a micro-cell service area with a radius of several hundred meters. Cell stations control most of the tasks concerning the air interface. These include *dynamic channel allocation* (DCA), *superframe establishment*, *diversity support*, and many more. What is unique is the use of diversity, both for the uplink (by means of post-detection selection) and for the downlink (by means of transmitter antenna selection), controlled by the CS. The average (peak) RF power depends on the kind of CS. An outdoor standard-power CS, for example, transmits at an RF power of 20 mW (160 mW).

12.3 PHS Radio Characteristics

12.3.1 Speech Coding

Similarly to DECT, a 64 kbit/s full-rate voice coding is employed first; then the signal is transcoded into 32 kbit/s ADPCM (*Adaptive Differential Pulse Code Modulation*) based on the ITU-T Recommendation G.726.

Half-rate and quarter-rate voice coding methods are not specified at present, but getting available, a superframe structure permits the use of multiple codecs with bit-rates down to 8 kbit/s.

The ADPCM compresses speech data without degrading speech quality. Performance of voiceband data transmission via modems is not significantly degraded, and non-speech service can be supported.

12.3.2 Modulation

Within PHS a $\pi/4$-shifted differential quadrature phase shift keying technique (DQPSK) with a raised cosine pulse shaping with a roll-off factor of 0.5 is used. Demodulation can be performed in several ways, using coherent or non-coherent detection, delay detection or frequency discrimination detection [5, 10].

Being band-limited, the $\pi/4$ DQPSK signal preserves the constant envelope property better than other modulation schemes such as QPSK. As non-coherent detection is possible, $\pi/4$ DQPSK is well suited for mobile communication since the receiver implementation is simplified. Further, $\pi/4$ DQPSK performance is comparatively better than other modulation schemes (such as OQPSK) in the presence of multipath spread and fading [4].

According to [11], the use of the DQPSK modulation method enables a higher spectrum utilization efficiency compared with GMSK modulation. For multicellular systems this proposition cannot be proofed by means of the total traffic capacity balance, since this advantage is nullified by the greater required *Channel Reuse Distance* in consequence of the higher required signal-to-noise ratio of 3–4 dB C/I. A confrontation of $\pi/4$-shifted DQPSK and GMSK is given by [12, 4].

12.3.3 Access Method

In common with DECT, the access method applied in PHS is hybrid time-division/frequency-division multiple-access (TDMA/FDMA) with time-division duplexing (TDD). The ARIB Standard, RCR STD-28 Version 3, issued in November 1997, introduced specifications for PHS 64 kbit/s digital data transmission. It enables high-speed wireless access to the ISDN network from the PS by assigning a second TCH in parallel.

A TDMA frame has a length of 5 ms and carries 8 slots. The first four slots are downlink; the other four slots are uplink slots (see Table 12.1 and Figure 12.3).

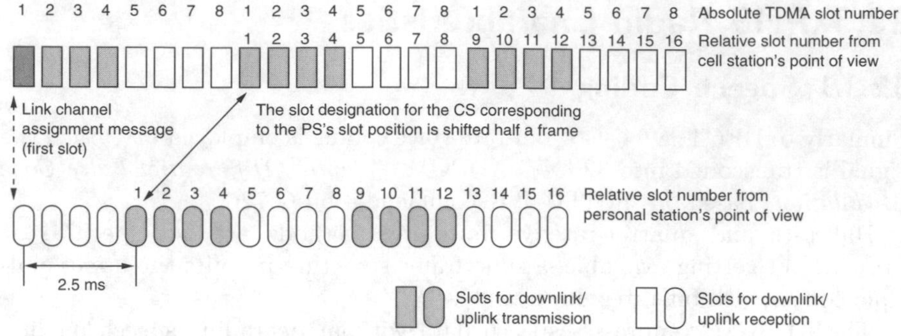

Figure 12.3: Example of relative slot numbers

The TDMA/TDD technology allows deviation from the allocation of paired transmit/receive channels, which is usually required in order to accomplish symmetric two-way communication and is able to support asymmetric communication relationships or to modify service bit rates. TDMA/TDD is flexible, because it does not need paired bands and both the lower and upper ends of the spectrum can easily be expanded to accommodate needs as done in RCR STD-28 Version 3.

12.3.3.1 Communication Physical Slot Designation Method

Designation of communication physical slots is performed by a signal (*link channel assignment* message) on a Signalling Control Channel (SCCH) (see Section 12.4.1.3), sent from CS. The slot designation position is indicated by the slot number, counted relatively to the first slot starting 2.5 ms after the signal (link channel assignment message) has been received by the PS.

An example is given in Figure 12.3: here the link channel assignment message is transmitted in the first TDMA slot. The PS waits 2.5 ms after the reception, and then starts to count the following slots until the relative value maps with that given in the message. As a bidirectional channel is always made up of a pair of slots, the second slot is defined to follow half a frame length (2.5 ms) later, corresponding to the TDD scheme.

Thus the physical time slot number is specified by a combination of absolute and relative slot number by the CS. The TDMA slot number (SN) of a communication carrier is obtained from the following equation:

$$\text{TDMA SN} = \{(\text{absolute SN} + \text{relative SN} - 2)\text{mod}4\} + 1 \qquad (12.1)$$

12.3.3.2 PHS Superframe Structure

The minimum cycle of the *downlink* logical control channel (LCCH) that specifies the slot position of the first repeated LCCH elements is specified as

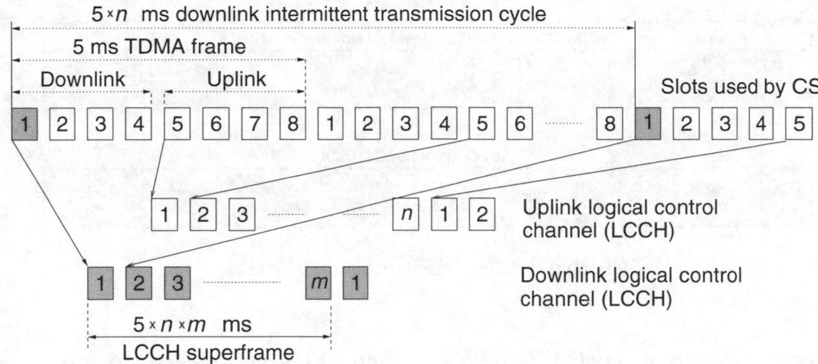

Figure 12.4: PHS superframe

the LCCH superframe. All transmission/reception timing of physical slots for controlling intermittent transmission and SCCH uplink slot designation is generated based on the superframe structure.

Elements (sub-channels) of the downlink LCCH (see Section 12.4.1 for a detailed description) are the Broadcast Control Channel (BCCH), Paging Channels (PCH, $P_1 - P_k$: number of paging groups $= k$), the Signalling Control Channel (SCCH) and an optional User-Specific Control Channel (USCCH).

The BCCH must be transmitted in the first slot of the LCCH superframe whereby the lead position of the superframe is reported. In detail, this is done by means of profile data contained in the *radio channel information broadcasting* message (see Section 12.5.5.1).

The downlink logical control channel (LCCH) has the superframe structure shown in Figure 12.4. After each n TDMA frames, the CS intermittently transmits an LCCH slot (as discussed in Section 12.4.1, the CS does not transmit control signals in each TDMA frame). The parameter m defines the number of LCCH elements that have to be transmitted until all kinds of information have been conveyed once.

For the uplink, no superframe structure is defined. Personal stations transmit their first signalling message by using the *Slotted Aloha* protocol (see Section 2.8.1), using an uplink SCCH slot, if they want a connection to be established.

12.3.4 Slot Structure

As a frame in PHS lasts 5 ms and each frame consists of 8 slots, each slot has a length of 625 µs in which 240 bits can be arranged. PHS specifies several time-slot formats corresponding to different logical channels. Basically, there are two categories: *control physical slots* used by *common control channels* (CCHs), and *communication physical slots* used by *traffic channels* (TCHs); see Figure 12.5.

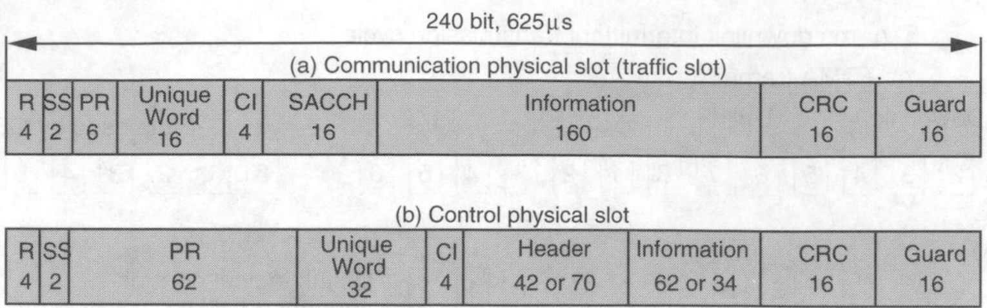

Figure 12.5: PHS time-slot formats

In contrast to the DECT standard, where, as a concession to more cost-effective hardware, certain slots (and related traffic channels) might not be used (see Figure 10.50), PHS makes use of all its time slots. Therefore all slot formats start with a 4-bit ramp time in which the PS or CS turns on its transmitter and a 2-bit start symbol for establishing the phase of the remote demodulator. At the end of each slot format, there is another common data field for performing an ITU-T 16-bit CRC. The start symbol is followed by a *preamble*, a layer-1 signal pattern used to establish bit synchronization. Its number of bits depends on the kind of slot format. Together, start symbol and preamble are repetitions of the pattern 1001. With control physical slots, a preamble of 62 bits is used to allow synchronization for each slot independently; with communication physical slots, a preamble of 6 bits serves to update the *synchronization* established in the previously transmitted slots.

Additionally, all time slots carry a *unique word*, known in advance by the receiver, which is different for the downlink and uplink channels. It also differs for control physical slots and communication physical slots, so that it helps to distinguish between them. Moreover, together with the CRC check, it is used for error detection. As it is known in advance, the receiving party listening for its dedicated slot either detects it or not (unique word detection error).

The *channel identifier* follows right after the unique word. It is similar to the flag bits in GSM bursts. For example, the channel identifier sequence 0000 indicates that the time slot carries user information (TCH), while 0001 indicates that the time slot carries a fast associated control channel (FACCH). Slow associated control channels (SACCH) do not have a special channel identifier since they are part of each communication physical slot, with the exception of an optional user-specific packet channel (USPCH), described in Section 12.4.2.4.

Major data field in Figure 12.5 (a) is the *information field*, which carries the user data. For telephone services such as voice transmission it consists of 160 bits from an adaptive differential pulse code modulation (*ADPCM*) encoder. Therefore, in common with DECT, the bit rate carried by communication physical slots (with the exception of USPCH) is

$$\frac{160 \text{ bit/frame}}{0.005 \text{ s/frame}} = 32 \text{ kbit/s} \qquad (12.2)$$

Control physical slots have headers containing addresses. Broadcasting point-to-multipoint channels such as the *broadcasting control channel* (BCCH) and the *paging channel* (PCH) only need to transmit a 42-bit CS identification code. Thus the information field contains 62 bits. On the other hand, bidirectional point-to-point channels such as the *signalling control channel* (SCCH) need to specify a CS identification code (42 bits) and a PS identification code (28 bits). In these channels the length of the information field is 34 bit. This implies that the information rate for logical control channels is either

$$\frac{62 \text{ bit/frame}}{0.005 \text{ s/frame}} = 12.4 \text{ kbit/s} \quad \text{or} \quad \frac{34 \text{ bit/frame}}{0.005 \text{ s/frame}} = 6.8 \text{ kbit/s} \qquad (12.3)$$

Finally, all slot formats have a *guard time* of 41.7 µs, corresponding to 16 bits, i.e., a burst carried in a slot is 224 bits in length.

12.3.5 Radio-Frequency Band

The PHS band spans 26.1 MHz from 1893.5–1919.6 MHz. Originally the radio-frequency band allocated for PHS service spanned 23.1 MHz in the range of 1895.0–1918.1 MHz. Because it was working close to capacity in hot-spot areas, an extension became necessary. There are 87 carriers in the PHS band, with a spacing of 300 kHz. Figure 12.6 shows the relationship between the PHS frequency band and the carrier numbers. Control carriers for private use are assigned to 1898.450 MHz and 1900.250 MHz for Japan, and 1903.850 MHz and 1905.650 MHz for other countries. In Japan, four control carriers are reserved for public use. One control channel is assigned to each of the three PHS operators, and the remaining one is set aside as a spare channel. In order to protect the public control channels from adjacent channel interferences, they are enclosed by guard channels. Currently the spectrum for public use is 15 MHz. The spectrum for private use is 11.1 MHz, which can be shared with public use. In addition, the first 10 carriers of private use (1895.15–1897.85 MHz) are also designated for direct communication between personal stations (Transceiver mode, Walkie Talkie).

12.3.6 Frequency Allocation

As Figure 12.6 shows, the PHS frequency band is not separately allocated for each operator, but is shared, with the exception of the channels that are pre-assigned as control channels dynamically by all PHS operators by means of *dynamic channel assignment* (DCA). This is an autonomous decentralized radio channel control technology, which enables efficient and flexible use of frequencies and time slots according to the local interference levels of CS and PS.

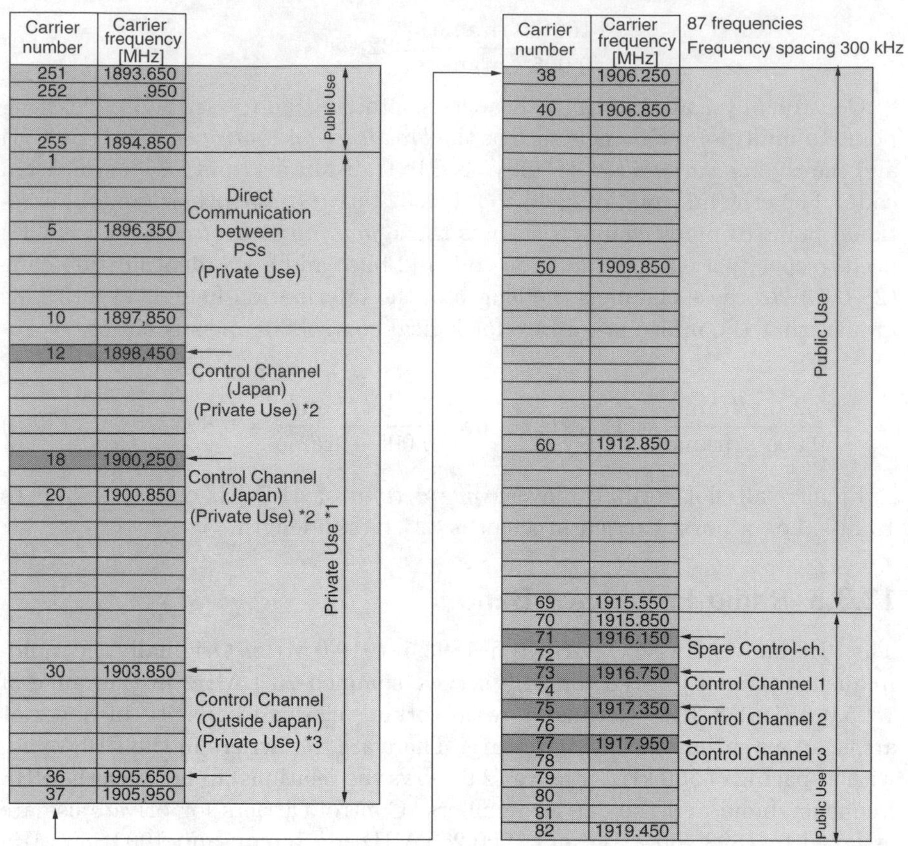

Figure 12.6: PHS frequency table

In contrast to the DECT system, where it is the mobile's task to select a suitable channel, in PHS this is done by the cell station. With a *link channel establishment request* or a *TCH switching channel request* (see Section 12.5.5.1), the PS asks for the assignment of a channel. The cell station can automatically pick up carriers at random and select an available carrier that has no interference problem. As a result of checking signal strength when a call is established, the CS renews an internal two-dimensional frequency–time matrix with available channels.

If no carrier is available, the CS refuses the request. The PS will then automatically request again. This can be done up to three times; then the PS has to wait a certain time before another try is possible.

12.3.7 Micro-cellular Architecture

PHS applies a micro-cellular architecture, thus the available spectrum is used more efficiently. PSs with low-power consumption can be used and long standby/talk times can be realized. The PHS system allows to incorporate a large number of cell stations without frequency planning. (In fact, this is only true for traffic channels. Concerning the fixed control channels, frequency planning still has to be done). For example, NTT launched a PHS service in July 1995 with some 25 000 standard cell stations in Tokyo, using a conventional small-cell structure, 150–200 m, with antennas mounted on public telephone boxes.

The micro-cellular technology allows easy addition and removal of CSs in the field, based on factors such as traffic demand and the presence of other operators serving the same area. Cell stations in city areas usually cover a micro-cell with a diameter of 200 m and transmit with an average power of 10 mW. Increasing cell size is possible but at the cost of higher-power transmitters and more accurate low-noise amplifiers.

12.3.8 Handover

In PHS, a handover is referred to as *channel switching*. During a call, both network elements—the cell station and the personal station—observe the channel quality. This is done by evaluating the *radio signal strength indicator* (RSSI) and the *frame-error ratio*. In response to deteriorating quality, either network element can initiate a handover.

PHS distinguishes between two types of handover: one is called *recalling-type* and the other *traffic channel switching-type*.

TCH switching-type This kind of handover is performed if a high frame-error ratio is detected but the received signal strength indicates that the terminal is still in the vicinity of the serving cell station; this is mostly due to rising interference. Remedial measures can be taken by switching to another slot or frequency. Thus the serving cell station does not need to be changed. To change the channel, the PS transmits a *TCH switching request* message on the fast associated control channel (FACCH). Thereupon the CS replies with a *TCH switching indication* message, assigning PS to a new physical channel. After successful exchange of some synchronization bursts on the new channel communication is resumed. In other systems this function is called intracell handover.

Recalling-type Within this inter-cell handover the PS establishes a connection in a new cell in the same way as a call originating in a new communication cell. This function allows the call to be maintained whilst the traffic channel is switched between cell stations or interfaces as the personal station moves during the call. The stimulus for this kind of handover had been that the received radio signal strength became too weak to maintain communications with the serving cell station. Measuring signal strengths on common control channels received from surrounding cells, the PS picks up a new CS. Therefore it transmits a *link channel establishment request* message to the chosen cell station on a signalling control channel (SCCH). In answer to this, the CS responds with a *link channel assignment* message, also transmitted on SCCH, and both stations start synchronizing their operations on the new channel.

To perform this kind of intercell handover, both PS and CS need access to a signalling control channel. For the uplink, a *Slotted-ALOHA* protocol is applied, and the downlink's access is restricted by the superframe structure. As a result, seamless handovers cannot take place owing to a short time gap up to 2 seconds.

According to [6], intercell handovers are performed to support connections at vehicular speed. However, this is not guaranteed by the providers. If a new cell station is not available when the handover is required, the communication channel is switched back to the previous CS. Therefore the communication channel is restored when returning quickly to the previous cell.

12.4 PHS Radio Channel Structures

The Personal Handyphone System supports traffic channels (TCHs) and control channels (CCHs), as shown in Figure 12.7.

12.4.1 Logical Control Channels (LCCH)

The basic structure for all kind of control slots, with the exception of the associated control channel (ACCH), is a control slot as shown in Figure 12.8.

12.4.1.1 Broadcast Control Channel (BCCH)

This is a one-way downlink channel for broadcasting control information from the cell station to the personal station. It transmits information related to

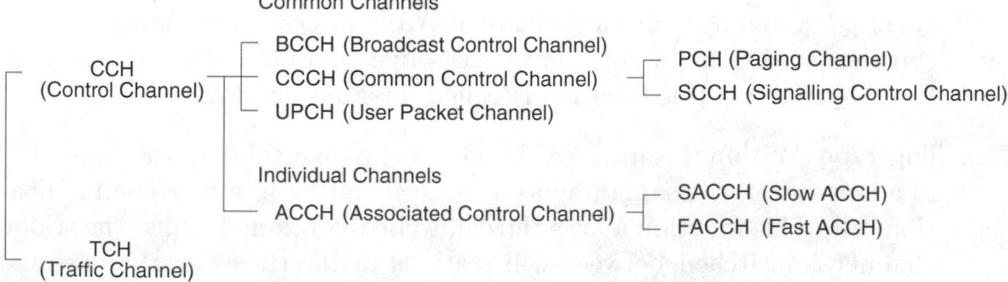

Figure 12.7: Function channel structure

channel structure, system information and restrictions. The BCCH carries general system information to all terminals in a cell. Cell stations establish a superframe structure (see Figure 12.4) and inform personal stations by means of messages transmitted on the BCCH (see Section 12.5.5.1). The transmission of these paging messages is scheduled in a manner that enables terminals to turn off their receivers for a high proportion of the time. Thus, corresponding to the sleep mode of digital cellular systems, PHS defines a *battery saving* mode of operation for PSs that do not have a call in progress.

12.4.1.2 Paging Channel (PCH)

This is a one-way downlink point-to-multipoint channel that simultaneously transmits identical information to individual cells or a wide area of multiple cells (the paging area) from the CS to PS (for more details see Section 12.5.5.1).

12.4.1.3 Signalling Control Channel (SCCH)

This is a bidirectional point-to-point channel that transmits information needed for call connection between CS and PS by means of messages (for detailed information see Section 12.5.5.1). It transmits independent information to each cell. Personal stations contend for access to the reverse direction randomly, by means of the *Slotted Aloha* protocol.

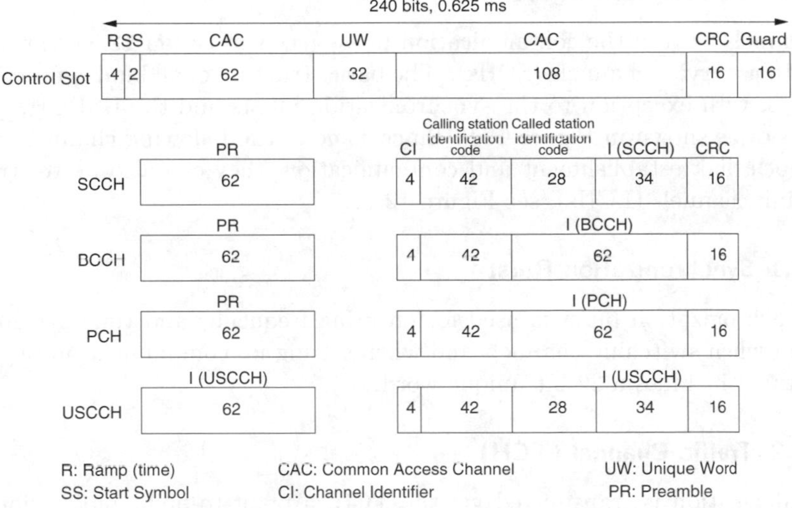

Figure 12.8: Control physical slot structures

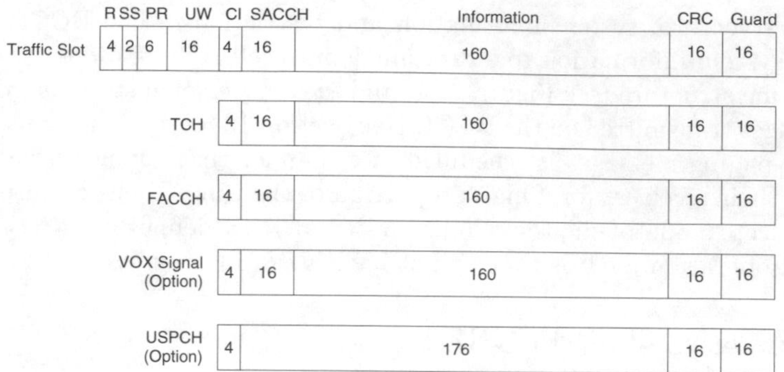

Figure 12.9: Communication physical slot structures

12.4.1.4 User-Specific Control Channel (USCCH)

A User Packet Channel (UPCH) that can be set up on the control physical slot is defined as the USCCH (see Figure 12.8). If it satisfies the specified items, the usage method is arbitrary. This two-way channel for packet data communication of user information can be used for transmitting short messages to and from personal stations that do not have calls in progress. The uplink channel is random-access.

12.4.2 Service Channels

All channels used in the communication phase (see subsection 12.5.6) are referred to as service channels (SCHs). The basic structure of all kinds of service channels, with exception to the synchronization bursts and the USPCH, is a traffic slot as shown in Figure 12.9. Since some of the following channels are used both, link establishment and communication, they can also be referred to as link channels (LCHs), see Figure 12.12.

12.4.2.1 Synchronization Burst

The synchronization burst is used for ensuring frequency and time synchronization when switching channels and when setting up communication physical slots. It includes a 32-bit unique word.

12.4.2.2 Traffic Channel (TCH)

User information is transmitted via the TCH, a point-to-point bidirectional channel.

12.4.2.3 Associated Control Channel (ACCH)

This is a two-way channel that is associated with a TCH. It carries out transmission of control information and user packet data necessary for call control. The channel that is ordinarily auxiliary to TCH is called SACCH (*Slow Associated Control Channel*), and the channel that temporarily steals TCH capacity and carries out high-speed data transmission is called FACCH (*Fast Associated Control Channel*). In case of a higher signalling demand, e.g. during a handover, this channel temporary uses the TCH capacity to provide fast data transfer. SACCH and FACCH are mutually independent, and may exist on the same physical slot.

Slow Associated Control Channel (SACCH) The SACCH conveys bidirectional control signals between a cell station and a personal station related to a call in progress.

Fast Associated Control Channel (FACCH) The FACCH can temporarily steal slots from the traffic channel (these are the slots carrying the user information), in order to exchange time-critical control information, e.g., for handover support.

12.4.2.4 User-Specific Packet Channel (USPCH)

The UPCH that can be set up on the communication physical slot is defined as USPCH. If it satisfies the specified items, the usage method is arbitrary. The USPCH uses a service channel assigned to a specific terminal. The time-slot structure of the USPCH (see Figure 12.9) resembles the communication physical slot structure, but with the difference that the USPCH does not contain a SACCH field. Instead, these 16 bits become part of the information field, now holding 176 bits.

12.4.2.5 Voice Activity Channel (VOX)

This is an optional channel that transmits VOX control start and background noise information. It operates based on a voice-activity detector that turns off the PS transmitter when a pause in the speech of the subscriber using the PS is detected. When this happens, the PS transmits a special message in VOX format in one out of every four frames describing the background noise at the PS. In such a way, this operation modus helps conserve battery power.

12.5 Network Operations

In a manner similar to DECT and GSM, PHS formally defines three network control protocols:

1. Radio-Frequency Transmission Management (RT)

2. Mobility Management (MM)
3. Call Control (CC)

These layer-3 standards specify procedures for establishing, maintaining, switching, releasing network connections, PS location and authentication at the radio interface of the Personal Handyphone System. These procedures apply to messages exchanged through link channels in the link channel establishment phase and service channels in the other phases. Their aim is to provide the environment, procedures and messages needed for radio control, mobility control and call control in link channels as in service channels.

12.5.1 Radio-Frequency Transmission Management (RT)

RT incorporates radio resources management functions and also signal encryption. The radio frequency transmission entity RT has functions related to management of radio resources. These functions include radio zone selection, radio line setup, maintenance, switching and disconnection functions.

12.5.2 Mobility Management (MM)

The mobility management entity has functions related to the mobility support of a PS. These functions include location registration, authentication functions and handover support.

For each network control protocol (RT, MM, CC), a state transition diagram can be set up. Examples of how this is done are shown in Figure 12.10 and Figure 12.11. Incoming and outgoing messages (the latter are in parentheses), which are described below, stimulate state transitions.

Messages for mobility management (MM), which are transmitted on SACCH/FACCH, are (see Figures 12.10 and 12.11):

Authentication Request (CS → PS, DL) This message is for confirming the legitimacy of PS, and it is transmitted from CS to PS.

Authentication Response (PS → CS, UL) This message is for reporting the response to the Authentication Request. It is transmitted from PS to CS.

Function Request (PS → CS, UL) This message is transmitted from PS to CS when PS requests MM functions from CS.

Function Request Response (CS → PS, DL) This message is transmitted from CS to PS for responding to the MM function request from PS.

Location Registration Acknowledge (CS → PS, DL) This message, transmitted from CS to PS, is for reporting that the location registration request was accepted.

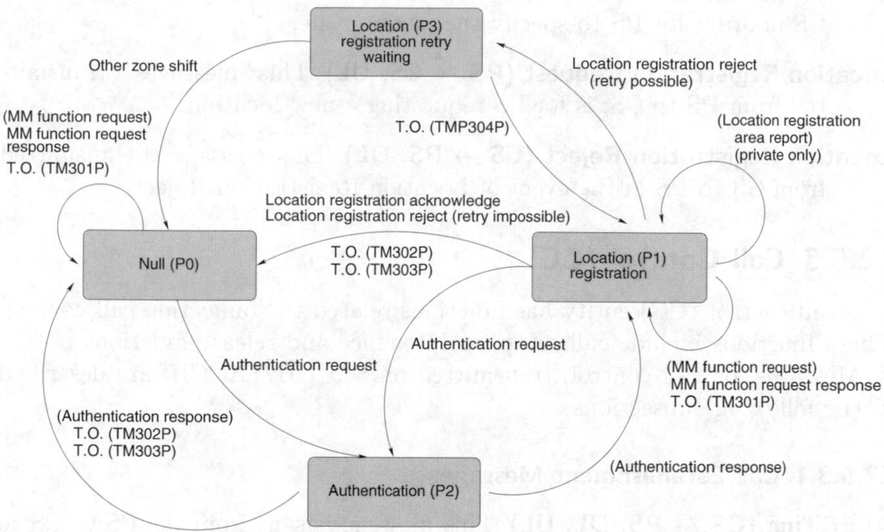

Figure 12.10: MM state transition diagram (PS side)

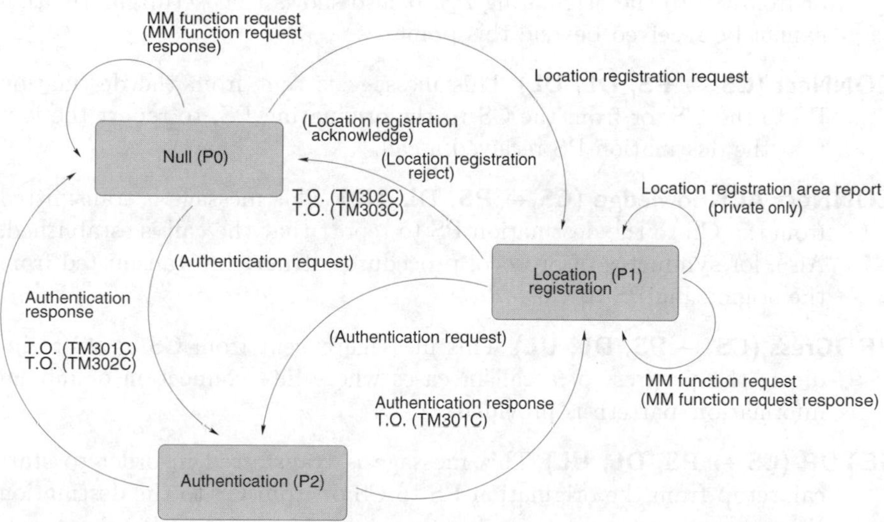

Figure 12.11: MM state transition diagram (CS side)

Location Registration Area Report (PS → CS, UL) This message is used to transmit from PS to CS the specified CS-ID and reception level from CS in order for PS to specify the paging zone.

Location Registration Request (PS → CS, UL) This messages, transmitted from PS to CS, is for PS requesting a new location.

Location Registration Reject (CS → PS, DL) This message is transmitted from CS to PS in the event of Location Registration Reject.

12.5.3 Call Control (CC)

The call control (CC) entity has functions related to connection call control. These functions include call setup, maintenance and release functions.

Messages for call control, transmitted on SACCH/FACCH, are described in the following subsections.

12.5.3.1 Call Establishment Messages

ALERTing (CS ↔ PS, DL, UL) This message is sent from the PS to CS to indicate the fact that the PS is in the process of calling. The CS sends this message to report to the PS that a call is initiated on the destination side.

CALL PROCeeding (CS ↔ PS, DL, UL) This message shows that the requested setup is started. It is transferred from the destination PS to CS or from CS to the originating PS. It also shows that setup information cannot be received beyond this point.

CONNect (CS ↔ PS, DL, UL) This message is sent from the destination PS to the CS, or from the CS to the originating PS, to report the fact that the destination PS received a call.

CONNect ACKnowledge (CS ↔ PS, DL, UL) This message is transmitted from the CS to the destination PS to report that the call is established. Also, for symmetry of protocol procedures, it may be transmitted from the originating PS to CS.

PROGress (CS ↔ PS, DL, UL) This message is sent from CS or PS to indicate the progress of a call in cases where interconnection or inband information/pattern is provided.

SETUP (CS ↔ PS, DL, UL) This message is transferred in order to start call setup from the origination PS to CS or from CS to the destination PS.

SETUP ACKnowledge (CS → PS, DL) This optional message is sent from CS to the origination PS to describe that call setup is initiated and to request further information. It can only be used in private systems.

12.5.3.2 Call Clearing Messages

DISConnect (CS $\leftrightarrow$ PS, DL, UL) This message is transmitted from PS to request call clearing from CS, or from CS in order to state that the call was disconnected.

RELease (CS $\leftrightarrow$ PS, DL, UL) This message is transmitted in one direction from either PS or CS. It shows that the equipment that transmitted this message has already disconnected the traffic channel. Further, it is transmitted to release the traffic channel and call reference.

RELease COMPlete (CS $\leftrightarrow$ PS, DL, UL) This message is transmitted in one direction from PS or CS. It shows that the equipment that transmitted this message has already released the traffic channel and call reference.

12.5.3.3 Other Messages

FACility (CS $\leftrightarrow$ PS, DL, UL) This message is sent in order to request or check supplementary services.

INFOrmation (CS $\leftrightarrow$ PS, DL, UL) This optional message is used when it is desired to transmit information from PS or CS.

NOTIFY (CS $\rightarrow$ PS, DL) This message is sent from CS to PS to indicate information pertaining to a call, such as user suspended or user resumed.

STATus (CS $\leftrightarrow$ PS, DL, UL) This message is transferred from either PS or CS at any time within the call for reporting error status or as a response to STATus ENQuiry message.

STATus ENQuiry (CS $\leftrightarrow$ PS, DL, UL) This optional message is transmitted at any time for requesting a status message from layer 3 to PS or CS. If this message is received, transmission of the STATus message is mandatory as a response.

12.5.4 Protocol Model

In general, communication is made up of three phases: call establishment, communication and call release. As shown in Figure 12.13, the *link channel establishment phase* is followed by the *service channel establishment phase*, and finally by the *communication phase*. Figure 12.12 shows what channels are being used as link, service and logical control channels respectively.

The link channel establishment and the service channel establishment phases use the control protocol stack, while the communication phase uses layers 1 to 2 of the user plane conformant to the ISO-OSI model (see Section 2.5). At the end of the communication phase, the control plane (management) is used to disconnect the call; see Figure 12.13.

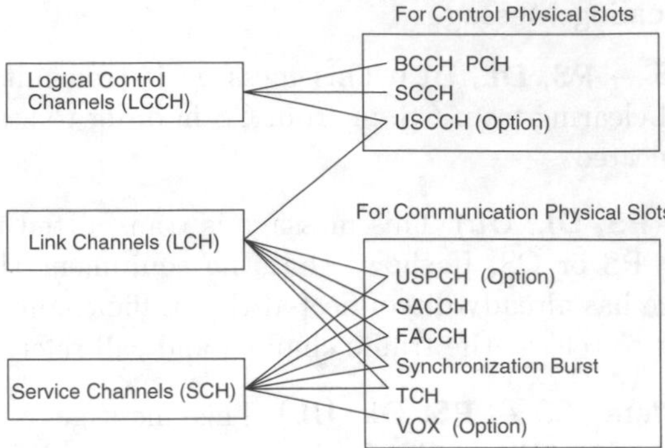

Figure 12.12: Correspondence between protocol phase channels and function channels

12.5.5 Call Establishment

The call connection phase consists of the *link channel establishment phase* for establishing the link with the radio interface, and the *service channel establishment phase* for establishing the radio link for telephone service such as voice transmission and non-telephone service such as ISDN.

12.5.5.1 Link Channel Establishment

The link channel establishment phase is defined as the stage of using control channel functions to select a channel (henceforth referred to as *link channel*) with the quality and capacity required for each service's call connection, and to select the protocol type required in the next phase of a call connection. For the purpose of assigning a dedicated channel for communication, personal stations and base stations exchange messages on common control channels, such as BCCH, SCCH and PCH. Figure 12.14 shows the functions of call control involved in this phase.

A possible start of a communication is described below: On receiving a paging message from the CS, or a stimulus to register the location of the PS or the user's wish to originate a call, the PS starts the process by transmitting a link channel establishment request message using a signalling control channel (SCCH). After receiving a request from the PS, the CS selects a service channel and transmits the chosen pair of slot and frequency, together with a link channel assignment message, to the personal station, again using the SCCH. In contrast to DECT, the CS is responsible for allocating a service channel. The PS then tunes to the assigned channel, and the two stations transmit synchronization bursts to establish communication.

Messages defined in the link channel establishment phase are as follows:

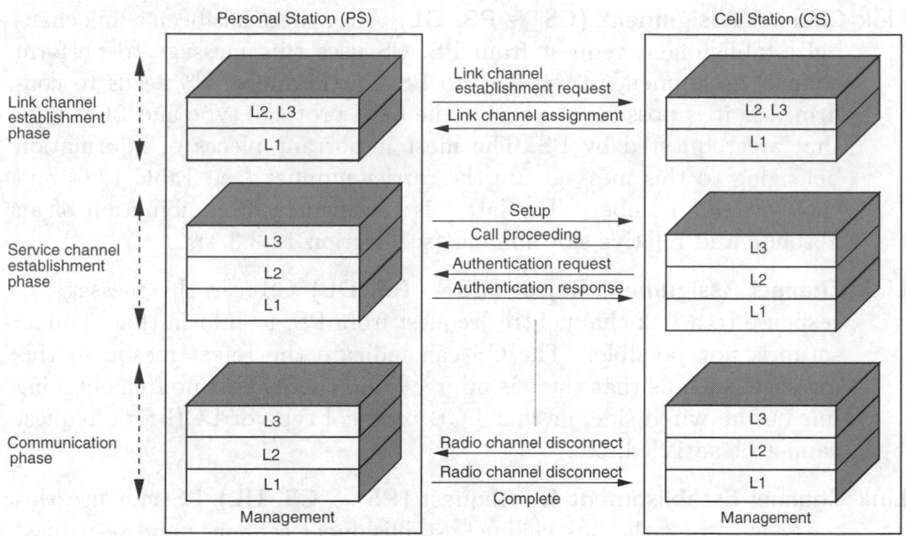

Figure 12.13: Basic structure of signals

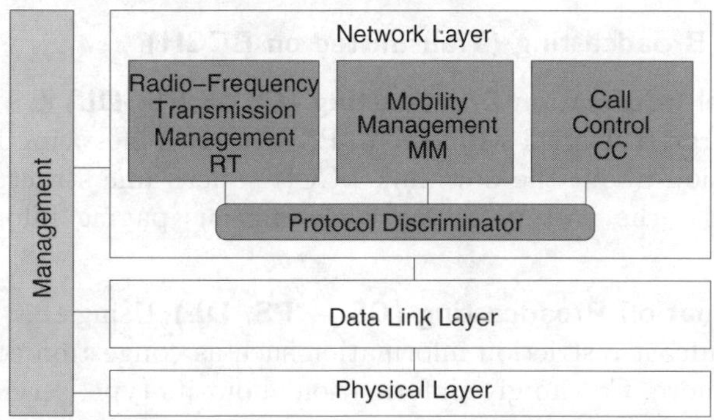

Figure 12.14: Hierarchical structure of service channel establishment phase and communication phase

Messages for Channel Setup (Transmitted on SCCH)

Idle (CS → PS, DL) In downlink SCCH timing, only when there is no significant signal to be transmitted.

Link Channel Establishment Request (PS → CS, UL) The PS carries out link set up request to CS using this message. The link set up request is only possible from PS to CS (uplink). PS designates the link channel (LCH) type needed for communication and the call connection protocol type, and transmits the link channel establishment request to CS.

Link Channel Assignment (CS $\rightarrow$ PS, DL) After there has been a link channel establishment request from PS, CS uses this message to perform channel assignment in response to PS. Furthermore, CS needs to confirm that it is possible to provide the LCH protocol type and LCH type that are requested by PS. The most important pieces of information belonging to this message are the carrier number (see Table 12.6) and the time slot number. The latter is specified by a combination of an absolute and relative slot number (see Section 12.3.3.1).

Link Channel Assignment Reject (CS $\rightarrow$ PS, DL) CS uses this message, in response to a link channel (re-)request from PS, to inform that channel setup is not possible. The CS can indicate the reject reason in this message, such as that there is no free radio channel, or no free outgoing line on the wired side, or that LCH protocol type or LCH type request cannot be satisfied, etc.

Link Channel Establishment Re-Request (PS $\rightarrow$ CS, UL) PS can use this message to halt the link channel establishment sequence and re-request CS for link channel establishment after a link channel assignment message has been received from CS.

Messages for Broadcasting (Transmitted on BCCH)

Radio Channel Information Broadcasting (CS $\rightarrow$ PS, DL) CS must broadcast the radio channel structure to PS using this message. This includes information about the downlink LCCH superframe structure (see Section 12.4), the battery saving cycle, various paging information and others.

System Information Broadcasting (CS $\rightarrow$ PS, DL) Using this message, CS can broadcast restriction information such as congestion control to PS. Furthermore, CS can give information about its type, service attributes and other.

2nd-System Information Broadcasting (CS $\rightarrow$ PS, DL) By using this message, CS can broadcast information such as country code and system type to PS.

3rd-System Information Broadcasting (CS $\rightarrow$ PS, DL) CS can broadcast 3rd-system information for a public system to PS, using this message.

Option Information Broadcasting (CS $\rightarrow$ PS, DL) Using this message, CS can broadcast private system option information to PS.

Message for Paging (Transmitted on PCH)

Paging (CS $\rightarrow$ PS, DL) Using this message, CS reports that PS received a call. When PS responds to the paging from CS, it is necessary to activate the link channel establishment request.

12.5.5.2 Service Channel Establishment

The service channel establishment phase is defined as the stage of using link channel functions obtained in the link channel establishment phase to select a channel (henceforth referred to as the *service channel*) with the capacity required for providing service and to select the protocol type required in the communication phase.

Ordinary call control (CC), higher-level mobility management (MM) and radio-frequency transmission management (RT) functions are performed in this phase. Messages are transmitted by means of an associated control channel (ACCH).

Messages Pertaining to Activation of Communication (Transmitted on SACCH/FACCH)

Definition Information Request (PS $\rightarrow$ CS, UL) This message is transmitted to CS for requesting notification of definition information, e.g. broadcast information like area information.

Definition Information Response (CS $\rightarrow$ PS, DL) This message is transmitted from CS to PS to report definition information including area information like a channel switching FER threshold value.

Encryption Key Set (PS $\rightarrow$ CS, UL) This message is used to report the encryption key, and is transmitted from PS to CS as necessary. It is applied in the service channel establishment phase. If the encryption key was set by this message, the encryption function is effective starting from the first TCH data of the communication phase.

Function Request (PS $\rightarrow$ CS, UL) This message is transmitted from PS to CS to perform an RT function request to CS.

Function Request Response (CS $\rightarrow$ PS, DL) This message is transmitted from CS to PS to notify PS of standards and to respond to an RT function request from PS.

Paging Response (PS $\rightarrow$ CS, UL) This message is transmitted from PS to CS to respond to a paging call from CS.

Zone Information Indication (CS $\rightarrow$ PS, DL) This message is transmitted from CS to PS to report zone information.

Messages Pertaining to Connection Release (Transmitted on SACCH/ FACCH)

PS Release (CS $\rightarrow$ PS, DL) This message is transmitted from CS to PS in order to unilaterally release the radio channel.

Radio-Channel Disconnect (CS → PS, DL) CS transmits this message to PS to release the radio channel.

Radio-Channel Disconnect Complete (PS → CS, UL) This message from PS to CS is used to indicate the fact that the radio channel was released and as a response to Radio-Channel Disconnect. After PS transmits this message, it enters standby state.

Messages Pertaining to Connection Establishment (Transmitted on SACCH/FACCH)

Condition Inquiry (CS → PS, DL) This message is transmitted from CS to PS to query the reception level of the local zone and peripheral zones.

Condition Report (CS ↔ PS, DL, UL) This message is transmitted from CS to PS to autonomously report a condition, and from PS to CS to respond to a condition inquiry.

Encryption Control (CS ↔ PS, DL, UL) This messages is transmitted in either direction to indicate operation or stopping of the encryption function.

Encryption Control Acknowledgement (CS ↔ PS, DL, UL) This message is transmitted in either direction as an acknowledgement of encryption control. Encryption control in the communication phase (start/stop of encryption process) is activated by transmission of this message.

TCH Switching Indication (CS → PS, DL) This message from CS to PS is used to indicate TCH switching. After PS has received this message, it immediately starts the operation of switching to the specified channel.

TCH Switching Request (PS → CS, UL) This message is transmitted from PS to CS to perform a TCH switching request from PS to CS.

TCH Switching Request Reject (CS → PS, DL) This message is transmitted from CS to PS when the TCH Switching Request to CS from PS is rejected.

TCH Switching Re-Request (PS → CS, UL) This message is transmitted from PS to CS to re-request TCH switching from PS to CS. It is used for channel switching during communication.

Transmission Power Control (CS ↔ PS, DL, UL) This message is transmitted in order to specify operation or prohibition of the Transmission Power Control function.

VOX Control (CS ↔ PS, DL, UL) This message is transmitted to indicate operation or prohibition of the VOX function.

PS-ID Notification (CS $\leftrightarrow$ PS, DL, UL) This message is transmitted in either direction to report the respective PS identification code (PS-ID) during communication.

12.5.6 Communication Phase

When both parties are connected, the call enters the communication phase which is characterized by the exchange of user information which can be both, voice and non-voice data. Additional control signals between peer-entities are transmitted by an associated control channel (ACCH).

For reception optimization and interference reduction during a call, either network element can advise the other to adjust its transmitted power by transmitting a transmission power control message on either the slow associated control channel (SACCH) or the fast associated control channel (FACCH). This message requests an adjustment between -32 dB and $+32$ dB relative to the current power. Other messages pertaining to connection establishment and connection release that are used in this phase have been presented in Section 12.5.5.2.

In the communication phase, the optimum channel and the optimum protocol for each service will be employed.

12.6 Network Interfaces/Technologies

12.6.1 Private Communication System Application

Home Digital Cordless Telephone The PHS terminal, together with a CS that includes a cordless base unit function, can be used as a home digital cordless telephone, thereby extending the access method of conventional wired service. The air interface between PSs and CSs complies with the previously described PHS air interface standard. To provide security, the standard specifies special features, e.g., scrambling.

Wireless PBX/LAN PHS can also function as a private communication system for office use, such as the digital wireless Private Branch eXchange (PBX) system and the wireless Local Area Network (LAN). The digital wireless PBX system consists of Personal Stations (PSs), Cell Stations (CSs) and a digital PBX, which is responsible for the overall system control. The air interface between PSs and CSs complies with the previously described PHS air interface standard.

One possible future development is the integrated services concept of the fixed and the mobile networks. The PHS MoU [9] group has specified the minimum common technical requirements for PHS/GSM dual-mode terminals, which realize the integration of the GSM service and the PHS home-office cordless telephone service.

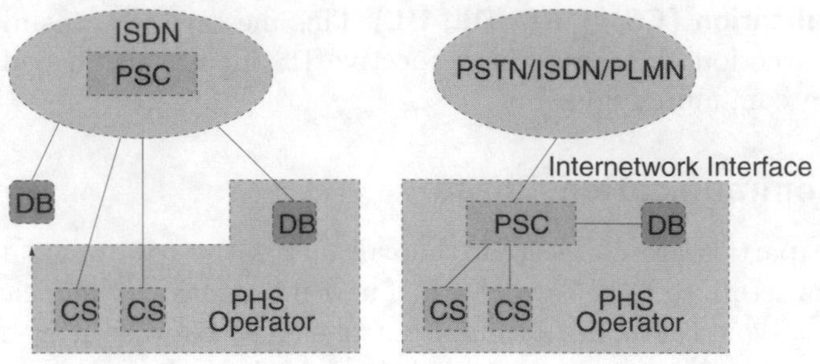

Figure 12.15: Types of public PHS networks

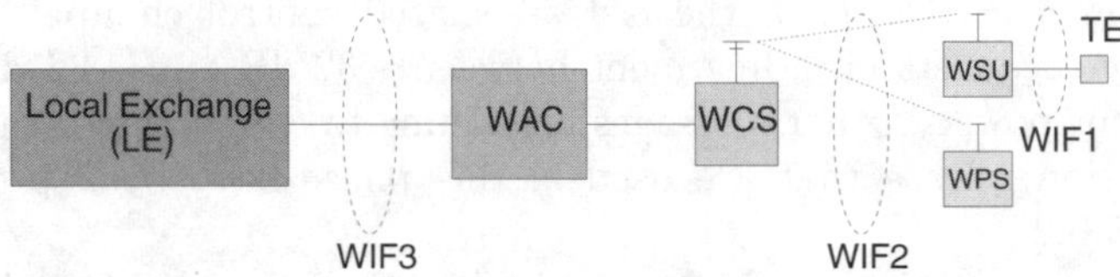

Figure 12.16: PHS Wireless Local Loop network model

12.6.2 Public PHS

This application of PHS technology allows a wide-area public communication service with moderate mobility. It provides coverage for urban and suburban areas, both outdoors and indoors. Public indoor spaces include underground, shopping centres, railway stations, etc.

The basic public PHS network consists of Personal Stations (PSs), public Cell Stations (CSs), a PHS Switching Centre (PSC), and a Service Control Point (SCP) that stores PHS subscriber data and location information (Independent Network Type). However, if an existing intelligent network (IN) already offers PSC and SCP facilities, they can be overtaken in order to realize PHS services (Public Network Utilizing type). In both networks, the core equipment is the PSC, which is a digital switching system with ISDN functions including PHS service software. The two types of public PHS networks are shown in Figure 12.15.

The network interface between a CS and the public PHS PSC conforms to an ISDN interface that is modified to carry out PHS-specific functions such as location registration, authentication and handover.

12.6.3 Wireless Local Loop and FWA (WLL/FWA)

The PHS-WLL System is a wireless access system that provides various services, such as plain old telephone service (POTS), high-speed data service, limited mobile service, etc. ISDN services will be supported in the future.

A PHS-WLL network consists of *WLL Access Controllers* (WACs), *WLL Cell Stations* (WCSs), *WLL Subscriber Units* (WSUs) and *WLL Personal Stations* (WPSs). WACs are located between the *Local Exchange* (LE) and the WCSs. The interface between the *WLL Access Controller* (WAC) and an LE, referred to in Figure 12.16 as *WLL Interface 1* (WIF1), can be either an analog or a digital interface, such as an analogue two-wire interface, a digital V5-interface, etc. In WLL systems, the PHS-specific functions such as location registration, handover (for WPS) and authentication are managed by the WAC. The *WLL Interface 2* (WIF2) is an air interface based on the standard described in this chapter.

12.6.3.1 Alteration of the term WLL to FWA

The International Telecommunication Union's Radio-Communication Sector (ITU-R) recently issued ITU-R Recommendation F.1399, which describes new wireless access terminology adopted by ITU-R. Based on this, the term WLL (*Wireless Local Loop*) in RCR STD-28 was replaced by FWA (*Fixed Wireless Access*).

In the previous RCR STD-28 Version 3.2, the public PHS air interface (used mainly for providing analogue telephony service) was fully applied to FWA system. The latest version 3.3 (released in spring 2000) introduces a new type of air interface that makes ISDN and leased line services available via the FWA system [3]. Thereby, the public PHS air interface is used to establish the communication channels. Hereafter respective control information of the layers 2 and 3 is transparently conveyed through these channels. An ISDN service, consisting of two 2B+D type, is realized by mapping the D channel on one 32 kbit/s communication channel and two 64 kbit/s communications channels are assigned to two B channels (B1 and B2).

12.7 Standards and References

The PHS technology was developed in Japan and standardized there. The *Association of Radio Industries and Businesses* (ARIB), formerly known as the Research and Development Centre for Radio Systems (RCR), drafted a standard for the PHS Common Air Interface, which was published as RCR STD-28 [2].

Meanwhile, the *Telecommunication Technology Committee* (TTC) has formulated standards based on ITU-T Recommendations for the PHS User–Network and Network–Network Roaming and Public Cell Station–PHS Service Control Procedures, which were published at the end of 1994. The internetworking standards (JT-Q761, 762 and 763) relate to the functions required to connect a PHS call between different PHS carrier networks. The roaming standard (JT-Q1218a), finalized in November 1994, defines the functionality needed to enable a PHS user to make calls via a PHS network with which the user has no contact.

Further extensions to the basic PHS standards are being developed. One such is aimed at Wireless Local Loop/Fixed Wireless Access (WLL/FWA) applications including support of ISDN services and another at PHS over cable TV (CATV) networks. Recently (December 2000) NTTDoCoMo announced the launch of a PHS video download service, the first in the world [8].

PHS has been developed in parallel with the Japanese Personal Digital Cellular (PDC) system, and both have been conceived within an overall evolutionary framework. Just as dual-mode DECT/GSM terminals have been announced in Europe, it is envisaged that dual-mode PHS/PDC terminals may also be developed in Japan.

Bibliography

[1] ARIB. *Personal Handy Phone System RCR Standard Version 1 (RCR-STD-28)*, Dec. 1993.

[2] ARIB. *ARIB STANDARD, RCR STD-28, Version 2*, Dec. 1995.

[3] ARIB. *ARIB STANDARD, RCR STD-28, Version 3.3*, March 2000.

[4] S. Ganu, S. Sripadham. *Comparison of modulation tecniques: GMSK and PI/4 DQPSK.* Project Report, Bradley Department of Electrical and Computer Engineering, Virginia Tech Blacksburg, Virginia Tech Blacksburg, USA.

[5] H.D. Lüke. *Signalübertragung.* Springer-Verlag, Berlin, 1995.

[6] MPT. *PHS Guidebook.* Technical Report, Ministry of Posts and Telecommunications, Japan, 1996.

[7] MPT. *Policy Reports.* Technical Report, Ministry of Posts and Telecommunications, Japan, Oct. 1998.

[8] NTT DoCoMo. *PHS Hot News.* `http://www.phsmou.or.jp/hotnews/docomoeggy.html`, December 2000.

[9] PHS MoU Group. *PHS MoU Group Technical Specifications.* `http://www.phsmou.or.jp/tech1.html`.

[10] Theodore S. Rappaport. *Wireless Communications.* Prentice-Hall PTR, Englewood Cliffs, New Jersey, 1996.

[11] Walter H.W. Tuttlebee. *Cordless Telecommunications Worldwide.* Springer-Verlag, Berlin, 1997.

[12] Ingo Willimowski. *Vergleich der Telekommunikationssysteme DECT und PHS unter besonderer Berücksichtigung zukünftiger Teilnehmerzugangsnetze.* In *IMST*, 1997.

13
Wireless Broadband Systems and Wireless ATM*

Broadband systems are generally those systems that provide a high transmission rate as well as other features like integration of services. An exact definition of this term can be found in ITU Rec. I.113, which characterizes broadband services as having a higher transmission rate than a primary multiplex connection in ISDN (2048 kbit/s).

A brief overview of the current state of development of wireless broadband systems, particularly in Europe, including the fundamentals of ATM in B-ISDN, is given below. This is followed by important aspects of development of wireless ATM for movable and mobile stations.

13.1 European Research on Broadband Systems

Until 1995, the EU research programmes *Research and Development in Advanced Communications Technologies in Europe* (RACE) I and RACE II were devoted to the development and testing of prototypes of systems with broadband radio transmission.

The importance of wireless broadband systems is evident from the number of projects carried out in this field, e.g., within the European research programme *Advanced Communication Technologies and Services* (ACTS) [3]. Example projects that prepared the standardization and development of wireless broadband systems are briefly referenced in the following subsections.

The follow-up (5th) framework programme of the EU, called *Information Society Technology* (IST), from 1999 onwards and others are devoted to the exploitation of the results gained during the programs RACE and ACTS to define private and public wireless broadband systems, integrated with 3G cellular mobile radio systems. The respective work can be seen as the first steps towards the definition of the *Next Generation* (NG) of wireless systems (see Chapter 18).

*With the collaboration of Arndt Kadelka, Andreas Krämling, Dietmar Petras**

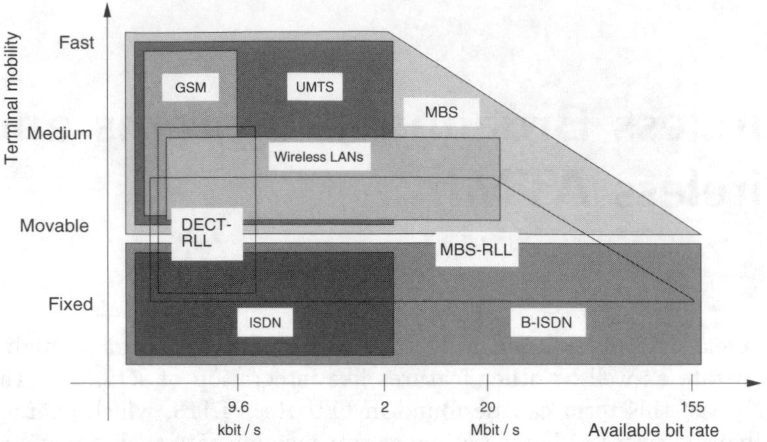

Figure 13.1: MBS and other data networks

13.1.1 MBS

The RACE II/MBS project *Mobile Broadband System* (MBS) undertook the development and testing of a technology and system concept for a wireless ATM system at 60 GHz that demonstrated the possibility of video transmission with a 16 Mbit/s transmission rate (net) at 50 km/h speed of movement of the terminal [9, 23]. Narrowband services were also to be provided. The MBS system made a particularly important impact, and convinced the professional world of the possibility of providing the services of broadband ISDN to mobile users through wireless ATM transmission [7, 15, 20, 22, 24].

In addition to providing a link-up to the broadband ISDN, the MBS concept also supports a cooperation with other systems such as UMTS. The type of network and level of integration can vary all the way from a privately operated MBS system with a low level of service integration and mobility up to a public MBS system with a high level of integration, extensive mobility and coverage of a wide area [6]. Figure 13.1 shows MBS in relationship to other systems with differing levels of mobility support for their terminals and transmission rates as seen from the year 1993 [22]. It can be seen that MBS combines the wide-ranging service spectrum of broadband ISDN with the mobility of mobile radio networks whilst offering the services of wideband and narrowband systems such as UMTS, WLAN, GSM, DECT and their derivatives for RLL applications.

Owing to the flexibility of MBS and the availability of the services of B-ISDN, a number of different applications are possible. These are indicated (with no claim to completeness) according to the data rates required and the mobility of their users in Figure 13.2.

The author and his research group were responsible for designing the radio and network protocols for MBS, which, although they were not implemented

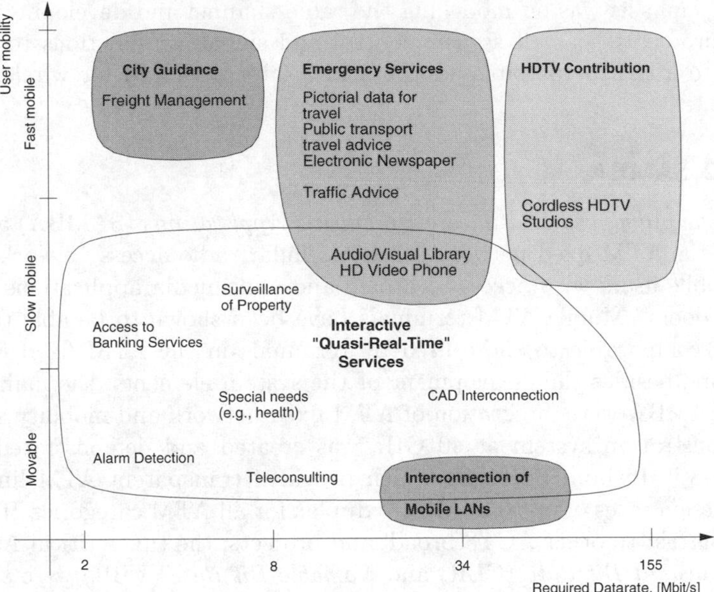

Figure 13.2: MBS applications and services

in the demonstrator system, were developed as part of the project and were incorporated into a number of successor projects in the ACTS programme, where they were developed further (see Section 13.1.2). For example, the MBS project as one of the first proposed an ATM-based radio interface for mobile use and specified it as part of the system [20, 21, 25].

13.1.2 Wireless Broadband Communications in the ACTS Programme

As the successor to RACE II, the ACTS research programme [2] of the EU from the year 1995 on was conducting field trials and demonstrations to monitor the developed systems for real applications.

Along with the development of UMTS, promising attributes of MBS have been developed further in the following ACTS projects.

13.1.2.1 Magic WAND

The *Wireless ATM Network Demonstrator* (WAND) extended the use of ATM technology to mobile users, and examined realistic user environments. The field of application covered Internet services over ATM in indoor areas with a 20 Mbit/s transmission rate at 5 GHz. The project realized an indoor wireless ATM demonstration network.

The emphasis was on modelling the radio channel and developing channel access protocols, as well as new control and signalling functions to be submitted to ETSI for possible adoption in a later standard for wireless ATM systems.

13.1.2.2 SAMBA

System for Advanced Multimedia Broadband Applications (SAMBA) aimed to expand the ATM fixed network using a cellular radio access network to provide mobile users with access to broadband multimedia applications indoors and outdoors. Mobile ATM terminals have been shown to be able to access services comparable to those used by terminals in the ATM fixed network. Therefore, besides the development of the system elements, the main priorities of SAMBA were integration of ATM fixed network and mobility support. A demonstration system at 40 GHz was created and demonstrated at the EXPO'98 in Lisbon, Portugal, which provided transparent ATM links with transmission rates of up to 34 Mbit/s duplex for all ATM categories of service.

In contrast to other ACTS broadband projects, the time-critical ATM services *Constant Bit Rate* (CBR) and *Variable Bit Rate* (VBR) were also supported and appropriate provisions made for radio protocols so that the radio channel offered a quality of service comparable to a fibre-optic transmission path (within the framework of ATM quality of service requirements).

The SAMBA project also developed a technology, not yet provided by the ATM Forum, for call handover between different ATM fixed network access points. The author and his research group were responsible for the implementation of the protocols of the radio interface and of the ATM network protocols, using their experience from MBS (see Section 13.1.1) [17, 18].

13.1.3 ATMmobil

This four year programme by the German Ministry of Research and Technology started in 1996 with the development of concepts and the corresponding demonstrators for four forms of wireless ATM systems.

1. The concept ATM-RLL introduced ATM for point-to-multipoint line-of-sight radio transmission at 26/40 GHz to bridge the last few miles with radio in a local loop (RLL) area.

2. The concept *Wireless ATM* (W-ATM) LAN examined the wireless connection of mobile computers to support multimedia applications at 5 and 19 GHz [8].

3. The concept cellular W-ATM linked mobile terminals with an ATM radio interface over a cellular network at 5 GHz to an ATM broadband network access point [5, 21].

4. The concept *Integrated Broadband Mobile System* (IBMS) studied the development of wireless transmission technology for indoors and outdoors. Along with infrared as the medium for indoor purposes, millimetre waves at 5, 17, 40 and 60 GHz have been used. Adaptive antennas, single-carrier transmission technology, radio interface, radio resources and mobility management were the focus of research [10].

The author proposed and headed the ATMmobil project, and members of his research group participated in the implementation work of the first, second and third concept mentioned above and were also involved in the ETSI/BRAN standardization of H/2 (see Section 13.1.6).

13.1.4 Wireless Broadband for Next-Generation Cellular

With the further development of UMTS it soon became clear that this system would not be able to meet the expectations raised, namely to support high-bitrate multimedia on the move. Consequently, the integration of two systems with quite different air interfaces has been started to study in the IST programme that are able to combine the high mobility support of WLAN systems based wireless ATM technology as defined by the ETSI/BRAN standard H/2 (see Section 14.2). There are two projects worth mentioning that followed this approach.

13.1.4.1 IST/BRAIN and IST/MIND

Broadband Radio Access to IP Networks (BRAIN) and its continuation since the year 2001, called *Mobile IP Network Simulator* (MIND) developed and demonstrated the IP core-network based integration of UMTS and H/2 systems with a common subscriber management and mobility support, including

- horizontal handover between calls providing the same air interface standard and
- vertical handover between base stations supporting different air interfaces (see Figure 18.9).

The main actors were the companies Ericsson, Nokia, Siemens and NTT DoCoMo.

13.1.4.2 IST/WINEGLASS

The project *Wireless IP Network as a Generic Plattform for Location Aware Services* (WINEGLASS) aims at integrating *UMTS Terrestrial Radio Access Network* (UTRAN) and WLANs to an IP core network. The access and core network architectures are explored to provide QoS and mobility support. Location-aware applications for business and public usage will be supported by a transparent solution for ubiquitous use of the IP-based core network. A testbed will be developed comprising wireless and fixed network technology and the related software to demonstrate the feasibility of the concept.

13.1.5 The Role of the ATM Forum in the Standardization of Wireless ATM Systems

Although the ATM Forum is not an official standards body, it is playing an important role in the quasi-standardization of certain forms of the ATM fixed network through its association with industry and its products. In June 1996 the ATM Forum became involved in WLAN standardization. The WLAN group originally wanted to focus its attention on mobility support by ATM fixed networks, a project that was supposed to run until the first quarter of 1999 [19].

13.1.6 The ETSI Contribution to W-ATM Standardization

The project ETSI/BRAN took-up the results of the European research projects mentioned above and specified in early 2000 a wireless ATM LAN called *HIPERLAN/2* (H/2) that is currently seen as one candidate standard for wireless broadband systems (see Section 14.2). This system is the only one known to date (2001) able to differentiate the QoS needs of applications competing for transmission capacity at the air interface of a base station.

13.2 Services in Broadband Networks

Multiplex data rates of up to 155 Mbit/s on the wireless user connection are necessary for the integration of wireless broadband applications in broadband fixed networks, say Internet or B-ISDN. These kinds of applications require services with constant and variable bitrates for non-interactive data as well as for bursty-type interactive data. Along with voice transmission, applications with continuous bit streams include video conferencing in which real-time requirements must also be strictly maintained. Interactive services are characterized by a wide fluctuation in the requirements for bit rates. Thus a short information request to a database can result in a very long response (counted in bit time) requiring a high transmission rate. A distinction is made between the following:

- Interactive services
  - Telefony
  - Video telephony
  - Broadband video conferencing

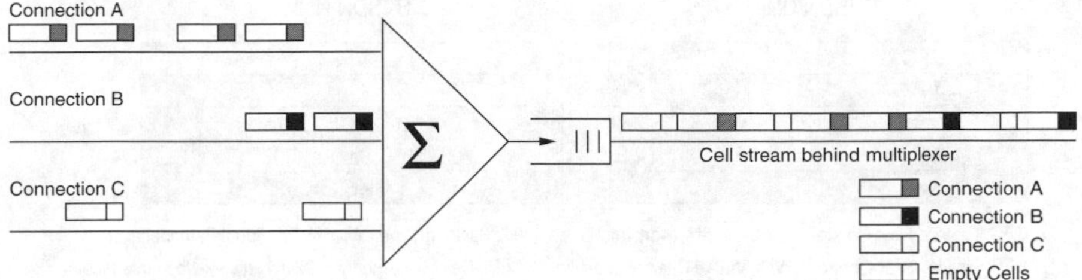

Figure 13.3: Statistical multiplexing of cells to a medium

- Inquiry services
  - Access to databases
  - Radio, TV, HDTV, Video-on-Demand
  - Electronic newspaper
  - Videopost
- Data communications
  - LAN links
  - File transfer
  - CAM links
  - High-resolution video transmission

Synchronous transmission methods like *Synchronous Time-Division Multiplexing* (STDM) have difficulty coping with the varying requirements of broadband services. Although an oversizing of the transmission capacity of synchronous channels reduces waiting times, it results in a poor utilization of capacity in the transmission medium. *Asynchronous Transfer Mode* (ATM) technology is better suited to dealing with the demands of broadband services.

13.2.1 ATM, the Transmission Technology in B-ISDN

ATM is the connection-oriented packet-switching method used in B-ISDN. ATM combines the advantages of connection and packet-oriented switching—specifically the statistical multiplcxing of data from different connections to one medium and the message switching of packets in the network nodes between the communicating terminals. The data streams to be transmitted are divided into short blocks of a fixed length, referred to as ATM cells. The cells of different connections are transmitted with time interleaving over a physical channel. Depending on their data rates, the connections are dynamically allocated varying amounts of transmission capacity, with some of them transmitting a large number of cells per time unit and others only very few. The cells in each connection are transmitted in the order of their arrival.

The ATM multiplexer adds empty cells to the multiplex data stream if none of the connections requires transmission capacity and a synchronous transmission method is being used (see Figure 13.3).

Elements of ATM have been chosen to form the basis of the wireless broadband air interface of H/2 to enable high speed wireless communication with

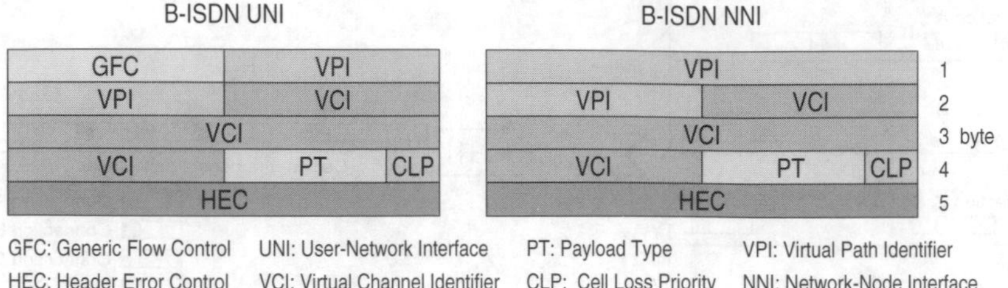

Figure 13.4: Header of an ATM cell

a QoS as defined for fixed ATM networks. This is the reason why the basics of ATM are described shortly in the following.

13.2.2 Structure of an ATM Cell

An ATM cell comprises 53 bytes, consisting of a 5-byte long header field and a 48-byte long information field containing data of the higher ATM protocol-stack layers and user data. The switching of the cells is connection-oriented. All cells in a virtual connection take the same transmission path, which was established when virtual channels were set up on different switching sections in the network during call setup. The cells are controlled through the network on the basis of the routing information stored in the cell header (see Figure 13.4).

Virtual Channel Identifier (VCI), 2 bytes An identification of the virtual channel differentiates between the different concurrent logical channels and their cells. The virtual channel number is always only assigned to one switching section.

Virtual Path Identifier (VPI), 8 or 12 bits A channel group is identified by the parameter VPI. A differentiation can be made between a large number of groups in the same direction, each containing several virtual channels. The cells of channels in the same group can be processed especially quickly by the switch-fabric of an ATM switch and then forwarded; *cross-connects* are used for this purpose.

Payload Type (PT), 3 bits This parameter describes the type of information field and provides a differentiation between user and signalling information. The signalling information is required, for example, to update the routing tables managed in the switching centres. The switching centre carries out updates by evaluating the header field as well as the information field of the ATM cell. User data being transmitted in the information field of an ATM cell is not taken into account in the switching.

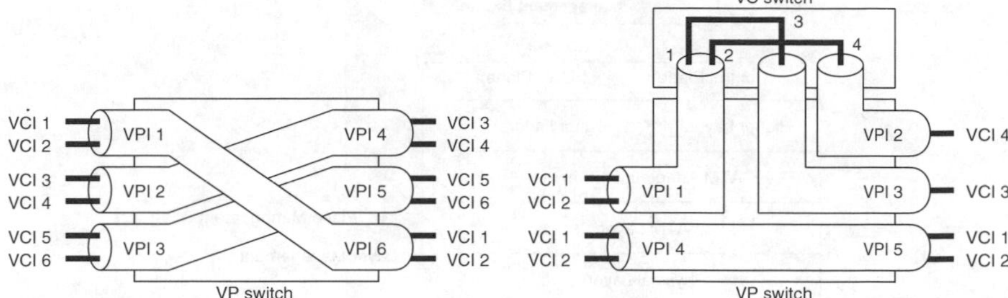

Figure 13.5: Virtual path switching und virtual channel switching; VCI=Virtual Channel Identifier, VPI=Virtual Path Identifier

Header Error Control (HEC), 1 byte Because the header of an ATM cell contains data that is vital to the transport of the cells, it is protected by a checksum. This permits the detection of transmission errors.

Cell Loss Priority (CLP), 1 bit This parameter identifies cells with a lower priority, which are discarded in the ATM switching centre when there is an overflow in the queue. The bit is also used by multiplex and switching nodes for flow control and traffic shaping.

13.2.3 ATM Switching Technology

As with other packet-switched methods, ATM cells are switched on the basis of the routing information contained in the cell header. The complete originating and target addresses are sent only during connection setup, so that the routing information can be kept as short as possible, thus increasing throughput. Identification of the logical channels is assigned for the different sections of a connection (VCI, VPI). During connection setup, the ATM switching centres enter relationships between input and output identification (line + logical channel identification) into their routing tables using incoming control information.

The switching of the cells is based on the information in these tables and takes place as follows. Switching control derives the logical channel identification (VCI, VPI) from the incoming cells. The identification VPI, VCI of the next connection section is obtained through the relationship of the incoming and outgoing pairs (VCI, VPI) in its routing table, and entered into the cell header, and the cell is then switched through to the appropriate exit of the switching network.

Figure 13.5 shows the switching elements used in the switching centres. A distinction is made between *VP switches* and *VC switches*. The *Virtual Path* (VP) switch only evaluates the VPI values of the cells, so that the cells can be switched quickly. The entries of the VCI fields remain unchanged.

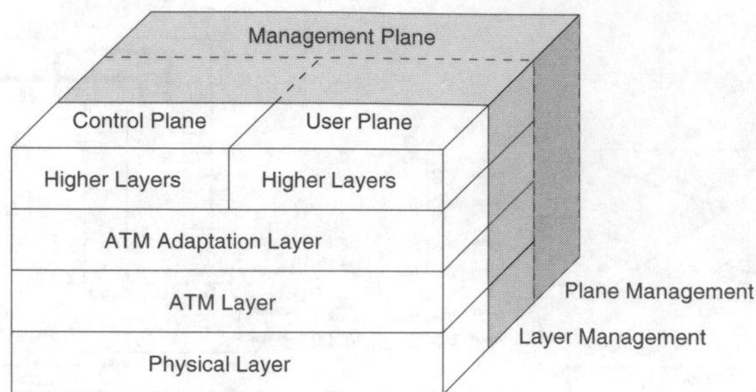

Figure 13.6: The ATM reference model

The VCI identities are changed only if transmission has occurred through a *Virtual Channel* (VC) switch.

13.2.4 ATM Reference Model

According to the recommendations of the OSI reference model, an ATM reference model with four layers can also be specified (see Figure 13.6). These layers include the *physical layer*, the *ATM layer*, the *ATM Adaptation Layer* (AAL)) and a layer that represents the functions of the *higher layers*. Since the physical layer corresponds to layer 1 of the ISO/OSI reference model and the ATM layer corresponds to layer 3 (network), the data link layer has not been taken into account for the ATM reference model. This is because ATM networks typically use fibre links that guarantee a very low *Bit Error Ratio* (BER) and therefore do not need any error control in layer 2.

Furthermore, three different planes are defined: the *user plane*, the *control plane* and the *management plane*. The management plane comprises the functions for *plane management* as well as the functions for *layer management*. Plane management is responsible for management of the entire system, whereas layer management controls the individual layers.

The services, protocols and interfaces of the physical layer depend on the transmission medium used, and contain all the functions required for the bit-by-bit transmission of information. The tasks of the ATM layer include:

- Multiplexing and demultiplexing cells of different connections
- Control of VCI and VPI-oriented functions
- Generation or evaluation of cell header information
- Generic flow control at the *User Network Interface* (UNI)
- Establishment, routing, operation and release of connections

The *ATM Adaptation Layer* adapts the higher ranking layers to the ATM layer. It carries out the necessary segmentation of data streams into cells on

Table 13.1: Classes of AAL Services

Class	1	2	3	4
Time relat.	Continuous	Continuous	Non-cont.	Non-cont.
Bit rate	Constant	Variable	Variable	Variable
Connection	Conn.-orient.	Conn.-orient.	Conn.-orient.	Conn.less
Examples	Emulation of sync. circuit switching (voice)	Variable bit rate (video)	Conn.-orient. data trans.	Conn.less data trans.

the transmitting side and their reassembly into messages on the receiving side, and ensures safe transmission. The AAL layer is divided into two sublayers:

1. The *Segmentation and Reassembly* (SAR) sublayer for mapping protocol data units of the higher layers to the ATM cells and the reverse.

2. The *Convergence Sublayer* (CS), which compensates for the undesirable effects caused by the different cell transit times of different services.

For example, the digital sample values used for voice transmission are consolidated over a certain period of time to enable the complete utilization of the capacity of an ATM cell. The receiver generates a continuous data stream again from the arriving ATM cells.

The different cell transit times through the network are offset by the ATM adaptation layer in the receiver through the addition of a constant time delay before the output.

13.2.5 ATM Classes of Service

The number of protocols required for the ATM adaptation layer has been kept to a minimum through the classification of services into four different groups according to the following parameters: time relationship between source and sink, bit rate and type of connection (see Table 13.1). Time-continuous services with constant or variable bit rates are differentiated according to those that are connection-oriented and those that are connectionless.

In accordance with the class of service, control data is also added to the user information. This is used to allow the restoration of user information that is divided among several cells. The control data is generated by the SAR sublayer when the data is divided into cells. In the receiver the corresponding SAR sublayer must join the data together in the correct sequence according to the control data.

For identification of the sequence of the individual ATM cells, each message contained in a cell is assigned a sequence number to enable the receiver to detect the loss of any cells. Whereas the sequence number is available with all the classes of service, additional backup data and different segment types

Table 13.2: Classes of service and their quality of service parameters

	ATM Layer Service Classes				
Attribute	CBR	VBR(RT)	VBR(NRT)	ABR	UBR
CLR	×	×	×	×	—
CTD	×	×	×	—	—
CDV	×	×	—	—	—

CBR	Constant Bit Rate	VBR	Variable Bit Rate	UBR	Unspecified Bit Rate
RT	Real-Time	NRT	Non-Real-Time	ABR	Available Bit Rate
×	Specified	—	Unspecified		

are being planned only for some of the classes. Owing to the varying control data part, the user data part is 44–48 bytes.

In its specification *Traffic Management (V 4.0)* [4] the ATM Forum differentiates between different classes of service that represent different applications. The specified QoS parameters for the following classes of service are listed in Table 13.2:

Unspecified Bit Rate (UBR) No QoS parameters are specified for the UBR class of service. No guarantees are provided for this class, which means that it only benefits from a *best-effort service.*

Available Bit Rate (ABR) The ABR class of service is particularly suitable for applications that are not real-time-oriented and also have no requirements in terms of transmission rate. Only the cell-loss rate is defined as a parameter for the quality of service.

Constant Bit Rate (CBR) The CBR class of service is planned for real-time-oriented services with constant bit rates that have a stringent requirement for *Cell Loss Ratio* (CLR), *Cell-Transfer Delay* (CTD) and *Cell-Delay Variance* (CDV).

Non-real-time Variable Bit Rate The VBR class of service is a compromise between the VBR and ABR classes of service. This is used for applications that require more quality-of-service guarantees than the ABR class but attach no importance to the assurance of a specific variance in cell transmission times (CDV).

Real-time Variable Bit Rate Real-time-oriented applications that place a stringent demand on delays and their variance as well as on cell loss ratios require a real-time VBR service.

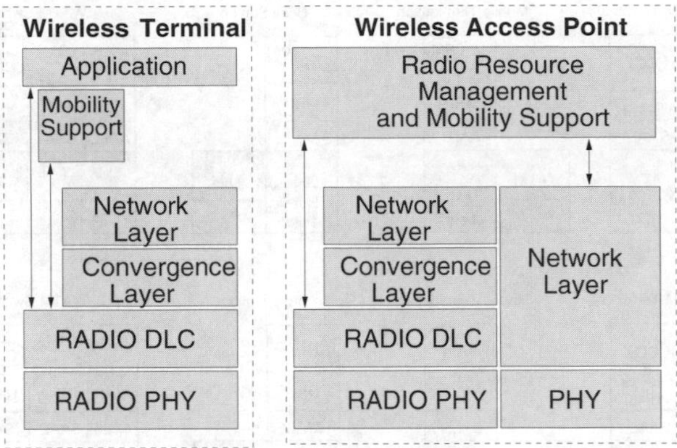

Figure 13.7: HIPERLAN/2 layers and architecture

13.2.6 Functions and Protocols of the AAL

Because transmission errors cannot be completely prevented even with fibre-optic technology, an end-to-end error correction procedure dependent on type of service is provided in the AAL layer.

The AAL protocols type 1 and type 2 are used for the real-time-oriented CBR and VBR services. They supply their PDUs with sequence numbers and check sums to aid in the detection of lost or incorrectly inserted ATM cells. An FEC procedure can be used as an option to correct bit errors [13]. If the bit-error ratio in the ATM layer exceeds the correction capabilities of the code used—something that may particularly occur from time to time on a radio transmission path—the QoS requested by the user cannot be guaranteed by this procedure.

An ARQ protocol based on the functions for detection of bit errors and cell loss in the lower AAL sublayers—*Common Part Convergence Sublayer* (CPCS) and *Segmentation and Reassembly* (SAR)—is provided in the highest sublayer of the AAL (*Service Specific Convergence Sublayer*, SSCS) in AAL protocols type 3/4 and type 5 [11]. According to [12], these ARQ protocols can be executed efficiently with a packet-loss probability of 10^{-3}. If the packet-loss probability for packets 1 kbyte in length is to be maintained then the bit-error probability 10^{-7} cannot be exceeded. The bit-error probability of a radio transmission path protected by an FEC procedure usually lies above that limit, and is thereforc too high for an ARQ procedure to be carried out efficiently in the AAL.

Figure 13.7 shows the design of the ETSI/BRAN protocol stack that is heavily supported by manufacturers aiming to provide wideband and broadband Internet services to their customers. Instead of a W-ATM terminal, any class of wireless broadband terminals can be supported by this protocol stack. This design favours the straightforward support of Internet applications by

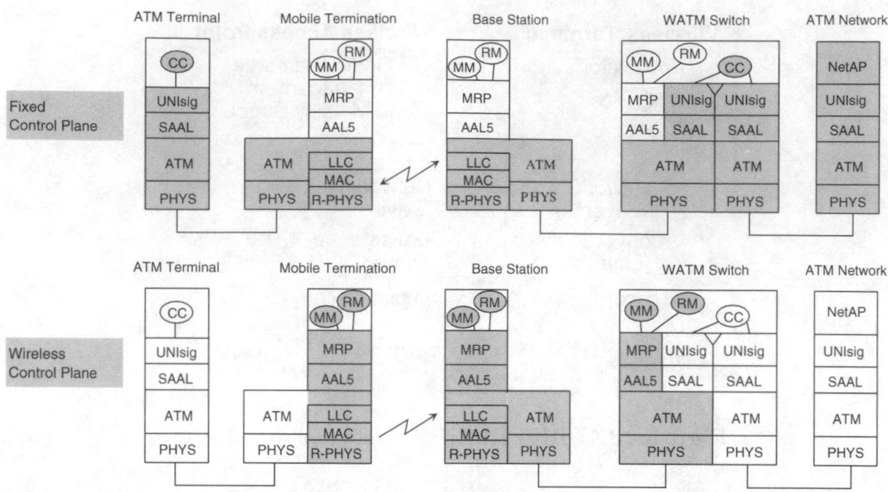

Figure 13.8: Protocol stack in wireless ATM

using the *Internet Protocol* (IP) in the network layer instead of ATM signalling. A convergence sublayer is provided in both protocol stacks to enable use of the air interface provided by H/2 by any network service and related protocols.

13.3 Mobility Support for W-ATM Systems

The architecture of a broadband mobile radio system might use an ATM or IP network as the transport platform (see Figure 6.1). Neither an ATM network nor an IP network contain mobility support functions such as location management of mobile users or handover support. Figure 13.8 presents a typical W-ATM protocol stack for the control plane [16]. Appropriately for an ATM fixed network, the standard protocol for user network signalling (UNIsig, [14]) is supported for *Call Control* (CC) in the *fixed control plane* (see shadowed areas). The radio access system is transparent for these protocols. Protocols for *Radio Resource Management* (RRM) and *Mobility Management* (MM) are supported in the *wireless control plane* within the radio access system (*Mobility and Resource Management Protocol*, MRP).

In the following, handover protocols for an assumed ATM fixed network are explained. The problems and solutions addressed would be very similar for an IP based fixed network.

13.3.1 Radio Handover

Radio handover takes place between two *Base Station Transceivers* (BSTs) of the same base station; see Figure 13.9. The virtual connections are switched

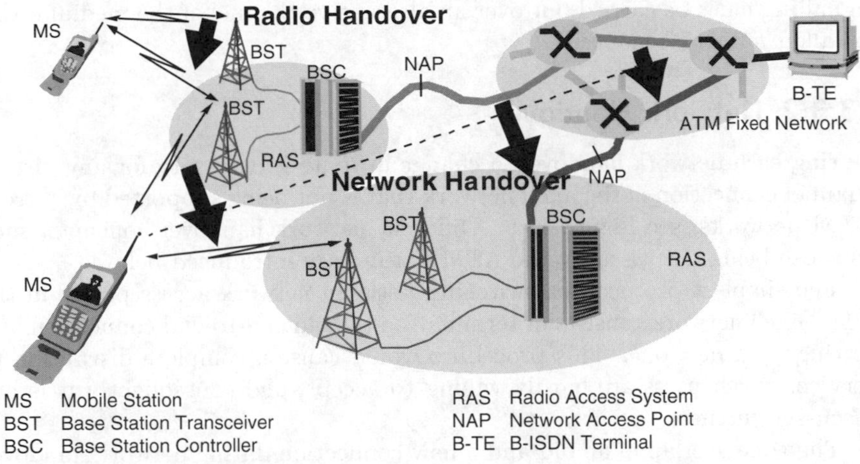

Figure 13.9: Radio and network handover in wireless ATM

over within the *Base Station Controller* (BSC), independently of the ATM fixed network.

A handover can be divided into three stages:

1. **Measurement of radio resources** A handover is normally executed because of poor transmission conditions and the subsequent unsatisfactory QoS of the radio channel. The conditions can be assessed only if measurements are conducted frequently enough. It is also necessary for the effects of radio propagation to be recognized and taken into account. This necessitates measurements not only of the actual connection but also those of the alternative radio channels.

2. **Handover decision** This is the stage when it is decided whether a handover should be excecuted. The relevant algorithm uses as its criteria the measurement values provided of the radio conditions and the capacity utilization of the radio cells.

3. **Handover execution** The signalling protocol is executed at this stage in a handover to enable the relevant MAC connection to be switched from one radio cell to another.

Wireless ATM systems have the advantage that terminals and base stations are able to initiate handovers independently of one other. A terminal can initiate a handover because of receiving conditions; a base station can force a terminal to execute a handover (*forced handover*) if the receiving conditions or cell load require it.

There are two types of handover initiated by a terminal. If the handover is carried out during the actual connection, it is called a *backward handover*. If the connection has already been interrupted or is very poor, the handover signalling must be carried out over an alternative channel of the medium; this is called *forward handover*.

13.3.2 Network Handover

During each network handover, a change in route is necessary for any virtual channel connection in the fixed network that is not being supported by current ATM networks; see Figure 13.9. Different network handover concepts, such as those being discussed by the ATM Forum, are introduced below.

The simplest protocol for switching between network access points in the ATM fixed network consists of terminating the old end-to-end connection and setting up a new one. This procedure would cause a complete disruption to service, which users are hardly willing to accept, and consequently it is not discussed further.

The time overlap of an old and a new connection during network handover would confront the corresponding terminal (in the fixed network) with two simultaneous connections. At times the user data would be divided between both connections in a way that would be confusing to the corresponding terminal, and it would not be possible to receive the user data without some modification to the terminal's protocols. Because of the distance involved and the subsequent difficulty in synchronizing the data streams of both connections, a dynamic modification of the protocols of terminals would result in a reduced quality of service. Furthermore, this kind of modification would affect all ATM terminals, and therefore is not workable.

Current mobile radio systems, such as GSM, have *Mobile-services Switching Centres* (MSCs) that supply the functions necessary for network handover between old and new connections without any loss of information. A similar solution is also possible with ATM networks. This involves breaking down a virtual end-to-end connection into two virtual channel connections, one each from the mobile switching centre to the mobile station and to the corresponding terminal (see Figure 13.10). If during the network handover a virtual channel connection is being set up between a mobile station and the mobile radio switching centre from the new *Network Access Point* (NAP) then this can under certain conditions through the use of a bridging function produce a *seamless* handover between the old and the new virtual channel connection at the mobile switching centre.

Virtual channel connections are set up between entities of the *ATM Adaptation Layer* (AAL). A separation of the end-to-end connection into two virtual channel connections as shown in Figure 13.10 requires a transition function between the terminating AALs of the two virtual connections in an interworking function. A switching function, as well as the bridging function as one of the main components of a network handover, is then carried out in the AAL.

A possible alternative is leaving the switching function to the ATM layer. The actual switching is then carried out by an enhanced ATM switching centre with mobility functions which is therefore also referred to as an *ATM Mobility-*

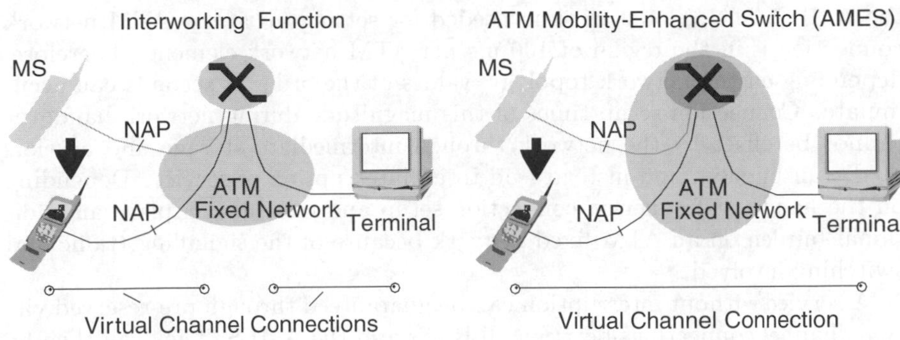

Figure 13.10: Network handover support in the AAL layer

Figure 13.11: Network handover support in the ATM layer

Enhanced Switch (AMES). In this case the virtual end-to-end channel connection remains intact (see Figure 13.11). Consequently only an end-to-end connection is available for call control in the network, and no interworking functions or adaptation functions are required between subnets. The role of the bridging function is more difficult to accomplish in the AMES, because the standard stipulates asynchronous transmission and the ATM cells are not numbered. Therefore with each network handover there is the danger of ATM cells being lost.

A network handover in the ATM layer requires the rerouting of the virtual end-to-end channel connection in the VP or the VC switch within the ATM layer of the ATM fixed network.

Two typical concepts for executing a network handover are presented below. They are based on the rerouting of a virtual channel connection.

13.3.2.1 Virtual Tree

There are different suggestions for executing a network handover. For example, a distinction must be made between backward and forward handover protocols when the handover is initiated by a mobile station (see Section 13.3.1). The suitability of the network handover protocol depends mainly on the choice of the radio handover protocol.

With *backward handover* protocols, in a network handover the mobile station deregisters from the old *Radio Access System* (RAS), which consists of a BSC and several BSTs, before it changes to the new RAS. Therefore in preparation a virtual channel connection is set up to form a new RAS over the AMES that is currently carrying out the connection. The end-to-end connection will then be switched to that system during the network handover.

With *forward handover* protocols, it is not possible for the AMES to be informed in advance. The mobile station registers over the radio interface in the new RAS, which must then set up a new virtual channel connection to the AMES. The amount of time needed for setting up a new ATM network connection is in the region of 100 ms per ATM network element. Therefore, depending on the network topology, values of the order of seconds can accumulate. Connection setup times of this magnitude during network handover cannot be offset by the network through intermediate storage and accelerated transmission, and instead lead to an interruption of service. Depending on the handover frequency, connection setup and termination place an additional burden on an ATM fixed network because of the signalling traffic and switching involved.

A service without interruption can be guaranteed through pre-reserved virtual channel connections between all RASs and the AMES. They can then be used immediately during network handover, and produce the respective connection setup. However, each RAS must then constantly maintain reserved virtual channel connections to the AMES. This produces a so-called virtual connection tree (see Figure 13.12) for each connection from a mobile station to AMES. Every virtual branch of this tree consists of two virtual channel connections (one for each transmission direction), which in turn consists of a chain of virtual channels between ATM network elements. In Figure 13.12 the virtual branches are abstracted as a direct line.

A virtual tree is set up for each call and continues to exist for the duration of an entire call. Only one virtual branch of a tree carries user data, whereas all the other virtual branches are seen as preparatory branches and remain unused until replacing a branch formerly being used. Although they are set up, i.e., all entries in the routing tables of the ATM network elements are available, they do not transmit ATM cells. In a network handover the virtual branch of the new radio access system is used, whereas the old virtual branch is idle but continues to exist.

In AMES this changeover must be supported by a switching function, which is carried out differently depending on transmission direction. There are various concepts for this [1, 25]. Virtual trees tie up additional capacity in a fixed network, because all branches must provide the quality of service required for an RAS.

13.3.2.2 Extension of Virtual Channel Connections between Radio Access Systems

Instead of the routing between different virtual branches being switched over, a virtual channel connection can be extended through the execution of a network handover that includes the extension of the virtual channel connection from the old to the new radio access system. The old radio access system then takes over the function of the AMES.

When a connection is being set up, the location area of the mobile station establishes the first RAS as the anchor for the entire connection. If a network handover occurs, the virtual channel connection is extended bidirectionally from the anchor to the new RAS. In each subsequent network

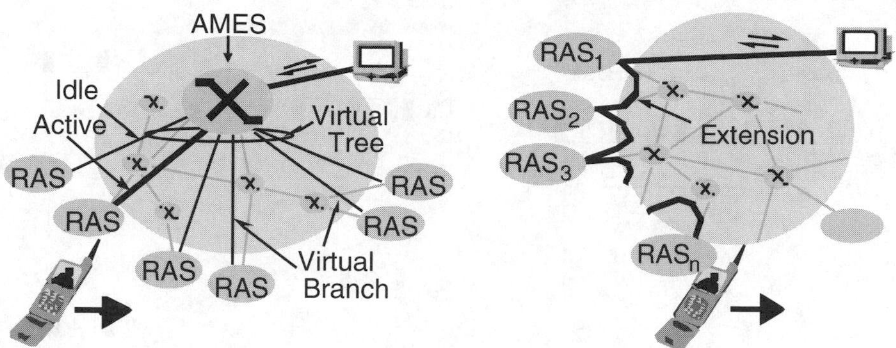

Figure 13.12: Virtual connection tree

Figure 13.13: Extension of virtual channel connections

handover another extension is added between each old RAS and the new one, thereby producing a chain of extensions such as that shown in Figure 13.13. The route can be shortened again in a network handover when the mobile station changes back to the previous RAS.

Here too a distinction has to be made between backward and forward handovers. In the first case the extension can be set up before the change in radio access system. In the second case the extension of the bidirectional virtual channel connection to the old RAS must be set up after the radio handover by the new RAS.

As discussed in Section 13.3.2.1, this kind of connection setup can result in an interruption to service, which explains why it makes sense to use reserved virtual channels between neighbouring RASs. Similarly to the virtual tree, as a precaution for each communicating mobile station a bidirectional virtual channel connection should be set up to each neighbouring RAS.

13.3.2.3 Guaranteed Quality of Service

For network handovers in wireless ATM systems it is essential that the *Quality of Service* (QoS) for the respective ATM connection be maintained. The procedures given as examples in the previous subsections allow the execution of an accelerated network handover through the use of preset channels. These concepts on their own, however, cannot guarantee that the QoS as agreed at connection setup is also maintainable during a network handover.

It is clear from Figure 13.14 that ATM cells in a virtual tree can be lost during a network handover. In a forward handover the AMES is notified by the new radio access system (RAS 2) that a handover has taken place. The AMES thereupon switches over the VCC to the new branch of the tree. ATM cells remaining on the old branch after a handover will be lost (cell 3 in this example). It is obvious that the number of lost cells is dependent on the data rate of the service and on the duration of the handover signalling.

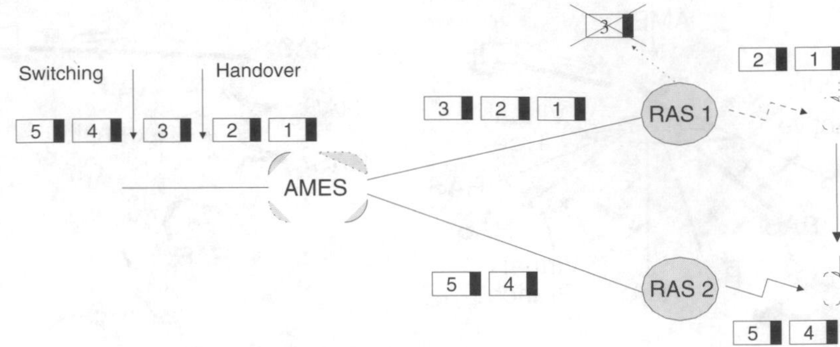

Figure 13.14: Cell loss during handover in a simple virtual tree

Protocols that have a minimal effect on quality of service during handover are currently at the specification stage. It is generally not possible for the QoS of the ATM fixed network to be guaranteed in all situations in a mobility-supporting wireless ATM system.

Bibliography

[1] A. Acampora, N. Naghshineh. *An architecture and methology for mobile-executed handoff in cellular ATM networks. IEEE Journal on Selected Areas in Communication*, Vol. 1294, No. 8, pp. 1365–1375, 1994.

[2] *Advanced Communications Technologies and Services (ACTS).* `http://slarti.ucd.ie/inttelec/acts/index.html`.

[3] Advanced Communications Technologies and Services (ACTS). *ACTS Information Window.* `http://www.infowin.org/ACTS/`.

[4] ATM Forum. *Traffic Management Specification—Version 4.0.* ATM Forum—Technical Committee, 1995.

[5] H. Bakker, W. Schoedl. *ATM up to the mobile terminal—Impact on the design of a cellular broadband Systeme.* In *Proceedings Int. Switching Symposium 1997, XVI World Telecommunications Congress*, pp. 379–385, September 1997.

[6] J. Brázio, C. Belo, S. Sveat, B. Langen, D. Plassmann, P. Roman, A. Cimmino. *MBS scenarios.* In *RACE Mobile Workshop*, pp. 194–197, Metz, France, June 1993.

[7] M. Chelouche, A. Plattner. *Mobile Broadband System (MBS): Trends and impact on 60 GHz band MMIC development. Electronics and Communication Engineering Journal*, pp. 187–197, June 1993.

[8] Y. Du, S. Hulyakar, D. Evans, D. Petras, C. Ngo, P. May, C. Herrman, M. Duque-Anton, R. Kraemer, R. Fifield, D. Verma. *System architecture of a home wireless ATM network.* In *Proceedings of 5th International Conference on Universal Personal Communications (ICUPC96)*, pp. 477–481, 1996.

[9] L. Fernandes. *R2067-MBS: A system concept and technologies for mobile broadband communication.* In *RACE Mobile Telecommunications Summit*, Cascais, Portugal, Nov. 1995.

[10] G. P. Fettweis, M. Bronzel, H. Schubert, V. Aue, A. Maempel, A. Wolisz, G. Walf, J. P. Ebert. *A closed solution for an integrated broadband mobile system.* In *Proceedings of 5th International Conference on Universal Personal Communications (ICUPC96)*, pp. 707–711, 1996.

[11] R. Händel, M.-N. Huber, S. Schröder. *ATM Networks: Concepts, Protocols, Applications.* Addison-Wesley, Reading, Massachusetts, 1994.

[12] T. Henderson. *Design principles and performance analysis of SSCOP: a new ATM adaptation protocol.* *Computer Communication Review*, Vol. 25, No. 2, pp. 47–59, 1995.

[13] ITU-T. *Recommendation I.363: B-ISDN ATM Adaptation Layer Specification*, 1993.

[14] ITU-T. *Recommendation Q.2931: B-ISDN Application Protocols For Access Signalling*, 1995.

[15] B. Jülich, D. Plassmann. *Protocol design and performance analysis of an intermediate-hop radio network architecture for MBS.* In *Proceedings of WCN'94*, pp. 1178–1182, The Hague, The Netherlands, Sep. 1994.

[16] A. Kadelka, N. Esseling, J. Zidbeck, J. Tulliluoto. *SAMBA Trial Platform—Mobility management and interconnection to B-ISDN.* In *ACTS Telecommunications Summit '97*, pp. 501–506, Aalborg (DK), Oct. 1997.

[17] H. Kist, D. Petras. *Service strategy for VBR services at an ATM air interface.* In *Proceedings of EPMCC' 97*, pp. 181–189, Bonn, Germany, Sept. 1997.

[18] A. Krämling, G. Seidel, M. Radimirsch, W. Detlefsen. *Performance evaluation of MAC schemes for wireless ATM Systems with centralised control considering processing delays.* In *Proceedings of EPMCC'97*, pp. 173–180, Bonn, Germany, Sept. 1997.

[19] J. Kruys, M. Niemi. *An overview of wireless ATM standardization.* In *ACTS Mobile Communications Summit*, pp. 250–255, Nov. 1996.

[20] D. Petras. *Performance evaluation of medium access control schemes for mobile broadband systems.* In *Proceedings of Sixth Nordic Seminar on Digital Mobile Radio Communications DMR VI*, pp. 255–261, Stockholm, Sweden, June 1994.

[21] D. Petras. *Medium Access Control Protocol for wireless, transparent ATM access.* In *IEEE Wireless Communication Systems Symposium*, pp. 79–84, Long Island, New York, Nov. 1995.

[22] C.-H. Rokitansky, H. Hussmann. *Mobile broadband system (MBS)—system architecture.* In *Fourth Winlab Workshop on Third Generation Wireless Information Networks*, pp. 309–316, Brunswick Hilton, East Brunswick, New Jersey, Oct. 1993.

[23] C.-H. Rokitansky, M. Scheibenbogen. *Mobile Broadband System: System Description Document.* CEC Deliverable 68, RACE, R2067/UA/WP215/DS/P/68.b1, 1995.

[24] H. Steffan, D. Plaßmann. *Mathematical performance analysis for broadband systems with beamforming antennas.* In *2. ITG-Fachtagung Mobile Kommunikation '95*, pp. 449–456, Neu-Ulm, Germany, September 1995.

[25] B. Walke, D. Petras, D. Plassmann. *Wireless ATM: Air interface and network protocols of the mobile broadband system. IEEE Personal Communications Magazine*, Vol. 3, No. 4, pp. 50–56, Aug. 1996.

14
Wireless Local Area Networks

Since the introduction of lightweight portable computers (laptops, notebooks), a great deal of attention has been focused on the development of wireless computer networks (*Wireless Local Area Network*, WLAN).

Thanks to standardization in the field of LANs, it is comparatively easy to find systems that will still be upgradable even in a few years' time. Around 70 % of all computers connected to networks are compliant with the *Institute of Electrical and Electronics Engineers* (IEEE) 802.3 (*Ethernet*) and IEEE 802.5 (*Token Ring*) standards. Connection is normally over a permanent wireline link. The problems that can occur are the surfacing of mechanical defects (corrosion) after a few years and violations of rules on radiated interference. It is difficult to adapt these networks to cope with changing office conditions. Mobile network nodes are not possible.

The obvious approach is to leave out the cable entirely. This idea is almost as old as the concept of the so-called ALOHA system, which used radio to connect terminals to their processing computers. The newer WLANs work with the most up-to-date radio technology. Data is encrypted and extensive error-protection mechanisms are available. Integrity of data is also guaranteed.

Just like wireline LANs, WLANs can be divided into different architectures and performance categories. Many companies offer products for wireless point-to-point connections, but only very few build LANs for multipoint communication. Today wireless networks use spread-spectrum, narrowband microwave or infrared signals for transmission (see Table 14.1). Because of legal regulations, networks using spread-spectrum and narrowband microwave cannot be operated in most countries unless special authorization has been given. The only exemption of this is operation in a licence exempt band, e.g., *Industrial, Scientific and Medical* (ISM) band, where under a set of given rules for channelization and emitted power operation of radio equipment is allowed, generally. An example of this is the 2.4 GHz band, where the first WLANs have been positioned.

Until now, WLANs have only had a very small share of the market. This is partly due to the higher costs per network node, but no doubt also because of the late standardization in this area. In spite of this, suppliers are projecting a growth in wireless networks over the next few years. Standards such as IEEE 802.11 or HIPERLAN discussed below will help to increase user acceptance of WLANs.

Table 14.1: Characteristics of different transmission techniques

	Spread spectrum	Microwave	Infrared
Frequency [GHz]	1–6	18.825–19.205	30 000
Range [m]	30–250	10–50	25
Power [W]	<1	0.025	—

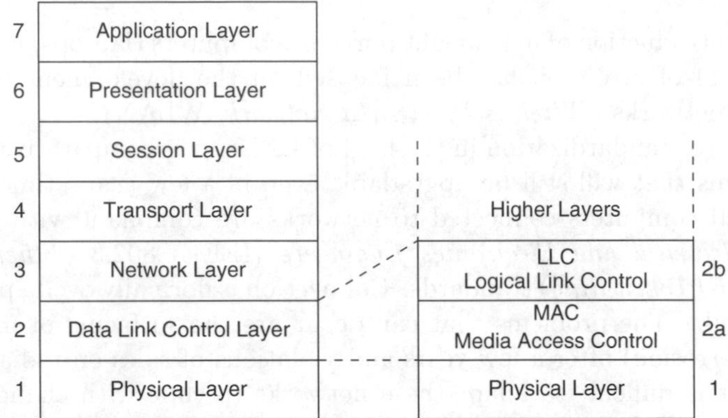

Figure 14.1: The IEEE 802/ISO 8802 standard

The diversity of LAN systems in terms of cabling, transmission techniques, transmission speeds, access procedures and variations thereof necessitated standardization in order to facilitate their acceptance and enable different LANs to work together.

Committee 802 of the *Institute of Electrical and Electronics Engineers* (IEEE) in the early 1980s had developed a standard for *Local Area Networks* (LANs) with speeds of up to 20 Mbit/s that offers security on the communications side for manufacturers as well as for users and has largely been accepted.

The standard mainly restricts itself to the lower two layers of the ISO/OSI reference model (see Figure 14.1). A separation is made between *Logical Link Control* (LLC) and *Medium Access Control* (MAC). The LLC layer upwardly offers all systems a standard interface for establishing logical connections. The MAC sublayer supports protocols such as Token Ring, Token Bus and *Carrier Sense Multiple Access with Collision Detection* (CSMA/CD) (*Ethernet*).

In Western Europe the standards for wireless radio LANs are specified by ETSI. The technical group *Radio-Equipment and Systems* (RES) 10 at ETSI has developed *High Performance Radio Local Area Network Type 1* (HIPERLAN/1), an European standard for LANs [13]. The frequency bands at 5.2 GHz and 17.1 GHz are being reserved throughout Europe for this WLAN.

High Performance Radio Local Area Network (HIPERLAN) Type 1 describes a WLAN for computer–computer/terminal communication. Since this standard has not found products so far, its description that was contained in the first edition of this book can be found now under `http://www.comnets.rwth-aachen.de/~mrn/FirstEdition/`. ETSI *Broadband Radio Access Network* (BRAN)—combining RES 10 and TM 4—also has workgroups that are developing specifications for wireless *Asynchronous Transfer Mode* (ATM) systems under the designation of HIPERLAN Type 2 (*Wireless-ATM LAN*), HIPERACCESS (*Wireless-ATM Remote Access*) and HIPERLINK (*Wireless-ATM Link*). The Standardization of HIPERLAN Type 2 was completed in March 2000 (see Section 14.2).

In parallel to HIPERLAN in Europe, the IEEE in the USA has specified 802.11 a standard for WLANs [20]. Both standards are equipped with an IEEE 802.2 or ISO 8802 compatible interface [22], allowing them to replace the wired LAN transmission systems described above. Because of the restriction of the radio medium (e.g., radio range), both standards must contain functions for the management and maintenance of the radio network that far exceed the normal tasks of the MAC sublayer. These are described along with the technical characteristics of *High Performance Radio Local Area Network Type 2* (HIPERLAN/2) and IEEE 802.11 in the Sections 14.2 and 14.3.

14.1 General Properties of a WLAN

A WLAN can be used as a universally accepted broadband and flexible ad hoc LAN and as such can be connected to other LANs.

WLAN terminals have to be small so they can be used in portable computers. Plans are underway to make them the size of a *Personal Computer Memory Card Interface Association* (PCMCIA) card with the dimensions 85×54×10.5 mm (excluding antenna system).

Since WLAN systems support applications for battery-operated systems, they must offer low power consumption of a few hundred mW and offer an energy-saving mode.

WLAN networks should support the mobility of terminals, e.g. stations should be designed to be able to exchange information with other stations at up to a speed of 10 m/s, which corresponds to 36 km/h, or up to a rotational speed of 360 °/s.

14.1.1 Network Environments for WLANs

The *ETSI Technical Report* (ETR) 069 [14] produced by the ETSI RES 10 committee is taken as an example here to define the services and possibilities required from a WLAN. Some of the applications that will benefit from new solutions and an overview of the WLAN network topologies are presented in the following [5].

14.1.1.1 WLAN Applications

Wireless offices A WLAN is a better option than a fixed network in buildings or in environments where constructional changes are so frequent that cabling cannot be installed, e.g., film and photographic studios. Furthermore, it should be possible, for example, to use portable computers in different locations and connect them easily to a network.

Ad hoc networks Ad hoc networks [35, 39] are radio networks without any kind of permanent communications infrastructure. A group of users can form its own closed complex. This means that at conferences, conventions, and large functions, or in the event of accidents or catastrophes, computers can communicate with each other without having been cabled together beforehand. Each user carries his part of the network with him in the form of his computer with a radio-LAN connection.

Medicine Within a radio-LAN, doctors would be able to have direct and interactive access to remote data such as X-rays when visiting their patients. This would make the work of doctors easier and produce better and faster diagnosis for patients.

Industrial applications More and more work in industry is being automated. In many cases the controlling computers are in a central facility managing a large number of machines that are restricted to use in their locations. A wireless connection to the network would allow these machines (e.g., industrial robots or unmanned vehicles) more freedom of movement and the possibility of being used with more flexibility. Maintenance personnel would use laptops to retrieve the data needed for diagnostic purposes.

14.1.1.2 Network Topologies

A WLAN need not to be organized centrally, and instead may have a completely distributed architecture with a dynamic allocation of network and network node identifiers. Each station (node) is differentiated from the other stations by a unique *Node Identifier* (NID). A number of stations are combined into a network with a shared *WLAN Identifier* (LID). This network forms a WLAN.

In contrast to a wireline network, WLANs using the same radio channel cannot be separated from one another. Overlapping can occur. Another problem with radio channels is the range restriction. Mobile WLAN nodes and unfavourable propagation characteristics can cause the fragmentation of a network.

The following network topologies exist as a result of the channel characteristics and the applications presented:

Independent WLANs Two WLANs, A and B, are considered to be independent of each other if no member from WLAN A is located in the trans-

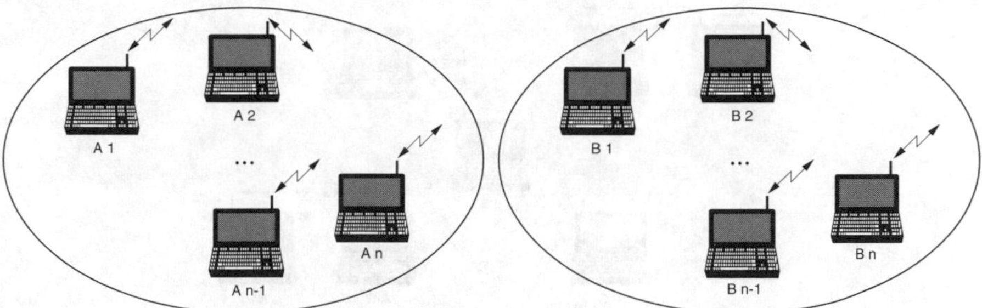

Figure 14.2: Independent WLANs

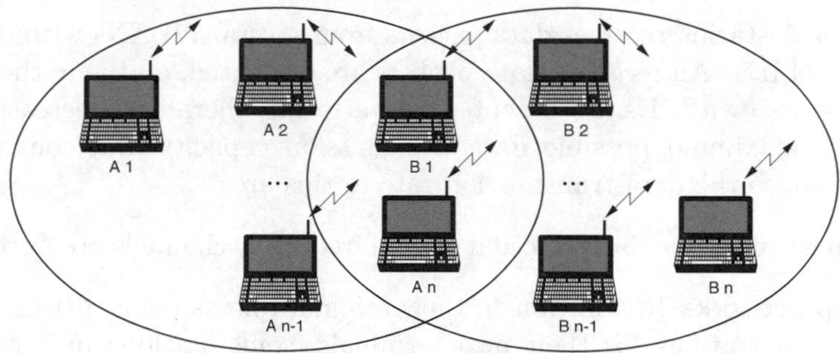

Figure 14.3: Overlapping WLANs

mission range of a member from network B (see Figure 14.2). Even if the same frequencies are used for transmission in both networks, it is assumed that subnetworks A and B do not share the communication medium and therefore also do not cause any interference in each other's network. WLAN A could be an ad hoc network set up for a conference of company X, and WLAN B an LAN used in the factory of company Y.

Overlapping WLANs If the radio range of some of the stations in network A should overlap with some of the stations in network B then these members share the communications medium and its transmission capacity in the area where the overlapping occurs (see Figure 14.3).

An overlapping of networks produces two effects:

- The senders in the different WLANs use the same frequency band, thereby increasing the occurrences of interference. As a result, optimal use of the frequency band is no longer possible because not all the stations are able to receive from each other (*hidden stations*) and therefore can cause interference to each other (see Section 14.2.10).

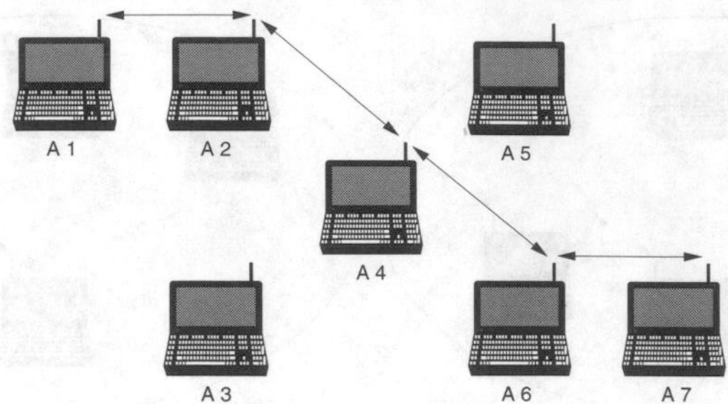

Figure 14.4: Communication in a multihop radio network

- A station receives data packets from several WLANs with different LIDs. All received data packets are evaluated, and only those with its own LIDs are accepted. As a result, there is a decrease in the maximum possible data transmission capacity and consequently also the data transmission rate in this area.

These effects can be reduced if several frequency channels are introduced.

Multihop networks In addition to their original role as transmitting and receiving stations for their own terminals, some stations in a multihop network also perform the function of relay stations. This allows data to be transmitted over larger distances despite the restricted reachable range of the radio medium. In Figure 14.4 the relay stations (*forwarders*) 2, 4 and 6 are forwarding the traffic of node 1 to destination 7.

Interworking Since most WLAN applications already exist, it must be possible for WLANs to be connected to the usual fixed networks (see Figure 14.5). This affects the network layer, and is not part of the WLAN standard.

Communication Security A radio channel can be listened for in a neighbouring area. An encryption algorithm with the appropriate cryptographic management is therefore needed. This protects confidential data from unauthorized eavesdropping, and also guarantees communication security for the radio network.

Typical WLAN encryption schemes provide for a common set of keys, one of which is used in the encryption operation. Each key has a number that is transmitted with the encrypted data to the receiver. In addition, a common initialization vector for encryption and decryption is required and if necessary transmitted.

The level of transmission security increases with the frequency of the change of key and initialization vectors.

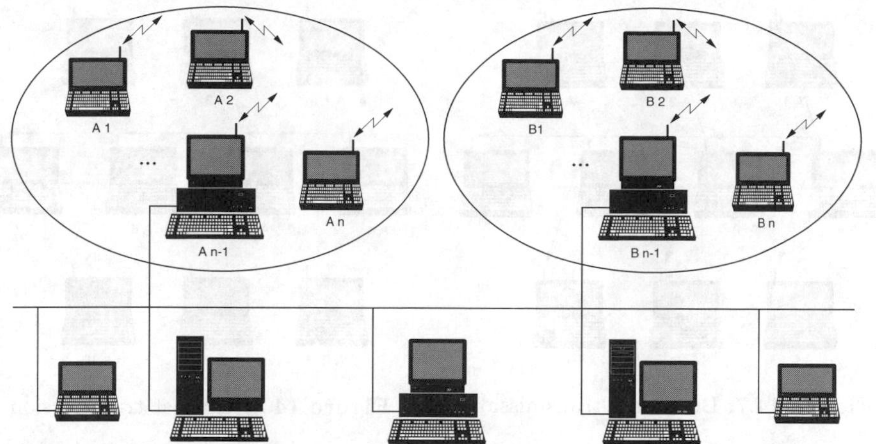

Figure 14.5: Connection to fixed network

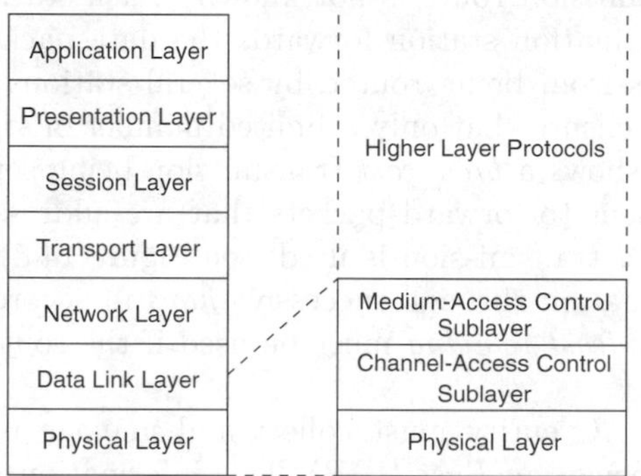

Figure 14.6: ISO/OSI and WLAN reference models

Addressing of Service Access Points For compatibility with the ISO-MAC service definition (see Figure 14.6) the MAC service of a WLAN typically uses 48-bit LAN-MAC addresses for identification of *MAC Service Access Points* (MSAPs). Individual MSAPs and group addresses for contacting several MSAPs are available.

Forwarding The WLAN-MAC protocol may incorporate a *multihop relaying* facility that enables the transmission of data beyond the boundaries of a station's sending area—in other words over several stations. A WLAN-MAC entity may either be a forwarder or a non-forwarder. Only forwarders carry out the forwarding of *MAC Service Data Units* (MSDUs) when required. Point-to-point (*unicast*) as well as broadcast transmission (*broadcast, multicast*) is possible for the transmission of packets.

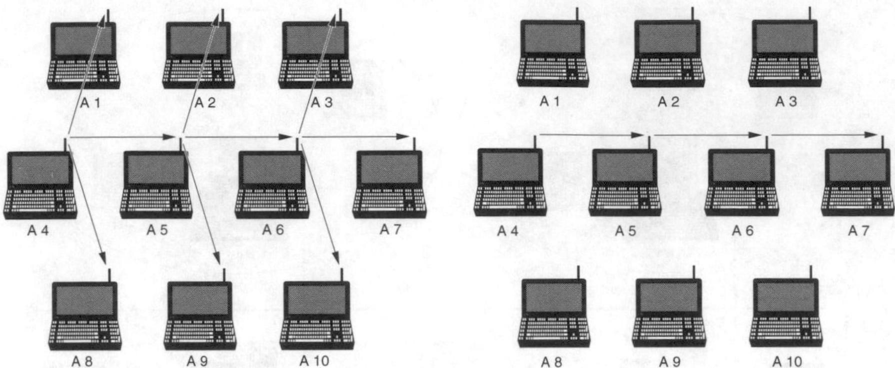

Figure 14.7: Broadcast transmission

Figure 14.8: Unicast transmission

Broadcast relaying is used to relay information to all WLAN-MAC entities or when the transmission route is not known. Each station that recognizes a route to the destination station forwards the data packet accordingly. To avoid data packets from being routed by several stations at the same time, the protocol must ensure that only a limited number of stations can forward data. Figure 14.7 shows a *broadcast* transmission being sent by station 4.

It is more efficient to forward packets that are addressed to a particular receiver if a *unicast* transmission is used (see Figure 14.8). A packet is then routed to its destination through successive *hops* in accordance with an (optimal) route. *Broadcast relaying* must be used if the route is not known, as explained above.

Each WLAN-MAC entity must collect and manage routing information in its *Routing Information Base* (RIB). This information is continuously updated. The data in the RIB becomes obsolete and is discarded once it has passed its validity period. Because of this constant updating of routing information, it is even possible to specify quasi-optimal paths for the forwarding of packets in continuously changing WLANs.

14.2 HIPERLAN/2*

High Performance Radio Local Area Network Type 2 (HIPERLAN/2) offers wireless broadband network access. Generally, broadband systems are those in which the transmission rate is higher than the primary multiplex rate of ISDN (2048 kbit/s). See ITU Rec. I. 113 for a more comprehensive definition of *broadband.*

HIPERLAN/2 is being standardized by the *European Telecommunications Standards Institute* (ETSI) as part of the BRAN project. An overview of ETSI

*__With the collaboration of Andreas Hettich, Arndt Kadelka, and Jörg Peetz__

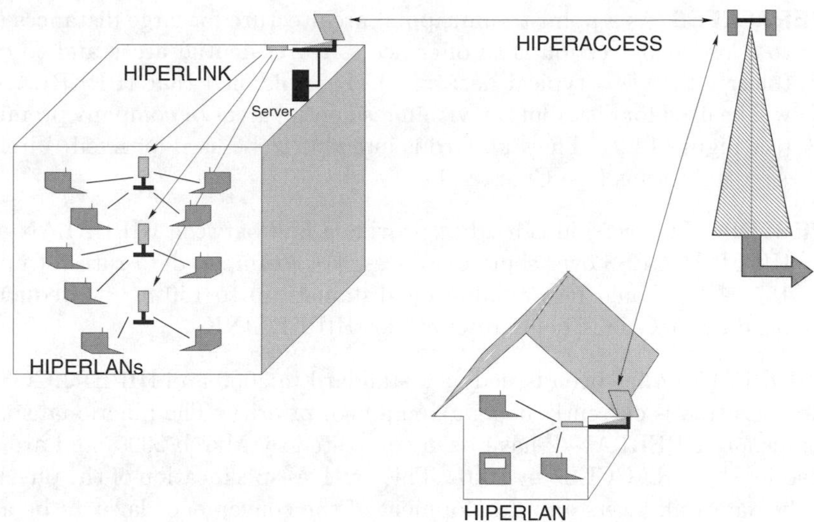

Figure 14.9: ETSI-BRAN systems: broadband radio access networks

BRAN is presented, followed by a detailed description of the HIPERLAN/2 protocol stack.

14.2.1 BRAN System Description

Starting in the year 2000 the ETSI project BRAN has provided services with bit rates of up to 54 MBit/s and will be providing service with bit rates of up to 155 Mbit/s for access to fixed networks in private and public environments. Wireless access networks are regarded as a competitive alternative to wired access systems. The advantages of the wireless networks are the speed with which they can be set up and the fact that they do not require an existing wired infrastructure but are able to expand wired LANs.

With the availability of the functional specifications for HIPERLAN/1 in 1998, three different BRAN standards have been or are now being developed (see Figure 14.9):

HIPERLAN/2 This version for short distances (up to 200 m) is being planned as a supplementary access mechanism for 3G systems and for private users as a wireless LAN. It offers high-speed access (27 Mbit/s, 54 Mbit/s optional, typical data rate) to different networks, including UMTS core networks, ATM networks and IP-based networks. HIPERLAN/2 is operated in the frequency band at around 5 GHz. The HiperLAN Global Forum is promoting the introduction of HIPERLAN/2 worldwide (see `www.hiperlan2.com`).

HIPERACCESS As a point-to-multipoint architecture for large distances (up to 5 km), this version is to offer access to residential areas and to customers (27 Mbit/s typical data rate). It is planned that HIPERLAN/2 will be used for distribution within residential areas or company premises (see Figure 14.9). The standard is intended to be used for fixed Wireless Access Systems (see Chapter 11).

HIPERLINK This version is used to provide a link between HIPERLAN and HIPERACCESS over short distances. For example, data rates of up to 155 Mbit/s should be available for distances up to 150 m. A frequency band at 17 GHz is being reserved for HIPERLINK.

HIPERLAN/2 has been issued as a standard in 2000 and HIPERACCESS standardization is currently being given a high priority. The functional specifications for HIPERLAN/2 have been completed by March 2000, and are expected for HIPERACCESS by 2001. This involves specification of the physical and the data link layers and development of the convergence layer to be able to serve Ethernet (IEEE 802.3), ATM and IEEE 1394.

14.2.2 System Description

The current most popular standard for wireless LANs is IEEE 802.11 issued by the *Institute of Electrical and Electronics Engineers*. It offers data rates of 1 Mbit/s and recently as high as up to 11 Mbit/s called 802.11b; a newer version 802.11a does offer rates up to 54 Mbit/s. However, a new class of WLAN standards that support quality of service, handover and data integrity are necessary to satisfy the requirements of wireless LANs. This demand was the motivation for the standardization of H/2.

14.2.2.1 Network Configuration

An H/2 network typically consists of many *Access Points* (APs) that together provide complete or partial radio coverage to a geographical area. Mobile terminals (*Mobile Termination*, MT) communicate with the AP over the H/2 radio interface. MTs are able to move freely within this area and are automatically supplied with a service by the most favourable access point. The AP in Figure 14.10 consists of one *Access Point Controller* (APC) and one or more *Access Point Transceiver* (APT)s. The APs automatically search for the most favourable frequency channel in order to avoid the need for frequency planning.

H/2 can be operated in two operating modes that may also be offered simultaneously:

Centralized Mode (CM) In CM all APs are linked to the fixed network and the MTs form an association with them. All usage data is transmitted over the AP, even if two MTs in the same cell are communicating with

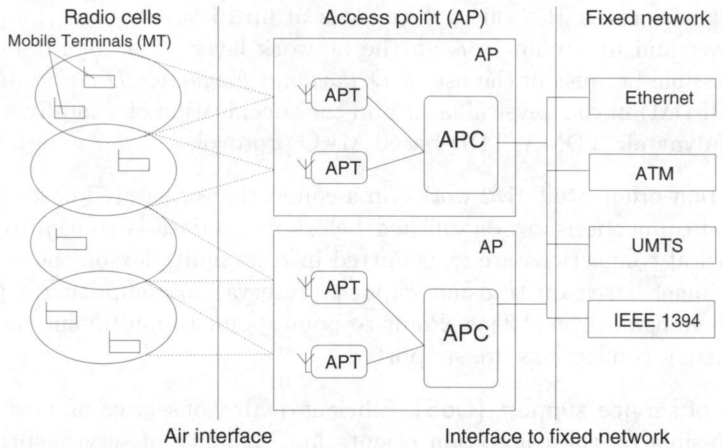

Figure 14.10: HIPERLAN/2 network configuration

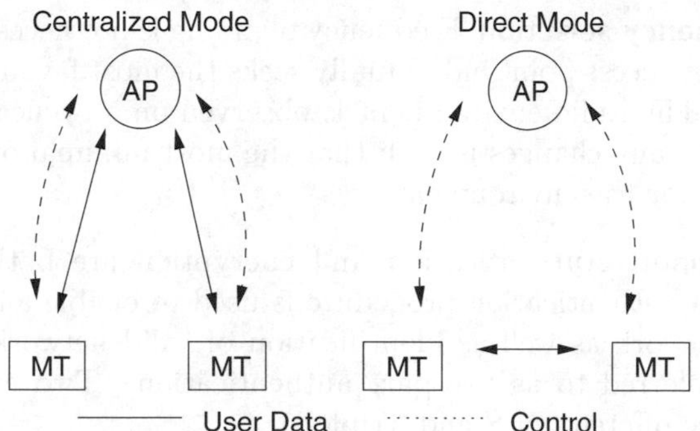

Figure 14.11: HIPERLAN/2 operating modes

one another (see Figure 14.11). CM must be supported by all MTs and APs.

Direct Mode (DM) In direct communication mode radio cells are managed by a central control entity, called *Central Controller* (CC). In this mode MTs that are located in radio range to one another can directly exchange usage data under the control of the CC. A CC or an MT can be connected to the fixed network in DM, thereby having access to the network. DM is optional.

14.2.2.2 Features

H/2 offers features that are designed to meet the requirements of future wireless multimedia services. These features are briefly summarized below:

High transfer rates H/2 offers data rates of up to 54 Mbit/s on the physical layer and up to 36 Mbit/s on the network layer. The high data rate is possible because of the use of *Orthogonal Frequency Division Multiplex* (OFDM) in the physical layer and the specification of a totally new type of dynamic TDMA/TDD-based MAC protocol.

Connection-orientated H/2 works on a connection-orientated basis, i.e., logical connections are established before usage data is transmitted. The logical connections are transmitted in time multiplex on the frequency channel based on transmit capacity reservation comprising a periodic planning horizon of 2 ms. Point-to-point, point-to-multipoint and broadcasting connections are supported.

Quality of service support (QoS) Efficient quality of service support is made possible by the connection orientation. Quality of service-specific parameters, such as data rate, delay, and variation of delay and loss rates, can be agreed individually for each connection.

Dynamic frequency selection Frequency planning is not necessary in H/2 because each access point individually seeks the most favourable frequency channel. The radio environment is observed on a cyclical basis and the reaction to any changes is such that the most minimal of interference is produced for the environment.

Encryption support Authentication and encryption are both supported in H/2. The authentication procedure is used to enable authorized access to the network as well as identification of valid network access points. This is referred to as two-part authentication. Two encryption algorithms are offered, DES and Triple-DES.

Mobility support The mobility of H/2 users is supported by the fact that handovers are possible from one network access point to a more favourable one. However, some data can be lost during this process.

Network and application independence H/2 has not been optimized for any specific network protocols or applications. Different network protocols, such as Ethernet, ATM and IEEE 1394, can use the WLAN HIPERLAN/2 because of the so-called convergence layers defined as part of the H/2 standard. The flexibility of the channel access protocol allows the use of any type of application, such as audio, video or LAN.

Energy-saving mode H/2 devices can agree their own individual sleep times. Different energy-saving modes can be implemented during these times, depending on the device.

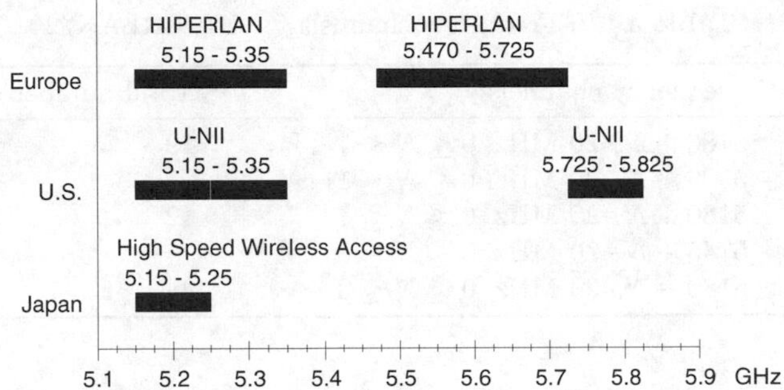

Figure 14.12: Frequency allocations for HIPERLAN/2 systems

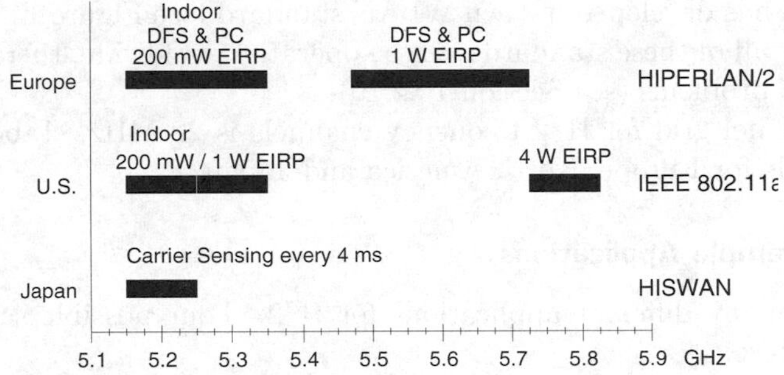

Figure 14.13: Regulatory restrictions for frequency bands

14.2.2.3 Spectrum for HIPERLAN/2

H/2 was designed for operation in the 5-GHz frequency band. The frequencies shown in Figure 14.12 are available for the core markets Europe, North America and Japan. In Europe 455 MHz of *licence-exempt* spectrum overall has been allocated [12]. In general, the spectrum is not exclusively available for WLANs but is also allocated for radar and satellite-based services. A total of 300 MHz of *unlicenced* spectrum is available for the USA, and only 100 MHz for Japan.

Specifications exist in Europe for the maintenance of spectrum masks as well as for the implementation of transmitter *Power Control* (PC) with an extensive average power reduction of 3 dB and *Dynamic Frequency Selection* (DFS) (see Section 14.2.5.3). The values given for *Equivalent Isotropic Radiated Power* (EIRP) in Figure 14.13 relate to the maximum average transmit power in Europe.

A need for PC or DFS does not exist in the USA. There the EIRP values relate to the maximum peak transmit power. All equipment adhering to the standards of the *Unlicensed National Information Infrastructure* (U-NII) band, particularly IEEE 802.11a, can operate within the spectrum for U-NII.

Table 14.2: Frequency channels for HIPERLAN/2

Region	Frequency channels	Total number of channels
Europe	$5180 + N \cdot 20$ MHz, $0 \leq N \leq 7$	19
	$5500 + N \cdot 20$ MHz, $0 \leq N \leq 11$	
USA	$5180 + N \cdot 20$ MHz, $0 \leq N \leq 7$	12
	$5745 + N \cdot 20$ MHz, $0 \leq N \leq 3$	
Japan	$5180 + N \cdot 20$ MHz, $0 \leq N \leq 3$	4

In Japan HIPERLAN/2 and IEEE 802.11a share the 100 MHz wide spectrum for the socalled *High-speed Wireless Access Network* (HISWAN). For frequency etiquette *carrier sensing* is prescribed for every 4 ms. Although each region has developed its own WLAN standard (see Figure 14.13) it is expected that all of these standards will be operated worldwide, thereby causing coexistence problems (see Section 14.2.10).

The channel grid for H/2 frequency channels is 20 MHz. Table 14.2 lists the channels for Europe, North America and Japan.

14.2.2.4 Sample Applications

There are many different applications for H/2. Four possible scenarios are presented here.

Corporate LAN An H/2 network can be operated in an office building as access to its Ethernet-LAN. The main application is the Internet. Mobile users within a company can move about freely since H/2 deals with handovers. For users leaving their own network areas IP mobility can enable a handover to a public network access point (AP).

Hot spots H/2 can be used in *hot spots*, such as airports, train stations and city centres. This facility provides business people with access to public services, to the Internet or, if the corresponding protection mechanisms are available, to their company networks.

Access to 3rd generation cellular networks H/2 will provide high-speed access and thus complementing the wide-area W-CDMA networks. This means that subscribers in areas with high traffic volumes and limited mobility *(hot spots)* will be able to benefit from both high data rates and quality of service of H/2 wide-area service of UMTS. The fixed network must be able to provide seamless transition between these different types of access networks. User terminals must also support both radio interfaces.

Networking in home locations There is a growing trend in the field of entertainment electronics to use IEEE 1394 buses *(fire wire)* to network together different equipment. Because of its high transmission speed

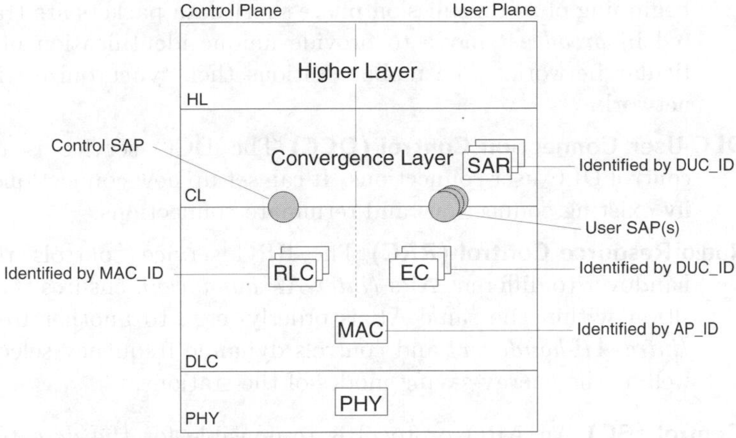

Figure 14.14: HIPERLAN/2 service model

and quality of service guarantees, H/2 can be used to provide a wireless connection of individual groups of equipment, e.g. plug-and-play multimedia clusters. A cable modem or an *Asynchronous Digital Subscriber Line* (ADSL) can also implement an access point to service providers.

14.2.3 Service Model

The specifications for ETSI-BRAN systems primarily relate to the lower three layers of the ISO/OSI reference model. The H/2 service model is shown in Figure 14.14. A *Data Link Control* (DLC) layer is inserted over the *Physical Layer* (PHY). A *Convergence Layer* (CL) is introduced to create a transition to the higher layers. This layer converts the data of the control and user planes of the higher layers to data of the DLC layer that are exchanged over service access points of the DLC layer. The convergence layer is also responsible for the *Segmentation and Reassembly* (SAR) of protocol data units (PDU) on the user side.

The DLC layer is organized into three functional units: RLC sublayer on the control plane, error control layer on the user plane and MAC sublayer. The following services are offered:

Radio Link Control (RLC) The RLC protocol handles most of the control tasks on the DLC plane. It is an asymmetrical protocol in which the AP resolves protocol queries by the MT or initiates RLC procedures on its own (*master–slave* principle). The tasks of the RLC include:

Association Control Function (ACF) The ACF service controls registration and authentication functions within a radio cell and provides the appropriate signalling using *beacon* signals. At the beginning of a transmission phase short data packets are transmitted in *broadcast* mode to provide unique identification of a particular network. The mobile stations then synchronize with this network.

DLC User Connection Control (DCC) The DCC service is used to control DLC user connections. It can set up new connections, modify existing connections and terminate connections.

Radio Resource Control (RRC) The RRC service controls the MT handover to different APs *(inter-AP handover)*, ensures that transition within the same AP is orderly, e.g., to another frequency *(intra-AP-handover)* and controls dynamic frequency selection as well as the energy-saving modes of the stations.

Error Control (EC) An ARQ protocol is responsible for the detection and correction of transmission errors on the user plane. EC is based on the sending of retransmission requests for erroneous packets, thereby enabling an enhance of the quality of service of a connection. ARQ protocols are symmetric, i.e., the entities in the AP and the MT have an identical structure.

Medium Access Control (MAC) The MAC sublayer is responsible for access control on a medium. The AP allocates capacity to the terminals that use this capacity to transfer usage and control data. Thus, like the RLC protocol, this also is a master–slave mode of operation. Capacity allocation uses a strategy that is designed to guarantee the quality of service of the connections competing for transmit capacity.

If the service model is considered from a standpoint that the AP and the MT are separate, it can be determined that in respect to the RLC, EC and SAR entities the AP always contains the partner entities of the entities in an MT. The number of RLC, EC and SAR entities in the AP is consequently the sum of these entities of all MTs.

The number of SAR entities depends on the number of connections requiring a segmentation and reassembly of data packets. The number of EC entities depends on the number of connections requiring error correction through packet retransmission.

Figure 14.15 (left) shows the entities of an AP. One RLC entity exists for each registered terminal. The corresponding partner entity is located in the MT (see Figure 14.15, right). The number of EC and SAR entities in a MT is considerably lower than in an AP, because only the connections of the MT are handled in the respective MT, whereas the connections of all MTs have to be handled in the AP. An AP and an MT each have one MAC entity. Only multi-frequency transmission requires a larger number of MAC entities.

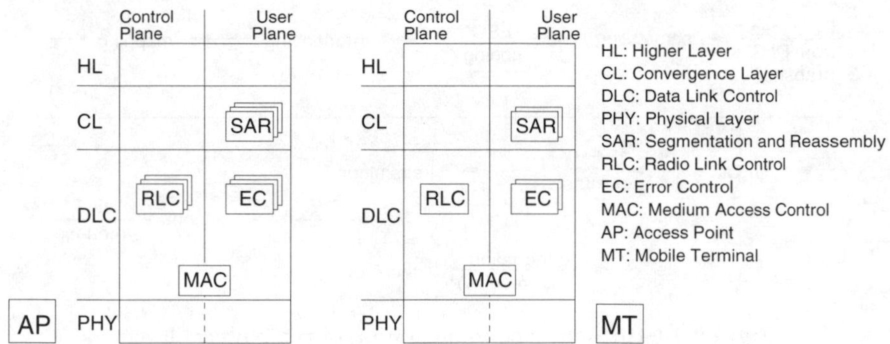

Figure 14.15: Service model for AP (left) and MT (right)

14.2.4 Physical Layer

H/2 and IEEE 802.11a both use *Orthogonal Frequency Division Multiplex* (OFDM) for bit transmission in the PHY. The advantage of OFDM over other multiplexing techniques is that no steep-edged filters are required because each OFDM subcarrier of a given channel, which consists of many simultaneously operated subcarriers, can utilize the entire spectrum. Orthogonal subcarriers ensure that the individual subcarriers can be separated again on the receiver side. Overall, the use of OFDM gives the assurance of increased spectral efficiency. This technique can be implemented cost-effectively because only *Fast Fourier Transformation* (FFT) or *Inverse Fast Fourier Transformation* (IFFT) is used for demodulation and modulation of the symbols to be transmitted, respectively, and no equalizers are required. See [28] for a more detailed explanation of OFDM.

In addition to transmitting and receiving data, layer 1 also carries out the following functions. These functions are listed but not explained in detail.

- Measurement of radio channel quality
- Antenna selection
- Transmitter power control

14.2.4.1 Reference Configuration

Figurc 14.16 shows the reference configuration for the functions of the physical layer. The standard only specifies the sending side; the receiving side is specified by performance characteristics only.

The elements in Figure 14.16 show the individual tasks of the physical layer. These functions are briefly explained below. Link adaptation, coding and modulation are covered in detail in Sections 14.2.4.3 and 14.2.4.4.

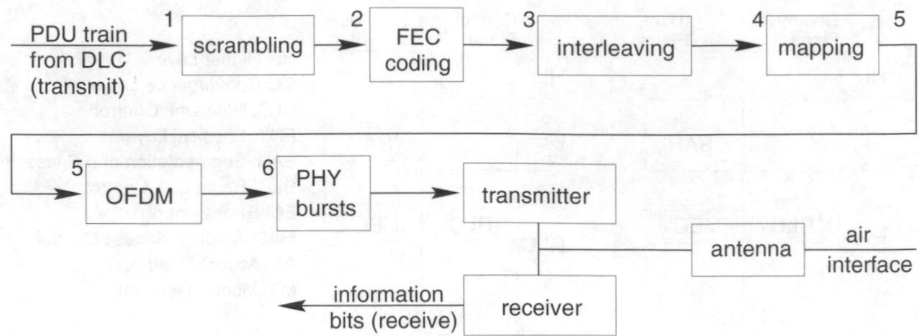

Figure 14.16: Reference configuration of the physical layer

Link adaptation The physical layer selects a suitable PHY mode for *information bits (1)* arriving (see Section 14.2.4.4).

Scrambling A scrambler is used to scramble the *information bits* byte-by-byte so that the same bits produce different *scrambled bits (2)* in the various MAC frames. Consequently, the power density spectrum of a data burst varies over time despite the same content. The generator polynom of the scrambler is $S(x) = x^7 + x^4 + 1$.

FEC coding For error correction a convolutional coder is used to code the scrambled bits. This technique is called forward-error correction (FEC). The coded scrambled bits are called *encoded bits (3)*.

Interleaving Interleaving increases the efficiency of the convolutional coder and reduces the influence of frequency-selective fading. The block length of the interleaver equals the number of codable bits per OFDM symbol N_{CBPS} (see Section 14.2.4.4). The interleaving can be specified through two permutations:

$$\begin{aligned}
i &= \frac{N_{CBPS}}{16} \cdot (k \bmod 16) + \left\lfloor \frac{k}{16} \right\rfloor, \\
k &= 0, 1, ..., N_{CBPS} - 1, \\
j &= s \cdot \left\lfloor \frac{i}{s} \right\rfloor + (i + N_{CBPS} - \left\lfloor 16 \cdot \frac{i}{N_{CBPS}} \right\rfloor) \bmod s, \\
i &= 0, 1, ..., N_{CBPS} - 1, \\
s &= max\left(\frac{N_{CBPS}}{2}, 1\right)
\end{aligned}$$

The result of interleaving is *interleaved bits (4)*.

Mapping Depending on the modulation technique used, the *interleaved bits* are broken down into groups of 1, 2, 4 or 6 bits (corresponding to BPSK,

Figure 14.17: Functional units of a FEC coder

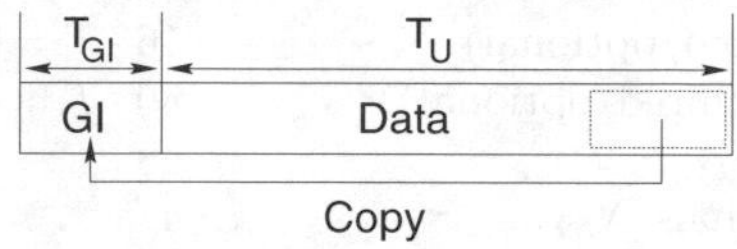

Figure 14.18: OFDM symbol with cyclical extension

QPSK, 16QAM and 64QAM) and allocated to the corresponding complex modulation symbol $d = (I + j \cdot Q) \cdot K_{MOD}$. The K_{MOD} stands for one of the normalization factors based on the modulation used. These mapped symbols are called *sub-carrier modulation symbols (5)*.

Modulation OFDM modulation transforms *sub-carrier modulation symbols* into baseband signals by means of IFFT. The symbols are then referred to as *complex-valued baseband OFDM modulation symbols (6)*.

PHY burst composition A PHY burst—the data structure for the physical symbol-by-symbol information transmission on the sub-carriers used—is set up through the addition of pilot sub-carriers and a preamble.

14.2.4.2 Coding

The coder shown in Figure 14.17 is used in the coding of scrambled bits. First, 6 bits are added to the *PDU train* for code termination. A punctured convolutional code with the code rate 1/2 and the influence length 7 is used to code the data. The puncturing works with code rates of 3/4 or 9/16.

14.2.4.3 Modulation

The complex-valued symbols from the allocation tables are then assigned to the sub-carriers. There are 48 sub-carriers for data and four for reference information. This reference is necessary in assessing channels for the coherent modulation being used. Each OFDM symbol therefore is transmitted on 52 sub-carriers with a length, including guard time, of T_U =4 µs. Other parameters of OFDM modulation can be derived from Table 14.3.

Figure 14.18 illustrates the structure of an OFDM symbol with cyclical extension through the guard time that protects against intersymbol interference.

Table 14.3: Transmission parameters of OFDM

Parameter	Value
Sample frequency $F_s = 1/T$	20 MHz
Symbol duration T_U	$64 \cdot T = 3{,}2$ µs
Guard time T_{GI} (required/optional)	$16 \cdot T = 0{,}8$ µs / $8 \cdot T = 0{,}4$ µs
Symbol interval T_S (required/optional)	$80 \cdot T = 4{,}0$ µs / $72 \cdot T = 3{,}6$ µs
Sub-carrier for data N_{SD}	48
Sub-carrier for pilot signals N_{SP}	4
Number of sub-carriers N_{ST}	52 $(N_{SD} + N_{SP})$
Sub-carrier interval δ_f	0,3125 MHz $(1/T_U)$
Interval of external sub-carriers	16,25 MHz $(N_{ST} \cdot \delta_f)$

Table 14.4: Transmission rates based on modulation technique and code rate

Modulation-technique	Code rate	Codable bits per symbol	Data bytes per symbol	Usable data rate (Mbit/s)
BPSK	1/2	48	3	6
BPSK	3/4	48	4,5	9
QPSK	1/2	96	6	12
QPSK	3/4	96	9	18
16QAM	9/16	192	13,5	27
16QAM	3/4	192	18	36
		optional		
64QAM	3/4	288	27	54

14.2.4.4 Link Adaptation

Transmission rates vary depending on the choice of modulation technique and on the code rate. Table 14.4 presents all the possible combinations of modulation techniques and code rates and the resulting transmission rates.

The combination of modulation and code rate is referred to as PHY mode and enables a layer 2 entity to select a suitable mode based on receiving conditions of the partner equipment that are characterized by the BER.

14.2.5 The Data Link Layer

The task of the *Data Link Control* (DLC) layer is to divide the joint shared physical medium, the radio channel, fairly and efficiently among the mobile stations. Another task of layer 2 is to protect transmission from errors on the radio channel for each individual connection. On the user plane the DLC layer is therefore broken down into two sublayers: *Medium Access Control* (MAC) and *Error Control* (EC). In the MAC sublayer a TDMA technique is used to subdivide the physical channel into time slots and transmission capacity is allocated to stations depending on their requirements. By means of an ARQ

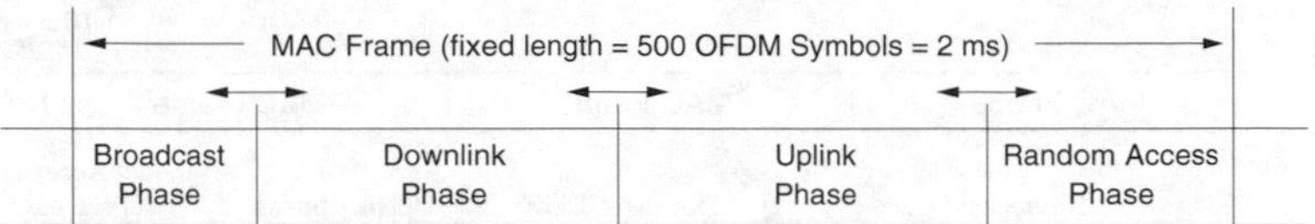

Figure 14.19: Transmission phases within a MAC frame

protocol (see Section 14.2.5.2) the EC sublayer, on the other hand, resolves transmission errors using data acknowledgement and packet retransmission requests.

14.2.5.1 The MAC Sublayer

The MAC sublayer divides the physical channel into frames of a constant length. The ETSI-BRAN standard defines a frame length of 2 ms. At an OFDM symbol length of 4 µs, this equates exactly to 500 OFDM symbols per frame. A frame is further subdivided into transmission phases with differing functions [36] (see Figure 14.19).

A central entity, the AP (called *Central Controller* (CC) in the *Home Environment Extension* (HEE) of H/2 (see Section 14.2.13) coordinates access to the shared physical channel. It allocates dynamic transmission capacity in the form of variable-sized time slots to the terminals. Through the introduction of priorities preference can be given to connections with high quality of service requirements. The AP therefore needs information about the number and the service class of packets waiting for transmission in the MTs associated to it.

Thus the following activities are carried out (see Figure 14.19):

Downlink phase User data is transmitted by the AP to the MT.

Uplink phase User data is transferred from the MT to the AP.

Random access phase The capacity requirements of registered MTs as well as the registration requests of MTs that are not yet registered are sent.

Structure of the MAC Frame The physical layer provides transport channels to handle the transmission of MAC protocol data units during the different phases. The transport channels of the different phases are shown in Figure 14.20. Depending on demand, the capacity of the transport channels' capacity changes in the successive MAC frames, controlled by the BCH and the *Frame Channel* (FCH).

Broadcast Channel (BCH) The BCH is the beacon channel where basic information about a radio cell, such as identification of the AP on the DLC layer and the current transmitter power of the AP, is transmitted to all MTs. In addition, the BCH contains pointers to the positions of

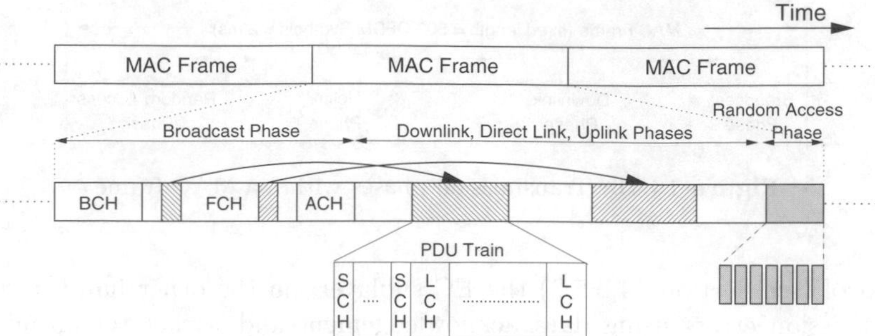

Figure 14.20: Allocation of channel capacity to a MT

the FCH and the *Random Channel* (RCH) in the frame and defines their lengths. This provides the entire MAC frame with a variable structure.

Frame Channel (FCH) The AP uses the FCH to deliver a description of contents of the slots for the downlink and the uplink phases.

Access Feedback Channel (ACH) The status of access to the RCH of the preceding MAC frame is described in the ACH. The ACH indicates the RCH time slots in which protocol data units have been received correctly by the AP.

Downlink phase The *Long Channel* (LCH) and the *Short Channel* (SCH) are transmitted in the *downlink phase.* Fixed-size packets used to send the usage and control data of the current MT and its connections are transmitted over these transport channels. These packets are organized into variable groups called PDU-*trains.* Each PDU-train contains the data belonging to one MT. The AP defines in the FCH the number of PDU-trains to be send in a frame and their starting points in time.

Uplink phase: The *uplink phase* essentially has the same structure as the *downlink phase.*

Direct Link Phase This phase has been defined for the direct communication between MTs under the control of the AP for the HEE of H/2.

Random Channel (RCH) The RCH is used for initial access to a network during a handover and for *Resource Requests* (RRqs). Up to 31 RCHs can be used in the random access phase.

The *Long-Channel-PDU* (L-PDU) and the *Short-Channel-PDU* (S-PDU) in a PDU train are of different lengths. Figure 14.21 shows how the two different types of PDU are constructed. L-PDUs mainly contain usage data from connections. The size of this PDU is 54 bytes of which 49.5 bytes (396 bits) are reserved for usage data; the remaining bytes are used for DLC information.

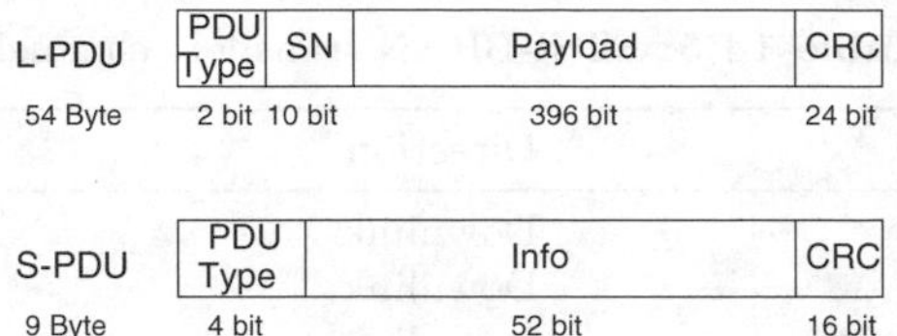

Figure 14.21: Structure of MAC protocol data units

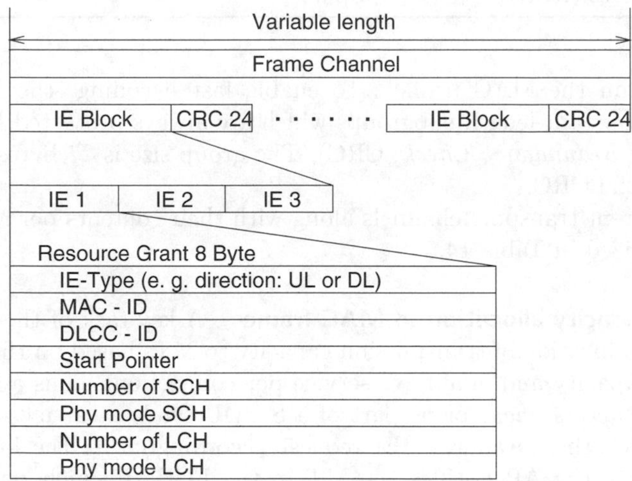

Figure 14.22: Principal structure of a *Frame Channel* (FCH)

S-PDUs with a size of 9 bytes carry DLC control information, such as ARQ acknowledgements. S-PDUs are also used on the uplink for *Resource Requests* (RRqs). These RRqs can therefore be transmitted either in the S-PDU or in the RCH in competition with other MTs.

The FCH specifies the contents of PDU trains and its position in the *uplink* and *downlink phases.* Consequently, these contents contain a list of allocation information (*Resource Grant Information Element*, RG-IE) that always describes a connection in terms of allocated transmission capacity and the timing of the respective PDU. The RG-IE has a constant length of 8 bytes and, using a *Medium Access Control Identifier* (MAC-ID) and a DLCC-ID, addresses the connection of a specific MT (see Figure 14.22). An indicator field describes the position of the connection within the MAC frame. The number of S- and L-PDUs and their PHY modes are also indicated. The modulation and the code rate of the PDU are allocated dynamically per connection per MAC frame. The FCH may contain more than one RG-IE per phase to enable more than one MT to send and receive on the uplink and downlink phase.

Because of this dynamic structure, the length of the FCH is variable. The length is determined by the number of connections being provided with re-

Table 14.5: ETSI-BRAN transport channels

Physical channel	Direction	Length (bytes)
Broadcast (BCH)	Downlink	15
Frame (FCH)	Downlink	$n \cdot 27$
Access feedback (ACH)	Downlink	9
Short transport (SCH)	Down-/Uplink	9
Long transport (LCH)	Down-/Uplink	54
Random access (RCH)	Uplink	9

sources within the MAC frame. To enable fast decoding, the FCH is divided into constant-length subgroups which are always protected by a checksum (*Cyclic Redundancy Check*, CRC). The group size is 27 bytes, including 3 bytes for the CRC.

The different transport channels along with their contents per MAC frame are summarized in Table 14.5.

Dynamic capacity allocation in MAC frames A key task of the MAC sublayer is the allocation of transmission capacity to MTs based on their requirements for capacity and quality of service per connection. Thus a MT within the *random access phase* or as part of a S-PDU sends a capacity request to the AP which then evaluates the request accordingly. At the beginning of a MAC frame the AP notifies the MT in the FCH at which point it is to start receiving data from the AP during the downlink phase and when it is allowed to transmit in the uplink phase. Also conveyed are the start time and the length of the random access phase during which MTs are permitted to transmit their *Resource Requests* (RRqs) to the AP using the slotted-ALOHA protocol. An overlapping of the transmission of access bursts of different MTs can occur at the AP in the RCH. A collision occurs when this causes a total or a partial disruption to a resource request. Due to the varying distances of *Mobile Terminal* (MT)s to the AP, the receive power of the different access bursts can vary so much that the most powerful burst in the AP can be decoded despite a collision (*signal capture*). A mechanism for the detection of possible collisions by the MTs is provided by feedback of the slot number in which a resource request was received by the AP in the ACH. To resolve collisions, H/2 uses the *Exponential-Binary-Backoff* algorithm familiar from the Ethernet in which the average waiting time until a retransmission attempt is doubled after each collision based on the RCH slot length as the time unit. This method does not totally exclude a repeated collision (*deadlock*). More effective methods are described in [40].

It is possible to reserve an RCH time slot to certain MTs, which did not have the opportunity to transmit their dynamic parameters in a S-PDU during the actual MAC frame, to poll their capacity requests there (see Figure 14.20). For a more detailed description of polling strategies refer to [18].

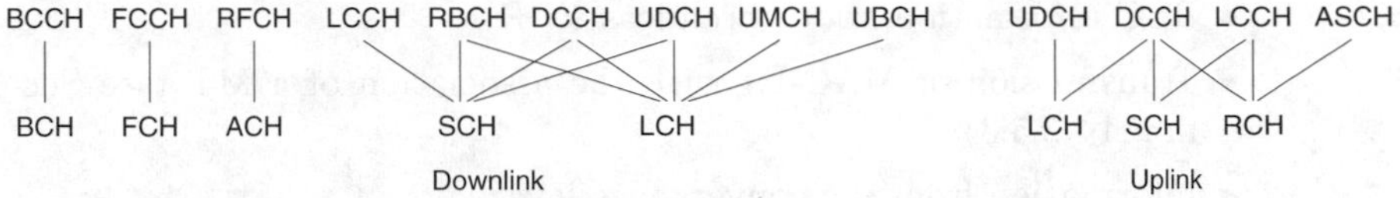

Figure 14.23: Mapping of logical channels onto transport channels

If an MT within a MAC frame has already been allocated transmitting capacity during the uplink phase, then the resource request for usage data to be sent is transmitted in a control PDU in the SCH. Alternatively, other control PDUs, such as ARQ information or RLC information, can also be transmitted in the SCH (see Sections 14.2.5.2 and 14.2.5.3).

The H/2 standard does not specify the allocation strategy that determines how many time slots are to be reserved for which MT. The reason is that it does not have any influence on the interaction between different H/2 devices.

Mapping logical channels onto transport channels The term *logical channel* describes a path that is allocated from the DLC layer. A group of logical channels is defined for the data transmission services that are used by the MAC entity. Each logical channel is defined by the type of information it transmits. Logical channels can be regarded as logical network endpoints between logical entities, such as RLC and EC. Figure 14.23 shows how logical channels are mapped onto transport channels.

Broadcast Control Channel (BCCH) The BCCH is used on the downlink and transmits general control information that is broadcast by the AP. It contains a constant amount of data. If the AP or the CC supports a number of antenna sectors, many BCCHs exist accordingly.

Frame Control Channel (FCCH) An FCCH is used on the downlink and transmits information that describes the other sections of the MAC frame. The following information can be transmitted in a FCCH:

Resource grants for SCH and LCH that are used for the transmission of information from RBCH, DCCH, LCCH, UDCH, UBCH and UMCH.

Announcement of sectors not being used in a MAC frame.

Random Feedback Channel (RFCH) The task of an RFCH is to inform the MTs that used the RCH in the last MAC frame about the results of the access attempts made. An RFCH is transmitted for each sector element in each MAC frame.

RLC Broadcast Channel (RBCH) The RBCH is used on the downlink and transmits general control information about a radio cell. The information is only transmitted when it is required. This information can include the following:

- RLC information about broadcasts.
- Transmission of MAC-ID with the association of a MT (see Section 14.2.5.3).
- Information from the convergence layer.

Dedicated Control Channel (DCCH) The DCCH transmits the RLC messages belonging to an MT. It is set up implicitly with the association of an MT (see Section 14.2.5.3).

User Broadcast Channel (UBCH) The UBCH transmits usage data from the convergence layer in broadcast mode. UBCH data cannot be protected through ARQ.

User Multicast Channel (UMCH) The UMCH transmits PTM usage data to all MTs belonging to the particular PTM group. UMCH data cannot be protected through ARQ.

User Data Channel (UDCH) This channel is used to transmit point-to-point usage data between MT and AP in CM or between two MTs in DM. UDCH data can be protected through ARQ (see Section 14.2.5.2).

Link Control Channel (LCCH) ARQ messages between EC entities can be transmitted over the LCCH. This channel can also be used to transmit *Resource Requests* (RRqs). The LCCH is assigned to a UDCH.

Association Control Channel (ASCH) An ASCH is only used when an MT makes its first contact with an AP, i.e., during an association or a handover. Therefore, it is only used by MTs that are not yet associated with an AP.

14.2.5.2 The EC Sublayer

The *Error Control* (EC) sublayer offers three modes of transport:

1. EC mode (see Section 14.2.5.2)
2. Repetition mode (see Section 14.2.5.2)
3. Unacknowledged mode

Either EC or unacknowledged mode can be selected for the transmission of usage data in a UDCH and either repetition or unacknowledged mode for a UBCH. UMCH, DCCH and RBCH always use unacknowledged mode. The transport mode is negotiated during connection setup. A bidirectional LCCH is set up for EC mode and a unidirectional LCCH is set up for repetition mode.

Irrespective of which transport mode has been selected, the sender must assign the *Sequence Number* (SN) of a PDU in the LCH. Since the sequence

Table 14.6: Contents of ARQ Feedback message

Name	Bits	Significance
FC	1	Flow Control
CAI	1	Cumulative Acknowledgement Indication
BMN1	7	Block Number of BMB1
BMB1	8	Bit Map Block 1
BMN2	5	Block Number of BMB2 relative to BMN1
BMB2	8	Bit Map Block 2
BMN3	5	Block Number of BMB3 relative to BMN2
BMB3	8	Bit Map Block 3

Table 14.7: Contents of Discard message

Name	Bits
Discard Sequence Number	10
Repeated Discard Sequence Number	10

numbers are 10 bits wide, an arithmetic modulo 2^{10} is used with all operations. The window size is negotiated between sender and receiver, with the valid values being 32, 64, 128, 256 and 512. The maximum window size corresponds to half of the SN number space. This restriction is necessary since a *Selective-Repeat-ARQ* protocol is used.

The tasks of the EC sublayer also include producing and evaluating the 24 bit long CRC field. If the receiver detects a checksum error, this entire L-PDU is discarded.

EC mode An ARQ protocol is used for error control in EC mode. This protocol is from the class of *Selective-Repeat*-ARQ protocols. Only those PDUs that the receiver has requested again are retransmitted. S-PDUs are used to exchange control data over the SCH. Two PDUs are defined: `ARQ Feedback` (see Table 14.6) and `Discard` (see Table 14.7). The receiver uses the message `ARQ Feedback` to inform the sender that L-PDUs from the LCH have either been received successfully or are missing and `Discard` to notify the sender of discarded L-PDUs.

H/2 uses *partial-bitmap* acknowledgements, i.e., positive and negative acknowledgements are coded in a bitmap. Table 14.6 presents three bitmap blocks each with a length of 8 bits. Within a *Bit Map Block* (BMB) a 0 means a negative acknowledgement, 1 is a positive acknowledgement. BMBs are *byte-aligned*, so 7 bits are sufficient for addressing. The positions of the second and the third BMB are described relative to the previous one, which means that 5 bits suffice. The BMB are ordered sequentially (BMN1 $\leq$ BMN2 $\leq$ BMN3), with BMNi being the block number of BMBi. The sender is not permitted to

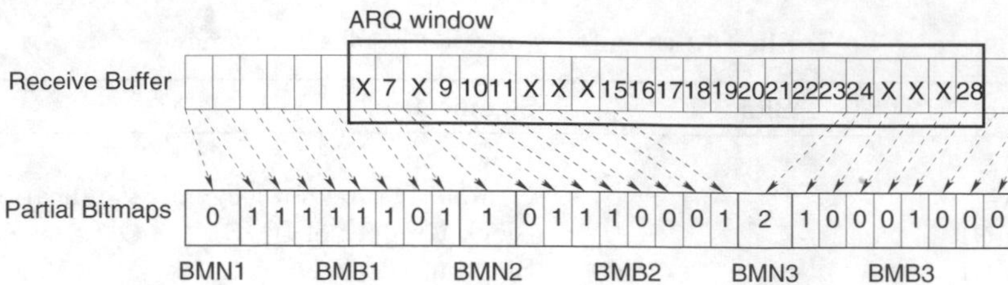

Figure 14.24: Example of a partial bitmap acknowledgement

shift its transmit window until a `cumulative acknowledgement` (CumAck) message has been received. All PDUs up to the first 0 in BMB1 are positively acknowledged through a CumAck, indicated by a `CAI` flag. Figure 14.24 presents an example of the coding of a partial bitmap acknowledgement.

The receive buffer is shown in the top part of Figure 14.24. PDUs that are received correctly are indicated by their sequence numbers; missing PDUs are identified by an X. The lower part shows three bitmap blocks. Since the BMBs are byte-aligned, they cover the sequence numbers SN 0–7, 8–15 and 24-31. However, since the receive buffer does not start until SN 6 (the first PDU not correctly received, the bits for SN_0 to SN_5 are coded with 1 (correctly received). At the same time the remaining bits in BMB3 are filled with 0 since SN_{29} to SN_{31} have not yet been received and possibly have not even yet been sent. BMN1 relates to the 7 significant bits of SN_0. BMN2 is the result of the 7 significant bits from SN_8 minus BMN1. At the same time BMN3 is relative to BMN2.

H/2 also supports *Flow Control* (FC), which enables the receiver to stop the sender if the receiver is temporarily not in a position to process all the PDUs of the sender. If an FC bit is set, the sender is only allowed to transmit PDUs up to the highest sequence number in BMB3. The receiver sets the FC bit for the duration of flow control.

The sender can use a `Discard` message to inform the receiver that it has stopped retransmitting PDUs because, for example, the remaining life of the PDUs has expired. The `Discard` message contains the sequence numbers up to the one when the PDUs were discarded. A copy of the sequence number is also entered into the `Discard` message to increase the robustness of the message. The receipt of a `Discard` message is acknowledged with a CumAck. Only then is the sender permitted to shift its window.

Repetition mode Repetition mode is only used in broadcast (UBCH) mode. In this mode the sender transmits each PDU several times but each repetition has the same SN as the original one. The receiver stores the PDUs according to SN and might fill the gaps with repeated PDUs. Correctly received PDUs are forwarded in the right sequence to the adaptation layer. The receiver does not generate any acknowledgements. If a *Protocol Data Unit* with an SN beyond the receive window but within an acceptable range of 512 SN is received, the receive window is shifted automatically. Because of the acceptance range the maximum window size is limited to 256. The sender can shift its window when it no longer wants to transmit the same PDUs.

This mechanism increases the receive probability in broadcast mode considerably, albeit at the expense of a loss in capacity.

14.2.5.3 The Radio Link Control Sublayer

The *Radio Link Control* (RLC) sublayer contains services for controlling the H/2-DLC layer. The tasks are organized into the following three groups:

- Functions for association (*Association Control Function*, ACF)
- Functions for the control of DLC user connections (*DLC User Connection Control*, DCC)
- Functions for the control of radio resources (*Radio Resource Control*, RRC)

The RLC protocol provides the respective service groups with appropriate signalling procedures. The three service groups of the RLC sublayer and the corresponding protocol are explained below.

Association control functions The *Association Control Function* (ACF) contains procedures for association, authentication, encryption and deassociation. An MT initiates the association procedure to make contact with an AP. Figure 14.25 presents the different options for executing this procedure. The entire association procedure is divided into many sub-procedures that either have to be executed or, alternatively, can be omitted.

In each case the MT must synchronize with the carrier signal of the radio cell. A *beacon* signal is sent in the BCCH of each MAC frame. In addition to other information, the BCCH contains the identification of the AP and the network on the DLC plane. Due to its length, the complete address of the AP (e.g., IEEE-802.2 address) cannot be transmitted in the BCH of each individual frame; instead it is transmitted periodically in the RBCH. The RBCH is part of the expanded broadcast phase and is disclosed in the FCCH (see Section 14.2.5.1). Therefore, MTs must wait for this AP address to determine whether access to a particular radio network is permitted. Alternatively, an MT may also ignore this information and initiate the association directly. For an association the MT requests a new MAC-ID from the AP over the ASCH. This MAC-ID is allocated by the AP over the RBCH, and it is used for the unique addressing of the MT during the entire time in which the MT is associated with the AP. A logical signalling channel (*Dedicated Control Channel*, DCCH) is implicitly set up over which all other RLC messages are transmitted on the uplink and the downlink.

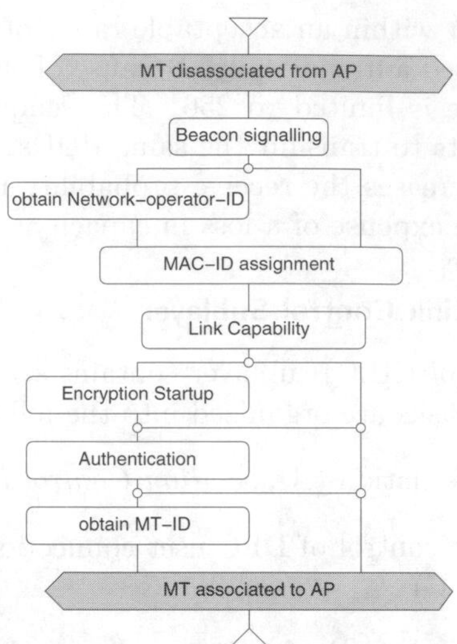

Figure 14.25: Association

In the next step, optional parameters are exchanged between MT and AP, thus establishing the working mode. Using the *link-capability* procedure, the MT sends the following parameters to the AP:

- Version of the DLC protocol used in the MT (currently only Phase 1).
- A flag indicating whether direct mode is supported.
- List of supported convergence layers.
- List of supported authentication and encryption procedures.

The AP makes its decision on the basis of this information and selects the suitable options. The AP also establishes which of the following procedures should be implemented during association:

Encryption startup Negotiation and startup of encryption.

Authentication Mutual authentication is carried out. The MT authentication checks terminal access to the fixed network. If the authentication fails, no access is granted. AP authentication enables an MT to detect false APs. Through the input during the *link-capability* procedure the AP decides whether authentication should be carried out. An MT can cancel the complete association at any time in case an authentication is still wanted.

Obtain MT-ID The AP can request the MT to make known its own address.

The MT is associated with the AP as soon as these procedures have been carried out successfully.

The MT or the AP can initiate a de-association at any time. There are two types of de-association: explicit and implicit. In the case of explicit de-association, a short exchange of signalling takes place between AP and MT, and AP and MT terminate all connections. Implicit de-association occurs when MT and AP can no longer maintain a radio link and therefore an exchange of signalling is no longer possible. The AP initiates the *MT-Alive* procedure when it determines that the MT is no longer transmitting and waits for a response. If the AP does not receive a response—especially after a number of attempts have been made—it releases the radio resources of the MT.

DLC user connection control These control functions (*DLC Connection*, DLCC) are used to set up, terminate and renegotiate *Unicast*-DLC user connections. All procedures in this group can be initiated either by the MT or the AP. If an MT initiates connection setup, it sends a recommendation for the characteristics of this connection, such as ARQ window size and *polling*. However, it is always the AP that establishes the connection parameters.

In addition to *unicast* connections, H/2 also supports *multicast* and broadcast connections. *Multicast* supports two operating modes: *NxUnicast* and MAC-*Multicast*. With *NxUnicast*, multicast transmission is carried out in the same way as unicast connections, i.e., through the use of a number of UDCH. So ARQ can be used within a multicast group. In the case of MAC-*Multicast*, a multicast group is allocated a unique MAC-ID (logical channel: UMCH). Each MT in this group therefore has at least two MAC-IDs: a *Unicast*-MAC-ID, which is assigned during association, and a MAC-ID for each multicast group. In this case, ARQ procedures cannot be used in the multicast group, because each protocol data unit may only be sent once to each member of this group.

Radio resource control RRC offers functions and procedures for radio measurements, handover, power saving, power control and dynamic frequency selection.

Radio measurements If an MT is associated, it continuously measures the quality of the current radio channel. These measurements are based on values such as *Received Signal Strength* (RSS). To find a candidate for a possible handover, the MT measures the frequencies of neighbouring APs at fixed (or variable) intervals. If the MT only has one transceiver, it must first synchronize with the respective frequency before it can carry out the measurements. No communication is possible with an AP during this time. To avoid unnecessary transmission, MT deregisters itself from the AP for this period (*MT-Absence* procedure).

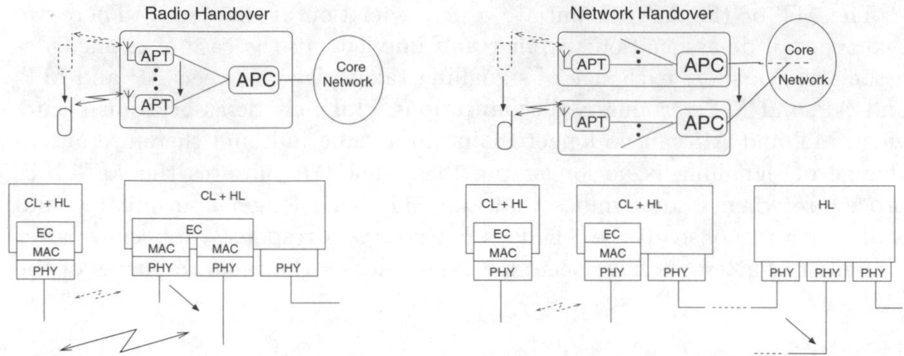

Figure 14.26: Radio and network handover scenarios

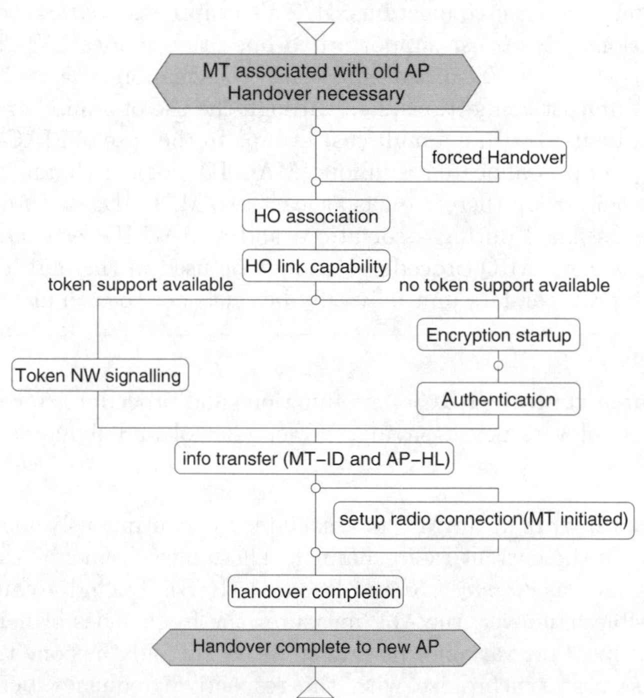

Figure 14.27: Network handover procedure

Handover H/2 supports three types of handover: *sector handover*, *radio handover* and *network handover*. These handovers take into account the current implementation of the AP. For instance, a sector handover can be implemented if the AP, i.e., the current *Access Point Transceiver* (APT), supports many antenna sectors. During this handover the MT changes the antenna sector. This handover can be carried out completely within the DLC layer so that higher protocol layers are not affected. In the case of a radio handover, the MT changes from the supply area of one APT to the supply area of another one, which, however, is controlled by the same *Access Point Controller* (APC) (see Figure 14.26). Here too it is possible for the handover to be executed within the DLC layer. All relevant information about existing connections, encryption algorithms and authentication is available in the AP, so there is no need to request this information. However, a renegotiation of connection parameters is possible if, for example, sufficient resources are not available in the new radio cell. In network handovers the MT changes over to a different APC. This type of handover also affects the higher protocol layers if no information about the MT is available in the new AP. The AP can request the information from the MT; the network handover procedure is then similar to the association procedure (see Figure 14.27, right path). If the handover is supported by special signalling in the fixed network, most of these parameters can be exchanged over the fixed network. Consequently, retransmission across the radio interface can be avoided. Specifications for this fixed network signalling are not included in the H/2 standard.

Dynamic channel selection To facilitate an independent selection of a suitable frequency the AP can request that the associated MTs measure the frequencies of neighbouring APs. On the basis of these and their own channel measurements, the AP can dynamically change its own frequency. The RLC protocol offers procedures that enable the associated MTs to be notified of the impending change.

Power control The transmit power is adapted to cope with the distance between AP and MT (i.e., path loss) and the quality of the radio channel (C/I). This mechanism is aimed at reducing co-channel interference from neighbouring cells.

Power saving This function allows an MT to enter into a power-saving mode or to leave it again. The power of the sender is controlled accordingly. No MAC frames are received in power-saving mode. The power-saving function is initiated by the MT. After negotiating the sleep time N, the MT can *put itself to sleep* for N MAC frames. The following four scenarios are possible after N frames:

- The AP *wakes up the MT* (possible reason: data from the AP to be transmitted).

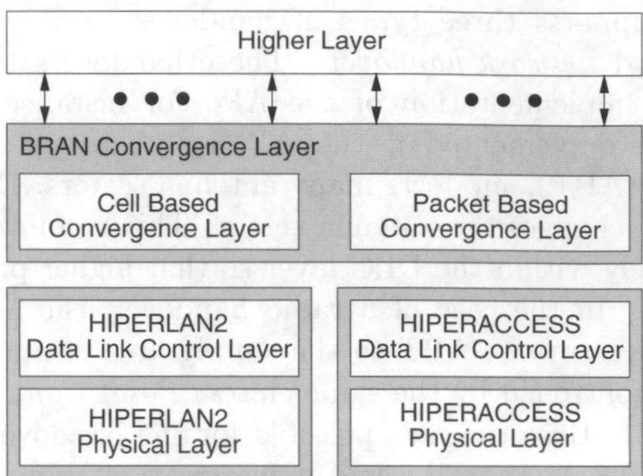

Figure 14.28: Convergence layers

- The MT *wakes up* (possible reason: data from MT to be transmitted).
- The AP arranges for the MT *to continue sleeping* (for another N frames).
- The MT misses the `Wake-Up` message from the AP. The MT then starts the *MT-Alive* procedure.

14.2.6 The Convergence Layer

The physical and the data link layers in H/2 were designed to function independently of any specific network protocols. The aim of H/2 is to support different network layers, and to this end the *Convergence Layer* (CL) was inserted between layers 2 and 3.

The task of the convergence layer is to provide the services used by the network layer, specifically to setup and to release data connections, and correspondingly to implement the quality of service requests of higher layers to the parameters of the H/2 layers. *Segmentation and Reassembly* (SAR) of U-PDUs (usage data) of higher layers also takes place in the convergence layer.

As shown in Figure 14.28, there are two groups of CL: *cell-based* and *packet-based.* A cell-based convergence layer processes fixed-sized ATM cells. A packet-oriented convergence layer processes variable-length packets (e.g., IEEE 802.3 (Ethernet), IP and IEEE 1394).

14.2.6.1 Packet-Oriented Convergence Layer

Figure 14.29 shows the user plane of the convergence layer. The user plane can be divided into two sublayers: a *Common Part Convergence Sublayer* (CPCS) and a *Service Specific Convergence Sublayer* (SSCS). The common part comprises functions for *padding* packets with filler bits as well as other important

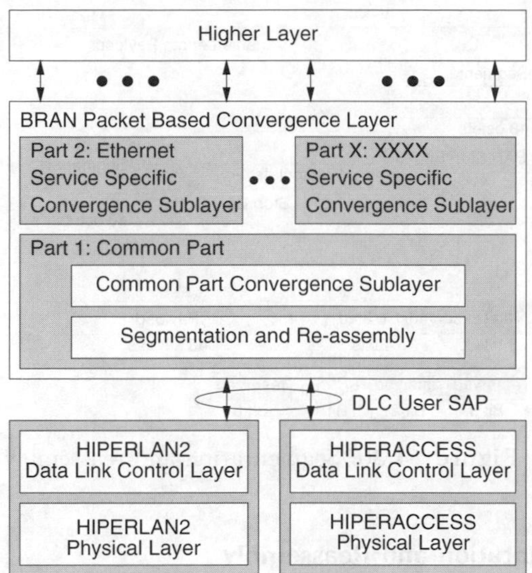

Figure 14.29: User plane of the packet-oriented convergence layer

information on the sender side along with the interpretation and resolution of the latter by the receiver side. Furthermore, the SAR function is a component of the common part. It is located directly on the DLC user SAP and communicates over this SAP using fixed-length PDUs (see Section 14.2.6.2).

The SSCS is specific to each CL and actually adapts the different data formats to the H/2-DLC format. Terminals with many SSCSs are conceivable. Currently, the SSCS is selected during association. Therefore, reregistration is required in the network in order to change the SSCS.

Packet-oriented convergence layers also have a control part. The control part is allocated to the SSCS, is specific to the different network layer protocols and interacts with the RLC sublayer (see Section 14.2.5.3) over the DLC control SAP. The tasks of the control part include:

- Negotiation of parameters
- Specification of quality of service models
- Monitoring of unicast, multicast and broadcast connections
- Handover control

The SAR function is one of the main tasks of the convergence layer and, therefore, is covered in detail in the next section.

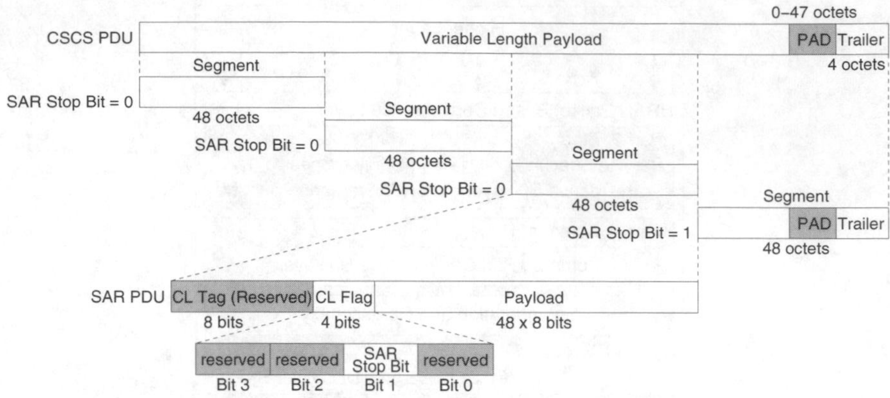

Figure 14.30: Segmentation and reassembly

14.2.6.2 Segmentation and Reassembly

Figure 14.30 presents the complete SAR process. First, a 4-byte trailer is attached. This trailer consists of 2 bytes for a length field, which indicates the length of the CPCS-SDU (CPCS payload) in bytes, and 2 bytes for future use. The CPCS-SDU together with the 4-byte trailer is then padded to a multiple of 48 bytes through padding bytes. The receiver uses the length field to calculate the number of padding bits so that it can remove them.

The CPCS-PDU is forwarded to the SAR sublayer, which splits it into segments of 48 bytes.

An 8-bit CL tag and 4 -bit CL flags are added to each segment. Only bit 1 of the CL flag is currently defined as a SAR stop bit; all other bits are reserved for the future. The SAR sublayer transfers a 49.5 byte-long CL-PDU on the DLC user SAP to the DLC layer and also receives the same from the DLC layer.

14.2.6.3 Convergence Layer for Ethernet

The convergence layer for Ethernet (IEEE 802.3 and 802.3ac) is specified as a *Service Specific Convergence Sublayer* (SSCS) and therefore also uses the *Common Part Convergence Sublayer* (CPCS) of the packet-oriented convergence layer. The following functions are carried out:

Preservation of the Ethernet SSCS-PDU This function protects the transparency of Ethernet SSCS PDUs.

Mapping of traffic classes according to 802.1p (optional) 802.1p supports the traffic classes listed in Table 14.8. These classes are mapped onto DLC connections. Depending on the number of supported DLC connections, several traffic classes can be multiplexed to one DLC connection.

Table 14.8: IEEE 802.1p traffic classes

Priority	Traffic class	Remarks
0 (Default)	Best effort	LAN traffic
1	Background	
2	—	Not used
3	Excellent effort	For special customers
4	Controlled load	Subject to connection acceptance control
5	Video	<100ms delay and variance in delay
6	Voice	<10ms delay and variance in delay
7	Network control	Control data

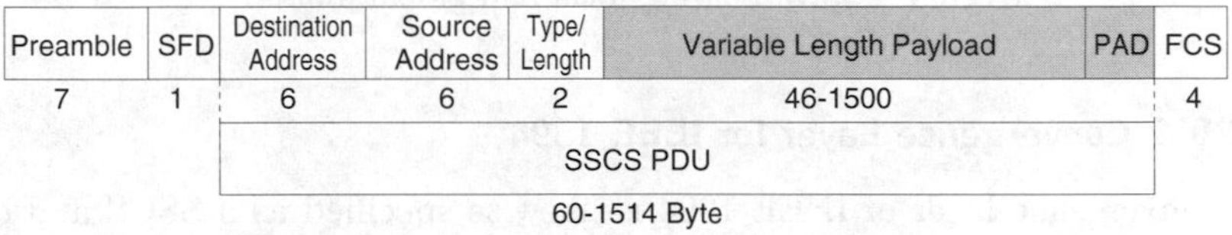

Preamble	SFD	Destination Address	Source Address	Type/ Length	Variable Length Payload	PAD	FCS
7	1	6	6	2	46-1500		4

SSCS PDU

60-1514 Byte

Figure 14.31: Coding of an SSCS-PDU

Mapping of Ethernet broadcast service This service is mapped onto the broadcast services of the data link layer.

Mapping of Ethernet PTM services These services are mapped onto the PTM service of the data link layer.

Emulation of collision domains All users associated with the same AP are treated as if they were all linked to the same LAN segment. Consequently, the AP must forward all frames of the Ethernet broadcast or PTM service to all MTs concerned. The AP must also forward Ethernet frames to a destination terminal. For their part, the MTs must filter out and discard the Ethernet frames that they themselves sent to the AP, which in turn had forwarded them in broadcast or PTP mode.

Figure 14.31 shows the coding of an *Service Specific Convergence Sublayer PDU* (SSCS-PDU). The preamble, the frame identification (*Start Frame Delimiter*, SFD) and the *Frame Check Sequence* (FCS) are not transmitted but added instead by the receiver of the SSCS-PDU.

The protocol of the Ethernet convergence layer is asymmetric, i.e., only the AP is aware of the allocation of IEEE-802-MAC addresses to the DLC-MAC-ID. Upon receipt of a valid SSCS-PDU, the convergence layer in the AP checks the destination address and forwards the SSCS-PDU to the appropriate terminal.

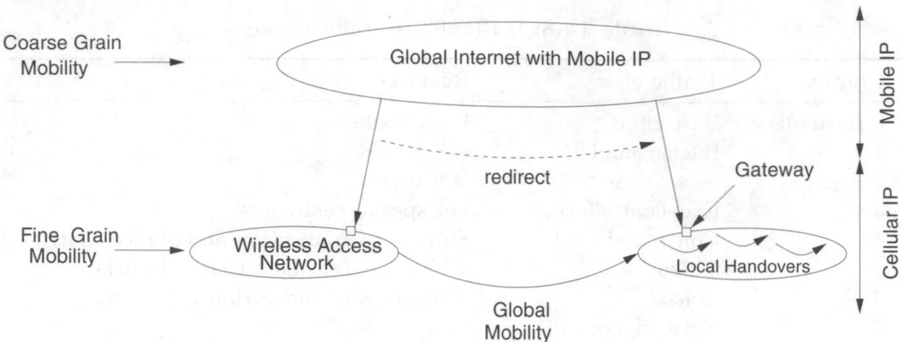

Figure 14.32: Serving local and global mobility

14.2.6.4 Convergence Layer for IEEE 1394

The convergence layer of IEEE 1394 is likewise specified as a SSCS and consequently also uses the CPCS of the packet-oriented convergence layer. The features of the convergence layer for IEEE 1394 include synchronization of IEEE 1394 timers across the air interface and processing of isochronous data streams with special requirements for the variance of transmission delay.

14.2.6.5 Convergence layer for ATM

In contrast to the other convergence layers, the one for ATM is cell-based rather than packet-oriented. Segmentation and reassembly are not necessary because the usage part of an ATM cell with 48 bytes fits into the usage part of a DLC-PDU.

The functions of the ATM convergence layer are therefore limited to administrating MAC-IDs and listing the *DLC User Connection Identifier* (DUC-ID)s used for the identification of terminals and connections.

14.2.6.6 IP Network Mobility

Serving mobility in an H/2-based access connected to the IP network requires to integrate H/2 and IP mobility functions. When an MT performs a network handover from one AP to another the access network is involved. The H/2 handover procedure has to be aligned to IP mobility functions by the *Convergence Layer* (CL). A new IP CL supporting these functions is currently under development.

Several concepts for IP mobility support exist and are discussed in the IETF. They can be roughly divided into two groups serving either *Global* or *Local Mobility* (see also Figure 14.32). For instance Mobile IP [34] is designed to support *Global Mobility*, when a MT moves from one IP-subnet to another. Analyses have shown that the signalling procedures defined for Mobile IP are very time-consuming, when large areas of the IP network have to be traversed

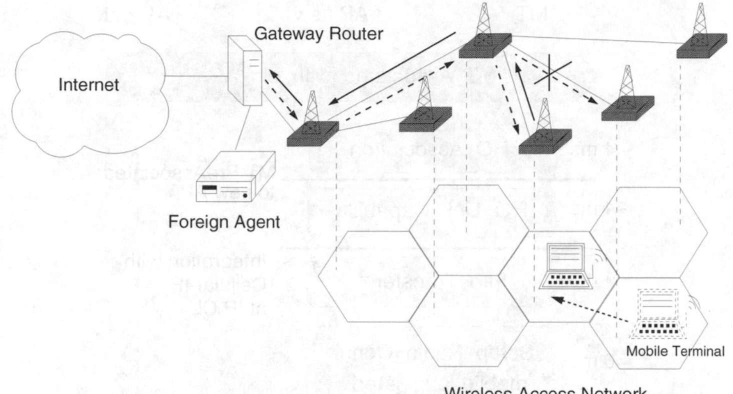

Figure 14.33: Local handover within access network

[47]. Fast and seamless handovers cannot be guaranteed. Therefore, *Local Mobility* concepts are proposed such as *Cellular IP* [11] and *HAWAII* [37] that allow fast handover execution since negotiation with the network takes place only locally (see Figure 14.33).

The H/2-based radio access is designed to serve in local areas rather than to cover large subnets. A fast handover execution is required to maintain QoS for MTs.

Serving Local Mobility with H/2 Serving *Local Mobility* in an H/2-based access network, where an MT changes its point of attachment to the IP network during a network handover, requires to address the following points of interworking:

- Re-registration to the IP network
- Authentication
- Maintaining on-going connections

H/2 RLC provides the basis to support the network handover execution at the radio access. Figure 14.34 shows the different sub-tasks to be performed by MT and target AP during the handover procedure. The boxes indicate sub-procedures and the time values the approximate time needed for their execution.

Similar to an initial association to an AP, in case of network handover the MT obtains the local identifier MAC-ID during `HO_association` in order to establish basic communication on DLC level with the target AP. After this link capabilities and further specific CL information will be negotiated. The `Info_Transfer` sub-precedure is used by the MT to provide the IP address of the old AP to the new AP. By this, APs maintain an internal data base

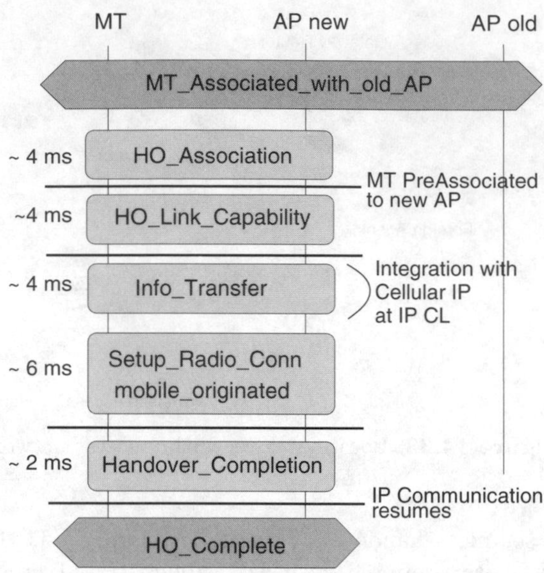

Figure 14.34: Slow RLC signalling procedure executed during network handover

mapping DLC layer AP addresses to their IP addresses. This data base stores the addresses of the neighbouring APs and is used for network handover optimization as described in Section 14.2.7.

Finally, the on-going DLC connections have to be re-established. Thereby, different mechanisms are supported at DLC level. Connections may be re-established via the air, which is the most simple case, but requires a rather long time for execution, since all relevant connection information has to be sent from MT to AP. Alternatively, the AP may retrieve this connection information from the network, e.g., from the former serving AP. However, this requires additional signalling support via the network.

14.2.7 Handover Integration with IP

As described in the previous section, several actions have to be performed during network handover to resume connectivity on DLC level. For re-establishment of IP connectivity at the new AP similar actions have to be performed on IP level:

- Gathering IP address information
- Address update for IP unicast (e.g., by using Mobile IP, Cellular IP)
- Address update for IP multicast (e.g., by using *Internet Group Management Protocol* (IGMP) [21])

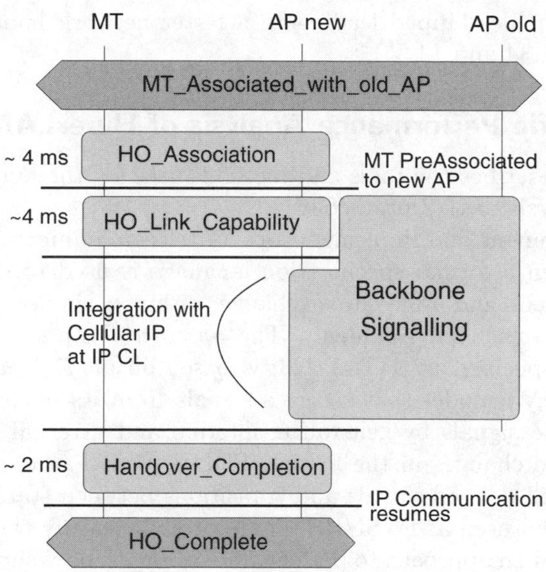

Figure 14.35: Fast RLC signalling procedure executed during network handover

Performing re-establishment on DLC and IP level in sequence is time-consuming and may lead to intensive packet loss during handover. Furthermore, these IP protocols in their conventional usage base on host-to-network communication relying on a stable link layer connection, i.e., *DLC User Connection* (DUC), that has to be established before.

To reduce handover latency the AP can serve as proxy and provides network-relevant information to the MT that is periodically broadcast via the IP subnet. For instance, advertisements of mobility capable IP routers will be provided by the AP only on request of the MT, i.e., conveyed by the RLC `Info_Transfer` procedure. This avoids the periodic broadcasting of this IP information via the radio link.

The IP CL provides the relevant information from the network to DLC and triggers the respective IP procedures on DLC indication.

Using this tight integration of DLC and IP protocols allows also to perform backbone signalling between the old and the new AP during network handover (see Figure 14.35).

On arrival of an MT at the new AP, the MT provides the DLC addresses of the old AP. The new AP retrieves the corresponding IP address of the old AP from its internal data base and requests the relevant information such as on-going DUCs from the old AP. This signalling used for information retrieval can be encapsulated in UDP datagrams using conventional IP transmission in the access network. IP packets mis-routed to the old AP during handover execution are forwarded to the new AP using IP tunnelling. According to this protocol integration the RLC procedures used for `INFO_TRANSFER` and

DUC setup can be skipped leading to a faster network handover execution (see Figures 14.34 and 14.35).

14.2.8 Traffic Performance Analysis of HiperLAN/2

This section describes the tools and methods used by the authors in the performance analysis of H/2 protocols.

The development and implementation of telecommunication protocols are facilitated when a formal specification language is used to describe the signalling sequences and the data exchange within a single system, between systems and to their environment. The *Specification and Description Language* (SDL) specified by ITU-T [23] is based on an *Extended Finite State Machine* (EFSM) model that receives signals from its environment and responds to these signals by generating internal and external signals that are related to state changes in the model. SDL depicts a protocol as an EFSM with states, well-defined signals and transitions between states.

SDL has been used at ETSI/BRAN to formally specify the H/2 protocols. The following is an approach to performance evaluation by simulation on basis of protocols formally specified in SDL.

14.2.8.1 HiperLAN/2 Specification in SDL

Our H/2 simulator is entirely specified in SDL completely reflecting the H/2 specifications of ETSI/BRAN. These specifications have been extended by algorithms, e.g., DLC scheduler and handover algorithms that are not part of the standard but vital to complete a H/2 system.

When simulating the traffic performance of a H/2 protocol stack, we in fact are executing the protocols in detail instead of studying a model of the protocols.

Figure 14.36 shows the highest level of the SDL system specification. It contains the four main blocks: `MT`, `AP`, `Channel` and `SimControl`. The blocks `MT` and `AP` contain the SDL specification of the respective H/2 protocols. Every AP and every MT in the simulation is represented by one block of the respective type. The blocks `Channel` and `SimControl` are introduced to complete the simulator and to place the H/2 system in a given operations and usage environment.

The main tasks of `SimControl` are the initialization, control and execution of the simulation. At start-up all `AP` and `MT` blocks log-on to the `SimControl`, which distributes unique `AP` and `Mobile IDs`. Other parameters specific to a simulation to be executed are read from parameter files. `SimControl` also contains the traffic load generators.

The task of the `Channel` block is to transfer and broadcast signals from APs to MTs and vice versa. It incorporates a channel and a mobility model. Together with the PHY-layer model being part of the blocks *MT* and *AP*,

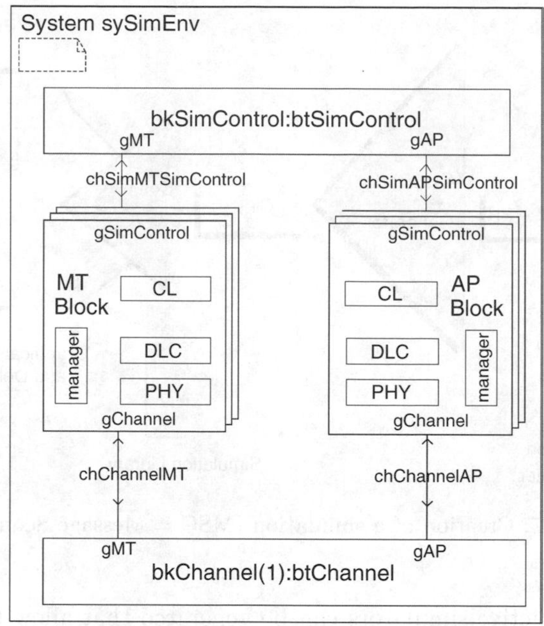

Figure 14.36: H/2 system environment in SDL

the actual *Carrier-to-Interference* (C/I) and the resulting *Packet Error Ratio* (PER) are calculated.

The blocks *AP* and *MT* are divided into sub-blocks representing the H/2 protocol layers PHY, DLC and CL. The `DLC` block combines the functionalities of the MAC, the RLC and the EC sub-layers (see Figure 14.15). Blocks also exist for the `DLC scheduler` and the `DLC layer manager`.

14.2.8.2 The SDL-Based H/2 Performance Simulator

Based on SDL-specified protocols a simulator for traffic performance analysis is generated using tools for automatic code generation plus a library for statistical analysis of protocol parameters [1]. Figure 14.37 shows the steps taken to produce a simulation program.

Based on an SDL system specified in the *SDL Graphical Representation* (SDL-GR), as shown in Figure 14.36, an *SDL Phrasal Representation* (SDL-PR) as defined in [23] is generated. *Abstract Data Types* (ADT) are used to integrate specific tools for performance analysis, e.g., traffic generators and statistical results analysis, into the block `SimControl`.

By means of a code generator the SDL specification is automatically translated into C++-code that is conventionally compiled and linked with a C++-based performance evaluation library to generate different types of simulators. Besides standalone simulator programs that run on UNIX or LINUX

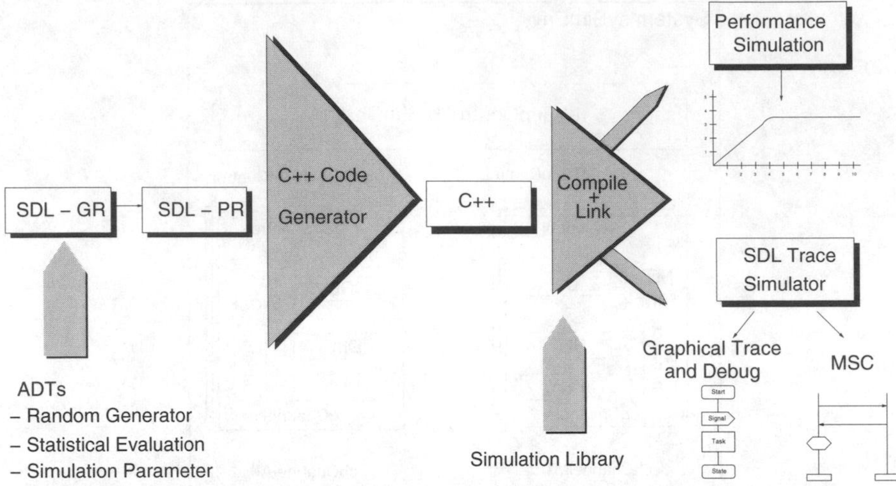

Figure 14.37: Creation of a simulation (MSC = Message Sequence Chart)

platforms, interactive simulators can be generated that allow the SDL specification to be traced and debugged.

14.2.8.3 Parameterization of the Simulator

Before an H/2 system can be studied under reproducible load and radio propagation conditions, a scenario including traffic generators, number of MTs and APs and a radio channel model has to be defined.

Modelling of the Arrival Process Five models of arrival processes (three based on actual applications and two theoretical ones) were used in the performance evaluation of the protocols.

The two theoretical arrival processes are used to analyse and develop an understanding of the protocol behaviour and are characterized as follows:

Poisson The Poisson arrival process produces events with a negative exponential distribution of inter-arrival times. Each event produces an ATM cell. Independent (Poisson) arrival of small packets (cells) is the most demanding situation for the scheduler at the air-interface and is used to evaluate an upper bound on the packet delay.

Saturation Saturation is used in our analytical study and produces a new packet if the previous packet was transmitted successfully. The arrival process therefore depends on the protocol. Saturation emulates a case in which the transmit queue is never empty without taking into account the waiting time in the transmit queue. Saturation is used to determine maximum traffic capacity and operating time.

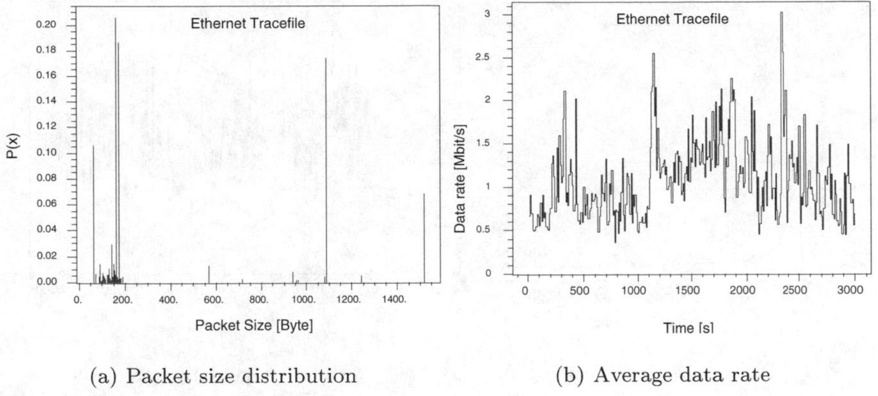

(a) Packet size distribution (b) Average data rate

Figure 14.38: Characteristics of Ethernet tracefiles

Three applications with very different requirements for quality of service support have been modelled:

N-ISDN A voice service with 64 kbit/s is modelled through the generation of fixed-length packets with a constant inter-arrival time. Based on the size of the ATM cells, 48 byte-sized packets are produced every 6 ms. A variance in inter-arrival time is not modelled; this is an ideal isochronous source.

LAN The LAN operation is modelled on the basis of typical Ethernet traffic. The measurements for the model were produced in 1994 by Bellcore Labs, which have recorded the traffic in its Intranet [3]. The characteristics of the model are presented in detail in the section on Ethernet traffic below.

Video Video transmission based on the MPEG-1 standard is considered in the evaluation of the support of real-time services. The film *Star Wars* was MPEG-coded by Bellcore Labs [15]. The resulting data was used in our simulation runs to generate the traffic load for video transmission. The characteristics of the model are explained in detail below.

Ethernet Figure 14.38 highlights the characteristics of the Ethernet tracefiles. Figure 14.38(a) shows the distribution of packet sizes from that tracefile. A review of the packet sizes 64 bytes (Ethernet acknowledgement), 162 bytes, 174 bytes, 1090 bytes and 1518 bytes (maximum size of an Ethernet frame) produces an interesting observation. 70 % of the packets are $\leq$ 174 bytes and 25 % are $\geq$ 1090 bytes, thus resulting in a mix of small and large packets. The average packet size of 425 bytes does not produce a statement about typical packet sizes because hardly any packets of this size exist.

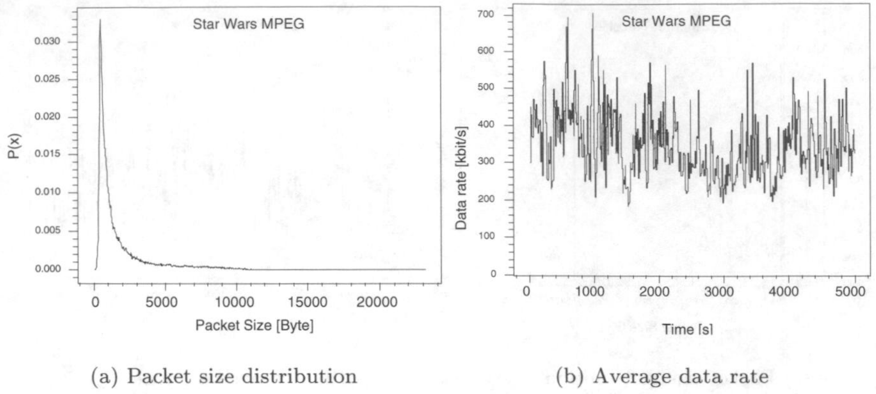

(a) Packet size distribution (b) Average data rate

Figure 14.39: Characteristics of *Star Wars* MPEG tracefiles

The course of the average data rate in Figure 14.38(b) is presented over time, with transmission taking place every ten seconds. What is evident is that the data rate fluctuates considerably. The inter-arrival time between the individual packets has been scaled to produce a variation in the average data rate. Each connection between AP and an MT in the simulation begins at a different randomly selected position in the tracefile.

MPEG The sequence of the *Moving Pictures Expert Group* (MPEG) frames is IBBPBBPBBPBB IBB..., so that 12 frames are in a *Group of Pictures* (GOP) [16, 46]. Figure 14.39 presents the characteristics of the *Star Wars* MPEG tracefiles.

The distribution of packet sizes in Figure 14.39(a) shows that the peak value is at about 500 bytes and then drops almost negative exponentially. The largest packets reach more than 20 kbytes and consequently are considerably larger than the largest packet permitted by IEEE 802.11 (4095 bytes). Segmentation was introduced for packets larger than 4095 bytes. The segments are then reassembled in the receiver. The measurements carried out all relate to MPEG packets and not to their segments.

The course of the mean data over time is shown in Figure 14.38(b), with transmission taking place every 10 seconds. It is evident that the mean data rate is subject to considerable fluctuation. The inter-arrival time is constant and was varied for different traffic loads.

Isolated H/2 System Figure 14.40 shows the reference scenario studied by simulation. It consists of an AP serving a video stream download (MPEG) and a duplex voice connection (N-ISDN), and 8 terminals exchanging Ethernet packets with their direct neighbouring MTs in a circular route. The data rate of the video traffic has been set to 5 Mbit/s whereas the mean data rate of the LAN traffic has been varied in the simulation runs to model different loads.

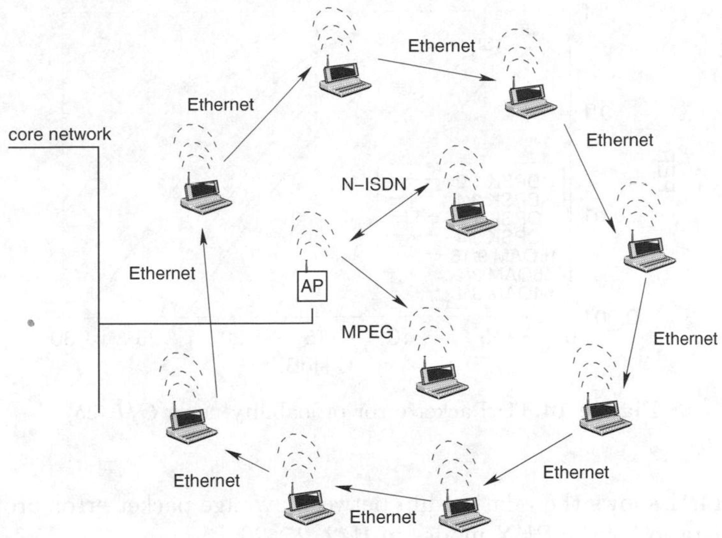

Figure 14.40: Simulation scenario

Cellular H/2 system We have also used simulation to study large cellular H/2 systems with 61 cells for the purposes of evaluating the performance of dynamic channel selection in conditions with strong interference and found that H/2 systems are still able to provide QoS guarantees such as in an isolated system [27].

Channel model The propagation of radio waves is modelled so that the receive power P_E is calculated based on transmit power P_S, wavelength λ, antenna gains g_S and g_E, the distance d [m] between sender and receiver, reference point $d_0 = 1$ m and the attenuation coefficient γ, as follows:

$$P_E(d) = \begin{cases} P_S \cdot g_S \cdot g_E \cdot \left(\frac{\lambda}{4\pi}\right)^2 & d \leq d_0 \\ P_S \cdot g_S \cdot g_E \cdot \left(\frac{\lambda}{4\pi}\right)^2 \cdot \frac{d_0}{d^\gamma} & d > d_0 \end{cases} \quad (14.1)$$

Only isotropic antennas were considered in the study so that antenna gains would equal one ($g_S = g_E = 1$). The propagation coefficient γ can be varied in the range of 2 (free space propagation) to 5 (high-density area). A γ of 3.5 was used for the indoor scenarios because of attenuation caused by walls.

The channel model used considers a carrier-to-interference ratio based on modulation type and code rate. Equation (14.1) calculates user signal C and interference I at a point in time at the position of the receiving station. The values of the interferer and ground noise N are added. Reliant on the $C/(I + N)$ (simplified: C/I) ratio established, a packet with a certain probability is received. This reliance is established through *link level* simulation.

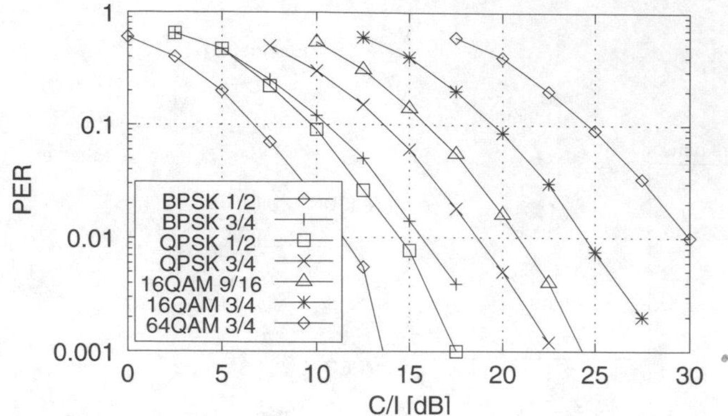

Figure 14.41: Packet error probability using C/I [25]

Figure 14.41 shows the relationship between average packet error probability and C/I ratio for the PHY modes in H/2 [25, 26].

Three error models are used in which an error is simulated through the signal fading of a user signal. This produces a C/I ratio per packet transmitted and a packet error occurs dependent on modulation and code rate, as shown in Figure 14.41. The depth of signal fading can be adjusted. The error models are:

None No signal fading occurs.

Uncorrelated Signal fading occurs with selectable probability. The occurrence of signal fading is uncorrelated.

Correlated The occurrence of signal fading is correlated. The Gilbert model is used [17]. In state *good* there is no signal fading; in state *bad* the signal fading has an adjustable depth. The length of the signal fading is on average 1 ms [19]. The average interval between two instances of signal fading is varied. This enables the simulation of different average packet error probabilities.

The simulator supports the use of mobility models. However, only stationary scenarios are presented in the following study.

14.2.9 Performance Characteristics of HiperLAN/2

14.2.9.1 Maximum Throughput Analysis of H/2

The H/2 MAC protocol provides the flexibility to accommodate a large variety of MTs and connections on the one hand and different QoS requirements on the other. The actual data rate supported by the MAC protocol can be defined by an AP before and during each MT connection by defining the size of a PDU train and the PHY modes.

The MAC layer contains some protocol overhead that is considered in the following to calculate the throughput available on top of the MAC layer assuming a number n_{MT} of MTs active in a radio cell. In this calculation the MTs use only one connection. The influence of a varying number of MTs and connections has been studied in [24].

By considering the static and dynamic lengths of the control channels and the PHY modes assigned (compare Table 14.4), the length L of the *MAC Frame* (MF) phases (see Figure 14.19) in number of OFDM symbols can be calculated:

BCH 15 bytes transmitted with the PHY mode BPSK 1/2 preceded by a preamble.

$$L_{BCH} = 15/3 + Preamble_{BCH} = 9 \tag{14.2}$$

FCH variable in length built up from n_{MT} *Resource Grant* (RG)s transmitted in 27 byte blocks containing 3 RGs and a CRC-24 checksum with the PHY mode BPSK 1/2.

$$L_{FCH} = 9 \cdot \lceil n_{MT}/3 \rceil \tag{14.3}$$

RCH variable and determined by the number of RCH slots ($Slots_{RCH}$). Each RCH slot has the length of an SCH transmitted with the PHY mode BPSK 1/2 plus the uplink PHY Preamble.

$$L_{RCH} = Slots_{RCH} \cdot 7 \tag{14.4}$$

ACH fixed with a length of 9 bytes transmitted with the PHY mode BPSK 1/2.

$$L_{ACH} = 3 \tag{14.5}$$

Uplink Phase One SCH per connection served in an MF is provided in order to allow piggy-backed resource requesting. Assuming BPSK 1/2 for SCH symbol transmission and a PHY preamble per PDU train:

$$L_{UL} = n_{MT} \cdot (Preamble_{UL} + 3) = n_{MT} \cdot 7. \tag{14.6}$$

From the Eqns. (14.2–14.6) the number of OFDM symbols per MF to transmit user data via LCH-PDUs can be calculated. With a length of 4 µs per OFDM symbol and a MF length of $t_{frame} = 2$ ms a number of 500 OFDM symbols are contained in total giving

$$L_{LCH} \quad = \quad 500 - L_{BCH,FCH,ACH,UL,RCH} \tag{14.7}$$

The total number of LCH-PDUs (N_{PDU}) per MF is

$$N_{PDU} = \lfloor L_{LCH} \cdot BpS/54 \rfloor, \tag{14.8}$$

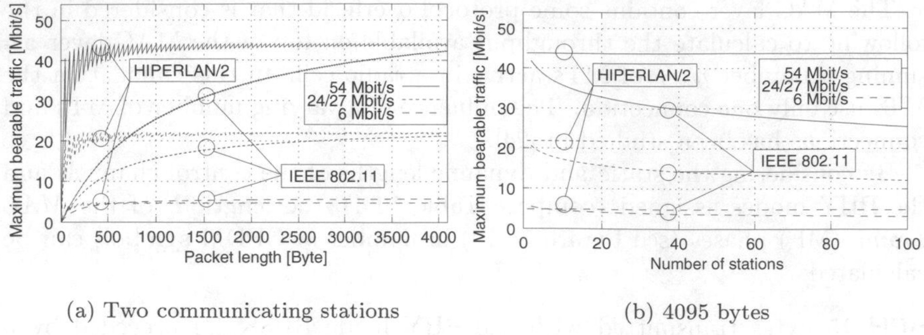

(a) Two communicating stations (b) 4095 bytes

Figure 14.42: Maximum traffic capacity

with *BpS* giving the number of bytes coded per OFDM symbol. The maximum throughput is then given by:

$$Throughput = N_{PDU} \cdot \frac{x}{\lceil \frac{x+4}{48} \rceil} \cdot \frac{8}{t_{frame}}, \tag{14.9}$$

where x is the length of the user data packet in bytes. N_{PDU} depends on the PHY mode chosen and thereby on the available data rate. In Eq. (14.9) the overhead introduced due to SAR to packets of 48 byte in the CL is already included.

Figure 14.42(b) shows the results of a 4095-byte packet length, a favourable packet length for IEEE 802.11, see Section 14.3, over a number of active stations. The maximum traffic capacity is not dependent on the number of active stations when an *Exhaustive Round Robin* (ERR) schedulers is used with H/2. On the other hand, IEEE 802.11 shows a high dependency, particularly in the case of high data rates. Although the maximum traffic capacity of 6 Mbit/s is about the same for both systems, H/2 clearly offers advantages when it comes to high data rates.

14.2.9.2 H/2 Performance in an Interference Limited Environment

The quality of a radio connection depends on position and movement of the mobile users, the radio environment (indoor or outdoor) and the burstiness of co-channel interference and may change either smoothly or even rapidly. In a multicellular environment the influence of co-channel interference becomes a major determining factor of system performance, where the DLC scheduler in the AP utilizes the radio channel capacity in an optimal way according to the QoS and capacity requirements of the individual connections served. The DLC algorithms have to take into account the various properties of the H/2 radio access that are mutually dependent:

ARQ is used to react on transmission errors by re-transmission. However, in case of a poor radio link the transmission delay increases. Re-

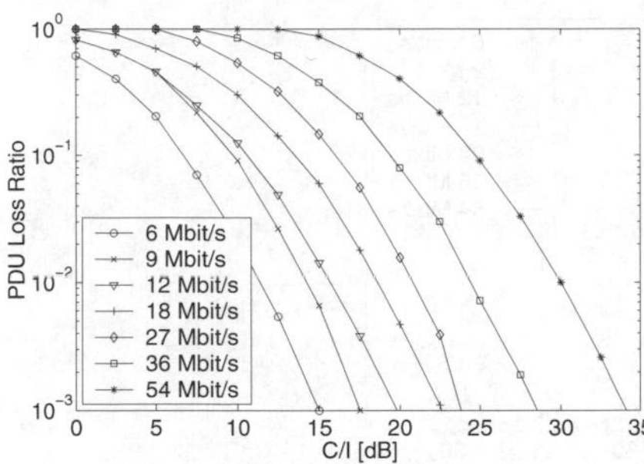

Figure 14.43: PER vs. C/I

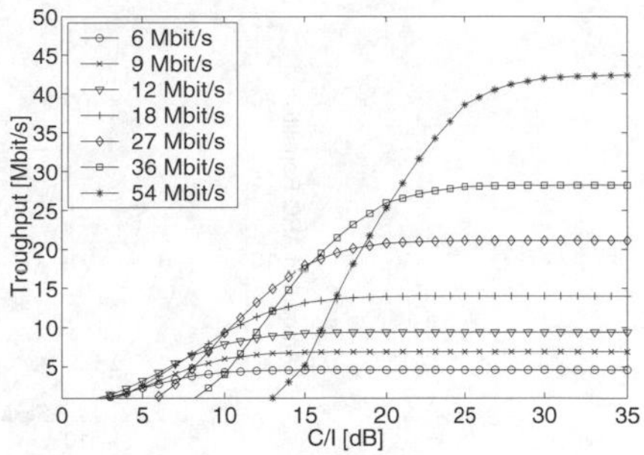

Figure 14.44: System throughput, including MAC signalling and ARQ re-transmission

transmissions have to be considered as additional DLC signalling overhead, so that the net system capacity will be reduced.

Discarding of PDUs allows to skip re-transmission of specific PDUs, e.g., when due dates expire.

Link Adaptation In case of a poor link quality the PHY mode chosen for LCH PDUs can be adapted to a more robust one, but requires more capacity in the MAC frame, resulting also in a degradation of system capacity. Figure 14.43 depicts this dependency of the *Packet Error Ratio* (PER) from the C/I.

Total system throughput, transmission delay and packet loss ratio of an individual connection are the most important parameters determining the performance of the H/2 radio access. The total system throughput can be calculated considering the protocol overhead introduced by MAC and ARQ re-transmissions.

Section 14.2.9.1 shows that the net MAC throughput is mainly determined by the PHY mode selected. ARQ retransmissions further reduce the net DLC throughput. This overhead can be estimated by assuming an ideal Selective Reject ARQ, that decreases the system throughput by $1 - PER_{PHYmode}$ (see Equation (2.52)), where $PER_{PHYmode}$ determines the current *Packet Error Ratio* experienced by the individual connection using a specific PHY mode.

$$Throughput_{DLC} = Throughput_{MAC} * (1 - PER_{PHYmode}) \quad (14.10)$$

Figure 14.44 shows the influence of link adaptation and ARQ on the total system throughput at the DLC layer.

Considering a high C/I value (> 30 dB) the PER can be neglected (see Figure 14.43) and ARQ has almost no impact. For example, in case the 36 Mbit/s

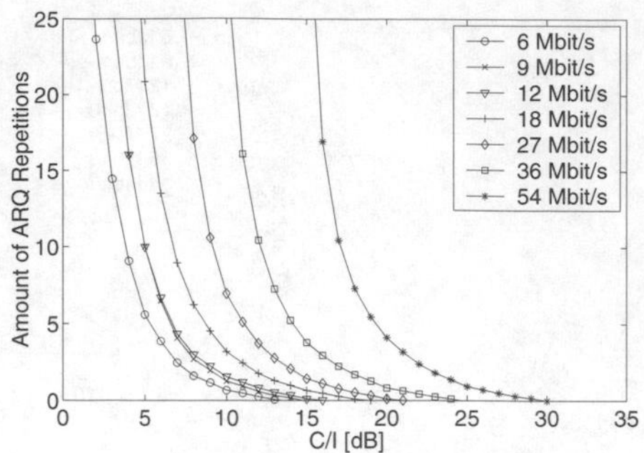

Figure 14.45: Amount of ARQ repetitions, with 1% rest error propability

PHY mode is used for the LCH PDU the system throughput is 28 Mbit/s. With decreasing C/I, i.e., increasing PER, the ARQ re-transmission overhead increases, so that the total system throughput is reduced. From a system throughput perspective link adaptation should switch the PHY mode at $C/I \approx 16$ dB to 27 Mbit/s PHY mode. At this point the PER increases to 25 % (see Figure 14.43) and, owing to ARQ re-transmissions, a high transmission delay will be experienced by the respective traffic. The amount of repetitions n_{rep} for a certain $PER_{PFY mode}$ can be estimated by

$$n_{rep} = \frac{p_{rest}}{PER_{PFY mode}} - 1, \quad (14.11)$$

where p_{rest} determines the rest error probability. To reach a low delay, the PER must be kept small to avoid too much repetition.

Figure 14.45 depicts the amount of re-transmissions to be performed for the various PHY modes, when $p_{rest} = 1$ % is considered.

Figure 14.46 shows results from simulations of a multicellular environment, where the usage ratio of the various PHY modes is analysed.

14.2.9.3 Dynamic Performance of the HiperLAN/2 DLC Layer

The results of Section 14.2.9.1 are derived for a quasi static system, where a fixed number of MTs and connections are served per MF. Due to the dynamic bandwidth requirements of broadband applications, the amount of data to be transmitted for a specific MT may change from frame to frame. Furthermore, the scheduling strategy used in the AP determines the number of PDU trains provided per MAC frame. To analyse the H/2 performance under dynamic conditions an event-driven computer simulation has been performed.

The scheduling algorithm considered here for dynamic performance analysis takes into account the different IP QoS service models (see Figure 14.47): Best Effort, Differentiated and Integrated Services. To accommodate the various QoS requirements a two-staged scheduling strategy is chosen. In the first

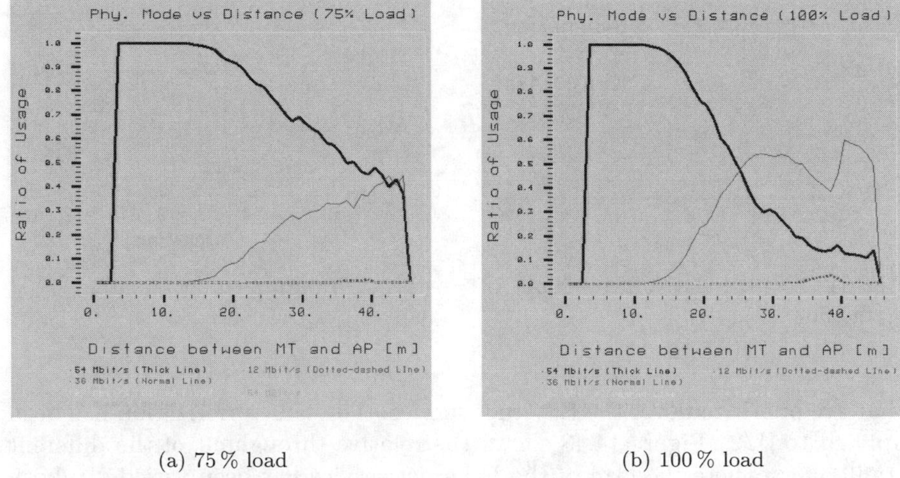

(a) 75 % load (b) 100 % load

Figure 14.46: AP-MT distance dependent PHY mode selection

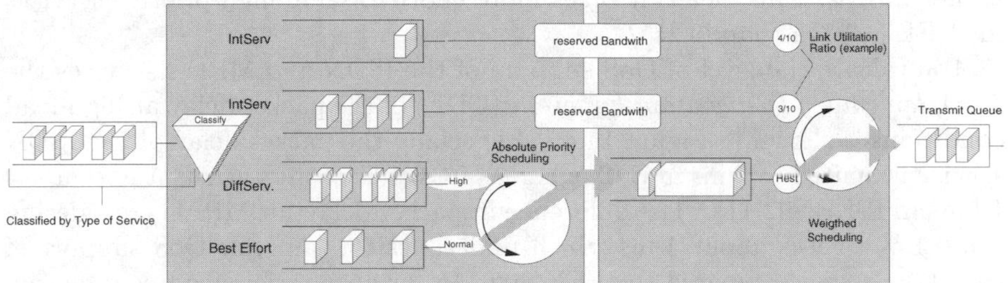

Figure 14.47: Concept of IP QoS model scheduling

stage *Priority Scheduling* is performed for Best Effort and Differentiated Services classes with higher priority for the Differentiated Services classes. In the second stage *Weighted Scheduling* is carried out where specific amounts of capacity are allocated for the Integrated Services traffic flows. The remaining capacity is used by the traffic resulting from the first stage. Call admission control ensures that sufficient resources are available for Differentiated Services traffic.

In the simulations the *Isolated HiperLAN/2 System* scenario described in Section 14.2.8.3 has been used. H/2 MAC signaling PDUs, i.e., BCH, FCH, ACH, SCH, and RCH, use the PHY mode BPSK 1/2, whereas data PDUs, i.e., LCH, are coded with 16QAM 9/16 (see Table 14.4). The effort for ARQ and SAR is included in all the results for H/2.

First, the performance of H/2 is compared with the one of IEEE 802.11a. The *Point Coordination Function* (PCF) of IEEE 802.11 with a *Contention free Repetition interval* of 10 ms is used to serve the ISDN and MPEG services

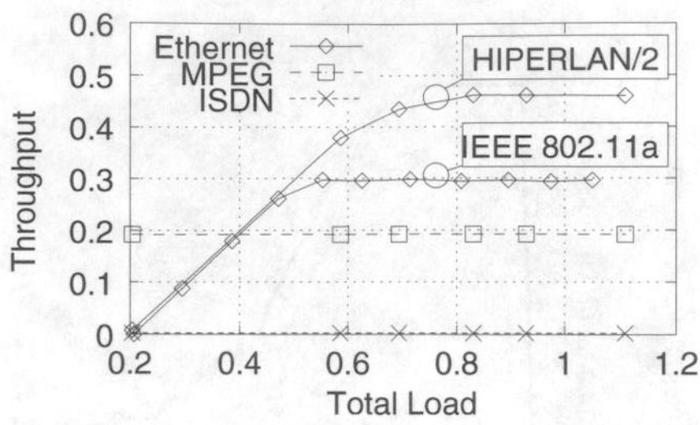

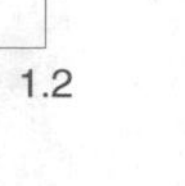

Figure 14.48: Relative throughput of traffic flows

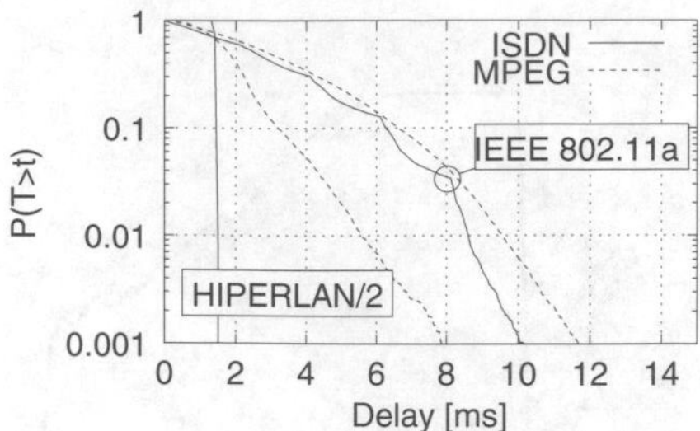

Figure 14.49: CDF of the user packet delay

that are prioritized over the Ethernet service. The same priorization has been applied to H/2. Figure 14.48 shows the relative throughput of the different traffic flows where the load of the Ethernet service has been varied. In both systems the prioritized services ISDN and MPEG are served comparably well under all load conditions, whereas the throughput for Ethernet is significantly higher in H/2. This means it takes more effort to serve high priority services in IEEE 802.11a than in H/2.

To evaluate the level of QoS support of the ISDN and MPEG services the *Complementary Distribution Function* (CDF) of the packet delay at high load conditions is shown in Figure 14.49. Especially the packet delay of the ISDN service is limited to 2 ms in H/2, whereas for 0.1 % of the packets it goes up to 10 ms in IEEE 802.11a. The difference in packet delay for MPEG considering the 0.1 % level is about 4 ms which underlines the superior QoS support of the H/2 system compared to IEEE 802.11a for an interference-free scenario. With co-channel interference H/2 performs comparably well [27].

14.2.9.4 Performance under Packet Errors

The simulations have been repeated for H/2 with an C/I value of 17.5 dB which results in an average PER of 5 % for LCH and no errors for the signalling PDUs (see Figure 14.41). The throughput for the different traffic flows for a PER of 0 % and 5 % is shown in Figure 14.50. It can be seen that the degradation in throughput is limited to the Ethernet service. It is in the order of the PER, i.e., 5 %.

Figure 14.51 shows the influence of packet errors on the packet delay. 5 % of the ISDN packets need more than one transmission for a successful reception. This is comparable to the performance of an ideal *Selective Repeat* ARQ protocol as it is limited only by the loss of the data packets (see Equation (2.52)). The additional delay for the MPEG packets mainly results from the re-sequencing in the receive buffer and it is limited to 3 ms. Even under hostile conditions with a PER of 5 % H/2 enables a good QoS support. It has

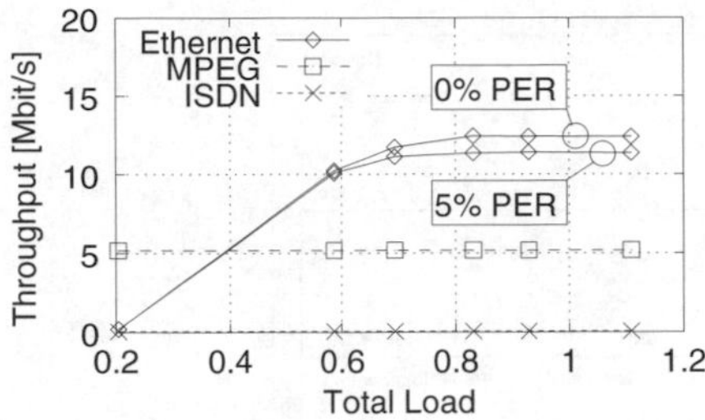

Figure 14.50: Throughput of the traffic flows, with PER = 5 %

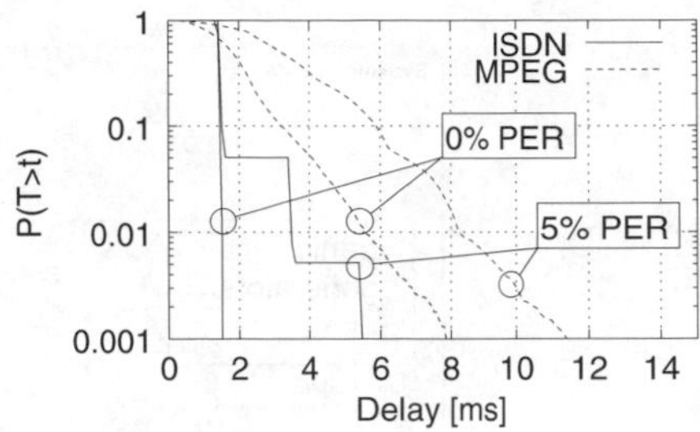

Figure 14.51: CDF of the packet delay, with PER = 5 %

to be noted that for delay sensitive services it may be better to choose a more robust PHY mode than used in the simulations to meet the required QoS.

14.2.10 Co-existence and Resource Sharing*

14.2.10.1 Frequency Sharing Rules for HiperLAN/2 and IEEE 802.11a

Figure 14.52 shows the basic architecture of the SDL-based simulator we have used to study the spectrum co-existence behaviour of H/2 and IEEE 802.11a. All relevant protocols are included. Accurate traffic models and radio channels will lead to reliable simulation results, but are at this point of time only roughly implemented.

Figure 14.53 illustrates a simulation scenario, where two H/2 mobile stations and one AP have to coexist with an IEEE 802.11a system including an AP and two mobile stations.

The distance between both systems is kept small. All stations are located no more than 5 m away from each other in order to simulate scenarios with harsh interference. Simple best-effort Poisson traffic is simulated. All terminals are transmitting packets with the power of 200 mW, TPC, LA and DFS are not applied. All mobile stations have duplex connections with their corresponding AP. The H/2 system transmits all BCH-, FCH-, and RCH-PDUs with BPSK modulation with a coding-rate of 1/2 (i.e., 6 Mbit/s), and all SCH- and LCH-PDUs are transmitted by using 16*Quadrature Amplitude Modulation* (QAM) modulation method with a coding rate of 9/16 so that a transmission rate of up to 27 Mbit/s is possible. The IEEE systems transmit the *Ready To Send* (RTS), *Clear To Send* (CTS) and ACK PDUs with BPSK 1/2 (6 Mbit/s) and the MAC frames including the data are sent with 16QAM 1/2 (i.e., 24 Mbit/s). Both systems transmit their packets at the same frequency using the same carrier.

*With the collaboration of Stefan Mangold

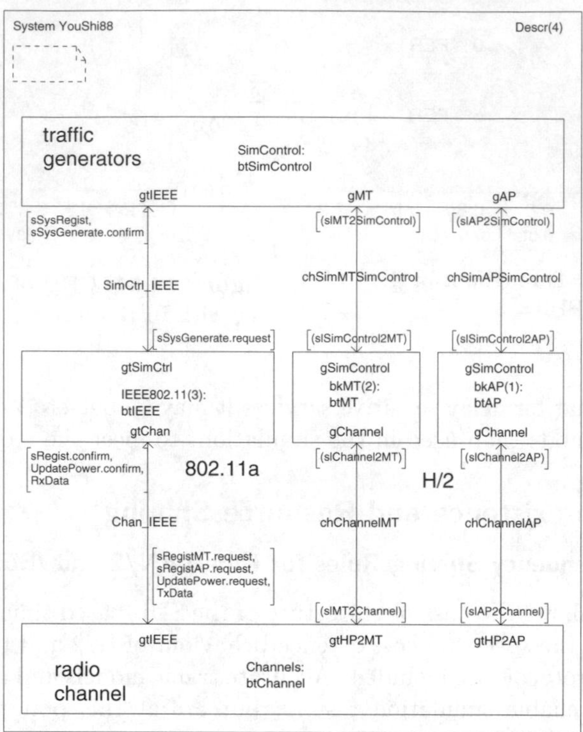

Figure 14.52: The simulation environment. The protocols are formally specified using SDL. As H/2 is centrally controlled, two different blocks are included.

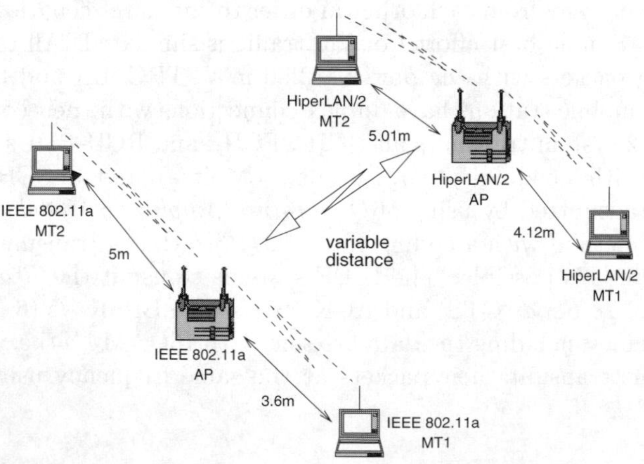

Figure 14.53: Typical scenario

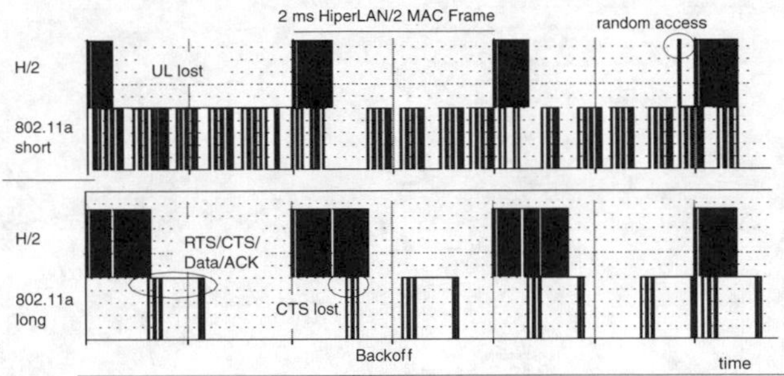

Figure 14.54: Simulated MAC frames with short and long data packets of IEEE 802.11a

14.2.10.2 Results

Results of coexistence simulations highly depend on the scenarios. The results found from our studies can only give a rough view on what problems may occur when *Frequency Sharing Rule* (FSR)s are not taken into account. For example, the following two figures show the H/2 throughput for two different configurations.

Figure 14.54 shows the records of two simulations. Results of a simple simulation scenario indicate the problems arising when HiperLAN/2 and 802.11a operate simultaneously without applying any sharing rule.

Both terminals have to carry the same loads of about 5 Mbit/s. Two configurations have been simulated, one in which the 802.11a is sending long packets of 1024 bytes without fragmentation (bottom) and one in which the 802.11a is sending short packets of 53 bytes, equivalent to the H/2 LCH PDUs (top). No DFS, LA, or TPC have been simulated. The simulation results indicate the following: with both configurations, with small to medium load the systems do not perform well. The 802.11a packets sent after carrier sensing and after RTS and CTS bursts very often interfere with the BCH PDU of HiperLAN/2 at the beginning of the MAC frame. Once the BCH of HiperLAN/2 is corrupted, the related MAC frame gets lost and no traffic can be carried in the UL (see Figure 14.54). This is due to the fact that the BCH of H/2 with this configuration is interfered by many 802.11a packets. Although the whole transmission period of an 802.11a packet fits into the not used parts of the HiperLAN/2 frame, at least with small loads, i.e., with longer periods in the HiperLAN/2 MAC frame which are not used, the BCH is very often corrupted. In contrast, if the traffic load of both systems is close to its maximum (in this scenario 20 Mbit/s), i.e., the H/2 frame is filled up well, the 802.11a system fails to operate and the HiperLAN/2 system reaches nearly its optimum.

Figure 14.55 shows the throughput over the offered traffic for both short and long packets.

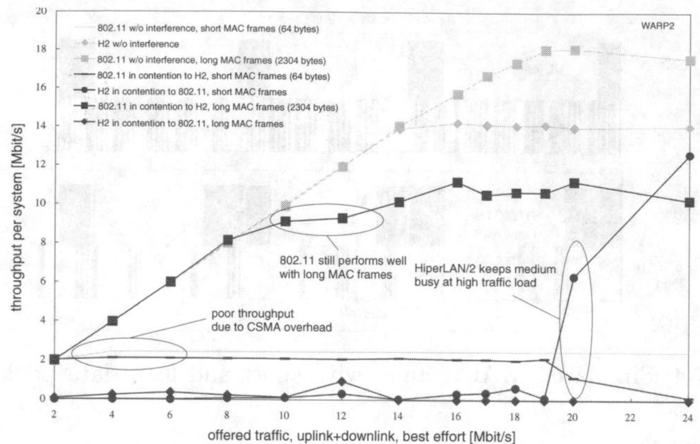

Figure 14.55: System throughput vs. traffic offer for the systems when sharing the spectrum

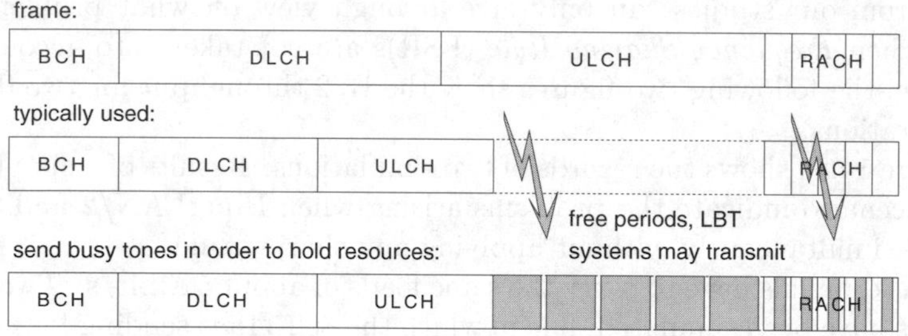

Figure 14.56: Interference protection by filling the H/2 MAC frame

14.2.10.3 A Way to guarantee QoS of an H/2 System interfered by an IEEE 802.11a System

One method to allow real-time traffic in a spectrum shared scenario with QoS guarantee is illustrated in Figure 14.56.

In order to avoid the transmission of a competing IEEE 802.11a terminal on the same carrier in not used parts of the H/2 MF, LA is applied and a MCS is selected that fills up the MF as much as possible. If this measure does not suffice to fill the MF completely, the AP would broadcast system-related management information in not used parts of the MF to fill it completely and avoid an IEEE 802.11a terminal sensing the channel as free and starting its transmission.

Since some random access slots of the RCH might be unused in H/2 and could therefore motivate a IEEE 802.11a terminal to start transmission, the AP transmits a NAK in an RCH slot as soon as it has detected that a slot is unused. No idle periods in the MF longer than the inter-frame space of 34 µs necessary for starting a transmission of an IEEE 802.11a terminal should occur (*Distributed Coordination Function IFS* (DIFS) = 34 µs) to avoid an IEEE 802.11a system to interfere H/2 during intervals where H/2 is aiming to guarantee QoS for real-time traffic.

It is obvious that a system that is contracting services to users with a QoS guarantee must be able to apply measures to keep a resource it had at the time when contracting a service. The only way to reach that goal is to occupy the resources for the duration of the service, e.g., to protect it against unpredictable interference of another system.

It has been found by simulation for the scenario of Figure 14.53 that H/2 remains unaffected and 802.11 does not reach any throughput if the method to guarantee QoS in H/2 proposed in this section is applied.

14.2.11 HiperLAN/2 and Adaptive Antennas*

The H/2 infrastructure based system as described in Section 14.2.2 also specifies sector antennas at an AP and specifies broadcast phases per sector in a common MAC frame as an option.

14.2.11.1 The Spatial Dimension

Cellular radio systems make use of the spatial dimension by re-using frequencies at geometric intervals determined by the propagation attenuation. The use of a smart array antenna opens up the spatial dimension within the single radio cell and permits, together with advanced signal processing techniques, the use of true SDMA. This access technique allows to serve different users on the same frequency at the same time. The ability to set up multiple antenna beams that can be directed to the MT locations adaptively, leads to an increase in capacity and reduces signal interference. In the following, we discuss the use of *Uniform Linear Array* (ULA) antennas consisting of M identical antenna elements.

Mainly two spatial signal processing algorithms are required to enable SDMA: Spatial filtering separates the signals impinging on the antenna array during receptions, and beamforming algorithms control the radiating directions of the array during transmission. Subsequently, we describe and evaluate a proposed concept for a joint TDMA-SDMA approach that nicely fits into the H/2 system:

- The 5 GHz frequency band permits compact construction of array antennas for AP deployment.
- With TDD the Information gathered at the uplink channel shortly later supports downlink beamforming on the same carrier.

***With the collaboration of Ulrich Vornefeld and Christoph Walke**

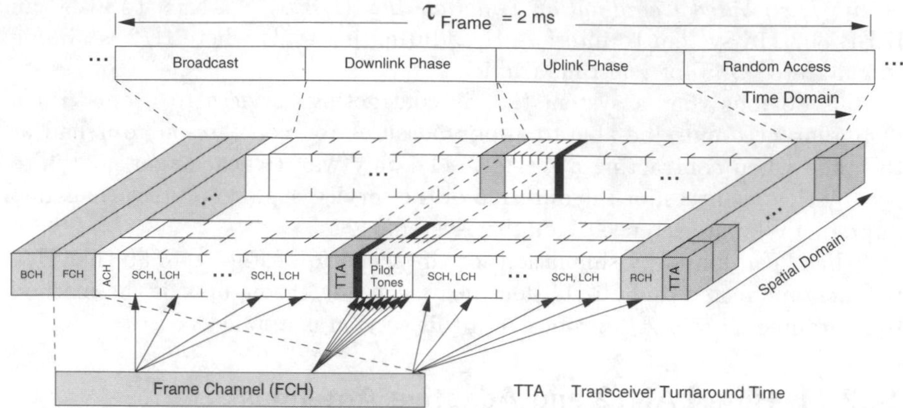

Figure 14.57: Spatially extended H/2 MAC protocol frame

- The MF duration (2 ms) is typically shorter than the channel coherence time. Spatial parameter estimation is needed less frequently that reduces the computational load.
- The H/2 reservation based MAC protocol permits a flexible transmit capacity allocation considering QoS requirements and the actual packets pending.
- ARQ error recovery at the radio interface provides fast retransmission in cases of transmission errors and inaccurate parameter estimation.

Owing to its flexible structure and its versatile signalling abilities the H/2 MAC-protocol can easily be extended to meet additional requirements imposed by space-time processing.

Figure 14.57 shows a possible MAC frame structure considering the spatial dimension, reflected by the parallel use of transport channels and a pilot tone phase introduced to support accurate uplink spatial parameter estimation [41]. The pilot tones might be omitted and parameter estimation might take place during concurrent uplink reception. This will reduce the probability of successful reception since estimation accuracy will degrade owing to increased interference power present during reception.

14.2.11.2 Space-Time Scheduling

For an SDMA system the scheduling is a two-dimensional problem. The scheduler has to determine the temporal transmission sequence of MAC PDUs as well as the MTs that can simultaneously be addressed or can be allowed to concurrently transmit their bursts. The applied algorithm is called space-time scheduling. Based on temporal input parameters like queue occupancy at AP

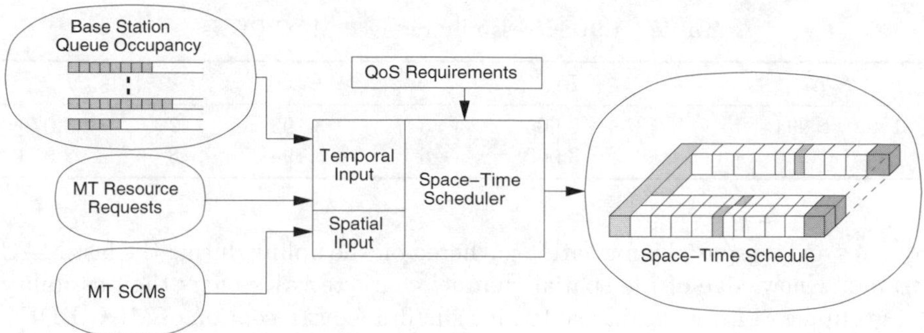

Figure 14.58: Space-time scheduling meeting QoS requirements (SCM: *Spatial Covariance Matrix*)

and RRqs received and on spatial input parameters, the algorithm determines the transmission and reception schedule obeying the agreed QoS requirements (see Figure 14.58).

Uplink scheduling The most intuitive approach to space-time scheduling on the uplink is to fix an upper threshold number of simultaneously transmittable MAC PDUs per time slot and to schedule the MAC PDUs with decreasing temporal priority. It is then the task of the receiver signal processing algorithms to spatially separate the concurrent signals. The maximum number of simultaneously transmittable MAC PDUs equals the number K of MTs operating in the system. The actual number of uplink MAC PDUs can be adjusted to meet the desired system dynamic complying with the downlink transmission needs and the MAC frame length of 2 ms.

A more sophisticated scheduling approach incorporates additional knowledge about the individual MT positions. By tracking these positions and monitoring the reception success, the scheduling algorithm can try to find spatially compatible groups of MTs that increase the mean number of successfully received bursts.

Both approaches are limited by the reception capabilities of the antenna system, i.e., it makes no sense to allow simultaneous transmission for more MTs than the spatial filtering algorithm is theoretically able to separate, while multipath propagation reduces the successful signal separation even further. Another limiting factor is the interference power. If the interference exceeds a certain threshold because too many MTs are transmitting simultaneously, it becomes impossible to separate any signal at all. Therefore, another objective of the scheduling algorithm is to split up transmission constellations on the time axis that cannot be separated in space.

Downlink scheduling Space-time scheduling on the downlink can be based on the due dates [42] or other parameters of MAC PDUs in the MT-specific AP

Table 14.9: Successfully received MAC PDUs

k	E_k	k	E_k	k	E_k	k	E_k
1	0.994	3	2.605	5	3.193	7	3.079
2	1.926	4	3.000	6	3.194	8	2.973

queues and on spatial information gathered on the uplink during the last MAC frame. Knowledge of the spatial channel characteristics offers the possibility of grouping MTs, which are suited for simultaneous reception of MAC-PDUs. Sets of MAC-PDUs being transmitted simultaneously on the same time slot in the MAC frame and transmitted into different directions have to be defined on the basis that the interference power has to be kept below a level γ. Thus, all sets of PDUs which elements cause less than γ as interference power at each other set element, are considered to be spatially compatible.

14.2.11.3 Random Access and SDMA

The number of simultaneously receivable signals is restricted by the antenna system and is interference-limited. With an increasing number of simultaneously transmitting MTs, the C/I decreases and a correct reception of a burst becomes less likely. The present interference situation depends on the number of simultaneous transmissions, the MTs' positions and the channel characteristics. Since no further restrictions can be imposed on the initial access to the RCH, i.e., the interference situation could not be taken into account, some MTs might not succeed in transmitting via the RCH. To control the retransmission attempts of these collided MTs, a collision resolution algorithm has to be applied that can make use of the enhanced reception capabilities. Thus, the transmission of RRqs will benefit from an increased throughput and a reduced delay. In H/2, MTs may use the RCH to transmit their RRqs to the AP. Especially for delay-sensitive services, this access should be carried out as fast as possible, i.e., the collision resolution algorithm has to be optimized for short delays, while throughput becomes a secondary optimization criterion. Collision resolution algorithms incorporating SDMA have been evaluated, e.g., in [43] (Slotted ALOHA) and [40] (splitting algorithms).

14.2.11.4 Performance Evaluation

Bit-level simulations were performed based on a channel model with stochastic directional scattering parameterized for a pico-cellular indoor multipath propagation environment. In all simulations, a ULA with 12 antenna elements and inter-element spacing of $\lambda/2$ was used. The unitary ESPRIT algorithm with Spatial Smoothing was used. Table 14.9 gives the mean E_k of the number of MAC PDUs successfully received by the AP on the uplink. A detailed description of the simulation scenario can be found in [41].

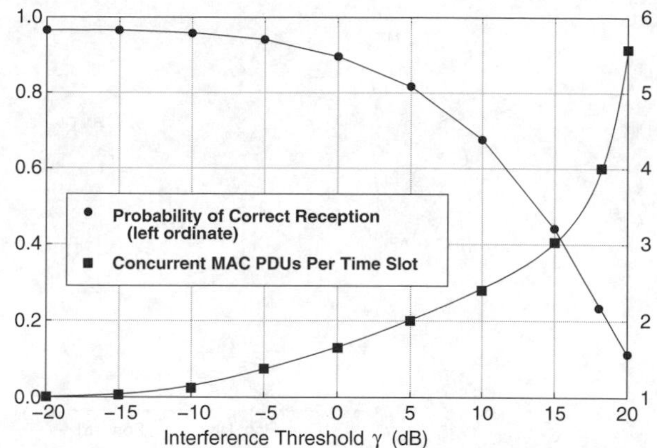

Figure 14.59: Successful reception and spatial exploitation for $k = 6$ MTs

The number of MTs allowed to transmit their bursts simultaneously has been set to k and the MT positions have been chosen at random, equally distributed within the coverage area of the AP antenna. No further spatial evaluation of the MT constellation was considered.

The values clearly show the limitation of the signal resolution capabilities of the antenna system, since the number of successfully received bursts only slightly increases for more than $k = 4$ MTs. For more than 6 MTs the values are decreasing despite the fact that the number of signals offered to detect has increased. The useful signal energy is corrupted by the additional intra cell interference. This effect shows the interference limitation. A performance evaluation of the space-time scheduling algorithm on the downlink was performed with $k = 6$ MTs under heavy traffic load. The probabilities of a successful reception and the capacity utilization of the available spatial channels have been evaluated as a function of the interference threshold γ. The logarithmic values of γ relate to the normalized transmission power $S = 1$.

Figure 14.59 clearly underlines the trade-off between large probability of correct MAC PDU reception and an efficient use of the $k = 6$ theoretically available spatial channels. These results encourage the idea to adapt the spatial exploitation of the available time slot to the QoS requirements of the service the PDUs belong to. On the uplink, a low threshold for the number of simultaneously transmitting MTs could be applied for time-critical PDUs, while for non-real-time services the schedule could be optimized for full spatial exploitation, i.e., throughput. The same applies for the downlink. In addition, here the parameter γ controls the interference conditions. A small value for γ, which stands for a nearly perfect spatial compatibility, is suitable for real-time oriented services, whereas a large γ leads to the desired high throughput especially of interest for non-real-time oriented services with the drawback of low reception probabilities. Aiming at very low delays, a very high probability

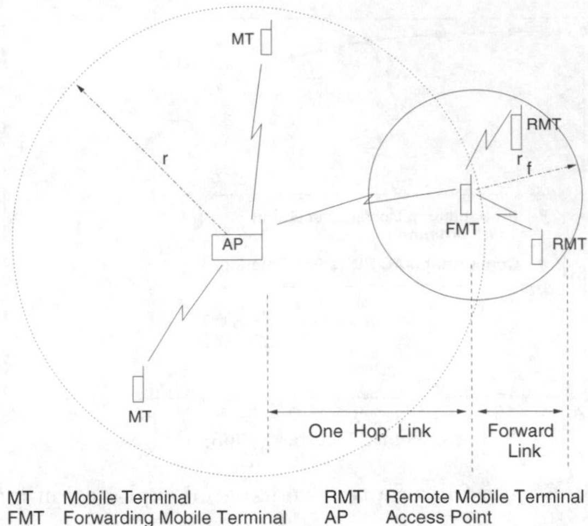

Figure 14.60: Wireless base station scenario

of a successful transmission is achieved by scheduling a MAC PDU in an exclusive time slot, although the additional capacity of the spatial dimension is then lost.

Together with position tracking of the MTs more elaborated scheduling algorithms will permit a further increase in efficiency, while obeying the delay constraints of real-time services.

14.2.12 HiperLAN/2 Wireless Base Station Concept*

In the following the so-called *Forwarding Mobile Terminal* (FMT), a H/2 relay, is presented as the key element to extend the communication range of an AP of the H/2 system [6, 7]. A typical forwarding scenario is presented in Figure 14.60. It generally applies to outdoor and indoor/office environments.

The *Remote Mobile Terminal* (RMT) is an MT that cannot communicate directly with the AP (one-hop) but needs a forwarding link for two-hop communication. The term *remote* differentiates it from a normal MT that is connected to the AP over the one-hop link. An MT associated to the AP via a one-hop link and located at the edge of the AP coverage area may perform the function of a forwarder and is thus called FMT. Since the boundary of the AP coverage area depends on the PHY mode [10] used, a new dimension is introduced into the H/2 world by the FMT to further complement the service coverage area.

---

*With the collaboration of Norbert Esseling**

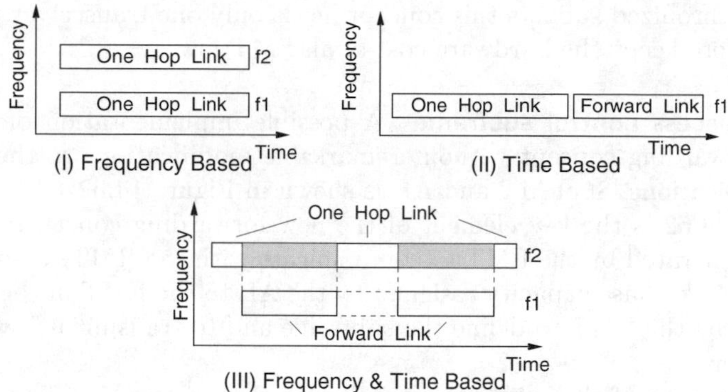

Figure 14.61: Forwarding concepts in time and frequency domains

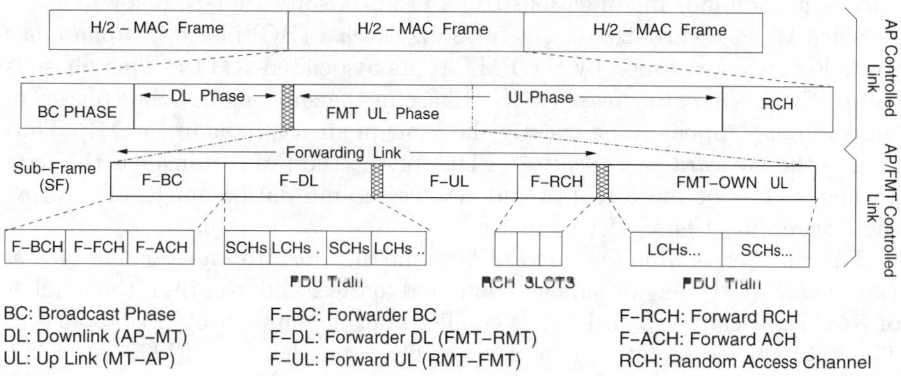

Figure 14.62: H/2 MAC frame forwarding subframe structures

14.2.12.1 Forwarding in H/2

There are a number of possibilities to employ forwarding for the H/2 system. The concepts are illustrated in Figure 14.61.

1. One-hop links (first hop) and forwarding links (second hop) operate on different frequencies.

2. The links operate on the same frequency but are separated in time on a time sharing basis.

3. A combination of 1.) and 2.). The FMT uses $f1$ for the second hop and $f2$ on a time sharing basis for the first hop.

FMTs may be fixed or mobile terminals. In the following the time-based concept 2. is further explained and investigated, as it enables the entire system to use synchronized subnets able to support well-defined QoS parameters. Due to the synchronized subnets this concept needs only one transceiver per FMT and therefore keeps the hardware cost of an FMT low.

Medium access control subframe A possible implementation of the time shared forwarding concept without remarkable modifications to the existing H/2 specifications [8] of MT and AP is shown in Figure 14.62.

Figure 14.62 is the key element of the new forwarding concept. The subframe is generated by the FMT to communicate with the RMTs associated to it. The uplink phase capacity assigned by the AP to the FMT in the H/2 MF is utilized by the FMT to define the subframe and to transmit its own uplink traffic if any.

The structure of the subframe is similar to that of the MF. The subframe is embedded into the MF so that an MT can operate on both frames when operating as an FMT. Further, an MT that operates temporarily as an RMT only requires minor modifications to its protocol software (see below).

Since MTs synchronize to APs in the *Broadcast* (BC) phase this phase has to be logically forwarded by the FMT to its associated RMTs. This phase is called *Forwarding Broadcast* (FBC) Phase in the subframe. Other phases of the subframe appear to be exactly the same in structure as in the MF. Thus during the *Forwarding Downlink* (FDL) phase, the FMT transmits the data packets to one or more RMTs and receives uplink data from RMTs during the *Forwarding Uplink* (FUL) phase.

The number of *Random Access Slots* in the *Forwarding Random Access Channel* (FRCH) may be smaller compared to the standard MF, if the number of RMTs associated to an FMT is smaller than the number of MTs to an AP. The subframe is monitored by a scheduler located in the FMT. An FMT transmitting the subframe will be received by the AP but only the FMT uplink phase of the subframe is accepted by the AP whilst the other parts are ignored. The data flow in the FMT is handled in queues that separate the data flow between the two links it is connecting.

Design aspects The forwarding concept is aimed to offer an acceptable performance for throughput, delay and other QoS parameters.

The radio resources requested by an FMT will naturally be more than that of an MT, as the FMT generates RRqs to the AP for the subframe including *Forwarding Broadcast Channel* (FBCH) and FRCH phases, its own uplink data and the data of the associated RMTs. This fact has to be reflected in the strategy adopted for the RGs in the AP.

In the FMT controlled concept, a standard AP is assumed to have no knowledge about an FMT and its subframe used for forwarding. This concept fits into the H/2 system with a minor modification to the specifications: The MT has to search for the BCH phase immediately after the end of the RCH phase even if there is a gap between RCH phase and the end of a MF.

As the subframe is controlled by the FMT it is acting like an AP for the RMTs, and for the AP both RMT as well as FMT are seen like MTs. The AP

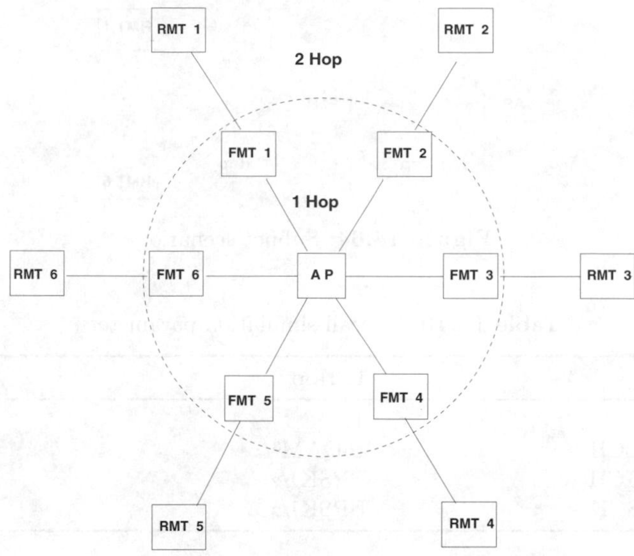

Figure 14.63: Star scenario

is not aware of the forwarding link. An RMT is connected to the AP via a two-hop link consuming approximately twice the MAC-capacity of a one-hop link.

The uplink phase granted to the FMT is *re-scheduled* by the FMT into a subframe and the FMT's uplink. The FMT has its own scheduler and other routines to handle the subframe management. The resource grants received for the subframe from the AP depend on the scheduling strategy in the AP.

The proposed forwarding scheme can be applied recursively. An RMT can also be used as a *Forwarder Remote Mobile Terminal* (FRMT) to connect a far remote terminal via serving FMT to the AP, and so forth. Besides point-to-point communication the forwarding concept seems to be especially beneficial for multicast and broadcast applications since no capacity appears to be wasted through the multi-hop transmission then.

14.2.12.2 Traffic Performance

Scenario To verify the FMT concept and to study the impact of FMTs onto the QoS parameters like transfer delay and system throughput simulation studies have been carried out.

In the simulation scenario of Figure 14.63 some MTs were placed out of the range of the AP to act as RMTs that are associated to FMTs. The FMTs are associated to the AP directly. Both FMTs and RMTs are assumed to have one MAC-connection each. Six FMTs and six RMTs are grouped together in the example scenario around the AP.

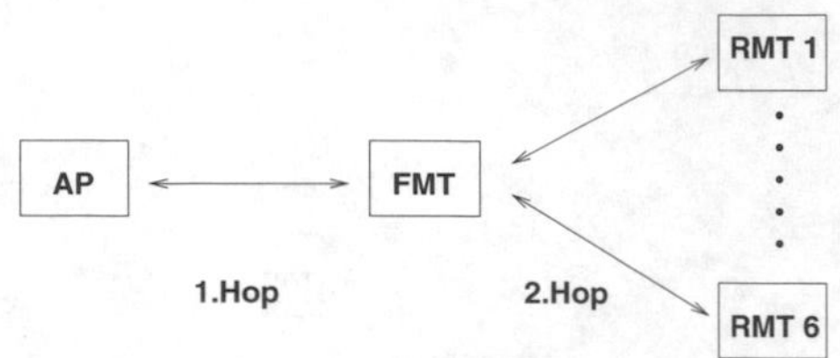

Figure 14.64: Subnet scenario

Table 14.10: Overall simulation parameters

Parameter	1. Hop	2. Hop
No. RCHs	4	1
PhyMode LCH	16QAM3/4	64QAM3/4
PhyMode FCH	BPSK1/2	BPSK3/4
PhyMode SCH	BPSK1/2	BPSK3/4

This scenario is called the *star scenario* in the following and is used as a reference scenario to investigate effects caused by a varying number of FMTs.

In contrast to the star scenario, the scenario in Figure 14.64 is used to study the effects of an FMT serving a varying number of RMTs associated to a single FMT. This set of served RMTs is a *subnet scenario* with respect to the serving FMT.

The simulation system (see Section 14.2.8.2) can be loaded with a mix of constant bit rates, Poisson shaped and video traffic sources. In the simulation results shown end-to-end user connections (i.e., connections between AP and RMT) loaded with Poisson traffic were studied. Each user connection was loaded with the same amount of traffic, and no additional user traffic between FMT and the AP was considered.

The most harmful effect of the presented concept is the reduction of capacity for the whole system, as the data has to travel two-hops instead of one. To keep the impact of forwarding on capacity as low as possible, higher modulation schemes were used on the second hop. The number of RCHs on the second hop was reduced to one to save capacity there. The resulting simulation parameters are summarized in Table 14.10.

Maximum system throughput The simulation studies on maximum system throughput have been performed for the star and the subnet scenarios. The total system throughput is defined as the sum of data of all active connections per hop and per direction. From the perspective of an RMT user in the star and subnet scenario, the available end-to-end throughput times the number of RMTs is studied.

Figure 14.65 shows the simulation results for the star scenario for a varying number of FMTs and one RMT associated to each FMT. A combination of one FMT and one RMT results in a maximum end-to-end (AP-to-RMT) throughput of approximately 14 Mbit/s using the PHY modes for LCH as given in Table 14.10. The maximum end-to-end throughput was derived from the calculated and simulated total system throughput, taking traffic on first and corresponding second hop and the overhead of the MF and the sub-frame into account in a way similar to Section 14.2.9.1.

An increasing number of FMTs each serving a single RMT strongly decreases the capacity available for AP-to-RMT connections as more and more capacity is needed for overhead to organize information transfer via the subframes. A value of approximately 3.5 Mbit/s is reached for the example scenario of six FMT as depicted in Figure 14.63. The simulated results are supplemented with the calculated throughput for the proposed subframe structure.

In Figure 14.66 the simulation for the subnet scenario is illustrated, where the number of RMTs connected to a single FMT is varied. As expected, the increasing number of RMTs does not reduce the total throughput as dramatically as observed in the star scenario. This becomes obvious keeping in mind that the support of an additional sub-frame (in the star scenario) needs much more resources (new FBC, FRCH, additional guardtimes for the whole subframe) than the support of an additional terminal in the subframe in the subnet scenario, where only small additional overhead in the *Forwarding Frame Channel* (FFCH) and small additional guardtimes for FDL and FUL are needed. Compared to a system without FMTs [24] the number of RMTs to be supported by an H/2 MF with comparable throughput is about half the number of MTs supported in a conventional setup. This result reflects the fact that end-to-end data for an RMT has to be transmitted twice (to and from the FMT) and therefore needs about double the system capacity on MAC level.

Delay In Figure 14.67 the CDF of the delay for one-hop and two-hop links are shown. The CDF allows a complete understanding of the delay behaviour since it gives the probability of a given delay time that can be guaranteed for a connection.

An additional delay of a little more than 2 ms in UL and DL directions for the two-hop forwarding link is observed compared to the one-hop link.

In downlink direction this is caused by the fact that the capacity for forwarding of a received downlink packet by the FMT is granted in the following MF. As the sub-frame is always scheduled after the first hop downlink of the MF the packet has to wait one complete MF and the additional time between it was sent to the FMT and its offset to the transmission in the subframe.

For similar reasons the uplink shows the same behaviour. In this case the position of the sub-frame is scheduled before the first hop uplink to keep the end-to-end downlink delay as low as possible. Unfortunately the packet

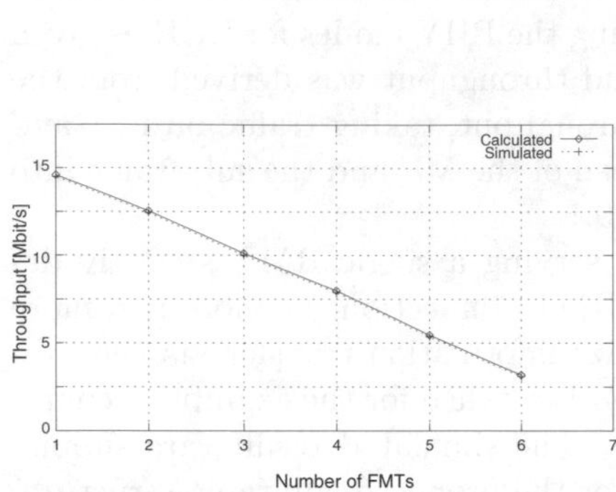

Figure 14.65: Maximum throughput vs. number of FMTs in the star scenario

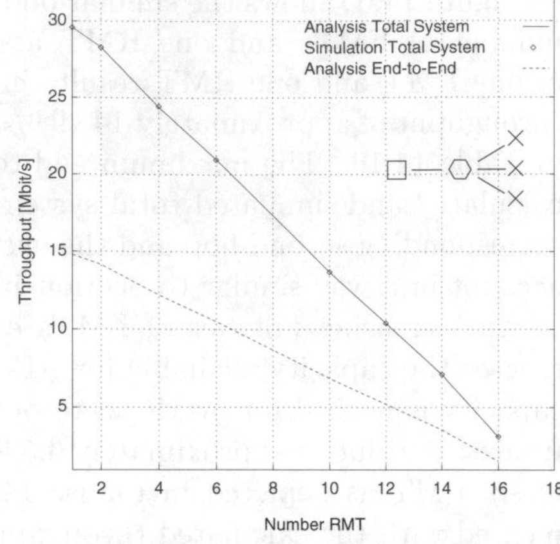

Figure 14.66: Maximum throughput vs. number of RMTs in the subnet scenario

cannot be forwarded without delay, as capacity first has to be requested for this packet on the uplink to the AP.

14.2.12.3 Applications of the Forwarding Concept

The total spectrum allocated to WLAN systems (see Figure 14.12) in Europe and the US is quite large compared to the spectrum allocated to cellular radio (see Figure 6.4) taking the short communication range and resulting possible traffic per square element into account. Since the transmit power is very limited in the 5 GHz band, the interference range is very limited and can be expected to be below 400 m outdoors. This will result in an excessive spectrum capacity of WLAN systems at 5 GHz.

It can be argued that under these conditions no care needs to be taken to efficiently use the frequency spectrum and multihop comunication would be an attractive choice to trade cost of infrastructure against system capacity. In fact, forwarding terminals help a lot to avoid the establishment of H/2 APs that each need access to a distribution system, e. g., a fixed network. Since the propagation characteristics of radio waves at 5 GHz tend to follow the laws of optical waves, it appears to be difficult to completely illuminate a service area containing obstructions. Forward mobile terminals placed at fixed locations are expected to substantially contribute to keep the number of APs small. It appears even possible to consider 3-hops (2 FMT in sequence) to be acceptable seen from the packet delay introduced then.

This philosophy reaches its limitation if the traffic generated per area element exceeds a value that can be served by multihop links. Then, FMTs will have to be replaced by APs to better exploit the spectrum capacity at 5 GHz by avoiding multihop communication. It is clear from these considerations

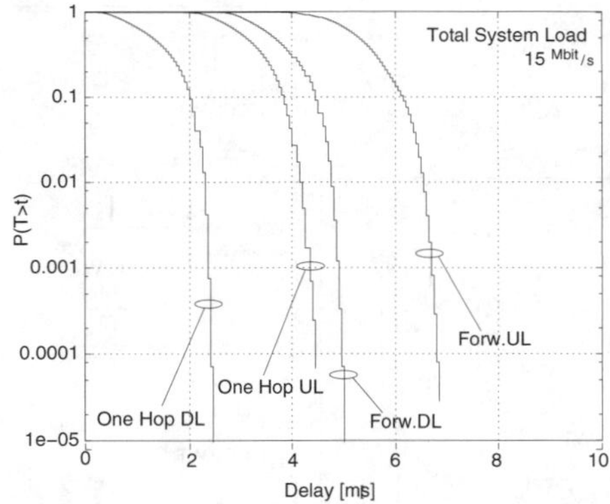

Figure 14.67: CDF of delay under forwarding

that the forwarder concept is best suited for an introduction of the H/2 system aiming at a complete area coverage and that forwarders will have to be replaced with growing acceptance of the system and a respective increase of the traffic load.

14.2.13 HiperLAN/2 Ad hoc Networking for Home Environments*

Beside infrastructure-based configurations, the self-organizing communication among home devices including the support of IEEE 1394 is another objective of H/2. The H/2 *Home Environment Extension* (HEE) specifies an ad-hoc network configuration to allow plug-and-play operation [9]. The concept of a central control of the air interface has been kept but is now performed by a *Central Controller* (CC), a terminal that has been assigned this role ad hoc in a decentral way. The direct communication called *Direct Link* (DiL) between terminals has been introduced to be the normal way to exchange data between fixed and mobile terminals in unicast, multicast or broadcast mode. Since a CC is controlling a cluster of terminals it is called a clusterhead.

14.2.13.1 Ad hoc Definition of Central Controllers

A H/2 HEE network instead of an AP needs a CC, a wireless terminal that controls access to the radio channel. The CC Selection process specifies a decentrally operating algorithm to identify a terminal best suited to coordinate transmission of terminals in its neighborhood. Figure 14.68 shows the corresponding possible scenarios. On the left-hand side a set of terminals is depicted before or during the CC Selection process whereby each terminal provides

***With the collaboration of Jörg Peetz and Andreas Hettich**

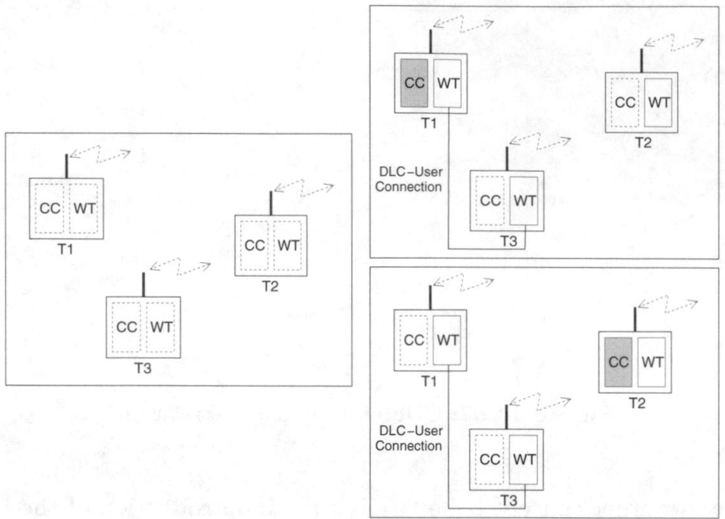

Figure 14.68: One-hop ad hoc scenarios

both a *Wireless Terminal* (WT) and the CC functionality that in general is available from so-called CC-capable terminals only. On the right-hand side the operation after network establishment is shown. From the upper right hand to the lower right-hand figure, the CC function has changed based on a CC Handover procedure.

The CC Selection algorithm [33] ensures that within one subnet only one CC is established. It is performed in a decentralized and autonomous way by each CC-capable terminal when powered on or when the CC has been lost. A CC-capable terminal in general aims to be an MT and to associate to an existing CC. Therefore it scans the available carrier frequencies to find an operational CC to associate to it. If the association procedure fails, e.g., because the detected CC belongs to a foreign subnet, scanning is continued. If no CC has been found the dynamic CC selection algorithm is started.

The CC Selection algorithm of a terminal lasts for a fixed duration T_{CC}. A CC probing terminal alternating probes and scans. These probing and scanning phases form together *probing periods*, see Figure 14.69. Each probing phase is subdivided into *probing frames* where a CC-capable terminal sends one so-called beacon signal per probing frame to inform other terminals about its existence. In order to avoid collisions, the transmit time of the beacon is determined randomly. Since a probing terminal operates on one frequency channel for a whole probing phase, the probability of two terminals probing on the same frequency and not hearing each other is close to zero [33].

During frequency scanning a terminal searches for further terminals probing on other frequency carriers. Figure 14.70 shows the frequency scanning phase T_S, see also Figure 14.69, performed within a probing period of length

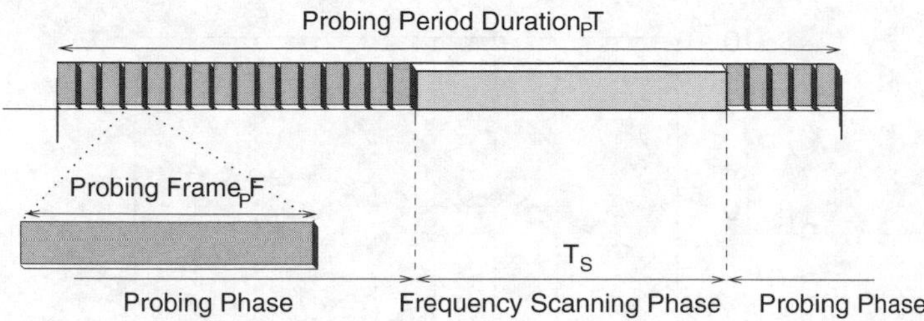

Figure 14.69: CC probing period of a CC-capable terminal

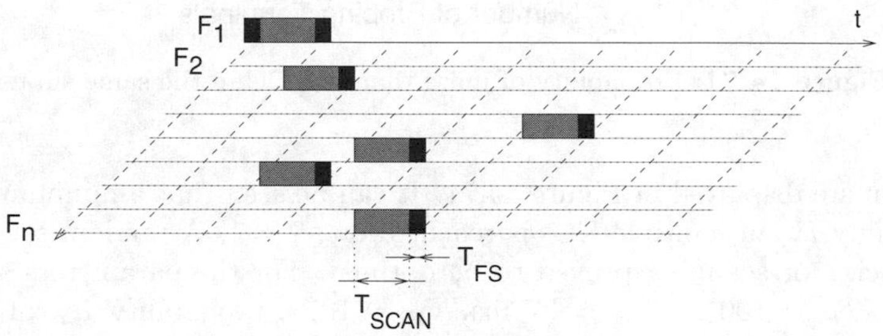

Figure 14.70: Frequency scanning

T_P. A CC-capable terminal scans all available frequencies for beacons. The *Frequency Switching Time* T_{FS} is the maximum time to change a frequency channel and is defined to be 1 ms [10] . The *Frequency Scanning Time* T_{SCAN} determines the duration a CC-capable terminal scans one frequency channel for a beacon. After having scanned a frequency the CC-capable terminal switches to the next frequency in random order.

The performance of the CC Selection process is characterized by the probability P_m that, after a random number of probing terminals have used the CC Selection Process for a duration T_{CC}, there are multiple CC-capable terminals deciding they have to be the CC. The probability P_m can be reduced significantly if probing and frequency scanning phases are repeated N_P times within T_{CC}. To avoid synchronized and steadily overlapping scanning phases of different probing terminals, the start of a frequency scanning phase is determined randomly.

Each CC-capable terminal that senses a beacon immediately withdraws from CC Selection, starts a back-off to await the result of the CC Selection process of the other terminals and then starts its initial scan process again to find an operational CC.

Performance of CC Selection The probability P_m has been calculated to evaluate an optimum ratio $\frac{T_P}{T_S}$ and the results have been validated by simulation [33]. The simulation results for a duration $T_{CC} = 500$ ms of the CC

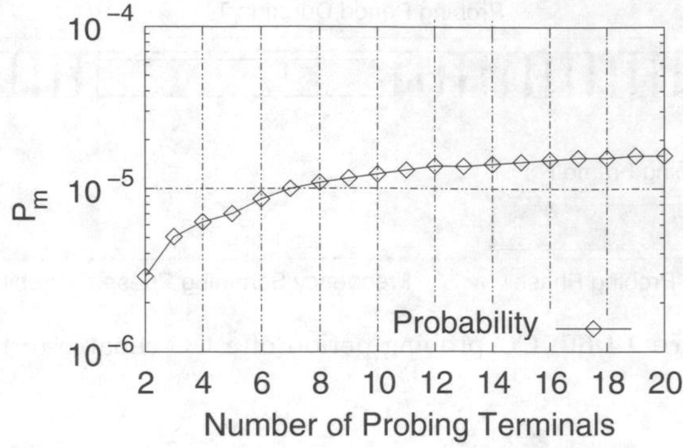

Figure 14.71: Probability of more than one CC in the same subnet

selection are displayed in Figure 14.71. It can be seen that a minimum of the probability P_m of multiple CCs is reached for $\frac{T_P}{T_S} = e$. A random selection of frequencies for scanning proved to be optimal. For the parameters specified in H/2 ($T_P = 100$ ms, $T_S = 37$ ms, $N_P = 10$) a probability P_m of 10^{-5} is achieved giving a quite good performance of the CC Selection process.

CC Handover To allow the current CC to switch off or to go into a sleep mode the CC function must be able to perform a dynamical handover between terminals. The CC Selection algorithm described before is too time consuming to serve for a CC Handover.

The CC Handover procedure as described is part of the H/2 HEE standard. It is transparent to the MTs in the network in the sense that the data transfer on the MAC layer goes on without interruption. The process only involves the control plane of H/2 in the RLC protocol. A CC Handover either is initiated by the current CC itself or proposed by any MT in the subnet. Following, the CC selects a terminal known to be CC-capable and sends a request to it. Upon a positive reply from the CC candidate all MT are informed about the forthcoming CC Handover and all other RLC procedures are stopped to freeze the RLC data in the CC and not allow connections to be established with the old CC.

The data to be transmitted from old CC to new CC comprise data related to all associated MTs (like MAC-ID, MT capabilities, etc.). To maintain ongoing DiL connections between MTs, all connection specific RLC parameters have to be transmitted, too.

The new CC will poll all MTs after CC Handover immediately to ask for their RRqs and in addition will assign them an LCH PDU to transmit user data. It has been found that the handover typically will interrupt the ongoing traffic by about 1 ms only [33].

14.2.13.2 Direct Link Communication

Unicast In infrastructure-based networks, data are exchanged between two WTs via Uplink and Downlink. For ad-hoc networks, however, DiL communication is preferred due to the self-organizing configuration. In case of a CC Handover, for example, Uplink and Downlink connections need to be re-established whereas all DiLs remain unchanged. Another advantage is the gain in capacity because a direct transfer between two WTs requires approximately half the spectrum resources than transmission via the CC. Just the scheduling of transmission capacity and the corresponding signaling is performed by the CC. Besides saving capacity, the DiL also reduces the delay of user data. Since there is no intermediate station, resource request and allocation delay is reduced. Thus the DiL communciation shows significant advantages regarding QoS especially for the use in a self-organizing HEE. Considering the radio link, another benefit of DiL might apply if there is a better direct radio connection between the involved WTs than via the CC. In this case a higher PHY mode may be chosen inducing a higher transmission rate. However, even the opposite is possible and then the CC may take over relais functionality resulting in conventional Uplink and Downlink conditions.

The signaling mechanism for DiL communication is described in detail in [9]. A WT with data pending for another WT indicates the required capacity to the CC by a `Resource Request` message sent in the Uplink. The CC responds using so-called `Resource Grants` during the Broadcast Phase and informs the WTs about the capacity that has actually been allocated in a MF. The transmission of user data is then started in the indicated slots whereby the CC is not involved except it is acting as relais node or if it takes an active part in the DiL as sender or receiver itself. DiL is the only mode of communication that also allows a duplex connection whereas Uplink and Downlink are simplex in any case. The specific extensions for requesting duplex capacity can be found in [9].

Multicast The objective of multicasting is to save capacity in the case that a WT intends to transmit PDUs not only to one destination WT, but to a group of WTs. The simplest, but most inefficient, solution for this task is the approach of multiple unicast DiL connections. Therefore, another specific mode of transmission is provided by the H/2 standard [9] reserving certain MAC *Infrastructure Domain* (ID)s for groups of terminals. During the establishment of such Multicast Groups as well as by the *Group Join Procedure*, the set of participating WTs is negotiated. By identifying the group MAC ID in `Resource Grants` transmitted within the broadcast phase of each MAC Frame, all terminals belonging to the Multicast group detect and receive the multicast data PDUs at the same time.

An acknowledgement by the receivers is not planned because unicast connections back to the sender would be needed that would introduce too much overhead. Furthermore, the coordination of several acknowledgements for the

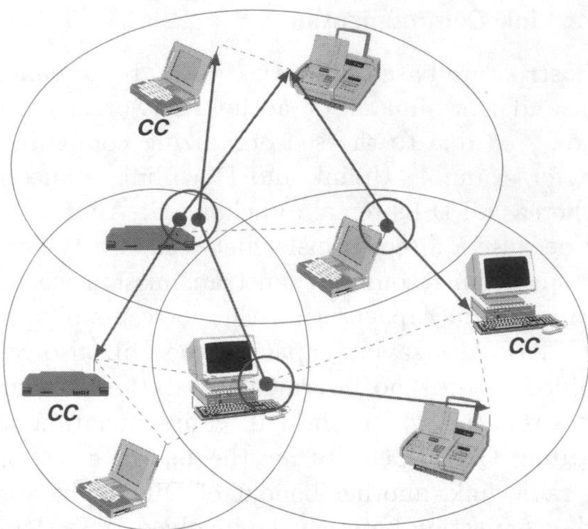

Figure 14.72: Multihop ad-hoc configuration

same PDUs would cause inconsistencies or at least high effort in evaluation. Thus, instead of an ARQ mechanism, the HEE envisions a repetetion mode and a second level FEC based on Reed Solomon error correction for DiL multicast data transmission. The described Multicast concept offers manifold fields of application, e.g., the distribution of video streams to several users.

14.2.13.3 The Multiple Subnets Model—First Solutions

Today's stage of the development of HEE considers only one hop networks whereby no communication between different subnets is specified. However, interconnection of subnets, meaning that WTs associated to any of the involved subnets can communicate to each other, is planned for the second phase of standardization. Since each subnet decides about its operation frequency channel according to interference minimization based on the *Dynamic Frequency Selection* (DFS), a forwarding concept is required that considers overlapping subnets on different frequency channels [31]. In Figure 14.72, an example for a corresponding multihop network configuration consisting of three interconnected subnets is depicted. The WTs operating as *Multiple-Frequency Forwarder WT* (MF-WT) are marked.

Multiple-Frequency Forwarding A new inter-subnet forwarding approach that is founded on an intermitted presence of forwarding WTs at each subnet has been presented in [31]. This approach enables *inter*-connection of subnets whereby only the RLC functions `MT_Absence` and `MT_Alive`, that are specified by the standard, are required. A terminal selected to be MF-WT, periodically withdraws from transmission for negotiated periods of $0 \leq mt - absence - time \leq 63$ MFs. During these absence periods, it continues operation in the

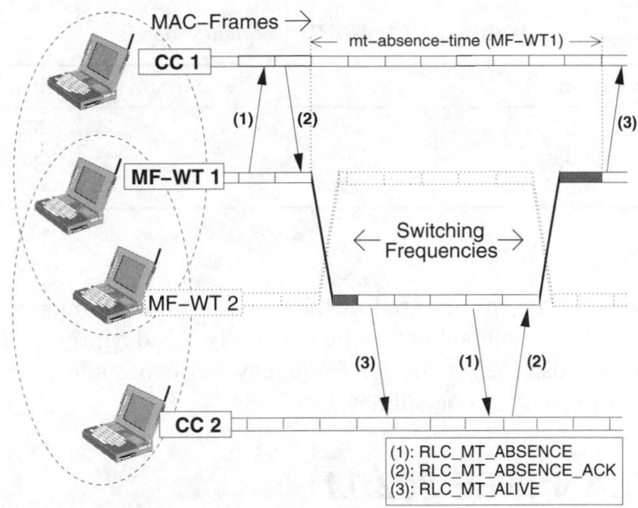

Figure 14.73: MF-WT operation in subnet 1 and 2

second subnet. Since an MF-WT has direct radio contact only to WTs of one subnet, it emulates the other subnets' destination WTs and caches all *inter*-subnet traffic. For inter-subnet communication, DiL is the only possible mode of transmission because the forwarding terminal is a WT. Therefore, it must not send data in the Downlink or receive data in the Uplink phase. This, however, would be the case if no direct link is applied, except forwarding only between the CCs leading to a severe throughput degradation and poor delay results.

Figure 14.73 illustrates the operation of one (two) MF-WT(s) successfully associated to the CCs of two subnets in detail. The MF-WT is present alternating either for CC1 or CC2. To leave the current CC, for example CC1, it transmits the `RLC_MT_ABSENCE` message and, when it receives the acknowledgement, the radio connection to CC1 is intermitted and the absence period timer is started. After switching frequency, that is expected to last 1 ms [10], the MF-WT synchronizes to CC2. If the MF-WT awakes in subnet 2 earlier than expected, its presence is signalled by transmitting an `RLC_MT_ALIVE` message via the RCH. Otherwise, it is scheduled by the CC automatically and just starts transmission. To return to subnet 1 the `MT_Absence` procedure is executed again. This sequence is repeated periodically whereby symmetric and asymmetric periods are adjustable according to the load situation.

As an MF-WT is only present during time intervals in one subnet, the *inter*-subnet-throughput is limited to less than 44 % of the maximum possible *intra*-subnet-throughput [31]. In case of several terminals located in the overlapping area of involved subnets, the *Alternating Forwarder* approach allows to enhance throughput and reduce delay. If, for example, two networks are coupled by two MF-WTs, as depicted in Figure 14.73, the absence periods of

Table 14.11: U–NII frequency bands

Band of operation [GHz]			Maximum transmit power [mW]
5.15	...	5.25	50
5.25	...	5.35	250
5.725	...	5.825	1000

both MF-WTs can be coordinated in the way that alternating one of the MF-WTs is available for one subnet. Thus a nearly steady inter-connection can be achieved, see also [32]. Only for frequency switching and synchronization no MF-WT is present at one subnet.

14.3 WLAN IEEE 802.11*

The purpose of the IEEE 802.11 standard is "To provide wireless connectivity to automatic machinery, equipment or stations that require rapid deployment, which may be portable or hand-held, or which may be mounted on moving vehicles within a local area" [20].

The standard describes the functions and services required to operate within IEEE 802.11 compliant ad hoc and infrastructure networks. It defines MAC procedures and PHY signalling techniques. Furthermore, it describes procedures to provide privacy of user information.

IEEE 802.11 has been designed for use in *Industrial, Scientific and Medical* (ISM) bands. The *Federal Communications Commission* (FCC) in the USA prescribed maximum power levels only, band edge interference, and the requirement to use spread spectrum in order to minimize interference with already existing communication systems. It designated the frequency bands 902–928 MHz, 2400–2483.5 MHz and 5725–5850 MHz to these ISM bands [45].

IEEE 802.11 has defined two PHY standards for the 2.4 GHz ISM band: one using the *Frequency Hopping Spread Spectrum* (FHSS) technique, and one using the *Direct Sequence Spread Spectrum* (DSSS) technique. An alternative is the specification of an *Infrared* (IR) physical layer. IEEE 802.11 stations using any of the three technologies operate at a data rate of 1 Mbit/s (optionally 2 Mbit/s) and at 11 MBit/s according to standard IEEE 802.11b. Recently work has been finished on the specification of a 20 Mbit/s PHY at 5 GHz. Target frequencies are those opened by the FCC in 1997 for *Unlicensed Information Infrastructure Networks* (U–NII); see Table 14.11.

A key issue of 802.11 is that a mobile station is able to communicate with any other mobile or wired station transparently, which means that above the MAC layer, 802.11 appears like any other 802.x LAN and offers comparable

***With the collaboration of Christian Plenge and Stefan Mangold**

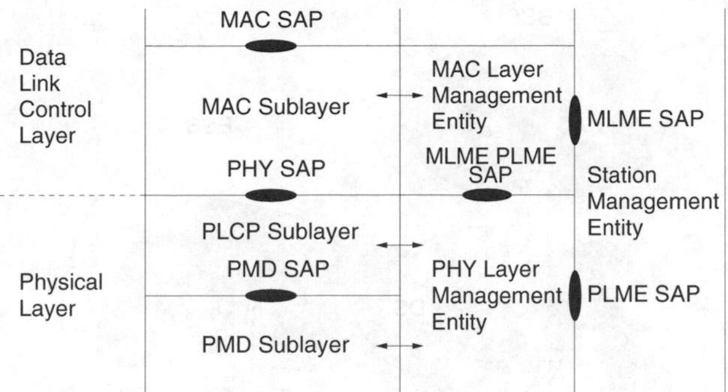

Figure 14.74: IEEE 802.11 reference model

services (see Figure 14.74). This implies that mobility is handled at MAC level [2].

The protocols are specified for slowly moving stations, usually indoors but not limited to this, communicating among each other (*Ad Hoc Mode*) or with stations beyond their direct communication range with the support of an infrastructure (*Infrastructure Mode*). The communication is packet-oriented [44].

14.3.1 Architecture of IEEE 802.11 Networks

The IEEE 802.11 network architecture is hierarchical [20]. Its basic element is the *Basic Service Set* (BSS), which is a set of IEEE 802.11 stations (STA) controlled by a single *Coordination Function* (CF). The *Distributed Coordination Function* (DCF) is mandatory for each BSS, whereas the *Point Coordination Function* (PCF) is optional [30].

An *Independent BSS* (IBSS) is the most basic IEEE 802.11 type of network. It is an ad hoc network consisting of a minimum of two stations. A BSS may also be part of a larger network, the so-called *Extended Service Set* (ESS). The ESS consists of multiple BSS interconnected by the *Distribution System* (DS). The ESS appears as a single BSS to the IEEE 802.2 LLC layer.

Logically the BSS and the DS work on different media. The BSS operates on the *Wireless Medium* (WM), whereas the DS uses the *Distribution System Media* (DSM). As the IEEE 802.11 architecture is specified independently of any specific media, the WM and DSM may or may not be the same.

Stations connected to the DS are called *Access Points* (AP), which provide the *Distribution System Services* (DSS) in order to enable the MAC to transport MAC SDUs between stations that cannot communicate over a single instance of the WM. A *Portal* is the logical point where a non IEEE 802.11 LAN is connected to the DS. This allows the communication across different types of LANs. Figure 14.75 comprises the components of the 802.11 architecture.

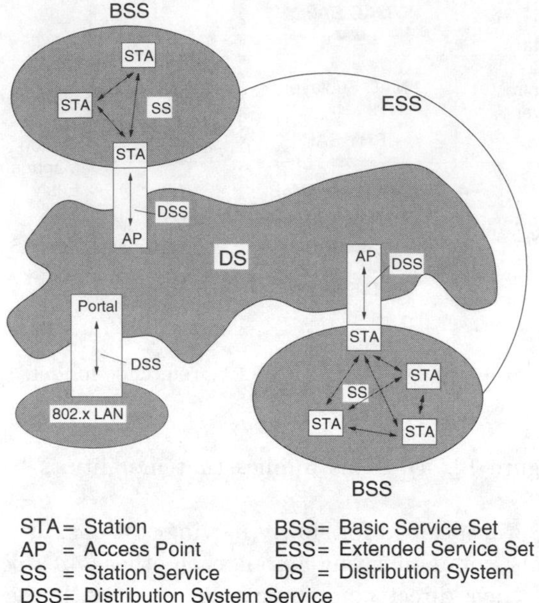

Figure 14.75: Components of the IEEE 802.11 architecture

IEEE 802.11 only specifies the use of the WM address space. Therefore the address space of the DSM and the integrated wired LAN may differ from the WM. IEEE 802.11 uses the 48-bit address space, so it is most likely that the address spaces of WM, DSM and wired LAN is the same.

14.3.2 Services of IEEE 802.11 Networks

The services of IEEE 802.11 are sorted into two categories: the *Station Services* (SS) and the *Distribution System Services* (DSS). The services are listed in Table 14.12.

Table 14.12: Services of IEEE 802.11 networks

Station services	Distribution system services
Authentication	Association
Deauthentication	Disassociation
Privacy	Distribution
MSDU delivery	Integration
	Reassociation

In an *Independent Basic Service Set* (IBSS) or ad hoc network, only the *Station Services* are available. Besides the *MSDU Delivery*, which will be explained in detail in Section 14.3.3, the station services provide access and confidentiality control [20]:

Authentication The identity of stations is delivered to each other. Here, only link level authentication is provided. End-to-End system or user-to-user authentication is outside the scope of IEEE 802.11. Nevertheless these services are supported.

Deauthentication An existing authentication is cancelled.

Privacy The contents of link level messages are protected from eavesdropping. IEEE 802.11 specifies an optional privacy algorithm: *Wired Equivalent Privacy* (WEP).

The *Distribution System Services* are used to distribute messages in the *Distribution System* and to support mobility.

The *Distribution* service delivers MSDUs within the DS, but it is not specified in IEEE 802.11. The standard provides the DS with sufficient information in order to be able to determine the target AP of an MSDU. This is done by the three association-related services (*Association*, *Reassociation* and *Disassociation*). The *Integration* service enables the delivery of MSDUs between non 802.11 LAN and DS via a *Portal*. The *Integration* service depends on the *Distribution* service, and performs media and address space translation if necessary. The *Integration* service is outside the scope of IEEE 802.11.

Mobility is supported by association services. Three types of mobility are distinguished in 802.11:

No-transition Stations are stationary or move within a BSS.

BSS-transition Stations move from one BSS to another BSS within the same ESS.

ESS-transition Stations move from one BSS in one ESS to another BSS in a different ESS. This kind of mobility is not supported by 802.11; running services will be disrupted.

The association services needed for mobility support are as follows:

Association makes an AP aware of a station in the BSS. It is always invoked by the station. A station cannot associate with more than one AP at the same time. This ensures that the *Distribution* service finds a unique AP for a message in the DS. A station selects the appropriate AP out of several possible APs by scanning. *Association* is sufficient to handle *No-transition* mobility and necessary but not sufficient to support *BSS-transition* mobility.

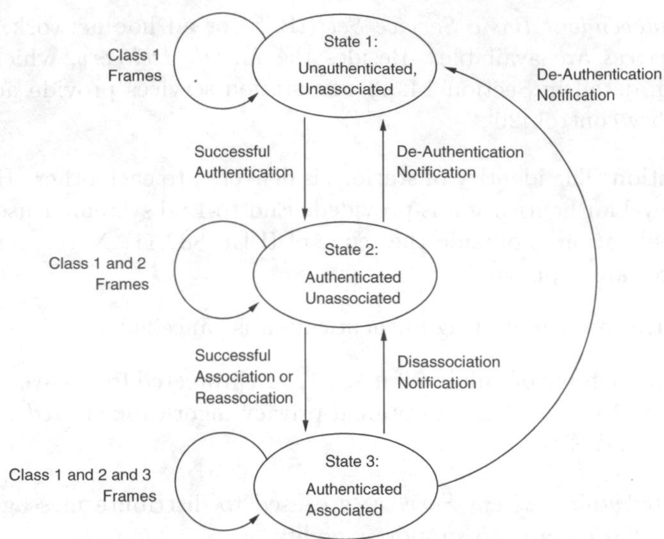

Figure 14.76: Relationship between IEEE 802.11 services

Disassociation cancels an existing *Association*. The DS is informed about the *Disassociation* by the related AP.

Reassociation transfer an existing *Association* from one AP to another, thus supporting *BSS-transition* mobility by informing the DS about the change of AP. In order to speed up the *Reassociation* procedure, a *Pre-authentication* may be invoked prior to the *Reassociation*.

The relationship between the services is shown in Figure 14.76. The frame types allowed to be transmitted between two stations are grouped in three classes corresponding to the state of the station. In an IBSS only class 1 and 2 frames are allowed, since there is no AP and therefore no *Association* is possible [20].

14.3.3 IEEE 802.11 MAC Sublayer

The IEEE 802.11 MAC protocol provides two types of service: asynchronous and contention-free. The asynchronous type of service is provided by the *Distributed Coordination Function* (DCF), which implements as the basic access method the *Carrier Sense Multiple Access with Collision Avoidance* (CSMA/CA) protocol. The contention-free type of service is provided by the *Point Coordination Function* (PCF), which basically implements a polling access method. Unlike the DCF, the implementation of the PCF is not mandatory. Furthermore, the PCF itself relies on the asynchronous service provided by the DCF as shown in Figure 14.77.

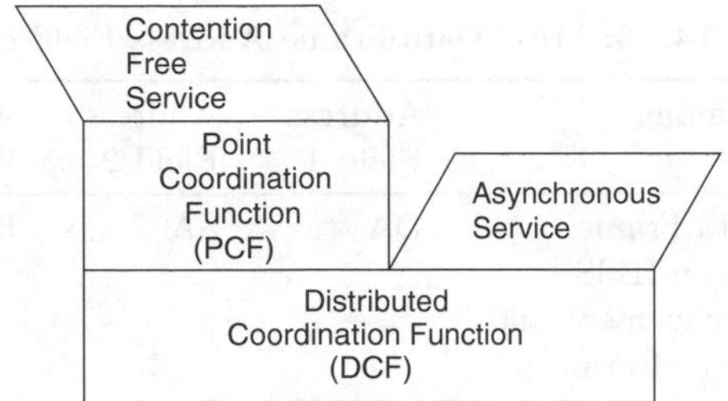

Figure 14.77: IEEE 802.11 MAC architecture

14.3.3.1 Address Mapping

The IEEE 802.11 standard uses 48-bit address representation. In order to avoid an additional address mapping, five types of addresses are used:

BSS Identifier (BSSID) The BSSID uniquely identifies each BSS. In an infrastructure BSS, the BSSID is the MAC address of the AP serving the BSS. In an independent BSS (IBSS), the BSSID is formed by a random number.

Destination Address (DA) The MAC address of the final recipient of an MSDU.

Source Address (SA) The MAC address of the originator of a MSDU.

Receiver Address (RA) The MAC address of the intended immediate receiver on the wireless medium for the MPDU.

Transmitter Address (TA) The MAC address of the recent station transmitting this MPDU on the wireless medium.

MAC Data Frames have four address fields. The type of addresses assigned to these fields depend on the *To DS* and *From DS* bit settings (see Table 14.13) [44].

14.3.3.2 MAC Services

As a service provider, the MAC sublayer provides its services to the LLC sublayer at the MAC SAP (see Figure 14.74).

The MAC offers the *Asynchronous Data Service*, which provides the LLC with the ability to exchange *MAC Service Data Units* (MSDU). This service is offered on a best-effort connectionless basis. There are no guarantees that the submitted MSDU will be delivered successfully. Two classes of services are offered: one allowing re-ordering of MSDUs inside the MAC, the other guaranteeing the strict order of MSDUs.

Table 14.13: MAC Data Frame Address Field content

To DS	From DS	Meaning	Address Field 1	Address Field 2	Address Field 3	Address Field 4
0	0	Data Frame within IBSS Management and Control Frames	DA	SA	BSSID	N/A
0	1	Data Frame to DS	DA	BSSID	SA	N/A
1	0	Data Frame from DS	BSSID	SA	DA	N/A
1	1	WDS Frame distributed from one AP to another AP	RA	TA	DA	SA

Table 14.14: 802.11 MAC primitives and their parameters

Primitive	Parameter
MA-UNITDATA.request	Source address, destination address, routing information, data, priority, service class
MA-UNITDATA.indication	Source address, destination address, routing information, data, reception status, priority, service class
MA-UNITDATA-STATUS.indication	Source address, destination address, transmission status, provided priority, provided service class

Broadcast and multicast transport is part of the asynchronous data service that will experience a lower quality of service compared to unicast MSDUs owing to the characteristics of the WM [20].

The 802.11 MAC offers a security service by using the WEP. This includes:

1. Confidentiality
2. Authentication
3. Access control in conjunction with layer management

MAC Service Primitives The MAC service primitives supported by the IEEE 802.11 MAC layer are taken from the ISO/IEC 8022-2 standard [22]. The service primitives at the MAC SAP and their parameters are listed in Table 14.14.

The meaning of the parameters listed in Table 14.14 are:

Source address (SA) Address of the MAC sublayer entity receiving the MA-UNITDATA.request primitive.

Destination address (DA) An individual or a group MAC sublayer entity address.

Routing information Must be null for 802.11.

Data MSDU that is transported without modification by the MAC between to LLC sublayer entities. The size of the MSDU must be less or equal to 2304 octets for 802.11.

Priority 802.11 offers two priority levels: *Contention* and *Contention-Free.*

Service class Two classes are defined: *Reorderable Multicast* and *Strictly Ordered.*

Reception status The result of the reception is reported, namely *Success* and *Failure.*

Transmission status Information about the transmission passed back to the local LLC sublayer entity. Possible values are:

1. Successful
2. Undeliverable
3. Excessive data length
4. Non-null source routing
5. Unsupported priority
6. Unsupported service class
7. Unavailable priority
8. Unavailable service class

14.3.3.3 MAC Protocol

The IEEE 802.11 MAC protocol is based on a *Carrier Sense Multiple Access with Collision Avoidance* (CSMA/CA) protocol. The time to sense the carrier is defined by the *Interframe Space* (IFS). The collision avoidance is done by a random backoff procedure [44].

Interframe Space The time between two frames is called *Interframe Space* (IFS). In order to determine whether the medium is free, a station has to use the carrier sense function for a specified IFS. The standard specifies four different IFSs, which represent three different priority levels for the channel-access. The shorter the IFS, the higher the priority. The IFSs are specified as time gaps on the medium and are independent of the channel data rate. Owing to the different characteristics of the different PHY specifications, the IFS time durations are specific for each PHY.

Some relations between the IFS are shown in Figure 14.78. The IFS are listed in order, from the shortest to the longest [20].

Short IFS (SIFS) The SIFS is used for the immediate acknowledgement (ACK frame) of a data frame, the answer (*Clear To Send* (CTS) frame)

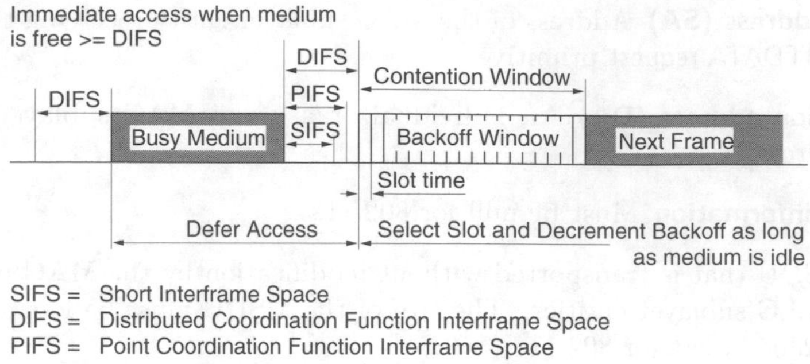

Figure 14.78: Interframe space relationship

to a *Request To Send* (RTS) frame, a subsequent MPDU of a fragmented MSDU, response to any polling by the PCF, and any frames of the AP during the *Contention-Free Period* (CFP).

Point Coordination Function IFS (PIFS) The PIFS is used by stations operating under the PCF to gain access to the medium at the start of the CFP.

Distributed Coordination Function IFS (DIFS) The DIFS is used by stations operating under the DCF to gain access to the medium to transmit data or management frames.

Extended IFS (EIFS) The EIFS is used by the DCF whenever the PHY indicates that a frame transmission did not result in a correct *Frame Check Sequence* (FCS). The EIFS allows another station to acknowledge what was, to this station, an incorrectly received frame.

Distributed Coordination Function (DCF) According to the DCF (see Figure 14.79), a station must sense the medium before initiating the transmission of a packet. If the medium is sensed as being idle for a time interval greater than a DIFS then the station transmits the packet. Otherwise, the transmission is deferred and the backoff process is started. Specifically, the station computes a random number uniformly distributed between zero and a maximum called *Contention Window* (CW). The random number is multiplied by the *slot time*, resulting in the backoff interval used to set the backoff timer. This timer is decremented only when the medium is idle, whereas it is frozen when another station is transmitting. Each time the medium becomes idle, the station waits for a DIFS and then periodically decrements the backoff timer.

As soon as the backoff timer expires, the station is authorized to access the medium. If two or more stations start transmission simultaneously, a collision occurs. Unlike wired networks (e.g. with CSMA/CD in IEEE 802.3), in a

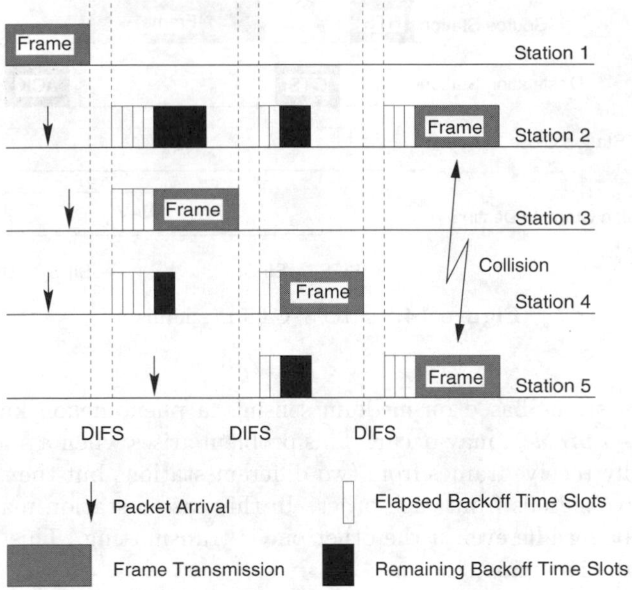

Figure 14.79: Basic access mechanism

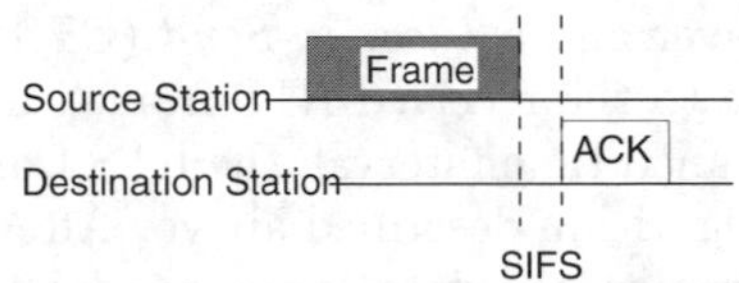

Figure 14.80: Acknowledgement mechanism

wireless environment collision detection is not possible. Hence, as shown in Figure 14.80, a positive acknowledgement is used to notify the sending station that the transmitted frame has been successfully received. The transmission of the acknowledgement is initiated at a time interval equal to the SIFS after the end of the reception of the previous frame.

If the acknowledgement is not received in a specified time interval, the station assumes that the transmitted frame was not successfully received, and hence schedules a retransmission and enters the backoff process again. However, to reduce the probability of collisions, after each unsuccessful transmission attempt the *Contention Window* is doubled until a predefined maximum (CW_{max}) is reached. After a successful transmission, the *Contention Window* is reset to CW_{min}.

After each frame transmission, a station must execute a new backoff process. Therefore at least one backoff is in between two transmissions of the same station [2].

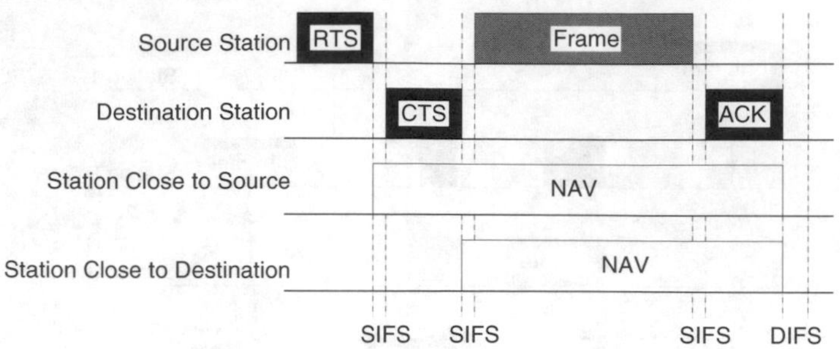

Figure 14.81: RTS/CTS mechanism

In radio systems based on medium sensing, a phenomenon known as the *hidden-station problem* may occur. This problem arises when a station is able to successfully receive frames from two different stations but the two stations cannot receive signals from each other. In this case a station may sense the medium as being idle even if the other one is transmitting. This results in a collision at the receiving station.

To deal with the hidden-station problem, the IEEE 802.11 MAC protocol includes a mechanism based on the exchange of two short control frames (see Figure 14.81): a *Request-to-Send* (RTS) frame that is sent by a potential transmitter to the receiver and a *Clear-to-Send* (CTS) frame that is sent by the receiver in response to the received RTS frame. If the CTS frame is not received within a predefined time interval, the RTS frame is retransmitted by executing the backoff algorithm described above. After a successful exchange of the RTS and CTS frames, the data frame can be sent by the transmitter after waiting for a SIFS. The implementation of the RTS packet is optional, whereas all stations must be able to answer to a RTS with the belonging CTS.

The RTS and CTS frames include a duration field that specifies the time interval necessary to completely transmit the data frame and the related acknowledgement. This information is used by stations that can hear either the transmitter or the receiver to update their *Net Allocation Vector* (NAV), a timer that, unlike the backoff timer, is continuously decremented irrespective of the status of the medium. Since stations that can hear either the transmitter or the receiver refrain from transmitting until their NAV has expired, the probability of a collision occurring because of a hidden station is reduced. Of course, the drawback of using the RTS/CTS mechanism is an increased overhead, which may be significant for short data frames.

Furthermore, the RTS/CTS mechanism can be regarded as a way to improve the MAC protocol performance. In fact, when the mechanism is enabled, collisions can obviously occur only during the transmission of the RTS frame. Since the RTS frame is usually much shorter than the data frame, the waste of bandwidth and time due to the collision is reduced.

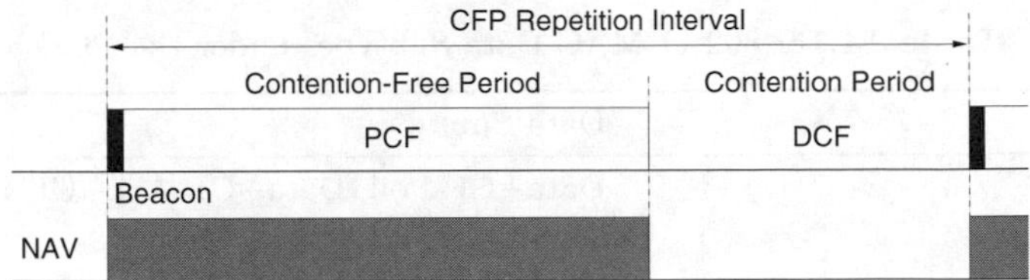

Figure 14.82: Relationship between CFP and CP

In both cases the effectiveness of the RTS/CTS mechanism depends upon the length of the data frame to be protected. Consequently, the RTS/CTS mechanism relies on a threshold, the *RTS threshold*: the mechanism is enabled for data frame sizes over the threshold and disabled for data frame sizes under the threshold [2].

The RTS/CTS is useful also while operating overlapping BSS or IBSS.

Point Coordination Function In order to support time-bounded services, the IEEE 802.11 standard defines the *Point Coordination Function* (PCF) to permit a *Point Coordinator* (PC) to have priority access to the medium. Usually an AP in an infrastructure based network acts as a PC.

Although PCF is optional, all stations are able to obey the medium access rules of the PCF, because it is based on the DCF. Stations that are able to respond to polls by the PC are called *Contention-Free-Pollable* (CF-Pollable). Only these stations are able to transmit frames according to the PCF (besides the AP).

The PCF controls the frame transfers during the so-called *Contention-Free Period* (CFP), which alterates with the *Contention Period* (CP) under the control of the DCF. The CFP is periodically repeated in time at the *Contention-Free Repetition Rate* (CFPRate) and starts with the transmission of a beacon (see Section 14.3.3.4). The beacon contains the maximum duration of the CFP (`CFPMaxDuration`), and all stations in the BSS except the PC set their NAV to CFPMaxDuration, thus guaranteeing the control of the PCF for this amount of time. Figure 14.82 shows these relations.

The PC gains control at the nominal beginning of the CFP by waiting PIFS after the medium is sensed idle instead of DIFS. It maintains control for the entire CFP by waiting a shorter time than stations using the DCF access procedure. The PC maintains a *Polling List*, which consists of the *Association Idcntificr* (AID) of the stations requesting polling. A *CF-Pollable* station may request to be added to the polling list during *Association* or *Reassociation*.

Special Data Subtypes are defined for the use during CFP in order to enable "piggybacking" of polls and acknowledgements (see Table 14.15). A *CF-Poll* is used by the PC to poll a station for the transmission of a data frame. *CF-Ack* is the acknowledgement to a successfully received frame under the PCF, either by a station or the PC. The *Null Function* is used to indicate that no

Table 14.15: 802.11 MAC Data Subtypes under the PCF

Station	Data Subtypes
PC	Data+CF-Poll, Data+CF-Ack+CF-Poll, CF-Poll, CF-Ack+CF-Poll, CF-End, CF-End+CF-Ack
PC and CF-Pollable	Data, Data+CF-Ack, CF-Ack, Null Function

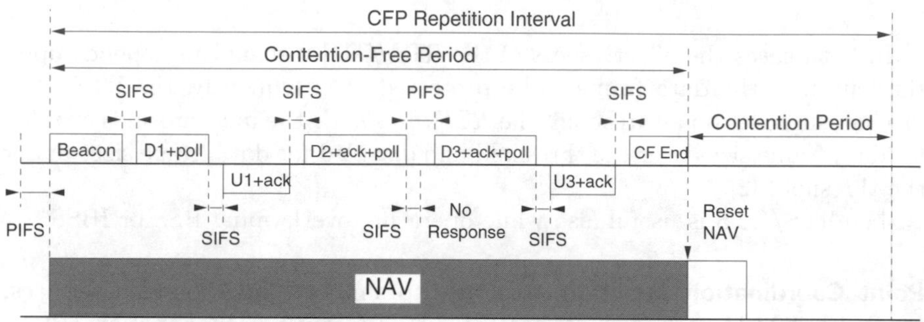

Figure 14.83: Example of PCF frame transfer

data has to be transmitted. If all stations on the polling list have been polled and no more data has to be transmitted by the PC during one CFP, the PC may prematurely stop the current CFP by sending a *CF-End*. On receiving a *CF-End*, all stations reset their NAV.

Figure 14.83 shows an example of a sequence of frame transmissions during a CFP. Usually the gap between two transmissions under the PCF is SIFS unless a station does not respond to a *CF-Poll*. In the later case the PC regains control of the medium after a PIFS.

14.3.3.4 Synchronization

All stations within a single BSS will be synchronized to a common clock while maintaining a local timer. This is done by performing a *Timing Synchronization Function* (TSF). The TSF is different for infrastructure BSS and IBSS.

The synchronization is maintained by broadcasting the TSF timer in a so-called *beacon*. Beacons are transmitted periodically at the *Target Beacon Transmission Times* (TBTT). In case the medium is sensed busy at TBTT, the transmission of the beacon is delayed as shown in Figure 14.84 [38].

Beacon Generation in Infrastructure Networks In an infrastructure network the access point is responsible for the transmission of the synchronization beacon. At each TBTT, the AP schedules a beacon as the next frame to be transmitted according to the PCF procedures described in Section 14.3.3.3.

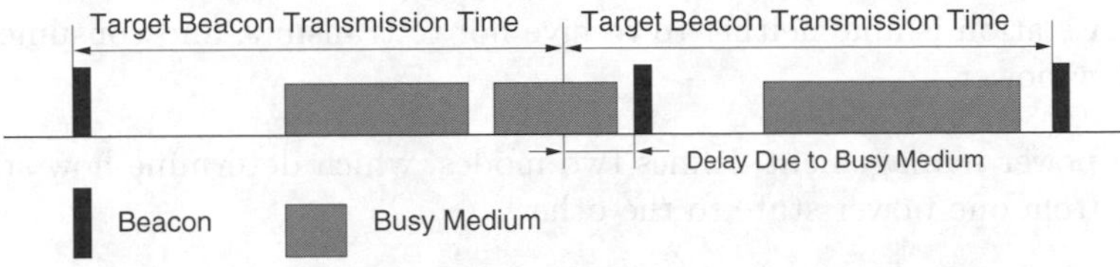

Figure 14.84: Delayed beacon transmission on a busy medium

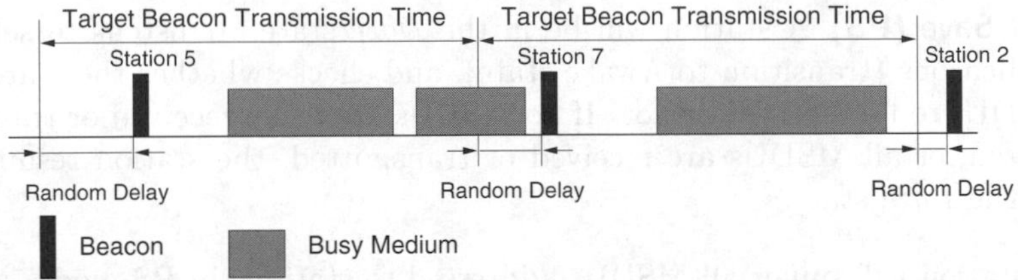

Figure 14.85: Beacon transmission in an IBSS

Beacon Generation in an IBSS In an independent BSS the beacon generation is distributed among all stations. At each TBTT all members of a IBSS perform the following procedure:

1. Suspend the decrementing of the backoff timer for any pending non-beacon transmission.
2. Calculate a random delay uniformly distributed in the range between zero and $2 \cdot CW_{min} \cdot slottime$.
3. Wait for the period of random delay similar to the backoff algorithm.
4. If a beacon arrives in between then the random delay timer and the beacon transmission are cancelled.
5. If the random delay timer expires, send a beacon.

Performing the aforementioned procedure, the transmission of the beacon in an IBSS is performed by different stations using different random delays as shown in Figure 14.85.

14.3.3.5 Power-Saving Mode

IEEE 802.11 compliant devices will most probably be battery-powered. Therefore the standard supports a power conservation mode. A station may be in one of two power states:

Awake A station is fully powered.

Doze A station is able neither to receive nor to transmit, and consumes very low power.

The power management defines two modes, which determine how stations transit from one power state to the other:

Active Mode (AM) A station will be in the *Awake* state. A station on the polling list of a PC will be in *Active Mode* for the duration of the CFP.

Power Save (PS) A station will be in the *Doze* state. It listens to selected beacons (transition to Awake state), and checks whether there are MSDUs to be received or not. If no MSDUs are to be received or transmitted, or all MSDUs are received or transmitted, the station resumes to the *Doze* state.

A station will buffer all MSDUs addressed to stations in PS mode. Multicast and broadcast MSDUs are announced by a field in the beacon called the *Delivery Traffic Indication Message* (DTIM). Only selected beacons contain a DTIM. The multicast and broadcast MSDUs are transmitted immediately after the DTIM beacon [38].

The procedures to announce unicast MSDUs to stations in PS mode are different for infrastructure and independent BSS.

Power Management in an Infrastructure BSS Stations changing power management modes have to inform the AP via a successful frame exchange. Stations having buffered cells within the AP are announced in a *Traffic Indication Map* (TIM), which is included in all beacons generated by the AP. A station receiving and interpreting the TIM may poll the AP for the buffered frames during the *Contention Period* using the *PS-Poll* frame; therefore it is not necessary that a station listens to each single beacon. CF-pollable stations do not poll for frames, since these are transmitted during the CFP.

Power Management in an IBSS As the beacons are generated by different stations (see Section 14.3.3.4) within the IBSS, the TIM is not sufficient to announce traffic to stations in PS mode. Instead, a station announcing unicast traffic to another station uses the *Ad Hoc Traffic Indication Message* (ATIM). The receiving station acknowledges the reception of the ATIM, and remains in the Awake state for the whole beacon interval, waiting for the announced MSDUs to be delivered.

All stations in PS mode have to wake up for a so-called *ATIM Window*, which follows the TBTT. During the *ATIM Window*, only beacons and ATIMs will be transmitted. After the ATIM, every station in PS mode knows whether it has to remain in the *Awake* state or is allowed to go to the *Doze* state. Figure 14.86 shows the basic operation of power management in an IBSS.

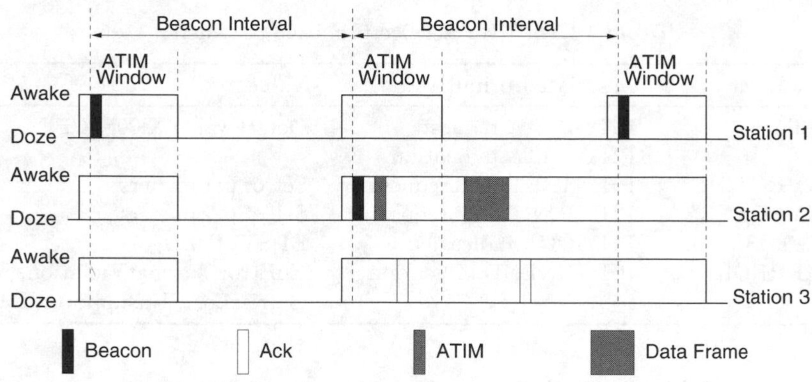

Figure 14.86: Power management in an IBSS

14.3.4 IEEE 802.11 Physical Layer Specification

14.3.4.1 Scope

The IEEE 802.11 *Physical Layer* (PHY) consists of two sublayers: the *Physical Layer Convergence Procedure* (PLCP) sublayer and the *Physical Medium-Dependent* (PMD) sublayer (see Figure 14.74). The PLCP sublayer maps the 802.11 MAC PDUs into a framing suitable for transmitting and receiving user data and management information between the associated PMD entities. The PMD system describes the method of transmitting and receiving data through a wireless medium.

14.3.4.2 Physical Layer Service Definition

Two groups of service primitives are distinguished: *service-user-to-service-provider* and *sublayer-to-sublayer* service primitives.

The *service-user-to-service-provider* service primitive is:

PHY-DATA This primitive is used to transfer octects of data between the PHY and MAC sublayers.

The *sublayer-to-sublayer* service primitives are:

PHY-TXSTART With this primitive, the start of the transmission of a MPDU is requested by the MAC layer.

PHY-TXEND The MAC layer requests the end of a transmission.

PHY-RXSTART The PHY layer indicates to the local MAC entity that the PLCP has received a valid *Start Frame Delimiter* (SFD) and PLCP header.

PHY-RXEND With this primitive, the end of a MPDU currently received is indicated to the local MAC entity.

Table 14.16: PHY service primitive parameters

Parameter	Associated primitive	Value
DATA	PHY-DATA.request PHY-DATA.indication	Octect value X'00'–X'FF'
TXVECTOR	PHY-TXSTART.request	Set of parameters
RXVECTOR	PHY-RXSTART.request	Set of parameters
STATUS	PHY-CCA.indication	BUSY, IDLE
RXERROR	PHY-RXEND.indication	NoError, FormatViolation, CarrierLost, UnsupportedRate

Table 14.17: FHSS specific service primitive parameters

Parameter	Associated parameter	Value
LENGTH	TXVECTOR, RXVECTOR	1–4095
DATARATE	TXVECTOR, RXVECTOR	1, 1.5, 2, 2.5, 3, 3.5, 4, 4.5
RSSI	RXVECTOR	0–RSSI Max

PHY-CCA The PHY layer indicates the state of the medium to the local MAC entity. Possible states are IDLE and BUSY. It is generated whenever the state of the medium changes.

PHY-CCARESET With this primitive, the MAC layer requests to reset the *Clear Channel Assessment* (CCA) state machine.

Table 14.16 lists the parameters belonging to the PHY service primitives.

The architecture of the IEEE 802.11 MAC is intended to be PHY-independent. However, some PHY implementations require the interaction with the *MAC Sublayer Management Entity* (MLME) through the PHY SAP using the PHY service primitives. Therefore the parameter lists *TXVECTOR* and *RXVECTOR* are PHY-dependent. Two fields are mandatory, namely *DATARATE*, specifying the transmit or receive data rate, and *LENGTH*, determining the length of an MPDU.

14.3.4.3 Frequency Hopping Spread Spectrum

FHSS Specific Service Parameters The *Frequency Hopping Spread Spectrum* (FHSS) dependent definitions of *TXVECTOR* and *RXVECTOR* are listed in Table 14.17.

The *Receive Signal Strength Indicator* (RSSI) is an optional parameter indicating the energy observed at the antenna used to receive the current PPDU. RSSI is intended to be used in a relative manner, since the absolute accuracy of the RSSI is not specified.

PLCP Preamble		PLCP Header			PSDU
Sync	SFD	PLW	PSF	HEC	
80 bits	16 bits	12 bits	4 bits	16 bits	Variable number of octects

Figure 14.87: FHSS frame format

Table 14.18: Frequency range, channels and hopping sets

Geographical region	Operating frequency [GHz]	Regulatory range [GHz]	Hopping set
North America	2.402 ... 2.480	2.400 ... 2.4835	79
Europe	2.402 ... 2.480	2.400 ... 2.4835	79
Japan	2.473 ... 2.495	2.471 ... 2.497	23
Spain	2.447 ... 2.473	2.445 ... 2.475	27
France	2.448 ... 2.482	2.4465 ... 2.4835	35

FHSS Frame Format The *Physical Layer Convergence Procedure* (PLCP) maps an MAC PDU into a frame format designed for the FHSS radio transceiver.

The frame format of the FHSS PLCP is shown in Figure 14.87. It consists of three parts: the PLCP Preamble, the PLCP Header and the PSDU.

The PLCP Preamble consists of a 80-bit SYNC field and a 16-bit *Start Frame Delimiter* (SFD). The SYNC field is used to detect a receivable signal, and to give frequency correction and synchronization with the packet timing. The SFD defines the frame timing.

The PLCP Header consists of a 12-bit *PSDU Length Word* (PLW), a 4-bit *PLCP Signalling Field* (PSF) and a 16-bit *Header Error Check* (HEC). The PLW determines the number of octects contained in the PSDU, the PSW defines the data rate of the PSDU, and the HEC provides an error detection facility for the PLCP Header.

FHSS Regulatory Requirements The regulatory requirements address the operating frequency range, the regulatory range and the hopping sequences, as listed in Table 14.18.

FHSS Modulation and data rates The FHSS PMD operates at the following data rates: 1 Mbit/s, 2 Mbit/s and 11 Mbit/s (called 802.11b). 1 Mbit/s is mandatory, whereas 2 Mbit/s is optional. For operation at 1 Mbit/s a two-level *Gaussion Frequency Shift Keying* (GFSK) modulation with a nominal bandwidth bit period (BT) = 0.5 is used. For operation at 2 Mbit/s a four-level GFSK modulation with a nominal bandwidth bit period (BT) = 0.5 is used.

PLCP Preamble		PLCP Header				PSDU
Sync	SFD	SIGNAL	SERVICE	LENGTH	CRC	
144 bits	16 bits	8 bits	8 bits	16 bits	16 bits	Variable number of octects

Figure 14.88: DSSS frame format

Table 14.19: Frequency range, channels and power levels

Geographical region	Frequency channels [GHz]	Number of channels	Transmit power
North America	2.412 ... 2.462	11	1000 mW
Europe	2.412 ... 2.472	13	100 mW (EIRP)
Japan	2.484	1	10 mW/MHz
Spain	2.457 ... 2.462	2	100 mW (EIRP)
France	2.457 ... 2.472	4	100 mW (EIRP)

FHSS Receiver Sensitivity The sensitivity is defined as the minimum signal level required for a *Frame Error Ratio* (FER) of 3% for PSDUs of 400 octects generated with pseudorandom data. For the 1 Mbit/s FHSS the sensitivity should be less than or equal to −80 dBm. For the 2 Mbit/s FHSS the sensitivity should be less than or equal to −75 dBm.

14.3.4.4 Direct Sequence Spread Spectrum

DSSS Frame Format The *Physical Layer Convergence Procedure* (PLCP) maps an MAC PDU into a frame format designed for the DSSS radio transceiver.

The frame format of the DSSS PLCP is shown in Figure 14.88. It consists of three parts: the PLCP Preamble, the PLCP Header and the PSDU.

The PLCP Preamble consists of a 128-bit SYNC field and a 16-bit *Start Frame Delimiter* (SFD). The SYNC field is used to detect a receivable signal, and to give frequency correction and synchronization with the packet timing. The SFD defines the frame timing.

The PLCP Header consists of an 8-bit SIGNAL field, an 8-bit SERVICE field, a 16-bit LENGTH field and a 16-bit CRC field. The SIGNAL field defines the data rate of the PSDU, the SERVICE field is reserved for future use, the LENGTH indicates the number of microseconds required to transmit the PSDU, and the CRC field allows error detection in the PLCP Header.

DSSS Regulatory Requirements The regulatory requirements address the operating frequency range, the number of channels operated and the maximum transmit power levels, as listed in Table 14.19.

PLCP Preamble		PLCP Header				PSDU
Sync	SFD	DR	DCLA	LENGTH	CRC	
57-73 slots	4 slots	3 slots	32 slots	16 bits	16 bits	Variable number of octects

Figure 14.89: IR frame format

DSSS Modulation and Data Rates The symbol duration is exactly 11 chips long. The following 11-chip Barker sequence is used as the PN code sequence:

$$+1, -1, +1, +1, -1, +1, +1, +1, -1, -1, -1$$

The *Direct Sequence Spread Spectrum* (DSSS) PMD operates at two data rates: a *basic access rate* and an *enhanced access rate.* The *basic access rate* is based on 1 Mbit/s *Differential Binary Phase Shift Keying* (DBPSK) modulation. The *enhanced access rate* is based on 2 Mbit/s DQPSK modulation.

Recently the transmission speed has been extended to up to 11 Mbit/s. This extension is defined in the IEEE 802.11 supplement standard *Higher-Speed Physical Layer Extension in the 2.4 GHz Band*, known as IEEE Standard 802.11b-1999. IEEE 802.11b defines the *High Rate Direct Sequence Spread Spectrum* (HR/DSSS) transmission mode with a chip rate of 11 Mchip/s, providing the same occupied channel bandwidth and channelization scheme as legacy DSSS. The higher data rate is achieved through a transmission mode based on 8-chip *Complementary Code Keying* (CCK) modulation. The code set of complementary codes is richer than the set of Walsh codes. At 11 Mbit/s, the spreading code length is 8 and the symbol duration is 8 rather than 11 chips. Data bits encode the symbols with DQPSK and QPSK.

DSSS Receiver Sensitivity The sensitivity is defined as the minimum signal level required for a *Frame-Error Ratio* (FER) of 8 % for PSDUs of 1024 bytes. The sensitivity is specified for the 2 Mbit/s DQPSK modulation, and should be less than or equal to −80 dBm.

14.3.4.5 Infrared

The infrared (IR) PHY uses light in the 850–950 nm range. The PHY IR is not directed, but uses both line-of-sight and reflected energy. The operation is limited to indoor environments and the typical range is around 10 m. IEEE 802.11 IR devices are able to deal with disturbances from other common consumer devices operating in the same spectral range, such as infrared remote controls.

IR Frame Format The *Physical Layer Convergence Procedure* (PLCP) maps an MAC PDU into a frame format designed for the IR radio transceiver.

The frame format of the IR PLCP is shown in Figure 14.89. It consists of three parts: the PLCP Preamble, the PLCP Header and the PSDU.

The PLCP Preamble consists of a SYNC field that is 57–73 slots long and a 4-slot *Start Frame Delimiter* (SFD). A slot corresponds to one of the L-positions of a symbol, and has a 250 ns duration. The SYNC field is used to perform clock recovery (slot synchronization), automatic gain control, signal–to–noise ratio estimation, and diversity selection. The SFD is provided to perform bit and symbol synchronization.

The PLCP Header consists of a 3-slot *Data Rate* (DR) field, a 32-slot *DC Level Adjustment* (DCLA) field, a 16-bit LENGTH field and a 16-bit CRC field. The DR field defines the data rate of the LENGTH, CRC and PSDU fields, the DCLA field is required to allow the receiver to stabilize to the dc level, the LENGTH indicates the number of microseconds required to transmit the PSDU, and the CRC field allows an error detection in the PLCP Header.

IR Regulatory Requirements Worldwide, there are currently no frequency and bandwidth allocation regulatory restrictions on infrared emissions.

IR Modulation and data rates The IR PMD operates at two data rates: a *basic access rate* and an *enhanced access rate*. The *basic access rate* is based on 1 Mbit/s 16-*Pulse Position Modulation* (PPM). A group of 4 data bits is mapped to one 16-PPM symbol. The *enhanced access rate* is based on 2 Mbit/s 4-PPM. A group of 2 data bits is mapped to one 4-PPM symbol.

IR Receiver Sensitivity The sensitivity is defined as the minimum irradiance at the photodetector plane required for a *Frame-Error Ratio* (FER) of $4 \cdot 10^{-5}$, with a PSDU of 512 octets and with an unmodulated background IR source between 800 nm and 1000 nm, with a level of 0.1 mW/cm^2. For 1 Mbit/s the sensitivity should be less than or equal to $2 \cdot 10^{-5}$ mW/cm^2. For 2 Mbit/s the sensitivity should be less than or equal to $8 \cdot 10^{-5}$ mW/cm^2.

14.3.5 Traffic Performance of WLAN IEEE 802.11a

For results of the analysis and simulation of the traffic performance of IEEE 802.11a please see Section 14.2.9 and Chapter 15.

14.4 Bluetooth*

In 1994 the Swedish company Ericsson Mobile Communications commissioned a study on the feasibility of a cost-effective low-energy radio interface between mobile telephones and their accessories. The objective of the study was to investigate the possibility of eliminating the need for connection cables. The company Nokia has been experimenting with infrared as a short-distance communication medium since 1990. Portable devices currently use the infrared

*With the collaboration of Ian Herwono

interface of the *Infrared Data Association* (IrDA) for transmission. Although this is a cost-effective option, it has some disadvantages:

- Restricted range (typically one or two metres).
- Direction-sensitive and dependent on direct line-of-sight link.
- In principle, can only be used between two devices.

In contrast, radio communication offers a wider range of coverage, can penetrate through objects and different materials and links many devices simultaneously.

Consequently, the *Bluetooth Special Interest Group* (BSIG), consisting of leading companies in the areas of mobile radio communication, laptops and DSP, was formed in 1998. The objective of this consortium is to develop a *de facto* standard for the interface and its control software to ensure the compatibility of the equipment of different manufacturers. The first Bluetooth devices have been shown at CeBit 2000.

14.4.1 Frequency Range and Transmit Power Control

The frequency band being used is the licence-free *Industrial, Scientific and Medical* (ISM) frequency band that falls in the 2.4–2.4835 GHz range in the USA and Europe (only partially available in France and Spain) and in the 2.471–2.497 GHz range in Japan. The frequency modulation is *Gaussian Frequency Shift Keying* (GFSK) with a modulation index between 0.28 and 0.35. Transmit power control is between 1 and 100 mW, the range at 10–100 m.

14.4.2 Transmission Channel

The Bluetooth system provides point-to-point as well as point-to-multipoint connections. In the case of the latter, many Bluetooth devices share one channel. In this channel, connecting a *piconet*, the initializing unit is the master and the other ones are slaves. A total of up to 7 active slaves is possible.

In Europe and in the USA a 83.5 MHz frequency bandwidth in which 79 channels at 1 MHz intervals are defined is available. In Japan, France and Spain 23 channels are defined. The transmission channel is derived from a pseudo-random *Frequency Hopping Spread Spectrum* (FHSS) method, spread over the 79, or 23, frequencies. The sequence is unique for each piconet, determined by the device address of the master. The Bluetooth timer of the master determines the cycle.

A channel is divided into consecutively-numbered time slots of 625 µs length, each slot having a different frequency, that is 1600 hops/s. This should ensure that simultaneous interference, such as from baby phones, garage door openers, cordless telephones and, particularly, microwave oven equipment [29],

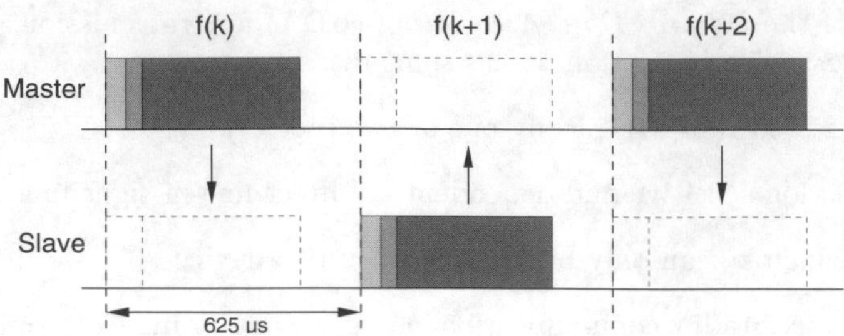

Figure 14.90: TDD method used with Bluetooth

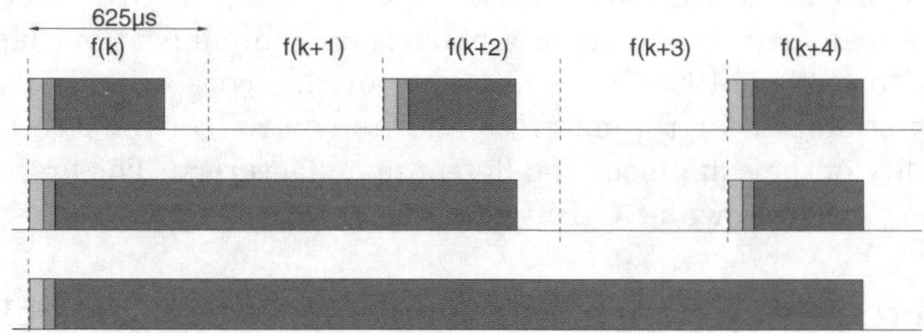

Figure 14.91: Multislot packets

does not remarkably affect transmission. All Bluetooth devices participating on the piconet are time and hop-synchronized. The TDD method is applied, i.e., master and slaves are alternating transmitting (see Figure 14.90).

The master sends in the even-numbered slots, the slaves in the odd-numbered ones. A packet starts with the beginning of the slot and can also spread to over five slots. During this time the frequency remains constant (see Figure 14.91).

14.4.3 Types of Connection

Two different types of connection can be established between master and slave(s):

- *Synchronous Connection-Oriented* (SCO)
- *Asynchronous Connectionless* (ACL) packet-oriented

14.4.3.1 SCO Connections

An SCO connection is a symmetric point-to-point connection between master and a specific slave. It typically transmits time-critical information, such as speech. A master can establish up to three SCO connections to one or different slaves. A slave can support up to three SCO connections from the same master and two from different masters.

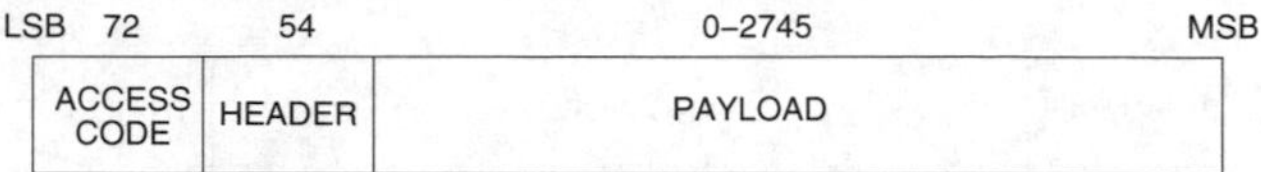

Figure 14.92: Standard packet format

The master sends SCO packets at regular intervals, called SCO intervals. The SCO slave is allowed to respond in the slave-to-master slots that follow unless another slave was addressed in the preceding master-to-slave packet. A slave is permitted to transmit in the reserved SCO slots even if it cannot read the address in the packet header.

The master uses an SCO-Setup message over the LM protocol to establish an SCO connection. This message contains time parameters.

14.4.3.2 ACL Connections

The master can exchange packets with any slave on a per-slot basis in the slots that are not reserved for SCO connections. ACL connections are packet-switched connections for the master and all slaves of a piconet. Only *one* ACL connection can exist between master and slave. Most ACL packet types use packet retransmission based on an ARQ protocol for data integrity purposes.

A slave is only permitted to send an ACL packet if it was addressed in the preceding master-to-slave packet. If it cannot read the address, it is not allowed to transmit. ACL packets that are not specifically addressed are broadcast packets and can be read by all slaves. No transmission takes place if no data is being sent or no polling signal has been received by a slave.

14.4.4 Packet Description

All packets have a standard format the details of which vary according to the different packet types involved (see Figure 14.92 and Section 14.4.4.2).

14.4.4.1 Packet Format

Each packet—with the exception of *Frequency Hop Synchronization* (FHS)—begins with a 72 bit wide `Access Code`. This code is used for synchronization, frequency offset compensation and identification. All packets in the same piconet use the same access code that identifies them.

With *inquiry* (see Section 14.4.8.2) and *paging* (see Section 14.4.8.3) the access code itself functions as a signalling message without a header and a payload. There are three different types of access code:

Channel Access Code, CAC The CAC code identifies a piconet and is contained in all packets that are exchanged over a piconet channel.

Device Access Code (DAC) This code is used for special signalling processes, e.g., paging and page response.

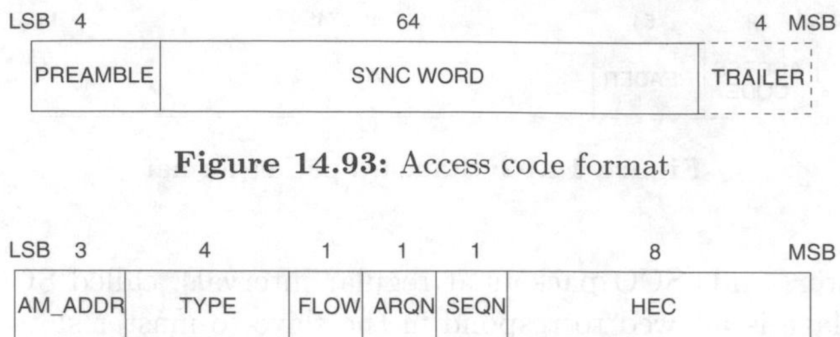

Figure 14.93: Access code format

Figure 14.94: Packet header format

Inquiry Access Code (IAC) There are two different IAC codes. One is the *General Inquiry Access Code* (GIAC) that is the same in all devices and can be used to determine which other devices are in range. The other one, the *Dedicated Inquiry Access Code* (DIAC), may only be used within a selected group of devices.

The 4-bit wide `Preamble` in the access code is used to compensate for the propagation delays (see Figure 14.93).

The 64-bit wide `Sync Word` is derived from the 24-bit wide *Lower Address Part* (LAP) from the Bluetooth device address: for the *Channel Access Code* (CAC) from the LAP of the master, for GIAC and DIAC from the LAP of selected devices and for the DAC from the LAP of the slave. The construction guarantees a large Hamming distance between Sync words.

The `Trailer` is attached as soon as the packet header follows the access code. This is typically the case with the CAC but the `Trailer` is also attached to the DAC or the IAC if these codes are used in FHS packets.

The packet header is shown in Figure 14.94. It contains *Link Control* (LC) information and consists of six fields. The entire header consists of 18 bits that are coded with a 1/3 FEC (see Figure 14.95) to give it a size of 54 bits.

The parameter *Active Member Address* (`AM_ADDR`) makes a distinction between the active mobile subscribers in a piconet. It is possible for 7 active slaves to be connected to a single master, with each individual slave identified by its temporary 3-bit wide `AM_ADDR`. The address is contained in the master-to-slave as well as the slave-to-master packets. Packets with a 0 address are broadcast packets from a master for all slaves or they are FHS packets. Slaves that leave the piconet or are parked (see Section 14.4.8.4) disclose their `AM_ADDR` and receive a new one when they deregister from the piconet.

The `TYPE` field distinguishes between the 16 different packet types. It also establishes the physical link (ACL or SCO) and the slot size of a packet.

The `FLOW` bit controls packets over the ACL link. If the memory of a receiver is full, the `FLOW` bit sends a 0 bit to it in order to stop transmission temporarily.

ARQN *(Automatic Repeat Request Indication)* informs the source whether a packet was received error-free. If errors occurred at the receiver, NAK is set (ARQN=0) and the packet is retransmitted as the next one.

The SEQN *(Sequential Numbering Scheme)* bit ensures that packets are numbered consecutively. The bit is inverted for each new packet transmitted with a CRC and compared with the bit of the next packet. A packet that arrives twice is detected.

Each header has an *Header Error Check* (HEC) to check integrity.

14.4.4.2 Packet Types

Packet types are differentiated according to the connections for which they are used. There are packet types for SCO and ACL connections and shared packet types.

A distinction is made between five shared packet types.

ID packets contain a DAC or an IAC and have a fixed length of 68 bits.

NULL packets have no payload and consist of 126 bits. The receiver sends the packet back to the sender with information about the success of the last transmission or the status of the receiver's memory.

POLL packets are similar to NULL packets but have no payload. However, these packets need an acknowledgement from the receiver. They can be used by the master to poll slaves in the piconet; the slaves have to respond even if they have no information to send.

FHS packets are special control packets that include the Bluetooth device address and the current time of the sender. The payload contains 144 information bits plus a 16-bit wide CRC code. These bits are then 2/3-FEC-coded again so that the overall size is 240 bits. FHS packets are used for frequency-hop synchronization when a piconet has not yet been set up or when an existing piconet is being changed.

DM1 packets support control messages for each type of connection. They can also transmit regular user data. All packets in a shared-packet type group are one slot large.

SCO packets have no CRC and are not retransmitted if an error occurs. They are typically used for voice transmission at 64 kbit/s. SCO packets likewise comprise one slot.

HV1 (high-quality voice) packets transmit 10 bytes of information which are protected by a 1/3-FEC. There is no CRC and the payload has a fixed size of 240 bits.

HV2 packets transmit 20 bytes of information which is protected by a 2/3-FEC. The payload is 240 bits in size.

Table 14.20: ACL packet types

Type	User payload (bytes)	FEC	CRC	Symmetric max. rate (kb/s)	Asymm. max. rate (kb/s) Forward	Reverse
DM1	0–17	2/3	yes	108.8	108.8	108.8
DH1	0–27	no	yes	172.8	172.8	172.8
DM3	0–121	2/3	yes	258.1	387.2	54.4
DH3	0–183	no	yes	390.4	585.6	86.4
DM5	0–224	2/3	yes	286.7	477.8	36.3
DH5	0–339	no	yes	433.9	723.2	57.6
AUX1	0–29	no	no	185.6	185.6	185.6

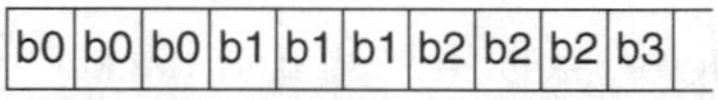

b0	b0	b0	b1	b1	b1	b2	b2	b2	b3

Figure 14.95: 1/3-FEC scheme for the header

HV3 packets are structured in a similar way but transmit 30 bytes of information and have no FEC.

DV (data voice) packets—a combination of data and voice information—are also defined.

14.4.5 Error Correction

Three error correction methods have been defined for Bluetooth:

- 1/3-FEC
- 2/3-FEC
- ARQ for data

The aim of using FEC for the payload is to reduce the number of repeated transmissions. However, in an interference-free environment FEC reduces throughput. Therefore, an option is available to eliminate FEC through the use of DH instead of DM packets for ACL connections (see Table 14.20) and instead of HV packets for SCO connections.

14.4.5.1 1/3-FEC

Irrespective of packet type, the packet header is always protected by a 1/3-FEC because it contains important information (see Figure 14.95).

14.4.5.2 2/3 FEC

The 2/3-FEC is used in DM packets, in the data field of DV packets, in FHS packets and in HV2 packets. A (15,10) Hamming code with 5 parity bits attached to each block of 10 information bits is used. This code can correct all simple errors in any code word and detect duplicate errors.

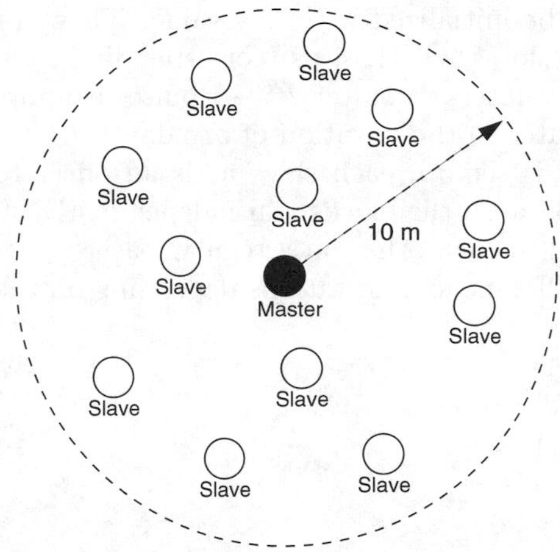

Figure 14.96: Piconet with master and slaves

14.4.5.3 ARQ

An ARQ protocol is used with DM and DH data packets and the data field of DV packets. It ensures that retransmission continues until an acknowledgement from the receiver confirms error-free receipt (or a time-out has been exceeded). The information is contained piggy-back in the packet header of the returned packet. A CRC is added to the packet to establish whether the payload is correct. This method only functions with payloads with an CRC and not with packet headers and payloads from voice packets.

14.4.6 Master–Slave Definition

A channel in a piconet is totally controlled by the master. Its Bluetooth device address (`BD_ADDR`) specifies the frequency-hopping sequence and the CAC. Its system clock determines the phase in the hopping sequence and sets the timing. Furthermore, the master controls the channel through polling. According to the definition, the master is the initializing unit that sets up a connection to one or more slaves. However, the master and the slaves have the same structure, i.e., any of them can be the master of a piconet. This master–slave role can even still be changed after the setup of a connection.

14.4.7 The Bluetooth System Clock

Each Bluetooth unit has an internal system clock that determines the timing and hopping of the sender. It is based on a separate independent clock (*native clock*) which is never set or stopped. For synchronization with other entities, offsets are simply added to produce temporarily synchronous Bluetooth clocks. These Bluetooth system clocks have no relationship to the time of the day;

they can therefore be initialized with any value. They are broken down into at least half of one slot, thus 312.5 µs (corresponding to 3.2 kHz). The clock runs with a 28-bit counter, i.e., after $2^{28} - 1$ pulses it jumps back to zero; this approximately equates to the duration of one day.

When a piconet is set up, each slave adds an offset to its own clock for synchronziation. Because the clocks run independently of one another, they have to be adjusted to each other on a regular basis.

The clock has different manifestations depending on the different states of a Bluetooth unit:

CLKN: Native clock

CLKE: Estimated clock

CLK: Master clock

The *Native Clock* (CLKN) runs independently and is a reference for the other forms. In states with heavy activity an oscillator with a high level of accuracy drives the clock; otherwise a less accurate oscillator is used.

Estimated Clock (CLKE) and *Native Clock* (CLK) are derived from the CLKN through the addition of an offset. A clock becomes a CLKE when a unit during paging estimates the clock of the receiver in order to run more closely with the CLKN of the receiver. This process aims at accelerating connection setup.

The CLK controls activities in the piconet. All Bluetooth units use this clock to control the transmission and receipt of data. Slaves add an appropriate offset in order to synchronize their CLKNs with the CLKN of the master. The master does not have to add anything because the CLK is identical to its CLKN.

14.4.8 The States of a Bluetooth Entity

Figure 14.97 presents an overview of the states in which a Bluetooth unit can find itself.

14.4.8.1 Standby

`Standby` is the normal state for a Bluetooth unit in low-power mode. Only the CLKN runs with the accuracy of the *Low Power Oscillator* (LPO).

The controller is allowed to exit from `Standby` state to scan page or inquiry messages or to enter the page or inquiry state itself. After it has responded to a page message, the unit enters the `Connection` state as a slave. After a successful paging attempt it enters this state as a master.

14.4.8.2 Inquiry

Inquiry and paging procedures are used for setting up new connections. The inquiry procedure enables a unit to determine which other communicating en-

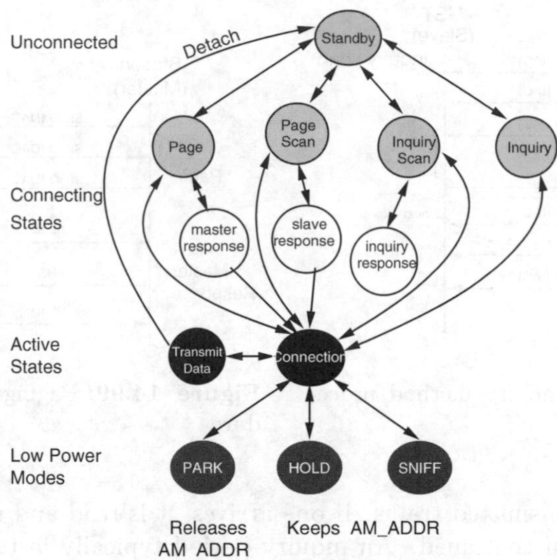

Figure 14.97: Overview of Bluetooth states

tities are operating in its range, along with their device addresses and CLKN. In this state it collects this information about all entities that are responding to its message.

In `Inquiry Scan` state a unit scans for the inquiry messages (IAC) of other units for at least 18 slots using a constant hop frequency. The inquiry procedure uses 32 different inquiry-hop frequencies equivalent to an *inquiry hopping sequence*. These frequencies are stipulated by the GIAC (or additionally by the DIAC). This phase originates from the CLKN of the initializing unit and changes every 1.28 s. If the unit was previously in `Connection` state, existing ACL connections may be interrupted. If the inquiry message of a sender is detected during an inquiry period, the Bluetooth unit moves into `Inquiry-Response` state.

In `Inquiry Response` state the Bluetooth unit responds to inquiry messages with a conventional FHS packet containing its own parameters. The problem could arise that several units in the immediate vicinity of the sender are replying at the same time. Although this is not likely since each unit has an independent clock and therefore uses a different phase for the inquiry-hopping sequence, the following protocol is also followed: Each unit waits for a random number of slots and then sends a response.

In `Inquiry` state the initializing unit transmits the IAC (GIAC or DIAC) at 10 ms intervals, called trains, to other feasible units in the range. A distinction is made between two trains, A and B, each containing 16 hops from the 32 inquiry-hop frequencies. The inquiry-hopping sequence is derived from the GIAC of the sender. The sending unit scans for inquiry-response messages

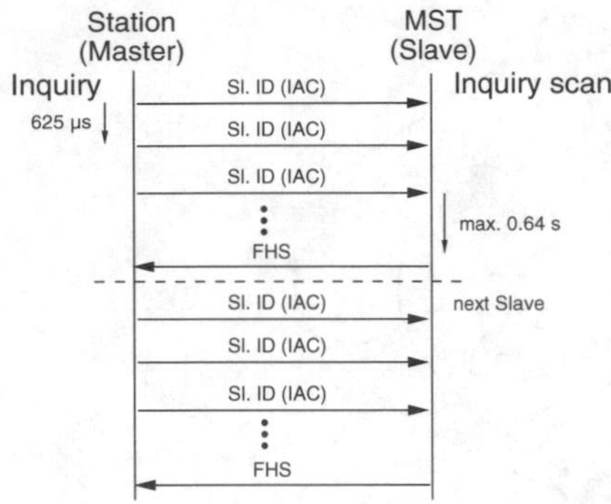

Figure 14.98: Inquiry method procedure

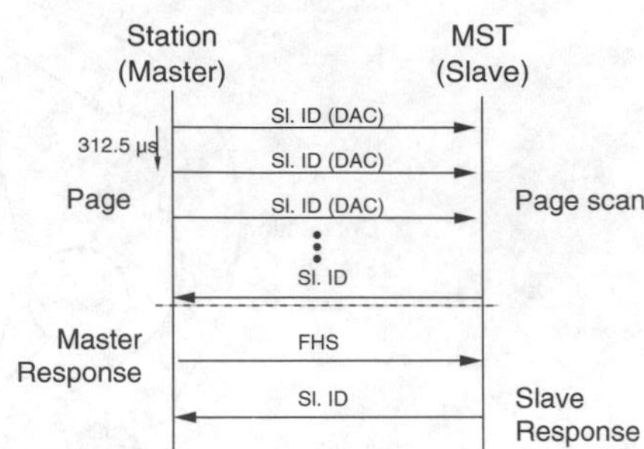

Figure 14.99: Paging method procedure

between the transmitted trains. If one arrives, it is read and transmission of the train is then continued. An inquiry period typically lasts 5.12 s. If the unit was previously in `Connection` state, existing ACL connections may be interrupted.

Inquiry messages are shown in Figure 14.98.

14.4.8.3 Paging

Paging is used to set up connections (see Figure 14.99). Different methods exist, and each Bluetooth device must support one of them. The paging function is applied when devices make initial contact with one another or when paging immediately follows an inquiry.

In `Page Scan` state a Bluetooth unit waits for its own DAC during an 18-slot *scan window*. The frequency that is selected to match the *page hopping sequence* remains constant. The sequence consists of 32 hops (or 16 in a limited hop system), with each frequency only occurring once. The hopping sequence is determined by the `BD_ADDR`, whereas the phase corresponds to the bits CLKN_{16-12} of the unit, i.e., other frequencies are used every 1.28 s. If a suitable DAC arrives, the unit moves into `Slave Response` state. If the unit was previously in `Connection` state, existing ACL connections could be interrupted.

A master enters `Page` state to activate a slave and to set up a connection to the slave. Therefore, it again transmits the DAC of the slave to various hop channels. Because the CLKN of the two units are not synchronized, the master does not know when and at which frequency the slave is being activated. Therefore, it sends a train of identical DACs on different frequencies and keeps scanning for responses until one arrives. The page-hopping sequence is derived from the `BD_ADDR` of the slave. For this phase the master calculates an estimate of the CLKN of the slave (CLKE) which it can sometimes derive from earlier connections or from the inquiry method. Although the hopping sequence of both units may match, the phases can be different enough that they do not meet up with each other. Therefore, the master also sends frequencies that are either immediately before or after the estimated one. The procedure is accelerated because two frequencies are used in each slot. Because the master does not know the time when the slave is moving into `Page-Scan` state, it first resends an A train (10 ms), waits for a response and then sends a B train, a procedure similar to the one for an inquiry. If the master receives a response in the meantime, it moves to `Master Response` state. Existing ACL connections could be interrupted for the paging.

In `Slave Response` state the slave responds exactly one slot (625 µs) after it has received a message from the master. For the response it sends its own DAC on the same frequency which also received message. It is activated one-half slot later and waits for the arrival of an FHS packet from the master. If it does not receive it, after two slots it selects the next master-to-slave frequency, applicable to the page-hopping sequence. If no response is received after by the time `pagerespTO`, the slave returns to `Page-Scan` state. If it receives an FHS packet from the master, it sends its DAC as an acknowledgement. It then converts to the values of the CAC and the CLK of the master that it received in the FHS packet. Finally, it moves into `CONNECTION` state.

Paging typically takes 0.64 s. See Figure 14.99 for an illustration of the paging process.

14.4.8.4 Connection

In this state connections are set up and packets can be sent back and forth. The CAC and the CLK of the master are used in the piconet. The hopping scheme is oriented towards a *channel hopping sequence*, i.e., 79 (23 in a reduced case) frequencies are used. The `Connection` state begins with a Poll packet sent by the master to check that a switchover has been made to its timing and *channel frequency hopping*. The slave may respond with any packet. If the slave does not receive the Poll packet or the master does not receive a response for a `newconnectionTO` number of slots, then both return to `Page/Page-Scan` state. The first information packets contain control messages characterizing the connection and providing more details about the communicating Bluetooth units. These messages are exchanged between the *Link Manager* (LM). The transmission of user data can now commence.

The `Connection` state is terminated by a Detach or a Reset command. Detach is used for terminating a connection in the normal way, i.e., all configuration files in the Bluetooth link controller remain valid. After a Reset, on the other hand, the controller must be reconfigured.

There are different ways in which Bluetooth units can be in `Connection` state:

- In `Active Mode` the Bluetooth unit is an active participant in the piconet, i.e., the master controls transmission based on the requirements of the slaves. In addition, it regularly transmits packets to the slaves to ensure that they remain synchronized. Any packet type suffices for this purpose because all that is required is the CAC. The slaves only read the master-to-slave slots that are addressed to them. However, if they participate in an ACL connection, they must listen in on each ACL slot. Based on the `Type Indication` in the packet header, they have information about how many slots a packet comprises.

- In `Sniff Mode` the slave no longer has to listen in on each ACL slot but only on certain slots, called sniff slots. Consequently the master only transmits to the slaves in these slots. The gap between sniff slots is called T_{sniff}, the number of them D_{sniff}. These parameters are provided over the *Link Manager Protocol* (LMP).

- The ACL connection of a slave can be transferred to `Hold Mode`, which means this connection is no longer supported (but the SCO connection continues to be so). The purpose of the change to this mode is to create capacity for *scanning*, *paging* and *inquiry*, or to enable the slave to take part in another piconet. The Bluetooth unit can also move into a Low-Power-Sleep state. Before a slave moves into `Hold Mode`, it comes to an agreement with the master on how long it will remain there. At the point in time `holdTO` it is activated again and synchronized and waits for further instructions from the master.

- If a slave does not have to participate in a piconet but wants to remain synchronized, it can change to `Park Mode`—a low-power state with little activity. In contrast to the states described above, in this state it provides its `AM_ADDR`. In return it receives two new addresses:

  - `PM_ADDR`: *Parked Member Address* (8 bit)
  - `AR_ADDR`: *Access Request Address* (8 bit)

  A `PM_ADDR` differentiates a parked slave from other parked slaves. It is required for master-initiated *unparking*. The slave uses `AR_ADDR` for the unparking initiated by itself. The parked slave wakes up at regular times so that it can listen in on the piconet in order to resynchronize and scan for broadcast messages. To make this possible, the master sets up a *beacon channel* (described in detail below). In addition to providing the possibility of low power usage, park mode is used to link more than seven slaves with the master. The number of parked and active slaves can be varied as needed, allowing the virtual connection of up to 255 slaves if `PM_ADDR` is used and even many more than that with `BD_ADDR`. There is no limit to the number of parked slaves possible.

A *beacon channel* consists of a beacon slot or a train of equally wide beacon slots that are transmitted periodically at fixed intervals. A beacon channel is illustrated in Figure 14.100.

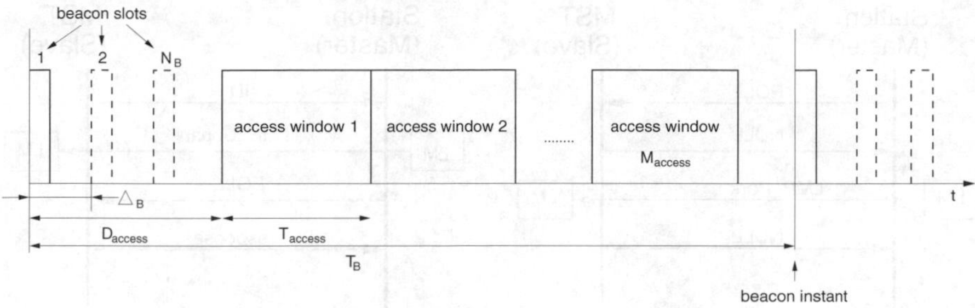

Figure 14.100: Beacon-channel

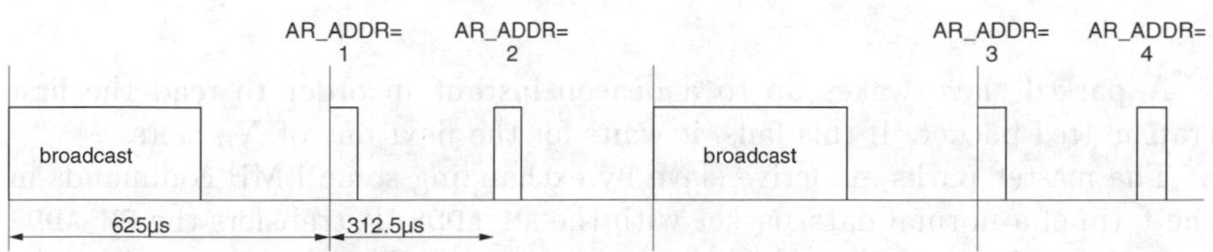

Figure 14.101: Beacon access window

The *beacon instant* serves as a *beacon timing* reference. N_B and T_B are selected so that sufficient beacon slots are available for a parked slave to synchronize at a specific time in an interference-prone environment. In `Park` state the slave receives the beacon parameters through an LMP command. In addition, the timing of the beacon instant is indicated by an offset Δ_B. A beacon channel has four aims:

- Transmission of master-to-slave packets for resynchronization
- Provision of commands to parked slaves for change in parameters
- Provision of general broadcast commands for parked slaves
- Reparking of one or more slaves

Because a slave can synchronize with any packet which has the appropriate CAC, the beacon slots do not need to contain any specific broadcast packets; any packet can be used, e.g., a NULL packet if no information is being transmitted. However, if actual broadcast information is being transmitted to the parked slaves, the first packet in the message must be repeated in each beacon slot.

Slaves can use the *beacon access window* to send their requirements for unparking to the master. For greater reliability the beacon access window is repeated M_{access} times. A beacon access window is shown in Figure 14.101.

A slave is allowed to transmit in the half slot applicable to its `AR_ADDR` if the master has transmitted a broadcast packet beforehand.

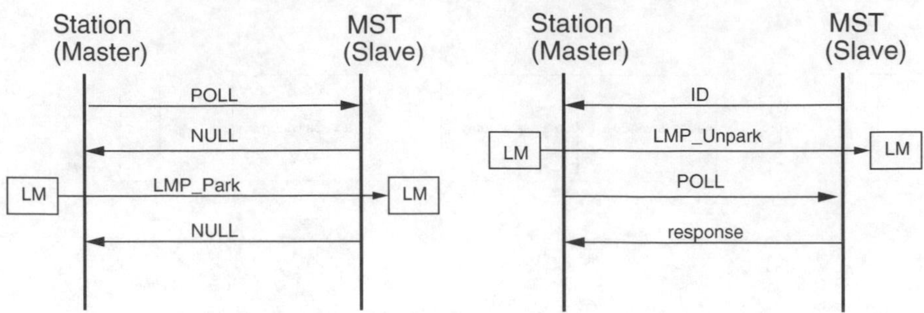

Figure 14.102: Parking procedure

Figure 14.103: Unparking procedure

A parked slave wakes up to a beacon instant in order to read the first transmitted packet. If this fails, it waits for the next one of N_B slots.

The master parks an active slave by exchanging some LMP commands in the form of a normal data packet with the AM_ADDR. It transfers the PM_ADDR and a AR_ADDR as well as the beacon parameters. The slave relinquishes its AM_ADDR and moves into Park state. A master can only park one slave at a time. A master or even a slave can activate unparking (see Figure 14.102).

In *master-activated unparking* the master sends an LMP-Unpark command with the new AM_ADDR in a broadcast packet to a specific slave. This broadcast packet can also contain different PM_ADDR, which means that many slaves can be activated simultaneously. The addressed slave moves into active state and waits for a packet containing its AM_ADDR from the master. This should be a Poll packet. The slave then responds sending any sort of packet. If a Poll packet is not received by the time newconnectionTO, the slave returns to Park state with the old beacon parameters. If a master does not receive a response to its Poll packet, it repeats the unparking process.

If a slave decides to become active in Park state, it sends an AccessRequest message (its own ID with the DAC of the master) in the half slot containing its AR_ADDR. It then waits for a LMP-Unpark command from the master. If one is not received, it repeats the process in the following access window so long until it receives an Unpark command. If a command is still not received after the last (M_{access}) access window and after an additional time N_{poll}, the slave is allowed to return to sleep mode. However, if it receives the command for unparking, it becomes active and waits for a Poll packet from the master.

14.4.9 Scatternet

Many piconets can provide coverage to the same area. Each piconet has a different master; consequently, each one has a different hopping sequence and phase, independent of the others. The more piconets that overlap, the greater is the probability of collisions. This results in reduced performance, which is the case with all FHSS systems.

Time-multiplexing enables a Bluetooth device to participate in two or more overlapping piconets. It should use the corresponding master device address for the applicable channel and a clock offset to ensure it has the correct phase. This enables the device to operate as a slave in many piconets but as master only in one. If two piconets are synchronized with the same master and use the same hopping sequence, they count as the same piconet. A group of piconets in which connections exist between the different piconets is called a *scatternet.* A master or a slave can become a slave in another piconet if it is paged by the master of the respective piconet. Because the paging unit is automatically the master, the master–slave role can, if necessary, be switched.

If only ACL connections exist, a unit can announce to the immediate piconet that it wants to change to `Hold` or `Park` mode so that it can be active in these modes in other piconets. To do so it merely has to change the piconet parameters. In `Sniff` mode the slave can be active between the sniff slots in other piconets. However, the Bluetooth unit must adapt its `CLKN` each time in the scatternet through the addition of an offset because the masters of the different piconets are not synchronized. Due to the independent drift of the CLKs of the masters, regular updates are also necessary.

14.4.10 Bluetooth Security

All Bluetooth units incorporate the same authentication and encryption procedures to protect against fraudulent use and to ensure data protection. These include a public accessible unique address for each user, two secret keys and a random number that differs for each new transaction.

If two Bluetooth units want to use encryption to communicate with one another, the connection between them must be set up in a special way. The prerequisite is an existing connection between both parties. If this is the first time both units are using encryption in their communication, an *initialization key* (K_{init}) derived from the device address (`BD_ADDR`) of the initializing unit (*claimant*), a PIN code and a random number from the verifying unit *(verifier)* need to be generated. This K_{init} is used in the authentication. If this is successful, a new key called a unit key (K_A) is generated, permanently stored and (usually) never changed. The unit key is then used in future connections set up between these units. The key is exchanged and the unit with limited memory capacity provides its key as the *link key.* The *encryption key* (K_C) is then derived from this key, a 96-bit wide *Ciphering Offset Number* (COF), and a 128-bit wide random number (see Figure 14.104).

14.5 Performance of Bluetooth

If Bluetooth is used for establishing (cost-effective) public wireless networks in form of pico- and scatternets, a number of Bluetooth base stations or access points—each of them having a coverage radius of up to 10 m—can be positioned in such areas like exhibition halls, supermarkets, or museums. By

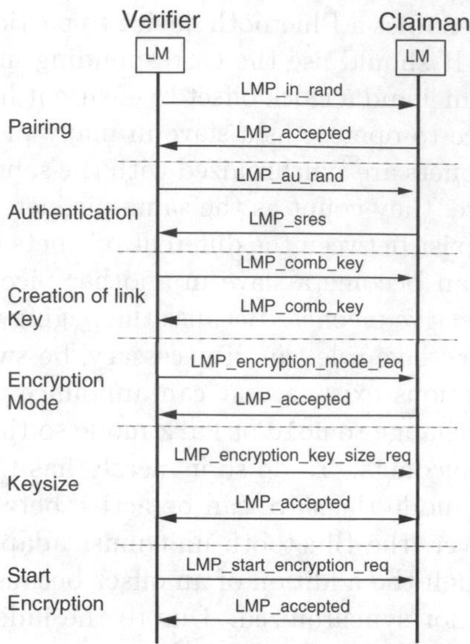

Figure 14.104: Encryption procedure

using a Bluetooth-capable mobile device and connecting wirelessly to one of the base stations, a visitor or customer may request data sheets of exhibited computers, test reports of products sold in supermarkets, or authors' biographies of the paintings hanging on the museum's walls, from a local content server that is connected to the base stations via LAN.

Since a single Bluetooth base station—acting as a master unit—can only serve up to seven active slave units simultaneously, and all of the slaves in a piconet must share the same radio channel (see Section 14.4.2), the capacity and throughput of such Bluetooth-based wireless networks become a critical issue. Consider a usage scenario, e. g., in a supermarket, wherein a Bluetooth network (see Figure 14.105) is used to provide customers with such information like product test reports, special offers, or product availability, anytime and anywhere they want. Each customer (with a Bluetooth-capable mobile device) moves with a typical behaviour towards a target point. After reaching the point, he or she tries to connect to a base station in range and perform data transmission for a specific duration. Depending on the current state of the user's Bluetooth device, e. g., whether the device has been already paged in the piconet or not, the connection setup time may vary; a maximum of 9.6 s has been assumed. After completing the transmission, the customer moves again towards the next target point.

Furthermore, only ACL connections are considered and DM5-packets (477.8 kbit/s) are used for downlink, and DM1-packets (108.8 kbit/s) for uplink transmissions. According to the positioning of the base stations shown in Figure 14.105, no coverage gap exists and up to four piconets may over-

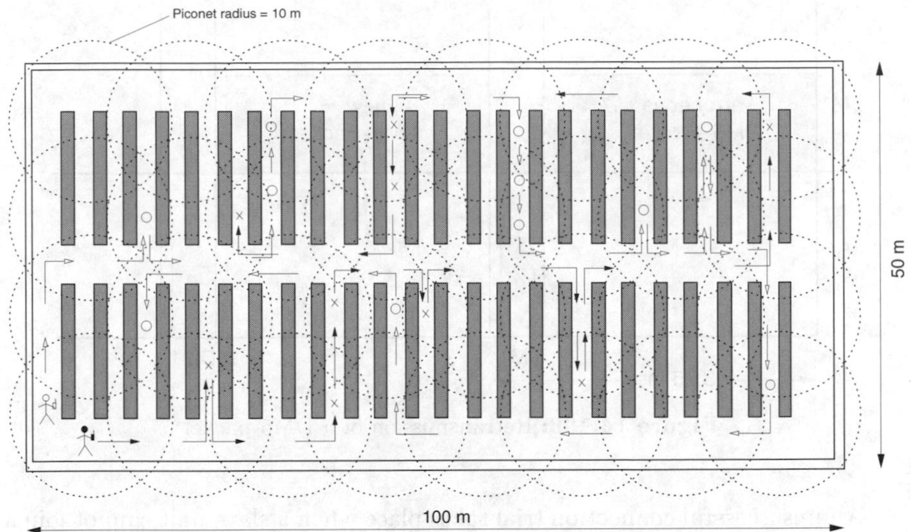

Figure 14.105: An example scenario for a Bluetooth network

lap. The probabilities that in these areas the same or adjacent frequencies are hopped by different (independent) piconets simultaneously, which lead to collisions (*jammed slots*) and packet retransmissions, are taken into account in terms of *Slot Error Rate* (SER_n), which is approximated for n overlapping piconets (79-hop systems) to:

$$SER_2 = 3.77\,\%, \qquad SER_3 = 7.39\,\%, \qquad SER_4 = 10.6\,\%. \tag{14.12}$$

In areas where no overlapping exists ($n = 1$) and in accordance with a maximum BER of 0.1 % required in the Bluetooth specification [4], the slot error rate of DM5-packets remains to: $SER_1 = 0.804$ %.

As illustrated in Figure 14.106 the transmission of a single DM5-packet according to the ARQ protocol including a DM1 acknowledgement packet ends after six slots $\cong$ 3.75 ms. Hence, the additional time Δt needed to retransmit DM5-packets on a piconet channel, e.g., due to piconet interferences, calculates to:

$$\Delta t = \lceil N_{DM5} \cdot SER_n \rceil \cdot 3.75 \text{ ms}, \tag{14.13}$$

$$N_{DM5} = \left\lceil \frac{\mathit{Data\ amount\ in\ bytes}}{224 \text{ byte}} \right\rceil. \tag{14.14}$$

The above mentioned scenario has been studied by simulation in order to evaluate the performance of Bluetooth regarding its system capacity in terms of *Blocking Probability* (P_B), which is defined by:

$$P_B = \frac{\{Sum\ of\ unsuccessful\ connection\ trials\}}{\{Sum\ of\ all\ connection\ trials\}} \tag{14.15}$$

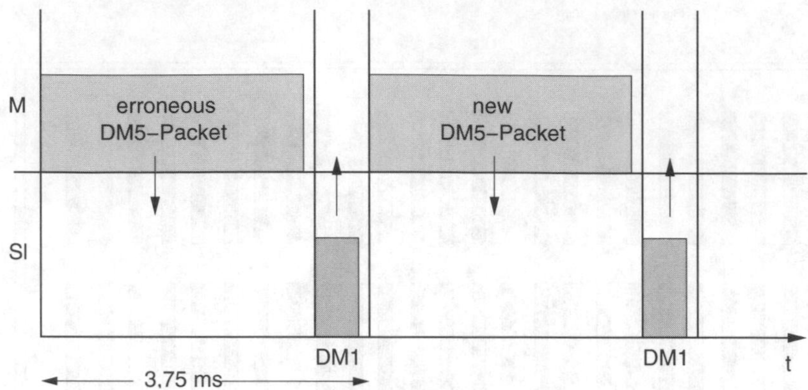

Figure 14.106: Retransmission of a DM5-packet

An unsuccessful connection trial takes place when a slave unit cannot join a piconet because all of the base stations in range have already served seven slave units simultaneously. A blocked slave unit retries to establish a connection until a free piconet's MAC-address can be allocated. The time, during which the slave unit has been blocked, is defined as the *Blocking Time.*

Other parameters used in the simulation are listed in Table 14.21 and described as follows:

Speed the constant speed of each slave unit (customer) while moving.

Idle time the time each customer spends at each target point without transmitting data.

Mean transmission time the average time, in which the data should be transmitted at a target point (without delays, e.g., caused by piconet interference); it is randomly determined according to the Rayleigh distribution function with the corresponding *variance.*

Simulation time the total simulated time with the corresponding *time interval.*

Slave number the constant number of existing slaves during a simulation, which represents a constant slave density.

Max. target points the maximum number of target points that are randomly positioned for each customer.

The blocking probabilities, that have been averaged from the blocking probabilities of each simulated slave unit at variable slave densities, are depicted in Figure 14.107 and the corresponding average blocking time in Figure 14.108. A mean number of 28 target points have been randomly positioned for each slave unit during the simulation. Since the ascertained additional time Δt due

Table 14.21: Simulation parameters

Parameter	Value
Speed v	3 km/h
Idle time t_{idle}	5 s
Mean transmission time t_{trans}	15 s
Trans. time variance t_{var}	22 s
Simulation time t_{sim}	up to 250,000 s
Time interval t_{int}	0.5 s
Slave number N_{slave}	50–300
Max. target points $N_{tp,max}$	150

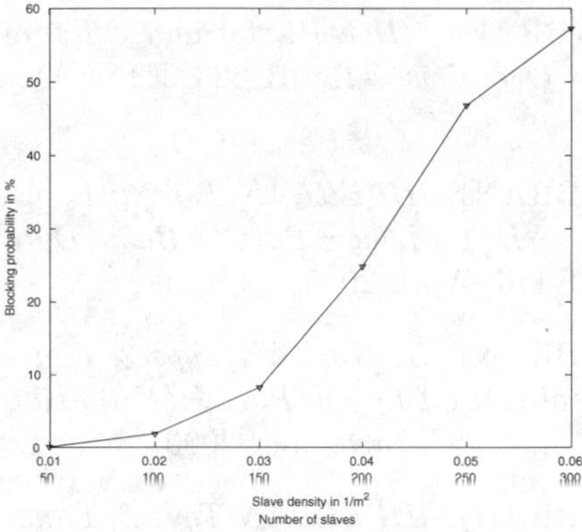

Figure 14.107: Blocking probability

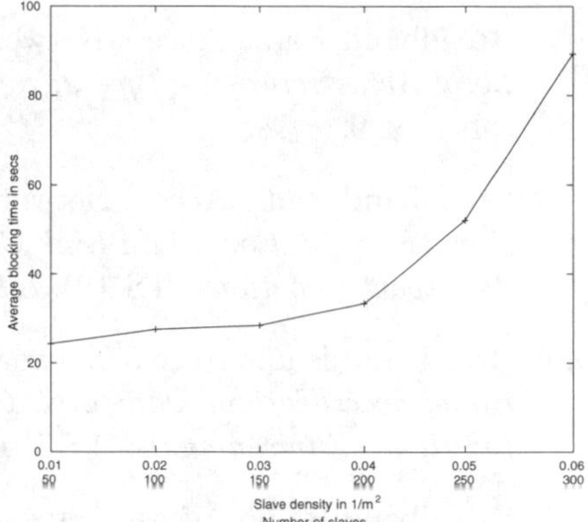

Figure 14.108: Blocking time

to piconet interference corresponds only to 3.4 % of the mean transmission time (t_{trans}), the relatively high blocking probability and blocking time have been mainly caused by lack of free piconet channels. It appears that Bluetooth in this scenario with only one *Transceiver* (TRX) per piconet (eight available MAC-addresses) is highly loaded already with a small number of about 100 slave units. This indicates a need for multi TRX operation in this scenario whereby, although the interference would be much higher, a substantial higher capacity could be gained.

Bibliography

[1] AixCom. *SDL Performance Evaluation Environment Toolkit SPEET, SPEET Class Library.* for a demo version please see `www.AixCom.de`.

[2] G. Anastasi, E. De Stefano, L. Lenzini. *QoS provided by the IEEE 802.11 wireless LAN to advanced data applications: A simulation analysis.* In *Workshop on Nomadic Computing*, Geneva, Switzerland, April 1997.

[3] Bellcore (now Telcordia). *Ethernet Tracefile pAug.TL.* Available via ftp://ftp.telcordia.com/pub/world/wel/lan_traffic/.

[4] Bluetooth SIG. *Specification of the Bluetooth System.* http://www.bluetooth.com, 1999. v1.0 B.

[5] B. Bourin. *HIPERLAN—Markets and Applications.* In *Proceedings of Wireless Networks (WCN), Catching the Mobile Future*, Vol. 3, pp. 863–868, Amsterdam, IOS Press, Sept. 1994.

[6] Broadband Radio Access Networks (BRAN). *HIgh PErformance Radio Local Area Network (HIPERLAN) Type 2; Requirements and architectures for wireless broadband access.* TR 101 031, ETSI, April 1997.

[7] Broadband Radio Access Networks (BRAN). *HIgh PErformance Radio Local Area Network—Type 2; System Overview.* TR 101 683, ETSI, October 1999.

[8] Broadband Radio Access Networks (BRAN). *HIPERLAN Type 2; Functional Specification; Data Link Control (DLC) Layer; Part 1: Basic Data Transport Functions.* TS 101 761-1, ETSI, April 2000.

[9] Broadband Radio Access Networks (BRAN). *HIPERLAN Type 2; Functional Specification; Data Link Control (DLC) Layer; Part 4: Extension for Home Environments.* TS 101 761-4, ETSI, November 2000.

[10] Broadband Radio Access Networks (BRAN). *HIPERLAN Type 2; Functional Specification; Part 2: Physical (PHY) layer.* DTS/BRAN 101 475, ETSI, April 2000.

[11] A. Campbell, et al. *Cellular IP, Internet Draft.* IETF, January 2000.

[12] European Radiocommunications Committee. *Decision on the harmonised frequency bands to be designated for the introduction of High Performance Radio Local Area Networks (HIPERLANs).* Technical Report ERC/DEC/(99)23, ERC, nov 1999.

[13] ETSI. *Radio Equipent and Systems (RES); HIPERLAN.* Functional specifications, 1995.

[14] ETSI. (TC RES10). *High Performance Radio Local Area Network (HIPERLAN); Services and Facilities—ETR 096.* Technical report, ETSI RES; Radio Equipment and Systems, Feb. 1993.

[15] M. W. Garrett. *MPEG Tracefile MPEG.data.* Available via anonymous ftp from ftp.telcordia.com/pub/vbr.video.trace/.

[16] M. W. Garrett. *Contributions Toward Real-Time Services on Packet Switched Networks.* PhD thesis, Columbia University, March 1993.

[17] E. N. Gilbert. *Capacity of a burst-noise channel. Bell System Techn. Journal*, Vol. 39, pp. 1253–1266, September 1960.

[18] A. Hettich, P. Sievering. *Introduction of a New Polling Strategy for a Wireless LAN at an ATM Radio Interface.* In *5th Workshop on Mobile Multimedia Communications* MoMuC'98, pp. 267–274, Oct. 1998.

[19] Andreas Hettich. *Leistungsbewertung der Standards HIPERLAN/2 und IEEE 802.11 für drahtlose lokale Netze.* Dissertation, Lehrstuhl für Kommunikationsnetze, RWTH Aachen, Germany, 2000. ISBN: 3-86073-824-0.

[20] IEEE. *Wireless LAN Medium Access Control (MAC) and Physical Layer (PHY) Specifications.* Standard IEEE 802.11, IEEE, New York, Nov. 1997.

[21] IETF. *RFC 2236, Internet group management protocol, version 2.*

[22] ISO. *ISO 8802-2, Information Processing Systems-Local Area Networks—-Part 2: Logical Link Control.* International standard, 1990.

[23] ITU-T. *Recommendation Z-100, Specification and Description Language SDL.* ITU-T.

[24] A. Kadelka, A. Hettich, S. Dick. *Performance Evaluation of the MAC Protocol of ETSI BRAN HiperLAN/2 Standard.* In *Proc. of the European Wireless'99, ISBN 3-8007-2490-1*, pp. 157–162, Munich, Germany, Oct. 1999.

[25] J. Khun-Jush, P. Schramm, U. Wachsmann, F. Wenger. *Link Performance of the HIPERLAN/2 Physical Layer in Fading Environments.* In *Proc. European Wireless'99*, pp. 145–149, Munich, Germany, October 1999.

[26] J. Khun-Jush, P. Schramm, U. Wachsmann, F. Wenger. *Structure and Performance of the HIPERLAN/2 Physical Layer.* In *50th IEEE Vehicular Technology Conference (VTC'99 fall)*, pp. 2667–2671, Amsterdam, Netherlands, September 1999.

[27] A. Krämling, M. Scheibenbogen. *Influence of Channel Allocation on the System Capacity of Wireless ATM Networks.* In *2nd International Workshop on Wireless Mobile ATM Implementation (wmATM'99)*, pp. 224–234, S.F. Bay Area, U.S.A., June 1999.

[28] M. Lott. *Comparison of Frequency and Time Domain Differential Modulation in an OFDM System for Wireless ATM.* pp. 877–883, Houston, Texas, USA, May 1999.

[29] K. Matheus, S. Zürbes. *On the Influence of Microwave Ovens on the Performance of Bluetooth Networks.* In *Proc. of the 4th European Personal Mobile Communications Conference*, Vienna, Austria, Feb. 2001.

[30] K. Pahlavan, A. Zahedi, P. Krishnamurthy. *Wideband local access: Wireless LAN and wireless ATM. IEEE Communications Magazine*, pp. 34–40, Nov. 1997.

[31] J. Peetz. *A Concept for Interconnecting HiperLAN/2 Ad Hoc Subnets Operating on Different Frequency Channels.* In *Proc. 4th European Personal Mobile Communications Conference (EPMCC'01)*, p. 21.3, Vienna, Austria, Feb. 2001.

[32] J. Peetz. *HiperLAN/2 Multihop Ad Hoc Communication by Multiple-Frequency Forwarding.* In *Proc. of Vehicular Technology Conference (VTC) Spring 2001*, Rhodes, Greece, May 2001.

[33] J. Peetz, A. Hettich, O. Klein. *HiperLAN/2 Ad Hoc Network Configuration By CC Selection.* In *Proc. European Wireless 2000 (EW2000)*, pp. 1–6, Dresden, Germany, Sept. 2000.

[34] C. E. Perkins. *Mobile IP—Design and Principles.* Addison-Wesley, Reading, Massachusetts, 1. edition, 1998.

[35] C.E. Perkins. *Ad-Hoc Networking.* Addison-Wesley, Boston, USA, 2001.

[36] D. Petras, A. Hettich, A. Krämling. *Design Principles for a MAC Protocol of an ATM Air Interface.* In *ACTS Mobile Summit 1996*, pp. 639–646, Granada, Spain, November 1996.

[37] R. Ramjee, et al. *IP micro mobility using HAWAII, Internet Draft.* IETF, January 2000.

[38] C. Röhl, H. Woesner, A. Wolisz. *A short look on power saving mechanisms in the wireless LAN standard IEEE 802.11. Advances in Wireless Communications*, pp. 219–226, 1998.

[39] C.K. Toh. *Wireless ATM and Ad-Hoc Networks.* Kluwer Academic, Dordrecht, 1997.

[40] U. Vornefeld, D. Schleimer, B. Walke. *Fast Collision Resolution for Real Time Services in SDMA based Wireless ATM Networks.* In *49th IEEE Vehicular Technology Conference (VTC'99)*, pp. 1151–1155, Houston, Texas, USA, May 1999.

[41] U. Vornefeld, Ch. Walke. *SDMA Techniques for Wireless ATM.* In *Proc. 2nd wmATM Workshop.*, pp. 107–117, San Francisco, CA, USA, June 1999.

[42] B. Walke, W. Rosenbohm. *Deadline-Oriented Servicing: Waiting-time Distributions. IEEE Trans. Software Engineering*, Vol. SE-6, No. 3, pp. 304–312, May 1980.

[43] J. Ward, R. T. Compton. *High Throughput Slotted ALOHA Packet Radio Networks with Adaptive Arrays. IEEE Transactions on Communications*, Vol. 41, No. 3, pp. 460–470, 1993.

[44] J. Weinmiller, M. Schläger, A. Festag, A. Wolisz. *Performance study of access control in wireless LANs—IEEE 802.11 DFWMAC and ETSI RES10 HIPERLAN. Mobile Networks and Applications*, Vol. 2, No. 1, pp. 55–76, 1997.

[45] E. K. Wesel. *Wireless Multimedia Communications: Networking Video, Voice, and Data.* Addison-Wesley, Reading, Massachusetts, 1998.

[46] A. Wong, C.-T. Chen, D. J. Le Gall, F.-C. Jeng, K. M. Uz. *MCPIC: A Video Coding Algorithm for Transmission and Storage Applications. IEEE Commununications Magazine*, Vol. 28, No. 11, pp. 24–32, November 1990.

[47] J. Wu. *An IP Mobility Support Architecture for 4GW Wireless Infrastructure.* In *Proc. of 1999 Personal Computing and Communication Workshop (PCC'99)*, Lund, Sweden, November 1999.

[illegible] *[illegible]* [illegible] Vol. 41, No. 3, pp. 408–420, [illegible]

[illegible]

[illegible]

[illegible] A. Wong, [illegible] Video Coding [illegible] Storage Applications [illegible] IEEE [illegible]

[illegible]

15 Self-organizing WLANs with QoS Guarantee

A particularly efficient form of media access control is required for packet-oriented radio interfaces, because this process is constantly needed for each packet and not only once per connection for setting up a channel as is usual with many second-generation mobile radio systems.

It is very easy for a central base station to reserve transmission capacity on the uplink; on the downlink it determines how it is used anyway. A base station either is connected directly to the fixed network or uses a wireless connection in order to reach a neighbouring base station on the route to the fixed network access point. One simple way of linking base stations (node) together for packet transmission is under the control of a central node (see Figure 15.1).

Current proposals for controlling access rights to the radio medium are therefore based on the use of central facilities [6, 8]. In simpler versions transmission capacity has to be requested packet-by-packet by means of the S-Aloha protocol, which is a very inefficient procedure. The best procedure entails using a transmitted packet to transmit a reservation request for the subsequent packet if the latter is already in the queue.

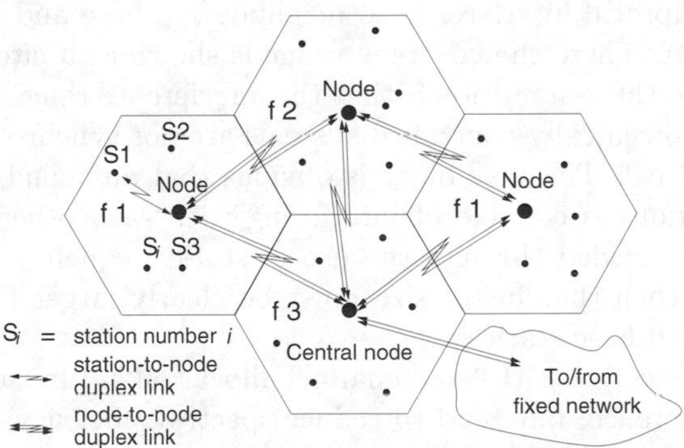

Figure 15.1: Cellular wireless LAN network with central control of stations by the node of the cell, where nodes use frequencies f_i in a 3-cluster and wireless communication between nodes is controlled by a central node.

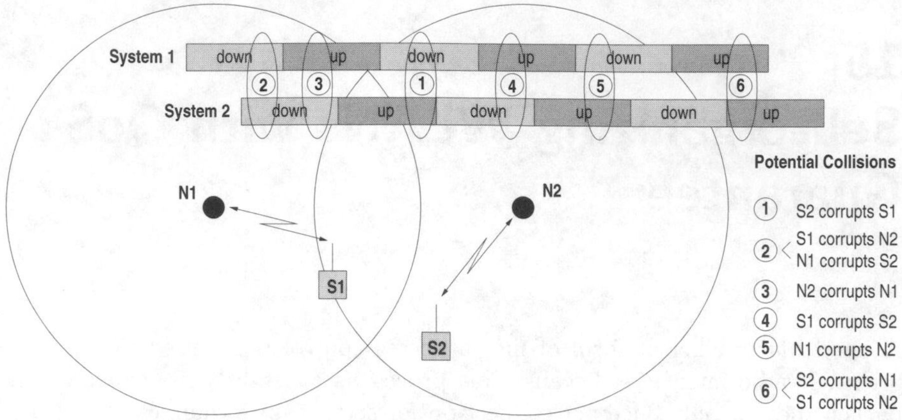

Figure 15.2: Unlike in cellular TDMA/FDD networks, with TDMA/TDD and non-synchronized nodes all the cells interfere with each other at random times.

Uplink use, centrally coordinated by a base station, will not be subject to any interference as long as no other base stations are active in its interference range. Otherwise the mobile stations controlled by different base stations would be transmitting at the same time or overlapping and possibly colliding.

Packets received from a mobile station's own base station could also be affected by interference from other base stations.

Cell clusters in which each cell has a fixed allocation of frequencies are used to avoid reciprocal interference from the base station supply areas (see Figure 15.1). Frequency etiquettes for coordinating the times when frequencies are used by a base station under a decentral control are also being discussed.

If clustering is not considered then a number of possibilities are available for avoiding reciprocal interference to neighbouring base and mobile stations (see Figure 15.2). There the coverage range is shown as a circle around each base station, and the assumption is that the interference range is clearly larger and that the two centrally controlled systems are not synchronized with base stations N1 and N2. The possibility is obvious that any combination of base and mobile stations is capable of interfering with each other. If this interference is to be avoided through the use of static frequency planning (with clusterization), then the cluster size must be clearly larger than is the case with synchronized base stations.

Frequency etiquettes with fixed channel allocation aid in the prevention of reciprocal interference, but tend to reduce spectral efficiency compared with systems with dynamic channel allocation. This is because capacity is allocated to stations and systems at fixed times and cannot be changed dynamically, e.g., when a system does not need the current capacity but a neighbouring system is experiencing a high load.

The prerequisite for dynamic flexible channel allocation under co-channel interference is a criteria that allows a projection of future usage based on current occupation of the medium.

15.1 Channel Concept for a Packet-Oriented Radio Interface

The following statements generally apply [2]:

1. A radio channel can be divided among several users if they require an already-known *Constant Bit Rate* (CBR).
2. During a sufficiently short period of time T_c, each communications relationship possesses a CBR characteristic.
3. The time needed for setting up a connection can be regarded as the practical lower limit for T_c.
4. If a terminal is multiplexing several services on a connection then the time needed is greater than T_c.
5. Multimedia terminals require protocols that support centrally controlled and ad hoc networks.

Channel-oriented reservation-based medium allocation is an interesting option if the following characteristics are required:

- Simple allocation of transmission capacity
- Support of terminals with highly variable bit rates
- Mechanisms for avoiding collisions
- Transmission over radio by terminals using different modulation techniques
- Transparent support for layer-3 protocols and applications based on them
- Compatibility with centrally and decentrally-controlled (ad-hoc) protocols

15.1.1 Bursty Sources of Information and Packet Trains

Bursty traffic sources can be described using the *packet train model* [5], which is a finite state machine model with at least two states in which "cars" (packets) of the "train" (close sequence of packets) are produced after a random short period of time in one state and a longer random period of time in the other

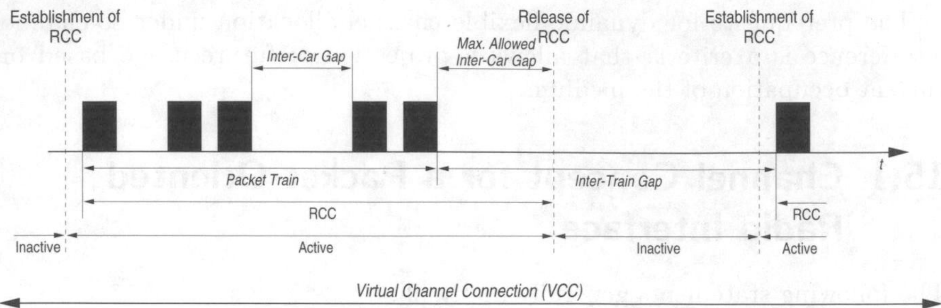

Figure 15.3: The packet train model (RCC = Real Channel Connection)

(see Figure 15.3). The longer intervals correspond to the inter-train times; the shorter intervals to the inter-car times.

If the inter-car gap T_i is regarded as a parameter then different train lengths can be defined for the same traffic source, thereby influencing the frequency of the trains [9]. Obviously T_i should correspond to T_c, the time needed to establish a channel, with T_i increasing with T_c.

According to Figure 15.3, an existing virtual connection is only supported with the transmission capacity of the radio channel (*Real Channel Connection*, RCC) for the duration of a train, and the capacity is withdrawn once the train has come to an end, i.e. the inter-car gap has exceeded.

The receiver uses a reverse channel with the correct transmission rate to communicate with the sender in order to acknowledge individual *cars* or groups of cars of the train, in other words to transmit its own data and send acknowledgements piggybacked. In rigid TDD systems with fixed time intervals for transmission facilities and the same transmission rates per direction, the occupancy of a TDD channel also leads to the occupancy of the one in the opposite direction. With asymmetrical allocation of channel capacity each receiver must indicate the receive time channel in its transmit channel so that other stations can be made aware of the state of occupancy.

This allows a bidirectional physical channel to be recognized as occupied and respected by all stations currently in the detection radius of the respective receiver during the duration of a train, thereby avoiding spontaneous interference by accessing stations, which are hidden stations as far as the receiver is concerned.

What remains to be determined is the amount of capacity to be allocated per physical channel for the duration of a train and how connection setup takes place (see the recommendations in [2, 9, 13]). Suggestions for this are also given in the DECT standard providing a multilink protocol; see Section 10.4.2.6.

15.1.2 Connection Admission Control for Packet-Based Radio Interfaces

If the packet train model is followed and physical connections are set up only for the duration of need then the connection admission controller must ensure that for each accepted virtual connection each new train is guaranteed to receive a physical channel with only a short delay. Therefore the mechanisms familiar from ATM networks appear to be directly transferrable.

If the decentral organization of an ad hoc network is selected, then the connection admission control entity is implemented on a distributed basis. A relevant proposal can be found in [9].

In the W-ATM applications being anticipated, VBR services requiring real-time support (see Table 13.2) will only make a relatively small demand on capacity utilization. If the plan is to interrupt the physical channels reserved for ABR services at any time despite the packet trains being run then it can be expected that the quality of service of these VBR services will be guaranteed.

In the following section a complete design of a decentrally organized W-ATM LAN that is based on channel reservation is introduced and its performance is analysed.

15.2 W-CHAMB: A Self-organizing W-LAN with QoS Guarantee*

Inspired from GPRS and ETSI/DECT concepts, we developed a channel-oriented solution—*Wireless CHannel-oriented Ad-hoc Multihop Broadband* (W-CHAMB) network—for a self-organizing wireless broadband network with multi-hop capability. W-CHAMB adopts the key idea of GPRS, that is statistical multiplexing of bursty traffic through packet reservation, and the most advanced feature of DECT, that is dynamic channel selection according to the measured signal level *Receive Signal Strength Indicator* (RSSI). It differs from GPRS and DECT with its ability to operate in a multihop environment and in a fully distributed manner. The most significant feature of W-CHAMB is that it meets QoS parameters for different classes of service and realizes statistical multiplexing of bursty traffic in a fully distributed and efficient manner.

W-CHAMB networks will be best suited to be operated in the 5–6 GHz licence-exempt spectrum because there is a high capacity of 455 MHz available in Europe and of 300 MHz in the US. The respective frequencies have very limited ability to penetrate obstructions and have very unpredictable propagation characteristics so that frequency planning is impossible there. Although the most common way to realize a wireless broadband network is to use a centrally controlled MAC protocol to realize statistical multiplexing of

---

*With the collaboration of Bangnan Xu. ©2001 IEEE. Reprinted, with permission, from [11]

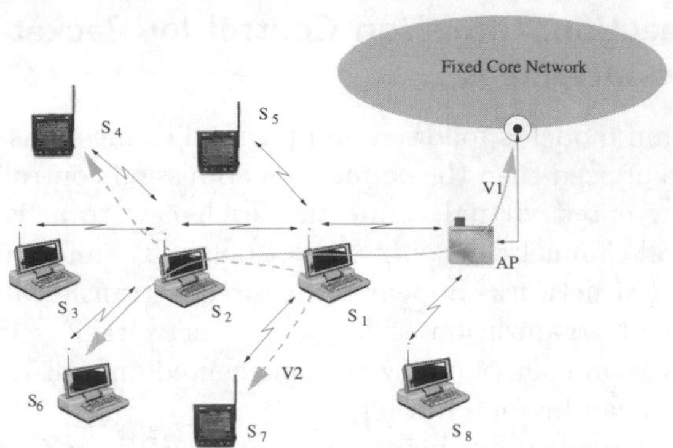

Figure 15.4: Network architecture

bursty traffic sources and to guarantee QoS (see, e.g., H/2), self-organization of a network appears to be not feasible, or even impossible, to implement based on a central controller. Multihop transmission must be considered as a means to achieve a reasonable radio coverage, e.g., to cover shadowed areas for signals at 5 GHz.

Figure 15.4 shows the network architecture of a W-CHAMB network. *Wireless Terminals* (WTs), e.g. Palm, Laptop etc., can communicate with each other directly or through one or more WTs acting as relays to reach their destinations. Some of the WTs might decide not to offer the relay function. If an *Access Point* (AP) to a fixed core network is available, WTs can access the core network. With a number of APs, a large communication zone can be achieved to provide a public/private radio access network with a full coverage of a large area. Association with an AP is necessary if a WT wants to be reachable globally.

15.2.1 W-CHAMB Channel structure

Transmission of packets in W-CHAMB networks is channel-oriented. The transmission time scale in W-CHAMB networks is organized in periodic frames, each containing a fixed number of time slots, see Figure 15.5. WTs are synchronized on a frame and slot basis. A time slot carries a burst containing fields for settling time for *Automatic Gain Control* (AGC), synchronization, *Error Control* (EC) and user data, and has some guard time, that is related to the signal propagation delay.

Exclusive transmit capacity without contention on the MAC layer for point-to-point communications as long as needed is defined by means of *Traffic Channels* (TCHs). TCHs are aquired by means of a random access scheme, which is described in Section 15.2.2, by contending on one of the *Access Channels* (ACHs). One or more slots of a MAC frame are used as ACH, where a number of energy signals (key) and an access signalling PDU (ACC PDU) can be transmitted. The other slots are used as TCHs, each for transmission of one data packet per frame. Each data packet has a user part, such as an ATM cell, and a packet header containing information, such as packet identifier, sequence number, time information, etc., see [10].

To use channel resources more flexibly for heterogeneous applications, periodic slots are used as physical channels to provide transmit capacity for several *Logical Channels* (LCHs). A LCH x/y is using a number of slots per frame, the duration being a multiple of the frame duration. x gives the number of slots used per frame, and y is the repetition period of these slots counted in frames. E.g., an LCH 1/2 defines a physical channel capacity according to one slot every second frame, whilst an LCH 2/1 uses two slots in every frame (see Figure 15.5). By means of defining different slot-to-frame relations sub-multiplexing of the capacity represented by one TCH becomes possible. The smallest traffic channel capacity available is LCH 1/y where y is a design parameter that is chosen according the traffic characteristic and QoS requirements. A number of LCHs can be grouped to realize a so-called *Real Channel Connection* (RCC). The notion of an RCC is used to define the transmit capacity assigned to a one hop link on the layer 2 of the ISO/OSI reference model between two communicating stations in terms of TCHs.

At the end of a MAC frame a number of minislots that carry energy signals (E-signals) may follow. Each minislot, defined as *Energy Channel* (ECH) is associated with a TCH. An E-signal is only a single on-off pulse of the unmodulated carrier, so that the related overhead is generally quite small. E-Signals could alternately be sent in parallel on a narrow band frequency channel to keep the overhead caused by E-Signals negligible. Two E-signals may be sent on one ECH. One is the Busy-E-signal to solve the hidden station problem. The receiving WT sends a Busy-E-signal to indicate to its neighbour WTs the corresponding TCH is its current reserved channel. The other one is the ACK-E-signal that is used to acknowledge a correctly received packet. A duplex channel can be realized either by TDD operation of one TCH or by two independent TCHs with related ECHs.

15.2.2 Dynamic Channel Reservation

An RCC must be established before data transfer between two WTs. As an example, we demonstrate the procedure of *Dynamic Channel Reservation* (DCR) for S_1 to send a PDU train to S_2, see Figure 15.4 and Figure 15.5. To establish an RCC an ACC PDU is transmitted via the ACH using a distributed access priority realized by a key. A priority mechanism is used in the W-CHAMB network to prioritize real-time oriented services, see [12]. The ACC PDU with a PDU *Identifier* (ID) defines the type of a PDU, e.g. access or network management, and containing the addresses of the transmitter and receiver of the one-hop connection. An abbreviated unique *Virtual Channel Identifier* (VCI), which reduces the MAC overhead, refers to a VC that has been previously established, and a channel list contains the proposed silent

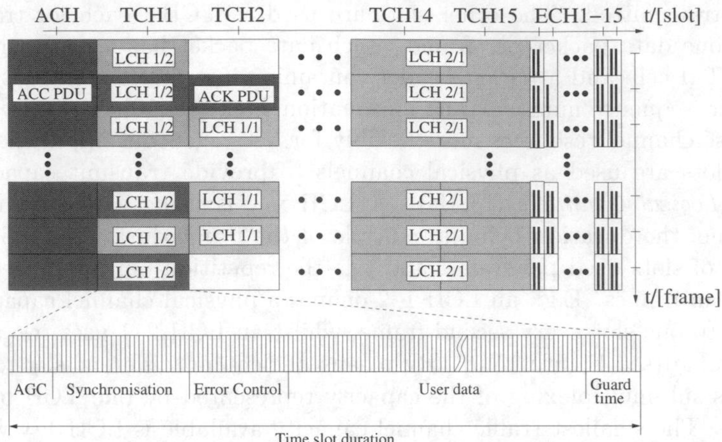

Figure 15.5: W-CHAMB channel structure

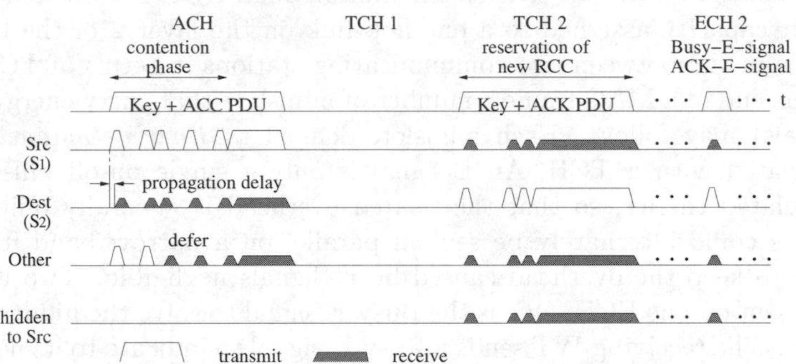

Figure 15.6: Connection set up reducing the collision with hidden stations

TCHs that can be used by the responding station to acknowledge a new RCC. Further, the PDU comprises signalling information of higher layer protocols, e.g., the QoS parameters demanded that inform the MAC layer of the throughput and delay requirements for the requested RCC.

The ACC PDU is preceded by a key. The key is a code representing priority values. In Figure 15.6 as an example a key is used with 8 binary symbols per code.

A 'one' bit symbol is generated by transmitting an energy burst whereby the 'zero' bit symbol defines an equivalent listening duration. In the example, station S_1 transmits 5 energy bursts and listens for 3 'zero' bit symbols. A competing station that senses another station transmitting during listening in the contention phase defers from transmitting an ACC PDU in the current ACH. This access protocol is known as the EY-NPMA protocol of ETSI/HIPERLAN/1 [3].

A surviving station, say S_1, sends its ACC PDU to the destination (Dest) S_2, and all other stations in the transmit range with radius R_{tx} of station S_1 mark the proposed TCHs contained in that message as reserved for a time duration T_{res}. The destination station S_2 if reached selects one out of the proposed TCHs according to a minimum required RSSI margin value out of its local channel occupancy list and responds to the calling station in the respective TCH, e.g. LCH 1/1 on TCH 2 in Figure 15.5 and 15.6, with an ACK PDU that is preceded by an abbreviated key to avoid collisions owing to hidden stations. By this procedure, it is guaranteed that S2 will reach S_1 safely and vice versa with a high probability. All stations in the receive range of S_1 and S_2 are aware of the TCH being in use as soon as stations S_1 and S_2 are transmitting on the respective TCH and ECH. In case none of the proposed TCHs is free at S_2, this station may either pre-empt a packet-train that has a low priority to free a proposed TCH or may block the channel requested by not answering to it. Station S_1 in turn may repeat its request, possibly proposing other suitable TCHs.

15.3 Data Transfer and Acknowledgement

After an RCC has been established, the data transfer of packets, that are transmitted transparently as payload of the data PDU, is started. All other stations in the receive range of the respective stations will recognize and respect the occupied TCHs through measuring the signal strength and a collision free transmission with guaranteed QoS becomes possible.

The receipt of a packet on the reserved TCH requires the receiving WT to respond with an acknowledgement. In the TDD operation of one TCH, the acknowledgement can be piggy-backed on the data PDU. Otherwise, the receiving WT in W-CHAMB network sends an ACK-E-signal on the respective ECH for the purpose of a positive acknowledgement if a packet is received error-free on the reserved TCH and no ACK-E-signal if an error occurred. The erroneous PDU will be repeated on the next transmission opportunity.

An ACK-E-signal on the ECH is uniquely associated with a reserved TCH by setting the decision signal level, called ACK signal level, for the ACK-E-signal appropriately. The ACK signal level should be much higher than the signal detection level, but smaller than or same as the required signal decode level of the WT receiver. Its power might even be controlled according to the actual needs. To describe the operation of ACK-E-signals, we assume that S_1 has sent a packet to S_2 on the reserved TCH 2. There are three possible results of this packet:

1. S_2 successfully receives the packet and sends an ACK-E-signal on the corresponding ECH. S_1 detects this ACK-E-signal with a sufficiently high signal level (larger than ACK signal level). The transmission of this packet on TCH 2 is acknowledged then.

2. S_2 successfully receives the packet and sends an ACK-E-signal on the corresponding ECH. But S_1 cannot detect the ACK-E-signal due to heavy fading. The packet will be repeated by S_1. Then S_2 will receive the correct packet twice. The duplicated packet is filtered out by the packet number. The probability that a packet is received twice can be kept reasonably low by selecting a suitable ACK signal level.
3. S_2 does not receive the packet. No corresponding ACK-E-signal is sent by S_2. The correct operation of the MAC level acknowledgement requires that S_1 shall not receive any ACK-E-signal for TCH 2 sent by another WT that would cause a false acknowledgement. As no WT in the detection range of S_1 may reserve TCH 2 for reception, no ACK-E-signal for TCH 2 shall be sent by other WTs except S_2 in the detection range of S_1. WSs which are out of the detection range of S_1 may send ACK-E-signals for TCH 2 due to the spatial reuse of TCH 2. But those ACK-E-signals will reach S_1 with a much lower signal level than the agreed ACK signal level. So no false acknowledgement shall be received by S_1. The errored packet sent on TCH 2 will be retransmitted until it is correctly received.

15.3.1 Release of RCC

Bursty traffic sources tend to use a channel as can be described by the packet train model, see Figure 15.3. If an inactive duration of the source exceeds the *Maximum Allowed Inter-Car Gap* (MAIG), S_1 stops transmitting on the reserved LCH and S_2 stops sending the Busy-E-signal on the corresponding ECH. WTs in the range of S_1 and/or S_2 that then detect the unused TCH 2 will mark it in their local channel occupancy list as free.

15.3.2 Performance of W-CHAMB in Comparison with IEEE 802.11a and HiperLAN/2

Intensive stochastic computer simulations based on SDL specified protocols have been used to evaluate and compare the traffic performance in *HIPERLAN/2* (H/2), IEEE 802.11a and W-CHAMB systems. The physical layer of W-CHAMB uses the same OFDM modulation scheme as standardized for IEEE 802.11a and H/2. The size of a data packet sent in one slot in W-CHAMB is 9 OFDM symbols (36 µs). Each packet contains a 6 bytes packet header that is not counted in the throughput performance. The slot duration is 45 µs including 9 µs for physical overhead. The number of slots of one W-CHAMB MAC frame is 16.

15.3.2.1 One hop scenario

Figure 14.40 shows a one hop scenario for a fully connected WLAN. It consists of an AP serving a video stream download (MPEG) and a duplex voice

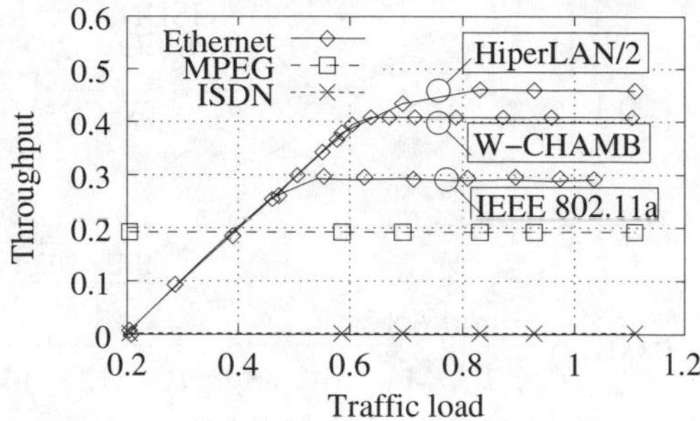

Figure 15.7: One hop throughput performance

connection (N-ISDN), and eight terminals exchanging Ethernet packets with their direct neighbour WTs in a circular way. The data rate of the Video traffic has been set to 5 Mbit/s whereas the mean data rate of the LAN traffic has been varied to model different loads. Traffic loads of MPEG and Ethernet are read from trace files [1, 4] (see Figure 14.38 and Figure 14.39).

In the simulation, the transmission rate of H/2 is set to 27 Mbit/s. The transmission rate of IEEE 802.11a is 24 Mbit/s. The PCF of IEEE 802.11 with Contention Free Repetition Interval of 10 ms length is used to serve the N-ISDN and MPEG services that have priority over the Ethernet service. The transmission rate of W-CHAMB is also 24 Mbit/s. Figure 15.7 shows the relative throughput of the different traffic flows under no bit errors where the load of the Ethernet service has been varied. In all systems the prioritized services N-ISDN and MPEG are served well under all load conditions, whereas the throughput for Ethernet is different. H/2 has the highest throughput for the Ethernet service. The throughput of W-CHAMB for Ethernet service is a little lower than that of H/2, but significantly higher than that of IEEE 802.11a.

The CDF of the packet delay at high load conditions is shown in Figure 15.8. Although W-CHAMB has neither a central controller like H/2 nor a point coordinator like IEEE 802.11, the packet delay of N-ISDN is limited to 6 ms and the delay of MPEG is limited to 13 ms at the total traffic load of 0.75, which meets the requirements of high performance video. In this one hop scenario, H/2 has the the best traffic performance for all services.

15.3.2.2 Two hop scenario

Figure 15.9 shows a two hop scenario with a central AP to a fixed network. Some MTs are placed out of the coverage range of the AP to act as *Remote Mobile Terminals* (RMTs). The end-to-end user connections (i.e. connections between AP and RMT) are loaded with Poisson traffic. We use this scenario to compare the throughput capacity of IEEE 802.11a, H/2 with *Forwarding Mobile Terminals* (FMTs) (see Section 14.2.12) and W-CHAMB.

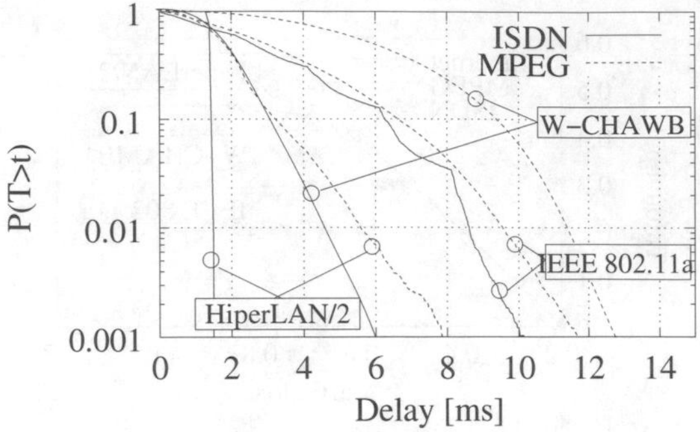

Figure 15.8: CDF of MPEG and N-ISDN at high traffic load conditions

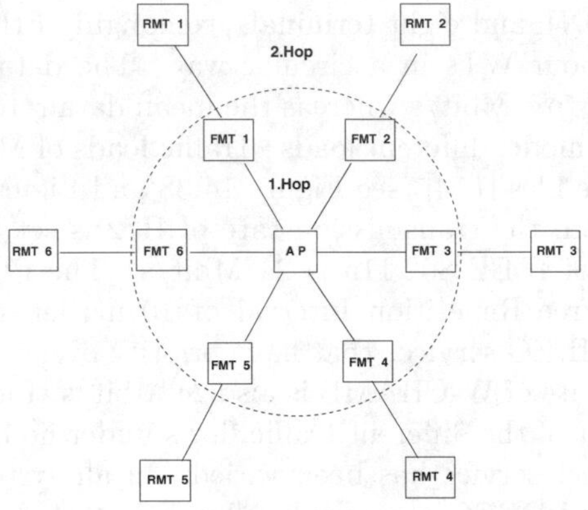

Figure 15.9: Two hop scenario

In Figure 15.10 the maximum system end-to-end throughput of the three systems under transmission rates of 18, 36 and 54 Mbit/s is shown. The packet size with W-CHAMB is 9 OFDM symbols, equivalent to 81, 162 and 243 bytes under transmission rates of 18, 36 and 54 Mbit/s, respectively. The respective packet sizes with IEEE 802.11a are 115, 196 and 277 bytes, each including a 34 byte packet header. The packet size with H/2 is standardized as 54 bytes including 6 byte packet header under all transmission rates. From Figure 15.10 we see that the throughput of IEEE 802.11a is very low due to the short packet size and t wo hop transmission. For H/2 with FMT, an increasing number of FMTs each serving one RMT strongly decrease the capacity available for AP-to-RMT connections as more and more capacity is needed for the overhead introduced by the sub-frames of the FMTs. In contrast to H/2 with FMT, an increasing number of FMTs increases the maximum throughput of W-CHAMB slightly because of the spatial frequency reuse of the decentralized MAC protocol.

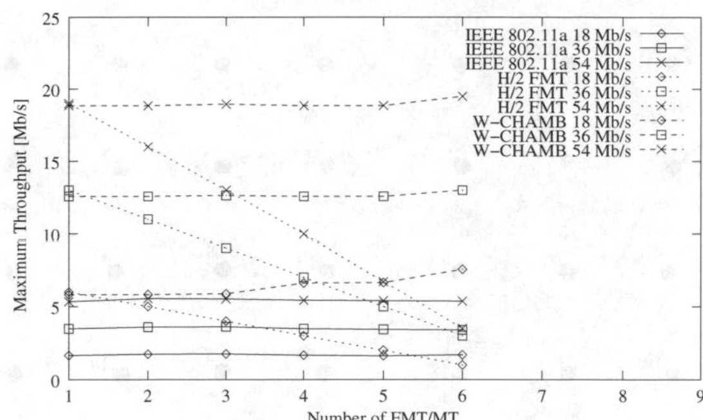

Figure 15.10: Maximum throughput performance

15.3.2.3 Multihop scenario

To study the traffic performance in a multihop scenario, a 5×5 square grid network with/without AP is used, see Figure 15.11(b). A desired network connectivity is achieved by adjusting the fixed transmit power of the wireless stations accordingly. The connectivity is defined as the mean number of neighbours to a WT, normalized by the number of the maximum possible number of neighbours $c = \frac{1}{N(N-1)} \sum_{i=1}^{N} n_i$, where n_i is the number of neighbours to station i, N is the number of stations in the network. This means that a fully connected network has a connectivity of 1. Different from the access scenario in Figure 15.11, no access to a fixed network is considered at the ad hoc scenario in Figure 15.11(a), resulting in an unbalanced traffic distribution to the WTs. Each wireless station randomly selects another station as its traffic sink. The Min-hop routing algorithm is used to establish a multihop connection.

The packet error rate depends on the *Carrier-to-Interference* (C/I) ratio and packet size. The link-level simulation results of [7] concerning the relation between C/I and the packet error rate are used as a basis here (see Figure 14.41). We assume that the power at the distance γ from the transmitter is $W = k\gamma^{-\alpha}$, where k is a constant for all stations. A typical value for WLAN environments is $\alpha = 4$.

To compare the traffic performance of W-CHAMB and IEEE 802.11a in the multihop environment with different network connectivities at the ad hoc scenario in Figure 15.11(a), realistic packet sizes read from an Ethernet trace file [1] are used for the traffic load. The transmission rate is 24 Mbit/s. Figure 15.12(a) shows the throughput of W-CHAMB and IEEE 802.11a with different network connectivities. For both systems, the smaller the network connectivity is, the lower is the throughput. This is because at a smaller network connectivity more hops are necessary for an end-to-end connection. It can also be seen in Figure 15.12(a) that the maximum throughput of W-CHAMB is much higher than that of IEEE 802.11a over all network connec-

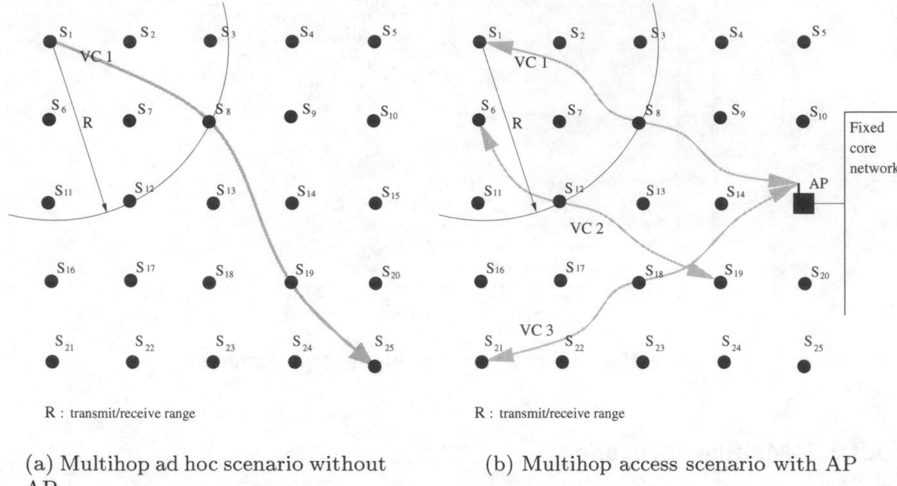

(a) Multihop ad hoc scenario without AP

(b) Multihop access scenario with AP

Figure 15.11: Multihop scenario

tivities. It appears that the channel-oriented MAC protocol in a multi-hop scenario is much more efficient than the packet-oriented one.

Figure 15.12(b) compares the mean delay achieved with the W-CHAMB and IEEE 802.11a systems at the ad hoc scenario in Figure 15.11(a). The impact of the network connectivity on the mean delay with W-CHAMB and IEEE 802.11a can clearly be seen. With a larger network connectivity, a better delay performance can be achieved. It is interesting to see that IEEE 802.11a has a better overall mean delay performance than W-CHAMB as long as it is not saturated. The reason for that is that IEEE 802.11a can transmit a packet as large as 2304 bytes, whereas W-CHAMB has to segment large packets into 102 byte fragments at the transmission rate of 24 Mbit/s. The fragments are transmitted over a number of W-CHAMB MAC frames as a packet train. It can be seen that IEEE 802.11a goes into saturation at much lower traffic load than W-CHAMB can carry. It is worth mentioning that IEEE 802.11a is not able to differentiate between service classes and the mean delay of all service classes together is a too rough measure to characterize a system.

The unique feature of W-CHAMB is QoS guarantee for real time traffic in a multihop network without any central control. This feature makes it best suited to be applied as a self-organizing wireless broadband multihop network. To evaluate the grade of the QoS support of real time traffic, we study the network shown in Figure 15.11 at a connectivity of 0.34. For the ad hoc scenario in Figure 15.11(a), five stations are loaded with real time (rt)-VBR traffic. All other WTs have one ABR connection each. For the access scenario with AP in Figure 15.11(b), there are 25 end-to-end virtual connections (VCs) in total that are established permanently during the simulation. Five rt-VBR end-to-end connections between the AP and 5 WTs are established to model

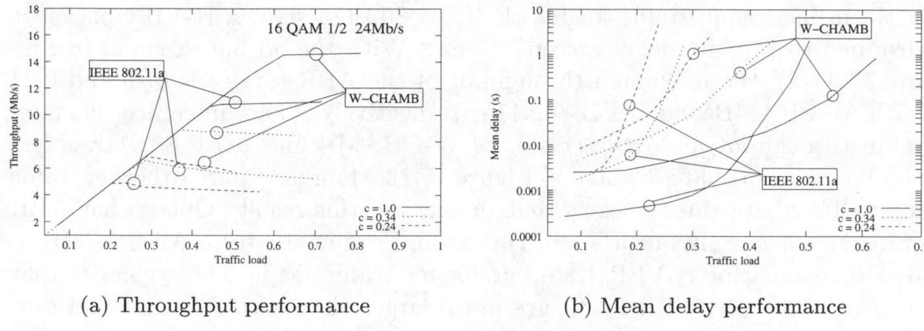

(a) Throughput performance (b) Mean delay performance

Figure 15.12: Traffic performance with different connectivities for the network in Figure 15.11(a)

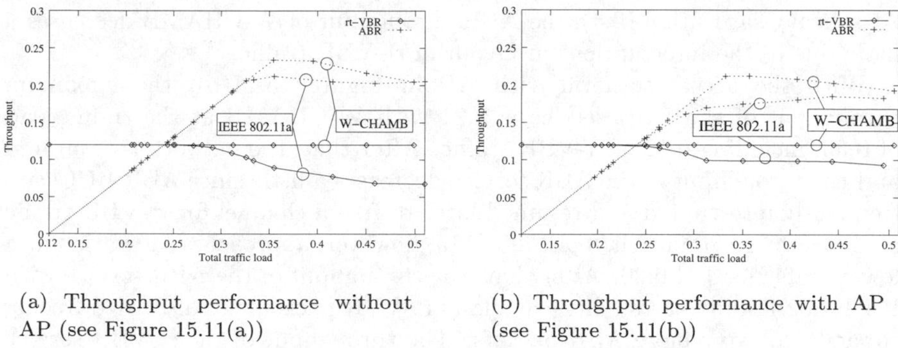

(a) Throughput performance without AP (see Figure 15.11(a))

(b) Throughput performance with AP (see Figure 15.11(b))

Figure 15.13: Throughput performance at $c = 0.34$

the download real time video stream. Another 10 WTs receive ABR traffic from the AP, resulting in 10 download ABR end-to-end connections. 5 upload ABR end-to-end connections are established to transport ABR traffic from five WTs to the AP. There are five WTs that randomly select another WT as ABR traffic sink, resulting in 5 direct link end-to-end connections. For both scenarios, Figure 15.11(a) and Figure 15.11(b), the resulting mean of the five rt-VBR end-to-end connections is 1.6 hops, while the resulting mean of all ABR end-to-end connections is 1.9 hops. The packet size of rt-VBR traffic is modelled by an autoregressive Markovian process, with a mean of 3060 bytes and a maximum of 6120 bytes, yielding a burstiness factor of 2. The rt-VBR traffic source produces 24 packets per second. The packet size of ABR traffic is read from an Ethernet trace file (see Figure 14.38). The interarrival time of Ethernet packets read from the trace file has been varied to model different loads.

Figure 15.13 shows that for both scenarios in Figure 15.11, the prioritized rt-VBR service is served with the same throughput under all load conditions with W-CHAMB, whereas the throughput of rt-VBR traffic decreases with an increasing traffic load with IEEE 802.11a since a rt-VBR packet is dropped there if its delay exceeds 30 ms. With the ad hoc scenario in Figure 15.11(a), the maximum throughput of the ABR service is 0.25 and 0.21 with W-CHAMB and IEEE 802.11a, respectively. The difference in maximum throughput for ABR services of W-CHAMB and IEEE 802.11a is less significant than that indicated in Figure 15.12(a) under a pure Ethernet traffic load. There are three reasons that do explain this result. One is that ABR transmissions are frequently interrupted and resumed later in W-CHAMB to free a channel for rt-VBR traffic at higher traffic load. The second is that most packets of rt-VBR traffic are much larger than with Ethernet and that IEEE 802.11a achieves a higher throughput with a larger packet size. The third reason is that beyond the point of saturation (0.27) with IEEE 802.11a more and more rt-VBR packets are discarded freeing capacity to carry ABR packets so that the ABR throughput remains high on cost of rt-VBR traffic. At a heavy saturation (0.4), the ABR throughput of W-CHAMB decreases a little due to the interruptions in favour of rt-VBR traffic.

With the access scenario with AP in Figure 15.11(b), the maximum throughput of the ABR service is 0.21 with W-CHAMB at the traffic load of 0.33, including 0.12 of rt-VBR traffic. After that, the network is saturated and the throughput of the ABR service decreases a little since ABR RCCs are frequently interrupted and resumed later to free a channel for rt-VBR traffic at a saturation condition. IEEE 802.11a, however, approaches saturation at a lower traffic load of 0.26. After that, the throughput of the ABR service with IEEE 802.11a increases a little as many rt-VBR packets are discarded freeing capacity to carry more ABR packets. The throughput of the rt-VBR service with IEEE 802.11a does not decrease further, as visible from Figure 15.13(a), after the load of 0.31 since a WS or the AP transports rt-VBR packets with preference over ABR packets.

Different from the ad hoc scenario in Figure 15.11(a), there are 5 rt-VBR VCs and 15 ABR VCs in the same AP at the access scenario in Figure 15.11(b) making it the highest loaded station. The maximum throughput of ABR traffic of 0.21 and 0.17 in Figure 15.13(b) instead of 0.25 and 0.21 in Figure 15.13(a) with W-CHAMB and IEEE 802.11a, respectively, is due to the bottleneck effect at the AP.

Figure 15.14 shows the CDF of the rt-VBR service with W-CHAMB and IEEE 802.11a under various traffic loads. For W-CHAMB, the delay distribution of the rt-VBR service is still under control even under a heavy overloaded condition (0.5) for both scenarios in Figure 15.11. With an increasing traffic load from 0.3 to 0.5, the packet delay of rt-VBR traffic increases several milliseconds because the probability that a station cannot find a free channel for a rt-VBR burst at its arrival increases. Several milliseconds are necessary to interrupt the transmission of an ABR packet train to free a channel for the rt-VBR traffic. IEEE 802.11a instead is not able to differentiate rt-VBR and ABR traffic. At a very low traffic load condition, rt-VBR service has a good delay performance. But the delay performance degrades rapidly

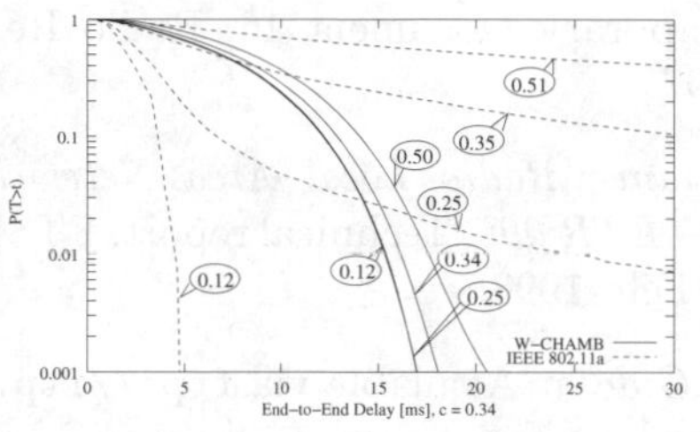

(a) CDF of rt-VBR traffic without AP

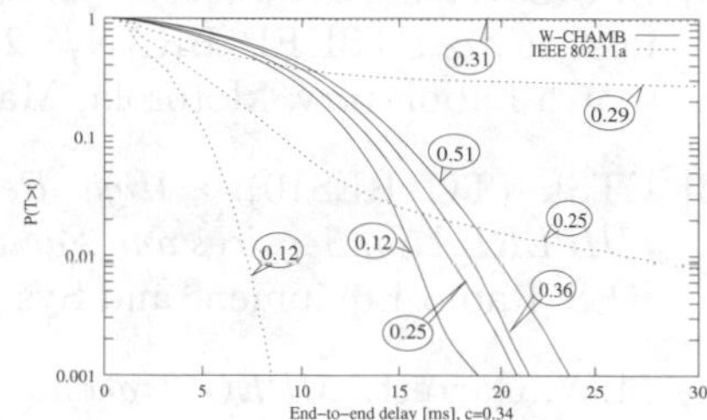

(b) CDF of rt-VBR traffic with AP

Figure 15.14: CDF of rt-VBR traffic at $c = 0.34$

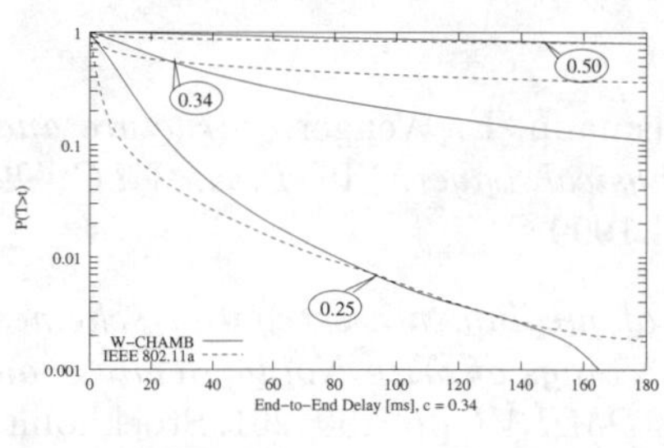

(a) CDF of ABR traffic without AP

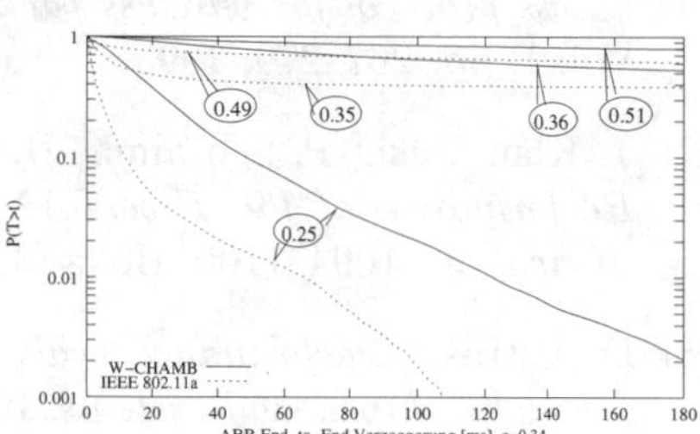

(b) CDF of ABR traffic with AP

Figure 15.15: CDF of ABR traffic at $c = 0.34$

with a increasing traffic load. With a moderate traffic load of 0.25, the delay performance of the rt-VBR service is no longer acceptable.

Figure 15.13 and 15.14 reveal that IEEE 802.11a cannot support a rt-VBR service in a multihop environment, whereas W-CHAMB is able to guarantee QoS even under a high traffic load for both scenarios, with and without AP. Figure 15.15 shows the CDF of the ABR traffic. The delay distribution of the ABR service significantly depends on the traffic load for both, W-CHAMB and IEEE 802.11a. In comparison with the ad hoc scenario in Figure 15.11(a), the access scenario in Figure 15.11(b) has a worse delay performance for both, ABR and rt-VBR services, due to the bottleneck effect at the AP.

Bibliography

[1] Bellcore. *Ethernet Tracefile pAug.TL.* Available via `ftp://ftp.telcordia.com/pub/world/wel/lan_traffic/`.

[2] L. Dellaverson. *An etiquette for sharing multi-media radio channels.* Submission to ETSI EP BRAN# 2, Temporary Document 16, Radio Research Laboratory, Motorola, May 1997.

[3] ETSI. (TC RES10). *High Performance Radio Local Area Network (HIPERLAN); Services and Facilities—ETR 096.* Technical report, ETSI RES; Radio Equipment and Systems, Feb. 1993.

[4] M.W. Garrett. *MPEG Tracefile MPEG.data.* Available via `ftp://ftp.telcordia.com/pub/vbr.video.trace/`.

[5] R. Jain, S. A. Routhier. *Packet trains. IEEE Journal on Selected Areas in Communication*, Vol. JSAC-4-86, No. 4, pp. 986–995, 1986.

[6] M. J. Karol, L. Zhao, K. Y. Eng. *An efficient demand-assignment multiple access protocol for wireless packet (ATM) networks. Wireless Networks*, Vol. 1, pp. 267–279, 1995.

[7] J. Khun-Jush, P. Schramm, U. Wachsmann, F. Wenger. *Structure and Performance of the HiperLAN/2 Physical Layer.* In *Proc. VTC '99 Spring*, pp. 1094–1100, Holland, April 1999.

[8] D. Petras. *Performance evaluation of medium access control schemes for mobile broadband systems.* In *Proceedings of Sixth Nordic Seminar on Digital Mobile Radio Communications DMR VI*, pp. 255–261, Stockholm, Sweden, June 1994.

[9] B. Walke, S. Böhmer, M. Lott. *Protocols for a wireless ATM multihop network.* In *Proceedings of International Zurich Seminar*, pp. 75–82, Zürich, 1998.

[10] B. Walke, S. Böhmer, M. Lott. *Protocols for a Wireless ATM Multihop Network.* In *Proc. 1998 International Zurich Seminar on Broadband Comm.*, pp. 75–82, ETH Zürich, Switzerland, Feb. 1998.

[11] B. Walke, B. Xu. *Self-Organising W-ATM Multihop Networks with QoS Guarantee.* In *Proc. International Symposium 3rd Generation Infrastructure and Services*, pp. 3–9, Athens, Greece, July 2001.

[12] B. Xu, B. Walke. *A Multiple Access Protocol for Ad-hoc Wireless ATM Multihop Networks.* In *Proc. IEEE VTC '99*, pp. 1141–1145, Houston, USA, May 1999.

[13] B. Xu, B. Walke. *Protocols and algorithms supporting QoS in ad-hoc wireless ATM multihop networks.* In *Proceedings of the European Personal and Mobile Communications Conference (EPMCC)*, pp. 79–84, Paris, Mar. 1999.

16 Mobile Satellite Communication*

16.1 Fundamentals

In principle communications satellites provide the same connectivity as terrestrial (wireless and wireline) networks. The advantages of satellites, such as fast wide-area coverage, flexible transmission parameters and cost independence due to distance, are compared with the disadvantages, such as restricted channel capacity because of the frequencies available, orbital positions, need for line-of-sight connectivity and high initial investment besides relatively long signal propagation times. As a result, only certain application areas have been developed for satellites in the past.

16.1.1 Application Areas

Satellites are being widely used for distribution functions, e.g., for transmitting television and radio programmes as well as for data. Existing communications networks can be totally bridged through the use of satellite systems. Satellite communication, which until recently was almost only exclusively used for navigation and aviation as well as in land vehicles, is a branch that opens up a totally new world of applications. Satellite paging along with GPS (*Global Positioning System*) and GLONAS have recently been introduced for civil use. The interest in global personal communications is leading to big efforts in the development of new satellite systems that operate at the low orbital heights.

Currently mobile satellite systems are being used mainly in areas where no other terrestrial communications systems are available (on the open seas, in the desert, in rural regions, etc.). These systems are also attractive to users who operate internationally and otherwise use different kinds of terrestrial mobile radio systems, requiring them to carry terminals with different standards.

A differentiation is made between worldwide, regional and national systems, depending on the coverage area of a satellite system. In terms of institutional and organizational structure, a distinction is made between international, national and private operators of satellite systems. Tables 16.1–16.5 present a compilation of the known parameters of all systems currently in development.

***With the collaboration of Branko Bjelajac and Alexander Guntsch**

Table 16.1: Narrowband satellite systems concentrating on telephony applications—Part 1

System	Globalstar	ICO	IRIDIUM	Odyssey	Ellipso	ECCO (later: Aries)
Company	Loral, Qualcomm, Alcatel Espace	ICO Global Communications	Motorola, Boeing	TRW, Teleglobe Canada	Mobile Communications Holdings	Constellation Inc., Telebras
Orbit	LEO (circular)	MEO (circular)	LEO (circular)	MEO (circular)	MEO (circ. + ell.)	LEO (circular)
Path altitude [km]	1414	10 390	780	10 354	520–7846 (ell.) 8040 (circ.)	2000
No. of sats. + spare sats.	$48+8$	$10+2$	$66+12$	$12+2$ on the Ground	$14+3$	$11+1$ (later add'l $35+7$)
No. of orbits	8	2	11	3	3 (2 ellipt. + 1 circular)	1 equat. (later add'l 7)
No. of ground stations	100	12	15-20	7	≥ 20	11 (more later)
Inclination [°]	52	45	86	50	116.5 (ellipt.)/ 0 (circ.)	0 (later 62)
Min. elevation [°]	10	10	10	20	25–30	5
Cells/sat.	16	163	48	37	61	32
ISL	—	—	4/sat.	—	—	—
Access methods	CDMA	FDMA/TDMA	FDMA/TDMA	CDMA	CDMA	CDMA
Duplex method		FDD	TDD			
Cluster size	6		180 (global) 5 (USA) 4 (Europa)	6.3		
Error handling	FEC $r=\frac{1}{3}$–$\frac{1}{2}$		FEC MS: $r=\frac{4}{3}$ Grnd.st.: $r=\frac{1}{2}$	FEC $r=\frac{1}{2}$		FEC Voice: $r=\frac{1}{2}$ Data: $r=\frac{1}{4}$

Table 16.2: Narrowband satellite systems concentrating on telephony applications—Part 2

System	Globalstar	ICO	IRIDIUM	Odyssey	Ellipso	ECCO (later: Aries)
Modulation	QPSK	QPSK	QPSK	QPSK		BPSK
No. of channels/sat.	2700/ 2.4 kbit/s	4500/ 4.8 kbit/s	4070/2.4 kbit/s	2800		1500
No. of channels in entire system	130 000	45 000	283 000	27 600		11 000 (later 46 000)
Channel bandwidth [kHz]		25.2				
Transmission rate [kbit/s]	2.4–9.6	up to 144	2.4	2.4–9.6	0.3-9.6	bis 9.6
Voice transmission rate [kbit/s]	2.4/4.8/9.6	4.8	2.4	9.6		
Bit-error ratio voice/data	$10^{-3}/10^{-6}$		$10^{-2}/10^{-3}$	$10^{-3}/10^{-5}$		$10^{-3}/10^{-6}$
Frequency UL [MHz]	1610–1621.35	1980–2010	1621.35–1626.5	1610–1621.35	1990–2025	1610–1621.35
Frequency DL [MHz]	2483.5–2500	2170–2200	1621.35–1626.5	2483.5–2500	2165–2200	2483.5–2500
Bandwidth UL+DL [MHz]	27.85[1]	70	5.15	27.85[1]	27.85[1]	27.85[1]
Frequency GW–Sat [GHz]	5.091–5.25	5.15–5.25	29.1–29.3	29.1–29.4	15.45–15.65	5.05–5.25
Frequency Sat–GW [GHz]	6.875–7.055	6.975–7.075	19.4–19.6	19.3–19.6	6.775–7.075	6.825–7.025
Satellite weight [kg]	450	2600	689	2000	689 & 877	425
Antenna type	Planar horn		Planar	Dual refl.		
Satellite intelligence	Available	Available	Available	No	No	No
Satellite transmission mode	Transparent	Regenerative	Regenerative, switching	Transparent	Transparent	Transparent
Transmitter power [W]	1100	5000	1430	4500	2300	815
Power reserve [dB]	3–10 dyn.		16	6		
Max. delay [ms]	11.5		8.22	44.3		
Eavesdropping security	Possible		High	Possible		

Table 16.2: Narrowband satellite systems concentrating on telephony applications—Part 2 (continued)

System	Globalstar	ICO	IRIDIUM	Odyssey	Ellipso	ECCO (later: Aries)
Terminal (mobile)	Dual-mode	Dual-mode	Dual-mode	Dual-mode	Dual-mode	Dual-mode
Terminal trans. power [W]		3.8				
Services	Voice, data, fax, RDSS, SMS	Voice, data, fax, RDSS, SMS	Voice, data, fax, RDSS, SMS	Voice, data, fax, RDSS, SMS	Voice, data, fax, RDSS, SMS	Voice, data, fax, RDSS, SMS
RDSS accuracy [km]	0.3–2		0.5			
Coverage area	74° S–74° N	Global	Global	Global	55° S–90° N	23° S–23° N (later global)
Availability [%]	90–95		90–95	99.5 (User) 99.9 (Grnd.st.)		
No. of users [Mill]	2–5		1.4	2.3	1.0	> 1.0
System cost [Mill. US$]	2.6	2.6	4.4	3.2	0.9	0.55 (later 1.7)
Price of terminal [US$]	750	1000	2000–3000	550	1000	1500
Cost/min [US$]	0.35–0.53	1–2	3	0.65	0.12–0.5	
Start comm. operations	1998	2000	2001	1998	2000	2000
Useful life [years]	10	12	5–10	15	5	5
German partners	DB Aerospace	Deutsche Telecom				
Licence	01/95 granted by FCC	10/95 frequency alloc. by ITU	01/95 granted by FCC	01/95 granted by FCC	06/97 granted by FCC	06/97 granted by FCC

[a]The UL and DL frequency bands are occupied by CDMA systems simultaneously

Table 16.3: Narrowband satellite systems concentrating on message transfer

System	Orbcomm	E-Sat	Faisat	GE Starsys	GEMnet	LEO One
Company	Magellan Systems, Teleglobe, Orbital Science	Echostar Communications	Final Analysis, Polyglott Enterp., VITA	GE American Comm., CLS North America	CTA Orbital Sciences	dBX Corp.
Orbit	LEO	LEO	LEO	LEO	LEO	LEO
Path altitude [km]	825	1260	1000	1067	1000	950
No. of satellites	48	6	32 + 6	24	38	48
No. of orbits	8		6	4		8
No. of ground stations	min. 1 per country					USA 3, others in other countries
Inclination	45°					50°
Intersatellite links	—	—	—	—	—	—
Access method	CDMA	CDMA		CDMA		
Transmission rate [kbit/s]	0.3–2.4		UL: 1.2–19.2 DL: 1.2–38.4			UL: 2.4–9.6 DL: up to 24
Frequency uplink [MHz]	148–149.9	148–148.905	Fin. An.: 455–456, 459–460 VITA: 148–149.9	148–149.9		148–150.05
Frequency downlink [MHz]	137–138	137.0725–137.9725	400–401 (Fin. An. and VITA)	137–138		137–138
Bandwidth UL+DL [MHz]	2.9			2.9		3.05
Frequency GW–Sat [MHz]	148–149.9					148–150.05
Frequency Sat–GW [MHz]	137–138					400.15–401
Satellite weight [kg]	46			80		192

Table 16.3: Narrowband satellite systems concentrating on message transfer (continued)

System	Orbcomm	E-Sat	Faisat	GE Starsys	GEMnet	LEO One
Terminal (communicator)	×	×	×	×	×	×
Services (Monitoring, control, message transfer)	×	×	×	×	×	×
Coverage area	Global	N. America	Global	Global	Global	65° S–65° N
System costs [Mill. US$]	0.35	0.05	0.25	0.17	0.16	0.25
Price of terminal [US$]						100–500
Start comm. operations	1998	1998	2002	1999	1999	2000
Useful life	4				5–7	5

Table 16.4: Broadband satellite systems concentrating on data transfer

System	Teledesic	Celestri LEO[a]	M-Star	SkyBridge
Company	Teledesic	Motorola	Motorola	Alcatel Espace, Loral Space & Comm.
Orbit	LEO	LEO	LEO	LEO
Path altitude [km]	1375	1400	1350	1469
No. of satellites	288	63	72	80
No. of orbits	12	7	12	2x20
No. of ground stations		2 control + 6 antenna centres, many GWs	6	200
Inclination	40°	48°	47°	53°
No. of cells per sat.	576	260		
Intersatellite links	8 per sat. (visual)	6 per sat. (visual)	4 per sat.	No
Access method	DL: Asynch. TDMA UL: MF-TDMA	DL: FDM/TDM UL: FDM/TDMA		CDMA/TDMA/FDMA
No. of channels per satellite	125 000 at 16 kbit/s			
No. of channels in total system	Simult. 36 Mill.	Simult. 395 000 at 64 kbit/s, in total 1.8 Mill. users at 64 kbit/s		
Transmission rate [Mbit/s]	UL: up to 2, DL: up to 64	Up to 155.52	2.048–51.84	0.016–60
Frequency UL [GHz]	28.6–29.1 and 27.6–28.4[a]	28.6–29.1 & 29.5–30	47.2–50.2	14–14.5
Frequency DL [GHz]	18.8–19.8 and 17.8–18.6[a]	18.8–19.8 & 19.7–20.2	37.5–40.5	11.7–12.7

[a]Motorola in 1998 has joined the Teledesic consortium

Table 16.4: Broadband satellite systems concentrating on data transfer (continued)

System	Teledesic	Celestri LEO	M-Star	SkyBridge
Bandwidth UL+DL [MHz]	2600	2000	6000	1500
Frequency GW–Sat [GHz]	In above band	In above band	In user-sat. band	12.75–13.25 and 13.75–14 and 17.3–17.8
Frequency Sat–GW [GHz]	In above band	In above band	In user-sat. band	10.7–11.7
Terminal	Fixed terminal (Ant.-Ø 0.16–1.8 m)	Fixed terminal	Terminal (Ant.-Ø 0.66–1.5 m)	Fixed terminal
Trans. power terminal [W]	0.01–4.7			
Services	Multimedia, video, data	Multimedia, video, data	Data, high-rate network connection	Multimedia, video, data
Coverage area	95 % of area and 100 % of population	70° S–70° N (with Celestri[b] global)	Global	Global
System costs [Mill. US$]	9	12.9	6.1	3.5
Start comm. operations	2005	2002	2002	2005
Useful life [years]	10	8		
Licence from FCC	Granted March 1997	Application accepted	Application accepted	Application accepted

[a] Gigalinks

[b] The Celestri system consists of the Celestri-LEO, M-Star and Millennium (GEO systems)

Table 16.5: Broadband GEO satellite systems concentrating on data transfer

System	Spaceway	Express-way	Millenium (Celestri-GEO)	Astrolink	GE Star	PAC 1–8 and Galaxy Sat.	Inmarsat-3
Company	Hughes	Hughes	Motorola, Vebacom	Lockheed Martin Comm.	GE Americom	PanAmSat	Inmarsat Contractors
Orbit	GEO/MEO	GEO	GEO	GEO	GEO	GEO	GEO
Path altitude [km]	10 352	35 786	35 786	35 786	35 786	35 786	35 786
No. of satellites	8 GEO (HS702) 20 MEO	14	4	9	9	16	5
Different sat. positions				5	5		
No. of ground stations						7	
No. of cells per sat.	48				44		
Intersatellite links	Yes			Yes	No		
Access method	UL: TDMA/ FDMA DL: TDM			UL: TDMA/ FDMA DL: TDM			
No. of channels per sat.	276 480 at 16 kbit/s			64 at 125 MHz, 4 at 250 MHz			
No. of channels in total system	248 832	588 000		576 at 125 MHz			
Transmission rate [kbit/s]	16–16000			16–9600	384–40 000		
Frequency uplink [GHz]	28.35–28.6 & 29.25–30	V- and Ku-band	29.5–30	28.35–28.6 & 29.25–30	28.35–28.6 & 29.25–30	5.925–6.425	
Frequency downlink [GHz]	19.7–20.2 + 5 GHz in-band 17.7–18.8	V- and Ku-band	18.55–18.8 and 19.7–20.2	19.7–20.2 + 5 GHz in-band 17.7–18.8	19.7–20.2 + 5 GHz in-band 17.7–18.8	3.7–4.2	

Table 16.5: Broadband GEO satellite systems concentrating on data transfer (continued)

System	Spaceway	Express-way	Millenium (Celestri-GEO)	Astrolink	GE Star	PAC 1-8 u. Galaxy Sat.	Inmarsat-3
Bandwidth UL+DL [GHz]	2[a]		1.25	2[a]	2[a]	1	
Frequency GW–Sat [GHz]	In-band user-sat.			In-band user-Sat.	In-band user-sat.	14.0–14.5	
Frequency Sat–GW [GHz]	In-band user-sat.			In-band user-sat.	In-band user-sat.	11.7–12.2	
Terminal	Terminal (Ant.-Ø 0.7 m)	Terminal	Terminal	Terminal	Terminal	Terminal	Terminal
Services	Multimedia, video, data, voice		Video, data broadcast	Data, video, Internet	Data, video, audio	Voice, data, fax, multi-media, RDSS	Voice, data, fax, Pos. det.
Coverage area	Continents ex. parts of Russia	Global	America	Global	America, Europe, Asia, West Pacific, Caribbean	Global	70°§–70° N
System costs [Mill. US$]	3.2	3.9	2.3	9	4		0.69
Price of terminal [US$]	1000			Several 100			
Start comm. operations	2002		2001	2000		In operation	since 1997
Useful life [years]	15					11	13
Licence from FCC	05/97			05/97	05/97		

[a]Several systems share frequency band

16.1.2 Satellite Organizations

Until now, the commercial operations of international satellite systems have almost exclusively been carried out by state-run operating companies such as Intelsat, Eutelsat and Inmarsat (see Appendix B). The primary responsibility of Intelsat is to provide regular radio services worldwide, namely telephony and data links as well as the transmission and distribution of TV and radio programmes. Satellites for regular radio services specifically for Europe are operated by Eutelsat *(European Telecommunications Satellite Organization).* The organization Inmarsat (*International Maritime Satellite Organization*) was set up to provide a worldwide maritime satellite radio service.

These operating organizations are all structured in a similar way. They are based on international agreements that were ratified as national law by the participating countries. These member countries establish the objectives of the organization and entrust national telecommunications authorities or operating companies with creating satellite earth stations. The satellite organizations of the countries have no customer contact, but instead provide satellite capacity and, among other things, specify the frequency ranges available for the operation of satellite systems. The operating companies design the satellite services for the customers.

State-run operators of satellite systems are competing against private operators, international companies and consortia, which mainly concentrate on implementing new concepts such as worldwide satellite-based mobile communications systems. These organizations focus on providing almost complete wide-area coverage of the earth's surface using *Low Earth Orbit* (LEO) and *Medium Earth Orbit* (MEO) satellite systems. Some of these satellite systems are introduced in the following sections, accompanied by advantages and disadvantages of geostationary and non-geostationary satellite systems.

16.1.3 Satellite Orbits

Path height is possibly the most important aspect of describing satellite systems. As shown in Figure 16.1, the categorization breaks down into LEO, MEO, *Highly Elliptical Orbit* (HEO) and *Geostationary Orbit* (GEO) systems. LEO system satellites are located between a height of 200 km and the inner Van Allen belt (named after James Alfred van Allen, b. 1914, an American physicist. The Van Allen belts are two radiation belts of the earth (zones with high-intensity ionizing radiation)) at a height of 1500 km. The satellites of the MEO systems are located between the two Van Allen belts, between 5000 and 13 000 km. LEO and MEO satellites are in circular orbits. In addition to the circular systems, there are also HEO systems that are able to achieve a better coverage of densely populated areas using elliptical orbits without having to forfeit the advantages of low orbits.

During the 1960s, developers of the first systems started discussing the advantages and disadvantages of LEO, MEO and GEO systems. The technical

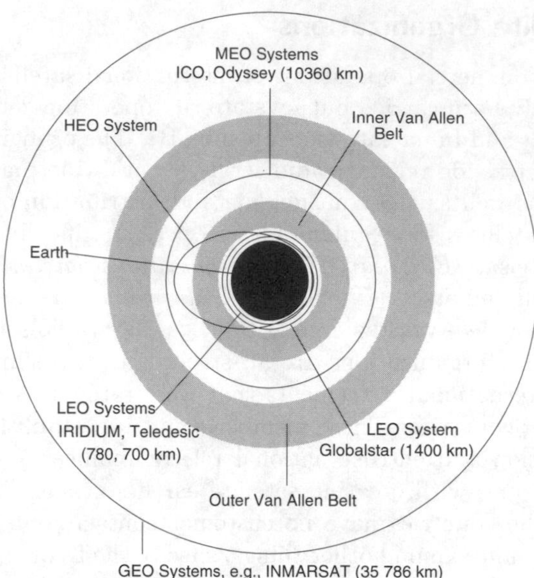

Figure 16.1: Orbits of satellite systems

advantages of LEO systems, such as lower signal propagation times and path loss, were weighed up against the practicability of geostationary systems. The first test satellites such as TELSTAR operated in the lower orbits. Before the mobile applications of satellites were considered as a possibility, GEO systems were viewed as being superior. The satellite launches, particularly during the 1960s, were still unreliable and global coverage was less important than it is today, so the GEO systems, which can manage with only one satellite, were favoured. With interest in mobile communications over satellites increasing, there was a renewed effort to speed up the development of systems with low earth orbits.

16.1.4 Elevation Angles and Coverage Zones

With LEO satellite systems the radio coverage zone on the ground is divided, following the usual cellular principles, into individual cells by a satellite to allow the reuse of frequency bands within the entire coverage zone. The size of the illuminated zone is established by the minimum elevation angle ϵ_{min}, which can be determined from the maximum possible distance between a mobile terminal and the satellite (see Figure 16.2).

The elevation ϵ is

$$\epsilon = \arccos\left(\frac{r_e + h}{d}\sin\phi\right) \tag{16.1}$$

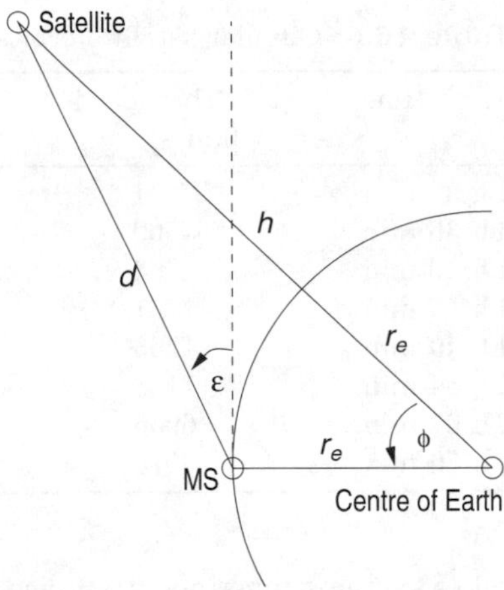

Figure 16.2: Elevation angle in LEO systems

Theoretically, elevation angles of 0° are possible. However, there is a tendency to maintain the minimum elevation angle ϵ_{min}, which for practical reasons is typically at 10° to avoid larger areas without radio coverage because of shadowing [12]. The radius of a coverage zone depends on the earth's radius r_e and the path height h of the satellite above the earth's surface:

$$r_{cov} = r_e \left[\arccos \left(\frac{r_e}{r_e + h} \cos \epsilon_{min} \right) - \epsilon_{min} \right] \tag{16.2}$$

With a given maximum distance d_{max} and a minimum elevation ϵ_{min}, the orbit height works out as

$$h = \sqrt{d_{max}^2 + 2dr_e \sin \epsilon_{min} + r_e^2} - r_e \tag{16.3}$$

The area covered by the satellite is then

$$A_{cov} = 2\pi r_e^2 (1 - \cos \phi) \tag{16.4}$$

These equations can be used to determine the radius of the coverage zone, the orbit speed of the satellite and the travel time around the earth. If the number of cells per coverage zone is available then the size of the cells can be established and an initial prediction made about the amount of time a user spends in a cell. The above values for some orbit heights are given in Table 16.6. For example, an IRIDIUM satellite (at a 780 km height) travels at approximately 28 500 km/h over a stationary earth point, and is only visible from there for about 10–12 min (see Figure 16.39).

16.1.5 Frequency Regulation for Mobile Satellites

Frequency allocation is an important aspect in the planning of satellite systems. Since it is the international authorities who allocate the frequencies,

Table 16.6: Circular satellite orbits

Orbit height [km]	Travel time	Orbit speed [km/s]	Radius of coverage zone [km]
400	1 h 32 min	7681	1333
600	1 h 36 min	7569	1737
800	1 h 40 min	7463	2065
1000	1 h 45 min	7361	2349
1200	1 h 49 min	7263	2592
1400	1 h 54 min	7168	2805
2000	2 h 07 min	6906	3320
35786	23 h 56 min	3075	6027

operators of global satellite systems must negotiate licensing agreements with all the countries in which they want to offer their services. Member states of the International Telecommunications Union (ITU) are committed to the resolutions of the World Radio Conference (WRC).

Commercial satellite radio is currently using frequencies mainly in the C- and Ku-bands. In the C-band these are 5.925–6.425 GHz range for the uplink and 3.7–4.2 GHz for the downlink; and in the Ku-band they are 11.7–12.2 GHz for the downlink and 14–14.5 GHz for the uplink. Because of the growing interest in satellite communications, additional frequencies in the S- and the K/Ka-bands were reserved for satellite radio at WRC 1992. The frequency bands 1980–2010 for the uplink and 2170–2200 MHz for the downlink were made available in the S-band for the satellite segment of UMTS. With the availability of 3.5 GHz of bandwidth each (27.5–31 GHz and 17.7–21.2 GHz), this means that considerably more bandwidth is now available in the Ka-band than was provided in the previous frequency bands.

Since the initial networks are mostly being built by American companies, applications for licences are first made to the American regulatory body, the *Federal Communications Commission* (FCC). These licences then establish the pattern for regulation in other countries. Accordingly, the first frequencies allocated by the FCC for mobile satellite systems were in the 1610–1626.5 MHz band. As is shown in Figure 16.3, 5.15 MHz was allocated to the IRIDIUM-TDMA system and 11.35 MHz to the CDMA systems to be shared by Odyssey and Globalstar. In case only one of the CDMA systems becomes operational, the frequency band 1616.25–1621.35 MHz will be renegotiated.

16.2 Geostationary Satellite Systems

The geostationary orbit is at an altitude of 35 786 km above the earth's surface. Satellites in this circular orbit have an angular velocity that is the same as the rotational speed of the earth. Therefore to an observer on the ground the satellite is at in a fixed position. Because a geostationary satellite can

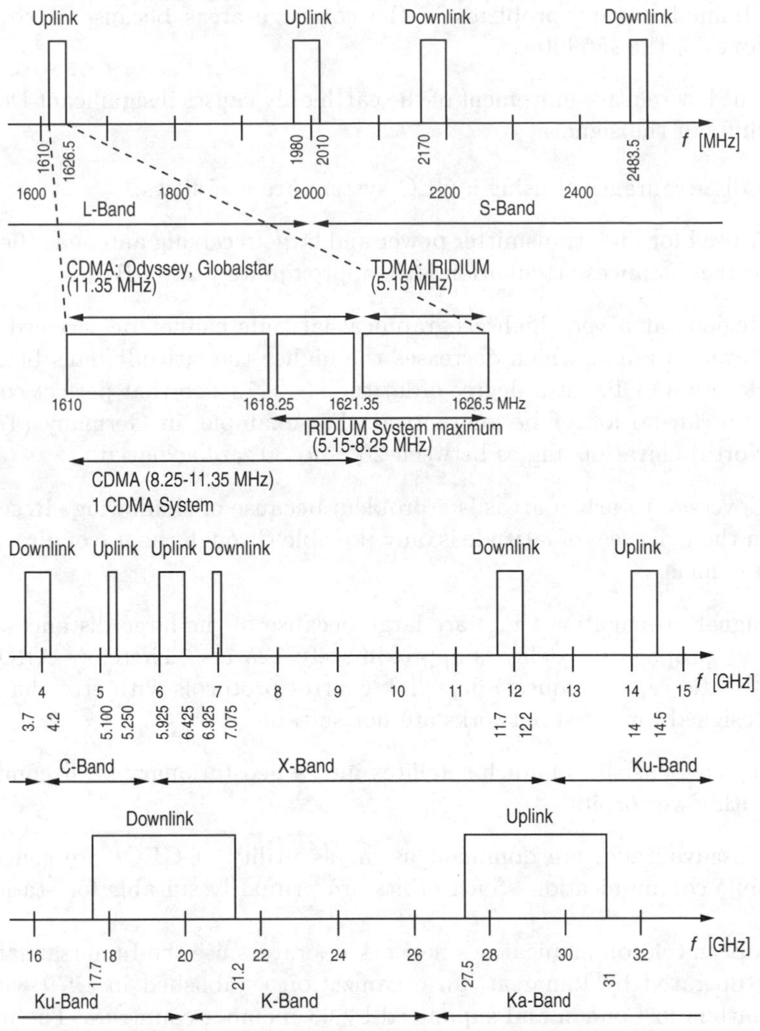

Figure 16.3: Frequencies for satellite radio

drift up to 60 km from its position owing to the influence of other celestrial bodies, positional corrections are essential for the satellite. Ground stations can follow these positional changes better than mobile stations.

The advantages of geostationary satellite systems compared with non-geostationary systems include:

- Simple configuration.
- Coverage of large areas with only one satellite. Large distances can easily be bridged.

- Minimal routing problems in the coverage areas because of coverage zones of the satellites.
- The low relative movement of the earth only causes insignificant Doppler shifts of the signals.

The disadvantages of using a GEO system are as follows:

- A need for high transmitter power and large receiving antennas (because of the distances attenuation is at approximately 200 dB).
- Regions at a very high geographical latitude cannot be serviced. The elevation angle, which decreases the higher the latitude, falls below an elevation of 10° at a degree of latitude $\varphi > 72°$, so that perfect connection can no longer be guaranteed. For example, in Germany (47°–55° North) elevation angles between 20° and 30° are achieved.
- Coverage to urban areas is a problem because of shadowing. Reception in these degrees of latitude is only possible through the use of directional antennas.
- Signal propagation times are large because of the large distances. One way propagation delay is approximately 125 ms. Therefore ARQ (automatic repeat request) data link control protocols with error handling designed for wired networks are not suitable.
- It is expensive to launch satellites into a geostationary orbit compared with lower orbits.

These disadvantages are dominant as far as utility of GEOs are concerned for mobile communication. Such orbits are primarily suitable for stationary services.

European telecommunications network operators use the Inmarsat satellite system operated by Inmarsat, an organization established in 1979 with its headquarters in London and supported by 63 member countries. The mobile radio services offered by Inmarsat were initially restricted to maritime applications. Inmarsat is still the only supplier today that offers global mobile satellite communications on the sea, on land and in the air.

The Inmarsat satellite system consists of 11 satellites: 4 operating satellites and 7 reserve satellites. Reserve satellites can be leased from different organizations.

All Inmarsat satellites are located in a geostationary orbit. The four operating satellites are distributed in the equatorial plane so that they can cover the entire globe, except for the polar regions. The reserve satellites are located in positions in the immediate vicinity of the operating satellites so that they can be replaced if necessary.

The four coverage zones of the operating satellites are the West Atlantic, the East Atlantic, the Indian Ocean and the Pacific. As can be seen in Figure 16.4,

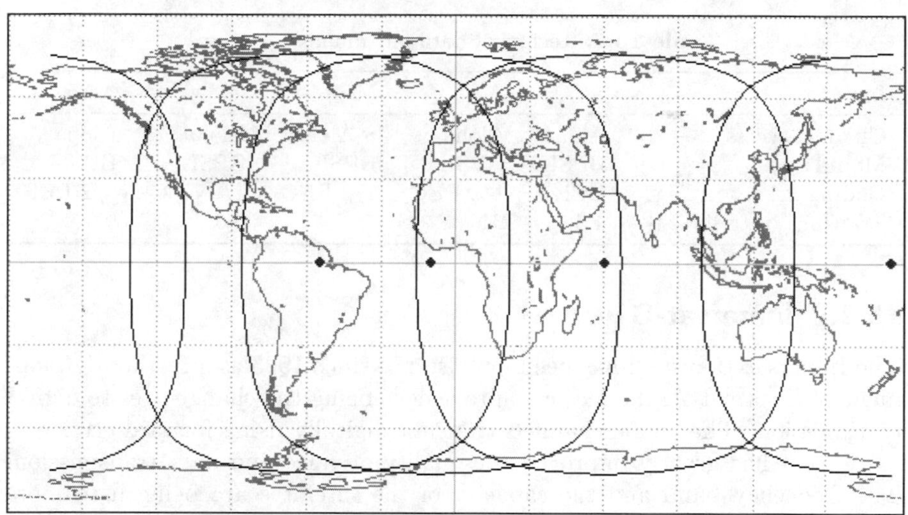

Figure 16.4: Configuration of Inmarsat operating satellites

there is a difference in how much the coverage zones overlap each other. The largest overlap is between the West Atlantic and the East Atlantic. The reason for this is to provide better servicing of the large traffic between America and Europe.

16.2.1 Inmarsat-A

The Inmarsat-A service caters for the following applications:

- Self-dialled telephone connections from/to user terminals.
- Self-dialled (half-duplex) telephone connections from/to user terminals.
- Calls to terminals and retrieval of data stored in terminals. The information is provided only if the caller provides unique identification.
- Use of telephone connections to transmit fax messages and data (up to 9600 bit/s with a modem).
- Data transfer from terminals at a data rate of 64 kbit/s.
- Special applications for broadcast transmission, video freezing frames and commentator transmission routes.

Inmarsat-A is the first of six systems being operated by Inmarsat. It went into operation in 1979, and transmits analogue modulated signals. Portable terminals require a parabolic antenna of approximately 1 m diameter and weighing 20–60 kg. With stationary terminals the antennas are adjusted by hand in the direction of the desired satellite.

Table 16.7: Technical data on Inmarsat systems

Inmarsat	A	B	M	C	Aero
Channel access	Aloha	Aloha	S-Aloha	Aloha	Aloha
Modulation	BPSK	QPSK	BPSK	BPSK	BPSK
Coding	BCH	1/2 FEC	1/2 FEC	1/2 FEC	1/2 FEC
Transmis. rate [kbit/s]	4.8	24	3	0.6	0.6

16.2.2 Inmarsat-B

The Inmarsat-B system has been in existence since 1993 as a further development of Inmarsat-A, the major improvement being the change over to digital transmission. The traffic channels transmit digitally using forward error correction. It has greatly improved the quality of voice and data transmission. The frequency band and the capacity of the satellites are being used more economically. Improvement in satellite technology have produced a reduction in the transmission costs for a call. The communications services offered by Inmarsat-B are comparable to those of Inmarsat-A (see Table 16.7).

16.2.3 Inmarsat-C

Inmarsat-C was developed for worldwide telex and data transmission (X.25) from small mobile terminals. The following applications are some of those available:

- Data transmission in X.25 mode from/to mobile terminals (*Mobile Earth Station*, MES) at data rates of up to 600 bit/s.
- Text transmission from the MES in X.25 mode and output of messages on fax equipment.
- Data transmission in X.400 mode to mailboxes.
- Position location of MES by a central switching centre.
- Formation of closed user groups through the allocation of appropriate telephone numbers.

The net bit rate transmitted in both directions is 600 bit/s. Dynamic intermediate storage (*store-and-forward* transmission) located in the earth station guarantees reliable data transfer. An earth station automatically makes another request for erroneous or missing data blocks. The data blocks are not forwarded to the receiver until the message has arrived error-free at the destination.

Signals are scrambled and spread by the sender to reduce tramsmission errors that occur owing to fast fading. The length of messages is limited to 32 kbytes, primarily because of the intermediate storage.

The terminals used in the Inmarsat-C service are relatively small and weigh approximately 5 kg. They consist of a transmit/receive antenna unit for all directions, a transmitter/receiver, an operating unit and a printer (if required). The power consumption in receiving mode is 25 W and in sending mode 100 W pulsed.

16.2.4 Inmarsat-Aero

The Inmarsat-Aero service, which was introduced in 1992, enables communication from and to airplanes (see also the TFTS service in Section 5.1). Two types of antenna are available:

- Low-gain antennas (0 dBi) for data transmission, mainly for internal operating purposes for airplanes.
- High-gain antennas (12 dBi), which allow higher data rates, thereby enabling voice links to the ground.

16.2.5 Inmarsat-M

The Inmarsat-M service, a digital expansion of the Inmarsat-A service, has been available since 1993. Inmarsat-M terminals are small and considerably lighter in weight than Inmarsat-A terminals. Mobile radio users can make use of the following communications services:

- Telephone
- Fax
- Data transfer (up to 4800 bit/s)
- Variations of the Inmarsat-C service

Because of the low transmission rate of 3 kbit/s, the voice quality with Inmarsat-M is poorer than in connections with Inmarsat-A and -B. Because of the terminal's smaller antenna (30–50 cm) and lower transmitter power, the gain in the Inmarsat-M system is 8 dB less than with Inmarsat-A. It was necessary to reduce the data rate to 3 kbit/s to achieve good transmission quality despite the poor power budget.

16.3 Non-Geostationary Satellite Systems

The advantages of LEO and MEO/ICO systems compared with GEO systems are:

- Lower transmitter power required because of low orbit heights.
- Higher average elevation angle. Because of the large number of satellites, the satellite with the shortest distance to the mobile radio user can always be selected.

- High operating reliability through increased redundancy.
- Good coverage of regions with highest geographical latitude (e.g., polar regions).
- Small signal propagation time.

However, there are also disadvantages due to the lower orbits:

- Short link duration to satellites due to changing elevation angles.
- Smaller coverage area per satellite.
- Complex system control.

At their nearest point to earth (*perigee*), HEO satellites approach up to a couple of hundred kilometres from the earth's surface in order to be able to reach the furthest distance (*apogee*) in the area of the geostationary orbit. They are used there for communication purposes because that is where their orbit speeds are the lowest and a satellite therefore remains visible for a long time from a certain point. Satellites in these orbits are always passing through the Van Allen belt and are consequently subject to increased radiation.

16.3.1 ICO

In September 1994 Inmarsat founded a company called ICO (Intermediate Circular Orbit) to introduce and operate the Inmarsat-P21 project (see Table 16.1). ICO's biggest investors are INMARSAT (US$ 150 million) and Hughes Space and Communications (US$ 94 million). The ICO system is the only mobile satellite system that has been initiated by an operator of an existing satellite system.

ICO will be using 10 satellites in two orbits with 45° inclination to the equatorial plane (see Figure 16.22; the coverage zones of the ICO system are shown in Figure 16.5). Maximum use will be made of path diversity. It is very probable that two satellites will be visible at the lower elevation, and the satellite with the best channel quality will be selected. It is anticipated that the greatest proportion of users of the ICO system will be accessing through GSM dual-mode terminals, and will use the satellite system only if the GSM network cannot guarantee an adequate radio connection.

ICO will be dispersing 12 earth stations (*Service Area Nodes*, SAN) across the globe (see Figure 16.6), and these will be interconnected over a broadband network (P-network). Because there are no plans for links between satellites, the P-network will forward wide-area connections to the nearest network gateway, which will create the link to other networks such as PSTN, PLMN and PSDN.

The *terrestrial terminal and control* (TT&C) stations are responsible for operation and maintenance of the satellites, as well as for control of the configuration through position tracking. Data (such as battery state and position)

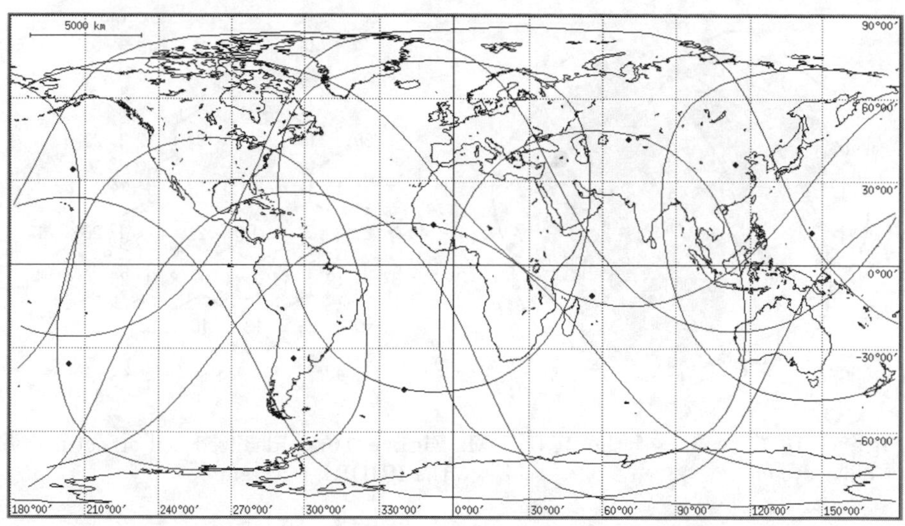

Figure 16.5: Coverage zones of the ICO system

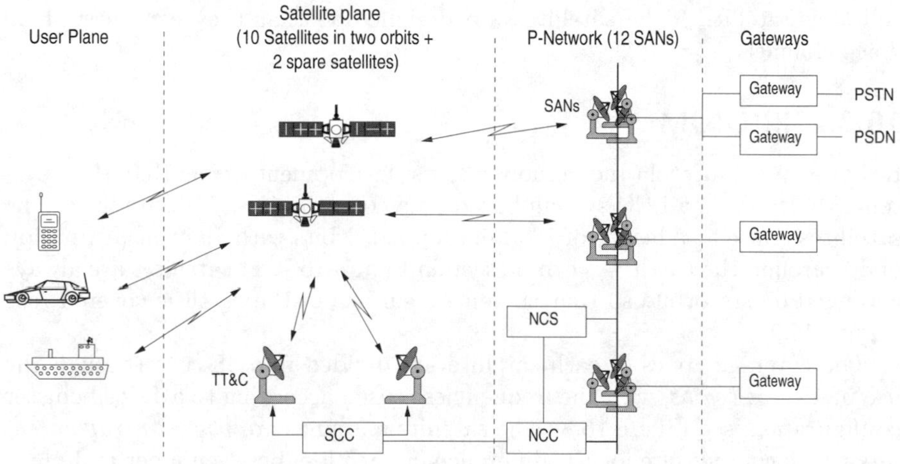

Figure 16.6: Architecture of the ICO system

are recorded. The *Network Control Stations* (NCS) check traffic loads and the connection quality of the different links, and are responsible for channel allocation. The *Network Control Centre* (NCC) carries out system management for the entire system [4, 9].

ICO is planning to use 10 MHz in each of the frequency bands 1980–2010 MHz and 2170–2200 MHz for the uplink and the downlink of satellite user connections (see Figure 16.3). Frequencies from the bands between 5100 and 5250 MHz as well as between 6925 and 7075 MHz are planned for the

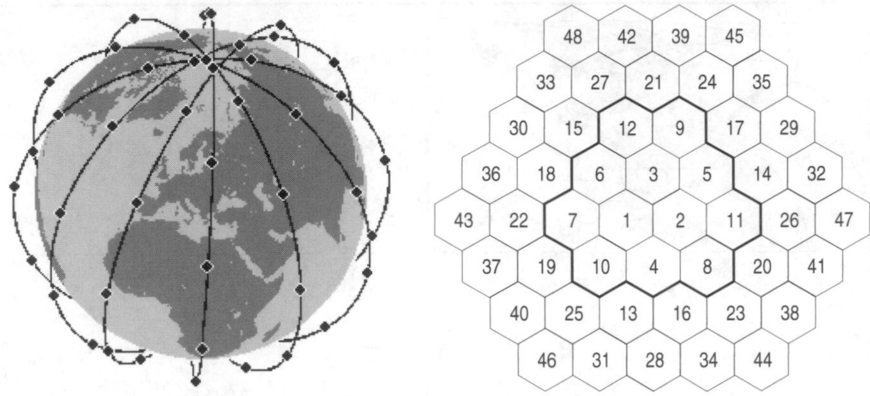

Figure 16.7: Orbits of the IRIDIUM satellite system

Figure 16.8: Idealized cell structure of the IRIDIUM system

satellite-to-SAN link; see Figure 16.6. A TDMA/FDMA method with six time slots per carrier will be used in the uplink and the downlink and 163 cells per satellite. The satellites are designed so that they can carry 4500 voice channels.

16.3.2 IRIDIUM

In June 1990 Motorola Inc. announced its development of the IRIDIUM system. IRIDIUM is a LEO system based on a constellation of 66 satellites. The satellites cruise at a height of 780 km on polar orbits with an orbit inclination of 86° around the earth. As can be seen in Figure 16.7, 11 satellites are always arranged on six orbits so that the entire surface of the earth is covered (see Figure 16.9).

The coverage zones of each satellite are divided into 48 cells through the use of 48 *spot beams*, with the frequencies reused according to a 12-cell cluster configuration; see Figure 16.8. The satellite antennas are *phased-array antennas* which compensate for the difference in path loss between inner and outer cells using the different antenna gain of the respective radiation diagram.

The first satellites were launched into orbit in July 1997. Commercial operation has been opened in November 1998. The IRIDIUM system uses *inter-satellite links* (ISL), for which 200 MHz are being provided in the Ka-band, as is also the case for the gateway satellite link (see Figure 16.3). The ISLs reduce the number of gateways required. Furthermore, coverage over oceans could not be guaranteed without ISLs. ISLs exist to each of the four adjacent satellites. The difficulty is with interorbital links, because the distances and directions for these links are constantly changing. The routing in the ISL-based satellite network is controlled from the NCSs on the ground.

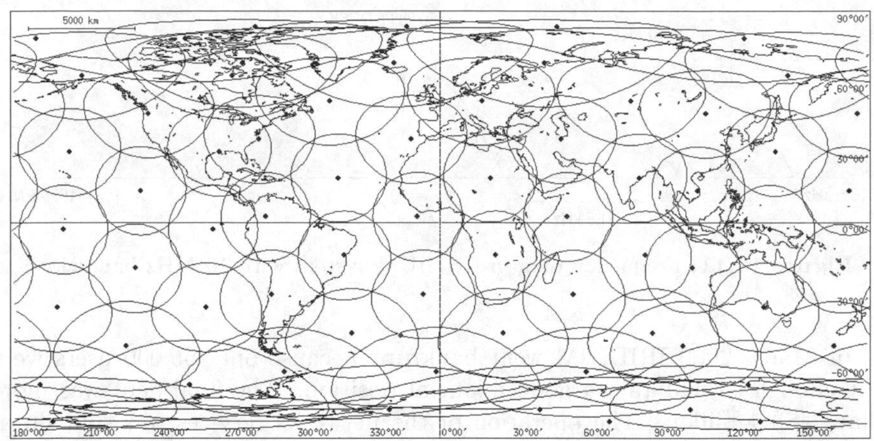

Figure 16.9: Coverage zones of the IRIDIUM system

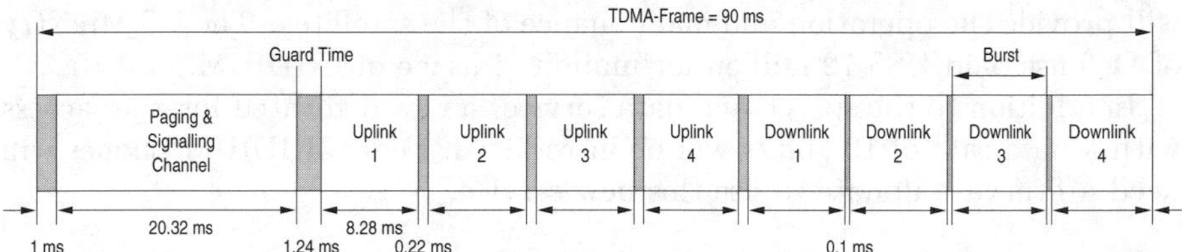

Figure 16.10: TDMA frame in IRIDIUM

The mobile terminals are pocket-sized dual-mode devices that enable connections to be established to satellites as well as to a terrestrial mobile radio network. The transmission rates are 4800 bit/s for speech and 2400 bit/s for data. Multiple access is through the use of a combined FDMA/TDMA/TDD procedure. The TDMA frame planned is 90 ms long, and contains four uplink and four downlink channels as well as a channel for signalling and paging (see Figure 16.10).

The IRIDIUM system uses a QPSK modulation technique with 50 kbit/s. Frequency planning for IRIDIUM before the allocation of an additional 5.15 MHz by the FCC is shown in Figure 16.11. The channel bandwidth is 31.5 kHz and the frequency spacing is 41.67 kHz [8].

The 5.15 MHz frequency band allocated by the FCC to the IRIDIUM system is divided into 124 carriers, each containing four duplex channels, thereby providing 496 channels. Since the frequencies are reused four times per satellite (12 cells per cluster, 48 cells per satellite) it is theoretically possible that 1984 connections could exist simultaneously per satellite because of the given cell structure. However, owing to the limited battery power of 1400 W available a satellite can only maintain 1100 connections at a time.

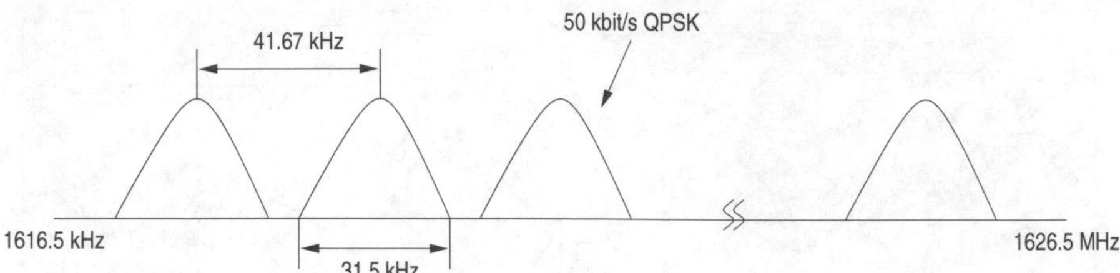

Figure 16.11: Frequency plan in IRIDIUM system with 10 MHz bandwidth

In March 2000 IRIDIUM went bankrupt because only 55,000 users were not enough to maintain a good financial position. The final debt was more than US$ 4 billion. The operation of the networks services was shut down and it was planned to destroy all 66 satellites after August 2000, because the operation and maintenance of the satellites cost several million US$ per month. In October 2000 new investors were found, among them Boeing that will provide the operation and maintenance of the satellites. The US Ministry of Defense paid US$ 72 million for unlimited usage of IRIDIUM.

In addition to the speech and data services a new dedicated Internet access with a data rate of 10 kbit/s will be introduced. Older IRIDIUM phones will need a firmware update to use this new service.

16.3.3 Globalstar

In contrast to IRIDIUM, which was developed for worldwide coverage, Globalstar, another representative of LEO systems, supplies coverage only to areas between a 70° northern and southern latitude (see Figure 16.12). Its main purpose is to provide coverage to the industrial countries. Direct (intersatellite) links between the satellites are not planned. The transmission and access technologies closely follow the standard TIA-IS95; see Section 5.3.

Forty-eight satellites circle eight circular orbits at a height of 1400 km (see Figure 16.13). Two satellites are constantly visible below an elevation angle of at least 10° in the coverage area of the system. The coverage area of a satellite is divided into 16 cells, which are grouped into six or nine segments in a concentric arrangement around the foot of the satellite (see Figure 16.14). A synchronous CDMA channel allocation method is used to facilitate handovers. Communication between mobile station and satellite is only possible if an earth station is located in the area being illuminated by the satellite, because the satellites have only a transponder function and no onboard switching at all. The signal of a connection between two mobile stations must therefore cover twice the distance (two hops). The signal propagation times are short anyway (approximately 70 ms) because of the low orbital height. For example, in North America six earth stations are required to cope with the traffic volume.

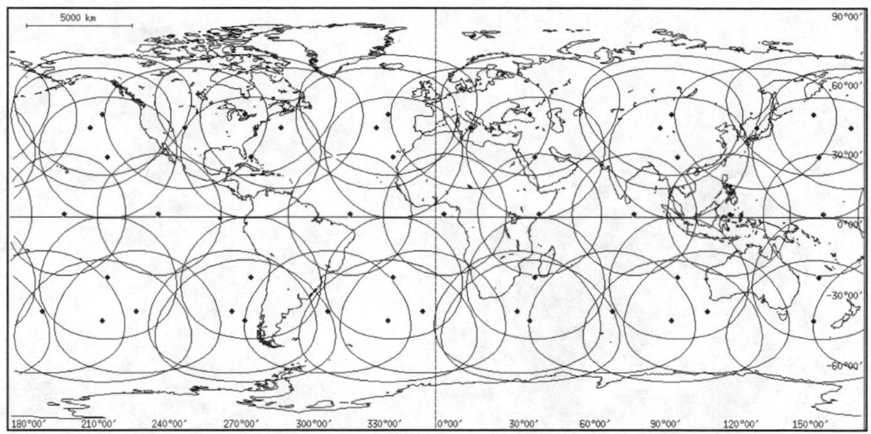

Figure 16.12: Coverage zones of the Globalstar system

16.3.4 TELEDESIC

There has been a sharp increase in the demand for broadband communications systems during the last few years. However, coverage with fibre optics will not be widespread in the world, even in the next few years. The LEO and MEO satellite systems offer services with up to a 9.6 kbit/s transmission rate only, and, this naturally does not meet the requirements for multimedia data transmission. Services such as video conferencing, telelearning and video telephony require a much larger bandwidth.

TELEDESIC, a company founded in June 1990 by Bill Gates, the owner of Microsoft, and Craig McCaw, one of the owners of the largest mobile radio network operators worldwide, is planning a satellite system that will be offering broadband services up to 2 Mbit/s transmission rate per terminal over satellite (see Table 16.4). The first satellites are to be launched into their orbits in 1999, and operations are to start in 2002.

The TELEDESIC system is supposed to be compatible with fibre-optic networks and to provide the quality and services of fibre-optic technology in remote, rural areas at the price of fibre optics in urban areas.

In contrast to other systems, TELEDESIC will operate in the Ka-band (30 GHz downlink, 20 GHz uplink) in order to provide broadband services. Communication in the Ka-band is heavily affected by attenuation caused by rain.In the TELEDESIC system 924 satellites (840 operating and 84 in reserve) operate in 21 orbits at 700 km above the earth's surface. The orbit inclination is 98° so that there is almost global coverage except for a small gap at the poles (see Figure 16.15), this also reduces the loss due to rain. Four reserve satellites are distributed evenly over each orbit. If one of the satellites malfunctions, the other operating satellites and the closest reserve satellite spread out over the respective orbit. Because of the large number of

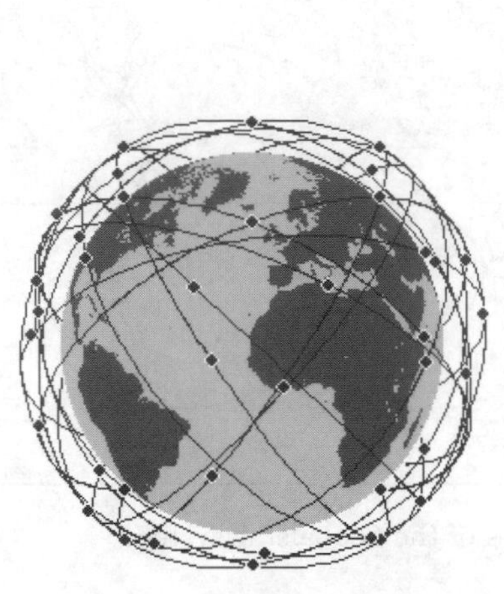

Figure 16.13: Orbits of the Globalstar satellite system

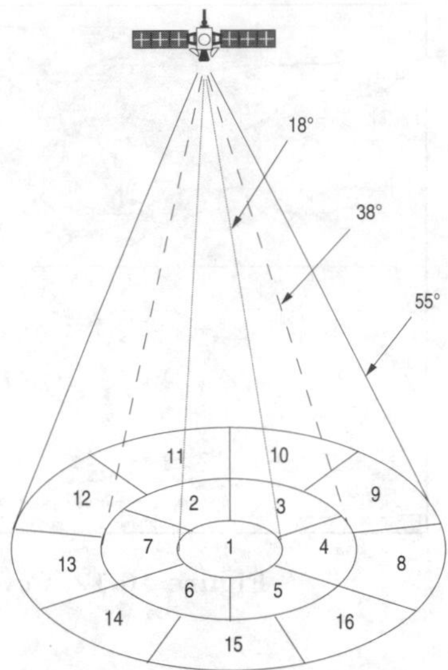

Figure 16.14: Cell layout on Globalstar downlink

satellites, there is a minimal elevation of 40°, which helps to reduce the effects of heavily elevation-dependent rain attenuation. The TELEDESIC system has been developed to offer 99.9 % availability.

Initially eight satellites are to be launched into orbit at the same time in order to keep the number of rocket launches to a minimum. Weighing 700 kg, the satellites will be much lighter in weight in comparison with other systems. Figure 16.16 shows a TELEDESIC satellite [13]. It consists of a quadratic *solar panel* of 12 m edge length, with a total of 24 plates arranged in a daisy shape with eight *daisy leaves*. The plates contain the phased-array antennas. The octagonal baseplate contains eight pairs of ISL antennas along with a drive unit. The power supply unit is attached to the top of the solar panel. The satellites are so designed that they can be launched by as many as 20 different types of rockets.

The TELEDESIC network uses links in the 60 GHz band for communication between satellites (ISL) in order to feed connections as closely as possible to the destination user in the fixed network; intersatellite links exist to eight neighbouring satellites. The network transmits in packet-switched mode and supports broadband communication based on the ATM standard (*Asynchronous Transfer Mode*). The data packets are 512 bits long. The packet header contains the address and the sequencing information, which are used to restore the sequence at the receiver, as well as fields for error protection and user information. Deviating from the ATM standard, TELEDESIC uses a connectionless protocol (see Figure 16.17). The packets that arrive at the receiver over different paths must therefore be buffered and restored again in

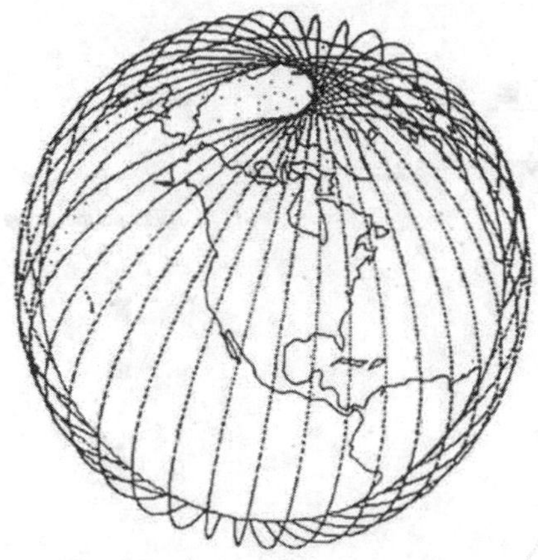

Figure 16.15: Satellite orbit in the TELEDESIC system

Figure 16.16: A TELEDESIC satellite

the right sequence. Each node independently routes the packet (*Distributed Adaptive Routing Algorithm*) over the path that is currently able to guarantee the minimum delay.

TELEDESIC is also breaking new ground in its approach to cell layout. Cells in the TELEDESIC system are ground-based, which makes cell planning easier but on the other hand requires more complicated antenna technology. The earth's surface is covered by 20 000 *supercells*. These supercells are divided into bands, which at the equator contain 250 supercells but towards the poles contain fewer supercells so that the North-South boundaries shift between the bands. As Figure 16.18 shows, a supercell consists of nine square small cells.

Ground-based cells require much less administration than satellite-based cells. The TELEDESIC system uses *steering phased-array antennas* (see Section 16.4.1.1). As illustrated in Figure 16.20, the satellite controls its spot beam and compensates for the movement of the satellite and earth when the satellite passes over the cell cluster. In contrast with satellite systems without ground-based cells, there is no *interbeam handover* between antennas of the same satellite.

A supercell is supplied by an antenna which serves the nine cells in consecutive order (SDMA). This cycle which is made up of transmit times of 2.276 ms and guard times of 0.292 ms is 23.111 ms long (see Figure 16.19). The cells are scanned according to the numbers given in Figure 16.18. The combination of a large number of satellites and the different multiplexing procedures produces a high level of system capacity. With the TELEDESIC system each frequency used is repeated 350 times. Each satellite can scan 64 supercells simultaneously. Each cell has 1800 channels with a 16 kbit/s transmission

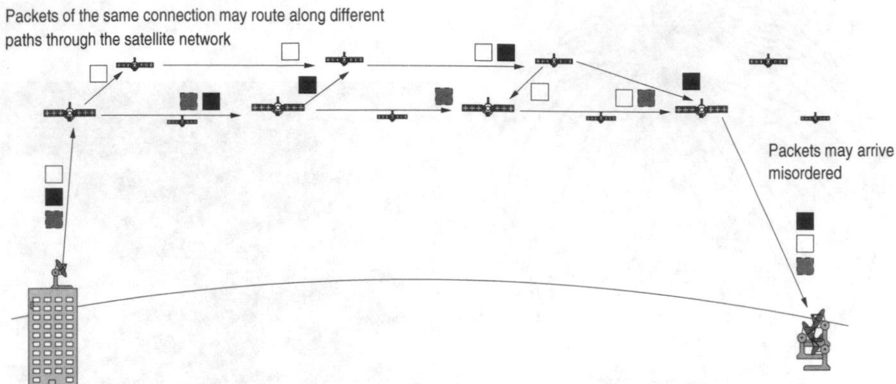

Figure 16.17: Packet-switched connectionless TELEDESIC network

rate. A 400 MHz bandwidth is being planned for both the uplink and the downlink. As illustrated in Figure 16.21, the 400 MHz on the uplink will be distributed over 1800 FDM channels. The time slot belongs to the 16 kbit/s channel, and a terminal totally occupies one or more channels for the duration of a call. On the other hand, an ATDMA ($A = adaptive$) method is used on the downlink. The channels are not allocated permanently, because of the connectionless transmission. A terminal must therefore read all the packets and check the headers to determine whether the packet belongs to its own connection before the sequence of the packets is restored. The packets are separated from one another through certain bit sequences to enable detection of the beginning and end of a packet. In this connection, the satellite always only transmits as long as packets are still in the queue.

A user connection can occupy up to 128 channels, so data rates between 16 and 2048 kbit/s are supported. However, data rates of 155.52 Mbit/s to 1.24416 Gbit/s are supported for special applications.

Fixed and mobile terminals are planned for the TELEDESIC system. The transmitter power of the terminals varies between 0.01 W and 4.7 W, and the antenna diameter between 16 cm and 1.8 m, depending on climatic conditions, requirements for availability and transmission rate [16].

16.3.5 Odyssey

A MEO system, the Odyssey satellite system consists of 12 satellites that travel around the earth at an altitude of 10 370 km on three orbits with an inclination of 55° (see Figure 16.22). The antennas of the satellites can be controlled dynamically and are mainly directed towards coastal areas and the interior. Similarly to the TELEDESIC system, the adaptive antennas are controlled to create ground-based cells (see Section 16.3.4). Because of

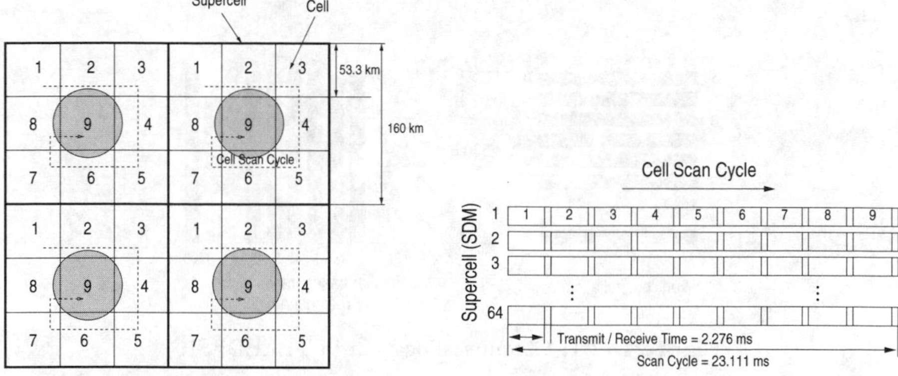

Figure 16.18: Cell structure of the TELEDESIC system

Figure 16.19: Cell scan cycle

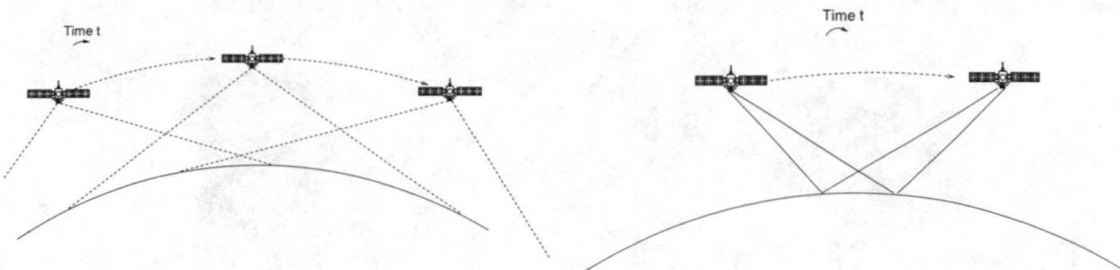

Figure 16.20: Left: Coverage zone which moves with the satellite. Right: Ground-based cells

the approximately six-hour earth circulation time and the relatively large illumination zones, handovers are normally not required during a connection. The coverage zones of the Odyssey system are shown in Figure 16.23.

Connections between users of the system are carried out over 1 of 10 to 11 earth stations distributed around the world. For example, two earth stations (one each on the West and East coast) are sufficient for good-quality operation in North America. A channel-oriented CDMA method is used as the access procedure. The first starts originally were planned for 1997, and initially six satellites were intended to be put into operation in early 1998. This schedule has been delayed, i. e., until end 1998 Odyssey has not been operational at all.

16.4 Antennas and Satellite Coverage Zones

There are many points to be considered in modelling antennas, because of the vital importance of the antenna characteristics to the performance of the system. Antennas determine the size of the cells and consequently the area of the satellite coverage zone. It is desirable that antennas have a high gain in the area of a cell but at the same time that the gain drop dramatically at

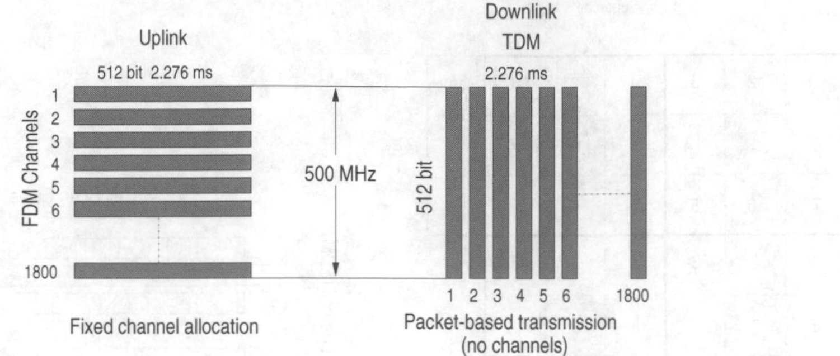

Figure 16.21: Channel allocation in TELEDESIC

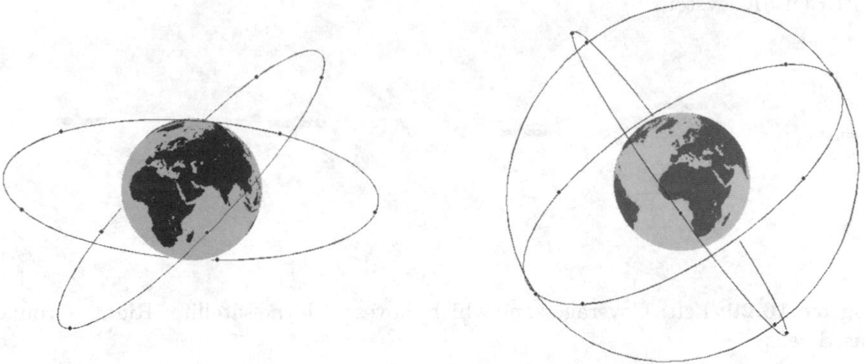

Figure 16.22: Orbits of the ICO (left) and Odyssey (right) satellite systems

a cell boundary in order to prevent undesirable interference. Splitting up the satellite coverage zones into cells provides a possibility for the multiple use of frequencies.

16.4.1 Antennas

16.4.1.1 Phased-Array Antennas (PAR)

The satellite systems planned for use in non-geostationary orbits may use *phased-array antennas*.

Individual dipoles act as omnidirectional beams. However, if groups of dipoles are used, a certain directional pattern is created through the phase difference at the receiver, which in a distant field is only dependent on the direction. The directional pattern of a dipole group is dependent on the layout, phases and amplitudes of the dipoles. This is exploited when phased-array antennas are used. The phases of the dipoles are then controlled electrically,

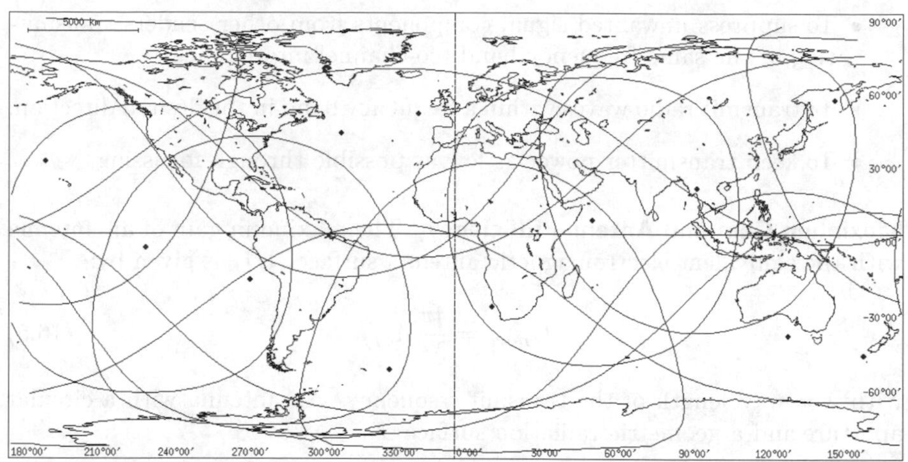

Figure 16.23: Coverage zones of the Odyssey system

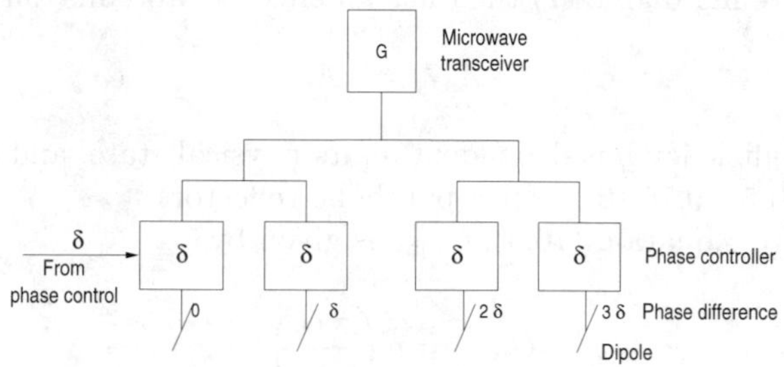

Figure 16.24: Principle of phased-array antennas

as shown in Figure 16.24 for one antenna with four dipoles. This provides for high speeds of beam steering.

Satellite systems with PAR are lighter in weight than mechanically controlled antennas. Phased-array antennas prove to be advantageous, even on the ground, compared with those that swivel mechanically, because with high-gain antennas two antennas are needed owing to the slowness of the swivelling when mechanically swivelled antennas are used in handovers between different satellites.

16.4.1.2 Parameters

The main tasks of a satellite antenna are:

- To receive the desired signal with a given polarization in a specific frequency band.

- To suppress unwanted signal components from other senders transmitting in the same frequency band (co-channel interference).
- To transmit radio waves within a frequency band in the desired direction.
- To keep transmitter power as low as possible through focussing.

Maximum Gain and Antenna Efficiency The maximum gain of an antenna with the equivalent electromagnetic antenna surface A_{eff} is given by:

$$G_{max} = \frac{4\pi}{\lambda^2} A_{eff} \tag{16.5}$$

with λ = wavelength of the transmit frequency. An antenna with a circular aperture and a geometric radiation surface of

$$A = \pi D^2/4 \tag{16.6}$$

(with D = antenna diameter) then has an effective antenna surface

$$A_{eff} = \eta A. \tag{16.7}$$

The antenna efficiency η is dependent on its physical state, and is typically in the region of 0.55–0.75 (0.55 for a parabolic reflector).

According to Equation (16.5), $G_{\max}$ is given by

$$G_{\max} = \eta \left(\frac{\pi D}{\lambda} \right)^2 \tag{16.8}$$

Depending on the angle θ, the antenna gain decreases compared with the maximum gain for $\theta = 0$.

Characteristics Figure 16.25 shows the modelled characteristics of satellite antennas in CCIR-Rep. 558-4. The characteristics of the steep drop are described by

$$G(\theta) = G_{\max} - 3 \left(\frac{\theta}{\theta_o} \right)^2 \quad \text{[dB]} \tag{16.9}$$

For an angle $\theta > \alpha_{SL}$ the assumption is that the antenna gain is very small (−100 dB). This means that signals incident on the antenna from an angle $\theta > \alpha_{SL}$ cannot be received. Another way of looking at this would be that the antenna has no transmitting effects for $\theta > \alpha_{SL}$.

Antennas with spherical characteristics are normally assumed for mobile users. The antenna gain is constant at 0 dB for all angles.

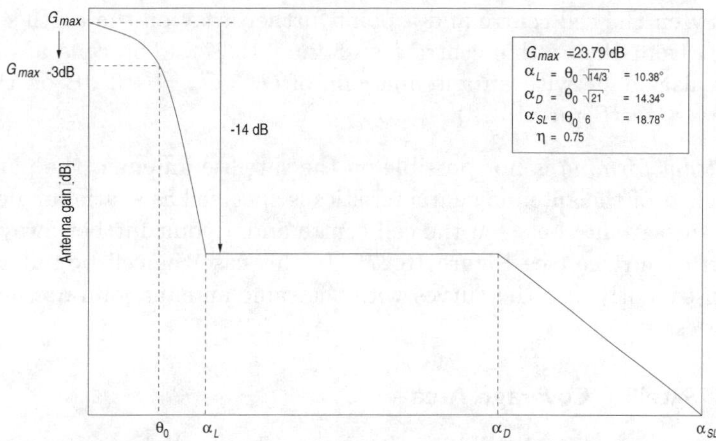

Figure 16.25: Characteristics of satellite antennas (values for IRIDIUM)

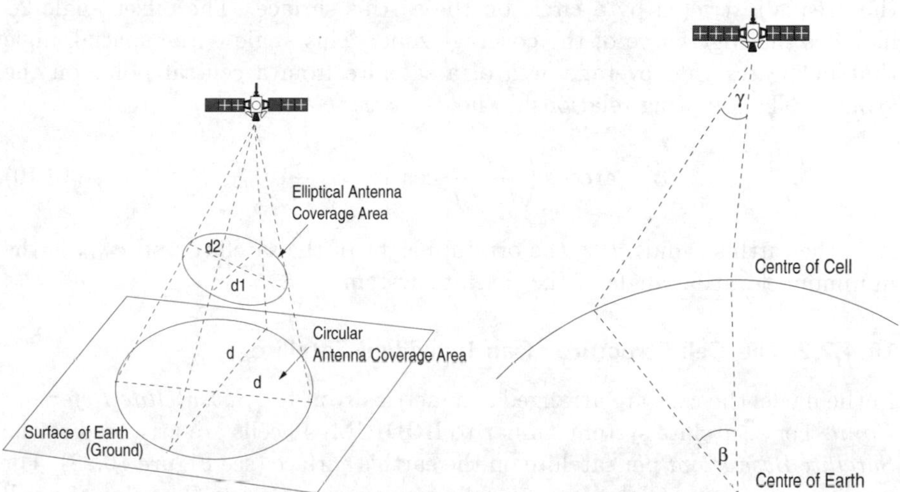

Figure 16.26: *Beam forming* by the satellite antenna

Figure 16.27: Definition of antenna parameter angles β and γ

16.4.2 Satellite Coverage Area and Cell Structure

The size and shape of cells produced by satellite antennas is determined by the characteristics of the antenna used. According to Figure 16.25, there are two approaches to defining the angle parameters α of the antenna characteristics [2]:

1. With a satellite antenna with *beam forming*, the parameter angle α of the antenna characteristics can be specified as the spatial angle β

between the cell centre and a point further away on the earth's surface seen from the earth's centre (see Figure 16.27). The cells are then of circular shape with an antenna gain of $G = G_{max} - 3$ dB on the edge curves (see Figure 16.26).

2. If *beam forming* is not possible on the satellite antenna, the parameter angle α of the antenna characteristics is specified as spatial angle γ seen by the satellite between the cell centre and a point further away on the earth's surface (see Figure 16.27). In this case the cell boundaries and consequently also the curves with the same antenna gain are no longer circles.

16.4.2.1 Satellite Coverage Area

The coverage area of a satellite is determined by its illumination zone. The coverage area can be of any shape. The assumption is made that a satellite can shape its coverage area (*beam forming*, Assumption 1). In the model the area is restricted by a circle on the earth's surface. The offset angle 2ϕ includes the edge curve of the coverage zone. This angle is the spatial angle that delineates the coverage area of a satellite from a central point on the ground. The following relationship holds for ϕ:

$$\phi = \arccos\left(\frac{r_e}{r_e + h} \cos\, \epsilon_{min}\right) - \epsilon_{min} \tag{16.10}$$

r_e is the earth's radius, h is the orbital height of the satellite and ϵ_{min} is the minimum elevation angle of the satellite system.

16.4.2.2 The Cell Structure of an Individual Satellite

In the model the cells are arranged symmetrically at the *Subsatellite Reference Point.* For a satellite system similar to IRIDIUM, 48 cells are arranged at the *Satellite Base Point* per satellite on the earth's surface (see Figure 16.28). For a system similar to Globalstar, 16 cells appear in two rings. The size of a cell is determined by the antenna chacteristics. The cell edge forms a closed curve on which the signalling of the user to his satellite is reduced by half (-3 dB) compared with the position in the cell centre.

16.4.2.3 The Cell Structure in a Satellite System

The overlapping of all the coverage areas of satellites is usually more distinct than the overlapping of the cells of an individual satellite. The overlapped areas of the coverage areas are particularly large at the earth's poles because of the predefined orbits. If the coverage zones of two satellites overlap near the equator then up to five satellites could be overlapping near the poles. In satellite systems this produces very complex cell structures, which are constantly changing because of the changing overlapping of the satellite coverage areas.

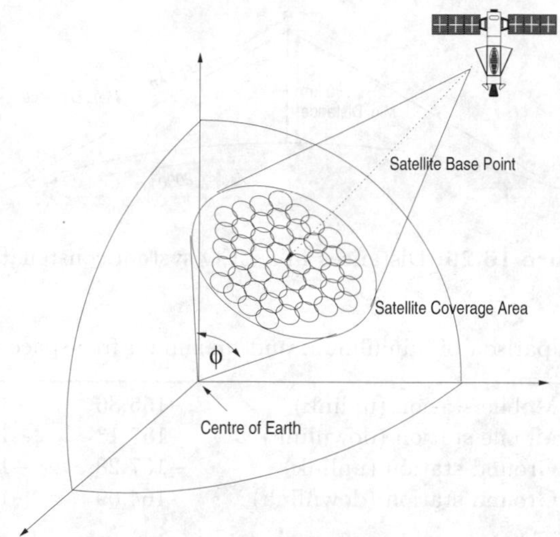

Figure 16.28: Position of cells in satellite coverage area

The cell structures are formed by the antenna characteristics and free-space attenuation.

16.4.3 Radio Propagation

Because of large distances between a satellite and a transmit/receive station, a large free-space attenuation L_0 undergo signals given by

$$L_0[\text{dB}] = 92.4 + 20 \log f_c[\text{GHz}] + 20 \log d[\text{km}] \qquad (16.11)$$

The *free-space attenuation* increases logarithmically with the path length, which corresponds to 6 dB with each doubling of the distance. Atmospheric and rain attenuation further increase the signal degradation. Though C-band losses due to atmospheric and rain attenuation are relatively small, this effect has to be taken into account at higher frequencies (over 10 GHz) (see Figure 2.3). The signal is also affected by the following factors:

1. Change in polarization and resulting additional interference
2. Increase in noise temperature

16.4.3.1 Path Loss in LEO and MEO Satellite Systems

As mentioned above, the free-space attenuation depends on the frequency and the distance of the mobile user or base stations from the ground to the satellite. The existence of different constellations of the systems produces different free-space attenuations. There is a big difference between maximum and minimum

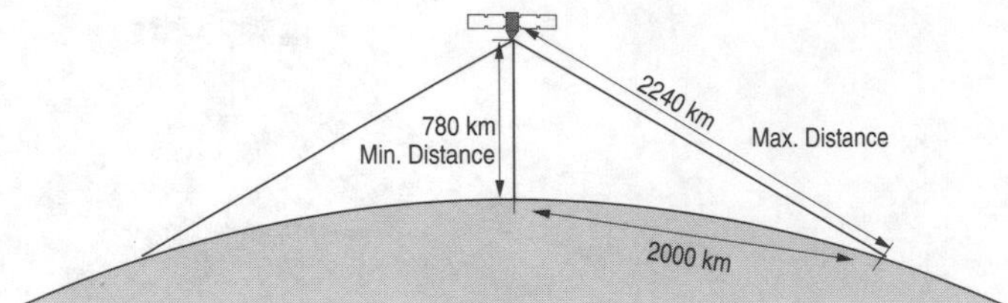

Figure 16.29: Distances in a LEO system constellation

Table 16.8: Comparison of mininimum and maximum free-space attenuation [dB]

IRIDIUM	Mobile station (uplink)	−156.36	−163.28	6.92
	Mobile station (downlink)	−157.13	−164.06	6.93
	Ground station (uplink)	−167.28	−174.21	6.93
	Ground station (downlink)	−164.69	−171.62	6.93
Globalstar	Mobile station (uplink)	−161.51	−167.44	5.93
	Mobile station (downlink)	−162.30	−168.23	5.93
	Ground station (uplink)	−172.45	−178.38	5.93
	Ground station (downlink)	−169.85	−175.78	5.93
Odyssey	Mobile station (uplink)	−178.81	−181.08	2.27
	Mobile station (downlink)	−179.59	−181.86	2.27
	Groud station (uplink)	−189.74	−192.01	2.27
	Ground station (downlink)	−187.15	−189.42	2.27

distance owing to the low flight paths and relatively large illumination zones, especially with the LEO and MEO constellations, illustrated in Figure 16.29 for a LEO constellation.

According to Equation (16.11), the free-space attenuation is dependent on the frequency used and the distance between satellite and mobile station. The different distances that can exist between users and their connecting satellites produce varying levels of free-space attenuation. The distance d between a user and a satellite depends on the *elevation angle*:

$$d = \sqrt{r_e^2 + h^2 - (r_e \cos \epsilon)^2} - r_e \sin \epsilon \tag{16.12}$$

r_e is the earth's radius, h is the height of the satellite and ϵ is the elevation angle.

Examples of minimum and maximum free-space attenuation resulting within a satellite illumination zone are given in Table 16.8 for three satellite systems with different orbit heights: IRIDIUM (780 km), Globalstar (1389 km) and Odyssey (10 354 km).

The difference between maximum and minimum free-space attenuation (last column in Table 16.8) in LEO system constellations is considerably greater than in the MEO constellation Odyssey, because the difference in dis-

Table 16.9: Rain rate statistics [mm/h] of different regions of Europe [NASA Reference Publication 1082(03)]

Percent annually	Zone C	Zone D_1	Zone B_2
0.001	78	90	70
0.01	28	35.5	23.5
0.1	7.2	9.8	6.1

tance between the satellite base point and the edge of the illumination zone is greater.

16.4.3.2 Attenuation Caused by Weather Conditions (3 ... 180 GHz)

With frequencies over 1 GHz, scattering and absorption of electromagnetic energy due to rain cause considerable attenuation to signal amplitude. These effects increase at higher frequency.

Statistics of Rain Rate The problems increase with the statistical characteristics of the rain rate. The maximum rain rates often exceed average values. Table 16.9 lists the rain rates that are reached 0.001 %, 0.01 % and 0.1 % of the time in zones C, D_1 and B_2. These zones cover most of the area of Europe. Germany, France and the British Isles belong to Zone C, large parts of Southern Europe belong to Zone D_1 and Eastern Europe to B_2 [15].

Calculation of Rain Attenuation Different methods are available for predicting rain attenuation. According to the CCIR recommendation [15], the attenuation A is given by

$$A = aR^b L_{eff} [\text{dB}] \tag{16.13}$$

The value aR^b is the specific attenuation [dB/km] and L_{eff} the effective path length due to rain. The coefficients a and b are based on frequency, raindrop size, elevation and other factors, but can be approximated as follows [14]:

$$a = \begin{cases} 4.21 \cdot 10^{-5} f^{2.42}, & 2.9 < f < 54 \text{ GHz} \\ 4.09 \cdot 10^{-2} f^{0.699}, & 54 < f < 180 \text{ GHz} \end{cases} \tag{16.14}$$

$$b = \begin{cases} 1.41 \cdot f^{-0.0779}, & 8.5 < f < 25 \text{ GHz} \\ 2.63 \cdot f^{-0.272}, & 25 < f < 164 \text{ GHz} \end{cases} \tag{16.15}$$

The rain height h_r, which depends on the degree of latitude ϕ, is required in order to calculate the effective path length based on the CCIR recommendation:

$$h_r = 5.1 \text{ km} - 2.15 \log(1 + 10^{(\phi - 27)/25}) \text{ km} \tag{16.16}$$

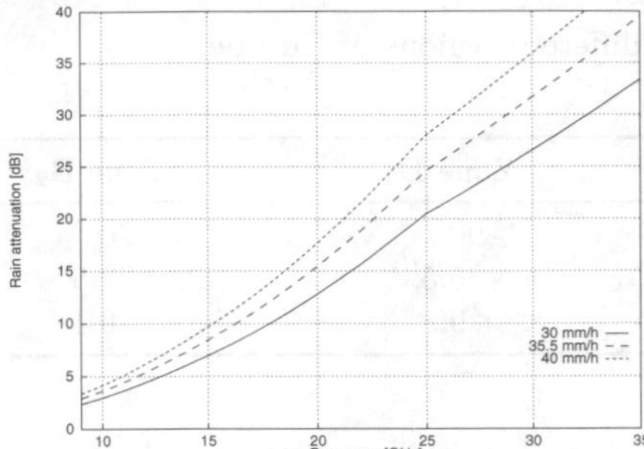

Figure 16.30: Rain attenuation versus frequency on the basis of different rain rates and 10° elevation

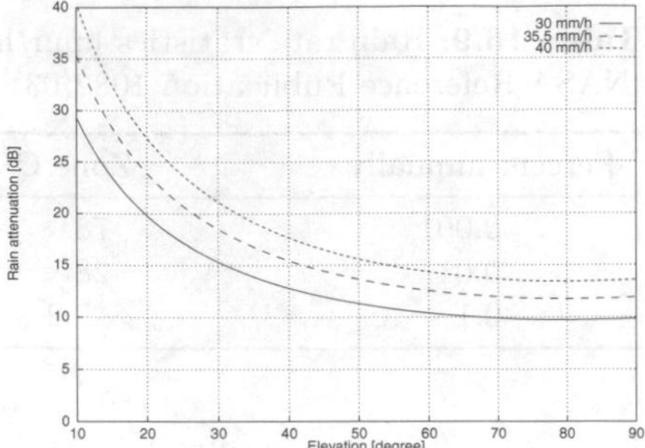

Figure 16.31: Rain attenuation versus elevation at 20 GHz

The path length L can then be calculated from h_r and the height of the receiver over NN, h_0, as well as from the elevation angle θ as follows [15]:

$$L = \begin{cases} \dfrac{2(h_r - h_0)}{(\sin^2(\theta) + 2(h_r - h_0)/8500)^{\frac{1}{2}} + \sin(\theta)} \text{ km}, & \theta < 10^\circ \\ \dfrac{h_r - h_0}{\sin(\theta)} \text{ km}, & \theta \geq 10^\circ \end{cases} \tag{16.17}$$

The attenuation is then:

$$A = aR^b L r_p, \qquad \text{with} \quad r_p = \frac{90}{90 + 4L\cos\theta} \tag{16.18}$$

The basis used in system design is normally the value $R_{0.01}$ [15], which indicates the rain rate that 99.99 % of the time is actually lower, and the required transmission quality guaranteed for this rain rate. As can be extrapolated from the equations, the rain attenuation depends on many different factors, such as rain rate, frequency, elevation, geographical latitude, height of earth station, as well as other values such as raindrop diameter.

At frequencies above 10 GHz rain attenuation has a strong effect on connection quality. The dependency of the frequency was determined through a simulation of the geographical latitude of 40° and an elevation of 10° and illustrated in Figure 16.30. The rain attenuation increases dramatically the higher the frequency. Frequency selection therefore has a great impact on transmission quality, thereby limiting suitable frequency bands.

Figure 16.31 shows the simulated rain attenuation over the elevation at 20 GHz. Below a 30° elevation, the rain attenuation rises sharply as the elevation decreases. The aim therefore is to strive for constellations with high elevations. This can only be achieved through increasing the number of satellites, which results in high costs and high risks or through high orbits which in turn increases the signal propagation times.

16.4.4 Power Control

The blocking probability of calls depends in part on signal strength, so power control can be used to compensate for path loss. Otherwise mobile stations located a minimal distance from the satellite would have a high signal strength at the antennas of the satellite and consequently cause comparatively strong co-channel interference.

In general, power control serves the following purpose in satellite systems:

- Compensation of propagation attenuation.
- Compensation of deterioration in radio propagation conditions (e.g., through meteorological influences) in order to maintain a good-quality link.
- Compensation of co-channel interference. This occurs when more than one mobile station is maintaining a connection on the same channel at the same time.
- Protection of mobile station and satellite batteries through reduction of emitted signal power.

A possible solution to this problem would be for mobile stations operating in the proximity of the satellite base point to reduce their power by the following in favour of mobile stations on the edge of the satellite radiation zone:

$$L = 20 \log \left(\frac{4\pi f d_{max}}{c} \right) \tag{16.19}$$

d_{max} is the maximum distance between a satellite and a mobile station on the edge of the illumination zone, c is the speed of light and f is the frequency.

16.5 Interference in the Satellite Radio Network

Besides attenuation, signals experience disturbance due to interference from other signals. There are two different types of interference:

1. Co-channel interference caused by the use of the same radio channel in a neighbouring cluster.
2. Adjacent-channel interference caused by the use of adjacent radio channels in the same cluster.

16.5.1 Co-Channel Interference

Channel allocation methods are based on an evaluation of the carrier-to-interference (C/I) ratio measured at the antennas of the satellite and of the mobile user. A simple model is used here to explain the level of interference.

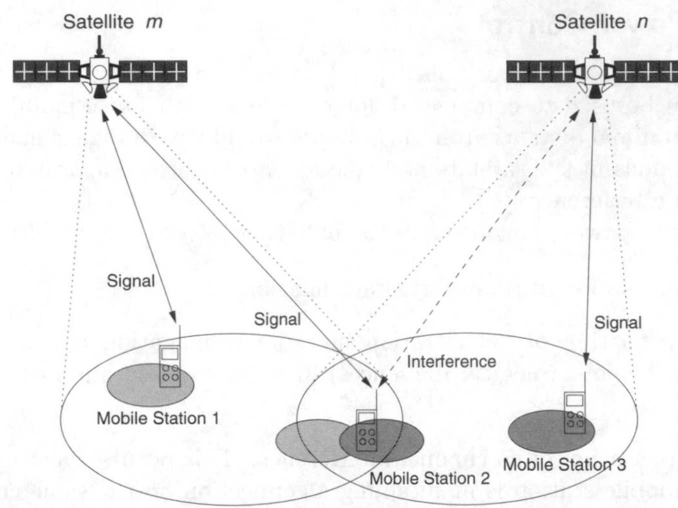

Figure 16.32: Co-channel interference caused by the use of the same channel

No influences caused by weather (attenuation, polarization) or environmental influences (shadowing, fading) are taken into account. The received and transmitted signal strength is determined solely on the basis of free-space attenuation, the characteristics of the antenna used and the power control of the mobile stations. Interference in the simplified model is therefore caused by the user signal of a call and the noise signals received from calls of other mobile stations. Interference can only occur if both connections are using the same physical channel.

Figure 16.32 shows that interference can occur in the uplink and in the downlink for various reasons, thereby necessitating a separate calculation of C/I_{Uplink} and the C/I_{Downlink}. The carrier-to-interference ratios will be calculated below.

16.5.2 Uplink Carrier-to-Interference Ratio

The carrier-to-interference ratio C/I_{Uplink} measured by the satellite is

$$C/I_{\text{Uplink}} = \frac{C_{\text{Uplink}}}{\sum_{x=1}^{N} I_x(\text{mobile station}_x)} \tag{16.20}$$

C_{Uplink} is for the user signal power of a mobile station received on the channel of a satellite. I is the interference power received on the same channel from mobile stations located in the coverage area of the satellite and maintaining a connection to this or to a neighbouring satellite. N is the number of co-channel mobile stations that are operating in the illumination zone of the satellite.

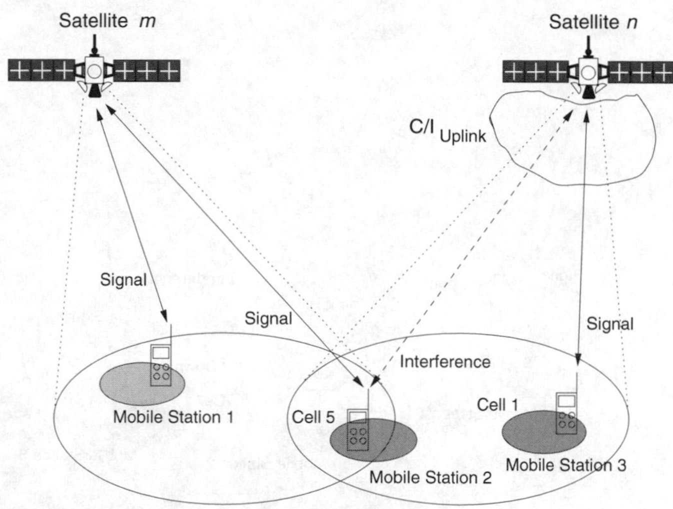

Figure 16.33: Development of uplink interference level

For example, in Figure 16.33 mobile station 3 has set up a connection to satellite n (antenna 1); mobile station 2, on the other hand, has a connection to satellite m (antenna 5). Since both mobile stations are transmitting on the same channel, mobile station 2 is interfering with the signal sent by mobile station 3. C/I_{Uplink} of the connection to mobile station 3 received at satellite n (antenna 1) is therefore given by

$$C/I_{\text{Uplink}}(\text{Sat } n,\ \text{Ant } 1, \text{MS } 3) = \frac{C(\text{Sat } n,\ \text{Ant } 1, \text{MS } 3)}{I(\text{Sat } n,\ \text{Ant } 1, \text{MS } 2)} \tag{16.21}$$

16.5.3 Downlink Carrier-to-Interference Ratio

The carrier-to-interference ratio C/I_{Downlink} at a mobile station is

$$C/I_{\text{Downlink}} = \frac{C_{\text{Downlink}}}{\sum_{x=1}^{N} \sum_{y=1}^{M_x} I_{xy}(\text{satellite antenna}_{xy})} \tag{16.22}$$

C_{Downlink} is the user signal level received on a channel at the mobile station. N is the number of satellites in whose coverage area a mobile station requesting a channel is located. M_x is the number of antennas for each of these satellites that is maintaining a connection to another mobile station on the same channel. I_{xy} is the interference level sent by the above-mentioned satellite antennas and received at the mobile station.

Figure 16.34 illustrates this example in more detail. Satellite n (antenna 1) has established a connection to mobile station 3. Satellite m (antenna 5), on the other hand, has a connection to mobile station 2. Since both satellites are transmitting on the same channel, satellite n (antenna 1) is interfering with

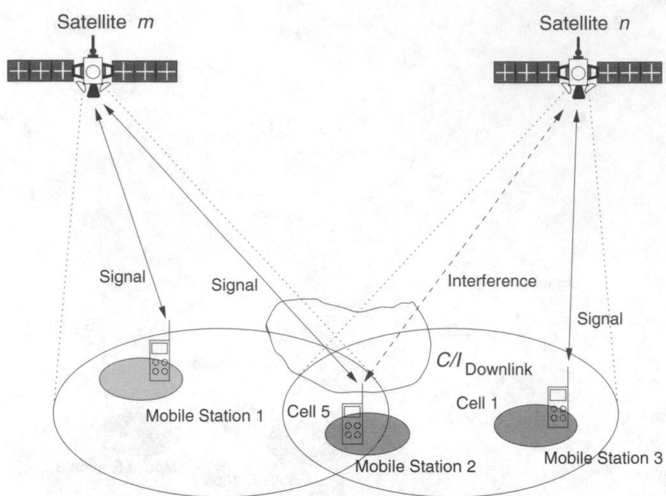

Figure 16.34: Development of downlink interference level

the user signal of the satellite (antenna 5) at mobile station 2. C/I_{Downlink} of the connection to satellite m (antenna 5) received at this mobile station is then given by

$$C/I_{\text{Downlink}}(\text{Sat } m,\ \text{Ant } 5,\ \text{MS } 2) = \frac{C(\text{Sat } m,\ \text{Ant } 5,\ \text{MS } 2)}{I(\text{Sat } n,\ \text{Ant } 1,\ \text{MS } 2)} \qquad (16.23)$$

16.5.4 Model of a Land Mobile Satellite Channel

A channel model is introduced to determine the bit error probability and incorporates a channel structure similar to the IRIDIUM system. Consideration will only be given to the communications link between mobile station and satellite. Because of the higher transmit power of the ground station with a higher antenna gain (see [5]) and constant line-of-sight path, the link between satellite and ground station (*gateway*) is not considered important for the discussion here.

As is the case with terrestrial mobile radio channels, land mobile satellite channels are also affected by the following interference effects [1]:

Fading due to shadowing Obstacles such as trees and buildings on the transmission path cause short-term shadowing and consequently produce considerable signal attenuation. The shadowed areas become larger as the satellite elevation drops, which results in a minimum elevation angle necessary for transmission.

Multipath fading Objects in the immediate vicinity of a mobile radio user can cause reflection of the signal with different propagation times of the resulting paths at the receiver. The superposition (interference) of the reflected signal with the signal transmitted over the line-of-sight produces short-term signal fading.

In contrast to terrestrial channels, where a mobile station's speed of movement has a major effect on the duration of fading, with satellites this hardly makes any difference. For example, a satellite in the IRIDIUM system has an orbital speed (orbit time: 1 h 40‘ 27“) of around 27 800 km/h. Taking into account the earth's rotation (1670 km/h at the equator) through addition of the two velocity vectors produces the relative speed to earth, which is higher than the mobile station speed by many magnitudes. The IRIDIUM satellite is approximately 150 times faster than an automobile travelling at 170 km/h. Consequently the fading duration is shorter by a factor of 50 than in terrestrial mobile radio networks.

Between 1984 and 1987, the German Aerospace Institute (DLR: Deutsche Forschungsanstalt für Luft- und Raumfahrt, Oberpfaffenhofen, Germany), studied channel behaviour for different elevation angles and in different environments over the geostationary MARCES satellite. A modulated carrier signal in the L-band (1.54 GHz) was beamed over satellite from a ground station in Spain (ESA in Villafrance) to a measuring vehicle in different locations in Europe (Stockholm, Copenhagen, Hamburg, Munich, Barcelona and Cadiz) [11].

The results show short-term fading signals in open areas, e.g., as caused by trees and bridges to a travelling automobile. Even stronger multipath and fading signals caused by shadowing were observed in urban areas. Shadowing is particularly more pronounced in cities, and leads to time durations that alternate between good and poor reception conditions.

The measurements taken were used to develop a model for a land mobile satellite channel that reproduces the stochastic behaviour of the signal carrier received. The basis of this model is formed by two channel states that mirror each other in good and poor reception conditions. A good channel state is represented by a direct line-of-sight connection to the satellite, whereas a poor state is characterized by shadowing of the direct transmission path. In both cases the satellite signal is reflected by a large number of objects located in the immediate vicinity of the mobile station. The respective signals are received by the mobile station with differing propagation times and amplitudes, and superpose themselves with those of the direct transmission path.

This so-called DLR model of the land mobile satellite channel appears in Figure 16.35. A distinction is made between the good channel state with Rice fading and the bad state with lognormally distributed Rayleigh fading and low receive power. The switch controlled through the Markov process reproduces a random succession of good and bad channel states. The transmit signal is multiplied with the fading signal $a(t)$ and an *AWGN signal* (white Gaussian noise) is then added.

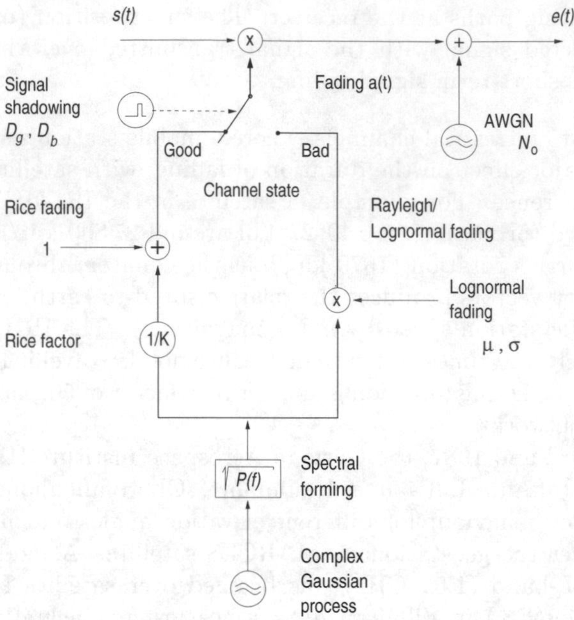

Figure 16.35: Analogue (DLR) model of land mobile satellite channel

The channel model is characterized by the Rice factor c, the mean value μ and the dispersion σ^2 of the lognormal attenuation, as well as by the average durations D_g and D_b of the good or bad channel states. The time part of bad channel states is indicated as shadowing factor A [10]. These parameters are given in [11].

The average duration of channel states depends on the following factors:

- Environment (*open, suburban, urban*)
- Elevation of visible satellite
- Speed at which mobile station is moving

These factors can be used to simulate the signal profile of a land mobile satellite channel for different receiving conditions. Figure 16.36 provides an example of simulation results for the *receive signal envelope*. The information digitally modulated to the carrier signal is recovered in the demodulator of the receiver, and the current channel quality is assessed through an evaluation of the error-correction unit in the receiver. The connection is jeopardized if the channel quality is not adequate, and must then be replaced by a connection with a better channel.

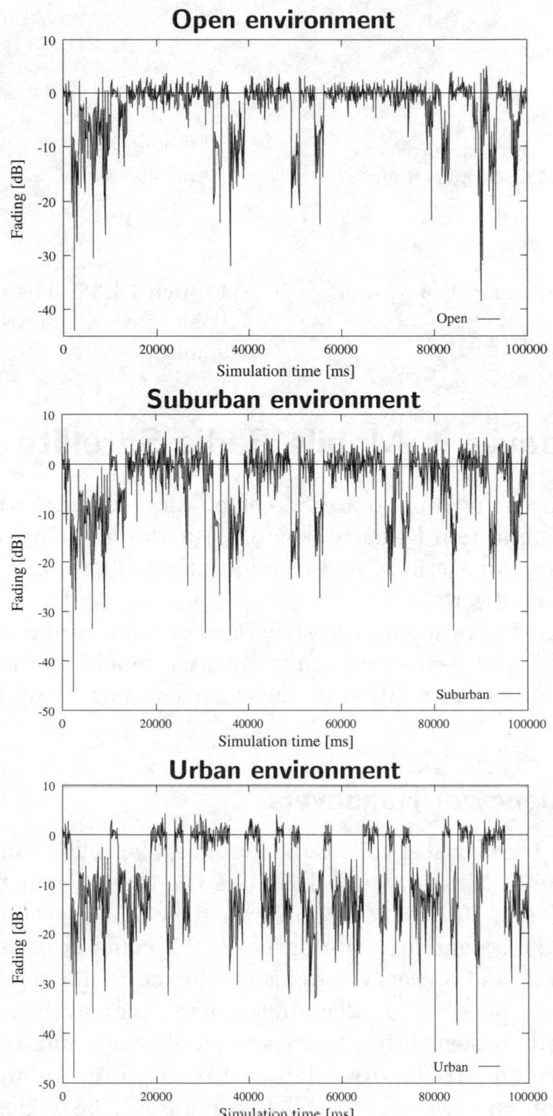

Figure 16.36: Signal profile in different environments, velocity $v = 108$ km/h, 21° satellite elevation

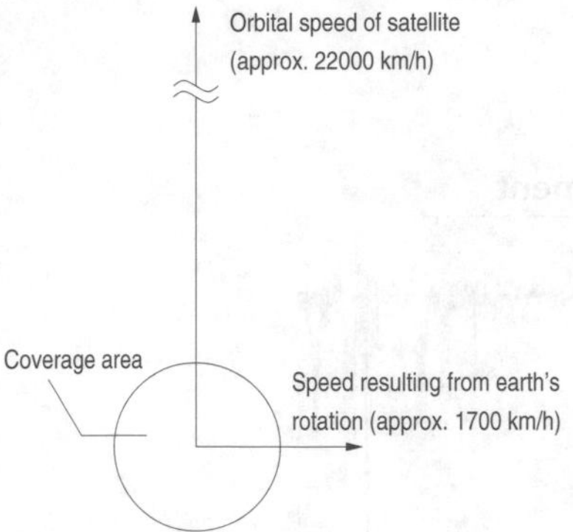

Figure 16.37: Movement of a coverage zone (spot)

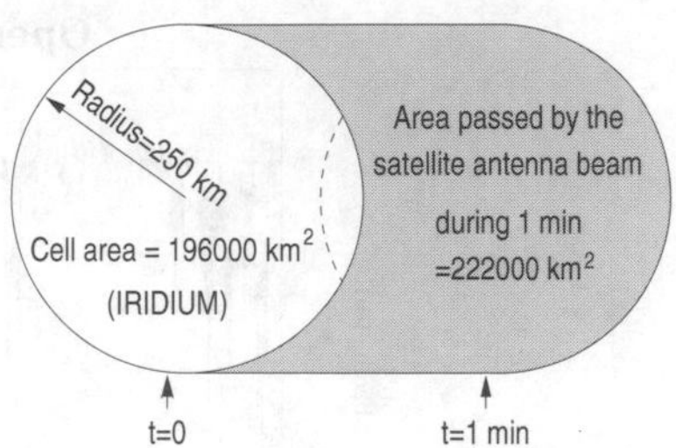

Figure 16.38: The area of an IRIDIUM system cell overflown in one minute

16.6 Handover in Mobile Radio Satellite Systems

A handover is a changeover to a different physical channel without the risk of the current connection being broken off. Handovers help to provide users with good-quality connections, and consequently are an important feature of the mobile radio network.

In contrast to the stationary cell structures of terrestrial cellular networks, where handovers are necessary because of user mobility, the cells in LEO satellite systems move in relation to the earth's surface at up to 7 km/s (see Figure 16.37).

16.6.1 Frequency of Handovers

Compared with terrestrial mobile radio systems, the cells of satellite systems are very large; for example, with IRIDIUM the diameter of the cells is approximately 500 km. If these cells were stationary on the ground, there would seldom be a need for a handover during a call. If it could be assumed that communicating mobile radio users are distributed evenly in a certain area of the world, it would be possible to calculate the new area travelled during a given time interval and consequently the average number of handovers executed.

In one minute an IRIDIUM satellite cell travels over an area that is 1.12 times larger than the area of the cell (see Figure 16.38). Based on the assumption that a cell has a traffic volume of 80 Erl., the number of handover occurrences would be on average $80 \cdot 1.12 = 90$ per minute.

Based on an average call duration of 2 min and 80 Erl. of traffic, the number of new calls would be 40 per minute. The number of handovers would then be greater by a factor of 2.25 than with terrestrial systems.

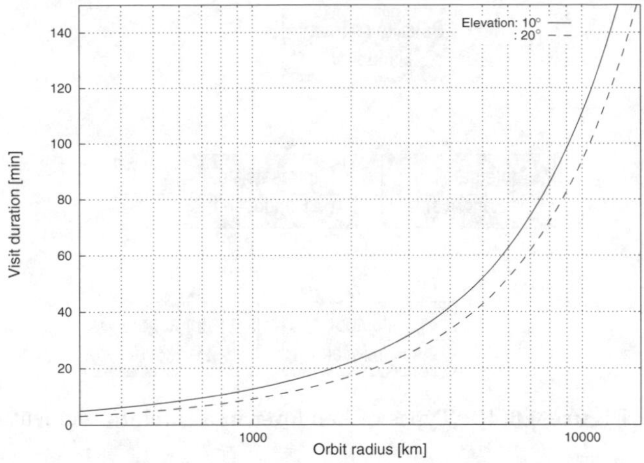

Figure 16.39: Maximum visit duration in a coverage zone

With satellite systems communication is only possible over a line-of-sight link. Therefore the time span of the line-of-sight link is extremely important. It depends on the orbital speed of the satellite and the size of the coverage zone (spot), as well as the position and direction of movement relative to earth. Studies have shown that the maximum duration of a visit in a coverage zone can be calculated as

$$t_{max}(r) = \frac{2\, r_{cov}}{v_g} \tag{16.24}$$

The assumption is that the mobile user is not moving, because his speed compared to the orbital speed of the satellite and the speed based on the earth's rotation is negligible. In Equation (16.24) r_{cov} is the radius of the coverage zone and v_g the velocity of the satellite base point relative to the earth's surface. For two elevation angles Figure 16.39 shows the maximum visit duration of a mobile user in a coverage zone over the orbit radius.

16.6.2 Types of Handover

A differentiation is made between intrasatellite handovers within the coverage zone of a satellite and intersatellite handovers between different satellites. Intrasatellite handovers are divided into intracell and intercell handovers (see Figure 16.40).

16.6.2.1 Intersatellite Handovers

Even stationary terminals are covered by the same satellite only for just a few minutes (see Figure 16.39).

When a connection is being established or a handover is being executed, the satellite that remains visible for as long a period of time as possible must

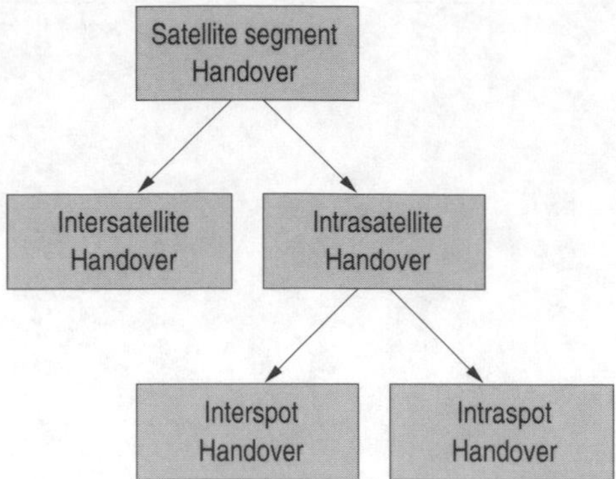

Figure 16.40: Types of handover in a satellite segment

be selected in order to keep the number of handovers required to a minimum. During this period of time, the channel parameters inevitably change considerably because unfavourable but still acceptable elevations occur. If a satellite is selected because of optimal channel parameters, it is more than likely that more handovers will have to be executed.

The objective is to keep the elevation angle at an acceptable level in order to guarantee satisfactory connection quality. There are two procedures known for this [3].

Maximization of current elevation angle With this strategy the satellite that is visible below the currently largest elevation angle is selected at all times during an existing connection. Handovers are also carried out if the connection quality is sufficient in order to maximize them. The shadowing probability is reduced because the connection is constantly being executed above the optimal elevation angle. The disadvantage of this procedure is that handovers could be executed unnecessarily because a satellite can provide sufficient connection quality even at a lower elevation.

Minimization of intersatellite handover rate This strategy allows handovers only when a given minimum elevation angle ϵ_{min} has been reached. For a new connection the satellite that is visible the longest is always selected on the condition $\epsilon \geq \epsilon_{min}$. This necessitates a position establishment by the mobile station.

With this procedure it is possible to keep the handover rate low, but in some cases the satellite selected cannot supply an acceptable connection quality.

Possible criteria for an intersatellite handover are as follows:

- The signal quality falls below a certain margin, i.e., the distance between satellite and user has become too great.

- The signal carrier-to-interference ratio is too low because other users are using the same channel and causing co-channel interference.

If a handover is initiated, the satellite that has the best signal quality or the best elevation angle is selected.

16.6.2.2 Intersegment Handover

The concept of an intersegment handover is used to describe an important function for integrating future LEO satellite systems into terrestrial cellular networks. An intersegment handover enables an active connection to be changed from a terrestrial cell to a satellite cell (*generated by a spotbeam antenna*) or the other way around.

A typical application would be the support of local cellular networks (e.g., of a city) by a satellite network in areas away from cities.

Intersegment handovers are discussed below based on the GSM system. A possibility for the integration of ground stations (*gateways*) into the GSM network is also introduced.

Integration at the network level This integration concept deals with the direct connection of satellite gateways to mobile switching centres (MSCs) of the GSM network (see Chapter 3). As illustrated in Figure 16.41, the gateway together with the *Controller* forms a *Satellite Relay Subsystem* (SRSS) which has a similar function as the *Base Station Subsystem* (BSS) of GSM. In addition, the SRSS has its own database, the *Satellite Visitor Location Register* (S-VLR), into which information on all active users located within the coverage zone (*footprint*) of the satellite is entered. Just like the VLR in GSM, this database can be used through the MSC or through the *Home Location Register* (HLR). The advantage of this kind of integration of satellite system and GSM is that it allows the use of existing protocols and interfaces for mobility management. It requires a close cooperation between satellite and GSM network operators [7].

Intersegment handovers in integrated GSM satellite systems Intersegment handovers, which correspond to intra-MSC handovers in GSM, are introduced below.

Figure 16.42 presents the handover protocol for changing to the GSM segment and Figure 16.43 that for the opposite direction. In both cases the relationship to the intra-MSC handover protocol described in Section 3.6.8 is apparent. Even the protocol elements are essentially identical. Changes to cell and channel identity appear to be necessary in the satellite segment to comply with the GSM handover protocol. The next paragraph deals with a peculiarity of satelllite channel access.

Random access to satellites Unlike terrestrial networks, in which mobile stations, except for the time-displacement factor (*timing advance*),

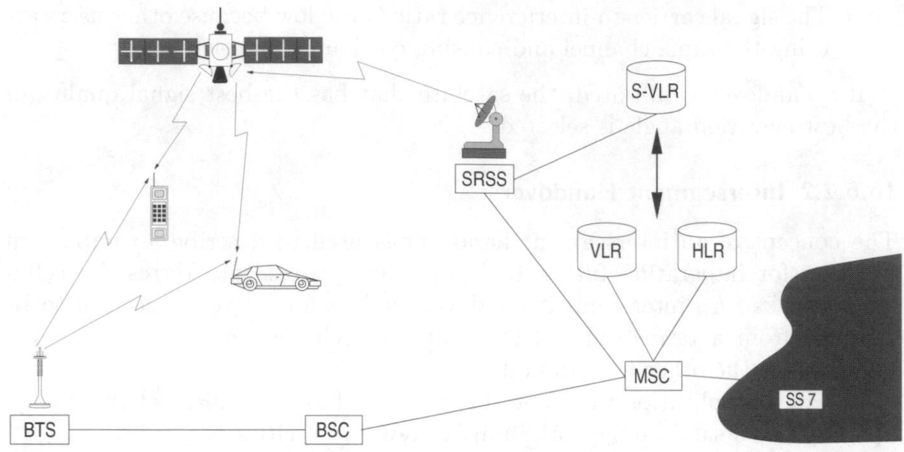

Figure 16.41: System level integration scenario in GSM systems

synchronize with the logical time-multiplex channel for multiple access before they access the channel, synchronization in the satellite segment is not always possible during a handover. Because the signal propagation times are not known and depend on the current distance between satellite and user or gateway, it is not possible for a mobile station to be notified before it accesses the satellite channel of the signal propagation time correction that it must undertake in order to maintain the frame structure with the satellite.

Separate channels therefore exist in the satellite segment for the transmission of channel allocation requests (Handover_Access, Channel_Request) that require no synchronization with the satellite or with other users. Each mobile station is allowed to transmit without considering whether the channel is already occupied. The pure ALOHA procedure is an example of this kind of *Time-Random Multiple-Access* protocol (TRMA protocol).

If several mobile users are transmitting at the same time on the Pure ALOHA channel, collisions will occur and collided data packets have to be retransmitted. The packets are not transmitted again until a random waiting time has passed to ensure that another collision may not occur when they are resent.

If the duration of a data packet is T, as illustrated in Figure 16.44, and it is assumed that all packets are of the same length, then no other stations are allowed to send in the time frame between $-T$ and $+T$ ($t = 0$ at end of transmission) so that the communication can be received correctly. The phase of conflict, i.e., the time during which no other station is allowed to send, is then $2T$. It is double the length of a packet. Equivalently to the analysis of the Slotted ALOHA protocol in Section 2.8.1, independent access by mobile stations can be modelled by a Poisson arrival process for packets. The probability of a successful transmission during the time T, which then

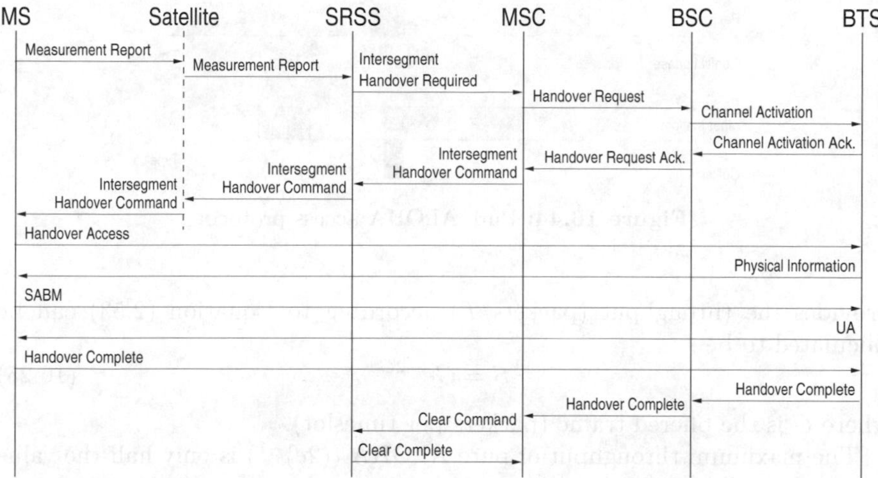

Figure 16.42: Intersegment handover protocol for changeover from satellite to GSM segment

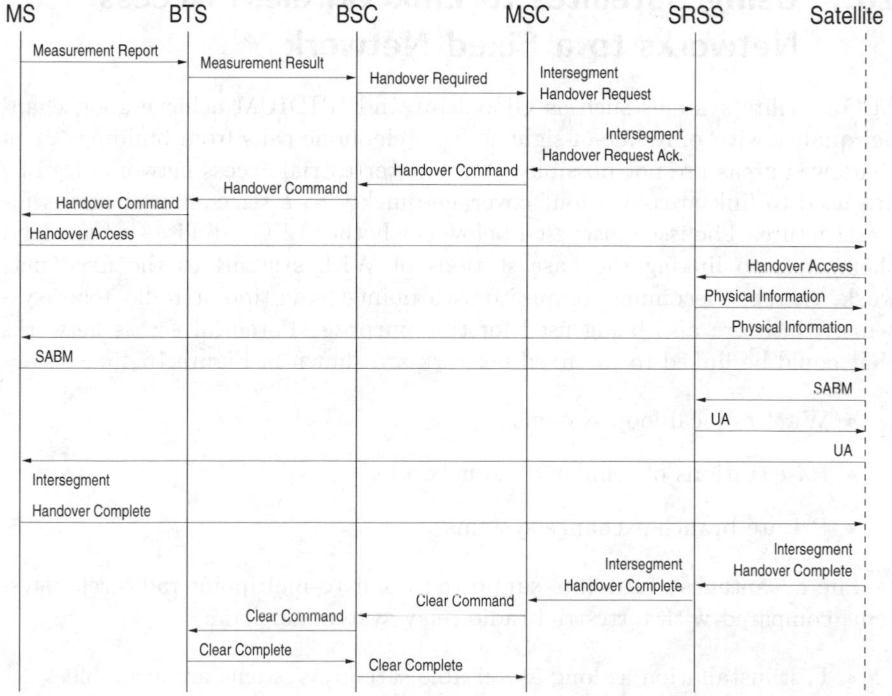

MS
BTS
BSC
MSC
SRSS
Satellite
Measurement Report
Measurement Result
Handover Required
Intersegment
Handover Request
Intersegment
Handover Request Ack.
Handover Command
Handover Command
Handover Command
Handover Access
Handover Access
Physical Information
Physical Information
SABM
SABM
UA
UA
Intersegment
Handover Complete
Intersegment
Handover Complete
Intersegment
Handover Complete
Clear Command
Clear Command
Clear Complete
Clear Complete

Figure 16.43: Intersegment handover protocol from GSM to satellite segment

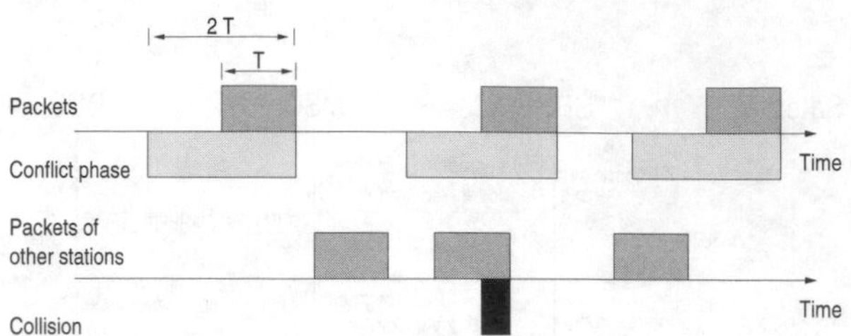

Figure 16.44: Pure ALOHA access protocol

provides the throughput (packets/T) according to Equation (2.53) can be calculated to be

$$S = Ge^{-2G} \tag{16.25}$$

where G is the offered traffic (packets per timeslot).

The maximum throughput of pure ALOHA ($(2e)^{-1}$) is only half the value known from S-ALOHA ($1/e$).

16.7 Using Satellites to Link Wireless Access Networks to a Fixed Network

LEO satellite systems such as Globalstar and IRIDIUM achieve good channel quality with only line-of-sight links. Telephone calls from buildings or in shadowed areas are not possible. Wireless terrestrial access networks (WLL) are used to link areas without coverage quickly to a telecommunications infrastructure. The issue discussed below is whether LEO satellite systems lend themselves to linking the base stations of WLL systems to the fixed network, thereby becoming competitors to point-to-multipoint radio relay systems, which are also being used for this purpose. Potential access networks that could be linked to the fixed network are shown in Figure 16.45:

- Wireless local-loop systems
- Base stations of cellular radio networks
- Private branch exchange systems

The advantages of satellite-supported point-to-multipoint radio relay systems compared with terrestrial radio relay systems include:

- Fast installation as long as suitable satellite systems are available.
- Low additional costs for infrastructure to handle new connections.

The disadvantages are:

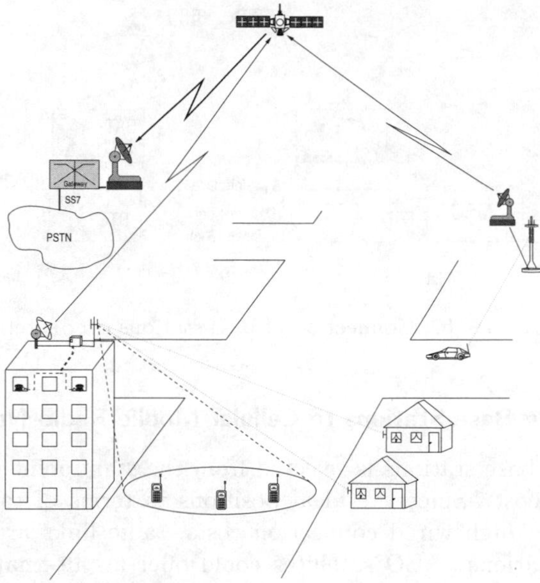

Figure 16.45: Scenarios showing the use of a satellite system as a transport network

- Higher signal propagation times.
- High system costs of satellite systems.

This raises the question of how many small or large cities could be supplied coverage by a given LEO system. The example considered involves a penetration of 1 % as well as of 10 %, with the WLL systems of a city being modelled as a point source/sink. With satellite cell sizes of 600 km this approximation is sufficiently exact. However, the area over which the WLL systems extend should not be neglected in the capacity study of the satellite system.

16.7.1 Simple Fictional WLL System

Since the first LEO satellite systems only support low transmission rates, WLL systems, e.g., according to the DECT standard, can only be linked through the trunking of several satellite channels per, say, DECT channel. However, this would considerably reduce the number of calls that could be carried. Therefore a fictional WLL system with a transmission rate of 4.8 kbit/s adapted to the satellite channel is assumed here for the capacity study. The voice quality will then be low, but it can be assumed that there will be a large market for services at a lower quality than is normally the case with telecommunications networks in places like Central Europe, as long as the price is sufficiently low. The Internet is an example of a data transfer system with currently a poor quality (in terms of throughput) but a high acceptance.

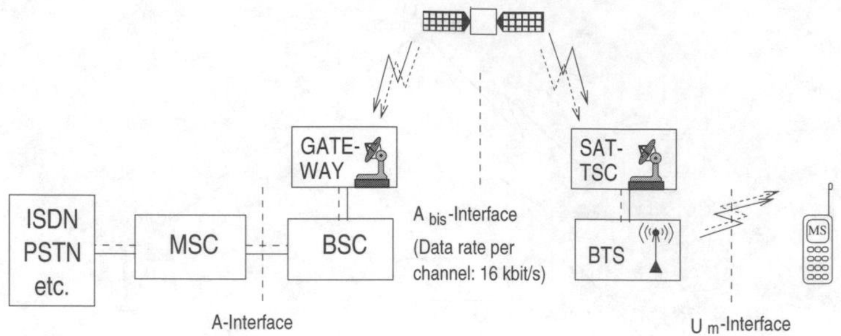

Figure 16.46: Connection of base stations over satellites

16.7.1.1 Linking Base Stations to Cellular Mobile Radio Networks

The location of base stations is selected from the standpoint of radio supply conditions and cost. Since the ideal positions in terms of radio supply frequently result in high wired connection costs, radio links are often used to link the base stations. LEO satellites could offer an alternative connection option.

Figure 16.46 illustrates the architecture of base stations linked over satellite. A new element that comes into play is the satellite-terminal controller (SAT-TSC) which is used to link base stations to the satellite system exploiting the higher gain of a directed antenna.

This kind of configuration places high demands on the satellite system in terms of delay time and bit-error ratio. There must be an assurance that no timers are running out in protocols and that messages are not retransmitted too often so that possible delay times do not become too large.

16.7.1.2 Potential Systems

Compared with the traffic to be carried, the capacity of the IRIDIUM system is too low even on the assumption that voice transmission will be at only 4.8 kbit/s (see scenario 1a in Figure 16.47). However, WLL connections could be implemented in the L-band for example (see scenario 1b in Figure 16.47) for linking GSM base stations.

The TELEDESIC system would be more appropriate as a transport network for WLL systems. Figure 16.48 presents a scenario with a DECT-WLL system. Since TELEDESIC has considerably more channels available than IRIDIUM, and DECT can be seen as representative of smaller WLL systems, this could be an interesting combination. Furthermore, the suitability for linking GSM base stations is being discussed. Figure 16.48 also shows the connection of larger users, e.g., GSM base stations in the Ka-band.

Rough calculations show that the IRIDIUM system would only be considered in special cases, and the TELEDESIC system only with major limitations, as the technology connecting WLL base stations to the fixed network if

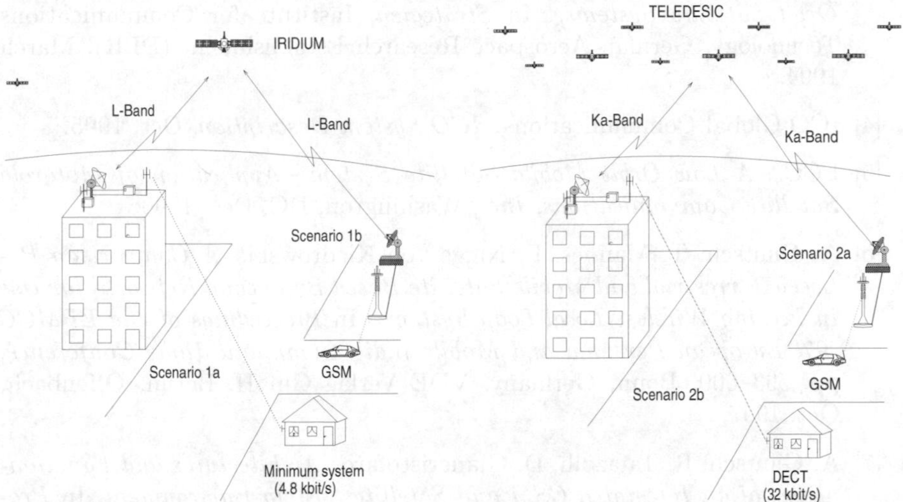

Figure 16.47: IRIDIUM for connection of a WLL system

Figure 16.48: TELEDESIC for connection of DECT-WLL and GSM systems

a considerable proportion (10 %) of the traffic volume is to be carried, e.g., in Europe or the USA. The satellite cells are still too large for this application. The traffic that occurs in this area exceeds the capacity of satellite systems many times over. So the choice is to use this option only in isolated cases or to wait for satellite systems with definitely smaller cell radii and/or clearly higher capacity.

The situation is totally different when it comes to linking large rural areas or third world countries with very low traffic volumes per square unit to telecommunications facilities. Even the IRIDIUM system with its small traffic capacity could be an interesting option economically. The TELEDESIC system could be used there as a transport network for WLL systems for large regions without any restrictions [6].

Bibliography

[1] H. Bischle, W. Schäfer, E. Lutz. *Modell für die Berechnung der Paketfehlerrate unter Berücksichtigung der Antennencharakteristik.* DLR Deutsche Forschungsanstalt für Luft und Raumfahrt e.V., Weßling, Germany.

[2] B. Bjelajac. *CIR based Hybrid and Borrowing Channel Allocation Schemes.* RACE R2117 SAINT, SAINT/A3300/AAU_003, 1994.

[3] A. Böttcher, M. Werner. *Strategies for Handover Control in Low Earth Orbit Satellite Systems.* In *Strategien*, Institute for Communications Technology, German Aerospace Research Establishment (DLR), March 1994.

[4] ICO Global Communications. *ICO System Description*, Oct. 1995.

[5] FCC. *A Low Orbit Mobile Satellite System—Application of Motorola Satellite Communications, Inc.*, Washington, DC, Dec. 1990.

[6] A. Guntsch, T. Mannes, F. Nigge von Kiedrowski. *A Comparison Between Terrestrial and Mobile Satellite Based Broadband Networks for Use in Feeding Wireless Local Loop Systems.* In *Proceedings of 2nd EPMCC '97, European Personal and Mobile radio Communications Conference*, pp. 593–600, Bonn, Germany, VDE Verlag GmbH, Berlin, Offenbach, Oct. 1997.

[7] A. Guntsch, R. Tafazolli, D. Giancristofaro. *Architectures and Functionalities of an Integrated GSM and Satellite System Environment.* In *Proceedings RACE Mobile Telecommunication Summit, Cascais, Portugal*, pp. 393–397, Nov. 1995.

[8] Iridium Inc. *Application for a Mobile Satellite System before the Federal Communications Commission*, Dec. 1990.

[9] K. G. Johannsen. *Mobile P-Service Satellite System Comparison. International Journal of Satellite Communications*, Vol. 13, pp. 453–71, 1995.

[10] E. Lutz. *Mobilkommunikation über geostationäre (GEO) und umlaufende (LEO) Satelliten*, 1993.

[11] E. Lutz, D. Cygan, M. Dippold, F. Dolainsky, W. Papke. *The Land Mobile Satellite Communication Channel—Recording, Statistics and Channel Model. IEEE Transactions on Vehicular Technology*, Vol. VT-40, No. 2, pp. 375–385, 1991.

[12] G. Maral, J. de Ridder. *Low Earth Orbit Satellite Systems for Communications. International Journal of Satellite Communications*, Vol. 9, pp. 209–225, 1991.

[13] *Mike's Spacecraft Library, Homepage of NASA.* http://leonardo.jpl.nasa.gov/msl/QuickLooks/teledesicQL.html, 1996.

[14] R. L. Olsen, D. V. Rogers, D. B. Hodge. *The aR^b Relation in the Calculation of Rain Attenuation. IEEE Transactions on Antennas and Propagation*, Vol. AP-26, pp. 318–329, 1978.

[15] W. K. Pratt. *Digital Image Processing*, 1991.

[16] M. A. Sturza. *Architecture of the TELEDESIC Satellite System.* In *International Mobile Satellite Conference—IMSC '95*, pp. 212–218, 1995.

17 UPT—Universal Personal Telecommunication*

People are becoming more mobile. At the same time they have a greater need to be reachable and to be able to reach others. A variety of different systems that address all the different categories of mobility already exist:

- Employees can be reached through a DECT terminal anywhere in their company.
- Tradesmen can be called over a *Digital Communication System* (DCS) within a city.
- Business people can be reached over the same telephone number via GSM anywhere in Europe, wherever they are located.
- Adventurous types are linked to civilization through the INMARSAT satellite system.

For each of these communications systems, the user requires a special terminal.

In contrast to these radio-based systems, wireline telecommunications systems capable of reaching almost every household in the industrialized countries are available. Fixed networks do not offer users mobility. To be reached at another terminal, the user has to inform the caller of his new location area in the form of a telephone number. With Universal Personal Telecommunication (UPT) the telecommunications network assumes responsibility for establishing a user's location through an interrogation of databases based on a personal telephone number. This provides mobility in telecommunications networks to a larger number of users without the need for special new terminals.

17.1 Classification of Telecommunications Services

The services of a telecommunications network first allow users to exchange information over the network in order to communicate. A service is what is made available to users by telecommunications authorities and private service providers for communication over public and private networks.

The ITU-T distinguishes [20] between two types of service:

***With the collaboration of Matthias Fröhlich**

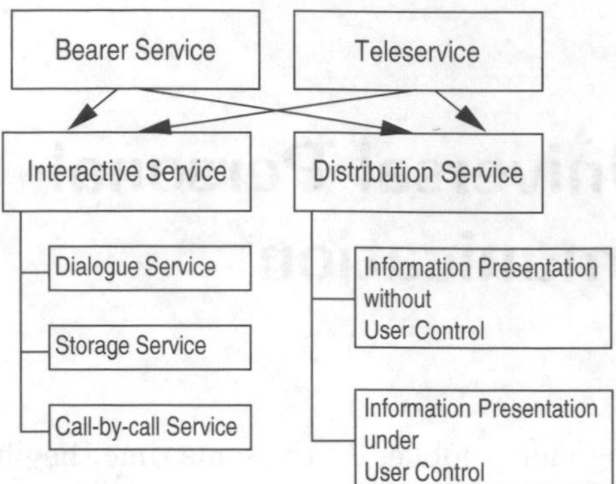

Figure 17.1: Classification of telecommunications services

Bearer services, also called communications services, are used to transfer data between precisely defined user–network interfaces. According to the ISO/OSI reference model, a bearer service is provided by layers 1–3 (the *Open Systems Interconnection* reference model (OSI-RM, [6, 25]) of the ISO (*International Organization for Standardization*) describes communication between open systems as seven interrelated layers). An example of this is the socalled B-channel in ISDN with 64 kbit/s of transmission capacity.

Teleservices offer users the possibility of communicating with one another over network terminals, and are provided by layers 1–7 of the OSI-RM. This also specifies the communications functions of the terminals. An example of this is the telephone service in ISDN.

With both services a separate distinction is made between interactive services and distribution services, which are subdivided further as shown in Figure 17.1.

Services are described according to the service features offered to the user. A differentiation is made between *general* connection features, *basic* service features and *extended* service features (see Figure 17.2).

Extended service features are only available in connection with a bearer or teleservice, and can be provided either outside of the network or within the network. Within the network they are subject to standardization and are specifically referred to as *supplementary services* (see Figure 17.3).

Value-added services (VAS) are extended service features that are provided by service nodes outside the network in OSI layers 4–7. These services usually contain storage, call-up or conversion functions for interfaces, protocols and bit-rate adaptation. They require an interface to a bearer or teleservice of the network.

Figure 17.4 presents a breakdown of the bearer services, teleservices, supplementary services and value-added services in a telecommunications net-

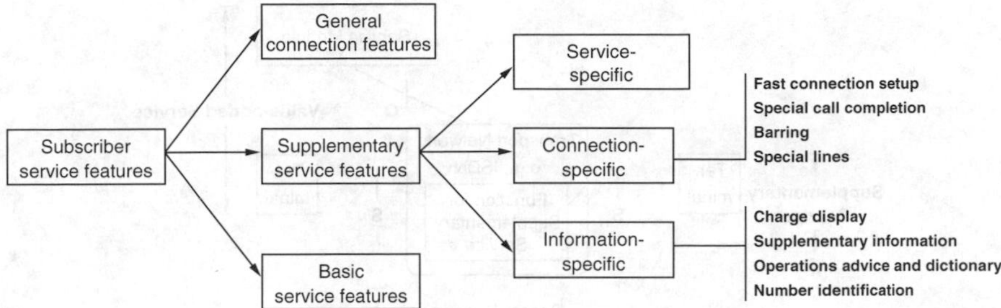

Figure 17.2: Breakdown of the service features of telecommunications services

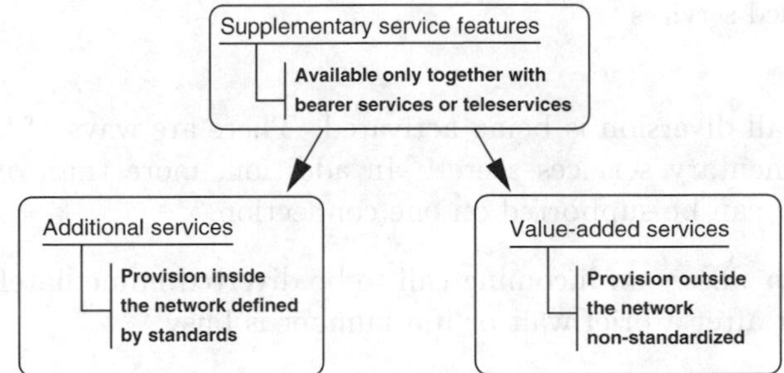

Figure 17.3: Separation of supplementary service features into additional and value-added services

work, with S representing the access interface of the user between terminal and network and Q the interface between value-added service and network.

17.2 Extended Service Features in ISDN and GSM

The current worldwide success of GSM is due to a great extent to the supplementary and value-added services offered. Owing to the harmonization between ISDN and GSM, most of the extended service features of ISDN are also available in GSM. The following section describes the most important supplementary and value-added services of ISDN. The service features offered additionally by GSM will be described in Section 17.2.2.

17.2.1 Supplementary and Value-Added Services in ISDN

Because of full digitalization of its network, ISDN offers a variety of supplementary services compared with the analogue telephone network, the most important of which can be summarized into seven groups:

Number identification displays the telephone number of the party calling or of the party being called. The latter is of interest, for example,

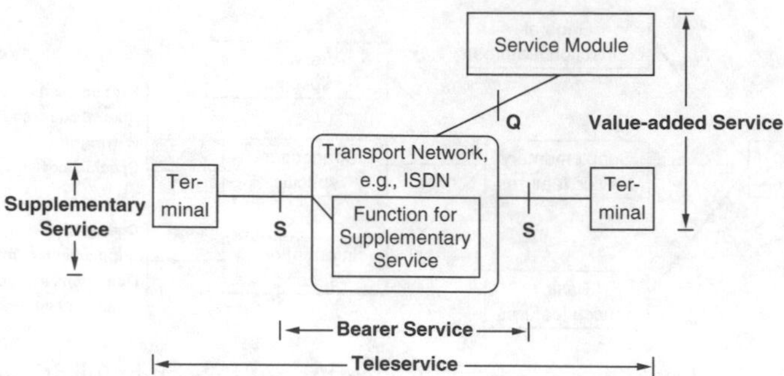

Figure 17.4: Breakdown of bearer services, teleservices, supplementary services and value-added services

when call diversion is being activated. There are ways of keeping these supplementary services secret. In addition, more than one telephone number can be supported on one connection.

Call diversion allows an incoming call to be diverted immediately to another number after a brief wait or if a number is busy.

Special call completion includes supplementary service features such as call holding for enquiries as well as automatic callback or call-waiting if the number called is busy.

Group connections include conference and three-party calls as well as calls for closed user groups.

Charge functions allow to bill all or some incoming calls to the called subscriber's account and to display the charges for these calls.

Supplementary information can be exchanged between the terminals using transparent user signalling.

Barring functions can be used for suppressing unwanted incoming calls, as well as for restricting specific or all outgoing calls.

For the first time, some of the supplementary and value-added services of ISDN are supporting the mobility of users, or, in other words, a personalization of the services is offered. Thus, for example, call diversion allows incoming calls to be routed to another location, thereby supporting the user's mobility. Personalization is provided to a limited degree through the exclusive assignment of multiple-access telephone numbers to specific groups of people or the barring of certain telephone numbers for incoming calls.

17.2.2 Supplementary and Value-Added Services in GSM

Depending on the system, GSM offers several extended service features that are not supported by ISDN. In particular, these are services that support mobility and personalization directly through the system. The main ones include the following:

Device mobility, i.e., the ability to move freely within the radio coverage area of a base station during a call. Because the radio connection is switched automatically (*handover*), it is even possible to change between any number of adjacent base stations without causing an existing connection to be broken off.

Authentication of the user, i.e., establishing and checking the identity independently of the terminal used. A *Personal Identification Module* (PIM), usually in the form of a *smart card*, is used along with a *Personal Identification Number* (`PIN`).

Localization describes the automatic identification and storage of a user's location area in the network. This allows the user to be reached under his GSM telephone number irrespective of his location.

Registration means making available the services subscribed to within the framework of the contract agreed between operator and user (service profile). The costs that are incurred are charged at the same time to the user in accordance with the terms of the agreement. The subscriber's identity must be clearly established through the authentication process in order to prevent fraudulent use.

Personalization of the service profile allows the user to configure certain service features individually within the framework of the agreed service agreement. Thus, for example, the user could activate or deactivate a voice mail service (automatic answering service in the network) or restrict his own accessibility by blocking incoming calls because of the extra cost these would occur or to avoid being disturbed.

The large number of users subscribing to the GSM network underline the demand for these extended service features, so an effort is being made to support them also in the fixed network—wherever this would be technically feasible. The first step in this direction is the service referred to as *Universal Personal Telecommunication* (UPT), which is introduced below.

17.3 The UPT Service for Universal Personal Telecommunication

The UPT service has been under development since the early 1990s, and was standardized in its outline form through international recommendations

[21, 24] of the ITU in 1993. The introduction of the UPT service will take place in several stages based on the so-called *Capability Sets* [26].

In wireline telecommunications networks such as ISDN the telephone number of a user is firmly linked to the network connection of the terminal. All services of the network accessed from this network connection are charged to the user. This fixed relationship between network connection and telephone number of the user will be eliminated with the UPT service.

Identification of UPT users is carried out independently of the terminal addressing and the access points of the network. A unique UPT number allows a user to make and receive calls at any access point and from any terminal in the network.

A telecommunications network must offer different supplementary services in order to support the UPT service:

Personal mobility, which allows a UPT user to make and receive calls using different terminals in accordance with his service agreement (*UPT User's Service Profile*).

Identification of UPT users on the basis of network-independent UPT numbers.

Charging and billing, which is based on the UPT number independently of the terminal.

Uniform access and authentication functions for UPT services in different networks.

Security functions for protecting the personal data of UPT users.

Configuration functions, which are used by the UPT user and the UPT subscriber to tailor the subscribed services to meet individual requirements.

The UPT service of the first phase offers these supplementary services in a limited form, and only supports the telephone service in the analogue telephone network, in ISDN and in mobile radio networks. Other restrictions affect the scope of the security functions as well as the access and authentication functions.

These supplementary services will be available during the second phase in a less restricted form and, e.g., data services will be supported. However, this phase has not yet been standardized.

17.3.1 Existing Studies of the UPT Service

The mobile radio system UMTS (*Universal Mobile Telecommunications System*) [2, 29] is being designed to support standardized personalized worldwide user telephone numbers and supplementary services similar to those of the UPT service. Concepts for user administration in UMTS using a distributed

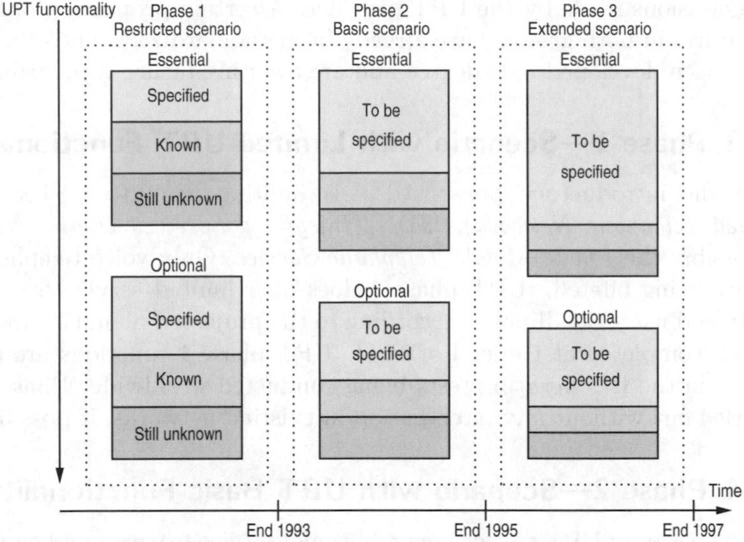

Figure 17.5: Timetable for the development and standardization of UPT service features

database are being developed [28] and quantitively evaluated on the basis of models. Flat directory hierarchies used in connection with the interrogation technique *passing* are proving to be particularly favourable.

The extended service features of ISDN such as call diversion and number identification enable services similar to those of UPT to be implemented through a PC connected to the network access interface of the user. This kind of system is being developed [27] and evaluated on a simulated basis. For comparison purposes, several alternatives to UPT implementations are being simulated and studied, e.g., in the Deutsche Telekom network. The implementation with a PC at the network access interface of the user is showing clear advantages in terms of waiting times for requested services. However, this does not offer a complete UPT service, because the system does not support UPT calls from non-UPT terminals.

17.3.2 Further Development of UPT

The introduction of UPT into existing telecommunications networks is a dynamic development process. The fundamental objectives mentioned in Section 17.3 are to be seen as guidelines that should be considered in any further development of UPT [30].

Figure 17.5 shows the phases projected in the development of UPT. These show a differentiation between essential and optional service features. Essential service features must be available at the outset of the corresponding development phase, whereas the offering of optional service features depends on the decisions made by the UPT provider. A certain proportion of the service features in both groups have already been standardized. The others have already been developed to a degree and are awaiting future standardization.

17.3.3 Phase 1—Scenario with Limited UPT Functionality

During the introductory phase, UPT is limiting itself to PSTN (*Public Switched Telephone Network*), ISDN (*Integrated Services Digital Network*) and possibly the *Public Mobile Telephone Service*. Only voice telephone services are being offered. UPT phase 1 does offer limited service features for security and user-friendliness. According to the projected plan, the first phase has been completed at the end of 1993. UPT phase 1 functions are already available in the IN operating tests being conducted worldwide. Phase 1 is to be carried out without any intervention in existing networks, if possible.

17.3.4 Phase 2—Scenario with UPT Basic Functionality

Phase 2 does offer UPT subscribers additional service features and supports a larger number of networks. During this phase, UPT also linked up with GSM networks and connection-oriented data networks. Smart cards and card readers offer users improved security against fraudulent use by outsiders. Phase 2, which was completed in 1995, is using the technology of intelligent networks.

17.3.5 Phase 3—Scenario with Extended UPT Functionality

Phase 3 is offering extended UPT functionality. This phase was not specified in detail, but only regarded as a target for UPT to allow a more effective integration of new technological developments and an opportunity to respond to the economic developments of UPT services. Phase 3 results were projected for late 1997, and are available since then.

17.3.6 Service Features of UPT in Phase 1 of its Introduction

As already explained, the ITU-T is considering some of the service features are being essential and others as being optional. Four service features in phase 1 are essential for the implementation of UPT [21]:

- *Authentication* of a UPT user's identity protects both the UPT provider and the user from unauthorized use of the UPT service.
- *In-call registration* enables a UPT user to notify the network of the location area where he can be reached through a terminal. This registration can be restricted to a specific period of time or revoked through an explicit deregistration or new registration. It is possible for several subscribers to be registered at the same terminal.

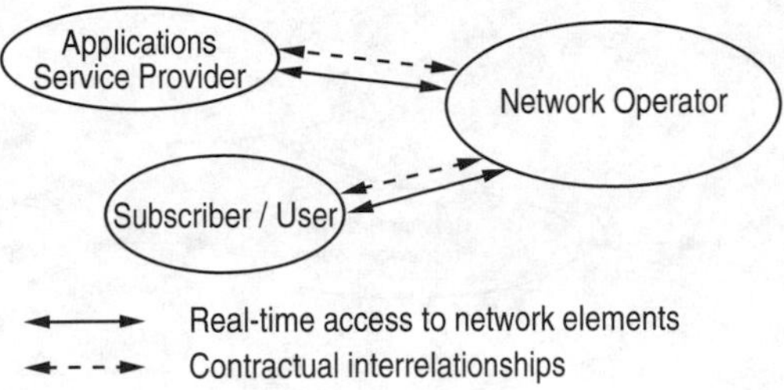

Figure 17.6: Business relationshipment between conventional telephone subscriber and system provider

- An *outgoing UPT call* allows a UPT user to use a terminal and have the call charged to his bill. So long as no *follow-on* service feature is available, an authentication must be carried out for each UPT call in order to prevent fraudulent use. The service feature *follow* indicates to the network that other calls are following the outcalls. New authentication is then not necessary.
- *In-call delivery* refers to the call forwarding service. The UPT subscriber makes his location known to the network beforehand through an in-call registration. The incoming call is then forwarded to this terminal. The subscriber to the network connection must be able to prevent registration and call forwarding at his terminal.

Access to UPT functions using registration is shown in Section 17.4.2.

17.4 Business Relationship between UPT Users and Providers

To be able to use the telecommunications services, users must enter into a business relationship with the provider.

Three different market participants can be identified in the conventional monopolistic telecommunications market. Figure 17.6 identifies these three business partners as:

- The network operator who provides network access and the communications network
- The subscriber to the service
- The service provider of the applications who contributes to the value-added service in the network

The user has a direct business relationship with the network operator.

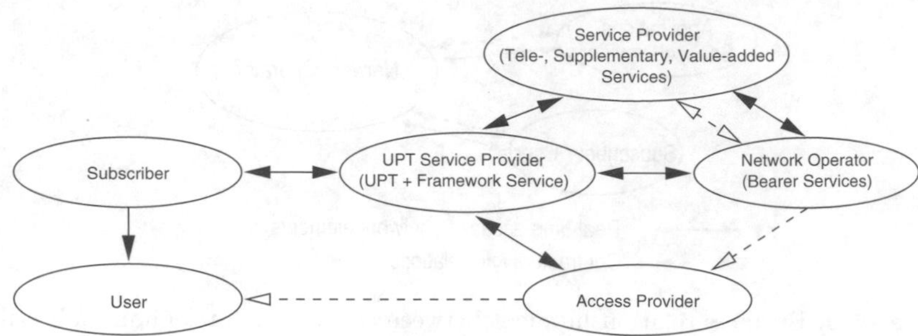

Figure 17.7: UPT business relationships

With UPT there is a complicated situation from the view of the user (see Figure 17.7). In addition to

- a network operator, a user and an applications service provider,

there are also

- a UPT service provider, a network access provider and a subscriber to the UPT services

who are involved in the UPT service arrangements.

A business relationship no longer exists between subscriber and network operator. Instead, a subscriber can order the UPT services from a UPT service provider. In the case of private individuals, this subscriber will be the user himself. With companies, the organization appears as the subscriber to the UPT services that it makes available to its employees through proper registration in the user service profile. Since users usually do not use their own terminals (mobility), it is important that a relationship exists with a network access provider who can provide access to the network.

A deregulation of the telecommunications market is necessary in order to enable the relationship between user and network operator to be split up. A similar situation was observed with the mobile radio networks in Europe, where users acquire mobile radio services from a service provider. The service provider is not automatically also the network operator.

17.4.1 Charging—New Concepts in the Introduction of UPT

In conventional telecommunications networks the person placing a call is charged for the duration of a successfully completed call, depending on time of day and distance between user A and user B.

With UPT, users do not have to be registered at their home locations, and instead may be linked to the network at location C. Users placing a call are not aware of the distance between home location B and visitor location C. Therefore they have no way of estimating how much a call will cost unless additional information is provided. The following scenarios are therefore possible

- User A absorbs all the costs of the call, in other words connections A–B as well as B–C.
- User A pays the cost of the connection A–B. User B pays the cost for his mobility and consequently the cost of connection B–C.
- User B absorbs all the costs.

The second solution is the one that would appear to be ideal, because the costs of the extended connection are charged to the person who is responsible for the higher cost. Other solutions are conceivable for determining the charges; however, cost transparency is important for the acceptance of new services so that users have an easy way of estimating the costs of the calls they initiate.

With UPT it may be necessary also to charge the UPT user for calls that were not switched successfully. The reason for this is the running costs and investment in processors and databases in the network. This measure could be justified at least for outgoing calls, where UPT calls have to be checked for approval.

17.4.2 Example of Registration of a UPT Subscriber

Figure 17.8 presents an example of what a communication between UPT subscriber and UPT service provider could look like with a dialogue control over the telephone.

This UPT function can also be carried out over analogue connections for terminals, e.g., using DTMF signalling.

First, the user dials the access code, which can be in the form of a previously unused dialling code (in the example #0185). After he has authenticated himself by entering his own UPT number and his personal identification number, he dials the menu for registration. He then requests registration for incoming calls. Now he can notify the UPT service to which terminal the calls should be directed and for how long. This completes the access procedure to the UPT service.

In man–machine communication, especially acoustic communication, it is particularly important to allow a person a chance to return to the previously processed dialogue levels to enable errors to be corrected. In addition, the user should be able to have personal contact with an operator if assistance is needed.

17.4.3 Options for Authentication

For security reasons a UPT user must be authenticated before UPT resources are accessed. This can be done in different ways.

- The easiest way, which requires no intervention in terminals or the network, is authentication using a DTMF transmitter or by dialling certain numbers at a terminal. These signals are picked up by an appropriate

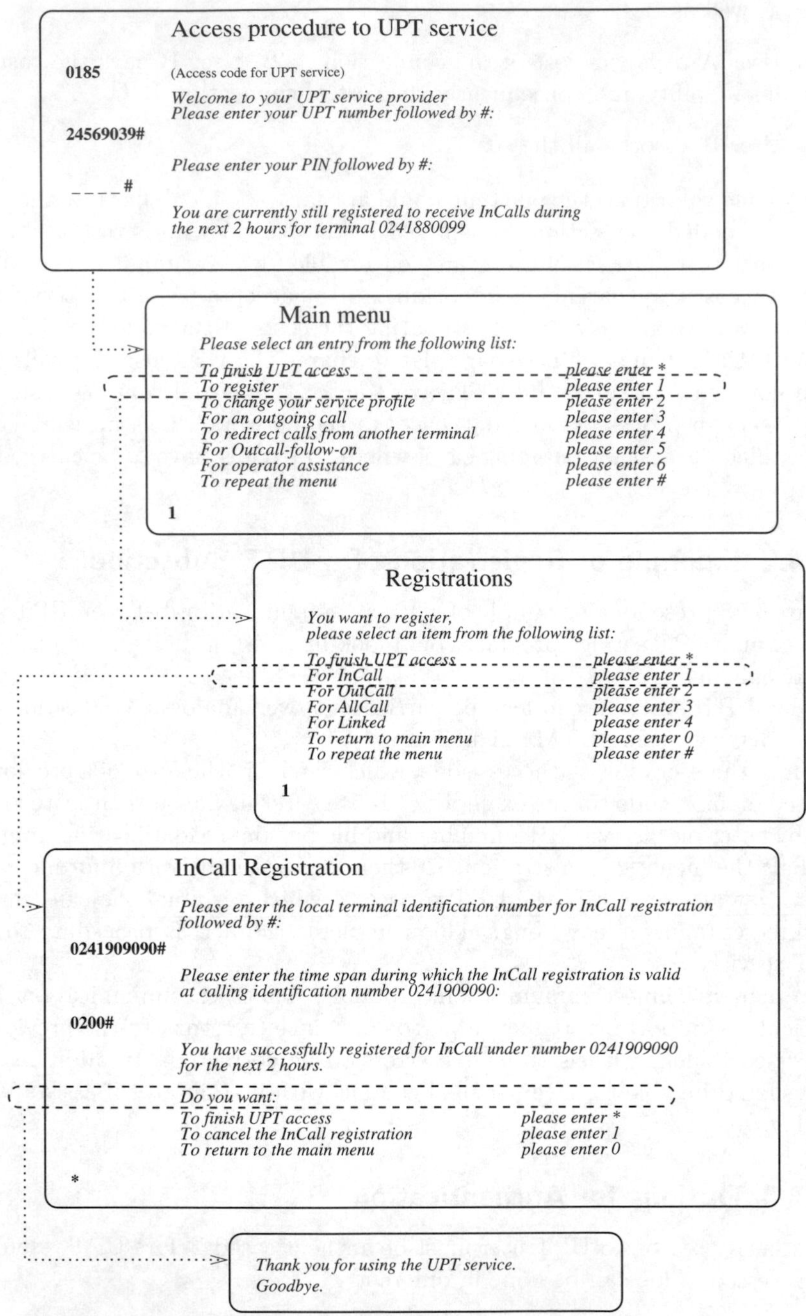

Figure 17.8: Example of an in-call registration with a UPT service provider

functional unit and forwarded for evaluation. This type of authentication is regarded as weak, because it does not provide any special security. However, this approach can be implemented in every network and at every terminal, which is also a requirement of the ITU-T [21]. With this model the authentication data travels over the traffic channel, which has already been switched and puts no strain on the signalling network.

- A variation of the above option involves using an intelligent terminal for an automatic execution of the authentication. The terminal is then responsible for transmitting the DTMF signals. This option does not require signalling functions either.

- Card readers and smart cards offer a higher level of security and are easier to use. As soon as authentication is required, the card-reading device reads the inserted smart card and carries out the authentication. The signalling can be implemented over traffic channels or over signalling channels.

Thus different versions/variations, which are also able to offer different levels of user comfort, are possible, depending on the terminal intelligence or authentication security required.

17.5 UPT Service Profile

One of the elements that determines the functionality of UPT is the *Flexible Service Profile* (FSP). This is part of a database, and is used to supply the data associated with UPT users to the *Service Control Functions* (SCFs). One of the most important entries in the service profile is the terminal identification that a user has registered for incoming calls for a specific period of time. With each call, irrespective of whether it is an incoming or an outgoing call, the FSP must be interrogated for certain data relating to the call. A differentiation is made between permanent and variable entries in the structure of the FSP. Permanent entries relate to data that is fixed when the UPT identification is established and seldom needs to be changed. Variable entries concern data with values that change owing to actions of the user, such as change of location.

Permanent entries that can be changed only by the service provider include:

- UPT number

- Identification of the home local terminal connection

- Basic services contracted by the subscriber

- Supplementary services contracted

- Number of maximum possible authentication attempts

- Type of authentication procedures

- Security options
- Barred destination numbers (e.g., police, speaking clock)

Permanent entries that can be changed by the subscriber include:

- Approved payment options (e.g., credit card calls)
- Maximum ceiling on calls
- Maximum number of accesses from other terminals
- Released user functions

The following variable entries relate to service control:

- Type of active authentication
- Active option for charging
- Active security mode
- Status of supplementary services

The following variable entries are used for mobility control and supervision:

- Relevant terminal number
- Fallback terminal for incoming calls
- Relevant terminal for incoming calls
- Standard number of outgoing calls per registration
- Destination for call forwarding—absolute
- Destination for call forwarding—when terminal is busy
- Destination for call forwarding—when there is no answer
- Destination for call forwarding—if called party cannot be reached

17.6 Requests to UPT-Supported Networks

A model for the functional architecture of UPT was developed in I.373 [14] to enable the formulation of requests to the UPT-supporting network. Figure 17.9 illustrates this model, which consists of five layers hierarchically positioned underneath each other, which will later be related to the network. This abstraction is desirable to avoid the need of having to specifiy any specific form of network. UPT can thus be implemented in fixed as well as in mobile networks.

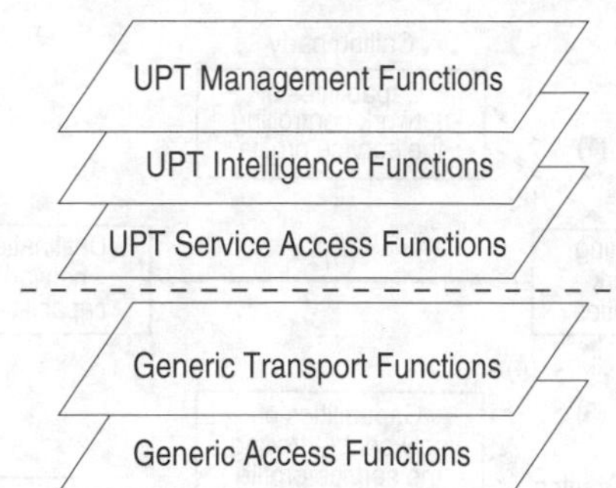

Figure 17.9: Functional grouping of UPT

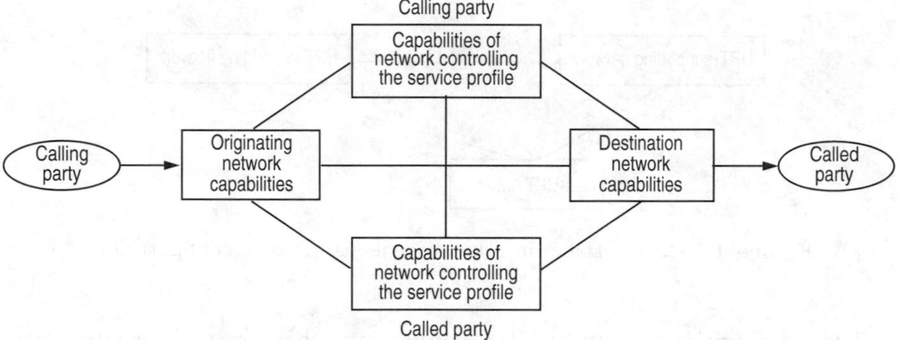

Figure 17.10: Reference model for UPT calls

A UPT call can cross different networks on its route from the person placing the call to the person receiving the call. These networks have different capabilities for supporting UPT. Connections for traffic channels and for signalling exist between the different network components such as the network and service profiles. I.373 provides a reference model for UPT calls to present an overview of the dependences between the individual components (see Figure 17.10). For each UPT call a different sequence of communication between the network components and the service profile is conceivable. Only an example of a call between two UPT users is shown here (see Figure 17.11).

When UPT is offered in heterogeneous network configurations, conflicts can occur because networks have different capabilities for supporting UPT.

Figure 17.12 shows where interfaces occur between networks supporting UPT and non-supporting UPT. Two possibilities for handling UPT connections across network boundaries are to be considered. For networks that do not support UPT, either no UPT service or only a limited UPT service can be offered.

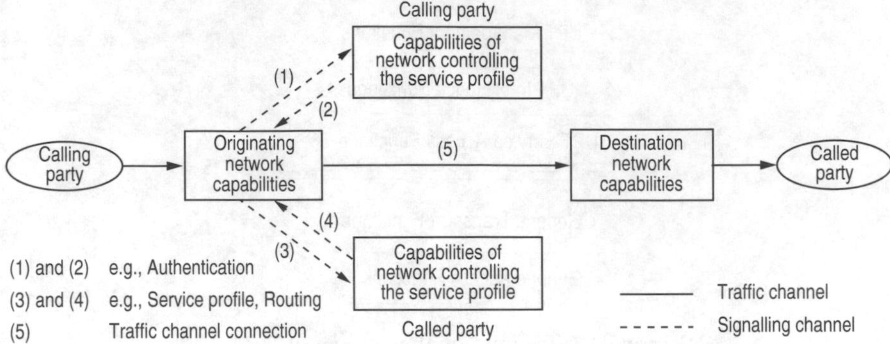

Figure 17.11: Example of a UPT call: call between two UPT users

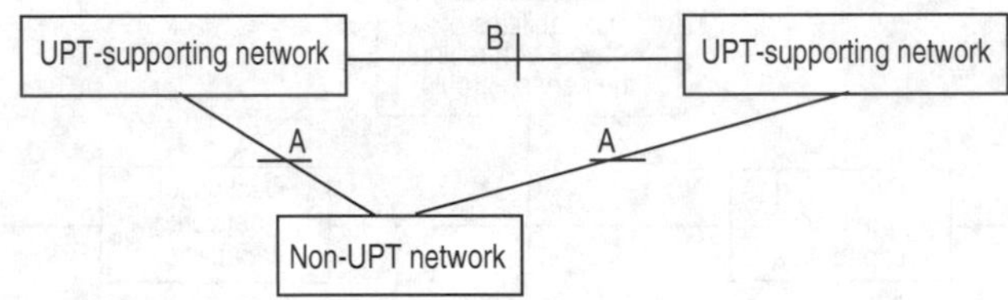

Figure 17.12: Interfaces in a heterogeneous network configuration

The same options apply to UPT connections that are to be routed over non-UPT-supporting networks. The first option is very unsatisfactory, which is why the second option should be taken into consideration even if additional costs are involved.

17.7 PSCS as a Further Development of UPT

Personal Services Communication Space (PSCS) is a service concept that was developed within the framework of the EU research programme RACE by the Mobilise project partners [5]. PSCS represents a further development of UPT phase 1, and is serving as the basis for projecting the services and functions of UPT phases 2 and 3. It enhances the UPT service in the following ways:

- User-friendliness during access
- Use of services beyond telephony, such as fax, e-mail and data services
- Easier access procedures to the service profile
- Flexible routing schemes, such as e-mail forwarding to a fax device

To achieve a certain level of user-friendliness, PSCS requires special terminals that have at least an alphanumeric display and a card reader for smart cards. Consequently, demands that UPT allow access from any terminal cannot be met, but what is gained is a far more flexible man–machine interface.

17.8 Numbering and Dialling

In telecommunications networks users are addressed by sequences of digits. By entering the telephone number of the party being called (user B), the calling party (user A) is instructing the network to set up a connection. For technical reasons, direct dialling was previously used to route a call across the network elements. On the basis of how it is structured, the telephone number can provide a direct indication of where user B is located with a direct dialling system. A telephone number therefore has a direct relationship to the location of user B, and also determines how a call is routed through the network. Because of the individual numbers that make up the telephone number, the switching centre is able to route the call to user B.

In addition to addressing users, the telephone number is used for the billing of telecommunications services. This gives the telephone number economic significance along with its technical importance.

This direct control has now been replaced by indirect dialling systems, which is why it is necessary to establish an association between telephone number and location of user B. In an indirect dialling system the routing of a call is largely separate from the telephone number. A path established from the routing tables filed in the switching centres is switched between the two users. In this case the telephone number only serves for the addressing, and does not contain any information about a subscriber's location.

However, users of telecommunications services have the knowledge to determine from the telephone number the calling area of user B, the charges and type of telecommunications services required. It is therefore important to retain this information function for the acceptance of new numbering systems.

Current and future numbering systems are explained in the following sections.

17.8.1 ISDN, PSTN

Telephone numbers in public telecommunications networks as well as in ISDN for a long time period consisted of 12 digits. Since the end of 1996, this number has been extended to 15 digits [10]. The telephone number structure is provided in ITU Recommendation E.163 [7]; see Figure 17.13.

In their structure, national destination codes indicate the direct dialling control of the network that existed until about 1977. Like the network, they are arranged hierarchically, with

- position 1 for the central switching centre
- position 2 for the home location switching centre

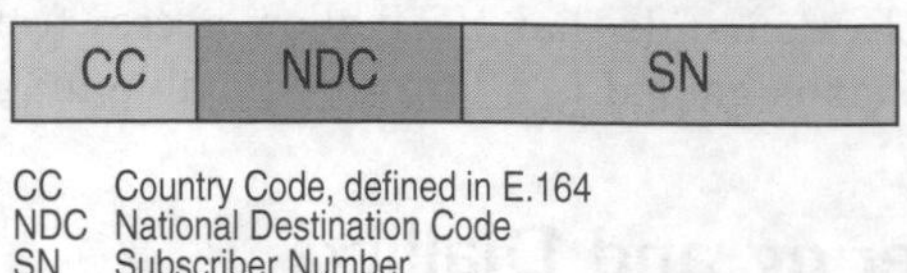

CC Country Code, defined in E.164
NDC National Destination Code
SN Subscriber Number

Figure 17.13: Structure of an ISDN telephone number

- position 3 for the customer switching centre
- position 4 for the subscriber switching centre

The national destination code is different for each local network, and typically unique for a whole country. It consists of two to four digits, depending on the size of the local network.

A user within the same local network can be reached through his respective telephone number. For example, parties in Germany beyond one's own local network can be reached through dialling the prefix 0 followed by the national destination code and the respective telephone number. Users outside a given country can be reached through dialling the international prefix, e.g. 00, the country code and the respective telephone number. The length of the telephone number to be dialled therefore depends on the distance between the two communicating parties.

Special services can be selected by dialling special prefixes before the actual telephone number of the person being called. A 1 is reserved as the first digit for this purpose.

Telecommunications connections are charged based on the national destination code. Knowledge of the structure of the national destination codes provides the user with an information source for determining the cost of a call. According to newer tariffing models there will be probably only two national tarif zones in the future: local and long distance. Then there will no longer be advantage for the user to differentiate from the destination code where the long-distance call party is located, and the acceptance of the indirect dialling system will grow.

17.8.2 Public Mobile Telephone Network—GSM

The numbering used in the public mobile telephone network is similar to that of PSTN. ITU Recs. E.212 [8] and E.213 [9] provide the relevant structures. The GSM numbering scheme is based on these recommendations.

Figure 17.14 shows how a public mobile telephone number is structured. It consists of the respective country code, the network code and the terminal identification number. A complete integration into the ISDN/PSTN numbering plan could have been possible, but was not implemented. Instead, as mentioned above, GSM networks are differentiated from the PSTN through separate dialling codes which function as a traffic discrimination feature.

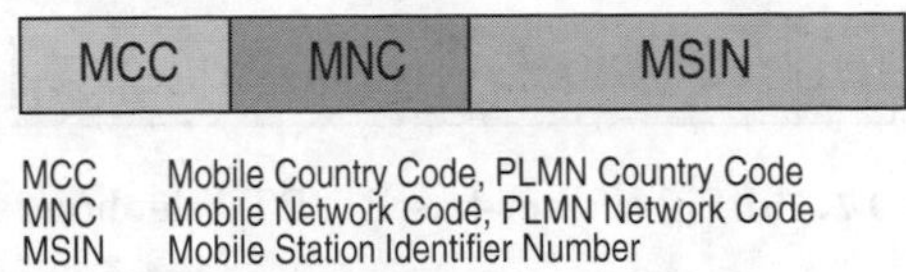

Figure 17.14: Numbering structure in GSM

17.8.3 UPT

The ITU presents proposals for UPT numbering in Rec. E.168 [13]. In these proposals a UPT number is defined as a user's unique identification, and is dialled by a caller to reach the user. The UPT indicator is the part of the UPT number that identifies the call as a UPT call, e.g., in Germany 0700 is used as the prefix.

From the standpoint of the user, the UPT number must meet the following requirements [12]:

- A UPT number must be recognizable as such by users, so that they have the opportunity to determine that a call is a UPT call and therefore will be handled and charged in a special way.
- A UPT number must be as short as possible to minimize the number of digits to be dialled.
- It must be possible to dial the UPT number from any terminal in the PSTN. Therefore digits have already been reserved (digits 0...9 and possibly the characters # and *).
- UPT users should be able to retain their UPT numbers if they change service provider.
- In the future the UPT number should be valid in all networks, at all terminals and for the use of any service.
- Any further developments and changes to the UPT numbering plan should have as little an effect as possible on existing UPT numbers.

Network operators have other requirements of UPT numbers:

- It must be easy for a network to recognize a UPT call.
- The numbering capacity must be protected even when UPT is introduced.
- UPT numbers should have no influence on routing.
- The administration of UPT numbers must be kept simple.
- UPT numbers must fit into existing numbering plans (see E.164) [11].

Figure 17.15: Subscriber-based UPT telephone number

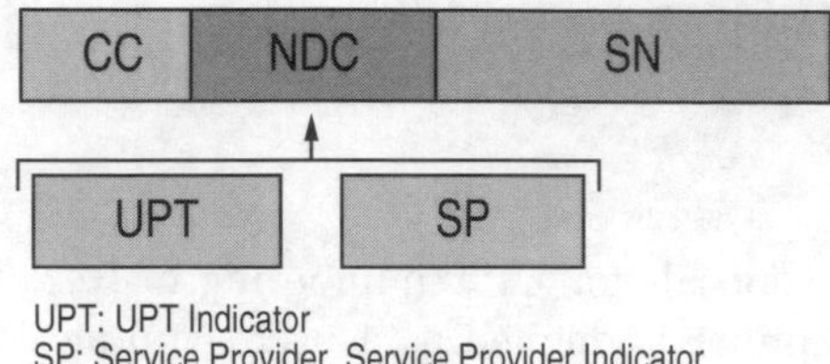

Figure 17.16: Country-based UPT telephone number

Recommendation E.168 differentiates between three scenarios that all comply with E.164 and therefore are aimed at meeting the requirements of both user and network operator.

17.8.3.1 Scenario 1—Subscriber-Based Concept

The subscriber-based concept gives no indication to the subscriber that a call is a UPT call (see Figure 17.15). All information concerning the UPT subscriber is administered by the *Flexible Service Profile* (FSP) at his home location. In scenario 1, `CC` remains the country code, `NDC` the national destination code and `SN` the subscriber's telephone number. `NDC` and `SN` together produce a unique telephone number for a country. Owing to the lack of other criteria, the entire number has to be evaluated and searched in a database to determine whether the call is to be recognized as a UPT call.

17.8.3.2 Scenario 2—Country-Based Concept

In the country-based concept, `CC` retains its significance as the country code (see Figure 17.16). Unlike scenario 1, however, `NDC` identifies a call as a UPT call. This applies to the party making the call as well as to the network. Special codes can be used to select UPT services or a UPT service provider instead of the local network. This concept has already be applied in the integration of GSM mobile radio networks. For example, the first NDC digit can select the UPT service while the other digits would be responsible for the selection of the UPT service provider.

17.8.3.3 Scenario 3—Country Code-Based Concept

The country code-based concept indicates `CC` as the UPT indicator (see Figure 17.17). There are two possible options for the NDC field:

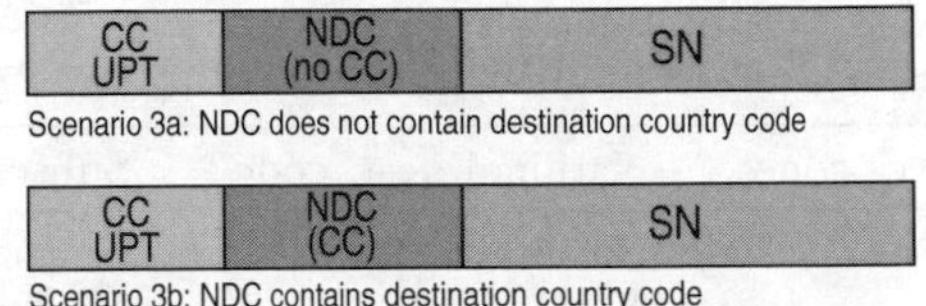

Figure 17.17: Country code-based concept for UPT telephone numbers

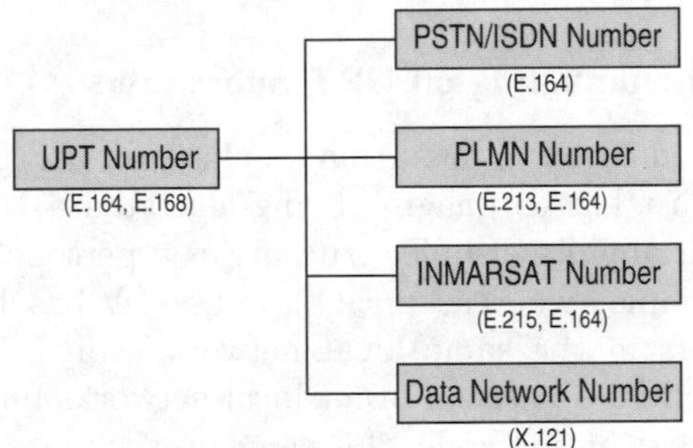

Figure 17.18: Relationship between UPT and network numbers

- NDC without country code or • NDC with country code

The first option (scenario a in Figure 17.17) does not give the user the possibility to establish the destination country of his call. The caller therefore has no way of estimating the cost of the call on the basis of the telephone number. In this case other sources of information must be used to obtain some indication of cost. Furthermore, with this numbering scheme the telephone number is distributed globally, whereas situation b in Figure 17.17 allows for a national administration of the telephone numbers.

17.8.3.4 Relationship Between UPT and Other Network Numbers

Figure 17.18 indicates which ITU recommendations could have an influence on possible UPT numbering scenarios [13].

The UPT number should provide the network with sufficient information for using databases to establish a subscriber's network number.

17.8.3.5 Responsibility for Number Allocation

The different scenarios produce different requirements for the administration of the numbers. Table 17.1 indicates who has responsibility for the allocation of numbers in each case. Because central allocation appears less flexible than decentral allocation, scenario 2 would be the preferred one.

Table 17.1: Responsibility for number allocation

Scenario	Country code	National dest. code	Subscriber telephone no.
1	ITU	National	National
2	ITU	National	National
3	ITU	(a) ITU	ITU
		(b) ITU	National

17.8.3.6 Effect of Numbering on UPT Subscribers

The numbering used in UPT has a major effect on acceptance by potential UPT users. One just has to imagine being allocated a 15-digit combination of numbers without any discernible structure as a personal telephone number for UPT. For example, whereas until now user A has been able to reach eight different users in the same local network using a six-digit telephone number and five different users in other local networks but with possibly the identical national destination code, this same user will now have to remember 13 different 15-digit UPT telephone numbers. This situation surely will not contribute towards the success of UPT.

Because a call has to be recognizable as a UPT call to the calling party, scenario 1 from E.168 would not appear to be practical. With scenario 3 a user would always have to dial the complete 15-digit telephone number in order to reach a UPT subscriber. This solution does not seem user-friendly either. In comparison, scenario 2 offers the possibility of omitting the country code with calls within a country (after all, only about 3 % of all calls made in Germany are international calls). Thus each country could create a numbering plan tailored to its own needs. However, there has to be an agreement on how the same UPT indication numbers are used.

According to the scenarios discussed, the UPT number would also have to be changed if there is a change in UPT service provider because the UPT indicator for the service provider would be different. There is a strong commitment to remaining with a provider because of the costs involved in making a number change. This bond to the provider can be broken if the *Personal User Identity* (`PUI`) can be switched as the indirect address between UPT number and the *Flexible Service Profile* of the party called. The `PUI` is contained in a database, which allocates a UPT number to the address of the FSP. This allows the UPT number to be retained if there is a change in the FSP's service provider or home location area. It then suffices for the address of the FSP to be changed in the `PUI`. The UPT number thereby has the significance of a true, possibly permanently valid, personal telephone number. For subscribers who have never changed their FSPs (this probably applies to most subscribers) the `PUI` would be a one-to-one duplication of the UPT number to the FSP address.

Although interrogation of a PUI database introduces some access delay, there are advantages to this approach that increase the acceptance of UPT:

- Users are assigned a true personal telephone number, which in turn is also easy to remember.
- The network operator is able to change the internal processes in his UPT network without having to change the UPT number. This allows him to change the structure and the location of his FSP databases without any problem.

The disadvantage when a UPT subscriber changes from provider A to provider B is that the `PUI` from provider A has to route every call to provider B. In this case other database distributions have to be taken into account:

1. A global database stores the `PUI` for all UPT subscribers.
2. A national database stores the `PUI` for all UPT subscribers whose first service provider was based in that country.
3. A database in the country of the party making the call holds the `PUI`.
4. As described above, the UPT provider holds the `PUI` for the subscribers who first used UPT with him.

All the possibilities have advantages and disadvantages. The second solution seems the most practical. Mobility between countries is not as great as between service providers. Every country or alliance of countries (e.g., the EU or ETSI) can therefore take responsibility for its own allocation of personal telephone numbers. The third solution creates problems in terms of data consistency, whereas the fourth solution saddles UPT provider A with tasks from UPT provider B. This leaves the former managing data on subscribers who have not been his customers for a long time.

From the standpoint of user-friendliness, priority is given to the concept in which a `PUI` guarantees the personal telephone number. The time delay associated with the additional database access can be alleviated through caching strategies.

17.9 Intelligent Networks and Their Value-Added Services

Now whenever a new service is installed in a telecommunications network, the control programs in each switching centre have to be updated, which is an enormous expense in terms of time and cost, and also involves operative risks that are difficult to assess.

Because of the growing emphasis on the ability to introduce services quickly and flexibly in the future, efforts were begun in the USA and in Europe in the

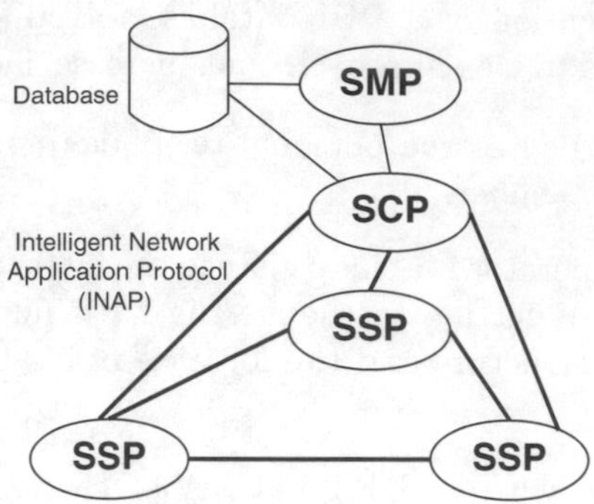

Figure 17.19: The principles of the functioning of an intelligent network

late 1980s to develop the concept of *Intelligent networks*, IN [1]. This is not a new telecommunications network, but instead an architecture designed for the further development of existing networks and new developments of future networks.

With INs certain network services that extend beyond pure switching and transmission functions will be concentrated in autonomous network nodes. This will allow new services to be implemented centrally rather than from each switching node, thereby accelerating the process of introducing these services into the network.

The concept of intelligent networks is being standardized in international recommendations of the ITU in several *capability sets* (CS) [22]. So far only the recommendations of the first capability set (CS-1) have been passed. This section provides an overview of intelligent networks, with special emphasis on their role in the implementation of UPT services.

17.9.1 The Functional Principle of an Intelligent Network

An important advantage of intelligent networks compared with a conventional network architecture is in the separation of switching and service-provisioning functions. Figure 17.19 clarifies this principle.

IN calls are recognized at digital *service switching points* (SSP) according to service codes (e.g., 0 800 for a no-charge call in the network of Deutsche Telekom AG). The Intelligent Network Application Protocol (INAP) is used to instruct the *service control point* (SCP) of the IN call. The SCP controls the execution of the IN service. The *service management point* (SMP) is responsible for setting up, changing, administering and monitoring the IN services. All necessary parameters of the IN services are stored in the database attached to the SMP, which the SCP also accesses when providing a service.

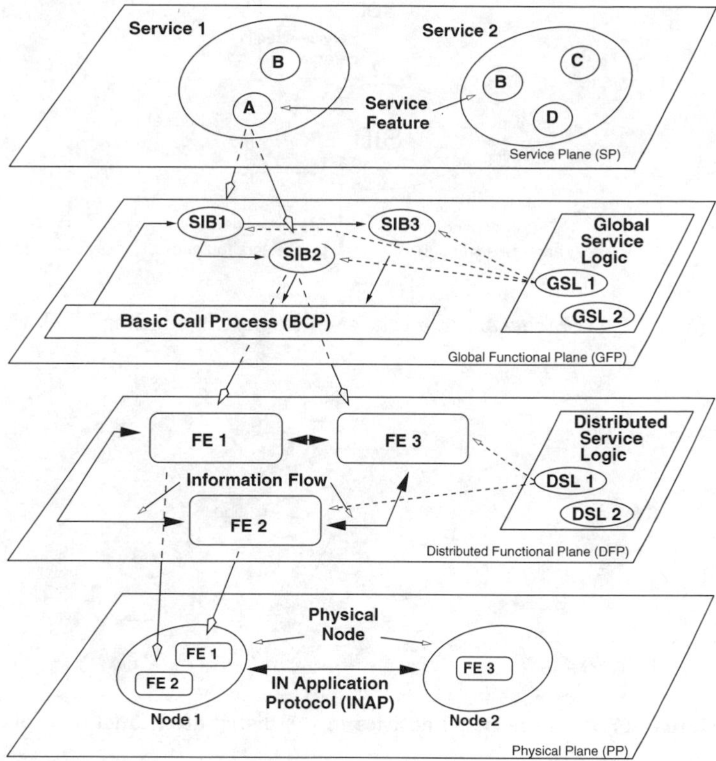

Figure 17.20: Layer model for description of services in intelligent networks

17.9.2 Description of Services in Intelligent Networks

IN services are described at four different planes in a *conceptual model* of the intelligent network IN conceptual model (INCM) (see Figure 17.20); these planes are interrelated with each one dealing with a different aspect of IN services.

Service plane (SP) At the service plane a service is described solely from the standpoint of the user on the basis of different *service features* (SF) and through usage scenarios [16].

Global functional plane (GFP) The global functional plane breaks down a service into *service-independent building blocks* (SIB), which are used in sequence to provide some service [16], based on the description in the service plane. Service support data (SSD) and call-instance data (CID) as well as a logical beginning and a logical end to the execution are defined for configuration of a SIB module (see Figure 17.21). The operational control process and the initialization of the service-independent and call-instance data is carried out by the *global service logic* (GSL).

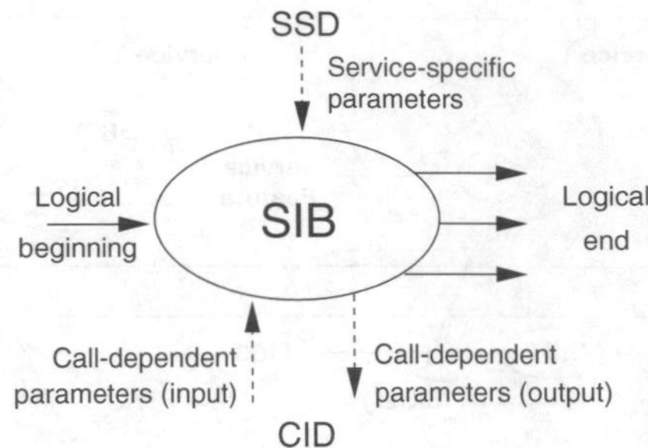

Figure 17.21: Structure of a service-independent building block (SIB)

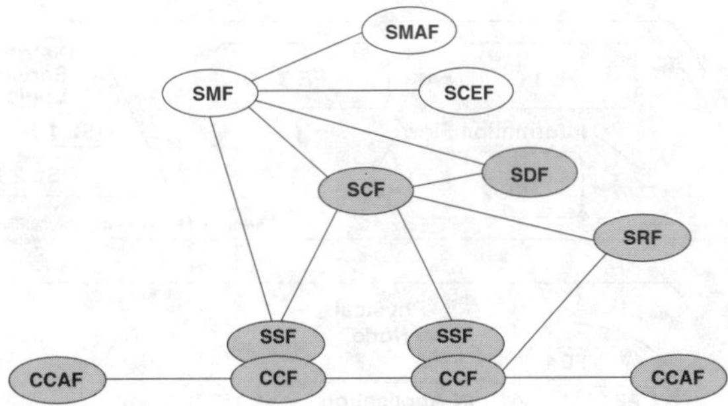

Figure 17.22: Functional entities in the distributed functional plane

A special role is played by the service-independent module BCP (*basic call process*, or basic call sequence), which functions as the central interface to the switching and transmission functions.

Distributed functional plane (DFP) A service-independent building block (SIB) is implemented by one or several *functional entities* (FE) in the distributed functional plane [17]. The interaction between two functional entities takes place through information flow (IF). Figure 17.22 shows the functional entities of the distributed functional plane and how they are linked.

Functional entities are divided into those that provide and those that manage IN services. The functions for providing IN services are shown in grey in Figure 17.22, and comprise the following:

Call control agent functions (CCAF) The call control agent function is the interface between the user and the switching functions of the network.

Call control functions (CCF) The call control function is responsible for controlling and setting up connections. The CCF also provides a mechanism for detecting so-called *trigger points* (TP) for IN services.

Service switching functions (SSF) The service switching function enables cooperation of the call control functions of a network (CCF) with the service control function (SCF).

Service control functions (SCF) All functions used for providing an IN service in the network are carried out as service control functions. To do so, the SCF has access to the data that is supplied by the SDF.

Service data functions (SDF) This function stores all data that is required to provide, execute and bill the IN services.

Specialized resource functions (SRF) A specialized resource function provides resources, such as functions for voice input and output, that can be used by the user to interact with the network.

The other functional entities serve in the management of the IN services, in other words in providing and managing the services, and consist of the following:

Service creation environment functions (SCEF) This function can be used to create new IN services or to change existing ones.

Service management functions (SMF) The service management function provides the functions necessary to activate services from the SCEF for availability in the network.

Service management access functions (SMAF) This access function provides the administrator of IN services with access to the SMF.

Physical plane (PP) The functional entities of the distributed functional planes are allocated to the nodes of the intelligent network at the physical plane [18]. A sample model of a physical node is illustrated in Figure 17.19. Protocols implement the information flow between different nodes. The Intelligent Network Application Protocol (INAP) [15, 19] is specifically discussed in this context.

17.9.3 The Intelligent Network Applications Protocol

In the capability set 1 (CS-1) the Intelligent Network Application Protocol (INAP) enables interaction between the following functional entities of the distributed functional plane:

- Service switching function (SSF)
- Service control function (SCF)
- Service data function (SDF)
- Specialized resource function (SRF)

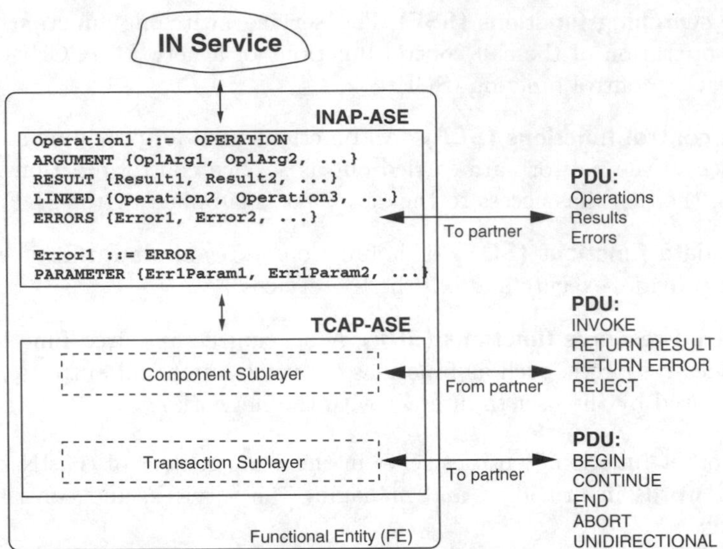

Figure 17.23: Definition of the INAP protocol data units with ASN-1 and transmission using the application service element TCAP

The *Remote Operations Service Element* (ROSE) of the OSI reference model is used in the *Intelligent Network Applications Protocol* (INAP) to transfer the protocol data units (PDU), thus defining the INAP as a client-server system. The protocol data units of the ROSE application protocol are in turn transmitted through messages in the *transaction capability application part* (TCAP). The transmission is carried out by the *Application Service Element* (ASE) of the application part for transaction processing (TCAP). The IN service in turn accesses the remote operations of the INAP using an ASE.

Abstract Syntax Notation No. 1, (ASN-1) [6, 23] is used in defining the protocol data units of the INAP. The macro definition `OPERATION` indicates the name of each remote operation as well as optionally the invoking parameters (`ARGUMENT`), return values (`RESULT`), error messages (`ERRORS`) along with linked operations (`LINKED`) (see Figure 17.23).

Reference [19] defines 58 remote operations for capability set 1 (CS-1) of an intelligent network. Only 34 remote operations have been accepted by ETSI in [3, 4] for Europe-wide standardization. Table 17.2 shows that most requests for remote operations take place between SCF and SSF.

17.9.4 UPT in the IN Layer Model

The functional architecture of UPT was presented in Section 17.6. The functional layers shown there can be assigned to the IN functional units in an

Table 17.2: Distribution of INAP signals over functional components

<table>
<tr><th colspan="3">Information flow</th><th>ITU</th><th>ETSI</th></tr>
<tr><td>SCF</td><td>→</td><td>SSF</td><td>27</td><td>19</td></tr>
<tr><td>SSF</td><td>→</td><td>SCF</td><td>23</td><td>7</td></tr>
<tr><td>SCF</td><td>→</td><td>SDF</td><td colspan="2">3</td></tr>
<tr><td>SDF</td><td>→</td><td>SCF</td><td colspan="2">2</td></tr>
<tr><td>SCF</td><td>→</td><td>SRF</td><td colspan="2">2</td></tr>
<tr><td>SRF</td><td>→</td><td>SCF</td><td colspan="2">1</td></tr>
</table>

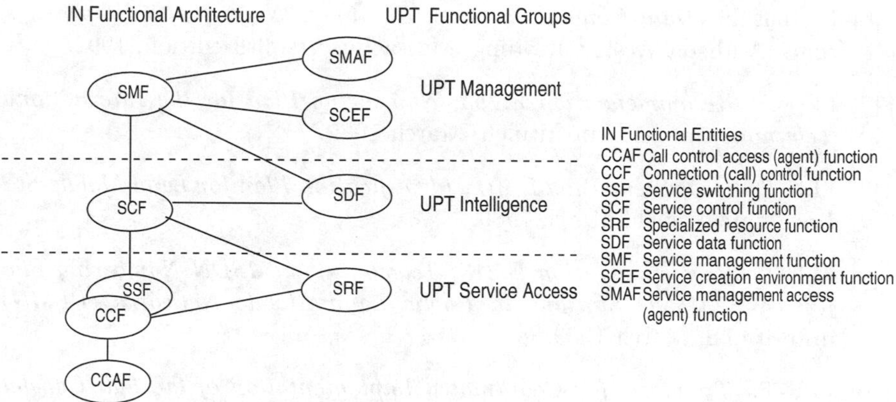

Figure 17.24: Mapping of UPT functional groups to the IN layer model

IN-based solution. Figure 17.24 shows that intelligent networks offer a good basis for the introduction of UPT.

Bibliography

[1] W. D. Ambrosch, A. Maher, B. Sasscer. *The Intelligent Network.* Springer, Berlin, Heidelberg, New York, 1989.

[2] S. Chia. *The Universal Mobile Telecommunications System. IEEE Communications Magazine*, Vol. 30, No. 12, pp. 54–62, 1992.

[3] ETSI-TC-SPS. *Intelligent Network(IN); Intelligent Network Capability Set 1 (CS1); Core Intelligent Network Application Protocol (INAP); Part 1: Protocol Specification.* European Telecommunication Standard ETS 300 374-1, European Telecommunications Standards Institute (ETSI), Sophia Antipolis, 1994.

[4] ETSI-TC-SPS. *Intelligent Network(IN); Intelligent Network Capability Set 1 (CS1); Core Intelligent Network Application Protocol (INAP); Part*

5: Protocol specification for the Service Control Function (SCF)—Service Data Function (SDF) interface. European Telecommunication Standard prETS 300 374-5, European Telecommunications Standards Institute (ETSI), Sophia Antipolis, 1996.

[5] C. Görg, S. Kleier, M. Guntermann, M. Fröhlich, H. Bisseling. *A European Solution for Advanced UPT: Integration of Services for Personal Communication.* In *Integrating Telecommunications and Distributed Computing—from Concept to Reality*, pp. 603–617, Melbourne, Australia, TINA '95 Conference, February 1995.

[6] F. Halsall. *Data Communications, Computer Networks and Open Systems.* Addison-Wesley, Reading, Massachusetts, 3rd edition, 1992.

[7] ITU-T. *Recommendation E.163, Numbering Plan for the International Telephone Service.* info@itu.ch, March 1989.

[8] ITU-T. *Recommendation E.212, Identification Plan for Land Mobile Stations.* info@itu.ch, March 1989.

[9] ITU-T. *Recommendation E.213, Telephone and ISDN Numbering Plan for Land Mobile Stations in Public Land Mobile Networks (PLMN).* info@itu.ch, March 1989.

[10] ITU-T. *Timetable for Coordinated Implementation of the Full Capability of the Numbering Plan for the ISDN Era (Recommendation E.164).* info@itu.ch, March 1989.

[11] ITU-T. *Recommendation E.164, Numbering Plan for the ISDN Era.* info@itu.ch, Aug. 1991.

[12] ITU-T. *Draft Recommendation F.851, Universal Personal Telecommunication (UPT)—Service Description.* info@itu.ch, July 1993.

[13] ITU-T. *Recommendation E.168, Application of E.164 Numbering Plan for UPT.* info@itu.ch, March 1993.

[14] ITU-T. *Recommendation I.373, Network Capabilities to Support Universal Personal Telecommunication (UPT).* info@itu.ch, March 1993.

[15] ITU. Telecommunication Standardization Sector (ITU-T). *General Aspects of the Intelligent Network Application Protocol.* ITU-T Recommendation Q.1208, International Telecommunication Union (ITU), Geneva, 1993.

[16] ITU. Telecommunication Standardization Sector (ITU-T). *Intelligent Network—Service Plane Architecture.* ITU-T Recommendation Q.1202, International Telecommunication Union (ITU), Geneva, 1993.

[17] ITU. Telecommunication Standardization Sector (ITU-T). *Intelligent Network Distributed Functional Plane Architecture.* ITU-T Recommendation Q.1204, International Telecommunication Union (ITU), Geneva, 1993.

[18] ITU. Telecommunication Standardization Sector (ITU-T). *Intelligent Network Physical Plane Architecture.* ITU-T Recommendation Q.1205, International Telecommunication Union (ITU), Geneva, 1993.

[19] ITU. Telecommunication Standardization Sector (ITU-T). *Interface recommendations for Intelligent Network CS-1.* ITU-T Recommendation Q.1218, International Telecommunication Union (ITU), Geneva, 1993.

[20] ITU. Telecommunication Standardization Sector (ITU-T). *Principles of Telecommunication Services Supported by an ISDN and the Means to Describe Them.* ITU-T Recommendation I.210, International Telecommunication Union (ITU), Geneva, 1993.

[21] ITU. Telecommunication Standardization Sector (ITU-T). *Principles of Universal Personal Telecommunication (UPT).* ITU-T Recommendation F.850, International Telecommunication Union (ITU), Geneva, 1993.

[22] ITU. Telecommunication Standardization Sector (ITU-T). *Q-Series Intelligent Network Recommendation Structure.* ITU-T Recommendation Q.1200, International Telecommunication Union (ITU), Geneva, 1993.

[23] ITU. Telecommunication Standardization Sector (ITU-T). *Specification of Abstract Syntax Notation One (ASN.1).* ITU-T Recommendation X.208, International Telecommunication Union (ITU), Geneva, 1993.

[24] ITU. Telecommunication Standardization Sector (ITU-T). *Vocabulary of Terms for Universal Personal Telecommunications.* ITU-T Recommendation I.114, International Telecommunication Union (ITU), Geneva, 1993.

[25] ITU. Telecommunication Standardization Sector (ITU-T). *Information Technology—Open Systems Interconnection (OSI)—Basic Reference Model: The Basic Model.* ITU-T Recommendation X.200, International Telecommunication Union (ITU), Geneva, 1994.

[26] ITU. Telecommunication Standardization Sector (ITU-T). *Universal Personal Telecommunication (UPT)—Service Description (Capability Set 1).* ITU-T Recommendation F.851, International Telecommunication Union (ITU), Geneva, 1995.

[27] S. Kleier. *Neue Konzepte zur Unterstützung von Mobilität in Telekommunikationsnetzen.* Dissertation, RWTH Aachen, Lehrstuhl für Kommunikationsnetze, Aachen, 1996. ISBN 3-86073-386-9.

[28] D. R. Lawniczak. *Modellierung und Bewertung der Datenverwaltungskonzepte in UMTS*, 1995. ISBN 3-86073-381-8.

[29] T. Norp, A. J. M. Roovers. *UMTS Integrated with B-ISDN. IEEE Communications Magazine*, Vol. 32, No. 11, pp. 60–65, 1994.

[30] J. Sundborg. *Universal Personal Telecommunication (UPT)—Concept and Standardisation. Ericsson Review*, Vol. 4, pp. 140–155, 1993.

18
Next Generation Systems

Wireless access to fixed networks is seen to be of increasing importance, since wired communications, i.e., via cable or fibre, severely limits the usage of network dependent services, e.g., Internet services. Wireless Internet access needs high bit-rate data transmission and a much larger capacity of the radio spectrum than is available to date with cellular systems to enable multiple users at the same time and same location to communicate simultaneously.

18.1 Applications and Services

Humans and machines will be transmitters and receivers to exchange real-time and non real-time application data of large volumes as indicated in Figure 18.1. The machine to machine wireless communication is expected to raise more traffic than that resulting from or destined for humans.

Today's restrictions of the *Quality of Service* (QoS) such as data throughput, delay, or *Bit Error Ratio* (BER), will be substantially reduced. By allowing a wideband or even broadband access to the fixed network, the user will also get access to various (high bit-rate) multimedia applications and services at the office, at home, at hot spots and even on the move. At the bottom of Figure 18.2 fixed and mobile radio networks, called transport platforms, are shown to access the example multi-media applications shown in the upper part of the figure that are running on workstations, called service platforms.

Our recent investigations have shown that the demand for the provision of MM services to moving and mobile users for many applications could be satisfied by a much more cost-effective system than cellular mobile radio, which would then adopt the function of back-up to provide the services that must be realized instantaneously to users without any cost concern, e.g., speech and urgent data. The *Next Generation* (NG) system foreseen and described later in this section is based on a hybrid system approach comprising *Third-Generation* (3G) cellular and wireless broadband. Apparently, multi-mode terminals will be needed to exploit the possibilities offered by that. This vision is going beyond the evolution scenario depicted in Figure 18.3 that was presented in a workshop on 'Visions of the Wireless World' [1].

Table 18.1 shows the times needed to transfer some exemplary media objects of different sizes, such as web pages, book content or video, over various mobile telecommunication systems. Accordingly, it is obvious that systems like (E)GPRS or UMTS on the move cannot support continuous media even

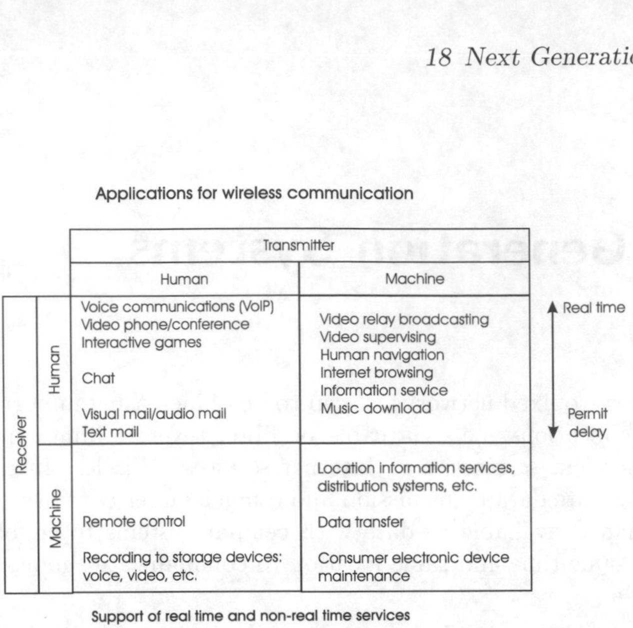

Applications for wireless communication

Receiver	Transmitter: Human	Transmitter: Machine
Human	Voice communications (VoIP) Video phone/conference Interactive games Chat Visual mail/audio mail Text mail	Video relay broadcasting Video supervising Human navigation Internet browsing Information service Music download
Machine	Remote control Recording to storage devices: voice, video, etc.	Location information services, distribution systems, etc. Data transfer Consumer electronic device maintenance

Real time ↔ Permit delay

Support of real time and non-real time services

Figure 18.1: Applications and users of future mobile and wireless services

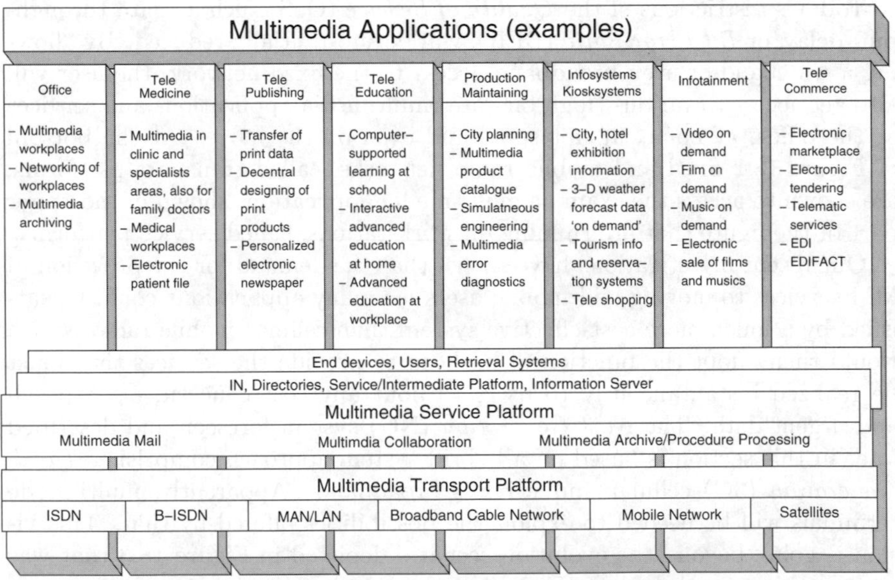

Multimedia Applications (examples)

Office	Tele Medicine	Tele Publishing	Tele Education	Production Maintaining	Infosystems Kiosksystems	Infotainment	Tele Commerce
– Multimedia workplaces – Networking of workplaces – Multimedia archiving	– Multimedia in clinic and specialists areas, also for family doctors – Medical workplaces – Electronic patient file	– Transfer of print data – Decentral designing of print products – Personalized electronic newspaper	– Computer–supported learning at school – Interactive advanced education at home – Advanced education at workplace	– City planning – Multimedia product catalogue – Simulatneous engineering – Multimedia error diagnostics	– City, hotel exhibition information – 3–D weather forecast data 'on demand' – Tourism info and reserva–tion systems – Tele shopping	– Video on demand – Film on demand – Music on demand – Electronic sale of films and musics	– Electronic marketplace – Electronic tendering – Telematic services – EDI EDIFACT

End devices, Users, Retrieval Systems

IN, Directories, Service/Intermediate Platform, Information Server

Multimedia Service Platform

Multimedia Mail | Multimdia Collaboration | Multimedia Archive/Procedure Processing

Multimedia Transport Platform

ISDN | B–ISDN | MAN/LAN | Broadband Cable Network | Mobile Network | Satellites

Figure 18.2: Multi-media (MM) applications based on MM-workstations connected via various MM-transport networks

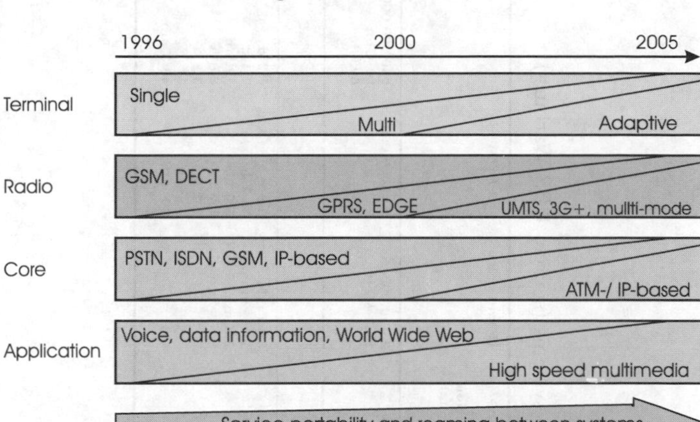

Figure 18.3: Evolution of terminals, network elements and applications towards NG systems

with the unrealistic high data rate assumed there. One example of this is the 15-minute video (300,000,000 bytes).

The expected evolution of terminals, radio interface standards, core networks and applications for mobile usage is shown schematically over time in Table 18.2. Terminals have evolved from single to multi-band and must further evolve to become multi-mode terminals to adapt to the various air interfaces expected for 3G and NG systems. The radio interfaces soon will see 3G+ systems, e.g., integrating cellular and digital broadcasting or modified 3G systems to be able to provide asymmetrical data transport. The core networks will evolve to provide packet-based high data-rate transmission via advanced Internet technology or ATM based networks. High speed multimedia applications will be served by combining the most effective radio access technologies. The services available from any type of mobile or wireless system will be portable across the various air interfaces to provide the feeling of an unified unrestricted connectivity across the radio space.

Broadband wireless access for mobile (moving) terminals, e.g., for terminals installed in vehicles, poses higher system requirements than for stationary terminals. High bit rate wireless access will be provided by the 3G mobile telecommunication systems, e.g., UMTS. Since UMTS will provide a gross transmission rate of about 256 kbit/s to mobile users in vehicles that will result in somewhat like 64 kbit/s net bit rate, only small displays for multimedia communication may be realizable with an accepted picture quality.

Table 18.2 shows the sequence of generations of radio systems extrapolated to *Fourth-Generation* (4G) systems expected to be launched in the year 2011.

Table 18.1: Required transfer times for example objects over various mobile radio systems (source: UBS Warburg)

Application		Page of text/E-mail	Picture on laptop	Web home page/ 30 pp simple colour presentation	E-mail + attachment	Book (300 pp)	30 pp simple colour presentation Professional quality photograph	15 min video	CD-ROM	
Size	(bytes)	3,125	50,000	100,000	750,000	937,500	2,000,000	2,400,000	300,000,000	650,000,000
Size	(bits)	25 kbit	400 kbit	800 kbit	6 Mbit	7.5 Mbit	16 Mbit	19.2 Mbit	24 Gbit	52 Gbit
Book equiv's		1 page	16 pages	32 pages	240 pages	1	2.1	2.6	320	693.3
Technology	data rate (kbit/s)	Time to transmit (s)								
GSM	9.6	2.6	41.7	1.4 min	10.4 min	13 min	27.8 min	33.3 min	2.9 d	6.3 d
HSCSD	28.8	0.87	13.9	27.8	3.5 min	4.3 min	9.3 min	11.1 min	23.1 h	2.1 d
GPRS	115	0.22	3.5	7	52.2	1.1 min	2.3 min	2.8 min	5.8 h	12.8 h
EDGE	384	0.07	1	2.1	15.6	19.5	41.7	50	1.7 h	3.8 h
UMTS moving	384	0.07	1	2.1	15.6	19.5	41.7	50	1.7 h	3.8 h
UMTS stationary	2000	0.01	0.2	0.4	3	3.8	8	9.6	20 min	43.3 min

Table 18.2: Generations of radio-based systems and services provided

Perspective	1G (1981)	2G (1991)	3G (2001)	4G (2011?)
Business model	Monopoly Network centric			Deregulated Client/Server
Application	Voice	Voice SMS	Voice Multimedia	Voice Virtual reality
Terminal	Phone	Phone w. message	Multimedia terminal	Virtual reality device
Bearer	Symmetric circuit	Symmetric packet (GPRS)		Asymmetric packet
Network	'Old' circuit	ISDN based	All IP (GPRS)	All IP
Access	Analog FDMA	Digital TDMA cdmaOne	Digital WCDMA EDGE cdma2001Xev	Multi access WCDMA/EDGE Bluetooth WLAN, MBS new air i/f?

Figure 18.4 shows the evolution from 2G to 4G cellular systems and underlines that multiple radio interfaces competing worldwide to serve the same applications will continue to exist in the future, although based on new technology compared to 3G systems. It becomes very clear also that

- The intention of the IMT-2000 initiative to end up with one universal 3G air interface standard has not been met and will not be met in the future.
- New applications and related frequency bands require air interfaces optimised for the specific task and will contribute to extent the multitude of radio interface standards.
- The spectrum range to be covered by 4G systems will make it difficult to implement the respective radio interfaces as part of one single terminal, since the size and energy consumption will then not fit the specific requirements.
- The 3G radio interface will be a temporary solution that will be replaced soon by much more advanced systems, probably based on OFDM or coded OFDM modulation instead of spread spectrum.
- Higher transmission speed per user is one of the most demanding parameters and that spectrum availability is a prerequisite to make this vision a reality.

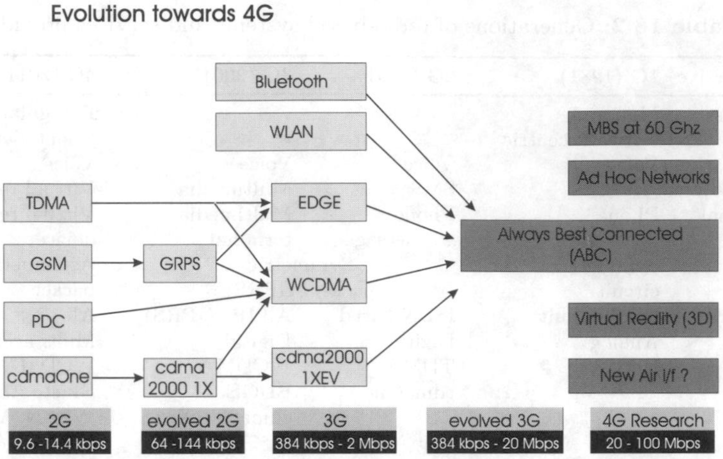

Figure 18.4: Evolution from 2G to 4G wireless systems [1]

18.2 Spectrum Issues for Next Generation Systems

In 2000 in the UK and Germany licence fees of about 5 and 8 billion Euro per operator were paid for 10 MHz of paired UMTS spectrum. The cost for infrastructure, even when limited to highly populated areas including subsistence costs for the new terminals will amount to another 5–10 billion Euro per operator. As a consequence of high licence, infrastructure and subsistence costs in Europe, it is expected that high bit-rate services will be too expensive to afford and the focus of UMTS operators will be on narrow-band data services, where not much has to be transmitted per event, but a charge comparable to that of a one-minute speech conversation can be charged. The short message service of GSM and I-mode of *Nippon Telegraph and Telephone Corporation* (NTT) DoCoMo are examples of this.

Current 3G systems with the spectrum available, in principle, can support high bit-rate Internet access for a very small number of concurrent users per cell only, but it has even been shown that with the standardized UMTS protocols in combination with TCP/IP only medium bit-rates will be reached for WWW Internet access (see Figure 6.50).

The current spectrum allocations of WARC-92 for 3G are unable to provide the spectrum capacity needed for wideband or even broadband data transmission [6].

To reach this capacity would require about 300 to 500 MHz of additional spectrum exploited in a pico-cellular network layout. Such an amount of new spectrum is available only beyond 2.69 GHz, i.e., beyond the IMT-2000 extension bands as defined by *World Radiocommunication Conference* (WRC)-

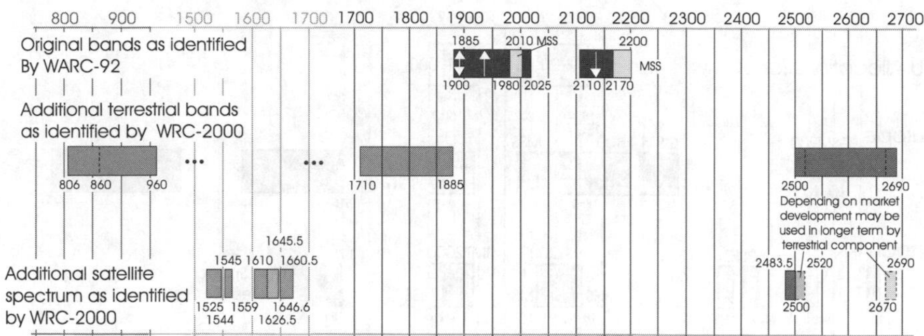

Figure 18.5: IMT-2000 extension bands

2000. Figure 18.5 shows a comprehensive overview on the frequency bands identified by WARC-92 and by WRC-2000 for IMT-2000 systems. ITU-R WP8F - Spectrums Group currently is discussing the usage of these bands. The bands shown are not available in all the member regions internationally and not early enough.

These bands consist of 169 MHz at 2.5 to 2.69 GHz of new spectrum plus the re-farming of 319 MHz of spectrum currently being used for cellular, e.g., GSM in Europe. Since the applications served in these 319 MHz already exist, the re-farming will not contribute to increase the capacity to serve more mobile services at all but only shift services from 2G to 3G systems.

Frequency bands in use today for second generation mobile radio services, e.g., 806 to 960 MHz dependent on region and country (GSM900) and 1710 to 1885 MHz (GSM1800) will be available for IMT-2000 after the respective licences will have expired or after the regulation conditions have been changed accordingly. New assignments of these bands will be possible, e.g., in Germany for the GSM bands from the year 2015 on.

The IMT-2000 extension band 2500 to 2690 MHz—devoted for terrestrial radio services in the range from 2520 to 2670 MHz—is not available in some countries of Asia and in North America. This band will become available in other countries between the years 2005 and 2010, e.g., in Germany the band will be available from January 2008 on.

In view of these facts, WRC-92 has also allocated spectrum between 5 and 6 GHz for wireless broadband systems that are also called *Wireless Local Area Network* (WLAN), see Figure 18.6. The spectrum is proposed to be licence-exempt and amounts to 455 MHz in Europe, 300 MHz in the USA and 100 MHz in Japan and will speed up the breakthrough of wireless systems that recently have been standardized as IEEE 802.11a and HiperLAN/2 systems. The WLAN spectrum partly has been allocated for indoors with 200 mW EIRP transmit power and partly outdoors with 1 W EIRP.

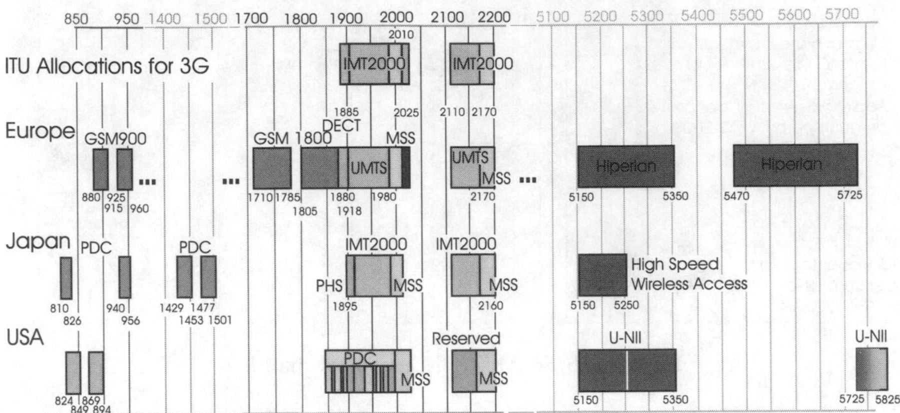

Figure 18.6: Spectrum allocations for some world regions for 3G and for WLANs

18.2.1 Asymmetric Traffic Characteristics of Uplink and Downlink Usage

The higher the transmit rate of a service, the higher is the expected asymmetry of usage of the uplink and downlink channels, making the downlink a bottleneck in IMT-2000 systems. Both UMTS Forum and ITU-R have published a projection of the future usage of IMT-2000 systems and have identified the spectrum needed for the specific services (see Figure 18.7). A substantial asymmetry of the expected average traffic has been predicted there especially for medium and high rate multimedia traffic. The grade of asymmetry dependent on the services used might change from cell to cell over time and should be taken into account when considering spectrum allocation for the extension of the currently available bands for IMT-2000 systems. It would be best to be able to adapt the asymmetry of the spectrum load to the occupancy of the spectrum dynamically, dependent on the current load situation in a cell and on the development of the usage of services in a mobile radio system.

This would require the further development of 3G systems to be able to support higher bit rate services on the downlink than on the uplink that could be realized by allowing a variable duplex spacing and multiple FDM channels for parallel usage on the downlink. The respective standardization is under way for UMTS.

One interesting approach under study is the combination of cellular and digital broadcast services in a way to integrate the high downlink capacity of digital TV-broadcast systems into a communications session supported via cellular radio. Different ways can be imagined to implement that:

1. Usage of *Digital Video Broadcasting Terrestrial* (DVB-T) on the downlink to complement, say, UMTS, as being studied in the IST/DRIVE

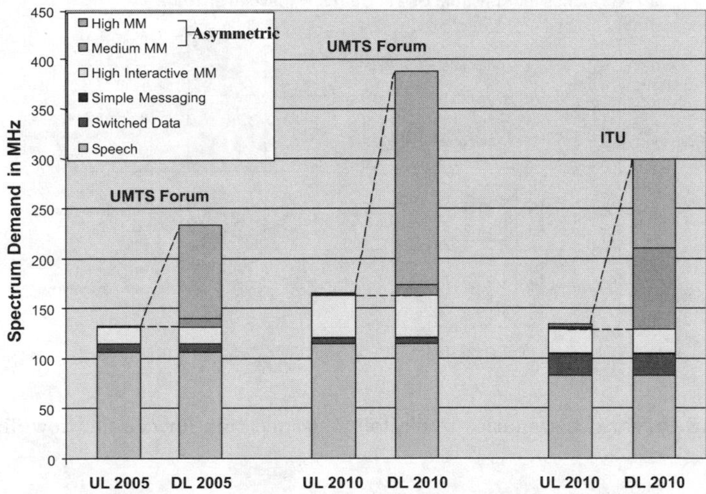

Figure 18.7: Projection of the future usage of IMT-2000 systems (Source: UMTS-Forum, Report No.6 and ITU-R Report M.[IMT.SPEC])

project, with a common control of the distribution of service data to the radio links of the two systems.

2. Time-shared usage of spectrum (spectrum co-farming) with defence organizations against leasing fees combined with well-established rules under what conditions to clear the band from public usage when needed for military purposes, see Figure 18.8 (Scenario 1).

3. Usage of some 8 MIIz wide channels from the broadcast spectrum to realize UMTS downlink channels with a dynamic channel selection to be able to react to the actual availability of channels in time at the respective location, see Figure 18.8 (Scenario 2).

4. Usage of some 8 MHz wide channels from the broadcast spectrum to realize UMTS fixed downlink channels plus dynamic channel allocation, see Figure 18.8 (Scenario 3).

The hybrid system resulting from the integration of UMTS and DVB-T systems described under item 1 of cause would require multi-mode terminals able to support both air interfaces and a common coordination channel in the core network to dynamically decide which system's downlink should be used actually during a communications session. Experiments have shown the design to be feasible but having a limited flexibility owing to the MPEG-2 container used for data transmission in the DVB-T standard. Per DVB-T (8 MHz) channel even at very high speed a total data rate of 12 Mbit/s would be possible. A capacity improvement of the combined system could be

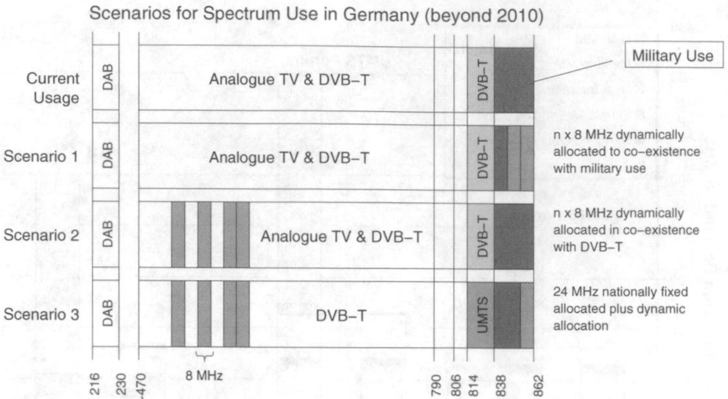

Figure 18.8: Various scenarios to exploit TV-spectrum for cellular downlink usage

realized, especially for broadcast applications in UMTS but not for point-to-point communications owing to the large TV broadcast cells and the resulting small capacity.

The homogeneous system resulting from scenarios 1 to 3 appears to be much more attractive, since the downlink in the TV-broadcast band could be provided from all the base stations of a cellular operator in the coverage area, instead of from only the TV towers, allowing a much higher capacity gain. Dynamic channel allocation to provide UMTS downlinks on TV channels needs a concept called co-farming of spectrum [6] to guarantee both, the broadcast operator and the cellular operator a mutual benefit from this. Further, the UMTS standard would need to be extended to allow various duplex spacing and multiple downlink channels.

18.3 Wireless LANs to Supplement Cellular Radio Networks

The services that particularly demand an asymmetrical air interface and substantial spectrum allocation in addition are termed high bit rate multimedia in Figure 18.7. The contents of these services are among others text, music and image data that are stored as conserves somewhere in the network, e.g., contents of a mailbox or a server data base, typically addressed via the Internet. The related mass data typically do not have the same value per bit transmitted as speech data and in many cases of access do not require immediate service as is typically for services provided by a cellular radio network. It appears convincing to consider alternate air interfaces to provide these services in a much more cost-efficient way, thereby unloading the cellular spectrum.

WLANs have been proposed to complement the service capacity of 3G systems and have been studied, e.g., in the IST/BRAIN project [3, 4]. The

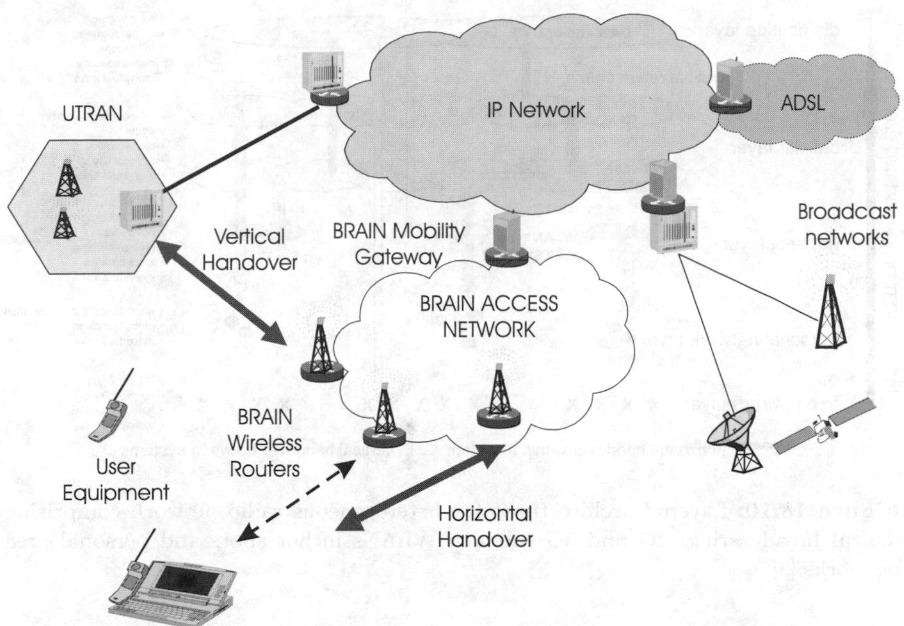

Figure 18.9: Hybrid network architecture studied in the BRAIN project

co-operation of cellular and WLAN is by means of access to a common core network using a multi-mode terminal controlled by a mobility management function comprising both access systems to provide the user with the best service possible at a given location, see Figure 18.9. The idea behind BRAIN is to provide broadband services in so-called hot spots where a WLAN-based pico-cellular coverage is provided, including horizontal handover across access points providing broadband access and vertical handover between the two different access networks.

The WLAN technology is expected to be based on ETSI/BRAN H/2 or IEEE 802.11a. Systems combining UMTS and H/2 are expected to come into operation in Europe in the year 2003, according to announcements in 2001 of some network manufacturers.

A further step ahead would be the integration of systems like *Digital Audio Broadcast* (DAB) and DVB-T, besides *Personal Area Networks* (PANs) like Bluetooth into a common network architecture as shown in Figure 18.7 [3]. Again, an architecture like this is close to implementation since the elements are available already and a combination of these into a heterogeneous system appears feasible. Then, radio interfaces developed specifically for the various speeds of mobility and classes of service will be combined in the form of separated access networks to a common core network to provide universal mobile access for all services known from fixed networks.

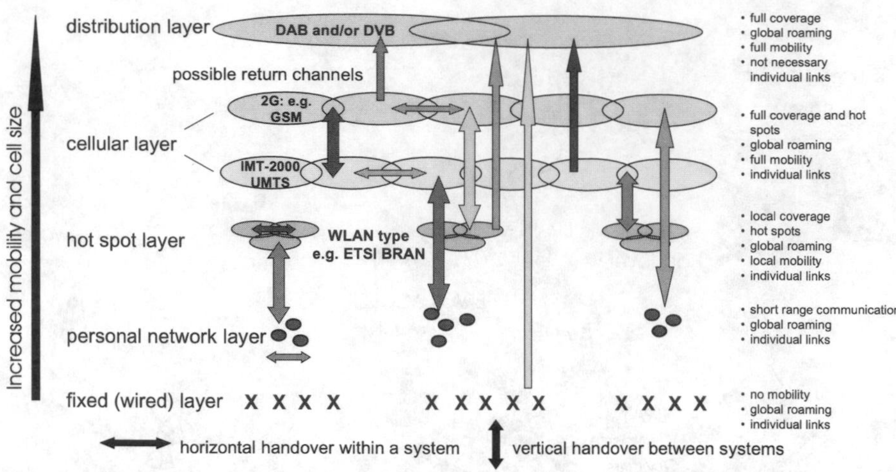

Figure 18.10: Layered architecture of a heterogeneous radio network comprising digital broadcasting, 2G and 3G cellular, WLANs in hot spots and personal area networks [3]

18.4 The Wireless Media System: A Candidate for Next Generation Systems

It is not a big step to imagine that WLAN access will not be limited to hot spots but provided by a sufficiently dense network of scattered cells providing a non-contiguous radio coverage in the area that are controlled in a way to provide virtually a continuous connectivity of mobile terminals to a wireless broadband network for access to multimedia contents. This concept of a *Wireless Media System* (WMS) [5] is similar to the Infostation concept [2] and would have a number of advantages over current designs as explained in Figure 18.11. The figure shows two buttons 'red' and 'black' of a terminal controlling the two radio interfaces to the service characteristics and costs of services available from cellular radio and wireless Internet access via WLAN. Accordingly, cellular is characterized as a narrowband service with high *Quality of Service* (QoS) and high cost of usage whilst WLANs provide broadband access to the Internet with an Internet typical QoS and low cost of usage. Taking the current licence fees and the costs to deploy the systems in the field into account it appears attractive for a subscriber to both types of systems, separately, to use as much as possible the WLAN-based system that would appear like a broadband wireless 'telephone booth' but without the need to approach the AP more than, say some hundred metres outdoors.

The modems currently under development for IEEE 802.11a and H/2 allow at 1 W EIRP outdoors to bridge a maximum distance of 2.2 km at 6 Mbit/s with line-of-sight connectivity and of 92 m with obstructions. This figures

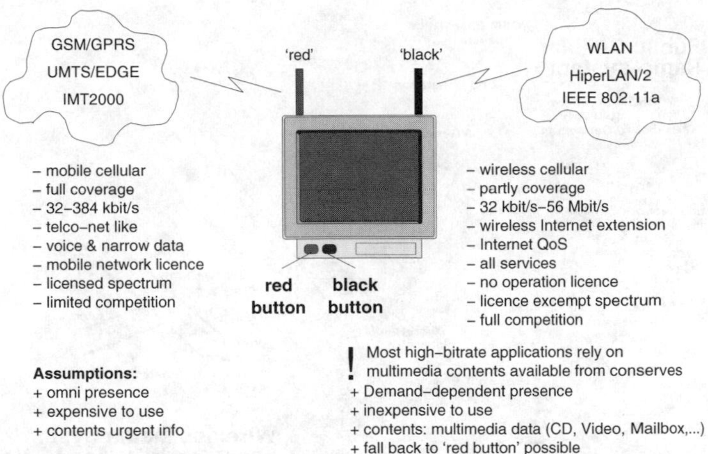

Figure 18.11: Terminal of the Wireless Media System

are reduced at 54 Mbit/s to 322 m and 32 m, respectively. It has also been estimated that a terminal speed of about 100 km/h would be possible at the lower bitrate PHY modes standardized for H/2. Transmission in the proposed WMS is bi-directional between a *Media Access Point* (MAP) connected to the fixed core network and moving or mobile *Media Terminals* (MTs) that are equipped with a large storage capacity. Of course, real-time oriented media data can also be supported if a MT does not move outside the radio coverage range when communicating. Although the WMS is able to operate on its own and does not require a close interworking with a cellular radio system, it would benefit from such an integration for some applications to a great extent, for others not.

Figure 18.12 shows the integration of 2G and 3G cellular radio systems via their *Internet Protocol* (IP) based core network with the *Wireless Media System* (WMS) that is based on its own Intranet. It is assumed that the mobile terminals at least are supporting one mobile radio and one WLAN air interface.

1. The integration might be tight if the Intranet of the WMS is part of the core network of the mobile radio network.

2. The integration might be loose if both fixed IP networks are connected to exchange control and signaling information.

3. It might also be that both systems are operated completely separated from each other. For example a mobile terminal might contain two *Subscriber Identity Modules* (SIMs) according to the two different contracts signed with the operators of the mobile radio network and WMS, respectively.

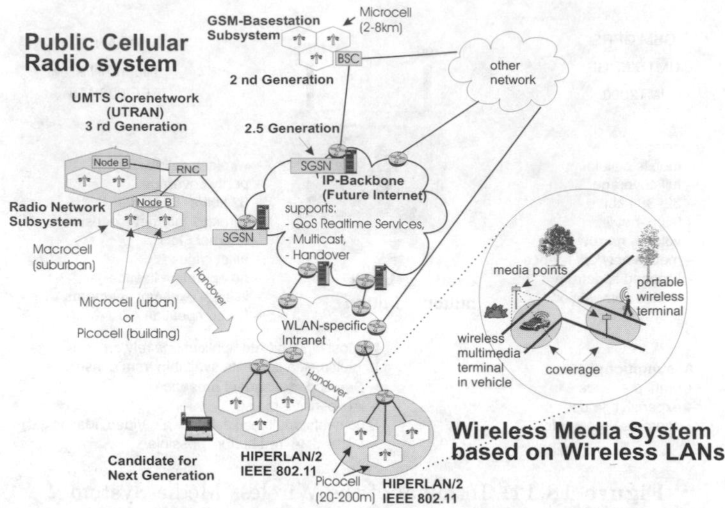

Figure 18.12: A NG system by example of the WMS

The system proposed in Figure 18.9 incorporates a vertical handover between the mobile radio network and WLAN-based WMS. One possibility is to rely instead on a fast re-establishment of a connection when switching between systems if both systems don't have a mobility management function in common, see variant 3 above.

A handover from the WMS to the mobile radio network anyway would result in a substantial degradation of quality of service for multimedia services and the subscriber would not be happy with a handover at all. A handover in the other direction should happen under the control of the current by communicating applications and could then avoid an interrupt to the ongoing session completely since the mobile radio networks can be assumed to cover the area where the WMS is available, in addition.

The following characteristics of the WMS are worth mentioning[7]:

- Operation in a licence-exempt (licence-free) or in a low-cost licensed band using a broadband standardized radio interface, e.g., H/2, IEEE 802.11(b/a), etc.

- High bit-rate wireless transmission of media data between an MT and the MAP that is connected via *Line Of Sight* (LOS) radio, cable or fibre to an Intranet containing WMS-specific multimedia servers.

- An MAP provides a certain radio coverage and is located in the service area in a number according to the expected MT density and service usage. In total the MAPs do not provide an ubiquitous but only a fragmented area radio coverage. An MAP is equipped with a large capacity storage (cache) to be able to download media contents with the full speed available at the air-interface.

- The MAPs altogether are not able to serve a contiguous area but only a scattered radio coverage of the service area. The service area might be quite large comprising motor ways, a city or highly populated area. Smaller service areas are sports arenas, air ports, railway stations, city centres, etc.
- Continuous availability of media contents is provided to the user of an MT, in spite of a discontinuous radio coverage, through a specific service control software running on the MT and on top of the Intranet that simulates a continuous connectivity to the WMS by buffering of content data in MAPs and MT. A spontaneous access request to new content data, typically, is executed with some delay only, since the respective MT must wait until it has radio connectivity to the WMS.
- At each MAP the MT receives media data to be stored in its local cache from the network in an amount that is required for the expected processing and consuming within a planned future time period, e.g., within an hour. On the uplink the MT transmits its cached data in the radio coverage range of an MAP.
- An MT when reaching an MAP relates to the association that it had established earlier with the WMS-service and then receives the data it had requested earlier with the high bitrate of the WLAN air-interface to be stored in the MT's local cache to be consumed or processed later. The amount of data transmitted is large enough to cover the expected time of processing (e.g., contents of mailbox) or to cover the duration of local consumption for a well defined time horizon, e.g., half an hour. Further, when reaching an MAP, the MT transmits all data waiting locally via the WMS to the final destination.
- An MT may use all services known from cellular radio networks including speech and interactive media data.
- Commands issued by the MT to the WMS will determine which media data should be transmitted next and the appropriate MAP to download the respective data will be determined either by the MT or the network itself, e.g., based on geographical information or routes of terminal movement known or estimated by the network. Before reaching the next MAP the data stored locally in the MT is consumed or processed without a link to the network, and the results may be transferred to the network at the next MAP.
- A cellular radio system might be used at any time as back-up and to order data to be cached at some MAPs in the vicinity of the MT for download, when it has arrived there.

- An AP may use a highly directive antenna to only cover areas where MTs may roam, e.g., streets where MTs are operated in vehicles or urban areas for portable MTs.

An example scenario is shown in Figure 18.13 (left), with two MAPs and their transmitting and receiving devices (transceivers) that are mounted on two gantries of a highway. An AP is equipped with a control unit that is connected to the Internet or local multimedia servers via a service control in the network.

As long as a vehicle resides in an MAP´s coverage area, an MT in the vehicle can set up a connection to request and download its application data wirelessly, or to send data that has been prepared meanwhile, e.g., emails or video recordings. When the vehicle leaves the coverage area, no radio link exists any longer.

In another scenario depicted in Figure 18.13 (right) the transceivers of the MAPs are mounted on the masts of street lighting within a residential area. There, a user using a portable MT may set up a radio link to an MAP and transmit as long as he/she stays within the corresponding coverage area.

One example of a content to be moved in a short while across the air interface of the WMS is the 15-minute video (300,000,000 bytes) of Table 18.1. The 20 min needed with UMTS for a radio terminal will be reduced to 1.6 min with 25 Mbit/s available from the MAP, e.g., at a crossing equipped with traffic lights. The video may be a part of a film that could be consumed (re-played) offline. The next part of the film (the next 15-minute video-scene) can then be downloaded at the next access point. If the download processes at several access points can be controlled (synchronized) perfectly, the user may enjoy seeing the film without any interruption (on the backseat of a car in the city).

In order to minimize the number of MAPs a trading of spectrum load against infrastructure costs is possible, e.g., with H/2 systems by extending the coverage area of an MAP through multi-hop communication using a wireless MAP working as a relay, see the *Wireless Base Station* (WBS) in Figure 18.14.

The MAP is realized there by an H/2 Access Point that is able to illuminate the whole crossing and in addition with line-of-sight coverage provides radio coverage in all the streets that go straight ahead form the MAP. Another WLAN system that is affixed at the next crossing and is operating as a relay (without a connection to the fixed communication network) enables to provide radio coverage into those streets in a city that are shaded with respect to the MAP. Communication between the vehicle shown and the MAP would then be via two hops, one from the vehicle to the relay (2. hop) and another from the relay to the MAP (1. hop) and vice versa. The relays shown are positioned at fixed locations but might be mobiles themselves. It might be advisable to use directional antennas (with a high gain) to connect an AP to its fixed relays.

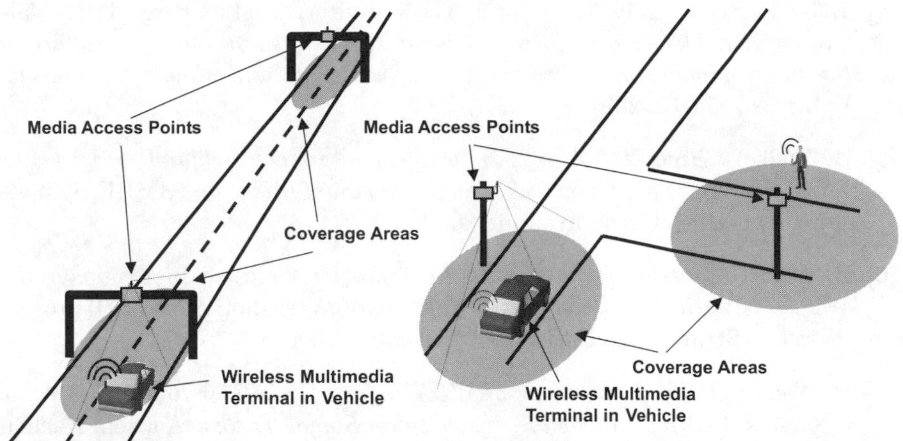

Figure 18.13: Wireless Media System scenarios with MTs on highway and in residential area

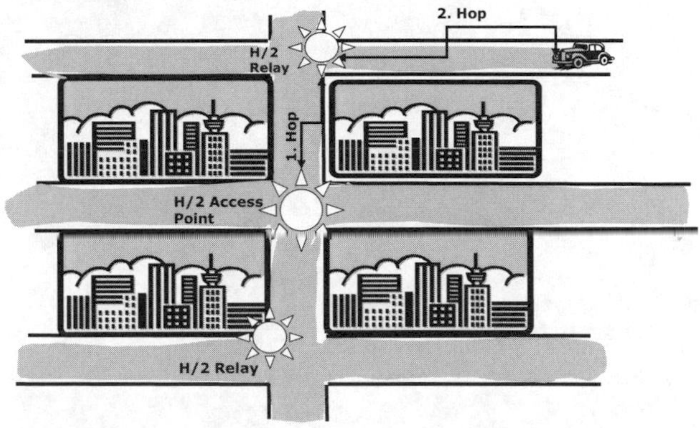

Figure 18.14: Multi-hop approach with HiperLAN/2 using wireless base stations

Bibliography

[1] H. Ericsson. *3G Services and the Roadmap Ahead.* In *Visions of the Wireless World*, Brussels, Belgium, Workshop of the Wireless Strategic Initiative, 12 December 2000.

[2] R. H. Frenkiel, B. R. Badrinath, J. Borres, R. D. Yates. *The Infostation Challenge: Balancing Cost and Ubiquity in Delivering Wireless Data. IEEE Personal Communications*, Vol. 7, No. 2, pp. 66–71, April 2000.

[3] W. Konhaeuser. *Innovating the Mobile World beyond the Third Generation.* In *Visions of the Wireless World*, Brussels, Belgium, Workshop of the Wireless Strategic Initiative, 12 December 2000.

[4] J. Urban, D. Wisely, E. Bolinth, G. Neureiter, M. Liljeberg, T. R. Valladares. *BRAIN—an architecture for a broadband radio access network of the nesxt generation. Wireless Communications and Mobile Computing*, Vol. 1, No. 55-75, 2001.

[5] B. Walke. *HiperLAN/2 – Ein weitgehend in Deutschland entwickelter Standard.* In *Zukunftsperspektiven Mobilkommunikation*, BMBF Symposium, Neu Ulm, Germany, June 2000.

[6] B. Walke. *Spectrum Issues for Next Generation Cellular.* In *Visions of the Wireless World*, Brussels, Belgium, Proc. of Workshop of the IST project 'Wireless Strategic Initiative', 12 December 2000.

[7] B. Walke. *On the Importance of WLANs for 3G Cellular Radio to Become a Success.* In *Proc. Aachen Symposium on Signal Theory*, Aachen, Aachen University of Technology, September 2001. Available from `http://www.comnets.rwth-aachen.de/~walke`.

Appendix A
Queuing and Loss Systems

Queuing systems model real systems and enable the calculation of their performance parameters.

A.1 The Queuing System M/M/n-∞

This system models the interarrival times T_A from randomly received jobs as well as the random service times T_B of the jobs, each through a negative exponential (*Markov*, M) distribution.

The n servers are assumed to be identical, and free servers can be occupied in any sequence with the arrival of a request; when all servers are occupied, the job waits in one of the s queuing positions in the queue until it is serviced. Waiting calls are serviced in any sequence, which is not allowed to be based on the servicing time (see Figure A.1).

The system is directly suitable for modelling mobile radio systems with a large number of independent users who occupy one of several channels available in a cell for their communication. Some base station systems have a queuing option for calls while waiting for the the availability of one out of several trunked channels.

A.1.1 State Process as Special Birth-and-Death Process

State Space and Transitions

A state $(= 0, 1, 2, \ldots \infty)$ defines the number of calls in a system (those being processed and those queuing). The state transition rates correspond to the arrival rate λ or the state-dependent servicing rate (see Figure A.2).

The traffic offered, which is

$$A = \frac{\lambda}{\epsilon} \tag{A.1}$$

is used to calculate the state probability for x calls in the system:

$$p_x = P(X = x) = \begin{cases} P_0 \left(\dfrac{A^x}{x!}\right) & \text{for } 0 \le x \le n \\ P_0 \left(\dfrac{A^n}{n!}\right)\left(\dfrac{A}{n}\right)^{(x-n)} & \text{for } x > n \end{cases} \tag{A.2}$$

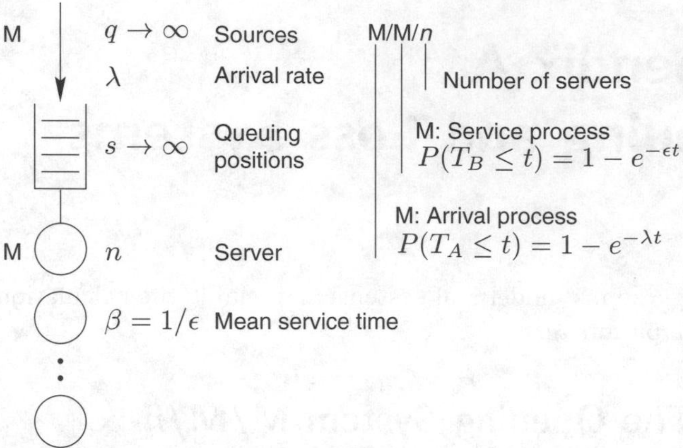

Figure A.1: Queuing system M/M/n-∞

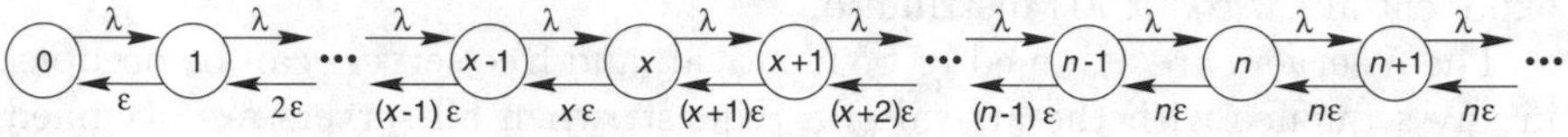

0, 1, …, x-1, x, x+1, …, n-1, n, n+1, …
λ λ λ λ λ λ λ λ λ λ
ε 2ε (x-1) ε xε (x+1)ε (x+2)ε (n-1) ε nε nε nε

Figure A.2: State diagram of M/M/n-∞ queuing system

$$P_0 = \left[\sum_{i=0}^{n-1} \frac{A^i}{i!} + \frac{A^n}{n!} \sum_{i=0}^{\infty} \left(\frac{A}{n} \right)^i \right]^{-1} = \left[\sum_{i=0}^{n-1} \frac{A^i}{i!} + \frac{nA^n}{n!(n-A)} \right]^{-1} \tag{A.3}$$

In pure queuing systems ($s \to \infty$):

- The sum in Equation A.3 only converges if the stationary condition $A < n$ is fulfilled.
- The traffic carried (also called the traffic value) Y equals the traffic offered A if $A \leq n$; otherwise the system will become unstable and the queue increases without a limit.

The *utilization level* is given by

$$\rho = \frac{A}{n} \tag{A.4}$$

A.1.2 Characteristic Performance Parameters

Queuing probability p_w

The probability of an arriving call having to queue is

$$p_w = \frac{p_n}{1 - \rho} \tag{A.5}$$

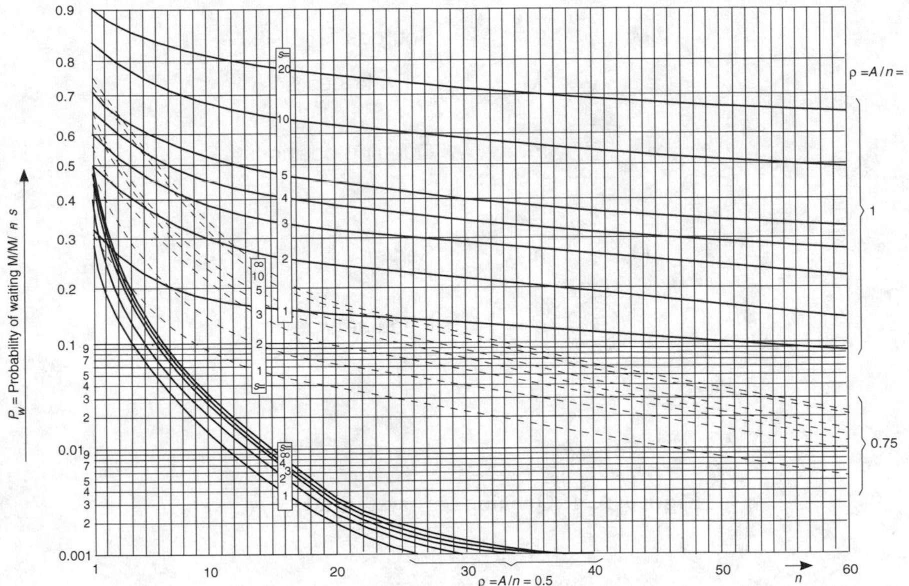

Figure A.3: Queuing probability in an M/M/n-s queuing system

Erlang queuing formula

$$p_w = P_0 \frac{A^n}{n!(1-\rho)} = \frac{\dfrac{A^n}{n!(1-\rho)}}{\displaystyle\sum_{i=0}^{n-1} \frac{A^i}{i!} + \frac{A^n}{n!(1-\rho)}} \tag{A.6}$$

Figure A.3 shows the queuing probability p_w as a function of the number of channels n, with the utilization level ρ as a parameter (see Equation A.4) for the models with $s = 1, 2, \ldots \infty$ queuing positions. $s = 0$ results in a loss system (see Section A.2).

Traffic value Y

Traffic value Y indicates the average number of servers occupied simultaneously.

$$Y = \sum_{x=0}^{n-1} x p_x + n \sum_{x=n}^{\infty} p_x = A \tag{A.7}$$

In pure queuing systems the offered traffic is the same as the carried traffic.

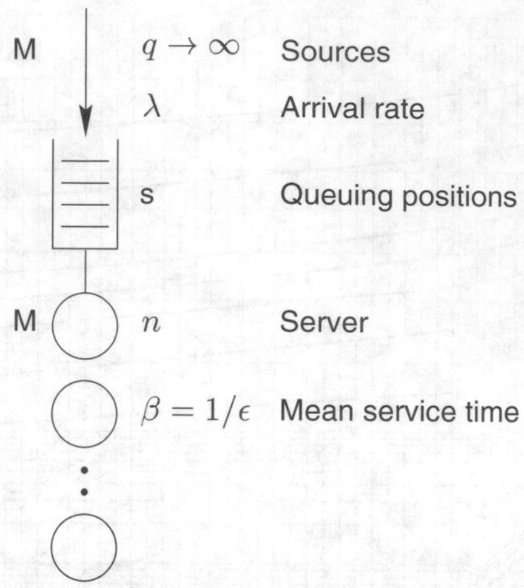

Figure A.4: Queuing-loss system M/M/$n - s$

Queuing load (average queue length)

L_Q is used to indicate the average number of queuing positions occupied. It is given by

$$L_Q = \sum_{x=n}^{\infty} (x-n)p_x = p_n \frac{\rho}{(1-\rho)^2} \tag{A.8}$$

Average queuing time W and W^-

- $W = E[T_W]$: average queuing time of all arriving calls:

$$W = \frac{L_Q}{\lambda} \tag{A.9}$$

- $W^- = E[T_W | T_W > 0]$: average waiting time for queuing calls only:

$$W^- = \frac{L_Q}{p_w \lambda} \tag{A.10}$$

The approximation of a stationary boundary, $A \to n$, makes $W^- \to \infty$.

A.2 The Queuing-Loss System M/M/n-s

Here the number of queuing positions is limited to $s < \infty$. If the traffic exceeds the number of servers available plus queuing positions then the calls above that number will not be processed (see Figure A.4). $s = 0$ equates to a pure-loss system.

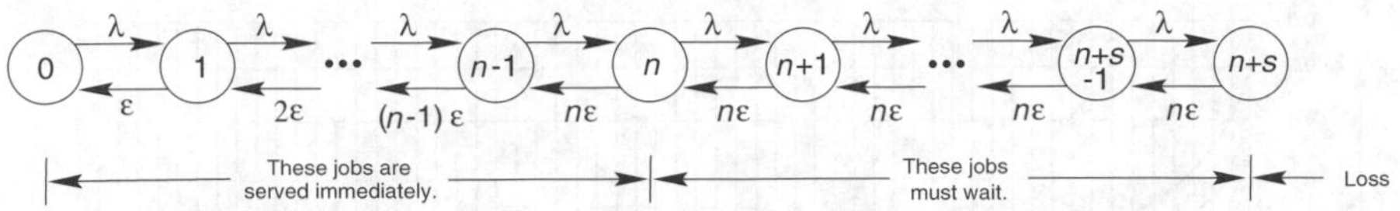

Figure A.5: State diagram of an M/M/$n - s$ queuing-loss system

A.2.1 State Process as a Special Birth-and-Death Process

State space and transitions

The state number ($= 0, 1, 2, \ldots, n + s$) of calls in a system (serviced and queuing) is

$$P(X = x) = \begin{cases} P_0 \left(\dfrac{A^x}{x!}\right) & \text{for } 0 \leq x \leq n \\ P_0 \left(\dfrac{A^n}{n!}\right)\left(\dfrac{A}{n}\right)^{(x-n)} & \text{for } n+1 \leq x \leq n+s \end{cases} \tag{A.11}$$

$$P_0 = \left(\sum_{i=0}^{n} \frac{A^i}{i!} + \sum_{i=n+1}^{n+s} \frac{A^i}{n!} \frac{1}{n^{i-n}}\right)^{-1} \tag{A.12}$$

See Figure A.5.

A.2.2 Characteristic Values

Loss probability

Calls are lost if no further ones can be accepted; thus the state probability for the state $n + s$ corresponds to the loss probability p_l:

$$p_l = P_{n+s} \tag{A.13}$$

Figure A.6 shows the loss probability p_l as a function of the number n of available parallel channels for different levels of utilization ρ of the channels and s queuing positions of the queuing-loss system. Figure A.7 shows the loss probability p_l as a function of the number n of channels for a pure-loss system, with the utilization level ρ as a curve parameter. The queuing probability is illustrated in Figure A.3.

Traffic value Y

$$Y = \sum_{k=1}^{n} k p_k + \sum_{k=n+1}^{n+s} n p_k \tag{A.14}$$

$$Y = A(1 - p_l) \tag{A.15}$$

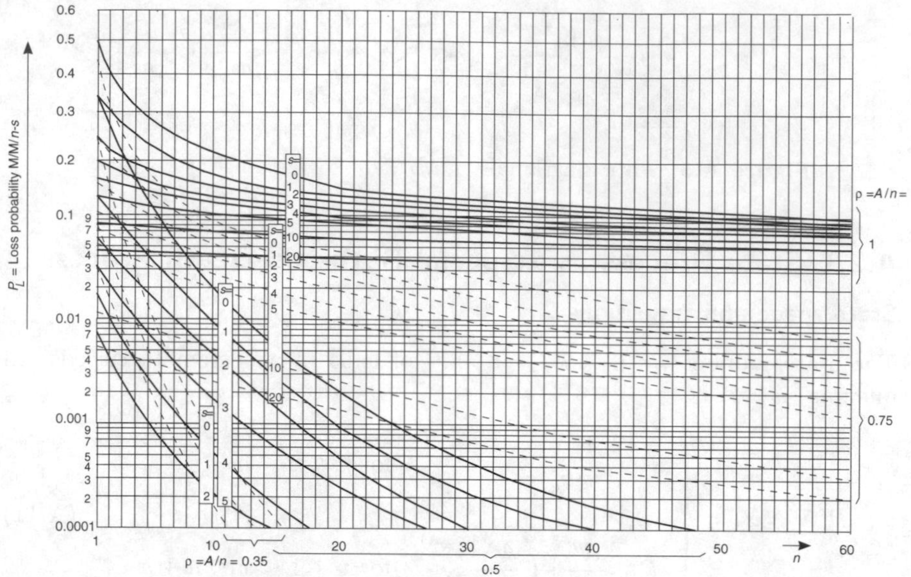

Figure A.6: M/M/n-s queuing-loss system

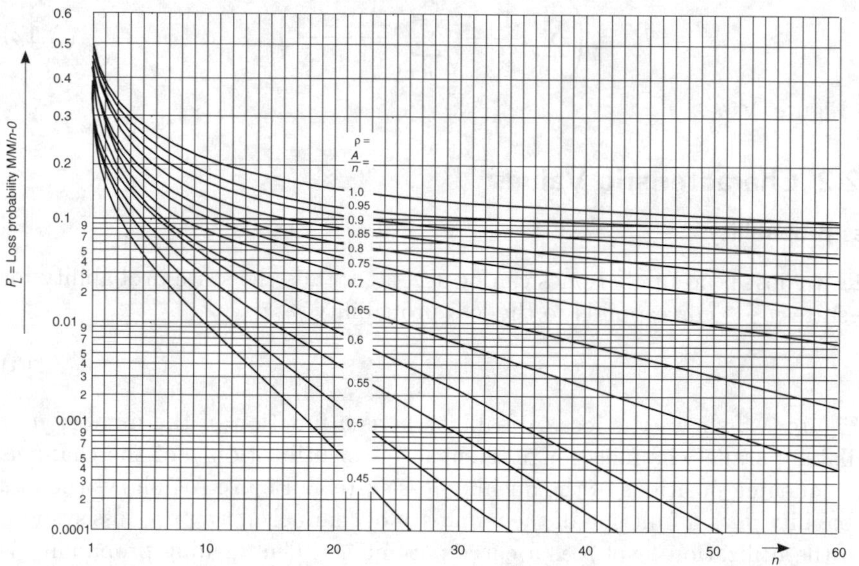

Figure A.7: M/M/n loss system

Queuing load L_Q

$$L_Q = \sum_{k=n+1}^{n+s} (k-n)p_k \tag{A.16}$$

Average Queuing Time Q and Q^-

- The average queuing time for all arriving calls is

$$W = \frac{L_Q}{\lambda} \tag{A.17}$$

- The average queuing time of the queuing calls only is

$$W^- = \frac{L_Q}{p_w \lambda} \tag{A.18}$$

Appendix B
Standards and Recommendations

As a result of the rapid developments in communications and the variety of communications systems developed by different manufacturers all over the world, agreements on international, regional and national standards or recommendations are essential. According to the ISO, the definitions of a standard and a recommendation are as follows:

Standard: "*A technical specification or other document available to the public, drawn up with the cooperation and consensus or general approval of all interests affected by it, based on the consolidated results of science, technology and experience, aimed at the promotion of optimum community benefits and approved by a body recognized on the national, regional or international level.*"

Recommendation: "*A binding document which contains legislative, regulatory or administrative rules and which is adopted and published by an authority legally vested with the necessary power.*"

Standardization (or the drawing up of recommendations) in the area of telecommunications is defining specifications within a task area to allow as many manufacturers as possible to develop their products accordingly. This results in the elimination of trade barriers, thereby resulting in an international harmonization of telecommunications infrastructure. Other objectives being pursued in the mobile radio area include noise-free message transfer, optimal use of frequency spectrum and minimization of electromagnetic incompatibility between radio services.

On the other hand, product development is restricted owing to standards and the resultant compliance with certain regulations. Another disadvantage is the long time span between the standardization of a new system and its commercial introduction. An effort is being made to consider future technological innovation in standards in order to minimize the problem.

B.1 International Standards Organizations

The most important international standards organizations in the telecommunications area include the following:

- ISO: International Standardization Organization

- ITU: International Telecommunications Union
- IEC: International Electrotechnical Commission
- INTELSAT/INMARSAT: International Telecommunications Satellite Organization/International Maritime Satellite Organization

B.1.1 ISO

The ISO was founded in 1946 as an institution under the umbrella of UNESCO, and therefore falls within the scope of the United Nations. Its members for the most part include national standards organizations, which meet at a general assembly every three years.

The ISO has very extensive responsibilities. Its standardization work is divided into *Technical Committees* (TC) according to the different technical fields, which, depending on the scope of the work involved, are subdivided into *Subcommittees* (SC) and then into *Working Groups* (WG) and other groups.

The actual work itself is carried out by 100 000 volunteers in WGs all over the world. Many of these volunteers are delegated to the ISO by their employers whose products are being standardized. Others who are involved include government officials who are promoting one of their country's standards for international acceptance. Scientific experts are also in many of the WGs.

In general, the ISO's involvement in telecommunications is more from the standpoint of message transfer between computer systems. For example, some of the standards that were developed for *Local Area Networks* (LAN) are also applicable for voice transmission.

B.1.2 ITU

The International Telecommunications Union (ITU), with its headquarters in Geneva, was founded in 1865 and is one of the oldest international organizations in existence. As a subsidiary organization of the UN, it is active in four different task areas today:

- International allocation and registration of transmit and receive frequencies
- Coordination of efforts in the elimination of interference in radio traffic
- Coordination of the development of telecommunications systems
- Preparation of agreements on performance guarantees and tariffs

The tasks, rights and duties of the member states of the ITU are regulated in the *International Telecommunication Convention* (ITC), the international telecommunications contract. Every three to four years, the proposals that have been drawn up are presented to the general assembly, and with a high enough quorum are published as recommendations. Internationally these recommendations have the weight of standards. This approach has now changed, and recommendations are now introduced on CD-ROM.

The *World [Administrative] Radio Conference* (W[A]RC) was convened by the ITU to revise the *Radio Regulations* to meet current needs.

Until the end of 1992, the standardization work of the ITU was carried out by five committees based in Geneva [3] and entrusted with the following areas of responsibility:

- CCITT: Consultative Committee for International Telephone and Telegraph
- CCIR: Consultative Committee for International Radiocommunication
- IFRB: International (Radio) Frequency Registration Board
- BDT: Telecommunications Development Bureau
- General Secretariat

B.1.2.1 New ITU Structure

As the result of a reorganization in 1993, the ITU now consists of four committees [6]:

- *Telecommunications standardization sector* (ITU-T), which is involved in the standardization activities that were previously carried out by the CCITT and partially by the CCIR.
- *Radio communications sector* (ITU-R), which carries out the remaining standardization activities of the CCIR along with the previous work of the IFRB.
- *Development sector* (ITU-D), which has taken over the functions of the BDT.
- *ITU General Secretariat*, which supports the activities of these committees.

B.1.2.2 CCITT

This international organization was under the control of the national PTTs (*Post, Telephone and Telegraph*) until a few years ago. As a result of deregulation and privatization of the PTT in many countries, its role has increasingly been taken over by new national institutions and ministries.

The CCITT produces and publishes interfaces, services and protocols as technical recommendations for telephony, telegraphy and data communications.

For example, the CCITT produced the recommendation for transmission over large distances in open data networks. It also developed standards for cooperation between mobile land, see and air communications systems, as well as standards for linking these different mobile telephone services with the international telegraph and public telephone network.

Another aspect of CCITT's standardization work is focused on integrated systems for the simultaneous transmission of voice and data (ISDN). Voice transmission quality, switching technology and tariff issues relating to mobile radio and telephone are other activity areas.

Depending on the subject area, the work of the CCITT is carried out by the following *Study Groups* [5]:

- SG I Service Definitions
- SG II Network Operations
- SG III Tariff Principles
- SG IV Maintenance
- SG V Safety, Protection and EMC
- SG VI Outside Plant
- SG VII Dedicated Networks
- SG VIII Terminal for Telematic Services
- SG IX Telegraphs
- SG X Software
- SG XI ISDN, Network Switching and Signalling
- SG XII Transmission and Performance
- SG XV Transmission Systems and Equipment
- SG XVII Data Transmission
- SG XVIII ISDN and Digital Communications

The CCITT acts in a consulting capacity to the ITU.

B.1.2.3 CCIR

Like the CCITT, the CCIR is a consultative committee of the ITU in the area of radio technology; its principles and regulations have developed over the decades, and have gradually been adapted to cover new aspects, applications, situations and requirements.

The recommendations of CCIR mainly relate to the planning and coordination of radio communications and radio services, the technical characteristics of systems, and the effective and efficient use of frequency spectrum. The CCIR has drawn up criteria for the prevention of noisy interference, and is working together with the IRFB to ensure that these are applied and adhered to.

The *Interim Working Party* (IWP) was founded through a resolution of CCIR for the purpose of designing standards for future mobile radio systems.

The work of CCIR is divided into study groups, which deal with the following technical areas:

- SG 1 Spectrum Management Techniques
- SG 4 Fixed Satellite Services
- SG 5 Radio Wave Propagation in Non-Ionized Media
- SG 6 Radio Wave Propagation in Ionized Media
- SG 7 Science Services
- SG 8 Mobile, Radio Determination and Amateur Services
- SG 9 Fixed Services
- SG 10 Broadcasting Services—Sound
- SG 11 Broadcasting Services—Television
- SG 12 Inter-Service Sharing and Compatibility

The work is further divided into individual groups, which help to alleviate the responsibilities of the study groups. The most important group for wireless communication—the SG 8, which deals with issues relating to mobile and radio communication—is divided into the following work groups:

- Land Mobile Services
- Maritime Mobile Services
- Aeronautical Mobile Services
- Amateur Services
- Maritime Satellite Services
- Land Mobile Satellite Services
- Future Global Maritime Distress and Safety System (FGMDSS)

The *Commission Mixte CCIR/CCITT pour les Transmissions Televisuelles et Sonares* (CMTT) [4] deals with the interests of both organizations CCIR/CCITT.

B.1.2.4 International Frequency Registration Board, IFRB

After clarification of all technical issues in accordance with CCIR regulations, the IFRB administers all radio transmission of commercial, military and scientific origin, including amateur radio, which is carried on earth or over satellite. The IFRB regulates the allocation of frequencies worldwide and specifies the maximum number of radio channels in each frequency spectrum where interference could occur. Notification of the IFBR is required if a frequency:

- could cause damaging interference to the services of other administrations
- is to be used for international radio connections
- is to be internationally recognized

After it has been checked by the IFRB (mainly because of the possibility of interference), the frequency is entered in the *International Frequency Master File.*

Based on the IFBR frequency allocation, the world was divided into three regions:

- Region 1: Europe, Middle East, former USSR and Africa
- Region 2: Greenland, North and South America
- Region 3: Far East, Australia and New Zealand

WARC (*World Administrative Radio Conference*)—recently abbreviated to WRC—is a periodic conference at which applications specifying the use of frequency spectrum are deliberated and formally agreed upon by organizations from the three regions.

B.1.2.5 BDT

The *Board of Directors for Telecommunications* (BDT), founded in 1989, has the same status as CCITT and CCIR. Its task is to guarantee technical cooperation and to raise funds for financing the development of telecommunications networks in less industrialized countries.

B.1.2.6 General Secretariat

This committee is responsible for administration, finance, publications and technical recommendations.

B.1.2.7 Cooperation Between ISO and CCITT

There are areas such as *Open Systems Interconnection* (OSI) where there is a considerable overlap in the work carried out by the ISO and CCITT. This previously led to a race between the two organizations, which almost resulted in standardization efforts keeping pace with technical progress. The *Joint Technical Committee* (JTC) was founded to enable cooperation between the two bodies.

B.1.3 IEC

The IEC has been in existence since 1906 and, like the ISO and the ITU, is a UN institution. It develops standards in the electric and electronic component area as well as for operational safety and for the environmental conditions of products.

Its suborganization, the *Committee International Special Perturbations Radio* (CISPR), studies radio interference. In 1988, the ISO and the IEC jointly formed the *Joint Technical Committee* (JTCI), which is drawing up standards for information technology.

B.1.4 INTELSAT/INMARSAT

These organizations are responsible for the development, procurement and management of international satellite services. They are developing their own satellite radio standards.

The largest operator of a satellite communications system is INTELSAT with 200 earth stations in over 100 countries. It was set up in 1964 to deal with the many technical and administrative problems associated with worldwide satellite communications systems [8]. Starting with 14 countries, the membership number has been increasing continuously. By joining INTELSAT, a member state undertakes to seek approval before operating its own satellites.

Founded in 1980, the INMARSAT organization operates a global communications satellite system for maritime radio traffic, and offers telephone and telegraph services along with data transfer between coastal ground stations and ships and production platforms, which are always equipped with mobile transmitting and receiving stations [7]. All services offered by national PTTs are accessible over the INMARSAT network. In a sense, INMARSAT could be regarded as an extension of the PTT organizations in the field of satellite communication.

B.1.5 ATM Forum

The ATM Forum is a honorary international organization based in Mountain View, California, USA, which has the aim of promoting the use of ATM

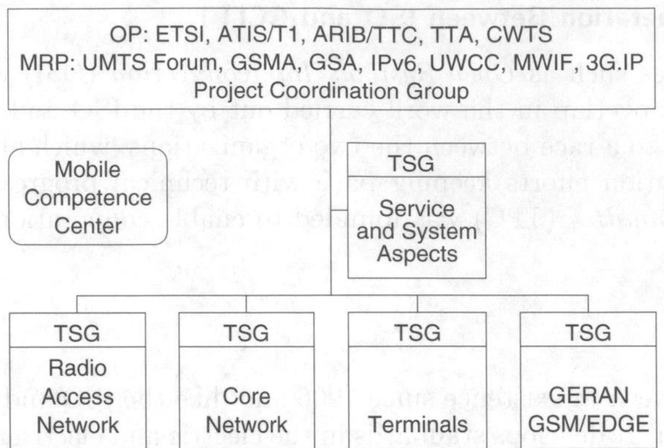

Figure B.1: Organization and responsibilities of 3GPP

products and services. It promotes the interoperability of products, particularly through the specification of interfaces.

The Forum maintains close contacts with other industrial associations. It consists of a worldwide technical committee (TC), three marketing committees in North America, Asia and Europe, and the *Enterprise Network Roundtable* on the Internet, which allows end users to exchange information.

B.1.6 Third Generation Partnership Project

The *3rd Generation Partnership Project* (3GPP) is big, with an organization inspired by the one setup by ETSI for GSM. It is estimated that more than thousand people are contributing in one way or another. This is an unprecedented number of experts working on the same project. Such an organization with its well defined procedures and rules is crucial for the success of 3G standardization. Surprisingly, it works! This large machine has delivered almost stable specifications, accepted by the majority of major industrial players, in only two years. Of course, this would not have been possible without the background work undertaken prior to the 3GPP within research projects carried out in all the countries involved.

The purpose of 3GPP is to prepare, approve and maintain global technical specifications for a 3G mobile system [1]. Figure B.1 shows the general organization of 3GPP, which includes [9, 2]:

Organizational Partners (OP) The OPs are open standards organizations which are officially recognized in their own countries, and which have the capability and authority to define, publish and set standards nationally or regionally, and have signed the partnership project agreement.

Mobility and Resource Management Protocol (MRP) These partners are responsible for identifying market requirements. Their role in 3GPP is to bring a consensus view on market requirements (e.g., services, features and functions). An MRP is invited by the OPs.

Individual Members (IM) IMs contribute technically or otherwise to the *Technical Specification Groups* (TSGs). They must be members of an OP.

Observers The status of *Observer* may be granted by the OPs to an entity that has the qualifications to become a future OP.

Detailed technical work is carried out in the TSGs. The tasks of these groups are to draw up, approve and maintain technical specifications and reports. The Service and System Aspect group is responsible for the technical coordination of work being undertaken within 3GPP, as well as for the integrity of the overall architecture and system.

As indicated in Figure B.1, there are five TSGs, each subdivided into subgroups:

Service and System Aspects TSG Responsible for service capability definition and development, stage one, as well as for the charging and accounting description, network management and security aspects, overall architecture definition evolution and maintenance, principles for the definition of end-to-end-transmission (codec aspects), and project coordination.

Radio Access Network TSG This TSG is the most relevant to CDMA technology. It is responsible for the layers 1 to 3 specifications, Iub specification, Iur and Iu interfaces, UTRAN operations and maintenance requirements, base station radio performance, and conformance tests specifications.

Core Network TSG This TSG is responsible for mobility management, call connection control signaling between the user equipment and the core network, definition of interworking functions between the core network and external networks, packet-related questions such as quality of service mapping, core network aspects of the Iu interface, and core network operations and maintenance requirements.

Terminal TSG In charge of service capability protocols, messaging, end-to-end interworking services, *User Services Identity Module* (USIM) to mobile terminal interface, model and framework for terminal interfaces and services execution, and conformance test specifications for terminals, including the radio aspects.

GERAN TSG Created in mid-2000 to ensure cohesion between GSM and the 3G specifications. The *GSM/EDGE Radio Access Network* (GERAN) TSG is responsible for maintaining and developing the GSM technical specifications and technical reports, including GSM-evolved radio access technologies such as GPRS and EDGE.

In addition, there is an *Mobile Competence Centre* (MCC), the role of which is to provide administrative and technical support for the management of work items and change requests.

Decision-making within the *Project Coordination Group* (PCG) and TSGs is based on consensus among the OPs for the PCG, or *Individual Members* (IM)s for the TSG, or by a vote where this is unavoidable. 3GPP very precisely defines the decision-making rules for each level of the organization.

IMs of the 3GPP are bound by the *Intellectual Property Rights* (IPRs) policies of their respective OPs. IMs are encouraged to declare, at the earliest opportunity, any IPRs they might have which they believe to be essential, or potentionally essential, to any work ongoing within 3GPP. After comparing their IPR policies, ARIB, ETSI, T1 TTA and TTC have agreed that these policies share common principles which are quite similar, and which have been established to maximize the success of the 3GPP. There is also a 3GPP2 with partly the same members as 3GPP, see `www.3gpp2.org`.

B.2 European Standards Organizations

The decision of the EU states within the framework of the Treaties of Rome (1983) to introduce the common European market in 1992, along with the emphasis on the important political, social and economic role to be played by telecommunications, has produced major changes in the European communications market since 1987 as a result of the publication of the so-called *Green Paper*.

Moreover, the EU states are directing, controlling and financing a large number of research programmes that, in addition to furthering their technical objectives, are also promoting cooperations between companies from different countries. The most important projects in the telecommunications area include: *Advanced Communication Technologies and Services* (ACTS), *Cooperation in the Field of Scientific and Technical Research* (COST) and *European Strategic Programme for Information Technology* (ESPRIT). Aside from a purely technical orientation, these programmes promote cooperation between the manufacturers of different countries, thereby playing an important role in the standardization process.

B.2.1 CEN/CENELEC

The *Comité Européene de Normalisation* (CEN) is the equivalent of the ISO in Europe, with the difference that the standards that it develops are binding.

Its sister organization, the *Comité Européene des Normalisation Electrotechniques* (CENELEC), develops standards in the electrical engineering area, and is therefore the European equivalent of the IEC. The standards that it develops with the label

- EN must be published and applied as national standards

- ENV can be applied voluntarily

B.2.2 CEPT

The *European Conference for Posts and Telecommunications* (CEPT), the members of which are the European PTTs (*Posts, Telephone and Telegraph*) and the public telephone network operators, was responsible for drawing up standards in the telephone, telegraph and data network area until 1988. Many of the recommendations it developed were taken over by the CCITT. A large proportion of CEPT's tasks were transferred to ETSI (*European Telecommunications Standards Institute*) when it was founded in 1988. CEPT is still involved in strategy and planning.

Three CEPT committees play a major role in the development of European standards:

Technical Recommendations Applications Committee (TRAC): This committee passes recommendations as *Normes Européennes de Télécommunication* (NET), which are binding for all PTT members.

Comité des Coordinations des Radiocommunications (CR): The CR's task is the development of stategies geared towards optimizing use of the frequency band and allocation of frequencies to the different services. The group CEPT/ERC (*European Radio Conference*) is responsible for the organizational aspects, and handles coordination between CEPT participants, with ETSI and on an international basis (see Figure B.2). CEPT/ERC corresponds to the international group ITU-T/WRC (*World Radio Conference*). According to Figure B.3, the *European Radio Office*, which has been part of CEPT since 1991, oversees the ongoing activities relating to frequency regulation and works together with the corresponding national organizations, e.g., RegTP in Germany.

Service and Facilities (SF) Committee: The SF Committee specifies service capabilities.

B.2.3 ETSI

The role of the *European Telecommunications Standards Institute* (ETSI), which was founded in 1988 in accordance with the requirements of the EU *Green Paper*, is similar to that of the CCITT, but with the main objective of developing telecommunications standards for Europe. ETSI membership is not only restricted to EU states but is open to all countries on the European continent.

ETSI recently began to include five categories of members with equal status, namely:

- National standards institutions and administrations

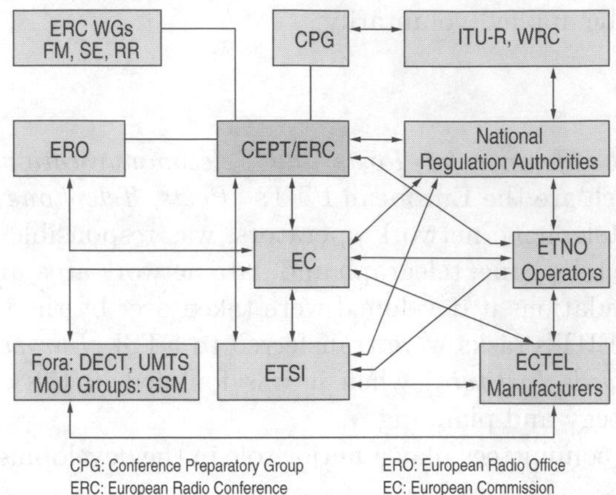

Figure B.2: European radio frequency planning coordinated by CEPT/ERC

- Public network operators
- Manufacturers
- User groups
- Private service providers, research institutes and consultants

ETSI has thus also extended its membership to manufacturers, which means that standards in the future will be developed in close cooperation between network operators and manufacturing firms.

Another new development at ETSI is that project teams are being formed with experts secunded and paid by the member organizations and responsible for the design of standards.

To detect future trends in the market and in telecommunications, ETSI has set up *Special Strategic Review Committees of Senior Experts*, which study the ETSI programme and the need for new standards in certain areas.

The former ETSI structure is illustrated in Figure B.4. The highest-ranking ETSI committee, the *General Assembly (GA)*, determines ETSI policies, elects new members and produces the budget. The administrative work is carried out by a secretariat based in Sophia-Antipolis, France, under the supervision of a director. The secretariat also oversees the work of the project teams (PT).

The *Technical Assembly* (TA) passes *European Telecommunications Standards* (ETS), which take effect with a majority vote. The TA appoints the *Technical Committees* (TC) and their chairmen. The standardization work is carried out by 12 TCs, which are divided into *Sub-Technical Committees* (STC). The Technical Subcommittees (STC) and the Project Teams (PT), which work under the direction of the respective Technical Committees (TC) (see Figure B.4), are listed in Table B.1.

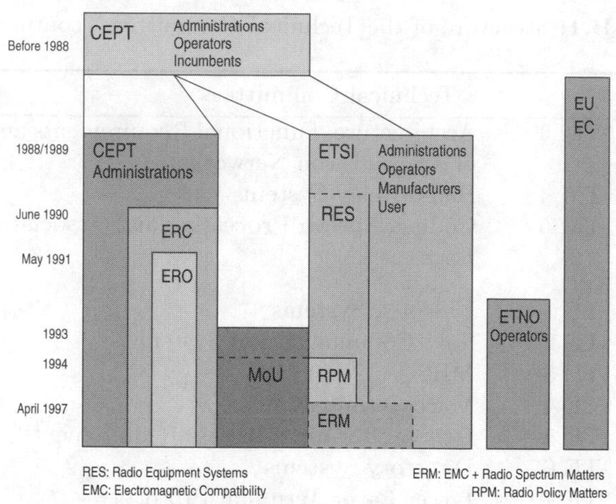

Figure B.3: CEPT, ERC and ETSI: Historical development

Table B.1: Structure of the Technical Committees

			Technical Committees
NA	STC	NA 1	User Interfaces Services and Charging
		NA 2	Numbering, Addressing, Routing and Interworking
		NA 4	Network Architecture Operations, Maintenance Principles and Performance
		NA 5	Broadband Networks
		NA 6	Intelligent Networks
		NA 7	Universal Personal Telecommunications
BT	STC	BT 1	Private Networking Aspects
		BT 2	Business Telecommunications Network Performance
	PT	PT 27	PTN Attendant Services
		PT 43	PTM Mobility
		PT 18V	PSTN Access PABX
		PT 22V	ONP Leased Lines
SPS	STC	SPS 1	Network Interconnection and Signalling
		SPS 2	Signalling Network and Mobility Applications
		SPS 3	Digital Switching
		SPS 5	Customer Access to the ISDN
	PT	PT 21V	DSS 1 PICS/PIXIT
TM	STC	TM 1	Transmission Equipment Fibres and Cables
		TM 2	Transmission Networks Management, Performance and Protection

Continued

Table B.1: Structure of the Technical Committees (continued)

		Technical Committees	
		TM 3	Architecture, Functional Requirements and Interfaces of Transmission Networks
		TM 4	Radio Relay Systems
		TM 5	Coding, Speech Processing and Associated Network Issues
TE	STC	TE 1	Videotex Systems
		TE 2	Text Communication Systems
		TE 3	MHS
		TE 4	Voice Terminals
		TE 5	General Terminal Access Requirements
		TE 6	Directory Systems
		TE 7	Lower Layer Terminal Protocols
		TE 8	Functional Standards
		TE 9	Card Terminal
		TE 10	Audio-Visual Management (AVM)
	PT	PT 34	Conformance Testing for Videophony
		PT 8V	Intelligent Caros
		PT 40	ISDN Terminals for ODA
		PT 17V	PSTN Access
		PT 19V	ISDN Programming Communications Interface
EE	STC	EE 1	Environment Conditions
		EE 2	Power Supply
		EE 3	Mechanical Structure
		EE 4	Electromagnetical Compatibility
HF	STC	HF 1	Telecommunication Services
		HF 2	People with Special Needs
		HF 3	Usability Evaluation
	PT	PT 16V	User Procedures for Videophony
		PT 36	Human Factor Guidelines for ISDN Equipment Design
RES	STC	RES 1	Maritime Mobile
		RES 2	Land Mobile
		RES 3	DECT
		RES 5	TFTS
		RES 6	Digital Trunking Systems (TETRA)
		RES 7	DSRR
		RES 8	Low Power Devices
		RES 9	EMC
		RES 10	Wireless LANs
	PT	PT 10	DECT

Continued

Table B.1: Structure of the Technical Committees (continued)

		Technical Committees	
		PT 19	TFTS
		PT 29	Trunked Mobile Systems
		PT 41	Radio LANs
SMG	STC	SMG 1	Services and Facilities
		SMG 2	Radio Aspects
		SMG 3	Network Aspects
		SMG 4	Data and Telematic Services
		SMG 5	UMTS
		SMG 6	Network Management Aspects
	PT	PT 12	Pan European Cellular Digital Radio Systems
		PT 12V	DCS1800 (PCN)
PS	STC	PS 2	ERMES Radio Aspects
		PS 3	ERMES Network Aspects
SES	STC	SES 1	General Systems Requirements
		SES 2	RF and IF Equipment
		SES 3	Interconnections to Terrestrial Networks Control and Monitoring Functions
		SES 4	TV and Sound Programmes Equipment
		SES 5	E.S for Mobile Services
	PT	PT 42	VSAT. ISDN
		PT 15V	Mobile Earth Stations for LMSS and RDSS
		PT 23V	News Gathering Earth Stations
ATM	STC	ATM 1	Conformance Testing Methodologies
		ATM 2	Conformance Testing Environment
	PT	PT 31	TTCN SDL
		PT 37	SDL Guide
		PT 38	Test Specification Handbook
		PT 39	Conformance Testing Consulting Group

ETSI has assigned a high priority to the mobile radio area. It has provided the specifications for systems such as GSM, DCS 1800, DECT, ERMES, TETRA and TFTS. The following committees are important in the mobile radio area:

- RES: Radio Equipment and Systems
- SMG: Special Mobile Group
- PS: Paging Systems

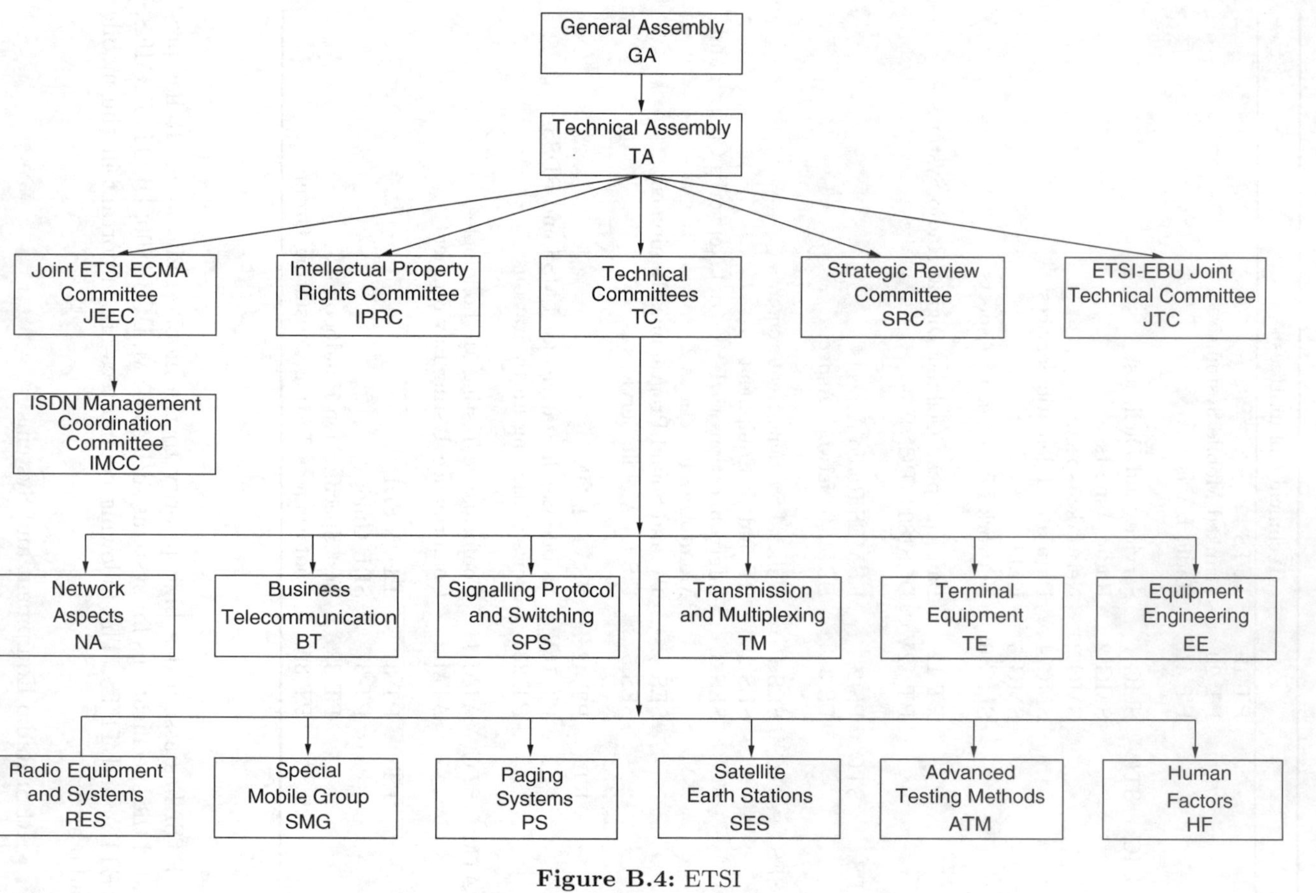

Figure B.4: ETSI

- SES: Satellite Earth Stations

ETSI is drawing up three types of standards:

- EN or ENV: European Norm
- ETS: European Telecommunications Standards
- NETS: Normes Européennes de Télécommunications

Owing to the overlap in the work being carried out by ETSI, CEN and CENELEC, the *Information Technology Steering Committee* (ITSTC) was set up to assume a coordinating role. The *European Radio Office* (ERO), with its headquarters in Copenhagen, Denmark, was established for the purposes of coordinating European frequency use.

Changes to the ETSI Organization

The reorganization implemented in 1997 had a major influence on the standardization process and associated documents. A brief overview of the old and new types of documents and the corresponding approval procedures is presented below. The *Radio Policy Matters* (RPM) group (see Figure B.3) has existed since 1994 and the *ETSI Radio Matters* ERM group since 1997. CEPT/ERC and ETSI work together in accordance with MoU 1994; RPM coordinates this cooperation on the ETSI side. The ERM group can be seen as a "horizontal technical committee" with responsibility for the coordination between different mobile radio systems. In future every mobile radio system will be dealt with in an ETSI project (EP) (see Figure B.5). Future cooperation between CEN/ERC and ETSI/ERM will be organized as shown in Figure B.6.

The labelling of the documents has changed as follows:

Old

ETSI Technical Report, ETR This provides an overview of the status of technology and/or requirements for a specific system.

ETSI Technical Specification, ETS This is the actual standard. It was also produced by STC and then presented to the *Public Enquiry* that incorporated the teams from the different countries who were involved in the standardization process.

New

Technical Report, TR This is similar to the reports previously prepared by ETR. It was released by the respective committee, and in that sense is more of a working paper.

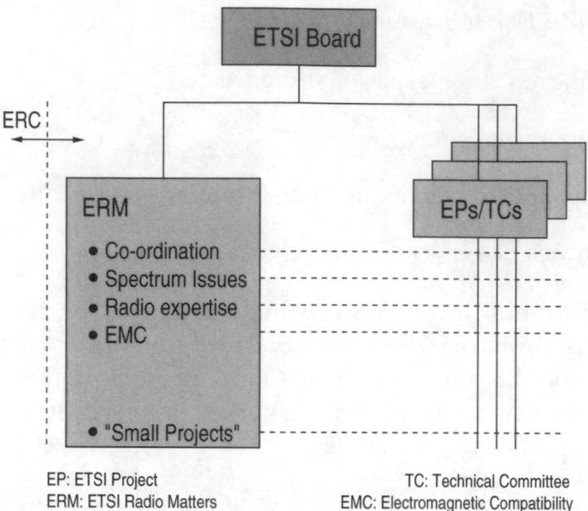

Figure B.5: Matrix organization for dealing with radio-related matters

Technical Specification, TS This is comparable to the previous ETS, but is not a standard because it is only released by the responsible committee.

ETSI Standard, ETS: This is a new type of paper comparable in content to the previous ETS. This is a de facto standard, which is ratified through an abbreviated procedure. According to the procedure, ETSI members are apprised of the standard (by a single contact person) over the Internet, and have 60 days in which to give their vote over the Internet. The experts from the different countries who were involved in the standardization work are not included.

ETSI Guide, EG The content is the same as that of TR, and the confirmation procedure is the same as with ETS.

Euro Norm, EN This is the de jure standard for which the national organizations are included in the ratification. The enquiry (and voting) procedure has also been shortened and now takes either 17 or 23 days.

B.2.4 ECMA

The *European Computer Manufacturers Associations* (ECMA) was established in 1960. Its members include approximately 20 leading computer manufacturers who on a volunteer basis develop standards in the data communications area, in particular *Open System Interconnection* (OSI), and the recently developed signalling system QSIG for private networks.

The most important ECMA committee is the *Communications, Networks and Interconnection* (TC32) Committee, which is divided into four working groups:

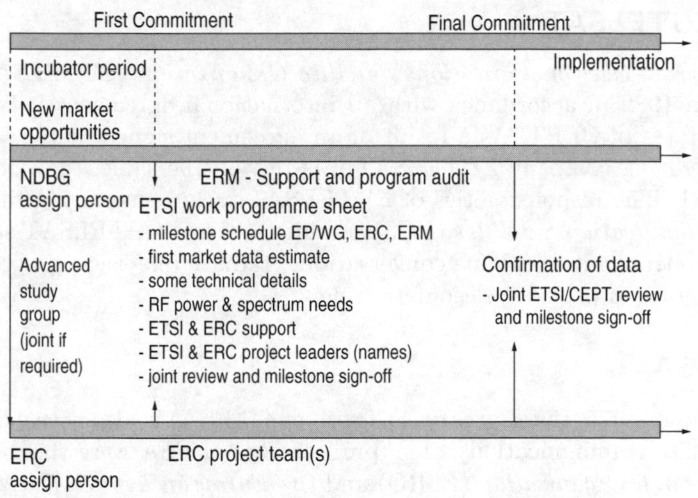

Figure B.6: Joint CEPT/ERC-ETSI implementation process

- TG11: Computer-Supported Telecommunications Applications
- TG12: PTN Managment
- TG13: PTN Networking
- TG14: PTN Signalling

The *Joint ETSI ECMA Committee* (JEEC) was founded in 1991 to coordinate standardization activities because of the overlapping of ETSI and ECMA in the area of private network standards. ETSI's work is directed towards the development of standards for interactive connections between public and private networks, whereas ECMA develops the standards for private networks. Information on ongoing work is exchanged between the two organizations. Furthermore, an opportunity exists for members of one organization to work with members of the other. The standards completed by ECMA are presented to ETSI and voted on for approval as an ETS.

B.2.5 EBU

The *European Broadcasting Union* (EBU), which is based in Geneva, is an association of European television and broadcasting corporations. The technical committee of EBU works on recommendations for norms and standards, such as the satellite transmission norm for television and broadcasting.

B.2.6 EUTELSAT

The *European Telecommunications Satellite Organization* (EUTELSAT) was founded in 1982 in accordance with an international agreement between 28 member states of CEPT. As a result of an agreement signed in 1983, the regional operating company EUTELSAT was put on the same legal footing as INTELSAT. The responsibilities of EUTELSAT include managing the *European Communication Satellites* (ECS) satellite system. EUTELSAT satellites are used to transmit telephone conversations, radiate television programmes, transfer data and provide teleconferencing.

B.2.7 ESA

Established in 1975, the *European Space Agency* (ESA) is also active in European satellite communications. The predecessors of ESA were the *European Space Research Organization* (ESRO) and the *European Launch Development Organization* (ELDO), along with the *Conferènce Européen pour des Télécommunications par Satellite* (CETS) [7].

ESA promotes cooperation between European countries in space research and technology, and develops operational space applications systems. An overview of international and European Standards organizations and their cooperations is presented in Figure B.7.

B.3 National Standards Organizations

National standardization facilities, organizations and professional associations concern themselves mainly with the specification of national standards, cooperation with international standards bodies and the translation of international standards into the local language. The best-known facilities are listed in Table B.2.

B.4 Quasi-Standards

B.4.1 Company Standards

Despite the numerous standards organizations in the world, many companies develop their own specifications, which can sometimes become de facto standards. These de facto standards are either accepted by the international and the national organizations or become serious competition to them for a long time.

Company standards are not published, and therefore play a major role in competition. Anyone who wants to use these standards in their own systems must apply for a licence. Examples of companies that have developed their own standards include:

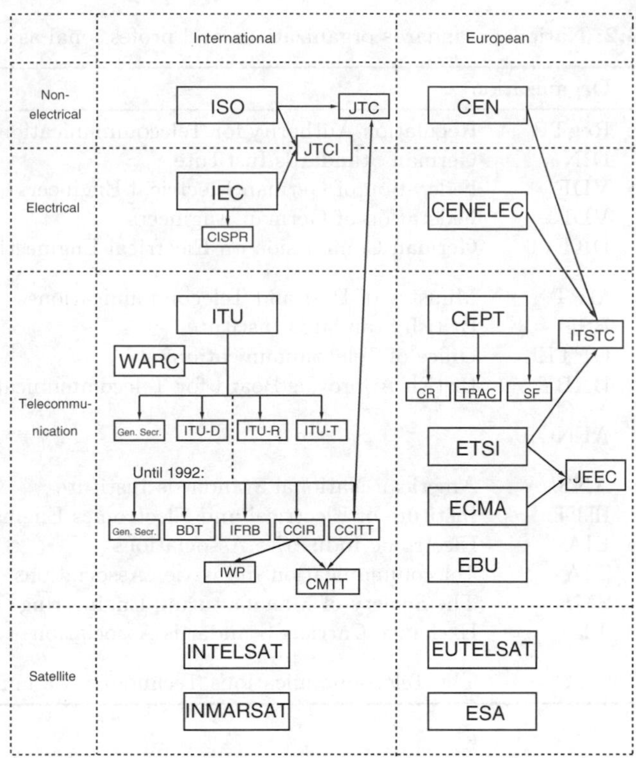

Figure B.7: Cooperation between international and European standards bodies

- Motorola: e.g., RD-LAP protocol, IRIDIUM, Altair WIN 802.11
- Ericsson: Mobitex
- Qualcomm: CDMA (IS-95)

B.4.2 User Standards

In addition to companies, other organizations or institutions such as universities, the military, the police and railways also develop their own de facto standards. These standards that are designed for internal use are called user standards. If there is an interest, they also can be used by outside users in their own systems.

The standards in this category include *Mobile IP*, which is a further development of TCP/IP (*Transmission Control Protocol/Internet Protocol*) of the Internet, which was developed by the *Defense Advanced Research Projects Agency* (DARPA).

Table B.2: National standards organizations and professional associations

Country	Organization	
Germany	RegTP	Regulation Authority for Telecommunications and Post
	DIN	German Standards Institute
	VDE	Federation of German Electrical Engineers
	VDI	Federation of German Engineers
	DKE	German Commission on Electrical Engineering
Great Britain	MPT	Ministry of Post and Telecommunications
	BSI	British Standards Institute
	OFTEL	Office of Telecommunications
	BABT	British Approvals Board for Telecommunications
France	AFNOR	
USA	ANSI	American National Standards Institute
	IEEE	Institute of Electrical and Electronics Engineers
	EIA	Electronic Industries Associations
	TIA	Telecommunication Industries Associations
	SME	The Society of Manufacturing Engineering
	T1	Exchange Carriers Standards Association
Japan	TTC	The Telecommunications Technology Committee

Bibliography

[1] 3GPP. *www.3gpp.org*. Web Pages.

[2] Stephan M. Blust. *Wireless Standards Development—A New Paradigm. IEEE Vehicular Technology Society News*, pp. 4–12, 11 2000.

[3] M. P. Clark. *Networks and Telecommunications: Design and Operation.* Wiley, Chichester, 1991.

[4] W. Heinrich. *Richtfunk-Technik*, 1988.

[5] R. J. Horrocks, R. W. A. Scarr. *Future Trends in Telecommunications.* Wiley, Chichester, 1993.

[6] ITU. DOC Server. *The ITU's structure.* Electronics Letters, August 1993.

[7] R. Kabel, T. Strätling. *Kommunikation per Satellit: Ein internationales Handbuch*, 1985.

[8] D. Roddy. *Satellitenkonnunikation: Grundlagen – Satelliten – Übertragungssysteme*, 1991.

[9] P. Sehier, J-M. Gabriagues, A. Urie. *Standardization of 3G mobile systems. Alcatel Telecommunications Review*, No. 1, pp. 11–18, 2001.

Appendix C
International Frequency Allocations

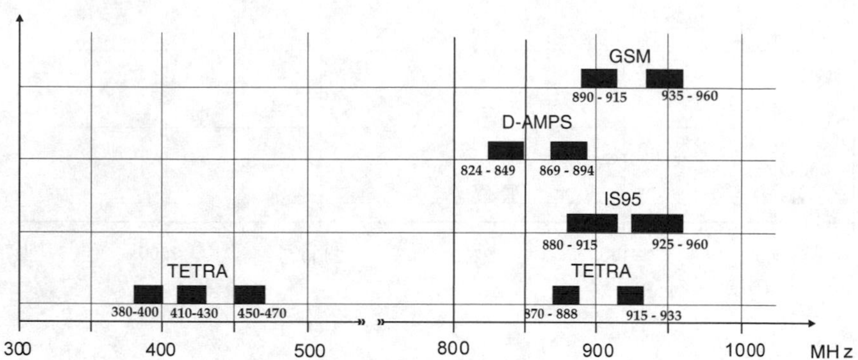

Figure C.1: Allocation of frequencies in accordance with Article 8 of the *Radio Regulations*

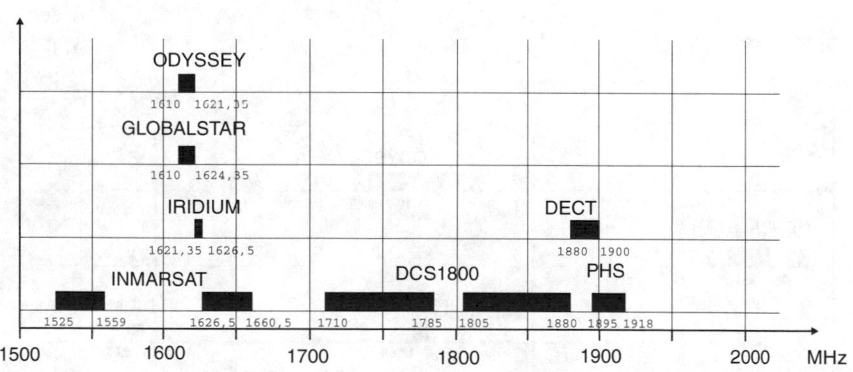

Figure C.2: Allocation of frequencies in accordance with Article 8 of the *Radio Regulations*

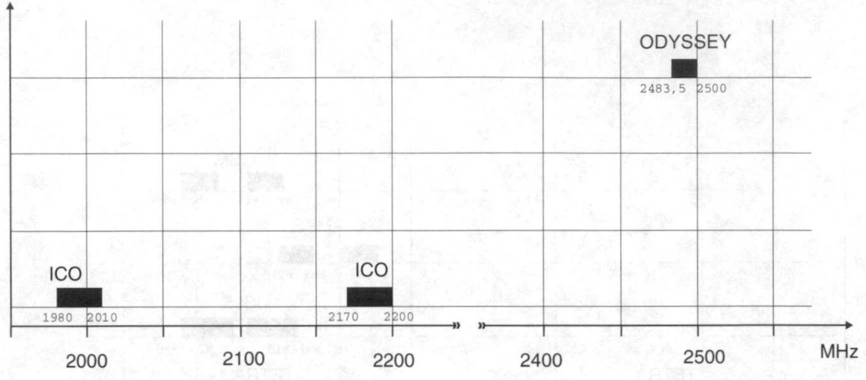

Figure C.3: Allocation of frequencies in accordance with Article 8 of the *Radio Regulations*

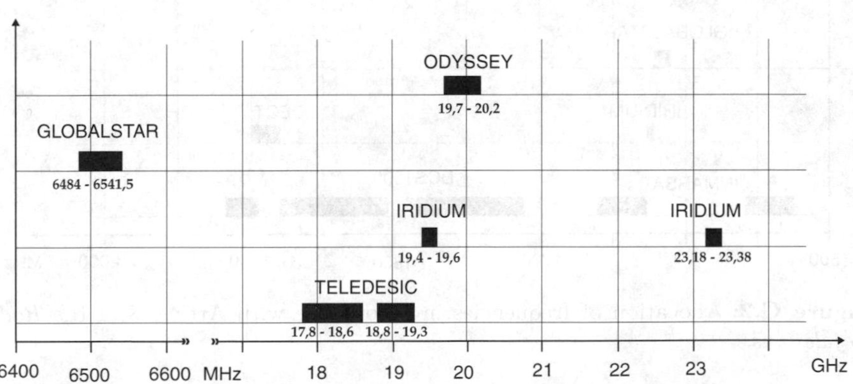

Figure C.4: Allocation of frequencies in accordance with Article 8 of the *Radio Regulations*

Appendix D The Frequencies of European Mobile Radio Systems

The latest data is available from the European Radio Office (ERO) under `http://www.ero.dk` (see Technical Report 25).

System	Freq. (lower band) [MHz]			Freq. (upper band) [MHz]		
			Cordless telephones			
CT1	914	...	915	959	...	960
CT1+	885	...	887	930	...	932
CT2	864.1	...	868.1			
DECT	1880	...	1900			
			Digital trunked radio (TETRA)			
			Frequency ranges based on CEPT recommendation, from which frequencies are to be selected on a national basis.			
	380	...	400			
	410	...	430			
	450	...	470			
	870	...	888	915	...	923
			Cellular mobile radio systems			
GSM	890	...	915	935	...	960
Extension band GSM	880	...	890	925	...	935
UIC (European orbits)	870	...	874	915	...	919
DCS1800	1710	...	1785	1805	...	1880
FPLMTS (UMTS)	1885	...	2010	2110	...	2200
FPLMTS satellite band	1980	...	2010	2170	...	2200
			Radio paging systems			
EUROSIGNAL (D, F, CH)	87.340					
	87.365					
	87.390					
	87.415					

Continued

System	Freq. (lower band) [MHz]	Freq. (upper band) [MHz]
Cityruf (D only)	465.970	
Euromessage	466.075	in D, F, I, GB
	466.230	
ERMES	169.4 ... 169.8	
Terrestrial flight telephone system		
TFTS	1670 ... 1675	1800 ... 1805
LEO satellite systems		
	1610 ... 1626.5	2483.5 ... 2500
INMARSAT systems		
Only sections of	1525 ... 1559	1626.5 ... 1660.5

Appendix E
The GSM Standard

Table E.1 provides an overview of the GSM standard. The standards for GSM1800 appear with the abbreviation DCS (Digital Communications System). The GSM/GPRS standards are listed seperately in Table E.2.

Table E.1: The GSM standard

GSM No.	GSM Title
01.02	General Description of a GSM PLMN
01.04 REP	Vocabulary in a GSM PLMN
01.06	Service Implementation Phases and Possible Further Phases in the GSM PLMN
01.78	Requirements for the CAMEL Feature
02.01	Principles of Telecommunication Services by a GSM PLMN
02.02	Bearer Services Supported by a GSM PLMN
02.03	Teleservices Supported by a GSM PLMN
02.04	General on Supplementary Services
02.05	Simultaneous and Alternate Use of Services
02.06	Types of Mobile Stations
02.06 DCS	Types of Mobile Stations
02.07	Mobile Station Features
02.08 REP	Report: Quality of Service
02.09	Security Aspects
02.10	Provision of Telecommunications Services
02.11	Service Accessibility
02.11 DCS	Service Accessibility
02.12	Licensing
02.13	Subscription to the Services of a GSM PLMN
02.14	Service Directory
02.15	Circulation of Mobile Stations
02.16	International MS Equipment Identities
02.17	Subscriber Identity Modules, Functional Characteristics
02.20	Collection Charges
02.24	Description of Advice of Charge
02.30	Man–Machine Interface of the Mobile Station
02.40	Procedures for Call Progress Indications
02.78	CAMEL Service Definition (Stage 1)
02.81	Number Identification Supplementary Services
02.82	Call Offering Supplementary Services

Continued

Table E.1: The GSM standard (continued)

GSM No.	GSM Title
02.83	Call Completion Supplementary Services
02.84	Multi-Party Supplementary Services
02.85	Community of Interest Supplementary Services
02.86	Charging Supplementary Services
02.87	Additional Information Transfer Supplementary Services
02.88	Call Restriction Supplementary Services
03.01	Network Functions
03.02	Network Architecture
03.03	Numbering, Addressing and Identification
03.04	Signalling Requirements Relating to Routing of Calls to Mobile Subscriber
03.05	Technical Performance Objectives
03.07	Restoration Procedures
03.08	Organization of Subscriber Data
03.09	Handover Procedures
03.10	GSM PLMN Connection Types
03.11	Technical Realization of Supplementary Services—General Aspects
03.12	Location Registration Procedures
03.12 DCS	Location Registration Procedures
03.13	Discontinuous Reception (DRX) in the GSM System
03.14	Support of DTMF via the GSM System
03.20	Security-Related Network Functions [A]
03.38	Alphabets and Language-Specific Information
03.40	Technical Realization of the Short Message Service Point-to-Point
03.41	Technical Realization of Short Message Service Cell Broadcast
03.42 REP	Report: Technical Realization of Advanced Data MHS Access
03.43	Technical Realization of Videotex
03.44	Support of Teletex in a GSM PLMN
03.45	Technical Realization of Facsimile Group 3 Service—Transparent
03.46	Technical Realization of Facsimile Group 3 Service—Non-Transparent
03.48 REP	Report: GSM Short Message Service—Cell Broadcast
03.50	Transmission Planning Aspects of the Speech Service in the GSM PLMN System
03.70	Routing of Calls to/from PDNs
03.78	Customized Applications for Mobile Network Enhanced Logic (CAMEL)
03.81	Technical Realization of Line Identification Supplementary Services
03.82	Technical Realization of Call Offering Supplementary Services

Continued

Table E.1: The GSM standard (continued)

GSM No.	GSM Title
03.83	Technical Realization of Call Completion Supplementary Services
03.84	Multi-Party Supplementary Services
03.86	Technical Realization of Charging Supplementary Services
03.88	Technical Realization of Call Restriction Supplementary Services
03.90	Unstructured Supplementary Service Data (USSD—Stage 2)
04.01	MS-BSS Interface—General Aspects and Principles
04.02	GSM PLMN Access Reference Configuration
04.03	MS-BSS Interface: Channel Structures and Access Capabilities
04.04	MS-BSS Layer 1—General Requirements
04.05	MS-BSS Data Link Layer—General Aspects
04.06	MS-BSS Data Link Layer Specification
04.07	Mobile Radio Interface Signalling Layer 3—General Aspects
04.08	Mobile Radio Interface—Layer 3 Specification
04.08 DCS	Mobile Radio Interface—Layer 3 Specification
04.10	Mobile Radio Interface Layer 3—Supplementary Services Specification—General Aspects
04.11	Point-to-Point Short Message Service Support on Mobile Radio Interface
04.12	Cell Broadcast Short Message Service Support on Mobile Radio Interface
04.21	Rate Adaption on MS–BSS Interface
04.22	Radio Link Protocol for Data and Telematic Services on the MS-BSS Interface
04.80	Mobile Radio Interface Layer 3—SS Specification—Formats and Coding
04.82	Mobile Radio Interface Layer 3—Call Offering SS Specification
04.88	Mobile Radio Interface Layer 3—Call Restriction SS Specification
04.90	Unstructured Supplementary Service Data (USSD)—Stage 3
05.01	Physical Layer on the Radio Path (General Description)
05.01 DCS	Physical Layer on the Radio Path (General Description)
05.02	Multiplexing and Multiple Access on the Radio Path
05.03	Channel Coding
05.04	Modulation
05.05	Radio Transmission and Reception
05.05 DCS	Radio Transmission and Reception
05.08	Radio Subsystem Link Control
05.08 DCS	Radio Subsystem Link Control
05.10	Radio Subsystem Synchronization
06.01	Speech Processing Functions: General Description
06.10	GSM Full-Rate Speech Transcoding

Continued

Table E.1: The GSM standard (continued)

GSM No.	GSM Title
06.11	Substitution and Muting of Lost Frames for Full-Rate Speech Traffic Channels
06.12	Comfort Noise Aspects for Full-Rate Speech Traffic Channels
06.31	Discontinuous Transmission (DTX) for Full-Rate Speech Traffic Channels
06.32	Voice Activity Detection
07.01	General on Terminal Adaptation Functions for MSs
07.02	Terminal Adaptation Functions for Services Using Asynchronous Bearer Capabilities
07.03	Terminal Adaptation Functions for Services Using Synchronous Bearer Capabilities
08.01	General Aspects on the BSS–MS Interface
08.02	BSS-MS Interface—Interface Principles
08.04	BSS-MSC Layer 1 Specification
08.06	Signalling Transport Mechanism for the BSS-MSC
08.08	BSS-MSC Layer 3 Specification
08.09	Network Management Signalling Support Related to BSS
08.20	Rate Adaption on the BSS–MSC Interface
08.51	BSC-BTS Interface, General Aspects
08.52	BSC-BTS Interface Principles
08.54	BSC-TRX Layer 1: Structure of Physical Circuit
08.56	BSC-BTS Layer 2 Specification
08.58	BSC-BTS Layer 3 Specification
08.58 DCS	BSC-BTS Layer 3 Specification
08.59	BSC-BTS C
08.60	Inband Control of Remote Transcoders and Rate Adaptors
09.01	General Network Interworking Scenarios
09.02	Mobile Application Part Specification
09.02 DCS	Mobile Application Part Specification
09.03	Requirements on Interworking Between the ISDN or PSIN and the PLMN
09.04	Interworking Between the PLMN and the CSPDN
09.05	Interworking Between the PLMN and the PSPDN for PAD Access
09.06	Interworking Between the PLMN and a PSPDN/ISDN for Support of Packet Switched Data
09.07	General Requirements on Interworking Between the PLMN and the ISDN or PSTN
09.09 REP	Detailed Signalling Interworking within the PLMN and with the PSIN/ISDN
09.10	Information Element Mapping Between MS-BS/BSS-MSC Signalling Procedures and MAP

Continued

Table E.1: The GSM standard (continued)

GSM No.	GSM Title
09.10 DCS	Information Element Mapping Between MS-BS/BSS-MSC Signalling Procedures and MAP
09.11	Signalling Interworking for Supplementary Services
11.10	Mobile Station Conformity Specifications
11.11	Specification of the SIM-ME Interface
11.11 DCS	Specification of the SIM-ME Interface
11.20	The GSM Base Station System: Equipment Specification
11.30 REP	Mobile Services Switching Centre
11.31 REP	Home Location Register Specification
11.32 REP	Visitor Location Register Specification
11.40	System Simulator Specification
12.00	Objectives and Structure of Network Managment
12.01	Common Aspects of GSM Network Managment
12.02	Subscriber, Mobile Equipment and Services Data Administration
12.03	Security Management
12.04	Performance Data Measurements
12.05	Subscriber Related Event and Call Data
12.06	GSM Network Change Control
12.07	Operations and Performance Management
12.10	Maintenance Provisions for Operational Integrity of MSs
12.11	Maintenance of the Base Station System
12.13	Maintenance of the Mobile-Services Switching Centre
12.14	Maintenance of Location Registers
12.20	Network Management Procedures and Messages
12.21	Network Management Procedures and Messages on the A_{bis}-Interface

Table E.2: The GPRS standards: an overview

Stage	GSM No.	GSM Title
1	02.60	GPRS Overview
2	03.60	System Description
	03.61	PTMM Service
	03.62	PTMG Service
	03.64	Radio Architecture
3 (new)	04.60	RLC/MAC Protocol (U_m-Interface)
	04.61	PTM-M Protocol
	04.62	PTM-G Protocol
	04.64	LAPG Protocol (MS-SGSN)
	04.65	SNDCP (MS-SGSN)
	07.60	User Interworking
	08.64	BSSGP (G_b-Interface)
	09.16	Network Service (G_s-Interface)
	09.18	Layer-3 Protocol (G_s-Interface)
	09.60	GTP (G_n-Interface)
	09.61	Network Interworking (G_i-Interface)
3 (modified)	03.20	Security Aspects
	04.03–04.07	System Scheduling
	04.08	Mobility Management
	05.xx	Radio Interface
	08.06	TRAU Frame Mod.
	08.08	BSSGP at G_b
	08.20	Physical Layer at G_b
	09.02	MAP at G_r and G_d
	11.11	SIM Additions
	11.10	MS Testing
	11.2x	BSS Testing
	12.xx	O & M Additions

Appendix F
The UMTS Standard

3G TS No.	Title
21.101	3rd Generation Mobile System Release 1999 Specifications
21.111	USIM and IC card Requirements
21.133	Security Threats and Requirements
21.900	3GPP Working Methods
21.904	UE Capability Requirements (UCR)
21.905	3G Vocabulary
21.910	Multi-Mode UE Issues
21.978	Feasibility Technical Report—CAMEL Control of VoIP Services
22.001	Principles of Telecommunication Services Supported by a GSM Public Land Mobile Network(PLMN)
22.002	Bearer Services Supported by a GSM PLMN
22.003	Teleservices Supported by a GSM PLMN
22.004	General on Supplementary Services
22.011	Service Accessibility
22.016	International Mobile Equipment Identities (IMEI)
22.022	Personalisation of GSM ME Mobile Functionality Specification; Stage 1
22.024	Description of Charge Advice Information (CAI)
22.030	Man-Machine Interface (MMI) of the Mobile Station (MS)
22.034	High-Speed Circuit-Switched Data (HSCSD); Stage 1
22.038	SIM Application Toolkit (SAT); Stage 1
22.041	Operator-Determined Call Barring
22.042	Network Identity and Time Zone (NITZ); Stage 1
22.043	Support of Localised Service Area (SoLSA); Stage 1
22.057	Mobile Station Application Execution Environment (MExE); Stage 1
22.060	General Packet Radio Service (GPRS); Stage 1
22.066	Support of Mobile Number Portability (MNP); Stage 1
22.067	Enhanced Multi-Level Precedence and Pre-Emption Service (eMLPP); Stage 1
22.071	Location Services (LCS); Stage 1 (T1P1)
22.072	Call Deflection (CD); Stage 1
22.078	CAMEL; Stage 1
22.079	Support of Optimal Routing; Stage 1
22.081	Line Identification Supplementary Services; Stage 1
22.082	Call Forwarding (CF) Supplementary Services; Stage 1
22.083	Call Waiting (CW) and Call Hold (HOLD) Supplementary Services; Stage 1

3G TS No.	Title
22.084	Multi-Party (MPTY) Supplementary Services; Stage 1
22.085	Closed User Group (CUG) Supplementary Services; Stage 1
22.086	Advice of Charge (AoC) Supplementary Services; Stage 1
22.087	User-to-User Signalling (UUS); Stage 1
22.088	Call Barring (CB) Supplementary Services; Stage 1
22.090	Unstructured Supplementary Service Data (USSD); Stage 1
22.091	Explicit Call Transfer (ECT) Supplementary Services; Stage 1
22.093	Call Completion to Busy Subscriber (CCBS); Stage 1
22.094	Follow-Me; Stage 1
22.096	Calling Name Presentation (CNAP); Stage 1 (T1P1)
22.097	Multiple Subscriber Profile (MSP); Stage 1
22.100	UMTS Phase 1
22.101	UMTS Service Principles
22.105	Services & Service Capabilities
22.115	Service Aspects Charging and Billing
22.121	Provision of Services in UMTS—The Virtual Home Environment
22.129	Handover Requirements between UMTS and GSM or other Radio Systems
22.135	Multicall Stage 1
22.140	Multimedia Messaging Service Stage 1
22.945	Study of Provision of Fax Services in GSM and UMTS
22.960	Mobile Multimedia Services
22.971	Automatic Establishment of Roaming Relationships
22.975	Advanced Addressing
23.002	Network Architecture
23.003	Numbering, Addressing and Identification
23.007	Restoration Procedures
23.008	Organisation of Subscriber Data
23.009	Handover Procedures
23.011	Technical Realization of Supplementary Services—General Aspects
23.012	Location Registration Procedures
23.014	Support of Dual-Tone Multi-Frequency (DTMF) Signalling
23.015	Technical Realisation of Operator-Determined Barring (ODB)
23.016	Subscriber Data Management; Stage 2
23.018	Basic Call Handling—Technical Realisation
23.032	Universal Geographical Area Description (GAD)
23.034	High-Speed Circuit-Switched Data (HSCSD); Stage 2
23.038	Alphabets & Language
23.039	Interface Protocols for the Connection of Short-Message Service Centers (SMSCs) to Short-Message Entities (SMEs)
23.040	Technical Realisation of Short-Message Service
23.041	Technical Realisation of Cell Broadcast Service
23.042	Compression Algorithm for SMS
23.054	Shared Interworking Functions; Stage 2
23.057	Mobile Station Application Execution Environment (MExE)

3G TS No.	Title
23.060	General Packet Radio Service (GPRS) Service Description; Stage 2
23.066	Support of GSM Mobile Number Portability (MNP); Stage 2
23.067	Enhanced Multi-Level Precedence and Preemption Service (EMLPP); Stage 2
23.072	Call Deflection Supplementary Services; Stage 2
23.073	Support of Localised Service Area (SoLSA); Stage 2
23.078	CAMEL Stage 2
23.079	Support of Optical Routing; Phase 1, Stage 2
23.081	Line Identification Supplementary Services; Stage 2
23.082	Call Forwarding (CF) Supplementary Services; Stage 2
23.083	Call Waiting (CW) and Call Hold (HOLD) Supplementary Services; Stage 2
23.084	MultiParty (MPTY) Supplementary Service; Stage 2
23.085	Closed User Group (CUG) Supplementary Service; Stage 2
23.086	Advice of Charge (AoC) Supplementary Service; Stage 2
23.087	User-to-User Signalling (UUS); Stage 2
23.088	Call Barring (CB) Supplementary Service; Stage 2
23.090	Unstructured Supplementary Service Data (USSD); Stage 2
23.091	Explicit Call Transfer (ECT) Supplementary Service; Stage 2
23.093	Call Completion to Busy Subscriber (CCBS); Stage 2
23.094	Follow-Me Stage 2
23.096	Name Identification Supplementary Service; Stage 2
23.097	Multiple Subscriber Profile (MSP); Stage 2
23.101	General UMTS Architecture
23.107	Quality of Service, Concept and Architecture
23.108	Mobile Radio Interface Layer 3 Specification Core Network Protocols Stage 2 (Structured Procedures)
23.110	UMTS Access Stratum Services and Functions
23.116	Super Charger; Stage 2
23.119	Gateway Location Register (GLR); Stage2
23.121	Architecture Requirements for Release 99
23.122	Non-Access Stratum Functions Related to Mobile Station (MS) in Idle Mode
23.127	Virtual Home Environment / Open Service Architecture
23.140	Multimedia Messaging Service (MMS)
23.146	Technical Realisation of Facsimile Group 3 Service—Non-Transparent
23.153	Out of Band Transcoder Control; Stage 2
23.171	Functional Stage 2 Description of Location Services in UMTS
23.908	Technical Report on Pre-Paging
23.909	Technical Report on the Gateway Location Register
23.910	Circuit-Switched Data Bearer Services
23.911	Technical Report on Out-of-Band Transcoder Control
23.912	Technical Report on Super-Charger
23.922	Architecture for an All IP network

3G TS No.	Title
23.923	Combined GSM and Mobile IP Mobility Handling in UMTS IP CN
23.925	UMTS Core Network-Based ATM Transport
23.927	VHE, Open Service Architecture
23.930	Iu Principles
23.972	Multimedia Telephony
24.007	Mobile Radio Interface Signalling Layer 3—General Aspects
24.008	Mobile Radio Interface Layer 3 Specification; Core Network Protocols-Stage 3
24.010	Mobile Radio Interface Layer 3 - Supplementary Services Specification - General Aspects
24.011	Point-to-Point (PP) Short-Message Service (SMS) Support on Mobile Radio Interface
24.012	Short Message Service Cell Broadcast (SMSCB) Support on the Mobile Radio Interface
24.022	Radio Link Protocol (RLP) for Data and Telematic Services on the (MS-BSS) Interface and the Base Station System—Mobile-Services Switching Centre (BSS-MSC) Interface
24.065	General Packet Radio Service (GPRS); Mobile Station (MS)—Serving GPRS Support Node (SGSN); Subnetwork-Dependent Convergence Protocol (SNDCP)
24.067	Enhanced Multi-Level Precedence and Pre-Emption service (eMLPP); Stage 3
24.072	Call Deflection Supplementary Service; Stage 3
24.080	Mobile Radio Layer 3 Supplementary Service Specification—Formats and Coding
24.081	Line Identification Supplementary Service; Stage 3
24.082	Call Forwarding Supplementary Service; Stage 3
24.083	Call Waiting (CW) and Call Hold (HOLD) Supplementary Services; Stage 3
24.084	Multi-Party (MPTY) Supplementary Service; Stage 3
24.085	Closed User Group (CUG) Supplementary Service; Stage 3
24.086	Advice of Charge (AoC) Supplementary Service; Stage 3
24.087	User-to-User Signalling (UUS); Stage 3
24.088	Call Barring (CB) Supplementary Service; Stage 3
24.090	Unstructured Supplementary Service Data (USSD); Stage 3
24.091	Explicit Call Transfer (ECT) Supplementary Service; Stage 3
24.093	Call Completion to Busy Subscriber (CCBS); Stage 3
24.096	Name Identification Supplementary Service; Stage 3
25.101	UE Radio Transmission and Reception (FDD)
25.102	UE Radio Transmission and Reception (TDD)
25.103	RF Parameters in Support of RRM
25.104	UTRA (BS) FDD; Radio transmission and Reception
25.105	UTRA (BS) TDD: Radio transmission and Reception
25.113	Base Station EMC
25.123	RF Parameters in Support of RRM (TDD)
25.133	RF Parameters in Support of RRM (FDD)

3G TS No.	Title
25.141	Base Station Conformance Testing (FDD)
25.142	Base Station Conformance Testing (TDD)
25.201	Physical Layer -General Description
25.211	Physical Channels and Mapping of Transport channels onto Physical Channels (FDD)
25.212	Multiplexing and Channel Coding (FDD)
25.213	Spreading and Modulation (FDD)
25.214	FDD: Physical Layer Procedures
25.215	Physical Layer: Measurements (FDD)
25.221	Physical Channels and Mapping of Transport Channels onto Physical Channels (TDD)
25.222	Multiplexing and Channel Coding (TDD)
25.223	Spreading and Modulation (TDD)
25.224	TDD: Physical Layer Procedures
25.225	Physical Layer: Measurements (TDD)
25.301	Radio Interface Protocol Architecture
25.302	Services Provided by the Physical Layer
25.303	UE Functions and Inter-Layer Procedures in Connected Mode
25.304	UE Procedures in Idle Mode and Procedures for Cell Reselection in Connected Mode
25.305	Stage 2 Functional Specification of Location Services in UTRAN (LCS)
25.321	Medium Access Control (MAC) Protocol Specification
25.322	Radio Link Control (RLC) Protocol Specification
25.323	Packet Data Convergence Protocol (PDCP) Protocol
25.324	Radio Interface for Broadcast/Multicast Services
25.331	Radio Resource Control (RRC) Protocol Specification
25.401	UTRAN Overall Description
25.402	Synchronisation in UTRAN; Stage 2
25.410	UTRAN Iu Interface: General Aspects and Principles
25.411	UTRAN Iu Interface Layer 1
25.412	UTRAN Iu Interface Signalling Transport
25.413	UTRAN Iu Interface RANAP Signalling
25.414	UTRAN Iu Interface Data Transport & Transport Signalling
25.415	UTRAN Iu Interface User Plane Protocols
25.419	UTRAN Iu Interface: Cell Broadcast Protocols Between SMS-CBC and RNC
25.420	UTRAN Iur Interface: General Aspects and Principles
25.421	UTRAN Iur Interface Layer 1
25.422	UTRAN Iur Interface signalling transport
25.423	UTRAN Iur Interface RNSAP signalling
25.424	UTRAN Iur Interface data transport & transport signalling for CCH data streams
25.425	UTRAN Iur interface user plane protocols for CCH data streams
25.426	UTRAN Iur and Iub interface data transport & transport signalling for DCH data streams

3G TS No.	Title
25.427	UTRAN Iur and Iub interface user plane protocols for DCH data streams
25.430	UTRAN Iub Interface: General Aspects and Principles
25.431	UTRAN Iub interface Layer 1
25.432	UTRAN Iub interface signalling transport
25.433	UTRAN Iub interface NBAP signalling
25.434	UTRAN Iub interface data transport & transport signalling for CCH data streams
25.435	UTRAN Iub interface user plane protocols for CCH data streams
25.442	UTRAN Implementation Specific O&M Transport
25.831	Study Items for future release
25.832	Manifestations of Handover and SRNS relocation
25.833	Physical layer items not for inclusion in Release 99
25.921	Guidelines and principles for protocol description and error handling
25.922	Radio Resource Management Strategies
25.925	Radio Interface for Broadcast/Multicast Services
25.926	UE Radio Access capabilities definition
25.931	UTRAN Functions, examples on signalling procedures
25.941	Document structure
25.942	RF system scenarios
25.943	Deployment aspects
25.944	Channel coding and multiplexing examples
25.990	Vocabulary for UTRAN
26.071	AMR speech Codec; General description
26.073	AMR speech Codec; C-source code
26.074	AMR speech Codec; Test sequences
26.090	AMR speech Codec; Transcoding Functions
26.091	AMR speech Codec; Error concealment of lost frames
26.092	AMR speech Codec; comfort noise for AMR Speech Traffic Channels
26.093	AMR speech Codec; Source Controlled Rate operation
26.094	AMR Speech Codec; Voice Activity Detector for AMR Speech Traffic Channels
26.101	AMR speech Codec; Frame Structure
26.102	AMR speech Codec; Interface to Iu and Uu
26.103	Codec lists
26.104	AMR speech Codec; Floating point C-Code
26.110	Codec for Circuit switched Multimedia Telephony Service; General Description
26.111	Codec for Circuit switched Multimedia Telephony Service; Modifications to H.324
26.131	Narrow Band (3.1kHz) Speech & Video Telephony Terminal Acoustic Characteristics
26.132	Narrow Band (3.1kHz) Speech & Video Telephony Terminal Acoustic Test Specification.

3G TS No.	Title
26.911	Codec for Circuit switched Multimedia Telephony Service; Terminal Implementor's Guide
26.912	Codec for Circuit switched Multimedia Telephony Service; Quantitative performance evaluation of H.324 Annex C over 3G
26.913	Quantitative performance evaluation of real-time packet switched multimedia services over 3G
26.915	QoS for Speech and Multimedia Codec; Quantitative performance evaluation of real-time packet switched multimedia services over 3G
26.975	AMR speech Codec; Performance Characterization of the GSM AMR Speech Codec
27.001	General on Terminal Adaptation Functions (TAF) for Mobile Stations (MS)
27.002	Terminal Adaptation Functions (TAF) for services using Asynchronous bearer capabilities
27.003	Terminal Adaptation Functions (TAF) for services using Synchronous bearer capabilities
27.005	Use of Data Terminal Equipment - Data Circuit terminating Equipment (DTE - DCE) interface for Short Message Service (SMS) and Cell Broadcast Service (CBS)
27.007	AT command set for 3G User Equipment (UE)
27.010	Terminal Equipment to User Equipment (TE-UE) multiplexer protocol User Equipment (UE)
27.060	GPRS Mobile Stations supporting GPRS
27.103	Wide Area Network Synchronisation
27.901	Report on Terminal Interfaces—An Overview
27.903	Discussion of Synchronisation Standards
29.002	Mobile Application Part (MAP)
29.007	General requirements on Interworking between the PLMN and the ISDN or PSTN
29.010	Information Element Mapping between Mobile Station—Base Station System (MS–BSS) and Base Station System—Mobile-services Switching Centre (BSS–MCS) Signalling Procedures and the Mobile Application Part (MAP)
29.011	Signalling Interworking for Supplementary Services
29.013	Signalling interworking between ISDN supplementary services Application Service Element (ASE) and Mobile Application Part (MAP) protocols
29.016	Serving GPRS Support Mode SGSN—Visitors Location Register (VLR); Gs Interface Network Service Specification
29.018	Serving GPRS Support Mode SGSN - Visitors Location Register (VLR); Gs Interface Layer 3 Specification
29.060	GPRS Tunnelling protocol (GPT) across the Gn and Gp interface
29.061	General Packet Radio Service (GPRS); Interworking between the Public Land Mobile Network (PLMN) supporting GPRS and Packet
29.078	CAMEL; Stage 3

3G TS No.	Title
29.119	GPRS Tunnelling Protocol (GTP) specification for Gateway Location Register (GLR)
29.120	Mobile Application Part (MAP) specification for Gateway Location Register (GLR); Stage 3
31.101	UICC / Terminal Interface; Physical and Logical Characteristics
31.102	Characteristics of the USIM Application
31.110	Numbering system for telecommunication IC card applications
31.111	USIM Application Toolkit (USAT)
31.120	Terminal tests for the UICC Interface
31.121	UICC Test Specification
32.005	GSM charging CS domain
32.015	GSM charging PS domain
32.101	3G Telecom Management principles and high level requirements
32.102	3G Telecom Management architecture
32.104	3G Performance Management
32.105	3G charging call event data
32.106	3G Configuration Management
32.111	3G Fault Management
33.102	Security Architecture
33.103	Security Integration Guidelines
33.105	Cryptographic Algorithm requirements
33.106	Lawful interception requirements
33.107	Lawful interception architecture and functions
33.120	Security Objectives and Principles
33.900	Guide to 3G security
33.901	Criteria for cryptographic Algorithm design process
33.902	Formal Analysis of the 3G Authentication Protocol with Modified Sequence number Management
34.108	Common Test Environments for User Equipment (UE) Conformance Testing
34.109	Logical Test Interface (TDD and FDD)
34.121	Terminal Conformance Specification, Radio Transmission and Reception (FDD)
34.122	Terminal Conformance Specification, Radio Transmission and Reception (TDD)
34.123-1	UE Conformance Specification, Part 1: Conformance Specification
34.123-2	UE Conformance Specification, Part 2: ICS
34.123-3	UE Conformance Specification, Part 3 - Abstract Test Suites
34.124	Electro-Magnetic Compatibility (EMC) for Terminal equipment; Stage 1
34.907	Report on electrical safety requirements and regulations
34.925	Specific Absorption Rate (SAR) requirements and regulations in different regions

Appendix G
Acronyms

2G	*Second-Generation*
3G	*Third-Generation*
3G.IP	*3G mobile Internet*
3GMSC	*Third-Generation MSC*
3GPP	*3rd Generation Partnership Project*
4G	*Fourth-Generation*
8-PSK	*8-Phase-Shift-Keying*
AACH	*Access Assignment Channel*
AAL	*ATM Adaptation Layer*
ABM	*Asynchronous Balanced Mode*
ABQP	*Aggregate BSS QoS Profile*
ABR	*Available Bit Rate*
AC	*Admission Control*
ACB	*Access Control Block*
ACC	*Area Communications Controller*
ACELP	*Algebraic Codebook Excited Linear Predictive*
ACF	*Association Control Function*
ACH	*Access Feedback Channel*
ACH	*Access Channel*
ACK	*Acknowledgement*
ACL	*Asynchronous Connectionless*
ACK	*Acknowledge*
ACS	*Access Channel*
ACTS	*Advanced Communication Technologies and Services*
ADA	*Alias Destination Address*
ADM	*Asynchronous Disconnected Mode*
ADPCM	*Adaptive Differential Pulse Code Modulation*
ADSL	*Asynchronous Digital Subscriber Line*
ADT	*Abstract Data Type*
AGC	*Automatic Gain Control*
AGCH	*Access Grant Channel*
AH	*Address Handling*
AI	*Air Interface*

AICH	*Acquisition Indication Channel*
AID	*Association Identifier*
AKF	*Autokorrelationsfunktion*
AL	*Advanced Link*
AL	*Ambience Listening*
AL	*Advanced Link*
AM	*Acknowledge Mode*
AM	*Active Mode*
AMES	*ATM Mobility-Enhanced Switch*
AND	*Access Network Domain*
ANSI	*American National Standards Institute*
AoC	*Advice of Charge*
AP-AICH	*Access Preamble—Acquisition Indication Channel*
AP	*Access Point*
APC	*Access Point Controller*
APCO	*American Public Safety Communications Officials*
APDU	*Application Protocol Data Units*
API	*Application Programmers Interface*
APT	*Access Point Transceiver*
AR	*Acknowledge Request*
ARIB	*Association of Industries and Businesses*
ARQ	*Automatic Repeat Request*
AS	*Access Slot*
AS	*Area Selection*
AS	*Access Stratum*
ASA	*Alias Source Address*
ASC	*Access Service Class*
ASCH	*Association Control Channel*
ASCI	*Advanced Speech Call Items*
ASE	*Application Service Elements*
ASP	*Active Server Page(s)*
ATDMA	*Adaptive Time-Division Multiple Access*
ATIM	*Ad Hoc Traffic Indication Message*
ATM	*Asynchronous Transfer Mode*
AuC	*Authentication Centre*
AUM	*Asynchronous Unbalanced Mode*
BBK	*Broadcast Block*
BC	*Broadcast*
BCH	*Broadcast Channel*
BCH	*Bose-Chaudhuri-Hocquenghem*
BCCH	*Broadcast Control Channel*

BCE	*Bearer Capability Element*
BCFE	*Broadcast Control Functional Entity*
BCS	*Block Check Sequence*
BCSM	*Basic Call State Model*
BEC	*Backward Error Correction*
BER	*Bit Error Ratio*
B-ISDN	*Broadband ISDN*
BKN	*BlocK Number*
BL	*Basic Link*
BLCH	*Base Station Linearization Channel*
BLE	*Base Link Entity*
BLEP	*Block Error Probability*
BLER	*Block Error Ratio*
BMB	*Bit Map Block*
BMC	*Broadcast/Multicast Control*
BMC	*Broadcast Message Control*
BMN	*Bit Map Number*
BN	*Bit Number*
BNCH	*Broadcast Network Channel*
BPSK	*Binary Phase Shift Keying*
BR	*Basic Rate*
BRAN	*Broadband Radio Access Network*
BRAIN	*Broadband Radio Access to IP Networks*
BS	*Base Station*
BSC	*Base Station Controller*
BSCH	*Broadcast Synchronization Channel*
BSIC	*Base Station Identification Code*
BSIG	*Bluetooth Special Interest Group*
BSN	*Block Sequence Number*
BSS	*Base Station Subsystem*
BSS	*Basic Service Set*
BSSAP	*Base Station Subsystem Application Part*
BSSID	*BSS Identifier*
BSSGP	*Base Station Subsystem GPRS Protocol*
BSSMAP	*Base Station Subsystem Mobile Application Part*
BST	*Base Station Transceiver*
BU	*Bad Urban*
BTS	*Base Transceiver Station*
BTSM	*BTS Management*
BVC	*BSSGP Virtual Connection*
BVCI	*BSSGP Virtual Connection Identifier*
C-RNTI	*Cell Radio Network Temporary Identity*

C-SAP	*Control Service Access Point*
C/I	*Carrier-to-Interference*
CAC	*Channel Access Control*
CAC	*Channel Access Code*
CAC	*Connection Admission Control*
CAD	*Call Authorized by Dispatcher*
CAMEL	*Customized Applications for Mobile network Enhanced Logic*
CAM	*Channel Access Mechanism*
CB	*Control Uplink Burst*
CB	*Cell Broadcast*
CBC	*Connectionless Bearer Control*
CBCH	*Cell Broadcast Channel*
CBR	*Constant Bit Rate*
CC	*Channel Combinations*
CC	*Central Controller*
CC	*Call Control*
CC	*Country Code*
CCA	*Clear Channel Assessment*
CCCH	*Common Control Channel*
CCF	*Call Control Functions*
CCH	*Control Channel*
CCIR	*Consultative Committee for International Radiocommunication*
CCITT	*Comité Consultatif International des Télégraphes et Téléphones*
CCK	*Complementary Code Keying*
CCPCH	*Common Control Physical Channel*
CCS7	*Common Channel Signalling System No. 7*
CCTrCH	*Coded Composite Transport Channel*
CDF	*Complementary Distribution Function*
CD/CA-ICH	*Collision Detection/Channel Assignment Indicator Channel*
CDM	*Code Division Multiplex*
CDMA	*Code Division Multiple Access*
CDV	*Cell-Delay Variance*
CEP	*Connection Endpoint*
CEPT	*Conférence Europeéne des Administrations des Postes et des Télécommunications*
CFPRate	*Contention-Free Repetition Rate*
CF	*Coordination Function*
CFP	*Contention-Free Period*

CGI	*Common Gateway Interface*
CI	*Cell Identity*
CIE	*Contents of Information Element*
CKSN	*Cyphering Key Sequence Number*
CL	*Convergence Layer*
CLCH	*Common Linearization Channel*
CLKE	*Estimated Clock*
CLK	*Native Clock*
CLKN	*Native Clock*
CLNP	*Connectionless Network Protocol*
CLNS	*Connectionless Network Service*
CL	*Convergence Layer*
CLT	*Common Linearization Time*
CLP	*Cell Loss Priority*
CLR	*Cell Loss Ratio*
CM	*Centralized Mode*
CM	*Call Management*
CMC	*Connectionless Message Control*
CMCE	*Circuit Mode Control Entity*
CN	*Core Network*
CNCL	*Communication Networks Class Library*
CND	*Core Network Domain*
COF	*Ciphering Offset Number*
CONP	*Connection Oriented Network Protocol*
CONS	*Connection Oriented Network Service*
COST	*European Cooperation in the Field of Scientific and Technical Research*
CP	*Contention Period*
CP	*Control Physical Channel*
CPCH	*Common Packet Channel*
CPCS	*Common Part Convergence Sublayer*
CPICH	*Common Pilot Channel*
CPU	*Central Processing Unit*
CR	*Call Report*
CRC	*Cyclic Redundancy Check*
CS	*Circuit-Switched*
CS	*Carrier Sensing*
CS	*Coding Scheme*
CS	*Convergence Sublayer*
CSCF	*Call State Control Function*
CSD	*Circuit Switched Domain*
CSE	*CAMEL Service Environment*
CSI	*CAMEL Subscription Information*

CSICH	*CPCH Status Indicator Channel*
CSMA/CA	*Carrier Sense Multiple Access with Collision Avoidance*
CSMA/CD	*Carrier Sense Multiple Access with Collision Detection*
CT	*Cordless Telephone*
CTCH	*Common Traffic Channel*
CTD	*Cell-Transfer Delay*
CTS	*Clear To Send*
CumAck	*Cumulative Acknowledgement*
CV	*Countdown Value*
CW	*Contention Window*
CW	*Call Waiting*
DA	*Destination Address*
DA	*Duplication Avoidance*
DAB	*Digital Audio Broadcast*
DAC	*Device Access Code*
DAWS	*Digital Advanced Wireless Services*
DBC	*Dummy Bearer Control*
DBPSK	*Differential Binary Phase Shift Keying*
DC	*Dedicated Control*
DCA	*Dynamic Channel Allocation*
DCC	*DLC User Connection Control*
DCH	*Dedicated Channel*
DCCH	*Dedicated Control Channel*
DCF	*Distributed Coordination Function*
DCFE	*Dedicated Control Functional Entity*
DCLA	*DC Level Adjustment*
DCR	*Dynamic Channel Reservation*
DCS	*Dynamic Channel Selection*
DCS	*Digital Cellular System*
DDE	*Dienstdateneinheit*
DECT	*Digital Enhanced Cordless Telecommunications*
DES	*Data Encryption Standard*
DFS	*Dynamic Frequency Selection*
DIAC	*Dedicated Inquiry Access Code*
DIFS	*Distributed Coordination Function IFS*
DiL	*Direct Link*
DISC	*Disconnect*
DL	*Downlink*
DLC	*Data Link Control*
DLCC	*DLC Connection*
DLCC-ID	*DLC Connection Identifier*

DLCI *Data Link Connection Identifier*

DLL *Data Link Layer*

DLR *German Aerospace Research Institute*

DM *Direct Mode*

DM *Disconnected Mode*

DMPD *Deferred Multicast Pattern Declaration*

DNC *Desired Number of Channels*

DoA *Directions of Arrival*

DP *Detection Point*

DPCH *Dedicated Physical Channel*

DPCCH *Uplink Dedicated Physical Control Channel*

DPCCH *Dedicated Physical Control Channel*

DPDCH *Uplink Dedicated Physical Data Channel*

DQPSK *Differential Quaternary Phase Shift Keying*

DR *Data Rate*

DRIVE *Dynamic Radio for IP-Services in Vehicular Environments*

DRNC *Drift Radio Network Controller*

DRNS *Drift Radio Network Subsystem*

DRX *Discontinuous Reception*

DS *Distribution System*

DSCH *Downlink Shared Channel*

DSCCH *DSCH Control Channel*

DSM *Distribution System Media*

DSMA *Data Sense Multiple Access*

DSMA *Slotted Digital Sense Multiple Access*

DSP *Digital Signal Processor*

DSS *Distribution System Services*

DSSS *Direct Sequence Spread Spectrum*

DSU *Data Service Unit*

DTCH *Dedicated Traffic Channel*

DTD *Document Type Definition*

DTE *Data Terminal Equipment*

DTIM *Delivery Traffic Indication Message*

DTMF *Dual-Tone Multiple Frequency*

DTX *Discontinuous Transmission*

DUC *DLC User Connection*

DUC-ID *DLC User Connection Identifier*

DVB-T *Digital Video Broadcasting Terrestrial*

DwPCH *Downlink Pilot Channel*

DwPTS	*Downlink Pilot Time Slot*
EC	*Error Control*
ECH	*Energy Channel*
ECSD	*Enhanced Circuit-Switched Data*
EDD	*Earliest Due Date*
EDGE	*Enhanced Data Rates for GSM Evolution*
EFR	*Enhanced Full Rate*
EFSM	*Extended Finite State Machine*
EGSM	*Extended GSM*
EGPRS	*Enhanced General Packet Radio Service*
EI	*Equipment Identity*
EIFS	*Extended IFS*
EIR	*Equipment Identity Register*
EIRP	*Equivalent Isotropic Radiated Power*
EL	*Event Label*
eMLPP	*enhanced Multi-Level Precedence and Pre-Emption*
ERO	*European Radio Office*
ERR	*Exhaustive Round Robin*
ESS	*Extended Service Set*
ETCS	*European Train Control System*
ETE	*Equivalent Telephony Erlang*
ETR	*ETSI Technical Report*
ETS	*European Telecommunication Standard*
ETSI	*European Telecommunications Standards Institute*
EU	*European Union*
EUTELSAT	*European Telecommunications Satellite Organization*
FACCH	*Fast Associated dedicated Control Channel*
FACH	*Forward Access Channel*
FAUSCH	*Fast Uplink Signalling Channel*
FBC	*Forwarding Broadcast*
FBCH	*Forwarding Broadcast Channel*
FBS	*Flexible Bearer Services*
FBI	*Feedback Information*
FC	*Flow Control*
FCA	*Fixed Channel Allocation*
FCB	*Frequency Correction Burst*
FCC	*Federal Communications Commission*
FCCH	*Frame Control Channel*
FCCH	*Frequency Correction Channel*
FCFS	*First Come First Serve*
FCH	*Frame Channel*

FCS	*Frame Check Sequence*
FDD	*Frequency Division Duplex*
FDL	*Forwarding Downlink*
FDM	*Frequency Division Multiplex*
FDMA	*Frequency Division Multiple Access*
FDT	*Formal Description Techniques*
FEC	*Forward Error Correction*
FER	*Frame Error Ratio*
FFCH	*Forwarding Frame Channel*
FFSK	*Fast-Frequency Shift Keying*
FFT	*Fast Fourier Transformation*
FH	*Frequency Hopping*
FHS	*Frequency Hop Synchronization*
FHSS	*Frequency Hopping Spread Spectrum*
FIFO	*First In First Out*
FMT	*Forwarding Mobile Terminal*
FN	*Frame Number*
FPGA	*Field Programmable Gate Array*
FPLMTS	*Future Public Land Mobile System*
FRCH	*Forwarding Random Access Channel*
FRMT	*Forwarder Remote Mobile Terminal*
FRS	*Fixed Relay Station*
FSM	*Finite State Maschine*
FSR	*Frequency Sharing Rule*
FT	*Fixed Radio Termination*
FTP	*File Transfer Protocol*
FUL	*Forwarding Uplink*
FUNET	*Finish University and Research Network*
FWA	*Fixed Wireless Access*
GC	*General Control*
GCR	*Group Call Register*
GEO	*Geostationary Orbit*
GERAN	*GSM/EDGE Radio Access Network*
GGSN	*Gateway GPRS Support Node*
GFSK	*Gaussian Frequency Shift Keying*
GIAC	*General Inquiry Access Code*
GIST	*Graphical Interactive Simulation Result Tool*
GLONAS	*Russian Global Positioning System*
GMM	*Global Multimedia Mobility*
GMM	*Global Mobility Management*

GMSC	*Gateway Mobile-services Switching Centre*
GMSK	*Gaussian Minimum Shift Keying*
GOP	*Group of Pictures*
GoS	*Grade of Service*
GP	*Guard Period*
GPRS	*General Packet Radio Service*
GPS	*Global Positioning System*
GR	*GPRS Register*
GRLP	*GPRS Radio Link Protocol*
GSM	*Global System for Mobile Communication*
GSM	*Groupe Spéciale Mobile*
GSMSCF	*GSM Service Control Function*
GSMSSF	*GSM Service Switching Function*
GSN	*GPRS Support Node*
GTP	*GPRS Tunnelling Protocol*
GTP-C	*GPRS Tunnelling Protocol (Control Plane)*
GTP-U	*GPRS Tunnelling Protocol (User Plane)*
GTSI	*Group TETRA Subscriber Identity*
GW	*Gateway to/from fixed network*
HBRBP	*High-Bit-Rate Bit Period*
HC	*Hop Counter*
HC-	*HIPERLAN-CAC-*
HCS	*HIPERLAN-CAC Service*
HCSAP	*HIPERLAN-CAC Service Access Point*
HDB	*Home Data Base*
HDLC	*High Level Data Link Control*
HDML	*Handheld Device Markup Language*
HEC	*Header Error Check*
HEC	*Header Error Control*
HEE	*Home Environment Extension*
HEO	*Highly Elliptical Orbit*
HID	*HIPERLAN Identifier*
HIPERLAN	*High Performance Radio Local Area Network*
H/1	*HIPERLAN/1*
H/2	*HIPERLAN/2*
HIPERLAN/1	*High Performance Radio Local Area Network Type 1*
HIPERLAN/2	*High Performance Radio Local Area Network Type 2*
HISWAN	*High-speed Wireless Access Network*
HLR	*Home Location Register*
HM-	*HIPERLAN-MAC*

HMPDU	*HIPERLAN-MAC Protocol Data Unit*
HMQoS	*HIPERLAN-MAC Quality of Service*
HMS	*HIPERLAN-MAC Service*
HND	*Home Network Domain*
HO	*Handover*
HR/DSSS	*High Rate Direct Sequence Spread Spectrum*
HSCSD	*High-Speed Circuit-Switched Data*
HSS	*Home Subscriber Server*
HTTP	*Hypertext Transfer Protocol*
HTML	*Hypertext Markup Language*
HTTP	*Hypertext Transfer Protocol*
I-CSCF	*Interrogating-CSCF*
IAC	*Inquiry Access Code*
IAM	*Initial Address Message*
IBMS	*Integrated Broadband Mobile System*
IBSS	*Independent Basic Service Set*
IC	*Integrated Circuit*
ICGW	*Incoming Call Gateway*
ICO	*Intermediate Circular Orbit*
ID	*Infrastructure Domain*
ID	*Identifier*
IDU	*Interface Data Unit*
IE	*Information Element*
IEC	*International Electrotechnical Commission*
IEEE	*Institute of Electrical and Electronics Engineers*
IEI	*Information Element Identifier*
IETF	*Internet Engineering Task Force*
IFFT	*Inverse Fast Fourier Transformation*
IFS	*Interframe Space*
IGMP	*Internet Group Management Protocol*
IM	*IP Multimedia Subsystem*
IM	*Individual Members*
IMEI	*International Mobile Equipment Identity*
IMGI	*International Mobile GPRS Identity*
IMM	*Interactive Multimedia*
IMSI	*International Mobile Subscriber Identity*
IMT-2000	*International Mobile Telecommunications at 2000 MHz*
IN	*Intelligent Network*
INMARSAT	*International Maritime Satellite Organization*

IP	*Internet Protocol*
IPR	*Intellectual Property Right*
IPv4	*IP Version 4*
IPv6	*IP Version 6*
IR	*Infrared*
IR	*Incremental Redundancy*
IRC	*Idle Receiver Control*
IrDA	*Infrared Data Association*
ISDN	*Integrated Services Digital Network*
ISL	*Inter Satellite Link*
ISM	*Industrial, Scientific and Medical*
ISO	*International Standards Organization*
ISP	*Intermediate Service Part*
IST	*Information Society Technology*
ITSI	*Individual TETRA Subscriber Identity*
ITTP	*Intelligent Terminal Transfer Protocol*
ITU	*International Telecommunication Union*
ITU-R	*Radiocommunication Standardization Sector of ITU*
ITU-T	*Telecommunication Standardization Sector of ITU*
IWF	*Interworking Function*
IWU	*Interworking Unit*
JVM	*Java Virtual Machine*
L2	*Layer 2*
L2R	*Layer-2 Relay*
L3	*Layer 3*
LA	*Location Area*
LA	*Link Adaptation*
LAC	*Location Area Code*
LAI	*Location Area Identifier*
LAN	*Local Area Network*
LAP	*Lower Address Part*
LAP.T	*Link Access Protocol for TETRA*
LAPD	*Link Access Procedure on the D-channel*
LB	*Linearization Uplink Burst*
LBT	*Listen Before Talk*
LC	*Link Control*
LCCH	*Link Control Channel*
LCE	*Link Control Entity*
LCH	*Long Channel*
LCH	*Linearization Channel*
LCH	*Logical Channel*
LE	*Late Entry*
LEO	*Low Earth Orbit*
LI	*Length Indicator*

LID	*WLAN Identifier*
LLC	*Logical Link Control*
LLME	*Lower-Layer Management Entity*
LM	*Link Manager*
LMP	*Link Manager Protocol*
LMS	*Least Mean Square*
LOS	*Line Of Sight*
L-PDU	*Long-Channel-PDU*
LPC	*Linear Predictive Coding*
LPO	*Low Power Oscillator*
LQC	*Link Quality Control*
LS	*Line Station*
LSC	*List Search Calls*
LTP	*Long-term Prediction*
LU	*Location Updating*
M3UA	*MTP3 User Adaptation Layer*
MAC	*Medium Access Control*
MAC-I	*Message Authentication Code Identifier*
MAC-ID	*Medium Access Control Identifier*
MANET	*Mobile Access Network*
MAHO	*Mobile-assisted Handover*
MAIG	*Maximum Allowed Inter-Car Gap*
MAP	*Mobile Application Part*
MAP	*Media Access Point*
MBC	*Multiple Bearer Control*
MBS	*Mobile Broadband System*
MC	*MultiCommunicator*
MCC	*Mobile Country Code*
MCC	*Mobile Competence Centre*
MCHO	*Mobile Controlled Handover*
MCS	*Modulation and Coding Scheme*
MDTRS	*Mobile Digital Trunked Radio System*
ME	*Mobile Equipment*
MED	*Mobile Equipment Domain*
MEO	*Medium Earth Orbit*
MER	*Message Error Ratio*
MES	*Mobile Earth Station*
MESA	*Mobility for Emergency and Safety Applications*
MExE	*Mobile Station Execution Environment*
MF	*MAC Frame*
MF-WT	*Multiple-Frequency Forwarder WT*
MGW	*Media Gateway*
MHCH	*Main HSCSD Subchannel*
MIME	*Multipurpose Internet Mail Extensions*

MIND	*Mobile IP Network Simulator*
MLE	*Mobile Link Entity*
MLME	*MAC Sublayer Management Entity*
MLPP	*Multi-Level Precedence and Pre-Emption*
MM	*Multimedia*
MM	*Mobility Management*
MMI	*Man–Machine Interface*
MMIC	*MMI Control*
MMS	*Multimedia Messaging Service*
MN	*Multiframe Number*
MNC	*Mobile Network Code*
MNI	*Mobile Network Identity*
MO-C	*Mobile Originated Call*
MPDCH	*Master Packet Data Channel*
MPDU	*MAC Protocol Data Unit*
MPEG	*Moving Pictures Expert Group*
MPT	*Ministry of Post and Telecommunication*
MRP	*Market Representation Partners*
MRSC-SN	*Multipoint Relay Set Change Sequence Number*
MRP	*Mobility and Resource Management Protocol*
MRT	*Multipoint Relay Table*
MS	*Mobile Station*
MSA	*Multi-Slot Assignment*
MSAP	*MAC Service Access Point*
MSC	*Mobile-services Switching Centre*
MSC	*Master System Controller*
MSC	*Multislot Capability*
MSDU	*MAC Service Data Unit*
MSIN	*Mobile Subscriber Identity*
MSISDN	*Mobile Station International ISDN Number*
MSRN	*Mobile Station Roaming Number*
MSS	*Maximum Segment Size*
MT	*Mobile Termination*
MT	*Mobile Terminal*
MT	*Media Terminal*
MT-C	*Mobile Terminated Call*
MTP	*Message Transfer Part*
MTU	*Maximum Transfer Unit*
MWIF	*Mobile Wireless Internet Forum*
N	*Network Layer*
N	*Network*
NAH	*Network Administration Host*
NAK	*No Acknowledge*

NAP	*Network Access Point*
NAS	*Non Access Stratum*
NAV	*Net Allocation Vector*
NBAP	*Node B Application Part*
NBS	*Narrow Band Sockets*
NCC	*Network Control Centre*
NCCH	*Notification Common Control Channel*
NCS	*Network Control Station*
NDB	*Normal Downlink Burst*
NDC	*National Destination Code*
NG	*Next Generation*
NID	*Node Identifier*
NL	*Network Layer*
NLOS	*Non-Line-of-Sight*
NNI	*Network-To-Network Interface*
NRML	*Normalized Residual MSDU Lifetime*
NRT	*Non Real-Time*
NSAP	*Network Layer SAP*
N-SAPI	*Network Layer Service Access Point Identifier*
NSH	*Non-Seamless Handover*
NSS	*Network and Switching Subsystem*
Nt	*Notification*
NT	*Network Termination Functional Group*
NTT	*Nippon Telegraph and Telephone Corporation*
NUB	*Normal Uplink Burst*
OAM	*Operating and Maintenance*
OACSU	*Off Air Call Setup*
OCCCH	*ODMA Common Control Channel*
ODCCH	*ODMA Dedicated Control Channel*
ODCH	*ODMA Dedicated Channel*
ODMA	*Opportunity Driven Multiple Access*
ODTCH	*ODMA Dedicated Traffic Channel*
OFDM	*Orthogonal Frequency Division Multiplex*
OHG	*Operator Harmonization Group*
OMC	*Operating and Maintenance Centre*
OP	*Organizational Partners*
ORACH	*ODMA Random Access Channel*
OSA	*Open Service Architecture*
OSI	*Open Systems Interconnection*
OSS	*Operation Subsystem*
O-SS	*Operator-Specific Services*
OTA	*Over-The-Air*

OVSF	*Orthogonal Variable Spreading Factor*
P-CCPCH	*Primary Common Control Physical Channel*
P-CSCF	*Proxy-CSCF*
P-CPICH	*Primary Common Pilot Channel*
P-SCH	*Primary Synchronization Channel*
PABX	*Private Automatic Branch Exchange*
PACCH	*Packet Associated Control Channel*
PAD	*Packet Assembly Disassembly*
PAGCH	*Packet Access Grant Channel*
PAM	*Pulse Amplitude Modulation*
PAN	*Personal Area Network*
PAP	*Push Access Protocol*
PBCCH	*Packet Broadcast Control Channel*
PBR	*Packet Bit Rate*
PBX	*Private Branch Exchange*
PC	*Point Coordinator*
PC	*Personal Computer*
PC	*Power Control*
PCCCH	*Packet Common Control Channel*
PCCH	*Paging Control Channel*
PCF	*Point Coordination Function*
PCG	*Project Coordination Group*
PCH	*Paging Channel*
PCI	*Protocol Control Information*
PCM	*Pulse Code Modulation*
PCMCIA	*Personal Computer Memory Card Interface Association*
PCN	*Personal Communication Network*
PCPCH	*Physical Common Packet Channel*
PCS	*Personal Communication System*
PD	*Protocol Discriminator*
PDA	*Personal Digital Assistent*
PDCCH	*Packet Dedicated Control Channel*
PDCH	*Packet Data Channel*
PDCP	*Packet Data Convergence Protocol*
PDN	*Public Data Network*
PDO	*Packet Data Optimized*
PDP	*Packet Data Protocol*
PDSCH	*Physical Downlink Shared Channel*
PDTCH	*Packet Data Traffic Channel*
PDU	*Protocol Data Unit*

PER	*Packet Error Ratio*
PFC	*Packet Flow Context*
PFI	*Packet Flow Identifier*
PHL	*Physical Layer*
PHP	*PHP Hypertext Protocol*
PHS	*Personal Handyphone System*
PHY	*Physical Layer*
PI	*Page Indicator*
PICH	*Paging Indication Channel*
PIFS	*Point Coordination Function IFS*
PIN	*Personal Identification Number*
PL	*Physical Layer*
PLL	*Physical Link Layer*
PLCP	*Physical Layer Convergence Procedure*
PLMN	*Public Land Mobile Network*
PLW	*PSDU Length Word*
PMD	*Physical Medium-Dependent*
PNCH	*Packet Notification Channel*
PP	*Portable Part*
PPG	*Push Proxy Gateway*
PT	*Portable Radio Termination*
PTMR	*Private Trunked Mobile Radio*
PNFE	*Paging and Notification Functional Entity*
PPCH	*Packet Paging Channel*
PPDU	*Physical PDU*
PPM	*Pulse Position Modulation*
PR	*Phrase Representation*
PR	*Primary Rate*
PRACH	*Physical Random Access Channel*
PS	*Packet-Switched*
PS	*Power Save*
PS	*Personal Station*
PSD	*Packet Switched Domain*
PSDU	*Physical SDU*
PSE	*Personalized Service Environment*
PSF	*PLCP Signalling Field*
PSK	*Phase Shift Keying*
PSDN	*Public Switched Data Network*
PSPDN	*Public Switched Packet Data Network*
PSPP	*Public Safety Partnership Project*
PSTN	*Public Switched Telephony Network*
PT	*Payload Type*
PTCH	*Packet Traffic Channel*
PTM	*Point-to-Multipoint*

PTM-B *PTM Broadcast*

PTM-G *PTM Group Call*

PTM-M *PTM Multi-cast*

PTM-S *PTM Single-cast*

PTM-SC *PTM Service Centre*

P-TMSI *Packet Temporary Mobile Subscriber Identity*

PTP *Point-To-Point*

PU *Payload Unit*

PUK *PIN Unblocking Key*

PUSCH *Physical Uplink Shared Channel*

PVC *Permanent Virtual Circuit*

PVCs *Permanent Virtual Circuits*

QAM *Quadrature Amplitude Modulation*

QN *Quarter-bit Number*

QoS *Quality of Service*

QPSK *Quaternary Phase Shift Keying*

RA *Receiver Address*

RA *Routing Area*

RA *Rate Adaptation*

RAB *Radio Access Bearer*

RACE *Research and Development in Advanced Communications Technologies in Europe*

RACH *Random Access Channel*

RAN *Radio Access Network*

RANAP *Radio Access Network Application Part*

RAS *Radio Access System*

RBB *Received Block Bitmap*

RBCH *RLC Broadcast Channel*

RCC *Real Channel Connection*

RCH *Random Channel*

RCPC *Rate-Compatible Punctured Convolutional code*

RD-LAP *Radio Data Link Access Procedure*

RDN *Radio Data Network*

RDSS *Radio Data Satellite Service*

RES *Radio-Equipment and Systems*

RFCH *Random Feedback Channel*

RF *Radio Frequency*

RFE *Routing Functional Entity*

RF Layer *Physical Radio Frequency Layer*

RFP *Radio Fixed Part*

RG *Resource Grant*

RG-IE *Resource Grant Information Element*

RHC *Robust Header Compression*

RIB	*Routing Information Base*
RLC	*Radio Link Control*
RLL	*Radio in the Local Loop*
RLP	*Radio Link Protocol*
RML	*Residual MSDU Lifetime*
RMT	*Remote Mobile Terminal*
RNC	*Radio Network Controller*
RNC	*Required Number of Channels*
RNR	*Receiver Not Ready*
RNS	*Radio Network Subsystem*
RNSAP	*Radio Network Subsystem Application Part*
RNTI	*Radio Network Temporary Identifier*
RPE	*Regular Pulse Excitation*
RR	*Radio Resource*
RRBP	*Relative Reserved Block Period*
RRC	*Radio Resource Control*
RRM	*Radio Resource Management*
RRq	*Resource Request*
RSS	*Received Signal Strength*
RSSI	*Receive Signal Strength Indicator*
RT	*Real-Time*
RTS	*Ready To Send*
RTT	*Round Trip Time*
S-ALOHA	*Slotted-ALOHA*
S-CCPCH	*Secondary Common Control Physical Channel*
S-CLNP	*TETRA Specific ConnectionLess Network Protocol*
S-CLNS	*TETRA Specific ConnectionLess Network Service*
S-CPICH	*Secondary Common Pilot Channel*
S-CSCF	*Serving-CSCF*
S-SCH	*Secondary Synchronization Channel*
S-PDU	*Short-Channel-PDU*
SA	*Source Address*
SABM	*Set Asynchronous Balance Mode*
SACCH	*Slow Associated dedicated Control Channel*
SAMBA	*System for Advanced Multimedia Broadband Applications*
SAN	*Service Area Node*
SAP	*Service Access Point*
SAPI	*Service Access Point Identifier*

SAR *Segmentation and Reassembly*

SAT *SIM Application Toolkit*

SB *Synchronization Downlink Burst*

SCAC *Service Creation and Accounting Centre*

SCB *Synchronization Burst*

SCCP *Signalling Connection Control Part*

SCF *Service Control Function*

SCH *Synchronization Channel*

SCH *Signalling Channel*

SCH *Short Channel*

SCH/F *Full-Size Signalling Downlink Burst*

SCH/HD *Half-Size Downlink Signalling Channel*

SCH/HU *Half-Size Uplink Signalling Channel*

SCCH *Synchronization Control Channel*

SCFE *Shared Control Functional Entity*

SCO *Synchronous Connection-Oriented*

SCP *Service Control Point*

SCR *Standard Context Routing*

SCS *Service Capability Server*

SCTP *Simple Control Transport Protocol*

SD *Send Duplicated*

SDCCH *Stand-Alone Dedicated Control Channel*

SDL *Specification and Description Language*

SDL-GR *SDL Graphical Representation*

SDL-PR *SDL Phrasal Representation*

SDMA *Space Division Multiple Access*

SDT *SDL Development Tool*

SDU *Service Data Unit*

SFD *Start Frame Delimiter*

SFN *System Frame Number*

SGSN *Serving GPRS Support Node*

SH *Seamless Handover*

SHCCH *Shared Channel Control Channel*

SI *Stall Indicator*

SIFS *Short Interframe Space*

SIM *Subscriber Identity Module*

SIP *Session Initiation Protocol*

SIR *Signal to Interference Ratio*

SJN *Shortest Job Next*

SM *Session Management*

SMG *Special Mobile Group*

SMS *Short Message Service*

SMSC *SMS Centre*

SMTP *Simple Mail Transfer Protocol*

SN *Sequence Number*

SN *Subscriber Number*

SNA *Short-number Addressing*

SNAF *Sub-Network Access Functions*

SND *Serving Network Domain*

SNDCP *Sub-Network Dependent Convergence Protocol*

SNR *Signal to Noise Ratio*

SNS *Sequence Number Space*

SOR *Start Of Reservation*

SP *Service Primitive*

SPD *Serving Profile Database*

SPEETCL *SDL Performance Evaluation Tool Class Library*

SRNC *Serving Radio Network Controller*

SRNS *Serving Radio Network Subsystem*

SS *Station Services*

SS *Supplementary Services*

SSCS *Service Specific Convergence Sublayer*

SSCS-PDU *Service Specific Convergence Sublayer PDU*

SSF *Service Switching Function*

SSG *Special Study Group*

SSI *Short Subscriber Identity*

SSL *Secure Socket Layer*

SSN *SubSlot Number*

STC *Standardization Technical Committee*

STCH *Stealing Channel*

STDM *Synchronous Time-Division Multiplexing*

STM *Synchronous Transfer Mode*

SVC *Switched Virtual Channel*

SVCs *Switched Virtual Circuits*

SwMI *Switching and Management Infrastructure*

T-AH *Alias Holding Timer*

T-PDU *Transport Layer PDU*

TA *Timing Advance*

TA *Terminal Adapter*

TA *Terminal Adapting*

TA *Transmitter Address*

TAF *Terminal Adaptation Function*

TBC *Traffic Bearer Control*

TBF	*Temporary Block Flow*
TBS	*Transport Block Set*
TBTT	*Target Beacon Transmission Times*
TC	*Transfer of Control*
TCAP	*Transaction Capability Application Part*
TCH	*Traffic Channel*
TCP	*Transport Control Protocol*
TCTF	*Target Channel Type Field*
TD-CDMA	*Time Division—Code Division Multiple Access*
TD-SCDMA	*Time Division—Synchronous Code Division Multiple Access*
TDA	*Target Destination Address*
TDD	*Time Division Duplex*
TDM	*Time Division Multiplex*
TDMA	*Time Division Multiple Access*
TE	*Terminal Equipment*
TEI	*TETRA Equipment Identity*
TEI	*Terminal Endpoint Identifier*
TETRA	*Terrestrial Trunked Radio*
TF	*Transport Format*
TFC	*Transport Format Combination*
TFCI	*Transport Format Combination Identifier*
TFCS	*Transport Format Combination Set*
TFI	*Transport Format Identifier*
TFS	*Transport Format Set*
TFT	*Traffic Flow Template*
TI	*Transaction Identifier*
TIA	*Telecommunication Industry Association*
TINA	*Telecommunication Infrastructure Networking Architecture*
TIM	*Traffic Indication Map*
TIPHON	*Telecommunications and Internet Protocol Harmonization Over Networks*
TL	*Transmission Line*
TLS	*Transport Layer Security*
TLLI	*Temporary Logical Link Identifier*
TLx-SAP	*TETRA LLC type x SAP*
TM	*Trunked Mode*
TME	*Transfer Mode Entity*
TMI	*TETRA Management Identity*
TMN	*Telecommunications Management Network*

TMSI	*Temporary Mobile Subscriber Identity*
TMV	*TETRA MAC Virtual*
TMV-SAP	*TETRA MAC Virtual SAP*
TMx-SAP	*TETRA MAC type x SAP*
TN	*Timeslot Number*
TND	*Transit Network Domain*
TP	*Traffic Physical Channel*
TPC	*Transmitter Power Control*
TRMA	*Time-Random Multiple Access*
TRAU	*Transcoding and Rate Adaption Unit*
TRX	*Transceiver*
TS	*Training Sequence*
TS	*Timeslot*
TSAP	*Transport Service Access Point*
TSC	*Trunked System Controller*
TSF	*Timing Synchronization Function*
TSG	*Technical Specification Group*
TSI	*TETRA Subscriber Identity*
TTC	*Terrestrial Terminal and Control*
TTI	*Transmission Time Interval*
TTML	*Tagged Text Markup Language*
TU	*Typical Urban*
U-NII	*Unlicensed National Information Infrastructure*
U-PDU	*User-PDU*
U-RNTI	*Cell Radio Network Temporary Identity*
UA	*Unnumbered Acknowledge*
UBCH	*User Broadcast Channel*
UBR	*Unspecified Bit Rate*
UDCH	*User Data Channel*
UDCP	*USSD Dialogue Control Protocol*
UDP	*User Datagram Protocol*
UE	*User Equipment*
UED	*User Equipment Domain*
UI	*Unacknowledged Information*
UIC	*Union Internationale des Chemins de fer*
UIM	*UMTS Subscriber Identity Module*
UL	*Uplink*
ULA	*Uniform Linear Array*
UMCH	*User Multicast Channel*

UMTS *Universal Mobile Telecommunication System*

UNI *User Network Interface*

UP *User Priority*

UP *Unallocated Physical Channel*

UpPCH *Downlink Pilot Channel*

UpPTS *Downlink Pilot Time Slot*

UPT *Universal Personal Telecommunication*

URL *Unified Resource Locator*

USCH *Uplink Shared Channel*

USF *Uplink State Flag*

USIM *User Services Identity Module*

USSD *Unstructured Supplementary Service Data*

USSD *Unspecified Service Signalling Data*

UTRA *UMTS Terrestrial Radio Access*

UTRAN *UMTS Terrestrial Radio Access Network*

US *United States*

USA *United States of America*

UWC *Universal Wireless Communications*

UWCC *Universal Wireless Communications Consortium*

VAD *Voice Activity Detection*

VBR *Variable Bit Rate*

VBS *Voice Broadcast Service*

VC *Virtual Channel*

VCC *Visited Country Code*

VCI *Virtual Channel Identifier*

VDB *Visited Data Base*

VEA *Very Early Assignment*

VGCS *Voice Group Call Service*

VHE *Virtual Home Environment*

VLR *Visitor Location Register*

VMSC *Visited MSC*

VNDC *Visited NDC*

VoIP *Voice over IP*

VP *Virtual Path*

VPI *Virtual Path Identifier*

VPN *Virtual Private Network*

VSN *Visited Subscriber Number*

VSSF *Visited Service Switching Function*

V+D *Voice plus Data*

W-ATM *Wireless ATM*

W-CDMA *Wideband Code Division Multiple Access*

W-CHAMB *Wireless CHannel-oriented Ad-hoc Multihop Broadband*

WAE *Wireless Application Environment*

WAN *Wide Area Network*

WAND *Wireless ATM Network Demonstrator*

WAP *Wireless Application Protocol*

WARC *World Administrative Radio Conference*

WBS *Wireless Base Station*

WBXML *WAP Binary Extensible Markup Language*

WCDMA *Wideband CDMA*

WDP *Wireless Datagram Protocol*

WEP *Wired Equivalent Privacy*

WG *Working Group*

WIM *WAP Identity Module*

WINEGLASS *Wireless IP Network as a Generic Plattform for Location Aware Services*

WLAN *Wireless Local Area Network*

WLL *Wireless Local Loop*

WM *Wireless Medium*

WMS *Wireless Media System*

WPD *Wake Pattern Declaration*

WRC *World Radiocommunication Conference*

WRS *Wireless Relay Station*

WS *Window Size*

WSP *Wireless Session Protocol*

WT *Wireless Terminal*

WTLS *Wireless Transport Layer Security*

WTP *Wireless Transaction Protocol*

WML *Wireless Markup Language*

WWW *World Wide Web*

XML *Extensible Markup Language*

Index

C

E

H

I

J

V

X